AC Circuits

- The Oscilloscope and Sine Wave Measurements
- Series RC Circuits
- Parallel RC Circuits
- Inductance Measurement
- Transformers
- RC and LC Filters
- Ceramic Band Pass Filter
- Fourier Theory
- Pulse Circuits and measurements
- Solar/Battery Charger and AC Inverter

Semiconductor/Linear Circuits

- Diodes (rectifier, Schottky, zener, LED)
- Rectifiers
- Build an AC-DC Power Supply (kit)
- Linear Regulators (3-terminal and LDO)
- Switching Regulators
- DC-DC Converters
- Bipolar Transistor
- MOSFET
- Basic Amplifiers: Gain and Frequency Response
- Power Amplifiers—Linear
- Power Amplifiers—Switching
- Build an Audio Amplifier (kit)
- 555 Timer and Applications
- Op Amps
- Instrument Amplifier with Sensor
- Comparators
- Active Filters
- RC Wein Bridge Oscillator
- Crystal Oscillator
- Transistor Switches
- Pulse Circuits and Bandwidth Measurement
- Pulse Width Modulation and Applications

MW00845459

Contemporary Electronics

Contemporary Electronics

Fundamentals, Devices, Circuits, and Systems

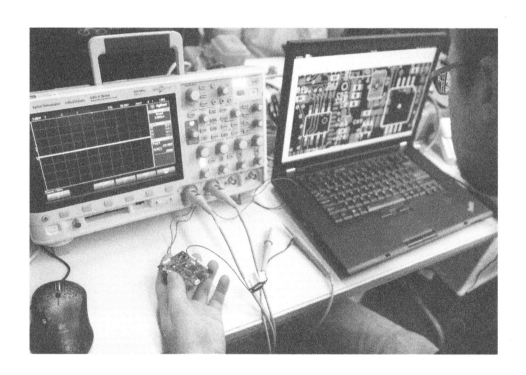

Louis E. Frenzel, Jr.

Connect
Learn
Succeed™

CONTEMPORARY ELECTRONICS: FUNDAMENTALS, DEVICES, CIRCUITS, AND SYSTEMS
Published by McGraw-Hill, a business unit of The McGraw-Hill Companies, Inc., 1221 Avenue of the
Americas, New York, NY, 10020. Copyright © 2014 by The McGraw-Hill Companies, Inc. All rights
reserved. Printed in the United States of America. No part of this publication may be reproduced or
distributed in any form or by any means, or stored in a database or retrieval system, without the prior written
consent of The McGraw-Hill Companies, Inc., including, but not limited to, in any network or other electronic
storage or transmission, or broadcast for distance learning.

Some ancillaries, including electronic and print components, may not be available to customers outside the
United States.

This book is printed on acid-free paper.

6 7 8 9 0 LKV 21 20

ISBN 978-0-07-337380-5
MHID 0-07-337380-X

Senior Vice President, Products & Markets: *Kurt L. Strand*
Vice President, General Manager, Products & Markets: *Martin J. Lange*
Vice President, Content Production & Technology Services: *Kimberly Meriwether David*
Managing Director: *Thomas Timp*
Director: *Michael D. Lange*
Executive Brand Manager: *Raghothaman Srinivasan*
Director of Development: *Rose Koos*
Development Editor: *Kelly H. Lowery*
Executive Marketing Manager: *Curt Reynolds*
Project Manager: *Jean R. Starr*
Senior Buyer: *Carol A. Bielski*
Lead Designer: *Matthew Baldwin*
Cover/Interior Designer: *Matthew Baldwin*
Cover Image: © *Adam Berry/Getty Images*
Content Licensing Specialist: *Shawntel Schmitt*
Photo Researcher: *Colleen Miller*
Media Project Manager: *Cathy L. Tepper*
Typeface: *11/13 Times*
Compositor: *MPS Limited*
Printer: *LSC Communications*

All credits appearing on page or at the end of the book are considered to be an extension of the
copyright page.

Library of Congress Cataloging-in-Publication Data

Frenzel, Louis E., Jr., 1938-
  Contemporary electronics : fundamentals, devices, circuits, and systems / Louis E. Frenzel, Jr.
    pages cm
  Includes index.
  ISBN 978-0-07-337380-5 (alk. paper)—ISBN 0-07-337380-X (alk. paper)
  1. Electronics. I. Title.
TK7816.F6786 2014
621.38—dc23
                            2012043572

The Internet addresses listed in the text were accurate at the time of publication. The inclusion of a website
does not indicate an endorsement by the authors or McGraw-Hill, and McGraw-Hill does not guarantee the
accuracy of the information presented at these sites.

www.mhhe.com

To Joan with love for her help, support, and patience.

Brief Contents

Contents

PART 2

Chapter 14   Alternating Voltage and Current   232

Chapter 15   Capacitance   257

Chapter 16   Capacitive Reactance   273

Chapter 17   Capacitive Circuits   282

Chapter 18   Inductance and Transformers   295

Chapter 19   Inductive Reactance   320

Chapter 20   Inductive Circuits   329

Chapter 21   RC and L/R Time Constants   344

Preface to the Instructor

Contemporary Electronics: Fundamentals, Devices, Circuits, and Systems offers a modern approach to fundamental courses for the electronics and electrical fields. It is designed for the first two or three electronics courses in the typical associate degree program in electronics technology. It includes both dc and ac circuits as well as semiconductor fundamentals and basic linear circuits. It addresses the numerous changes that have taken place over the past years in electronics technology, industry, jobs, and the knowledge and skills required by technicians and other technical workers. It can be used in separate dc and ac courses but also in a combined dc/ac course that some schools have adopted in the past years. *Contemporary Electronics* offers the student the benefit of being able to use a single text in two or three courses, minimizing expenses.

Goals of the Book

Contemporary Electronics has the following goals:

1. Address major changes in electronics technology to ensure courses and curricula are up to date.
2. Provide the knowledge and skills that industry wants in a technician.
3. Give students the core fundamentals on which to build future knowledge.
4. Provide the right balance between the fundamentals and current practical systems and test, measurement, and troubleshooting needs.
5. Provide a single text that can be used in two or three courses, thereby helping reduce student text costs.

Major Changes in Technology

While electronics fundamentals do not change, the technology does. All courses and curricula ideally cover the fundamentals but in the context of the current technology and practices. Here are the major changes that impact electronics education:

1. **Advances in semiconductor technology have now made it possible to package most electronic circuits into integrated circuit form.** Less need for knowing and understanding complex circuitry that cannot be accessed. Fewer components and greater reliability, so less need for troubleshooting to the component level.

2. **Changes in the way electronics circuits and equipment are designed.** Today, engineers design is at a higher level with software. Technicians do not design, calculate, or analyze circuits. Little or no need for engineering technicians.

3. **Most technicians now work at a higher systems level.** Because of the significant changes in electronic hardware, technicians work more with the equipment and its subassemblies, such as modules and printed circuit boards that make up a product or system rather than components and circuits. The technician works at the block diagram level where signal flow and interfaces, specifications, and standards are the most important considerations. Typically, the technician only has access to power connections, input, and output and, as a result, is more concerned about input and output impedances, frequency response, and output voltage or power levels.

4. **Most technician work today involves troubleshooting, repair, and servicing.** Technicians also install and operate equipment. A major function of most technicians is testing and measuring, using modern test equipment and performing a variety of calibration duties or testing to meet defined specifications or electronics standards.

What Industry Wants

The predominant employers of electronics technicians today are not in the electronics field. These employers are those that are heavy users of electronic equipment. The largest segment is manufacturing of all types, including automotive, pharmaceuticals, food processing, metal working, and others. Many technicians work for chemical plants, petroleum plants, and other systems using process control and instrumentation. The electric power industry is a huge employer. And there are too many others to include. Feedback from such employers generally indicates that they want technicians who know the fundamentals, how to use test equipment to make measurements, and how to troubleshoot at the systems level.

To further build enrollments and ensure job placement, all colleges need to survey the needs of local employers and adjust their curriculum accordingly.

What This New Text Offers

The core of this book is derived from two widely used and excellent textbooks. These are *Grob's Basic Electronics* by Mitchel Shultz and *Electronic Principles* by Albert Malvino and David Bates. The core fundamental chapters have been retained, while other more detailed analysis and design material has been eliminated. All this material has been updated to include the most recent information on components and circuits with particular emphasis on integrated circuits and the most recent technology. Here are the highlights:

1. Teach a "systems" approach to electronics that is more in keeping with what technicians actually do today, rather than in the past where component-level circuit work dominated.

2. Reduce the amount of detailed circuit analysis and design that technicians rarely if ever use on the job, yet ensure the retention of essential fundamentals and theories. Elimination of circuit analysis procedures such as branch, mesh, and nodal analysis and the superposition and Miller's theorems. Elimination of complex *RLC* circuit analysis.

3. Ensure that all chapters reflect current components, technologies, methodologies, and applications.

4. Minimize magnetic circuit theory, as technicians do not analyze or calculate magnetic products or circuits.

5. Extensive reduction of bipolar junction transistor (BJT) biasing and circuit analysis.

6. Increased coverage of MOSFETs, since over 90% of all electronic circuitry is made with MOSFETs instead of BJTs.

7. Increased coverage of ICs and their specifications and application. Addition of widely used ICs not previously covered, such as class D switching power amplifiers, increased coverage of switch mode power supply circuits, and phase-locked loops.

8. Added coverage of ac power wiring, as well as other wiring and cabling in which technicians are always involved.

9. A light and subtle introduction to digital and binary concepts, circuits, and applications. Given that most electronic equipment today is digital, the student needs an early look at this technology in the context of the electrical and electronics fundamentals.

10. Increase coverage of troubleshooting, testing, and measuring, with a systems emphasis that is more in alignment with current technician work practices.

Unique Features

- Two chapters on systems concepts, including block diagram, signal flow, and the big picture to understanding electronics.

- Three chapters on troubleshooting, including dc, ac and linear systems.

- System sidebars. These sidebars are chapter supplements that cover practical applications and related systems to tie the theory and concepts being learned to actual products and systems.

- A companion lab manual has been created to present fundamentals in a modern way that is more in keeping with what technicians will encounter in their jobs.

What's in It for You and the Student?

This textbook gives you the opportunity to update your first two or three electronics fundamentals courses. The result will be not only more modern and current course content but also an introduction to the concepts and approaches that students will need for available jobs.

The student will learn material relevant to available jobs and what industry wants. More importantly, the student can focus on significant concepts rather that waste time on excessive and unnecessary math and analysis. The result should be higher student interest and greater retention in these earlier courses.

Note to the Student

As you begin to study electronics in class and in this textbook, you are bound to ask, "Why do I need to learn this?" There are several answers. For one, the content of a course and book are typically dictated by the course syllabus, the course description in the college catalog, and, to some extent, accrediting body requirements. Sometimes, the content is based on what the instructors in the department think should be taught. In a number of enlightened schools, even local industry has had an input. In any case, the content is commonly greater in depth and breadth than necessary for the jobs in industry.

Course and textbook content is designed to teach you the fundamentals of the technology, so that you may understand the operation and application of the devices and equipment you will be working with. These fundamentals often do not seem to bear any relationship to the job you will do in industry, but they will serve you well later as you do the troubleshooting, testing, and other work. The objective of this text is to reduce the depth of coverage of the fundamentals to the only those that you need to know. Other "nice to know" fundamentals and procedures, as well as many analysis and design techniques, are either eliminated or minimized to make room for more real-world content like troubleshooting, system operation and more of the big picture of electronics.

Finally, pay particular attention to the lab work. Learn to use the test equipment and develop your troubleshooting skills on the lab experiments. And get as much hands-on practice as possible by building kits, experimenting on your own, and becoming more involved with real electronics equipment. Then in each case try to match the fundamentals with each practical project or learning activity you encounter. In other words, apply what you are learning as soon as you can.

Acknowledgments

My special thanks to the following people who helped with this project:

Mitchel Schultz, Dave Bates, and Albert Malvino for contributing their books to this project.

Raghu Srinivasan, McGraw Hill Global Publisher–Engineering/CS/Trade&Tech, for supporting the project.

Kelly Lowery for her help in keeping me organized, motivated, and on schedule.

Jean Starr for her patience and skill with the manuscript.

Chet Gottfried for making the book as error-free as possible.

Colleen Miller for her photo work.

Bill Hessmiller for his help with the tests, PowerPoints, and other ancillary materials.

Mike Lesiecki and Tom McGlew of the Maricopa Advanced Technology Education Center in Phoenix for their original NSF grant work and support with the "systems approach" to electronics technology education.

I would like to thank the following reviewers for taking their valuable time to review this textbook and provide valuable feedback and suggestions.

Norman Ahlhelm
Central Texas College

Michael Beavers
Lake Land College

David Becker
Pittsburgh Technical Institute

Randolph Blatt
Reading Area Community College

Stephen J. Cole
Computer Network Technology

Cory Cooksey
North Central Missouri College

Toby Cumberbatch
The Cooper Union

Robert M. Deeb
Southeastern Louisiana University

Christine Delahanty
Bucks County Community College

William I. Dolan
Kennebec Valley Community College

Roger Eddy
Forsyth Technical Community College

Dan Fergen
Mitchell Technical Institute

Richard Fornes
Johnson College

Gary George
American River College

Chris Haley
North Georgia Technical College

James W. Heffernan
Quinsigamond Community College

Paul Hollinshead
Cochise Community College

Patrick Hoppe
Gateway Technical Co

David V Jones
Lenoir Community College

M. Hugh King
Johnson County Community College

Stan Kohan
Richland College

Richard Le Blanc
Benjamin Franklin Institute of Boston

Dr. Shueh-Ji Lee
Grambling State University

Joe Morales
Dona Ana Community College

Jerry Morehouse
Career College of Northern Nevada

Yehuda Nishli
Career Institute of Health & Technology

Chrys Panayiotou
Indian River State College

Thomas Patton
The Community Colleges of Baltimore County

Philip Regalbuto
Trident Technical College

Mike M Samadi
Dekalb Technical College

Jess C. Sandoval
Oxnard Community College

Scott Segalewitz
University of Dayton

Guru Subramanyam
University of Dayton

David Turbeville
Angelina College

Charlie Williams
Cleveland Institute of Electronics

Thomas Zach
Southwestern Illinois College

Robert Zbikowski
Hibbing Community College

Walkthrough

ontemporary Electronics: Fundamentals, Devices, Circuits, and Systems provides the essential information students need to become successful technicians in the modern world. The content is illustrated by the accessible design with vivid diagrams and illustrations and review problems that help reinforce the material that is presented throughout each chapter.

LEARNING OUTCOMES provide an overview of the chapter to help students focus on the main concepts and retain key information.

· ▶

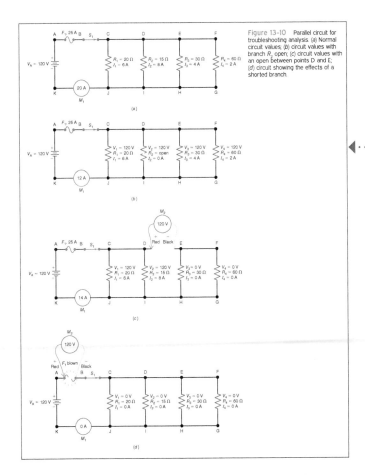

Figure 13-10 Parallel circuit for troubleshooting analysis. (a) Normal circuit values; (b) circuit values with branch R_2 open; (c) circuit values with an open between points D and E; (d) circuit showing the effects of a shorted branch.

Learning Outcomes

After studying this chapter, you should be able to:

> *Use* a multimeter in measuring voltage, current, and resistance in dc circuits.

> *Define* an *open* and a *short*.

> *Identify* when opens and shorts occur.

> *Identify* defective components and/or connections in dc circuits.

> *Test* for continuity in components, wires, cables, and connectors.

> *Troubleshoot* wires, cables, and connectors.

> *Troubleshoot* a dc system.

DIAGRAMS AND PHOTOS illustrate concepts and aid student understanding of new material.

◀ ·

CHAPTER 13 SYSTEM SIDEBAR

DC Power System

A good example of a dc electrical system is that used in an automobile or boat. Both are powered by batteries with subsystems for charging, lights, motors, controls, electronics, and other accessories. This section describes a simplified generic boat electrical system as an example. This hypothetical system is typical of what you may find on a small (<35-ft)

powerboat with inboard or outboard motors or a sailboat with an auxiliary motor. There are significant variations and a wide range of electrical features and accessories, but this system describes all the most common equipment. Refer to Fig. S13-1.

224   Chapter 13

SYSTEM SIDEBARS cover practical applications to the related systems and help students connect the theory and concepts being learned to actual products and systems. There are twenty-six system sidebars integrated throughout the text, so students can apply what they have learned.

EXAMPLE 13-1

Assume that the series circuit in Fig. 13-5 has failed. A technician troubleshooting the circuit used a voltmeter to record the following resistor voltage drops.

$$V_1 = 0 \text{ V}$$
$$V_2 = 0 \text{ V}$$
$$V_3 = 24 \text{ V}$$
$$V_4 = 0 \text{ V}$$

Based on these voltmeter readings, which component is defective and what type of defect is it? (Assume that only one component is defective.)

Answer:
To help understand which component is defective, let's calculate what the values of V_1, V_2, V_3, and V_4 are supposed to be. Begin by calculating R_T and I.

$$R_T = R_1 + R_2 + R_3 + R_4$$
$$= 150 \ \Omega + 120 \ \Omega + 180 \ \Omega + 150 \ \Omega$$
$$R_T = 600 \ \Omega$$

$$I = \frac{V_T}{R_T}$$
$$= \frac{24 \text{ V}}{600 \ \Omega}$$
$$I = 40 \text{ mA}$$

Next,

$$V_1 = I \times R_1$$
$$= 40 \text{ mA} \times 150 \ \Omega$$
$$V_1 = 6 \text{ V}$$
$$V_2 = I \times R_2$$
$$= 40 \text{ mA} \times 120 \ \Omega$$
$$V_2 = 4.8 \text{ V}$$

$$V_3 = I \times R_3$$
$$= 40 \text{ mA} \times 180 \ \Omega$$
$$V_3 = 7.2 \text{ V}$$
$$V_4 = I \times R_4$$
$$= 40 \text{ mA} \times 150 \ \Omega$$
$$V_4 = 6 \text{ V}$$

Next, compare the calculated values with those measured in Fig. 13-5. When the circuit is operating normally, V_1, V_2, and V_4 should measure 6 V, 4.8 V, and 6 V, respectively. Instead, the measurements made in Fig. 13-5 show that each of these voltages is 0 V. This indicates that the current I in the circuit must be zero, caused by an open somewhere in the circuit. The reason that V_1, V_2, and V_4 are 0 V is simple: $V = I \times R$. If $I = 0$ A, then each good resistor must have a voltage drop of 0 V. The measured value of V_3 is 24 V, which is considerably higher than its calculated value of 7.2 V. Because V_3 is dropping the full value of the applied voltage, it must be open. The reason the open R_3 will drop the full 24 V is that it has billions of ohms of resistance and, in a series circuit, the largest resistance drops the most voltage. Since the open resistance of R_3 is so much higher than the values of R_1, R_2, and R_4, it will drop the full 24 V of applied voltage.

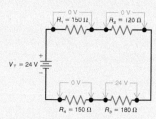

Figure 13-5   Series circuit for Example 13-1.

EXAMPLES illustrate comprehensive concepts and allow the students to visualize how they can apply what they have learned.

Chapter **13**

DC Troubleshooting

TROUBLESHOOTING is the focus of this textbook, in order to provide students with the information they need to know to help prepare them for their careers as technicians. In addition to the system sidebars that provide a practical application to the related systems, the author has provided three chapters on troubleshooting: Chapter 13, "DC Troubleshooting," and Chapter 26, "AC Testing and Troubleshooting," and Chapter 41, "Electronics Systems Troubleshooting."

REVIEW QUESTIONS are intended to reinforce the chapter learning outcomes. The questions test students' knowledge of the material as they read, helping them identify areas that may need further study.

1. The main troubleshooting test instrument for dc systems is a(n)
   a. oscilloscope.
   b. multimeter.
   c. clamp-on current meter.
   d. neon test bulb.

2. The very first step in any troubleshooting procedure is to
   a. acquire all necessary documentation.
   b. measure all source voltages.
   c. check the circuit breakers.
   d. verify that a problem actually exists.

3. The basic approach in dc troubleshooting could be described as
   a. signal injection.
   b. effect to cause reasoning.
   c. signal tracing.
   d. continuity checking.

4. Most dc system problems, excluding the voltage source, are
   a. shorts or opens.
   b. defective parts.
   c. corrosion.
   d. ground faults.

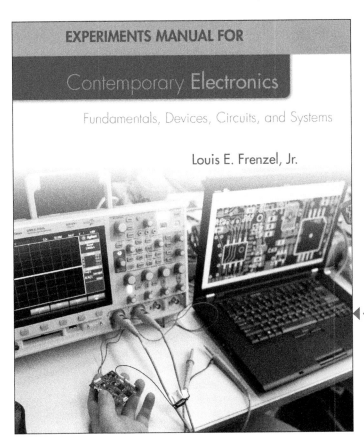

EXPERIMENTS MANUAL FOR
Contemporary Electronics
Fundamentals, Devices, Circuits, and Systems
Louis E. Frenzel, Jr.

THE EXPERIMENTS MANUAL for use with *Contemporary Electronics* provides practical, hands-on knowledge in the form of lab experiments and projects that will give students valuable practice in handling components, building and testing circuits, and using common test instruments.

This manual is divided into four major sections. Part 1 describes the equipment students will need and a basic procedure for running the experiments. The other three sections are divided into experiments associated with most major courses, including material on dc circuits, ac circuits, and solid-state devices with linear circuits.

THE ONLINE LEARNING Center features an outstanding student and instructor support package:

- Online quizzes for each chapter.
- ExamView and EZ Test Questions for instructors for each chapter.
- Instructor PowerPoint slides.
- An Instructor's Manual with answers to chapter questions and problems as well as guidelines for the lab experiments and materials.

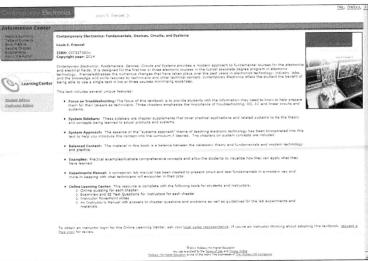

Visit the Online Learning Center for this textbook at **www.mhhe.com/frenzelle**.

Survey of Electronics

Learning Outcomes

After studying this chapter, you should be able to:

> *List* the five major sectors of the electronics industry and *name* three examples of each.

> *Name* the two major types of jobs in the electronics industry.

> *Describe* the kind of education that you will need for each of the major job types.

> *Name* the one thing that will keep your knowledge and skills current during your career.

> *List* the major players in the electronics industry and how they interact with one another.

> *Explain* two ways in which all electronic circuits can be represented.

With this book, you are about to begin your study of electronics. Your goal is no doubt a job and career in electronics. You could not have chosen a better career path. Not only is the electronics industry one of the largest, most dynamic and exciting, but also it can be one of the most lucrative. Jobs continue to be plentiful, and there is a lifetime of interesting jobs to be had. The electronics industry changes daily with a blizzard of innovative new products, components, technologies, and applications. This industry is one that will challenge you as well as keep you interested. This chapter introduces you to the industry, as well as to the jobs and the education you will need to succeed.

1.1 Life Impact

Just so you understand how important electronics is to our lives, take a minute and think about how electronics affects you personally. As a starting point, do the following:

- Make a comprehensive list of all the electronic products and services you own and use daily. Do it now.
- Make an hour-by-hour diary especially identifying things you do with electronics daily. Again, do it now.

Now, study your results. Are you astonished? Electronics is so pervasive that we simply take most of it for granted. It just is. We are not surprised or amazed anymore by even the most sophisticated electronic devices although we use them and perhaps covet some of them. And even common everyday appliances are loaded with electronics.

Now think about where all that equipment comes from. Someone has to design and build it, sell it, install it, and otherwise support it. And it must usually be operated, maintained, and serviced later. Lots of job opportunities are involved. Electronics is fun and interesting, and many engage in electronics as a hobby. Maybe that is how you became interested in electronics. Electronics is a great hobby because it helps you learn while having fun. Table 1-1 lists the most popular electronic hobbies. If you do not have an electronics hobby, you should consider starting one, as it is engaging, challenging, and educational. And more often than not your hobby becomes your career.

1.2 Major Segments of the Electronics Industry

The electronics industry is enormous and diverse. One estimate has the total of all electronic goods sold annually worldwide at over $1.5 trillion. And the industry continues to grow under the toughest economic conditions, attesting to its diversity and importance in our lives. To get a handle on how the industry is structured, it is best to divide it into segments, or areas of specialization. Then you can see how they are all interrelated. The five major divisions are components, communications, computers, control, and instrumentation. See Fig. 1-1. All electronic equipment and applications fall into one of these sectors if not several.

Components

Components are the individual parts that make up all circuits and equipment. These include resistors, capacitors, inductors, transformers, connectors, wire and cable, and printed circuit boards. The largest segment of the components field is semiconductors like integrated circuits (chips), transistors, diodes, solar cells, and many other specialty parts. These parts are used by the engineers to design all types of electronic equipment.

Communications

The oldest and one of the largest segments of electronics is communications. It began with the telegraph and telephone in the mid- to late 1800s. In the early 1900s, radio was developed, then the vacuum tube came along and the rest as they say is history. It was specifically the vacuum tube that created the electronics industry as we know it today. The vacuum tube brought us amplification and electronic switching, neither of which existed in the telegraph and telephone. Later on we got TV, satellites, and many other communications applications. Broadcast radio and TV dominated the early years, then two-way radio became commonplace, and radar was invented during World War II.

Table 1-1 Popular Electronic Hobbies

1. *Amateur radio.* Amateur radio operators, or hams, build and operate radio equipment to make contact with other hams to exchange signal reports, technical information, and personal experiences. A Federal Communications Commission (FCC) license is required.

2. *Computers.* Computer hobbyists build personal computers, write programs, work with peripheral equipment, interface computers to other gadgets, and boost computer performance. Software and programming are a major part of this hobby.

3. *Robots.* Building and experimenting with robots has become a huge hobby in the past few years.

4. *Model radio control.* Building model airplanes, boats, cars, and other objects that can be controlled remotely by radio will also get your outdoors.

5. *Audio.* Building and experimenting with high-fidelity stereo and surround sound equipment, speakers, and music is popular. Electronic music instruments and sound systems are also a part of this hobby.

6. *Home entertainment centers.* Building and using high-definition television and audio systems are indoor hobbies. Cable TV, satellite TV, Internet TV, wireless connectivity, gaming, and 3D TV are popular elements of this hobby.

7. *Home monitoring and control.* Electronic components and systems to monitor and control heating and air conditioning, lighting, appliances, and electrical systems (sprinklers, garage doors, security, etc.) provide energy savings, safety, and convenience.

8. *General experimenting.* Curiosity can lead to kit building, miscellaneous projects, and experimentation with various electronic gadgets and equipment.

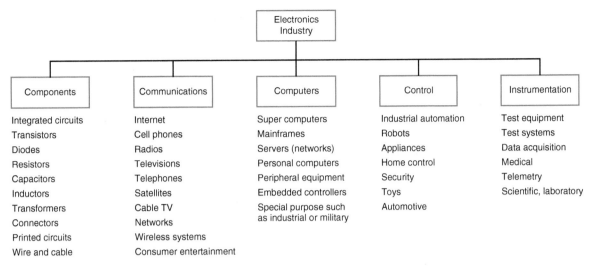

Figure 1-1 The major sectors of the electronics industry.

Communications refers to all the various types of wired and wireless technologies you use every day. Typical wired technologies include cable TV, computer networks, the telephone system, and the Internet. Typical wireless communication technologies include radio of all types, including broadcast and two-way radio, cell phones, satellites, and wireless networks. Today the communications industry continues to dominate with the huge telecommunications system, the Internet and networking, and of course the huge wireless industry with its cell phones, wireless networks, and links of all kinds. Leading the industry is the smartphone like the Apple iPhone and others that are not only phones but also full-blown computers in a handset with Internet access and other communications features.

Computers

Computers didn't really come along until World War II and later. And these were the huge vacuum tube monsters called mainframes. Transistors and integrated circuits made them smaller, faster, and better. During the 1970s, thanks to digital integrated circuits (ICs), we got the minicomputers. Then later in the 1970s the microcomputer was created. This put most of a computer's circuitry on a single chip of silicon. Called a microprocessor or central processing unit (CPU), it formed the basis for newer personal computers. They created a whole new industry making computers available and affordable for everyone. Today, PCs, laptops, and tablets are as common as the TV set.

The computer segment encompasses an enormous range of different types of computers. Computers process data. The largest and most powerful (meaning fast with lots of storage) computers are known as supercomputers that solve massively difficult scientific, engineering, and mathematical problems. Other large powerful computers are the mainframes that still serve the data processing needs of large business and government. Small but very powerful computers known as servers are the workhorse of our computer networks, from the Internet to local area networks to which most computers are connected today. The personal computer is probably the most visible and widespread. Laptop computers have now passed desktop personal computers in total volume of computers shipped. The tablet market is also growing, taking market share from laptops.

But the real breakthrough was the single-chip computer with the processor, memory, and input/output circuitry in one integrated circuit. This device, called an embedded controller or microcontroller, permits the power of a computer to be packaged into other electronic devices, expanding their functionality, versatility, and power. Today, virtually every electronic product in existence contains an embedded controller as its central control point. These small computers handle all monitoring and control functions in cell phones, TV sets, DVD players, and MP3 music players. In fact, you can say that every electronic piece of equipment made today is simply an embedded microcontroller surrounded by peripheral devices that perform the functions of the equipment.

This puts computers everywhere, in our cars, consumer electronics products, and appliances. It is impossible to name an electronic product that does not contain one. As you will find out, all electronic products are mainly an embedded controller surrounded by other circuits that customize it to the specific application.

The computer part of the electronics industry is also huge but dispersed. And with computers readily available in all forms from mainframes to PCs and to micros on a chip, the focus in the industry has turned to software. *Software* is the term used to describe the programs that a computer uses to perform as desired. There continues to be a great demand for people who can program computers.

Control

Control is a huge and diverse part of electronics. Think of electronics as that field of science that is used to monitor and control physical functions. Monitoring means observing and measuring physical things like temperature, pressure, mechanical position, liquid level, or light intensity. Sensors turn these physical characteristics into electrical signals that can be processed by electronic circuits. We may want to record the physical phenomena or, better still, use the information they provide as signals to tell electronic circuits what to do. That is the control part.

Control is simply the execution of various duties with electronic circuits. Some common control functions are turning lights off or on, turning motors off or on, operating pumps or valves, or controlling the transmission of data over a network.

Electronic controls are everywhere, in home appliances, cars and trucks, vending machines, military weapons, aircraft of all types, and most factories. Think of robots, garage door openers, traffic lights, remote keyless entry on a car, and the autopilot on a unmanned aerial vehicle (UAV). The examples are broad and diverse.

Monitoring and control is a very large segment of electronics involved in performing monitoring operations of various physical characteristics. Components called sensors or transducers are used to measure temperature, light intensity, pressure, and literally hundreds of other physical characteristics. These measurements are then used in control systems to activate appliances, robots, chemical plants, automotive systems, and hundreds of other devices. The signals being monitored are processed in various ways, and embedded controllers or computers produce outputs that control other devices, such as factory automation, manufacturing plants, security systems, appliances, and toys.

Instrumentation

Instrumentation is that segment of electronics involved with testing electronic circuits and equipment or other equipment using electronics and in making precise measurements of electrical and electronic signals. Working in electronics, you will eventually use a wide variety of electronic test instruments like multimeters, oscilloscopes, signal generators, and analyzers of all sorts. You cannot design, build, troubleshoot, or service electronic equipment without the need to measure voltage, current, power, frequency, or other electronic characteristics.

Instrumentation and measurement relate to precisely and accurately measuring electronic characteristics. One of the largest categories is test equipment, such as meters, oscilloscopes, signal generators, spectrum analyzers, and other general-purpose instruments that are used to test and measure all other electronic equipment. Test systems used for automated component or equipment testing for complete systems also fall into this category.

Instrumentation and measurement also include the category known as data acquisition, by which systems are used to collect data from a variety of sensors and other sources. A major segment of instrumentation and measurement is medical diagnosis and testing. Medical instruments measure EEG, EKG, temperature, blood characteristics, and chemical compositions and include MRI, CT, and x-ray machines.

Some examples of instrumentation besides generic test equipment are the instruments in a jet aircraft, the electronics in a chemical or process control plant, or an automated test system for cell phones.

1.3 A Converged Industry

As you can see, the electronics industry is enormous and diverse. Yet all these segments of electronics have a significant impact on our lives. They provide us with instantaneous information and communications, speed up and simplify our work with computers, and protect us at home and office.

While we still view electronics as being comprised of these basic sectors, as you can easily see, there are lots of crossovers and overlaps. The different segments converge in different devices and applications. Tablet computers contain wireless transceivers to connect to hot spots or the cellular network, and iPods and MP3 music players contain a control computer and lots of memory to store songs. Factory automation and control systems contain instrumentation, computers, and networking for communications. And almost everything contains an embedded controller. Home appliances like washers, dryers, refrigerators, dishwashers, blenders, toasters, coffeemakers, and most others are all loaded with electronic controls. Our consumer entertainment equipment like HD TV sets, DVD players, cable and satellite TV sources, stereo audio systems, and others are totally electronic. The modern automobile contains an ever-increasing number of electronic components, control systems, and safety features. It is difficult to name something today that does not include some electronic segment. Nevertheless, it is still best to keep these divisions separate in your mind as you decide what interests you most and how you wish to focus in your electronics career.

1.4 Jobs and Careers in the Electronics Industry

As you saw in the previous section, the electronics industry is roughly divided into five major specializations. The largest in terms of people employed and the dollar value of equipment purchased is the communications field, closely followed by the computer field. The components, industrial control, and instrumentation fields are considerably smaller. Hundreds of thousands of people are employed in these fields, and billions of dollars' worth of equipment is purchased each year. The growth rate varies from year to year depending on the economy, technological developments, and other factors. All segments of electronics have grown steadily over the years, creating a relatively constant opportunity for employment. If your interests lie in electronics, you will be glad to know that there are many opportunities for long-term jobs and careers. This section outlines the types of jobs available and the major kinds of employers.

The two major types of technical positions available in the electronics field are engineer and technician.

Engineers

Engineers design electronic components, equipment, and systems. Engineers work from specifications and create new components, equipment, or systems that are then manufactured. For example, some engineers specialize in integrated circuit design. They use sophisticated computer software called electronic design automation (EDA) to create the detailed circuits that ultimately become the chips making up the electronic equipment we use. Other engineers use the chips and other components to design the electronic end products like cell phones, DVD players, network routers, industrial controllers, or medical instruments like a pacemaker.

But while most engineers specialize in design, others work in manufacturing, testing, quality control, and management, among other areas. Engineers may also serve as field service personnel, installing and maintaining complex equipment and systems. If your interest lies in the design of electronic equipment, then an engineering position may be for you.

It is important to note that what engineers do is design and analyze. They use their heavy math and science knowledge to model electronic circuits and systems and use computer software to simulate the behavior of circuits, equipment, and systems. While engineers do indeed work with the end products they design and analyze, mostly they work at this higher abstract level.

Technicians

Technicians are most often employed in service jobs. The work typically involves equipment installation, troubleshooting and repair, testing and measuring, maintenance and calibration, or operation. Technicians in such positions are sometimes called field service technicians, field service engineers, or customer representatives. Today, the jobs for technicians are so diverse that the generic term *electronic technician* is rarely used. When you are looking for a job as a technician, you need to look at not only the job titles given above but also those using the terms mechanic, installer, associate, assistant, assembler, manufacturing tester, troubleshooter, and similar titles.

Technicians can also be involved in engineering. Engineers may use one or more technicians to assist in the design of equipment. They build and troubleshoot prototypes and in many cases actually participate in equipment design. A great deal of the work involves testing and measuring. In this capacity, the technician is known as an engineering technician, lab technician, engineering assistant, or associate engineer. Engineering technician positions were once widely available, but because of the widespread use of ICs and design software, engineers rarely need technicians to the extent they once did. Engineering technician jobs are rarely available today.

Technicians are also employed in manufacturing. They may be involved in the actual building and assembling of the equipment but, more typically, are concerned with final testing, measurement, and quality assessment of the finished products. Other positions involve quality control or repair of defective units. Online and telephone help and support is another common opportunity.

Technicians are the hands-on electronic workers. Their duties involve the equipment and systems and their service, installation, maintenance, calibration, and repair. Technicians do not design or do any significant amount of analysis. Therefore, their knowledge of math and science does not have to be as great as that of an engineer. Practical hands-on job training, work experience, and specific equipment or system training are far more important.

Other Technical Positions

There are many jobs in the electronics industry other than those of engineer or technician. For example, there are many outstanding jobs in technical sales. Selling complex electronic equipment or systems usually requires a strong technical education and background. The work may involve determining customer needs and related equipment specifications, writing technical proposals, making sales presentations to customers, and attending conferences and exhibits where equipment is sold. The pay potential in sales is generally much higher than in the engineering or service positions.

Another position is that of technical writer. Technical writers generate the technical documentation for electronic equipment and systems, producing installation and service manuals, maintenance procedures, and customer operations manuals. Most of this material is web-based today. This important task requires considerable depth of education and experience as well as a knack for writing, organizing, and categorization.

Finally, there is the position of trainer. Engineers and technicians are often used to train other engineers and technicians or customers. With the high degree of complexity that exists in electronic equipment today, there is a major need for training. Many individuals find education and training positions to be very desirable and satisfying. The work typically involves development of curriculum and programs, generating the necessary training manuals, presentation materials and lab exercises, creating online training, and conducting classroom training sessions in-house, online, or at a customer site.

1.5 Engineering and Technology Education

To get a good job in electronics today, you need some formal postsecondary education at a college or university. This education varies widely with the type of job but can be categorized as engineering education or technology education. They are similar because they both involve electronics, but they are not the same simply because the jobs they prepare the graduates for are so vastly different and require different levels of knowledge and skills. Education is essential to being successful in getting a job in electronics, but it is also the one single ingredient of continuing success in the field. Continuing personal education is the key to staying on the top of your job and field.

Engineering Education

Engineers need a bachelor's (BSEE), master's (MSEE), or doctoral (PhD) degree in electrical or electronic engineering. Figure 1-2 shows the general paths through college to a job. Such an education starts with a strong science and mathematics background, including physics, chemistry, calculus, statistics, and other advanced math courses. This is followed by specialized education in electronic circuits and equipment. The education is largely design- and analysis-oriented with an emphasis on computer simulation and software development.

Some jobs require additional education beyond the bachelor's degree. Figure 1-2 shows the path to graduate school that may include more advanced electrical or electronic courses leading to a master of science in electrical engineering (MSEE) degree. This prepares you for more advanced jobs with better pay. An alternative path is to pursue a master in business administration (MBA) degree. Some BSEE graduates find that their greater interest lies in the business side of the industry, such as finance, economics, marketing and sales, or management.

A doctorate degree or doctor of philosophy (PhD) is an advanced degree usually with a specialization in one specific area of electronics. It is the path to take to emphasize research or teaching.

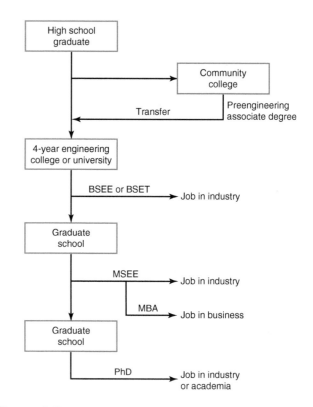

Figure 1-2 Educational paths for engineers.

Some engineers have a bachelor's degree in electronics technology from a technical college or university. Some typical degree titles are bachelor of technology (BT), bachelor of engineering technology (BET), and bachelor of science in engineering technology (BSET.)

Bachelor of technology degrees often begin with a two-year associate degree program followed by two additional years required for a bachelor of technology degree. During those last two years, the student takes more complex electronics courses along with additional science, math, and humanities courses. The main difference between the BT graduate and the BSEE engineering graduate is that the technologist usually takes courses that are more practical and hands on than engineering courses. Holders of BT degrees can generally design electronic equipment and systems, but do not typically have the depth of background in analytical mathematics or science that is required for complex design jobs. However, BT graduates are generally employed as engineers. Although many do design work, others are employed in engineering positions in manufacturing and field service rather than design.

Although a degree in electrical engineering is generally the minimum entrance requirement for engineering jobs in most organizations, people with other educational backgrounds (e.g., physics and math) also become engineers. Technicians who obtain sufficient additional education and appropriate experience may go on to become engineers as well.

Technology Education

Technology education is usually less stringent in the areas of math and science and more practical than engineering education. Usually less education is required to be a technician than an engineer. Engineers do far more math, design, and analysis, making their jobs far more mental. Technician jobs do not require the heavy math, science, and analytical background, but they do require clear logical thinking as well as good hands-on skills.

Technicians have some kind of postsecondary education in electronics, from a vocational or technical school, a community college, or a technical institute. The typical technology education paths are illustrated in Fig. 1-3. Many technicians are also educated in military training programs. Most technicians have an average of two years of formal post-high school education and an associate degree. Common degrees are associate in arts (AA), associate in science (AS), or associate of science in engineering technology (ASET) or associate of science in electronic engineering technology (ASEET), and associate in applied science (AAS). The AAS degree tends to cover more occupational and job-related subjects; the AA and AS degrees are more general and are designed to provide a foundation for transfer to a bachelor degree program. The math level is typically algebra and trigonometry, rather than calculus although some AAS programs may require an introduction to calculus. As for science, AAS programs do not commonly require engineering-level physics or chemistry although some introductory courses may be included.

Technicians with an associate degree from a community college can usually transfer to a bachelor of technology program and complete the bachelor's degree in another two years. Just keep in mind that associate degree holders are usually not able to transfer to an engineering degree program. If you decide to become an engineer, you must literally start over at an engineering school because of the big differences in math and science backgrounds needed. That is a choice few make because the BSET degree is far faster to achieve and the chance of working as an engineer just as good in most cases.

Many BSET graduates do go on to some kind of engineering job. Another path is an MBA degree if a business slant to your education is desired. Another alternative is a master in technology (MT) degree, which is available at a limited number of colleges and universities. Such degrees generally focus on teaching or some specialty subject.

Continuing Education

Continuing education refers to the education you obtain after graduating from college. And don't think it is not necessary. You should know up front that you cannot survive in the field of electronics without a continuous process of personal self-education. Electronics changes fast and furiously. New components, products, technologies, and methods are developed daily, and all have an impact on how products are designed and used. Usually what was current yesterday is often obsolete tomorrow. You always need to know the latest products and techniques to stay competitive in your job. When you enter the electronics field, consider the fact that you will immediately need to engage in some form of continuing education as soon as possible. The half-life of an engineering or technology degree is only a few years today, meaning that within those few years, half of what you learned will be obsolete or irrelevant. That may be depressing to some extent, but think of the bright side. Learning new stuff is fun, and that is half the excitement with electronics.

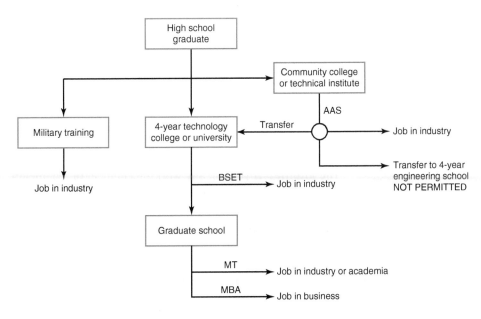

Figure 1-3 Educational paths for technicians.

There is always something new, interesting, and exciting to learn and get involved with. Besides learning more usually means earning more.

Where do you get continuing education? Outlined below are the most common sources that most engineers and technicians use today.

Advanced Degrees

If you have an AAS degree, think of going back to school to get a BSET degree. If you live near a college or university that will accept your previous college work, you are generally halfway to the bachelor's degree anyway. Maybe you can even complete the degree at night, and amazingly many employers will help you pay for that.

If you already have a bachelor's degree, think of going for a master's degree. Your BSEE or BSET will lead you to an MSEE or MT as described earlier.

A good option for both BSEE and BSET degree holders is a master of business administration (MBA). If you decide you like the business side of electronics more than the technical side, this is a good choice. You can parlay that degree into very high end marketing, financial, or management positions.

A PhD is the ultimate degree in engineering, but it is rarely worth the long process and very high cost. If you plan to teach engineering or do advanced research, then you will need it, of course. For most good jobs a master's is more than enough today.

Alternative Forms of Education

In addition to adding a degree, there are other methods for continuing education or staying current.

University Courses Sometimes you only need to take a course here or there to learn what you need. You can take regular university courses toward a master's degree or university-sponsored continuing education courses.

Seminars Many companies offer specialized seminars. Such seminars last from two days to a week and focus on a specialty like programming skills, RF design, or computer networking. You will have to ferret these out for yourself with Internet searches or magazine announcements, but often they will be just what you need on the job. Many employers will pay for them.

Company-Sponsored Classes Many larger companies offer internal courses for their employees. Training employees helps the company. You should take as many of these as you can, depending how relevant they are to your situation and available time.

Webinars Webinars are online seminars. They consist of a presentation over the Internet via PC and sometimes a telephone connection for audio. They are like a lecture and typically run about an hour. The subject is very focused, but there are lots of them online. Many are sponsored by companies which want to promote the use of their products in new designs. These are mostly free, so do as many as time allows, assuming they are relevant.

Books Books are still a good choice for self-education. You rarely need to read the entire book anyway. Mostly if you can find the books related to the subject you want to read, you will buy them for reference and specific knowledge. Check out your local bookstore and especially the college bookstore for relevant materials. Also search for the books you want online by going to sites like Amazon or Barnes & Noble. Go directly to technical book publishers websites to see what is available. Some good technical book publishers are Artech House, Elsevier, McGraw-Hill, Morgan Kauffman, Newnes, Prentice Hall, and Wiley. Also look for good used books to save a few dollars.

Magazines There are lots of electronic magazines written for engineers and technicians. These are what are called controlled circulation magazines or business-to-business (B2B) magazines. They are free to the subscriber and are paid for by advertisers who want to get their products in front of a targeted audience like engineers. These magazines come out at least monthly and sometimes two times a month. Many also include online newsletters weekly. These magazines have in-depth technical articles, new product information, and a whole range of business- and technology-related information. Subscribe and read them regularly. Some popular electronic magazines for engineers are *Electronic Design*, *EDN*, *EETimes*, and *Electronic Products*. The *IEEE Spectrum* is a great magazine, but it does require that you be a member of the Institute of Electrical and Electronic Engineers (IEEE), a professional association you should eventually look into after graduation.

Don't forget the hobby-oriented electronic magazines. These are excellent on the practical side as they cover not only theory and practice but often include construction projects. A couple of popular electronic magazines are *Circuit Cellar*, *Elector*, *Make*, and *Nuts & Volts*. Amateur radio magazines like *CQ*, *QEX*, and *QST* are also excellent sources of new learning. These are paid subscriptions but worth the price.

The Internet If you ask any working engineer or technician how and where he or she learns new stuff, their first answer is usually, "The Internet." The Internet offers a huge worldwide source of information and learning materials. And it is free in most cases. All you have to do is search for it. You have probably already done some of this so it may already be second nature. Just type in what you want to know; do a Google, Yahoo, or Bing search; and within seconds you will have at your fingertips just what you want. Maybe. You may have to refine your search to zero in, but it is most likely there in the form of a Wikipedia report, magazine article, technical paper, company white paper, or application note or other. It could even be a free online tutorial or website. Start using the Internet now while you are in school to reinforce what you are learning or get a different perspective or confirm how something works from another source.

Manufacturers Resources Component and equipment manufacturers want you to buy their products, so they offer tons of literature to help sell them. Most of this is on the Internet, but some is available in hard copy like data sheets, brochures, reference manuals, books, or other literature, usually free for the asking. Then there are the usual massive databases on their websites. Most offer product data sheets, application notes, white papers, and tutorials—all available for free. Here is one other resource you can begin using to educate yourself.

Licensing and Certification Programs There are numerous programs that offer to prepare you for a wide range of certifications, licenses or registration. Engineers can seek registration in their state as a professional engineer (PE). You have to be a BSEE graduate, have a certain number of years of experience as an engineer, and pass a rigorous exam. It is a tough process, but you learn a great deal in the process of preparing. Having a PE license opens new doors of employment and higher pay. For most jobs a PE license is not required.

There are similar programs for technicians. Numerous organizations offer certification programs that examine your knowledge and experience to certify you as knowledgeable and proficient in your field. Generic certifications are offered by organizations like the International Society for the Certification of Electronic Technicians (ISCET) and the Electronic Technicians Association–International (ETA-I).

The Federal Communications Commission (FCC) also offers its popular general radiotelephone operators license (GROL) that is obtained by passing a comprehensive exam on electronic fundamentals, communications techniques, and FCC rules and regulations. It is required for working on certain types of radio equipment but is also useful as a job-getting credential. There are also many specialty certifications in the fields of communications and industrial control. After you get your AAS degree, a good next step is a license or certification that will go a long way to giving you new knowledge as you prepare for the exams but also a great credential that is appreciated by many employers.

Hobby and Personal Experimentation Finally, don't forget that just enjoying electronics as a hobby can lead to learning and experience. While you cannot really document such involvement or claim it as experience, the knowledge and skills you gain are invaluable and show up in your work. So don't hesitate to build your own lab bench and build kits and things of your own design. Become involved with embedded controllers, personal computers, robots, radio, audio, video, or whatever interests you.

1.6 The Major Employers

The overall structure of the electronic industry is shown in Fig. 1-4. The four major segments of the industry are manufacturers, resellers, service organizations, and end users.

Manufacturers

It all begins, of course, with customer needs. Manufacturers translate customer needs into products and purchase components and materials from other companies to use in creating the products. Note that there are three types of manufacturers: component, equipment, and system manufacturers. Component manufacturers buy the raw materials like copper and other metals, plastic, and chemicals to create the

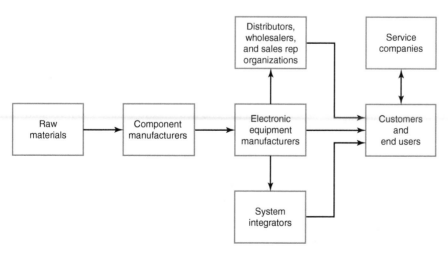

Figure 1-4 The structure of the electronics industry.

various resistors, capacitors, inductors, and transformers. Semiconductor manufacturers buy silicon and other materials like germanium, gallium, arsenic, phosphorus, indium, and other chemicals to make the transistors, diodes, and integrated circuits.

Then there are the equipment manufacturers that make complete products, which may be computers, cell phones, TV sets, cable boxes, military radios and radars, or automobile components like ignitions, fuel injectors, and fuel control computers. You will often hear these companies referred to as original equipment manufacturers (OEMs). Engineers design the products and manufacturing produces them. There are jobs for engineers, technicians, production workers, salespeople, field service personnel, technical writers, and trainers.

The final category is system manufacturers or integrators that put together larger, more complex systems like processes controls for petroleum processing, military aircraft, or satellite systems like GPS. Other system examples are air traffic control, broadband Internet access, wireless base stations, and electrical utility monitoring and control. Again, there are many jobs for engineers and technicians.

Resellers

Manufacturers that do not sell products directly to the end users sell the products to reselling organizations, which in turn sell them to the end user. For example, a manufacturer of marine communication equipment may not sell directly to a boat owner but instead to a regional distributor or marine electronics store or shop. This shop not only sells the equipment but also takes care of installation, service, and repairs. A cellular telephone or copy machine manufacturer also typically sells to a distributor or dealer that takes care of sales and service. Most of the jobs available in the reselling segment of the industry are in sales, service, and training.

Other sales organizations are sales representatives that sell components or equipment, or system integrators that buy equipment from others and assemble it into a more complex system for a specific application.

Service Organizations

These companies usually perform some kind of service, such as repair, installation, or maintenance. One example is an avionics company that does installation or service work on electronic equipment for private planes. Another is a systems integrator, a company that designs and assembles a piece of communication equipment or more often an entire system by using the products of other companies. Systems integrators put together systems to meet special needs and customize existing systems for particular jobs. Best Buy and similar retail organizations also perform service and repair.

End Users

The end user is the ultimate customer—and a major employer. Today, almost every person and organization is an end user of electronic equipment. The major categories of end users are:

- Consumers
- Government (national, state, county, and city)
- Military
- Transportation (airlines, railroads, trucking, and shipping)
- Education (schools, colleges, and universities)
- Hospitals and health care organizations
- General business
- Industry, manufacturing, process control, and automation.
- Telecommunications (telephone, broadcast, satellite, cellular, and networking)

What you will most likely find is that most of the good electronic technician jobs are not directly in the electronics industry itself but in the end user category.

1.7 Where Are You Headed?

Hopefully this chapter has given you some ideas and at least a working knowledge of the industry. And if you did not have any idea of what you wanted to do in electronics, perhaps you now have a better feel for what is available and what you may be doing on the job. If not, then continue to explore the field on your own, so when you do graduate from college, you will have a starting point. Once you get some real-world experience, you may want to change and you will readily identify new jobs and opportunities as you work.

One key to success in electronics is to find the most promising emerging technologies and find or prepare for jobs in those areas where the growth will be greatest. When growth is fast, many jobs open up as do opportunities for learning and advancement. Some of the more promising areas offering potential future growth are:

- *Alternative energy.* Solar, wind, geothermal, and other so-called green energy jobs are sparse now but slowly increasing and will ultimately offer some interesting new opportunities.
- *Biomedical.* The health care industry is huge and still growing. Electronic equipment is an enormous part of this industry.
- *Wireless.* The cellular industry continues its amazing growth with many opportunities.
- *Broadband Internet connectivity.* Cable TV and broadband wireless services are also growing. Fiber optics is a continuing growth area.

- *Electric utilities.* Retiring baby-boom generation personnel are leaving this legacy field wide open. The emergence of the smart grid and alternative energy sources makes electric utility jobs more exciting than ever.

1.8 How Electronic Equipment and Circuits Work

The study of electronics is one of learning electrical theories, electronic device characteristics, and circuit operation. But before venturing into those details, here is a simplified overview of how all electronic gadgets work.

Figure 1-5 shows the big picture. It is a relatively simple concept overall but becomes more complex as you dig deeper into the various elements. Inputs that are electrical signals representing some type of information—such as voice, video, sensor data, computer data, or other intelligence—are applied to circuits or equipment to be processed. These signals are voltages. A voltage is an electrical quantity that causes current to flow. The overall goal in electronics is to create the input voltages and then process them into other voltages called outputs. The result is some useful end result.

The processing takes many forms. Some common processes are amplification, attenuation, filtering, computation, conversion, decision making, interpretation, or translation. The process then generates new output signals that do something useful.

A simple example is given in Fig. 1-6. This public address system allows sound to be distributed over a wider area than

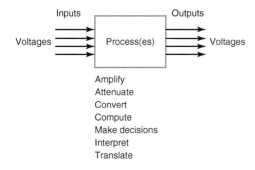

Figure 1-5   A model of all electronic circuits or equipment.

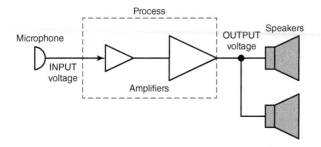

Figure 1-6   A public address system.

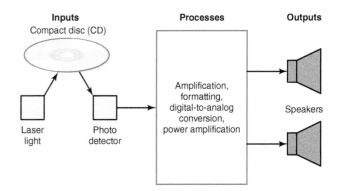

Figure 1-7   A compact disc (CD) player.

that normally covered by a human voice. A person speaks into a microphone. The microphone is a sensor that generates electric voltage that represents the voice. The voice voltage is amplified by several amplifier circuits, and a larger stronger voice signal is generated. This output signal is applied to one or more speakers. The speaker is a transducer that converts the signal into sound waves.

Another example is shown in Fig. 1-7. A compact disc (CD) player gets its inputs from a compact disk which has embedded music or other sounds. A laser light is shined on the bottom of the spinning CD, and reflections from the embedded music produce digital or pulse signals in the photo detector. These signals are then amplified, converted, and translated into the audio signals that drive the speakers or headphones. A DVD player works the same way, although the information on the DVD disc includes video and audio. The outputs are audio to speakers and video to a liquid-crystal display (LCD) or other TV screen.

A computer or laptop is a good example. See Fig. 1-8. The inputs to the computer are voltages developed from keyboards, mouse, disk drives, digital cameras, microphones, video cameras, or the Internet. These inputs are stored in a memory and processed by the computer in some way. The processing is defined by software. Software consists of many programs that define how to process the inputs and create new outputs. These programs are also stored in a memory. The computer then generates output signals that drive the LCD screen, speakers, disk drives, a printer, or other peripheral device. The computer also works with modems and interfaces to connect to networks and the Internet. The diagram in Fig. 1-8 is also representative of the new tablet computers.

An industrial control example is given in Fig. 1-9. A tank holds a liquid for some type of chemical process that is part of manufacturing an end product. The liquid must be kept at a specific temperature, so a heating element is attached to the bottom of the tank and a temperature sensor is used to

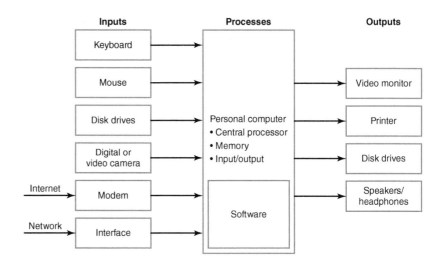

Figure 1-8 A complete example of the electronics model.

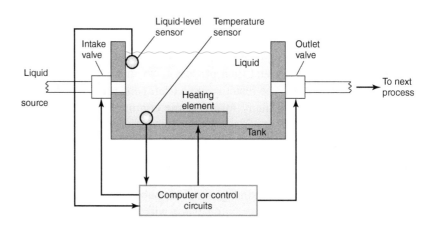

Figure 1-9 An industrial control example of the electronics model.

measure the temperature. The tank also has an output valve that can be opened to allow the liquid to pass on to the next process. An input valve lets more liquid into the tank as it drains. A liquid-level sensor is used to detect when the tank is full.

In this example, the inputs come from the temperature sensor and liquid-level sensor. The outputs are the heating element, input valve, and the output valve. These inputs and outputs attach to a computer or some specialized control circuit. If the liquid-level sensor detects that the tank is not full, it tells the control circuit to open the input valve and let the liquid in. The input valve is closed when the tank is full. The process here is decision making based on the liquid level.

When the tank is full, the controller next reads the temperature. If the liquid is not hot enough, it generates a signal that turns on the heating element until the desired higher temperature is reached. The control circuits turn off the heater. Finally, a signal to the output valve opens it to let the liquid flow to the next stage of the process. All the processing is built into a program in a computer or an electronic circuit.

1.9 An Electrical View of Electronics

Another way to view electronic circuits is shown in Fig. 1-10. It begins with a voltage source. Remember that a voltage is a form of electric energy that causes current to flow. The

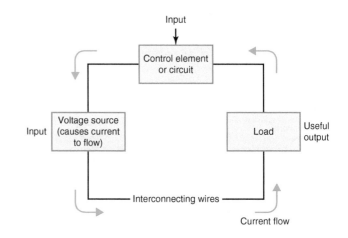

Figure 1-10 An electric current model of electronics.

voltage is an input. Current is electrons, subatomic particles that move through wires and electric components. The current flows through a load that produces the desired output. Some form of control element or circuit is used to vary the current in some way to produce the desired output. Another input causes the desired control. As indicated earlier, the whole objective of electronic circuits is to use a voltage to create current that is then controlled in a specific way to produce an output in the load.

Some simple examples are given in Fig. 1-11. Figure 1-11a shows a flashlight. The voltage comes from a battery. The load is a light-emitting diode (LED). A simple ON-OFF switch is the control element. In Fig. 1-11b, the standard alternating current (ac) voltage from a wall outlet is the voltage source. The load is the motor on an electric drill. The speed is controlled by an electric circuit that varies the current in the motor.

While these are simple examples, they illustrate the concept. Remember, all electronic circuits operate this way. And electronic equipment is made up of many such circuits operating concurrently to do more complex things.

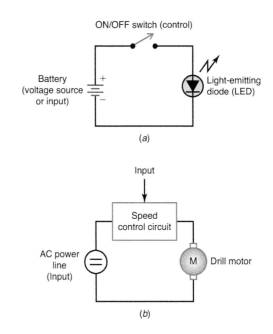

Figure 1-11 Sample circuit examples. (a) Flashlight. (b) Electric drill.

CHAPTER 1 REVIEW QUESTIONS

1. Which of the following is not one of the major segments of electronics?
   a. Medical electronics.
   b. Communications.
   c. Instrumentation.
   d. Computers.

2. Which of the following is the oldest segment of electronics?
   a. Control.
   b. Communications.
   c. Instrumentation.
   d. Computers.

3. Which of the following is the largest segment of electronics?
   a. Components.
   b. Communications.
   c. Instrumentation.
   d. Control.

4. Which of the following is not used in military electronics?
   a. Computers.
   b. Communications.
   c. Control.
   d. All are used.

5. An integrated circuit is a component.
   a. True.
   b. False.

6. The duties of a technician do not usually involve
   a. troubleshooting.
   b. installation.
   c. analysis and design.
   d. equipment testing.

7. The main duties of an engineer are
   a. equipment maintenance.
   b. troubleshooting.
   c. design and circuit analysis.
   d. equipment operation.

8. An engineer requires at least which degree for a job?
   a. Associate's.
   b. Bachelor's.
   c. Master's.
   d. PhD.

9. The primary degree of technician jobs is the
   a. associate's.
   b. bachelor's.
   c. master's.
   d. high school's.

10. A core difference between technician and engineer education is primarily
    a. the humanities.
    b. management training.
    c. math and science.
    d. electronics.

11. A graduate with an AAS degree in electronics technology can transfer directly to a BSEE program.
    a. True.
    b. False.

12. What must you do to stay competent and employable electronics?
    a. Get a master's degree.
    b. Work more than three jobs in your career.
    c. Find a mentor.
    d. Engage in some form of continuing education.

13. Which of the following in not a type of process that an electronic input signal may encounter?
    a. Stretching.
    b. Amplification.
    c. Filtering.
    d. Conversion.

14. The result of electronic processing is
    a. new inputs.
    b. removal of inputs.
    c. new outputs.
    d. a change in process.

15. What causes current to flow?
    a. Electric power.
    b. Voltage.
    c. Electrons.
    d. Magnetism.

16. Current flow is
    a. like atoms in motion.
    b. molecules.
    c. liquid atoms.
    d. moving electrons.

17. Which is not a major part of a simple electrical system?
    a. Load.
    b. Control element.
    c. Voltage source.
    d. Protection device.

18. What is a good supplement to an AAS degree in getting a job in electronics?
    a. A second AAS degree.
    b. A license or certification.
    c. A bachelor's degree.
    d. Any job experience.

19. A BSET degree is considered as part of which field of education?
    a. Technology.
    b. Engineering.
    c. Business.
    d. Science.

20. What is your best immediate source of learning in electronics?
    a. Books.
    b. Magazines.
    c. The Internet.
    d. Other people.

 ## CHAPTER 1 ESSAY QUESTIONS

1. What do you think is the *single* most important electronic invention?

2. What is your favorite electronic product?

3. What electronic product could you not do without?

4. Of the major segments of electronics, which interests you the most?

5. Would you rather be an engineer or a technician? Why?

6. Do you prefer hands-on work with electronic equipment or more abstract thinking about and analyzing of electronic products?

7. Does the business side of electronics (finance, accounting, economics, marketing, management, etc.) interest you? Why?

8. Which of the newer growth segments of electronics interests you most? Why?

9. Name the inputs, outputs, and main processes that take place in a smartphone like an Apple iPhone.

10. What is your main goal in pursuing a career in electronics? A good job, money, interest, security, fascination, contribute to society, or what?

11. What is your electronic hobby?

12. What hobby would you like to pursue?

Chapter 2

Electricity

Learning Outcomes

After studying this chapter, you should be able to:

> *List* the two basic particles of electric charge.

> *Describe* the basic structure of the atom.

> *Define* the terms *conductor, insulator,* and *semiconductor* and give examples of each.

> *Define* the coulomb unit of electric charge.

> *Define* potential difference and voltage and list the unit of each.

> *Define* current and list its unit of measure.

> *Define* resistance and conductance and list the unit of each.

> *List* three important characteristics of an electric circuit.

> *Define* the difference between electron flow and conventional current.

> *Describe* the difference between direct and alternating current.

We see applications of electricity all around us, especially in the electronic products we own and operate every day. For example, we depend on electricity for lighting, heating, and air-conditioning and for the operation of our vehicles, cell phones, appliances, computers, and home entertainment systems to name a few. The applications of electricity are extensive and almost limitless to the imagination.

Although there are many applications of electricity, electricity itself can be explained in terms of electric charge, voltage, and current. In this chapter, you will be introduced to the basic concepts of electricity which include a discussion of the following topics: basic atomic structure, the coulomb unit of electric charge, the volt unit of potential difference, the ampere unit of current, and the ohm unit of resistance. You will also be introduced to conductors, semiconductors, insulators, and the basic characteristics of an electric circuit

2.1 Negative and Positive Polarities

We see the effects of electricity in a battery, static charge, lightning, radio, television, and many other applications. What do they all have in common that is electrical in nature? The answer is basic particles of electric charge with opposite *polarities*. All the materials we know, including solids, liquids, and gases, contain two basic particles of electric charge: the *electron* and the *proton*. An electron is the smallest amount of electric charge having the characteristic called *negative polarity*. The proton is a basic particle with *positive polarity*.

The negative and positive polarities indicate two opposite characteristics that seem to be fundamental in all physical applications. Just as magnets have north and south poles, electric charges have the opposite polarities labeled negative and positive. The opposing characteristics provide a method of balancing one against the other to explain different physical effects.

It is the arrangement of electrons and protons as basic particles of electricity that determines the electrical characteristics of all substances. As an example, this paper has electrons and protons in it. There is no evidence of electricity, though, because the number of electrons equals the number of protons. In that case, the opposite electrical forces cancel, making the paper electrically neutral. The neutral condition means that opposing forces are exactly balanced, without any net effect either way.

When we want to use the electrical forces associated with the negative and positive charges in all matter, work must be done to separate the electrons and protons. Changing the balance of forces produces evidence of electricity. A battery, for instance, can do electrical work because its chemical energy separates electric charges to produce an excess of electrons at its negative terminal and an excess of protons at its positive terminal. With separate and opposite charges at the two terminals, electric energy can be supplied to a circuit connected to the battery. Fig. 2-1 shows a battery with its negative (−) and positive (+) terminals marked to emphasize the two opposite polarities.

2.2 Electrons and Protons in the Atom

Although there are any number of possible methods by which electrons and protons might be grouped, they assemble in specific atomic combinations for a stable arrangement. (An atom is the smallest particle of the basic elements which forms the physical substances we know as solids, liquids, and gases.) Each stable combination of electrons and protons makes one particular type of atom. For example, Fig. 2-2 illustrates the electron and proton structure of one atom of the gas, hydrogen. This atom consists of a central mass called the *nucleus* and one electron outside. The proton in the nucleus makes it the massive and stable part of the atom because a proton is 1840 times heavier than an electron.

In Fig. 2-2, the one electron in the hydrogen atom is shown in an orbital ring around the nucleus. To account for the electrical stability of the atom, we can consider the electron as spinning around the nucleus, as planets revolve around the sun. Then the electrical force attracting the electrons in toward the nucleus is balanced by the mechanical force outward on the rotating electron. As a result, the electron stays in its orbit around the nucleus.

In an atom that has more electrons and protons than hydrogen, all protons are in the nucleus, and all the electrons are in one or more outside rings. For example, the carbon atom illustrated in Fig. 2-3a has six protons in the nucleus and six electrons in two outside rings. The total number of electrons in the outside rings must equal the number of protons in the nucleus in a neutral atom.

The distribution of electrons in the orbital rings determines the atom's electrical stability. Especially important is the number of electrons in the ring farthest from the nucleus. This outermost ring requires eight electrons for stability, except when there is only one ring, which has a maximum of two electrons.

In the carbon atom in Fig. 2-3a, with six electrons, there are just two electrons in the first ring because two is its maximum number. The remaining four electrons are in the second ring, which can have a maximum of eight electrons.

As another example, the copper atom in Fig. 2-3b has only one electron in the last ring, which can include eight electrons. Therefore, the outside ring of the copper atom is less stable than the outside ring of the carbon atom.

Positive (+)   Negative (−)

Figure 2-1   Positive and negative polarities for the voltage output of a typical battery.

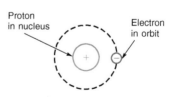

Proton in nucleus   Electron in orbit

Figure 2-2   Electron and proton in hydrogen (H) atom.

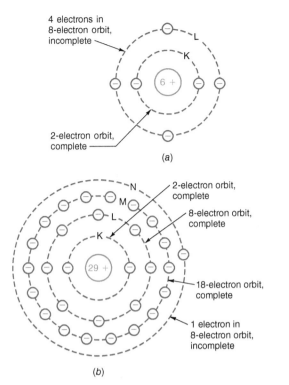

4 electrons in
8-electron orbit,
incomplete

L

K

6 +

2-electron orbit,
complete

(a)

N
M
L
K

29 +

2-electron orbit,
complete

8-electron orbit,
complete

18-electron orbit,
complete

1 electron in
8-electron orbit,
incomplete

(b)

Figure 2-3 Atomic structure showing the nucleus and its orbital rings of electrons. (a) Carbon (C) atom has six orbital electrons to balance six protons in nucleus. (b) Copper (Cu) atom has 29 protons in nucleus and 29 orbital electrons.

When many atoms are close together in a copper wire, the outermost orbital electron of each copper atom can easily break free from its home or parent atom. These electrons then can migrate easily from one atom to another at random. Such electrons that can move freely from one atom to the next are called *free electrons*. This freedom accounts for the ability of copper to conduct electricity very easily. It is the movement of free electrons that provides electric current in a metal conductor.

The net effect in the wire itself without any applied voltage, however, is zero because of the random motion of the free electrons. When voltage is applied, it forces all the free electrons to move in the same direction to produce electron flow, which is an electric current.

Conductors, Insulators, and Semiconductors

When electrons can move easily from atom to atom in a material, the material is a *conductor*. In general, all metals are good conductors, with silver the best and copper second. Their atomic structure allows free movement of the outermost orbital electrons. Copper wire is generally used for practical conductors because it costs much less than silver. The purpose of using conductors is to allow electric current to flow with minimum opposition.

The wire conductor is used only to deliver current produced by the voltage source to a device that needs the current to function. As an example, a bulb lights only when current flows through the filament.

A material with atoms in which the electrons tend to stay in their own orbits is an *insulator* because it cannot conduct electricity very easily. However, insulators can hold or store electricity better than conductors. An insulating material, such as glass, plastic, rubber, paper, air, or mica, is also called a *dielectric,* meaning it can store electric charge.

Insulators can be useful when it is necessary to prevent current flow. In addition, for applications requiring the storage of electric charge, as in capacitors, a dielectric material must be used because a good conductor cannot store any charge.

Carbon can be considered a semiconductor, conducting less than metal conductors but more than insulators. In the same group are germanium and silicon, which are commonly used for transistors and other semiconductor components. Practically all transistors are made of silicon.

Elements

The combinations of electrons and protons forming stable atomic structures result in different kinds of elementary substances having specific characteristics. A few familiar examples are the elements hydrogen, oxygen, carbon, copper, and iron. An *element* is defined as a substance that cannot be decomposed any further by chemical action. The atom is the smallest particle of an element that still has the same characteristics as the element. *Atom* is a Greek word meaning a "particle too small to be subdivided." As an example of the fact that atoms are too small to be visible, a particle of carbon the size of a pinpoint contains many billions of atoms. The electrons and protons within the atom are even smaller.

Table 2-1 lists some more examples of elements. These are just a few out of a total of 112. Notice how the elements are grouped. The metals listed across the top row are all good conductors of electricity. Each has an atomic structure with an unstable outside ring that allows many free electrons.

Semiconductors have four electrons in the outermost ring. This means that they neither gain nor lose electrons but share them with similar atoms. The reason is that four is exactly halfway to the stable condition of eight electrons in the outside ring.

The inert gas neon has a complete outside ring of eight electrons, which makes it chemically inactive. Remember that eight electrons in the outside ring is a stable structure.

Table 2-1 Examples of the Chemical Elements

Group	Element	Symbol	Atomic Number	Electron Valence
Metal conductors, in order of conductance	Silver	Ag	47	+1
	Copper	Cu	29	+1*
	Gold	Au	79	+1*
	Aluminum	Al	13	+3
	Iron	Fe	26	+2*
Semiconductors	Carbon	C	6	±4
	Silicon	Si	14	±4
	Germanium	Ge	32	±4
Active gases	Hydrogen	H	1	±1
	Oxygen	O	8	−2
Inert gases	Helium	He	2	0
	Neon	Ne	10	0

*Some metals have more than one valence number in forming chemical compounds. Examples are cuprous or cupric copper, ferrous or ferric iron, and aurous or auric gold.

Molecules and Compounds

A group of two or more atoms forms a molecule. For instance, two atoms of hydrogen (H) form a hydrogen molecule (H_2). When hydrogen unites chemically with oxygen, the result is water (H_2O), which is a compound. A compound, then, consists of two or more elements. The molecule is the smallest unit of a compound with the same chemical characteristics. We can have molecules for either elements or compounds. However, atoms exist only for elements.

2.3 Structure of the Atom

As illustrated in Figs. 2-2 and 2-3, the nucleus contains protons for all the positive charge in the atom. The number of protons in the nucleus is equal to the number of planetary electrons. Then the positive and negative charges are balanced because the proton and electron have equal and opposite charges. The orbits for the planetary electrons are also called *shells* or *energy levels*.

Atomic Number

The atomic number gives the number of protons required in the atom for each element. For the hydrogen atom in Fig. 2-2, the atomic number is 1, which means that the nucleus has one proton (balanced by one orbital electron). Similarly, the carbon atom in Fig. 2-3 with atomic number 6 has six protons in the nucleus (and six orbital electrons). The copper atom has 29 protons (and 29 electrons) because its atomic number is 29. The atomic number listed for each of the elements in Table 2-1 indicates the atomic structure.

Orbital Rings

The planetary electrons are in successive shells, or orbital rings, called K, L, M, N, O, P, and Q at increasing distances outward from the nucleus. Each shell has a maximum

Table 2-2 Shells of Orbital Electrons in the Atom

Shell	Maximum Electrons	Inert Gas
K	2	Helium
L	8	Neon
M	8 (up to calcium) or 18	Argon
N	8, 18, or 32	Krypton
O	8 or 18	Xenon
P	8 or 18	Radon
Q	8	—

number of electrons for stability. As indicated in Table 2-2, these stable shells correspond to inert gases, such as helium and neon.

The K shell, closest to the nucleus, is stable with two electrons, corresponding to the atomic structure for the inert gas, helium. Once the stable number of electrons has filled a shell, it cannot take any more electrons. The atomic structure with all its shells filled to the maximum number for stability corresponds to an inert gas.

Elements with a higher atomic number have more planetary electrons. These are in successive shells, tending to form the structure of the next inert gas in the periodic table. (The periodic table is a very useful grouping of all elements according to their chemical properties.) After the K shell has been filled with two electrons, the L shell can take up to eight electrons. Ten electrons filling the K and L shells is the atomic structure for the inert gas, neon.

The maximum number of electrons in the remaining shells can be 8, 18, or 32 for different elements, depending on their place in the periodic table. The maximum for an outermost shell, though, is always eight.

To illustrate these rules, we can use the copper atom in Fig. 2-3b as an example. There are 29 protons in the nucleus balanced by 29 planetary electrons. This number of electrons fills the K shell with two electrons, corresponding to the helium atom, and the L shell with eight electrons. The 10 electrons in these two shells correspond to the neon atom, which has an atomic number of 10. The remaining 19 electrons for the copper atom then fill the M shell with 18 electrons and 1 electron in the outermost N shell. These values can be summarized as follows:

K shell = 2 electrons
L shell = 8 electrons
M shell = 18 electrons
N shell = 1 electron
   Total = 29 electrons

For most elements, we can use the rule that the maximum number of electrons in a filled inner shell equals $2n^2$, where n is the shell number in sequential order outward from the nucleus. Then the maximum number of electrons in the first shell is $2 \times 1 = 2$; for the second shell $2 \times 2^2 = 8$, for the third shell $2 \times 3^2 = 18$, and for the fourth shell $2 \times 4^2 = 32$. These values apply only to an inner shell that is filled with its maximum number of electrons.

Electron Valence

The electron valence is the value of the number of electrons in an incomplete outermost shell (valence shell). A completed outer shell has a valence of zero. Copper, for instance, has a valence of one, as there is one electron in the last shell, after the inner shells have been completed with their stable number. Similarly, hydrogen has a valence of one, and carbon has a valence of four. The number of outer electrons is considered positive valence because these electrons are in addition to the stable shells.

Except for H and He, the goal of valence is eight for all atoms, as each tends to form the stable structure of eight electrons in the outside ring. For this reason, valence can also be considered the number of electrons in the outside ring needed to make eight. This value is the negative valence. As examples, the valence of copper can be considered +1 or −7; carbon has the valence of ±4. The inert gases have zero valence because they all have complete outer shells.

The valence indicates how easily the atom can gain or lose electrons. For instance, atoms with a valence of +1 can lose this one outside electron, especially to atoms with a valence of +7 or −1, which need one electron to complete the outside shell with eight electrons.

Subshells

Although not shown in the illustrations, all shells except K are divided into subshells. This subdivision accounts for different types of orbits in the same shell. For instance, electrons in one subshell may have elliptical orbits, and other electrons in the same main shell have circular orbits. The subshells indicate magnetic properties of the atom.

Particles in the Nucleus

A stable nucleus (that is, one that is not radioactive) contains protons and sometimes neutrons. The neutron is electrically neutral (it has no net charge). Its mass is almost the same as that of a proton.

A proton has the positive charge of a hydrogen nucleus. The charge is the same as that of an orbital electron but of opposite polarity. There are no electrons in the nucleus.

2.4 The Coulomb Unit of Electric Charge

If you rub a hard rubber pen or comb on a sheet of paper, the rubber will attract a corner of the paper if it is free to move easily. The paper and rubber then give evidence of a static electric charge. The work of rubbing resulted in separating electrons and protons to produce a charge of excess electrons on the surface of the rubber and a charge of excess protons on the paper.

Because paper and rubber are dielectric materials, they hold their extra electrons or protons. As a result, the paper and rubber are no longer neutral, but each has an electric charge. The resultant electric charges provide the force of attraction between the rubber and the paper. This mechanical force of attraction or repulsion between charges is the fundamental method by which electricity makes itself evident.

Any charge is an example of *static electricity* because the electrons or protons are not in motion. There are many examples. When you walk across a wool rug, your body becomes charged with an excess of electrons. Similarly, silk, fur, and glass can be rubbed to produce a static charge. This effect is more evident in dry weather, because a moist dielectric does not hold its charge so well. Also, plastic materials can be charged easily, which is why thin, lightweight plastics seem to stick to everything.

The charge of many billions of electrons or protons is necessary for common applications of electricity. Therefore, it is convenient to define a practical unit called the *coulomb* (C) as equal to the charge of 6.25×10^{18} electrons or protons stored in a dielectric (see Fig. 2-4). The analysis of static charges and their forces is called *electrostatics*.

The symbol for electric charge is Q or q, standing for quantity. For instance, a charge of 6.25×10^{18} electrons is stated as $Q = 1$ C. This unit is named after Charles A. Coulomb (1736–1806), a French physicist, who measured the force between charges.

Negative and Positive Polarities

Historically, negative polarity has been assigned to the static charge produced on rubber, amber, and resinous materials in

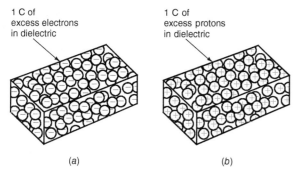

(a)   (b)

Figure 2-4 The coulomb (C) unit of electric charge. (a) Quantity of 6.25×10^{18} excess electrons for a negative charge of 1 C. (b) Same amount of protons for a positive charge of 1 C, caused by removing electrons from neutral atoms.

general. Positive polarity refers to the static charge produced on glass and other vitreous materials. On this basis, the electrons in all atoms are basic particles of negative charge because their polarity is the same as the charge on rubber. Protons have positive charge because the polarity is the same as the charge on glass.

Charges of Opposite Polarity Attract

If two small charged bodies of light weight are mounted so that they are free to move easily and are placed close to each other, one can be attracted to the other when the two charges have opposite polarity (Fig. 2-5a). In terms of electrons and protons, they tend to be attracted to each other by the force of attraction between opposite charges. Furthermore, the weight of an electron is only about $\frac{1}{1840}$ the weight of a proton. As a result, the force of attraction tends to make electrons move to protons.

Charges of the Same Polarity Repel

In Fig. 2-5b and c, it is shown that when the two bodies have an equal amount of charge with the same polarity, they repel each other. The two negative charges repel in Fig. 2-5b, and two positive charges of the same value repel each other in Fig. 2-5c.

Polarity of a Charge

An electric charge must have either negative or positive polarity, labeled $-Q$ or $+Q$, with an excess of either electrons or protons. A neutral condition is considered zero charge.

Note that we generally consider that the electrons move, rather than heavier protons. However, a loss of a given number of electrons is equivalent to a gain of the same number of protons.

Charge of an Electron

The charge of a single electron, designated Q_e, is 0.16×10^{-18} C. This value is the reciprocal of 6.25×10^{18} electrons, which is the number of electrons in 1 coulomb of charge. Expressed mathematically,

$$-Q_e = 0.16 \times 10^{-18} \text{ C}$$

($-Q_e$ denotes that the charge of the electron is negative.)

It is important to note that the charge of a single proton, designated Q_P, is also equal to 0.16×10^{-18} C. However, its polarity is positive instead of negative.

In some cases, the charge of a single electron or proton will be expressed in scientific notation. In this case, $-Q_e = 1.6 \times 10^{-19}$ C. It is for convenience only that Q_e or Q_P is sometimes expressed as 0.16×10^{-18} C instead of 1.6×10^{-19} C. The convenience lies in the fact that 0.16 is the reciprocal of 6.25 and 10^{-18} is the reciprocal of 10^{18}.

The Electric Field of a Static Charge

The ability of an electric charge to attract or repel another charge is a physical force. To help visualize this effect, lines of force are used, as shown in Fig. 2-6. All the lines form the electric field. The lines and the field are imaginary, since they cannot be seen. Just as the field of the force of gravity is not visible, however, the resulting physical effects prove that the field is there.

Each line of force in Fig. 2-6 is directed outward to indicate repulsion of another charge in the field with the same polarity as Q, either positive or negative. The lines are shorter

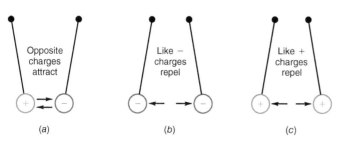

(a)   (b)   (c)

Figure 2-5 Physical force between electric charges. (a) Opposite charges attract. (b) Two negative charges repel each other. (c) Two positive charges repel.

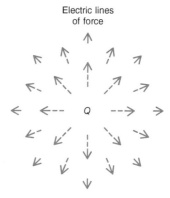

Figure 2-6 Arrows to indicate electric field around a stationary charge Q.

farther away from Q to indicate that the force decreases inversely as the square of the distance. The larger the charge, the greater the force. These relations describe Coulomb's law of electrostatics.

2.5 The Volt Unit of Potential Difference

Potential refers to the possibility of doing work. Any charge has the potential to do the work of moving another charge by either attraction or repulsion. When we consider two unlike charges, they have a *difference of potential.*

A charge is the result of work done in separating electrons and protons. Because of the separation, stress and strain are associated with opposite charges, since normally they would be balancing each other to produce a neutral condition. We could consider that the accumulated electrons are drawn tight and are straining themselves to be attracted toward protons to return to the neutral condition. Similarly, the work of producing the charge causes a condition of stress in the protons, which are trying to attract electrons and return to the neutral condition. Because of these forces, the charge of electrons or protons has potential because it is ready to give back the work put into producing the charge. The force between charges is in the electric field.

Potential between Different Charges

When one charge is different from the other, there must be a difference of potential between them. For instance, consider a positive charge of 3 C, shown at the right in Fig. 2-7a. The charge has a certain amount of potential, corresponding to the amount of work this charge can do. The work to be done is moving some electrons, as illustrated.

Assume that a charge of 1 C can move three electrons. Then the charge of +3 C can attract nine electrons toward the right. However, the charge of +1 C at the opposite side can attract three electrons toward the left. The net result, then, is that six electrons can be moved toward the right to the more positive charge.

In Fig. 2-7b, one charge is 2 C, and the other charge is neutral with 0 C. For the difference of 2 C, again 2 × 3 or 6 electrons can be attracted to the positive side.

In Fig. 2-7c, the difference between the charges is still 2 C. The +1 C attracts three electrons to the right side. The −1 C repels three electrons to the right side also. This effect is really the same as attracting six electrons.

Therefore, the net number of electrons moved in the direction of the more positive charge depends on the difference of potential between the two charges. This potential difference is the same for all three cases in Fig. 2-7. Potential difference is often abbreviated PD.

The only case without any potential difference between charges occurs when they both have the same polarity and are equal in amount. Then the repelling and attracting forces cancel, and no work can be done in moving electrons between the two identical charges.

The Volt Unit

The *volt unit* of potential difference is named after Alessandro Volta (1745–1827). Fundamentally, the volt is a measure of the amount of work or energy needed to move an electric charge. By definition, when 0.7376 foot-pound (ft · lb) of work is required to move 6.25×10^{18} electrons between two points, the potential difference between those two points is one volt. (Note that 6.25×10^{18} electrons make up one coulomb of charge.) The metric unit of work or energy is the joule (J). One joule is the same amount of work or energy as 0.7376 ft · lb. Therefore, we can say that the potential difference between two points is one volt when one joule of energy is expended in moving one coulomb of charge between those two points. Expressed as a formula, $1 \text{ V} = \dfrac{1 \text{ J}}{1 \text{ C}}$.

In electronics, potential difference is commonly referred to as voltage, with the symbol V. Remember, though, that voltage is the potential difference between two points and that two terminals are necessary for a potential difference to exist. A potential difference cannot exist at only one point!

Consider the 2.2-V lead-acid cell in Fig. 2-8a. Its output of 2.2 V means that this is the amount of potential difference between the two terminals. The lead-acid cell, then, is a voltage source, or a source of electromotive force (emf). The schematic symbol for a battery or dc voltage source is shown in Fig. 2-8b.

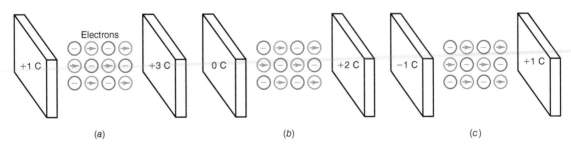

(a)   (b)   (c)

Figure 2-7 The amount of work required to move electrons between two charges depends on their difference of potential. This potential difference (PD) is equivalent for the examples in (a), (b), and (c).

$V = 2.2\ V$

(a)   (b)

Figure 2-8 Chemical cell as a voltage source. (a) Voltage output is the potential difference between the two terminals. (b) Schematic symbol of any dc voltage source with constant polarity. Longer line indicates positive side.

Sometimes the symbol E is used for emf, but the standard symbol V represents any potential difference. This applies either to the voltage generated by a source or to the voltage drop across a passive component such as a resistor.

It may be helpful to think of voltage as an electrical pressure or force. The higher the voltage, the more electrical pressure or force. The electrical pressure of voltage is in the form of the attraction and repulsion of an electric charge such as an electron.

The general equation for any voltage can be stated as

$$V = \frac{W}{Q} \qquad (2\text{-}1)$$

where V is the voltage in volts, W is the work or energy in joules, and Q is the charge in coulombs.

Let's take a look at an example.

EXAMPLE 2-1

What is the output voltage of a battery that expends 3.6 J of energy in moving 0.5 C of charge?

Answer: Use Equation (2-1).

$$V = \frac{W}{Q}$$

$$= \frac{3.6\ J}{0.5\ C}$$

$$= 7.2\ V$$

2.6 Charge in Motion Is Current

When the potential difference between two charges forces a third charge to move, the charge in motion is an *electric current*. To produce current, therefore, charge must be moved by a potential difference.

In solid materials, such as copper wire, free electrons are charges that can be forced to move with relative ease by a potential difference, since they require relatively little work to be moved. As illustrated in Fig. 2-9, if a potential difference is connected across two ends of a copper wire, the applied voltage forces the free electrons to move. This current is a drift of electrons, from the point of negative charge at one end, moving through the wire, and returning to the positive charge at the other end.

To illustrate the drift of free electrons through the wire shown in Fig. 2-9, each electron in the middle row is numbered, corresponding to a copper atom to which the free electron belongs. The electron at the left is labeled S to indicate that it comes from the negative charge of the source of potential difference. This one electron S is repelled from the negative charge $-Q$ at the left and is attracted by the positive charge $+Q$ at the right. Therefore, the potential difference of the voltage source can make electron S move toward atom 1. Now atom 1 has an extra electron. As a result, the free electron of atom 1 can then move to atom 2. In this way, there is a drift of free electrons from atom to atom. The final result is that the one free electron labeled 8 at the extreme right in Fig. 2-9 moves out from the wire to return to the positive charge of the voltage source.

Considering this case of just one electron moving, note that the electron returning to the positive side of the voltage source is not the electron labeled S that left the negative side. All electrons are the same, however, and have the same charge. Therefore, the drift of free electrons resulted in the charge of one electron moving through the wire. This charge in motion is the current. With more electrons drifting through the wire, the charge of many electrons moves, resulting in more current.

The current is a continuous flow of electrons. Only the electrons move, not the potential difference. For ordinary

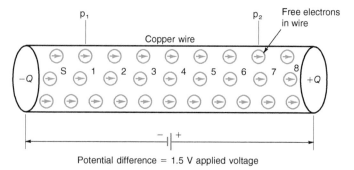

Potential difference = 1.5 V applied voltage

Figure 2-9 Potential difference across two ends of wire conductor causes drift of free electrons throughout the wire to produce electric current.

applications, where the wires are not long lines, the potential difference produces current instantaneously through the entire length of wire. Furthermore, the current must be the same at all points of the wire at any time.

Potential Difference Is Necessary to Produce Current

The number of free electrons that can be forced to drift through a wire to produce a moving charge depends on the amount of potential difference across the wire. With more applied voltage, the forces of attraction and repulsion can make more free electrons drift, producing more charge in motion. A larger amount of charge moving during a given period of time means a higher value of current. Less applied voltage across the same wire results in a smaller amount of charge in motion, which is a smaller value of current. With zero potential difference across the wire, there is no current.

Two cases of zero potential difference and no current can be considered to emphasize that potential difference is needed to produce current. Assume that the copper wire is by itself, not connected to any voltage source, so that there is no potential difference across the wire. The free electrons in the wire can move from atom to atom, but this motion is random, without any organized drift through the wire. If the wire is considered as a whole, from one end to the other, the current is zero.

As another example, suppose that the two ends of the wire have the same potential. Then free electrons cannot move to either end because both ends have the same force and there is no current through the wire. A practical example of this case of zero potential difference would be to connect both ends of the wire to just one terminal of a battery. Each end of the wire would have the same potential, and there would be no current. The conclusion, therefore, is that two connections to two points at different potentials are needed to produce a current.

The Ampere of Current

Since current is the movement of charge, the unit for stating the amount of current is defined in rate of flow of charge. When the charge moves at the rate of 6.25×10^{18} electrons flowing past a given point per second, the value of the current is one *ampere* (A). This is the same as one coulomb of charge per second. The *ampere unit* of current is named after André M. Ampère (1775–1836).

Referring back to Fig. 2-9, note that if 6.25×10^{18} free electrons move past p_1 in 1 second (s), the current is 1 A. Similarly, the current is 1 A at p_2 because the electron drift is the same throughout the wire. If twice as many electrons moved past either point in 1 s, the current would be 2 A.

The symbol for current is I or i for intensity, since the current is a measure of how intense or concentrated the electron flow is. Two amperes of current in a copper wire is a higher intensity than one ampere; a greater concentration of moving electrons results because of more electrons in motion. Sometimes current is called *amperage*. However, the current in electronic circuits is usually in smaller units, milliamperes and microamperes.

How Current Differs from Charge

Charge is a quantity of electricity accumulated in a dielectric, which is an insulator. The charge is static electricity, at rest, without any motion. When the charge moves, usually in a conductor, the current I indicates the intensity of the electricity in motion. This characteristic is a fundamental definition of current:

$$I = \frac{Q}{T} \qquad (2\text{-}2)$$

where I is the current in amperes, Q is in coulombs, and time T is in seconds. It does not matter whether the moving charge is positive or negative. The only question is how much charge moves and what its rate of motion is.

In terms of practical units,

$$1\ A = \frac{1\ C}{1\ s} \qquad (2\text{-}3)$$

One ampere of current results when one coulomb of charge moves past a given point in 1 s. In summary, Q represents a specific amount or quantity of electric charge, whereas the current I represents the rate at which the electric charge, such as electrons, is moving. The difference between electric charge and current is similar to the difference between miles and miles per hour.

EXAMPLE 2-2

The charge of 12 C moves past a given point every second. How much is the intensity of charge flow?

Answer:

$$I = \frac{Q}{T} = \frac{12\ C}{1\ s}$$
$$I = 12\ A$$

The fundamental definition of current can also be used to consider the charge as equal to the product of the current multiplied by the time. Or

$$Q = I \times T \qquad (2\text{-}4)$$

In terms of practical units,

$$1\ C = 1\ A \times 1\ s \qquad (2\text{-}5)$$

One coulomb of charge results when one ampere of current accumulates charge during one second. The charge is generally accumulated in the dielectric of a capacitor or at the electrodes of a battery.

For instance, we can have a dielectric connected to conductors with a current of 0.4 A. If the current can deposit electrons for 0.2 s, the accumulated charge in the dielectric will be

$$Q = I \times T = 0.4 \text{ A} \times 0.2 \text{ s}$$
$$Q = 0.08 \text{ C}$$

The formulas $Q = IT$ for charge and $I = Q/T$ for current illustrate the fundamental nature of Q as an accumulation of static charge in an insulator, whereas I measures the intensity of moving charges in a conductor. Furthermore, current I is different from voltage V. You can have V without I, but you cannot have current without an applied voltage.

The General Nature of Current

The moving charges that provide current in metal conductors such as copper wire are the free electrons of the copper atoms. In this case, the moving charges have negative polarity. The direction of motion between two terminals for this *electron current,* therefore, is toward the more positive end. It is important to note, however, that there are examples of positive charges in motion. Common applications include current in liquids, gases, and semiconductors. For the current resulting from the motion of positive charges, its direction is opposite from the direction of electron flow. Whether negative or positive charges move, though, the current is still defined fundamentally as Q/T. Note also that the current is provided by free charges, which are easily moved by an applied voltage.

2.7 Resistance Is Opposition to Current

The fact that a wire conducting current can become hot is evidence that the work done by the applied voltage in producing current must be accomplished against some form of opposition. This opposition, which limits the amount of current that can be produced by the applied voltage, is called *resistance.* Conductors have very little resistance; insulators have a large amount of resistance.

The atoms of a copper wire have a large number of free electrons, which can be moved easily by a potential difference. Therefore, the copper wire has little opposition to the flow of free electrons when voltage is applied, corresponding to low resistance.

Carbon, however, has fewer free electrons than copper. When the same amount of voltage is applied to carbon as to copper, fewer electrons will flow. Just as much current can be produced in carbon by applying more voltage. For the same current, though, the higher applied voltage means that more work is necessary, causing more heat. Carbon opposes the current more than copper, therefore, and has higher resistance.

The Ohm

The practical unit of resistance is the *ohm.* A resistance that develops 0.24 calorie of heat when one ampere of current flows

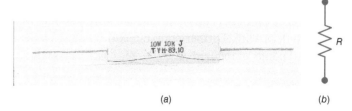

Figure 2-10   (a) Wire-wound type of resistor with cement coating for insulation. (b) Schematic symbol for any type of fixed resistor.

through it for one second has one ohm of opposition. The symbol for resistance is R. The abbreviation used for the ohm unit is the Greek letter *omega,* written as Ω. As an example of a low resistance, a good conductor such as copper wire can have a resistance of 0.01 Ω for a 1-ft length. The resistance-wire heating element in a 600-W 120-V toaster has a resistance of 24 Ω, and the tungsten filament in a 100-W 120-V lightbulb has a resistance of 144 Ω. The ohm unit is named after Georg Simon Ohm (1787–1854), a German physicist.

Figure 2-10a shows a wire-wound resistor. Resistors are also made with powdered carbon. They can be manufactured with values from a few ohms to millions of ohms.

In diagrams, resistance is indicated by a zigzag line, as shown by R in Fig. 2-10b.

Conductance

The opposite of resistance is *conductance.* The lower the resistance, the higher the conductance. Its symbol is G, and the unit is the *siemens* (S), named after Ernst von Siemens (1816–1892), a German inventor. (The old unit name for conductance is *mho,* which is *ohm* spelled backward.)

Specifically, G is the reciprocal of R, or $G = \frac{1}{R}$. Also, $R = \frac{1}{G}$.

EXAMPLE 2-3

Calculate the resistance for the following conductance values:
(a) 0.05 S (b) 0.1 S

Answer:

a.
$$R = \frac{1}{G}$$
$$= \frac{1}{0.05 \text{ S}}$$
$$= 20 \ \Omega$$

b.
$$R = \frac{1}{G}$$
$$= \frac{1}{0.1 \text{ S}}$$
$$= 10 \ \Omega$$

Notice that a higher value of conductance corresponds to a lower value of resistance.

EXAMPLE 2-4

Calculate the conductance for the following resistance values:
(a) 1 kΩ (b) 5 kΩ.

Answer:

a.
$$G = \frac{1}{R}$$
$$= \frac{1}{1000 \ \Omega}$$
$$= 0.001 \text{ S or } 1 \text{ mS}$$

b.
$$G = \frac{1}{R}$$
$$= \frac{1}{5000 \ \Omega}$$
$$= 0.0002 \text{ S or } 200 \ \mu\text{S}$$

Notice that a higher value of resistance corresponds to a lower value of conductance.

2.8 The Closed Circuit

In applications requiring current, the components are arranged in the form of a *circuit*, as shown in Fig. 2-11. A circuit can be defined as a path for current flow. The purpose of this circuit is to light the incandescent bulb. The bulb lights when the tungsten-filament wire inside is white hot, producing an incandescent glow.

The tungsten filament cannot produce current by itself. A source of potential difference is necessary. Since the battery produces a potential difference of 1.5 V across its two output terminals, this voltage is connected across the filament of the bulb by the two wires so that the applied voltage can produce current through the filament.

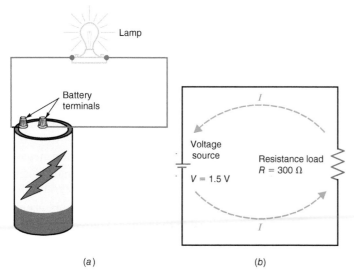

(a)   (b)

Figure 2-11 Example of an electric circuit with a battery as a voltage source connected to a lightbulb as a resistance. (*a*) Wiring diagram of the closed path for current. (*b*) Schematic diagram of the circuit.

In Fig. 2-11*b*, the schematic diagram of the circuit is shown. Here the components are represented by shorthand symbols. Note the symbols for the battery and resistance. The connecting wires are shown simply as straight lines because their resistance is small enough to be neglected. A resistance of less than 0.01 Ω for the wire is practically zero compared with the 300-Ω resistance of the bulb. If the resistance of the wire must be considered, the schematic diagram includes it as additional resistance in the same current path.

Note that the schematic diagram does not look like the physical layout of the circuit. The schematic shows only the symbols for the components and their electrical connections.

Any electric circuit has three important characteristics:

1. There must be a source of potential difference. Without the applied voltage, current cannot flow.

2. There must be a complete path for current flow, from one side of the applied voltage source, through the external circuit, and returning to the other side of the voltage source.

3. The current path normally has resistance. The resistance is in the circuit either to generate heat or limit the amount of current.

How the Voltage Is Different from the Current

It is the current that moves through the circuit. The potential difference does not move.

In Fig. 2-11, the voltage across the filament resistance makes electrons flow from one side to the other. While the current is flowing around the circuit, however, the potential difference remains across the filament to do the work of moving electrons through the resistance of the filament.

The circuit is redrawn in Fig. 2-12 to emphasize the comparison between *V* and *I*. The voltage is the potential difference across the two ends of the resistance. If you want to measure the PD, just connect the two leads of a voltmeter across the resistor. However, the current is the intensity of the electron flow past any one point in the circuit. Measuring the current is not as easy. You would have to break open the path at any point and then insert the current meter to complete the circuit.

The word *across* is used with voltage because it is the potential difference between two points. There cannot be a PD at one point. However, current can be considered at one point, as the motion of charges through that point.

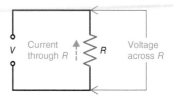

Figure 2-12 Comparison of voltage (*V*) across a resistance and the current (*I*) through *R*.

To illustrate the difference between *V* and *I* in another way, suppose that the circuit in Fig. 2-11 is opened by disconnecting the bulb. Now no current can flow because there is no closed path. Still, the battery has its potential difference. If you measure across the two terminals, the voltmeter will read 1.5 V even though the current is zero. This is like a battery sitting on a store shelf. Even though the battery is not producing current in a circuit, it still has a voltage output between its two terminals. This brings us to a very important conclusion: **Voltage can exist without current, but current cannot exist without voltage.**

The Voltage Source Maintains the Current

As current flows in a circuit, electrons leave the negative terminal of the cell or battery in Fig. 2-11, and the same number of free electrons in the conductor are returned to the positive terminal. As electrons are lost from the negative charge and gained by the positive charge, the two charges tend to neutralize each other. The chemical action inside the battery, however, continuously separates electrons and protons to maintain the negative and positive charges on the outside terminals that provide the potential difference. Otherwise, the current would neutralize the charges, resulting in no potential difference, and the current would stop. Therefore, the battery keeps the current flowing by maintaining the potential difference across the circuit. The battery is the voltage source for the circuit.

The Circuit Is a Load on the Voltage Source

We can consider the circuit as a means whereby the energy of the voltage source is carried by the current through the filament of the bulb, where the electric energy is used in producing heat energy. On this basis, the battery is the *source* in the circuit, since its voltage output represents the potential energy to be used. The part of the circuit connected to the voltage source is the *load resistance,* since it determines how much work the source will supply. In this case, the bulb's filament is the load resistance on the battery.

The current that flows through the load resistance is the *load current*. Note that a lower value of ohms for the load resistance corresponds to a higher load current. Unless noted otherwise, the term *load* by itself can be assumed generally to mean the load current. Therefore, a heavy or big load electrically means a high current load, corresponding to a large amount of work supplied by the source.

In summary, we can say that the closed circuit, normal circuit, or just a circuit is a closed path that has *V* to produce *I* with *R* to limit the amount of current. The circuit provides a means of using the energy of the battery as a voltage source. The battery has its potential difference *V* with or without the circuit. However, the battery alone is not doing

any work in producing load current. The bulb alone has resistance, but without current, the bulb does not light. With the circuit, the voltage source is used to produce current to light the bulb.

Open Circuit

When any part of the path is open or broken, the circuit is incomplete because there is no conducting path. The *open circuit* can be in the connecting wires or in the bulb's filament as the load resistance. The resistance of an open circuit is infinitely high. The result is no current in an open circuit.

Short Circuit

In this case, the voltage source has a closed path across its terminals, but the resistance is practically zero. The result is too much current in a *short circuit*. Usually, the short circuit is a bypass around the load resistance. For instance, a short across the tungsten filament of a bulb produces too much current in the connecting wires but no current through the bulb. Then the bulb is shorted out. The bulb is not damaged, but the connecting wires can become hot enough to burn unless the line has a fuse as a safety precaution against too much current.

2.9 The Direction of Current

Just as a voltage source has polarity, current has a direction. The reference is with respect to the positive and negative terminals of the voltage source. The direction of the current depends on whether we consider the flow of negative electrons or the motion of positive charges in the opposite direction.

Electron Flow

As shown in Fig. 2-13*a*, the direction of electron drift for the current *I* is out from the negative side of the voltage source. The *I* flows through the external circuit with *R* and returns to the positive side of *V*. Note that this direction from the negative terminal applies to the external circuit connected to the output terminals of the voltage source. *Electron flow* is also shown in Fig. 2-13*c* with reversed polarity for *V*.

Inside the battery, the electrons move to the negative terminal because this is how the voltage source produces its potential difference. The battery is doing the work of separating charges, accumulating electrons at the negative terminal and protons at the positive terminal. Then the potential difference across the two output terminals can do the work of moving electrons around the external circuit. For the circuit outside the voltage source, however, the direction of the electron flow is from a point of negative potential to a point of positive potential.

Conventional Current

A motion of positive charges, in the opposite direction from electron flow, is considered *conventional current*.

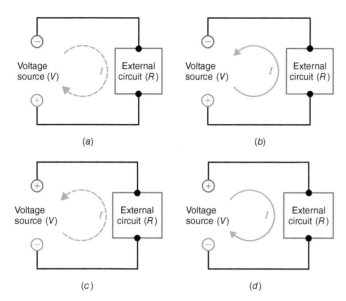

Figure 2-13 Direction of I in a closed circuit, shown for electron flow and conventional current. The circuit works the same way no matter which direction you consider. (a) Electron flow indicated with dashed arrow in diagram. (b) Conventional current indicated with solid arrow. (c) Electron flow as in (a) but with reversed polarity of voltage source. (d) Conventional I as in (b) but reversed polarity for V.

This direction is generally used for analyzing circuits in electrical engineering. The reason is based on some traditional definitions in the science of physics. By the definitions of force and work with positive values, a positive potential is considered above a negative potential. Then conventional current corresponds to a motion of positive charges "falling downhill" from a positive to a negative potential. The conventional current, therefore, is in the direction of positive charges in motion. An example is shown in Fig. 2-13b. The conventional I is out from the positive side of the voltage source, flows through the external circuit, and returns to the negative side of V. Conventional current is also shown in Fig. 2-13d, with the voltage source in reverse polarity.

Examples of Mobile Positive Charges

An ion is an atom that has either lost or gained one or more valence electrons to become electrically charged. For example, a positive ion is created when a neutral atom loses one or more valence electrons and thus becomes positively charged. Similarly, a negative ion is created when a neutral atom gains one or more valence electrons and thus becomes negatively charged. Depending on the number of valence electrons that have been added or removed, the charge of an ion may equal the charge of one electron (Q_e), two electrons (2 Q_e), three electrons (3 Q_e), etc. Ions can be produced by applying voltage to liquids and gases to ionize the atoms. These ions are mobile charges that can provide an electric current. Positive or negative ions are much less mobile than electrons, however, because an ion includes a complex atom with its nucleus.

An example of positive charges in motion for conventional current, therefore, is the current of positive ions in either liquids or gases. This type of current is referred to as ionization current. The positive ions in a liquid or gas flow in the direction of conventional current because they are repelled by the positive terminal of the voltage source and attracted to the negative terminal. Therefore, the mobile positive ions flow from the positive side of the voltage source to the negative side.

Another example of a mobile positive charge is the hole. Holes exist in semiconductor materials such as silicon and germanium. A hole possesses the same amount of charge as an electron but instead has positive polarity. Although the details of the hole charge are beyond the scope of this discussion, you should be aware that in semiconductors, the movement of hole charges are in the direction of conventional current.

It is important to note that protons themselves are not mobile positive charges because they are tightly bound in the nucleus of the atom and cannot be released except by nuclear forces. Therefore, a current of positive charges is a flow of either positive ions in liquids and gases or positive holes in semiconductors.

In this book, the current is considered as electron flow in the applications where electrons are the moving charges. A dotted or dashed arrow, as in Fig. 2-13a and c, is used to indicate the direction of electron flow for I. In Fig. 2-13b and d, the solid arrow means the direction of conventional current. These arrows are used for the unidirectional current in dc circuits. For ac circuits, the direction of current can be considered either way because I reverses direction every half-cycle with the reversals in polarity for V.

2.10 Direct Current (DC) and Alternating Current (AC)

The electron flow illustrated for the circuit with a bulb in Fig. 2-11 is *direct current (dc)* because it has just one direction. The reason for the unidirectional current is that the battery maintains the same polarity of output voltage across its two terminals.

The flow of charges in one direction and the fixed polarity of applied voltage are the characteristics of a dc circuit. The current can be a flow of positive charges, rather than electrons, but the conventional direction of current does not change the fact that the charges are moving only one way.

Furthermore, the dc voltage source can change the amount of its output voltage, but, with the same polarity, direct current still flows only in one direction. This type of source provides a fluctuating or pulsating dc voltage. A battery is a

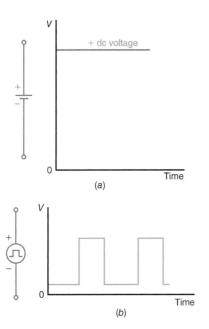

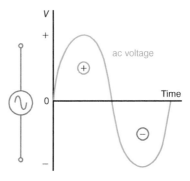

Figure 2-15 Sine wave ac voltage with alternating polarity, such as from an ac generator. Note schematic symbol at left. The ac line voltage in your home has this waveform.

Figure 2-14 (*a*) Steady dc voltage of fixed polarity, such as the output of a battery. Note schematic symbol at left. (*b*) DC pulses as used in binary digital circuits.

steady dc voltage source because it has fixed polarity and its output voltage is a steady value.

An alternating voltage source periodically reverses or alternates in polarity. The resulting *alternating current (ac)*, therefore, periodically reverses in direction. In terms of electron flow, the current always flows from the negative terminal of the voltage source, through the circuit, and back to the positive terminal, but when the generator alternates in polarity, the current must reverse its direction. The 60-cycle ac power line used in most homes is a common example. This frequency means that the voltage polarity and current direction go through 60 cycles of reversal per second.

The unit for 1 cycle per second is 1 hertz (Hz). Therefore 60 cycles per second is a frequency of 60 Hz.

The details of ac circuits are explained in Chap. 14. Direct-current circuits are analyzed first because they usually are simpler. However, the principles of dc circuits also apply to ac circuits. Both types are important because most electronic circuits include ac voltages and dc voltages. A comparison of dc and ac voltages and their waveforms is illustrated in Figs. 2-14 and 2-15. Their uses are compared in Table 2-3.

2.11 Sources of Electricity

There are electrons and protons in the atoms of all materials, but to do useful work, the charges must be separated to produce a *potential difference* that can make current flow. Some of the more common methods of providing electrical effects are listed here.

Static Electricity by Friction

Static electricity by friction is a method by which electrons in an insulator are separated by the work of rubbing to produce opposite charges that remain in the dielectric. Examples of how *static electricity* can be generated include combing your hair, walking across a carpeted room, or sliding two pieces of plastic across each other. An *electrostatic discharge (ESD)* occurs when one of the charged objects comes into contact with another dissimilarly charged object. The electrostatic discharge is in the form of a spark. The current from the discharge lasts for only a very short time but can be very large.

Table 2-3 Comparison of DC Voltage and AC Voltage	
DC Voltage	**AC Voltage**
Fixed polarity	Reverses in polarity
Can be steady or vary in magnitude	Varies between reversals in polarity
Steady value cannot be stepped up or down by a transformer	Can be stepped up or down for electric power distribution
Terminal voltages for transistor amplifiers	Signal input and output for amplifiers
Easier to measure	Easier to amplify
Heating effect is the same for direct or alternating current	

Conversion of Chemical Energy

Wet or dry cells, batteries, and fuel cells generate voltage. Here a chemical reaction produces opposite charges on two dissimilar metals, which serve as the negative and positive terminals.

Electromagnetism

Electricity and magnetism are closely related. Any moving charge has an associated *magnetic field;* also, any changing magnetic field can produce current. A motor is an example showing how current can react with a magnetic field to produce motion; a generator produces voltage by means of a conductor rotating in a magnetic field.

Photoelectricity

Some materials are photoelectric; that is, they can emit electrons when light strikes the surface. The element cesium is often used as a source of *photoelectrons*. Also, photovoltaic cells or solar cells use silicon to generate output voltage from the light input. In another effect, the resistance of the element selenium changes with light. When this is combined with a fixed voltage source, wide variations between *dark current* and *light current* can be produced. Such characteristics are the basis of many photoelectric devices, including television camera tubes, photoelectric cells, and phototransistors.

2.12 The Digital Multimeter

As an electronics technician, you can expect to encounter many situations where it will be necessary to measure the voltage, current, or resistance in a circuit. When this is the case, a technician will most likely use a digital multimeter (DMM)

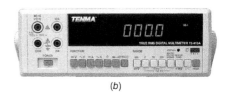

(a) (b)

Figure 2-16 Typical digital multimeters (DMMs) (*a*) Handheld DMM (*b*) Benchtop DMM.

to make these measurements. A DMM may be either a handheld or benchtop unit. Both types are shown in Fig. 2-16. All digital meters have numerical readouts that display the value of voltage, current, or resistance being measured.

Measuring Voltage

Figure. 2-17*a* shows a typical DMM measuring the voltage across the terminals of a battery. To measure any voltage, the meter leads are connected directly across the two points where the potential difference or voltage exists. For dc voltages, the red lead of the meter is normally connected to the positive (+) side of the potential difference, whereas the black lead is normally connected to the negative (−) side. When measuring an alternating (ac) voltage, the orientation of the meter leads does not matter since the voltage periodically reverses polarity anyway.

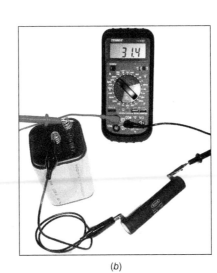
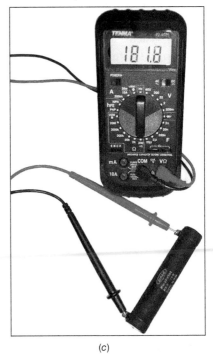

(a) (b) (c)

Figure 2-17 DMM measurements (*a*) Measuring voltage (*b*) Measuring current (*c*) Measuring resistance.

Measuring Current

Figure 2-17*b* shows the DMM measuring the current in a simple dc circuit consisting of a battery and a resistor. Notice that the meter is connected between the positive terminal of the battery and the right lead of the resistor. Unlike voltage measurements, current measurements must be made by placing the meter in the path of the moving charges. To do this, the circuit must be broken open at some point, and then the leads of the meter must be connected across the open points to recomplete the circuit. When measuring the current in a dc circuit, the black lead of the meter should be connected to the point that traces directly back to the negative side of the potential difference. Likewise, the red lead of the meter should be connected to the point that traces directly back to the positive side of the potential difference. When measuring ac currents, the orientation of the meter leads is unimportant.

Measuring Resistance

Figure 2-17*c* shows the DMM measuring the ohmic value of a single resistor. Note that the orientation of the meter leads is unimportant when measuring resistance. What is important is that no voltage is present across the resistance being measured, otherwise the meter could be damaged. Also, make sure that no other components are connected across the resistance being measured. If there are, the measurement will probably be both inaccurate and misleading.

CHAPTER 2 SYSTEM SIDEBAR

A Solar-Powered Home Electrical System

Virtually all homes are powered by the standard electrical grid that distributes 120 and 240 volts of alternating current (ac). The house wiring distributes the power to the lights and outlets that power our many appliances and devices. More and more, solar power systems are being deployed in homes to lighten the load on the utility system. Such solar power systems can eventually pay for themselves in energy usage and minimize the amount of energy consumption at the utility. The solar power system may power the entire house or provide a supplement to the grid power. Small portable solar power systems are also available to provide emergency power during a storm or other loss of grid power.

The typical solar power systems consist of the major components shown in Fig. S2-1. It begins with the solar panel that is made up of a large number of solar cells. These semiconductor cells generate about 0.5 volt of direct current (dc) when light shines on them. Multiple cells are connected in series and parallel to achieve a voltage of 12 or 24 volts of a desired power rating.

This voltage is connected to a charge controller. This is an electronic circuit that conditions the panel voltage into a voltage level suitable for charging a battery or bank of batteries. The battery stores energy to be used at night when the panels do not produce voltage. The charge controller parcels out exact amount of voltage and current to keep the batteries charged but prevents overcharging.

The battery voltage is sent to an inverter. An inverter is an electronic circuit that converts dc into ac. The ac output is usually a sine wave of 120 volts at 60 Hz or 60 cycles per second. This voltage is distributed to the house wiring.

Some systems have a special inverter called a grid tied inverter that lets the inverter output to be connected to the ac line from the utility. Extra power from the solar power system can then be sold back to the utility in some cases to further reduce the electrical bill.

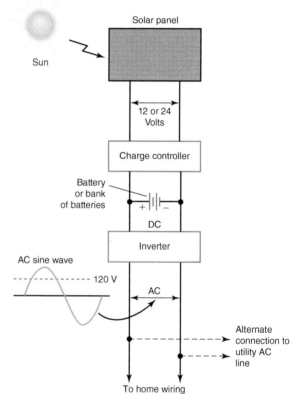

Figure S2-1   Home solar power system.

1. The most basic particle of negative charge is the
   a. coulomb.
   b. electron.
   c. proton.
   d. neutron.

2. The coulomb is a unit of
   a. electric charge.
   b. potential difference.
   c. current.
   d. voltage.

3. Which of the following is not a good conductor?
   a. copper.
   b. silver.
   c. glass.
   d. gold.

4. The electron valence of a neutral copper atom is
   a. +1.
   b. 0.
   c. ±4.
   d. −1.

5. The unit of potential difference is the
   a. volt.
   b. ampere.
   c. siemens.
   d. coulomb.

6. Which of the following statements is true?
   a. Unlike charges repel each other.
   b. Like charges repel each other.
   c. Unlike charges attract each other.
   d. Both b and c.

7. In a metal conductor, such as a copper wire,
   a. positive ions are the moving charges that provide current.
   b. free electrons are the moving charges that provide current.
   c. there are no free electrons.
   d. none of the above.

8. A 100-Ω resistor has a conductance, G, of
   a. 0.01 S.
   b. 0.1 S.
   c. 0.001 S.
   d. 1 S.

9. The most basic particle of positive charge is the
   a. coulomb.
   b. electron.
   c. proton.
   d. neutron.

10. If a neutral atom loses one of its valence electrons, it becomes a(n)
    a. negative ion.
    b. electrically charged atom.
    c. positive ion.
    d. both b and c.

11. The unit of electric current is the
    a. volt.
    b. ampere.
    c. coulomb.
    d. siemens.

12. A semiconductor, such as silicon, has an electron valence of
    a. ±4.
    b. +1.
    c. −7.
    d. 0.

13. Which of the following statements is true?
    a. Current can exist without voltage.
    b. Voltage can exist without current.
    c. Current can flow through an open circuit.
    d. Both b and c.

14. The unit of resistance is the
    a. volt.
    b. coulomb.
    c. siemens.
    d. ohm.

15. Except for hydrogen (H) and helium (He) the goal of valence for an atom is
    a. 6.
    b. 1.
    c. 8.
    d. 4.

16. One ampere of current corresponds to
    a. $\frac{1\text{ C}}{1\text{ s}}$.
    b. $\frac{1\text{ J}}{1\text{ C}}$.
    c. 6.25×10^{18} electrons.
    d. 0.16×10^{-18} C/s.

17. Conventional current is considered
    a. the motion of negative charges in the opposite direction of electron flow.
    b. the motion of positive charges in the same direction as electron flow.
    c. the motion of positive charges in the opposite direction of electron flow.
    d. none of the above.

18. When using a DMM to measure the value of a resistor
   a. make sure that the resistor is in a circuit where voltage is present.
   b. make sure there is no voltage present across the resistor.
   c. make sure there is no other component connected across the leads of the resistor.
   d. both b and c.

19. In a circuit, the opposition to the flow of current is called
   a. conductance.
   b. resistance.
   c. voltage.
   d. current.

20. Aluminum, with an atomic number of 13, has
   a. 13 valence electrons.
   b. 3 valence electrons.
   c. 13 protons in its nucleus.
   d. both b and c.

21. The nucleus of an atom is made up of
   a. electrons and neutrons.
   b. ions.
   c. neutrons and protons.
   d. electrons only.

22. How much charge is accumulated in a dielectric that is charged by a 4-A current for 5 seconds?
   a. 16 C.
   b. 20 C.
   c. 1.25 C.
   d. 0.8 C.

23. A charge of 6 C moves past a given point every 0.25 second. How much is the current flow in amperes?
   a. 24 A.
   b. 2.4 A.
   c. 1.5 A.
   d. 12 A.

24. What is the output voltage of a battery that expends 12 J of energy in moving 1.5 C of charge?
   a. 18 V.
   b. 6 V.
   c. 125 mV.
   d. 8 V.

25. Which of the following statements is false?
   a. The resistance of an open circuit is practically zero.
   b. The resistance of a short circuit is practically zero.
   c. The resistance of an open circuit is infinitely high.
   d. There is no current in an open circuit.

 CHAPTER 2 PROBLEMS

SECTION 2.5 The Volt Unit of Potential Difference

2.1 What is the output voltage of a battery if 10 J of energy is expended in moving 1.25 C of charge?

2.2 How much energy is expended, in joules, if a voltage of 12 V moves 1.25 C of charge between two points?

SECTION 2.6 Charge in Motion Is Current

2.3 A charge of 2 C moves past a given point every 0.5 s. How much is the current?

2.4 If a current of 500 mA charges a dielectric for 2 s, how much charge is stored in the dielectric?

SECTION 2.7 Resistance Is Opposition to Current

2.5 Calculate the resistance value in ohms for the following conductance values: (a) 0.002 S (b) 0.004 S (c) 0.00833 S (d) 0.25 S.

2.6 Calculate the conductance value in siemens for each of the following resistance values: (a) 200 Ω (b) 100 Ω (c) 50 Ω (d) 25 Ω.

Chapter 3

Resistors

Learning Outcomes

After studying this chapter, you should be able to:

> *List* several different types of resistors and *describe* the characteristics of each type.

> *Interpret* the resistor color code to determine the resistance and tolerance of a resistor.

> *Explain* the difference between a potentiometer and a rheostat.

> *Explain* the significance of a resistor's power rating.

Resistors are used in a wide variety of applications in all types of electronic circuits. Their main function in any circuit is to limit the amount of current or to produce a desired drop in voltage. Resistors are manufactured in a variety of shapes and sizes and have ohmic values ranging from a fraction of an ohm to several megohms. The power or wattage rating of a resistor is determined mainly by its physical size. There is, however, no direct correlation between the physical size of a resistor and its resistance value.

In this chapter, you will be presented with an in-depth discussion of the following resistor topics: resistor types, resistor color coding, potentiometers and rheostats, and power ratings.

3.1 Types of Resistors

The two main characteristics of a resistor are its resistance R in ohms and its power rating in watts (W). Resistors are available in a very wide range of R values, from a fraction of an ohm to many kilohms ($k\Omega$) and megohms ($M\Omega$). One kilohm is 1000 Ω and one megohm is 1,000,000 Ω. The power rating for resistors may be as high as several hundred watts or as low as $\frac{1}{10}$ W.

The R is the resistance value required to provide the desired current or voltage. Also important is the wattage rating because it specifies the maximum power the resistor can dissipate without excessive heat. *Dissipation* means that the power is wasted, since the resultant heat is not used. Too much heat can make the resistor burn. The wattage rating of the resistor is generally more than the actual power dissipation, as a safety factor.

Most common in electronic equipment are carbon resistors with a power rating of 1 W or less. The construction is illustrated in Fig. 3-1a. The leads extending out

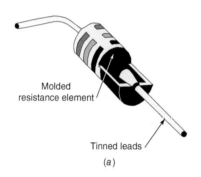

Molded
resistance element

Tinned leads

(a)

Figure 3-1 Carbon-composition resistor. (a) Internal construction. Length is about ¼ in. without leads for ¼-W power rating. Color stripes give R in ohms. Tinned leads have coating of solder. (b) Resistors mounted on printed circuit (PC) board.

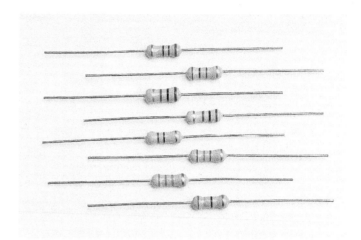

Figure 3-2 Carbon resistors with same physical size but different resistance values. The physical size indicates a power rating of ½ W.

from the resistor body can be inserted through the holes on a printed circuit (PC) board for mounting as shown in Fig. 3-1b. The resistors on a PC board are often inserted automatically by machine. Note that resistors are not polarity-sensitive devices. This means that it does not matter which way the leads of a resistor are connected in a circuit.

Resistors with higher R values usually have lower wattage ratings because they have less current. As an example, a common value is 1 $M\Omega$ at ¼ W, for a resistor only ¼ in. long. The lower the power rating, the smaller the actual size of the resistor. However, the resistance value is not related to physical size. Figure 3-2 shows several carbon resistors with the same physical size but different resistance values. The different color bands on each resistor indicate a different ohmic value. The carbon resistors in Fig. 3-2 each have a power rating of ½ W, which is based on their physical size.

Wire-Wound Resistors

For wire-wound resistors, a special type of wire called *resistance wire* is wrapped around an insulating core. The length of wire and its specific resistivity determine the R of the unit. Types of resistance wire include tungsten and manganin, as explained in Chap. 9, "Conductors, Insulators, and Semiconductors." The insulated core is commonly porcelain, cement, or just plain pressed paper. Bare wire is used, but the entire unit is generally encased in an insulating material. Typical fixed and variable wire-wound resistors are shown in Fig. 3-3.

Since they are generally used for high-current applications with low resistance and appreciable power, *wire-wound resistors* are available in wattage ratings from 1 W up to 100 W or more. The resistance can be less than 1 Ω up to several thousand ohms. For 2 W or less, carbon

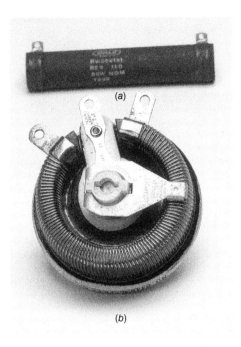

Figure 3-3 Large wire-wound resistors with 50-W power rating. (*a*) Fixed *R*, length of 5 in. (*b*) Variable *R*, diameter of 3 in.

resistors are preferable because they are generally smaller and cost less.

In addition, wire-wound resistors are used where accurate, stable resistance values are necessary. Examples are precision resistors for the function of an ammeter shunt or a precision potentiometer to adjust for an exact amount of *R*.

Carbon-Composition Resistors

Carbon-composition resistors are made of finely divided carbon or graphite mixed with a powdered insulating material as a binder in the proportions needed for the desired *R* value. As shown in Fig. 3-1*a*, the resistor element is enclosed in a plastic case for insulation and mechanical strength. Joined to the two ends of the carbon resistance element are metal caps with leads of tinned copper wire for soldering the connections into a circuit. These are called *axial leads* because they come straight out from the ends. Carbon-composition resistors normally have a brown body and are cylindrical.

Carbon-composition resistors are commonly available in *R* values of 1 Ω to 20 MΩ. Examples are 10 Ω, 220 Ω, 4.7 kΩ, and 68 kΩ. The power rating is generally $\frac{1}{10}$, $\frac{1}{8}$, $\frac{1}{4}$, $\frac{1}{2}$, 1, or 2 W.

Film-Type Resistors

There are two kinds of film-type resistors: *carbon-film* and *metal-film resistors*. The carbon-film resistor, whose

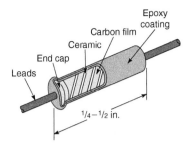

Figure 3-4 Construction of carbon-film resistor.

construction is shown in Fig. 3-4, is made by depositing a thin layer of carbon on an insulated substrate. The carbon film is then cut in the form of a spiral to form the resistive element. The resistance value is controlled by varying the proportion of carbon to insulator. Compared to carbon-composition resistors, carbon-film resistors have the following advantages: tighter tolerances, less sensitivity to temperature changes and aging, and they generate less noise internally.

Metal-film resistors are constructed in a manner similar to the carbon-film type. However, in a metal-film resistor, a thin film of metal is sprayed onto a ceramic substrate and then cut in the form of a spiral. The construction of a metal-film resistor is shown in Fig. 3-5. The length, thickness, and width of the metal spiral determine the exact resistance value. Metal-film resistors offer more precise *R* values than carbon-film resistors. Like carbon-film resistors, metal-film resistors are affected very little by temperature changes and aging. They also generate very little noise internally. In overall performance, metal-film resistors are the best, carbon-film resistors are next, and carbon-composition resistors are last. Both carbon- and metal-film resistors can be distinguished from carbon-composition resistors by the fact that the diameter of the ends is a little larger than that of the body. Furthermore, metal-film resistors are almost always coated with a blue, light green, or red lacquer which provides electrical, mechanical, and climate protection. The body color of carbon-film resistors is usually tan.

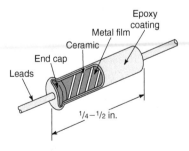

Figure 3-5 Construction of metal-film resistor.

Figure 3-6 Typical chip resistors.

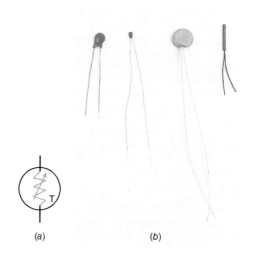

(a)   (b)

Figure 3-7 (a) Thermistor schematic symbol. (b) Typical thermistor shapes and sizes.

Surface-Mount Resistors

Surface-mount resistors, also called *chip resistors,* are constructed by depositing a thick carbon film on a ceramic base. The exact resistance value is determined by the composition of the carbon itself, as well as by the amount of trimming done to the carbon deposit. The resistance can vary from a fraction of an ohm to well over a million ohms. Power dissipation ratings are typically ⅛ to ¼ W. Figure 3-6 shows typical chip resistors. Electrical connection to the resistive element is made via two lead-less solder end electrodes (terminals). The end electrodes are C-shaped. The physical dimensions of a ⅛-W chip resistor are 0.125 in. long by 0.063 in. wide and approximately 0.028 in. thick. This is many times smaller than a conventional resistor having axial leads. Chip resistors are very temperature-stable and also very rugged. The end electrodes are soldered directly to the copper traces of a circuit board, hence the name *surface-mount.*

Thermistors

A *thermistor* is a thermally sensitive resistor whose resistance value changes with changes in operating temperature. Because of the self-heating effect of current in a thermistor, the device changes resistance with changes in current. Thermistors, which are essentially semiconductors, exhibit either a *positive temperature coefficient (PTC)* or a *negative temperature coefficient (NTC).* If a thermistor has a PTC, its resistance increases as the operating temperature increases. Conversely, if a thermistor has an NTC, its resistance decreases as its operating temperature increases. How much the resistance changes with changes in operating temperature depends on the size and construction of the thermistor. Note that the resistance does not undergo instantaneous changes with changes in the

operating temperature. A certain time interval, determined by the thermal mass (size) of the thermistor, is required for the resistance change. A thermistor with a small mass will change more rapidly than one with a large mass. Carbon- and metal-film resistors are different: their resistance does not change appreciably with changes in operating temperature.

Figure 3-7a shows the standard schematic symbol for a thermistor. Notice the arrow through the resistor symbol and the letter T within the circle. The arrow indicates that the resistance is variable as the temperature T changes. As shown in Fig. 3-7b, thermistors are manufactured in a wide variety of shapes and sizes. The shapes include beads, rods, disks, and washers.

Thermistors are frequently used in electronic circuits in which it is desired to provide temperature measurement, temperature control, and temperature compensation.

Photoresistors

A photoresistor or photocell is a light-sensitive resistor. Also known as a photoconductor or light-dependent resistor, the photoresistor is made of materials like cadmium sulphide, lead sulphide, or indium antimonide. Their primary characteristic is a negative-resistance variation, by which the device has its highest resistance in the dark and a decreasing resistance with light intensity.

Figure 3-8 shows the schematic symbols of a photoresistor, a photo of a typical device, and a generic response curve. Note that its resistance decreases with an increasing light intensity.

Photoresistors are used as sensors to detect the presence or absence of light or the intensity of the ambient light.

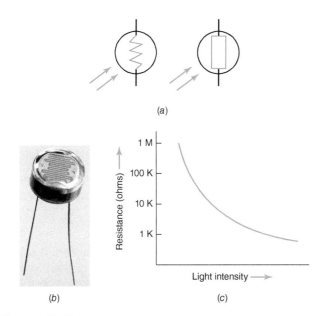

(a)

(b)   (c)

Figure 3-8 Photoresistors. (a) Schematic symbols.
(b) One type of photoresistor (c) Common response curve.

3.2 Resistor Color Coding

Because carbon resistors are small, they are *color-coded* to mark their R value in ohms. The basis of this system is the use of colors for numerical values, as listed in Table 3-1. In memorizing the colors, note that the darkest colors, black and brown, are for the lowest numbers, zero and one, whereas white is for nine. The color coding is standardized by the Electronic Industries Alliance (EIA).

Resistance Color Stripes

The use of colored bands or stripes is the most common system for color-coding resistors, as shown in Fig. 3-9. The colored

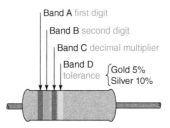

Figure 3-9 How to read color stripes on carbon resistors for R in ohms.

bands or stripes completely encircle the body of the resistor and are usually crowded toward one end. Reading from left to right, the first band closest to the edge gives the first digit in the numerical value of R. The next band indicates the second digit. The third band is the decimal multiplier, which tells us how many zeros to add after the first two digits.

In Fig. 3-10a, the first stripe is red for 2 and the next stripe is green for 5. The red multiplier in the third stripe means add two zeros to 25, or "this multiplier is 10^2." The result can be illustrated as follows:

Red	Green		Red	
↓	↓		↓	
2	5	×	100	= 2500

Therefore, this R value is 2500 Ω or 2.5 kΩ.

The example in Fig. 3-10b illustrates that black for the third stripe just means "do not add any zeros to the first two digits." Since this resistor has red, green, and black stripes, the R value is 25 Ω.

Table 3-1 Color Code	
Color	**Numerical Value**
Black	0
Brown	1
Red	2
Orange	3
Yellow	4
Green	5
Blue	6
Violet	7
Gray	8
White	9

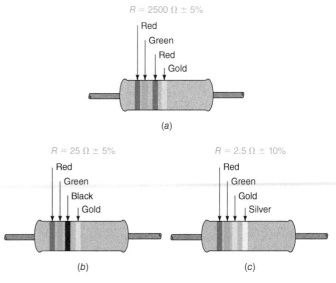

Figure 3-10 Examples of color-coded R values with percent tolerance.

Resistors under 10 Ω

For these values, the third stripe is either gold or silver, indicating a fractional decimal multiplier. When the third stripe is gold, multiply the first two digits by 0.1. In Fig. 3-10c, the R value is

$$25 \times 0.1 = 2.5 \ \Omega.$$

Silver means a multiplier of 0.01. If the third band in Fig. 3-9c were silver, the R value would be

$$25 \times 0.01 = 0.25 \ \Omega.$$

It is important to realize that the gold and silver colors represent fractional decimal multipliers only when they appear in the third stripe. Gold and silver are used most often however as a fourth stripe to indicate how accurate the R value is. The colors gold and silver will never appear in the first two color stripes.

Resistor Tolerance

The amount by which the actual R can differ from the color-coded value is the *tolerance*, usually given in percent. For instance, a 2000-Ω resistor with ±10% tolerance can have resistance 10% above or below the coded value. This R, therefore, is between 1800 and 2200 Ω. The calculations are as follows:

$$10\% \text{ of } 2000 \text{ is } 0.1 \times 2000 = 200.$$

For +10%, the value is

$$2000 + 200 = 2200 \ \Omega.$$

For −10%, the value is

$$2000 - 200 = 1800 \ \Omega.$$

As illustrated in Fig. 3-9, silver in the fourth band indicates a tolerance of ±10%, gold indicates ±5%. If there is no color band for tolerance, it is ±20%. The inexact value of carbon-composition resistors is a disadvantage of their economical construction. They usually cost only a few cents each, or less in larger quantities. In most circuits, though, a small difference in resistance can be tolerated.

Five-Band Color Code

Precision resistors (typically metal-film resistors) often use a five-band color code rather than the four-band code shown in Fig. 3-9. The purpose is to obtain more precise R values. With the five-band code, the first three color stripes indicate the first three digits, followed by the decimal multiplier in the fourth stripe and the tolerance in the fifth stripe. In the fifth stripe, the colors brown, red, green, blue, and violet represent the following tolerances:

Brown	±1	%
Red	±2	%
Green	±0.5	%
Blue	±0.25	%
Violet	±0.1	%

EXAMPLE 3-1

What is the resistance indicated by the five-band color code in Fig. 3-11? Also, what ohmic range is permissible for the specified tolerance?

Answer:

The first stripe is orange for the number 3, the second stripe is blue for the number 6, and the third stripe is green for the number 5. Therefore, the first three digits of the resistance are 3, 6, and 5, respectively. The fourth stripe, which is the multiplier, is black, which means add no zeros. The fifth stripe, which indicates the resistor tolerance, is green for ±0.5%. Therefore R = 365 Ω ± 0.5%. The permissible ohmic range is calculated as 365 × 0.005 = ±1.825 Ω, or 363.175 to 366.825 Ω.

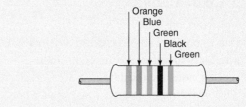

Orange
Blue
Green
Black
Green

Figure 3-11 Five-band code.

Wire-Wound-Resistor Marking

Usually, wire-wound resistors are big enough to have the R value printed on the insulating case. The tolerance is generally ±5% except for precision resistors, which have a tolerance of ±1% or less.

Some small wire-wound resistors may be color-coded with stripes, however, like carbon resistors. In this case, the first stripe is double the width of the others to indicate a wire-wound resistor. Wire-wound resistors that are color-coded generally have a power rating of 4 W or less.

Preferred Resistance Values

To minimize the problem of manufacturing different R values for an almost unlimited variety of circuits, specific values are made in large quantities so that they are cheaper and more easily available than unusual sizes. For resistors of ±10% the *preferred values* are 10, 12, 15, 18, 22, 27, 33, 39, 47, 56, 68, and 82 with their decimal multiples. As examples, 47, 470, 4700, and 47,000 are preferred values. In this way, there is a preferred value available within 10% of any R value needed in a circuit.

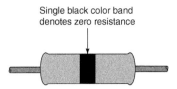

Figure 3-12 A zero-ohm resistor is indicated by a single black color band around the body of the resistor.

Zero-Ohm Resistors

Believe it or not, there is such a thing as a *zero-ohm resistor*. In fact, zero-ohm resistors are quite common. The zero-ohm value is denoted by the use of a single black band around the center of the resistor body, as shown in Fig. 3-12. Zero-ohm resistors are available in ⅛- or ¼-W sizes. The actual resistance of a so-called ⅛-W zero-ohm resistor is about 0.004 Ω, whereas a ¼-W zero-ohm resistor has a resistance of approximately 0.003 Ω.

But why are zero-ohm resistors used in the first place? The reason is that for most printed circuit boards, the components are inserted by automatic insertion machines (robots) rather than by human hands. In some instances, it may be necessary to short two points on the printed circuit board, in which case a piece of wire has to be placed between the two points. Because the robot can handle only components such as resistors, and not wires, zero-ohm resistors are used. Before zero-ohm resistors were developed, jumpers had to be installed by hand, which was time-consuming and expensive. Zero-ohm resistors may be needed as a result of an after-the-fact design change that requires new point-to-point connections in a circuit.

Chip Resistor Coding System

The chip resistor, shown in Fig. 3-13a, has the following identifiable features:

Body color: white or off-white

Dark film on one side only (usually black, but may also be dark gray or green)

End electrodes (terminals) that are C-shaped

Three- or four-digit marking on either the film or the body side (usually the film)

The resistance value of a chip resistor is determined from the three-digit number printed on the film or body side of the component. The three digits provide the same information as the first three color stripes on a four-band resistor. This is shown in Fig. 3-13b. The first two digits indicate the first two numbers in the numerical value of the resistance; the third digit indicates the multiplier. If a four-digit number is used, the first three digits indicate the first three numbers in the numerical value of the resistance, and the fourth digit indicates the multiplier. The letter R is used to signify a decimal point for values

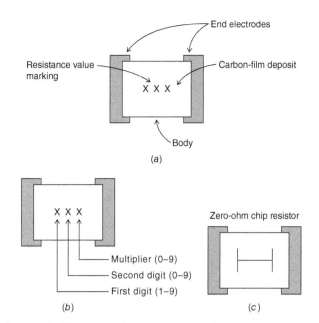

Figure 3-13 Typical chip resistor coding system.

between 1 and 10 ohms as in $2R7 = 2.7\ \Omega$. Figure 3-13c shows the symbol used to denote a zero-ohm chip resistor. Chip resistors are typically available in tolerances of $\pm1\%$ and $\pm5\%$. It is important to note, however, that the tolerance of a chip resistor is not indicated by the three- or four-digit code.

EXAMPLE 3-2

Determine the resistance of the chip resistor in Fig. 3-14.

Answer:
The first two digits are 5 and 6, giving 56 as the first two numbers in the resistance value. The third digit, 2, is the multiplier, which means add 2 zeros to 56 for a resistance of 5600 Ω or 5.6 kΩ.

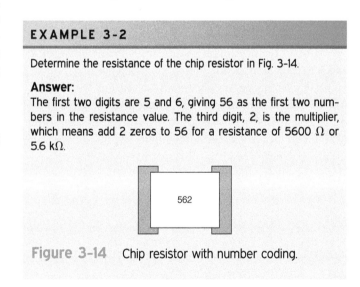

Figure 3-14 Chip resistor with number coding.

Thermistor Values

Thermistors are normally rated by the value of their resistance at a reference temperature T of 25°C. The value of R at 25°C is most often referred to as the zero-power resistance and is designated R_0. The term *zero-power resistance* refers to the resistance of the thermistor with zero-power dissipation. Thermistors normally do not have a code or marking system to indicate their resistance value in ohms. In rare cases, however, a three-dot code is used to indicate the value of R_0. In this case, the first and second dots indicate the first two significant digits, and the third dot is the multiplier. The colors used are the same as those for carbon resistors.

3.3 Variable Resistors

Variable resistors can be wire-wound, as in Fig. 3-3*b*, or carbon type, illustrated in Fig. 3-15. Inside the metal case of Fig. 3-15*a*, the control has a circular disk, shown in Fig. 3-15*b*, that is the carbon-composition resistance element. It can be a thin coating on pressed paper or a molded carbon disk. Joined to the two ends are the external soldering-lug terminals 1 and 3. The middle terminal is connected to the variable arm that contacts the resistor element by a metal spring wiper. As the shaft of the control is turned, the variable arm moves the wiper to make contact at different points on the resistor element. The same idea applies to the slide control in Fig. 3-16, except that the resistor element is straight instead of circular.

When the contact moves closer to one end, the R decreases between this terminal and the variable arm. Between the two ends, however, R is not variable but always has the maximum resistance of the control.

Carbon controls are available with a total R from 1000 Ω to 5 MΩ, approximately. Their power rating is usually ½ to 2 W.

Tapered Controls

The way R varies with shaft rotation is called the *taper* of the control. With a linear taper, a one-half rotation changes R by one-half the maximum value. Similarly, all values of R change in direct proportion to rotation. For a nonlinear taper, though, R can change more gradually at one end with bigger changes at the opposite end. This effect is accomplished by different densities of carbon in the resistance element. For a volume control, its audio taper allows smaller changes in R at low settings. Then it is easier to make changes without having the volume too loud or too low.

Figure 3-16   Slide control for variable *R*. Length is 2 in.

3.4 Rheostats and Potentiometers

Rheostats and potentiometers are variable resistances, either carbon or wire-wound, used to vary the amount of current or voltage in a circuit. The controls can be used in either dc or ac applications.

A *rheostat* is a variable R with two terminals connected in series with a load. The purpose is to vary the amount of current.

A *potentiometer,* generally called a *pot* for short, has three terminals. The fixed maximum R across the two ends is connected across a voltage source. Then the variable arm is used to vary the voltage division between the center terminal and the ends. This function of a potentiometer is compared with that of a rheostat in Table 3-2.

Rheostat Circuit

The function of the rheostat R_2 in Fig. 3-17 is to vary the amount of current through R_1. For instance, R_1 can be a small lightbulb that requires a specified I. Therefore, the two terminals of the rheostat R_2 are connected in series with R_1 and the source V to vary the total resistance R_T in the circuit. When R_T changes, I changes, as read by the meter.

In Fig. 3-17*b*, R_1 is 5 Ω and the rheostat R_2 varies from 0 to 5 Ω. With R_2 at its maximum of 5 Ω, then R_T equals 5 + 5 = 10 Ω. I equals 0.15 A or 150 mA. (The method for calculating I given R and V is covered in Chap. 4, "Ohm's Law.")

When R_2 is at its minimum value of 0 Ω R_T equals 5 Ω. Then I is 0.3 A or 300 mA for the maximum current. As a result, varying the rheostat changes the circuit resistance to vary the current through R_1. I increases as R decreases.

It is important that the rheostat have a wattage rating high enough for maximum I when R is minimum. Rheostats are often wire-wound variable resistors used to control relatively

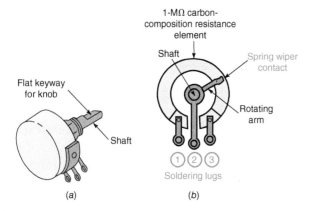

1-MΩ carbon-composition resistance element

Shaft

Spring wiper contact

Flat keyway for knob

Shaft

Rotating arm

① ② ③
Soldering lugs

(a)   (b)

Figure 3-15   Construction of variable carbon resistance control. Diameter is ¾ in. (*a*) External view. (*b*) Internal view of circular resistance element.

Table 3-2 Potentiometers and Rheostats	
Rheostat	**Potentiometer**
Two terminals	Three terminals
In series with load and *V* source	Ends are connected across *V* source
Varies the *I*	Taps off part of *V*

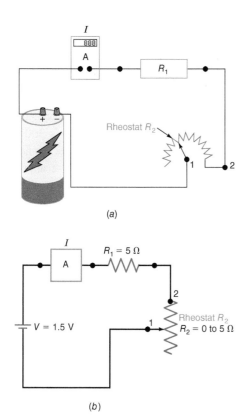

(a)

(b)

Figure 3-17 Rheostat connected in series circuit to vary the current *I*. Symbol for current meter is A for amperes. (*a*) Wiring diagram with digital meter for *I*. (*b*) Schematic diagram.

large values of current in low-resistance circuits for ac power applications.

Potentiometer Circuit

The purpose of the circuit in Fig. 3-18 is to tap off a variable part of the 100 V from the source. Consider this circuit in two parts:

1. The applied *V* is input to the two end terminals of the potentiometer.
2. The variable *V* is output between the variable arm and an end terminal.

Two pairs of connections to the three terminals are necessary, with one terminal common to the input and output. One pair connects the source *V* to the end terminals 1 and 3. The other pair of connections is between the variable arm at the center terminal and one end. This end has double connections for input and output. The other end has only an input connection.

When the variable arm is at the middle value of the 500-kΩ *R* in Fig. 3-18, the 50 V is tapped off between terminals 2 and 1 as one-half the 100-V input. The other 50 V is between terminals 2 and 3. However, this voltage is not used for output.

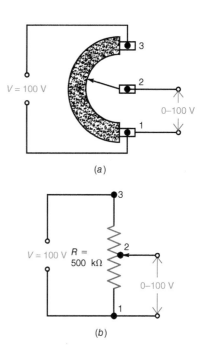

(a)

(b)

Figure 3-18 Potentiometer connected across voltage source to function as a voltage divider. (*a*) Wiring diagram. (*b*) Schematic diagram.

As the control is turned up to move the variable arm closer to terminal 3, more of the input voltage is available between 2 and 1. With the control at its maximum *R*, the voltage between 2 and 1 is the entire 100 V. Actually, terminal 2 is then the same as 3.

When the variable arm is at minimum *R*, rotated to terminal 1, the output between 2 and 1 is zero. Now all the applied voltage is across 2 and 3 with no output for the variable arm. It is important to note that the source voltage is not short-circuited. The reason is that the maximum *R* of the potentiometer is always across the applied *V*, regardless of where the variable arm is set. Typical examples of small potentiometers used in electronic circuits are shown in Fig. 3-19.

Figure 3-19 Small potentiometers and trimmers often used for variable controls in electronic circuits. Terminal leads are formed for insertion into a PC board.

Potentiometer Used as a Rheostat

Commercial rheostats are generally wire-wound, high-wattage resistors for power applications. However, a small, low-wattage rheostat is often needed in electronic circuits. One example is a continuous tone control in a receiver. The control requires the variable series resistance of a rheostat but dissipates very little power.

A method of wiring a potentiometer as a rheostat is to connect just one end of the control and the variable arm, using only two terminals. The third terminal is open, or floating, not connected to anything.

Another method is to wire the unused terminal to the center terminal. When the variable arm is rotated, different amounts of resistance are short-circuited. This method is preferable because there is no floating resistance.

Either end of the potentiometer can be used for the rheostat. The direction of increasing R with shaft rotation reverses, though, for connections at opposite ends. Also, the taper is reversed on a nonlinear control.

The resistance of a potentiometer is sometimes marked on the enclosure that houses the resistance element. The marked value indicates the resistance between the outside terminals.

3.5 Power Rating of Resistors

When current flows in a resistance, work is done and power is generated. This power is manifest as heat. As a result, a resistor must be able to withstand heat. Power is expressed in terms of watts. All resistors have a power rating. Power will be covered in more detail in the next chapter. In addition

to having the required ohms value, a resistor has a wattage rating high enough to dissipate the power produced by the current flowing through the resistance without becoming too hot. Carbon resistors in normal operation often become warm, but they should not get so hot that they "sweat" beads of liquid on the insulating case. Wire-wound resistors operate at very high temperatures; a typical value is 300°C for the maximum temperature. If a resistor becomes too hot because of excessive power dissipation, it can change appreciably in resistance value or burn open.

The power rating is a physical property that depends on the resistor construction, especially physical size. Note the following:

1. A larger physical size indicates a higher power rating.
2. Higher-wattage resistors can operate at higher temperatures.
3. Wire-wound resistors are larger and have higher wattage ratings than carbon resistors.

For approximate sizes, a 2-W carbon resistor is about 1 in. long with a ¼-in. diameter; a ¼-W resistor is about 0.25 in. long with a diameter of 0.1 in.

For both types, a higher power rating allows a higher voltage rating. This rating gives the highest voltage that may be applied across the resistor without internal arcing. As examples for carbon resistors, the maximum voltage is 500 V for a 1-W rating, 350 V for ½-W, 250 V for ¼-W, and 150 V for ⅛-W. In wire-wound resistors, excessive voltage can produce an arc between turns; in carbon-composition resistors, the arc is between carbon granules.

CHAPTER 3 SYSTEM SIDEBAR

Memristors

A memristor is a new type of resistor whose value of resistance varies as a measure of the applied voltage, its polarity, and the time of application. When a voltage is applied, the resistance increases over time. Reversing the polarity of the voltage causes the resistance to decrease. If the voltage is removed, the resistance simply remains constant at that point. The memristor remembers the resistance that exists when the current stops. Figure S3-1 shows the schematic symbol.

Memristors are wires made of titanium dioxide. They were developed by Hewlett-Packard Labs. Figure S3-2 shows what they look like. Memristors are not available as individual components like standard resistors shown in this chapter.

Figure S3-1   Schematic symbol of a memristor.

Instead, they are usually a part of a larger integrated circuit with multiple memristors.

Practical memristors are used primarily as computer memory storage cells. Since the memristor actually remembers its last resistance value when current ceases, it makes an ideal memory device. And it is a nonvolatile memory device because it remembers its state even when power is removed. Memory in computers today is largely

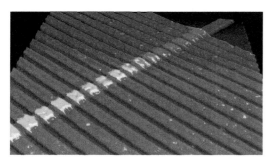

Figure S3-2 An image of a circuit with 17 memristors captured by an atomic force microscope. Each memristor is composed of two layers of titanium dioxide sandwiched between two wires. When a voltage is applied to the top wire of a memristor, the electrical resistance of the titanium dioxide layers is changed, which can be used as a method to store a bit of data.

dynamic random access memory (DRAM) that uses small capacitors as storage cells. These volatile devices lose their charge, and all storage is lost when power is removed from a DRAM.

Nonvolatile memory cells called flash memory use a special transistor to store data. However, they are larger, more expensive, and slower than DRAM. Most computers today use a combination of both DRAM and flash memory.

Using memristors as memory cells offers many benefits over DRAM and flash. Memristors have the following qualities:

- Nonvolatile
- Faster
- Smaller
- Uses less energy
- Immune to radiation

Practical memory devices using memristors are being developed. One promising version is called a nanostore. A nanostore is the combination of a memristor memory and a special processor. These can be combined in a variety of ways to process data and implement special computers for different applications.

Other applications for memristors are also being developed.

❓ CHAPTER 3 REVIEW QUESTIONS

1. A carbon composition resistor having only three color stripes has a tolerance of
   a. ±5%.
   b. ±20%.
   c. ±10%.
   d. ±100%.

2. A resistor with a power rating of 25 W is most likely a
   a. carbon-composition resistor.
   b. metal-film resistor.
   c. surface-mount resistor.
   d. wire-wound resistor.

3. When checked with an ohmmeter, an open resistor measures
   a. infinite resistance.
   b. its color-coded value.
   c. zero resistance.
   d. less than its color-coded value.

4. One precaution to observe when checking resistors with an ohmmeter is to
   a. check high resistances on the lowest ohms range.
   b. check low resistances on the highest ohms range.
   c. disconnect all parallel paths.

   d. make sure your fingers are touching each test lead.

5. A chip resistor is marked 394. Its resistance value is
   a. 39.4 Ω.
   b. 394 Ω.
   c. 390,000 Ω.
   d. 39,000 Ω.

6. A carbon-film resistor is color-coded with red, violet, black, and gold stripes. What are its resistance and tolerance?
   a. 27 Ω ± 5%.
   b. 270 Ω ± 5%.
   c. 270 Ω ± 10%.
   d. 27 Ω ± 10%.

7. A potentiometer is a
   a. three-terminal device used to vary the voltage in a circuit.
   b. two-terminal device used to vary the current in a circuit.
   c. fixed resistor.
   d. two-terminal device used to vary the voltage in a circuit.

8. A metal-film resistor is color-coded with brown, green, red, brown, and blue stripes. What are its resistance and tolerance?
   a. 1500 Ω ± 1.25%.
   b. 152 Ω ± 1%.
   c. 1521 Ω ± 0.5%.
   d. 1520 Ω ± 0.25%.

9. Which of the following resistors has the smallest physical size?
   a. wire-wound resistors.
   b. carbon-composition resistors.
   c. surface-mount resistors.
   d. potentiometers.

10. Which of the following statements is true?
    a. Resistors always have axial leads.
    b. Resistors are always made from carbon.
    c. There is no correlation between the physical size of a resistor and its resistance value.
    d. The shelf life of a resistor is about 1 year.

11. If a thermistor has a negative temperature coefficient (NTC), its resistance
    a. increases with an increase in operating temperature.
    b. decreases with a decrease in operating temperature.
    c. decreases with an increase in operating temperature.
    d. is unaffected by its operating temperature.

12. With the four-band resistor color code, gold in the third stripe corresponds to a
    a. fractional multiplier of 0.01.
    b. fractional multiplier of 0.1.
    c. decimal multiplier of 10.
    d. resistor tolerance of ±10%.

13. Which of the following axial-lead resistor types usually has a blue, light green, or red body?
    a. wire-wound resistors.
    b. carbon-composition resistors.
    c. carbon-film resistors.
    d. metal-film resistors.

14. A surface-mount resistor has a coded value of 4R7. This indicates a resistance of
    a. 4.7 Ω.
    b. 4.7 kΩ.
    c. 4.7 MΩ.
    d. none of the above.

15. Reading from left to right, the colored bands on a resistor are yellow, violet, brown and gold. If the resistor measures 513 Ω with an ohmmeter, it is
    a. well within tolerance.
    b. out of tolerance.
    c. right on the money.
    d. close enough to be considered within tolerance.

16. As the light shining on a photoresistor decreases, the resistance
    a. increases.
    b. decreases.
    c. remains the same.
    d. drops to zero.

17. The unique feature of a memristor is its
    a. variable resistance.
    b. power tolerance.
    c. memory ability.
    d. voltage sensitivity.

 CHAPTER 3 PROBLEMS

Answers to odd-numbered problems at back of book.

SECTION 3.2 Resistor Color Coding

3.1 Indicate the resistance and tolerance for each resistor shown in Fig. 3-20.

3.2 Indicate the resistance for each chip resistor shown in Fig. 3-21.

3.3 Calculate the permissible ohmic range of a resistor whose resistance value and tolerance are (a) 3.9 kΩ ± 5% (b) 100 Ω ± 10% (c) 120 kΩ ± 2% (d) 2.2 Ω ± 5% (e) 75 Ω ± 1%.

3.4 Using the four-band code, indicate the colors of the bands for each of the following resistors: (a) 10 kΩ ± 5% (b) 2.7 Ω ± 5% (c) 5.6 kΩ ± 10% (d) 1.5 MΩ ± 5% (e) 0.22 Ω ± 5%.

3.5 Using the five-band code, indicate the colors of the bands for each of the following resistors: (a) 110 Ω ± 1% (b) 34 kΩ ± 0.5% (c) 82.5 kΩ ± 2% (d) 62.6 Ω ± 1% (e) 105 kΩ ± 0.1%.

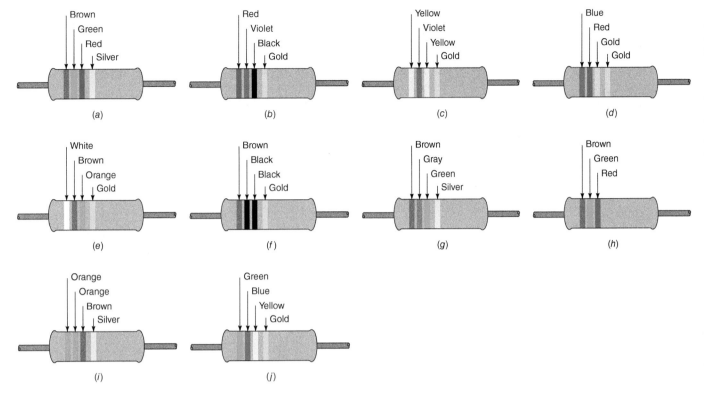

Figure 3-20 Resistors for Prob. 3.1.

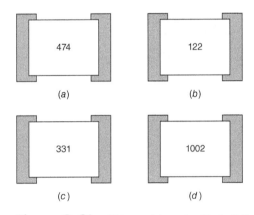

Figure 3-21 Chip resistors for Prob. 3.2.

SECTION 3.4 Rheostats and Potentiometers

3.6 Show two different ways to wire a potentiometer so that it will work as a rheostat.

Chapter 4

Ohm's Law

The mathematical relationship between voltage, current, and resistance was discovered in 1826 by Georg Simon Ohm. The relationship, known as Ohm's law, is the basic foundation for all circuit analysis in electronics. Ohm's law, which is the basis of this chapter, states that the amount of current, I, is directly proportional to the voltage, V, and inversely proportional to the resistance, R. Expressed mathematically, Ohm's law is stated as

$$I = \frac{V}{R}$$

Besides the coverage of Ohm's law, this chapter also introduces you to the concept of power. Power can be defined as the time rate of doing work. The symbol for power is P and the unit is the watt. All the mathematical relationships that exist between V, I, R, and P are covered in this chapter.

Learning Outcomes

After studying this chapter, you should be able to:

> *List* the three forms of Ohm's law.

> *Use* Ohm's law to calculate the current, voltage, or resistance in a circuit.

> *List* the multiple and submultiple units of voltage, current, and resistance.

> *Explain* the difference between a linear and a nonlinear resistance.

> *Explain* the difference between work and power and *list* the units of each.

> *Calculate* the power in a circuit when the voltage and current, current and resistance, or voltage and resistance are known.

> *Determine* the required resistance and appropriate wattage rating of a resistor.

4.1 The Current $I = V/R$

If we keep the same resistance in a circuit but vary the voltage, the current will vary. The circuit in Fig. 4-1 demonstrates this idea. The applied voltage V can be varied from 0 to 12 V, as an example. The bulb has a 12-V filament, which requires this much voltage for its normal current to light with normal intensity. The meter I indicates the amount of current in the circuit for the bulb.

With 12 V applied, the bulb lights, indicating normal current. When V is reduced to 10 V, there is less light because of less I. As V decreases, the bulb becomes dimmer. For zero volts applied, there is no current and the bulb cannot light. In summary, the changing brilliance of the bulb shows that the current varies with the changes in applied voltage.

For the general case of any V and R, Ohm's law is

$$I = \frac{V}{R} \qquad \text{(4-1)}$$

where I is the amount of current through the resistance R connected across the source of potential difference V. With volts as the practical unit for V and ohms for R, the amount of current I is in amperes. Therefore,

$$\text{Amperes} = \frac{\text{volts}}{\text{ohms}}$$

This formula says simply to divide the voltage across R by the ohms of resistance between the two points of potential difference to calculate the amperes of current through R. In Fig. 4-2, for instance, with 6 V applied across a 3-Ω resistance, by Ohm's law, the amount of current I equals $\frac{6}{3}$ or 2 A.

High Voltage but Low Current

It is important to realize that with high voltage, the current can have a low value when there is a very high resistance in the circuit. For example, 1000 V applied across 1,000,000 Ω results in a current of only $\frac{1}{1000}$ A. By Ohm's law,

$$I = \frac{V}{R}$$

$$= \frac{1000 \text{ V}}{1,000,000 \ \Omega} = \frac{1}{1000}$$

$$I = 0.001 \text{ A}$$

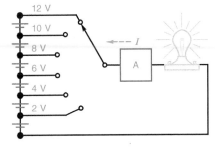

Figure 4-1 Increasing the applied voltage V produces more current I to light the bulb with more intensity.

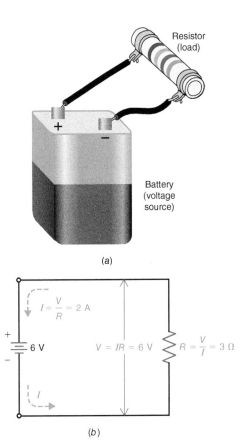

(a)

(b)

Figure 4-2 Example of using Ohm's law. (a) Wiring diagram of a circuit with a 6-V battery for V applied across a load R. (b) Schematic diagram of the circuit with values for I and R calculated by Ohm's law.

The practical fact is that high-voltage circuits usually do have small values of current in electronic equipment. Otherwise, tremendous amounts of power would be necessary.

Low Voltage but High Current

At the opposite extreme, a low value of voltage in a very low resistance circuit can produce a very high current. A 6-V battery connected across a resistance of 0.01 Ω produces 600 A of current:

$$I = \frac{V}{R}$$

$$= \frac{6 \text{ V}}{0.01 \ \Omega}$$

$$I = 600 \text{ A}$$

Less I with More R

Note the values of I in the following two examples also.

EXAMPLE 4-1

A heater with the resistance of 8 Ω is connected across the 120-V power line. How much is current I?

Answer:

$$I = \frac{V}{R} = \frac{120 \text{ V}}{8 \text{ } \Omega}$$

$$I = 15 \text{ A}$$

EXAMPLE 4-2

A small lightbulb with a resistance of 2400 Ω is connected across the same 120-V power line. How much is current I?

Answer:

$$I = \frac{V}{R} = \frac{120 \text{ V}}{2400 \text{ } \Omega}$$

$$I = 0.05 \text{ A}$$

Although both cases have the same 120 V applied, the current is much less in Example 4-2 because of the higher resistance.

Typical V and I

Transistors and integrated circuits generally operate with a dc supply of 5, 6, 9, 12, 15, 24, or 50 V. The current is usually in millionths or thousandths of one ampere up to about 5 A.

4.2 The Voltage V = IR

Referring back to Fig. 4-2, the voltage across R must be the same as the source V because the resistance is connected directly across the battery. The numerical value of this V is equal to the product $I \times R$. For instance, the IR voltage in Fig. 4-2 is 2 A $\times$ 3 Ω, which equals the 6 V of the applied voltage. The formula is

$$V = IR \qquad (4\text{-}2)$$

EXAMPLE 4-3

If a 12-Ω resistor is carrying a current of 2.5 A, how much is its voltage?

Answer:

$$V = IR$$

$$= 2.5 \text{ A} \times 12 \text{ } \Omega$$

$$= 30 \text{ V}$$

With I in amperes and R in ohms, their product V is in volts. Actually, this must be so because the I value equal to V/R is the amount that allows the IR product to be the same as the voltage across R.

Beside the numerical calculations possible with the IR formula, it is useful to consider that the IR product means

voltage. Whenever there is current through a resistance, it must have a potential difference across its two ends equal to the IR product. If there were no potential difference, no electrons could flow to produce the current.

4.3 The Resistance R = V/I

As the third and final version of Ohm's law, the three factors V, I, and R are related by the formula

$$R = \frac{V}{I} \qquad (4\text{-}3)$$

In Fig. 4-2, R is 3 Ω because 6 V applied across the resistance produces 2 A through it. Whenever V and I are known, the resistance can be calculated as the voltage across R divided by the current through it.

Physically, a resistance can be considered some material whose elements have an atomic structure that allows free electrons to drift through it with more or less force applied. Electrically, though, a more practical way of considering resistance is simply as a V/I ratio. Anything that allows 1 A of current with 10 V applied has a resistance of 10 Ω. This V/I ratio of 10 Ω is its characteristic. If the voltage is doubled to 20 V, the current will also double to 2 A, providing the same V/I ratio of a 10-Ω resistance.

Furthermore, we do not need to know the physical construction of a resistance to analyze its effect in a circuit, so long as we know its V/I ratio. This idea is illustrated in Fig. 4-3. Here, a box with some unknown material in it is connected in a circuit where we can measure the 12 V applied across the box and the 3 A of current through it. The resistance is 12V/3A, or 4 Ω. There may be liquid, gas, metal, powder, or any other material in the box; but electrically the box is just a 4-Ω resistance because its V/I ratio is 4.

EXAMPLE 4-4

How much is the resistance of a lightbulb if it draws 0.16 A from a 12-V battery?

Answer:

$$R = \frac{V}{I}$$

$$= \frac{12 \text{ V}}{0.16 \text{ A}}$$

$$= 75 \text{ } \Omega$$

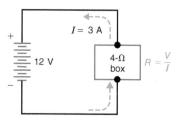

Figure 4-3 The resistance R of any component is its V/I ratio.

4.4 Practical Units

The three forms of Ohm's law can be used to define the practical units of current, potential difference, and resistance as follows:

$$1 \text{ ampere} = \frac{1 \text{ volt}}{1 \text{ ohm}}$$

$$1 \text{ volt} = 1 \text{ ampere} \times 1 \text{ ohm}$$

$$1 \text{ ohm} = \frac{1 \text{ volt}}{1 \text{ ampere}}$$

One **ampere** is the amount of current through a one-ohm resistance that has one volt of potential difference applied across it.

One **volt** is the potential difference across a one-ohm resistance that has one ampere of current through it.

One **ohm** is the amount of opposition in a resistance that has a V/I ratio of 1, allowing one ampere of current with one volt applied.

In summary, the circle diagram in Fig. 4-4 for $V = IR$ can be helpful in using Ohm's law. Put your finger on the unknown quantity and the desired formula remains. The three possibilities are

Cover V and you have IR.
Cover I and you have V/R.
Cover R and you have V/I.

4.5 Multiple and Submultiple Units

The basic units—ampere, volt, and ohm—are practical values in most electric power circuits, but in many electronics applications, these units are either too small or too big. As

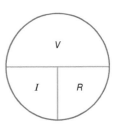

Figure 4-4 A circle diagram to help in memorizing the Ohm's law formulas $V = IR$, $I = V/R$, and $R = V/I$. The V is always at the top.

examples, resistances can be a few million ohms, the output of a high-voltage supply in a computer monitor is about 20,000 V, and the current in transistors is generally thousandths or millionths of an ampere.

In such cases, it is often helpful to use multiples and submultiples of the basic units. These multiple and submultiple values are based on the metric system of units discussed earlier, and a complete listing of all metric prefixes is in Table 4-1.

EXAMPLE 4-5

The I of 8 mA flows through a 5-kΩ R. How much is the IR voltage?

Answer:

$$V = IR = 8 \times 10^{-3} \times 5 \times 10^{3} = 8 \times 5$$
$$V = 40 \text{ V}$$

In general, milliamperes multiplied by kilohms results in volts for the answer, as 10^{-3} and 10^{3} cancel.

Table 4-1 Multiples and Submultiples of Units*			
Value	**Prefix**	**Symbol**	**Example**
$1\ 000\ 000\ 000\ 000 = 10^{12}$	tera	T	THz $= 10^{12}$ Hz
$1\ 000\ 000\ 000 = 10^{9}$	giga	G	GHz $= 10^{9}$ Hz
$1\ 000\ 000 = 10^{6}$	mega	M	MHz $= 10^{6}$ Hz
$1\ 000 = 10^{3}$	kilo	k	kV $= 10^{3}$ V
$100 = 10^{2}$	hecto	h	hm $= 10^{2}$ m
$10 = 10$	deka	da	dam $= 10$ m
$0.1 = 10^{-1}$	deci	d	dm $= 10^{-1}$ m
$0.01 = 10^{-2}$	centi	c	cm $= 10^{-2}$ m
$0.001 = 10^{-3}$	milli	m	mA $= 10^{-3}$ A
$0.000\ 001 = 10^{-6}$	micro	μ	μV $= 10^{-6}$ V
$0.000\ 000\ 001 = 10^{-9}$	nano	n	ns $= 10^{-9}$ s
$0.000\ 000\ 000\ 001 = 10^{-12}$	pico	p	pF $= 10^{-12}$ F

* Additional prefixes are exa $= 10^{18}$, peta $= 10^{15}$, femto $= 10^{-15}$, and atto $= 10^{-18}$.

EXAMPLE 4-6

How much current is produced by 60 V across 12 kΩ?

Answer:

$$I = \frac{V}{R} = \frac{60}{12 \times 10^3}$$

$$= 5 \times 10^{-3} = 5 \text{ mA}$$

Note that volts across kilohms produces milliamperes of current. Similarly, volts across megohms produces microamperes.

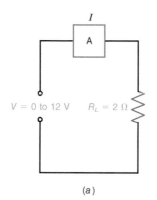

(a)

Volts V	Ohms Ω	Amperes A
0	2	0
2	2	1
4	2	2
6	2	3
8	2	4
10	2	5
12	2	6

(b)

In summary, common combinations to calculate the current I are

$$\frac{V}{k\Omega} = mA \quad \text{and} \quad \frac{V}{M\Omega} = \mu A$$

Also, common combinations to calculate IR voltage are

$$mA \times k\Omega = V$$
$$\mu A \times M\Omega = V$$

These relationships occur often in electronic circuits because the current is generally in units of milliamperes or microamperes. A useful relationship to remember is that 1 mA is equal to 1000 μA.

4.6 The Linear Proportion between V and I

The Ohm's law formula $I = V/R$ states that V and I are directly proportional for any one value of R. This relation between V and I can be analyzed by using a fixed resistance of 2 Ω for R_L, as in Fig. 4-5. Then when V is varied, the meter shows I values directly proportional to V. For instance, with 12 V, I equals 6 A; for 10 V, the current is 5 A; an 8-V potential difference produces 4 A.

All the values of V and I are listed in the table in Fig. 4-5b and plotted in the graph in Fig. 4-5c. The I values are one-half the V values because R is 2 Ω. However, I is zero with zero volts applied.

Plotting the Graph

The voltage values for V are marked on the horizontal axis, called the x axis or abscissa. The current values I are on the vertical axis, called the y axis or ordinate.

Because the values for V and I depend on each other, they are variable factors. The independent variable here is V because we assign values of voltage and note the resulting current. Generally, the independent variable is plotted on the x axis, which is why the V values are shown here horizontally and the I values are on the ordinate.

The two scales need not be the same. The only requirement is that equal distances on each scale represent equal

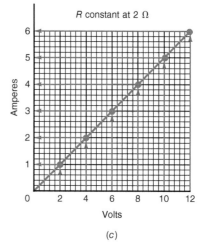

(c)

Figure 4-5 Experiment to show that I increases in direct proportion to V with the same R. (a) Circuit with variable V but constant R. (b) Table of increasing I for higher V. (c) Graph of V and I values. This is a linear volt-ampere characteristic. It shows a direct proportion between V and I.

changes in magnitude. On the x axis here, 2-V steps are chosen, whereas the y axis has 1-A scale divisions. The zero point at the origin is the reference.

The plotted points in the graph show the values in the table. For instance, the lowest point is 2 V horizontally from the origin, and 1 A up. Similarly, the next point is at the intersection of the 4-V mark and the 2-A mark.

A line joining these plotted points includes all values of I, for any value of V, with R constant at 2 Ω. This also applies to values not listed in the table. For instance, if we take the value of 7 V up to the straight line and over to the y axis, the graph shows 3.5 A for I.

Volt-Ampere Characteristic

The graph in Fig. 4-5c is called the volt-ampere characteristic of R. It shows how much current the resistor allows for different voltages. Multiple and submultiple units of V and I

can be used, though. For transistors, the units of I are often milliamperes or microamperes.

Linear Resistance

The *straight-line (linear) graph* in Fig. 4-5 shows that R is a linear resistor. A linear resistance has a constant value of ohms. Its R does not change with the applied voltage. Then V and I are directly proportional. Doubling the value of V from 4 to 8 V results in twice the current, from 2 to 4 A. Similarly, three or four times the value of V will produce three or four times I, for a proportional increase in current.

Nonlinear Resistance

Nonlinear resistance has a nonlinear volt-ampere characteristic. As an example, the resistance of the tungsten filament in a lightbulb is nonlinear. The reason is that R increases with more current as the filament becomes hotter. Increasing the applied voltage does produce more current, but I does not increase in the same proportion as the increase in V. Another example of a nonlinear resistor is a thermistor.

Inverse Relation between I and R

Whether R is linear or not, the current I is less for more R, with the applied voltage constant. This is an inverse relation; that is, I goes down as R goes up. Remember that in the formula $I = V/R$, the resistance is in the denominator. A higher value of R actually lowers the value of the complete fraction.

As an example, let V be constant at 1 V. Then I is equal to the fraction $1/R$. As R increases, the values of I decrease. For R of 2 Ω, I is ½ or 0.5 A. For a higher R of 10 Ω, I will be lower at ⅒ or 0.1 A.

4.7 Electric Power

The unit of electric *power* is the *watt* (W), named after James Watt (1736–1819). One watt of power equals the work done in one second by one volt of potential difference in moving one coulomb of charge.

Remember that one coulomb per second is an ampere. Therefore power in watts equals the product of volts times amperes.

$$\text{Power in watts} = \text{volts} \times \text{amperes}$$
$$P = V \times I \qquad \textbf{(4-4)}$$

When a 6-V battery produces 2 A in a circuit, for example, the battery is generating 12 W of power.

The power formula can be used in three ways:

$$P = V \times I$$
$$I = P \div V \quad \text{or} \quad \frac{P}{V}$$
$$V = P \div I \quad \text{or} \quad \frac{P}{I}$$

Which formula to use depends on whether you want to calculate P, I, or V. Note the following examples.

EXAMPLE 4-7

A toaster takes 10 A from the 120-V power line. How much power is used?

Answer:

$$P = V \times I = 120\,\text{V} \times 10\,\text{A}$$
$$P = 1200\,\text{W} \quad \text{or} \quad 1.2\,\text{kW}$$

EXAMPLE 4-8

How much current flows in the filament of a 300-W bulb connected to the 120-V power line?

Answer:

$$I = \frac{P}{V} = \frac{300\,\text{W}}{120\,\text{V}}$$
$$I = 2.5\,\text{A}$$

EXAMPLE 4-9

How much current flows in the filament of a 60-W bulb connected to the 120-V power line?

Answer:

$$I = \frac{P}{V} = \frac{60\,\text{W}}{120\,\text{V}}$$
$$I = 0.5\,\text{A} \quad \text{or} \quad 500\,\text{mA}$$

Note that the lower-wattage bulb uses less current.

Work and Power

Work and energy are essentially the same with identical units. Power is different, however, because it is the time rate of doing work.

As an example of work, if you move 100 lb a distance of 10 ft, the work is 100 lb × 10 ft or 1000 ft·lb, regardless of how fast or how slowly the work is done. Note that the unit of work is foot-pounds, without any reference to time.

However, power equals the work divided by the time it takes to do the work. If it takes 1 s, the power in this example is 1000 ft·lb/s; if the work takes 2 s, the power is 1000 ft·lb in 2 s, or 500 ft·lb/s.

Similarly, electric power is the rate at which charge is forced to move by voltage. This is why power in watts is the product of volts and amperes. The voltage states the amount of work per unit of charge; the current value includes the rate at which the charge is moved.

Watts and Horsepower Units

A further example of how electric power corresponds to mechanical power is the fact that

$$746 \text{ W} = 1 \text{ hp} = 550 \text{ ft·lb/s}$$

This relation can be remembered more easily as 1 hp equals approximately ¾ kilowatt (kW). One kilowatt = 1000 W.

Practical Units of Power and Work

Starting with the watt, we can develop several other important units. The fundamental principle to remember is that power is the time rate of doing work, whereas work is power used during a period of time. The formulas are

$$\text{Power} = \frac{\text{work}}{\text{time}} \qquad \text{(4-5)}$$

and

$$\text{Work} = \text{power} \times \text{time} \qquad \text{(4-6)}$$

With the watt unit for power, one watt used during one second equals the work of one joule. Or one watt is one joule per second. Therefore, 1 W = 1 J/s. The **joule** is a basic practical unit of work or energy.

To summarize these practical definitions,

$$1 \text{ joule} = 1 \text{ watt · second}$$
$$1 \text{ watt} = 1 \text{ joule/second}$$

In terms of charge and current,

$$1 \text{ joule} = 1 \text{ volt · coulomb}$$
$$1 \text{ watt} = 1 \text{ volt · ampere}$$

Remember that the ampere unit includes time in the denominator, since the formula is 1 ampere = 1 coulomb/second.

Kilowatt-Hours

The kilowatt-hour (kWh) is a unit commonly used for large amounts of electrical work or energy. The amount is calculated simply as the product of the power in kilowatts multiplied by the time in hours during which the power is used. As an example, if a lightbulb uses 300 W or 0.3 kW for 4 hours (h), the amount of energy is 0.3 × 4, which equals 1.2 kWh.

We pay for electricity in kilowatt-hours of energy. The power-line voltage is constant at 120 V. However, more appliances and lightbulbs require more current because they all add in the main line to increase the power.

Suppose that the total load current in the main line equals 20 A. Then the power in watts from the 120-V line is

$$P = 120 \text{ V} \times 20 \text{ A}$$
$$P = 2400 \text{ W} \quad \text{or} \quad 2.4 \text{ kW}$$

If this power is used for 5 h, then the energy or work supplied equals 2.4 × 5 = 12 kWh. If the cost of electricity is 6¢/kWh, then 12 kWh of electricity will cost 0.06 × 12 = 0.72 or 72¢. This charge is for a 20-A load current from the 120-V line during the time of 5 h.

EXAMPLE 4-10

Assuming that the cost of electricity is 6 ¢/kWh, how much will it cost to light a 100-W lightbulb for 30 days?

Answer:

The first step in solving this problem is to express 100 W as 0.1 kW. The next step is to find the total number of hours in 30 days. Since there are 24 hours in a day, the total number of hours for which the light is on is calculated as

$$\text{Total hours} = \frac{24 \text{ h}}{\text{day}} \times 30 \text{ days} = 720 \text{ h}$$

Next, calculate the number of kWh as

$$\begin{aligned}
\text{kWh} &= \text{kW} \times \text{h} \\
&= 0.1 \text{ kW} \times 720 \text{ h} \\
&= 72 \text{ kWh}
\end{aligned}$$

And finally, determine the cost. (Note that 6¢ = $0.06.)

$$\begin{aligned}
\text{Cost} &= \text{kWh} \times \frac{\text{cost}}{\text{kWh}} \\
&= 72 \text{ kWh} \times \frac{0.06}{\text{kWh}} \\
&= \$4.32
\end{aligned}$$

4.8 Power Dissipation in Resistance

When current flows in a resistance, heat is produced because friction between the moving free electrons and the atoms obstructs the path of electron flow. The heat is evidence that power is used in producing current. This is how a fuse opens, as heat resulting from excessive current melts the metal link in the fuse.

The power is generated by the source of applied voltage and consumed in the resistance as heat. As much power as the resistance dissipates in heat must be supplied by the voltage source; otherwise, it cannot maintain the potential difference required to produce the current.

The correspondence between electric power and heat is indicated by the fact that 1 W used during 1 s is equivalent to 0.24 calorie of heat energy. The electric energy converted to heat is considered dissipated or used up because the calories of heat cannot be returned to the circuit as electric energy.

Since power is dissipated in the resistance of a circuit, it is convenient to express the power in terms of the resistance R. The formula $P = V \times I$ can be rearranged as follows:

Substituting IR for V,

$$P = V \times I = IR \times I$$
$$P = I^2R \qquad \text{(4-7)}$$

This is a common form of the power formula because of the heat produced by current in a resistance.

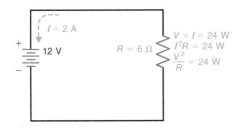

Figure 4-6 Calculating the electric power in a circuit as $P = V \times I$, $P = I^2R$, or $P = V^2/R$.

For another form, substitute V/R for I. Then

$$P = V \times I = V \times \frac{V}{R}$$
$$P = \frac{V^2}{R} \tag{4-8}$$

In all the formulas, V is the voltage across R in ohms, producing the current I in amperes, for power in watts.

Any one of the three formulas (4-4), (4-7), and (4-8) can be used to calculate the power dissipated in a resistance. The one to be used is a matter of convenience, depending on which factors are known.

In Fig. 4-6, for example, the power dissipated with 2 A through the resistance and 12 V across it is $2 \times 12 = 24$ W.

Or, calculating in terms of just the current and resistance, the power is the product of 2 squared, or 4, times 6, which equals 24 W.

Using the voltage and resistance, the power can be calculated as 12 squared, or 144, divided by 6, which also equals 24 W.

No matter which formula is used, 24 W of power is dissipated as heat. This amount of power must be generated continuously by the battery to maintain the potential difference of 12 V that produces the 2-A current against the opposition of 6 Ω.

EXAMPLE 4-11

Calculate the power in a circuit where the source of 100 V produces 2 A in a 50-Ω R.

Answer:
$$P = I^2R = 2 \times 2 \times 50 = 4 \times 50$$
$$P = 200 \text{ W}$$
This means that the source delivers 200 W of power to the resistance and the resistance dissipates 200 W as heat.

EXAMPLE 4-12

Calculate the power in a circuit in which the same source of 100 V produces 4 A in a 25-Ω R.

Answer:
$$P = I^2R = 4^2 \times 25 = 16 \times 25$$
$$P = 400 \text{ W}$$
Note the higher power in Example 4-12 because of more I, even though R is less than that in Example 4-11.

In some applications, electric power dissipation is desirable because the component must produce heat to do its job. For instance, a 600-W toaster must dissipate this amount of power to produce the necessary amount of heat. Similarly, a 300-W lightbulb must dissipate this power to make the filament white-hot so that it will have the incandescent glow that furnishes the light. In other applications, however, the heat may be just an undesirable by-product of the need to provide current through the resistance in a circuit. In any case, though, whenever there is current I in a resistance R, it dissipates the amount of power P equal to I^2R.

Components that use the power dissipated in their resistance, such as lightbulbs and toasters, are generally rated in terms of power. The power rating is given at normal applied voltage, which is usually the 120 V of the power line. For instance, a 600-W, 120-V toaster has this rating because it dissipates 600 W in the resistance of the heating element when connected across 120 V.

Note this interesting point about the power relations. The lower the source voltage, the higher the current required for the same power. The reason is that $P = V \times I$. For instance, an electric heater rated at 240 W from a 120-V power line takes 240 W/120 V = 2 A of current from the source. However, the same 240 W from a 12-V source, as in a car or boat, requires 240 W/12 V = 20 A. More current must be supplied by a source with lower voltage, to provide a specified amount of power.

4.9 Power Formulas

To calculate I or R for components rated in terms of power at a specified voltage, it may be convenient to use the power formulas in different forms. There are three basic power formulas, but each can be in three forms for nine combinations.

$$P = VI \qquad P = I^2R \qquad P = \frac{V^2}{R}$$

$$\text{or} \quad I = \frac{P}{V} \qquad \text{or} \quad R = \frac{P}{I^2} \qquad \text{or} \quad R = \frac{V^2}{P}$$

$$\text{or} \quad V = \frac{P}{I} \qquad \text{or} \quad I = \sqrt{\frac{P}{R}} \qquad \text{or} \quad V = \sqrt{PR}$$

EXAMPLE 4-13

How much current is needed for a 600-W, 120-V toaster?

Answer:
$$I = \frac{P}{V} = \frac{600}{120}$$
$$I = 5 \text{ A}$$

EXAMPLE 4-14

How much is the resistance of a 600-W, 120-V toaster?

Answer:

$$R = \frac{V^2}{P} = \frac{(120)^2}{600} = \frac{14,400}{600}$$
$$I = 24\ \Omega$$

EXAMPLE 4-15

How much current is needed for a 24-Ω R that dissipates 600 W?

Answer:

$$I = \sqrt{\frac{P}{R}} = \sqrt{\frac{600\ W}{24\ \Omega}} = \sqrt{25}$$
$$I = 5\ A$$

Note that all these formulas are based on Ohm's law $V = IR$ and the power formula $P = VI$. The following example with a 300-W bulb also illustrates this idea. Refer to Fig. 4-7. The bulb is connected across the 120-V line. Its 300-W filament requires a current of 2.5 A, equal to P/V. These calculations are

$$I = \frac{P}{V} = \frac{300\ W}{120\ V} = 2.5\ A$$

The proof is that the VI product is 120 × 2.5, which equals 300 W.

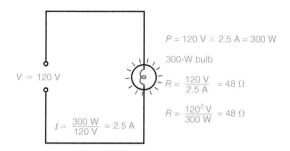

$P = 120\ V \times 2.5\ A = 300\ W$

300-W bulb

$V = 120\ V$

$R = \frac{120\ V}{2.5\ A} = 48\ \Omega$

$R = \frac{120^2\ V}{300\ W} = 48\ \Omega$

$I = \frac{300\ W}{120\ V} = 2.5\ A$

Figure 4-7   All formulas are based on Ohm's law and the power formula.

Furthermore, the resistance of the filament, equal to V/I, is 48 Ω. These calculations are

$$R = \frac{V}{I} = \frac{120\ V}{2.5\ A} = 48\ \Omega$$

If we use the power formula $R = V^2/P$, the answer is the same 48 Ω. These calculations are

$$R = \frac{V^2}{P} = \frac{120^2}{300}$$
$$R = \frac{14,400}{300} = 48\ \Omega$$

In any case, when this bulb is connected across 120 V so that it can dissipate its rated power, the bulb draws 2.5 A from the power line and the resistance of the white-hot filament is 48 Ω.

CHAPTER 4 SYSTEM SIDEBAR

The Binary Number System

Most electronic systems today are digital. Examples are TV sets, PCs and laptops, cell phones, CD/DVD players, electronic games, audio players, tablets, satellites, cable TV, and wireless hot spots. The inputs to be processed are not continuous analog signals but binary signals. The outputs are also in binary form.

Number Conversions

Binary signals represent numbers and codes. They are voltages that represent two basic digits, 0 and 1. By using these two numbers, you can represent any numerical value or any letter, symbol, or code. The binary number system uses the digits or bits 0 and 1 in a system like our decimal number system. It is based on the position of the digit or bit and the weight given to that position.

For example, take the number 7308. Remember from your basic math that the positions from right to left have the weights of units, tens, hundreds, and thousands, or expressed mathematically 10^0, 10^1, 10^2, and 10^3. There are 8 units, 0 tens, 3 hundreds and 7 thousands. The value of the number is computed by adding these values.

$$7000 + 300 + 0 + 8 = 7308$$

The binary number system works like that, but it uses positions with weights that are a power of 2 (*binary* means 2). The weights of a binary number from right to left are 2^0, 2^1, 2^2, 2^3 or 1, 2, 4, 8, and so on.

To evaluate a binary number, you add the weights of those bits that are binary 1. An example is the binary number 1110. The decimal value is

$$8 + 4 + 2 + 0 = 14$$

Decimal	Binary
0	0000
1	0001
2	0010
3	0011
4	0100
5	0101
6	0110
7	0111
8	1000
9	1001
10	1010
11	1011
12	1100
13	1101
14	1110
15	1111

The bit on the far left is the most significant bit (MSB), and the bit on the far right is the least significant bit (LSB). The table below shows the 16 values possible with a 4-bit number and the binary equivalent.

Using this system, any value can be represented. The binary value can mean anything assigned to it, such as a numerical value, a print character, a special code like a PIN number, or a control function. Such binary values are often called words or codes. Collectively, all this information is called data.

Electronic Representation

To implement the binary system in electronics, we use two different voltage levels to represent the 1 and 0. In some systems a binary 1 is represented by +5 volts and 0 by zero volts or ground. There are many other possibilities like binary 1 = +1.8 volts and binary 0 = +0.2 volts. Figure S4-1 shows what binary input signals might look like on a digital circuit. In this instance, the number 10011010 represents the decimal value 154. Here all the bits appear in parallel at the same time to the circuits. The output is shown on the right in Figure S4-1. Here the bits occur one after the other. Each bit has a specific time interval. This is referred to as *serial binary data*. The circuit shown is simply a digital parallel to serial data converter.

The ASCII System

Instead of representing numerical values, binary can be used to represent any character, like letters of the alphabet, punctuation, and special symbols. One widely used code is the American Standard Code for Information Interchange (ASCII). It is usually pronounced "ask key." This is an 8-bit code used in computers, printers, data communications in networks, the Internet, and in thousands of other places.

For example, the lowercase letter *j* in ASCII is 01101010. The capital letter *F* is 01000110. The symbol *?* is 00111111. Special values represent information or control functions. The end-of-text (ETX) function is 00000011 and signals that this is the end of the text being transmitted. The code 00000111 is BEL, a control code used to ring a bell or buzzer or some similar function. With 8 bits you can represent $2^8 = 256$ characters. In the standard ASCII code only half the characters (128) are commonly used.

You will learn more about the binary number system in a later course, and in the meantime you can translate between binary and decimal with almost any scientific calculator, which does the conversions for you automatically.

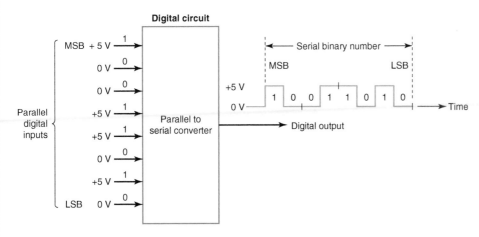

Figure S4-1 A parallel to serial data converter. Binary inputs and outputs represent numbers, characters, and codes using voltages.

1. With 24 V across a 1-kΩ resistor, the current, *I*, equals
   a. 0.24 A.
   b. 2.4 mA.
   c. 24 mA.
   d. 24 μA.

2. With 30 μA of current in a 120-kΩ resistor, the voltage, *V*, equals
   a. 360 mV.
   b. 3.6 kV.
   c. 0.036 V.
   d. 3.6 V.

3. How much is the resistance in a circuit if 15 V of potential difference produces 500 μA of current?
   a. 30 kΩ.
   b. 3 MΩ.
   c. 300 kΩ.
   d. 3 kΩ.

4. A current of 1000 μA equals
   a. 1 A.
   b. 1 mA.
   c. 0.01 A.
   d. none of the above.

5. One horsepower equals
   a. 746 W.
   b. 550 ft·lb/s.
   c. approximately ¾ kW.
   d. all of the above.

6. With *R* constant
   a. *I* and *P* are inversely related.
   b. *V* and *I* are directly proportional.
   c. *V* and *I* are inversely proportional.
   d. none of the above.

7. One watt of power equals
   a. 1 V × 1 A.
   b. $\dfrac{1\ J}{s}$
   c. $\dfrac{1\ C}{s}$
   d. both a and b.

8. A 10-Ω resistor dissipates 1 W of power when connected to a dc voltage source. If the value of dc voltage is doubled, the resistor will dissipate
   a. 1 W.
   b. 2 W.
   c. 4 W.
   d. 10 W.

9. If the voltage across a variable resistance is held constant, the current, *I*, is
   a. inversely proportional to resistance.
   b. directly proportional to resistance.
   c. the same for all values of resistance.
   d. both a and b.

10. A resistor must provide a voltage drop of 27 V when the current is 10 mA. Which of the following resistors will provide the required resistance and appropriate wattage rating?
    a. 2.7 kΩ, ⅛ W.
    b. 270 Ω, ½ W.
    c. 2.7 kΩ, ½ W.
    d. 2.7 kΩ, ¼ W.

11. The resistance of an open circuit is
    a. approximately 0 Ω.
    b. infinitely high.
    c. very low.
    d. none of the above.

12. The current in an open circuit is
    a. normally very high because the resistance of an open circuit is 0 Ω.
    b. usually high enough to blow the circuit fuse.
    c. zero.
    d. slightly below normal.

13. Power in a resistive circuit shows itself as
    a. light emissions.
    b. heat.
    c. physical vibrations.
    d. a magnetic field.

14. How much current does a 75-W lightbulb draw from the 120-V power line?
    a. 625 mA.
    b. 1.6 A.
    c. 160 mA.
    d. 62.5 mA.

15. The resistance of a short circuit is
    a. infinitely high.
    b. very high.
    c. usually above 1 kΩ.
    d. approximately zero.

16. Which of the following is considered a linear resistance?
    a. Lightbulb.
    b. Thermistor.
    c. 1-kΩ, ½-W carbon-film resistor.
    d. Both a and b.

17. How much will it cost to operate a 4-kW air-conditioner for 12 hours if the cost of electricity is 7¢/kWh?
a. $3.36.
b. 33¢.
c. $8.24.
d. $4.80.

18. What is the maximum voltage a 150-Ω, ⅛-W resistor can safely handle without exceeding its power rating? (Assume no power rating safety factor.)
a. 18.75 V.
b. 4.33 V.
c. 6.1 V.
d. 150 V.

19. Power is proportional to the
a. voltage.
b. current.
c. resistance.
d. square of the voltage or current.

 CHAPTER 4 PROBLEMS

SECTION 4.1 The Current $I = V/R$

In Probs. 4.1 to 4.4, solve for the current, I, when V and R are known. As a visual aid, it may be helpful to insert the values of V and R into Fig. 4-8 when solving for I.

4.1 a. $V = 10$ V, $R = 5$ Ω, $I = ?$

b. $V = 9$ V, $R = 3$ Ω, $I = ?$

c. $V = 36$ V, $R = 9$ Ω, $I = ?$

4.2 a. $V = 15$ V, $R = 3,000$ Ω, $I = ?$

b. $V = 120$ V, $R = 6,000$ Ω, $I = ?$

c. $V = 150$ V, $R = 10,000$ Ω, $I = ?$

4.3 If a 100-Ω resistor is connected across the terminals of a 12-V battery, how much is the current, I?

4.4 If one branch of a 120-V power line is protected by a 20-A fuse, will the fuse carry an 8-Ω load?

SECTION 4.2 The Voltage $V = IR$

In Probs. 4.5 to 4.9, solve for the voltage, V, when I and R are known. As a visual aid, it may be helpful to insert the values of I and R into Fig. 4-9 when solving for V.

4.5 a. $I = 9$ A, $R = 20$ Ω, $V = ?$

b. $I = 4$ A, $R = 15$ Ω, $V = ?$

4.6 a. $I = 4$ A, $R = 2.5$ Ω, $V = ?$

b. $I = 1.5$ A, $R = 5$ Ω, $V = ?$

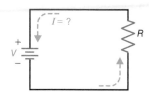

Figure 4-8   Figure for Probs. 4.1 to 4.4.

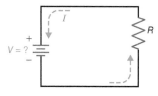

Figure 4-9   Figure for Probs. 4.5 to 4.9.

4.7 a. $I = 0.05$ A, $R = 1200$ Ω, $V = ?$

b. $I = 0.006$ A, $R = 2200$ Ω, $V = ?$

4.8 How much voltage is developed across a 1000-Ω resistor if it has a current of 0.01 A?

4.9 A lightbulb drawing 1.25 A of current has a resistance of 96 Ω. How much is the voltage across the lightbulb?

SECTION 4.3 The Resistance $R = V/I$

In Probs. 4.10 to 4.14, solve for the resistance, R, when V and I are known. As a visual aid, it may be helpful to insert the values of V and I into Fig. 4-10 when solving for R.

4.10 a. $V = 6$ V, $I = 1.5$ A, $R = ?$

b. $V = 24$ V, $I = 4$ A, $R = ?$

4.11 a. $V = 36$ V, $I = 9$ A, $R = ?$

b. $V = 240$ V, $I = 20$ A, $R = ?$

4.12 a. $V = 12$ V, $I = 0.002$ A, $R = ?$

b. $V = 45$ V, $I = 0.009$ A, $R = ?$

4.13 How much is the resistance of a motor if it draws 2 A of current from the 120-V power line?

4.14 If a CD player draws 1.6 A of current from a 13.6-Vdc source, how much is its resistance?

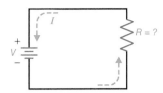

Figure 4.10   Figure for Probs. 4.10 to 4.14.

SECTION 4.5 Multiple and Submultiple Units

In Probs. 4.15 to 4.17, solve for the unknowns listed.

4.15 a. $V = 10$ V, $R = 100$ kΩ, $I = $?

    b. $I = 200$ μA, $R = 3.3$ MΩ, $V = $?

4.16 How much is the current, I, in a 470-kΩ resistor if its voltage is 23.5 V?

4.17 How much voltage will be dropped across a 40-kΩ resistance whose current is 250 μA?

SECTION 4.6 The Linear Proportion between *V* and *I*

4.18 Refer to Fig. 4-11. Draw a graph of the I and V values if (*a*) $R = 2.5$ Ω; (*b*) $R = 5$ Ω; (*c*) $R = 10$ Ω. In each case, the voltage source is to be varied in 5-V steps from 0 to 30 V.

SECTION 4.7 Electric Power

In Probs. 4.19 to 4.26, solve for the unknowns listed.

4.19 a. $V = 120$ V, $I = 625$ mA, $P = $?

    b. $P = 1.2$ kW, $V = 120$ V, $I = $?

4.20 a. $P = 6$ W, $V = 12$ V, $I = $?

    b. $P = 10$ W, $I = 100$ mA, $V = $?

4.21 a. $V = 75$ mV, $I = 2$ mA, $P = $?

    b. $P = 20$ mW, $I = 100$ μA, $V = $?

4.22 How much current do each of the following lightbulbs draw from the 120-V power line?

    a. 60-W bulb

    b. 100-W bulb

    c. 300-W bulb

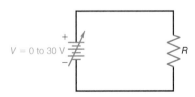

Figure 4-11   Circuit diagram for Prob. 4.18.

4.23 How much is the output voltage of a power supply if it supplies 75 W of power while delivering a current of 5 A?

4.24 How much power is consumed by a 12-V incandescent lamp if it draws 150 mA of current when lit?

4.25 How much will it cost to operate a 1500-W quartz heater for 48 h if the cost of electricity is 7¢/kWh?

4.26 How much will it cost to run an electric motor for 10 days if the motor draws 15 A of current from the 240-V power line? The cost of electricity is 7.5¢/kWh.

SECTION 4.8 Power Dissipation in Resistance

In Probs. 4.27 to 4.30, solve for the power, P, dissipated by the resistance, R.

4.27 How much power is dissipated by a 5.6-kΩ resistor whose current is 9.45 mA?

4.28 How much power is dissipated by a 50-Ω load if the voltage across the load is 100 V?

4.29 How much power is dissipated by a 600-Ω load if the voltage across the load is 36 V?

4.30 How much power is dissipated by an 8-Ω load if the current in the load is 200 mA?

SECTION 4.9 Power Formulas

In Probs. 4.31 to 4.39, solve for the unknowns listed.

4.31 a. $P = 2$ W, $R = 2$ kΩ, $V = $?

    b. $P = 50$ mW, $R = 312.5$ kΩ, $I = $?

4.32 Calculate the maximum current that a 1-kΩ, 1-W carbon resistor can safely handle without exceeding its power rating.

4.33 Calculate the maximum current that a 22-kΩ, ⅛-W resistor can safely handle without exceeding its power rating.

4.34 What is the hot resistance of a 60-W, 120-V lightbulb?

4.35 A 50-Ω load dissipates 200 W of power. How much voltage is across the load?

4.36 Calculate the maximum voltage that a 390-Ω, ½-W resistor can safely handle without exceeding its power rating.

4.37 How much current does a 960-W coffeemaker draw from the 120-V power line?

4.38 If a 4-Ω speaker dissipates 15 W of power, how much voltage is across the speaker?

4.39 What is the resistance of a 20-W, 12-V halogen lamp?

Series Circuits

Learning Outcomes

After studying this chapter, you should be able to:

> *Explain* why the current is the same in all parts of a series circuit.
> *Calculate* the total resistance of a series circuit.
> *Calculate* the current in a series circuit.
> *Determine* the individual resistor voltage drops in a series circuit.
> *Apply* Kirchhoff's voltage law to series circuits.
> *Determine* the polarity of a resistor's *IR* voltage drop.
> *Calculate* the total power dissipated in a series circuit.
> *Determine* the net voltage of series-aiding and series-opposing voltage sources.
> *Define* the terms *earth ground* and *chassis ground*.
> *Calculate* the voltage at a given point with respect to ground in a series circuit.
> *Describe* the effect of an open in a series circuit.
> *Describe* the effect of a short in a series circuit.

A series circuit is any circuit that provides only one path for current flow. An example of a series circuit is shown in Fig. 5-1. Here two resistors are connected end to end with their opposite ends connected across the terminals of a voltage source. Figure 5-1a shows the pictorial wiring diagram, and Fig. 5-1b shows the schematic diagram. The small dots in Fig. 5-1b represent free electrons. Notice that the free electrons have only one path to follow as they leave the negative terminal of the voltage source, flow through resistors R_2 and R_1, and return to the positive terminal. Since there is only one path for electrons to follow, the current, *I*, must be the same in all parts of a series circuit. To solve for the values of voltage, current, or resistance in a series circuit, we can apply Ohm's law. This chapter covers all of the characteristics of series circuits.

5.1 Why *I* Is the Same in All Parts of a Series Circuit

An electric current is a movement of charges between two points, produced by the applied voltage. When components are connected in successive order, as in Fig. 5-1, they form a series circuit. The resistors R_1 and R_2 are in series with each other and the battery.

In Fig. 5-2a, the battery supplies the potential difference that forces free electrons to drift from the negative terminal at A, toward B, through the connecting wires and resistances R_3, R_2, and R_1, back to the positive battery terminal at J. At the negative battery terminal, its negative charge repels electrons. Therefore, free electrons in the atoms of the wire at this terminal are repelled from A toward B. Similarly, free electrons at point B can then repel adjacent electrons, producing an electron drift toward C and away from the negative battery terminal.

At the same time, the positive charge of the positive battery terminal attracts free electrons, causing electrons to drift toward I and J. As a result, the free electrons in R_1, R_2, and R_3 are forced to drift toward the positive terminal.

The positive terminal of the battery attracts electrons just as much as the negative side of the battery repels electrons. Therefore, the motion of free electrons in the circuit starts at the same time and at the same speed in all parts of the circuit.

The electrons returning to the positive battery terminal are not the same electrons as those leaving the negative terminal. Free electrons in the wire are forced to move to the positive terminal because of the potential difference of the battery.

The free electrons moving away from one point are continuously replaced by free electrons flowing from an adjacent point in the series circuit. All electrons have the same speed as those leaving the battery. In all parts of the circuit, therefore, the electron drift is the same. An equal number of electrons move at one time with the same speed. That is why the current is the same in all parts of the series circuit.

In Fig. 5-2b, when the current is 2 A, for example, this is the value of the current through R_1, R_2, R_3, and the battery at the same instant. Not only is the amount of current the same throughout, but the current in all parts of a series circuit cannot differ in any way because there is just one current path for the entire circuit.

The order in which components are connected in series does not affect the current. In Fig. 5-3b, resistances R_1 and R_2 are connected in reverse order compared with Fig. 5-3a, but in both cases they are in series. The current through each is the same because there is only one path for the electron

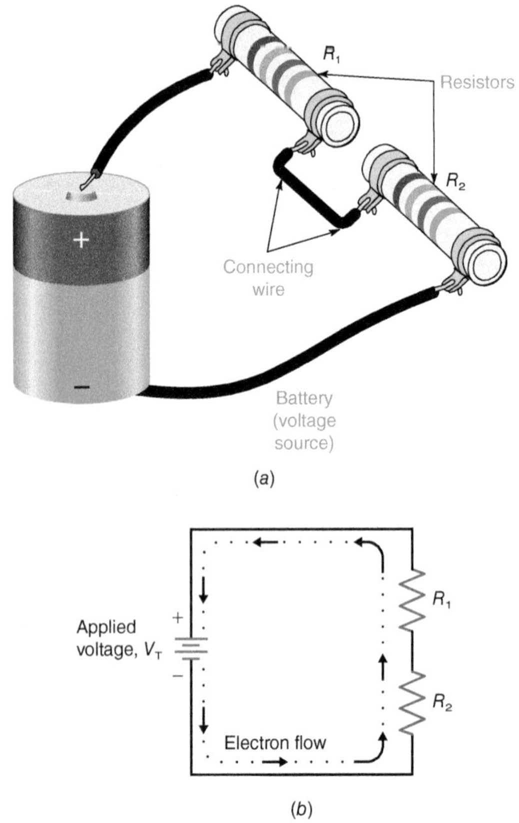

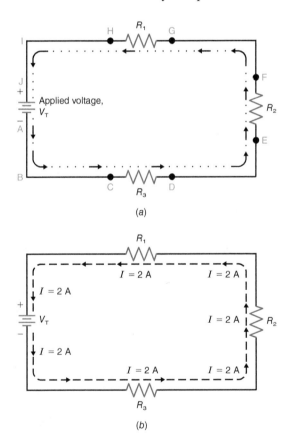

(a)

(b)

Figure 5-1 A series circuit. (a) Pictorial wiring diagram. (b) Schematic diagram.

Figure 5-2 There is only one current through R_1, R_2, and R_3 in series. (a) Electron drift is the same in all parts of a series circuit. (b) Current *I* is the same at all points in a series circuit.

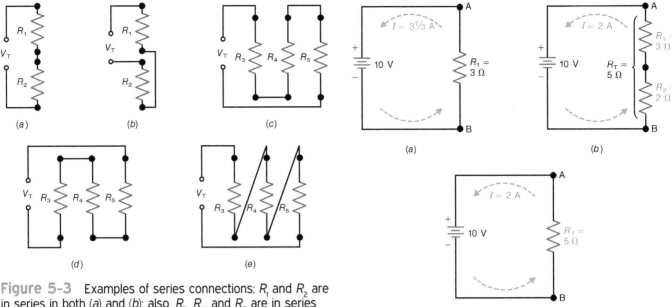

Figure 5-3 Examples of series connections: R_1 and R_2 are in series in both (a) and (b); also, R_3, R_4, and R_5 are in series in (c), (d), and (e).

Figure 5-4 Series resistances are added for the total R_T. (a) R_1 alone is 3 Ω. (b) R_1 and R_2 in series total 5 Ω. (c) The R_T of 5 Ω is the same as one resistance of 5 Ω between points A and B.

flow. Similarly, R_3, R_4, and R_5 are in series and have the same current for the connections shown in Fig. 5-3c, d, and e. Furthermore, the resistances need not be equal.

The question of whether a component is first, second, or last in a series circuit has no meaning in terms of current. The reason is that I is the same amount at the same time in all series components.

In fact, **series components** can be defined as those in the same current path. The path is from one side of the voltage source, through the series components, and back to the other side of the applied voltage. However, the series path must not have any point at which the current can branch off to another path in parallel.

5.2 Total R Equals the Sum of All Series Resistances

When a series circuit is connected across a voltage source, as shown in Fig. 5-3, the free electrons forming the current must drift through all the series resistances. This path is the only way the electrons can return to the battery. With two or more resistances in the same current path, therefore, the total resistance across the voltage source is the opposition of all the resistances.

Specifically, the total resistance R_T of a series string is equal to the sum of the individual resistances. This rule is illustrated in Fig. 5-4. In Fig. 5-4b, 2 Ω is added in series with the 3 Ω of Fig. 5-4a, producing the total resistance of 5 Ω. The total opposition of R_1 and R_2 limiting the amount of current is the same as though a 5-Ω resistance were used, as shown in the equivalent circuit in Fig. 5-4c.

Series String

A combination of series resistances is often called a **string.** The string resistance equals the sum of the individual resistances. For instance, R_1 and R_2 in Fig. 5-4b form a series string having an R_T of 5 Ω. A string can have two or more resistors.

By Ohm's law, the amount of current between two points in a circuit equals the potential difference divided by the resistance between these points. Because the entire string is connected across the voltage source, the current equals the voltage applied across the entire string divided by the total series resistance of the string. Between points A and B in Fig. 5-4, for example, 10 V is applied across 5 Ω in Fig. 5-4b and c to produce 2 A. This current flows through R_1 and R_2 in one series path.

Series Resistance Formula

In summary, the *total resistance* of a series string equals the sum of the individual resistances. The formula is

$$R_T = R_1 + R_2 + R_3 + \cdots + \text{etc.} \tag{5-1}$$

where R_T is the total resistance and R_1, R_2, and R_3 are individual series resistances.

This formula applies to any number of resistances, whether equal or not, as long as they are in the same series string. Note that R_T is the resistance to use in calculating the current in a series string. Then Ohm's law is

$$I = \frac{V_T}{R_T} \tag{5-2}$$

where R_T is the sum of all the resistances, V_T is the voltage applied across the total resistance, and I is the current in all parts of the string.

Note that adding series resistance reduces the current. In Fig. 5-4*a* the 3-Ω R_1 allows 10 V to produce $3\frac{1}{3}$ A. However, I is reduced to 2 A when the 2-Ω R_2 is added for a total series resistance of 5 Ω opposing the 10-V source.

EXAMPLE 5-1

Two resistances R_1 and R_2 of 5 Ω each and R_3 of 10 Ω are in series. How much is R_T?

Answer:

$$R_T = R_1 + R_2 + R_3 = 5 + 5 + 10$$

$$R_T = 20\ \Omega$$

EXAMPLE 5-2

With 80 V applied across the series string of Example 5-1, how much is the current in R_3?

Answer:

$$I = \frac{V_T}{R_T} = \frac{80\ \text{V}}{20\ \Omega}$$

$$I = 4\ \text{A}$$

This 4-A current is the same in R_3, R_2, R_1, or any part of the series circuit.

5.3 Series *IR* Voltage Drops

With current I through a resistance, by Ohm's law, the voltage across R is equal to $I \times R$. This rule is illustrated in Fig. 5-5 for a string of two resistors. In this circuit, I is 1 A because the applied V_T of 10 V is across the total R_T of 10 Ω, equal to the 4-Ω R_1 plus the 6-Ω R_2. Then I is 10 V/10 Ω = 1 A.

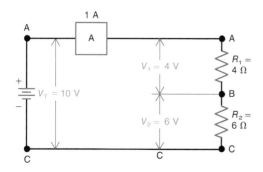

Figure 5-5 An example of *IR* voltage drops V_1 and V_2 in a series circuit.

For each *IR* voltage in Fig. 5-5, multiply each R by the 1 A of current in the series circuit. Then

$$V_1 = IR_1 = 1\ \text{A} \times 4\ \Omega = 4\ \text{V}$$
$$V_2 = IR_2 = 1\ \text{A} \times 6\ \Omega = 6\ \text{V}$$

The V_1 of 4 V is across the 4 Ω of R_1. Also, the V_2 of 6 V is across the 6 Ω of R_2. The two voltages V_1 and V_2 are in series.

The *IR* voltage across each resistance is called an *IR drop*, or a *voltage drop*, because it reduces the potential difference available for the remaining resistances in the series circuit. Note that the symbols V_1 and V_2 are used for the voltage drops across each resistor to distinguish them from the source V_T applied across both resistors.

In Fig. 5-5, the V_T of 10 V is applied across the total series resistance of R_1 and R_2. However, because of the *IR* voltage drop of 4 V across R_1, the potential difference across R_2 is only 6 V. The positive potential drops from 10 V at point A, with respect to the common reference point at C, down to 6 V at point B with reference to point C. The potential difference of 6 V between B and the reference at C is the voltage across R_2.

Similarly, there is an *IR* voltage drop of 6 V across R_2. The positive potential drops from 6 V at point B with respect to point C, down to 0 V at point C with respect to itself. The potential difference between any two points on the return line to the battery must be zero because the wire has practically zero resistance and therefore no *IR* drop.

Note that voltage must be applied by a source of *potential difference* such as the battery to produce current and have an *IR* voltage drop across the resistance. With no current through a resistor, the resistor has only resistance. There is no potential difference across the two ends of the resistor.

The *IR* drop of 4 V across R_1 in Fig. 5-5 represents that part of the applied voltage used to produce the current of 1 A through the 4-Ω resistance. Also, the *IR* drop across R_2 is 6 V because this much voltage allows 1 A in the 6-Ω resistance. The *IR* drop is more in R_2 because more potential difference is necessary to produce the same amount of current in the higher resistance. For series circuits, in general, the highest R has the largest *IR* voltage drop across it.

EXAMPLE 5-3

In Fig. 5-6, solve for R_T, I, and the individual resistor voltage drops.

Answer:

First, find R_T by adding the individual resistance values.

$$R_T = R_1 + R_2 + R_3$$
$$= 10\ \Omega + 20\ \Omega + 30\ \Omega$$
$$= 60\ \Omega$$

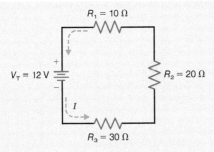

Figure 5-6 Circuit for Example 5-3.

Next, solve for the current, I.

$$I = \frac{V_T}{R_T}$$
$$= \frac{12 \text{ V}}{60 \text{ }\Omega}$$
$$= 200 \text{ mA}$$

Now we can solve for the individual resistor voltage drops.

$$V_1 = I \times R_1$$
$$= 200 \text{ mA} \times 10 \text{ }\Omega$$
$$= 2 \text{ V}$$
$$V_2 = I \times R_2$$
$$= 200 \text{ mA} \times 20 \text{ }\Omega$$
$$= 4 \text{ V}$$
$$V_3 = I \times R_3$$
$$= 200 \text{ mA} \times 30 \text{ }\Omega$$
$$= 6 \text{ V}$$

Notice that the individual voltage drops are proportional to the series resistance values. For example, because R_3 is three times larger than R_1, V_3 will be three times larger than V_1. With the same current through all the resistors, the largest resistance must have the largest voltage drop.

5.4 Kirchhoff's Voltage Law (KVL)

Kirchhoff's voltage law (KVL) states that the sum of all resistor voltage drops in a series circuit equals the **applied voltage.** Expressed as an equation, Kirchhoff's voltage law is

$$V_T = V_1 + V_2 + V_3 + \cdots + \text{etc.} \qquad \textbf{(5-3)}$$

where V_T is the applied voltage and V_1, V_2, V_3 . . . are the individual IR voltage drops.

EXAMPLE 5-4

A voltage source produces an IR drop of 40 V across a 20-Ω R_1, 60 V across a 30-Ω R_2, and 180 V across a 90-Ω R_3, all in series. According to Kirchhoff's voltage law, how much is the applied voltage V_T?

Answer:
$$V_T = 40 \text{ V} + 60 \text{ V} + 180 \text{ V}$$
$$V_T = 280 \text{ V}$$

Note that the IR drop across each R results from the same current of 2 A, produced by 280 V across the total R_T of 140 Ω.

EXAMPLE 5-5

An applied V_T of 120 V produces IR drops across two series resistors R_1 and R_2. If the voltage drop across R_1 is 40 V, how much is the voltage drop across R_2?

Answer:
Since V_1 and V_2 must total 120 V and V_1 is 40 V, the voltage drop across R_2 must be the difference between 120 V and 40 V, or

$$V_2 = V_T - V_1 = 120 \text{ V} - 40 \text{ V}$$
$$V_2 = 80 \text{ V}$$

It is logical that V_T is the sum of the series IR drops. The current I is the same in all series components. For this reason, the total of all series voltages V_T is needed to produce the same I in the total of all series resistances R_T as the I that each resistor voltage produces in its R.

A practical application of voltages in a series circuit is illustrated in Fig. 5-7. In this circuit, two 120-V lightbulbs

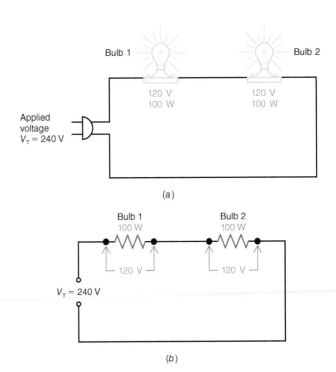

Figure 5-7 Series string of two 120-V lightbulbs operating from a 240-V line. (*a*) Wiring diagram. (*b*) Schematic diagram.

are operated from a 240-V line. If one bulb were connected to 240 V, the filament would burn out. With the two bulbs in series, however, each has 120 V for proper operation. The two 120-V drops across the bulbs in series add to equal the applied voltage of 240 V.

Note: A more detailed explanation of Kirchhoff's voltage law is provided in Chap. 8 (Sec. 8-2).

5.5 Polarity of *IR* Voltage Drops

When a voltage drop exists across a resistance, one end must be either more positive or more negative than the other end. Otherwise, without a potential difference no current could flow through the resistance to produce the voltage drop. The *polarity* of this *IR* voltage drop can be associated with the direction of *I* through *R*. In brief, electrons flow into the negative side of the *IR* voltage and out the positive side (see Fig. 5-8*a*).

If we want to consider conventional current, with positive charges moving in the opposite direction from electron flow, the rule is reversed for the positive charges. See Fig. 5-8*b*. Here the positive charges for *I* are moving into the positive side of the *IR* voltage.

However, for either electron flow or conventional current, the actual polarity of the *IR* drop is the same. In both *a* and *b* of Fig. 5-8, the top end of *R* in the diagrams is positive since this is the positive terminal of the source producing the current. After all, the resistor does not know which direction of current we are thinking of.

A series circuit with two *IR* voltage drops is shown in Fig. 5-9. We can analyze these polarities in terms of electron flow. The electrons move from the negative terminal of the source V_T through R_2 from point C to D. Electrons move into C and out from D. Therefore C is the negative side of the voltage drop across R_2. Similarly, for the *IR* voltage drop across R_1, point E is the negative side, compared with point F.

A more fundamental way to consider the polarity of *IR* voltage drops in a circuit is the fact that between any two points the one nearer to the positive terminal of the voltage source is more positive; also, the point nearer to the negative terminal of

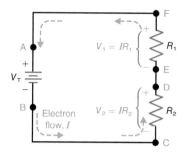

Figure 5-9 Example of two *IR* voltage drops in series. Electron flow shown for direction of *I*.

the applied voltage is more negative. A point nearer the terminal means that there is less resistance in its path.

In Fig. 5-9 point C is nearer to the negative battery terminal than point D. The reason is that C has no resistance to B, whereas the path from D to B includes the resistance of R_2. Similarly, point F is nearer to the positive battery terminal than point E, which makes F more positive than E.

Notice that points D and E in Fig. 5-9 are marked with both plus and minus polarities. The plus polarity at D indicates that it is more positive than C. This polarity, however, is shown just for the voltage across R_2. Point D cannot be more positive than points F and A. The positive terminal of the applied voltage must be the most positive point because the battery is generating the *positive potential* for the entire circuit.

Similarly, points B and C must have the most negative potential in the entire string, since point B is the negative terminal of the applied voltage. Actually, the plus polarity marked at D means only that this end of R_2 is less negative than C by the amount of voltage drop across R_2.

Consider the potential difference between E and D in Fig. 5-9, which is only a piece of wire. This voltage is zero because there is no resistance between these two points. Without any resistance here, the current cannot produce the *IR* drop necessary for a difference in potential. Points E and D are, therefore, the same electrically since they have the same potential.

When we go around the external circuit from the negative terminal of V_T, with electron flow, the voltage drops are drops in *negative potential*. For the opposite direction, starting from the positive terminal of V_T, the voltage drops are drops in positive potential. Either way, the voltage drop of each series *R* is its proportional part of the V_T needed for the one value of current in all resistances.

5.6 Total Power in a Series Circuit

The power needed to produce current in each series resistor is used up in the form of heat. Therefore, the *total power* used is the sum of the individual values of power dissipated in each part of the circuit. As a formula,

$$P_T = P_1 + P_2 + P_3 + \cdots + \text{etc.} \qquad (5\text{-}4)$$

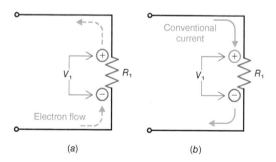

(a)   (b)

Figure 5-8 The polarity of *IR* voltage drops. (a) Electrons flow into the negative side of V_1 across R_1. (b) The same polarity of V_1 with positive charges into the positive side.

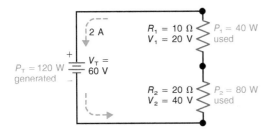

Figure 5-10 The sum of the individual powers P_1 and P_2 used in each resistance equals the total power P_T produced by the source.

As an example, in Fig. 5-10, R_1 dissipates 40 W for P_1, equal to 20 V × 2 A for the VI product. Or, the P_1 calculated as I^2R is $(2 × 2) × 10 = 40$ W. Also, P_1 is V^2/R, or $(20 × 20)/10 = 40$ W.

Similarly, P_2 for R_2 is 80 W. This value is $40 × 2$ for VI, $(2 × 2) × 20$ for I^2R, or $(40 × 40)/20$ for V^2/R. Thus, P_2 must be more than P_1 because R_2 is more than R_1 with the same current.

The total power dissipated by R_1 and R_2, then, is $40 + 80 = 120$ W. This power is generated by the source of applied voltage.

The total power can also be calculated as $V_T × I$. The reason is that V_T is the sum of all series voltages and I is the same in all series components. In this case, then, $P_T = V_T × I = 60 × 2 = 120$ W.

The total power here is 120 W, calculated either from the total voltage or from the sum of P_1 and P_2. This is the amount of power produced by the battery. The voltage source produces this power, equal to the amount used by the resistors.

5.7 Series-Aiding and Series-Opposing Voltages

Series-aiding voltages are connected with polarities that allow current in the same direction. In Fig. 5-11a, the 6 V of V_1 alone could produce a 3-A electron flow from the negative terminal, with the 2-Ω R. Also, the 8 V of V_2 could produce 4 A in the same direction. The total I then is 7 A.

Instead of adding the currents, however, the voltages V_1 and V_2 can be added, for a V_T of $6 + 8 = 14$ V. This 14 V produces 7 A in all parts of the series circuit with a resistance of 2 Ω. Then I is $14/2 = 7$ A.

Voltages are connected series-aiding when the plus terminal of one is connected to the negative terminal of the next. They can be added for a total equivalent voltage. This idea applies in the same way to voltage sources, such as batteries, and to voltage drops across resistances. Any number of voltages can be added, as long as they are connected with series-aiding polarities.

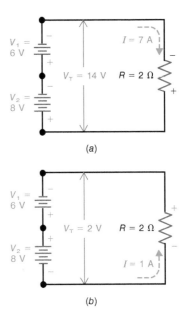

(a)

(b)

Figure 5-11 Example of voltage sources V_1 and V_2 in series. (a) Note the connections for series-aiding polarities. Here $8 V + 6 V = 14 V$ for the total V_T. (b) Connections for series-opposing polarities. Now $8 V − 6 V = 2 V$ for V_T.

Series-opposing voltages are subtracted, as shown in Fig. 5-11b. Notice here that the positive terminals of V_1 and V_2 are connected. Subtract the smaller from the larger value, and give the net V the polarity of the larger voltage. In this example, V_T is $8 − 6 = 2$ V. The polarity of V_T is the same as V_2 because its voltage is higher than V_1.

If two series-opposing voltages are equal, the net voltage will be zero. In effect, one voltage balances out the other. The current I also is zero, without any net potential difference.

5.8 Analyzing Series Circuits with Random Unknowns

Refer to Fig. 5-12. Suppose that the source V_T of 50 V is known, with a 14-Ω R_1 and 6-Ω R_2. The problem is to find R_T, I, the individual voltage drops V_1 and V_2 across each resistor, and the power dissipated.

We must know the total resistance R_T to calculate I because the total applied voltage V_T is given. This V_T is applied across the total resistance R_T. In this example, R_T is $14 + 6 = 20$ Ω.

Now I can be calculated as V_T/R_T, or $50/20$, which equals 2.5 A. This 2.5-A I flows through R_1 and R_2.

The individual voltage drops are

$$V_1 = IR_1 = 2.5 × 14 = 35 \text{ V}$$
$$V_2 = IR_2 = 2.5 × 6 = 15 \text{ V}$$

Note that V_1 and V_2 total 50 V, equal to the applied V_T.

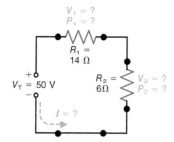

Figure 5-12 Analyzing a series circuit to find I, V_1, V_2, P_1, and P_2. See text for solution.

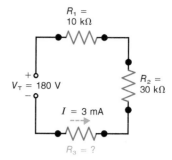

Figure 5-13 Find the resistance of R_3. See text for the analysis of this series circuit.

The calculations to find the power dissipated in each resistor are as follows:

$$P_1 = V_1 \times I = 35 \times 2.5 = 87.5 \text{ W}$$
$$P_2 = V_2 \times I = 15 \times 2.5 = 37.5 \text{ W}$$

These two values of dissipated power total 125 W. The power generated by the source equals $V_T \times I$ or 50×2.5, which is also 125 W.

General Methods for Series Circuits

For other types of problems with series circuits, it is useful to remember the following:

1. When you know the I for one component, use this value for I in all components, for the current is the same in all parts of a series circuit.

2. To calculate I, the total V_T can be divided by the total R_T, or an individual IR drop can be divided by its R. For instance, the current in Fig. 5-12 could be calculated as V_2/R_2 or $15/6$, which equals the same 2.5 A for I. However, do not mix a total value for the entire circuit with an individual value for only part of the circuit.

3. When you know the individual voltage drops around the circuit, these can be added to equal the applied V_T. This also means that a known voltage drop can be subtracted from the total V_T to find the remaining voltage drop.

These principles are illustrated by the problem in Fig. 5-13. In this circuit, R_1 and R_2 are known but not R_3. However, the current through R_3 is given as 3 mA.

With just this information, all values in this circuit can be calculated. The I of 3 mA is the same in all three series resistances. Therefore,

$$V_1 = 3 \text{ mA} \times 10 \text{ k}\Omega = 30 \text{ V}$$
$$V_2 = 3 \text{ mA} \times 30 \text{ k}\Omega = 90 \text{ V}$$

The sum of V_1 and V_2 is $30 + 90 = 120$ V. This 120 V plus V_3 must total 180 V. Therefore, V_3 is $180 - 120 = 60$ V.

With 60 V for V_3, equal to IR_3, then R_3 must be 60/0.003, equal to 20,000 Ω or 20 kΩ. The total circuit resistance is

60 kΩ, which results in the current of 3 mA with 180 V applied, as specified in the circuit.

Another way of doing this problem is to find R_T first. The equation $I = V_T/R_T$ can be inverted to calculate R_T.

$$R_T = \frac{V_T}{I}$$

With a 3-mA I and 180 V for V_T, the value of R_T must be 180 V/3 mA = 60 kΩ. Then R_3 is 60 kΩ − 40 kΩ = 20 kΩ.

The power dissipated in each resistance is 90 mW in R_1, 270 mW in R_2, and 180 mW in R_3. The total power is 90 + 270 + 180 = 540 mW.

Series Voltage-Dropping Resistors

A common application of series circuits is to use a resistance to drop the voltage from the source V_T to a lower value, as in Fig. 5-14. The load R_L here represents a radio that operates normally with a 9-V battery. When the radio is on, the dc load current with 9 V applied is 18 mA. Therefore, the requirements are 9 V at 18 mA as the load.

To operate this radio from 12.6 V, the voltage-dropping resistor R_S is inserted in series to provide a voltage drop V_S that will make V_L equal to 9 V. The required voltage drop for V_S is the difference between V_L and the higher V_T. As a formula,

$$V_S = V_T - V_L = 12.6 - 9 = 3.6 \text{ V}$$

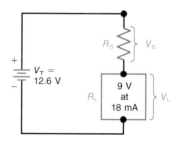

Figure 5-14 Example of a series voltage-dropping resistor R_S used to drop V_T of 12.6 V to 9 V for R_L. See text for calculations.

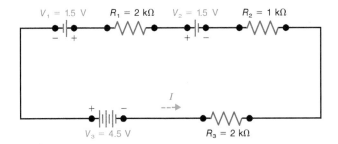

Figure 5-15 Finding the *I* for this series circuit with three voltage sources. See text for solution.

Furthermore, this voltage drop of 3.6 V must be provided with a current of 18 mA, for the current is the same through R_S and R_L. To calculate R_S,

$$R_S = \frac{3.6 \text{ V}}{18 \text{ mA}} = 0.2 \text{ k}\Omega = 200 \ \Omega$$

Circuit with Voltage Sources in Series

See Fig. 5-15. Note that V_1 and V_2 are series-opposing, with + to + through R_1. Their net effect, then, is 0 V. Therefore, V_T consists only of V_3, equal to 4.5 V. The total R is 2 + 1 + 2 = 5 kΩ for R_T. Finally, I is V_T/R_T or 4.5 V/5 kΩ, which is equal to 0.9 mA, or 900 μA.

5.9 Ground Connections in Electrical and Electronic Systems

In most electrical and electronic systems, one side of the voltage source is connected to ground. For example, one side of the voltage source of the 120-Vac power line in residential wiring is connected directly to *earth ground*. The reason for doing this is to reduce the possibility of electric shock. The connection to earth ground is usually made by driving copper rods into the ground and connecting the ground wire of the electrical system to these rods. The schematic symbol used for earth ground is shown in Fig. 5-16. In electronic circuits, however, not all ground connections are necessarily earth ground connections. The pitchforklike symbol shown in Fig. 5-16 is considered by many people to be the most appropriate symbol for a metal chassis or copper foil ground on printed-circuit boards. This *chassis ground symbol* represents a common return path for current and may or may not be connected to an actual earth ground. Another ground symbol, common ground, is shown in Fig. 5-16. This is just another symbol used to represent a common return path for current in a circuit. In all cases, ground is assumed to be at a potential of 0 V, regardless of which symbol is shown. Some schematic diagrams may use two or all three of the ground symbols shown in Fig. 5-16. In this type of circuit, each ground represents a common return path for only those circuits using the same ground symbol. When more than one type of ground symbol is shown on a schematic diagram, it is important to realize that each one is electrically isolated from the other. The term *electrically isolated* means that the resistance between each ground or common point is infinite ohms.

Although standards defining the use of each ground symbol in Fig. 5-16 have been set, the use of these symbols in the electronics industry seems to be inconsistent with their definitions. In other words, a schematic may show the earth ground symbol, even though it is a chassis ground connection. Regardless of the symbol used, the main thing to remember is that the symbol represents a common return path for current in a given circuit. In this text, the earth ground symbol shown in Fig. 5-16 has been arbitrarily chosen as the symbol representing a common return path for current.

Figure 5-17 shows a series circuit employing the earth ground symbol. Since each ground symbol represents the same electrical potential of 0 V, the negative terminal of V_T and the bottom end of R_3 are actually connected to the same point electrically. Electrons leaving the bottom of V_T flow through the common return path represented by the ground symbol and return to the bottom of R_3, as shown in the figure. One of the main reasons for using ground connections in electronic circuits is to simplify the wiring.

Voltages Measured with Respect to Ground

When a circuit has a ground as a common return, we generally measure the voltages with respect to this ground. The circuit in Fig. 5-18a is called a *voltage divider*. Let us consider this circuit without any ground, and then analyze the effect of grounding different points on the divider. It is important to realize that this circuit operates the same way with or without the ground. The only factor that changes is the reference point for measuring the voltages.

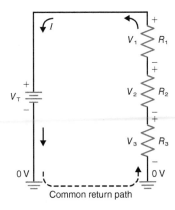

Figure 5-17 Series circuit using earth ground symbol to represent common return path for current.

Figure 5-16 Ground symbols.

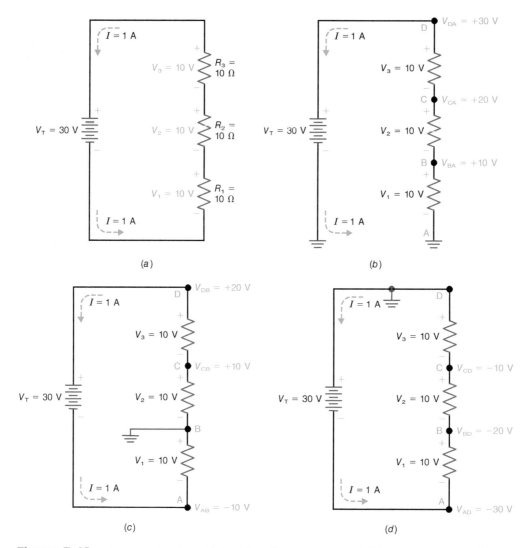

Figure 5-18 An example of calculating dc voltages measured with respect to ground. (a) Series circuit with no ground connection. (b) Negative side of V_T grounded to make all voltages positive with respect to ground. (c) Positive and negative voltages with respect to ground at point B. (d) Positive side of V_T grounded; all voltages are negative to ground.

In Fig. 5-18a, the three 10-Ω resistances R_1, R_2, and R_3 divide the 30-V source equally. Then each voltage drop is $30/3 = 10$ V for V_1, V_2, and V_3. The polarity of each resistor voltage drop is positive at the top and negative at the bottom, the same as V_T. As you recall, the polarity of a resistor's voltage drop is determined by the direction of current flow.

If we want to consider the current, I is $30/30 = 1$ A. Each IR drop is $1 \times 10 = 10$ V for V_1, V_2, and V_3.

Positive Voltages to Negative Ground

In Fig. 5-18b, the negative side of V_T is grounded, and the bottom end of R_1 is also grounded to complete the circuit. The ground is at point A. Note that the individual voltages V_1, V_2, and V_3 are still 10 V each. Also, the current is still 1 A. The direction of current is also the same, from the negative side of V_T, through the common ground, to the bottom end of R_1. The

only effect of the ground here is to provide a conducting path from one side of the source to one side of the load.

With the ground in Fig. 5-18b, though, it is useful to consider the voltages with respect to ground. In other words, the ground at point A will now be the reference for all voltages. When a voltage is indicated for only one point in a circuit, generally the other point is assumed to be ground. We must have two points for a potential difference.

Let us consider the voltages at points B, C, and D. The voltage at B to ground is V_{BA}. This *double subscript notation* indicates that we measure at B with respect to A. In general, the first letter indicates the point of measurement and the second letter is the reference point.

Then V_{BA} is +10 V. The positive sign is used here to emphasize the polarity. The value of 10 V for V_{BA} is the same as V_1 across R_1 because points B and A are across R_1. However,

V_1 as the voltage across R_1 cannot be given any polarity without a reference point.

When we consider the voltage at C, then, V_{CA} is +20 V. This voltage equals $V_1 + V_2$. Also, for point D at the top, V_{DA} is +30 V for $V_1 + V_2 + V_3$.

Positive and Negative Voltages to a Grounded Tap

In Fig. 5-18c point B in the divider is grounded. The purpose is to have the divider supply negative and positive voltages with respect to ground. The negative voltage here is V_{AB}, which equals −10 V. This value is the same 10 V as V_1, but V_{AB} is the voltage at the negative end A with respect to the positive end B. The other voltages in the divider are $V_{CB} = $ +10 V and $V_{DB} = $ +20 V.

We can consider the ground at B as a dividing point for positive and negative voltages. For all points toward the positive side of V_T, any voltage is positive to ground. Going the other way, at all points toward the negative side of V_T, any voltage is negative to ground.

Negative Voltages to Positive Ground

In Fig. 5-18d, point D at the top of the divider is grounded, which is the same as grounding the positive side of the source V_T. The voltage source here is *inverted,* compared with Fig. 5-18b, as the opposite side is grounded. In Fig. 5-18d, all voltages on the divider are negative to ground. Here, $V_{CD} = $ −10 V, $V_{BD} = $ −20 V, and $V_{AD} = $ −30 V. Any point in the circuit must be more negative than the positive terminal of the source, even when this terminal is grounded.

CHAPTER 5 SYSTEM SIDEBAR

Light-Emitting Diodes

Light-emitting diodes (LEDs) are everywhere in electronics. Almost all electronic devices have some form of indicator light or display, and all use LEDs. These red, green, blue, and yellow lights provide a visual indication of electronic or electric functions. LEDs in the form of seven-segment displays are still widely used in digital clocks. Even the newer large-screen HDTV sets use LEDs. Other uses are lighted signs and automobile taillights. And white LEDs are changing the way we light our world. You will learn more about these common devices in a later chapter, but for now here is a quickie introduction.

An LED is a solid-state or semiconductor device called a *diode*. A diode is a two-element component that allows current to flow through it in one direction only. It blocks current in the opposite direction. Diodes are widely used in electronics for rectification (converting ac to dc), circuit protection, voltage stabilization, and waveshaping. LEDs are special diodes that emit light when current is flowing through them. Figure S5-1 shows several common LEDs.

The schematic symbol for an LED is shown in Fig. S5-2. It has two components, the cathode (K) and the anode (A). When voltage is applied to an LED, we say that it is biased. If the anode is made positive with respect to the cathode, electrons will flow from cathode to anode. The arrow shows the direction of electron flow. If the polarity of the voltage is reversed, no current flows.

Note in Fig. S5-3 that a resistor is connected in series with the LED. This series voltage-dropping resistor or current-limiting resistor protects the diode from excessive voltage and current.

When current flows through an LED, a voltage drop appears across it. This voltage varies with the type and color of the LED. Red LEDs usually have a voltage drop of about 2 volts. Blue and white LEDs have a voltage drop of over 3 volts. Green and yellow LEDs have a voltage in the 2- to 3-volt range. Current in an LED is usually limited to 50 mA or so, depending on the application. For simple indicator functions, the current

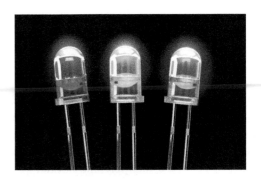

Figure S5-1   Several common LEDs.

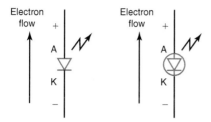

Figure S5-2   Schematic symbols for a light-emitting diode.

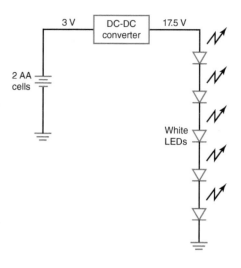

Figure S5-4   A dc-dc converter generates the higher dc voltage needed for 5 series LEDs.

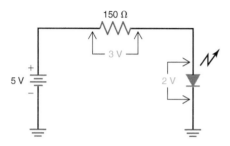

Figure S5-3   A series resistor provides the correct voltage and current for the LED.

is set to about 15–20 mA. For extra bright lighting, a current of 100 mA may be used. The amount of current determines the actual brightness.

In Fig. S5-3, the red LED needs 2 volts. The 5-volt dc power supply is too much, so the series resistor is used to drop 3 volts. If the desired current is 20 mA, the series resistor value must be

$$R = \frac{V_{drop}}{I} = \frac{3}{0.02} = 150\ \Omega$$

When multiple LEDs are used as in flashlights or LED lightbulbs, they are usually connected in series, as shown in Fig. S5-4. Five white LEDs in series will require about 3.5 volts per device for a total 17.5 volts. If the LEDs are to be powered by AA, C, or D cells that deliver about 1.5 volts, you would need 12 cells in series, which is monumentally impractical and expensive. The solution to this is to use an integrated circuit (IC) called a dc-dc converter. It can take the input from two 1.5-volt cells in series to deliver 3 volts to the dc-dc converter. The converter then puts out the desired 17.5 volts.

The forthcoming LED lightbulb replacements use an ac-dc converter called a *rectifier*. The normal 120-volt ac

has to be converted to dc first. The rectifier uses diodes and can be made to provide dc of a level suitable for strings of LEDs. Resistors are then used to adjust the current and brightness level. Some circuitry may also use a dc-dc converter or a regulator circuit to ensure that the LEDs get the correct voltage and current. Figure S5-5 shows one possible circuit. LED lightbulbs are expensive because they contain lots of related circuitry. But they are cool, efficient, and bright.

LEDs are also widely used in TV and other remote controls. These LEDs generate infrared (IR) light. Our eyes cannot see it, but it is light nonetheless. Digital codes representing the button pushes on the remote control switch the LED off and on at high speed to signal the TV set or other consumer device.

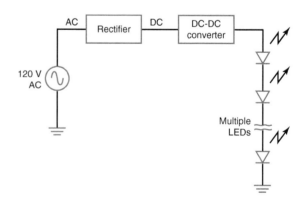

Figure S5-5   An LED lightbulb contains a complex electronic power supply circuit.

1. Three resistors in series have individual values of 120 Ω, 680 Ω, and 1.2 kΩ. How much is the total resistance, R_T?
   a. 1.8 kΩ.
   b. 20 kΩ.
   c. 2 kΩ.
   d. None of the above.

2. In a series circuit, the current, I, is
   a. different in each resistor.
   b. the same everywhere.
   c. the highest near the positive and negative terminals of the voltage source.
   d. different at all points along the circuit.

3. In a series circuit, the largest resistance has
   a. the largest voltage drop.
   b. the smallest voltage drop.
   c. more current than the other resistors.
   d. both a and c.

4. The polarity of a resistor's voltage drop is determined by
   a. the direction of current through the resistor.
   b. how large the resistance is.
   c. how close the resistor is to the voltage source.
   d. how far away the resistor is from the voltage source.

5. A 10-Ω and 15-Ω resistor are in series across a dc voltage source. If the 10-Ω resistor has a voltage drop of 12 V, how much is the applied voltage?
   a. 18 V.
   b. 12 V.
   c. 30 V.
   d. It cannot be determined.

6. How much is the voltage across a shorted component in a series circuit?
   a. The full applied voltage, V_T.
   b. The voltage is slightly higher than normal.
   c. 0 V.
   d. It cannot be determined.

7. How much is the voltage across an open component in a series circuit?
   a. The full applied voltage, V_T.
   b. The voltage is slightly lower than normal.
   c. 0 V.
   d. It cannot be determined.

8. A voltage of 120 V is applied across two resistors, R_1 and R_2, in series. If the voltage across R_2 equals 90 V, how much is the voltage across R_1?
   a. 90 V.
   b. 30 V.
   c. 120 V.
   d. It cannot be determined.

9. If two series-opposing voltages each have a voltage of 9 V, the net or total voltage is
   a. 0 V.
   b. 18 V.
   c. 9 V.
   d. none of the above.

10. On a schematic diagram, what does the chassis ground symbol represent?
    a. Hot spots on the chassis.
    b. The locations in the circuit where electrons accumulate.
    c. A common return path for current in one or more circuits.
    d. None of the above.

11. The notation, V_{BG}, means
    a. the voltage at point G with respect to point B.
    b. the voltage at point B with respect to point G.
    c. the battery (b) or generator (G) voltage.
    d. none of the above.

12. If a resistor in a series circuit is shorted, the series current, I,
    a. decreases.
    b. stays the same.
    c. increases.
    d. drops to zero.

13. A 6-V and 9-V source are connected in a series-aiding configuration. How much is the net or total voltage?
    a. −3 V.
    b. +3 V.
    c. 0 V.
    d. 15 V.

14. A 56-Ω and 82-Ω resistor are in series with an unknown resistor. If the total resistance of the series combination is 200 Ω, what is the value of the unknown resistor?
    a. 138 Ω.
    b. 62 Ω.
    c. 26 Ω.
    d. It cannot be determined.

15. How much is the total resistance, R_T, of a series circuit if one of the resistors is open?
   a. infinite (∞) Ω.
   b. 0 Ω.
   c. R_T is much lower than normal.
   d. None of the above.

16. If a resistor in a series circuit becomes open, how much is the voltage across each of the remaining resistors that are still good?
   a. Each good resistor has the full value of applied voltage.
   b. The applied voltage is split evenly among the good resistors.
   c. 0 V.
   d. It cannot be determined.

17. A 5-Ω and 10-Ω resistor are connected in series across a dc voltage source. Which resistor will dissipate more power?
   a. The 5-Ω resistor.
   b. The 10-Ω resistor.
   c. It depends on how much the current is.
   d. They will both dissipate the same amount of power.

18. Which of the following equations can be used to determine the total power in a series circuit?
   a. $P_T = P_1 + P_2 + P_3 + \cdots +$ etc.
   b. $P_T = V_T \times I$.
   c. $P_T = I^2 R_T$.
   d. All of the above.

19. Using electron flow, the polarity of a resistor's voltage drop is
   a. positive on the side where electrons enter and negative on the side where they leave.
   b. negative on the side were electrons enter and positive on the side where they leave.
   c. opposite to that obtained with conventional current flow.
   d. both b and c.

20. The schematic symbol for earth ground is
   a. .
   b. .
   c. .
   d. .

CHAPTER 5 PROBLEMS

SECTION 5.1 Why I Is the Same in All Parts of a Series Circuit

5.1 In Fig. 5-19, how much is the current, I, at each of the points A through F?

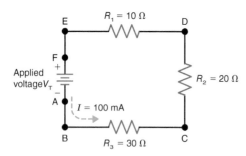

Figure 5-19

5.2 If R_1 and R_3 are interchanged in Fig. 5-19, how much is the current, I, in the circuit?

SECTION 5.2 Total R Equals the Sum of All Series Resistances

5.3 In Fig. 5-20, solve for R_T and I.

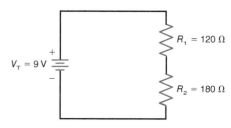

Figure 5-20

5.4 In Fig. 5-21 solve for R_T and I.

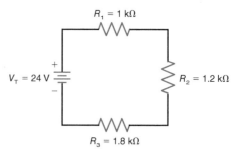

Figure 5-21

5.5 What are the new values for R_T and I in Fig. 5-21 if a 2-kΩ resistor, R_4, is added to the series circuit?

5.6 In Fig. 5-22, solve for R_T and I.

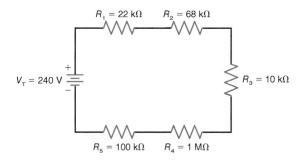

Figure 5-22

SECTION 5.3 Series *IR* Voltage Drops

5.7 In Fig. 5-21, find the voltage drops across R_1 and R_2.

5.8 In Fig. 5-20, find the voltage drops across R_1, R_2, and R_3.

5.9 In Fig. 5-22, find the voltage drops across R_1, R_2, R_3, R_4, and R_5.

5.10 In Fig. 5-23, solve for R_T, I, V_1, V_2, and V_3.

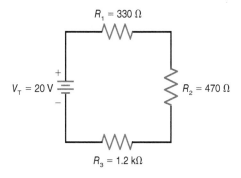

Figure 5-23

5.11 In Fig. 5-24, solve for R_T, I, V_1, V_2, V_3, and V_4.

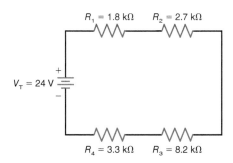

Figure 5-24

SECTION 5.4 Kirchhoff's Voltage Law (KVL)

5.12 Using Kirchhoff's voltage law, determine the value of the applied voltage, V_T, in Fig. 5-25.

5.13 If $V_1 = 2$ V, $V_2 = 6$ V, and $V_3 = 7$ V in Fig. 5-25, how much is V_T?

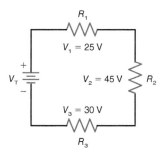

Figure 5-25

5.14 Determine the voltage, V_2, in Fig. 5-26.

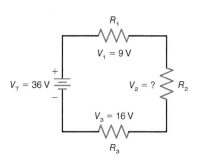

Figure 5-26

5.15 In Fig. 5-27, solve for the individual resistor voltage drops. Then, using Kirchhoff's voltage law, find V_T.

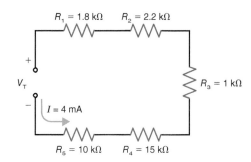

Figure 5-27

5.16 An applied voltage of 15 V is connected across resistors R_1 and R_2 in series. If $V_2 = 3$ V, how much is V_1?

SECTION 5.5 Polarity of *IR* Voltage Drops

5.17 In Fig. 5-28,
   a. Solve for R_T, I, V_1, V_2, and V_3.
   b. Indicate the direction of electron flow through R_1, R_2, and R_3.
   c. Write the values of V_1, V_2, and V_3 next to resistors R_1, R_2, and R_3.
   d. Indicate the polarity of each resistor voltage drop.

5.18 In Fig. 5-28, indicate the polarity for each resistor voltage drop using conventional current flow. Are the

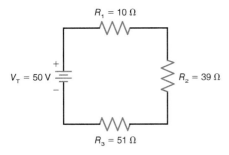

Figure 5-28

polarities opposite to those obtained with electron flow or are they the same?

5.19 If the polarity of V_T is reversed in Fig. 5-28, what happens to the polarity of the resistor voltage drops and why?

SECTION 5.6 Total Power in a Series Circuit

5.20 In Fig. 5-20, calculate P_1, P_2, and P_T.

5.21 In Fig. 5-21, calculate P_1, P_2, P_3, and P_T.

SECTION 5.7 Series–Aiding and Series–Opposing Voltages

5.22 In Fig. 5-29,
    a. How much is the net or total voltage, V_T across R_1?
    b. How much is the current, I, in the circuit?
    c. What is the direction of electron flow and conventional current through R_1?

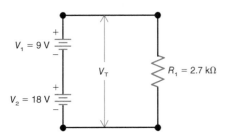

Figure 5-29

5.23 In Fig. 5-30,
    a. How much is the net or total voltage, V_T, across R_1?
    b. How much is the current, I, in the circuit?
    c. What is the direction of conventional current flow through R_1?

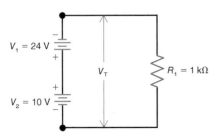

Figure 5-30

5.24 In Fig. 5-30, assume that V_2 is increased to 30 V. What is
    a. The net or total voltage, V_T, across R_1?
    b. The current, I, in the circuit?
    c. The direction of electron flow through R_1?

5.25 In Fig. 5-31,
    a. How much is the net or total voltage, V_T, across R_1 and R_2 in series?
    b. How much is the current, I, in the circuit?
    c. What is the direction of electron flow through R_1 and R_2?
    d. Calculate the voltage drops across R_1 and R_2.

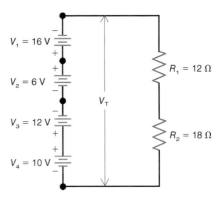

Figure 5-31

SECTION 5.8 Analyzing Series Circuits with Random Unknowns

5.26 In Fig. 5-32, calculate the value for the series resistor, R_s that will allow a 12-V, 150-mA CD player to be operated from a 30-V supply.

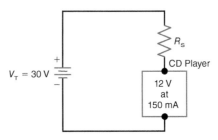

Figure 5-32

5.27 In Fig. 5-33, solve for R_T, V_3, V_4, R_3, P_T, P_1, P_2, P_3, and P_4.

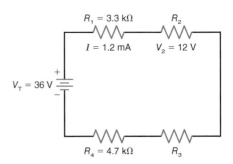

Figure 5-33

5.28 In Fig. 5-34, solve for I, R_T, V_T, V_2, and P_1.

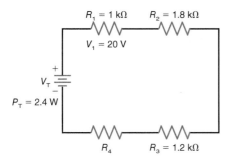

Figure 5-34

5.29 In Fig. 5-35, solve for R_3, I, V_T, and P_T.

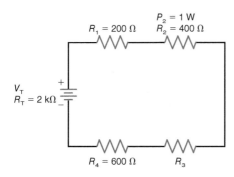

Figure 5-35

5.30 A 1.5-kΩ resistor is in series with an unknown resistance. The applied voltage, V_T, equals 36 V and the series current is 14.4 mA. Calculate the value of the unknown resistor.

5.31 How much resistance must be added in series with a 6.3-V, 150-mA lightbulb if the bulb is to be operated from a 12-V source?

5.32 A 22-Ω resistor is in series with a 12-V motor that is drawing 150 mA of current. How much is the applied voltage, V_T?

SECTION 5.9 Ground Connections in Electrical and Electronic Systems

5.33 In Fig. 5-36, solve for V_{AG}, V_{BG}, and V_{CG}.

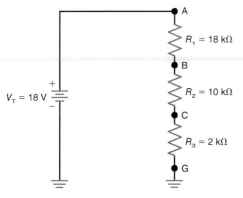

Figure 5-36

5.34 In Fig. 5-37, solve for V_{AG}, V_{BG}, and V_{CG}.

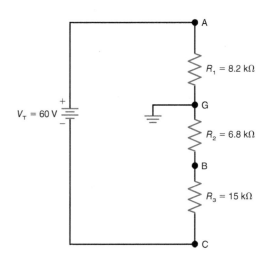

Figure 5-37

5.35 In Fig. 5-38, solve for V_{AG}, V_{BG}, V_{CG}, and V_{DG}.

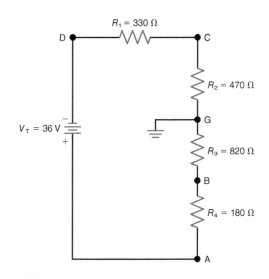

Figure 5-38

5.36 In Fig. 5-39, solve for V_{AG}, V_{BG}, and V_{CG}.

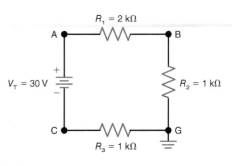

Figure 5-39

Parallel Circuits

A parallel circuit is any circuit that provides one common voltage across all components. Each component across the voltage source provides a separate path or branch for current flow. The individual branch currents are calculated as $\dfrac{V_A}{R}$ where V_A is the applied voltage and R is the individual branch resistance. The total current, I_T, supplied by the applied voltage, must equal the sum of all individual branch currents.

The equivalent resistance of a parallel circuit equals the applied voltage, V_A, divided by the total current, I_T. The term equivalent resistance refers to a single resistance that would draw the same amount of current as all the parallel-connected branches. The equivalent resistance of a parallel circuit is designated R_{EQ}.

This chapter covers all the characteristics of parallel circuits.

Learning Outcomes

After studying this chapter, you should be able to:

> *Explain* why voltage is the same across all branches in a parallel circuit.

> *Calculate* the individual branch currents in a parallel circuit.

> *Calculate* the total current in a parallel circuit using Kirchhoff's current law.

> *Calculate* the equivalent resistance of two or more resistors in parallel.

> *Calculate* the total conductance of a parallel circuit.

> *Calculate* the total power in a parallel circuit.

> *Solve* for the voltage, current, power, and resistance in a parallel circuit having random unknowns.

> *Describe* the effects of an open and short in a parallel circuit.

6.1 The Applied Voltage V_A Is the Same across Parallel Branches

A parallel circuit is formed when two or more components are connected across a voltage source, as shown in Fig. 6-1. In this figure, R_1 and R_2 are in parallel with each other and a 1.5-V battery. In Fig. 6-1b, points A, B, C, and E are equivalent to a direct connection at the positive terminal of the battery because the connecting wires have practically no resistance. Similarly, points H, G, D, and F are the same as a direct connection at the negative battery terminal. Since R_1 and R_2 are directly connected across the two terminals of the battery, both resistances must have the same potential difference as the battery. It follows that the voltage is the same across components connected in parallel. The parallel circuit arrangement is used, therefore, to connect components that require the same voltage.

A common application of parallel circuits is typical house wiring to the power line, with many lights and appliances connected across the 120-V source (Fig. 6-2). The wall receptacle has a potential difference of 120 V across each pair of terminals. Therefore, any resistance connected to an outlet has an applied voltage of 120 V. The lightbulb is connected to one outlet and the toaster to another outlet, but both have the same applied voltage of 120 V. Therefore, each operates independently of any other appliance, with all the individual branch circuits connected across the 120-V line.

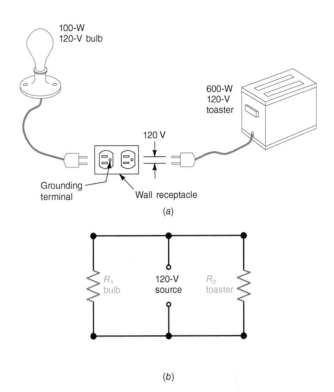

(a)

(b)

Figure 6-2 Lightbulb and toaster connected in parallel with the 120-V line. (a) Wiring diagram. (b) Schematic diagram.

6.2 Each Branch I Equals V_A/R

In applying Ohm's law, it is important to note that the current equals the voltage applied across the circuit divided by the resistance between the two points where that voltage is applied. In Fig. 6-3, 10 V is applied across the 5 Ω of R_2, resulting in the current of 2 A between points E and F through R_2. The battery voltage is also applied across the parallel resistance of R_1, applying 10 V across 10 Ω. Through R_1, therefore, the current is 1 A between points C and D. The current has a different value through R_1, with the same applied voltage, because the resistance is different. These values are calculated as follows:

$$I_1 = \frac{V_A}{R_1} = \frac{10}{10} = 1 \text{ A}$$

$$I_2 = \frac{V_A}{R_2} = \frac{10}{5} = 2 \text{ A}$$

Just as in a circuit with one resistance, any branch that has less R allows more I. If R_1 and R_2 were equal, however, the two branch currents would have the same value. For instance, in Fig. 6-1b each branch has its own current equal to 1.5 V/5 Ω = 0.3 A.

The I can be different in parallel circuits that have different R because V is the same across all the branches. Any voltage source generates a potential difference across its two terminals. This voltage does not move. Only I flows around

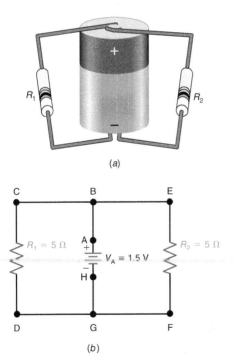

Figure 6-1 Example of a parallel circuit with two resistors. (a) Wiring diagram. (b) Schematic diagram.

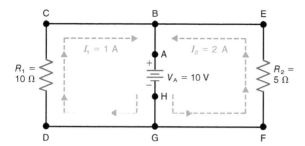

Figure 6-3 Parallel circuit. The current in each parallel branch equals the applied voltage V_A divided by each branch resistance R.

the circuit. The source voltage is available to make electrons move around any closed path connected to the terminals of the source. The amount of I in each separate path depends on the amount of R in each branch.

EXAMPLE 6-1

In Fig. 6-4, solve for the branch currents I_1 and I_2.

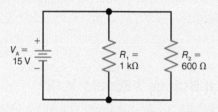

Figure 6-4 Circuit for Example 6-1.

Answer:

The applied voltage, V_A, of 15 V is across both resistors R_1 and R_2. Therefore, the branch currents are calculated as $\frac{V_A}{R}$, where V_A is the applied voltage and R is the individual branch resistance.

$$I_1 = \frac{V_A}{R_1}$$
$$= \frac{15\text{ V}}{1\text{ k}\Omega}$$
$$= 15\text{ mA}$$

$$I_2 = \frac{V_A}{R_2}$$
$$= \frac{15\text{ V}}{600\ \Omega}$$
$$= 25\text{ mA}$$

6.3 Kirchhoff's Current Law (KCL)

Components to be connected in parallel are usually wired directly across each other, with the entire parallel combination connected to the voltage source, as illustrated in Fig. 6-5.

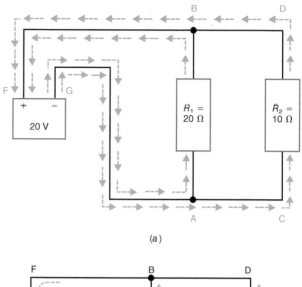

(a)

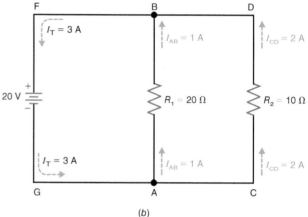

(b)

Figure 6-5 The current in the main line equals the sum of the branch currents. Note that from G to A at the bottom of this diagram is the negative side of the main line, and from B to F at the top is the positive side. (a) Wiring diagram. Arrows inside the lines indicate current in the main line for R_1; arrows outside indicate current for R_2. (b) Schematic diagram. I_T is the total line current for both R_1 and R_2.

This circuit is equivalent to wiring each parallel branch directly to the voltage source, as shown in Fig. 6-1, when the connecting wires have essentially zero resistance.

The advantage of having only one pair of connecting leads to the source for all the parallel branches is that usually less wire is necessary. The pair of leads connecting all the branches to the terminals of the voltage source is the **main line**. In Fig. 6-5, the wires from G to A on the negative side and from B to F in the return path form the main line.

In Fig. 6-5*b*, with 20 Ω of resistance for R_1 connected across the 20-V battery, the current through R_1 must be 20 V/20 Ω = 1 A. This current is electron flow from the negative terminal of the source, through R_1, and back to the positive battery terminal. Similarly, the R_2 branch of 10 Ω across the battery has its own branch current of 20 V/10 Ω = 2 A. This current flows from the negative terminal of the source, through R_2, and back to the positive terminal, since it is a separate path for electron flow.

All current in the circuit, however, must come from one side of the voltage source and return to the opposite side for a complete path. In the main line, therefore, the amount of current is equal to the total of the branch currents.

For example, in Fig. 6-5b, the total current in the line from point G to point A is 3 A. The total current at branch point A subdivides into its component branch currents for each of the branch resistances. Through the path of R_1 from A to B the current is 1 A. The other branch path ACDB through R_2 has a current of 2 A. At the branch point B, the electron flow from both parallel branches combines, so that the current in the main-line return path from B to F has the same value of 3 A as in the other side of the main line.

Kirchhoff's current law (KCL) states that the total current I_T in the main line of a parallel circuit equals the sum of the individual branch currents. Expressed as an equation, Kirchhoff's current law is

$$I_T = I_1 + I_2 + I_3 + \cdots + \text{etc.} \qquad (6\text{-}1)$$

where I_T is the total current and I_1, I_2, $I_3 \ldots$ are the individual branch currents. Kirchhoff's current law applies to any number of **parallel branches,** whether the resistances in the branches are equal or unequal.

EXAMPLE 6-2

An R_1 of 20 Ω, an R_2 of 40 Ω, and an R_3 of 60 Ω are connected in parallel across the 120-V power line. Using Kirchhoff's current law, determine the total current I_T.

Answer:

Current I_1 for the R_1 branch is 120/20 or 6 A. Similarly, I_2 is 120/40 or 3 A, and I_3 is 120/60 or 2 A. The total current in the main line is

$$I_T = I_1 + I_2 + I_3 = 6 + 3 + 2$$
$$I_T = 11 \text{ A}$$

EXAMPLE 6-3

Two branches R_1 and R_2 across the 120-V power line draw a total line current I_T of 15 A. The R_1 branch takes 10 A. How much is the current I_2 in the R_2 branch?

Answer:

$$I_2 = I_T - I_1 = 15 - 10$$
$$I_2 = 5 \text{ A}$$

With two branch currents, one must equal the difference between I_T and the other branch current.

EXAMPLE 6-4

Three parallel branch currents are 0.1 A, 500 mA, and 800 μA. Using Kirchhoff's current law, calculate I_T.

Answer:

All values must be in the same units to be added. In this case, all units will be converted to milliamperes: 0.1 A = 100 mA and 800 μA = 0.8 mA. Applying Kirchhoff's current law

$$I_T = 100 + 500 + 0.8$$
$$I_T = 600.8 \text{ mA}$$

You can convert the currents to A, mA, or μA units, as long as the same unit is used for adding all currents.

6.4 Resistances in Parallel

The combined equivalent resistance across the main line in a parallel circuit can be found by Ohm's law: *Divide the common voltage across the parallel resistances by the total current of all the branches.* Referring to Fig. 6-6a, note that the parallel resistance of R_1 with R_2, indicated by the equivalent resistance R_{EQ}, is the opposition to the total current in the main line. In this example, V_A/I_T is 60 V/3 A = 20 Ω for R_{EQ}.

The total load connected to the source voltage is the same as though one equivalent resistance of 20 Ω were connected across the main line. This is illustrated by the equivalent

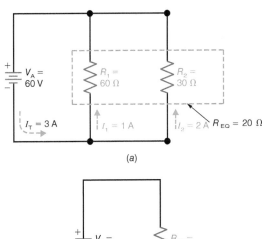

(a)

(b)

Figure 6-6 Resistances in parallel. (a) Combination of R_1 and R_2 is the total R_{EQ} for the main line. (b) Equivalent circuit showing R_{EQ} drawing the same 3-A I_T as the parallel combination of R_1 and R_2 in (a).

circuit in Fig. 6-6b. For any number of parallel resistances of any value, use the following equation,

$$R_{EQ} = \frac{V_A}{I_T} \qquad (6\text{-}2)$$

where I_T is the sum of all the branch currents and R_{EQ} is the equivalent resistance of all parallel branches across the applied voltage source V_A.

The first step in solving for R_{EQ} is to add all the parallel branch currents to find the I_T being delivered by the voltage source. The voltage source thinks that it is connected to a single resistance whose value allows I_T to flow in the circuit according to Ohm's law. This single resistance is R_{EQ}. An illustrative example of a circuit with two parallel branches will be used to show how R_{EQ} is calculated.

EXAMPLE 6-5

Two branches, each with a 5-A current, are connected across a 90-V source. How much is the equivalent resistance R_{EQ}?

Answer:

The total line current I_T is $5 + 5 = 10$ A. Then,

$$R_{EQ} = \frac{V_A}{I_T} = \frac{90}{10}$$
$$R_{EQ} = 9\ \Omega$$

Parallel Bank

A combination of parallel branches is often called a **bank.** In Fig. 6-6, the bank consists of the 60-Ω R_1 and 30-Ω R_2 in parallel. Their combined parallel resistance R_{EQ} is the bank resistance, equal to 20 Ω in this example. A bank can have two or more parallel resistors.

When a circuit has more current with the same applied voltage, this greater value of I corresponds to less R because of their inverse relation. Therefore, the combination of parallel resistances R_{EQ} for the bank is always less than the smallest individual branch resistance. The reason is that I_T must be more than any one branch current.

Why R_{EQ} Is Less than Any Branch R

It may seem unusual at first that putting more resistance into a circuit lowers the equivalent resistance. This feature of parallel circuits is illustrated in Fig. 6-7. Note that equal resistances of 30 Ω each are added across the source voltage, one branch at a time. The circuit in Fig. 6-7a has just R_1, which allows 2 A with 60 V applied. In Fig. 6-7b, the R_2 branch is added across the same V_A. This branch also has 2 A. Now the parallel circuit has a 4-A total line current because of $I_1 + I_2$. Then the third branch, which also takes 2 A for I_3, is added in Fig. 6-7c. The combined circuit with three branches, therefore, requires a total load current of 6 A, which is supplied by the voltage source.

The combined resistance across the source, then, is V_A/I_T, which is 60/6, or 10 Ω. This equivalent resistance R_{EQ}, representing the entire load on the voltage source, is shown in Fig. 6-7d. More resistance branches reduce the combined resistance of the parallel circuit because more current is required from the same voltage source.

Reciprocal Resistance Formula

We can derive the **reciprocal resistance formula** from the fact that I_T is the sum of all the branch currents, or,

$$I_T = I_1 + I_2 + I_3 + \cdots + \text{etc.}$$

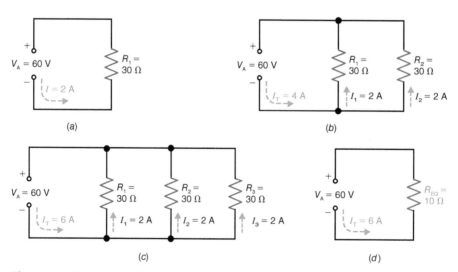

Figure 6-7 How adding parallel branches of resistors increases I_T but decreases R_{EQ}. (a) One resistor. (b) Two branches. (c) Three branches. (d) Equivalent circuit of the three branches in (c).

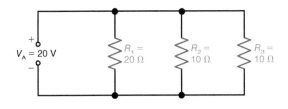

$$R_{EQ} = \cfrac{1}{\cfrac{1}{R_1} + \cfrac{1}{R_2} + \cfrac{1}{R_3}}$$

$R_{EQ} = 4\ \Omega$

(a)

$$R_{EQ} = \frac{V_A}{I_T} = \frac{20\text{ V}}{5\text{ A}}$$

$R_{EQ} = 4\ \Omega$

(b)

Figure 6-8 Two methods of combining parallel resistances to find R_{EQ}. (a) Using the reciprocal resistance formula to calculate R_{EQ} as 4 Ω. (b) Using the total line current method with an assumed line voltage of 20 V gives the same 4 Ω for R_{EQ}.

However, $I_T = V/R_{EQ}$. Also, each $I = V/R$. Substituting V/R_{EQ} for I_T on the left side of the formula and V/R for each branch I on the right side, the result is

$$\frac{V}{R_{EQ}} = \frac{V}{R_1} + \frac{V}{R_2} + \frac{V}{R_3} + \cdots + \text{etc.}$$

Dividing by V because the voltage is the same across all the resistances gives us

$$\frac{1}{R_{EQ}} = \frac{1}{R_1} + \frac{1}{R_2} + \frac{1}{R_3} + \cdots + \text{etc.}$$

Next, solve for R_{EQ}.

$$R_{EQ} = \frac{1}{\frac{1}{R_1} + \frac{1}{R_2} + \frac{1}{R_3} + \cdots + \text{etc.}} \qquad \textbf{(6-3)}$$

This reciprocal formula applies to any number of parallel resistances of any value. Using the values in Fig. 6-8a as an example,

$$R_{EQ} = \frac{1}{\frac{1}{20} + \frac{1}{10} + \frac{1}{10}} = 4\ \Omega$$

Total-Current Method

It may be easier to work without fractions. Figure 6-8b shows how this same problem can be calculated in terms of total current instead of by the reciprocal formula. Although the applied voltage is not always known, any convenient value can be assumed because it cancels in the calculations. It is usually simplest to assume an applied voltage of the same numerical value as the highest resistance. Then one assumed branch current will automatically be 1 A and the other branch currents will be more, eliminating fractions less than 1 in the calculations.

In Fig. 6-8b, the highest branch R is 20 Ω. Therefore, assume 20 V for the applied voltage. Then the branch currents are 1 A in R_1, 2 A in R_2, and 2 A in R_3. Their sum is $1 + 2 + 2 = 5$ A for I_T. The combined resistance R_{EQ} across the main line is V_A/I_T, or 20 V/5 A = 4 Ω. This is the same value calculated with the reciprocal resistance formula.

Special Case of Equal R in All Branches

If R is equal in all branches, the combined R_{EQ} equals the value of one branch resistance divided by the number of branches.

$$R_{EQ} = \frac{R}{n}$$

where R is the resistance in one branch and n is the number of branches.

This rule is illustrated in Fig. 6-9, where three 60-kΩ resistances in parallel equal 20 kΩ.

The rule applies to any number of parallel resistances, but they must all be equal. As another example, five 60-Ω resistances in parallel have the combined resistance of 60/5, or 12 Ω. A common application is two equal resistors wired in a parallel bank for R_{EQ} equal to one-half R.

Special Case of Only Two Branches

When there are two parallel resistances and they are not equal, it is usually quicker to calculate the combined resistance by the method shown in Fig. 6-10. This rule says that the combination of two parallel resistances is their product divided by their sum.

$$R_{EQ} = \frac{R_1 \times R_2}{R_1 + R_2} \qquad \textbf{(6-4)}$$

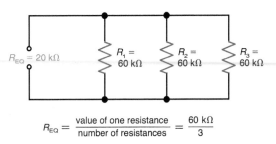

$$R_{EQ} = \frac{\text{value of one resistance}}{\text{number of resistances}} = \frac{60\text{ k}\Omega}{3}$$

Figure 6-9 For the special case of all branches having the same resistance, just divide R by the number of branches to find R_{EQ}. Here, $R_{EQ} = 60$ kΩ/3 = 20 kΩ.

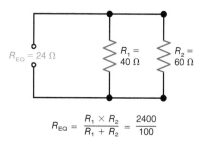

$$R_{EQ} = \frac{R_1 \times R_2}{R_1 + R_2} = \frac{2400}{100}$$

Figure 6-10 For the special case of only two branch resistances of any values R_{EQ} equals their product divided by the sum. Here, $R_{EQ} = 2400/100 = 24\ \Omega$.

where R_{EQ} is in the same units as all the individual resistances. For the example in Fig. 6-10,

$$R_{EQ} = \frac{R_1 \times R_2}{R_1 + R_2} = \frac{40 \times 60}{40 + 60} = \frac{2400}{100}$$
$$R_{EQ} = 24\ \Omega$$

Each R can have any value, but there must be only two resistances.

Short-Cut Calculations

Figure 6-11 shows how these special rules can help reduce parallel branches to a simpler equivalent circuit. In Fig. 6-11a, the 60-Ω R_1 and R_4 are equal and in parallel. Therefore, they are equivalent to the 30-Ω R_{14} in Fig. 6-11b. Similarly, the 20-Ω R_2 and R_3 are equivalent to the 10 Ω of R_{23}. The circuit in Fig. 6-11a is equivalent to the simpler circuit in Fig. 6-11b with just the two parallel resistances of 30 and 10 Ω.

Finally, the combined resistance for these two equals their product divided by their sum, which is 300/40 or 7.5 Ω, as shown in Fig. 6-11c. This value of R_{EQ} in

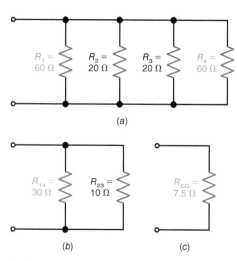

(a)

(b)   (c)

Figure 6-11 An example of parallel resistance calculations with four branches. (a) Original circuit. (b) Resistors combined into two branches. (c) Equivalent circuit reduces to one R_{EQ} for all the branches.

Fig. 6-11c is equivalent to the combination of the four branches in Fig. 6-11a. If you connect a voltage source across either circuit, the current in the main line will be the same for both cases.

The order of connections for parallel resistances does not matter in determining R_{EQ}. There is no question as to which is first or last because they are all across the same voltage source and receive their current at the same time.

Finding an Unknown Branch Resistance

In some cases with two parallel resistors, it is useful to be able to determine what size R_X to connect in parallel with a known R to obtain a required value of R_{EQ}. Then the factors can be transposed as follows:

$$R_X = \frac{R \times R_{EQ}}{R - R_{EQ}} \tag{6-5}$$

This formula is just another way of writing Formula (6-4).

EXAMPLE 6-6

What R_X in parallel with 40 Ω will provide an R_{EQ} of 24 Ω?

Answer:

$$R_X = \frac{R \times R_{EQ}}{R - R_{EQ}} = \frac{40 \times 24}{40 - 24} = \frac{960}{16}$$
$$R_X = 60\ \Omega$$

This problem corresponds to the circuit shown before in Fig. 6-10.

Note that Formula (6-5) for R_X has a product over a difference. The R_{EQ} is subtracted because it is the smallest R. Remember that both Formulas (6-4) and (6-5) can be used with only two parallel branches.

EXAMPLE 6-7

What R in parallel with 50 kΩ will provide an R_{EQ} of 25 kΩ?

Answer:

$$R = 50\ \text{k}\Omega$$

Two equal resistances in parallel have R_{EQ} equal to one-half R.

6.5 Conductances in Parallel

Since conductance G is equal to $1/R$, the reciprocal resistance formula (6-3) can be stated for conductance as $R_{EQ} = \frac{1}{G_T}$ where G_T is calculated as

$$G_T = G_1 + G_2 + G_3 + \cdots + \text{etc.} \tag{6-6}$$

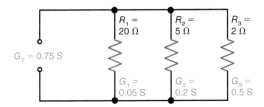

Figure 6-12 Conductances G_1, G_2, and G_3 in parallel are added for the total G_T.

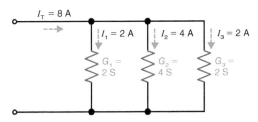

Figure 6-13 Example of how parallel branch currents are directly proportional to each branch conductance G.

With R in ohms, G is in siemens. For the example in Fig. 6-12, G_1 is $1/20 = 0.05$, G_2 is $1/5 = 0.2$, and G_3 is $1/2 = 0.5$. Then

$$G_T = 0.05 + 0.2 + 0.5 = 0.75 \text{ S}$$

Notice that adding the conductances does not require reciprocals. Each value of G is the reciprocal of R.

The reason why *parallel conductances* are added directly can be illustrated by assuming a 1-V source across all branches. Then calculating the values of $1/R$ for the conductances gives the same values as calculating the branch currents. These values are added for the total I_T or G_T.

Working with G may be more convenient than working with R in parallel circuits, since it avoids the use of the reciprocal formula for R_{EQ}. Each branch current is directly proportional to its conductance. This idea corresponds to the fact that each voltage drop in series circuits is directly proportional to each of the series resistances. An example of the currents for parallel conductances is shown in Fig. 6-13. Note that the branch with G of 4 S has twice as much current as the 2-S branches because the branch conductance is doubled.

6.6 Total Power in Parallel Circuits

Since the power dissipated in the branch resistances must come from the voltage source, the **total power** equals the sum of the individual values of power in each branch. This rule is illustrated in Fig. 6-14. We can also use this circuit as an example of applying the rules of current, voltage, and resistance for a parallel circuit.

The applied 10 V is across the 10-Ω R_1 and 5-Ω R_2 in Fig. 6-14. The branch current I_1 then is V_A/R_1 or $10/10$, which equals 1 A. Similarly, I_2 is $10/5$, or 2 A. The total I_T is $1 + 2 = 3$ A. If we want to find R_{EQ}, it equals V_A/I_T or $10/3$, which is $3\frac{1}{3}$ Ω.

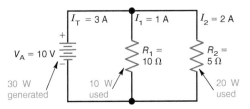

Figure 6-14 The sum of the power values P_1 and P_2 used in each branch equals the total power P_T produced by the source.

The power dissipated in each branch R is $V_A \times I$. In the R_1 branch, I_1 is $10/10 = 1$ A. Then P_1 is $V_A \times I_1$ or $10 \times 1 = 10$ W.

For the R_2 branch, I_2 is $10/5 = 2$ A. Then P_2 is $V_A \times I_2$ or $10 \times 2 = 20$ W.

Adding P_1 and P_2, the answer is $10 + 20 = 30$ W. This P_T is the total power dissipated in both branches.

This value of 30 W for P_T is also the total power supplied by the voltage source by means of its total line current I_T. With this method, the total power is $V_A \times I_T$ or $10 \times 3 = 30$ W for P_T. The 30 W of power supplied by the voltage source is dissipated or used up in the branch resistances.

It is interesting to note that in a parallel circuit, the smallest branch resistance will always dissipate the most power. Since $P = \dfrac{V^2}{R}$ and V is the same across all parallel branches, a smaller value of R in the denominator will result in a larger amount of power dissipation.

Note also that in both parallel and series circuits, the sum of the individual values of power dissipated in the circuit equals the total power generated by the source. This can be stated as a formula

$$P_T = P_1 + P_2 + P_3 + \cdots + \text{etc.} \qquad \textbf{(6-7)}$$

The series or parallel connections can alter the distribution of voltage or current, but power is the rate at which energy is supplied. The circuit arrangement cannot change the fact that all the energy in the circuit comes from the source.

6.7 Analyzing Parallel Circuits with Random Unknowns

For many types of problems with parallel circuits, it is useful to remember the following points.

1. When you know the voltage across one branch, this voltage is across all the branches. There can be only one voltage across branch points with the same potential difference.

2. If you know I_T and one of the branch currents I_1, you can find I_2 by subtracting I_1 from I_T.

The circuit in Fig. 6-15 illustrates these points. The problem is to find the applied voltage V_A and the value of R_3. Of the three branch resistances, only R_1 and R_2 are known. However, since I_2 is given as 2 A, the I_2R_2 voltage must be $2 \times 60 = 120$ V.

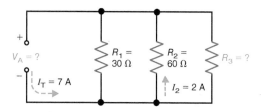

Figure 6-15 Analyzing a parallel circuit. What are the values for V_A and R_3. See solution in text.

Although the applied voltage is not given, this must also be 120 V. The voltage across all the parallel branches is the same 120 V that is across the R_2 branch.

Now I_1 can be calculated as V_A/R_1. This is $120/30 = 4$ A for I_1.

Current I_t is given as 7 A. The two branches take $2 + 4 = 6$ A. The third branch current through R_3 must be $7 - 6 = 1$ A for I_3.

Now R_3 can be calculated as V_A/I_3. This is $120/1 = 120\ \Omega$ for R_3.

CHAPTER 6 SYSTEM SIDEBAR

Christmas Lights

The small incandescent lights we put on Christmas trees are a good example of a series circuit. Also known as minilights, these strings of lights are multiple bulbs, all connected in series with the 120-volt power line. Figure S6-1 shows the basic connections of a typical string. Each bulb has a rating of 2.5 volts at 200 mA. If we put 50 of them in series, we will need a total voltage of $50 \times 2.5 = 125$ volts. This is close enough to the standard power-line voltage for proper operation. With 120 volts applied, each bulb will get 2.4 volts.

Note the unique three-wire configuration. The wiring looks complex since the wires are all twisted together. Nevertheless, the wiring is rather simple. Two of the wires pass directly through from the ac plug to the ac socket at the end for the string. This allows the string to be extended with another string plugged in parallel with the first string. Some longer strings have 100 lights. In this case, 2 strings of 50 lights are connected in parallel to the 120 volts using the same three-wire arrangement. The lower group of 50 bulbs is in parallel with the upper group. A feature of most strings of lights is having fuses in the ac plug for safety. If a short occurs somewhere in the string, the fuses will blow, protecting the home ac wiring and preventing a fire.

As you may have already guessed, if one bulb burns out, it will open the string and all the lights will go out. However, that does not actually occur because each bulb has a built-in shunt to keep the string unbroken. Each bulb has a small wire or shunt wrapped around the connections to the filament at its base that would appear to short-out the bulb. A special insulated coating on the wire prevents the short from occurring. Now if the filament in the bulb should open, the entire 120 volts from the line will appear across the shunt, causing the coating to burn away and a connection to be formed to keep the circuit intact. When this happens, the voltage across each bulb will rise slightly as the ac line voltage is now divided among fewer bulbs. The bulbs can handle a little more than their rated voltage and will glow slightly brighter. If too many bulbs burn out, the voltage across each bulb may reach the point where it will shorten the life of all the other bulbs, making burnouts more frequent.

Many strings of lights contain a flasher bulb. It looks like a standard bulb, but instead of a filament, it is actually a switch that repeatedly turns all the lights on and off. The switch is made of a bimetallic strip that bends when heated. Normally the strip keeps the circuit closed so the lights turn on. When current flows through, the strip bends and opens the switch contacts, turning the lights off. When the strip cools, it again closes the contacts, turning the lights on. This process continues as the lights flash off and on at a rate of about once every 1 or 2 seconds.

Figure S6-1 A series string of Christmas lights.

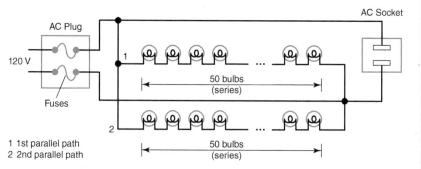

1 1st parallel path
2 2nd parallel path

1. A 120-kΩ resistor, R_1, and a 180-kΩ resistor, R_2, are in parallel. How much is the equivalent resistance, R_{EQ}?
   a. 72 kΩ.
   b. 300 kΩ.
   c. 360 kΩ.
   d. 90 kΩ.

2. A 100-Ω resistor, R_1, and a 300-Ω resistor, R_2, are in parallel across a dc voltage source. Which resistor dissipates more power?
   a. The 300-Ω resistor.
   b. Both resistors dissipate the same amount of power.
   c. The 100-Ω resistor.
   d. It cannot be determined.

3. Three 18-Ω resistors are in parallel. How much is the equivalent resistance, R_{EQ}?
   a. 54 Ω.
   b. 6 Ω.
   c. 9 Ω.
   d. none of the above.

4. Which of the following statements about parallel circuits is false?
   a. The voltage is the same across all branches in a parallel circuit.
   b. The equivalent resistance, R_{EQ}, of a parallel circuit is always smaller than the smallest branch resistance.
   c. In a parallel circuit the total current, I_T, in the main line equals the sum of the individual branch currents.
   d. The equivalent resistance, R_{EQ}, of a parallel circuit decreases when one or more parallel branches are removed from the circuit.

5. Two resistors, R_1 and R_2, are in parallel with each other and a dc voltage source. If the total current, I_T, in the main line equals 6 A and I_2 through R_2 is 4 A, how much is I_1 through R_1?
   a. 6 A.
   b. 2 A.
   c. 4 A.
   d. It cannot be determined.

6. How much resistance must be connected in parallel with a 360-Ω resistor to obtain an equivalent resistance, R_{EQ}, of 120 Ω?
   a. 360 Ω.
   b. 480 Ω.
   c. 1.8 kΩ.
   d. 180 Ω.

7. If one branch of a parallel circuit becomes open,
   a. all remaining branch currents increase.
   b. the voltage across the open branch will be 0 V.
   c. the remaining branch currents do not change in value.
   d. the equivalent resistance of the circuit decreases.

8. If a 10-Ω R_1, 40-Ω R_2, and 8-Ω R_3 are in parallel, calculate the total conductance, G_T, of the circuit.
   a. 250 mS.
   b. 58 S.
   c. 4 Ω.
   d. 0.25 μS.

9. Which of the following formulas can be used to determine the total power, P_T, dissipated by a parallel circuit.
   a. $P_T = V_A \times I_T$.
   b. $P_T = P_1 + P_2 + P_3 + \cdots + $ etc.
   c. $P_T = \dfrac{V^2_A}{R_{EQ}}$.
   d. All of the above.

10. A 20-Ω R_1, 50-Ω R_2, and 100-Ω R_3 are connected in parallel. If R_2 is short-circuited, what is the equivalent resistance, R_{EQ}, of the circuit?
    a. approximately 0 Ω.
    b. infinite (∞) Ω.
    c. 12.5 Ω.
    d. It cannot be determined.

11. If the fuse in the main line of a parallel circuit opens,
    a. the voltage across each branch will be 0 V.
    b. the current in each branch will be zero.
    c. the current in each branch will increase to offset the decrease in total current.
    d. both a and b.

12. A 100-Ω R_1 and a 150-Ω R_2 are in parallel. If the current, I_1, through R_1 is 24 mA, how much is the total current, I_T?
    a. 16 mA.
    b. 40 mA.
    c. 9.6 mA.
    d. It cannot be determined.

13. A 2.2-kΩ R_1 is in parallel with a 3.3-kΩ R_2. If these two resistors carry a total current of 7.5 mA, how much is the applied voltage, V_A?
    a. 16.5 V.
    b. 24.75 V.
    c. 9.9 V.
    d. 41.25 V.

14. How many 120-Ω resistors must be connected in parallel to obtain an equivalent resistance, R_{EQ}, of 15 Ω?
   a. 15.
   b. 8.
   c. 12.
   d. 6.

15. A 220-Ω R_1, 2.2-kΩ R_2, and 200-Ω R_3 are connected across 15 V of applied voltage. What happens to R_{EQ} if the applied voltage is doubled to 30 V?
   a. R_{EQ} doubles.
   b. R_{EQ} cuts in half.
   c. R_{EQ} does not change.
   d. R_{EQ} increases but is not double its original value.

16. If one branch of a parallel circuit opens, the total current, I_T,
   a. does not change.
   b. decreases.
   c. increases.
   d. goes to zero.

17. In a normally operating parallel circuit, the individual branch currents are
   a. independent of each other.
   b. not affected by the value of the applied voltage.

c. larger than the total current, I_T.
d. none of the above.

18. If the total conductance, G_T, of a parallel circuit is 200 μS, how much is R_{EQ}?
   a. 500 Ω.
   b. 200 kΩ.
   c. 5 kΩ.
   d. 500 kΩ.

19. If one branch of a parallel circuit is short-circuited,
   a. the fuse in the main line will blow.
   b. the voltage across the short-circuited branch will measure the full value of applied voltage.
   c. all the remaining branches are effectively short-circuited as well.
   d. both a and c.

20. Two lightbulbs in parallel with the 120-V power line are rated at 60 W and 100 W, respectively. What is the equivalent resistance, R_{EQ}, of the bulbs when they are lit?
   a. 144 Ω.
   b. 90 Ω.
   c. 213.3 Ω.
   d. It cannot be determined.

CHAPTER 6 PROBLEMS

SECTION 6.1 The Applied Voltage V_A Is the Same across Parallel Branches

6.1 In Fig. 6-16, how much voltage is across points
   a. A and B?
   b. C and D?
   c. E and F?
   d. G and H?

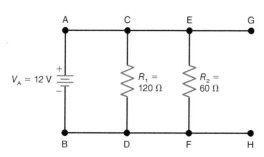

Figure 6-16

6.2 In Fig. 6-16, how much voltage will be measured across points C and D if R_1 is removed from the circuit?

SECTION 6.2 Each Branch I Equals $\dfrac{V_A}{R}$

6.3 In Fig. 6-16, solve for the branch currents, I_1 and I_2.

6.4 In Fig. 6-16, explain why I_2 is double the value of I_1.

6.5 In Fig. 6-16, assume a 10-Ω resistor, R_3, is added across points G and H.
   a. Calculate the branch current, I_3.
   b. Explain how the branch currents, I_1 and I_2 are affected by the addition of R_3.

6.6 In Fig. 6-17, solve for the branch currents I_1, I_2, and I_3.

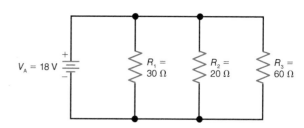

Figure 6-17

6.7 In Fig. 6-17, do the branch currents I_1 and I_3 remain the same if R_2 is removed from the circuit? Explain your answer.

6.8 In Fig. 6-18, solve for the branch currents I_1, I_2, I_3, and I_4.

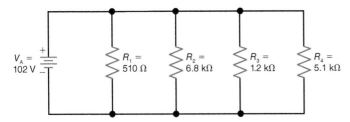

Figure 6-18

SECTION 6.3 Kirchhoff's Current Law (KCL)

6.9 In Fig. 6-16, solve for the total current, I_T.

6.10 In Fig. 6-17, solve for the total current, I_T.

6.11 In Fig. 6-18, solve for the total current, I_T.

6.12 In Fig. 6-19, solve for I_1, I_2, I_3, and I_T.

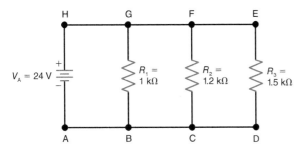

Figure 6-19

6.13 In Fig. 6-19, how much is the current in the wire between points
  a. A and B?
  b. B and C?
  c. C and D?
  d. E and F?
  e. F and G?
  f. G and H?

6.14 In Fig. 6-20, solve for I_1, I_2 I_3, and I_T.

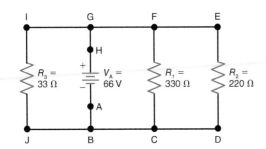

Figure 6-20

6.15 In Fig. 6-20, how much is the current in the wire between points
  a. A and B?
  b. B and C?
  c. C and D?
  d. E and F?
  e. F and G?
  f. G and H?
  g. G and I?
  h. B and J?

6.16 In Fig. 6-21, apply Kirchhoff's current law to solve for the unknown current, I_3.

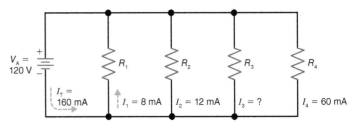

Figure 6-21

6.17 Two resistors R_1 and R_2 are in parallel with each other and a dc voltage source. How much is I_2 through R_2 if $I_T = 150$ mA and I_1 through R_1 is 60 mA?

SECTION 6.4 Resistances in Parallel

6.18 In Fig. 6-16, solve for R_{EQ}.

6.19 In Fig. 6-17, solve for R_{EQ}.

6.20 In Fig. 6-18, solve for R_{EQ}.

6.21 In Fig. 6-20, solve for R_{EQ}.

6.22 In Fig. 6-22, how much is R_{EQ} if $R_1 = 100\ \Omega$ and $R_2 = 25\ \Omega$?

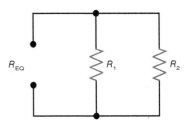

Figure 6-22

6.23 In Fig. 6-22, how much is R_{EQ} if $R_1 = 1.5$ MΩ and $R_2 = 1$ MΩ?

6.24 In Fig. 6-22, how much is R_{EQ} if $R_1 = 2.2$ kΩ and $R_2 = 220\ \Omega$?

6.25 In Fig. 6-22, how much is R_{EQ} if $R_1 = R_2 = 10$ kΩ?

6.26 In Fig. 6-22, how much resistance, R_2, must be connected in parallel with a 750 Ω R_1 to obtain an R_{EQ} of 500 Ω?

6.27 How much is R_{EQ} in Fig. 6-23 if $R_1 = 1.5$ kΩ, $R_2 = 1$ kΩ, $R_3 = 1.8$ kΩ, and $R_4 = 150$ Ω?

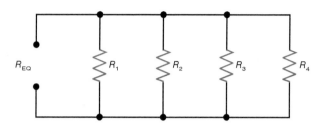

Figure 6-23

6.28 How much is R_{EQ} in Fig. 6-23 if $R_1 = R_2 = R_3 = R_4 = 2.2$ kΩ?

6.29 A technician is using an ohmmeter to measure a variety of different resistor values. Assume the technician has a body resistance of 750 kΩ. How much resistance will the ohmmeter read if the fingers of the technician touch the leads of the ohmmeter when measuring the following resistors:
a. 270 Ω.
b. 390 kΩ.
c. 2.2 MΩ.
d. 1.5 kΩ.
e. 10 kΩ.

SECTION 6.5 Conductances in Parallel

6.30 In Fig. 6-24, solve for G_1, G_2, G_3, G_T, and R_{EQ}.

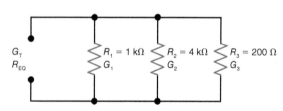

Figure 6-24

6.31 Find the total conductance, G_T for the following branch conductances; $G_1 = 100$ mS, $G_2 = 66.67$ mS, $G_3 = 250$ mS, and $G_4 = 83.33$ mS. How much is R_{EQ}?

SECTION 6.6 Total Power in Parallel Circuits

6.32 In Fig. 6-18, solve for P_1, P_2, P_3, P_4, and P_T.

6.33 In Fig. 6-20, solve for P_1, P_2, P_3, and P_T.

SECTION 6.7 Analyzing Parallel Circuits with Random Unknowns

6.34 In Fig. 6-25, solve for V_A, R_1, I_2, R_{EQ}, P_1, P_2, and P_T.

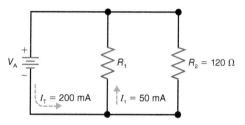

Figure 6-25

6.35 In Fig. 6-26, solve for V_A, I_1, I_2, R_2, I_T, P_2, and P_T.

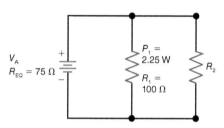

Figure 6-26

6.36 In Fig. 6-27, solve for R_3, V_A, I_1, I_2, I_T, P_1, P_2, P_3, and P_T.

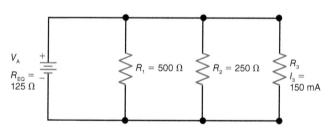

Figure 6-27

6.37 In Fig. 6-28, solve for I_T, I_1, I_2, R_1, R_2, R_3, P_2, P_3, and P_T.

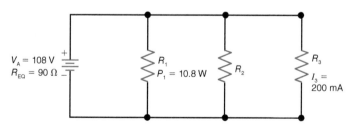

Figure 6-28

Chapter **7**

Series-Parallel Circuits and Voltage Dividers

Learning Outcomes

After studying this chapter, you should be able to:

> *Determine* the total resistance of a series-parallel circuit.

> *Calculate* the voltage, current, resistance, and power in a series-parallel circuit.

> *Calculate* the voltage, current, resistance, and power in a series-parallel circuit having random unknowns.

> *Explain* how a Wheatstone bridge can be used to determine the value of an unknown resistor.

> *List* other applications of balanced bridge circuits.

> *Calculate* the voltage drops in an unloaded voltage divider.

> *Explain* why resistor voltage drops are proportional to the resistor values in a series circuit.

> *Explain* why the branch currents are inversely proportional to the branch resistances in a parallel circuit.

> *Define* what is meant by the term *loaded voltage divider.*

A series-parallel circuit is one that combines both series and parallel connections. As you can imagine, there is an infinite variety of ways to interconnect resistors and other loads. In this chapter you will learn the process of analyzing and calculating any series-parallel combination using the techniques you have already learned.

This chapter also introduces several special forms of these combination circuits. Specifically, you will learn about the Wheatstone bridge and voltage dividers. These special circuits are all widely used throughout electronics. You will learn how to analyze these circuits and examine some common applications.

7.1 Finding R_T for Series-Parallel Resistances

In Fig. 7-1, R_1 is in series with R_2. Also, R_3 is in parallel with R_4. However, R_2 is *not* in series with either R_3 or R_4. The reason is that the current through R_2 is equal to the sum of the branch currents I_3 and I_4 flowing into and away from point A (see Fig. 7-1b).

The wiring is shown in Fig. 7-1a and the schematic diagram in Fig. 7-1b. To find R_T, we add the series resistances and combine the parallel resistances.

In Fig. 7-1c, the 0.5-kΩ R_1 and 0.5-kΩ R_2 in series total 1 kΩ for R_{1-2}. The calculations are

$$0.5 \text{ k}\Omega + 0.5 \text{ k}\Omega = 1 \text{ k}\Omega$$

Also, the 1-kΩ R_3 in parallel with the 1-kΩ R_4 can be combined, for an equivalent resistance of 0.5 kΩ for R_{3-4}, as in Fig. 7-1d. The calculations are

$$\frac{1 \text{ k}\Omega}{2} = 0.5 \text{ k}\Omega$$

This parallel R_{3-4} combination of 0.5 kΩ is then added to the series R_{1-2} combination for the final R_T value of 1.5 kΩ. The calculations are

$$0.5 \text{ k}\Omega + 1 \text{ k}\Omega = 1.5 \text{ k}\Omega$$

The 1.5 kΩ is the R_T of the entire circuit connected across the V_T of 1.5 V.

With R_T known to be 1.5 kΩ, we can find I_T in the main line produced by 1.5 V. Then

$$I_T = \frac{V_T}{R_T} = \frac{1.5 \text{ V}}{1.5 \text{ k}\Omega} = 1 \text{ mA}$$

This 1-mA I_T is the current through resistors R_1 and R_2 in Fig. 7-1a and b or R_{1-2} in Fig. 7-1c.

At branch point B, at the bottom of the diagram in Fig. 7-1b, the 1 mA of electron flow for I_T divides into two branch currents for R_3 and R_4. Since these two branch resistances are equal, I_T divides into two equal parts of 0.5 mA each. At branch point A at the top of the diagram, the two 0.5-mA branch currents combine to equal the 1-mA I_T in the main line, returning to the source V_T.

7.2 Resistance Strings in Parallel

More details about the voltages and currents in a series-parallel circuit are illustrated in Fig. 7-2, which shows two identical series **strings in parallel.** Suppose that four 120-V, 100-W lightbulbs are to be wired with a voltage source that produces 240 V. Each bulb needs 120 V for normal brilliance.

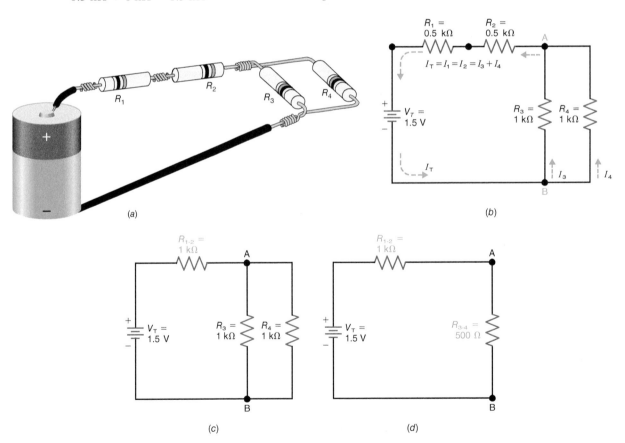

(a)

(b)

(c)

(d)

Figure 7-1 Example of a series-parallel circuit. (a) Wiring of a series-parallel circuit. (b) Schematic diagram of a series-parallel circuit. (c) Schematic with R_1 and R_2 in series added for R_{1-2}. (d) Schematic with R_3 and R_4 in parallel combined for R_{3-4}.

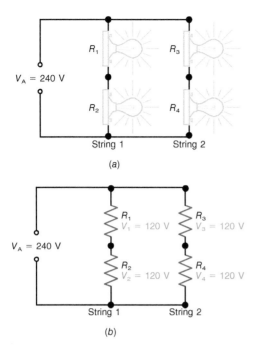

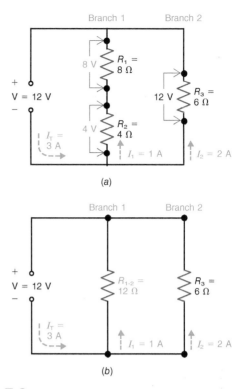

Figure 7-2 Two identical series strings in parallel. All bulbs have a 120-V, 100-W rating. (*a*) Wiring diagram. (*b*) Schematic diagram.

Figure 7-3 Series string in parallel with another branch. (*a*) Schematic diagram. (*b*) Equivalent circuit.

If the bulbs were connected directly across the source, each would have the applied voltage of 240 V. This would cause excessive current in all the bulbs that could result in burned-out filaments.

If the four bulbs were connected in series, each would have a potential difference of 60 V, or one-fourth the applied voltage. With too low a voltage, there would be insufficient current for normal operation, and the bulbs would not operate at normal brilliance.

However, two bulbs in series across the 240-V line provide 120 V for each filament, which is the normal operating voltage. Therefore, the four bulbs are wired in strings of two in series, with the two strings in parallel across the 240-V source. Both strings have 240 V applied. In each string, two series bulbs divide the 240 V equally to provide the required 120 V for normal operation.

Another example is illustrated in Fig. 7-3. This circuit has just two parallel branches. One branch includes R_1 in series with R_2. The other branch has just the one resistance R_3. Ohm's law can be applied to each branch.

Branch Currents I_1 and I_2

In Fig. 7-3*a*, each branch current equals the voltage applied across the branch divided by the total resistance in the branch. In branch 1, R_1 and R_2 total $8 + 4 = 12\ \Omega$. With 12 V applied, this branch current I_1 is $12/12 = 1$ A. Branch 2 has only the 6-Ω R_3. Then I_2 in this branch is $12/6 = 2$ A.

Series Voltage Drops in a Branch

For any one resistance in a string, the current in the string multiplied by the resistance equals the *IR* voltage drop across that particular resistance. Also, the sum of the series *IR* drops in the string equals the voltage across the entire string.

Branch 1 is a string with R_1 and R_2 in series. The I_1R_1 drop equals 8 V, whereas the I_1R_2 drop is 4 V. These drops of 8 and 4 V add to equal the 12 V applied. The voltage across the R_3 branch is also the same 12 V.

Calculating I_T

The total line current equals the sum of the branch currents for all parallel strings. Here I_T is 3 A, equal to the sum of 1 A in branch 1 and 2 A in branch 2.

Calculating R_T

The resistance of the total series-parallel circuit across the voltage source equals the applied voltage divided by the total line current. In Fig. 7-3*a*, $R_T = 12$ V/3 A, or 4 Ω. This resistance can also be calculated as 12 Ω in parallel with 6 Ω. Fig. 7-3*b* shows the equivalent circuit. Using the product divided by the sum formula, $72/18 = 4\ \Omega$ for the equivalent combined R_T.

Applying Ohm's Law

There can be any number of parallel strings and more than two series resistances in a string. Still, Ohm's law can be used in the same way for the series and parallel parts of the circuit. The series parts have the same current. The parallel parts have the same voltage. Remember that for V/R the R must include all the resistance across the two terminals of V.

7.3 Resistance Banks in Series

In Fig. 7-4a, the group of parallel resistances R_2 and R_3 is a bank. This is in series with R_1 because the total current of the bank must go through R_1.

The circuit here has R_2 and R_3 in parallel in one bank so that these two resistances will have the same potential difference of 20 V across them. The source applies 24 V, but there is a 4-V drop across R_1.

The two series voltage drops of 4 V across R_1 and 20 V across the bank add to equal the applied voltage of 24 V. The purpose of a circuit like this is to provide the same voltage for two or more resistances in a bank, where the bank voltage must be less than the applied voltage by the amount of the IR drop across any series resistance.

To find the resistance of the entire circuit, combine the parallel resistances in each bank and add the series resistance. As shown in Fig. 7-4b, the two 10-Ω resistances, R_2 and R_3 in parallel, are equivalent to 5 Ω. Since the bank resistance of 5 Ω is in series with 1 Ω for R_1, the total resistance is

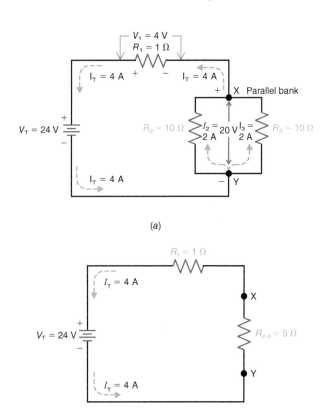

(a)

(b)

Figure 7-4 Parallel bank of R_2 and R_3 in series with R_1. (a) Original circuit. (b) Equivalent circuit.

6 Ω across the 24-V source. Therefore, the main-line current is 24 V/6 Ω, which equals 4 A.

The total line current of 4 A divides into two parts of 2 A each in the parallel resistances R_2 and R_3. Note that each branch current equals the bank voltage divided by the branch resistance. For this bank, 20/10 = 2 A for each branch.

The branch currents, I_2 and I_3, are combined in the main line to provide the total 4 A in R_1. This is the same total current flowing in the main line, in the source, into the bank, and out of the bank.

There can be more than two parallel resistances in a bank and any number of **banks in series.** Still, Ohm's law can be applied in the same way to the series and parallel parts of the circuit. The general procedure for circuits of this type is to find the equivalent resistance of each bank and then add all series resistances.

7.4 Resistance Banks and Strings in Series Parallel

In the solution of such circuits, the most important fact to know is which components are in series with each other and which parts of the circuit are parallel branches. The series components must be in one current path without any branch points. A branch point such as point A or B in Fig. 7-5 is common to two or more current paths. For instance, R_1 and R_6 are *not* in series with each other. They do not have the same current because the current through R_1 equals the sum of the branch currents, I_5 and I_6, flowing into and away from point A. Similarly, R_5 is not in series with R_2 because of the branch point B.

To find the currents and voltages in Fig. 7-5, first find R_T to calculate the main-line current I_T as V_T/R_T. In calculating R_T, start reducing the branch farthest from the source and work toward the applied voltage. The reason for following this order is that you cannot tell how much resistance is in series with R_1 and R_2 until the parallel branches are reduced to their equivalent resistance. If no source voltage is shown, R_T can still be calculated from the outside in toward the open terminals where a source would be connected.

To calculate R_T in Fig. 7-5, the steps are as follows:

1. The bank of the 12-Ω R_3 and 12-Ω R_4 in parallel in Fig. 7-5a is equal to the 6-Ω R_7 in Fig. 7-5b.
2. The 6-Ω R_7 and 4-Ω R_6 in series in the same current path total 10 Ω for R_{13} in Fig. 7-5c.
3. The 10-Ω R_{13} is in parallel with the 10-Ω R_5, across the branch points A and B. Their equivalent resistance, then, is the 5-Ω R_{18} in Fig. 7-5d.
4. Now the circuit in Fig. 7-5d has just the 15-Ω R_1, 5-Ω R_{18}, and 30-Ω R_2 in series. These resistances total 50 Ω for R_T, as shown in Fig. 7-5e.
5. With a 50-Ω R_T across the 100-V source, the line current I_T is equal to 100/50 = 2 A.

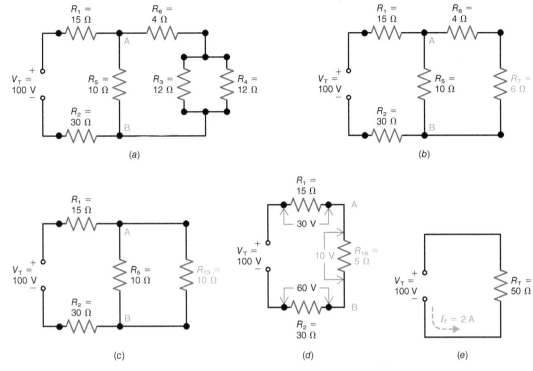

Figure 7-5 Reducing a series-parallel circuit to an equivalent series circuit to find the R_T. (a) Actual circuit. (b) R_3 and R_4 in parallel combined for the equivalent R_7. (c) R_7 and R_6 in series added for R_{13}. (d) R_{13} and R_5 in parallel combined for R_{18}. (e) The R_{18}, R_1, and R_2 in series are added for the total resistance of 50 Ω for R_T.

To see the individual currents and voltages, we can use the I_T of 2 A for the equivalent circuit in Fig. 7-5d. Now we work from the source V out toward the branches. The reason is that I_T can be used to find the voltage drops in the main line. The IR voltage drops here are

$$V_1 = I_T R_1 = 2 \times 15 = 30 \text{ V}$$
$$V_{18} = I_T R_{18} = 2 \times 5 = 10 \text{ V}$$
$$V_2 = I_T R_2 = 2 \times 30 = 60 \text{ V}$$

The 10-V drop across R_{18} is actually the potential difference between branch points A and B. This means 10 V across R_5 and R_{13} in Fig. 7-5c. The 10 V produces 1 A in the 10-Ω R_5 branch. The same 10 V is also across the R_{13} branch.

Remember that the R_{13} branch is actually the string of R_6 in series with the R_3-R_4 bank. Since this branch resistance is 10 Ω with 10 V across it, the branch current here is 1 A. The 1 A through the 4 Ω of R_6 produces a voltage drop of 4 V. The remaining 6-V IR drop is across the R_3-R_4 bank. With 6 V across the 12-Ω R_3, its current is ½ A; the current is also ½ A in R_4.

Tracing all the current paths from the voltage source in Fig. 7-5a, the main-line current, I_T, through R_1 and R_2 is 2 A. The 2-A I_T flowing into point B subdivides into two separate branch currents: 1 A of the 2-A I_T flows up through resistor, R_5. The other 1 A flows into the branch containing resistors

R_3, R_4, and R_6. Because resistors R_3 and R_4 are in parallel, the 1-A branch current subdivides further into ½ A for I_3 and ½ A for I_4. The currents I_3 and I_4 recombine to flow up through resistor R_6. At the branch point A, I_5 and I_6 combine resulting in the 2-A total current, I_T, flowing through R_1 back to the positive terminal of the voltage source.

7.5 Analyzing Series-Parallel Circuits with Random Unknowns

The circuits in Figs. 7-6 to 7-8 will be solved now. The following principles are illustrated:

1. With parallel strings across the main line, the branch currents and I_T can be found without R_T (see Fig. 7-6).

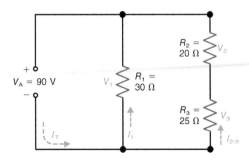

Figure 7-6 Finding all currents and voltages by calculating the branch currents first. See text for solution.

2. When parallel strings have series resistance in the main line, R_T must be calculated to find I_T, assuming no branch currents are known (see Fig. 7-8).

3. The source voltage is applied across the R_T of the entire circuit, producing an I_T that flows only in the main line.

4. Any individual series R has its own IR drop that must be less than the total V_T. In addition, any individual branch current must be less than I_T.

Solution for Figure 7-6

The problem here is to calculate the branch currents I_1 and $I_{2\text{-}3}$, total line current I_T, and the voltage drops V_1, V_2, and V_3. This order will be used for the calculations because we can find the branch currents from the 90 V across the known branch resistances.

In the 30-Ω branch of R_1, the branch current is $90/30 = 3$ A for I_1. The other branch resistance, with a 20-Ω R_2 and a 25-Ω R_3, totals 45 Ω. This branch current then is $90/45 = 2$ A for $I_{2\text{-}3}$. In the main line, I_T is 3 A $+ 2$ A, which is equal to 5 A.

For the branch voltages, V_1 must be the same as V_A, equal to 90 V, or $V_1 = I_1 R_1$, which is $3 \times 30 = 90$ V.

In the other branch, the 2-A $I_{2\text{-}3}$ flows through the 20-Ω R_2 and the 25-Ω R_3. Therefore, V_2 is $2 \times 20 = 40$ V. Also, V_3 is $2 \times 25 = 50$ V. Note that these 40-V and 50-V series IR drops in one branch add to equal the 90-V source.

If we want to know R_T, it can be calculated as V_A/I_T. Then 90 V/5 A equals 18 Ω. Or R_T can be calculated by combining the branch resistances of 30 Ω in parallel with 45 Ω. Then, using the product-divided-by-sum formula, R_T is $(30 \times 45)/(30 + 45)$ or 1350/75, which equals the same value of 18 Ω for R_T.

Solution for Figure 7-7

The division of branch currents also applies to Fig. 7-7, but the main principle here is that the voltage must be the same across R_1 and R_2 in parallel. For the branch currents, I_2 is 2 A, equal to the 6-A I_T minus the 4-A I_1. The voltage across the 10-Ω R_1 is 4×10, or 40 V. This same voltage is also across R_2. With 40 V across R_2 and 2 A through it, R_2 equals 40/2 or 20 Ω.

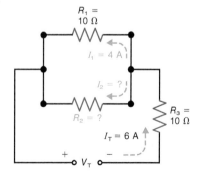

Figure 7-7 Finding R_2 in the parallel bank and its I_2. See text for solution.

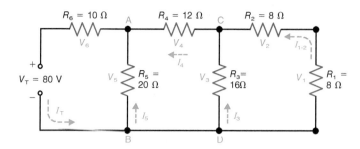

Figure 7-8 Finding all currents and voltages by calculating R_T and then I_T to find V_6 across R_6 in the main line.

We can also find V_T in Fig. 7-7 from R_1, R_2, and R_3. The 6-A I_T through R_3 produces a voltage drop of 60 V for V_3. Also, the voltage across the parallel bank with R_1 and R_2 has been calculated as 40 V. This 40 V across the bank in series with 60 V across R_3 totals 100 V for the applied voltage.

Solution for Figure 7-8

To find all currents and voltage drops, we need R_T to calculate I_T through R_6 in the main line. Combining resistances for R_T, we start with R_1 and R_2 and work in toward the source. Add the 8-Ω R_1 and 8-Ω R_2 in series with each other for 16 Ω. This 16 Ω combined with the 16-Ω R_3 in parallel equals 8 Ω between points C and D. Add this 8 Ω to the series 12-Ω R_4 for 20 Ω. This 20 Ω with the parallel 20-Ω R_5 equals 10 Ω between points A and B. Add this 10 Ω in series with the 10-Ω R_6, to make R_T of 20 Ω for the entire series-parallel circuit.

Current I_T in the main line is V_T/R_1, or 80/20, which equals 4 A. This 4-A I_T flows through the 10-Ω R_6, producing a 40-V IR drop for V_6.

Now that we know I_T and V_6 in the main line, we use these values to calculate all other voltages and currents. Start from the main line, where we know the current, and work outward from the source. To find V_5, the IR drop of 40 V for V_6 in the main line is subtracted from the source voltage. The reason is that V_5 and V_6 must add to equal the 80 V of V_T. Then V_5 is $80 - 40 = 40$ V.

Voltages V_5 and V_6 happen to be equal at 40 V each. They split the 80 V in half because the 10-Ω R_6 equals the combined resistance of 10 Ω between branch points A and B.

With V_5 known to be 40 V, then I_5 through the 20-Ω R_5 is $40/20 = 2$ A. Since I_5 is 2 A and I_T is 4 A, I_4 must be 2 A also, equal to the difference between I_T and I_5. The current flowing into point A equals the sum of the branch currents I_4 and I_5.

The 2-A I_4 through the 12-Ω R_4 produces an IR drop equal to $2 \times 12 = 24$ V for V_4. Note now that V_4 and V_3 must add to equal V_5. The reason is that both V_5 and the path with V_4 and V_3 are across the same two points AB or AD. Since the potential difference across any two points is the same regardless of the path, $V_5 = V_4 + V_3$. To find V_3 now, we can subtract the 24 V of V_4 from the 40 V of V_5. Then $40 - 24 = 16$ V for V_3.

With 16 V for V_3 across the 16-Ω R_3, its current I_3 is 1 A. Also, I_{1-2} in the branch with R_1 and R_2 is equal to 1 A. The 2-A I_4 consists of the sum of the branch currents, I_3 and I_{1-2}, flowing into point C.

Finally, with 1 A through the 8-Ω R_2 and 8-Ω R_1, their voltage drops are $V_2 = 8$ V and $V_1 = 8$ V. Note that the 8 V of V_1 in series with the 8 V of V_2 add to equal the 16-V potential difference V_3 between points C and D.

All answers for the solution of Fig. 7-8 are summarized below:

$R_T = 20\ \Omega$	$I_T = 4$ A	$V_6 = 40$ V
$V_5 = 40$ V	$I_5 = 2$ A	$I_4 = 2$ A
$V_4 = 24$ V	$V_3 = 16$ V	$I_3 = 1$ A
$I_{1-2} = 1$ A	$V_2 = 8$ V	$V_1 = 8$ V

7.6 The Wheatstone Bridge

A **Wheatstone bridge** is a circuit that is used to determine the value of an unknown resistance.* A typical Wheatstone bridge is shown in Fig. 7-9. Notice that four resistors are configured in a diamondlike arrangement, which is typically how the Wheatstone bridge is drawn. In Fig. 7-9, the applied voltage V_T is connected to terminals A and B, which are considered the input terminals to the Wheatstone bridge. A very sensitive zero-centered current meter M_1, called a *galvonometer*, is connected between terminals C and D, which are considered the output terminals.

As shown in Fig. 7-9, the unknown resistor R_X is placed in the same branch as a variable **standard resistor** R_S. It is important to note that the standard resistor R_S is a precision resistance variable from 0 to 9999 Ω in 1-Ω steps. In the other branch, resistors R_1 and R_2 make up what is known as the **ratio arm.** Resistors R_1 and R_2 are also precision resistors having very tight resistance tolerances. To determine the value of an unknown resistance R_X, adjust the standard resistor R_S until the current in M_1 reads exactly 0 μA. With zero current in M_1 the Wheatstone bridge is said to be balanced. But how does the balanced condition provide the value of the unknown resistance R_X? Good question. With zero current in

M_1, the voltage division among resistors R_X and R_S is identical to the voltage division among the ratio arm resistors R_1 and R_2. When the voltage division in the R_X-R_S branch is identical to the voltage division in the R_1-R_2 branch, the potential difference between points C and D will equal 0 V. With a potential difference of 0 V across points C and D, the current in M_1 will read 0 μA, which is the balanced condition. At balance, the equal voltage ratios can be stated as

$$\frac{I_1 R_X}{I_1 R_S} = \frac{I_2 R_1}{I_2 R_2}$$

Since I_1 and I_2 cancel in the equation, this yields

$$\frac{R_X}{R_S} = \frac{R_1}{R_2}$$

Solving for R_X gives us

$$R_X = R_S \times \frac{R_1}{R_2} \tag{7-1}$$

The ratio arm R_1/R_2 can be varied in most cases, typically in multiples of 10, such as 100/1, 10/1, 1/1, 1/10, and 1/100. However, the bridge is still balanced by varying the standard resistor R_S. The placement accuracy of the measurement of R_X is determined by the R_1/R_2 ratio. For example, if $R_1/R_2 = 1/10$, the value of R_X is accurate to within $\pm 0.1\ \Omega$. Likewise, if $R_1/R_2 = 1/100$, the value of R_X will be accurate to within $\pm 0.01\ \Omega$. The R_1/R_2 ratio also determines the maximum unknown resistance that can be measured. Expressed as an equation

$$R_{X(max)} = R_{S(max)} \times \frac{R_1}{R_2} \tag{7-2}$$

EXAMPLE 7-1

In Fig. 7-10, the current in M_1 reads 0 μA with the standard resistor R_S adjusted to 5642 Ω. What is the value of the unknown resistor R_X?

Answer:

Using Formula (7-1), R_X is calculated as follows:

$$R_X = R_S \times \frac{R_1}{R_2}$$
$$= 5642\ \Omega \times \frac{1\ k\Omega}{10\ k\Omega}$$
$$R_X = 564.2\ \Omega$$

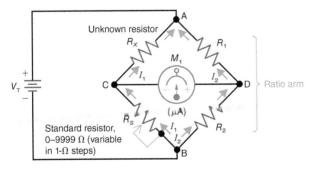

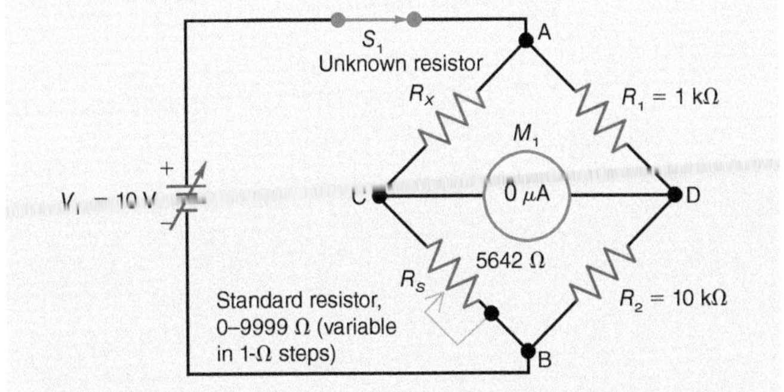

Figure 7-9 Wheatstone bridge.

Figure 7-10 Wheatstone bridge. See Examples 7-1 and 7-2.

* Sir Charles Wheatstone (1802–1875), English physicist and inventor.

EXAMPLE 7-2

In Fig. 7-10, what is the maximum unknown resistance R_x that can be measured for the ratio arm values shown?

Answer:

$$R_{X(max)} = R_{S(max)} \times \frac{R_1}{R_2}$$

$$= 9999\ \Omega \times \frac{1\ k\Omega}{10\ k\Omega}$$

$$R_{X(max)} = 999.9\ \Omega$$

If R_x is larger than 999.9 Ω, the bridge cannot be balanced because the voltage division will be greater than 1/10 in this branch. In other words, the current in M_1 cannot be adjusted to 0 μA. To measure an unknown resistance whose value is greater than 999.9 Ω, you would need to change the ratio arm fraction to 1/1, 10/1, or something higher.

Note that when the Wheatstone bridge is balanced, it can be analyzed simply as two series strings in parallel. The reason is that when the current through M_1 is zero, the path between points C and D is effectively open.

Other Balanced Bridge Applications

There are many other applications in electronics for **balanced bridge** circuits. For example, a variety of sensors is used in bridge circuits for detecting changes in pressure, flow, light, temperature, etc. These sensors are used as one of the resistors in a bridge circuit. Furthermore, the bridge can be balanced or zeroed at some desired reference level of pressure, flow, light, or temperature. Then, when the condition being sensed changes, the bridge becomes unbalanced

and causes a voltage to appear at the output terminals (C and D). This output voltage is then fed to the input of an amplifier or other device that modifies the condition being monitored, thus bringing the system back to its original preset level.

Consider the temperature control circuit in Fig. 7-11. In this circuit, a variable resistor R_3 is in the same branch as a negative temperature coefficient (NTC) thermistor whose resistance value at 25°C (R_0) equals 5 kΩ as shown. Assume that R_3 is adjusted to provide balance when the ambient (surrounding) temperature T_A equals 25°C. Remember, when the bridge is balanced, the output voltage across terminals C and D is 0 V. This voltage is fed to the input of an amplifier as shown. With 0 V into the amplifier, 0 V comes out of the amplifier.

Now let's consider what happens when the ambient temperature T_A increases above 25°C, say to 30°C. The increase in temperature causes the resistance of the thermistor to decrease, since it has an NTC. With a decrease in the thermistor's resistance, the voltage at point C decreases. However, the voltage at point D does not change because R_1 and R_2 are ordinary resistors. The result is that the output voltage V_{CD} goes negative. This negative voltage is fed into the amplifier, which in turn produces a positive output voltage. The positive output voltage from the amplifier turns on a cooling fan or air-conditioning unit. The air-conditioning unit remains on until the ambient temperature decreases to its original value of 25°C. As the temperature drops back to 25°C, the resistance of the thermistor increases to its original value, thus causing the voltage V_{CD} to return to 0 V. This shuts off the air conditioner.

Next, let's consider what happens when the ambient temperature T_A decreases below 25°C, say to 20°C. The decrease in temperature causes the resistance of the thermistor

Figure 7-11 Temperature control circuit using the balanced bridge concept.

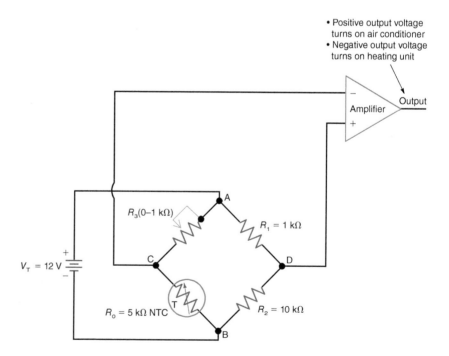

to increase, thus making the voltage at point C more positive. The result is that V_{CD} goes positive. This positive voltage is fed into the amplifier, which in turn produces a negative output voltage. The negative output voltage from the amplifier turns on a heating unit, which remains on until the ambient temperature returns to its original value of 25°C. Although the details of the temperature control circuit in Fig. 7-11 are rather vague, you should get the idea of how a balanced bridge circuit containing a thermistor could be used to control the temperature in a room. There are almost unlimited applications for balanced bridge circuits in electronics.

7.7 Series Voltage Dividers

The current is the same in all resistances in a series circuit. Also, the voltage drops equal the product of I times R. Therefore, the IR voltages are proportional to the series resistances. A higher resistance has a greater IR voltage than a lower resistance in the same series circuit; equal resistances have the same amount of IR drop. If R_1 is double R_2, then V_1 will be double V_2.

The series string can be considered a *voltage divider*. Each resistance provides an IR drop V equal to its proportional part of the applied voltage. Stated as a formula,

$$V = \frac{R}{R_T} \times V_T \qquad (7\text{-}3)$$

EXAMPLE 7-3

Three 50-kΩ resistors R_1, R_2, and R_3 are in series across an applied voltage of 180 V. How much is the IR voltage drop across each resistor?

Answer:

The voltage drop across each R is 60 V. Since R_1, R_2, and R_3 are equal, each has one-third the total resistance of the circuit and one-third the total applied voltage. Using the formula,

$$V = \frac{R}{R_T} \times V_T = \frac{50 \text{ k}\Omega}{150 \text{ k}\Omega} \times 180 \text{ V}$$
$$= \frac{1}{3} \times 180 \text{ V}$$
$$= 60 \text{ V}$$

Note that R and R_T must be in the same units for the proportion. Then V is in the same units as V_T.

Typical Circuit

Figure 7-12 illustrates another example of a proportional voltage divider. Let the problem be to find the voltage across R_3. We can either calculate this voltage V_3 as IR_3 or determine its proportional part of the total applied voltage V_T. The answer is the same both ways. Note that R_T is $20 + 30 + 50 = 100$ kΩ.

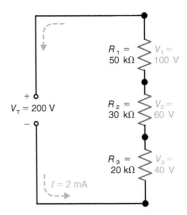

Figure 7-12 Series string of resistors as a proportional voltage divider. Each V_R is R/R_T fraction of the total source voltage V_T.

Proportional Voltage Method

Using Formula (7-3), V_3 equals 20/100 of the 200 V applied for V_T because R_3 is 20 kΩ and R_T is 100 kΩ. Then V_3 is 20/100 of 200 or ⅕ of 200, which is equal to 40 V. The calculations are

$$V_3 = \frac{R_3}{R_T} \times V_T = \frac{20}{100} \times 200 \text{ V}$$
$$V_3 = 40 \text{ V}$$

In the same way, V_2 is 60 V. The calculations are

$$V_2 = \frac{R_2}{R_T} \times V_T = \frac{30}{100} \times 200 \text{ V}$$
$$V_2 = 60 \text{ V}$$

Also, V_1 is 100 V. The calculations are

$$V_1 = \frac{R_1}{R_T} \times V_T = \frac{50}{100} \times 200 \text{ V}$$
$$V_1 = 100 \text{ V}$$

The sum of V_1, V_2, and V_3 in series is $100 + 60 + 40 = 200$ V, which is equal to V_T.

Method of IR Drops

If we want to solve for the current in Fig. 7-12, $I = V_T/R_T$ or 200 V/100 kΩ = 2 mA. This I flows through R_1, R_2, and R_3 in series. The IR drops are

$$V_1 = I \times R_1 = 2 \text{ mA} \times 50 \text{ k}\Omega = 100 \text{ V}$$
$$V_2 = I \times R_2 = 2 \text{ mA} \times 30 \text{ k}\Omega = 60 \text{ V}$$
$$V_3 = I \times R_3 = 2 \text{ mA} \times 20 \text{ k}\Omega = 40 \text{ V}$$

These voltages are the same values calculated by Formula (7-3) for proportional voltage dividers.

Two Voltage Drops in Series

For this case, it is not necessary to calculate both voltages. After you find one, subtract it from V_T to find the other.

As an example, assume that V_T is 48 V across two series resistors R_1 and R_2. If V_1 is 18 V, then V_2 must be 48 − 18 = 30 V.

The Largest Series R Has the Most V

The fact that series voltage drops are proportional to the resistances means that a very small R in series with a much larger R has a negligible IR drop. An example is shown in Fig. 7-13a. Here the 1 kΩ of R_1 is in series with the much larger 999 kΩ of R_2. The V_T is 1000 V.

The voltages across R_1 and R_2 in Fig. 7-13a can be calculated by the voltage divider formula. Note that R_T is 1 + 999 = 1000 kΩ.

$$V_1 = \frac{R_1}{R_T} \times V_T = \frac{1}{1000} \times 1000 \text{ V} = 1 \text{ V}$$

$$V_2 = \frac{R_2}{R_T} \times V_T = \frac{999}{1000} \times 1000 \text{ V} = 999 \text{ V}$$

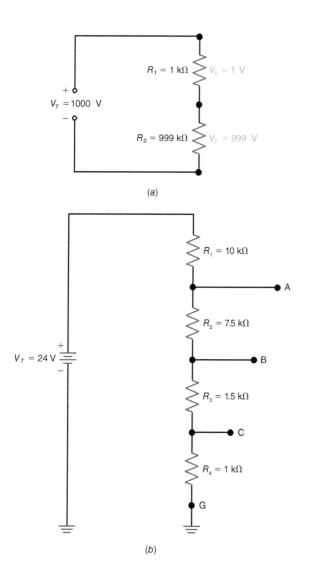

The 999 V across R_2 is practically the entire applied voltage. Also, the very high series resistance dissipates almost all the power.

The current of 1 mA through R_1 and R_2 in Fig. 7-13a is determined almost entirely by the 999 kΩ of R_2. The I for R_T is 1000 V/1000 kΩ, which equals 1 mA. However, the 999 kΩ of R_2 alone would allow 1.001 mA of current, which differs very little from the original I of 1 mA.

Voltage Taps in a Series Voltage Divider

Consider the series voltage divider with **voltage taps** in Fig. 7-13b, where different voltages are available from the tap points A, B, and C. Note that the total resistance R_T is 20 kΩ, which can be found by adding the individual series resistance values. The voltage at each tap point is measured with respect to ground. The voltage at tap point C, designated V_{CG}, is the same as the voltage across R_4. The calculations for V_{CG} are as follows:

$$V_{CG} = \frac{R_4}{R_T} \times V_T$$

$$= \frac{1 \text{ k}\Omega}{20 \text{ k}\Omega} \times 24 \text{ V}$$

$$V_{CG} = 1.2 \text{ V}$$

The voltage at tap point B, designated V_{BG}, is the sum of the voltages across R_3 and R_4. The calculations for V_{BG} are

$$V_{BG} = \frac{R_3 + R_4}{R_T} \times V_T$$

$$= \frac{1.5 \text{ k}\Omega \times 1 \text{ k}\Omega}{20 \text{ k}\Omega} \times 24 \text{ V}$$

$$V_{BG} = 3 \text{ V}$$

The voltage at tap point A, designated V_{AG}, is the sum of the voltages across R_2, R_3, and R_4. The calculations are

$$V_{AG} = \frac{R_2 + R_3 + R_4}{R_T} \times V_T$$

$$= \frac{7.5 \text{ k}\Omega + 1.5 \text{ k}\Omega + 1 \text{ k}\Omega}{20 \text{ k}\Omega} \times 24 \text{ V}$$

$$V_{AG} = 12 \text{ V}$$

Notice that the voltage V_{AG} equals 12 V, which is one-half of the applied voltage V_T. This makes sense, since $R_2 + R_3 + R_4$ make up 50% of the total resistance R_T. Similarly, since $R_3 + R_4$ make up 12.5% of the total resistance, the voltage V_{BG} will also be 12.5% of the applied voltage, which is 3 V in this case. The same analogy applies to V_{CG}.

Each tap voltage is positive because the negative terminal of the voltage source is grounded.

Advantage of the Voltage Divider Method

Using Formula (7-3), we can find the proportional voltage drops from V_T and the series resistances without knowing the amount of I. For odd values of R, calculating the I may

Figure 7-13 (a) Example of a very small R_1 in series with a large R_2; V_2 is almost equal to the whole V_T. (b) Series voltage divider with voltage taps.

be more troublesome than finding the proportional voltages directly. Also, in many cases, we can approximate the voltage division without the need for any written calculations.

7.8 Series Voltage Divider with Parallel Load Current

The voltage dividers shown so far illustrate just a series string without any branch currents. However, a voltage divider is often used to tap off part of the applied voltage for a load that needs less voltage than V_T. Then the added load is a parallel branch across part of the divider, as shown in Fig. 7-14. This example shows how the **loaded voltage** at the tap F is reduced below the potential it would have without the branch current for R_L.

Why the Loaded Voltage Decreases

We can start with Fig. 7-14a, which shows an R_1-R_2 voltage divider alone. Resistances R_1 and R_2 in series simply form a proportional divider across the 60-V source for V_T.

For the resistances, R_1 is 40 kΩ and R_2 is 20 kΩ, making R_T equal to 60 kΩ. Also, the current $I = V_T/R_T$, or 60 V/60 kΩ = 1 mA. For the divided voltages in Fig. 7-14a,

$$V_1 = \frac{40}{60} \times 60 \text{ V} = 40 \text{ V}$$

$$V_2 = \frac{20}{60} \times 60 \text{ V} = 20 \text{ V}$$

Note that $V_1 + V_2$ is 40 + 20 = 60 V, which is the total applied voltage.

However, in Fig. 7-14b, the 20-kΩ branch of R_L changes the equivalent resistance at tap F to ground. This change in the proportions of R changes the voltage division. Now the resistance from F to G is 10 kΩ, equal to the 20-kΩ R_2 and R_L in parallel. This equivalent bank resistance is shown as the 10-kΩ R_E in Fig. 7-14c.

Resistance R_1 is still the same 40 kΩ because it has no parallel branch. The new R_T for the divider in Fig. 7-14c is

40 kΩ + 10 kΩ = 50 kΩ. As a result, V_E from F to G in Fig. 7-14c becomes

$$V_E = \frac{R_E}{R_T} \times V_T = \frac{10}{50} \times 60 \text{ V}$$

$$V_E = 12 \text{ V}$$

Therefore, the voltage across the parallel R_2 and R_L in Fig. 7-14b is reduced to 12 V. This voltage is at the tap F for R_L.

Note that V_1 across R_1 increases to 48 V in Fig. 7-14c. Now V_1 is 40/50 × 60 V = 48 V. The V_1 increases here because there is more current through R_1.

The sum of $V_1 + V_E$ in Fig. 7-14c is 12 + 48 = 60 V. The IR drops still add to equal the applied voltage.

Path of Current for R_L

All current in the circuit must come from the source V_T. Trace the electron flow for R_L. It starts from the negative side of V_T, through R_L, to the tap at F, and returns through R_1 in the divider to the positive side of V_T. This current I_L goes through R_1 but not R_2.

Bleeder Current

In addition, both R_1 and R_2 have their own current from the source. This current through all the resistances in the divider is called the **bleeder current** I_B. The electron flow for I_B is from the negative side of V_T, through R_2 and R_1, and back to the positive side of V_T.

In summary, then, for the three resistances in Fig. 7-14b, note the following currents:

1. Resistance R_L has just its load current I_L.
2. Resistance R_2 has only the bleeder current I_B.
3. Resistance R_1 has both I_L and I_B.

Note that only R_1 is in the path for both the bleeder current and the load current.

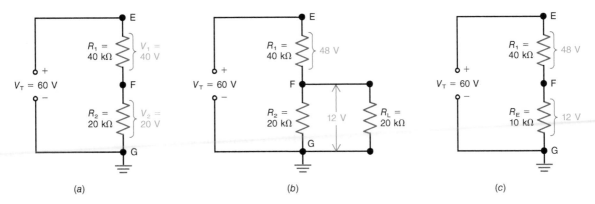

Figure 7-14 Effect of a parallel load in part of a series voltage divider. (a) R_1 and R_2 in series without any branch current. (b) Reduced voltage across R_2 and its parallel load R_L. (c) Equivalent circuit of the loaded voltage divider.

Analyzing Circuits with Simulation Software

By using network theorems and techniques along with the necessary mathematics, almost any electric or electronic circuit can be analyzed and designed. As you have seen in this chapter, such analysis requires knowledge of the various theorems and rules not to mention the mathematical procedures. However, such analysis takes time and a significant amount of calculation. Electronic calculators help considerably to simplify and speed up the analysis, but any such analysis is still a major task. Computers can be programmed to execute the math of the various theorems, further speeding up the calculations. Today there is an even better way to automate, simplify, and facilitate circuit analysis: simulation.

Circuit Simulation

Circuit simulation is the process of using a computer to simulate the behavior of all the various components like resistors, capacitors, diodes, transistors, and even some integrated circuits. Programs are written to emulate the various components. Then these components can be "interconnected" in software to form a circuit to be analyzed. By giving the components values, the simulation software can replicate in software what the real circuit will do. The software actually builds a mathematical model of the circuit. The simulation allows you to create, experiment with, analyze, and design circuits without actually physically building them. All the basic network theorems and methods as well as the related math are all implemented in software and automatically applied, providing a near instantaneous analysis with minimal work.

SPICE

Simulation did not become popular or useful until computers and software became affordable and easier to use. One of the first practical usable circuit simulation programs was called Simulation Program with Integrated Circuit Emphasis (SPICE). It was developed at the University of California, Berkley, in the 1970s. Numerous continuously updated and improved versions were implemented over the years. This was an open source program that anyone could acquire free and enhance. SPICE inspired the development of other simulation programs, and numerous commercial simulation software products started as SPICE derivatives.

Circuit simulation software literally changed the way that electronic circuits were analyzed and designed. They offered a way to test and prove the performance of a circuit before actually building one. It not only accelerated the design process but also made it far easier. Multiple iterations of a design could be accomplished in a short time. Simulation did not replace actual hardware prototypes but provided a way to shorten the actual hardware building and testing cycle by eliminating bad designs in the simulation process and saving testing and troubleshooting time. Today, simulation is the way most circuits are designed.

Multisim

One of the most refined and widely used simulation software products is National Instruments' Multisim. It is based on the SPICE software, but it has been highly enhanced with a friendly user interface as well as many useful analysis and design features. Multisim is widely used in electronic design work but is also used by some colleges and universities to teach circuit analysis and design. Multisim runs on almost any personal computer or laptop using the Windows operating system. Its primary application is to analyze and design analog or linear circuits.

To use Multisim, you start by entering the details of a circuit you wish to analyze. This is done with a schematic capture program. Using the mouse on the computer, you select components from a schematic data base and place them on the screen. All the various regular components are available, including resistors, capacitors, inductors, transformers, diodes, transistors, and a wide variety of common integrated circuits like op amps. Once all the desired components are present, you can connect them to form the circuit. Next you give each component a value and specify any dc or ac voltages to be connected. The computer builds the circuit in software.

Once the circuit has been built, you can save it as a file. Then you can analyze it. Multisim provides a dc analysis, an ac analysis, a transient analysis, and a variety of other test simulations. Just click on the simulation button, and the various voltages and/or currents are displayed.

An important feature is the simulated test instruments. You can connect a signal generator, digital multimeter, oscilloscope, and spectrum analyzer to show actual measurements of the circuit as it is being simulated.

Figure S7-1 shows a simple example of a basic dc circuit. Here a 12-volt dc source is connected to a 1-kΩ resistor. A multimeter is connected in series with the circuit to measure the current. Figure S7-2 shows an AC signal generator supplying a 1-V, 1-kHz sine wave to a 1-kΩ resistor. An oscilloscope is connected to show the voltage across the resistor. You can make actual measurements on the simulated oscilloscope.

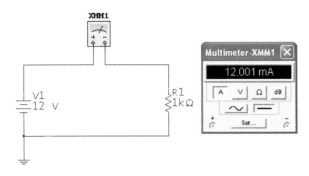

Figure S7-1 DC current measurement with a generic multimeter.

Figure S7-3 shows one unique feature of Multisim. Here a resistor-capacitor (*RC*) low-pass filter is being analyzed. It is connected to a Bode plotter. This is a feature that applies a continuingly increasing sine-wave frequency and then plots the output voltage response. In this way frequency-sensitive circuits like filters can be analyzed by looking at their output frequency response curves.

Any further analysis is done by manipulating circuit values, frequencies, and voltages so the circuit can be optimized. Multisim also has a feature that facilitates the printed circuit board layout of the circuit.

Most technicians do not generally perform analysis or design as part of their normal work. Doing analysis and design is the primary job of an engineer. Multisim is mostly used by engineers, but technicians can use it to further their knowledge of electronic circuits. It is a great learning tool. A student version is available for experimenting. Go to **www .ni.com/multisim** to learn more.

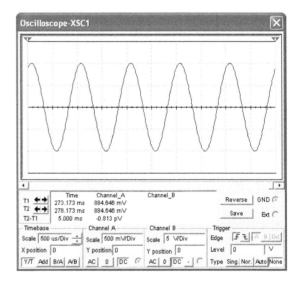

Figure S7-2 Voltage measurement with a generic oscilloscope.

Figure S7-3 (*a*) Bode plotter. (*b*) Bode plotter display.

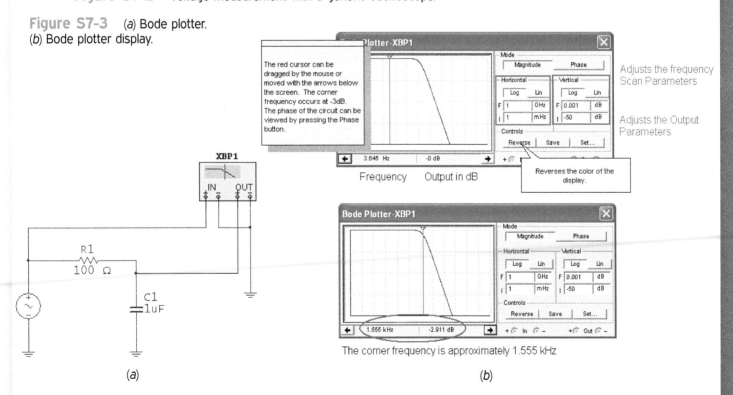

The red cursor can be dragged by the mouse or moved with the arrows below the screen. The corner frequency occurs at -3dB. The phase of the circuit can be viewed by pressing the Phase button.

Adjusts the frequency Scan Parameters

Adjusts the Output Parameters

Frequency   Output in dB

Reverses the color of the display.

The corner frequency is approximately 1.555 kHz

(*a*)

(*b*)

QUESTIONS 1–12 Refer to Fig. 7-15.

1. In Fig. 7-15,
   a. R_1 and R_2 are in series.
   b. R_3 and R_4 are in series.
   c. R_1 and R_4 are in series.
   d. R_2 and R_4 are in series.

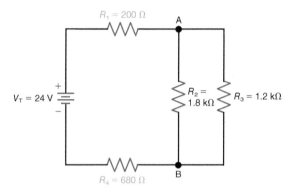

Figure 7-15

2. In Fig. 7-15,
   a. R_2, R_3, and V_T are in parallel.
   b. R_2 and R_3 are in parallel.
   c. R_2 and R_3 are in series.
   d. R_1 and R_4 are in parallel.

3. In Fig. 7-15, the total resistance, R_T, equals
   a. 1.6 kΩ.
   b. 3.88 kΩ.
   c. 10 kΩ.
   d. none of the above.

4. In Fig. 7-15, the total current, I_T, equals
   a. 6.19 mA.
   b. 150 mA.
   c. 15 mA.
   d. 25 mA.

5. In Fig. 7-15, how much voltage is across points A and B?
   a. 12 V.
   b. 18 V.
   c. 13.8 V.
   d. 10.8 V.

6. In Fig. 7-15, how much is I_2 through R_2?
   a. 9 mA.
   b. 15 mA.
   c. 6 mA.
   d. 10.8 mA.

7. In Fig. 7-15, how much is I_3 through R_3?
   a. 9 mA.
   b. 15 mA.

   c. 6 mA.
   d. 45 mA.

8. If R_4 shorts in Fig. 7-15, the voltage, V_{AB}
   a. increases.
   b. decreases.
   c. stays the same.
   d. increases to 24 V.

9. If R_2 becomes open in Fig. 7-15,
   a. the voltage across points A and B will decrease.
   b. the resistors R_1, R_3, and R_4 will be in series.
   c. the total resistance, R_T, will decrease.
   d. the voltage across points A and B will measure 24 V.

10. If R_1 opens in Fig. 7-15,
    a. the voltage across R_1 will measure 0 V.
    b. the voltage across R_4 will measure 0 V.
    c. the voltage across points A and B will measure 0 V.
    d. both b and c.

11. If R_3 becomes open in Fig. 7-15, what happens to the voltage across points A and B?
    a. It decreases.
    b. It increases.
    c. It stays the same.
    d. None of the above.

12. If R_2 shorts in Fig. 7-15,
    a. the voltage, V_{AB}, decreases to 0 V.
    b. the total current, I_T, flows through R_3.
    c. the current, I_3, in R_3 is zero.
    d. both a and c.

QUESTIONS 13–20 Refer to Fig. 7-16.

13. In Fig. 7-16, how much voltage exists between terminals C and D when the bridge is balanced?
    a. 0 V.
    b. 10.9 V.
    c. 2.18 V.
    d. 12 V.

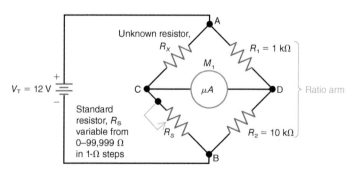

Figure 7-16

14. In Fig. 7-16, assume that the current in M_1 is zero when R_S is adjusted to 55,943 Ω. What is the value of the unknown resistor, R_X?
 a. 55,943 Ω.
 b. 559.43 Ω.
 c. 5,594.3 Ω.
 d. 10 kΩ.

15. In Fig. 7-16, assume that the bridge is balanced when R_S is adjusted to 15,000 Ω. How much is the total current, I_T, flowing to and from the terminals of the voltage source, V_T?
 a. Zero.
 b. Approximately 727.27 μA.
 c. Approximately 1.09 mA.
 d. Approximately 1.82 mA.

16. In Fig. 7-16, what is the maximum unknown resistor, $R_{X(max)}$, that can be measured for the resistor values shown in the ratio arm?
 a. 99.99 Ω.
 b. 9,999.9 Ω.
 c. 99,999 Ω.
 d. 999,999 Ω.

17. In Fig. 7-16, the ratio R_1/R_2 determines
 a. the placement accuracy of the measurement of R_X.
 b. the maximum unknown resistor, $R_{X(max)}$, that can be measured.
 c. the amount of voltage available across terminals A and B.
 d. both *a* and *b*.

18. In Fig. 7-16, assume that the standard resistor, R_S, has been adjusted so that the current in M_1 is exactly 0 μA. How much voltage exists at terminal C with respect to terminal B?
 a. 1.1 V.
 b. 0 V.
 c. 10.9 V.
 d. None of the above.

19. In Fig. 7-16, assume that the ratio arm resistors, R_1 and R_2, are interchanged. What is the value of the unknown resistor, R_x, if R_S equals 33,950 Ω when the bridge is balanced?
 a. 339.5 kΩ.
 b. 3.395 kΩ.
 c. 33,950 Ω.
 d. None of the above.

20. In Fig. 7-16, assume that the standard resistor, R_S, cannot be adjusted high enough to provide a balanced condition. What modification must be made to the circuit?
 a. Change the ratio arm fraction R_1/R_2, from $\frac{1}{10}$ to $\frac{1}{100}$ or something less.

b. Change the ratio arm fraction, R_1/R_2 from $\frac{1}{10}$ to $\frac{1}{1}$, $\frac{10}{1}$ or something greater.
 c. Reverse the polarity of the applied voltage, V_T.
 d. None of the above.

21. In a series circuit, the individual resistor voltage drops are
 a. inversely proportional to the series resistance values.
 b. proportional to the series resistance values.
 c. unrelated to the series resistance values.
 d. none of the above.

22. In a parallel circuit, the individual branch currents are
 a. not related to the branch resistance values.
 b. directly proportional to the branch resistance values.
 c. inversely proportional to the branch resistance values.
 d. none of the above.

23. Three resistors R_1, R_2, and R_3 are connected in series across an applied voltage, V_T, of 24 V. If R_2 is one-third the value of R_T, how much is V_2?
 a. 8 V.
 b. 16 V.
 c. 4 V.
 d. It is impossible to determined.

24. Two resistors R_1 and R_2 are in parallel. If R_1 is twice the value of R_2, how much is I_2 in R_2 if I_T equals 6 A?
 a. 1 A.
 b. 2 A.
 c. 3 A.
 d. 4 A.

QUESTIONS 25–29 Refer to Fig. 7-17.

25. In Fig. 7-17, how much is I_1 in R_1?
 a. 400 mA.
 b. 300 mA.
 c. 100 mA.
 d. 500 mA.

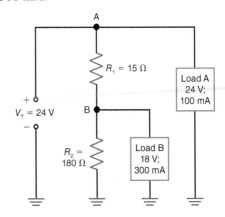

Figure 7-17

26. In Fig. 7-17, how much is the bleeder current, I_B?
  a. 500 mA.
  b. 400 mA.
  c. 100 mA.
  d. 300 mA.

27. In Fig. 7-17, how much is the total current, I_T?
  a. 500 mA.
  b. 400 mA.
  c. 100 mA.
  d. 300 mA.

28. In Fig. 7-17, what is the voltage, V_{BG}, if load B becomes open?
  a. 18 V.
  b. 19.2 V.
  c. 6 V.
  d. 22.15 V.

29. In Fig. 7-17, what happens to the voltage, V_{BG}, if load A becomes open?
  a. It increases.
  b. It decreases.
  c. It remains the same.
  d. It cannot be determined.

CHAPTER 7 PROBLEMS

SECTION 7.1 Finding R_T for Series-Parallel Resistances

7.1 In Fig. 7-18, identify which components are in series and which ones are in parallel.

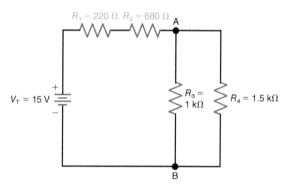

Figure 7-18

7.2 In Fig. 7-18,
  a. how much is the total resistance of just R_1 and R_2?
  b. what is the equivalent resistance of R_3 and R_4 across points A and B?
  c. how much is the total resistance, R_T, of the entire circuit?
  d. how much is the total current, I_T, in the circuit?
  e. how much current flows into point B?
  f. how much current flows away from point A?

7.3 In Fig. 7-18, solve for the following: I_1, I_2, V_1, V_2, V_3, V_4, I_3, and I_4.

7.4 In Fig. 7-19, identify which components are in series and which ones are in parallel.

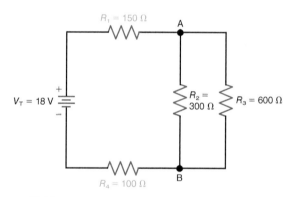

Figure 7-19

7.5 In Fig. 7-19,
  a. how much is the total resistance of just R_1 and R_4?
  b. what is the equivalent resistance of R_2 and R_3 across points A and B?
  c. how much is the total resistance, R_T, of the entire circuit?
  d. how much is the total current, I_T, in the circuit?
  e. how much current flows into point B and away from point A?

7.6 In Fig. 7-19, solve for the following: I_1, V_1, V_2, V_3, I_2, I_3, I_4, V_4, P_1, P_2, P_3, P_4, and P_T.

7.7 In Fig. 7-20,
  a. what is the total resistance of branch 1?
  b. what is the total resistance of branch 2?
  c. how much are the branch currents I_1 and I_2?
  d. how much is the total current, I_T, in the circuit?

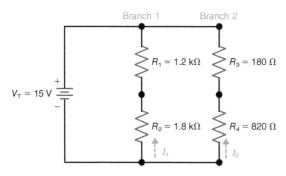

Figure 7-20

    e. how much is the total resistance, R_T, of the entire circuit?

    f. what are the values of V_1, V_2, V_3, and V_4?

7.8 In Fig. 7-21, solve for the following:

    a. branch currents I_1, I_2, I_3 and the total current, I_T

    b. R_T

    c. V_1, V_2, V_3, V_4, and V_5

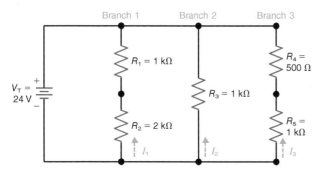

Figure 7-21

SECTION 7.3 Resistance Banks in Series

7.9 In Fig. 7-22,

    a. what is the equivalent resistance of R_1 and R_2 in parallel across points A and B?

    b. what is the total resistance, R_T, of the circuit?

    c. what is the total current, I_T, in the circuit?

    d. how much voltage exists across points A and B?

    e. how much voltage is dropped across R_3?

    f. solve for I_1 and I_2.

    g. how much current flows into point B and away from point A?

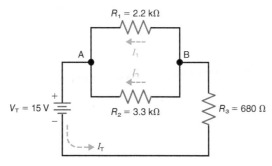

Figure 7-22

7.10 In Fig. 7-23,

    a. what is the equivalent resistance of R_2 and R_3 in parallel across points A and B?

    b. what is the total resistance, R_T, of the circuit?

    c. what is the total current, I_T, in the circuit?

    d. how much voltage exists across points A and B?

    e. how much voltage is dropped across R_1?

    f. solve for I_2 and I_3.

    g. how much current flows into point B and away from point A?

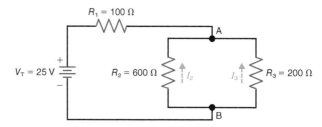

Figure 7-23

SECTION 7.4 Resistance Banks and Strings in Series Parallel

7.11 In Fig. 7-24, solve for R_T, I_T, V_1, V_2, V_3, V_4, I_1, I_2, I_3, and I_4.

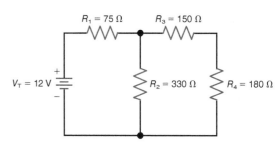

Figure 7-24

7.12 In Fig. 7-25, solve for R_T, I_T, V_1, V_2, V_3, V_4, V_5, V_6, I_1, I_2, I_3, I_4, I_5, and I_6.

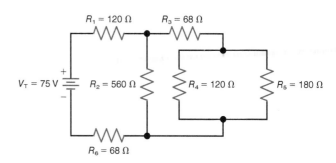

Figure 7-25

7.13 In Fig. 7-26, solve for R_T, I_1, I_2, I_3, V_1, V_2, V_3, and the voltage, V_{AB}.

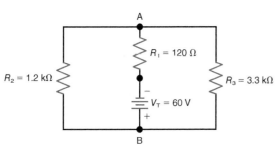

Figure 7-26

7.14 In Fig. 7-27, solve for R_T, I_T, V_1, V_2, V_3, V_4, V_5, V_6, I_1, I_2, I_3, I_4, I_5, and I_6.

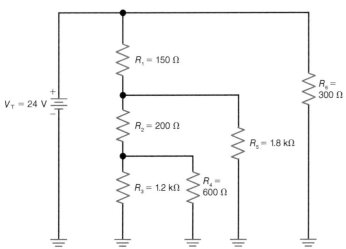

Figure 7-27

SECTION 7.5 Analyzing Series-Parallel Circuits with Random Unknowns

7.15 In Fig. 7-28 solve for V_1, V_2, V_3, I_2, R_3, R_T, and V_T.

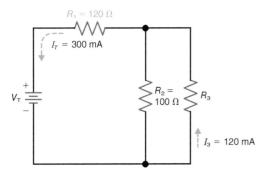

Figure 7-28

7.16 In Fig. 7-29, solve for R_2, V_1, V_2, V_3, I_3, I_T, and V_T.

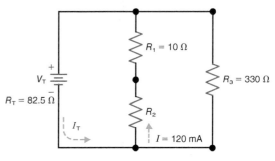

Figure 7-29

7.17 In Fig. 7-30, solve for R_T, I_T, V_1, V_2, V_3, V_4, V_5, V_6, I_1, I_2, I_3, I_4, I_5, and I_6.

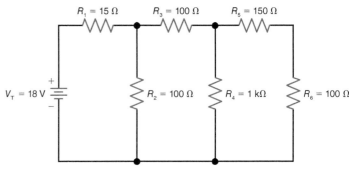

Figure 7-30

SECTION 7.6 The Wheatstone Bridge

Problems 18–22 refer to Fig. 7-31.

7.18 In Fig. 7-31,
    a. how much current flows through M_1 when the Wheatstone bridge is balanced?
    b. how much voltage exists between points C and D when the bridge is balanced?

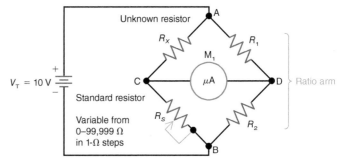

Figure 7-31

7.19 In Fig. 7-31, assume that the bridge is balanced when $R_1 = 1$ kΩ, $R_2 = 5$ kΩ, and $R_S = 34{,}080\ \Omega$. Determine
    a. the value of the unknown resistor, R_X.
    b. the voltages V_{CB} and V_{DB}.
    c. the total current, I_T, flowing to and from the voltage source, V_T.

7.20 In reference to Prob. 19, which direction (C to D or D to C) will electrons flow through M_1 if
a. R_S is reduced in value.
b. R_S is increased in value.

7.21 In Fig. 7-31, calculate the maximum unknown resistor, $R_{X(max)}$, that can be measured for the following ratio arm values:

a. $\dfrac{R_1}{R_2} = \dfrac{1}{1000}$

b. $\dfrac{R_1}{R_2} = \dfrac{100}{1}$

7.22 Assume that the same unknown resistor, R_X, is measured using different ratio arm fractions in Fig. 7-31. In each case, the standard resistor, R_S, is adjusted to provide the balanced condition. The values for each measurement are

a. $R_S = 123 \ \Omega$ and $\dfrac{R_1}{R_2} = \dfrac{1}{1}$.

b. $R_S = 1232 \ \Omega$ and $\dfrac{R_1}{R_2} = \dfrac{1}{10}$.

c. $R_S = 12{,}317 \ \Omega$ and $\dfrac{R_1}{R_2} = \dfrac{1}{100}$.

Calculate the value of the unknown resistor, R_X, for each measurement. Which ratio arm fraction provides the greatest accuracy?

PROBLEMS 23–25 Refer to Fig. 7-32.

7.23 In Fig. 7-32, to what value must R_3 be adjusted to provide zero volts across terminals C and D when the ambient temperature, T_A, is 25°C? (Note: R_0 is the resistance of the thermistor at an ambient temperature, T_A, of 25°C.)

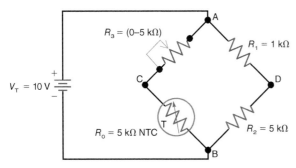

Figure 7-32

7.24 In Fig. 7-32, assume that R_S is adjusted to provide zero volts across terminals C and D at an ambient temperature, T_A, of 25°C. What happens to the polarity of the output voltage, V_{CD}, when
a. the ambient temperature, T_A, increases above 25°C?
b. the ambient temperature, T_A, decreases below 25°C?

7.25 In Fig. 7-32, assume that R_3 has been adjusted to 850 Ω to provide zero volts across the output terminals C and D. Determine
a. the resistance of the thermistor.
b. whether the ambient temperature, T_A, has increased or decreased from 25°C.

SECTION 7.7 Series Voltage Dividers

7.26 A 100-Ω R_1 is in series with a 200-Ω R_2 and a 300-Ω R_3. The applied voltage, V_T, is 18 V. Calculate V_1, V_2, and V_3.

7.27 A 10-kΩ R_1 is in series with a 12-kΩ R_2, a 4.7-kΩ R_3, and a 3.3-kΩ R_4. The applied voltage, V_T, is 36 V. Calculate V_1, V_2, V_3, and V_4.

7.28 In Fig. 7-33, calculate V_1, V_2, and V_3.

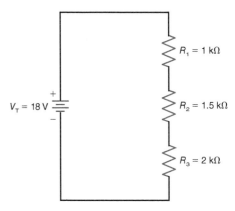

Figure 7-33

7.29 In Fig. 7-34, calculate V_1, V_2, and V_3. Note that resistor R_2 is three times the value of R_1 and resistor R_3 is two times the value of R_2.

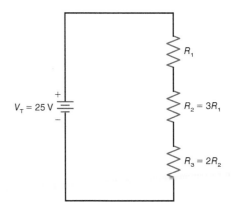

Figure 7-34

7.30 In Fig. 7-35, calculate
a. V_1, V_2, and V_3.
b. V_{AG}, V_{BG}, and V_{CG}.

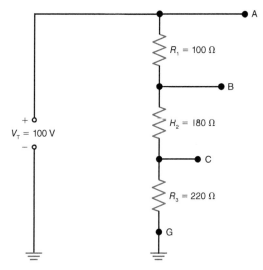

Figure 7-35

SECTION 7.8 Series Voltage Divider with
Parallel Load Current

7.31 In Fig. 7-36, calculate I_1, I_2, I_L, V_{BG}, and V_{AG} with
  a. S_1 open.
  b. S_1 closed.

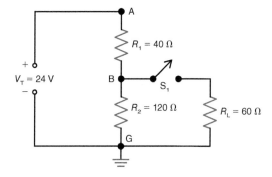

Figure 7-36

7.32 In Fig. 7-36, explain why the voltage, V_{BG}, decreases when the switch, S_1 is closed.

7.33 In Fig. 7-37, calculate I_1, I_2, I_L, V_{BG}, and V_{AG} with
  a. S_1 open.
  b. S_1 closed.

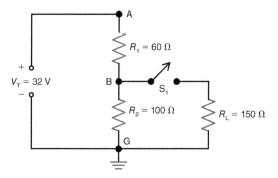

Figure 7-37

7.34 With S_1 closed in Fig. 7-37, which resistor has only the bleeder current, I_B, flowing through it?

7.35 In Fig. 7-38, calculate I_1, I_2, I_L, V_{BG}, and V_{AG} with
  a. S_1 open.
  b. S_1 closed.

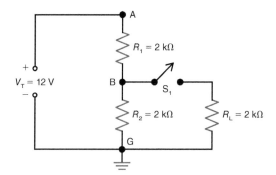

Figure 7-38

Network Theorems

Learning Outcomes

After studying this chapter, you should be able to:

> *State* and *apply* Kirchhoff's current law.

> *State* and *apply* Kirchhoff's voltage law.

> *Determine* the Thevenin equivalent circuits with respect to any pair of terminals in a complex network.

> *Apply* Thevenin's theorem in solving for an unknown voltage or current.

> *State* the internal source conditions for ideal and practical voltage and current sources.

A network is a combination of components, such as resistances and voltage sources, all interconnected to achieve a particular end result. Occasionally there is a need to analyze such a network to determine how it works and performs. In many cases, Kirchhoff's laws can be used to analyze the circuit, as you saw in a previous chapter. However, in some cases more sophisticated methods are needed to speed up and simplify the analysis. Good examples are circuits with multiple voltage sources or a bridge circuit.

This chapter delves a bit deeper into Kirchhoff's laws to show you some alternative approaches to circuit analysis. Then several new techniques are introduced. These new methods convert the circuit to some equivalent circuit that can be more easily solved with basic series and parallel circuit methods.

Specifically, these techniques are Thevenin's theorem and the maximum power transfer theorem. These methods are introduced as basic resistive circuits with dc voltages sources. Voltage and current sources are analyzed. Just remember, as you learn these methods, they can also be applied to circuits with ac voltages sources as well as with inductors and capacitors.

8.1 Kirchhoff's Current Law (KCL)

The algebraic sum of the currents entering and leaving any point in a circuit must equal zero. Or stated another way, *the algebraic sum of the currents into any point of the circuit must equal the algebraic sum of the currents out of that point.* Otherwise, charge would accumulate at the point, instead of having a conducting path. An *algebraic sum* means combining positive and negative values.

Algebraic Signs

In using Kirchhoff's laws to solve circuits, it is necessary to adopt conventions that determine the algebraic signs for current and voltage terms. A convenient system for currents is to *consider all currents into a branch point as positive and all currents directed away from that point as negative.*

As an example, in Fig. 8-1 we can write the currents as

$$I_A + I_B - I_C = 0$$

or

$$5\,A + 3\,A - 8\,A = 0$$

Currents I_A and I_B are positive terms because these currents flow into P, but I_C, directed out, is negative.

Current Equations

For a circuit application, refer to point C at the top of the diagram in Fig. 8-2. The 6-A I_T into point C divides into the 2-A I_3 and 4-A I_{4-5}, both directed out. Note that I_{4-5} is the current through R_4 and R_5. The algebraic equation is

$$I_T - I_3 - I_{4-5} = 0$$

Substituting the values for these currents,

$$6\,A - 2\,A - 4\,A = 0$$

For the opposite directions, refer to point D at the bottom of Fig. 8-2. Here the branch currents into D combine to equal the main-line current I_T returning to the voltage source. Now I_T is directed out from D with I_3 and I_{4-5} directed in. The algebraic equation is

$$I_3 + I_{4-5} - I_T = 0$$

or

$$2\,A + 4\,A - 6\,A = 0$$

$$I_{in} = I_{out}$$

Note that at either point C or point D in Fig. 8-2, the sum of the 2-A and 4-A branch currents must equal the 6-A total line current. Therefore, Kirchhoff's current law can also be stated as $I_{in} = I_{out}$. For Fig. 8-2, the equations of current can be written:

| At point C: | $6\,A = 2\,A + 4\,A$ |
| At point D: | $2\,A + 4\,A = 6\,A$ |

Kirchhoff's current law is the basis for the practical rule in parallel circuits that the total line current must equal the sum of the branch currents.

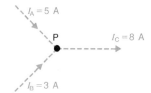

Figure 8-1 Current I_C out from point P equals 5 A + 3 A into P.

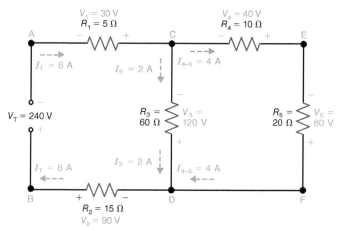

Figure 8-2 Series-parallel circuit illustrating Kirchhoff's laws. See text for voltage and current equations.

EXAMPLE 8-1

In Fig. 8-3, apply Kirchhoff's current law to solve for the unknown current, I_3.

Answer:

In Fig. 8-3, the currents I_1, I_2, and I_3 flowing into point X are considered positive, whereas the currents I_4 and I_5 flowing away from point X are considered negative. Expressing the currents as an equation gives us,

$$I_1 + I_2 + I_3 - I_4 - I_5 = 0$$

or

$$I_1 + I_2 + I_3 = I_4 + I_5$$

Inserting the values from Fig. 8-3,

$$2.5\,\text{A} + 8\,\text{A} + I_3 = 6\,\text{A} + 9\,\text{A}$$

Solving for I_3 gives us

$$I_3 = 6\,\text{A} + 9\,\text{A} - 2.5\,\text{A} - 8\,\text{A}$$
$$= 4.5\,\text{A}$$

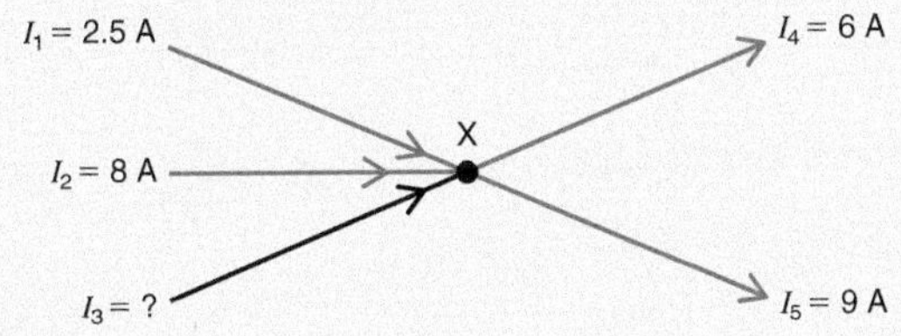

Figure 8-3

8.2 Kirchhoff's Voltage Law (KVL)

The algebraic sum of the voltages around any closed path is zero. If you start from any point at one potential and come back to the same point and the same potential, the difference of potential must be zero.

Algebraic Signs

In determining the algebraic signs for voltage terms in a KVL equation, first mark the polarity of each voltage as shown in Fig. 8-2. A convenient system is to *go around any closed path and consider any voltage whose negative terminal is reached first as a negative term and any voltage whose positive terminal is reached first as a positive term.* This method applies to *IR* voltage drops and voltage sources. The direction can be clockwise or counterclockwise.

Remember that electrons flowing into a resistor make that end negative with respect to the other end. For a voltage source, the direction of electrons returning to the positive terminal is the normal direction for electron flow, which means that the source should be a positive term in the voltage equation.

When you go around the closed path and come back to the starting point, the algebraic sum of all the voltage terms

must be zero. There cannot be any potential difference for one point.

If you do not come back to the start, then the algebraic sum is the voltage between the start and finish points.

You can follow any closed path because the voltage between any two points in a circuit is the same regardless of the path used in determining the potential difference.

Loop Equations

Any closed path is called a *loop*. A loop equation specifies the voltages around the loop.

Figure 8-2 has three loops. The outside loop, starting from point A at the top, through CEFDB, and back to A, includes the voltage drops V_1, V_4, V_5, and V_2 and the source V_T.

The inside loop ACDBA includes V_1, V_3, V_2, and V_T. The other inside loop, CEFDC with V_4, V_5, and V_3, does not include the voltage source.

Consider the voltage equation for the inside loop with V_T. In the clockwise direction starting from point A, the algebraic sum of the voltages is

$$-V_1 - V_3 - V_2 + V_T = 0$$

or

$$-30\,\text{V} - 120\,\text{V} - 90\,\text{V} + 240\,\text{V} = 0$$

Voltages V_1, V_3, and V_2 have negative signs, because the negative terminal for each of these voltages is reached first. However, the source V_T is a positive term because its plus terminal is reached first, going in the same direction.

For the opposite direction, going counterclockwise in the same loop from point B at the bottom, V_2, V_3, and V_1 have positive values and V_T is negative. Then

$$V_2 + V_3 + V_1 - V_T = 0$$

or

$$90\,\text{V} + 120\,\text{V} + 30\,\text{V} - 240\,\text{V} = 0$$

When we transpose the negative term of $-240\,\text{V}$, the equation becomes

$$90\,\text{V} + 120\,\text{V} + 30\,\text{V} = 240\,\text{V}$$

This equation states that the sum of the voltage drops equals the applied voltage.

$\Sigma V = V_T$

The Greek letter Σ means "sum of." In either direction, for any loop, the sum of the *IR* voltage drops must equal the applied voltage V_T. In Fig. 8-2, for the inside loop with the source V_T, going counterclockwise from point B,

$$90\,\text{V} + 120\,\text{V} + 30\,\text{V} = 240\,\text{V}$$

This system does not contradict the rule for algebraic signs. If 240 V were on the left side of the equation, this term would have a negative sign.

Stating a loop equation as $\Sigma V = V_T$ eliminates the step of transposing the negative terms from one side to the other to make them positive. In this form, the loop equations show that Kirchhoff's voltage law is the basis for the practical rule in series circuits that the sum of the voltage drops must equal the applied voltage.

When a loop does not have any voltage source, the algebraic sum of the IR voltage drops alone must total zero.

For instance, in Fig. 8-2, for the loop CEFDC without the source V_T, going clockwise from point C, the loop equation of voltages is

$$-V_4 - V_5 + V_3 = 0$$
$$-40 \text{ V} - 80 \text{ V} + 120 \text{ V} = 0$$
$$0 = 0$$

Notice that V_3 is positive now, because its plus terminal is reached first by going clockwise from D to C in this loop.

EXAMPLE 8-2

In Fig. 8-4a, apply Kirchhoff's voltage law to solve for the voltages V_{AG} and V_{BG}.

Answer:

In Fig. 8-4a, the voltage sources V_1 and V_2 are connected in a series-aiding fashion since they both force electrons to flow through the circuit in the same direction. The earth ground connection at the junction of V_1 and V_2 is used simply for a point of reference. The circuit is solved as follows:

$$V_T = V_1 + V_2$$
$$= 18 \text{ V} + 18 \text{ V}$$
$$= 36 \text{ V}$$

$$R_T = R_1 + R_2 + R_3$$
$$= 120 \text{ }\Omega + 100 \text{ }\Omega + 180 \text{ }\Omega$$
$$= 400 \text{ }\Omega$$

$$I = \frac{V_T}{R_T}$$

$$= \frac{36 \text{ V}}{400 \text{ }\Omega}$$

$$= 90 \text{ mA}$$

$$V_{R_1} = I \times R_1$$
$$= 90 \text{ mA} \times 120 \text{ }\Omega$$
$$= 10.8 \text{ V}$$

$$V_{R_2} = I \times R_2$$
$$= 90 \text{ mA} \times 100 \text{ }\Omega$$
$$= 9 \text{ V}$$

$$V_{R_3} = I \times R_3$$
$$= 90 \text{ mA} \times 180 \text{ }\Omega$$
$$= 16.2 \text{ V}$$

Figure 8-4b shows the voltage drops across each resistor. Notice that the polarity of each resistor voltage drop is negative at the end where the electrons enter the resistor and positive at the end where they leave.

Next, we can apply Kirchhoff's voltage law to determine if we have solved the circuit correctly. If we go counterclockwise (CCW) around the loop, starting and ending at the positive (+) terminal of

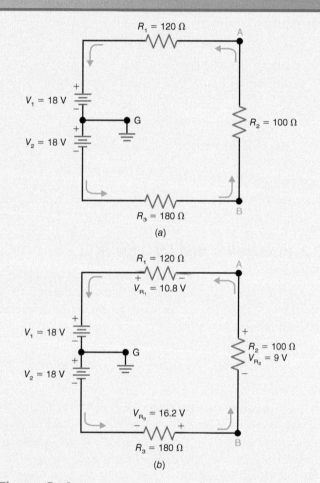

(a)

(b)

Figure 8-4

V_1, we should obtain an algebraic sum of 0 V. The loop equation is written as

$$V_1 + V_2 - V_{R_3} - V_{R_2} - V_{R_1} = 0$$

Notice that the voltage sources V_1 and V_2 are considered positive terms in the equation because their positive (+) terminals were reached first when going around the loop. Similarly, the voltage drops V_{R_1}, V_{R_2}, and V_{R_3} are considered negative terms because the negative (−) end of each resistor's voltage drop is encountered first when going around the loop.

Substituting the values from Fig. 8-4b gives us

$$18\text{ V} + 18\text{ V} - 16.2\text{ V} - 9\text{ V} - 10.8\text{ V} = 0$$

It is important to realize that the sum of the resistor voltage drops must equal the applied voltage, V_T, which equals $V_1 + V_2$ or 36 V in this case. Expressed as an equation,

$$V_T = V_{R_1} + V_{R_2} + V_{R_3}$$
$$= 10.8\text{ V} + 9\text{ V} + 16.2\text{ V}$$
$$= 36\text{ V}$$

It is now possible to solve for the voltages V_{AG} and V_{BG} by applying Kirchhoff's voltage law. To do so, simply add the voltages algebraically between the start and finish points, which are points A and G for V_{AG} and points B and G for V_{BG}. Using the values from Figure 8-4b,

$$V_{AG} = -V_{R_1} + V_1 \qquad \text{(CCW from A to G)}$$
$$= -10.8\text{ V} + 18\text{ V}$$
$$= 7.2\text{ V}$$

Going clockwise (CW) from A to G produces the same result.

$$V_{AG} = V_{R_2} + V_{R_3} - V_2 \qquad \text{(CW from A to G)}$$
$$= 9\text{ V} + 16.2\text{ V} - 18\text{ V}$$
$$= 7.2\text{ V}$$

Since there are fewer voltages to add going counterclockwise from point A, it is the recommended solution for V_{AG}.

The voltage, V_{BG}, is found by using the same technique.

$$V_{BG} = V_{R_3} - V_2 \qquad \text{(CW from B to G)}$$
$$= 16.2\text{ V} - 18\text{ V}$$
$$= -1.8\text{ V}$$

Going around the loop in the other direction gives us

$$V_{BG} = -V_{R_2} - V_{R_1} + V_1 \qquad \text{(CCW from B to G)}$$
$$= -9\text{ V} - 10.8\text{ V} + 18\text{ V}$$
$$= -1.8\text{ V}$$

Since there are fewer voltages to add going clockwise from point B, it is the recommended solution for V_{BG}.

8.3 Thevenin's Theorem

Named after M. L. Thevenin, a French engineer, Thevenin's theorem is very useful in simplifying the process of solving for the unknown values of voltage and current in a network. By Thevenin's theorem, many sources and components, no matter how they are interconnected, can be represented by an equivalent series circuit with respect to any pair of terminals in the network. In Fig. 8-5, imagine that the block at the left contains a network connected to terminals A and B. Thevenin's theorem states that the *entire* network connected to A and B can be replaced by a single voltage source V_{TH} in series with a single resistance R_{TH}, connected to the same two terminals.

Voltage V_{TH} is the open-circuit voltage across terminals A and B. This means finding the voltage that the network produces across the two terminals with an open circuit between A and B. The polarity of V_{TH} is such that it will produce current from A to B in the same direction as in the original network.

Resistance R_{TH} is the open-circuit resistance across terminals A and B, but with all sources killed. This means finding the resistance looking back into the network from terminals A and B. Although the terminals are open, an ohmmeter across AB would read the value of R_{TH} as the resistance of the remaining paths in the network without any sources operating.

Thevenizing a Circuit

As an example, refer to Fig. 8-6a, where we want to find the voltage V_L across the 2-Ω R_L and its current I_L. To use Thevenin's theorem, mentally disconnect R_L. The two open ends then become terminals A and B. Now we find the Thevenin equivalent of the remainder of the circuit that is still connected to A and B. In general, open the part of the circuit to be analyzed and "thevenize" the remainder of the circuit connected to the two open terminals.

Our only problem now is to find the value of the open-circuit voltage V_{TH} across AB and the equivalent resistance R_{TH}. The Thevenin equivalent always consists of a single voltage source in series with a single resistance, as in Fig. 8-6d.

The effect of opening R_L is shown in Fig. 8-6b. As a result, the 3-Ω R_1 and 6-Ω R_2 form a series voltage divider without R_L.

Furthermore, the voltage across R_2 now is the same as the open-circuit voltage across terminals A and B. Therefore VR_2 with R_L open is V_{AB}. This is the V_{TH} we need for the Thevenin equivalent circuit. Using the voltage divider formula,

$$V_{R_2} = \frac{6}{9} \times 36\text{ V} = 24\text{ V}$$
$$V_{R_2} = V_{AB} = V_{TH} = 24\text{ V}$$

This voltage is positive at terminal A.

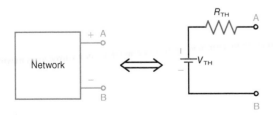

Figure 8-5 Any network in the block at the left can be reduced to the Thevenin equivalent series circuit at the right.

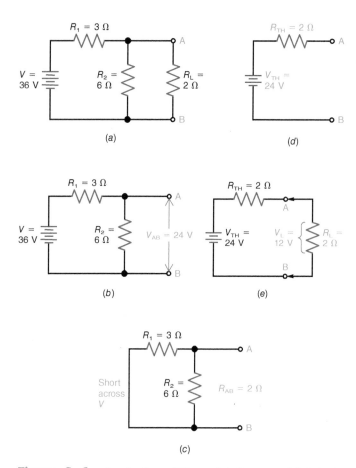

Figure 8-6 Application of Thevenin's theorem. (a) Actual circuit with terminals A and B across R_L. (b) Disconnect R_L to find that V_{AB} is 24 V. (c) Short-circuit V to find that R_{AB} is 2 Ω. (d) Thevenin equivalent circuit. (e) Reconnect R_L at terminals A and B to find that V_L is 12 V.

To find R_{TH}, the 2-Ω R_L is still disconnected. However, now the source V is short-circuited. So the circuit looks like Fig. 8-6c. The 3-Ω R_1 is now in parallel with the 6-Ω R_2 because both are connected across the same two points. This combined resistance is the product over the sum of R_1 and R_2.

$$R_{TH} = \frac{18}{9} = 2 \ \Omega$$

Again, we assume an ideal voltage source whose internal resistance is zero.

As shown in Fig. 8-6d, the Thevenin circuit to the left of terminals A and B then consists of the equivalent voltage V_{TH}, equal to 24 V, in series with the equivalent series resistance R_{TH}, equal to 2 Ω. This Thevenin equivalent applies for any value of R_L because R_L was disconnected. We are actually thevenizing the circuit that feeds the open AB terminals.

To find V_L and I_L, we can finally reconnect R_L to terminals A and B of the Thevenin equivalent circuit, as shown in Fig. 8-6e. Then R_L is in series with R_{TH} and V_{TH}. Using the

voltage divider formula for the 2-Ω R_{TH} and 2-Ω R_L, $V_L = 1/2 \times 24$ V $= 12$ V. To find I_L as V_L/R_L, the value is 12 V/2 Ω, which equals 6 A.

These answers of 6 A for I_L and 12 V for V_L apply to R_L in both the original circuit in Fig. 8-6a and the equivalent circuit in Fig. 8-6e. Note that the 6-A I_L also flows through R_{TH}.

The same answers could be obtained by solving the series-parallel circuit in Fig. 8-6a, using Ohm's law. However, the advantage of thevenizing the circuit is that the effect of different values of R_L can be calculated easily. Suppose that R_L is changed to 4 Ω. In the Thevenin circuit, the new value of V_L would be 4/6 × 24 V = 16 V. The new I_L would be 16 V/4 Ω, which equals 4 A. If we used Ohm's law in the original circuit, a complete, new solution would be required each time R_L was changed.

Looking Back from Terminals A and B

The way we look at the resistance of a series-parallel circuit depends on where the source is connected. In general, we calculate the total resistance from the outside terminals of the circuit in toward the source as the reference.

When the source is short-circuited for thevenizing a circuit, terminals A and B become the reference. Looking back from A and B to calculate R_{TH}, the situation becomes reversed from the way the circuit was viewed to determine V_{TH}.

For R_{TH}, imagine that a source could be connected across AB, and calculate the total resistance working from the outside in toward terminals A and B. Actually, an ohmmeter placed across terminals A and B would read this resistance.

This idea of reversing the reference is illustrated in Fig. 8-7. The circuit in Fig. 8-7a has terminals A and B open, ready to be thevenized. This circuit is similar to that in Fig. 8-6 but with the 4-Ω R_3 inserted between R_2 and terminal A. The interesting point is that R_3 does not change the value of V_{AB} produced by the source V, but R_3 does increase the value of R_{TH}. When we look back from terminals A and B, the 4 Ω of R_3 is in series with 2 Ω to make R_{TH} 6 Ω, as shown for R_{AB} in Fig. 8-7b and R_{TH} in Fig. 8-7c.

Let us consider why V_{AB} is the same 24 V with or without R_3. Since R_3 is connected to the open terminal A, the source V cannot produce current in R_3. Therefore, R_3 has no IR drop. A voltmeter would read the same 24 V across R_2 and from A to B. Since V_{AB} equals 24 V, this is the value of V_{TH}.

Now consider why R_3 does change the value of R_{TH}. Remember that we must work from the outside in to calculate the total resistance. Then, A and B are like source terminals. As a result, the 3-Ω R_1 and 6-Ω R_2 are in parallel, for a combined resistance of 2 Ω. Furthermore, this 2 Ω is in series with the 4-Ω R_3 because R_3 is in the main line from terminals A and B. Then R_{TH} is 2 + 4 = 6 Ω. As shown in Fig. 8-7c, the Thevenin equivalent circuit consists of V_{TH} = 24 V and R_{TH} = 6 Ω.

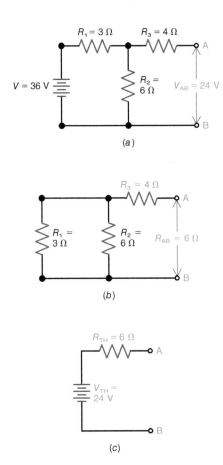

(a)

(b)

(c)

Figure 8-7 Thevenizing the circuit of Fig. 8–6 but with a 4-Ω R_3 in series with the A terminal. (a) V_{AB} is still 24 V. (b) Now the R_{AB} is 2 + 4 = 6 Ω. (c) Thevenin equivalent circuit.

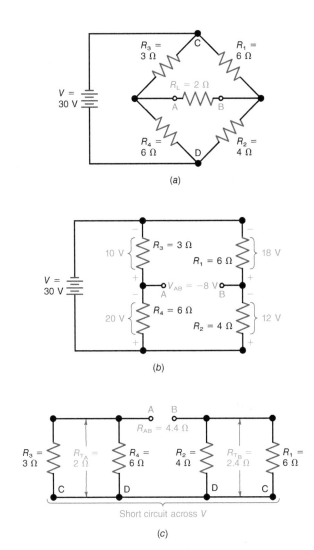

(a)

(b)

(c)

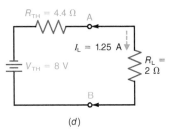

(d)

Figure 8-8 Thevenizing a bridge circuit. (a) Original circuit with terminals A and B across middle resistor R_L. (b) Disconnect R_L to find V_{AB} of −8 V. (c) With source V short-circuited, R_{AB} is 2 + 2.4 = 4.4 Ω. (d) Thevenin equivalent with R_L reconnected to terminals A and B.

8.4 Thevenizing a Bridge Circuit

As another example of Thevenin's theorem, we can find the current through the 2-Ω R_L at the center of the bridge circuit in Fig. 8-8a. When R_L is disconnected to open terminals A and B, the result is as shown in Fig. 8-8b. Notice how the circuit has become simpler because of the open. Instead of the unbalanced bridge in Fig. 8-8a which would require Kirchhoff's laws for a solution, the Thevenin equivalent in Fig. 8-8b consists of just two voltage dividers. Both the R_3-R_4 divider and the R_1-R_2 divider are across the same 30-V source.

Since the open terminal A is at the junction of R_3 and R_4, this divider can be used to find the potential at point A. Similarly, the potential at terminal B can be found from the R_1 R_2 divider. Then V_{AB} is the difference between the potentials at terminals A and B.

Note the voltages for the two dividers. In the divider with the 3-Ω R_3 and 6-Ω R_4, the bottom voltage V_{R_4} is 6/9 × 30 = 20 V. Then V_{R_3} at the top is 10 V because both must add up to equal the 30-V source. The polarities

are marked negative at the top, the same as the source voltage V.

Similarly, in the divider with the 6-Ω R_1 and 4-Ω R_2, the bottom voltage V_{R_2} is 4/10 × 30 = 12 V. Then V_{R_1} at the top is 18 V because the two must add up to equal the 30-V source. The polarities are also negative at the top, the same as V.

Now we can determine the potentials at terminals A and B with respect to a common reference to find V_{AB}. Imagine that the positive side of the source V is connected to a chassis ground. Then we would use the bottom line in the diagram as our reference for voltages. Note that V_{R_4} at the bottom of the R_3-R_4 divider is the same as the potential of terminal A with respect to ground. This value is -20 V, with terminal A negative.

Similarly, V_{R_2} in the R_1-R_2 divider is the potential at B with respect to ground. This value is -12 V with terminal B negative. As a result, V_{AB} is the difference between the -20 V at A and the -12 V at B, both with respect to the common ground reference.

The potential difference V_{AB} then equals

$$V_{AB} = -20 - (-12) = -20 + 12 = -8 \text{ V}$$

Terminal A is 8 V more negative than B. Therefore, V_{TH} is 8 V, with the negative side toward terminal A, as shown in the Thevenin equivalent in Fig. 8-8d.

The potential difference V_{AB} can also be found as the difference between V_{R_3} and V_{R_1} in Fig. 8-8b. In this case, V_{R_3} is 10 V and V_{R_1} is 18 V, both positive with respect to the top line connected to the negative side of the source V. The potential difference between terminals A and B then is $10 - 18$, which also equals -8 V. Note that V_{AB} must have the same value no matter which path is used to determine the voltage.

To find R_{TH}, the 30-V source is short-circuited while terminals A and B are still open. Then the circuit looks like Fig. 8-8c. Looking back from terminals A and B, the 3-Ω R_3 and 6-Ω R_4 are in parallel, for a combined resistance R_{T_A} of 18/9 or 2 Ω. The reason is that R_3 and R_4 are joined at terminal A, while their opposite ends are connected by the short circuit across the source V. Similarly, the 6-Ω R_1 and 4-Ω R_2 are in parallel for a combined resistance R_{T_B} of 24/10 = 2.4 Ω. Furthermore, the short circuit across the source now provides a path that connects R_{T_A} and R_{T_B} in series. The entire resistance is 2 + 2.4 = 4.4 Ω for R_{AB} or R_{TH}.

The Thevenin equivalent in Fig. 8-8d represents the bridge circuit feeding the open terminals A and B with 8 V for V_{TH} and 4.4 Ω for R_{TH}. Now connect the 2-Ω R_L to terminals A and B to calculate I_L. The current is

$$I_L = \frac{V_{TH}}{R_{TH} + R_L} = \frac{8}{4.4 + 2} = \frac{8}{6.4}$$

$$I_L = 1.25 \text{ A}$$

This 1.25 A is the current through the 2-Ω R_L at the center of the unbalanced bridge in Fig. 8-8a. Furthermore, the amount of I_L for any value of R_L in Fig. 8-8a can be calculated from the equivalent circuit in Fig. 8-8d.

8.5 Voltage Sources

An *ideal dc voltage source* produces a load voltage that is constant. The simplest example of an ideal dc voltage source is a perfect battery, one whose internal resistance is zero. Figure 8-9a shows an ideal voltage source connected to a variable load resistance of 1 Ω to 10 MΩ. The voltmeter reads 10 V, exactly the same as the source voltage.

Figure 8-9b shows a graph of load voltage versus load resistance. As you can see, the load voltage remains fixed at 10 V when the load resistance changes from 1 Ω to 1 MΩ. In other words, an ideal dc voltage source produces a constant load voltage, regardless of how small or large the load resistance is. With an ideal voltage source, only the load current changes when the load resistance changes.

Second Approximation

An ideal voltage source is a theoretical device; it cannot exist in nature. Why? When the load resistance approaches zero, the load current approaches infinity. No real voltage source can produce infinite current because a real voltage source always has some internal resistance. The second approximation of a dc voltage source includes this internal resistance.

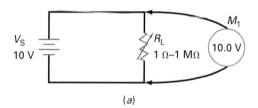

(a)

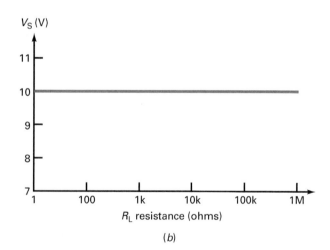

(b)

Figure 8-9 (a) Ideal voltage source and variable load resistance; (b) load voltage is constant for all load resistances.

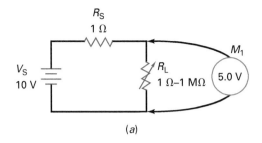

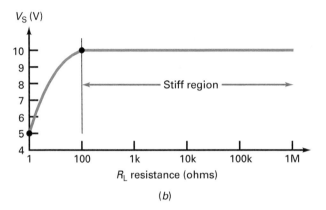

(b)

Figure 8-10 (a) Second approximation includes source resistance; (b) load voltage is constant for large load resistances.

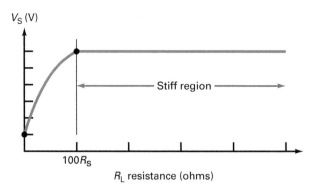

Figure 8-11 Stiff region occurs when load resistance is large enough.

Figure 8-10a illustrates the idea. A source resistance R_S of 1 Ω is now in series with the ideal battery. The voltmeter reads 5 V when R_L is 1 Ω. Why? Because the load current is 10 V divided by 2 Ω, or 5 A. When 5 A flows through the source resistance of 1 Ω, it produces an internal voltage drop of 5 V. This is why the load voltage is only half of the ideal value, with the other half being dropped across the internal resistance.

Figure 8-10b shows the graph of load voltage versus load resistance. In this case, the load voltage does not come close to the ideal value until the load resistance is much greater than the source resistance. But what does *much greater* mean? In other words, when can we ignore the source resistance?

Stiff Voltage Source

Now is the time when a new definition can be useful. So, let us invent one. We can ignore the source resistance when it is at least 100 times smaller than the load resistance. Any source that satisfies this condition is a *stiff voltage source*. As a definition,

$$\text{Stiff voltage source: } R_S < 0.01R_L \qquad \text{(8-1)}$$

This formula defines what we mean by a *stiff voltage source*. The boundary of the inequality (where < is changed to =) gives us the following equation:

$$R_S = 0.01R_L$$

Solving for load resistance gives the minimum load resistance we can use and still have a stiff source:

$$R_{L(min)} = 100R_S \qquad \text{(8-2)}$$

In words, the minimum load resistance equals 100 times the source resistance.

Equation (8-2) is a derivation. We started with the definition of a stiff voltage source and rearranged it to get the minimum load resistance permitted with a stiff voltage source. As long as the load resistance is greater than $100R_S$, the voltage source is stiff. When the load resistance equals this worst-case value, the calculation error from ignoring the source resistance is 1%, small enough to ignore in a second approximation.

Figure 8-11 visually summarizes a stiff voltage source. The load resistance has to be greater than $100R_S$ for the voltage source to be stiff.

EXAMPLE 8-3

The definition of a stiff voltage source applies to ac sources as well as to dc sources. Suppose an ac voltage source has a source resistance of 50 Ω. For what load resistance is the source stiff?

Solution:

Multiply by 100 to get the minimum load resistance:

$$R_L = 100R_S = 100(50 \text{ Ω}) = 5 \text{ kΩ}$$

As long as the load resistance is greater than 5 kΩ, the ac voltage source is stiff and we can ignore the internal resistance of the source.

A final point: Using the second approximation for an ac voltage source is valid only at low frequencies. *At high frequencies, additional factors such as lead inductance and stray capacitance come into play.* We will deal with these high-frequency effects in a later chapter.

Voltage Sources and Voltage Dividers

Voltage dividers show up in practically every electronic circuit or system. They occur when a load is connected to a voltage source. The internal resistance of the voltage source and the load resistance form a voltage divider that literally prevents the entire voltage source output from appearing across the load. It is critical that you recognize the presence of a voltage divider in all of these situations.

EXAMPLE 8-4

An amplifier acts as a generator when it is driving a load. The amplifier has an internal resistance also called the output impedance (Z_o) of 50 Ω. Refer to Figure 8-12. The diamond symbol is the equivalent generator. Its output voltage with no load is 3 volts. A load (R_L) of 300 Ω is connected to the amplifier. Is the load resistance stiff? What is the voltage across the load?

Answer:

Multiply R_L by 0.01 to get the stiff source resistance.

$$R_S = 0.01(300) = 3\ \Omega$$

The 50-Ω source is greater than 3 Ω so is not "stiff."

The output voltage is the source voltage without load or 3 volts multiplied by the voltage divider formula:

$$\text{Load voltage} = 3(300)/(300 + 50) = 2.57 \text{ volts}$$

Lost voltage (0.43 V) appears across the internal resistance of the amplifier. The amplifier output impedance dissipates power and generates heat as a result.

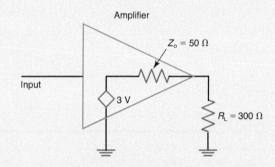

Figure 8-12 Amplifier acting as a generator.

Practical Voltage Sources

There are no ideal voltage sources as all practical devices and circuits have some internal resistance. However, some sources come close to ideal. The best examples of near ideal voltage sources are batteries and regulated power supplies. Both have extremely low internal series equivalent resistances.

When the voltage source is a circuit like an amplifier or oscillator, it acts as a voltage source but in some cases with a considerable internal resistance. This internal resistance is referred to as the output impedance or resistance of the circuit. When a load is connected to it, a voltage divider is formed between the internal resistance and the load. With a load, the output voltage is less than the open-circuit or no-load condition because of this voltage divider action.

8.6 Maximum Power Transfer Theorem

When transferring power from a voltage source to a load, it is often necessary to transfer maximum possible power rather than to optimize the amount of voltage across the load. The condition for that to happen is when the load resistance (R_L) equals the internal resistance of the generator (r_i).

In the diagram in Fig. 8-13, when R_L equals r_i, the load and generator are matched. The matching is significant because the generator then produces maximum power in R_L, as verified by the values listed in Table 8-1.

Maximum Power in R_L

When R_L is 100 Ω to match the 100 Ω of r_i, maximum power is transferred from the generator to the load. With higher resistance for R_L, the output voltage V_L is higher, but the current is reduced. Lower resistance for R_L allows more current, but V_L is less. When r_i and R_L both equal 100 Ω, this

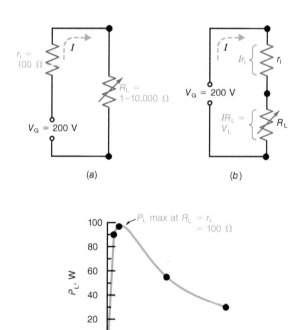

Figure 8-13 Circuit for varying R_L to match r_i. (a) Schematic diagram. (b) Equivalent voltage divider for voltage output across R_L. (c) Graph of power output P_L for different values of R_L. All values are listed in Table 8-1.

Table 8-1 Effect of Load Resistance on Generator Output*

	R_L, Ω	$I = V_G/R_T,$ A	$Ir_i,$ V	$IR_L,$ V	$P_L,$ W	$P_i,$ W	$P_T,$ W	Efficiency = $P_L/P_T,$ %
	1	1.98	198	2	4	392	396	1
	50	1.33	133	67	89	178	267	33
$R_L = r_i \rightarrow$	100	1	100	100	100	100	200	50
	500	0.33	33	167	55	11	66	83
	1,000	0.18	18	180	32	3.24	35.24	91
	10,000	0.02	2	198	4	0.04	4.04	99

* Values calculated approximately for circuit in Fig. 8-13, with $V_G = 200$ V and $r_i = 100 \ \Omega$.

combination of current and voltage produces the maximum power of 100 W across R_L.

Maximum Voltage across R_L

If maximum voltage, rather than power, is desired, the load should have as high a resistance as possible. Note that R_L and r_i form a voltage divider for the generator voltage, as illustrated in Fig. 8-13b. The values for IR_L listed in Table 8-1 show how the output voltage V_L increases with higher values of R_L.

Maximum Efficiency

Note also that the efficiency increases as R_L increases because there is less current, resulting in less power lost in r_i. When R_L equals r_i, the efficiency is only 50%, since one-half the total generated power is dissipated in r_i, the internal resistance of the generator. In conclusion, then, matching the load and generator resistances is desirable when the load requires maximum power rather than maximum voltage or efficiency, assuming that the match does not result in excessive current.

CHAPTER 8 SYSTEMS SIDEBAR

Digital-to-Analog Converter

A digital-to-analog converter (DAC) is a circuit that takes a series of binary numbers and generates an analog signal proportional to the binary values. A good example is the DAC that converts the digital music on a CD into the analog sound you hear. A DAC converts the digital music in an MP3 player or iPod into analog music. The analog video you see on you HDTV screen is created by a DAC. And a DAC also converts the digital voice you are receiving on your cell phone into the analog sound you hear in the earphone.

In most electronic systems today, the analog signals they process are in digital form. These analog signals like voice, video, or sensor voltages are digitized into a sequence of binary numbers proportional to the voltage by an analog-to-digital converter (ADC). To recover the original analog signals or the processed versions requires a DAC.

The Conversion Process

Figure S8-1 shows the analog-to-digital (A/D) and digital-to-analog (D/A) conversion process. Note the symbols for the ADC and DAC. Now refer to Fig. S8-2. The smooth curve is the analog signal. The ADC samples or measures the analog signal at specific time intervals. At each sampling point, the ADC produces a proportional binary number. Note for example, at sample time 4 the voltage is 14 volts so the ADC generates 1110 or the binary value for 14. That value is stored in a memory. The binary values for each analog sample are shown along the top of the plot. They are stored in sequence as the memory content in Fig. S8-1 shows.

Recovering a Digitized Signal

You recover the analog signal with a DAC. The binary number sequence is stored in a memory and then transferred at a fixed rate to the DAC. Each binary value produces a fixed level at the DAC output. Fig. S8-2 illustrates the original analog input and the DAC output. This results in a stair-step output, but it is a good representation of the original analog signal. While the stepped approximation of

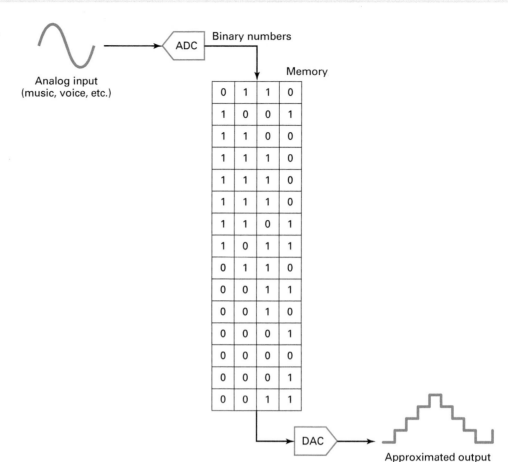

Figure S8-1   The A/D and D/A process.

Analog input
(music, voice, etc.)

Binary numbers

Memory

Approximated output

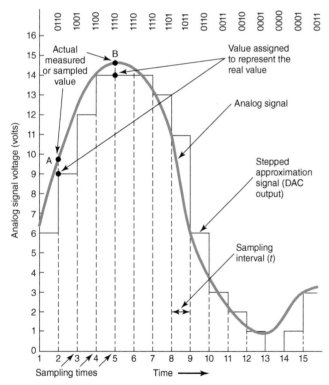

Figure S8-2   Original analog signal and recovered stepped approximation.

the analog signal may appear crude, it can be smoothed to look more like the original by feeding the signal through a filter. In some cases, as in digital voice or video, a person cannot tell the stepped version from the original. By using a binary number with more bits and by sampling more frequently, the stepped approximation will become much closer to the original.

DAC Operation

One common way to make a DAC is with the circuit shown in Fig. S8-3a. The network, which is made up of resistors R_1 through R_8 and the related switches, generates a current proportional to the binary value represented by bits D_0 through D_3. The bit value is set by the switches to either ground or zero volts (binary 0) or −5 volts (binary 1). This is a 4-bit DAC. This network is referred to as an R2R network because it only uses two values of resistor, in this case 10 and 20 kΩ. The switches themselves are actually transistors that turn off or on like a mechanical switch, only faster. There are 16, possible input values from 0000 to 1111, with D_0 the LSB and D_3 the MSB.

The current produced by the network at point D in Fig. S8-3 is fed to an operational amplifier (op amp) that

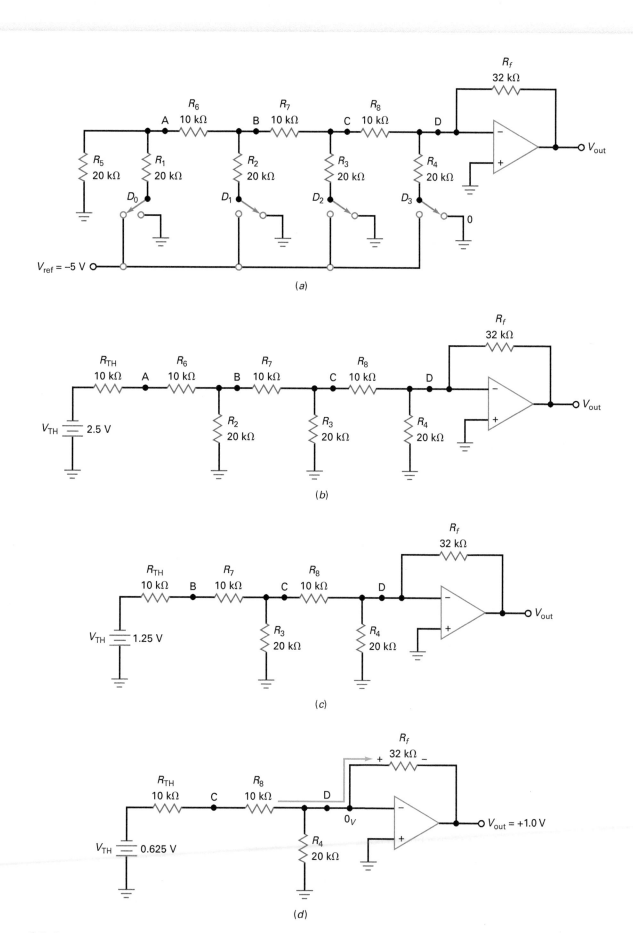

Figure S8-3 (a) Original circuit; (b) thevenized at point A; (c) thevenized at point B; (d) thevenized at point C.

converts the current into a proportional voltage. The op amp output (V_{out}) is this input current (I_{in}) multiplied by the value of feedback resistor R_f. The op amp also inverts the voltage, and that is indicated by the minus sign in the expression below.

$$V_{out} = -(I_{in})(R_f)$$

One big issue with this circuit is how it is analyzed so you can see in detail how it works. One good approach is to thevenize the circuit for each setting of the switches from 0000 to 1111. Remember that D_0 is the least significant bit and D_3 is the most significant bit. To see the LSB value, we can analyze the circuit with an input of 0001. This sets D_0 to -5 V and all the other switches to ground or 0 V. Now you can use Thevenin's to find the equivalent circuit and calculate the output.

With the switches D_0 to D_4 connected as in Fig. S8-3a, the binary input is $D_0 = 1$, $D_1 = 0$, $D_2 = 0$, and $D_3 = 0$. First, thevenize the circuit from point A, looking toward D_0. When doing so, R_5 (20 kΩ) becomes in parallel with R_1 (20 kΩ) and the equivalent equals 10 kΩ. The thevenized voltage at point A is one-half of V_{ref} and is equal to $+2.5$ V. This equivalent is shown in Fig. S8-3b.

Next, thevenize Fig. S8-3b from point B. Notice how R_{TH} (10 kΩ) is in series with R_6 (10 kΩ). This 20 kΩ

value is in parallel with R_2 (20 kΩ) and again gives us 10 kΩ. The thevenized voltage seen from point B is again reduced by half to 1.25 V. This equivalent is shown in Fig. S8-3c.

Now, thevenize Fig. S8-3c from point C. Again, R_{TH} (10 kΩ) is in series with R_7 (10 kΩ), and this 20 kΩ value becomes in parallel with R_3 (10 kΩ). V_{TH} is equal to 0.625 V. Notice how the V_{TH} values have been cut in half at each step. The Thevenin equivalent has been reduced to that of Fig. S8-3d.

In Fig. S8-3d, the op amp's inverting input and the top of R_4 (20 kΩ) is at a virtual ground. The voltage is equal to zero volts at this point. This places the entire 0.625 V of V_{TH} across R_{TH} and R_8 (10 kΩ). This results in an input current I_{in} of

$$I_{in} = \frac{0.625 \text{ V}}{20 \text{ k}\Omega} = -31.25 \ \mu A$$

Again, because of the virtual ground, this input current is forced to flow through R_f (32 kΩ) and produces an output voltage of

$$V_{out} = -(-I_{in})(R_f) = -(-31.25 \ \mu A)(32 \text{ k}\Omega) = +1.0 \text{ V}$$

This output voltage is the smallest output increment above 0 V and is referred to as the circuit's output resolution.

CHAPTER 8 SELF-TEST

1. Kirchhoff's current law states that
   a. the algebraic sum of the currents flowing into any point in a circuit must equal zero.
   b. the algebraic sum of the currents entering and leaving any point in a circuit must equal zero.
   c. the algebraic sum of the currents flowing away from any point in a circuit must equal zero.
   d. the algebraic sum of the currents around any closed path must equal zero.

2. When applying Kirchhoff's current law,
   a. consider all currents flowing into a branch point positive and all currents directed away from that point negative.
   b. consider all currents flowing into a branch point negative and all currents directed away from that point positive.
   c. remember that the total of all the currents entering a branch point must always be greater than the sum of the currents leaving that point.

   d. the algebraic sum of the currents entering and leaving a branch point does not necessarily have to be zero.

3. If a 10-A I_1 and a 3-A I_2 flow into point X, how much current must flow away from point X?
   a. 7 A.
   b. 30 A.
   c. 13 A.
   d. It cannot be determined.

4. Three currents I_1, I_2, and I_3 flow into point X, whereas current I_4 flows away from point X. If $I_1 = 2.5$ A, $I_3 = 6$ A, and $I_4 = 18$ A, how much is current I_2?
   a. 21.5 A.
   b. 14.5 A.
   c. 26.5 A.
   d. 9.5 A.

5. When applying Kirchhoff's voltage law, a closed path is commonly referred to as a
   a. node.
   b. principal node.
   c. loop.
   d. branch point.

6. Kirchhoff's voltage law states that
   a. the algebraic sum of the voltage sources and *IR* voltage drops in any closed path must total zero.
   b. the algebraic sum of the voltage sources and *IR* voltage drops around any closed path can never equal zero.
   c. the algebraic sum of all the currents flowing around any closed loop must equal zero.
   d. none of the above.

7. When applying Kirchhoff's voltage law,
   a. consider any voltage whose positive terminal is reached first as negative and any voltage whose negative terminal is reached first as positive.
   b. always consider all voltage sources as positive and all resistor voltage drops as negative.
   c. consider any voltage whose negative terminal is reached first as negative and any voltage whose positive terminal is reached first as positive.
   d. always consider all resistor voltage drops as positive and all voltage sources as negative.

8. The algebraic sum of +40 V and −30 V is
   a. −10 V.
   b. +10 V.
   c. +70 V.
   d. −70 V.

9. A resistor is an example of a(n)
   a. bilateral component.
   b. active component.
   c. passive component.
   d. both *a* and *c*.

10. When solving for the Thevenin equivalent resistance, R_{TH},
   a. all voltage sources must be opened.
   b. all voltage sources must be short-circuited.
   c. all voltage sources must be converted to current sources.
   d. none of the above.

11. Thevenin's theorem states that an entire network connected to a pair of terminals can be replaced with
   a. a single current source in parallel with a single resistance.
   b. a single voltage source in parallel with a single resistance.
   c. a single voltage source in series with a single resistance.
   d. a single current source in series with a single resistance.

12. With respect to terminals A and B in a complex network, the Thevenin voltage, V_{TH}, is
   a. the voltage across terminals A and B when they are short-circuited.
   b. the open-circuit voltage across terminals A and B.
   c. the same as the voltage applied to the complex network.
   d. none of the above.

13. An ideal voltage source has
   a. zero internal resistance.
   b. infinite internal resistance.
   c. a load-dependent voltage.
   d. a load-dependent current.

14. A real voltage source has
   a. zero internal resistance.
   b. infinite internal resistance.
   c. a small internal resistance.
   d. a large internal resistance.

15. If a load resistance is 100 Ω, a stiff voltage source has a resistance of
   a. less than 1 Ω.
   b. at least 10 Ω.
   c. more than 10 kΩ.
   d. less than 10 kΩ.

16. The Thevenin voltage is the same as the
   a. shorted-load voltage.
   b. open-load voltage.
   c. ideal source voltage.
   d. Norton voltage.

17. The Thevenin resistance is equal in value to the
   a. load resistance.
   b. half the load resistance.
   c. internal resistance of a Norton circuit.
   d. open-load resistance.

18. To get the Thevenin voltage, you have to
   a. short the load resistor.
   b. open the load resistor.
   c. short the voltage source.
   d. open the voltage source.

19. The voltage out of an ideal voltage source
   a. is zero.
   b. is constant.
   c. depends on the value of load resistance.
   d. depends on the internal resistance.

20. Thevenin's theorem replaces a complicated circuit facing a load by an
   a. ideal voltage source and parallel resistor.
   b. ideal current source and parallel resistor.
   c. ideal voltage source and series resistor.
   d. ideal current source and series resistor.

21. A cercuit consists of a generator with an internal resistance of 75 Ω and a load. Which value of load resistance will produce maximum output voltage?
a. 52 Ω.
b. 75 Ω.
c. 200 Ω.
d. 1 kΩ.

22. Which load resistance will allow a 50-ohm generator to produce maximum power?
a. 25 Ω.
b. 50 Ω.
c. 75 Ω.
d. 100 Ω.

CHAPTER 8 PROBLEMS

SECTION 8.1 Kirchhoff's Current Law (KCL)

8.1 If a 5-A I_1 and a 10-A I_2 flow into point X, how much is the current, I_3, directed away from that point?

8.2 Applying Kirchhoff's current law, write an equation for the currents directed into and out of point X in Prob. 8-1.

8.3 In Fig. 8-14, solve for the unknown current, I_3.

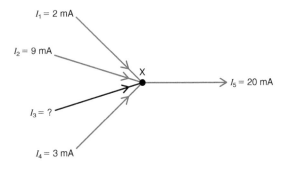

Figure 8-14

8.4 In Fig. 8-15, solve for the following unknown currents: I_3, I_5, and I_8.

8.5 Apply Kirchhoff's current law in Fig. 8-15 by writing an equation for the currents directed into and out of the following points:
a. Point X
b. Point Y
c. Point Z

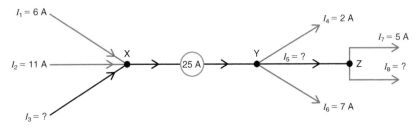

Figure 8-15

SECTION 8.2 Kirchoff's Voltage Law (KVL)

8.6 In Fig. 8-16,
a. Write a KVL equation for the loop CEFDC going clockwise from point C.
b. Write a KVL equation for the loop ACDBA going clockwise from point A.
c. Write a KVL equation for the loop ACEFDBA going clockwise from point A.

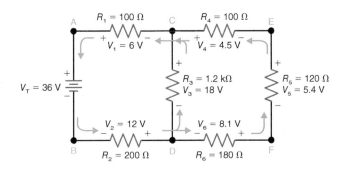

Figure 8-16

8.7 In Fig. 8-16,
a. Determine the voltage for the partial loop CEFD going clockwise from point C. How does your answer compare to the voltage drop across R_3?
b. Determine the voltage for the partial loop ACDB going clockwise from point A. How does your answer compare to the value of the applied voltage, V_T, across points A and B?
c. Determine the voltage for the partial loop ACEFDB going clockwise from point A. How does your answer compare to the value of the applied voltage, V_T, across points A and B?
d. Determine the voltage for the partial loop CDFE going counterclockwise from point C. How does your answer compare to the voltage drop across R_4?

8.8 In Fig. 8-17, solve for the voltages V_{AG} and V_{BG}. Indicate the proper polarity for each voltage.

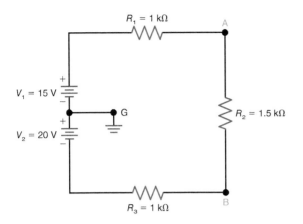

Figure 8-17

8.9 In Fig. 8-18, solve for the voltages V_{AG} and V_{BG}. Indicate the proper polarity for each voltage.

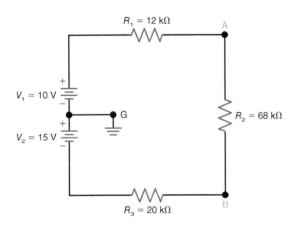

Figure 8-18

SECTION 8.3 Thevenin's Theorem

8.10 In Fig. 8-19, draw the Thevenin equivalent circuit with respect to terminals A and B (mentally remove R_L).

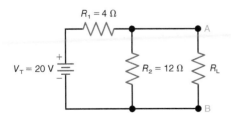

Figure 8-19

8.11 In Fig. 8-19, use the Thevenin equivalent circuit to calculate I_L and V_L for the following values of R_L; $R_L = 3\ \Omega$ and $R_L = 12\ \Omega$.

8.12 In Fig. 8-20, use the Thevenin equivalent circuit to calculate I_L and V_L for the following values of R_L: $R_L = 100\ \Omega$ and $R_L = 5.6\ k\Omega$.

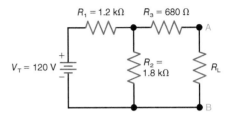

Figure 8-20

8.13 In Fig. 8-21, use the Thevenin equivalent circuit to solve for I_L and R_L.

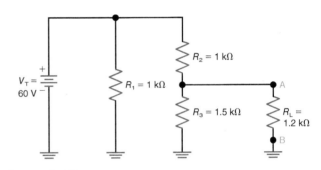

Figure 8-21

SECTION 8.4 Thevenizing a Bridge Circuit

8.14 In Fig. 8-22, draw the Thevenin equivalent circuit with respect to terminals A and B (mentally remove R_L).

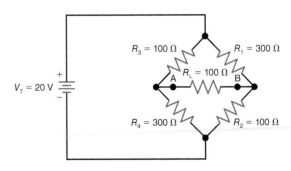

Figure 8-22

8.15 Using the Thevenin equivalent circuit for Fig. 8-22, calculate the values for I_L and V_L.

8.16 In Fig. 8-23, draw the Thevenin equivalent circuit with respect to terminals A and B (mentally remove R_L).

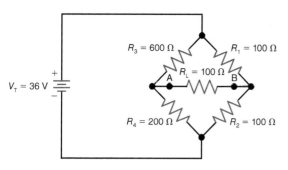

Figure 8-23

8.17 Using the Thevenin equivalent circuit for Fig. 8-23, calculate the values for I_L and V_L.

SECTION 8.5 Voltage Sources

8.18 A given voltage source has an ideal voltage of 12 V and an internal resistance of 0.1 Ω. For what values of load resistance will the voltage source appear stiff?

8.19 A load resistance may vary from 270 Ω to 100 kΩ. For a stiff voltage source to exist, what is the largest internal resistance the source can have?

8.20 The internal output resistance of a function generator is 50 Ω. For what values of load resistance does the generator appear stiff?

8.21 The internal resistance of a voltage source equals 0.05 Ω. How much voltage is dropped across this internal resistance when the current through it equals 2 A?

Chapter 9

Conductors, Insulators, and Semiconductors

Learning Outcomes

After studying this chapter, you should be able to:

> *Explain* the main function of a conductor in an electric circuit.

> *Calculate* the cross-sectional area of round wire when the diameter is known.

> *List* the advantages of using stranded wire versus solid wire.

> *List* common types of connectors used with wire conductors.

> *Define* the terms *pole* and *throw* as they relate to switches.

> *Explain* how fast-acting and slow-blow fuses differ.

> *Calculate* the resistance of a wire conductor whose length, cross-sectional area, and specific resistance are known.

Conductors are materials that pass electrons with minimal resistance. Conductors are usually wires but also take other forms like conducting paths on printed circuit boards.

An insulator is any material that resists or prevents the flow of electric charge, such as electrons. The resistance of an insulator is very high, typically several hundreds of megohms or more. An insulator provides the equivalent of an open circuit with practically infinite resistance and almost zero current.

In this chapter, you will be introduced to a variety of topics that includes wire conductors, insulators, sensors, connectors, mechanical switches and, fuses. All of these topics relate to the discussion of conductors and insulators since they either pass or prevent the flow of electricity, depending on their condition or

> *Explain* the meaning of *temperature coefficient of resistance.*

> *Name* two common types of resistive sensors and *explain* how they work.

> *Explain* why insulators are sometimes called *dielectrics.*

> *Explain* what is meant by the *corona effect.*

> *Explain* the operation of RTD and strain gauge sensors.

> *Define* the term *semiconductor.*

state. Finally, you will be introduced to semiconductors, materials that can be conductors or insulators. Today, semiconductors are at the heart of electronics as they form all of our diodes, transistors, and integrated circuits.

9.1 Function of the Conductor

In Fig. 9-1, the resistance of the two 10-ft lengths of copper wire conductor is 0.08 Ω. This R is negligibly small compared with the 144-Ω R of the tungsten filament in the lightbulb. When the current of 0.833 A flows in the bulb and the series conductors, the IR voltage drop of the conductors is only 0.07 V, with 119.93 V across the bulb. Practically all the applied voltage is across the bulb filament. Since the

bulb then has its rated voltage of 120 V, approximately, it will dissipate the rated power of 100 W and light with full brilliance.

The current in the wire conductors and the bulb is the same, since they are in series. However, the IR voltage drop in the conductor is practically zero because its R is almost zero.

Also, the I^2R power dissipated in the conductor is negligibly small, allowing the conductor to operate without becoming hot. Therefore, the conductor delivers energy from the source to the load with minimum loss by electron flow in the copper wires.

Although the resistance of wire conductors is very small, for some cases of high current, the resultant IR drop can be appreciable. For example, suppose that the 120-V power line is supplying 30 A of current to a load through two conductors, each of which has a resistance of 0.2 Ω. In this case, each conductor has an IR drop of 6 V, calculated as 30 A $\times$ 0.2 Ω = 6 V. With each conductor dropping 6 V, the load receives a voltage of only 108 V rather than the full 120 V. The lower-than-normal load voltage could result in the load not operating properly. Furthermore, the I^2R power dissipated in each conductor equals 180 W, calculated as $30^2 \times 0.2 \Omega$ = 180 W. The I^2R power loss of 180 W in the conductors is considered excessively high.

9.2 Standard Wire Gage Sizes

Table 9-1 lists the standard wire sizes in the system known as the American Wire Gage (AWG) or Brown and Sharpe (B&S) gage. The gage numbers specify the size of round wire in terms of its diameter and cross-sectional area. Note the following three points:

1. As the gage numbers increase from 1 to 40, the diameter and circular area decrease. Higher gage numbers indicate thinner wire sizes.

2. The circular area doubles for every three gage sizes. For example, No. 10 wire has approximately twice the area of No. 13 wire.

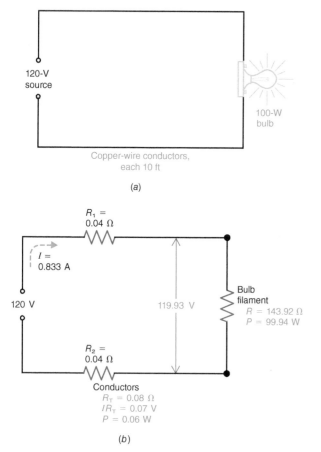

Figure 9-1 The conductors should have minimum resistance to light the bulb with full brilliance. (*a*) Wiring diagram. (*b*) Schematic diagram. R_1 and R_2 represent the very small resistance of the wire conductors.

Table 9-1 Standard Copper Wire Sizes

Gage No.	Diameter, Mils	Area, Circular Mils	Ohms per 1000 ft of Copper Wire at 25°C*	Gage No.	Diameter, Mils	Area, Circular Mils	Ohms per 1000 ft of Copper Wire at 25°C*
1	289.3	83,690	0.1264	21	28.46	810.1	13.05
2	257.6	66,370	0.1593	22	25.35	642.4	16.46
3	229.4	52,640	0.2009	23	22.57	509.5	20.76
4	204.3	41,740	0.2533	24	20.10	404.0	26.17
5	181.9	33,100	0.3195	25	17.90	320.4	33.00
6	162.0	26,250	0.4028	26	15.94	254.1	41.62
7	144.3	20,820	0.5080	27	14.20	201.5	52.48
8	128.5	16,510	0.6405	28	12.64	159.8	66.17
9	114.4	13,090	0.8077	29	11.26	126.7	83.44
10	101.9	10,380	1.018	30	10.03	100.5	105.2
11	90.74	8234	1.284	31	8.928	79.70	132.7
12	80.81	6530	1.619	32	7.950	63.21	167.3
13	71.96	5178	2.042	33	7.080	50.13	211.0
14	64.08	4107	2.575	34	6.305	39.75	266.0
15	57.07	3257	3.247	35	5.615	31.52	335.0
16	50.82	2583	4.094	36	5.000	25.00	423.0
17	45.26	2048	5.163	37	4.453	19.83	533.4
18	40.30	1624	6.510	38	3.965	15.72	672.6
19	35.89	1288	8.210	39	3.531	12.47	848.1
20	31.96	1022	10.35	40	3.145	9.88	1069

* 20° to 25°C or 68° to 77°F is considered average room temperature.

3. The higher the gage number and the thinner the wire, the greater the resistance of the wire for any given length.

In typical applications, hookup wire for electronic circuits with current of the order of milliamperes is generally about No. 22 gage. For this size, 0.5 to 1 A is the maximum current the wire can carry without excessive heating.

House wiring for circuits where the current is 5 to 15 A is usually No. 14 gage. Minimum sizes for house wiring are set by local electrical codes, which are usually guided by the National Electrical Code published by the National Fire Protection Association.

Circular Mils

The cross-sectional area of round wire is measured in circular mils, abbreviated cmil. A mil is one-thousandth of an inch, or 0.001 in. One circular mil is the cross-sectional area of a wire with a diameter of 1 mil. The number of circular mils in any circular area is equal to the square of the diameter in mils or cmil = d^2(mils).

EXAMPLE 9-1

What is the area in circular mils of a wire with a diameter of 0.005 in.?

Answer:

We must convert the diameter to mils. Since 0.005 in. equals 5 mil,

$$\text{Circular mil area} = (5 \text{ mil})^2$$
$$\text{Area} = 25 \text{ cmil}$$

Note that the circular mil is a unit of area, obtained by squaring the diameter, whereas the mil is a linear unit of length, equal to one-thousandth of an inch. Therefore, the circular-mil area increases as the square of the diameter. As illustrated in Fig. 9-2, doubling the diameter quadruples the area. Circular mils are convenient for round wire because the cross section is specified without using the formula πr^2 or $\pi d^2/4$ for the area of a circle.

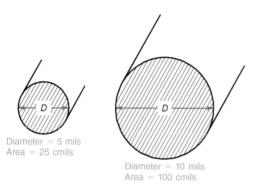

Figure 9-2 Cross-sectional area for round wire. Doubling the diameter increases the circular area by four times.

Diameter = 5 mils
Area = 25 cmils

Diameter = 10 mils
Area = 100 cmils

9.3 Types of Wire Conductors

Most wire conductors are copper due to its low cost, although aluminum and silver are also used sometimes. The copper may be tinned with a thin coating of solder, which gives it a silvery appearance. Tinned wire is easier to solder for connections. The wire can be solid or stranded, as shown in Fig. 9-3a and b. Solid wire is made of only one conductor. If bent or flexed repeatedly, solid wire may break. Therefore solid wire is used in places where bending and flexing are not encountered. House wiring is a good example of the use of solid wire. Stranded wire is made up of several individual strands put together in a braid. Some uses for stranded wire include telephone cords, extension cords, and speaker wire.

Stranded wire is flexible, easier to handle, and less likely to develop an open break. Sizes for stranded wire are equivalent to the sum of the areas for the individual strands. For instance, two strands of No. 30 wire correspond to solid No. 27 wire.

EXAMPLE 9-2

A stranded wire is made up of 16 individual strands of No. 27 gage wire. What is its equivalent gage size in solid wire?

Answer:

The equivalent gage size in solid wire is determined by the total circular area of all individual strands. Referring to Table 9-1, the circular area for No. 27 gage wire is 201.5 cmils. Since there are 16 individual strands, the total circular area is calculated as follows:

Total cmil area = 16 strands × 201.5 cmils per strand
= 3224 cmils

Referring to Table 9-1, we see that the circular area of 3224 cmils corresponds very closely to the cmil area of No. 15 gage wire. Therefore, 16 strands of No. 27 gage wire is roughly equivalent to No. 15 gage solid wire.

Very thin wire, such as No. 30, often has an insulating coating of enamel or shellac. It may look like copper, but the

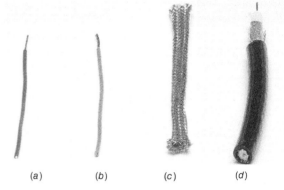

(a)   (b)   (c)   (d)

Figure 9-3 Types of wire conductors. (a) Solid wire. (b) Stranded wire. (c) Braided wire for very low R. (d) Coaxial cable. Note braided wire for shielding the inner conductor.

coating must be scraped off the ends to make a good connection. This type of wire is used for small coils.

Heavier wires generally are in an insulating sleeve, which may be rubber or one of many plastic materials. General-purpose wire for connecting electronic components is generally plastic-coated hookup wire of No. 20 gage. Hookup wire that is bare should be enclosed in a hollow insulating sleeve called *spaghetti.*

The braided conductor in Fig. 9-3c is used for very low resistance. It is wide for low R and thin for flexibility, and the braiding provides many strands. A common application is a grounding connection, which must have very low R.

Transmission Lines

Constant spacing between two conductors through the entire length provides a transmission line. A common example is the coaxial cable in Fig. 9-3d.

Coaxial cable with an outside diameter of ¼ in. is generally used for the signals in cable television. In construction, there is an inner solid wire, insulated from metallic braid that serves as the other conductor. The entire assembly is covered by an outer plastic jacket. In operation, the inner conductor has the desired signal voltage with respect to ground, and the metallic braid is connected to ground to shield the inner conductor against interference. Coaxial cable, therefore, is a shielded type of transmission line.

You will learn more about transmission lines in Chap. 25.

Wire Cable

Two or more conductors in a common covering form a cable. Each wire is insulated from the others. Cables often consist of two, three, ten, or many more pairs of conductors, usually color-coded to help identify the conductors at both ends of a cable.

The ribbon cable in Fig. 9-4, has multiple conductors but not in pairs. This cable is used for multiple connections to a computer and associated equipment.

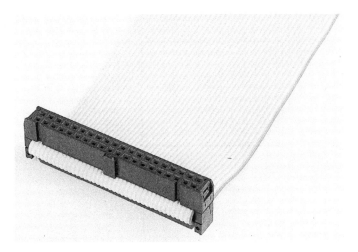

Figure 9-4   Ribbon cable with multiple conductors.

9.4 Connectors

Refer to Fig. 9-5 for different types of connectors. The spade lug in Fig. 9-5*a* is often used for screw-type terminals. The alligator clip in Fig. 9-5*b* is convenient for a temporary connection. Alligator clips come in small and large sizes. The banana pins in Fig. 9-5*c* have spring-type sides that make a tight connection. The terminal strip in Fig. 9-5*d* provides a block for multiple solder connections.

The RCA-type plug in Fig. 9-5*e* is commonly used for shielded cables with audio equipment. The inner conductor of the shielded cable is connected to the center pin of the plug, and the cable braid is connected to the shield. Both connections must be soldered.

The phone plug in Fig. 9-5*f* is still used in many applications but usually in a smaller size. The ring is insulated from the sleeve to provide for two connections. There may be a separate tip, ring, and sleeve for three connections. The sleeve is usually the ground side.

The plug in Fig. 9-5*g* is called an *F connector*. It is universally used in cable television because of its convenience. The center conductor of the coaxial cable serves as the center pin of the plug, so that no soldering is needed. Also, the shield on the plug is press-fit onto the braid of the cable underneath the plastic jacket.

Figure 9-5*h* shows a USB connector. This type of connector is often used to connect the components of a computer system, such as a printer and a keyboard, to the computer.

Figure 9-5*i* shows a spring-loaded metal hook as a grabber for a temporary connection to a circuit. This type of connector is often used with the test leads of a VOM or a DMM.

9.5 Printed Wiring

Most electronic circuits are mounted on a plastic or fiberglass insulating board with printed wiring, as shown in Fig. 9-6. This is a printed circuit (PC) or printed-wiring (PW) board. One side has the components, such as resistors, capacitors,

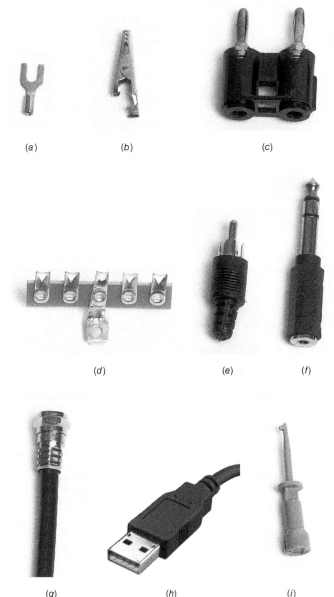

Figure 9-5   Common types of connectors for wire conductors. (*a*) Spade lug. (*b*) Alligator clip. (*c*) Double banana-pin plug. (*d*) Terminal strip. (*e*) RCA-type plug for audio cables. (*f*) Phone plug. (*g*) F-type plug for cable TV. (*h*) Multiple-pin connector plug. (*i*) Spring-loaded metal hook as grabber for temporary connection in testing circuits.

coils, transistors, diodes, and integrated circuit (IC) units. The other side has the conducting paths printed with silver or copper on the board, instead of using wires. On a double-sided board, the component side also has printed wiring. Sockets, small metal eyelets, or holes in the board are used to connect the components to the wiring.

With a bright light on one side, you can see through to the opposite side to trace the connections. However, the circuit may be drawn on the PC board. Modern printed circuit boards (PCBs) frequently have multiple layers. The PCB is made up

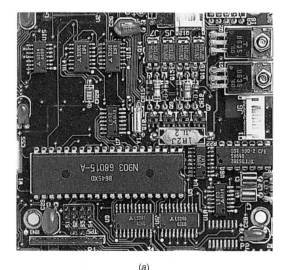

(a)

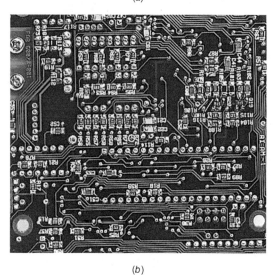

(b)

Figure 9-6 Printed-wiring board. (a) Component side with resistors, capacitors, transistors, and integrated circuits. (b) Side with printed wiring for the circuit.

of alternating layers of insulating fiberglass material and copper wiring patterns. The multiple layers are connected by short vertical conductors called *vias*. Such boards are widely used today as connecting points for small but very complex integrated circuits with hundreds of input and output connections.

9.6 Switches

A switch is a component that allows us to control whether the current is ON or OFF in a circuit. A closed switch has practically zero resistance, whereas an open switch has nearly infinite resistance.

Figure 9-7 shows a switch in series with a voltage source and a lightbulb. With the switch closed, as in Fig. 9-7*a*, a complete path for current is provided and the light is ON. Since the switch has very low resistance when it is closed, all of the source voltage is across the load, with 0 V across the closed contacts of the switch. With the switch open, as

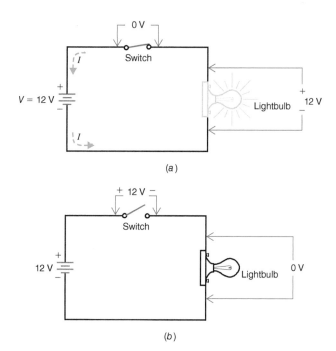

Figure 9-7 A series switch used to open or close a circuit. (a) With the switch closed, current flows to light the bulb. The voltage drop across the closed switch is 0 V. (b) With the switch open, the light is OFF. The voltage drop across the open switch is 12 V.

in Fig. 9-7*b*, the path for current is interrupted and the bulb does not light. Since the switch has very high resistance when it is open, all of the source voltage is across the open switch contacts, with 0 V across the load.

Switch Ratings

All switches have a current rating and a voltage rating. The current rating corresponds to the maximum allowable current that the switch can carry when closed. The current rating is based on the physical size of the switch contacts as well as the type of metal used for the contacts. Many switches have gold- or silver-plated contacts to ensure very low resistance when closed.

The voltage rating of a switch corresponds to the maximum voltage that can safely be applied across the open contacts without internal arcing. The voltage rating does not apply when the switch is closed, since the voltage drop across the closed switch contacts is practically zero.

Switch Definitions

Toggle switches are usually described as having a certain number of poles and throws. For example, the switch in Fig. 9-7 is described as a *single-pole, single-throw (SPST)* switch. Other popular switch types include the *single-pole, double-throw (SPDT), double-pole, single-throw (DPST),* and *double-pole, double-throw (DPDT)*. The schematic symbols for each type are shown in Fig. 9-8. Notice that the SPST switch has two connecting terminals, whereas the SPDT has three, the DPST has four, and the DPDT has six.

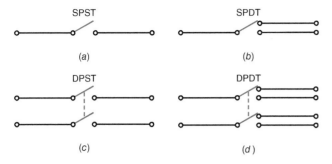

(a) SPST   *(b)* SPDT

(c) DPST   *(d)* DPDT

Figure 9-8   Switches. (*a*) Single-pole, single-throw (SPST). (*b*) Single-pole, double-throw (SPDT). (*c*) Double-pole, single-throw (DPST). (*d*) Double-pole, double-throw (DPDT).

The term *pole* is defined as the number of completely isolated circuits that can be controlled by the switch. The term *throw* is defined as the number of closed contact positions that exist per pole. The SPST switch in Fig. 9-7 can control the current in only one circuit, and there is only one closed contact position, hence the name single-pole, single-throw.

Figure 9-9 shows a variety of switch applications. In Fig. 9-9*a*, an SPDT switch is being used to switch a 12-Vdc source

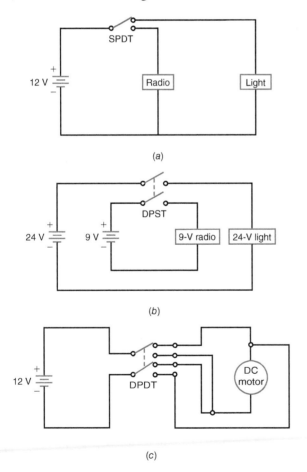

(a)

(b)

(c)

Figure 9-9   Switch applications. (*a*) SPDT switch used to switch a 12-V source between one of two different loads. (*b*) DPST switch controlling two completely isolated circuits simultaneously. (*c*) DPDT switch used to reverse the polarity of voltage across a dc motor.

between one of two different loads. In Fig. 9-9*b*, a DPST switch is being used to control two completely separate circuits simultaneously. In Fig. 9-9*c*, a DPDT switch is being used to reverse the polarity of voltage across the terminals of a dc motor. (Reversing the polarity reverses the direction of the motor.) Note that the dashed lines shown between the poles in Fig. 9-9*b* and 9-9*c* indicate that both sets of contacts within the switch are opened and closed simultaneously.

Switch Types

Figure 9-10 shows a variety of toggle switches. Although the toggle switch is a very popular type of switch, several other types are found in electronic equipment. Additional types include push-button switches, rocker switches, slide switches, rotary switches, and DIP switches.

Push-button switches are often spring-loaded switches that are either normally open (NO) or normally closed (NC). Figure 9-11 shows the schematic symbols for both types. For the normally open switch in Fig. 9-11*a*, the switch contacts remain open until the push button is depressed. When the push button is depressed, the switch closes, allowing current to pass. The normally closed switch in Fig. 9-11*b* operates opposite the normally open switch in Fig. 9-11*a*. When the push button is depressed, the switch contacts open to interrupt current in the circuit. A typical push-button switch is shown in Fig. 9-12.

Figure 9-13 shows a DIP (dual in-line package) switch. It consists of eight miniature rocker switches, where each switch can be set separately. A DIP switch has pin connections that fit into a standard IC socket.

Figure 9-14 shows another type of switch known as a rotary switch. As shown, it consists of three wafers or decks mounted on a common shaft.

Figure 9-10   A variety of toggle switches.

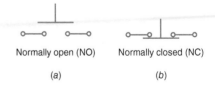

Normally open (NO)   Normally closed (NC)

(a)   *(b)*

Figure 9-11   Push-button switch schematic symbols. (*a*) Normally open (NO) push-button switch. (*b*) Normally closed (NC) push-button switch.

Figure 9-12   Typical push-button switch.

Figure 9-13   Dual in-line package (DIP) switch.

Figure 9-14   Rotary switch.

9.7 Fuses

Many circuits have a fuse in series as a protection against an overload from a short circuit. Excessive current melts the fuse element, blowing the fuse and opening the series circuit. The purpose is to let the fuse blow before the components and wiring are damaged. The blown fuse can easily be replaced by a new one after the overload has been eliminated. A glass-cartridge fuse with holders is shown in Fig. 9-15. This is a type 3AG fuse with a diameter of ¼ in. and length of 1¼ in. *AG* is an abbreviation of "automobile glass," since that was one of the first applications of fuses in a glass holder to make the wire link visible. The schematic symbol for a fuse is —⌒⌒— as shown in Fig. 9-17*a*.

The metal fuse element may be made of aluminum, tin-coated copper, or nickel. Fuses are available with current ratings from ⅟₅₀₀ A to hundreds of amperes. The thinner the wire element in the fuse, the smaller its current rating. For example, a 2-in. length of No. 28 wire can serve as a 2-A fuse. As typical applications, the rating for fuses in each branch of older house wiring is often 15 A; the high-voltage circuit in a television receiver is usually protected by a ¼-A glass-cartridge fuse. For automobile fuses, the ratings are generally 10 to 30 A because of the higher currents needed with a 12-V source for a given amount of power.

Slow-Blow Fuses

Slow-blow fuses have coiled construction. They are designed to open only on a continued overload, such as a short circuit. The purpose of coiled construction is to prevent the fuse from blowing on a temporary current surge. As an example, a slow-blow fuse will hold a 400% overload in current for up to 2 s. Typical ratings are shown by the curves in Fig. 9-16. Circuits with an electric motor use slow-blow fuses because the starting current of a motor is much more than its running current.

Circuit Breakers

A circuit breaker can be used in place of a fuse to protect circuit components and wiring against the high current caused by a short circuit. It is constructed of a thin bimetallic strip that expands with heat and in turn trips open the circuit. The advantage of a circuit breaker is that it can be reset once the bimetallic strip cools down and the short circuit has been removed. Because they can be reset, almost all new residential house wiring is protected by circuit breakers rather than fuses. The schematic symbol for a circuit breaker is often shown as ＿⌒＿.

Testing Fuses

In glass fuses, you can usually see whether the wire element inside is burned open. When measured with an ohmmeter, a good fuse has practically zero resistance. An open fuse reads

(a)   (b)   (c)

Figure 9-15   (*a*) Glass-cartridge fuse. (*b*) Fuse holder. (*c*) Panel-mounted fuse holder.

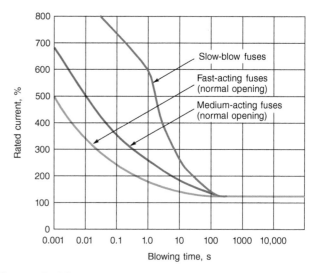

Figure 9-16   Chart showing percentage of rated current vs. blowing time for fuses.

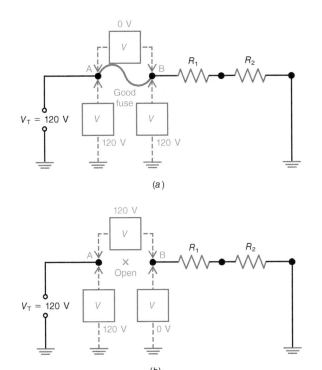

(a)

(b)

Figure 9-17 When a fuse opens, the applied voltage is across the fuse terminals. (a) Circuit closed with good fuse. Note schematic symbol for any type of fuse. (b) Fuse open. Voltage readings are explained in the text.

infinite ohms. Power must be off or the fuse must be out of the circuit to test a fuse with an ohmmeter.

When you test with a voltmeter, a good fuse has zero volts across its two terminals (Fig. 9-17a). If you read appreciable voltage across the fuse, this means that it is open. In fact, the full applied voltage is across the open fuse in a series circuit, as shown in Fig. 9-17b. This is why fuses also have a voltage rating, which gives the maximum voltage without arcing in the open fuse.

Referring to Fig. 9-17, notice the results when measuring the voltages to ground at the two fuse terminals. In Fig. 9-17a, the voltage is the same 120 V at both ends because there is no voltage drop across the good fuse. In Fig. 9-17b, however, terminal B reads 0 V because this end is disconnected from V_T by the open fuse. These tests apply to either dc or ac voltages.

9.8 Wire Resistance

The longer a wire, the higher its resistance. More work must be done to make electrons drift from one end to the other. However, the greater the diameter of the wire, the less the resistance, since there are more free electrons in the cross-sectional area. As a formula,

$$R = \rho\frac{l}{A} \tag{9-1}$$

where R is the total resistance, l the length, A the cross-sectional area, and ρ the specific resistance or *resistivity* of the conductor. The factor ρ then enables the resistance of

different materials to be compared according to their nature without regard to different lengths or areas. Higher values of ρ mean more resistance. Note that ρ is the Greek letter *rho*.

Specific Resistance

Specific resistance (ρ) is the rating in circular mil ohms per foot of a wire with a cross-sectional area of 1 cmil and a length of 1 ft. For example, the specific resistance of copper is 10.4, and for aluminum, it is 17. Since silver, copper, gold, and aluminum are the best conductors, they have the lowest values of specific resistance. Tungsten and iron have much higher resistance.

EXAMPLE 9-3

How much is the resistance of 100 ft of No. 20 gage copper wire?

Answer:

Note that from Table 9-1, the cross-sectional area for No. 20 gage wire is 1022 cmil; the ρ for copper is 10.4. Using Formula (9-1) gives

$$R = \rho\frac{l}{A}$$

$$= 10.4 \times \frac{100 \text{ ft}}{1022 \text{ cmil}}$$

$$R = 1.02 \ \Omega$$

EXAMPLE 9-4

How much is the resistance of a 100-ft length of No. 23 gage copper wire?

Answer:

$$R = \rho\frac{l}{A}$$

$$= 10.4 \times \frac{100 \text{ ft}}{509.5 \text{ cmil}}$$

$$R = 2.04 \ \Omega$$

Types of Resistance Wire

For applications in heating elements, such as a toaster, an incandescent lightbulb, or a heater, it is necessary to use wire that has more resistance than good conductors like silver, copper, or aluminum. Higher resistance is preferable so that the required amount of I^2R power dissipated as heat in the wire can be obtained without excessive current. Typical materials for resistance wire are the elements tungsten, nickel, or iron and alloys* such as manganin, Nichrome, and constantan. These types are generally called *resistance wire* because R is greater than that of copper wire for the same length.

---

* An *alloy* is a fusion of elements without chemical action between them. Metals are commonly alloyed to alter their physical characteristics.

9.9 Temperature Coefficient of Resistance

This factor with the symbol alpha (α) states how much the resistance changes for a change in temperature. A positive value for α means that R increases with temperature; with a negative α, R decreases; zero for α means that R is constant.

Positive α

All metals in their pure form, such as copper and tungsten, have positive temperature coefficients. The α for tungsten, for example, is 0.005.

In practical terms, a positive α means that heat increases R in wire conductors. Then I is reduced for a specified applied voltage.

Negative α

Note that carbon has a negative temperature coefficient. In general, α is negative for all semiconductors, including germanium and silicon. Also, all electrolyte solutions, such as sulfuric acid and water, have a negative α.

A negative value of α means less resistance at higher temperatures. The resistance of semiconductor diodes and transistors, therefore, can be reduced appreciably when they become hot with normal load current.

Zero α

Zero α means that R is constant with changes in temperature. The metal alloys constantan and manganin, for example, have a value of zero for α. They can be used for precision wire-wound resistors that do not change resistance when the temperature increases.

Hot Resistance

Because resistance wire is made of tungsten, Nichrome, iron, or nickel, there is usually a big difference in the amount of resistance the wire has when hot in normal operation and when cold without its normal load current. The reason is that the resistance increases with higher temperatures, since these materials have a positive temperature coefficient.

As an example, the tungsten filament of a 100-W, 120-V incandescent bulb has a current of 0.833 A when the bulb lights with normal brilliance at its rated power, since $I = P/V$. By Ohm's law, the hot resistance is V/I, or 120 V/0.833 A, which equals 144 Ω. If, however, the filament resistance is measured with an ohmmeter when the bulb is not lit, the cold resistance is only about 10 Ω.

The Nichrome heater elements in appliances and the tungsten heaters in vacuum tubes also become several hundred degrees hotter in normal operation. In these cases, only the cold resistance can be measured with an ohmmeter. The hot resistance must be calculated from voltage and current measurements with the normal value of load current. As a practical rule, the cold resistance is generally about one-tenth the hot resistance. In troubleshooting, however, the approach is usually just to check whether the heater element is open. Then it reads infinite ohms on the ohmmeter.

Superconductivity

The effect opposite to hot resistance occurs when cooling a metal down to very low temperatures to reduce its resistance. Near absolute zero, 0 K or –273°C, some metals abruptly lose practically all their resistance. As an example, when cooled by liquid helium, the metal tin becomes superconductive at 3.7 K. Tremendous currents can be produced, resulting in very strong electromagnetic fields. Such work at very low temperatures, near absolute zero, is called *cryogenics*.

New types of ceramic materials have been developed and are stimulating great interest in superconductivity because they provide zero resistance at temperatures much above absolute zero. One type is a ceramic pellet, with a 1-in. diameter, that includes yttrium, barium, copper, and oxygen atoms. The superconductivity occurs at a temperature of 93 K, equal to –160°C. This value is still far below room temperature, but the cooling can be done with liquid nitrogen, which is much cheaper than liquid helium. As research continues, it is likely that new materials will be discovered that are superconductive at even higher temperatures.

9.10 Resistive Sensors

A sensor is an electrical device that converts some physical characteristic like temperature, pressure or light into an electrical characteristic. Sensors are components that allow physical variables to be sensed and even measured accurately. Resistive sensors are very common. Two of these are the resistance temperature detector and the strain gauge.

Resistance Temperature Detectors

A resistance temperature detector (RTD), also called a resistive thermal device, is a sensor made of metal wire whose resistance changes with temperature. A wire with a specific length and diameter is wound on a form and housed in a protective container that has good heat transfer. While different types of wire are used, the most common is platinum. RTDs are available with resistances of ten to several thousand ohms but by far the most common in use is 100 ohms at 0°C temperature. The temperature variation is almost perfectly linear with a positive temperature coefficient. The resistance sensitivity of platinum α is 0.00383, or for a 100-Ω RTD this translates to a resistance variation of 0.385 ohm per degree Celsius (Ω/°C). Knowing this sensitivity value allows you to determine temperature by measuring the resistance. RTDs are useful in making temperature measurements in the -200 to $+500$°C range.

While most RTDs are of the wire type, a newer type uses a thin film of platinum on a substrate to form the resistive element. These RTDs are usually smaller and faster-acting but are not nearly as accurate as the wire version. Its upper temperature range is also less.

There are some critical factors to consider in making accurate temperature measurements. First is response time. In almost any temperature-related measurement, it takes time for an RTD to respond to a temperature change. That response time varies with the specific RTD sensor but is generally in the 0.5 to 5-s range.

Second, there is a self-heating factor that can skew measurement accuracy. To use an RTD, it must be connected into a circuit, such as a voltage divider or bridge, so that the resistance variation can be converted to a proportional voltage variation that can be easily measured. That means current must flow through the RTD. This current flow produces a heating effect that will affect the temperature variation. This variation is small, but must sometimes be compensated for to extract the best measurement accuracy. The goal is to keep the current as small as possible to minimize the heating effect.

The most common way of translating the resistance variation into a voltage is to connect the RTD into a bridge circuit, as shown in Fig. 9-18. Three of the resistors are fixed with the RTD providing the variation to produce unbalance. The bridge is initially balanced at some temperature for reference. As the temperature varies from the calibrated value, the RTD resistance changes and the bridge becomes unbalanced, producing a voltage across the bridge points A and B. This voltage change is very small, usually in the millivolt range. Therefore, amplification is usually required to boost the signal into a range that can be easily measured. A special amplifier called an *instrumentation amplifier* is used because of its high gain and high input resistance.

A common problem with the bridge circuit is that the RTD is normally located at some remote point where it is directly attached to the device or assembly whose temperature

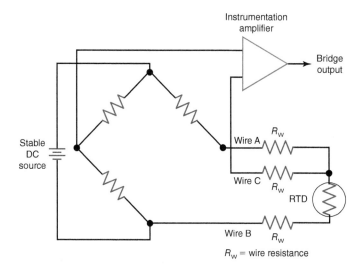

Figure 9-19 The three-wire connection that compensates for wire length resistance in connecting the remote sensor.

is to be measured. That means long wires are used to connect the RTD to the bridge and other circuitry. Typically the resistance of the wires is high enough to introduce substantial error into the measurement. The most common way to compensate for this error is to use a three-wire connection, as shown in Fig. 9-19. The assumption is that the wires are all the same length and have the same resistance. Wires A and B are in the opposite sides of the bridge, and cancel or balance one another. Wire C is the sense wire and essentially carries little or no current if the amplifier or measuring device has a very high resistance, like most do.

The output of the instrumentation amplifier connected to the bridge is a voltage proportional to the RTD resistance. The variation is nonlinear because of the bridge connection. Therefore the voltage reading must be corrected to provide a precise temperature equivalent. This can be done with circuitry or in a microcomputer if the bridge output voltage is first digitized into an equivalent binary number that can be manipulated by an algorithm implementing the compensation equation.

Strain Gauge

A strain gauge is a sensor used to measure force or stress. It is based on the principle of resistance change for a variation of the dimensions of a resistive element. The resistive element may be wire or a type of metal foil. As you saw earlier in this chapter, resistance will change if the physical dimensions of a wire are changed. For example, stretching a wire will cause its length to increase and its diameter to decrease. Both variations are small but do cause a measurable increase in resistance. In addition, compressing the wire causes its length to decrease and its diameter to increase slightly. This causes the resistance of the wire to decrease. A strain

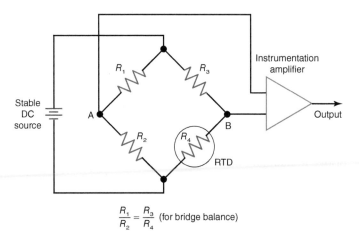

$$\frac{R_1}{R_2} = \frac{R_3}{R_4} \text{ (for bridge balance)}$$

Figure 9-18 R_1, R_2 and R_3 are fixed values. The RTD could also be placed in the R_3 position.

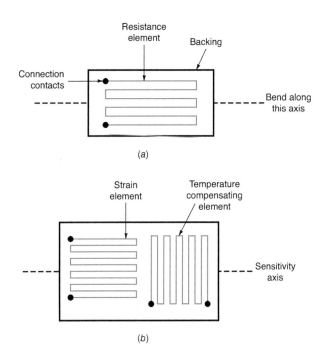

(a)

(b)

Figure 9-20 (*a*) Strain gauge element and (*b*) with temperature compensating resistance.

gauge is a sensor made of wire or foil pattern mounted on a paper or plastic backing. One example is given in Fig. 9-20. To measure stress, the gauge is physically attached with a strong adhesive to an object whose stress is to be measured. Figure 9-21 shows a strain gauge mounted on a metal plate that is to be put under force. As the metal piece is bent down, the resistance of gauge pattern increases. If an upward force is applied, the gauge pattern is compressed, causing a resistance decrease.

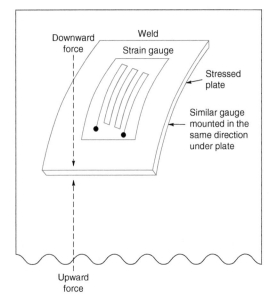

Figure 9-21 The strain gauge is glued to the stressed plate. Its resistance varies as the plate bends.

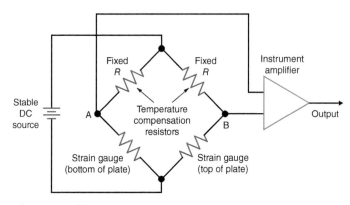

Figure 9-22 Using a strain gauge on each side of the bridge increases the sensitivity of the bridge, producing a greater output.

The resistance variation is converted to a voltage variation by connecting the strain gauge into a bridge circuit. As with the RTD, the resistance variation is very small, resulting in a very small resistance variation. That translates into a microvolt variation that is amplified, measured, and used to determine stress in pounds or tons.

One way to increase the output of the bridge is to put a similar strain gauge on the other side of the member being stressed. See Figs. 9-21 and 9-22. As the member is bent down, the upper gauge increases in resistance while the lower-gauge resistance decreases. Putting the strain elements on opposite sides of the bridge doubles the output variation.

Strain gauges come in a variety of sizes typically from 60 to 1000 Ω with 120 and 240 being the most common values. Each gauge has a gauge factor (GF) rating that essentially tells how sensitive the gauge is to force. The variation is usually specified in terms of microinches or micrometers of variation. This translates into resistance variations well below an ohm. Special semiconductor strain gauges are also available. While they have a greater resistance variation for a given force, their variation is very nonlinear and more difficult to translate into an accurate measurement. Metal foil and wire are still the best for accuracy.

Some strain gauges have a matching resistive element on the same backing but at a right angle to the main stressed resistive element. See Fig. 8-20. This second resistive element is insensitive to the force, so its dimensions do not change. However, it does provide temperature compensation when connected into the bridge circuit. As temperature varies, both resistive elements change the same amount, causing the bridge to maintain balance.

Figure 8-22 shows how temperature compensation strain gauges are connected into a bridge circuit. The fixed resistors are or can be the temperature-compensating elements. As the temperature changes, both resistances will vary, but by their being in opposite sides of the bridge, balance will be maintained.

Like RTDs, strain gauges are also usually mounted at some distance from the measuring equipment. Long wires connect the gauge elements to the bridge. For that reason, wire resistance must be taken into account. This is done using the same three-wire technique described earlier. Refer back to Fig. 9-19. Note that a high-gain instrumentation amplifier is used to boost the microvolt signal level to a higher level, making the measurement easier and more accurate.

In most modern applications, the instrument amplifier output is digitized in an analog-to-digital converter (ADC) and sent to a microcomputer, where the value is used in a calculation to determine the force in pounds or some other unit.

9.11 Insulators

Substances that have very high resistance, of the order of many megohms, are classed as insulators. With such high resistance, an insulator cannot conduct appreciable current when voltage is applied. As a result, insulators can have either of two functions. One is to isolate conductors to eliminate conduction between them. The other is to store an electric charge when voltage is applied.

An insulator maintains its charge because electrons cannot flow to neutralize the charge. The insulators are commonly called *dielectric materials,* which means that they can store a charge.

Among the best insulators, or dielectrics, are air, vacuum, rubber, wax, shellac, glass, mica, porcelain, oil, dry paper, textile fibers, and plastics such as Bakelite, Formica, and polystyrene. Pure water is a good insulator, but saltwater is not. Moist earth is a fairly good conductor, and dry, sandy earth is an insulator.

For any insulator, a high enough voltage can be applied to break down the internal structure of the material, forcing the dielectric to conduct. This dielectric breakdown is usually the result of an arc, which ruptures the physical structure of the material, making it useless as an insulator. Table 9-2 compares several insulators in terms of dielectric strength, which is the voltage breakdown rating. The higher the dielectric strength, the better the insulator, since it is less likely to break down at a high value of applied voltage. The breakdown voltages in Table 9-2 are approximate values for the standard thickness of 1 mil, or 0.001 in. More thickness allows a higher breakdown-voltage rating. Note that the value of 20 V/mil for air or vacuum is the same as 20 kV/in.

Insulator Discharge Current

An insulator in contact with a voltage source stores charge, producing a potential on the insulator. The charge tends to

Table 9-2 Voltage Breakdown of Insulators

Material	Dielectric Strength, V/mil	Material	Dielectric Strength, V/mil
Air or vacuum	20	Paraffin wax	200–300
Bakelite	300–550	Phenol, molded	300–700
Fiber	150–180	Polystyrene	500–760
Glass	335–2000	Porcelain	40–150
Mica	600–1500	Rubber, hard	450
Paper	1250	Shellac	900
Paraffin oil	380		

remain on the insulator, but it can be discharged by one of the following methods:

1. **Conduction through a conducting path.** For instance, a wire across the charged insulator provides a discharge path. Then the discharged dielectric has no potential.

2. **Brush discharge.** As an example, high voltage on a sharp pointed wire can discharge through the surrounding atmosphere by ionization of the air molecules. This may be visible in the dark as a bluish or reddish glow, called the *corona effect.*

3. **Spark discharge.** This is a result of breakdown in the insulator because of a high potential difference that ruptures the dielectric. The current that flows across the insulator at the instant of breakdown causes the spark.

A corona is undesirable because it reduces the potential by brush discharge into the surrounding air. In addition, the corona often indicates the beginning of a spark discharge. A potential of the order of kilovolts is usually necessary for a corona because the breakdown voltage for air is approximately 20 kV/in. To reduce the corona effect, conductors that have high voltage should be smooth, rounded, and thick. This equalizes the potential difference from all points on the conductor to the surrounding air. Any sharp point can have a more intense field, making it more susceptible to a corona and eventual spark discharge.

9.12 Semiconductors

Semiconductors are materials whose resistance is some value between that of an insulator and that of a conductor. A semiconductor may be made with just about any resistance value between a fraction of an ohm and many megohms. Semiconductors are used primarily to create diodes, transistors, and integrated circuits.

Semiconductor are elements or atoms that have four valence (outer-shell) electrons. Examples are carbon (C), silicon (Si), and germanium (Ge). Figure 9-23*a* show a carbon

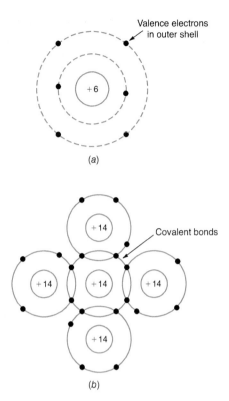

Valence electrons in outer shell

(a)

Covalent bonds

(b)

Figure 9-23 Semiconductors have four valence electrons, (a) A carbon atom. (b) A crystal structure formed with silicon atoms showing only the valence shells.

atom with four electrons in the outer shell. These atoms easily combine to create what are called *covalent bonds* with adjacent atoms. This produces a solid called a *crystal lattice structure*. Figure 9-23b has a two-dimensional portion of such a structure, showing only the valence

shell of silicon atoms. In reality, of course, the structure is three-dimensional.

In a crystal semiconductor structure all the valence atoms are fully involved with one another, meaning that they are strongly attached to each other and are not available to support electron flow. This is called an *intrinsic or pure semiconductor* whose resistance is very high. It is actually an insulator.

By adding other elements called *dopants* to the crystal structure, in a process called doping, extra electrons are added or an electron deficiency, or "hole," is created. Typical dopant atoms are boron, phosphorus, arsenic, gallium, and aluminum. This allows the semiconductor to conduct because it has resistance. The amount of resistance depends on the amount of dopant atoms added. The greater the number of dopant atoms added, the lower the resistance.

The doping process produces two types of semiconductors, *n*-type and *p*-type. The *n*-type semiconductors have an excess of electrons. The *p*-type semiconductors have a shortage of electrons in the form of vacant places in the valence shells of some atoms. These vacancies are called *holes* and have a positive charge. They attract electrons. Current flow is by electrons in *n*-type material and by holes in *p*-type materials.

By themselves the *p*- and *n*-type materials can be used to form resistors, capacitors, and other passive components. When combined, the *p*- and *n*-type materials are used to form diodes and transistors. Using special techniques, the semiconductors can be formed to make complete circuits called integrated circuits. Most electronic circuitry today is made in this way, using silicon semiconductor materials.

You will learn more about semiconductors in Chap. 8.

CHAPTER 9 SYSTEM SIDEBAR

Lightbulbs

A lightbulb seems like such a simple device. A filament of tungsten inside an evacuated glass bulb is connected to 120 or 240 volts, and the filament heats up and emits an enormous amount of light. Such lightbulbs have been around for about 130 years since Edison (and Swan in England) invented them in the late 1800s. We are still using them today. However, things are changing. With increased emphasis on energy efficiency, the U.S. government has passed rules and regulations that will change how we light our homes and offices. Furthermore, the latest electronic technology has brought us some alternative lighting choices.

The U.S. law is the Energy Independence and Security Act, passed by Congress in 2007. It provides a plan to phase out standard incandescent bulbs in place of more efficient bulbs.

The phaseout started in 2012 and will continue through 2014. With 20 to 25% of a home's energy consumption, lighting can provide a significant energy savings nationwide if consumers begin to replace their standard lightbulbs with some of the more efficient alternatives like compact fluorescent lights (CFL) or light-emitting diode (LED) lights. The result will be less demand on the utilities and greater savings of energy and, hopefully, some improvement in environmental conditions.

The basic standard lightbulb works great but is horribly inefficient. What we want from the bulb is light, but what we get most of is heat. Heat is produced by the power being generated to produce the light. Only about 5% of the power goes to operate the filament, while 95% of the power is wasted as heat. The goal of regulation is to save energy by forcing

people to use more efficient bulbs. Up until a few years ago, good practical alternatives didn't exist. Now, thanks to electronics, we have multiple options, as shown in Fig. S9-1, that are more efficient but more costly.

Halogen Bulb

The halogen bulb is just like the standard light bulb but contains halogen gas, such as iodine or bromine. This makes the bulb emit more light with less power consumption. In a normal incandescent bulb, the tungsten filament gets so hot that it vaporizes the tungsten, which is then deposited inside on the glass bulb, darkening it. Eventually the tungsten is used up, and the filament "burns out." The typical life of an incandescent bulb is in the 750- to 1000-hour range. By adding halogen gas, the halogen combines with the tungsten vapor so that it is not deposited on the bulb. Instead, it is essentially returned to the filament. Without the darkening, the light stays brighter longer. In addition, the life of the bulb is considerable extended to as many as 2500 hours. The result is an improvement in the efficiency of roughly 28% over a standard incandescent. It is estimated that a halogen bulb can produce the light output of a 60 watt bulb with only 43 W of power consumption. A 72-W halogen bulb produces approximately the same amount of light as a 100-W bulb. The main disadvantage is that the halogen bulb runs considerably hotter.

Halogen bulbs are readily available now with the standard Edison E27 screw base (also called the A19 standard screw base). The cost is slightly higher but is typically offset with the energy savings and the longer life. Halogen bulbs have been around for years in the form of automobile headlights and specialty lighting apparatus. Most automobile bulbs are halogen today.

Compact Fluorescent Lamps

Fluorescent bulbs have also been around for years in the form of long white tubes found mostly in offices, factories, and businesses rather than homes. Now they are available in a compact spiral or helical format that can be screwed into a standard lamp base. Compact fluorescent lamps (CFLs) operate like a standard fluorescent. Small filament electrodes at

each end of the tube are powered up, and they heat the mercury gas in the tube. Then a high voltage is applied between the two electrodes. The high voltage initiates an arc of electron current through the gas. That ionizes the mercury gas. A high-voltage, high-frequency ac signal keeps the current flowing and the gas ionized. The mercury atoms are temporarily put into a higher-energy state. Then they return their energy by emitting ultraviolet (UV) light, which we cannot see. However, the UV stimulates the phosphor coating inside the glass, converting it into visible light.

To generate the excitation for the mercury gas, the 120-Vac input is first rectified into dc and then used to power a high-frequency oscillator that generates the high voltage between the electrodes to stimulate the gas. All CFLs therefore have a rather sophisticated electronic circuit in their base, making them more expensive than a standard incandescent bulb.

The energy savings of a CFL over an incandescent is significant and in the 75% range. A CFL with the light output of a standard 60-W incandescent uses only about 13–14 W of energy and runs considerably cooler. CFL bulb life is in the 8000- to 10,000-hour range. The primary downsides are light color and lack of a dimming ability. The fluorescent light color is less appealing to most people who describe it as "colder." Some of the newer CFLs now produce light more the color of the familiar incandescent. Some of the newer CFLs are also dimmable. For those consumers who do not like the spiral shape, there is a new CFL that encases the spiral in a standard glass bulb. CFLs also emit high-frequency radio energy that can interfere with some sensitive shortwave and high-frequency radios.

Light-Emitting Diodes

A light-emitting diode (LED) is a semiconductor component called a *diode* that emits light when current passes through it. Most LEDs emit red, yellow, or green light depending on the type of semiconductor used. However, using special semiconductor material combinations, it is now possible to produce very bright white light.

LEDs are diodes and operate from dc. The 120 Vac from the screw socket has to be first rectified into dc and then conditioned by an integrated circuit to operate multiple LEDs. The result is a very efficient bulb with energy savings of 75–80% over an incandescent. An LED bulb with the light output of a 60-W incandescent consumes only 13–14 W. And the bulb life is an estimated 25,000 hours.

The primary disadvantage of the LED bulb is its very high cost. LED bulbs cost in the $20 to $50 range, which is excessive for most consumers. Even when considering the energy savings and the long life, it is still a great deal to pay for a lightbulb that would otherwise cost a little over a dollar. However, as progress is made with the electronics and the manufacturing, costs will decline further, making the LED bulb the light source of the future.

1. A closed switch has a resistance of approximately
   a. infinity.
   b. zero ohms.
   c. 1 MΩ.
   d. none of the above.

2. An open fuse has a resistance that approaches
   a. infinity.
   b. zero ohms.
   c. 1 to 2 Ω.
   d. none of the above.

3. How many connecting terminals does an SPDT switch have?
   a. 2.
   b. 6.
   c. 3.
   d. 4.

4. The voltage drop across a closed switch equals
   a. the applied voltage.
   b. zero volts.
   c. infinity.
   d. none of the above.

5. For round wire, as the gage numbers increase from 1 to 40,
   a. the diameter and circular area increase.
   b. the wire resistance decreases for a specific length and type.
   c. the diameter increases but the circular area remains constant.
   d. the diameter and circular area decrease.

6. The circular area of round wire, doubles for
   a. every 2 gage sizes.
   b. every 3 gage sizes.
   c. each successive gage size.
   d. every 10 gage sizes.

7. Which has more resistance, a 100-ft length of No. 12 gage copper wire or a 100-ft length of No. 12 gage aluminum wire?
   a. The 100-ft length of No. 12 gage aluminum wire.
   b. The 100-ft length of No. 12 gage copper wire.
   c. They both have exactly the same resistance.
   d. It cannot be determined.

8. In their pure form, all metals have a
   a. negative temperature coefficient.
   b. temperature coefficient of zero.
   c. positive temperature coefficient.
   d. very high resistance.

9. The current rating of a switch corresponds to the maximum current the switch can safely handle when it is
   a. open.
   b. either open or closed.
   c. closed.
   d. none of the above.

10. How much is the resistance of a 2000-ft length of No. 20 gage aluminum wire?
    a. less than 1 Ω.
    b. 20.35 Ω.
    c. 3.33 kΩ.
    d. 33.27 Ω.

11. How many completely isolated circuits can be controlled by a DPST switch?
    a. 1.
    b. 2.
    c. 3.
    d. 4.

12. Which of the following metals is the best conductor of electricity?
    a. steel.
    b. aluminum.
    c. silver.
    d. gold.

13. What is the area in circular mils (cmils) of a wire whose diameter, d, is 0.01 in.?
    a. 0.001 cmil.
    b. 10 cmil.
    c. 1 cmil.
    d. 100 cmil.

14. The term *pole* as it relates to switches is defined as
    a. the number of completely isolated circuits that can be controlled by the switch.
    b. the number of closed contact positions that the switch has.
    c. the number of connecting terminals the switch has.
    d. none of the above.

15. An RTD measures what physical characteristic?
    a. Pressure.
    b. Temperature.
    c. Light intensity.
    d. Liquid level.

16. A strain gauge measures what physical characteristic?
    a. Pressure.
    b. Temperature.
    c. Light intensity.
    d. Liquid level.

17. In Fig. 9-18, resistors R_1 and R_2 are 220 Ω. The RTD resistance is 100 Ω. What value should R_3 be to balance the bridge?
   a. 50 Ω.
   b. 100 Ω.
   c. 200 Ω.
   d. 220 Ω.

18. If the bridge in Fig. 9-18 is initially balanced and the temperature increases, how does the voltage across points A and B vary?
   a. Voltage at A decreases with respect to B.
   b. Voltage does not change.
   c. Voltage at B increases with respect to A.
   d. Voltage remains at zero.

19. What is the main purpose of the three wire bridge arrangement normally used with RTDs and strain gauges?
   a. To ensure the sensor resistance remains constant.
   b. For temperature compensation.
   c. To bring the bridge into balance.
   d. To compensate for the resistance of connecting wires.

CHAPTER 9 PROBLEMS

SECTION 9.1 Function of the Conductor

9.1 In Fig. 9-24, an 8-Ω heater is connected to the 120-Vac power line by two 50-ft lengths of copper wire. If each 50-ft length of wire has a resistance of 0.08 Ω, then calculate the following:
   a. The total length of copper wire that connects the 8-Ω heater to the 120-Vac power line.
   b. The total resistance, R_T, of the circuit.
   c. The current, I, in the circuit.
   d. The voltage drop across each 50-ft length of copper wire.
   e. The voltage across the 8-Ω heater.
   f. The I^2R power loss in each 50-ft length of copper wire.
   g. The power dissipated by the 8-Ω heater.
   h. The total power, P_T, supplied to the circuit by the 120-Vac power line.
   i. The percentage of the total power, P_T, dissipated by the 8-Ω heater.

SECTION 9.2 Standard Wire Gage Sizes

9.2 Determine the area in circular mils for a wire if its diameter, d, equals
   a. 0.005 in.
   b. 0.021 in.
   c. 0.032 in.
   d. 0.05 in.
   e. 0.1 in.
   f. 0.2 in.

9.3 What is the approximate AWG size of a wire whose diameter, d, equals 0.072 in.?

9.4 Using Table 9-1, determine the resistance of a 1000-ft length of copper wire for the following gage sizes:
   a. No. 10 gage.
   b. No. 13 gage.
   c. No. 16 gage.
   d. No. 24 gage.

9.5 Which would you expect to have more resistance, a 1000-ft length of No. 14 gage copper wire or a 1000-ft length of No. 12 gage copper wire?

9.6 Which would you expect to have more resistance, a 1000-ft length of No. 23 gage copper wire or a 100-ft length of No. 23 gage copper wire?

SECTION 9.3 Types of Wire Conductors

9.7 If an extension cord is made up of 65 strands of No. 28 gage copper wire, what is its equivalent gage size in solid wire?

9.8 What is the gage size of the individual strands in a No. 10 gage stranded wire if there are eight strands?

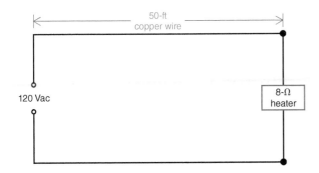

Figure 9-24

SECTION 9.6 Switches

9.9 With the switch, S_1, closed in Fig. 9-25,
  a. How much is the voltage across the switch?
  b. How much is the voltage across the lamp?
  c. Will the lamp light?
  d. What is the current, I, in the circuit based on the specifications of the lamp?

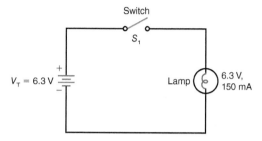

Figure 9-25

9.10 With the switch, S_1, open in Fig. 9-25,
  a. How much is the voltage across the switch?
  b. How much is the voltage across the lamp?
  c. Will the lamp light?
  d. What is the current, I, in the circuit based on the specifications of the lamp?

9.11 Draw a schematic diagram showing how an SPDT switch can be used to supply a resistive heating element with either 6 V or 12 V.

9.12 Draw a schematic diagram showing how a DPDT switch can be used to
  a. allow a stereo receiver to switch between two different speakers.
  b. reverse the polarity of voltage across a dc motor to reverse its direction.

9.13 An SPST switch is rated at 10 A/250 V. Can this switch be used to control a 120-V, 1000-W appliance?

SECTION 9.8 Wire Resistance

9.14 Calculate the resistance of the following conductors:
  a. 250 ft of No. 20 gage copper wire.
  b. 250 ft of No. 20 gage aluminum wire.

9.15 What is the resistance for each conductor of a 50-ft extension cord made of No. 14 gage copper wire?

9.16 A 100-ft extension cord uses No. 14 gage copper wire for each of its conductors. If the extension cord is used to connect a 10-A load to the 120-Vac power line, how much voltage is available at the load?

SECTION 9.9 Temperature Coefficient of Resistance

9.17 A tungsten wire has a resistance, R, of 20 Ω at 20°C. Calculate its resistance at 70°C.

9.18 The resistance of a Nichrome wire is 1 kΩ at 20°C. Calculate its resistance at 220°C.

Batteries

Learning Outcomes

After studying this chapter, you should be able to:

> *Explain* the difference between primary and secondary cells.

> *Define* what is meant by the *internal resistance of a cell*.

> *List* several different types of voltaic cells.

> *Explain* how cells can be connected to increase either the current capacity or voltage output of a battery.

> *Explain* why the terminal voltage of a battery drops with more load current.

> *Explain* the difference between voltage sources and current sources.

A *battery* is a group of cells that generate energy from an internal chemical reaction. The cell itself consists of two different conducting materials as the electrodes that are immersed in an electrolyte. The chemical reaction between the electrodes and the electrolyte results in a separation of electric charges as ions and free electrons. Then the two electrodes have a difference of potential that provides a voltage output from the cell.

A battery provides a source of steady dc voltage of fixed polarity and is a good example of a generator or energy source. Batteries are the oldest form of voltage source, and today they are more important than ever because of the huge demand for portable and mobile electronic devices like cell phones, laptop computers, tablets, and hybrid/electric vehicles. This chapter summarizes the types of batteries and how they are used. The battery supplies voltage to a circuit as the load to produce the desired load current. An important factor is the internal resistance, r_i of the source, which affects the output voltage when a load is connected. A low r_i means that the source can maintain a constant output voltage for different values of load current. For the opposite case, a high r_i makes the output voltage drop, but a constant value of load current can be maintained.

10.1 Introduction to Batteries

We rely on batteries to power an almost unlimited number of electronic products available today. For example, batteries are used in cars, personal computers (PCs), handheld radios, laptops, cameras, MP3 players, and cell phones, to name just a few of the more common applications. Batteries are available in a wide variety of shapes and sizes and have many different voltage and current ratings. The different sizes and ratings are necessary to meet the needs of the vast number of applications. Regardless of the application, however, all batteries are made up of a combination of individual voltaic cells. Together, the cells provide a steady dc voltage at the output terminals of the battery. The voltage output and current rating of a battery are determined by several factors, including the type of elements used for the electrodes, the physical size of the electrodes, and the type of electrolyte.

As you know, some batteries become exhausted with use and cannot be recharged. Others can be recharged hundreds or even thousands of times before they are no longer able to produce or maintain the rated output voltage. Whether a battery is rechargeable or not is determined by the type of cells that make up the battery. There are two types, primary cells and secondary cells.

Primary Cells

Primary cells cannot be recharged. After it has delivered its rated capacity, the primary cell must be discarded because the internal chemical reaction cannot be restored. Figure 10-1 shows a variety of dry cells and batteries, all of which are of the primary type. In Table 10-1 several different

Table 10-1 Cell Types and Open-Circuit Voltage

Cell Name	Type	Nominal Open-Circuit* Voltage, Vdc
Carbon-zinc	Primary	1.5
Zinc chloride	Primary	1.5
Manganese dioxide (alkaline)	Primary or secondary	1.5
Mercuric oxide	Primary	1.35
Silver oxide	Primary	1.5
Lithium	Primary	3.0
Lithium-ion	Secondary	3.7
Lead-acid	Secondary	2.1
Nickel-cadmium	Secondary	1.2
Nickel-metal-hydride	Secondary	1.2
Nickel-iron (Edison) cell	Secondary	1.2
Nickel-zinc	Secondary	1.6
Solar	Secondary	0.5

* Open-circuit V is the terminal voltage without a load.

cells are listed by name. Each of the cells is listed as either the primary or the secondary type. Notice the open-circuit voltage for each of the cell types listed.

Secondary Cells

Secondary cells can be recharged because the chemical action is reversible. When it supplies current to a load resistance, the cell is *discharging* because the current tends to neutralize the separated charges at the electrodes. For the opposite case, the current can be reversed to re-form the electrodes as the chemical action is reversed. This action is *charging* the cell. The charging current must be supplied by an external dc voltage source, with the cell serving as a load resistance. The discharging and recharging is called *cycling* of the cell. Since a secondary cell can be recharged, it is also called a *storage cell*. The most common type is the lead-acid cell generally used in automotive batteries (Fig. 10-2). In addition, the list in Table 10-1 indicates which are secondary cells.

Figure 10-1 Typical dry cells and batteries. These primary types cannot be recharged.

Figure 10-2 Example of a 12-V auto battery using six lead-acid cells in series. This is a secondary type, which can be recharged.

Dry Cells

What we call a *dry cell* really has a moist electrolyte. However, the electrolyte cannot be spilled, and the cell can operate in any position.

Sealed Rechargeable Cells

A sealed rechargeable cells is a secondary cell that can be recharged, but it has a sealed electrolyte that cannot be refilled. These cells are capable of charge and discharge in any position.

10.2 The Voltaic Cell

When two different conducting materials are immersed in an electrolyte, as illustrated in Fig. 10-3*a* , the chemical action of forming a new solution results in the separation of charges. This device for converting chemical energy into electric energy is a voltaic cell. It is also called a *galvanic cell,* named after Luigi Galvani (1737–1798).

In Fig. 10-3*a*, the charged conductors in the electrolyte are the electrodes or plates of the cell. They are the terminals that connect the voltage output to an external circuit, as shown in Fig. 10-3*b*. Then the potential difference resulting from the separated charges enables the cell to function as a source of applied voltage. The voltage across the cell's terminals forces current to flow in the circuit to light the bulb.

Current Outside the Cell

Electrons from the negative terminal of the cell flow through the external circuit with R_L and return to the positive terminal. The chemical action in the cell separates charges continuously to maintain the terminal voltage that produces current in the circuit.

The current tends to neutralize the charges generated in the cell. For this reason, the process of producing load current is considered discharging of the cell. However, the internal chemical reaction continues to maintain the separation of charges that produces the output voltage.

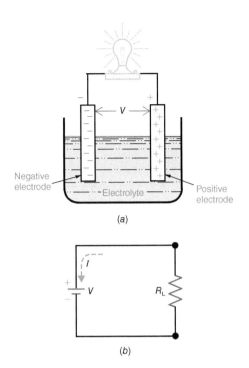

(a)

(b)

Figure 10-3 How a voltaic cell converts chemical energy into electric energy. (*a*) Electrodes or plates in liquid electrolyte solution. (*b*) Schematic of a circuit with a voltaic cell as a dc voltage source *V* to produce current in load R_L, which is the lightbulb.

Current Inside the Cell

The current through the electrolyte is a motion of ion charges. Notice in Fig. 10-3*b* that the current inside the cell flows from the positive terminal to the negative terminal. This action represents the work being done by the chemical reaction to generate the voltage across the output terminals.

The negative terminal in Fig. 10-3*a* is considered the anode of the cell because it forms positive ions in the electrolyte. The opposite terminal of the cell is its cathode.

Internal Resistance, r_i

Any practical voltage source has internal resistance, indicated as r_i, which limits the current it can deliver. For a chemical cell, as in Fig. 10-3, the r_i is mainly the resistance of the electrolyte. For a good cell, r_i is very low, with typical values less than 1Ω. As the cell deteriorates, though, r_i increases, preventing the cell from producing its normal terminal voltage when load current is flowing because the internal voltage drop across r_i opposes the output terminal voltage. This is why you can often measure the normal voltage of a dry cell with a voltmeter, which drains very little current, but the terminal voltage drops when the load is connected.

The voltage output of a cell depends on the elements used for the electrodes and the electrolyte. The current rating depends mostly on the size. Larger batteries can supply more current. Dry cells are generally rated up to 250 mA, and the lead-acid wet cell can supply current up to 300 A or more. Note that a smaller r_i allows a higher current rating.

10.3 Common Types of Primary Cells

In this section, you will be introduced to several different types of primary cells in use today.

Carbon-Zinc

The carbon-zinc dry cell is a very common type because of its low cost. It is also called the *Leclanché cell,* named after its inventor. The voltage output of the carbon-zinc cell is 1.4 to 1.6 V, with a nominal value of 1.5 V. The suggested current range is up to 150 mA for the D size, which has a height of 2¼ in. and volume of 3.18 in.³. The C, AA, and AAA sizes are smaller, with lower current ratings.

The electrochemical system consists of a zinc anode and a manganese dioxide cathode in a moist electrolyte. The electrolyte is a combination of ammonium chloride and zinc chloride dissolved in water. For the round-cell construction, a carbon rod is used down the center. The rod is chemically inert. However, it serves as a current collector for the positive terminal at the top. The path for current inside the cell includes the carbon rod as the positive terminal, the manganese dioxide, the electrolyte, and the zinc can which is the negative electrode. The carbon rod also prevents leakage of the electrolyte but is porous to allow the escape of gases which accumulate in the cell.

In operation of the cell, the ammonia releases hydrogen gas which collects around the carbon electrode. This reaction is called *polarization,* and it can reduce the voltage output. However, the manganese dioxide releases oxygen, which combines with the hydrogen to form water. The manganese dioxide functions as a *depolarizer.* Powdered carbon is also added to the depolarizer to improve conductivity and retain moisture.

Carbon-zinc dry cells are generally designed for an operating temperature of 70°F. Higher temperatures will enable the cell to provide greater output. However, temperatures of 125°F or more will cause rapid deterioration of the cell.

The chemical efficiency of the carbon-zinc cell increases with less current drain. Stated another way, the application should allow for the largest battery possible, within practical limits. In addition, performance of the cell is generally better with intermittent operation. The reason is that the cell can recuperate between discharges, probably by depolarization.

As an example of longer life with intermittent operation, a carbon-zinc D cell may operate for only a few hours with a continuous drain at its rated current. Yet the same cell could be used for a few months or even a year with intermittent operation of less than 1 hour at a time with smaller values of current.

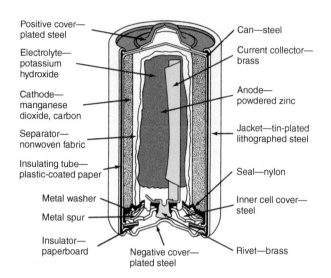

Figure 10-4 Construction of the alkaline cell.

Alkaline Cell

Another popular type is the manganese-zinc cell shown in Fig. 10-4, which has an alkaline electrolyte. It is available as either a primary or a secondary cell, but the primary type is more common.

The electrochemical system consists of a powdered zinc anode and a manganese dioxide cathode in an alkaline electrolyte. The electrolyte is potassium hydroxide, which is the main difference between the alkaline and Leclanché cells. Hydroxide compounds are alkaline with negative hydroxyl (OH) ions, whereas an acid electrolyte has positive hydrogen (H) ions. The voltage output from the alkaline cell is 1.5 V.

The alkaline cell has many applications because of its ability to work at high efficiency with continuous, high discharge rates. Depending on the application, an alkaline cell can provide up to seven times the service of a carbon-zinc cell. As examples, in a portable CD player, an alkaline cell will normally have twice the service life of a general-purpose carbon-zinc cell; in toys, the alkaline cell typically provides about seven times more service.

The outstanding performance of the alkaline cell is due to its low internal resistance. Its r_i is low because of the dense cathode material, the large surface area of the anode in contact with the electrolyte, and the high conductivity of the electrolyte. In addition, alkaline cells perform satisfactorily at low temperatures.

Additional Types of Primary Cells

The miniature button construction shown in Fig. 10-5 is often used for the mercury cell and the silver oxide cell. The cell diameter is ⅜ to 1 in.

The labels in Figure 10-4:
Positive cover—plated steel
Electrolyte—potassium hydroxide
Cathode—manganese dioxide, carbon
Separator—nonwoven fabric
Insulating tube—plastic-coated paper
Metal washer
Metal spur
Insulator—paperboard
Can—steel
Current collector—brass
Anode—powdered zinc
Jacket—tin-plated lithographed steel
Seal—nylon
Inner cell cover—steel
Rivet—brass
Negative cover—plated steel

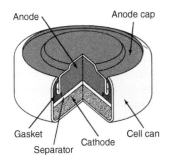

Anode — Anode cap

- Anodes are a gelled mixture of amalgamated zinc powder and electrolyte.
- Cathodes
  Silver cells: AgO_2, MnO_2, and conductor
  Mercury cells: HgO and conductor (may contain MnO_2)
  Manganese dioxide cells: MnO_2 and conductor

Gasket — Separator — Cathode — Cell can

Figure 10-5 Construction of miniature button type of primary cell. Diameter is $3/8$ to 1 in. Note the chemical symbols AgO_2 for silver oxide, HgO for mercuric oxide, and MnO_2 for manganese dioxide.

Mercury Cell

The electrochemical system consists of a zinc anode, a mercury compound for the cathode, and an electrolyte of potassium or sodium hydroxide. Mercury cells are available as flat, round cylinders and miniature button shapes. Note, though, that some round mercury cells have the top button as the negative terminal and the bottom terminal positive. The open-circuit voltage is 1.35 V when the cathode is mercuric oxide (HgO) and 1.4 V or more with mercuric oxide/manganese dioxide. The 1.35-V type is more common.

The mercury cell is used where a relatively flat discharge characteristic is required with high current density. Its internal resistance is low and essentially constant. These cells perform well at elevated temperatures, up to 130°F continuously or 200°F for short periods. One drawback of the mercury cell is its relatively high cost compared with a carbon-zinc cell. Mercury cells are becoming increasingly unavailable due to the hazards associated with proper disposal after use.

Silver Oxide Cell

The electrochemical system consists of a zinc anode, a cathode of silver oxide (AgO_2) with small amounts of manganese dioxide, and an electrolyte of potassium or sodium hydroxide. It is commonly available in the miniature button shape shown in Fig. 10-5. The open-circuit voltage is 1.6 V, but the nominal output with a load is considered 1.5 V. Typical applications include hearing aids, cameras, and electronic watches, which use very little current.

Summary of the Most Common Types of Dry Cells

The most common types of dry cells include carbon-zinc, zinc chloride (heavy duty), and manganese-zinc (alkaline). It should be noted that the alkaline cell is better for heavy-duty use than the zinc chloride type. They are available in the round, cylinder types, listed in Table 10-2, for the D, C, AA, and AAA sizes. The small button cells generally use either mercury or silver oxide. All these dry cells are the primary type and cannot be recharged. Each has an output of 1.5 V except for the 1.35-V mercury cell.

Table 10-2 Sizes for Popular Types of Dry Cells*		
Size	**Height, in.**	**Diameter, in.**
D	$2\frac{1}{4}$	$1\frac{1}{4}$
C	$1\frac{3}{4}$	1
AA	$1\frac{7}{8}$	$\frac{9}{16}$
AAA	$1\frac{3}{4}$	$\frac{3}{8}$

* Cylinder shape shown in Fig. 10-1.

Any dry cell loses its ability to produce output voltage even when it is not being used. The shelf life is about 2 years for the alkaline type, but much less with the carbon-zinc cell, especially for small sizes and partially used cells. The reasons are self-discharge within the cell and loss of moisture from the electrolyte. Therefore, dry cells should be used fresh from the manufacturer. It is worth noting, however, that the shelf life of dry cells is steadily increasing due to advances in battery technology.

Note that shelf life can be extended by storing the cell at low temperatures, about 40 to 50°F. Even temperatures below freezing will not harm the cell. However, the cell should be allowed to return to normal room temperature before being used, preferably in its original packaging, to avoid condensation.

The alkaline type of dry cell is probably the most cost-efficient. It costs more but lasts much longer, besides having a longer shelf life. Compared with size-D batteries, the alkaline type can last about 10 times longer than the carbon-zinc type in continuous operation, or about seven times longer for typical intermittent operation. The zinc chloride heavy-duty type can last two or three times longer than the general-purpose carbon-zinc cell. For low-current applications of about 10 mA or less, however, there is not much difference in battery life.

Lithium Cell

The lithium cell is a relatively new primary cell. However, its high output voltage, long shelf life, low weight, and small

Figure 10-6 Lithium battery.

Figure 10-7 Common 12-V lead-acid battery used in automobiles.

volume make the lithium cell an excellent choice for special applications. The open-circuit output voltage is 3 V. Figure 10-6 shows an example of a lithium battery with a 6-V output.

A lithium cell can provide at least 10 times more energy than the equivalent carbon-zinc cell. However, lithium is a very active chemical element. Many of the problems in construction have been solved, though, especially for small cells delivering low current. One interesting application is a lithium cell as the dc power source for a cardiac pacemaker. The long service life is important for this use.

Two forms of lithium cells are in widespread use, the lithium–sulfur dioxide ($LiSO_2$) type and the lithium–thionyl chloride type. Output is approximately 3 V.

In the $LiSO_2$ cell, the sulfur dioxide is kept in a liquid state by using a highpressure container and an organic liquid solvent, usually methyl cyanide. One problem is safe encapsulation of toxic vapor if the container should be punctured or cracked. This problem can be significant for safe disposal of the cells when they are discarded after use.

The shelf life of the lithium cell, 10 years or more, is much longer than that of other types.

10.4 Lead-Acid Wet Cell

Where high load current is necessary, the lead-acid cell is the type most commonly used. The electrolyte is a dilute solution of sulfuric acid (H_2SO_4). In the application of battery power to start the engine in an automobile, for example, the load current to the starter motor is typically 200 to 400 A. One cell has a nominal output of 2.1 V, but lead-acid cells are often used in a series combination of three for a 6-V battery and six for a 12-V battery. Examples are shown in Figs. 10-2 and 10-7.

The lead-acid type is a secondary cell or storage cell, which can be recharged. The charge and discharge cycle can be repeated many times to restore the output voltage, as long as the cell is in good physical condition. However, heat with excessive charge and discharge currents shortens the useful life to about 3 to 5 years for an automobile battery. The lead-acid type has a relatively high output voltage, which allows fewer cells for a specified battery voltage.

Construction

Inside a lead-acid battery, the positive and negative electrodes consist of a group of plates welded to a connecting strap. The plates are immersed in the electrolyte, consisting of eight parts of water to three parts of concentrated sulfuric acid. Each plate is a grid or framework, made of a lead-antimony alloy. This construction enables the active material, which is lead oxide, to be pasted into the grid. In manufacture of the cell, a forming charge produces the positive and negative electrodes. In the forming process, the active material in the positive plate is changed to lead peroxide (PbO_2). The negative electrode is spongy lead (Pb).

Automobile batteries are usually shipped dry from the manufacturer. The electrolyte is put in at installation, and then the battery is charged to form the plates. With maintenance-free batteries, little or no water need be added in normal service. Some types are sealed, except for a pressure vent, without provision for adding water.

Current Ratings

Lead-acid batteries are generally rated in terms of the amount of discharge current they can supply for a specified period of time. The output voltage should be maintained above a minimum level, which is 1.5 to 1.8 V per cell. A common rating is ampere-hours (A · h) based on a specific discharge time, which is often 8 h. Typical A · h ratings for automobile batteries are 100 to 300 A · h.

As an example, a 200-A · h battery can supply a load current of $^{200}/_8$ or 25 A, based on an 8-h discharge. The battery can supply less current for a longer time or more current for a shorter time. Automobile batteries may be rated in "cold cranking amps" (CCAs), which is related to the job of starting the engine. The CCA rating specifies the amount of current, in amperes, the battery can deliver at 0°F for 30 seconds while maintaining an output voltage of 7.2 V for a 12-V battery. The higher the CCA rating, the greater the starting power of the battery.

Note that the ampere-hour unit specifies coulombs of charge. For instance, 200 A · h corresponds to 200 A × 3600 s (1 h = 3600 s). This equals 720,000 A · s, or coulombs. One ampere-second is equal to one coulomb. Then the charge equals 720,000 or 7.2×10^5 C. To put this much charge back into the battery would require 20 h with a charging current of 10 A.

The ratings for lead-acid batteries are given for a temperature range of 77 to 80°F. Higher temperatures increase the chemical reaction, but operation above 110°F shortens the battery life.

Low temperatures reduce the current capacity and voltage output. The amperehour capacity is reduced approximately 0.75% for each decrease of 1°F below the normal temperature rating. At 0°F, the available output is only 40% of the amperehour battery rating. In cold weather, therefore, it is very important to have an automobile battery up to full charge. In addition, the electrolyte freezes more easily when diluted by water in the discharged condition.

Specific Gravity

The state of discharge for a lead-acid cell is generally checked by measuring the specific gravity of the electrolyte. Specific gravity is a ratio comparing the weight of a substance with the weight of water. For instance, concentrated sulfuric acid is 1.835 times as heavy as water for the same volume. Therefore, its specific gravity equals 1.835. The specific gravity of water is 1, since it is the reference.

In a fully charged automotive cell, the mixture of sulfuric acid and water results in a specific gravity of 1.280 at room temperatures of 70 to 80°F. As the cell discharges, more water is formed, lowering the specific gravity. When the specific gravity is below about 1.145, the cell is considered completely discharged.

Specific-gravity readings are taken with a battery hydrometer, such as the one in Fig. 10-8. With this type of hydrometer, the state of charge of a cell within the battery is indicated by the number of floating disks. For example, one floating disk indicates the cell is at 25% of full charge. Two floating disks indicate the cell is at 50% of full charge. Similarly, three floating disks indicate 75% of full charge, whereas four floating disks indicate the cell is at 100% of full charge. The number of floating disks is directly correlated with the value of the specific gravity. As the specific gravity increases, more disks will float. Note that all cells within the battery must be tested for full charge.

The importance of the specific gravity can be seen from the fact that the opencircuit voltage of the lead-acid cell is approximately equal to

$$V = \text{specific gravity} + 0.84$$

Figure 10-8 Hydrometer to check specific gravity of lead-acid battery.

For the specific gravity of 1.280, the voltage is 1.280 + 0.84 = 2.12 V, as an example. These values are for a fully charged battery.

Charging the Lead-Acid Battery

The requirements are illustrated in Fig. 10-9. An external dc voltage source is necessary to produce current in one direction. Also, the charging voltage must be more than the battery emf. Approximately 2.5 V per cell is enough to overcome the cell emf so that the charging voltage can produce current opposite to the direction of the discharge current.

Note that the reversal of current is obtained by connecting the battery V_B and charging source V_G with + to + and

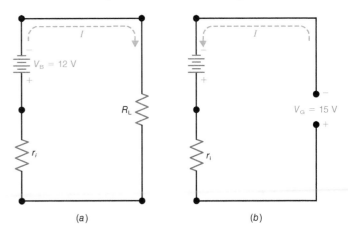

Figure 10-9 Reversed directions for charge and discharge currents of a battery. The r_i is internal resistance. (a) The V_B of the battery discharges to supply the load current for R_L. (b) The battery is the load resistance for V_G, which is an external source of charging voltage.

Figure 10-10   Charger for auto batteries.

− to −, as shown in Fig. 10-9*b*. The charging current is reversed because the battery effectively becomes a load resistance for V_G when it is higher than V_B. In this example, the net voltage available to produce a charging current is $15 - 12 = 3$ V.

A commercial charger for automobile batteries is shown in Fig. 10-10. This unit can also be used to test batteries and jump-start cars. The charger is essentially a dc power supply, rectifying input from an ac power line to provide dc output for charging batteries.

Float charging refers to a method in which the charger and the battery are always connected to each other to supply current to the load. In Fig. 10-11, the charger provides current for the load and the current necessary to keep the battery fully charged. The battery here is an auxiliary source for dc power.

It may be of interest to note that an automobile battery is in a floating-charge circuit. The battery charger is an ac generator or alternator with rectifier diodes, driven by a belt from the engine. When you start the car, the battery supplies the cranking power. Once the engine is running, the alternator charges the battery. It is not necessary for the car to be moving. A voltage regulator is used in this system to maintain the output at approximately 13 to 15 V.

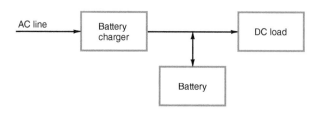

Figure 10-11   Circuit for battery in float-charge application.

10.5 Additional Types of Secondary Cells

A secondary cell is a storage cell that can be recharged by reversing the internal chemical reaction. A primary cell must be discarded after it has been completely discharged. The lead-acid cell is the most common type of storage cell. However, other types of secondary cells are available. Some of these are described next.

Nickel–Cadmium (NiCd) Cell

The nickel-cadmium (NiCd) cell is popular because of its ability to deliver high current and to be cycled many times for recharging. Also, the cell can be stored for a long time, even when discharged, without any damage. The NiCd cell is available in both sealed and nonsealed designs, but the sealed construction shown in Fig. 10-12 is common. Nominal output voltage is 1.2 V per cell. Applications include portable power tools, alarm systems, and portable radio or video equipment.

The NiCd cell is a true storage cell with a reversible chemical reaction for recharging that can be cycled up to 1000 times. Maximum charging current is equal to the 10-h discharge rate. Note that a new NiCd battery may need charging before use. A disadvantage of NiCd batteries is that they can develop a memory whereby they won't accept full charge if they are routinely discharged to the same level and then charged. For this reason, it is a good idea to discharge them to different levels and occassionally to discharge them completely before recharging to erase the memory.

Figure 10-12   Examples of nickel-cadmium cells. The output voltage for each is 1.2 V.

Nickel-Metal-Hydride (NiMH) Cell

Nickel-metal-hydride (NiMH) cells are currently finding widespread application in those high-end portable electric and electronic products and hybrid/electric vehicles where battery performance parameters, notably run-time, are of major concern. NiMH cells are an extension of the proven, sealed, NiCd cells discussed previously. A NiMH cell has about 40% more capacity than a comparably sized NiCd cell, however. In other words, for a given weight and volume, a NiMH cell has a higher A·h rating than a NiCd cell.

With the exception of the negative electrode, NiMH cells use the same general types of components as a sealed NiCd cell. As a result, the nominal output voltage of a NiMH cell is 1.2 V, the same as the NiCd cell. In addition to having higher A·h ratings compared to NiCd cells, NiMH cells also do not suffer nearly as much from the memory effect. As a result of the advantages offered by the NiMH cell, they are finding widespread use in the power-tool market where additional operating time and higher power are of major importance.

The disadvantage of NiMH cells versus NiCd cells is their higher cost. Also, NiMH cells self-discharge much more rapidly during storage or nonuse than NiCd cells. Furthermore, NiMH cells cannot be cycled as many times as their NiCd counterparts. NiMH cells are continually being improved, and it is foreseeable that they will overcome, at least to a large degree, the disadvantages listed here.

Lithium-Ion (Li-Ion) Cell

Lithium-ion cells (and batteries) are extremely popular and have found widespread use in today's consumer electronics market. They are commonly used in laptop computers, cell phones, handheld radios, and iPods, to name a few of the more common applications. The electrodes of a lithium-ion cell are made of lightweight lithium and carbon. Since lithium is a highly reactive element, the energy density of lithium-ion batteries is very high. Their high energy density makes lithium-ion batteries significantly lighter than other rechargeable batteries of the same size. The nominal open-circuit output voltage of a single lithium-ion cell is approximately 3.7 V.

Unlike NiMH and NiCd cells, lithium-ion cells do not suffer from the memory effect. They also have a very low self-discharge rate of approximately 5% per month compared with approximately 30% or more per month with NiMH cells and 10% or more per month with NiCd cells. In addition, lithium-ion cells can handle several hundred charge-discharge cycles in their lifetime.

Lithium-ion batteries do have a few disadvantages, however. For one thing, they are more expensive than similar-capacity NiMH or NiCd batteries because the manufacturing process is much more complex. Also, they begin degrading the instant they leave the factory and last only about 2 or 3 years whether they are used or not. Another disadvantage is that higher temperatures cause them to degrade much more rapidly than normal.

Lithium-ion batteries are not as durable as NiMH and NiCd batteries either. In fact, they can be extremely dangerous if mistreated. For example, they may explode if they are overheated or charged to an excessively high voltage. Furthermore, they may be irreversibly damaged if discharged below a certain level. To avoid damaging a lithium-ion battery, a special circuit monitors the voltage output from the battery and shuts it down when it is discharged below a certain threshold level (typically 3 V) or charged above a certain limit (typically 4.2 V). This special circuitry makes lithium-ion batteries even more expensive than they already are.

Nickel-Iron (Edison) Cell

Developed by Thomas Edison, the nickel-iron cell was once used extensively in industrial truck and railway applications. However, it has been replaced almost entirely by the lead-acid battery. New methods of construction with less weight, though, are making this cell a possible alternative in some applications.

The Edison cell has a positive plate of nickel oxide, a negative plate of iron, and an electrolyte of potassium hydroxide in water with a small amount of lithium hydroxide added. The chemical reaction is reversible for recharging. Nominal output is 1.2 V per cell.

Nickel-Zinc Cell

The nickel-zinc cell has been used in limited railway applications. There has been renewed interest in it for use in electric cars because of its high energy density. However, one drawback is its limited cycle life for recharging. Nominal output is 1.6 V per cell.

Fuel Cells

A fuel cell is a chemical source of direct current (dc) like a battery but does not contain the chemicals. Fuel cells develop the voltage across two electrodes using hydrogen and oxygen gases. The output voltage is low in the 0.3- to 1.2-V range, so many are usually connected together to get higher voltage. The current capability is a function of the size of the electrodes and the volume of oxygen and hydrogen supplied.

Figure 10-13 shows the basic construction of a fuel cell. The two electrodes are called the anode (+) and the cathode (−). Hydrogen is supplied to the anode, and oxygen or just plain air is supplied to the cathode. Between the anode and cathode is a thin membrane that contains the electrolyte. As the oxygen and hydrogen are supplied, a chemical reaction takes place, freeing up electrons that can pass through the external load. By-products of the chemical reaction are the production of water and heat.

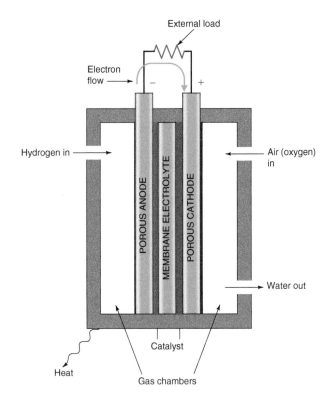

Figure 10-13 Construction of a fuel cell.

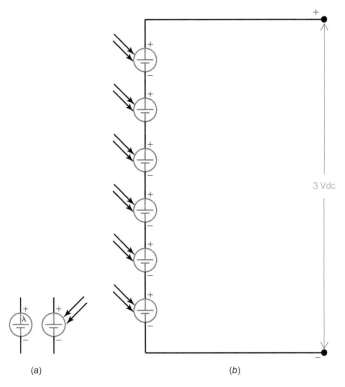

Figure 10-14 Solar cells. (*a*) Schematic symbols. Arrows indicate light. λ is lamda, the Greek symbol for light and wavelength. (*b*) Cells are connected in series to produce more voltage.

As long as the gases are supplied to the cell, voltage is produced. It does not wear out and never needs recharging.

Fuel cells are generally impractical as they are large and require storage for the gases. This often means large tanks. More practical fuel cells have been created that use less expensive gases. For example, a practical fuel cell can extract the oxygen from air which contains about 21% oxygen. The hydrogen can be derived from a methanol fuel that is less expensive. Methanol is basically just wood alcohol. Natural gas may also be used.

Fuel cells are still impractical for most applications because of their size and cost. However they have been widely used in spacecraft like the NASA shuttle and in the space station. They are also used as backup power for hospitals, military installations, cell phone base stations, and radio/TV broadcast stations. Large units can produce many hundreds of kilowatts of power. Their predicted use in automobiles has never developed.

Solar Cells

A solar cell converts the sun's light energy directly into electric energy. The cells are made of semiconductor materials, which generate voltage output with light input. Silicon, with an output of 0.5 V per cell, is mainly used now. Research is continuing, however, on other materials, such as cadmium sulfide and gallium arsenide, that might provide more output. In practice, the cells are arranged in modules that are assembled into a large solar array for the required power. Figure 10-14*a* shows the schematic symbols used to

represent a solar cell. Figure 10-14*b* shows a series and parallel connection of cells to provide more voltage and current.

In most applications, the solar cells are used in combination with a lead-acid cell that store the electricity for use when the light disappears. When there is sunlight, the solar cells charge the battery and supply power to the load. When there is no light, the battery supplies the required power. Solar power is expensive compared to other power sources but is clean and useful in various applications. Along with wind power, solar sources are gradually being adopted as major power sources to replace conventional ones. As prices decline popularity will increase.

10.6 Series-Connected and Parallel-Connected Cells

An applied voltage higher than the voltage of one cell can be obtained by connecting cells in series. The total voltage available across the battery of cells is equal to the sum of the individual values for each cell. Parallel cells have the same voltage as one cell but have more current capacity. The combination of cells is called a *battery*.

Series Connections

Figure 10-15 shows series-aiding connections for three dry cells. Here the three 1.5-V cells in series provide a total

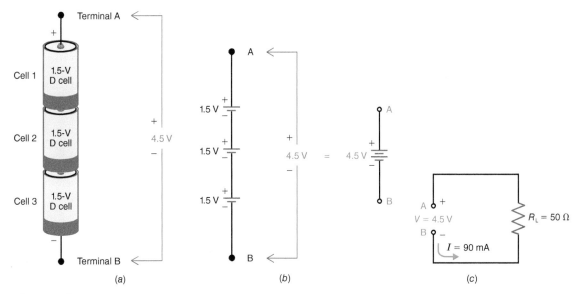

(a)   *(b)*   *(c)*

Figure 10-15   Cells connected in series for higher voltage. Current rating is the same as for one cell. (*a*) Wiring. (*b*) Schematic symbol for battery with three series cells. (*c*) Battery connected to load resistance R_L.

battery voltage of 4.5 V. Notice that the two end terminals, A and B, are left open to serve as the plus and minus terminals of the battery. These terminals are used to connect the battery to the load circuit, as shown in Fig. 10-15c.

In the lead-acid battery in Fig. 10-2, short, heavy metal straps connect the cells in series. The current capacity of a battery with cells in series is the same as that for one cell because the same current flows through all series cells.

Parallel Connections

For more current capacity, the battery has cells in parallel, as shown in Fig. 10-16. All positive terminals are strapped together, as are all the negative terminals. Any point on the positive side can be the plus terminal of the battery, and any point on the negative side can be the negative terminal.

The parallel connection is equivalent to increasing the size of the electrodes and electrolyte, which increases the current capacity. The voltage output of the battery, however, is the same as that for one cell.

Identical cells in parallel supply equal parts of the load current. For example, with three identical parallel cells producing a load current of 300 mA, each cell has a drain of 100 mA. Bad cells should not be connected in parallel with good cells, however, since the cells in good condition will supply more current, which may overload the good cells. In addition, a cell with lower output voltage will act as a load resistance, draining excessive current from the cells that have higher output voltage.

Series-Parallel Connections

To provide a higher output voltage and more current capacity, cells can be connected in series-parallel combinations.

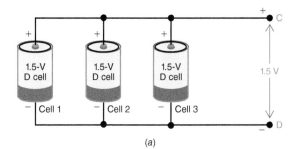

(a)

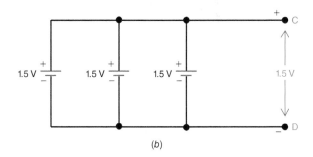

(b)

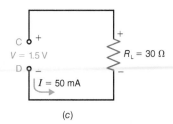

(c)

Figure 10-16   Cells connected in parallel for higher current rating. (*a*) Wiring. (*b*) Schematic symbol for battery with three parallel cells. (*c*) Battery connected to load resistance R_L.

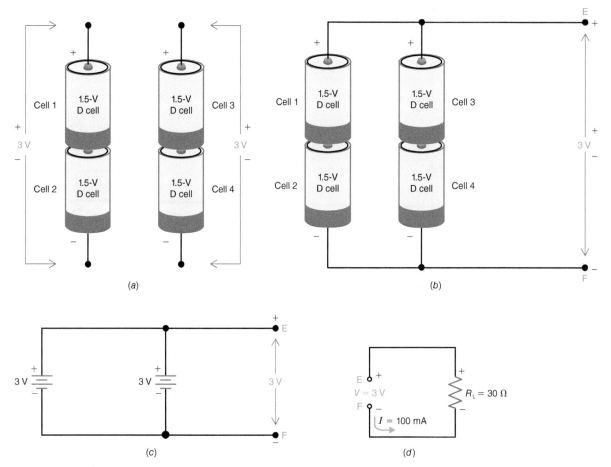

Figure 10-17 Cells connected in series-parallel combinations. (*a*) Wiring two 3-V strings, each with two 1.5-V cells in series. (*b*) Wiring two 3-V strings in parallel. (*c*) Schematic symbol for the battery in (*b*) with output of 3 V. (*d*) Equivalent battery connected to load resistance R_L.

Figure 10-17 shows four D cells connected in series-parallel to form a battery that has a 3-V output with a current capacity of ½ A. Two of the 1.5-V cells in series provide 3 V total output voltage. This series string has a current capacity of ¼ A, however, assuming this current rating for one cell.

To double the current capacity, another string is connected in parallel. The two strings in parallel have the same 3-V output as one string, but with a current capacity of ½ A instead of the ¼ A for one string.

10.7 Current Drain Depends on Load Resistance

It is important to note that the current rating of batteries, or any voltage source, is only a guide to typical values permissible for normal service life. The actual amount of current produced when the battery is connected to a load resistance is equal to $I = V/R$ by Ohm's law.

Figure 10-18 illustrates three different cases of using the applied voltage of 1.5 V from a dry cell. In Fig. 10-18*a*, the load resistance R_1 is 7.5 Ω. Then I is 1.5/7.5 = ⅕ A or 200 mA.

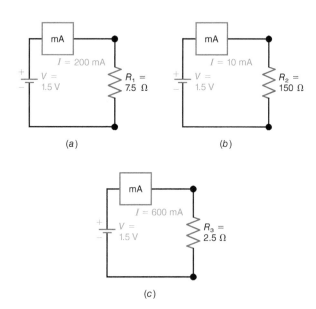

Figure 10-18 An example of how current drain from a battery used as a voltage source depends on the R of the load resistance. Different values of I are shown for the same V of 1.5 V. (*a*) V/R_1 equals I of 200 mA. (*b*) V/R_2 equals I of 10 mA. (*c*) V/R_3 equals I of 600 mA.

A No. 6 carbon-zinc cell with a 1500 mA·h rating could supply this load of 200 mA continuously for about 7.5 h at a temperature of 70°F before dropping to an end voltage of 1.2 V. If an end voltage of 1.0 V could be used, the same load would be served for a longer period of time.

In Fig. 10-18b, a larger load resistance R_2 is used. The value of 150 Ω limits the current to $1.5/150 = 0.01$ A or 10 mA. Again using the No. 6 carbon-zinc cell at 70°F, the load could be served continuously for 150 h with an end voltage of 1.2 V. The two principles here are

1. The cell delivers less current with higher resistance in the load circuit.

2. The cell can deliver a smaller load current for a longer time.

In Fig. 10-18c, the load resistance R_3 is reduced to 2.5 Ω. Then I is $1.5/2.5 = 0.6$ A or 600 mA. The No. 6 cell could serve this load continuously for only 2.5 h for an end voltage of 1.2 V. The cell could deliver even more load current, but for a shorter time. The relationship between current and time is not linear. For any one example, though, the amount of current is determined by the circuit, not by the current rating of the battery.

10.8 Internal Resistance of a Generator

Any source that produces voltage output continuously is a generator. It may be a cell separating charges by chemical action or a rotary generator converting motion and magnetism into voltage output, for common examples. In any case, all generators have internal resistance, which is labeled r_i in Fig. 10-19. This internal resistance is generally known as the equivalent series resistance (ESR).

The internal resistance, r_i, is important when a generator supplies load current because its internal voltage drop, Ir_i, subtracts from the generated emf, resulting in lower voltage across the output terminals. Physically, r_i may be the resistance of the wire in a rotary generator, or r_i is the resistance of the electrolyte between electrodes in a chemical cell. More generally, the internal resistance r_i is the opposition to load current inside the generator.

Since any current in the generator must flow through the internal resistance, r_i is in series with the generated voltage, as shown in Fig. 10-19c. It may be of interest to note that, with just one load resistance connected across a generator, they are in series with each other because R_L is in series with r_i.

If there is a short circuit across the generator, its r_i prevents the current from becoming infinitely high. As an example, if a 1.5-V cell is temporarily short-circuited, the short-circuit current I_{sc} could be about 15 A. Then r_i is V/I_{sc}, which equals 1.5/15, or 0.1 Ω for the internal resistance. These are typical values for a carbon-zinc D-size cell. (The value of r_i would be lower for a D-size alkaline cell.)

Why Terminal Voltage Drops with More Load Current

Figure 10-20 illustrates how the output of a 100-V source can drop to 90 V because of the internal 10-V drop across r_i. In Fig. 10-20a, the voltage across the output terminals is equal to the 100 V of V_G because there is no load current in an open circuit. With no current, the voltage drop across r_i is zero. Then the full generated voltage is available across the output terminals. This value is the generated emf, *open-circuit voltage,* or *no-load voltage.*

We cannot connect the test leads inside the source to measure V_G. However, measuring this no-load voltage without any load current provides a method of determining the internally generated emf. We can assume that the voltmeter draws practically no current because of its very high resistance.

In Fig. 10-20b with a load, however, current of 0.1 A flows to produce a drop of 10 V across the 100 Ω of r_i. Note that R_T is $900 + 100 = 1000$ Ω. Then I_L equals 100/1000, which is 0.1 A.

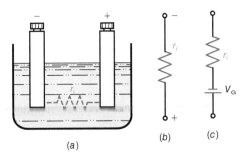

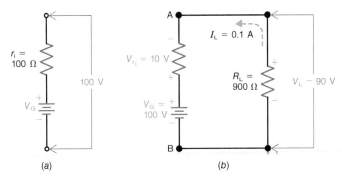

Figure 10-19 Internal resistance r_i is in series with the generator voltage V_G. (a) Physical arrangement for a voltage cell. (b) Schematic symbol for r_i. (c) Equivalent circuit of r_i in series with V_G.

Figure 10-20 Example of how an internal voltage drop decreases voltage at the output terminal of the generator. (a) Open-circuit voltage output equals V_G of 100 V because there is no load current. (b) Terminal voltage V_L between points A and B is reduced to 90 V because of 10-V drop across 100-Ω r_i with 0.1-A I_L.

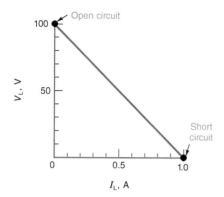

Figure 10-21 How terminal voltage V_L drops with more load current.

As a result, the voltage output V_L equals $100 - 10 = 90$ V. This terminal voltage or load voltage is available across the output terminals when the generator is in a closed circuit with load current. The 10-V internal drop is subtracted from V_G because they are series-opposing voltages.

The graph in Fig. 10-21 shows how the terminal voltage V_L drops with increasing load current I_L.

The lower the internal resistance of a generator, the better it is in producing full output voltage when supplying current for a load. For example, the very low r_i, about 0.01 Ω, for a 12-V lead-acid battery, is the reason it can supply high values of load current and maintain its output voltage.

For the opposite case, a higher r_i means that the terminal voltage of a generator is much less with load current. As an example, an old dry battery with r_i of 500 Ω would appear normal when measured by a voltmeter but be useless because of low voltage when normal load current flows in an actual circuit.

How to Measure r_i

The internal resistance of any generator can be measured indirectly by determining how much the output voltage drops for a specified amount of load current. The difference between the no-load voltage and the load voltage is the amount of internal voltage drop $I_L r_i$. Dividing by I_L gives the value of r_i. As a formula,

$$r_i = \frac{V_{NL} - V_L}{I_L} \qquad \textbf{(10-1)}$$

A convenient technique for measuring r_i is to use a variable load resistance R_L. Vary R_L until the load voltage is one-half the no-load voltage. This value of R_L is also the value of r_i, since they must be equal to divide the generator voltage equally. For the same 100-V generator with the 10-Ω r_i used in Example 10-1, if a 10-Ω R_L were used, the load voltage would be 50 V, equal to one-half the no-load voltage.

You can solve this circuit by Ohm's law to see that I_L is 5 A with 20 Ω for the combined R_T. Then the two voltage drops of 50 V each add to equal the 100 V of the generator.

EXAMPLE 10-1

Calculate r_i if the output of a generator drops from 100 V with zero load current to 80 V when $I_L = 2$ A.

Answer:

$$r_i = \frac{100 - 80}{2}$$
$$= \frac{20}{2}$$
$$r_i = 10\ \Omega.$$

10.9 Constant-Voltage and Constant-Current Sources

A generator with very low internal resistance is considered a constant-voltage source. Then the output voltage remains essentially the same when the load current changes. This idea is illustrated in Fig. 10-22a for a 6-V lead-acid battery with an r_i of 0.005 Ω. If the load current varies over the wide range of 1 to 100 A, the internal Ir_i drop across 0.005 Ω is less than 0.5 V for any of these values.

Constant-Current Generator

A constant-current generator has very high resistance, compared with the external load resistance, resulting in constant current, although the output voltage varies.

The constant-current generator shown in Fig. 10-23 has such high resistance, with an r_i of 0.9 MΩ, that it is the main factor determining how much current can be produced by V_G. Here R_L varies in a 3:1 range from 50 to 150 kΩ. Since the current is determined by the total resistance of R_L and r_i in series, however, I is essentially constant at 1.05 to 0.95 mA, or approximately 1 mA. This relatively constant I is shown by the graph in Fig. 10–23b.

Note that the terminal voltage V_L varies in approximately the same 3:1 range as R_L. Also, the output voltage is much

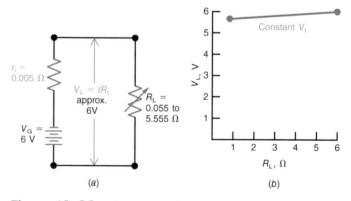

Figure 10-22 Constant-voltage generator with low r_i. The V_L stays approximately the same 6 V as I varies with R_L. (a) Circuit. (b) Graph for V_L.

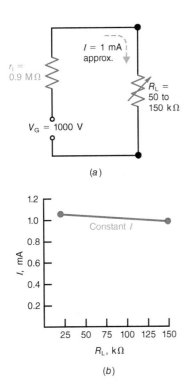

(a)

(b)

Figure 10-23 Constant-current generator with high r_i. The I stays approximately the same 1 mA as V_L varies with R_L. (a) Circuit. (b) Graph for I.

less than the generator voltage because of the high internal resistance compared with R_L. This is a necessary condition, however, in a circuit with a constant-current generator.

A common practice is to insert a series resistance to keep the current constant, as shown in Fig. 10-24a. Resistance R_1

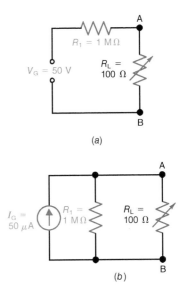

(a)

(b)

Figure 10-24 Voltage source in (a) equivalent to current source in (b) for load resistance R_L across terminals A and B.

must be very high compared with R_L. In this example, I_L is 50 μA with 50 V applied, and R_T is practically equal to the 1 MΩ of R_1. The value of R_L can vary over a range as great as 10:1 without changing R_T or I appreciably.

A circuit with an equivalent constant-current source is shown in Fig. 10-24b. Note the arrow symbol for a current source. As far as R_L is concerned, its terminals A and B can be considered as receiving either 50 V in series with 1 MΩ or 50 μA in a shunt with 1 MΩ.

CHAPTER 10 SYSTEM SIDEBAR

Battery Charging

Since secondary cells and batteries can be recharged, they are widely used in all types of equipment. Most cars and trucks still use lead-acid batteries, as do golf carts and similar vehicles. Hybrid and electric cars use nickel-metal-hydride (NiMH) and lithium-ion batteries. All types of portable power tools use NiMH, nickel-cadmium (NiCd) or lithium batteries. The highest volume of rechargeable batteries are used in the billions of cell phones, portable music players, personal navigation devices, laptop and tablet computers, and other mobile devices. Most of these use a form of lithium-ion cell or battery. How many battery chargers do you personally own? If you said several, you are typical of the modern electronics user. You should know a bit more

about chargers, because they are so ubiquitous. They are a good example of a small system.

Charging Basics

The simplest and most fundamental way to charge a secondary cell or battery is simply to apply to it a dc voltage somewhat higher than the final open-circuit voltage of the battery. For a lead-acid battery this is 12.6 V or for a lithium battery 4.2 V. The dc source can be another battery, but more usually it is an ac-to-dc power supply, such as the widely use wall chargers we all use.

Refer to Fig. S10-1. Note the polarity of the external source is such that it will reverse the current flow in the

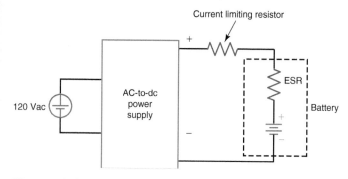

Figure S10-1   The basic battery charging circuit.

battery. The external voltage is higher than the voltage of the discharged battery, so the two voltages will oppose one another. The larger voltage ensures that the current is reversed, so recharging takes place. When a battery is discharged, its equivalent series resistance (ESR) is much higher than when the battery is fully charged. The discharged resistance depends on the battery chemistry, age, temperature, and other factors. Nevertheless, the ESR is still low and will cause a very high current to flow as the battery is charging. During charging, ESR will decrease, but the battery voltage will increase, offering more opposition to the external voltage and thereby gradually decreasing the current. The external resistor will limit the current to some maximum value as specified by the battery manufacturer to prevent damage and overheating. The dc power supply must be able to supply this current value.

The reverse current in the battery will reverse the chemical process and cause the battery voltage to rise. As it rises over time, it will offer more and more opposition to the external charging voltage. As a result the charging current will gradually decrease until the battery is fully charged and nearly equal to the external charging voltage.

While such a simple charging circuit will theoretically work for any battery, it is rarely used. Modern batteries are complex, sensitive, and dangerous. It is easy to damage a battery if it not charged using a formula or procedure recommended by the battery manufacturer. The battery could overheat and even catch fire or explode under the wrong conditions. Leaving the charging source connected for too long can also ruin the battery. The chemistry can be permanently altered under the wrong conditions, making it impossible to recharge it again. With the high cost of batteries and the danger they pose, virtually all batteries are recharged under very specific conditions as determined by the battery manufacturer. This is accomplished by an electronic charging circuit designed for the battery. At the heart of all modern chargers is an integrated circuit (IC) and numerous discrete components, making it a small system.

A Representative Charging System

Figure S10-2 shows a typical charging circuit. The 120-Vac input is applied to an ac-to-dc power supply. It may be a linear type that uses a transformer to step the voltage down into the range required by the battery. One or more diodes are then used to rectify the ac into pulsating dc. A large capacitor used as a filter smoothes the pulses into a more constant dc voltage. This dc voltage is applied to the charging IC.

Alternatively, the ac-to-dc power supply may be of the switching type that uses a diode rectifier and filter capacitor to produce dc from the high-voltage ac and a switching transistor that chops the dc into pulses that are averaged in a filter at a lower level. This lower-voltage dc is applied to the charging IC.

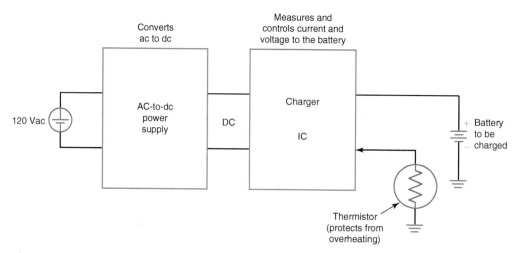

Figure S10-2   The charger IC implements a specific procedure of voltage and current control as specified by the battery type and manufacturer.

The charging IC is a complex electronic circuit designed to provide a constant current and then a constant voltage to the battery over specific period of time. The circuit implements an algorithm or procedure defined by the battery manufacturer. It varies depending on the battery type. For a common lithium-ion battery, the procedure usually begins by applying a small current to start and then steps it up gradually to some constant value until the voltage rises to a desired level. The charging current is usually based on the capacity (C) of the battery in ampere-hours (A·h) or milliampere hours (mA·h). For example, a smartphone battery may have a capacity of 1800 mA·h. An initial charging current may only be $0.1C$ or 180 mA. It is then stepped up according to some plan over time.

When the battery voltage rises to the recommended level, the constant current is turned off and a constant voltage is then applied to continue the charging process. All during this time, the battery current is being monitored. Should the recommended levels be exceeded, the IC adjusts automatically to maintain the suggested current/voltage limits. Some chargers even monitor battery temperature, which is typically sensed by a thermistor mounted next to the battery. Should temperature limits be exceeded, the IC automatically cuts off the power to the battery, preventing damage.

Figure S10-3 shows a typical charging cycle for a lithium-ion battery. Note the discharged battery voltage is initially low in the 2.0-V range or less. An initial current of $0.02C$ is applied for a short time and then stepped up to $0.1C$ for another period. Then a constant current of $1C$ is applied until the battery voltage reaches the desired final voltage of 4.2 V. Note the rise of the battery voltage over this constant current phase of the charging. When the desired end voltage is reached, the current is decreased over time but the charger output voltage remains at 4.2 V. The amount of charging time is usually several hours. When charging has been completed as determined by a specific time period and battery condition, the charger cuts off.

Other Charging Systems

The charging system in an automobile uses a mechanical generator driven by the engine. It is called an *alternator* because it generates ac. This is rectified into dc by diodes and then used to operate a charging circuit. It usually supplies 13 to 15 V to the lead-acid battery with current control.

Solar-powered chargers are also popular for boats and portable equipment. A solar panel producing from 14 to 20 V is applied to a charge controller that contains the charging circuit. Most of these systems use lead-acid batteries.

Hybrid/electric vehicles use very high-voltage lithium-ion or NiMH batteries. Voltages are in the 300- to 400-V range. Special chargers operated by the small gasoline engine keep the battery charged during operation. Electric-only vehicles rely on an external charger located at your home or a public charging station. The at-home systems use the 240-Vac line and a special charging circuit that connects by cable and connector to the vehicle.

There are several important points to remember about battery charging.

- The charging IC may be inside the device with the battery, as in a cell phone or portable music player. The ac-to-dc power supply is external and usually the "wall wart" or cube you are familiar with. Some products may have the charging IC inside the charger itself.

- Each charger is designed for a specific battery type. You cannot and must not use the charger for anything other than charging batteries of this particular size and type. For example, you cannot use a lithium-ion charger on a NiMH battery or any other type. Severe damage to the battery or charger or both will most likely result.

- Chargers do not have an ON-OFF switch and constantly draw current from the ac line. The current will be high initially but will drop off to a trickle once the battery is fully charged. Nevertheless, since it constantly draws current, the charger should be unplugged to save energy. It has been estimated that because so many people leave their multiple chargers constantly plugged in, millions of kilowatts of power are wasted annually.

- All secondary batteries have a finite life and must eventually be replaced. They can only be recharged so many times before the chemical process or physical state of the materials is permanently ruined. Longest life is achieved by following the strict recharging process. The actual battery life will depend more on the usage time and pattern, so the life of a battery could be years or considerably less.

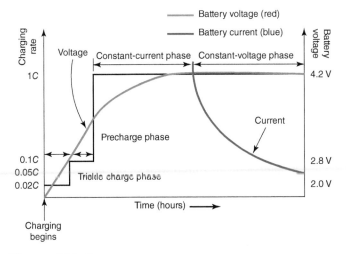

Figure S10-3 The charging profile of a lithium-ion battery.

1. Which of the following cells is not a primary cell?
   a. Carbon-zinc.
   b. Alkaline.
   c. Zinc chloride.
   d. Lead-acid.

2. The dc output voltage of a C-size alkaline cell is
   a. 1.2 V.
   b. 1.5 V.
   c. 2.1 V.
   d. about 3 V.

3. Which of the following cells is a secondary cell?
   a. Silver oxide.
   b. Lead-acid.
   c. Nickel-cadmium.
   d. Both *b* and *c*.

4. What happens to the internal resistance, r_i, of a voltaic cell as the cell deteriorates?
   a. It increases.
   b. It decreases.
   c. It stays the same.
   d. It usually disappears.

5. The dc output voltage of a lead-acid cell is
   a. 1.35 V.
   b. 1.5 V.
   c. 2.1 V.
   d. about 12 V.

6. Cells are connected in series to
   a. increase the current capacity.
   b. increase the voltage output.
   c. decrease the voltage output.
   d. decrease the internal resistance.

7. Cells are connected in parallel to
   a. increase the current capacity.
   b. increase the voltage output.
   c. decrease the voltage output.
   d. decrease the current capacity.

8. Five D-size alkaline cells in series have a combined voltage of
   a. 1.5 V.
   b. 5.0 V.
   c. 7.5 V.
   d. 11.0 V.

9. The main difference between a primary cell and a secondary cell is that
   a. a primary cell can be recharged and a secondary cell cannot.
   b. a secondary cell can be recharged and a primary cell cannot.
   c. a primary cell has an unlimited shelf life and a secondary cell does not.
   d. primary cells produce a dc voltage and secondary cells produce an ac voltage.

10. A constant-voltage source
    a. has very high internal resistance.
    b. supplies constant-current to any load resistance.
    c. has very low internal resistance.
    d. none of the above.

11. A constant-current source
    a. has very low internal resistance.
    b. supplies constant current to a wide range of load resistances.
    c. has very high internal resistance.
    d. both *b* and *c*.

12. The output voltage of a battery drops from 6.0 V with no load to 5.4 V with a load current of 50 mA. How much is the internal resistance, r_i?
    a. 12 Ω.
    b. 108 Ω.
    c. 120 Ω.
    d. It cannot be determined.

13. The output of a single solar cell is approximately
    a. 0.5 V.
    b. 1.2 V.
    c. 1.5 V.
    d. 3.0 V.

14. Another name for the internal resistance of a battery or cell is
    a. generator impedance.
    b. output impedance.
    c. equivalent series resistance.
    d. electrolyte resistance.

15. The internal resistance of a battery
    a. cannot be measured with an ohmmeter.
    b. can be measured with an ohmmeter.
    c. can be measured indirectly by determining how much the output voltage drops for a given load current.
    d. both *a* and *c*.

SECTION 10.6 Series-Connected and Parallel-Connected Cells

In Probs. 10-1 to 10-5, assume that each individual cell is identical and that the current capacity for each cell is not being exceeded for the load conditions presented.

10.1 In Fig. 10-25, solve for the load voltage, V_L, the load current, I_L, and the current supplied by each cell in the battery.

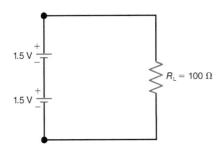

Figure 10-25

10.2 Repeat Prob. 10-1 for the circuit in Fig. 10-26.

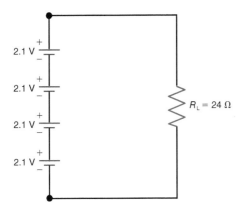

Figure 10-26

10.3 Repeat Prob. 10-1 for the circuit in Fig. 10-27.

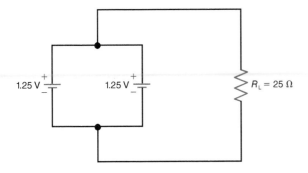

Figure 10-27

10.4 Repeat Prob. 10-1 for the circuit in Fig. 10-28.

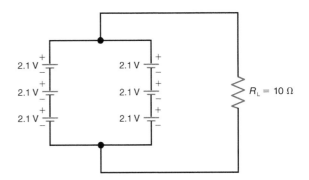

Figure 10-28

10.5 Repeat Prob. 10-1 for the circuit in Fig. 10-29.

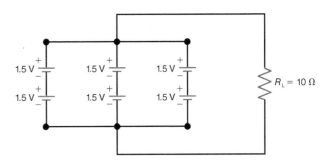

Figure 10-29

SECTION 10.8 Internal Resistance of a Generator

10.6 With no load, the output voltage of a battery is 9 V. If the output voltage drops to 8.5 V when supplying 50 mA of current to a load, how much is its internal resistance?

10.7 The output voltage of a battery drops from 6 V with no load to 5.2 V with a load current of 400 mA. Calculate the internal resistance, r_i.

10.8 A 9-V battery has an internal resistance of 0.6 Ω. How much current flows from the 9-V battery in the event of a short circuit?

10.9 A 1.5-V AA alkaline cell develops a terminal voltage of 1.35 V while delivering 25 mA to a load resistance. Calculate r_i.

10.10 Refer to Fig. 10-30. With S_1 in position 1, $V = 50$ V. With S_1 in position 2, $V = 37.5$ V. Calculate r_i.

10.11 A generator has an open-circuit voltage of 18 V. Its terminal voltage drops to 15 V when a 75-Ω load is connected. Calculate r_i.

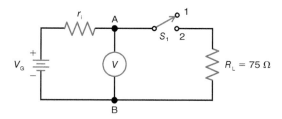

Figure 10-30

SECTION 10.9 Constant-Voltage and Constant-Current Sources

10.12 Refer to Fig. 10-31. If $r_i = 0.01\ \Omega$, calculate I_L and V_L for the following values of load resistance:
   a. $R_L = 1\ \Omega$.
   b. $R_L = 10\ \Omega$.
   c. $R_L = 100\ \Omega$.

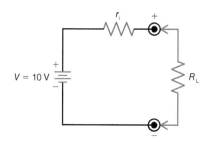

Figure 10-31

10.13 Refer to Fig. 10-31. If $r_i = 10\ M\Omega$, calculate I_L and V_L for the following values of load resistance:
   a. $R_L = 0\ \Omega$.
   b. $R_L = 100\ \Omega$.
   c. $R_L = 1\ k\Omega$.
   d. $R_L = 100\ k\Omega$.

10.14 Redraw the circuit in Fig. 10-32 using the symbol for a current source.

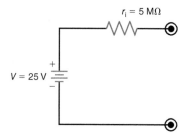

Figure 10-32

Chapter 11
Magnetism

Learning Outcomes

After studying this chapter, you should be able to:

> *Define* the terms and units of magnetic flux, flux density, magnetomotive force, and field intensity.

> *Define* the term *relative permeability*.

> *List* three classifications of magnetic materials.

> *Describe* the Hall effect.

> *Explain* the B-H magnetization curve.

> *Define* the terms *satiration* and *hysteresis*.

> *Determine* the magnetic polarity of a solenoid using the left-hand rule.

> *State* Lenz's and Faraday's laws.

> *Define* the term *saturation* as it relates to a magnetic core.

> *Explain* what is meant by magnetic hysteresis.

Magnetism is that invisible force field that is produced by certain materials and by current flowing in a conductor. This mysterious phenomenon is closely tied to electricity and electronics and is an essential part of your understanding of circuits and systems. In fact, you can't have electricity and electronics without magnetism. This chapter introduces you to the strange world of magnetic fields and how they affect us in many ways. You will learn how magnetic fields are expressed and measured. Even more significant you will learn about electromagnetism and electromagnetic induction. Learning about magnetism completes your understanding of electrical fundamentals.

Think about this. Without magnetism there would be no electric generators, no motors, no electric power distribution, no magnetic recording, or no radio waves. Electric components like inductors, transformers, relays, solenoids and microphones would not exist without magnetism. That is how important this chapter really is.

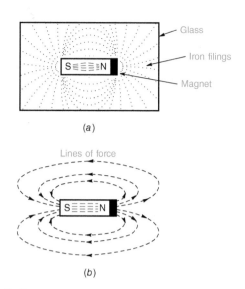

Figure 11-2 Magnetic field of force around a bar magnet. (*a*) Field outlined by iron filings. (*b*) Field indicated by lines of force.

11.1 The Magnetic Field

A magnetic field is produced by an object called a *magnet*. Magnets are made from metals like iron, iron alloys, and other materials generally known as ferromagnetic materials. They are formed into many different shapes depending on their use. A common form used primarily to explain the action of the magnet is a rectangular bar. The two ends of the bar are called the poles of the magnet. They are referred to as the north (N) and south (S) poles. Figure 11-1 shows two bar magnets, one a permanent magnet (PM) that is always producing a magnetic field and an electromagnet (EM) that produces a magnetic field when current flows through the coil of wire around it.

As shown in Figs. 11-1 and 11-2, the north and south poles of a magnet are the points of concentration of magnetic strength. The practical effects of this ferromagnetism result from the magnetic field of force between the two poles at opposite ends of the magnet. Although the magnetic field is invisible, evidence of its force can be seen when small

iron filings are sprinkled on a glass or paper sheet placed over a bar magnet (Fig. 11-2*a*). Each iron filing becomes a small bar magnet. If the sheet is tapped gently to overcome friction so that the filings can move, they become aligned by the magnetic field.

Many filings cling to the ends of the magnet, showing that the magnetic field is strongest at the poles. The field exists in all directions but decreases in strength with increasing distance from the poles of the magnet.

Field Lines

To visualize the magnetic field without iron filings, we show the field as lines of force, as in Fig. 11-2*b*. The direction of the lines outside the magnet shows the path a north pole would follow in the field, repelled away from the north pole of the magnet and attracted to its south pole. Although we cannot actually have a unit north pole by itself, the field can be explored by noting how the north pole on a small compass needle moves.

The magnet can be considered the generator of an external magnetic field, provided by the two opposite magnetic poles at the ends. This idea corresponds to the two opposite terminals on a battery as the source of an external electric field provided by opposite charges.

Magnetic field lines are unaffected by nonmagnetic materials such as air, vacuum, paper, glass, wood, or plastics. When these materials are placed in the magnetic field of a magnet, the field lines are the same as though the material were not there.

However, the magnetic field lines become concentrated when a magnetic substance such as iron is placed in the field. Inside the iron, the field lines are more dense, compared with the field in air.

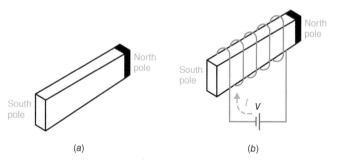

Figure 11-1 Poles of a magnet. (*a*) Permanent magnet (PM). (*b*) Electromagnet (EM) produced by current from a battery.

North and South Magnetic Poles

The earth itself is a huge natural magnet, with its greatest strength at the North and South Poles. Because of the earth's magnetic poles, if a small bar magnet is suspended so that it can turn easily, one end will always point north. This end of the bar magnet is defined as the *north-seeking pole,* as shown in Fig. 11-3a. The opposite end is the *south-seeking pole.* When polarity is indicated on a magnet, the north-seeking end is the north pole (N) and the opposite end is the south pole (S). It is important to note that the earth's geographic North Pole has south magnetic polarity and the geographic South Pole has north magnetic polarity. This is shown in Fig. 11-3b.

Similar to the force between electric charges is the force between magnetic poles causing attraction of opposite poles and repulsion between similar poles:

1. A north pole (N) and a south pole (S) tend to attract each other.

2. A north pole (N) tends to repel another north pole (N), and a south pole (S) tends to repel another south pole (S).

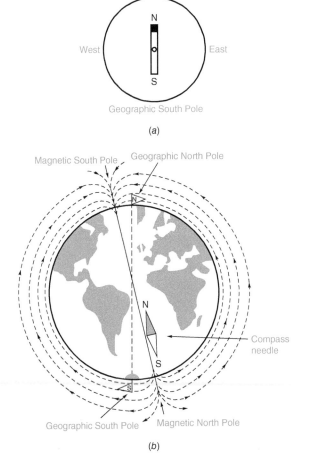

(a)

(b)

Figure 11-3 Definition of north and south poles of a bar magnet. (a) North pole on bar magnet points to geographic North Pole of the earth. (b) Earth's magnetic field.

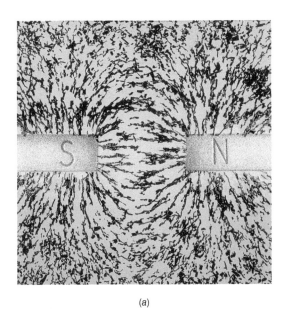

(a)

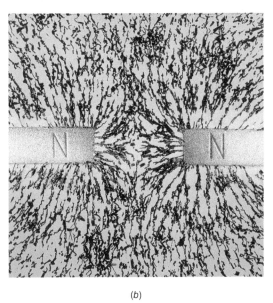

(b)

Figure 11-4 Magnetic field patterns produced by iron filings. (a) Field between opposite poles. The north and south poles could be reversed. (b) Field between similar poles. The two north poles could be south poles.

These forces are illustrated by the fields of iron filings between opposite poles in Fig. 11-4a and between similar poles in Fig. 11-4b.

11.2 Magnetic Units of Measurement

The entire group of magnetic field lines, which can be considered flowing outward from the north pole of a magnet, is called *magnetic flux.* Its symbol is the Greek letter ϕ (phi). A strong magnetic field has more lines of force and more flux than a weak magnetic field.

There are two basic ways of expressing and measuring magnetic fields: field strength and flux density.

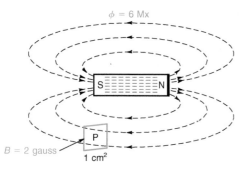

Figure 11-5 Total flux ϕ is six lines or 6 Mx. Flux density B at point P is two lines per square centimeter or 2 G.

The Maxwell

Field strength is expressed in lines of force. One maxwell (Mx) unit equals one magnetic field line. In Fig. 11-5, as an example, the flux illustrated is 6 Mx because there are six field lines flowing in or out for each pole. A 1-lb magnet can provide a magnetic flux ϕ of about 5000 Mx.

The Weber

The weber is a larger unit of magnetic flux. One weber (Wb) equals 1×10^8 lines or maxwells. Since the weber is a large unit for typical fields, the microweber unit can be used. Then $1\ \mu\text{Wb} = 10^{-6}\ \text{Wb}$.

Conversion between Units

Converting from maxwells (Mx) to webers (Wb) or vice versa, is easier if you use the following conversion formulas:

$$\#\text{Wb} = \#\text{Mx} \times \frac{1\ \text{Wb}}{1 \times 10^8\ \text{Mx}}$$

$$\#\text{Mx} = \#\text{Wb} \times \frac{1 \times 10^8\ \text{Mx}}{1\ \text{Wb}}$$

In practice it is rare ever to have to measure magnetic field strength or to make conversions between maxwells and webers. However, it is useful to know the units of measurement.

Systems of Magnetic Units

The basic units in metric form can be defined in two ways:

1. The centimeter-gram-second system defines small units. This is the cgs system.
2. The meter-kilogram-second system is for larger units of a more practical size. This is the mks system.

Furthermore, the Système International (SI) units provide a worldwide standard in mks dimensions. They are practical values based on the ampere of current.

For magnetic flux ϕ, the maxwell (Mx) is a cgs unit, and the weber (Wb) is an mks or SI unit. The SI units are preferred for science and engineering, but the cgs units are still used in many practical applications of magnetism.

Flux Density *B*

Another unit of magnetic strength is flux density. As shown in Fig. 11-5, the *flux density* is the number of magnetic field lines per unit area of a section perpendicular to the direction of flux. As a formula,

$$B = \frac{\phi}{A} \qquad \textbf{(11-1)}$$

where ϕ is the flux through an area A and the flux density is B.

The Gauss

In the cgs system, the gauss (G) is one line per square centimeter, or $1\ \text{Mx/cm}^2$. As an example, in Fig. 11-5, the total flux ϕ is six lines, or 6 Mx. At point P in this field, however, the flux density B is 2 G because there are two lines per square centimeter. The flux density is higher close to the poles, where the flux lines are more crowded. As an example of flux density, B for a 1-lb magnet would be 1000 G at the poles.

As typical values, B for the earth's magnetic field can be about 0.2 G; a large laboratory magnet produces B of 50,000 G. Since the gauss is so small, kilogauss units are often used, where $1\ \text{kG} = 10^3\ \text{G}$.

The Tesla

In SI, the unit of flux density B is webers per square meter (Wb/m^2). One weber per square meter is called a *tesla,* abbreviated T. The tesla is a larger unit than the gauss, as $1\ \text{T} = 1 \times 10^4\ \text{G}$.

Conversion between Units

Converting from teslas (T) to gauss (G), or vice versa, is easier if you use the following conversion formulas:

$$\#\text{G} = \#\text{T} \times \frac{1 \times 10^4\ \text{G}}{1\ \text{T}}$$

$$\#\text{T} = \#\text{G} \times \frac{1\ \text{T}}{1 \times 10^4\ \text{G}}$$

Again, it is rare to have to make such conversions on the job.

11.3 Induction by the Magnetic Field

The electric or magnetic effect of one body on another without any physical contact between them is called *induction.* For instance, a permanent magnet can induce an unmagnetized iron bar to become a magnet without the two touching. The iron bar then becomes a magnet, as shown in Fig. 11-6. What happens is that the magnetic lines of force generated by the permanent magnet make the internal molecular magnets in the iron bar line up in the same direction, instead of the random directions in unmagnetized iron. The magnetized iron bar then has magnetic poles at the ends, as a result of magnetic induction.

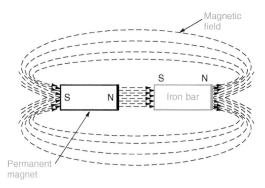

Figure 11-6 Magnetizing an iron bar by induction.

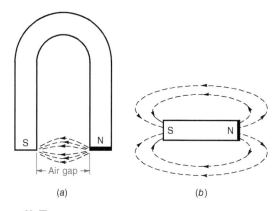

(a)   (b)

Figure 11-7 The horseshoe magnet in (a) has a smaller air gap than the bar magnet in (b).

Note that the induced poles in the iron have polarity opposite from the poles of the magnet. Since opposite poles attract, the iron bar will be attracted. Any magnet attracts to itself all magnetic materials by induction.

Although the two bars in Fig. 11-6 are not touching, the iron bar is in the magnetic flux of the permanent magnet. It is the invisible magnetic field that links the two magnets, enabling one to affect the other. Actually, this idea of magnetic flux extending outward from the magnetic poles is the basis for many inductive effects in ac circuits. More generally, the magnetic field between magnetic poles and the electric field between electric charges form the basis for wireless radio transmission and reception.

Polarity of Induced Poles

Note that the north pole of the permanent magnet in Fig. 11-6 induces an opposite south pole at this end of the iron bar. If the permanent magnet were reversed, its south pole would induce a north pole. The closest induced pole will always be of opposite polarity. This is the reason why either end of a magnet can attract another magnetic material to itself. No matter which pole is used, it will induce an opposite pole, and opposite poles are attracted.

Relative Permeability

Soft iron, as an example, is very effective in concentrating magnetic field lines by induction in the iron. This ability to concentrate magnetic flux is called *permeability*. Any material that is easily magnetized has high permeability, therefore, because the field lines are concentrated by induction.

Numerical values of permeability for different materials compared with air or vacuum can be assigned. For example, if the flux density in air is 1 G but an iron core in the same position in the same field has a flux density of 200 G, the relative permeability of the iron core equals 200/1, or 200.

The symbol for relative permeability is μ_r (mu), where the subscript r indicates relative permeability. Typical values for μ_r are 100 to 9000 for iron and steel. There are no units because μ_r is a comparison of two flux densities and the units cancel.

Air Gap

As shown in Fig. 11-7, the air space between the poles of a magnet is its air gap. The shorter the air gap, the stronger the field in the gap for a given pole strength. Since air is not magnetic and cannot concentrate magnetic lines, a larger air gap provides additional space for the magnetic lines to spread out.

Referring to Fig. 11-7a, note that the horseshoe magnet has more crowded magnetic lines in the air gap, compared with the widely separated lines around the bar magnet in Fig. 11-7b. Actually, the horseshoe magnet can be considered a bar magnet bent around to place the opposite poles closer. Then the magnetic lines of the poles reinforce each other in the air gap. The purpose of a short air gap is to concentrate the magnetic field outside the magnet for maximum induction in a magnetic material placed in the gap.

11.4 Types of Magnets

The two broad classes are permanent magnets and electromagnets. An electromagnet needs current from an external source to maintain its magnetic field. With a permanent magnet, not only is its magnetic field present without any external current but the magnet can maintain its strength indefinitely.

Permanent Magnets

Permanent magnets are made of hard magnetic materials, such as cobalt steel, magnetized by induction in the manufacturing process. A very strong field is needed for induction in these materials. When the magnetizing field is removed, however, residual induction makes the material a permanent magnet. A common PM material is *alnico*, a commercial alloy of aluminum, nickel, and iron, with cobalt, copper, and titanium added to produce about 12 grades. The Alnico V grade is often used for PM loudspeakers (Fig. 11-8). In this application, a typical size of PM slug for a steady magnetic field is a few ounces to about 5 lb, with a flux ϕ of 500 to

Figure 11-8 Example of a PM loudspeaker.

25,000 lines or maxwells. One advantage of a PM loudspeaker is that only two connecting leads are needed for the voice coil because the steady magnetic field of the PM slug is obtained without any field-coil winding.

Commercial permanent magnets will last indefinitely if they are not subjected to high temperatures, physical shock, or a strong demagnetizing field. If the magnet becomes hot, however, the molecular structure can be rearranged, resulting in loss of magnetism that is not recovered after cooling. A permanent magnet does not become exhausted with use because its magnetic properties are determined by the structure of the internal atoms and molecules.

Electromagnets

Current in a wire conductor has an associated magnetic field. If the wire is wrapped in the form of a coil, as in Fig. 11-9, the current and its magnetic field become concentrated in a smaller space, resulting in a stronger field. With the length much greater than its width, the coil is called a *solenoid*. It acts like a bar magnet, with opposite poles at the ends.

More current and more turns make a stronger magnetic field. Also, the iron core concentrates magnetic lines inside the coil. Soft iron is generally used for the core because it is easily magnetized and demagnetized.

The coil in Fig. 11-9, with the switch closed and current in the coil, is an electromagnet that can pick up the steel nail shown. If the switch is opened, the magnetic field is reduced to zero, and the nail will drop off. This ability of an

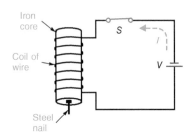

Figure 11-9 Electromagnet holding a nail when switch *S* is closed for current in the coil.

electromagnet to provide a strong magnetic force of attraction that can be turned on or off easily has many applications in lifting magnets, buzzers, bells or chimes, and relays. A *relay* is a switch with contacts that are opened or closed by an electromagnet.

Another common application is magnetic tape or disk recording. The tape or disk is coated with fine particles of iron oxide. The recording head is a coil that produces a magnetic field in proportion to the current. As the disk passes through the air gap of the head, small areas of the coating become magnetized by induction. On playback, the moving magnetic disk produces variations in electric current.

Classification of Magnetic Materials

When we consider materials simply as either magnetic or nonmagnetic, this division is based on the strong magnetic properties of iron. However, weak magnetic materials can be important in some applications. For this reason, a more exact classification includes the following three groups:

1. *Ferromagnetic materials.* These include iron, steel, nickel, cobalt, and commercial alloys such as alnico and Permalloy. Certain types of ceramics are also used to produce very strong magnets. They become strongly magnetized in the same direction as the magnetizing field, with high values of permeability from 50 to 5000. Permalloy has a μ_r of 100,000 but is easily saturated at relatively low values of flux density.

2. *Paramagnetic materials.* These include aluminum, platinum, manganese, and chromium. Their permeability is slightly more than 1. They become weakly magnetized in the same direction as the magnetizing field.

3. *Diamagnetic materials.* These include bismuth, antimony, copper, zinc, mercury, gold, and silver. Their permeability is less than 1. They become weakly magnetized but in the direction opposite from the magnetizing field.

The basis of all magnetic effects is the magnetic field associated with electric charges in motion. Within the atom, the motion of its orbital electrons generates a magnetic field. There are two kinds of electron motion in the atom. First is the electron revolving in its orbit. This motion provides a diamagnetic effect. However, this magnetic effect is weak because thermal agitation at normal room temperature results in random directions of motion that neutralize each other.

More effective is the magnetic effect from the motion of each electron spinning on its own axis. The spinning electron serves as a tiny permanent magnet. Opposite spins provide opposite polarities. Two electrons spinning in opposite directions form a pair, neutralizing the magnetic fields. In the atoms of ferromagnetic materials, however, there are many unpaired electrons with spins in the same direction, resulting in a strong magnetic effect.

In terms of molecular structure, iron atoms are grouped in microscopically small arrangements called *domains.* Each domain is an elementary *dipole magnet,* with two opposite poles. In crystal form, the iron atoms have domains parallel to the axes of the crystal. Still, the domains can point in different directions because of the different axes. When the material becomes magnetized by an external magnetic field, though, the domains become aligned in the same direction. With PM materials, the alignment remains after the external field is removed.

Ferrites

Ferrite is the name for nonmetallic materials that have the ferromagnetic properties of iron. Ferrites have very high permeability, like iron. However, a ferrite is a nonconducting ceramic material, whereas iron is a conductor. The permeability of ferrites is in the range of 50 to 3000. The specific resistance is 10^5 $\Omega\cdot$cm, which makes a ferrite an insulator.

A common application is a ferrite core, usually adjustable, in the coils of radio frequency transformers. The ferrite core is much more efficient than iron when the current alternates at high frequency. The reason is that less I^2R power is lost by eddy currents in the core because of its very high resistance.

A ferrite core is used in small coils and transformers for signal frequencies up to 100 MHz, approximately. The high permeability means that the transformer can be very small. However, ferrites are easily saturated at low values of magnetizing current. This disadvantage means that ferrites are not used for power transformers.

Another application is in ferrite beads (Fig. 11-10). A bare wire is used as a string for one or more beads. The bead concentrates the magnetic field of the current in the wire. This construction serves as a simple, economical RF choke, instead of a coil. The purpose of the choke is to reduce the current just for an undesired radio frequency.

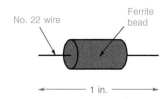

Figure 11-10   Ferrite bead equivalent of an inductor.

11.5 Magnetic Shielding

The idea of preventing one component from affecting another through their common electric or magnetic field is called *shielding.* Examples are the braided copper wire shield around the inner conductor of a coaxial cable, a metal shield can that encloses an RF coil, or a shield of magnetic material enclosing a cathode-ray tube.

The problem in shielding is to prevent one component from inducing an effect in the shielded component. The shielding materials are always metals, but there is a difference between using good conductors with low resistance, such as copper and aluminum, and using good magnetic materials such as soft iron.

A good conductor is best for two shielding functions. One is to prevent induction of static electric charges. The other is to shield against the induction of a varying magnetic field. For static charges, the shield provides opposite induced charges, which prevent induction inside the shield. For a varying magnetic field, the shield has induced currents that oppose the inducing field. Then there is little net field strength to produce induction inside the shield.

The best shield for a steady magnetic field is a good magnetic material of high permeability. A steady field is produced by a permanent magnet, a coil with steady direct current, or the earth's magnetic field. A magnetic shield of high permeability concentrates the magnetic flux. Then there is little flux to induce poles in a component inside the shield. The shield can be considered a short circuit for the lines of magnetic flux.

11.6 The Hall Effect

An interesting and useful magnetic phenomenon is the Hall effect. It has been observed that a small voltage is generated across a conductor carrying current in an external magnetic field. The Hall voltage was very small with typical conductors, and little use was made of this effect. However, with the development of semiconductors, larger values of Hall voltage can be generated. The semiconductor material indium arsenide (InAs) is generally used. As illustrated in Fig. 11-11, the InAs element inserted in a magnetic

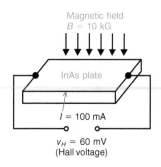

Figure 11-11   The Hall effect. Hall voltage V_H generated across the element is proportional to the perpendicular flux density B.

field can generate 60 mV with B equal to 10 kG and an I of 100 mA. The applied flux must be perpendicular to the direction of the current. With current in the direction of the length of conductor, the generated voltage is developed across the width.

The amount of Hall voltage V_H is directly proportional to the value of flux density B. This means that values of B can be measured by V_H. This value of V_H is then read by the meter, which is calibrated in gauss. The original calibration is made in terms of a reference magnet with a specified flux density.

11.7 Ampere-Turns of Magnetomotive Force (mmf)

The strength of the magnetic field of a coil magnet depends on how much current flows in the turns of the coil. The more current, the stronger the magnetic field. Also, more turns in a specific length concentrate the field. The coil serves as a bar magnet with opposite poles at the ends, providing a magnetic field proportional to the ampere-turns. As a formula,

$$\text{Ampere-turns} = I \times N = \text{mmf} \qquad \textbf{(11-2)}$$

where I is the current in amperes multiplied by the number of turns N. The quantity IN specifies the amount of *magnetizing force* or *magnetic potential*, which is the *magnetomotive force (mmf)*.

The practical unit is the ampere-turn. The SI abbreviation for ampere-turn is A, the same as for the ampere, since the number of turns in a coil usually is constant but the current can be varied. However, for clarity we shall use the abbreviation A·t.

As shown in Fig. 11-12, a solenoid with 5 turns and 2 amperes has the same magnetizing force as one with 10 turns and 1 ampere, as the product of the amperes and turns is 10 for both cases. With thinner wire, more turns can be placed in a given space. The amount of current is determined by the resistance of the wire and the source voltage. The number of ampere-turns necessary depends on the magnetic field strength required.

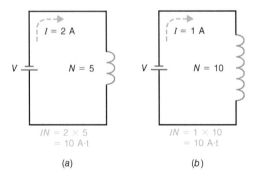

IN = 2 × 5
   = 10 A·t

IN = 1 × 10
   = 10 A·t

(a)   (b)

Figure 11-12 Two examples of equal ampere-turns for the same mmf. (a) *IN* is 2 × 5 = 10. (b) *IN* is 1 × 10 = 10.

EXAMPLE 11-1

Calculate the ampere-turns of mmf for a coil with 2000 turns and a 5-mA current.

Answer:

$$\text{mmf} = I \times N = 2000 \times 5 \times 10^{-3}$$
$$= 10 \text{ A·t}$$

EXAMPLE 11-2

A coil with 4 A is to provide a magnetizing force of 600 A·t. How many turns are necessary?

Answer:

$$N = \frac{\text{A·t}}{I} = \frac{600}{4}$$
$$= 150 \text{ turns}$$

EXAMPLE 11-3

The wire in a solenoid of 250 turns has a resistance of 3 Ω. (a) How much is the current when the coil is connected to a 6-V battery? (b) Calculate the ampere-turns of mmf.

Answer:

a. $I = \dfrac{V}{R} = \dfrac{6 \text{ V}}{3 \text{ Ω}}$
   $= 2 \text{ A}$

b. $\text{mmf} = I \times N = 2 \text{ A} \times 250 \text{ t}$
   $= 500 \text{ A·t}$

The ampere-turn A·t, is an SI unit. It is calculated as IN with the current in amperes.

The cgs unit of mmf is the *gilbert*, abbreviated Gb. One ampere-turn equals 1.26 Gb. The number 1.26 is approximately $4\pi/10$, derived from the surface area of a sphere, which is $4\pi r^2$.

11.8 Field Intensity (*H*)

The ampere-turns of mmf specify the magnetizing force, but the intensity of the magnetic field depends on the length of the coil. At any point in space, a specific value of ampere-turns must produce less field intensity for a long coil than for a short coil that concentrates the same mmf. Specifically, the field intensity H in mks units is

$$H = \frac{\text{ampere-turns of mmf}}{l \text{ meters}} \qquad \textbf{(11-3)}$$

This formula is for a solenoid. The field intensity H is at the center of an air core. For an iron core, H is the intensity through the entire core. By means of units for H, the

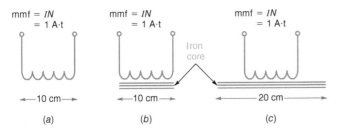

mmf = IN
= 1 A·t

mmf = IN
= 1 A·t

mmf = IN
= 1 A·t

Iron core

←10 cm→   ←10 cm→   ←20 cm→

(a)   (b)   (c)

Figure 11-13 Relation between ampere-turns of mmf and the resultant field intensity *H* for different cores. Note that *H* = mmf/length. (a) Intensity *H* is 10 A·t/m with an air core. (b) *H* = 10 A·t/m in an iron core of the same length as the coil. (c) *H* is 1/0.2 = .5 A·t/m in an iron core twice as long as the coil.

magnetic field intensity can be specified for either electromagnets or permanent magnets, since both provide the same kind of magnetic field.

The length in Formula (11-3) is between poles. In Fig. 11-13*a*, the length is 0.1 m between the poles at the ends of the coil. In Fig. 11-13*b*, *l* is also 0.1 m between the ends of the iron core. In Fig. 11-13*c*, though, *l* is 0.2 m between the poles at the ends of the iron core, although the winding is only 0.2 m long.

The examples in Fig. 11-13 illustrate the following comparisons:

1. In all three cases, the mmf is 1 A·t for the same value of *IN*.

2. In Fig. 11-13*a* and *b*, *H* equals 10 A·t/m. In *a*, this *H* is the intensity at the center of the air core; in *b*, this *H* is the intensity through the entire iron core.

3. In Fig. 11-13*c*, because *l* is 0.2 m, *H* is 1/0.2, or 5 A·t/m. This *H* is the intensity in the entire iron core.

Units for *H*

The field intensity is basically mmf per unit of length. In practical units, *H* is ampere-turns per meter. The cgs unit for *H* is the *oersted*, abbreviated Oe, which equals one gilbert of mmf per centimeter.

Permeability (*μ*)

Whether we say *H* is 1000 A·t/m or 12.6 Oe, these units specify how much field intensity is available to produce magnetic flux. However, the amount of flux produced by *H* depends on the material in the field. A good magnetic material with high relative permeability can concentrate flux and produce a large value of flux density *B* for a specified *H*. These factors are related by the formula:

$$B = \mu \times H \qquad (11\text{-}4)$$

or

$$\mu = \frac{B}{H} \qquad (11\text{-}5)$$

Using SI units, *B* is the flux density in webers per square meter, or teslas; *H* is the field intensity in ampere-turns per

meter. In the cgs system, the units are gauss for *B* and oersted for *H*. The factor *μ* is the absolute permeability, not referred to any other material, in units of *B/H*.

In the cgs system, the units of gauss for *B* and oersteds for *H* have been defined to give *μ* the value of 1 G/Oe, for vacuum, air, or space. This simplification means that *B* and *H* have the same numerical values in air and in vacuum. For instance, the field intensity *H* of 12.6 Oe produces a flux density of 12.6 G in air.

Furthermore, the values of relative permeability *μ*ᵣ are the same as those for absolute permeability in *B/H* units in the cgs system. The reason is that *μ* is 1 for air or vacuum, used as the reference for the comparison. As an example, if *μ*ᵣ for an iron sample is 600, the absolute *μ* is also 600 G/Oe.

In SI, however, the permeability of air or vacuum is not 1. This value is $4\pi \times 10^{-7}$, or 1.26×10^{-6}, with the symbol μ_0. Therefore, values of relative permeability *μ*ᵣ must be multiplied by 1.26×10^{-6} for μ_0 to calculate *μ* as *B/H* in SI units.

For an example of $\mu_r = 100$, the SI value of *μ* can be calculated as follows:

$$\mu = \mu_r \times \mu_0$$
$$= 100 \times 1.26 \times 10^{-6} \frac{\text{T}}{\text{A·t/m}}$$
$$\mu = 126 \times 10^{-6} \frac{\text{T}}{\text{A·t/m}}$$

In practice, such conversions are rarely needed.

11.9 *B-H* Magnetization Curve

The *B-H* curve in Fig. 11-14 is often used to show how much flux density *B* results from increasing the amount of field intensity *H*. This curve is for soft iron. Similar curves can be obtained for all magnetic materials.

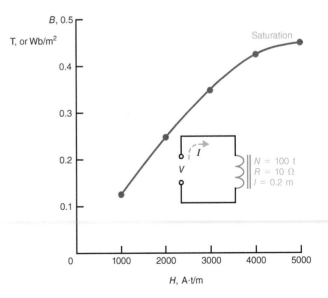

Figure 11-14 *B-H* magnetization curve for soft iron. No values are shown near zero, where *μ* may vary with previous magnetization.

Calculating H and B

The values are calculated as follows:

1. The current I in the coil equals V/R. For a 10-Ω coil resistance with 20 V applied, I is 2 A. Increasing values of V produce more current in the coil.

2. The ampere-turns IN of magnetizing force increase with more current. Since the turns are constant at 100, the values of IN increase from 200 for 2 A to 1000 for 10 A.

3. The field intensity H increases with higher IN. The values of H are in mks units of ampere-turns per meter. These values equal $IN/0.2$ because the length is 0.2 m. Therefore, each IN is divided by 0.2, or multiplied by 5, for the corresponding values of H. Since H increases in the same proportion as I, sometimes the horizontal axis on a B-H curve is given only in amperes, instead of in H units.

4. The flux density B depends on the field intensity H and the permeability of the iron. The values of B are obtained by multiplying $\mu \times H$. However, for SI units, the values of μ_r must be multiplied by 1.26×10^{-6} to obtain $\mu \times H$ in teslas.

Saturation

Note that the permeability decreases for the highest values of H. With less μ, the iron core cannot provide proportional increases in B for increasing values of H. In Fig. 11-14, for values of H above 4000 A·t/m, approximately, the values of B increase at a much slower rate, making the curve relatively flat at the top. The effect of little change in flux density when the field intensity increases is called *saturation*.

Iron becomes saturated with magnetic lines of induction. After most of the molecular dipoles and the magnetic domains are aligned by the magnetizing force, very little additional induction can be produced. When the value of μ is specified for a magnetic material, it is usually the highest value before saturation.

11.10 Magnetic Hysteresis

Hysteresis means "lagging behind." With respect to the magnetic flux in an iron core of an electromagnet, the flux lags the increases or decreases in magnetizing force. Hysteresis results because the magnetic dipoles are not perfectly elastic. Once aligned by an external magnetizing force, the dipoles do not return exactly to their original positions when the force is removed. The effect is the same as if the dipoles were forced to move against internal friction between molecules. Furthermore, if the magnetizing force is reversed in direction by reversal of the current in an electromagnet, the flux produced in the opposite direction lags behind the reversed magnetizing force.

Hysteresis Loss

When the magnetizing force reverses thousands or millions of times per second, as with rapidly reversing alternating current, hysteresis can cause a considerable loss of energy. A large part of the magnetizing force is then used to overcome the internal friction of the molecular dipoles. The work done by the magnetizing force against this internal friction produces heat. This energy wasted in heat as the molecular dipoles lag the magnetizing force is called the *hysteresis loss*. For steel and other hard magnetic materials, hysteresis losses are much higher than in soft magnetic materials like iron.

When the magnetizing force varies at a slow rate, hysteresis losses can be considered negligible. An example is an electromagnet with direct current that is simply turned on and off or the magnetizing force of an alternating current that reverses 60 times per second or less. The faster the magnetizing force changes, however, the greater the hysteresis effect.

Hysteresis Loop

To show the hysteresis characteristics of a magnetic material, its values of flux density B are plotted for a periodically reversing magnetizing force. See Fig. 11-15. This curve is the hysteresis loop of the material. The larger the area enclosed by the curve, the greater the hysteresis loss. The hysteresis loop is actually a B-H curve with an ac magnetizing force.

Values of flux density B are indicated on the vertical axis. The units can be gauss or teslas.

The horizontal axis indicates values of field intensity H. On this axis, the units can be oersteds, ampere-turns per

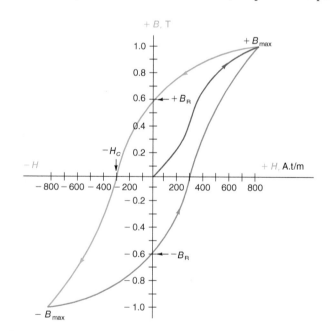

Figure 11-15 Hysteresis loop for magnetic materials. This graph is a B-H curve like Fig. 11-14, but H alternates in polarity with alternating current.

meter, ampere-turns, or magnetizing current because all factors are constant except *I*.

Opposite directions of current result in opposite directions of +*H* and −*H* for the field lines. Similarly, opposite polarities are indicated for flux density as +*B* or −*B*.

The current starts from zero at the center, when the material is unmagnetized. Then positive *H* values increase *B* to saturation at +B_{max}. Next *H* decreases to zero, but *B* drops to the value B_R, instead of to zero, because of hysteresis. When *H* becomes negative, *B* drops to zero and continues to −B_{max}, which is saturation in the opposite direction from +B_{max} because of the reversed magnetizing current.

Then, as the −*H* values decrease, the flux density is reduced to −B_R. Finally, the loop is completed; positive values of *H* produce saturation at B_{max} again. The curve does not return to the zero origin at the center because of hysteresis. As the magnetizing force periodically reverses, the values of flux density are repeated to trace out the hysteresis loop.

The value of either +B_R or −B_R, which is the flux density remaining after the magnetizing force has been reduced to zero, is the *residual induction* of a magnetic material, also called its *retentivity*. In Fig. 11-15, the residual induction is 0.6 T, in either the positive or the negative direction.

The value of −H_C, which equals the magnetizing force that must be applied in the reverse direction to reduce the flux density to zero, is the *coercive force* of the material. In Fig. 11-15, the coercive force −H_C is 300 A·t/m.

Demagnetization

To demagnetize a magnetic material completely, the residual induction B_R must be reduced to zero. This usually cannot be accomplished by a reversed dc magnetizing force because the material then would become magnetized with opposite polarity. The practical way is to magnetize and demagnetize the material with a continuously decreasing hysteresis loop. This can be done with a magnetic field produced by alternating current. Then, as the magnetic field and the material are moved away from each other or the current amplitude is reduced, the hysteresis loop becomes smaller and smaller. Finally, with the weakest field, the loop collapses practically to zero, resulting in zero residual induction.

This method of demagnetization is also called *degaussing*. One application is degaussing the metal electrodes in a color picture tube with a deguassing coil providing alternating current from the power line. Another example is erasing the recorded signal on magnetic tape by demagnetizing with an ac bias current. The average level of the erase current is zero, and its frequency is much higher than the recorded signal.

11.11 Magnetic Field around an Electric Current

In Fig. 11-16, the iron filings aligned in concentric rings around the conductor show the magnetic field of the current in the wire. The iron filings are dense next to the conductor, showing that the field is strongest at this point. Furthermore, the field strength decreases inversely as the square of the distance from the conductor. It is important to note the following two factors about the magnetic lines of force:

1. The magnetic lines are circular because the field is symmetrical with respect to the wire in the center.
2. The magnetic field with circular lines of force is in a plane perpendicular to the current in the wire.

From points C to D in the wire, the circular magnetic field is in the horizontal plane because the wire is vertical. Also, the vertical conductor between points EF and AB has the associated magnetic field in the horizontal plane. Where the conductor is horizontal, as from B to C and D to E, the magnetic field is in a vertical plane.

These two requirements of a circular magnetic field in a perpendicular plane apply to any charge in motion. Whether electron flow or motion of positive charges is considered, the associated magnetic field must be at right angles to the direction of current.

In addition, the current need not be in a wire conductor. As an example, the beam of moving electrons in the vacuum of a cathode-ray tube has an associated magnetic field. In all cases, the magnetic field has circular lines of force in a plane perpendicular to the direction of motion of the electric charges.

Clockwise and Counterclockwise Fields

With circular lines of force, the magnetic field would tend to move a magnetic pole in a circular path. Therefore, the direction of the lines must be considered either clockwise or counterclockwise. This idea is illustrated in

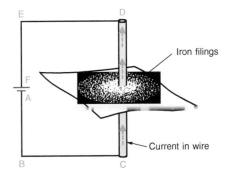

Figure 11-16 How iron filings can be used to show the invisible magnetic field around the electric current in a wire conductor.

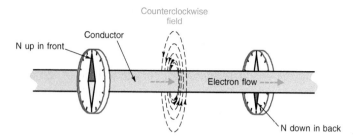

Figure 11-17 Rule for determining direction of circular field around a straight conductor. Field is counterclockwise for direction of electron flow shown here. Circular field is clockwise for reversed direction of electron flow.

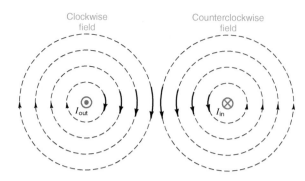

Figure 11-18 Magnetic fields aiding between parallel conductors with opposite directions of current.

Fig. 11-17, showing how a north pole would move in the circular field.

The directions are tested with a magnetic compass needle. When the compass is in front of the wire, the north pole on the needle points up. On the opposite side, the compass points down. If the compass were placed at the top, its needle would point toward the back of the wire; below the wire, the compass would point forward. When all these directions are combined, the result is the circular magnetic field shown with counterclockwise lines of force. (The counterclockwise direction of the magnetic field assumes that you are looking into the end of the wire, in the same direction as electron flow.)

Instead of testing every conductor with a magnetic compass, however, we can use the following rule for straight conductors to determine the circular direction of the magnetic field: *If you grasp the conductor with your left hand so that the thumb points in the direction of electron flow, your fingers will encircle the conductor in the same direction as the circular magnetic field lines.* In Fig. 11-17, the direction of electron flow is from left to right. Facing this way, you can assume that the circular magnetic flux in a perpendicular plane has lines of force in the counterclockwise direction.

The opposite direction of electron flow produces a reversed field. Then the magnetic lines of force rotate clockwise. If the charges were moving from right to left in Fig. 11-17, the associated magnetic field would be in the opposite direction with clockwise lines of force.

Note: If you are assuming conventional current flow from positive to negative, you can define the direction of the magnetic field with your right hand.

Fields Aiding or Canceling

When the magnetic lines of two fields are in the same direction, the lines of force aid each other, making the field stronger. When magnetic lines are in opposite directions, the fields cancel.

In Fig. 11-18, the fields are shown for two conductors with opposite directions of electron flow. The dot in the middle of the field at the left indicates the tip of an arrowhead to show current up from the paper. The cross symbolizes the back of an arrow to indicate electron flow into the paper.

Notice that the magnetic lines *between the conductors* are in the same direction, although one field is clockwise and the other counterclockwise. Therefore, the fields aid here, making a stronger total field. On either side of the conductors, the two fields are opposite in direction and tend to cancel each other. The net result, then, is to strengthen the field in the space between the conductors.

11.12 Magnetic Polarity of a Coil

Bending a straight conductor into a loop, as shown in Fig. 11-19, has two effects. First, the magnetic field lines are more dense inside the loop. The total number of lines is the same as those for the straight conductor, but the lines inside the loop are concentrated in a smaller space. Furthermore, all lines inside the loop are aiding in the same direction. This makes the loop field effectively the same as a bar magnet with opposite poles at opposite faces of the loop.

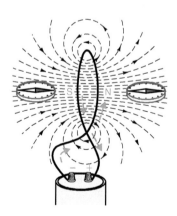

Figure 11-19 Magnetic poles of a current loop.

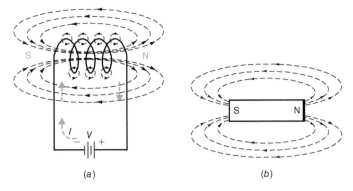

(a)   (b)

Figure 11-20 Magnetic poles of a solenoid. (a) Coil winding. (b) Equivalent bar magnet.

Solenoid as a Bar Magnet

A coil of wire conductor with more than one turn is generally called a *solenoid*. An ideal solenoid, however, has a length much greater than its diameter. Like a single loop, the solenoid concentrates the magnetic field inside the coil and provides opposite magnetic poles at the ends. These effects are multiplied, however, by the number of turns as the magnetic field lines aid each other in the same direction inside the coil. Outside the coil, the field corresponds to a bar magnet with north and south poles at opposite ends, as illustrated in Fig. 11-20.

Magnetic Polarity

To determine the magnetic polarity of a solenoid, use the *left-hand rule* illustrated in Fig. 11-21: *If the coil is grasped with the fingers of the left hand curled around the coil in the direction of electron flow, the thumb points to the north pole of the coil.* The left hand is used here because the current is electron flow. Use your right hand for conventional current flow.

The solenoid acts like a bar magnet, whether or not it has an iron core. Adding an iron core increases the flux density inside the coil. In addition, the field strength is uniform for the entire length of the core. The polarity is the same, however, for air-core and iron-core coils.

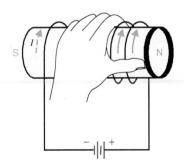

Figure 11-21 Left-hand rule for north pole of a coil with current *I*. The *I* is electron flow.

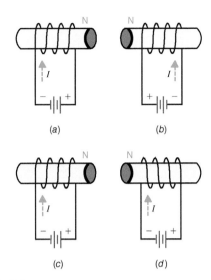

(a)   (b)

(c)   (d)

Figure 11-22 Examples for determining the magnetic polarity of a coil with direct current *I*. The *I* is electron flow. The polarities are reversed in (a) and (b) because the battery is reversed to reverse the direction of current. Also, (d) is the opposite of (c) because of the reversed winding.

The magnetic polarity depends on the direction of current flow and the direction of winding. The current is determined by the connections to the voltage source. Electron flow is from the negative side of the voltage source, through the coil, and back to the positive terminal.

The direction of winding can be over and under, starting from one end of the coil, or under and over with respect to the same starting point. Reversing either the direction of winding or the direction of current reverses the magnetic poles of the solenoid. See Fig. 11-22. When both are reversed, though, the polarity is the same.

11.13 Motor Action between Two Magnetic Fields

The physical motion from the forces of magnetic fields is called *motor action*. One example is the simple attraction or repulsion between bar magnets.

We know that like poles repel and unlike poles attract. It can also be considered that fields in the same direction repel and opposite fields attract.

Consider the repulsion between two north poles, illustrated in Fig. 11-23. Similar poles have fields in the same direction. Therefore, the similar fields of the two like poles repel each other.

A more fundamental reason for motor action, however, is the fact that the force in a magnetic field tends to produce motion from a stronger field toward a weaker field. In Fig. 11-23, note that the field intensity is greatest in the space between the two north poles. Here the field lines of similar poles in both magnets reinforce in the same direction. Farther away the field intensity is less, for essentially one magnet only. As

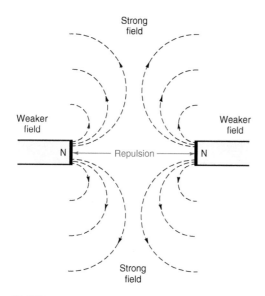

Figure 11-23 Repulsion between similar poles of two bar magnets. The motion is from the stronger field to the weaker field.

a result, there is a difference in field strength, providing a net force that tends to produce motion. The direction of motion is always toward the weaker field.

To remember the directions, we can consider that the stronger field moves to the weaker field, tending to equalize field intensity. Otherwise, the motion would make the strong field stronger and the weak field weaker. This must be impossible because then the magnetic field would multiply its own strength without any work added.

Force on a Straight Conductor in a Magnetic Field

Current in a conductor has its associated magnetic field. When this conductor is placed in another magnetic field from a separate source, the two fields can react to produce motor action. The conductor must be perpendicular to the magnetic field, however, as shown in Fig. 11-24. This way,

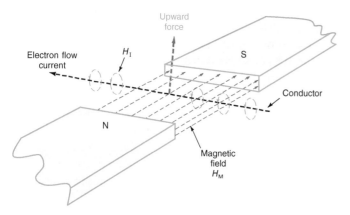

Figure 11-24 Motor action of current in a straight conductor when it is in an external magnetic field. The H_I is the circular field of the current. The H_M indicates field lines between the north and south poles of the external magnet.

the perpendicular magnetic field produced by the current is in the same plane as the external magnetic field.

Unless the two fields are in the same plane, they cannot affect each other. In the same plane, however, lines of force in the same direction reinforce to make a stronger field, whereas lines in the opposite direction cancel and result in a weaker field.

To summarize these directions:

1. When the conductor is at 90°, or perpendicular to the external field, the reaction between the two magnetic fields is maximum.
2. When the conductor is at 0°, or parallel to the external field, there is no effect between them.
3. When the conductor is at an angle between 0 and 90°, only the perpendicular component is effective.

In Fig. 11-24, electrons flow in the wire conductor in the plane of the paper from the bottom to the top of the page. This flow provides the counterclockwise field H_I around the wire in a perpendicular plane cutting through the paper. The external field H_M has lines of force from left to right in the plane of the paper. Then lines of force in the two fields are parallel above and below the wire.

Below the conductor, its field lines are left to right in the same direction as the external field. Therefore, these lines reinforce to produce a stronger field. Above the conductor, the lines of the two fields are in opposite directions, causing a weaker field. As a result, the net force of the stronger field makes the conductor move upward out of the page toward the weaker field.

If electrons flow in the reverse direction in the conductor or if the external field is reversed, the motor action will be in the opposite direction. Reversing both the field and the current, however, results in the same direction of motion.

Rotation of a Conductor Loop in a Magnetic Field

When a loop of wire is in the magnetic field, opposite sides of the loop have current in opposite directions. Then the associated magnetic fields are opposite. The resulting forces are upward on one side of the loop and downward on the other side, making it rotate. This effect of a force in producing rotation is called *torque.*

The principle of motor action between magnetic fields producing rotational torque is the basis of all electric motors. Since torque is proportional to current, the amount of rotation indicates how much current flows through the coil.

11.14 Induced Current

Just as electrons in motion provide an associated magnetic field, when magnetic flux moves, the motion of magnetic lines cutting across a conductor forces free electrons in the

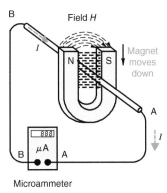

Figure 11-25 Induced current produced by magnetic flux cutting across a conductor. Direction of *I* here is for electron flow.

conductor to move, producing current. This action is called *induction* because there is no physical connection between the magnet and the conductor. The induced current is a result of generator action as the mechanical work put into moving the magnetic field is converted into electric energy when current flows in the conductor.

Referring to Fig. 11-25, let the conductor AB be placed at right angles to the flux in the air gap of the horseshoe magnet. Then, when the magnet is moved up or down, its flux cuts across the conductor. The action of magnetic flux cutting across the conductor generates current. The fact that current flows is indicated by the microammeter.

When the magnet is moved downward, current flows in the direction shown. If the magnet is moved upward, current will flow in the opposite direction. Without motion, there is no current.

Direction of Motion

Motion is necessary for the flux lines of the magnetic field to cut across the conductor. This cutting can be accomplished by motion of either the field or the conductor. When the conductor is moved upward or downward, it cuts across the flux. The generator action is the same as moving the field, except that the relative motion is opposite. Moving the conductor upward, for instance, corresponds to moving the magnet downward.

Conductor Perpendicular to External Flux

To have electromagnetic induction, the conductor and the magnetic lines of flux must be perpendicular to each other. Then the motion makes the flux cut through the cross-sectional area of the conductor. As shown in Fig. 11-25, the conductor is at right angles to the lines of force in the field *H*.

The reason the conductor must be perpendicular is to make its induced current have an associated magnetic field in the same plane as the external flux. If the field of the induced current does not react with the external field, there can be no induced current.

How Induced Current Is Generated

The induced current can be considered the result of motor action between the external field *H* and the magnetic field of free electrons in every cross-sectional area of the wire. Without an external field, the free electrons move at random without any specific direction, and they have no net magnetic field. When the conductor is in the magnetic field *H*, there still is no induction without relative motion, since the magnetic fields for the free electrons are not disturbed. When the field or conductor moves, however, there must be a reaction opposing the motion. The reaction is a flow of free electrons resulting from motor action on the electrons.

Referring to Fig. 11-25, for example, the induced current must flow in the direction shown because the field is moved downward, pulling the magnet away from the conductor. The induced current of electrons then has a clockwise field; lines of force aid *H* above the conductor and cancel *H* below. When motor action between the two magnetic fields tends to move the conductor toward the weaker field, the conductor will be forced downward, staying with the magnet to oppose the work of pulling the magnet away from the conductor.

The effect of electromagnetic induction is increased when a coil is used for the conductor. Then the turns concentrate more conductor length in a smaller area. As illustrated in Fig. 11-26, moving the magnet into the coil enables the flux to cut across many turns of conductors.

Lenz's Law

Lenz's law is the basic principle for determining the direction of an induced voltage or current. Based on the principle of conservation of energy, the law simply states that the direction of the induced current must be such that its own magnetic field will oppose the action that produced the induced current.

In Fig. 11-26, for example, the induced current has the direction that produces a north pole at the left to oppose the motion by repulsion of the north pole being moved in. This is why it takes some work to push the permanent magnet into the coil. The work expended in moving the permanent magnet is the source of energy for the current induced in the coil.

Using Lenz's law, we can start with the fact that the left end of the coil in Fig. 11-26 must be a north pole to oppose

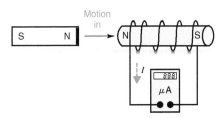

Figure 11-26 Induced current produced by magnetic flux cutting across turns of wire in a coil. Direction of *I* here is for electron flow.

the motion. Then the direction of the induced current is determined by the left-hand rule for electron flow. If the fingers coil around the direction of electron flow shown, under and over the winding, the thumb will point to the left for the north pole.

For the opposite case, suppose that the north pole of the permanent magnet in Fig. 11-26 is moved away from the coil. Then the induced pole at the left end of the coil must be a south pole by Lenz's law. The induced south pole will attract the north pole to oppose the motion of the magnet being moved away. For a south pole at the left end of the coil, then, the electron flow will be reversed from the direction shown in Fig. 11-26. We could generate an alternating current in the coil by moving the magnet periodically in and out.

11.15 Generating an Induced Voltage

Consider a magnetic flux cutting a conductor that is not in a closed circuit, as shown in Fig. 11-27. The motion of flux across the conductor forces free electrons to move, but in an open circuit, the displaced electrons produce opposite electric charges at the two open ends.

For the directions shown, free electrons in the conductor are forced to move to point A. Since the end is open, electrons accumulate here. Point A then develops a negative potential.

At the same time, point B loses electrons and becomes charged positively. The result is a potential difference across the two ends, provided by the separation of electric charges in the conductor.

The potential difference is an electromotive force (emf), generated by the work of cutting across the flux. You can measure this potential difference with a voltmeter. However, a conductor cannot store electric charge. Therefore, the voltage is present only while the motion of flux cutting across the conductor is producing the induced voltage.

Induced Voltage across a Coil

For a coil, as in Fig. 11-28a, the induced emf is increased by the number of turns. Each turn cut by flux adds to the induced voltage, since each turn cut forces free electrons to accumulate at the negative end of the coil with a deficiency of electrons at the positive end.

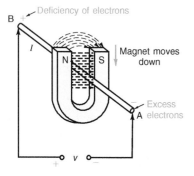

Figure 11-27   Voltage induced across open ends of conductor cut by magnetic flux.

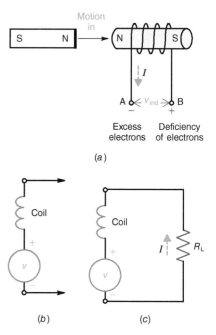

Figure 11-28   Voltage induced across coil cut by magnetic flux. (a) Motion of flux generating voltage across coil. (b) Induced voltage acts in series with the coil. (c) The induced voltage is a source that can produce current in an external load resistor R_L connected across the coil.

The polarity of the induced voltage follows from the direction of induced current. The end of the conductor to which the electrons go and at which they accumulate is the negative side of the induced voltage. The opposite end, with a deficiency of electrons, is the positive side. The total emf across the coil is the sum of the induced voltages, since all the turns are in series.

Furthermore, the total induced voltage acts in series with the coil, as illustrated by the equivalent circuit in Fig. 11-28b, showing the induced voltage as a separate generator. This generator represents a voltage source with a potential difference resulting from the separation of charges produced by electromagnetic induction. The source v then can produce current in an external load circuit connected across the negative and positive terminals, as shown in Fig. 11-28c.

The induced voltage is in series with the coil because current produced by the generated emf must flow through all the turns. An induced voltage of 10 V, for example, with R_L equal to 5 Ω, results in a current of 2 A, which flows through the coil, the equivalent generator v, and the load resistance R_L.

The direction of current in Fig. 11-28c shows electron flow around the circuit. Outside the source v, the electrons move from its negative terminal, through R_L, and back to the positive terminal of v because of its potential difference.

Inside the generator, however, the electron flow is from the + terminal to the − terminal. This direction of electron flow results from the fact that the left end of the coil in Fig. 11-28a must be a north pole, by Lenz's law, to oppose the north pole being moved in.

Notice how motors and generators are similar in using the motion of a magnetic field, but with opposite applications. In a motor, current is supplied so that an associated magnetic field can react with the external flux to produce motion of the conductor. In a generator, motion must be supplied so that the flux and conductor can cut across each other to induce voltage across the ends of the conductor.

Faraday's Law of Induced Voltage

The voltage induced by magnetic flux cutting the turns of a coil depends on the number of turns and how fast the flux moves across the conductor. Either the flux or the conductor can move. Specifically, the amount of induced voltage is determined by the following three factors:

1. *Amount of flux.* The more magnetic lines of force that cut across the conductor, the higher the amount of induced voltage.
2. *Number of turns.* The more turns in a coil, the higher the induced voltage. The v_{ind} is the sum of all individual voltages generated in each turn in series.
3. *Time rate of cutting.* The faster the flux cuts a conductor, the higher the induced voltage. Then more lines of force cut the conductor within a specific period of time.

These factors are fundamental in many applications. Any conductor with current will have voltage induced in it by a change in current and its associated magnetic flux.

The amount of induced voltage can be calculated by Faraday's law:

$$v_{ind} = N \frac{d\phi \,(\text{webers})}{dt \,(\text{seconds})} \qquad \textbf{(11-6)}$$

where N is the number of turns and $d\phi/dt$ specifies how fast the flux ϕ cuts across the conductor. With $d\phi/dt$ in webers per second, the induced voltage is in volts.

As an example, suppose that magnetic flux cuts across 300 turns at the rate of 2 Wb/s.

To calculate the induced voltage,

$$v_{ind} = N \frac{d\phi}{dt}$$
$$= 300 \times 2$$
$$v_{ind} = 600 \text{ V}$$

It is assumed that all flux links all turns, which is true for an iron core.

Rate of Change

The symbol d in $d\phi$ and dt is an abbreviation for *change*. The $d\phi$ means a change in the flux ϕ, and dt means a change in time. In mathematics, dt represents an infinitesimally small change in time, but in this book we are using the d to mean rate of change in general. The results are exactly the same for the practical changes used here because the rate of change is constant.

As an example, if the flux ϕ is 4 Wb one time but then changes to 6 Wb, the change in flux $d\phi$ is 2 Wb. The same idea applies to a decrease as well as an increase. If the flux changed from 6 to 4 Wb, $d\phi$ would still be 2 Wb. However, an increase is usually considered a change in the positive direction, with an upward slope, whereas a decrease has a negative slope downward.

Similarly, dt means a change in time. If we consider the flux at a time 2 s after the start and at a later time 3 s after the start, the change in time is $3 - 2$, or 1 s for dt. Time always increases in the positive direction.

Combining the two factors of $d\phi$ and dt, we can say that for magnetic flux increasing by 2 Wb in 1 s, $d\phi/dt$ equals 2/1, or 2 Wb/s. This states the rate of change of the magnetic flux.

As another example, suppose that the flux increases by 2 Wb in 0.5 s. Then

$$\frac{d\phi}{dt} = \frac{2 \text{ Wb}}{0.5 \text{ s}} = 4 \text{ Wb/s}$$

Polarity of the Induced Voltage

The polarity is determined by Lenz's law. Any induced voltage has the polarity that opposes the change causing the induction. Sometimes this fact is indicated by using a negative sign for v_{ind} in Formula (11-6). However, the absolute polarity depends on whether the flux is increasing or decreasing, the method of winding, and which end of the coil is the reference.

When all these factors are considered, v_{ind} has polarity such that the current it produces and the associated magnetic field oppose the change in flux producing the induced voltage. If the external flux increases, the magnetic field of the induced current will be in the opposite direction. If the external field decreases, the magnetic field of the induced current will be in the same direction as the external field to oppose the change by sustaining the flux. In short, the induced voltage has polarity that opposes the change.

11.16 Relays

A *relay* is an electromechanical device that operates by electromagnetic induction. It uses either an ac- or a dc-actuated electromagnet to open or close one or more sets of contacts. Relay contacts that are open when the relay is not energized are called *normally open* (NO) contacts. Conversely, relay contacts that are closed when the relay is not energized are called *normally closed* (NC) contacts. Relay contacts are held in their resting or normal position either by a spring or by some type of gravity-actuated mechanism. In most cases, an adjustment of the spring tension is provided to set the restraining force on the normally open and normally closed contacts to some desired level based on predetermined circuit conditions.

Figure 11-29 shows the schematic symbols that are commonly used to represent relay contacts. Figure 11-29a shows the symbols used to represent normally open contacts, and

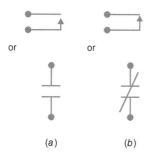

(a)   (b)

Figure 11-29 Schematic symbols commonly used to represent relay contacts. (a) Symbols used to represent normally open (NO) contacts. (b) Symbols used to represent normally closed (NC) contacts.

Fig. 11-29b shows the symbols used to represent normally closed contacts. When normally open contacts close, they are said to *make,* whereas when normally closed contacts open, they are said to *break.* Like mechanical switches, the switching contacts of a relay can have any number of poles and throws.

Figure 11-30 shows the basic parts of an SPDT armature relay. Terminal connections 1 and 2 provide connection to the electromagnet (relay coil), and terminal connections 3, 4, and 5 provide connections to the SPDT relay contacts that open or close when the relay is energized. A relay is said to be *energized* when NO contacts close and NC contacts open. The movable arm of an electromechanical relay is called the *armature.* The armature is magnetic and has contacts that make or break with other contacts when the relay is energized. For example, when terminals 1 and 2 in Fig. 11 30 are connected to a dc source, current flows in the relay coil and an electromagnet is formed. If there is sufficient current in the relay coil, contacts 3 and 4 close (make) and contacts 4 and 5 open (break). The armature is attracted whether the electromagnet produces a north or a south pole on the end adjacent to the armature. Figure 11-31 is a photo of a typical relay.

Reed Relay

Another popular type of relay is the *reed relay.* See Fig. 11-32. It uses two thin reeds of magnetic material sealed in a glass

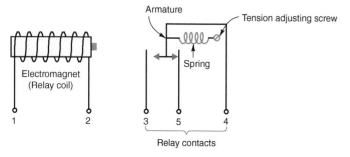

Figure 11-30 Basic parts of an SPDT armature relay. Terminal connections 1 and 2 provide connection to the electromagnet, and terminal connections 3, 4, and 5 provide connections to the SPDT relay contacts which open or close when the relay is energized.

Figure 11-31 Typical relay.

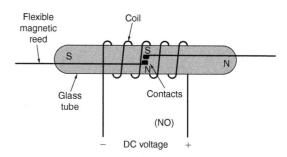

Figure 11-32 A read relay.

housing. The two reeds overlap, and this region forms the contacts. A coil is placed around the glass tube. When the coil is energized by an external dc voltage, current flows and an electromagnet is formed producing a magnetic field with the polarity shown in the figure. This strong field magnetizes the two reeds, forming two miniature bar magnets as shown. Because the contacts have opposite magnetic polarities, they attract each other and thereby contact. These NO contacts remain touching until the external dc voltage is removed. With no dc voltage, there is no longer a magnetic field and the two reeds become demagnetized. Because the reeds are made of a springlike material, they spring apart, opening the contacts.

Reed relays are used in lower-voltage, lower-power applications. They also work with a permanent bar magnet instead of an electromagnet. A reed relay without a coil is sometimes used as a switch sensor that is activated by a permanent bar magnet. Such a sensor may be mounted on a window sill. A bar magnet on the window nearby keeps the contacts closed. If the window is opened, the bar magnet moves away and the contacts open. This condition is used in a security system to sense if a window is opened. Such a sensor is called a *proximity sensor.*

Relay Specifications

Manufacturers of electromechanical relays always supply a specification sheet for each of their relays. The specification sheet contains voltage and current ratings for both the relay coil and its switch contacts. The specification sheet also includes information regarding the location of the relay coil and switching contact terminals. And finally, the specification sheet will indicate whether the relay can be energized from either an ac or a dc source. The following is an explanation of a relay's most important ratings.

Pickup voltage. The minimum amount of relay coil voltage necessary to energize or operate the relay.

Pickup current. The minimum amount of relay coil current necessary to energize or operate the relay.

Holding current. The minimum amount of current required to keep a relay energized or operating. (The holding current is less than the pickup current.)

Dropout voltage. The maximum relay coil voltage at which the relay is no longer energized.

Contact voltage rating. The maximum voltage the relay contacts can switch safely.

Contact current rating. The maximum current the relay contacts can switch safely.

Contact voltage drop. The voltage drop across the closed contacts of a relay when operating.

Insulation resistance. The resistance measured across the relay contacts in the open position.

Relay Applications

Figure 11-33 shows schematic diagrams for two relay systems. The diagram in Fig. 11-33a represents an open-circuit system. With the control switch S_1 open, the SPST relay contacts are open and the load is inoperative. Closing S_1 energizes the relay. This closes the NO relay contacts and makes the load operative.

Figure 11-33b represents a closed-circuit system. In this case, the relay is energized by the control switch S_1, which is closed during normal operation. With the relay energized, the normally closed relay contacts are open and the load is inoperative. When it is desired to operate the load, the control switch S_1 is opened. This returns the relay contacts to their normally closed position, thereby activating the load.

It is important to note that a relay can be energized using a low-voltage, low-power source. However, the relay contacts can be used to control a circuit whose load consumes much more power at a much higher voltage than the relay coil circuit. In fact, one of the main advantages of using a relay is its ability to switch or control very high power loads with a relatively low amount of input power. In remote-control applications, a relay can control a high-power load a long distance away much more efficiently than a mechanical switch can. When a mechanical switch is used to control a high-power load a long distance away, the I^{2}R power loss in the conductors carrying current to the load can become excessive.

A common example of such remote control is the starting system in a car or truck. The voltage source is the 12-V battery. A low-current, key-operated ignition switch is used to

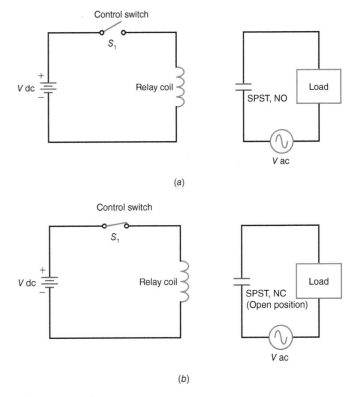

Figure 11-33 Schematic diagrams for two relay systems. (a) Open-circuit system. (b) Closed-circuit system.

energize a remote starter relay located near the starter motor. The starter motor is a powerful dc motor with a gear assembly that spins the engine to give it a start. Such motors require very high current, upwards of 100 A. The starter relay contacts can handle such current, whereas a manual ignition switch could not.

Common Relay Troubles

If a relay coil develops an open, the relay cannot be energized. The reason is simple. With an open relay coil, the current is zero and no magnetic field is set up by the electromagnet to attract the armature. An ohmmeter can be used to check for the proper relay coil resistance. An open relay coil measures infinite (∞) resistance. Since it is usually not practical to repair an open relay coil, the entire relay must be replaced.

A common problem with relays is dirty or corroded switch contacts. The switch contacts develop a thin carbon coating after extended use from arcing across the contact terminals when they are opened and closed. Dirty switch contacts usually produce intermittent operation of the load being controlled—for example, a motor. In some cases, the relay contacts may chatter (vibrate) if they are dirty.

One final point: The manufacturer of a relay usually indicates its life expectancy in terms of the number of times the relay can be energized (operated). A typical value is 5 million operations.

DC Motors

A motor is one of those electrical devices we take for granted. They are mostly invisible but everywhere. There are literally billions of motors in use around us. As an introductory exercise to this sidebar, take an inventory of the motors you own and use daily. Write down all the motors you can find. Include those in your home, car, and, if applicable, work/office environment. You will be totally surprised. Do it now.

Here is just a summary of what you should have found:

- **Home Motors**
  Refrigerator, washing machine, dryer, dishwasher, can opener, blender, mixer. Air conditioner and heating system, vacuum cleaner, analog clocks and watches, tooth brush, shaver, hair dryer, vent fan, ceiling fans, hard disk and DVD drives in computers, DVD/CD/VCR drive motors, PC cooling fans, printer, power tools (drill, saw, etc.), garage door. Golf carts.

- **Work Office Motors**
  Copiers, fax machine, printers, hard disk and DVD drives in PCs, industrial machines in factories, forklifts, robots, electric or hybrid car.

- **Car/Truck Motors**
  Starter motor, windshield wipers, power windows, power seats, cooling fans, air-conditioning and heating fan motors, CD drive in radio/audio system, sun roof.

  Many if not most of these motors have electronic circuits to control them.

  This sidebar introduces you to the most common types of dc motors. A later sidebar will cover ac motors.

Permanent Magnet Motor

A common type of dc motor is the permanent magnet (PM) motor. It uses a permanent magnet to produce the field. The rotating element called the *armature* is a coil of many turns of copper wire. When voltage is applied to the armature coil, a magnetic field is produced. That field interacts with the PM field as shown in Fig. S11-1. Only one turn of wire is shown to simplify the illustration and explanation. The armature coil rotates inside the magnetic field between the PM poles. The attraction and repulsion of the magnetic poles produce rotation.

Note the cylindrical segments attached to the armature coil ends. These rotating electrical contacts are called the *commutator*. They act like a switch that constantly reverses

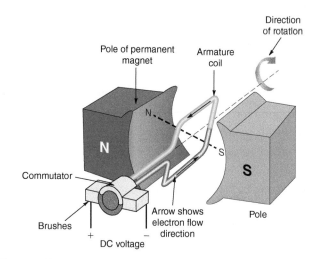

Figure S11-1 Permanent Magnet dc motor.

the dc voltage polarity to the armature to ensure that the magnetic field is in the right direction to keep it rotating in one direction. The commutator segments make contact with fixed connections called brushes that connect to the dc voltage source.

In practical motors, there are many rotor windings and commutator segments that provide a smoother continuous rotation.

Wound-Field DC Motors

PM dc motors are used primarily for low-power applications where the amount of horsepower and torque are relatively low. For higher-power applications, a dc motor with a wound field is used. A *wound field* refers to an electromagnetic field produced by a coil instead of a PM.

The magnetic field is produced by a winding on a set of magnetic poles. This winding is connected to the dc voltage source along with the armature coil. The connection may be in series or parallel, as shown in Fig. S11-2. The series connection (*a*) produces very high starting torque, but its speed is harder to control. The parallel connection (*b*) produces better speed control. Both series and parallel field coils can also be used together for improved speed and torque control.

The speed of a dc motor is a function of the armature current in a dc motor. The higher the current, the stronger the magnetic field and the stronger the interaction with the fixed field. This increases speed. Controlling speed in electromagnetic fields is determined by both the armature

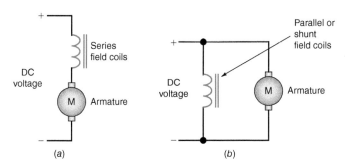

Figure S11-2 Wound-field dc motors produce more horsepower and torque. (*a*) Series-connected dc motor. (*b*) Shunt-controlled dc motor.

current and the field current. The higher the current, the greater the speed.

Direction of rotation of the motor shaft is determined by the direction of current flow in the armature. Reversing the polarity of the applied voltage reverses the direction of rotation.

Stepper Motors

A stepper motor is a dc motor that works on the same magnetic principles as other dc motors except that the motor does not rotate in a smooth continuous motion. Instead, the motor rotates in steps or increments. To make this happen, the motor uses a special structure and pulses of dc instead of a steady continuously applied dc voltage.

Figure S11-3 shows a simple stepper motor design to illustrate the principle. The motor is made up of fixed stator coils that are electromagnets. The rotating element is the rotor, which is a permanent magnet. The stator coils are energized by a sequence of dc voltage pulses, as shown in Fig. S11-4. When the A voltage is on (Fig. S11-3*a*), the upper coil A is energized, creating a magnet that attracts the rotor bar magnet. All other coils are off at this time. The A pulse turns off, and the B voltage energizes the B coil, pulling the rotor around 90° clockwise (Fig. S11-3*b*). The sequence then continues as B goes off and C turns

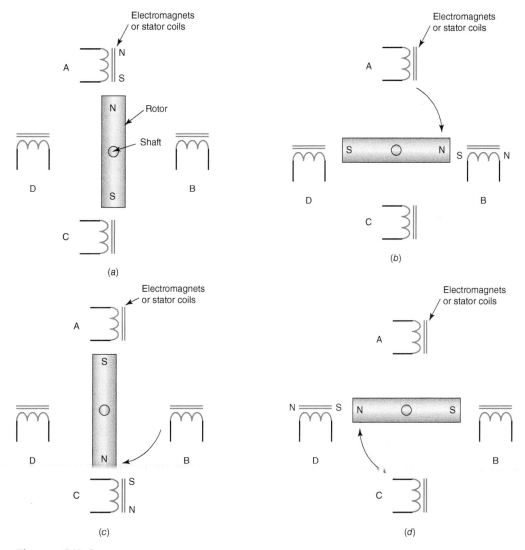

Figure S11-3 Stepper motor operation. (*a*) 0° position: A coil energized; all others off. (*b*) 90° position: B coil energized; all others off. (*c*) 180° position: C coil energized; all others off. (*d*) 270° position: D coil energized; all others off.

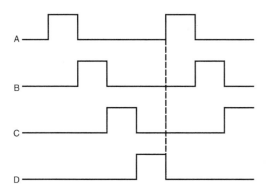

Figure S11-4 DC voltage pulses used to operate the stepper motor stator coils.

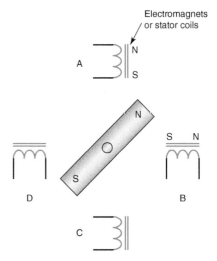

Figure S11-5 Half-stepping; 45° position. Both coils A and B energized.

on, moving the rotor another 90° clockwise (Fig. S11-3c). The C coil is deenergized, and D is turned on again, moving the rotor another 90° (Fig. S11-3d). Then the sequence repeats.

The speed of rotation is based on how fast the dc pulses occur or the pulse frequency. The width or duration of each pulse determines how long the rotor stays in that position.

A 90° stepper motor is impractical, as it is desirable to have many smaller steps for smoother operation. One way to produce more steps is the use what is called half-stepping, where two of the coils are energized at the same time. Figure S11-5 shows how coils A and B can be energized simultaneously. Each coil attracts the rotor equally, so the rotor positions itself halfway between the A and B coils, creating 45° steps. The stator coil pulse sequences are modified to produce the desired half-step operation.

Modern stepper motors use more stator coils and have a rotor that is more like a gear where the teeth are permanent magnets. With this arrangement, step increments of 1.8, 3.6, and 7.5° increments are possible.

The pulse sequences are usually generated by a programmed microcontroller chip or a special stepper motor sequencer chip. Transistors operated by the microcontroller or stepper chip turn the stator coils off and on. Figure S11-6 shows typical circuitry.

Brushless DC Motors

A brushless dc (BLDC) motor is a variation of the stepper motor that produces a continuous rotation rather than

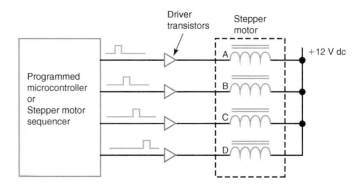

Figure S11-6 Stepper motor control circuits.

steps. By using multiple stator coils, multiple rotor magnets, and unique pulse sequences, the motor can be made to rotate continuously. In some variations, the permanent magnet rotor is outside or surrounding the stator coils. Special electronic circuitry in the form of programmable microcontroller and driver circuits are needed, but these are inexpensive and commonly available. The BLDC motor does not have a commutator and brushes that produce sparking and electrical noise, and the brushes do not wear out. BLDC motors are used in computer disk drives, fans, and other places where standard dc motors are used.

❓ CHAPTER 11 REVIEW QUESTIONS

1. The maxwell (Mx) is a unit of
   a. flux density.
   b. permeability.
   c. magnetic flux.
   d. field intensity.

2. With bar magnets,
   a. like poles attract each other and unlike poles repel each other.
   b. unlike poles attract each other and like poles repel each other.

c. there are no north or south poles on the ends of the magnet.

d. none of the above.

3. The tesla (T) is a unit of
   a. flux density.
   b. magnetic flux.
   c. permeability.
   d. magnetomotive force.

4. 1 maxwell (Mx) is equal to
   a. 1×10^8 Wb.
   b. $\dfrac{1 \text{ Wb}}{\text{m}^2}$.
   c. 1×10^4 G.
   d. one magnetic field line.

5. 1 Wb is equal to
   a. 1×10^8 Mx.
   b. one magnetic field line.
   c. $\dfrac{1 \text{ Mx}}{\text{cm}^2}$.
   d. 1×10^4 kG.

6. The electric or magnetic effect of one body on another without any physical contact between them is called
   a. its permeability.
   b. induction.
   c. the Hall effect.
   d. hysteresis.

7. A commercial permanent magnet will last indefinitely if it is not subjected to
   a. a strong demagnetizing field.
   b. physical shock.
   c. high temperatures.
   d. all of the above.

8. What is the name for a nonmetallic material that has the ferromagnetic properties of iron?
   a. lodestone.
   b. toroid.
   c. ferrite.
   d. solenoid.

9. One tesla (T) is equal to
   a. $\dfrac{1 \text{ Mx}}{\text{m}^2}$.
   b. $\dfrac{1 \text{ Mx}}{\text{cm}^2}$.
   c. $\dfrac{1 \text{ Wb}}{\text{m}^2}$.
   d. $\dfrac{1 \text{ Wb}}{\text{cm}^2}$.

10. The ability of a material to concentrate magnetic flux is called its
    a. induction.
    b. permeability.

c. Hall effect.
d. diamagnetic.

11. If the north (N) pole of a permanent magnet is placed near a piece of soft iron, what is the polarity of the nearest induced pole?
    a. South (S) pole.
    b. North (N) pole.
    c. It could be either a north (N) or a south (S) pole.
    d. It is impossible to determine.

12. A magnet that requires current in a coil to create the magnetic field is called a(n)
    a. permanent magnet.
    b. electromagnet.
    c. solenoid.
    d. either *b* or *c*.

13. The point at which a magnetic material loses its ferromagnetic properties is called the
    a. melting point.
    b. freezing point.
    c. Curie temperature.
    d. leakage point.

14. A material that becomes strongly magnetized in the same direction as the magnetizing field is classified as
    a. diamagnetic.
    b. ferromagnetic.
    c. paramagnetic.
    d. toroidal.

15. Which of the following materials are nonmagnetic?
    a. Air.
    b. Wood.
    c. Glass.
    d. All of the above.

16. The gauss (G) is a unit of
    a. flux density.
    b. magnetic flux.
    c. permeability.
    d. none of the above.

17. One gauss (G) is equal to
    a. $\dfrac{1 \text{ Mx}}{\text{m}^2}$.
    b. $\dfrac{1 \text{ Wb}}{\text{cm}^2}$.
    c. $\dfrac{1 \text{ Mx}}{\text{cm}^2}$.
    d. $\dfrac{1 \text{ Wb}}{\text{m}}$.

18. 1 μWb equals
    a. 1×10^8 Mx.
    b. 10,000 Mx.
    c. 1×10^{-8} Mx.
    d. 100 Mx.

19. A toroid
  a. is an electromagnet.
  b. has no magnetic poles.
  c. uses iron for the core around which the coil is wound.
  d. all of the above.

20. When a small voltage is generated across the width of a conductor carrying current in an external magnetic field, the effect is called
  a. the Doppler effect.
  b. the Miller effect.
  c. the Hall effect.
  d. the Schultz effect.

21. The weber (Wb) is a unit of
  a. magnetic flux.
  b. flux density.
  c. permeability.
  d. none of the above.

22. The flux density in the iron core of an electromagnet is 0.25 T. When the iron core is removed, the flux density drops to 62.5×10^{-6} T. What is the relative permeability of the iron core?
  a. $\mu_r = 4$.
  b. $\mu_r = 250$.
  c. $\mu_r = 4000$.
  d. It is impossible to determine.

23. What is the flux density, B, for a magnetic flux of 500 Mx through an area of 10 cm²?
  a. 50×10^{-3} T.
  b. 50 G.
  c. 5000 G.
  d. Both a and b.

24. The geographic North Pole of the earth has
  a. no magnetic polarity.
  b. south magnetic polarity.
  c. north magnetic polarity.
  d. none of the above.

25. With an electromagnet,
  a. more current and more coil turns mean a stronger magnetic field.
  b. less current and fewer coil turns mean a stronger magnetic field.
  c. if there is no current in the coil, there is no magnetic field.
  d. Both a and c.

26. A current of 20 mA flowing through a coil with 500 turns produces an mmf of
  a. 100 A·t.
  b. 1 A·t.
  c. 10 A·t.
  d. 7.93 A·t.

27. A coil with 1000 turns must provide an mmf of 50 A·t. The required current is
  a. 5 mA.
  b. 0.5 A.
  c. 50 μA.
  d. 50 mA.

28. The left-hand rule for solenoids states that
  a. if the fingers of the left hand encircle the coil in the same direction as electron flow, the thumb points in the direction of the north pole.
  b. if the thumb of the left hand points in the direction of current flow, the fingers point toward the north pole.
  c. if the fingers of the left hand encircle the coil in the same direction as electron flow, the thumb points in the direction of the south pole.
  d. if the thumb of the right hand points in the direction of electron flow, the fingers point in the direction of the north pole.

29. The physical motion resulting from the forces of two magnetic fields is called
  a. Lenz's law.
  b. motor action.
  c. the left-hand rule for coils.
  d. integration.

30. Motor action always tends to produce motion from
  a. a stronger field toward a weaker field.
  b. a weaker field toward a stronger field.
  c. a north pole toward a south pole.
  d. none of the above.

31. A conductor will have an induced current or voltage only when there is
  a. a stationary magnetic field.
  b. a stationary conductor.
  c. relative motion between the wire and magnetic field.
  d. both a and b.

32. The polarity of an induced voltage is determined by
  a. motor action.
  b. Lenz's law.
  c. the number of turns in the coil.
  d. the amount of current in the coil.

33. For a relay, the pickup current is defined as
  a. the maximum current rating of the relay coil.
  b. the minimum relay coil current required to keep a relay energized.
  c. the minimum relay coil current required to energize a relay.
  d. the minimum current in the switching contacts.

34. The movable arm of an attraction-type relay is called the
    a. contacts.
    b. relay coil.
    c. terminal.
    d. armature.

35. For a conductor being moved through a magnetic field, the amount of induced voltage is determined by
    a. the rate at which the conductor cuts the magnetic flux.
    b. the number of magnetic flux lines that are cut by the conductor.
    c. the time of day during which the conductor is moved through the field.
    d. both *a* and *b*.

36. Degaussing is done with
    a. strong permanent magnets.
    b. alternating current.
    c. static electricity.
    d. direct current.

37. Hysteresis losses
    a. increase with higher frequencies.
    b. decrease with higher frequencies.
    c. are greater with direct current.
    d. increase with lower frequencies.

38. The saturation of an iron core occurs when
    a. all of the molecular dipoles and magnetic domains are aligned by the magnetizing force.
    b. the coil is way too long.
    c. the flux density cannot be increased in the core when the field intensity is increased.
    d. both *a* and *c*.

39. The unit of field intensity is the
    a. oersted.
    b. gilbert.
    c. A·t/m.
    d. both *a* and *c*.

40. For a single conductor carrying an alternating current, the associated magnetic field is
    a. only on the top side.
    b. parallel to the direction of current.
    c. at right angles to the direction of current.
    d. only on the bottom side.

41. A coil with 200 mA of current has an mmf of 80 A·t. How many turns does the coil have?
    a. 4000 turns.
    b. 400 turns.
    c. 40 turns.
    d. 16 turns.

42. The magnetic field surrounding a solenoid is
    a. like that of a permanent magnet.
    b. unable to develop north and south poles.
    c. one without magnetic flux lines.
    d. unlike that of a permanent magnet.

43. For a relay, the holding current is defined as
    a. the maximum current the relay contacts can handle.
    b. the minimum amount of relay coil current required to keep a relay energized.
    c. the minimum amount of relay coil current required to energize a relay.
    d. the maximum current required to operate a relay.

44. A vertical wire with electron flow into this page has an associated magnetic field which is
    a. clockwise.
    b. counterclockwise.
    c. parallel to the wire.
    d. none of the above.

45. How much is the induced voltage when a magnetic flux cuts across 150 turns at the rate of 5 Wb/s?
    a. 7.5 kV.
    b. 75 V.
    c. 750 V.
    d. 750 mV.

 CHAPTER 11 ESSAY QUESTIONS

1. Name two magnetic materials and three nonmagnetic materials.

2. Explain the difference between a permanent magnet and an electromagnet.

3. Draw a horseshoe magnet and its magnetic field. Label the magnetic poles, indicate the air gap, and show the direction of flux.

4. Define relative permeability, shielding, induction, and Hall voltage.

5. Give the symbols, cgs units, and SI units for magnetic flux and for flux density.

6. How are the north and south poles of a bar magnet determined with a magnetic compass?

7. What is the difference between flux ϕ and flux density B?

8. Draw a diagram showing two conductors connecting a battery to a load resistance through a closed switch. (a) Show the magnetic field of the current

in the negative side of the line and in the positive side. (b) Where do the two fields aid? Where do they oppose?

9. State the rule for determining the magnetic polarity of a solenoid. (a) How can the polarity be reversed? (b) Why are there no magnetic poles when the current through the coil is zero?

10. Why does the motor action between two magnetic fields result in motion toward the weaker field?

11. Why does current in a conductor perpendicular to this page have a magnetic field in the plane of the paper?

12. Why must the conductor and the external field be perpendicular to each other to have motor action or to generate induced voltage?

13. Explain briefly how either motor action or generator action can be obtained with the same conductor in a magnetic field.

14. Assume that a conductor being cut by the flux of an expanding magnetic field has 10 V induced with the top end positive. Now analyze the effect of the following changes: (a) The magnetic flux continues to expand, but at a slower rate. How does this affect the amount of induced voltage and its polarity? (b) The magnetic flux is constant, neither increasing nor decreasing. How much is the induced voltage? (c) The magnetic flux contracts, cutting across the conductor with the opposite direction of motion. How does this affect the polarity of the induced voltage?

15. Assume that you have a relay whose pickup and holding current values are unknown. Explain how you can determine their values experimentally.

16. List two factors that determine the strength of an electromagnet.

17. What is meant by magnetic hysteresis?

18. What is meant by the saturation of an iron core?

Chapter 12

DC Testing with Analog and Digital Multimeters

Learning Outcomes

After studying this chapter, you should be able to:

> *Explain* the difference between analog and digital meters.

> *Explain* the construction and operation of a moving-coil meter.

> *Calculate* the value of shunt resistance required to extend the current range of a basic moving-coil meter.

> *Calculate* the value of multiplier resistance required to make a basic moving-coil meter capable of measuring voltage.

> *Explain* the ohms-per-volt rating of a voltmeter.

> *Explain* what is meant by *voltmeter loading*.

> *Explain* how a basic moving-coil meter can be used with a battery to construct an ohmmeter.

> *List* the main features of a digital multimeter (DMM)

The digital multimeter (DMM) is the most common measuring instrument used by electronics technicians. All DMMs can measure dc and ac voltage, current, and resistance, and some can even measure and test electronic components such as capacitors, diodes, and transistors. A DMM uses a numeric display to indicate the value of the measured quantity.

An analog multimeter uses a moving pointer and a printed scale. Like a DMM, an analog multimeter can measure voltage, current, and resistance. One disadvantage of an analog multimeter, however, is that the meter reading must be interpreted based on where the moving pointer rests along the printed scale. Although analog multimeters find somewhat limited use in electronics, there is great value in understanding their basic construction and operation. One of the main reasons for covering analog meters is that the concepts of series, parallel, and series-parallel circuits learned earlier are applied. In this chapter, therefore, you will be provided with a basic overview of the construction and operation of an analog meter as well as a concept called voltmeter loading. You will also learn about the main features of a DMM.

12.1 Moving-Coil Meter

Figure 12-1 shows two types of multimeters used by electronics technicians. Figure 12-1a shows an analog volt-ohm-milliammeter (VOM), and Fig. 12-1b shows a digital multimeter (DMM). Both types are capable of measuring voltage, current, and resistance.

A moving-coil meter movement is generally used in an analog VOM. The construction consists of a coil of fine wire wound on a drum mounted between the poles of a permanent magnet. When direct current flows in the coil, the magnetic field of the current reacts with the magnetic field of the permanent magnet.* The resultant force turns the drum with its pointer, winding up the restoring spring. When the current is removed, the pointer returns to zero. The amount of deflection indicates the amount of current in the coil. When the polarity is connected correctly, the pointer will read up-scale, to the right; the incorrect polarity forces the pointer off-scale, to the left.

The pointer deflection is directly proportional to the amount of current in the coil. If 100 μA is the current needed for full-scale deflection, 50 μA in the coil will produce half-scale deflection. The accuracy of the moving-coil meter mechanism is 0.1–2%.

Values of I_M

The full-scale deflection current I_M is the amount needed to deflect the pointer all the way to the right to the last mark on the printed scale. Typical values of I_M are from about 10 μA to 30 mA. In a VOM, the I_M is typically either 50 μA or 1 mA.

Refer to the analog VOM in Fig. 12-1a. The mirror along the scale is used to eliminate reading errors. The meter is read when the pointer and its mirror reflection appear as one. This eliminates the optical error called *parallax* caused by looking at the meter from the side.

Values of r_M

The internal resistance of the wire of the moving coil is represented by r_M. Typical values range from 1.2 Ω for a 30-mA movement to 2000 Ω for a 50-μA movement. A movement

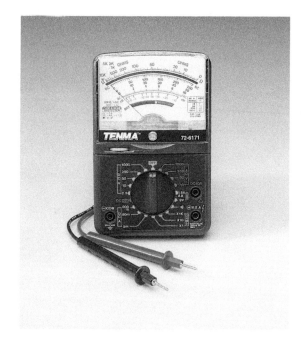

(a)

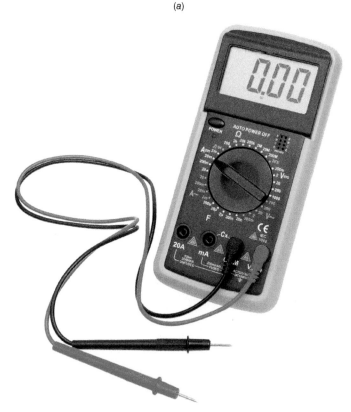

Figure 12–1   Typical multimeters used for measuring V, I, and R. (a) Analog VOM. (b) DMM.

---

* For more details on the interaction between two magnetic fields, see Chap. 11, "Magnetism."

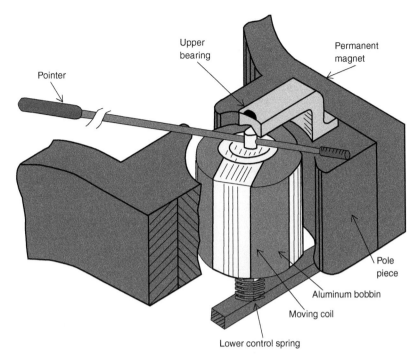

Figure 12-2 Close-up view of a moving-coil meter movement.

with a smaller I_M has a higher r_M because many more turns of fine wire are needed. An average value of r_M for a 1-mA movement is about 50 Ω. Figure 12-2 provides a close-up view of the basic components contained within a meter movement. Notice that the moving coil is wound around a drum which rotates when direct current flows through the wire of the moving coil.

12.2 Meter Shunts

A *meter shunt* is a precision resistor connected across the meter movement for the purpose of shunting, or bypassing, a specific fraction of the circuit's current around the meter movement. The combination then provides a current meter with an extended range. The shunts are usually inside the meter case. However, the schematic symbol for the current meter usually does not show the shunt.

In current measurements, the parallel bank of the movement with its shunt is connected as a current meter in series in the circuit (Fig. 12-3). Note that the scale of a meter with an internal shunt is calibrated to take into account the current through both the shunt and the meter movement. Therefore, the scale reads total circuit current.

Resistance of the Meter Shunt

In Fig. 12-3*b*, the 1-mA movement has a resistance of 50 Ω, which is the resistance of the moving coil r_M. To double the range, the shunt resistance R_S is made equal to the 50 Ω of the movement. When the meter is connected in series in a circuit

where the current is 2 mA, this total current into one terminal of the meter divides equally between the shunt and the meter movement. At the opposite meter terminal, these two branch currents combine to provide the 2 mA of the circuit current.

Inside the meter, the current is 1 mA through the shunt and 1 mA through the moving coil. Since it is a 1-mA movement, this current produces full-scale deflection. The scale is doubled, however, reading 2 mA, to account for the additional 1 mA through the shunt. Therefore, the scale reading indicates total current at the meter terminals, not just coil current. The movement with its shunt, then, is a 2-mA meter. Its internal resistance is $50 \times \frac{1}{2} = 25$ Ω.

Another example is shown in Fig. 12-4. In general, the shunt resistance for any range can be calculated with Ohm's law from the formula

$$R_S = \frac{V_M}{I_S} \qquad \textbf{(12-1)}$$

where R_S is the resistance of the shunt and I_S is the current through it.

Voltage V_M is equal to $I_M \times r_M$. This is the voltage across both the shunt and the meter movement, which are in parallel.

Calculating I_S

This current through the shunt alone is the difference between the total current I_T through the meter and the divided current I_M through the movement or

$$I_S = I_T - I_M \qquad \textbf{(12-2)}$$

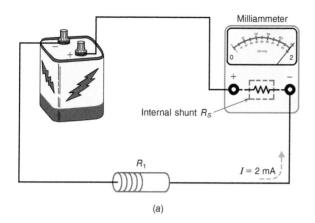

(a)

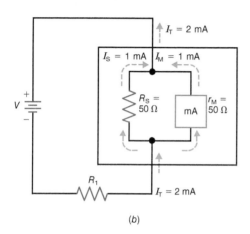

(b)

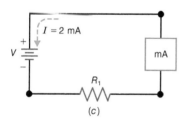

(c)

Figure 12-3 Example of meter shunt R_S in bypassing current around the movement to extend the range from 1 to 2 mA. (a) Wiring diagram. (b) Schematic diagram showing the effect of the shunt. With $R_S = r_M$ the current range is doubled. (c) Circuit with 2-mA meter to read the current.

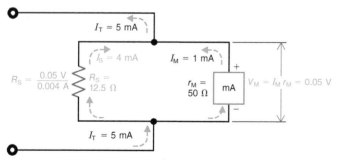

Figure 12-4 Calculating the resistance of a meter shunt. R_S is equal to V_M/I_S. See text for calculations.

Use the values of current for full-scale deflection, as these are known. In Fig. 12-4,

$$I_S = 5 - 1 = 4 \text{ mA} \quad \text{or} \quad 0.004 \text{ A}$$

Calculating R_S

The complete procedure for using the formula $R_S = V_M/I_S$ can be as follows:

1. Find V_M. Calculate this for full-scale deflection as $I_M \times r_M$. In Fig. 12-4, with a 1-mA full-scale current through the 50-Ω movement,

$$V_M = 0.001 \times 50 = 0.05 \text{ V} \quad \text{or} \quad 50 \text{ mV}$$

2. Find I_S. For the values that are shown in Fig. 12-4,

$$I_S = 5 - 1 = 4 \text{ mA} = 0.004 \text{ A} \quad \text{or} \quad 4 \text{ mA}$$

3. Divide V_M by I_S to find R_S. For the final result,

$$R_S = 0.05/0.004 = 12.5 \text{ }\Omega$$

This shunt enables the 1-mA movement to be used for an extended range from 0 to 5 mA.

Note that R_S and r_M are inversely proportional to their full-scale currents. The 12.5 Ω for R_S equals one-fourth the 50 Ω of r_M because the shunt current of 4 mA is four times the 1 mA through the movement for full-scale deflection.

The shunts usually are precision wire-wound resistors. For very low values, a short wire of precise size can be used.

Since the moving-coil resistance, r_M, is in parallel with the shunt resistance, R_S, the resistance of a current meter can be calculated as

$$R_M = \frac{R_S \times r_M}{R_S + r_M}.$$

In general, a current meter should have very low resistance compared with the circuit where the current is being measured. As a general rule, the current meter's resistance should be no greater than $\frac{1}{100}$ of the circuit resistance. The higher the current range of a meter, the lower its shunt resistance, R_S, and in turn the overall meter resistance, R_M.

EXAMPLE 12-1

A shunt extends the range of a 50-μA meter movement to 1 mA. How much is the current through the shunt at full-scale deflection?

Answer:

All currents must be in the same units for Formula (12-2). To avoid fractions, use 1000 μA for the 1-mA I_T. Then

$$I_S = I_T - I_M = 1000 \text{ }\mu\text{A} - 50 \text{ }\mu\text{A}$$
$$I_S = 950 \text{ }\mu\text{A}$$

EXAMPLE 12-2

A 50-μA meter movement has an r_M of 1000 Ω. What R_s is needed to extend the range to 500 μA?

Answer:

The shunt current I_s is $500 - 50$, or 450 μA. Then

$$R_S = \frac{V_M}{I_S}$$

$$= \frac{50 \times 10^{-6}\,\text{A} \times 10^3\,\Omega}{450 \times 10^{-6}\,\text{A}} = \frac{50{,}000}{450}$$

$$R_S = 111.1\ \Omega.$$

12.3 Voltmeters

Although a meter movement responds only to current in the moving coil, it is commonly used for measuring voltage by the addition of a high resistance in series with the movement (Fig. 12-5). The series resistance must be much higher than the coil resistance to limit the current through the coil. The combination of the meter movement with this added series resistance then forms a voltmeter. The series resistor, called a *multiplier,* is usually connected inside the voltmeter case.

Since a voltmeter has high resistance, it must be connected in parallel to measure the potential difference across two points in a circuit. Otherwise, the high-resistance multiplier would add so much series resistance that the current in the circuit would be reduced to a very low value. Connected in parallel, though, the high resistance of the voltmeter is an advantage. The higher the voltmeter resistance, the smaller the effect of its parallel connection on the circuit being tested.

The circuit is not opened to connect the voltmeter in parallel. Because of this convenience, it is common practice to make voltmeter tests in troubleshooting. The voltage measurements apply the same way to an *IR* drop or a generated emf.

The correct polarity must be observed in using a dc voltmeter. Connect the negative voltmeter lead to the negative side of the potential difference being measured and the positive lead to the positive side.

Multiplier Resistance

Figure 12-5 illustrates how the meter movement and its multiplier R_1 form a voltmeter. With 10 V applied by the battery in Fig. 12-5a, there must be 10,000 Ω of resistance to limit the current to 1 mA for full-scale deflection of the meter movement. Since the movement has a 50-Ω resistance, 9950 Ω is added in series, resulting in a 10,000-Ω total resistance. Then *I* is 10 V/10 kΩ = 1 mA.

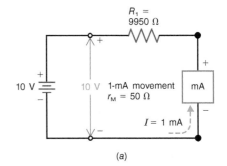

(a)

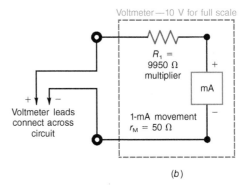

(b)

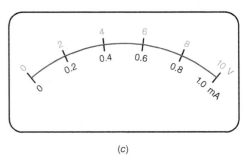

(c)

Figure 12-5 Multiplier resistor R_1 added in series with meter movement to form a voltmeter. (a) Resistance of R_1 allows 1 mA for full-scale deflection in 1-mA movement with 10 V applied. (b) Internal multiplier R_1 forms a voltmeter. The test leads can be connected across a potential difference to measure 0 to 10 V. (c) 10-V scale of voltmeter and corresponding 1-mA scale of meter movement.

With 1 mA in the movement, the full-scale deflection can be calibrated as 10 V on the meter scale, as long as the 9950-Ω multiplier is included in series with the movement. It doesn't matter to which side of the movement the multiplier is connected.

If the battery is taken away, as in Fig. 12-5b, the movement with its multiplier forms a voltmeter that can indicate a potential difference of 0 to 10 V applied across its terminals. When the voltmeter leads are connected across a potential difference of 10 V in a dc circuit, the resulting 1-mA current through the meter movement produces full-scale deflection, and the reading is 10 V. In Fig. 12-5c, the 10-V scale is shown corresponding to the 1-mA range of the movement.

If the voltmeter is connected across a 5-V potential difference, the current in the movement is ½ mA, the deflection is one-half of full scale, and the reading is 5 V. Zero voltage across the terminals means no current in the movement, and the voltmeter reads zero. In summary, then, any potential difference up to 10 V, whether an IR voltage drop or a generated emf, can be applied across the meter terminals. The meter will indicate less than 10 V in the same ratio that the meter current is less than 1 mA.

The resistance of a multiplier can be calculated from the formula

$$R_{mult} = \frac{\text{full-scale } V}{\text{full-scale } I} - r_M \qquad (12\text{-}3)$$

Applying this formula to the example of R_1 in Fig. 12-5 gives

$$R_{mult} = \frac{10 \text{ V}}{0.001 \text{ A}} - 50 \text{ } \Omega = 10{,}000 - 50$$

$$R_{mult} = 9950 \text{ } \Omega \quad \text{or} \quad 9.95 \text{ k}\Omega$$

We can take another example for the same 10-V scale but with a 50-μA meter movement, which is commonly used. Now the multiplier resistance is much higher, though, because less I is needed for full-scale deflection. Let the resistance of the 50-μA movement be 2000 Ω. Then

$$R_{mult} = \frac{10 \text{ V}}{0.000 \, 050 \text{ A}} - 2000 \text{ } \Omega = 200{,}000 - 2000$$

$$R_{mult} = 198{,}000 \text{ } \Omega \quad \text{or} \quad 198 \text{ k}\Omega$$

Typical Multiple Voltmeter Circuit

An example of a voltmeter with multiple voltage ranges is shown in Fig. 12-6. Resistance R_1 is the series multiplier for the lowest voltage range of 2.5 V. When higher resistance is needed for the higher ranges, the switch adds the required series resistors.

The meter in Fig. 12-6 requires 50 μA for full-scale deflection. For the 2.5-V range, a series resistance of $2.5/(50 \times 10^{-6})$, or 50,000 Ω, is needed. Since r_M is 2000 Ω,

the value of R_1 is 50,000 − 2000, which equals 48,000 Ω or 48 kΩ.

For the 10-V range, a resistance of $10/(50 \times 10^{-6})$, or 200,000 Ω, is needed. Since $R_1 + r_M$ provides 50,000 Ω, R_2 is made 150,000 Ω, for a total of 200,000-Ω series resistance on the 10-V range. Similarly, additional resistors are switched in to increase the multiplier resistance for the higher voltage ranges. Note the separate jack and extra multiplier R_6 on the highest range for 5000 V. This method of adding series multipliers for higher voltage ranges is the circuit generally used in commercial multimeters.

Voltmeter Resistance

The high resistance of a voltmeter with a multiplier is essentially the value of the multiplier resistance. Since the multiplier is changed for each range, the voltmeter resistance changes.

Ohms-per-Volt Rating

To indicate the voltmeter's resistance independently of the range, analog voltmeters are generally rated in ohms of resistance needed for 1 V of deflection. This value is the ohms-per-volt rating of the voltmeter. If a meter needs 50,000-Ω R_V for 2.5 V of full-scale deflection, the resistance per 1 V of deflection then is 50,000/2.5, which equals 20,000 Ω/V.

The ohms-per-volt value is the same for all ranges because this characteristic is determined by the full-scale current I_M of the meter movement. To calculate the ohms-per-volt rating, take the reciprocal of I_M in ampere units. For example, a 1-mA movement results in 1/0.001 or 1000 Ω/V; a 50-μA movement allows 20,000 Ω/V, and a 20-μA movement allows 50,000 Ω/V. The ohms-per-volt rating is also called the *sensitivity* of the voltmeter.

A higher ohms-per-volt rating means a higher voltmeter resistance R_V. The R_V can be calculated as the product of the ohms-per-volt rating and the full-scale voltage of each range.

Usually the ohms-per-volt rating of a voltmeter is printed on the meter face.

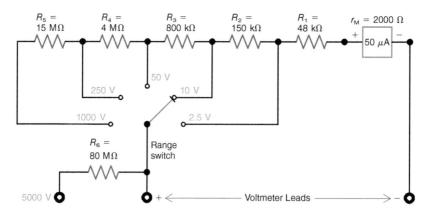

Figure 12-6 A typical voltmeter circuit with multiplier resistors for different ranges.

12.4 Loading Effect of a Voltmeter

When the voltmeter resistance is not high enough, connecting it across a circuit can reduce the measured voltage, compared with the voltage present without the voltmeter. This effect is called *loading down* the circuit, since the measured voltage decreases because of the additional load current for the meter.

Loading Effect

Voltmeter loading can be appreciable in high-resistance circuits, as shown in Fig. 12-7. In Fig. 12-7a, without the voltmeter, R_1 and R_2 form a voltage divider across the applied

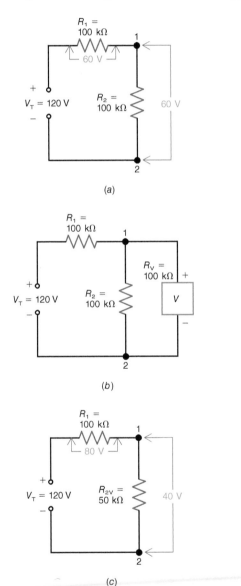

(a)

(b)

(c)

Figure 12-7 How the loading effect of the voltmeter can reduce the voltage reading. (a) High-resistance series circuit without voltmeter. (b) Connecting voltmeter across one of the series resistances. (c) Reduced R and V between points 1 and 2 caused by the voltmeter as a parallel branch across R_2. The R_{2V} is the equivalent of R_2 and R_V in parallel.

voltage of 120 V. The two equal resistances of 100 kΩ each divide the applied voltage equally, with 60 V across each.

When the voltmeter in Fig. 12-7b is connected across R_2 to measure its potential difference, however, the voltage division changes. The voltmeter resistance R_V of 100 kΩ is the value for a 1000-ohms-per-volt meter on the 100-V range. Now the voltmeter in parallel with R_2 draws additional current, and the equivalent resistance between the measured points 1 and 2 is reduced from 100,000 to 50,000 Ω. This resistance is one-third the total circuit resistance, and the measured voltage across points 1 and 2 drops to 40 V, as shown in Fig. 12-7c.

As additional current drawn by the voltmeter flows through the series resistance R_1, this voltage goes up to 80 V.

Similarly, if the voltmeter were connected across R_1, this voltage would go down to 40 V, with the voltage across R_2 rising to 80 V. When the voltmeter is disconnected, the circuit returns to the condition in Fig. 12-7a, with 60 V across both R_1 and R_2.

The loading effect is minimized by using a voltmeter with a resistance much greater than the resistance across which the voltage is measured. As shown in Fig. 12-8, with a voltmeter resistance of 10 MΩ, the loading effect is negligible. Because R_V is so high, it does not change the voltage division in the circuit. The 10 MΩ of the meter in parallel with the

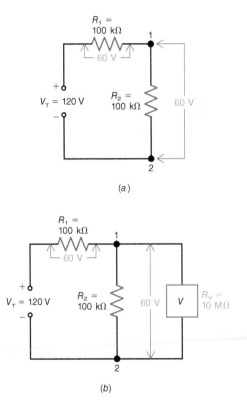

(a)

(b)

Figure 12-8 Negligible loading effect with a high-resistance voltmeter. (a) High-resistance series circuit without voltmeter, as in Fig. 12-7a. (b) Same voltages in circuit with voltmeter connected because R_V is so high.

100,000 Ω for R_2 results in an equivalent resistance practically equal to 100,000 Ω.

With multiple ranges on a VOM, the voltmeter resistance changes with the range selected. Higher ranges require more multiplier resistance, increasing the voltmeter resistance for less loading. As examples, a 20,000-ohms-per-volt meter on the 250-V range has an internal resistance R_V of 20,000 × 250, or 5 MΩ. However, on the 2.5-V range, the same meter has an R_V of 20,000 × 2.5, which is only 50,000 Ω.

On any one range, though, the voltmeter resistance is constant whether you read full-scale or less than full-scale deflection. The reason is that the multiplier resistance set by the range switch is the same for any reading on that range.

Correction for Loading Effect

The following formula can be used to correct for a loading effect:

Actual reading + correction

$$V = V_M + \frac{R_1 R_2}{R_V(R_1 + R_2)} V_M \qquad \textbf{(12-4)}$$

Voltage V is the corrected reading the voltmeter would show if it had infinitely high resistance. Voltage V_M is the actual voltage reading. Resistances R_1 and R_2 are the voltage-dividing resistances in the circuit without the voltmeter resistance R_V. As an example, in Fig. 12-7,

$$V = 40 \text{ V} + \frac{100 \text{ k}\Omega \times 100 \text{ k}\Omega}{100 \text{ k}\Omega \times 200 \text{ k}\Omega} \times 40\text{V}$$

$$= 40 + \frac{1}{2} \times 40 = 40 + 20$$

$$V = 60 \text{ V}$$

The loading effect of a voltmeter causes too low a voltage reading because R_V is too low as a parallel resistance. This corresponds to the case of a current meter reading too low because R_M is too high as a series resistance. Both of these effects illustrate the general problem of trying to make any measurement without changing the circuit being measured.

Note that the digital multimeter (DMM) has practically no loading effect as a voltmeter. The input resistance is usually 10 MΩ or 20 MΩ, the same on all ranges.

12.5 Ohmmeters

An ohmmeter consists of an internal battery, the meter movement, and a current-limiting resistance, as illustrated in Fig. 12-9. For measuring resistance, the ohmmeter leads are connected across the external resistance to be measured. Power in the circuit being tested must be off. Then only the ohmmeter battery produces current for deflecting the meter movement. Since the amount of current through the meter depends on the external resistance, the scale can be calibrated in ohms.

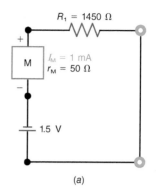

(a)

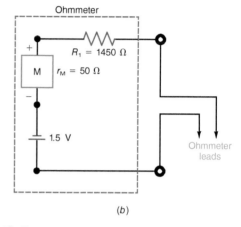

(b)

Figure 12-9 How meter movement M can be used as an ohmmeter with a 1.5-V battery. (a) Equivalent closed circuit with R_1 and the battery when ohmmeter leads are short-circuited for zero ohms of external R. (b) Internal ohmmeter circuit with test leads open, ready to measure an external resistance.

The amount of deflection on the ohms scale indicates the measured resistance directly. The ohmmeter reads up-scale regardless of the polarity of the leads because the polarity of the internal battery determines the direction of current through the meter movement.

Series Ohmmeter Circuit

In Fig. 12-9a, the circuit has 1500 Ω for $(R_1 + r_M)$. Then the 1.5-V cell produces 1 mA, deflecting the moving coil full-scale. When these components are enclosed in a case, as in Fig. 12-9b, the series circuit forms an ohmmeter. Note that M indicates the meter movement.

If the leads are short-circuited together or connected across a short circuit, as in Fig. 12-9a, 1 mA flows. The meter movement is deflected full-scale to the right. This ohmmeter reading is 0 Ω.

When the ohmmeter leads are open, not touching each other, the current is zero. The ohmmeter indicates infinitely high resistance or an open circuit across its terminals.

Therefore, the meter face can be marked zero ohms at the right for full-scale deflection and infinite ohms at the left for no deflection. In-between values of resistance

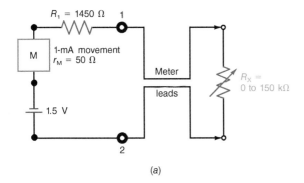

(a)

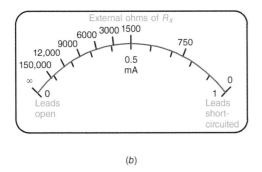

(b)

Figure 12-10 Back-off ohmmeter scale with R readings increasing from right to left. (a) Series ohmmeter circuit for the unknown external resistor R_X to be measured. (b) Ohms scale has higher R readings to the left of the scale as more R_X decreases I_M.

result when less than 1 mA flows through the meter movement. The corresponding deflection on the ohms scale indicates how much resistance is across the ohmmeter terminals.

Back-Off Ohmmeter Scale

Figure 12-10 illustrates the calibration of an ohmmeter scale in terms of meter current. The current equals V/R_T. Voltage V is the fixed applied voltage of 1.5 V supplied by the internal battery. Resistance R_T is the total resistance of R_X and the ohmmeter's internal resistance. Note that R_X is the external resistance to be measured.

The ohmmeter's internal resistance r_i is constant at $50 + 1450$, or $1500\ \Omega$ here. If R_X also equals $1500\ \Omega$, for example, R_T equals $3000\ \Omega$. The current then is 1.5 V/3000 Ω, or 0.5 mA, resulting in half-scale deflection for the 1-mA movement. Therefore, the center of the ohms scale is marked for $1500\ \Omega$. Similarly, the amount of current and meter deflection can be calculated for any value of the external resistance R_X.

Note that the ohms scale increases from right to left. This arrangement is called a *back-off scale,* with ohm values increasing to the left as the current backs off from full-scale deflection. The back-off scale is a characteristic of

any ohmmeter where the internal battery is in series with the meter movement. Then more external R_X decreases the meter current.

A back-off ohmmeter scale is expanded at the right near zero ohms and crowded at the left near infinite ohms. This nonlinear scale results from the relation $I = V/R$ with V constant at 1.5 V. Recall that with V constant, I and R are inversely related.

The highest resistance that can be indicated by the ohmmeter is about 100 times its total internal resistance. Therefore, the infinity mark on the ohms scale, or the "lazy eight" symbol ∞ for infinity, is only relative. It just means that the measured resistance is infinitely greater than the ohmmeter resistance.

It is important to note that the ohmmeter circuit in Fig. 12-10 is not entirely practical. The reason is that if the battery voltage, V_b, is not exactly 1.5 V, the ohmmeter scale will not be calibrated correctly. Also, a single ohmmeter range is not practical when it is necessary to measure very small or large resistance values. Without going into the circuit detail, you should be aware of the fact that commercially available ohmmeters are designed to provide multiple ohmmeter ranges as well as compensation for a change in the battery voltage, V_b.

Multiple Ohmmeter Ranges

Commercial multimeters provide for resistance measurements from less than 1 Ω up to many megohms in several ranges. The range switch in Fig. 12-11 shows the multiplying factors for the ohms scale. On the $R \times 1$ range, for low-resistance measurements, read the ohms scale directly. In the example here, the pointer indicates 12 Ω. When the range switch is on $R \times 100$, multiply the scale reading by 100; this reading would then be 12×100 or 1200 Ω. On the $R \times 10,000$ range, the pointer would indicate 120,000 Ω.

A multiplying factor, instead of full-scale resistance, is given for each ohm range because the highest resistance is

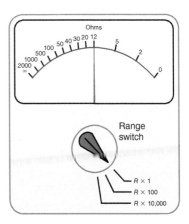

Figure 12-11 Multiple ohmmeter ranges with just one ohms scale. The ohm reading is multiplied by the factor set on the range switch.

infinite on all ohm ranges. This method for ohms should not be confused with full-scale values for voltage ranges. For the ohmmeter ranges, always multiply the scale reading by the $R \times$ factor. On voltage ranges, you may have to multiply or divide the scale reading to match the full-scale voltage with the value on the range switch.

Zero-Ohms Adjustment

To compensate for lower voltage output as the internal battery ages, an ohmmeter includes a variable resistor to calibrate the ohms scale. A back-off ohmmeter is always adjusted for zero ohms. With the test leads short-circuited, vary the ZERO OHMS control on the front panel of the meter until the pointer is exactly on zero at the right edge of the ohms scale. Then the ohm readings are correct for the entire scale.

This type of ohmmeter must be zeroed again every time you change the range because the internal circuit changes.

When the adjustment cannot deflect the pointer all the way to zero at the right edge, it usually means that the battery voltage is too low and it must be replaced. Usually, this trouble shows up first on the $R \times 1$ range, which takes the most current from the battery.

12.6 Multimeters

Multimeters are also called *multitesters,* and they are used to measure voltage, current, or resistance. Table 12-1 compares the features of the main types of multimeters: first, the volt-ohm-milliammeter (VOM) in Fig. 12-12, and next the digital multimeter (DMM) in Fig. 12-13. The DMM is explained in more detail in the next section.

Besides its digital readout, an advantage of the DMM is its high input resistance R_V as a dc voltmeter. The R_V is usually 10 MΩ, the same on all ranges, which is high enough to prevent any loading effect by the voltmeter in most circuits. Some types have an R_V of 22 MΩ. Many modern DMMs are autoranging; that is, the internal circuitry selects the proper range for the meter and indicates the range as a readout.

Figure 12-12 Analog VOM that combines a function selector and range switch.

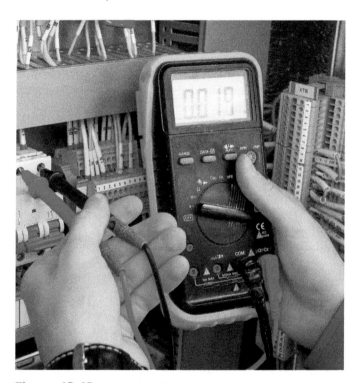

Figure 12-13 Portable digital multimeter (DMM).

For either a VOM or a DMM, it is important to have a low-voltage dc scale with resolution good enough to read 0.2 V or less. The range of 0.2 to 0.6 V, or 200 to 600 mV, is needed for measuring dc bias voltages in transistor circuits.

Table 12-1 **VOM Compared to DMM**	
VOM	**DMM**
Analog pointer reading	Digital readout
DC voltmeter R_V changes with range	R_V is 10 or 22 MΩ, the same on all ranges
Zero-ohms adjustment changed for each range	Zero-ohms adjustment unnecessary
Ohm ranges up to $R \times 10,000\ \Omega$, as a multiplying factor	Ohm ranges up to 20 MΩ; each range is the maximum

Low-Power Ohms (LPΩ)

Another feature needed for transistor measurements is an ohmmeter that does not have enough battery voltage to bias a semiconductor junction into the ON or conducting state. The limit is 0.2 V or less. The purpose is to prevent any parallel conduction paths in the transistor amplifier circuit that can lower the ohmmeter reading.

Decibel Scale

Most analog multimeters have an ac voltage scale calibrated in decibel (dB) units, for measuring ac signals. The decibel is a logarithmic unit used for comparison of power levels or voltage levels. The mark of 0 dB on the scale indicates the reference level, which is usually 0.775 V for 1 mW across 600 Ω. Positive decibel values above the zero mark indicate ac voltages above the reference of 0.775 V; negative decibel values are less than the reference level.

Amp-Clamp Probe

The problem of opening a circuit to measure I can be eliminated by using a probe with a clamp that fits around the current-carrying wire. Its magnetic field is used to indicate the amount of current. An example is shown in Fig. 12-14. The clamp probe measures just ac amperes, generally for the 60-Hz ac power line.

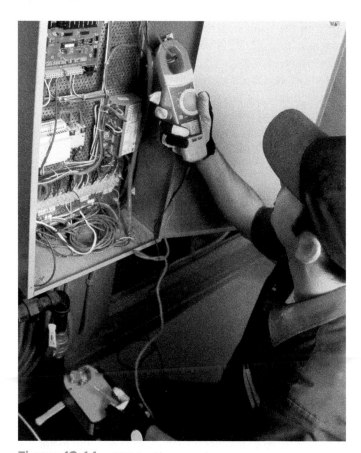

Figure 12-14 DMM with amp-clamp accessory.

High-Voltage Probe

An accessory probe can be used with a multimeter to measure dc voltages up to 30 kV. This probe is often referred to as a *high-voltage probe*. One application is measuring the high voltage of 20 to 30 kV at the anode of the color picture tube in a television receiver. The probe is just an external multiplier resistance for the dc voltmeter. The required R for a 30-kV probe is 580 MΩ with a 20-kΩ/V meter on the 1000-V range.

12.7 Digital Multimeter (DMM)

The digital multimeter has become a very popular test instrument because the digital value of the measurement is displayed automatically with decimal point, polarity, and the unit for V, A, or Ω. Digital meters are generally easier to use because they eliminate the human error that often occurs in reading different scales on an analog meter with a pointer. Examples of the portable DMM are shown in Fig. 12-13.

The basis of the DMM operation is an analog-to-digital (A/D) converter circuit. It converts analog voltage values at the input to an equivalent binary form. These values are processed by digital circuits to be shown on a liquid-crystal display (LCD) as decimal values.

Voltage Measurements

The A/D converter requires a specific range of voltage input; typical values are −200 mV to +200 mV. For DMM input voltages that are higher, the voltage is divided down. When the input voltage is too low, it is increased by a dc amplifier circuit. The measured voltage can then be compared to a fixed reference voltage in the meter by a comparator circuit. Actually, all functions in the DMM depend on the voltage measurements by the converter and comparator circuits.

The input resistance of the DMM is in the range of 10 to 20 MΩ. This R is high enough to eliminate the problem of voltmeter loading in most transistor circuits. Not only does the DMM have high input resistance, but the R is the same on all ranges.

With ac measurements, the ac input is converted to dc voltage for the A/D converter. The DMM has an internal diode rectifier that serves as an ac converter.

R Measurement

As an ohmmeter, the internal battery supplies I through the measured R for an IR drop measured by the DMM. The battery is usually the small 9-V type commonly used in portable equipment. A wide range of R values can be measured from a fraction of an ohm to more than 30 MΩ. Remember that power must be off in the circuit being tested with an ohmmeter.

A DMM ohmmeter usually has an open-circuit voltage across the meter leads, which is much too low to turn on a semiconductor junction. The result is low-power ohms operation.

I Measurements

To measure current, internal resistors provide a proportional *IR* voltage. The display shows the *I* values. Note that the DMM still must be connected as a series component in the circuit when current is measured.

Diode Test

The DMM usually has a setting for testing semiconductor diodes, either silicon or germanium. Current is supplied by the DMM for the diode to test the voltage across its junction. Normal values are 0.7 V for silicon and 0.3 V for germanium. A short-circuited junction will read 0 V. The voltage across an open diode reads much too high. Most diodes are silicon.

Resolution

This term for a DMM specifies how many places can be used to display the digits 0 to 9, regardless of the decimal point. For example, 9.99 V is a three-digit display; 9.999 V would be a four-digit display. Most portable units, however, compromise with a 3½-digit display. This means that the fourth digit at the left for the most significant place can only be a 1. If not, then the display has three digits. As examples, a 3½-digit display can show 19.99 V, but 29.99 V would be read as 30.0 V. Note that better resolution with more digits can be obtained with more expensive meters, especially the larger DMM units for bench mounting. Actually, though, 3½-digit resolution is enough for practically all measurements made in troubleshooting electronic equipment.

Range Overload

The DMM selector switch has specific ranges. Any value higher than the selected range is an overload. An indicator on the display warns that the value shown is not correct. Then a higher range is selected. Some units have an *autorange function* that shifts the meter automatically to a higher range as soon as an overload is indicated.

12.8 Meter Applications

Table 12-2 summarizes the main points to remember when using a voltmeter, ohmmeter, or milliammeter. These rules apply whether the meter is a single unit or one function on a multimeter. The voltage and current tests also apply to either dc or ac circuits.

To avoid excessive current through the meter, it is good practice to start on a high range when measuring an unknown value of voltage or current. It is very important not to make the mistake of connecting a current meter in parallel, because **usually this mistake ruins the meter.** The mistake of connecting a voltmeter in series does not damage the meter, but the reading will be wrong.

If the ohmmeter is connected to a circuit in which power is on, the meter can be damaged, beside giving the wrong

Table 12-2 Direct-Current Meters

Voltmeter	Milliammeter or Ammeter	Ohmmeter
Power on in circuit	Power on in circuit	Power off in circuit
Connect in parallel	Connect in series	Connect in parallel
High internal *R*	Low internal *R*	Has internal battery
Has internal series multipliers; higher *R* for higher ranges	Has internal shunts; lower resistance for higher current ranges	Higher battery voltage and more sensitive meter for higher ohm ranges

reading. An ohmmeter has its own internal battery, and the power must be off in the circuit being tested. When *R* is tested with an ohmmeter, it may be necessary to disconnect one end of *R* from the circuit to eliminate parallel paths.

Connecting a Current Meter in the Circuit

In a series-parallel circuit, the current meter must be inserted in a branch to read branch current. In the main line, the meter reads the total current. These different connections are illustrated in Fig. 12-15. The meters are shown by dashed lines to illustrate the different points at which a meter could be connected to read the respective currents.

If the circuit is opened at point A to insert the meter in series in the main line here, the meter will read total line current I_T through R_1. A meter at B or C will read the same line current.

To read the branch current through R_2, this R must be disconnected from its junction with the main line at either end. A meter inserted at D or E, therefore, will read the R_2 branch current I_2. Similarly, a meter at F or G will read the R_3 branch current I_3.

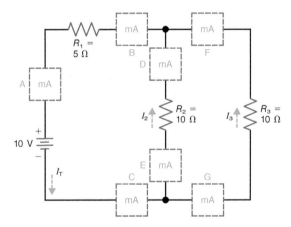

Figure 12-15 How to insert a current meter in different parts of a series-parallel circuit to read the desired current *I*. At point A, B, or C, the meter reads I_T; at D or E, the meter reads I_2; at F or G, the meter reads I_3.

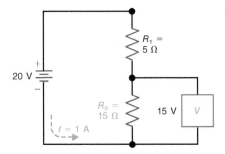

Figure 12-16 With 15 V measured across a known R of 15 Ω, the I can be calculated as V/R or 15/15 Ω = 1 A.

Calculating I from Measured Voltage

The inconvenience of opening the circuit to measure current can often be eliminated by the use of Ohm's law. The voltage and resistance can be measured without opening the circuit, and the current calculated as V/R. In the example in Fig. 12-16, when the voltage across R_2 is 15 V and its resistance is 15 Ω, the current through R_2 must be 1 A. When values are checked during troubleshooting, if the voltage and resistance are normal, so is the current.

This technique can also be convenient for determining I in low-resistance circuits where the resistance of a microammeter may be too high. Instead of measuring I, measure V and R and calculate I as V/R.

Furthermore, if necessary, we can insert a known resistance R_S in series in the circuit, temporarily, just to measure V_S. Then I is calculated as V_S/R_S. The resistance of R_S, however, must be small enough to have little effect on R_T and I in the series circuit.

This technique is often used with oscilloscopes to produce a voltage waveform of IR which has the same waveform as the current in a resistor. The oscilloscope must be connected as a voltmeter because of its high input resistance.

Checking Fuses

Turn the power off or remove the fuse from the circuit to check with an ohmmeter. A good fuse reads 0 Ω. A blown fuse is open, which reads infinity on the ohmmeter.

A fuse can also be checked with the power on in the circuit by using a voltmeter. Connect the voltmeter across the two terminals of the fuse. A good fuse reads 0 V because there is practically no IR drop. With an open fuse, though, the voltmeter reading is equal to the full value of the applied voltage. Having the full applied voltage seems to be a good idea, but it should not be across the fuse.

Voltage Tests for an Open Circuit

Figure 12-17 shows four equal resistors in series with a 100-V source. A ground return is shown here because voltage measurements are usually made with respect to chassis or earth

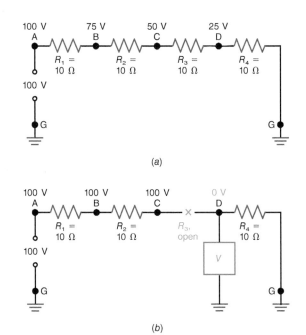

(a)

(b)

Figure 12-17 Voltage tests to localize an open circuit. (a) Normal circuit with voltages to chassis ground. (b) Reading of 0 V at point D shows R_3 is open.

ground. Normally, each resistor would have an IR drop of 25 V. Then, at point B, the voltmeter to ground should read $100 - 25 = 75$ V. Also, the voltage at C should be 50 V, with 25 V at D, as shown in Fig. 12-17a.

However, the circuit in Fig. 12-17b has an open in R_3 toward the end of the series string of voltages to ground. Now when you measure at B, the reading is 100 V, equal to the applied voltage. This full voltage at B shows that the series circuit is open without any IR drop across R_1. The question is, however, which R has the open? Continue the voltage measurements to ground until you find 0 V. In this example, the open is in R_3 between the 100 V at C and 0 V at D.

The points that read the full applied voltage have a path back to the source of voltage. The first point that reads 0 V has no path back to the high side of the source. Therefore, the open circuit must be between points C and D in Fig. 12-17b.

12.9 Checking Continuity with the Ohmmeter

A wire conductor that is continuous without a break has practically zero ohms of resistance. Therefore, the ohmmeter can be useful in testing for continuity. This test should be done on the lowest ohm range. There are many applications. A wire conductor can have an internal break which is not visible because of the insulated cover, or the wire can have a bad connection at the terminal. Checking for zero ohms between any two points along the conductor tests continuity. A break in the conducting path is evident from a reading of infinite resistance, showing an open circuit.

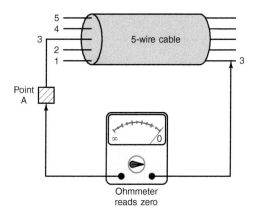

Figure 12-18 Continuity testing from point A to wire 3 shows that this wire is connected.

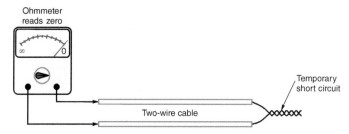

Figure 12-19 Temporary short circuit at one end of a long two-wire line to check continuity from the opposite end.

As another application of checking continuity, suppose that a cable of wires is harnessed together, as illustrated in Fig. 12-18, where the individual wires cannot be seen, but it is desired to find the conductor that connects to terminal A. This is done by checking continuity for each conductor to point A. The wire that has zero ohms to A is the one connected to this terminal. Often the individual wires are color-coded, but it may be necessary to check the continuity of each lead.

An additional technique that can be helpful is illustrated in Fig. 12-19. Here it is desired to check the continuity of the two-wire line, but its ends are too far apart for the ohmmeter leads to reach. The two conductors are temporarily short-circuited at one end, however, so that the continuity of both wires can be checked at the other end.

In summary, the ohmmeter is helpful in checking the continuity of any wire conductor. This includes resistance-wire heating elements, such as the wires in a toaster or the filament of an incandescent bulb. Their cold resistance is normally just a few ohms. Infinite resistance means that the wire element is open. Similarly, a good fuse has practically zero resistance. A burned-out fuse has infinite resistance; that is, it is open. Any coil for a transformer, solenoid, or motor will also have infinite resistance if the winding is open.

12.10 Alternating Current Measurements

All VOMs and DMMs can also measure alternating current (ac) and ac voltage. The meters are calibrated to measure ac with a sine-wave shape and will display the value of the root-mean-square (rms) or the effective value. The meters are usually only accurate for ac frequencies below about 5 kHz. Check the meter specifications for the exact figure. You will learn more about ac measurements in the next part of this book.

CHAPTER 12 SYSTEM SIDEBAR

Data Acquisition System

A data acquisition system, sometimes called a DAS or DAQ, is a collection of electronic components and equipment used to gather, store, process, and display data. The whole idea is to gather physical information (usually at a remote location), capture it, and ultimately use it to make decisions about the object or environment being investigated. Such systems are used in a huge variety of scientific and engineering endeavors, especially research and development. Here are a few examples:

- Medical patient monitoring
- Technical stress measurements on a new aircraft or missile in flight
- Monitoring the process flow in a chemical or petroleum plant

- Gathering data on the performance of a new automotive engine
- A weather station
- A satellite used for observation

System Components

A DAS is made up of the following components and equipment. These are illustrated in Fig. S12-1.

- Sensors. These are transducers that convert physical characteristics like temperature, pressure, light level, position, or fluid level/flow rate into an electric signal that can be captured and measured. Some sensors are analog devices, but others may be digital, like switch closures.

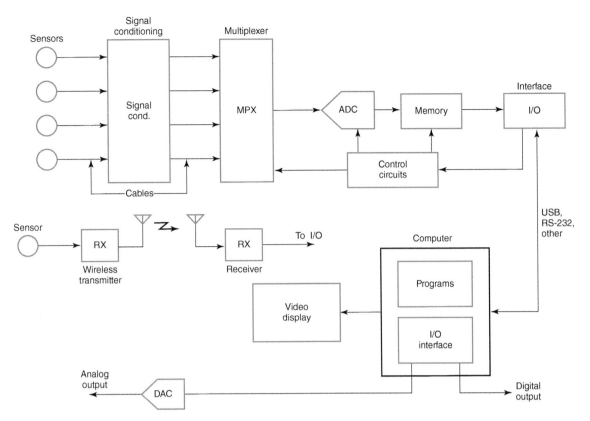

Figure S12-1   A block diagram of a generic data acquisition system.

- Signal conditioning. The signals from transducers may be small and need amplification or must be adjusted in some way. DC power may have to be applied or some network used to linearize the output. Filtering to remove noise is another signal-conditioning function.

- Connection medium. This is usually the cable that connects the sensor to the signal-conditioning equipment. Most cables are twisted pair conductors, and some may be shielded to protect against noise. The medium could also be a wireless link.

- Multiplexer. A multiplexer is a multi-input high-speed electronic switch that selects one of the sensors to be measured and captured. The multiplexer accepts an address from a computer that tells it which sensor to read.

- Analog-to-digital converter (ADC). The ADC digitizes the analog sensor information. It generates a sequence of binary numbers representing the sensor signal.

- Memory. This is an electronic memory that stores binary data for use later. It can be separate memory or part of a related computer. Different types of memory can be used, including flash, DRAM, hard disk drive, or magnetic tape.

- Input/output (I/O) interface. This is an electronic connection that sends the sensor data from the ADC to the

computer. It is a digital interface common to a computer, such as a USB port.

- Outputs. Some DAQ systems have outputs. These may be analog output voltages or digital signals. Outputs are generated by the computer to provide control signals back to the system being monitored. The control signals are usually feedback signals that respond to conditions in the system. Analog outputs are generated by a digital-to-analog converter (DAC) with input from the computer via the I/O interface.

- Computer. This is usually a laptop or desktop computer, but it could also be an embedded microcontroller dedicated to the DAS. The computer usually automates the reading of the sensors and the collection of all the data. It organizes the data into files.

- Real-time clock (RTC). An RTC is an electronic clock that keeps the time and date in digital form that can be attached to and stored with the data. In some applications it is useful to know exactly when the data were captured. The RTC may be part of the computer.

- Software. The programs that process and analyze the data may convert measurement units from the sensors, make calculations, scale the readings, compute new data from the measurements, and otherwise process the data as needed by the application. The software converts the raw data into information that can be used to indicate the status of the tested device or make decisions about what to change. The software helps analyze and make sense of the data so knowledge of how to respond can be developed.

- Display. This is usually the computer display screen. The data can be displayed directly or converted into more usable and understandable charts, graphs, curves, or other visuals.

A Commercial Product

Figure S12-2 shows a commercial DAQ. It is the National Instruments NI USB-6008. It has eight analog inputs that can range from ±1 V to ±10 V. The ADC samples the signals at a maximum rate of 10k samples per second (S/s) and generates 12-bit binary numbers. There is a small internal memory of 512 bytes (a byte is 8 bits). The unit connects to an external computer with a standard universal serial bus (USB) cable and connectors. The external laptop or PC hosts the unit and provides additional storage for the data. Multiple types of software are available to automate the system and analyze and display the data.

This is only one of many different configurations. Other DAQ systems are contained on a printed circuit board that plugs into a PC bus. Others are in the form of modules that

Figure S12-2 The NI USB 6008 DAQ. Its dimensions are 8.5 by 8.2 by 2.3 cm. The screw terminals on both sides are used for connecting the external sensor cables from transducers or signal-conditioning equipment. The USB connector is on the top.

plug into a special chassis. You can even get a complete DAQ on a single chip for small applications. The different blocks in Fig. 12-1 are partitioned and packaged differently in a variety of products.

❓ CHAPTER 12 REVIEW QUESTIONS

1. For a moving-coil meter movement, I_M is
   a. the amount of current needed in the moving-coil to produce full-scale deflection of the meter's pointer.
   b. the value of current flowing in the moving-coil for any amount of pointer deflection.
   c. the amount of current required in the moving-coil to produce half-scale deflection of the meter's pointer.
   d. none of the above.

2. For an analog VOM with a mirror along the printed scale,
   a. the pointer deflection will be magnified by the mirror when measuring small values of voltage, current, and resistance.
   b. the meter should always be read by looking at the meter from the side.
   c. the meter is read when the pointer and its mirror reflection appear as one.
   d. both *a* and *b*.

3. A current meter should have a(n)
   a. very high internal resistance.
   b. very low internal resistance.
   c. infinitely high internal resistance.
   d. none of the above.

4. A voltmeter should have a
   a. resistance of about 0 Ω.
   b. very low resistance.
   c. very high internal resistance.
   d. none of the above.

5. Voltmeter loading is usually a problem when measuring voltages in
   a. parallel circuits.
   b. low-resistance circuits.
   c. a series circuit with low-resistance values.
   d. high-resistance circuits.

6. To double the current range of a 50-μA, 2-kΩ moving-coil meter movement, the shunt resistance, R_S, should be

a. 2 kΩ.
b. 1 kΩ.
c. 18 kΩ.
d. 50 kΩ.

7. A voltmeter using a 20-μA meter movement has an Ω/V rating of
   a. $\dfrac{20 \text{ k}\Omega}{\text{V}}$.
   b. $\dfrac{50 \text{ k}\Omega}{\text{V}}$.
   c. $\dfrac{1 \text{ k}\Omega}{\text{V}}$.
   d. $\dfrac{10 \text{ M}\Omega}{\text{V}}$.

8. As the current range of an analog meter is increased, the overall meter resistance, R_M,
   a. decreases.
   b. increases.
   c. stays the same.
   d. none of the above.

9. As the voltage range of an analog VOM is increased, the total voltmeter resistance, R_V,
   a. decreases.
   b. increases.
   c. stays the same.
   d. none of the above.

10. An analog VOM has an Ω/V rating of 10 kΩ/V. What is the voltmeter resistance, R_V, if the voltmeter is set to the 25-V range?
    a. 10 kΩ.
    b. 10 MΩ.
    c. 25 kΩ.
    d. 250 kΩ.

11. What shunt resistance, R_S, is needed to make a 100-μA, 1-kΩ meter movement capable of measuring currents from 0 to 5 mA?
    a. 25 Ω.
    b. 10.2 Ω.
    c. 20.41 Ω.
    d. 1 kΩ.

12. For a 30-V range, a 50-μA, 2-kΩ meter movement needs a multiplier resistor of
    a. 58 kΩ.
    b. 598 kΩ.
    c. 10 MΩ.
    d. 600 kΩ.

13. When set to any of the voltage ranges, a typical DMM has an input resistance of
    a. about 0 Ω.
    b. 20 kΩ.

c. 10 MΩ.
d. 1 kΩ.

14. When using an ohmmeter to measure resistance in a circuit,
    a. the power in the circuit being tested must be off.
    b. the power in the circuit being tested must be on.
    c. the power in the circuit being tested may be on or off.
    d. the power in the circuit being tested should be turned on after the leads are connected.

15. Which of the following voltages cannot be displayed by a DMM with a 3½-digit display?
    a. 7.64 V.
    b. 13.5 V.
    c. 19.98 V.
    d. 29.98 V.

16. What type of meter can be used to measure ac currents without breaking open the circuit?
    a. An analog VOM.
    b. An amp-clamp probe.
    c. A DMM.
    d. There isn't such a meter.

17. Which of the following measurements is usually the most inconvenient and time-consuming when troubleshooting?
    a. Resistance measurements.
    b. DC voltage measurements.
    c. Current measurements.
    d. AC voltage measurements.

18. An analog ohmmeter reads 18 on the $R \times 10k$ range. What is the value of the measured resistance?
    a. 180 kΩ.
    b. 18 kΩ.
    c. 18 Ω.
    d. 180 Ω.

19. Which meter has a higher resistance, a DMM with 10 MΩ of resistance on all dc voltage ranges or an analog VOM with a 50 kΩ/V rating set to the 250-V range?
    a. The DMM.
    b. The analog VOM.
    c. They both have the same resistance.
    d. It cannot be determined.

20. When using an ohmmeter to measure the continuity of a wire, the resistance should measure
    a. about 0 Ω if the wire is good.
    b. infinity if the wire is broken (open).
    c. very high resistance if the wire is good.
    d. both *a* and *b*.

SECTION 12.2 Meter Shunts

12.1 Calculate the value of the shunt resistance, R_S, needed to extend the range of the meter movement in Fig. 12-20 to (a) 10 mA; (b) 25 mA.

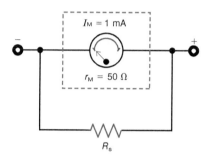

Figure 12-20

12.2 Calculate the value of the shunt resistance, R_S, needed to extend the range of the meter movement in Fig. 12-21 to (a) 100 μA; (b) 1 mA; (c) 5 mA; (d) 10 mA; (e) 50 mA; (f) 100 mA.

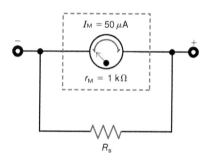

Figure 12-21

12.3 Why is it desirable for a current meter to have very low internal resistance?

SECTION 12.3 Voltmeters

12.4 Calculate the required multiplier resistance, R_{mult}, in Fig. 12-22 for each of the following voltage ranges: (a) 10 V; (b) 500 V.

12.5 What is the Ω/V rating of the voltmeter in Prob. 12-4?

12.6 Calculate the required multiplier resistance, R_{mult}, in Fig. 12-23 for each of the following voltage ranges: (a) 3 V; (b) 100 V.

12.7 What is the Ω/V rating of the voltmeter in Prob. 12-6?

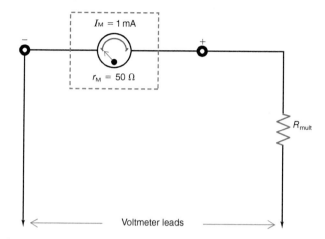

Figure 12-22

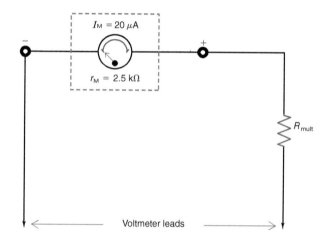

Figure 12-23

SECTION 12.4 Loading Effect of a Voltmeter

12.8 Refer to Fig. 12-24. (a) Calculate the dc voltage that should exist across R_2 without the voltmeter present. (b) Calculate the dc voltage that would be measured across R_2 using a 10 kΩ/V analog voltmeter set to the 10-V range. (c) Calculate the dc voltage that would be measured across R_2 using a DMM having an R_V of 10 MΩ on all dc voltage ranges.

12.9 Refer to Fig. 12-25. (a) Calculate the dc voltage that should exist across R_2 without the voltmeter present. (b) Calculate the dc voltage that would be measured across R_2 using a 100 kΩ/V analog voltmeter set to the 10-V range. (c) Calculate the dc voltage that would be measured across R_2 using a DMM with an R_V of 10 MΩ on all dc voltage ranges.

12.10 In Prob. 12-9, which voltmeter produced a greater loading effect? Why?

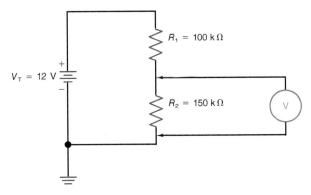

Figure 12-24

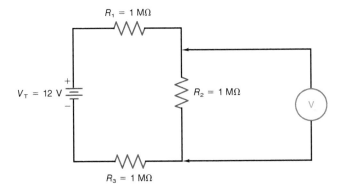

Figure 12-25

SECTION 12.5 Ohmmeters

12.11 Figure 12-26 shows a series ohmmeter and its corresponding meter face. How much is the external resistance, R_X, across the ohmmeter leads for (a) full-scale deflection; (b) three-fourths full-scale deflection; (c) one-half-scale deflection; (d) one-fourth full-scale deflection; (e) no deflection?

12.12 For the series ohmmeter in Fig. 12-26, is the orientation of the ohmmeter leads important when measuring the value of a resistor?

12.13 Analog multimeters have a zero-ohm adjustment control for the ohmmeter portion of the meter. What purpose does it serve and how is it used?

SECTION 12.8 Meter Applications

12.14 On what range should you measure an unknown value of voltage or current? Why?

12.15 What might happen to an ohmmeter if it is connected across a resistor in a live circuit?

12.16 Why is one lead of a resistor disconnected from the circuit when measuring its resistance value?

12.17 Is a current meter connected in series or in parallel? Why?

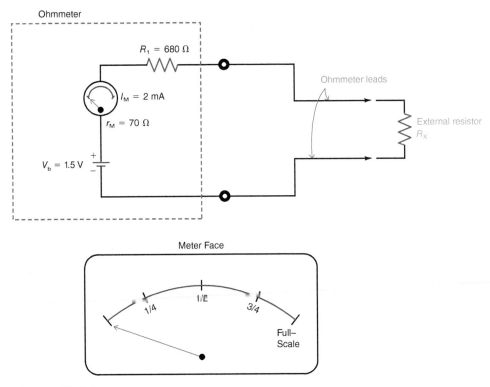

Figure 12-26

12.18 How can the inconvenience of opening a circuit to measure current be eliminated in most cases?

12.19 What is the resistance of a

  a. good fuse?

  b. blown fuse?

12.20 In Fig. 12-27, list the voltages at points A, B, C, and D (with respect to ground) for each of the following situations:

  a. All resistors normal.

  b. R_1 open.

  c. R_2 open.

  d. R_3 open.

  e. R_4 open.

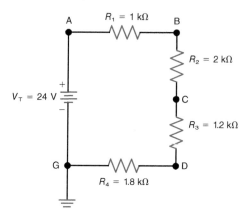

Figure 12-27

DC Troubleshooting

Learning Outcomes

After studying this chapter, you should be able to:

> *Use* a multimeter in measuring voltage, current, and resistance in dc circuits.

> *Define* an *open* and a *short*.

> *Identify* when opens and shorts occur.

> *Identify* defective components and/or connections in dc circuits.

> *Test* for continuity in components, wires, cables, and connectors.

> *Troubleshoot* wires, cables, and connectors.

> *Troubleshoot* a dc system.

The one activity that all technicians and engineers face most often is troubleshooting. This of course is the logical process of tracking down defects, failures, and operational problems in electric and electronic equipment. Troubleshooting is primarily a learned skill, but there are basic fundamentals that all technicians and engineers should know. With these basic fundamentals and some real-world, hands-on experience, you can become an expert troubleshooter. This chapter brings together all the various rules and techniques for finding problems in almost any dc circuit.

13.1 Introduction to Troubleshooting

Troubleshooting begins when a component, circuit, piece of equipment, or system is not working as it should. That implies you know how the device under consideration actually works so that you can recognize the malfunction or improper operation. Once you realize something is wrong, there are basic steps that will lead you to the problem. Once the problem has been discovered, repairing the defect is relatively easy. Finding the problem is always the challenge. It is clearly a puzzle to solve.

A Basic Troubleshooting Procedure

Here is a list of steps that will help you zero in on any problem. Of course, you may modify the steps or their sequence as information becomes available to you.

- Recognize that the problem exists. Recognition assumes that you know what to expect from a specific component, circuit, or piece of equipment. Perhaps you are expecting to measure a specific voltage or see a visible response, like a light turning on, an audible response, or a physical occurrence, like a relay being actuated.

- Obtain any available documentation. If any schematics, manuals, or other sources are available, acquire them. These may be print, online, or simply derived by tracing existing circuits.

- Verify that power is available. All dc circuits and systems have a dc power source, such as a battery or power supply. Make sure that source is providing the desired voltage. Measure it first. Be sure to check for blown fuses or tripped circuit breakers if they are present.

- Make a visual inspection. Look especially for broken wires, poor solder connections, poorly fitting or dirty connectors, or burned resistors or wires. Smell is another sense to use as it may detect something burned.

- Trace the voltage through the circuit from source to the ultimate load.

- Check the load to see that it is good. Is a light burned out, a motor defective, a relay damaged, or is there another problem?

- Use the techniques to be described in the following sections to identify the specific problem.

- Repair the circuit.

- Test to ensure proper operation is restored.

Test Instruments

Your main aid in the dc troubleshooting process is your multimeter. It can measure voltage, current, and resistance. Either a digital or analog multimeter will do, but be sure

to refer to Chap. 12 for usage information. Also have the meter instruction manual handy in case you have operational issues.

13.2 Resistor Troubles

The most common trouble in resistors is an open. When the open resistor is a series component, there is no current in the entire series path.

Noisy Controls

In applications requiring polentiometers, carbon controls are preferred because the smoother change in resistance results in less noise when the variable arm is rotated. With use, however, the resistance element becomes worn by the wiper contact, making the *control noisy*. An example is when a volume or tone control makes a scratchy noise as the shaft is rotated. That indicates either a dirty or worn-out resistance element. If the control is just dirty, it can be cleaned by spraying the resistance element (if accessible) with a special contact cleaner. If the resistance element is worn out, the control must be replaced.

Checking Resistors with an Ohmmeter

Resistance is measured with an ohmmeter. The ohmmeter has its own voltage source so that it is always used without any external power applied to the resistance being measured. Separate the resistance from its circuit by disconnecting one lead of the resistor or remove it completely. Then connect the ohmmeter leads across the resistance to be measured.

An open resistor reads infinitely high ohms. For some reason, infinite ohms is often confused with zero ohms. Remember, though, that infinite ohms means an open circuit. The current is zero, but the resistance is infinitely high. Furthermore, it is practically impossible for a resistor to become short-circuited in itself. The resistor may be short-circuited by some other part of the circuit. However, the construction of resistors is such that the trouble they develop is an open circuit with infinitely high ohms.

The ohmmeter must have an ohms scale capable of reading the resistance value, or the resistor cannot be checked. In checking a 10-MΩ resistor, for instance, if the highest R the ohmmeter can read is 1 MΩ, it will indicate infinite resistance, even if the resistor has its normal value of 10 MΩ. An ohms scale of 100 MΩ or more should be used for checking such high resistances.

To check resistors of less than 10 Ω, a low-ohms scale of about 100 Ω or less is necessary. Center scale should be 6 Ω or less. Otherwise, the ohmmeter will read a normally low resistance value as zero ohms.

When checking resistance in a circuit, it is important to be sure there are no parallel resistance paths. Otherwise, the measured resistance can be much lower than the actual resistor value, as illustrated in Fig. 13-1a. Here, the ohmmeter

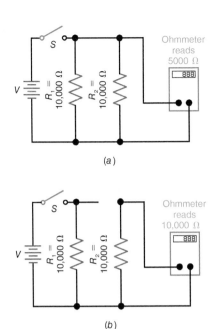

(a)

(b)

Figure 13-1 Parallel R_1 can lower the ohmmeter reading for testing R_2. (a) The two resistances R_1 and R_2 are in parallel. (b) R_2 is isolated by disconnecting one end of R_1.

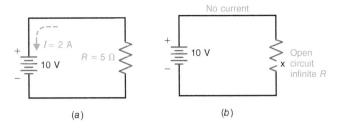

(a)   (b)

Figure 13-2 Effect of an open circuit. (a) Normal circuit with current of 2 A for 10 V across 5 Ω. (b) Open circuit with no current and infinitely high resistance.

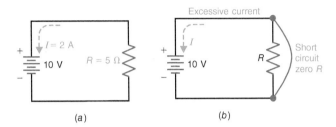

(a)   (b)

Figure 13-3 Effect of a short circuit. (a) Normal circuit with current of 2 A for 10 V across 5 Ω. (b) Short circuit with zero resistance and excessively high current.

reads the resistance of R_2 in parallel with R_1. To check across R_2 alone, one end is disconnected, as in Fig. 13-1b.

For very high resistances, it is important not to touch the ohmmeter leads. There is no danger of shock, but the body resistance of about 50,000 Ω as a parallel path will lower the ohmmeter reading.

Changed Value of R

In many cases, the value of a carbon-composition resistor can exceed its allowed tolerance; this is caused by normal resistor heating over a long period of time. In most instances, the value change is seen as an increase in R. This is known as *aging*. As you know, carbon-film and metal-film resistors age very little. A surface-mount resistor should never be rubbed or scraped because this will remove some of the carbon deposit and change its resistance.

13.3 Open-Circuit and Short-Circuit Troubles

Ohm's law is useful for calculating I, V, and R in a closed circuit with normal values. However, an open circuit or a short circuit causes trouble that can be summarized as follows: An open circuit (Fig. 13-2) has zero I because R is infinitely high. It does not matter how much the V is. A short circuit has zero R, which causes excessively high I in the short-circuit path because of no resistance (Fig. 13-3).

In Fig. 13-2a, the circuit is normal with I of 2 A produced by 10 V applied across R of 5 Ω. However, the resistor is shown open in Fig. 13-2b. Then the path for current has

infinitely high resistance and there is no current in any part of the circuit. The trouble can be caused by an internal open in the resistor or a break in the wire conductors.

In Fig. 13-3a, the same normal circuit is shown with I of 2 A. In Fig. 13-3b, however, there is a short-circuit path across R with zero resistance. The result is excessively high current in the short-circuit path, including the wire conductors. It may be surprising, but there is no current in the resistor itself because all the current is in the zero-resistance path around it.

Theoretically, the amount of current could be infinitely high with no R, but the voltage source can supply only a limited amount of I before it loses its ability to provide voltage output. The wire conductors may become hot enough to burn open, which would open the circuit. Also, if there is any fuse in the circuit, it will open because of the excessive current produced by the short circuit.

Note that the resistor itself is not likely to develop a short circuit because of the nature of its construction. However, the wire conductors may touch, or some other component in a circuit connected across the resistor may become short-circuited.

13.4 Troubleshooting: Opens and Shorts in Series Circuits

In many cases, electronics technicians are required to repair a piece of equipment that is no longer operating properly. The technician is expected to troubleshoot the equipment and restore it to its original operating condition. To *troubleshoot* means "to diagnose or analyze." For example, a technician may diagnose a failed electronic circuit by using a

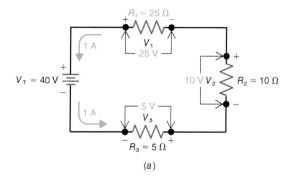

(a)

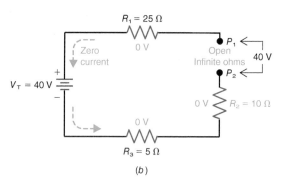

(b)

Figure 13-4 Effect of an open in a series circuit. (a) Normal circuit with 1-A series current. (b) Open path between points P_1 and P_2 results in zero current in all parts of the circuit.

digital multimeter (DMM) to make voltage, current, and resistance measurements. Once the defective component has been located, it is removed and replaced with a good one. But here is one very important point that needs to be made about troubleshooting: To troubleshoot a defective circuit, you must understand how the circuit is supposed to work in the first place. Without this knowledge, your troubleshooting efforts could be nothing more than guesswork. What we will do next is analyze the effects of both opens and shorts in series circuits.

The Effect of an Open in a Series Circuit

An *open circuit* is a break in the current path. The resistance of an open circuit is extremely high because the air between the open points is a very good insulator. Air can have billions of ohms of resistance. For a series circuit, a break in the current path means zero current in all components.

Figure 13-4a, shows a series circuit that is operating normally. With 40 V of applied voltage and 40 Ω of total resistance, the series current is 40 V/40 Ω = 1 A. This produces the following *IR* voltage drops across R_1, R_2, and R_3: $V_1 = 1 A \times 25 \Omega = 25 V$, $V_2 = 1 A \times 10 \Omega = 10 V$, and $V_3 = 1 A \times 5 \Omega = 5 V$.

Now consider the effect of an open circuit between points P_1 and P_2 in Fig. 13-4b. Because there is practically infinite

resistance between the open points, the current in the entire series circuit is zero. With zero current throughout the series circuit, each resistor's *IR* voltage will be 0 V even though the applied voltage is still 40 V. To calculate V_1, V_2, and V_3 in Fig. 13-4b, simply use 0 A for *I*. Then, $V_1 = 0 A \times 25 \Omega = 0 V$, $V_2 = 0 A \times 10 \Omega = 0 V$, and $V_3 = 0 A \times 5 \Omega = 0 V$. But how much voltage is across points P_1 and P_2? The answer is 40 V. This might surprise you, but here's the proof: Let's assume that the resistance between P_1 and P_2 is $40 \times 10^9 \Omega$, which is 40 GΩ (40 gigohms). Since the total resistance of a series circuit equals the sum of the series resistances, R_T is the sum of 25 Ω, 15 Ω, 10 Ω, and 40 GΩ. Since the 40 GΩ of resistance between P_1 and P_2 is so much larger than the other resistances, it is essentially the total resistance of the series circuit. Then the series current *I* is calculated as 40 V/40 GΩ = 1×10^{-9} A = 1 nA. For all practical purposes, the current *I* is zero. This is the value of current in the entire series circuit. This small current produces about 0 V across R_1, R_2, and R_3, but across the open points P_1 and P_2, where the resistance is high, the voltage is calculated as $V_{open} = 1 \times 10^{-9}$ A $\times 40 \times 10^9 \Omega = 40 V$.

In summary, here is the effect of an open in a series circuit:

1. The current *I* is zero in all components.
2. The voltage drop across each good component is 0 V.
3. The voltage across the open points equals the applied voltage.

The Applied Voltage V_T Is Still Present with Zero Current

The open circuit in Fig. 13-4b is another example of how voltage and current are different. There is no current with the open circuit because there is no complete path for current flow between the two battery terminals. However, the battery still has its potential difference of 40 V across the positive and negative terminals. In other words, the applied voltage V_T is still present with or without current in the external circuit. If you measure V_T with a voltmeter, it will measure 40 V regardless of whether the circuit is closed, as in Fig. 13-4a, or open, as in Fig. 13-4b.

The same idea applies to the 120-Vac voltage from the power line in our homes. The 120 V potential difference is available from the terminals of the wall outlet. If you connect a lamp or appliance to the outlet, current will flow in those circuits. When there is nothing connected, though, the 120 V potential is still present at the outlet. If you accidentally touch the metal terminals of the outlet when nothing else is connected, you will get an electric shock. The power company is maintaining the 120 V at the outlets as a source to produce current in any circuit that is plugged into the outlet.

EXAMPLE 13-1

Assume that the series circuit in Fig. 13-5 has failed. A technician troubleshooting the circuit used a voltmeter to record the following resistor voltage drops.

$$V_1 = 0 \text{ V}$$
$$V_2 = 0 \text{ V}$$
$$V_3 = 24 \text{ V}$$
$$V_4 = 0 \text{ V}$$

Based on these voltmeter readings, which component is defective and what type of defect is it? (Assume that only one component is defective.)

Answer:
To help understand which component is defective, let's calculate what the values of V_1, V_2, V_3, and V_4 are supposed to be. Begin by calculating R_T and I.

$$R_T = R_1 + R_2 + R_3 + R_4$$
$$= 150 \text{ }\Omega + 120 \text{ }\Omega + 180 \text{ }\Omega + 150 \text{ }\Omega$$
$$R_T = 600 \text{ }\Omega$$

$$I = \frac{V_T}{R_T}$$
$$= \frac{24 \text{ V}}{600 \text{ }\Omega}$$
$$I = 40 \text{ mA}$$

Next,

$$V_1 = I \times R_1$$
$$= 40 \text{ mA} \times 150 \text{ }\Omega$$
$$V_1 = 6 \text{ V}$$
$$V_2 = I \times R_2$$
$$= 40 \text{ mA} \times 120 \text{ }\Omega$$
$$V_2 = 4.8 \text{ V}$$

$$V_3 = I \times R_3$$
$$= 40 \text{ mA} \times 180 \text{ }\Omega$$
$$V_3 = 7.2 \text{ V}$$
$$V_4 = I \times R_4$$
$$= 40 \text{ mA} \times 150 \text{ }\Omega$$
$$V_4 = 6 \text{ V}$$

Next, compare the calculated values with those measured in Fig. 13-5. When the circuit is operating normally, V_1, V_2, and V_4 should measure 6 V, 4.8 V, and 6 V, respectively. Instead, the measurements made in Fig. 13-5 show that each of these voltages is 0 V. This indicates that the current I in the circuit must be zero, caused by an open somewhere in the circuit. The reason that V_1, V_2, and V_4 are 0 V is simple: $V = I \times R$. If $I = 0$ A, then each good resistor must have a voltage drop of 0 V. The measured value of V_3 is 24 V, which is considerably higher than its calculated value of 7.2 V. Because V_3 is dropping the full value of the applied voltage, it must be open. The reason the open R_3 will drop the full 24 V is that it has billions of ohms of resistance and, in a series circuit, the largest resistance drops the most voltage. Since the open resistance of R_3 is so much higher than the values of R_1, R_2, and R_4, it will drop the full 24 V of applied voltage.

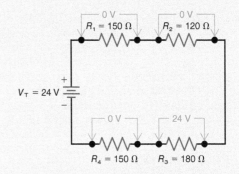

Figure 13-5 Series circuit for Example 13-1.

The Effect of a Short in a Series Circuit

A **short circuit** is an extremely low resistance path for current flow. The resistance of a short is assumed to be 0 Ω. This is in contrast to an open, which is assumed to have a resistance of infinite ohms. Let's reconsider the circuit in Fig. 13-4 with R_2 shorted. The circuit is redrawn for your convenience in Fig. 13-6. Recall from Fig. 13-4a that the normal values of V_1, V_2, and V_3 are 25 V, 10 V, and 5 V, respectively. With the 10-Ω R_2 shorted, the total resistance R_T will decrease from 40 Ω to 30 Ω. This will cause the series current to increase from 1 A to 1.33 A. This is calculated as 40 V/30 Ω = 1.33 A. The increase in current will cause the voltage drop across resistors R_1 and R_3 to increase from their normal values. The new voltage drops across R_1 and R_3 with R_2 shorted are calculated as follows:

$$V_1 = I \times R_1 = 1.33 \text{ A} \times 25 \text{ }\Omega \qquad V_3 = I \times R_3 = 1.33 \text{ A} \times 5 \text{ }\Omega$$
$$= 33.3 \text{ V} \qquad\qquad\qquad = 6.67 \text{ V}$$

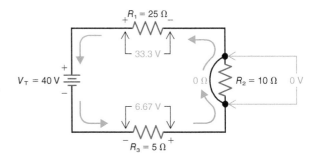

Figure 13-6 Series circuit of Fig. 13-4 with R_2 shorted.

The voltage drop across the shorted R_2 is 0 V because the short across R_2 effectively makes its resistance value 0 Ω. Then,

$$V_2 = I \times R_2 = 1.33 \text{ A} \times 0 \text{ }\Omega$$
$$= 0 \text{ V}$$

In summary, here is the effect of a short in a series circuit:

1. The current I increases above its normal value.
2. The voltage drop across each good component increases.
3. The voltage drop across the shorted component drops to 0 V.

EXAMPLE 13-2

Assume that the series circuit in Fig. 13-7 has failed. A technician troubleshooting the circuit used a voltmeter to record the following resistor voltage drops:

$$V_1 = 8 \text{ V}$$
$$V_2 = 6.4 \text{ V}$$
$$V_3 = 9.6 \text{ V}$$
$$V_4 = 0 \text{ V}$$

Based on the voltmeter readings, which component is defective and what type of defect is it? (Assume that only one component is defective.)

Answer:
This is the same circuit used in Example 13-1. Therefore, the normal values for V_1, V_2, V_3, and V_4 are 6 V, 4.8 V, 7.2 V, and 6 V, respectively. Comparing the calculated values with those measured in Fig. 13-7 reveals that V_1, V_2, and V_3 have increased from their normal values. This indicates that the current has increased, which is why we have a larger voltage drop across these resistors. The measured value of 0 V for V_4 shows a significant drop from its normal value of 6 V. The only way this resistor can have 0 V, when all other resistors show an increase in voltage, is if R_4 is shorted. Then $V_4 = I \times R_4 = I \times 0 \ \Omega = 0$ V.

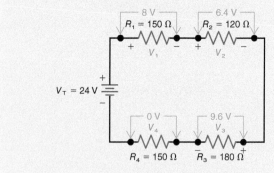

Figure 13-7 Series circuit for Example 13-2.

General Rules for Troubleshooting Series Circuits

When troubleshooting a series circuit containing three or more resistors, remember this important rule: The defective component will have a voltage drop that will change in the opposite direction as compared to the good components. In other words, in a series circuit containing an open, all the good components will have a voltage decrease from their normal value to 0 V. The defective component will have a voltage increase from its normal value to the full applied voltage. Likewise, in a series circuit containing a short, all good components will have a voltage increase from their normal values and the defective component's voltage drop will decrease from its normal value to 0 V. The point to be made here is simple: The component whose voltage changes in the opposite direction of the other components is the defective component. In the case of an open resistor, the voltage drop increases to the value of the applied voltage and all other resistor voltages decrease to 0 V. In the case of a short, all good components show their voltage drops increasing, whereas the shorted component shows a voltage decrease to 0 V. This same general rule applies to a series circuit that has components whose resistances have increased or decreased from their normal values but are neither open or shorted.

13.5 Troubleshooting: Opens and Shorts in Parallel Circuits

In a parallel circuit, the effect of an open or a short is much different from that in a series circuit. For example, if one branch of a parallel circuit opens, the other branch currents remain the same. The reason is that the other branches still have the same applied voltage even though one branch has effectively been removed from the circuit. Also, if one branch of a parallel circuit becomes shorted, all branches are effectively shorted. The result is excessive current in the shorted branch and zero current in all other branches. In most cases, a fuse will be placed in the main line that will burn open (blow) when its current rating is exceeded. When the fuse blows, the applied voltage is removed from each of the parallel-connected branches. The effects of opens and shorts are examined in more detail in the following paragraphs.

The Effect of an Open in a Parallel Circuit

An open in any circuit is an infinite resistance that results in no current. However, in parallel circuits there is a difference between an open circuit in the main line and an open circuit in a parallel branch. These two cases are illustrated in Fig. 13-8. In Fig. 13-8a the open circuit in the main line prevents any electron flow in the line to all the branches. The current is zero in every branch, therefore, and none of the bulbs can light.

However, in Fig. 13-8b the open is in the branch circuit for bulb 1. The **open branch** circuit has no current, then, and this bulb cannot light. The current in all the other parallel branches is normal, though, because each is connected to the voltage source. Therefore, the other bulbs light.

These circuits show the advantage of wiring components in parallel. An open in one component opens only one branch, whereas the other parallel branches have their normal voltage and current.

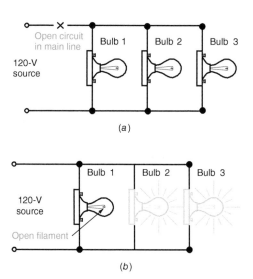

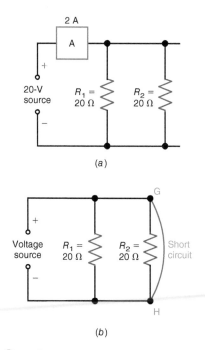

Figure 13-8 Effect of an open in a parallel circuit. (a) Open path in the main line—no current and no light for all bulbs. (b) Open path in any branch—bulb for that branch does not light, but the other two bulbs operate normally.

The Effect of a Short in a Parallel Circuit

A **short circuit** has practically zero resistance. Its effect, therefore, is to allow excessive current in the shorted circuit. Consider the example in Fig. 13-9. In Fig. 13-9a, the circuit is normal, with 1 A in each branch and 2 A for the total line current. However, suppose that the conducting wire at point G accidentally makes contact with the wire at point H,

Figure 13-9 Effect of a short circuit across parallel branches. (a) Normal circuit. (b) Short circuit across points G and H shorts out all the branches.

as shown in Fig. 13-9b. Since the wire is an excellent conductor, the short circuit results in practically zero resistance between points G and H. These two points are connected directly across the voltage source. Since the short circuit provides practically no opposition to current, the applied voltage could produce an infinitely high value of current through this current path.

The Short-Circuit Current

Practically, the amount of current is limited by the small resistance of the wire. Also, the source usually cannot maintain its output voltage while supplying much more than its rated load current. Still, the amount of current can be dangerously high. For instance, the short-circuit current might be more than 100 A instead of the normal line current of 2 A in Fig. 13-9a. Because of the short circuit, excessive current flows in the voltage source, in the line to the short circuit at point H, through the short circuit, and in the line returning to the source from G. Because of the large amount of current, the wires can become hot enough to ignite and burn the insulation covering the wire. There should be a fuse that would open if there is too much current in the main line because of a short circuit across any of the branches.

The Short-Circuited Components Have No Current

For the short circuit in Fig. 13-9b, the I is 0 A in the parallel resistors R_1 and R_2. The reason is that the short circuit is a parallel path with practically zero resistance. Then all the current flows in this path, bypassing the resistors R_1 and R_2. Therefore R_1 and R_2 are short-circuited or *shorted out* of the circuit. They cannot function without their normal current. If they were filament resistances of lightbulbs or heaters, they would not light without any current.

The short-circuited components are not damaged, however. They do not even have any current passing through them. Assuming that the short circuit has not damaged the voltage source and the wiring for the circuit, the components can operate again when the circuit is restored to normal by removing the short circuit.

All Parallel Branches Are Short-Circuited

If there were only one R in Fig. 13-9 or any number of parallel components, they would all be shorted out by the short circuit across points G and H. Therefore, a short circuit across one branch in a parallel circuit shorts out all parallel branches.

This idea also applies to a short circuit across the voltage source in any type of circuit. Then the entire circuit is shorted out.

Troubleshooting Procedures for Parallel Circuits

When a component fails in a parallel circuit, voltage, current, and resistance measurements can be made to locate the defective component. To begin our analysis, let's refer to the parallel circuit in Fig. 13-10a, which is normal. The individual branch currents I_1, I_2, I_3, and I_4 are calculated as follows:

$$I_1 = \frac{120 \text{ V}}{20 \text{ }\Omega} = 6 \text{ A}$$

$$I_2 = \frac{120 \text{ V}}{15 \text{ }\Omega} = 8 \text{ A}$$

$$I_3 = \frac{120 \text{ V}}{30 \text{ }\Omega} = 4 \text{ A}$$

$$I_4 = \frac{120 \text{ V}}{60 \text{ }\Omega} = 2 \text{ A}$$

By Kirchhoff's current law, the total current I_T equals $6 \text{ A} + 8 \text{ A} + 4 \text{ A} + 2 \text{ A} = 20 \text{ A}$. The total current I_T of 20 A is indicated by the ammeter M_1, which is placed in the main line between points J and K. The fuse F_1 between points A and B in the main line can safely carry 20 A, since its maximum rated current is 25 A, as shown.

Now consider the effect of an open branch between points D and I in Fig. 13-10b. With R_2 open, the branch current I_2 is 0 A. Also, the ammeter M_1 shows a total current I_T of 12 A, which is 8 A less than its normal value. This makes sense because I_2 is normally 8 A. Notice that with R_2 open, all other branch currents remain the same. This is because each branch is still connected to the applied voltage of 120 V. It is important to realize that voltage measurements across the individual branches would not help determine which branch is open because even the open branch between points D and I will measure 120 V.

In most cases, the components in a parallel circuit provide a visual indication of failure. If a lamp burns open, it doesn't light. If a motor opens, it stops running. In these cases, the defective component is easy to spot.

In summary, here is the effect of an open branch in a parallel circuit.

1. The current in the open branch drops to 0 A.
2. The total current I_T decreases by an amount equal to the value normally drawn by the now open branch.
3. The current in all remaining branches remains the same.
4. The applied voltage remains present across all branches whether they are open or not.

Next, let's consider the effect of an open between two branch points such as points D and E in Fig. 13-10c. With an open between these two points, the current through branch resistors R_3 and R_4 will be 0 A. Since $I_3 = 4 \text{ A}$ and $I_4 = 2 \text{ A}$ normally, the total current indicated by M_1 will drop from 20 A to 14 A as shown. The reason that I_3 and I_4 are now 0 A is that the applied voltage has effectively been removed from these two branches. If a voltmeter were placed across either points E and H or F and G, it would read 0 V. A voltmeter placed across points D and E would measure 120 V, however.

This is indicated by the voltmeter M_2 as shown. The reason M_2 measures 120 V between points D and E is explained as follows: Notice that the positive (red) lead of M_2 is connected through S_1 and F_1 to the positive side of the applied voltage. Also, the negative (black) lead of M_2 is connected to the top of resistors R_3 and R_4. Since the voltage across R_3 and R_4 is 0 V, the negative lead of M_2 is in effect connected to the negative side of the applied voltage. In other words, M_2 is effectively connected directly across the 120-V source.

EXAMPLE 13-3

In Fig. 13-10a, suppose that the ammeter M_1 reads 16 A instead of 20 A as it should. What could be wrong with the circuit?

Answer:
Notice that the current I_3 is supposed to be 4 A. If R_3 is open, this explains why M_1 reads a current that is 4 A less than its normal value. To confirm that R_3 is open, open S_1 and disconnect the top lead of R_3 from point E. Next place an ammeter between the top of R_3 and point E. Now, close S_1. If I_3 measures 0 A, you know that R_3 is open. If I_3 measures 4 A, you know that one of the other branches is drawing less current than it should. In this case, the next step would be to measure each of the remaining branch currents to find the defective component.

Consider the circuit in Fig. 13-10d. Notice that the fuse F_1 is blown and the ammeter M_1 reads 0 A. Notice also that the voltage across each branch measures 0 V and the voltage across the blown fuse measures 120 V as indicated by the voltmeter M_2. What could cause this? The most likely answer is that one of the parallel-connected branches has become short-circuited. This would cause the total current to rise well above the 25-A current rating of the fuse, thus causing it to blow. But how do we go about finding out which branch is shorted? There are at least three different approaches. Here's the first one: Start by opening switch S_1 and replacing the bad fuse. Next, with S_1 still open, disconnect all but one of the four parallel branches. For example, disconnect branch resistors R_1, R_2, and R_3 along the top (at points C, D, and E). With R_4 still connected, close S_1. If the fuse blows, you know R_4 is shorted! If the fuse does not blow, with only R_4 connected, open S_1 and reconnect R_3 to point E. Then, close S_1 and see if the fuse blows.

Repeat this procedure with branch resistors R_1 and R_2 until the shorted branch is identified. The shorted branch will blow the fuse when it is reconnected at the top (along points C, D, E, and F) with S_1 closed. Although this troubleshooting procedure is effective in locating the shorted branch, another fuse has been blown and this will cost you or the customer money.

Here's another approach to finding the shorted branch. Open S_1 and replace the bad fuse. Next, measure the resistance of

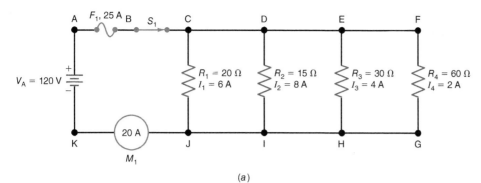

Figure 13-10 Parallel circuit for troubleshooting analysis. (a) Normal circuit values; (b) circuit values with branch R_2 open; (c) circuit values with an open between points D and E; (d) circuit showing the effects of a shorted branch.

(a)

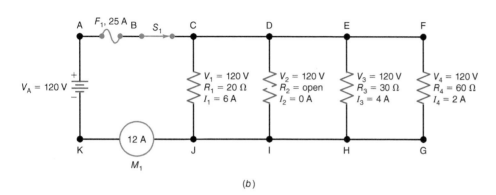

(b)

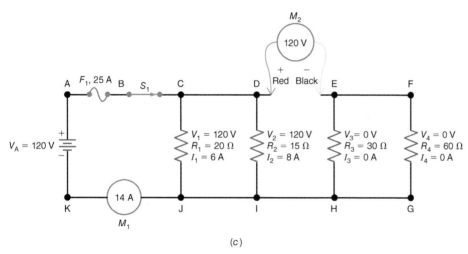

(c)

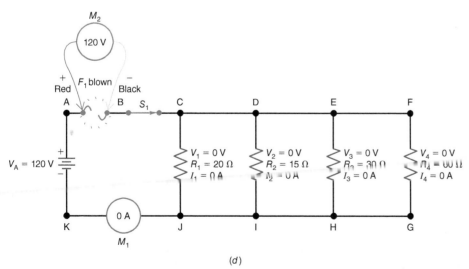

(d)

each branch separately. It is important to remember that when you make resistance measurements in a parallel circuit, one end of each branch must be disconnected from the circuit so that the rest of the circuit does not affect the individual branch measurement. The branch that measures 0 Ω is obviously the shorted branch. With this approach, another fuse will not get blown.

Here is yet another approach that could be used to locate the shorted branch in Fig. 13-10d. With S_1 open, place an ohmmeter across points C and J. With a shorted branch, the ohmmeter will measure 0 Ω. To determine which branch is shorted, remove one branch at a time until the ohmmeter shows a value other than 0 Ω. The shorted component is located when removal of a given branch causes the ohmmeter to show a normal resistance.

In summary, here is the effect of a shorted branch in a parallel circuit:

1. The fuse in the main line will blow, resulting in zero current in the main line as well as in each parallel-connected branch.

2. The voltage across each branch will equal 0 V, and the voltage across the blown fuse will equal the applied voltage.

3. With power removed from the circuit, an ohmmeter will measure 0 Ω across all the branches.

Before leaving the topic of troubleshooting parallel circuits, one more point should be made about the fuse F_1 and the switch S_1 in Fig. 13-10a: The resistance of a good fuse and the resistance across the closed contacts of a switch are practically 0 Ω. Therefore, the voltage drop across a good fuse or a closed switch is approximately 0 V. This can be proven with Ohm's law, since $V = I \times R$. If $R = 0\ \Omega$, then $V = I \times 0\ \Omega = 0\ V$. When a fuse blows or a switch opens, the resistance increases to such a high value that it is considered infinite. When used in the main line of a parallel circuit, the voltage across an open switch or a blown fuse is the same as the applied voltage. One way to reason this out logically is to treat all parallel branches as a single equivalent resistance R_{EQ} in series with the switch and fuse. The result is a simple series circuit. Then, if either the fuse or the switch opens, apply the rules of an open to a series circuit. As you recall from your study of series circuits, the voltage across an open equals the applied voltage.

13.6 Troubleshooting: Opens and Shorts in Series-Parallel Circuits

A short circuit has practically zero resistance. Its effect, therefore, is to allow excessive current. An open circuit has the opposite effect because an open circuit has infinitely high resistance with practically zero current. Furthermore, in series-parallel circuits, an open or short circuit in one path changes the circuit for the other resistances. For example, in Fig. 13-11,

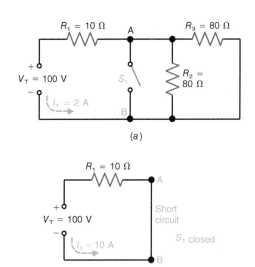

(a)

(b)

Figure 13-11   Effect of a short circuit with series-parallel connections. (a) Normal circuit with S_1 open. (b) Circuit with short between points A and B when S_1 is closed; now R_2 and R_3 are short-circuited.

the series-parallel circuit in Fig. 13-11a becomes a series circuit with only R_1 when there is a short circuit between terminals A and B. As an example of an open circuit, the series-parallel circuit in Fig. 13-12a becomes a series circuit with just R_1 and R_2 when there is an open circuit between terminals C and D.

Effect of a Short Circuit

We can solve the series-parallel circuit in Fig. 13-11a to see the effect of the short circuit. For the normal circuit with S_1 open, R_2 and R_3 are in parallel. Although R_3 is drawn

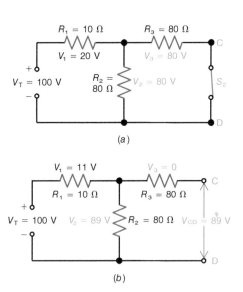

(a)

(b)

Figure 13-12   Effect of an open path in a series-parallel circuit. (a) Normal circuit with S_2 closed. (b) Series circuit with R_1 and R_2 when S_2 is open. Now R_3 in the open path has no current and zero IR voltage drop.

horizontally, both ends are across R_2. The switch S_1 has no effect as a parallel branch here because it is open.

The combined resistance of the 80-Ω R_2 in parallel with the 80-Ω R_3 is equivalent to 40 Ω. This 40 Ω for the bank resistance is in series with the 10-Ω R_1. Then R_T is 40 + 10 = 50 Ω.

In the main line, I_T is 100/50 = 2 A. Then V_1 across the 10-Ω R_1 in the main line is 2 × 10 = 20 V. The remaining 80 V is across R_2 and R_3 as a parallel bank. As a result, V_2 = 80 V and V_3 = 80 V.

Now consider the effect of closing switch S_1. A closed switch has zero resistance. Not only is R_2 short-circuited, but R_3 in the bank with R_2 is also short-circuited. The closed switch short-circuits everything connected between terminals A and B. The result is the series circuit shown in Fig. 13-11b.

Now the 10-Ω R_1 is the only opposition to current. I equals V/R_1, which is 100/10 = 10 A. This 10 A flows through the closed switch, through R_1, and back to the positive terminal of the voltage source. With 10 A through R_1, instead of its normal 2 A, the excessive current can cause excessive heat in R_1. There is no current through R_2 and R_3, as they are short-circuited out of the path for current.

Effect of an Open Circuit

Figure 13-12a shows the same series-parallel circuit as Fig. 13-11a, except that switch S_2 is used now to connect R_3 in parallel with R_2. With S_2 closed for normal operation, all currents and voltages have the values calculated for the series-parallel circuit. However, let us consider the effect of opening S_2, as shown in Fig. 13-12b. An open switch has infinitely high resistance. Now there is an open circuit between terminals C and D. Furthermore, because R_3 is in the open path, its 80 Ω cannot be considered in parallel with R_2.

The circuit with S_2 open in Fig. 13-12b is really the same as having only R_1 and R_2 in series with the 100-V source. The open path with R_3 has no effect as a parallel branch because no current flows through R_3.

We can consider R_1 and R_2 in series as a voltage divider, where each IR drop is proportional to its resistance. The total series R is 80 + 10 = 90 Ω. The 10-Ω R_1 is 10/90 or ⅑ of the total R and the applied V_T. Then V_1 is ⅑ × 100 V = 11 V and V_2 is ⅚ × 100 V = 89 V, approximately. The 11-V drop for V_1 and 89-V drop for V_2 add to equal the 100 V of the applied voltage.

Note that V_3 is zero. Without any current through R_3, it cannot have any voltage drop.

Furthermore, the voltage across the open terminals C and D is the same 89 V as the potential difference V_2 across R_2. Since there is no voltage drop across R_3, terminal C has the same potential as the top terminal of R_2. Terminal D is directly connected to the bottom end of resistor R_2. Therefore, the potential difference from terminal C to terminal D is the same 89 V that appears across resistor R_2.

Troubleshooting Procedures for Series-Parallel Circuits

The procedure for troubleshooting series-parallel circuits containing opens and shorts is a combination of the procedures used to troubleshoot individual series and parallel circuits. Figure 13-13a shows a series-parallel circuit with its normal operating voltages and currents. Across points A and B, the equivalent resistance R_{EQ} of R_2 and R_3 in parallel is calculated as

$$R_{EQ} = \frac{R_2 \times R_3}{R_2 + R_3}$$
$$= \frac{100\ \Omega \times 150\ \Omega}{100\ \Omega + 150\ \Omega}$$
$$R_{EQ} = 60\ \Omega$$

Since R_2 and R_3 are in parallel across points A and B, this equivalent resistance is designated R_{AB}. Therefore, R_{AB} = 60 Ω. The total resistance, R_T, is

$$R_T = R_1 + R_{AB} + R_4$$
$$= 120\ \Omega + 60\ \Omega + 180\ \Omega$$
$$= 360\ \Omega$$

The total current, I_T, is

$$I_T = \frac{V_T}{R_T}$$
$$= \frac{36\ V}{360\ \Omega}$$
$$= 100\ mA$$

The voltage drops across the individual resistors are calculated as

$$V_1 = I_T \times R_1$$
$$= 100\ mA \times 120\ \Omega$$
$$= 12\ V$$

$$V_2 = V_3 = V_{AB} = I_T \times R_{AB}$$
$$= 100\ mA \times 60\ \Omega$$
$$= 6\ V$$

$$V_4 = I_T \times R_4$$
$$= 100\ mA \times 180\ \Omega$$
$$= 18\ V$$

The current in resistors R_2 and R_3 across points A and B can be found as follows:

$$I_2 = \frac{V_{AB}}{R_2}$$
$$= \frac{6\ V}{100\ \Omega}$$
$$= 60\ mA$$

$$I_3 = \frac{V_{AB}}{R_3}$$
$$= \frac{6\ V}{150\ \Omega}$$
$$= 40\ mA$$

Figure 13-13 Series-parallel circuit for troubleshooting analysis. (*a*) Normal circuit voltages and currents; (*b*) circuit voltages with R_3 open between points A and B; (*c*) circuit voltages with R_2 or R_3 shorted between points A and B.

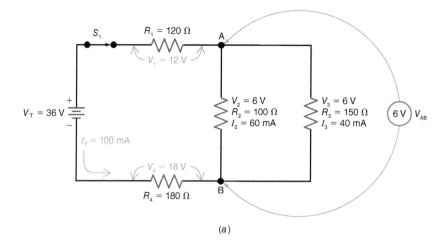

(*a*)

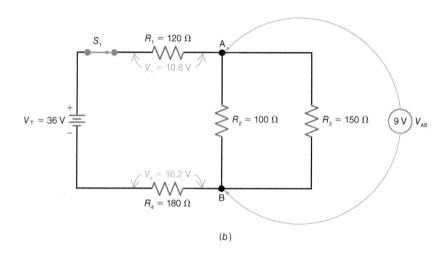

(*b*)

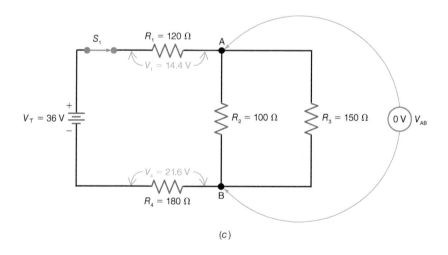

(*c*)

EXAMPLE 13-4

Assume that the series-parallel circuit in Fig. 13–13a has failed. A technician troubleshooting the circuit has measured the following voltages:

$$V_1 = 10.8 \text{ V}$$
$$V_{AB} = 9 \text{ V}$$
$$V_4 = 16.2 \text{ V}$$

These voltage readings are shown in Fig. 13–13b. Based on the voltmeter readings shown, which component is defective and what type of defect does it have?

Answer:
If we consider the resistance between points A and B as a single resistance, the circuit can be analyzed as if it were a simple series circuit. Notice that V_1 and V_4 have decreased from their normal values of 12 V and 18 V respectively, whereas the voltage V_{AB} across R_2 and R_3 has increased from 6 V to 9 V.

Recall that in a series circuit containing three or more components, the voltage across the defective component changes in a direction that is opposite to the direction of the change in voltage across the good components. Since the voltages V_1 and V_4 have decreased and the voltage V_{AB} has increased, the defective component must be either R_2 or R_3 across points A and B.

The increase in voltage across points A and B tells us that the resistance between points A and B must have increased. The increase in the resistance R_{AB} could be the result of an open in either R_2 or R_3.

But how do we know which resistor is open? At least three approaches may be used to find this out. One approach would be to calculate the resistance across points A and B. To do this, find the total current in either R_1 or R_4. Let's find I_T in R_1.

$$I_T = \frac{V_1}{R_1}$$
$$= \frac{10.8 \text{ V}}{120 \text{ }\Omega}$$
$$= 90 \text{ mA}$$

Next, divide the measured voltage V_{AB} by I_T to find R_{AB}.

$$R_{AB} = \frac{V_{AB}}{I_T}$$
$$= \frac{9 \text{ V}}{90 \text{ mA}}$$
$$= 100 \text{ }\Omega$$

Notice that the value of R_{AB} is the same as that of R_2. This means, of course, that R_3 must be open.

Another approach to finding which resistor is open would be to open the switch S_1 and measure the resistance across points A and B. This measurement would show that the resistance R_{AB} equals 100 Ω, again indicating that the resistor R_3 must be open.

The only other approach to determine which resistor is open would be to measure the currents I_2 and I_3 with the switch S_1 closed. In Fig. 13–13b, the current I_2 would measure 90 mA, whereas the current I_3 would measure 0 mA. With $I_3 = 0$ mA, R_3 must be open.

EXAMPLE 13-5

Assume that the series-parallel circuit in Fig. 13–13a has failed. A technician troubleshooting the circuit has measured the following voltages:

$$V_1 = 14.4 \text{ V}$$
$$V_{AB} = 0 \text{ V}$$
$$V_4 = 21.6 \text{ V}$$

These voltage readings are shown in Fig. 13–13c. Based on the voltmeter readings shown, which component is defective and what type of defect does it have?

Answer:
Since the voltages V_1 and V_4 have both increased, and the voltage V_{AB} has decreased, the defective component must be either R_2 or R_3 across points A and B. Because the voltage V_{AB} is 0 V, either R_2 or R_3 must be shorted.

But how can we find out which resistor is shorted? One way would be to measure the currents I_2 and I_3. The shorted component is the one with all the current.

Another way to find out which resistor is shorted would be to open the switch S_1 and measure the resistance across points A and B. Disconnect one lead of either R_2 or R_3 from point A while observing the ohmmeter. If removing the top lead of R_3 from point A still shows a reading of 0 Ω, then you know that R_2 must be shorted. Similarly, if removing the top lead of R_2 from point A (with R_3 still connected at point A) still produces a reading of 0 Ω, then you know that R_3 is shorted.

13.7 Troubleshooting Hints for Wires and Connectors

For all types of electronic equipment, a common problem is an open circuit in the wire conductors, the connectors, and the switch contacts.

You can check continuity of conductors, both wires and printed wiring, with an ohmmeter. A good conductor reads 0 Ω for continuity. An open reads infinite ohms.

A connector can also be checked for continuity between the wire and the connector itself. Also, the connector may be tarnished, oxide-coated, or rusted. Then it must be cleaned with either fine sandpaper or emery cloth. Sometimes, it helps just to pull out the plug and reinsert it to make sure of tight connections.

With a plug connector for cable, make sure the wires have continuity to the plug. Except for the F-type connector, most plugs require careful soldering to the center pin.

A switch with dirty or pitted contacts can produce intermittent operation. In most cases, the switch cannot be disassembled for cleaning. Therefore, the switch must be replaced with a new one.

13.8 Troubleshooting a DC System

The basic troubleshooting procedure for any dc system is similar to that as outlined earlier in Sec. 13-1. In most cases all you will need is a good digital (or analog) multimeter to measure dc voltage, current, and resistance and to check continuity.

The examples given here use the boat dc electrical system described in the System Sidebar. See Fig. S13-1.

Defective Running Lights

Assume that one of the running lights (red) does not light.

If only one of the lights does not work and the other running lights (green and white) are working, you can probably assume that 12 V is available since the bulbs are all in parallel. The OFF-ON switch SW_5 is probably OK. The most likely defect is a bad bulb. Remove the bulb, and check it visually to see if the filament is present. If not, or if the bulb is black, the bulb is probably burned out. If you are unsure, check its continuity with an ohmmeter. Then replace it if defective. Be sure to use the same type and wattage.

Another possibility is that the bulb socket or wiring is bad. Sockets exposed to salt water or air and a highly humid environment can corrode and produce an open or a high-resistance connection between the bulb and socket. A visual inspection will tell you. You may have to clean the bulb or socket contacts. Also check the connections from the power wiring to the socket for broken or corroded connections.

Cabin Light Failure

The problem has been determined as none of the cabin lights work.

Logical thinking will lead you to the conclusion that all the cabin lightbulbs and switches are good as the chances of all of them being bad is very low. As a starting point, you might go to the circuit breaker box and check the condition of the breaker. If it is off, turn it back on. That should fix the problem. However, if the breaker immediately trips again, then there is probably a short in one of the light fixtures or in the wiring. A visual inspection of each cabin light and its socket is in order. Corrosion may have produced a short across the socket that is creating the high current that trips the breaker. Cleaning or replacing the socket may be necessary. If the circuit breaker is not tripped, there may not be voltage at the breaker panel to distribute to the lights. Use your multimeter to measure the voltage at the breaker. Of course, it should be 12 V or close to it.

If no voltage is present, trace the wiring back to the batteries. Check the position of switch SW_4. If it is in position B, switch it to position A. It could be that battery 2 is not charged but battery 1 is. Measure the voltages at both batteries. A common problem is that the batteries have not been recharged. You will need to charge them, probably from the ac dockside charger or the solar charger. Set SW_1 to position B, and turn on the ac charger. Check to see that the charger voltage to the batteries is in the 13- to 15-V range. Note that SW_2 selects whether you charge $BATT_1$ or $BATT_2$. Close SW_3 if you wish to charge both at the same time.

AC Appliances Not Working

A cell phone charger and a microwave oven connected to the ac outlets are not working.

First, check for the presence of 120 Vac at the inverter outlet with your multimeter. Be careful in measuring this high voltage. If is not present, check to see that the inverter is turned on. It may or may not have an ON-OFF switch. If the unit is on and still no voltage is present, check the related circuit breaker. If it is OK, verify that 12 V is present at the breaker panel and the inverter input. If voltage is present, the inverter itself may be defective. It must be replaced.

DC troubleshooting is basically simple, at least electrically. The hard part is the actual physical wiring and mechanics of mounting and attaching the various electric components. Mechanical devices tend to go bad more often than electric devices so always expect problems there first. Visual inspections are a must.

The hardest part of troubleshooting a dc system is the wiring. Tracing the wires from source to load is a problem because the wiring is often hidden behind walls or bulkheads or otherwise buried in pipes or conduit. You must have the system wiring diagram for fastest diagnosis. Use any available cable identification, like color coding, to help you trace the wiring.

CHAPTER 13 SYSTEM SIDEBAR

DC Power System

A good example of a dc electrical system is that used in an automobile or boat. Both are powered by batteries with subsystems for charging, lights, motors, controls, electronics, and other accessories. This section describes a simplified generic boat electrical system as an example. This hypothetical system is typical of what you may find on a small (<35-ft) powerboat with inboard or outboard motors or a sailboat with an auxiliary motor. There are significant variations and a wide range of electrical features and accessories, but this system describes all the most common equipment. Refer to Fig. S13-1.

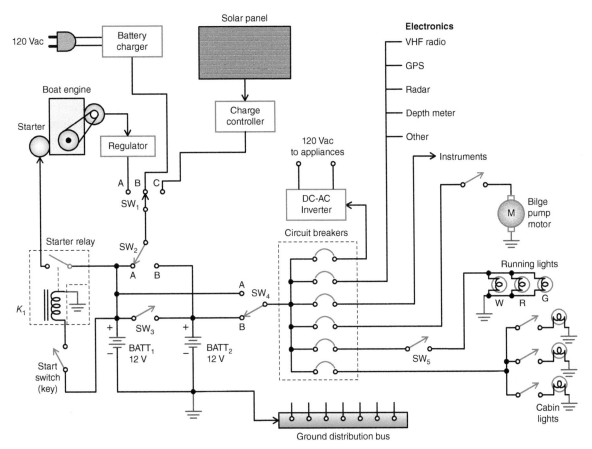

Figure S13-1 Boat dc electrical system.

Batteries

Most small boats use lead-acid 12-V batteries as the main power source. These are similar to standard 12-V car batteries, but special high-capacity marine batteries are available for larger boats and loads. Typically two batteries are used, $BATT_1$ and $BATT_2$, one for backup as it would be dangerous to lose battery power while a long way from dock or shore. One of these batteries would be used to start the engine, although some larger boats have a separate engine-starting battery. In this system, battery 1 is used for starting but can also be used to power the other subsystems in the boat. Battery 2 is the main or "house" battery for all subsystems. On larger boats, 24- or 48-V battery systems are used to handle the extra power for larger appliances and more accessories.

Starting System

The boat engine has a dc starter motor like a car engine. This motor requires a very high starting current of about 100 A or so; therefore, a simple starter switch cannot be used. Instead a starter relay or solenoid is used. This is K_1 in the figure. The starter switch is usually a key-operated, low-current switch that applies power to the relay. The relay or solenoid magnetically engages the heavy-duty contacts that connect the battery to the starter motor.

Charging System

In some systems, the boat engine also has an alternator or generator, as do all cars. This alternator produces ac, but that is rectified into dc by some internal diodes. The resulting dc is then fed to a regulator circuit that delivers the correct amount of voltage and current to the batteries to be charged.

Alternatively, an ac-operated battery charger can be used. In this system, a dockside ac outlet supplies 120-V, 60-Hz ac to the on-board battery charger that then charges the batteries. Some boats also have a solar panel and charge controller to recharge the batteries if they get too low while at sea.

Switching

You will note in Fig. S13-1 that many switches are used. These permit you to make the necessary connections between the batteries and charging sources. Switch SW_1 selects the charging source. When the engine is running, the batteries can be recharged by the alternator. At the dock, the ac battery charger is selected. At sea or anchor, the solar panel can be used for recharging. Switch SW_2 lets you select which battery to recharge. If you close SW_3, both batteries are connected in parallel and you can charge them at the same time. Switch SW_3 can also be closed if more power is needed to carry heavier accessory loads.

Switch SW_4 selects which battery is used to power all the various loads.

The switching arrangements vary widely from boat to boat. Some kind of master switching panel is provided near the batteries to select which configuration is needed.

Wiring

The battery and charging wiring require very heavy cables as they must carry many tens or even hundreds of amperes. Large wires ensure minimal voltage drop loss and overheating. Stranded wiring is the most common choice for flexibility in routing. Load wires use smaller cables as less amperage is needed. There are safety standards for determining the minimal wire sizes for different cable lengths and amperage.

Load Distribution

The 12 V from the batteries are distributed about the boat in much the same way ac house power is distributed. The battery voltage is connected to power bus bars that provide a way to connect the wiring to the various electrical devices needing power. A set of fuses or circuit breakers are used to divide the loads into logical subsystems or circuits. Loads connect to the power by a two-wire cable, with one connection to the positive bus and another to the negative ground bus.

In Fig. S13-1, note that a separate circuit breaker is used for each separate circuit. For example, one breaker is used to send power to the internal cabin lights. These are either 12-V incandescent bulbs in fixtures or newer lower-power LED lights. A switch is usually provided for each individual light.

A separate circuit controls the nighttime running lights. These include the red (left or port) and green (right or starboard) lights required by law. Often separate white transom or mast lights are also used.

The bilge pumps have a separate circuit. These pumps use dc motors to power them. Separate ON-OFF switches are used, and multiple bilge pumps are common.

Another circuit is provided for such instruments as speedometer, engine tachometer, engine temperature, oil indicator, depth indicator, and others including their lights. One or more circuits are also provided for electronics, including the mandatory VHF radio, GPS navigation system, radar, fish finder, and other equipment (long-range SSB radio, weather fax, etc.).

Additional circuits may be used for a horn, an anchor winch motor, or a small refrigerator, depending on how the boat is outfitted.

Note that many boats now have an inverter to power ac appliances and devices. An *inverter* is an electronic circuit that converts dc into 120-V, 60-Hz ac. These come in different sizes from a few hundred watts to several kilowatts or more, depending on the battery capacity. The inverter lets the boat owner use a microwave oven, TV set, radio/stereo, or another low-power ac device. It also serves to recharge cell phones, music players, and all the other gadgets a person uses.

A dc power system like this is relatively simple in concept because it only uses batteries, switches, and simple loads. What makes it complex is the extensive wiring and switching that is difficult to install and troubleshoot. Operational details differ, depending on the equipment involved. A wiring diagram and operations manual are essential to troubleshooting such a system.

❓ CHAPTER 13 REVIEW QUESTIONS

1. The main troubleshooting test instrument for dc systems is a(n)
   a. oscilloscope.
   b. multimeter.
   c. clamp-on current meter.
   d. neon test bulb.

2. The very first step in any troubleshooting procedure is to
   a. acquire all necessary documentation.
   b. measure all source voltages.
   c. check the circuit breakers.
   d. verify that a problem actually exists.

3. The basic approach in dc troubleshooting could be described as
   a. signal injection.
   b. effect to cause reasoning.
   c. signal tracing.
   d. continuity checking.

4. Most dc system problems, excluding the voltage source, are
   a. shorts or opens.
   b. defective parts.
   c. corrosion.
   d. ground faults.

5. In testing a resistor with an ohmmeter, you must
   a. remove the dc power source.
   b. make sure there are no other resistors or components in parallel with it.
   c. remove the resistor or at least disconnect one lead.
   d. all of the above.

6. The resistance of a short is essentially
   a. zero.
   b. less than 100 Ω.
   c. depends on the circuit.
   d. infinity.

7. The resistance of an open is essentially
   a. zero.
   b. over 100,000 Ω.
   c. depends on the circuit.
   d. infinity.

8. In measuring the resistance across a resistor in a series circuit and you measure the supply voltage, the problem is probably a(n)
   a. shorted resistor.
   b. resistor whose value has changed.
   c. open resistor.
   d. wiring problem.

9. An open in a series circuit causes the current to be
   a. maximum.
   b. zero.
   c. higher.
   d. lower.

10. A shorted component in a series circuit causes the current to be
   a. maximum.
   b. zero.
   c. higher.
   d. lower.

11. To measure the current in a series circuit, the ammeter is connected
   a. across one of the resistors.
   b. across the voltage source.
   c. in the negative lead of the source.
   d. in series with any part of the circuit.

12. An open branch in a parallel circuit can be detected by
   a. a lower total source current.
   b. a higher total source current.
   c. complete circuit failure.
   d. a lower source voltage.

13. A short across a resistor in a parallel circuit will cause the source current to be
   a. zero.
   b. some lower value.
   c. some extremely high value.
   d. infinite.

14. A common source of shorts and opens is
   a. defective connectors.
   b. defective wiring.
   c. defective components.
   d. all of the above.

15. If a resistor in a series circuit becomes open, how much is the voltage across each of the remaining resistors that are still good?
   a. Each good resistor has the full value of applied voltage.
   b. The applied voltage is split evenly among the good resistors.
   c. Zero.
   d. This is not possible to determine.

 CHAPTER 13 PROBLEMS

SECTION 13.4 Troubleshooting: Opens and
Shorts in Series Circuits

13.1 In Fig. 13-14, assume R_1 becomes open. How much is
   a. the total resistance, R_T?
   b. the series current, I?
   c. the voltage across each resistor, R_1, R_2, and R_3?

13.2 In Fig. 13-14, assume R_3 shorts. How much is
   a. the total resistance, R_T?
   b. the series current, I?
   c. the voltage across each resistor, R_1, R_2, and R_3?

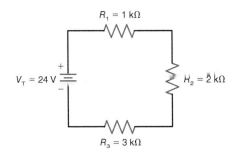

$R_1 = 1$ kΩ

$V_T = 24$ V

$R_2 = 2$ kΩ

$R_3 = 3$ kΩ

Figure 13-14

13.3 In Fig. 13-14, assume that the value of R_2 has increased but is not open. What happens to
  a. the total resistance, R_T?
  b. the series current, I?
  c. the voltage drop across R_2?
  d. the voltage drops across R_1 and R_3?

13.4 Table 13-1 shows voltage measurements taken in Fig. 13-15. The first row shows the normal values that exist when the circuit is operating properly. Rows 2 to 15 are voltage measurements taken when one component in the circuit has failed. For each row, identify which component is defective and determine the type of defect that has occurred in the component.

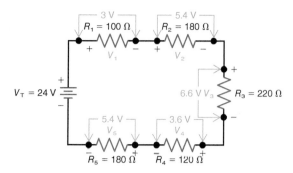

Figure 13-15 Circuit diagram. Normal values for V_1, V_2, V_3, V_4, and V_5 are shown on schematic.

Table 13-1 Voltage Measurements Taken in Fig. 13-15

	V_1	V_2	V_3	V_4	V_5	Defective Component
			Volts			
1 Normal values	3	5.4	6.6	3.6	5.4	None
2 Trouble 1	0	24	0	0	0	
3 Trouble 2	4.14	7.45	0	4.96	7.45	
4 Trouble 3	3.53	6.35	7.76	0	6.35	
5 Trouble 4	24	0	0	0	0	
6 Trouble 5	0	6.17	7.54	4.11	6.17	
7 Trouble 6	0	0	0	24	0	
8 Trouble 7	3.87	0	8.52	4.64	6.97	
9 Trouble 8	0	0	24	0	0	
10 Trouble 9	0	0	0	0	24	
11 Trouble 10	2.4	4.32	5.28	2.88	9.12	
12 Trouble 11	4	7.2	0.8	4.8	7.2	
13 Trouble 12	3.87	6.97	8.52	4.64	0	
14 Trouble 13	15.6	2.16	2.64	1.44	2.16	
15 Trouble 14	3.43	6.17	7.55	0.68	6.17	

SECTION 13.5 Troubleshooting: Opens and Shorts in Parallel Circuits

13.5 Figure 13-16 shows a parallel circuit with its normal operating voltages and currents. Notice that the fuse in the main line has a 25-A rating. What happens to the circuit components and their voltages and currents if
  a. the appliance in branch 3 shorts?
  b. the motor in branch 2 burns out and becomes an open?
  c. the wire between points C and E develops an open?
  d. the motor in branch 2 develops a problem and begins drawing 16 A of current?

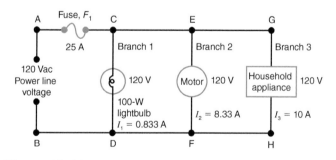

Figure 13-16

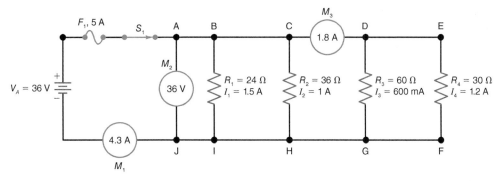

Figure 13-17 Circuit diagram. Normal values for I_1, I_2, I_3, and I_4 are shown on schematic.

Figure 13-17 shows a parallel circuit with its normal operating voltages and currents. Notice the placement of the meters M_1, M_2, and M_3 in the circuit. M_1 measures the total current I_T, M_2 measures the applied voltage V_A, and M_3 measures the current between points C and D. The following problems deal with troubleshooting the parallel circuit in Fig. 13-17.

13.6 If M_1 measures 2.8 A, M_2 measures 36 V, and M_3 measures 1.8 A, which component has most likely failed? How is the component defective?

13.7 If M_1 measures 2.5 A, M_2 measures 36 V, and M_3 measures 0 A, what is most likely wrong? How could you isolate the trouble by making voltage measurements?

13.8 If M_1 measures 3.3 A, M_2 measures 36 V, and M_3 measures 1.8 A, which component has most likely failed? How is the component defective?

13.9 If the fuse F_1 is blown, (a) how much current will be measured by M_1 and M_3? (b) How much voltage will be measured by M_2? (c) How much voltage will be measured across the blown fuse? (d) What is most likely to have caused the blown fuse? (e) Using resistance measurements, outline a procedure for finding the defective component.

13.10 If M_1 and M_3 measure 0 A but M_2 measures 36 V, what is most likely wrong? How could you isolate the trouble by making voltage measurements?

13.11 If the fuse F_1 has blown because of a shorted branch, how much resistance would be measured across points B and I? Without using resistance measurements, how could the shorted branch be identified?

13.12 If the wire connecting points F and G opens, (a) how much current will M_3 show? (b) How much voltage would be measured across R_4? (c) How much voltage would be measured across points D and E? (d) How much voltage would be measured across points F and G?

13.13 Assuming that the circuit is operating normally, how much voltage would be measured across (a) the fuse F_1; (b) the switch S_1?

13.14 If the branch resistor R_3 opens, (a) how much voltage would be measured across R_3? (b) How much current would be indicated by M_1 and M_3?

13.15 If the wire between points B and C breaks open, (a) how much current will be measured by M_1 and M_3? (b) How much voltage would be measured across points B and C? (c) How much voltage will be measured across points C and H?

SECTION 13.6 Troubleshooting: Opens and Shorts in Series-Parallel Circuits

13.16 Refer to Fig. 13-18. How much voltage will be indicated by the voltmeter when the wiper arm of the linear potentiometer R_2 is set (a) to point A; (b) to point B; (c) midway between points A and B.

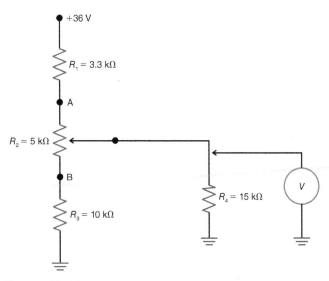

Figure 13-18

Table 13-2 Voltage Measurements Taken in Fig. 13-19

	V_1	V_2	V_3	V_4	V_5	V_6	Defective Component
				Volts			
1 Normal values	5	36	6.6	24	5.4	9	None
2 Trouble 1	0	0	0	0	0	50	
3 Trouble 2	17.86	0	0	0	0	32.14	
4 Trouble 3	3.38	40.54	0	40.54	0	6.08	
5 Trouble 4	3.38	40.54	0	0	40.54	6.08	
6 Trouble 5	6.1	43.9	8.05	29.27	6.59	0	
7 Trouble 6	2.4	43.27	7.93	28.85	6.49	4.33	
8 Trouble 7	50	0	0	0	0	0	
9 Trouble 8	7.35	29.41	16.18	0	13.24	13.24	
10 Trouble 9	0	40	7.33	26.67	6	10	
11 Trouble 10	3.38	40.54	40.54	0	0	6.08	
12 Trouble 11	5.32	35.11	0	28.67	6.45	9.57	
13 Trouble 12	5.25	35.3	7.61	27.7	0	9.45	

13.17 Table 13-2 shows voltage measurements taken in Fig. 13-19. The first row shows the normal values that exist when the circuit is operating normally. Rows 2 to 13 are voltage measurements taken when one component in the circuit has failed. For each row in Table 13-2, identify which component is defective and determine the type of defect that has occurred in the component.

13.18 Table 13-3 shows voltage measurements taken in Fig. 13-20. The first row shows the normal values when the circuit is operating properly. Rows 2 to 9 are voltage measurements taken when one component in the circuit has failed. For each row,

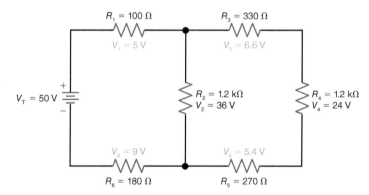

Figure 13-19 Circuit diagram. Normal operating voltages are shown.

Table 13-3 Voltage Measurements Taken in Fig. 13-20

	V_{AG}	V_{BG}	V_{CG}	V_{DG}	Defective Component
			Volts		
1 Normal values	25	20	7.5	2.5	None
2 Trouble 1	25	25	0	0	
3 Trouble 2	25	0	0	0	
4 Trouble 3	25	18.75	3.125	3.125	
5 Trouble 4	25	25	25	25	
6 Trouble 5	25	15	15	5	
7 Trouble 6	25	25	9.375	3.125	
8 Trouble 7	25	25	25	0	
9 Trouble 8	25	19.4	5.56	0	

identify which component is defective and determine the type of defect that has occurred in the component.

Table 13-4 shows voltage measurements taken in Fig. 13-21. The first row shows the normal values when the circuit is operating properly. Rows 2 to 9 are voltage measurements taken when one component in the circuit has failed. For each row, identify which component is defective and determine the type of defect that has occurred in the component.

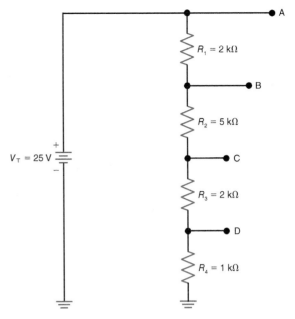

Figure 13-20   Series voltage divider.

Table 13-4 Voltage Measurements Taken in Fig. 13-21

	V_A	V_B	V_C	V_D	Comments	Defective Component
			Volts			
1 Normal values	36	24	15	6	—	None
2 Trouble 1	36	32	0	0	—	
3 Trouble 2	36	16.94	0	0	R_2 Warm	
4 Trouble 3	36	25.52	18.23	0	—	
5 Trouble 4	36	0	0	0	R_1 Hot	
6 Trouble 5	36	0	0	0	—	
7 Trouble 6	36	27.87	23.23	9.29	—	
8 Trouble 7	36	23.25	13.4	0	—	
9 Trouble 8	36	36	22.5	9	—	

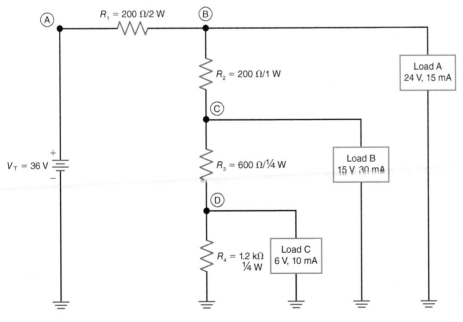

Figure 13-21   Loaded voltage divider.

Chapter 14

Alternating Voltage and Current

This chapter covers the analysis of alternating voltage. An alternating voltage is a voltage that continuously varies in amplitude and periodically reverses in polarity. One cycle includes two alternations in polarity. The number of cycles per second is the frequency measured in hertz (Hz). Every ac voltage has both amplitude variations and polarity reversals. The amplitude values and rate of polarity reversal, however, vary from one ac waveform to the next. This chapter covers the theory, the terminology, and the measurements of alternating voltage and current. The emphasis is on an naturally occurring ac waveform called the *sine wave*. Other forms of ac include square waves, triangular waves, and other nonsinusoidal shapes. The Fourier theory and the frequency domain are introduced as a way to analyze and view nonsinusoidal waves.

Learning Outcomes

After studying this chapter, you should be able to:

> *Describe* how a sine wave of alternating voltage is generated.

> *Calculate* the instantaneous value of a sine wave of alternating voltage or current.

> *Define* the following values for a sine wave: peak, peak-to-peak, root-mean-square, and average.

> *Calculate* the rms, average, and peak-to-peak values of a sine wave when the peak value is known.

> *Define frequency* and *period* and *list* the units of each.

> *Calculate* the wavelength when the frequency is known.

> *Explain* the concept of phase angles.

> *Describe* the makeup of a nonsinusoidal waveform.

> *Define* the term *harmonics*.

> *Define* the *Fourier theory*.

14.1 Alternating Current Applications

Figure 14-1 shows the output from an ac voltage generator with the reversals between positive and negative polarities and the variations in amplitude. In Fig. 14-1a, the waveform shown simulates an ac voltage as it would appear on the screen of an oscilloscope, which is an important test instrument for ac voltages. The oscilloscope shows a picture of any ac voltage connected to its input terminals. It also indicates the amplitude. The details of how to use the oscilloscope for ac voltage measurements are explained in Chap. 26.

In Fig. 14-1b, the graph of the ac waveform shows how the output from the generator in Fig. 14-1c varies with respect to time. Assume that this graph shows V at terminal 2 with respect to terminal 1. Then the voltage at terminal 1 corresponds to the zero axis in the graph as the reference level. At terminal 2, the output voltage has positive amplitude variations from zero up to the peak value and down to zero. All these voltage values are with respect to terminal 1. After a half-cycle, the voltage at terminal 2 becomes negative, still with respect to the other terminal. Then the same voltage variations are repeated at terminal 2, but they have negative polarity compared to the reference level. Note that if we take the voltage at terminal 1 with terminal 2 as the reference, the waveform in Fig. 14-1b would have the same shape but be inverted in polarity. The negative half-cycle would come first, but it does not matter which is first or second.

The characteristic of varying values is the reason that ac circuits have so many uses. For instance, a transformer can operate only with alternating current to step up or step down an ac voltage. The reason is that the changing current produces changes in its associated magnetic field. This application is just an example of inductance L in ac circuits, where the changing magnetic flux of a varying current can produce induced voltage.

A similar but opposite effect in ac circuits is capacitance C. The capacitance is important with the changing electric field of a varying voltage. Just as L has an effect with alternating current, C has an effect that depends on alternating voltage.

The L and C are additional factors, beside resistance R, in the operation of ac circuits.

In general, electronic circuits are combinations of R, L, and C, with both direct current and alternating current. Audio, video, and radio signals are ac voltages and currents. However, amplifiers that use transistors and integrated circuits need dc voltages to conduct any current at all. The resulting output of an amplifier circuit, therefore, consists of direct current with a superimposed ac signal.

14.2 Alternating-Voltage Generator

An alternating voltage is a voltage that continuously varies in magnitude and periodically reverses in polarity. In Fig. 14-1, the variations up and down on the waveform show the changes in magnitude. The zero axis is a horizontal line across the center. Then voltages above the center have positive polarity, and values below center are negative.

Figure 14-2 shows how such a voltage waveform is produced by a rotary generator. The conductor loop rotates through the magnetic field to generate the induced ac voltage across its open terminals. The magnetic flux shown here is vertical, with lines of force in the plane of the paper.

In Fig. 14-2a, the loop is in its horizontal starting position in a plane perpendicular to the paper. When the loop rotates counterclockwise, the two longer conductors move around a circle. Note that in the flat position shown, the two long conductors of the loop move vertically up or down but parallel to the vertical flux lines. In this position, motion of the loop does not induce a voltage because the conductors are not cutting across the flux.

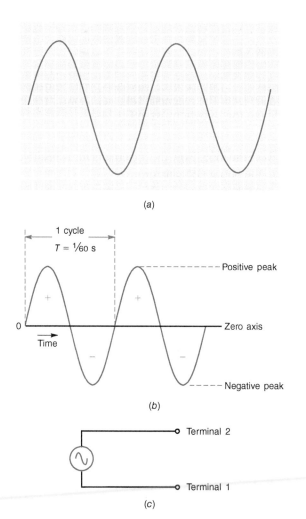

(a)

(b)

(c)

Figure 14-1 Waveform of ac power-line voltage with frequency of 60 Hz. Two cycles are shown. (a) Oscilloscope display. (b) Details of waveform and alternating polarities. (c) Symbol for an ac voltage source.

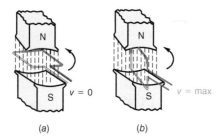

(a)   (b)

Figure 14-2 A loop rotating in a magnetic field to produce induced voltage *v* with alternating polarities. (*a*) Loop conductors moving parallel to magnetic field results in zero voltage. (*b*) Loop conductors cutting across magnetic field produce maximum induced voltage.

When the loop rotates through the upright position in Fig. 14-2*b*, however, the conductors cut across the flux, producing maximum induced voltage. The shorter connecting wires in the loop do not have any appreciable voltage induced in them.

Each of the longer conductors has opposite polarity of induced voltage because the conductor at the top is moving to the left while the bottom conductor is moving to the right. The amount of voltage varies from zero to maximum as the loop moves from a flat position to upright, where it can cut across the flux. Also, the polarity at the terminals of the loop reverses as the motion of each conductor reverses during each half-revolution.

With one revolution of the loop in a complete circle back to the starting position, therefore, the induced voltage provides a potential difference *v* across the loop, varying in the same way as the wave of voltage shown in Fig. 14-1. If the loop rotates at the speed of 60 revolutions per second, the ac voltage has a frequency of 60 Hz.

The Cycle

One complete revolution of the loop around the circle is a *cycle*. In Fig. 14-3, the generator loop is shown in its position at each quarter-turn during one complete cycle. The corresponding wave of induced voltage also goes through one cycle. Although not shown, the magnetic field is from top to bottom of the page, as in Fig. 14-2.

At position A in Fig. 14-3, the loop is flat and moves parallel to the magnetic field, so that the induced voltage is zero. Counterclockwise rotation of the loop moves the dark conductor to the top at position B, where it cuts across the field to produce maximum induced voltage. The polarity of the induced voltage here makes the open end of the dark conductor positive. This conductor at the top is cutting across the flux from right to left. At the same time, the opposite conductor below is moving from left to right, causing its induced voltage to have opposite polarity. Therefore, maximum induced voltage is produced at this time across the two

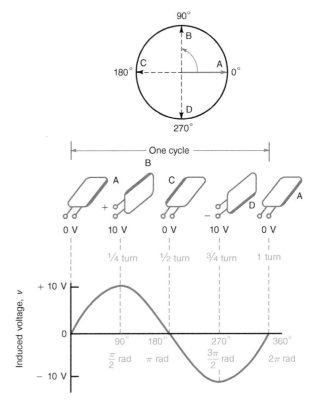

Figure 14-3 One cycle of alternating voltage generated by a rotating loop. The magnetic field, not shown here, is directed from top to bottom, as in Fig. 14-2.

open ends of the loop. Now the top conductor is positive with respect to the bottom conductor.

In the graph of induced voltage values below the loop in Fig. 14-3, the polarity of the dark conductor is shown with respect to the other conductor. Positive voltage is shown above the zero axis in the graph. As the dark conductor rotates from its starting position parallel to the flux toward the top position, where it cuts maximum flux, more and more induced voltage is produced with positive polarity.

When the loop rotates through the next quarter-turn, it returns to the flat position shown in C, where it cannot cut across flux. Therefore, the induced voltage values shown in the graph decrease from the maximum value to zero at the half-turn, just as the voltage was zero at the start. The half-cycle of revolution is called an *alternation*.

The next quarter-turn of the loop moves it to the position shown at D in Fig. 14-3, where the loop cuts across the flux again for maximum induced voltage. Note, however, that here the dark conductor is moving left to right at the bottom of the loop. This motion is reversed from the direction it had when it was at the top, moving right to left. Because the direction of motion is reversed during the second half-revolution, the induced voltage has opposite polarity with the dark conductor negative. This polarity is shown as negative voltage below the zero axis. The maximum value of induced voltage at the third quarter-turn is the same as at the first quarter-turn but with opposite polarity.

When the loop completes the last quarter-turn in the cycle, the induced voltage returns to zero as the loop returns to its flat position at A, the same as at the start. This cycle of values of induced voltage is repeated as the loop continues to rotate with one complete cycle of voltage values, as shown, for each circle of revolution.

Note that zero at the start and zero after the half-turn of an alternation are not the same. At the start, the voltage is zero because the loop is flat, but the dark conductor is moving upward in the direction that produces positive voltage. After one half-cycle, the voltage is zero with the loop flat, but the dark conductor is moving downward in the direction that produces negative voltage. After one complete cycle, the loop and its corresponding waveform of induced voltage are the same as at the start. *A cycle can be defined, therefore, as including the variations between two successive points having the same value and varying in the same direction.*

Angular Measure

Because the cycle of voltage in Fig. 14-3 corresponds to rotation of the loop around a circle, it is convenient to consider parts of the cycle in angles. The complete circle includes 360°. One half-cycle, or one alternation, is 180° of revolution. A quarter-turn is 90°. The circle next to the loop positions in Fig. 14-3 illustrates the angular rotation of the dark conductor as it rotates counterclockwise from 0 to 90 to 180° for one half-cycle, then to 270°, and returning to 360° to complete the cycle. Therefore, one cycle corresponds to 360°.

Radian Measure

In angular measure it is convenient to use a specific unit angle called the *radian* (abbreviated *rad*), which is an angle equal to 57.3°. Its convenience is due to the fact that a radian is the angular part of the circle that includes an arc equal to the radius r of the circle, as shown in Fig. 14-4. The circumference

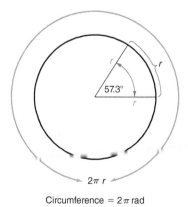

Circumference = 2π rad

Figure 14-4 One radian (rad) is the angle equal to 57.3°. The complete circle of 360° includes 2π rad.

around the circle equals $2\pi r$. A circle includes 2π rad, then, as each radian angle includes one length r of the circumference. Therefore, one cycle equals 2π rad.

As shown in the graph in Fig. 14-3, divisions of the cycle can be indicated by angles in either degrees or radians. The comparison between degrees and radians can be summarized as follows:

Zero degrees is also zero radians
$360° = 2\pi$ rad
$180° = \frac{1}{2} \times 2\pi$ rad $= \pi$ rad
$90° = \frac{1}{2} \times \pi$ rad $= \pi/2$ rad
$270° = 180° + 90°$ or π rad $+ \pi/2$ rad $= 3\pi/2$ rad

The constant 2π in circular measure is numerically equal to 6.2832. This is double the value of 3.1416 for π. The Greek letter π (pi) is used to represent the ratio of the circumference to the diameter for any circle, which always has the numerical value of 3.1416. The fact that 2π rad is 360° can be shown as $2 \times 3.1416 \times 57.3° = 360°$ for a complete cycle. To convert from degrees to radians or vice versa, use the following conversion formulas:

$$\#deg = \#rad \times \frac{180°}{\pi \text{ rad}}$$

$$\#rad = \#deg \times \frac{\pi \text{ rad}}{180°}$$

14.3 The Sine Wave

The voltage waveform in Figs. 14-1 and 14-3 is called a *sine wave, sinusoidal wave,* or *sinusoid* because the amount of induced voltage is proportional to the sine of the angle of rotation in the circular motion producing the voltage. The sine is a trigonometric function of an angle; it is equal to the ratio of the opposite side to the hypotenuse in a right triangle. This numerical ratio increases from zero for 0° to a maximum value of 1 for 90° as the side opposite the angle becomes larger.

The voltage waveform produced by the circular motion of the loop is a sine wave because the induced voltage increases to a maximum at 90°, when the loop is vertical, in the same way that the sine of the angle of rotation increases to a maximum at 90°. The induced voltage and sine of the angle correspond for the full 360° of the cycle.

The sine wave reaches one-half its maximum value in 30°, which is only one-third of 90°. This fact means that the sine wave has a sharper slope of changing values when the wave is near the zero axis, compared with more gradual changes near the maximum value.

The instantaneous value of a sine-wave voltage for any angle of rotation is expressed by the formula

$$v = V_P \sin \theta \tag{14-1}$$

where θ (Greek letter *theta*) is the angle, sin is the abbreviation for its sine, V_P is the maximum or peak voltage value, and v is the instantaneous value of voltage at angle θ.

EXAMPLE 14-1

A sine wave of voltage varies from zero to a peak of 100 V. How much is the voltage at the instant of 30° of the cycle? 45°? 90°? 270°?

Answer:

$$v = V_P \sin \theta = 100 \sin \theta$$

$$\begin{aligned}
\text{At } 30°: \quad v &= V_P \sin 30° = 100 \times 0.5 \\
&= 50 \text{ V} \\
\text{At } 45°: \quad v &= V_P \sin 45° = 100 \times 0.707 \\
&= 70.7 \text{ V} \\
\text{At } 90°: \quad v &= V_P \sin 90° = 100 \times 1 \\
&= 100 \text{ V} \\
\text{At } 270°: \quad v &= V_P \sin 270° = 100 \times -1 \\
&= -100 \text{ V}
\end{aligned}$$

The value of −100 V at 270° is the same as that at 90° but with opposite polarity.

To do the problems in Example 14-1, you must either refer to a table of trigonometric functions or use a scientific calculator that has trig functions.

Between zero at 0° and maximum at 90°, the amplitudes of a sine wave increase exactly as the sine value of the angle of rotation. These values are for the first quadrant in the circle, that is, 0 to 90°. From 90 to 180° in the second quadrant, the values decrease as a mirror image of the first 90°. The values in the third and fourth quadrants, from 180 to 360°, are exactly the same as 0 to 180° but with opposite sign. At 360°, the waveform is back to 0° to repeat its values every 360°.

In summary, the characteristics of the sine-wave ac waveform are:

1. The cycle includes 360° or 2π rad.
2. The polarity reverses each half-cycle.
3. The maximum values are at 90 and 270°.
4. The zero values are at 0 and 180°.
5. The waveform changes its values fastest when it crosses the zero axis.
6. The waveform changes its values slowest when it is at its maximum value. The values must stop increasing before they can decrease.

A perfect example of the sine-wave ac waveform is the 60-Hz power-line voltage in Fig. 14-1.

14.4 Alternating Current

When a sine wave of alternating voltage is connected across a load resistance, the current that flows in the circuit is also a sine wave. In Fig. 14-5, let the sine-wave voltage at the left in the diagram be applied across R of 100 Ω. The resulting

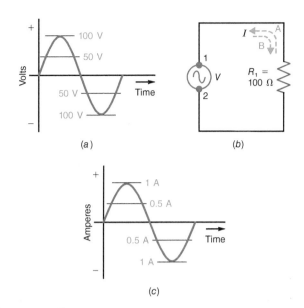

(a)

(b)

(c)

Figure 14-5 A sine wave of alternating voltage applied across R produces a sine wave of alternating current in the circuit. (a) Waveform of applied voltage. (b) AC circuit. Note the symbol for sine-wave generator V. (c) Waveform of current in the circuit.

sine wave of alternating current is shown at the right in the diagram. Note that the frequency is the same for v and i.

During the first alternation of v in Fig. 14-5, terminal 1 is positive with respect to terminal 2. Since the direction of electron flow is from the negative side of v, through R, and back to the positive side of v, current flows in the direction indicated by arrow A for the first half-cycle. This direction is taken as the positive direction of current in the graph for i, corresponding to positive values of v.

The amount of current is equal to v/R. If several instantaneous values are taken, when v is zero, i is zero; when v is 50 V, i equals 50 V/100, or 0.5 A; when v is 100 V, i equals 100 V/100, or 1 A. For all values of applied voltage with positive polarity, therefore, the current is in one direction, increasing to its maximum value and decreasing to zero, just like the voltage.

In the next half-cycle, the polarity of the alternating voltage reverses. Then terminal 1 is negative with respect to terminal 2. With reversed voltage polarity, current flows in the opposite direction. Electron flow is from terminal 1 of the voltage source, which is now the negative side, through R, and back to terminal 2. This direction of current, as indicated by arrow B in Fig. 14-5, is negative.

The negative values of i in the graph have the same numerical values as the positive values in the first half-cycle, corresponding to the reversed values of applied voltage. As a result, the alternating current in the circuit has sine-wave variations corresponding exactly to the sine-wave alternating voltage.

Only the waveforms for v and i can be compared. There is no comparison between relative values because the current and voltage are different quantities.

It is important to note that the negative half-cycle of applied voltage is just as useful as the positive half-cycle in producing current. The only difference is that the reversed polarity of voltage produces the opposite direction of current.

Furthermore, the negative half-cycle of current is just as effective as the positive values when heating the filament to light a bulb. With positive values, electrons flow through the filament in one direction. Negative values produce electron flow in the opposite direction. In both cases, electrons flow from the negative side of the voltage source, through the filament, and return to the positive side of the source. For either direction, the current heats the filament. The direction does not matter, since it is the motion of electrons against resistance that produces power dissipation. In short, resistance R has the same effect in reducing I for either direct current or alternating current.

14.5 Voltage and Current Values for a Sine Wave

Since an alternating sine wave of voltage or current has many instantaneous values through the cycle, it is convenient to define specific magnitudes to compare one wave with another. The peak, average, and root-mean-square (rms) values can be specified, as indicated in Fig. 14-6. These values can be used for either current or voltage.

Peak Value

This is the maximum value V_P or I_P. For example, specifying that a sine wave has a peak value of 170 V states the highest value the sine wave reaches. All other values during the cycle follow a sine wave. The peak value applies to either the positive or the negative peak.

To include both peak amplitudes, the *peak-to-peak (p-p) value* may be specified. For the same example, the peak-to-peak value is 340 V, double the peak value of 170 V, since the positive and negative peaks are symmetrical. Note that the two opposite peak values cannot occur at the same time. Furthermore, in some waveforms, the two peaks are not equal.

Average Value

This is an arithmetic average of all values in a sine wave for one alternation, or half-cycle. The half-cycle is used for the average because over a full cycle the average value is zero, which is useless for comparison. If the sine values for all angles up to 180° for one alternation are added and then divided by the number of values, this average equals 0.637.

Since the peak value of the sine function is 1 and the average equals 0.637, then

$$\text{Average value} = 0.637 \times \text{peak value} \qquad \textbf{(14-2)}$$

With a peak of 170 V, for example, the average value is 0.637 × 170 V, which equals approximately 108 V.

Root-Mean-Square, or Effective, Value

The most common method of specifying the amount of a sine wave of voltage or current is by relating it to the dc voltage and current that will produce the same heating effect. This is called its *root-mean-square* value, abbreviated *rms*. The formula is

$$\text{rms value} = 0.707 \times \text{peak value} \qquad \textbf{(14-3)}$$

or

$$V_{\text{rms}} = 0.707 V_P$$

and

$$I_{\text{rms}} = 0.707 I_P$$

With a peak of 170 V, for example, the rms value is 0.707 × 170, or 120 V, approximately. This is the voltage of the commercial ac power line, which is always given in rms value.

It is often necessary to convert from rms to peak value. This can be done by inverting Formula (14-3), as follows:

$$\text{Peak} = \frac{1}{0.707} \times \text{rms} = 1.414 \times \text{rms} \qquad \textbf{(14-4)}$$

or

$$V_P = 1.414 V_{\text{rms}}$$

and

$$I_P = 1.414 I_{\text{rms}}$$

Dividing by 0.707 is the same as multiplying by 1.414.

For example, commercial power-line voltage with an rms value of 120 V has a peak value of 120 × 1.414, which equals 170 V, approximately. Its peak-to-peak value is 2 × 170, or 340 V, which is double the peak value. As a formula,

$$\text{Peak-to-peak value} = 2.828 \times \text{rms value} \qquad \textbf{(14-5)}$$

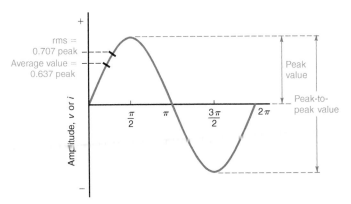

Figure 14-6 Definitions of important amplitude values for a sine wave of voltage or current.

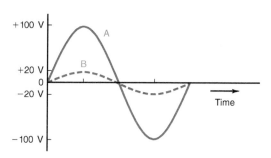

Figure 14-7 Waveforms A and B have different amplitudes, but they are both sine waves.

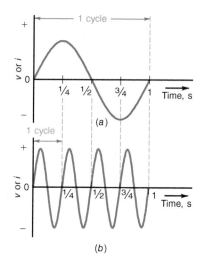

Figure 14-8 Number of cycles per second is the frequency in hertz (Hz) units. (a) $f = 1$ Hz. (b) $f = 4$ Hz.

The factor 0.707 for the rms value is derived as the square root of the average (mean) of all the squares of the sine values. If we take the sine for each angle in the cycle, square each value, add all the squares, divide by the number of values added to obtain the average square, and then take the square root of this mean value, the answer is 0.707.

The advantage of the rms value derived in terms of the squares of the voltage or current values is that it provides a measure based on the ability of the sine wave to produce power, which is I^2R or V^2/R. As a result, the rms value of an alternating sine wave corresponds to the same amount of direct current or voltage in heating power. An alternating voltage with an rms value of 120 V, for instance, is just as effective in heating the filament of a lightbulb as 120 V from a steady dc voltage source. For this reason, the rms value is also called the *effective* value.

Unless indicated otherwise, all sine-wave ac measurements are in rms values. The capital letters V and I are used, corresponding to the symbols for dc values. As an example, $V = 120$ V for ac power-line voltage.

Note that sine waves can have different amplitudes but still follow the sinusoidal waveform. Figure 14-7 compares a low-amplitude voltage with a high-amplitude voltage. Although different in amplitude, they are both sine waves. In each wave, the rms value = 0.707 × peak value.

14.6 Frequency

The number of cycles per second is the *frequency,* with the symbol f. In Fig. 14-3, if the loop rotates through 60 complete revolutions, or cycles, during 1 s, the frequency of the generated voltage is 60 cps, or 60 Hz. You see only one cycle of the sine waveform, instead of 60 cycles, because the time interval shown here is $\frac{1}{60}$ s. Note that the factor of time is involved. More cycles per second means a higher frequency and less time for one cycle, as illustrated in Fig. 14-8. Then the changes in values are faster for higher frequencies.

A complete cycle is measured between two successive points that have the same value and direction. In Fig. 14-8, the cycle is between successive points where the waveform

is zero and ready to increase in the positive direction. Or the cycle can be measured between successive peaks.

On the time scale of 1 s, waveform a goes through one cycle; waveform b has much faster variations, with four complete cycles during 1 s. Both waveforms are sine waves, even though each has a different frequency.

In comparing sine waves, the amplitude has no relation to frequency. Two waveforms can have the same frequency with different amplitudes (Fig. 14-7), the same amplitude but different frequencies (Fig. 14-8), or different amplitudes and frequencies. The amplitude indicates the amount of voltage or current, and the frequency indicates the rate of change of amplitude variations in cycles per second.

Frequency Units

The unit called the *hertz (Hz),* named after Heinrich Hertz, is used for cycles per second. Then 60 cps = 60 Hz. All metric prefixes can be used. As examples

$$1 \text{ kilohertz per second} = 1 \times 10^3 \text{ Hz} = 1 \text{ kHz}$$
$$1 \text{ megahertz per second} = 1 \times 10^6 \text{ Hz} = 1 \text{ MHz}$$
$$1 \text{ gigahertz per second} = 1 \times 10^9 \text{ Hz} = 1 \text{ GHz}$$
$$1 \text{ terahertz per second} = 1 \times 10^{12} \text{ Hz} = 1 \text{ THz}$$

Audio and Radio Frequencies

The entire frequency range of alternating voltage or current from 1 Hz to many megahertz can be considered in two broad groups: audio frequencies (AF) and radio frequencies (RF). *Audio* is a Latin word meaning "I hear." The audio range includes frequencies that can be heard as sound waves by the human ear. This range of audible frequencies is approximately 20 to 20,000 Hz.

The higher the frequency, the higher the pitch or tone of the sound. High audio frequencies, about 3000 Hz and above, provide *treble* tone. Low audio frequencies, about 300 Hz and below, provide *bass* tone.

Loudness is determined by amplitude. The greater the amplitude of the AF variation, the louder its corresponding sound.

Alternating current and voltage above the audio range provide RF variations, since electrical variations of high frequency can be transmitted by electromagnetic radio waves. See the System Sidebar, "The Frequency Spectrum," in this chapter for details.

Sonic and Ultrasonic Frequencies

Sonic and ultrasonic frequencies refer to sound waves, which are variations in pressure generated by mechanical vibrations, rather than electrical variations. The velocity of sound waves through dry air at 20°C equals 1130 ft/s. Sound waves above the audible range of frequencies are called *ultrasonic* waves. The range of frequencies for ultrasonic applications, therefore, is from 16,000 Hz up to several megahertz. Sound waves in the audible range of frequencies below 16,000 Hz can be considered *sonic* or sound frequencies. The term *audio* is reserved for electrical variations that can be heard when converted to sound waves.

14.7 Period

The amount of time it takes for one cycle is called the *period*. Its symbol is T for time. With a frequency of 60 Hz, as an example, the time for one cycle is $\frac{1}{60}$ s. Therefore, the period is $\frac{1}{60}$ s. Frequency and period are reciprocals of each other:

$$T = \frac{1}{f} \quad \text{or} \quad f = \frac{1}{T} \quad \text{(14-6)}$$

The higher the frequency, the shorter the period. In Fig. 14-8a, the period for the wave with a frequency of 1 Hz is 1 s, and the higher frequency wave of 4 Hz in Fig. 14-8b has a period of ¼ s for a complete cycle.

Units of Time

The second is the basic unit of time, but for higher frequencies and shorter periods, smaller units of time are convenient. Those used most often are:

$$T = 1 \text{ millisecond} = 1 \text{ ms} = 1 \times 10^{-3} \text{ s}$$
$$T = 1 \text{ microsecond} = 1 \text{ } \mu\text{s} = 1 \times 10^{-6} \text{ s}$$
$$T = 1 \text{ nanosecond} = 1 \text{ ns} = 1 \times 10^{-9} \text{ s}$$
$$T = 1 \text{ picosecond} = 1 \text{ ps} = 1 \times 10^{-12} \text{ s}$$

These units of time for a period are reciprocals of the corresponding units for frequency. The reciprocal of frequency in kilohertz gives the period T in milliseconds; the reciprocal of megahertz is microseconds; the reciprocal of gigahertz is nanoseconds.

EXAMPLE 14-2

An alternating current varies through one complete cycle in $\frac{1}{1000}$ s. Calculate the period and frequency.

Answer:

$$T = \frac{1}{1000} \text{ s}$$

$$f = \frac{1}{T} = \frac{1}{\frac{1}{1000}}$$

$$= \frac{1000}{1} = 1000$$

$$= 1000 \text{ Hz or 1 kHz}$$

EXAMPLE 14-3

Calculate the period for the two frequencies of 1 MHz and 2 MHz.

Answer:

a. For 1 MHz,

$$T = \frac{1}{f} = \frac{1}{1 \times 10^6}$$

$$= 1 \times 10^{-6} \text{ s} = 1 \text{ } \mu\text{s}$$

b. For 2 MHz,

$$T = \frac{1}{f} = \frac{1}{2 \times 10^6}$$

$$= 0.5 \times 10^{-6} \text{ s} = 0.5 \text{ } \mu\text{s}$$

To do these problems on a calculator, you need the reciprocal key, usually marked (1/x). Keep the powers of 10 separate, and remember that the reciprocal has the same exponent with opposite sign. With f of 2×10^6, for 1/f just punch in 2 and then press (2ndF) and the (1/x) key to see 0.5 as the reciprocal. The 10^6 for f becomes 10^{-6} for T so that the answer is 0.5×10^{-6} s or 0.5 μs.

14.8 Wavelength

When a periodic variation is considered with respect to distance, one cycle includes the *wavelength*, which is the length of one complete wave or cycle (Fig. 14-9). For example, when a radio wave is transmitted, variations in the electromagnetic field travel through space. Also, with sound waves, the variations in air pressure corresponding to the sound wave move through air. In these applications, the distance traveled by the wave in one cycle is the wavelength. The wavelength depends on the frequency of the variation and its velocity of transmission:

$$\lambda = \frac{\text{velocity}}{\text{frequency}} \quad \text{(14-7)}$$

where λ (the Greek letter lambda) is the symbol for one complete wavelength.

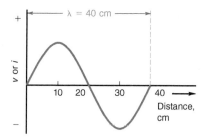

Figure 14-9 Wavelength λ is the distance traveled by the wave in one cycle.

Wavelength of Radio Waves

The velocity of electromagnetic radio waves in air or vacuum is 186,000 mi/s, or 3×10^{10} cm/s, which is the speed of light. Therefore,

$$\lambda\text{(cm)} = \frac{3 \times 10^{10}\text{ cm/s}}{f\text{(Hz)}} \qquad \textbf{(14-8)}$$

Note that the higher the frequency, the shorter the wavelength. For instance, the shortwave radio broadcast band of 5.95 to 26.1 MHz includes frequencies higher than the standard AM radio broadcast band of 535 to 1705 kHz.

EXAMPLE 14-4

Calculate λ for a radio wave with *f* of 30 GHz.

Answer:

$$\lambda = \frac{3 \times 10^{10}\text{ cm/s}}{30 \times 10^9\text{ Hz}} = \frac{3}{30} \times 10\text{ cm}$$
$$= 0.1 \times 10$$
$$= 1\text{ cm}$$

Such short wavelengths are called *microwaves*. This range includes λ of 1 m or less for frequencies of 300 MHz or more.

EXAMPLE 14-5

The length of a TV antenna is λ/2 for radio waves with *f* of 60 MHz. What is the antenna length in centimeters and feet?

Answer:

a.
$$\lambda = \frac{3 \times 10^{10}\text{ cm/s}}{60 \times 10^6\text{ Hz}} = \frac{1}{20} \times 10^4\text{ cm}$$
$$= 0.05 \times 10^4$$
$$= 500\text{ cm}$$

Then, λ/2 = ⁵⁰⁰/₂ = **250 cm.**

b. Since 2.54 cm = 1 in.,

$$\lambda/2 = \frac{250\text{ cm}}{2.54\text{ cm/in.}} = 98.4\text{ in.}$$
$$= \frac{98.4\text{ in}}{12\text{ in./ft}} = 8.2\text{ ft}$$

EXAMPLE 14-6

For the 6-m band used in amateur radio, what is the corresponding frequency?

Answer:

The formula λ = *v*/*f* can be inverted

$$f = \frac{v}{\lambda}$$

Then

$$f = \frac{3 \times 10\text{ cm/s}}{6\text{ m}} = \frac{3 \times 10^{10}\text{ cm/s}}{6 \times 10^2\text{ cm}}$$
$$= \frac{3}{6} \times 10^8 = 0.5 \times 10^8\text{ Hz}$$
$$= 50 \times 10^6\text{ Hz} \qquad \text{or} \qquad 50\text{ MHz}$$

Wavelength of Sound Waves

The velocity of sound waves is much lower than that of radio waves because sound waves result from mechanical vibrations rather than electrical variations. In average conditions, the velocity of sound waves in air equals 1130 ft/s. To calculate the wavelength, therefore,

$$\lambda = \frac{1130\text{ ft/s}}{f\text{ Hz}} \qquad \textbf{(14-9)}$$

This formula can also be used for ultrasonic waves. Although their frequencies are too high to be audible, ultrasonic waves are still sound waves rather than radio waves.

EXAMPLE 14-7

What is the wavelength of the sound waves produced by a loudspeaker at a frequency of 100 Hz?

Answer:

$$\lambda = \frac{1130\text{ ft/s}}{100\text{ Hz}}$$
$$\lambda = 11.3\text{ ft}$$

EXAMPLE 14-8

For ultrasonic waves at a frequency of 34.44 kHz, calculate the wavelength in feet and in centimeters.

Answer:

$$\lambda = \frac{1130}{34.44 \times 10^3}$$
$$= 32.8 \times 10^{-3}\text{ ft}$$
$$= 0.0328\text{ ft}$$

To convert to inches,

$$0.0328\text{ ft} \times 12 = 0.3936\text{ in.}$$

To convert to centimeters,

$$0.3936\text{ in.} \times 2.54 = 1\text{ cm} \quad \text{approximately}$$

Note that the 34.44-kHz sound waves in this example have the same wavelength (1 cm) as the 30-GHz radio waves in Example 14-4. The reason is that radio waves have a much higher velocity than sound waves.

14.9 Phase Angle

Referring back to Fig. 14-3, suppose that the generator started its cycle at point B, where maximum voltage output is produced, instead of starting at the point of zero output. If we compare the two cases, the two output voltage waves would be as shown in Fig. 14-10. Each is the same waveform of alternating voltage, but wave B starts at maximum, and wave A starts at zero. The complete cycle of wave B through 360° takes it back to the maximum value from which it started. Wave A starts and finishes its cycle at zero. With respect to time, therefore, wave B is ahead of wave A in values of generated voltage. The amount it leads in time equals one quarter-revolution, which is 90°. This angular difference is the phase angle between waves B and A. Wave B leads wave A by the phase angle of 90°.

The 90° phase angle between waves B and A is maintained throughout the complete cycle and in all successive cycles, as long as they both have the same frequency. At any instant, wave B has the value that A will have 90°

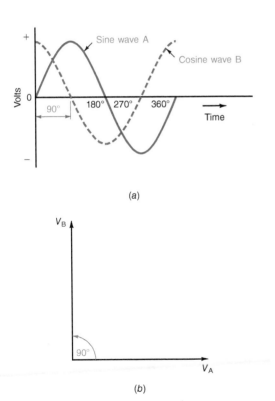

(a)

(b)

Figure 14-10 Two sine-wave voltages 90° out of phase. (a) Wave B leads wave A by 90°. (b) Corresponding phasors V_B and V_A for the two sine-wave voltages with phase angle $\theta = 90°$. The right angle shows quadrature phase.

later. For instance, at 180° wave A is at zero, but B is already at its negative maximum value, where wave A will be later at 270°.

To compare the phase angle between two waves, they must have the same frequency. Otherwise, the relative phase keeps changing. Also, they must have sine-wave variations because this is the only kind of waveform that is measured in angular units of time. The amplitudes can be different for the two waves, although they are shown the same here. We can compare the phases of two voltages, two currents, or a current with a voltage.

The 90° Phase Angle

The two waves in Fig. 14-10 represent a sine wave and a cosine wave 90° out of phase with each other. The 90° phase angle means that one has its maximum amplitude when the other is at zero value. Wave A starts at zero, corresponding to the sine of 0°, has its peak amplitude at 90 and 270°, and is back to zero after one cycle of 360°. Wave B starts at its peak value, corresponding to the cosine of 0°, has its zero value at 90 and 270°, and is back to the peak value after one cycle of 360°.

However, wave B can also be considered a sine wave that starts 90° before wave A in time. This phase angle of 90° for current and voltage waveforms has many applications in sine-wave ac circuits with inductance or capacitance.

The sine and cosine waveforms have the same variations but displaced by 90°. Both waveforms are called *sinusoids*. The 90° angle is called *quadrature phase*.

Phase-Angle Diagrams

To compare phases of alternating currents and voltages, it is much more convenient to use phasor diagrams corresponding to the voltage and current waveforms, as shown in Fig. 14-10b. The arrows here represent the phasor quantities corresponding to the generator voltage.

A phasor is a quantity that has magnitude and direction. The length of the arrow indicates the magnitude of the alternating voltage in rms, peak, or any ac value, as long as the same measure is used for all phasors. The angle of the arrow with respect to the horizontal axis indicates the phase angle.

The terms *phasor* and *vector* are used for a quantity that has direction, requiring an angle to specify the value completely. However, a vector quantity has direction in space, whereas a phasor quantity varies in time. As an example of a vector, a mechanical force can be represented by a vector arrow at a specific angle, with respect to either the horizontal or the vertical direction.

For phasor arrows, the angles shown represent differences in time. One sinusoid is chosen as the reference. Then the timing of the variations in another sinusoid can be compared to the reference by means of the angle between the phasor arrows.

The phasor corresponds to the entire cycle of voltage, but is shown only at one angle, such as the starting point, since the complete cycle is known to be a sine wave. Without the extra details of a whole cycle, phasors represent the alternating voltage or current in a compact form that is easier for comparing phase angles.

In Fig. 14-10b, for instance, the phasor V_A represents the voltage wave A with a phase angle of 0°. This angle can be considered the plane of the loop in the rotary generator where it starts with zero output voltage. The phasor V_B is vertical to show the phase angle of 90° for this voltage wave, corresponding to the vertical generator loop at the start of its cycle. The angle between the two phasors is the phase angle.

The symbol for a phase angle is θ (the Greek letter theta). In Fig. 14-10, as an example, $\theta = 90°$.

Phase–Angle Reference

The phase angle of one wave can be specified only with respect to another as reference. How the phasors are drawn to show the phase angle depends on which phase is chosen as the reference. Generally, the reference phasor is horizontal, corresponding to 0°. Two possibilities are shown in Fig. 14-11. In Fig. 14-11a, the voltage wave A or its phasor V_A is the reference. Then the phasor V_B is 90° counterclockwise. This method is standard practice, using counterclockwise rotation as the positive direction for angles. Also, a leading angle is positive. In this case, then, V_B is 90° counterclockwise from the reference V_A to show that wave B leads wave A by 90°

However, wave B is shown as the reference in Fig. 14-11b. Now V_B is the horizontal phasor. To have the same phase angle, V_A must be 90° clockwise, or −90° from V_B. This arrangement shows that negative angles, clockwise from the 0° reference, are used to show lagging phase angles. The reference determines whether the phase angle is considered leading or lagging in time.

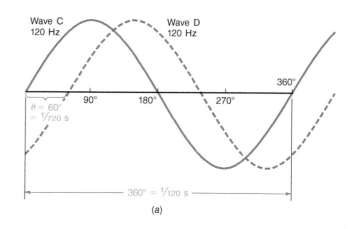

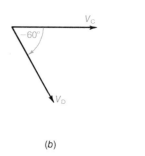

(b)

Figure 14-12   Phase angle of 60° is the time for ⁶⁰/₃₆₀ or ¹/₆ of the cycle. (a) Waveforms. (b) Phasor diagram.

The phase is not actually changed by the method of showing it. In Fig. 14-11, V_A and V_B are 90° out of phase, and V_B leads V_A by 90° in time. There is no fundamental difference whether we say V_B is ahead of V_A by +90° or V_A is behind V_B by −90°.

Two waves and their corresponding phasors can be out of phase by any angle, either less or more than 90°. For instance, a phase angle of 60° is shown in Fig. 14-12. For the waveforms in Fig. 14-12a, wave D is behind C by 60° in time. For the phasors in Fig. 14-12b, this lag is shown by the phase angle of −60°.

In-Phase Waveforms

A phase angle of 0° means that the two waves are in phase (Fig. 14-13).

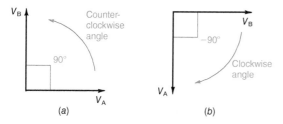

Figure 14-11   Leading and lagging phase angles for 90°. (a) When phasor V_A is the horizontal reference, phasor V_B leads by 90°. (b) When phasor V_B is the horizontal reference, phasor V_A lags by −90°.

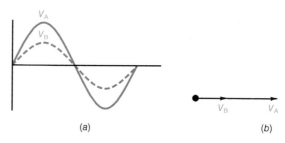

Figure 14-13   Two waveforms in phase, or the phase angle is 0°. (a) Waveforms. (b) Phasor diagram.

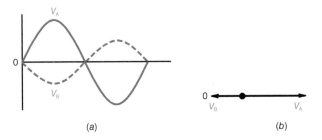

(a)   *(b)*

Figure 14-14 Two waveforms out of phase or in opposite phase with phase angle of 180°. (*a*) Waveforms. (*b*) Phasor diagram.

Out-of-Phase Waveforms

An angle of 180° means opposite phase, or that the two waveforms are exactly out of phase (Fig. 14-14). Then the amplitudes are opposing.

14.10 The Time Factor in Frequency and Phase

It is important to remember that the waveforms we are showing are graphs drawn on paper. The physical factors represented are variations in amplitude, usually on the vertical scale, with respect to equal intervals on the horizontal scale, which can represent either distance or time. To show wavelength, as in Fig. 14-9, the cycles of amplitude variations are plotted against distance or length. To show frequency, the cycles of amplitude variations are shown with respect to time in angular measure. The angle of 360° represents the time for one cycle, or the period *T*.

As an example of how frequency involves time, a waveform with stable frequency is actually used in electronic equipment as a clock reference for very small units of time. Assume a voltage waveform with the frequency of 10 MHz. The period *T* is 0.1 μs. Therefore, every cycle is repeated at 0.1-μs intervals. When each cycle of voltage variations is used to indicate time, then, the result is effectively a clock that measures 0.1-μs units. Even smaller units of time can be measured with higher frequencies. In everyday applications, an electric clock connected to the power line keeps correct time because it is controlled by the exact frequency of 60 Hz.

Furthermore, the phase angle between two waves of the same frequency indicates a specific difference in time. As an example, Fig. 14-12 shows a phase angle of 60° with wave C leading wave D. Both have the same frequency of 120 Hz. The period *T* for each wave then is $\frac{1}{120}$ s. Since 60° is one-sixth of the complete cycle of 360°, this phase angle represents one-sixth of the complete period of $\frac{1}{120}$ s. If we multiply $\frac{1}{6} \times \frac{1}{120}$, the answer is $\frac{1}{720}$ s for the time corresponding to the phase angle of 60°. If we

consider wave D lagging wave C by 60°, this lag is a time delay of $\frac{1}{720}$ s.

More generally, the time for a phase angle θ can be calculated as

$$t = \frac{\theta}{360} \times \frac{1}{f} \qquad \textbf{(14-10)}$$

where *f* is in Hz, θ is in degrees, and *t* is in seconds.

The formula gives the time of the phase angle as its proportional part of the total period of one cycle. For the example of θ equal to 60° with *f* at 120 Hz,

$$t = \frac{\theta}{360} \times \frac{1}{f}$$

$$= \frac{60}{360} \times \frac{1}{120} = \frac{1}{6} \times \frac{1}{120}$$

$$= \frac{1}{720} \text{ s}$$

14.11 Alternating Current Circuits with Resistance

An ac circuit has an ac voltage source. Note the symbol in Fig. 14-15 used for any source of sine-wave alternating voltage. This voltage connected across an external load resistance produces alternating current of the same waveform, frequency, and phase as the applied voltage.

The amount of current equals *V/R* by Ohm's law. When *V* is an rms value, *I* is also an rms value. For any instantaneous value of *V* during the cycle, the value of *I* is for the corresponding instant.

In an ac circuit with only resistance, the current variations are in phase with the applied voltage, as shown in Fig. 14-15*b*. This in-phase relationship between *V* and *I* means that such an ac circuit can be analyzed by the same methods used for dc circuits, since there is no phase angle to consider. Circuit components that have *R* alone include resistors, the filaments of lightbulbs, and heating elements.

The calculations in ac circuits are generally in rms values, unless noted otherwise. In Fig. 14-15*a*, for example, the

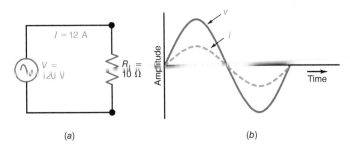

(a)   *(b)*

Figure 14-15 An ac circuit with resistance *R* alone. (*a*) Schematic diagram. (*b*) Waveforms.

120 V applied across the 10-Ω R_L produces rms current of 12 A. The calculations are

$$I = \frac{V}{R_L} = \frac{120 \text{ V}}{10 \text{ } \Omega} = 12 \text{ A}$$

Furthermore, the rms power dissipation is I^2R, or

$$P = 144 \times 10 = 1440 \text{ W}$$

14.12 Nonsinusoidal AC and DC Waveforms

The sine wave is the basic waveform for ac variations for several reasons. This waveform is produced by a rotary generator; the output is proportional to the angle of rotation. In addition, electronic oscillator circuits with inductance and capacitance naturally produce sine-wave variations.

Because of its derivation from circular motion, any sine wave can be analyzed in terms of angular measure, either in degrees from 0 to 360° or in radians from 0 to 2π rad.

Another feature of a sine wave is its basic simplicity; the rate of change of the amplitude variations corresponds to a cosine wave that is similar but 90° out of phase. The sine wave is the only waveform that has this characteristic of a rate of change with the same waveform as the original changes in amplitude.

In many electronic applications, however, other waveshapes are important. Any waveform that is not a sine or cosine wave is a *nonsinusoidal waveform*. Common examples are the square wave and sawtooth wave in Fig. 14-16.

For either voltage or current nonsinusoidal waveforms, there are important differences and similarities to consider. Note the following comparisons with sine waves.

1. In all cases, the cycle is measured between two points having the same amplitude and varying in the same direction. The period is the time for one cycle. In Fig. 14-16, T for any of the waveforms is 4 μs and the corresponding frequency is $1/T$, equal to 0.25 MHz.

2. Peak amplitude is measured from the zero axis to the maximum positive or negative value. However, peak-to-peak amplitude is better for measuring nonsinusoidal waveshapes because they can have unsymmetrical peaks, as in Fig. 14-16d. For all waveforms shown here, though, the peak-to-peak (p-p) amplitude is 20 V.

3. The rms value 0.707 of maximum applies only to sine waves because this factor is derived from the sine values in the angular measure used only for the sine waveform.

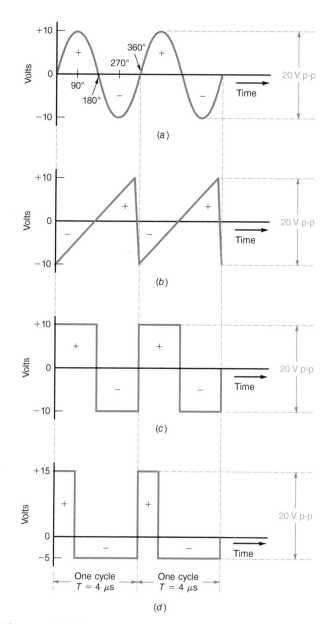

Figure 14-16 Comparison of sine wave with nonsinusoidal waveforms. Two cycles shown. (a) Sine wave. (b) Sawtooth wave. (c) Symmetrical square wave. (d) Unsymmetrical rectangular wave or pulse waveform.

4. Phase angles apply only to sine waves because angular measure is used only for sine waves. Note that the horizontal axis for time is divided into angles for the sine wave in Fig. 14-16a, but there are no angles shown for the nonsinusoidal waveshapes.

5. All waveforms represent ac voltages. Positive values are shown above the zero axis, and negative values below the axis.

The sawtooth wave in Fig. 14-16b represents a voltage that slowly increases to its peak value with a uniform or linear rate of change and then drops sharply to its starting value. This waveform is also called a *ramp voltage*. It is also often referred to as a *time base* because of its constant rate of change.

Note that one complete cycle includes a slow rise and a fast drop in voltage. In this example, the period T for a complete cycle is 4 μs. Therefore, these sawtooth cycles are repeated at the frequency of 0.25 MHz.

The square wave in Fig. 14-16c represents a switching voltage. First, the 10-V peak is instantaneously applied in positive polarity. This voltage remains on for 2 μs, which is one half-cycle. Then the voltage is instantaneously switched to −10 V for another 2 μs. The complete cycle then takes 4 μs, and the frequency is 0.25 MHz.

The rectangular waveshape in Fig. 14-16d is similar, but the positive and negative half-cycles are not symmetrical either in amplitude or in time. However, the frequency is the same 0.25 MHz and the peak-to-peak amplitude is the same 20 V, as in all the waveshapes. This waveform shows pulses of voltage or current, repeated at a regular rate.

DC Pulse Waveforms

The majority of nonsinusoidal waveforms are dc pulses. These have the appearance of the rectangular waveforms shown in Fig. 14-16c and d, but the voltage is of a single polarity, either positive or negative with respect to ground. When the voltage is present, the pulse is said to be ON. When the voltage is not present, the pulse is said to be OFF. Figure 14-17 shows three examples. Such waveforms are typically generated by digital circuits in which most signals have two distinct values. They are referred to as *rectangular waves.*

Some dc pulses also have a fixed dc component, that is a constant dc value on which the pulses vary. This is illustrated in Fig. 14-17c. Here the constant dc value is +0.2 V. The pulse varies from 0.2 to 3.5 V. The pulse amplitude is 3.3 V.

DC pulses are covered here simply because their variable nature makes them similar to ac voltages in that they are affected by inductors, transformers, and capacitors as ac waves are. Furthermore, because most electronic circuitry is digital today, it is essential to learn about dc or digital pulses as early as possible.

The most important characteristics of dc pulses are illustrated in Fig. 14-17. One cycle of a pulse is one interval at one level plus another interval at the second level or one OFF period and one ON period. The time between consecutive positive rises or consecutive negative falls is one period (T). The frequency is as with all other waves, the reciprocal of the period

$$f = \frac{1}{T}$$

The frequency is sometimes referred to as the pulse repetition rate (prr) or the pulse repetition frequency (prf).

The amplitude of the pulse is designated V_p as shown.

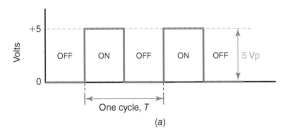

(a)

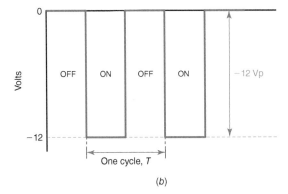

(b)

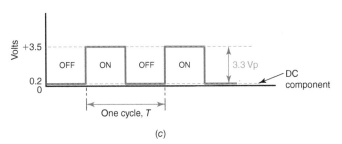

(c)

Figure 14-17 DC pulses. (a) positive. (b) negative. (c) positive with dc component.

Another important characteristic of dc pulses is that their ON and OFF times are often different. Figure 14-18 shows several examples. In Fig. 14-18a, the pulse ON time is very short while the OFF time is long. Figure 14-18b shows a pulse with a long ON time and a short OFF time. This characteristic is referred to as the pulse duty cycle (D). Duty cycle is defined as the ratio of the pulse on time (t_p) to the period.

$$D = \frac{t_p}{T} \tag{14-11}$$

The D value is a fraction that is often expressed as a percentage by simply multiplying it by 100. For instance in Fig. 14-18a, the duty cycle is

$$D = 5/40 = 0.125 \quad \text{or} \quad 12.5\%$$

The duty cycle indicates that the pulse is ON for 12.5% of the period.

In Fig. 14-18b, the duty cycle is

$$D = 60/80 = 0.75 \quad \text{or} \quad 75\%$$

The pulse is ON for three-quarters of the cycle.

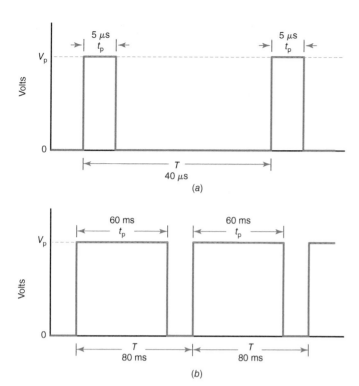

Figure 14-18   Duty cycles. (a) Short. (b) Long.

A pulse waveform with equal ON and OFF times has a 50% duty cycle and is called a *square wave*. Otherwise, pulse waveforms are referred to as rectangular waves.

The pulse waveforms in Figs. 14-17 and 14-18 appear to indicate that the voltage switches instantaneously from one level to the next. In reality, it takes a finite period of time for the pulse to switch from it low value to its high value and vice versa. A more realistic pulse is shown in Fig. 14-19. Note that the pulse rises gradually from its low value to its

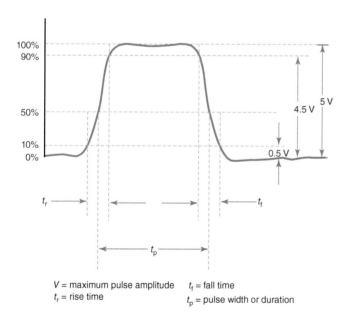

V = maximum pulse amplitude t_f = fall time
t_r = rise time t_p = pulse width or duration

Figure 14-19   Pulse rise and fall times and pulse width.

high value. This is called the leading edge of the pulse and the rise time. In a similar way, the pulse gradually changes from its high value to its lower value. This is the trailing edge of a pulse and the fall time.

Actually the rise and fall times are more precisely defined. The rise time (t_r) is the time from the 10% to the 90% value of the pulse amplitude. The fall time (t_f) is the time from the 90% value to the 10% value of the maximum pulse amplitude. For the pulse in Fig. 14-19, the peak pulse amplitude (V) is 5 V. The 10 and 90% values are 0.5 and 4.5 V, respectively. The rise and fall times are measured between these two values. In modern digital circuitry, typical rise and fall times are in the nanosecond and picosecond region.

Another key pulse characteristic is pulse width or pulse duration. This is the time duration that the pulse is on or t_p, as indicated earlier. That value is measured between the 50% amplitude points on the leading and trailing edges of the pulse. See Fig. 14-19.

EXAMPLE 14-9

A pulse waveform with a frequency of 10 MHz has a pulse duration (t_p) of 25 nanoseconds (ns). What is the period and duty cycle?

Answer:

$$T = 1/f = 1/10,000,000 \text{ Hz}$$
$$= 0.0000001 \text{ s or } 0.1 \text{ } \mu\text{s} \quad \text{or} \quad 100 \text{ ns}$$
$$D = t_p/T = 25/100 = 0.25 \quad \text{or} \quad 25\%$$

EXAMPLE 14-10

A pulse waveform has a peak value of 2.5 V. At what levels are the rise and fall times measured?

Answer:

10% of 2.5 V is 0.25 V; 90% of 2.5 V is 2.25 V.

14.13 The Frequency Domain and the Fourier Theory

There two different ways of looking at and examining electronic signals: the time domain and the frequency domain. We normally think about or analyze electronic signals as voltage variations occurring over time. The amplitude of the signal is plotted versus time. For example, Fig. 14-20a shows a 1-MHz sine wave with a peak amplitude of 2 V and peak to peak of 4 V. Such signals are said to be in the time domain. An oscilloscope is an example of an instrument that displays signals in the time domain.

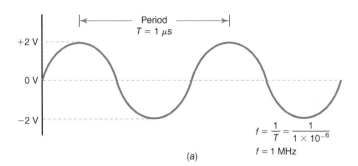

$$f = \frac{1}{T} = \frac{1}{1 \times 10^{-6}}$$

$$f = 1\ \text{MHz}$$

(a)

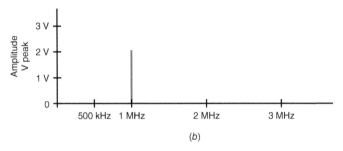

(b)

Figure 14-20 Time and frequency domain displays. (a) Time domain display of 1-MHz sine wave (4 V$_{p-p}$). (b) Frequency domain display.

It is also possible to display a signal in the frequency domain. Figure 14-20b shows the same signal in a frequency domain plot. The horizontal axis is frequency instead of time, and the signal amplitude is simply plotted as a straight vertical line proportional in length to the amplitude. The amplitude may be rms or peak voltage. Signal amplitudes may also be plotted in terms of power levels or decibels (dB). In communications electronics it is often more useful to work with a frequency domain display than a time domain display. With this arrangement, all the frequency components making up a complex analog or digital signal can be shown. The electronic test instrument known as a spectrum analyzer provides a frequency domain display of an input signal.

Fourier Analysis

The French mathematician Fourier discovered that any complex nonsinusoidal signal can be represented as a fundamental sine wave to which has been added harmonic sine waves. A Fourier analysis of any complex waveform will result in a mathematical expression made up of sine and/or cosine waves harmonically related. It is known as the *Fourier series*.

A *harmonic* is a sine wave whose frequency is some integer multiple of a fundamental sine-wave frequency. Consider a fundamental sine wave of 50 MHz. The second harmonic is simply 2 times the fundamental, or 100 MHz. The third and fourth harmonics are 150 and 200 MHz, respectively.

A common unit for frequency multiples is the *octave*, which is a range of 2:1. Doubling the frequency range—from 100 to 200 Hz, from 200 to 400 Hz, and from 400 to 800 Hz,

as examples—raises the frequency by one octave. The reason for this name is that an octave in music includes eight consecutive tones, for double the frequency. One-half the frequency is an octave lower. Another unit for representing frequency multiples is the decade. A decade corresponds to a 10:1 range in frequencies such as 100 Hz to 1 kHz and 30 kHz to 300 kHz.

EXAMPLE 14-11

What is the fifth harmonic of 7 MHz?

Answer:

$$7\ \text{MHz} \times 5 = 35\ \text{MHz}$$

EXAMPLE 14-12

What is the fundamental frequency related to a third harmonic of 81 MHz?

Answer:

$$81\ \text{MHz} \div 3 = 27\ \text{MHz}$$

An easy-to-understand example is the Fourier analysis of a square wave. Figure 14-21 shows an ac square wave with a peak voltage amplitude, V_p, of 1 volt. The waveform switches between +1 V and −1 V, having a peak-to-peak value of 2 V. The frequency is 1 kHz, having a period of 1 ms with positive and negative alternation times of 0.5 ms each. In a square wave, the positive, or on times, and the negative, or OFF times, are equal, meaning the duty cycle is 50%. A Fourier analysis of this square wave results in the mathematical expression

$$v = \frac{4V_p}{\pi} \sum_{n=1}^{\infty} \frac{\sin n\omega t}{n} \qquad (14\text{-}12)$$

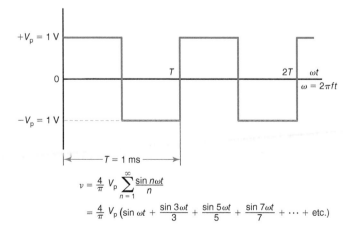

Figure 14-21 A square wave and its fourier expression.

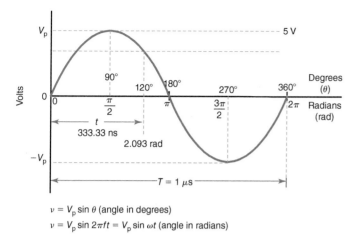

$v = V_p \sin \theta$ (angle in degrees)

$v = V_p \sin 2\pi ft = V_p \sin \omega t$ (angle in radians)

Figure 14-22 Sine wave expressed mathematically in radian measure.

The term v is the instantaneous voltage for given values of time. The uppercase Greek sigma, Σ, means summation of terms of the form $(\sin n\omega t)/n$, where n represents any integer from 1 to ∞.

The expression $(\sin n\omega t)/n$ is just another way to write the sine wave as a mathematical expression. As you have seen you can represent a sine wave with the basic trigonometric equation

$$v = V_p \sin \theta$$

The value v is the instantaneous voltage that occurs at the phase angle θ. Assume a sine wave with a peak voltage V_p of 5 V. See Fig. 14-22. At the phase angle of 120°, the value of the sine voltage is

$$v = 5 \sin 120° = 5(0.866) = 4.33 \text{ V}$$

Another way to write the sine expression is

$$v = V_p \sin 2\pi ft \qquad \text{(14-13)}$$

This is sometimes written as

$$v = V_p \sin \omega t \qquad \text{(14-14)}$$

The term ω (Greek lowercase omega) is equal to $2\pi f$. It is called the angular velocity, and it defines how fast the phasor representing the sine wave rotates. In either case, $2\pi ft$ or ωt is the phase angle in radians (rad).

Here, v is the instantaneous value of the sine voltage, f is the frequency, and t is the time duration from zero rads or degrees.

Assume a frequency f of 1 MHz or 1×10^6 Hz. The period T of course is 1 microsecond (1 μs). The time from 0 rad is 333.33 ns or 333.33×10^{-9} s.

With a peak value of 5 V the instantaneous value is

$$v = V_p \sin 2\pi ft = 5 \sin 2(3.14159)(1 \times 10^6)(333.33 \times 10^{-9})$$
$$= 5 \sin 2.093$$

In this expression, the value 2.093 is the angle in radians.

To compute the instantaneous value, find the sine of 2.093 by setting your calculator to radian measure instead of degrees and pressing the SIN key. The result is 0.866.

$$v = 5(0.866) = 4.33 \text{ V}$$

The angle of 120° is the same as 2.093 radians. Again refer to Fig. 14-22. Recall that one radian is 57.3°. Dividing 120° by 57.3 produces 2.093 radians.

In the expression $(\sin n\omega t)/n$, n is just the harmonic value. A value of 2 means the second harmonic of frequency f, 7 is the seventh harmonic of f, and so on. The n value in the denominator modifies the amplitude of the harmonic sine wave.

If you expand the expression it appears as shown below:

$$v = 4V_p/\pi \, \Sigma \, (\sin n\omega t)/n$$
$$= 4V_p/\pi \, [(\sin \omega t) + (\sin 3\omega t)/3 + (\sin 5\omega t)/5 + (\sin 7\omega t)/7 + \cdots + \text{etc.}]$$

All terms are multiplied by a coefficient of $4V_p/\pi$. Note that only odd values of integers show up as terms because terms with even integers become zero and drop out. What you have then is an expression that shows the addition of sine waves of different frequencies. There is a fundamental sine wave $(\sin \omega t)$ and odd harmonic sine waves $(\sin 3\omega t, \sin 5\omega t, \text{etc.})$.

Figure 14-23 shows the result of adding the sine waves. As you can see, the square wave consists of a sine wave at the fundamental frequency of 1 kHz plus additional sine waves at higher harmonic frequencies. The third harmonic is 3 kHz, and so on. In the case of the square wave, only the odd harmonics (third, fifth, seventh, etc.) exist. In Fig. 14-23, the fundamental, third, and fifth harmonics are added together, producing a composite wave that is beginning to look like a square wave. If you continue to add in the higher harmonics, the resulting wave will more closely approximate a pure square wave with vertical sides and a flat top.

The Fourier analysis produces an infinite number of harmonic components. Most of them are at such low amplitude levels as to be negligible in any practical analysis. An excellent square wave results from adding harmonics up to the 9th or 11th. The higher harmonics are the ones responsible for the steep or fast rise and fall times.

A frequency domain plot of the square wave is a group of vertical lines whose amplitude represents the fundamental sine wave of 1 kHz and the odd harmonics. See Fig. 14-24. This is what a spectrum analyzer would show if the square wave were connected to its input.

Figure 14-25 shows the relationship between the time and frequency domains. The two domains are at a right angle to one another. The peak values of the sine waves are plotted in the frequency domain.

As indicated earlier, any nonsinusoidal wave can be analyzed by the Fourier method. For example, the square wave

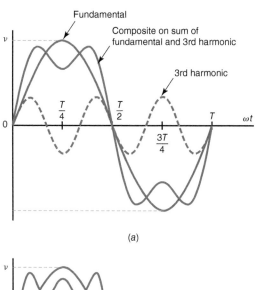

(a)

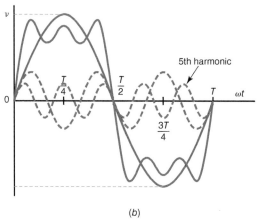

(b)

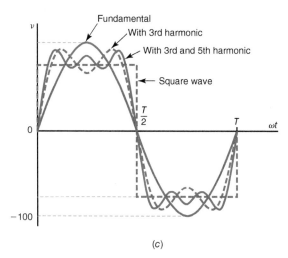

(c)

Figure 14-23 A square wave is the algebraic sum of a fundamental sine wave and all odd harmonics.

shown in Fig. 14-21 is an ac square wave with equal positive and negative voltage components. This resulted in fundamental and odd harmonic sine waves. If that same 1-kHz square wave were a pulsating dc signal for which the levels switched between 0 and, say, +3.3 V, the Fourier analysis would be identical in terms of the fundamental and odd harmonic sine waves; however, the mathematical expression

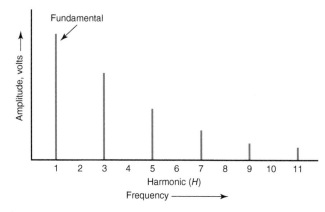

Figure 14-24 Frequency domain plot of a square wave.

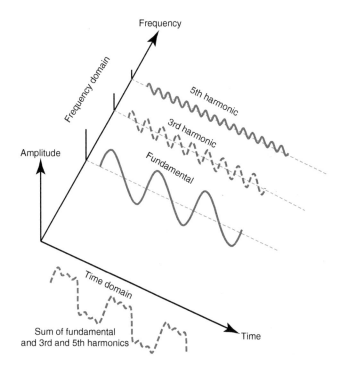

Figure 14-25 Relationship between the time and frequency domains.

would show a dc component equal to the average voltage over time. See Fig. 14-26. The average dc would be

$$V_p/2 = 3.3/2 = 1.15 \text{ V}$$

Any repeating nonsinusoidal wave can be analyzed with a Fourier series. Figure 14-27 shows the Fourier analysis of rectified sine waves. These waves are common in many fields of electronics, including power supplies, signal processing, and communications. Rectification is the process of converting an ac wave into pulses of dc. Figure 14-27a shows a half-wave rectified sine wave that is a series of positive half-sine pulses. A full-wave rectified signal is given in Fig. 14-27b. Both waveforms are made up of a fundamental and even harmonics only. These signals also contain a dc component equal to the average voltage value over time.

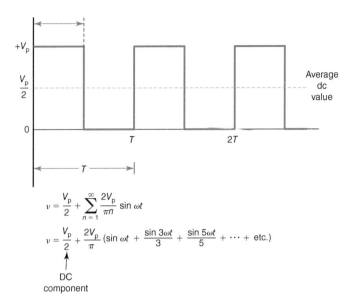

$$v = \frac{V_p}{2} + \sum_{n=1}^{\infty} \frac{2V_p}{\pi n} \sin \omega t$$

$$v = \frac{V_p}{2} + \frac{2V_p}{\pi} (\sin \omega t + \frac{\sin 3\omega t}{3} + \frac{\sin 5\omega t}{5} + \cdots + \text{etc.})$$

↑
DC
component

Figure 14-26   Fourier expression of a dc square wave.

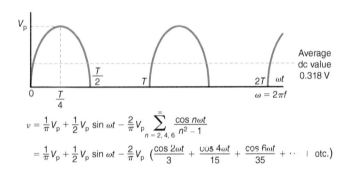

$$v = \frac{1}{\pi} V_p + \frac{1}{2} V_p \sin \omega t - \frac{2}{\pi} V_p \sum_{n=2,4,6}^{\infty} \frac{\cos n\omega t}{n^2 - 1}$$

$$= \frac{1}{\pi} V_p + \frac{1}{2} V_p \sin \omega t - \frac{2}{\pi} V_p (\frac{\cos 2\omega t}{3} + \frac{\cos 4\omega t}{15} + \frac{\cos 6\omega t}{35} + \cdots + \text{etc.})$$

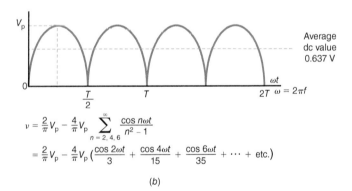

$$v = \frac{2}{\pi} V_p - \frac{4}{\pi} V_p \sum_{n=2,4,6}^{\infty} \frac{\cos n\omega t}{n^2 - 1}$$

$$= \frac{2}{\pi} V_p - \frac{4}{\pi} V_p (\frac{\cos 2\omega t}{3} + \frac{\cos 4\omega t}{15} + \frac{\cos 6\omega t}{35} + \cdots + \text{etc.})$$

(b)

Figure 14-27   Fourier analysis of rectified sine waves.
(a) half-wave. (b) full-wave.

Since most electronics is digital by way of binary dc pulses, a Fourier analysis of this type of waveform reveals more about the nature of the signal and how to deal with it. Let's examine a dc pulse wave train with a duty cycle something other than the 50% duty cycle of the square wave. Refer to the pulse wave train shown in Fig. 14-28. The pulse has a peak voltage amplitude of V, a period of T, a frequency

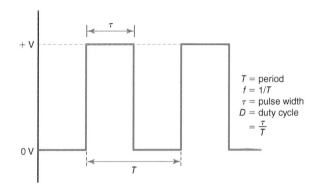

Figure 14-28   Rectangular pulse train.

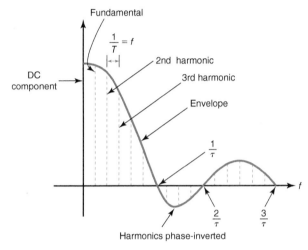

Figure 14-29   Plot of spectrum of a rectangular pulse.

of $f = 1/T$, and a pulse duration or width of τ (the Greek lowercase tau). The duty cycle (D) is the ratio of the pulse width to the period, or $D = \tau/T$.

If you were to work your way through the Fourier analysis, you would end up with the expression that contains sine wave of both odd and even harmonics and a dc component.

If you plot the fundamental and harmonic amplitudes on a frequency domain display, the resulting plot would look like that shown in Fig. 14-29. This is a generalized plot rather than one for a specific pulse train. The horizontal axis is frequency, and the vertical lines represent the peak values of the harmonics. The spacing between the vertical lines is, of course, the fundamental frequency of the pulse wave train, or $1/T$. If the duty cycle is anything besides 50%, which is indeed a special case, the resulting spectrum contains both odd and even harmonics. The amplitudes of the harmonics decease with frequency. At some higher frequency, the harmonic amplitudes become negative. What this means is that they are reversed in phase, that is 180° out of phase with the harmonic that starts in phase with the fundamental.

Continuing to refer to Fig. 14-29, notice that if you connect the peaks of the lines representing the harmonic amplitudes, you will trace out a curve which is called the *envelope*. This envelope curve can be expressed mathematically as $(\sin x)/x$. This is called the sinc function, or

$$\text{sinc}\,(x) = \frac{\sin x}{x} \qquad \textbf{(14-15)}$$

This particular expression shows up in many mathematical analyses in signal spectra. Using this expression, you can determine useful data from the plot of the harmonics. The most useful of these is the point where the sinc (x) curve first crosses the zero-amplitude line. Refer back to Fig. 14-29. As it turns out, that zero-crossing point is at a frequency of

$$f = 1/\tau$$

Remember τ is the pulse duration. The other zero-crossing points of the sinc (x) envelope are at integer multiples of the first zero-crossing point at $1/\tau$. The second zero-crossing point then is a $2/\tau$ and so on.

Why is the first zero-crossing point important? Mainly, for most pulse trains, the first zero-crossing point defines the range or band of frequencies where most of the energy in the wave is contained. The lower harmonics between the origin and $f = 1/\tau$ have the highest amplitudes and contribute the most to the shape and energy in the wave than do the higher harmonics. This, in turn, defines the minimum bandwidth or range of frequencies of the medium or circuits required to pass the signal with minimum distortion and attenuation. You will learn more about this in later chapters.

Just remember that you should try to think in terms of both time and frequency domains when you observe varying ac or dc signals. As you will discover, for some circuit analysis time domain analysis is best. For others a frequency domain view is a better explanation.

CHAPTER 14 SYSTEMS SIDEBAR

The Frequency Spectrum

The sine wave is the basic shape of many if not most electronic voltages. Examples are the ac power-line voltage at 60 Hz, pure audio tones in the 20- to 20,000-Hz range, and all radio waves. In addition, the sinusoidal shape can also apply to magnetic fields, electric fields, and even light waves. A radio wave is actually an electric field that occurs at a right angle to a magnetic wave with both varying sinusoidally and traveling together in a direction perpendicular to the two fields. You will hear this referred to as an *electromagnetic wave*.

Radio waves exist in free space. The range of electromagnetic waves runs continuously from 0 Hz to many THz and beyond. This range of frequencies is called the *electromagnetic frequency spectrum*. It can be pictured as shown in Fig. S14-1. Note several things. The wavelength is given as well as the frequency. In addition, the spectrum is usually divided into bands or segments for identification and classification of the different uses of the spectrum. The table below summarized these different ranges and classifications.

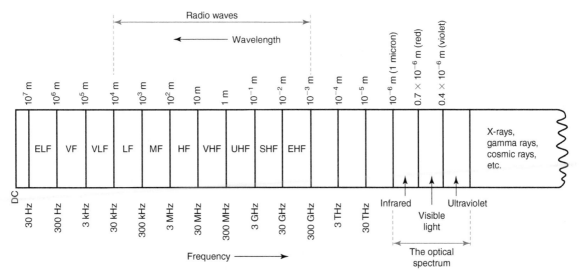

Figure S14-1   The electromagnetic frequency spectrum.

Frequency Band	Frequency Range	Application
Extremely low frequencies (ELF)	30 to 300 Hz	Power line, low audio frequencies
Voice frequencies (VF)	300 to 3000 Hz	Human voice range
Very low frequencies (VLF)	3 kHz to 30 kHz	Upper audio frequencies
Low frequencies (LF)	30 kHz to 300 kHz	Ultrasonic frequencies, sonar
Medium frequencies (MF)	300 kHz to 3 MHz	Marine, aircraft navigation, AM radio band
High frequencies (HF)	3 MHz to 30 MHz	Shortwave radio, ham, CB
Very high frequencies (VHF)	30 MHz to 300 MHz	TV, FM, mobile radio
Ultrahigh frequencies (UHF)	300 MHz to 3000 MHz	Cellular phone bands, short-range networks
Superhigh frequencies (SHF)	3 GHz to 30 GHz	Microwaves, satellites, radar
Extremely high frequencies (EHF)	30 GHz to 300 GHz	Millimeter waves, satellites, radar
Optical frequencies	1000 nm to 400 nm (wavelength)	Infrared (IR), visible, and ultraviolet (UV) light waves

Frequencies above 1 GHz are generally called *microwaves*. In the 30- to 300-GHz range, the signals are referred to as *millimeter waves* because of their wavelength.

Note the optical frequency range, which is expressed in wavelength rather than frequency. The low-frequency end of the infrared (IR) band is 1 micrometer (10^{-6} meter), also called a micron. This can also be expressed in nanometers (10^{-9} meter) as 1000 nm. Examples of IR waves are those emitted by most TV remote controls and light pulses that carry data over fiber-optic cables. IR is light but just not visible. Visible light is 400 nm (violet) to 700 nm (red). Ultraviolet is a shorter wavelength of <400 nm and not visible. Light waves are also sine waves.

While we normally think of the frequency spectrum as free-space electromagnetic waves, remember the spectrum or a part of it can exist on a cable like coaxial (cable TV) or a fiber-optic cable.

Finally, the free-space spectrum is shared by everyone. For that reason, there is potential for one signal interfering with another. As a result, the frequency spectrum is managed by the federal governments of most countries. In the United States, that job falls to the Federal Communications Commission (FCC) and the National Telecommunications and Information Administration (NTIA). In the United States most of the available useful spectrum has been assigned and is used. There is a spectrum shortage for future services. Think of the spectrum like land, for which there is only a finite amount available, and when that is all occupied, some different and perhaps unusual methods must be used.

? CHAPTER 14 REVIEW QUESTIONS

1. An alternating voltage is one that
   a. varies continuously in magnitude.
   b. reverses periodically in polarity.
   c. never varies in amplitude.
   d. both *a* and *b*.

2. One complete revolution of a conductor loop through a magnetic field is called a(n)
   a. octave.
   b. decade.
   c. cycle.
   d. alternation.

3. For a sine wave, one-half cycle is often called a(n)
   a. alternation.
   b. harmonic.
   c. octave.
   d. period.

4. One cycle includes
   a. 180°.
   b. 360°.
   c. 2π rad.
   d. both *b* and *c*.

5. In the United States, the frequency of the ac power-line voltage is
   a. 120 Hz.
   b. 60 Hz.
   c. 50 Hz.
   d. 100 Hz.

6. For a sine wave, the number of complete cycles per second is called the
   a. period.
   b. wavelength.
   c. frequency.
   d. phase angle.

7. A sine wave of alternating voltage has its maximum values at
   a. 90° and 270°.
   b. 0° and 180°.
   c. 180° and 360°.
   d. 30° and 150°.

8. To compare the phase angle between two waveforms, both must have
   a. the same amplitude.
   b. the same frequency.
   c. different frequencies.
   d. both *a* and *b*.

9. A 2-kHz sine wave has a period, T, of
   a. 0.5 μs.
   b. 50 μs.
   c. 500 μs.
   d. 2 ms.

10. If a sine wave has a period, T, of 40 μs, its frequency, f, equals
   a. 25 kHz.
   b. 250 Hz.
   c. 40 kHz.
   d. 2.5 kHz.

11. What is the wavelength of a radio wave whose frequency is 15 MHz?
   a. 20 m.
   b. 15 m.
   c. 0.753 ft.
   d. 2000 m.

12. The value of alternating current or voltage that has the same heating effect as a corresponding dc value is known as the
   a. peak value.
   b. average value.
   c. rms value.
   d. peak-to-peak value.

13. The wavelength of a 500-Hz sound wave is
   a. 60 km.
   b. 2.26 ft.

c. 4.52 ft.
   d. 0.226 ft.

14. How many radians are in one cycle of a sine wave?
   a. 1.
   b. π.
   c. 2π.
   d. 4.

15. A sine wave with a peak value of 20 V has an rms value of
   a. 28.28 V.
   b. 14.14 V.
   c. 12.74 V.
   d. 56.6 V.

16. A sine wave whose rms voltage is 25.2 V has a peak value of approximately
   a. 17.8 V.
   b. 16 V.
   c. 50.4 V.
   d. 35.6 V.

17. The unit of frequency is the
   a. hertz.
   b. maxwell.
   c. radian.
   d. second.

18. For an ac waveform, the period, T, refers to
   a. the number of complete cycles per second.
   b. the length of time required to complete one cycle.
   c. the time it takes for the waveform to reach its peak value.
   d. none of the above.

19. The wavelength of a radio wave is
   a. inversely proportional to its frequency.
   b. directly proportional to its frequency.
   c. inversely proportional to its amplitude.
   d. unrelated to its frequency.

20. Exact multiples of the fundamental frequency are called
   a. ultrasonic frequencies.
   b. harmonic frequencies.
   c. treble frequencies.
   d. resonant frequencies.

21. Raising the frequency of 500 Hz by two octaves corresponds to a frequency of
   a. 2 kHz.
   b. 1 kHz.
   c. 4 kHz.
   d. 250 Hz.

22. The Fourier theory is used to analyze
   a. sine waves.
   b. square waves.
   c. triangular waves.
   d. any nonsinusoidal wave.

23. The second harmonic of 7 MHz is
   a. 3.5 MHz.
   b. 28 MHz.
   c. 14 MHz.
   d. 7 MHz.

24. A sine wave has a peak voltage of 170 V. What is the instantaneous voltage at an angle of 45°?
   a. 240 V.
   b. 85 V.
   c. 0 V.
   d. 120 V.

25. Unless indicated otherwise, all sine-wave ac measurements are in
   a. peak-to-peak values.
   b. peak values.
   c. rms values.
   d. average values.

26. The duty cycle of a rectangular wave is the ratio of the
   a. OFF time to period.
   b. ON time to period.
   c. ON time to OFF time.
   d. OFF time to ON time.

SECTION 14.2 Alternating-Voltage Generator

14.1 For a sine wave of alternating voltage, how many degrees are included in
   a. ¼ cycle?
   b. ½ cycle?
   c. ¾ cycle?
   d. 1 complete cycle?

14.2 For a sine wave of alternating voltage, how many radians are included in
   a. ¼ cycle?
   b. ½ cycle?
   c. ¾ cycle?
   d. 1 complete cycle?

14.3 At what angle does a sine wave of alternating voltage
   a. reach its maximum positive value?
   b. reach its maximum negative value?
   c. cross the zero axis?

14.4 One radian corresponds to how many degrees?

SECTION 14.3 The Sine Wave

14.5 The peak value of a sine wave equals 20 V. Calculate the instantaneous voltage of the sine wave for the phase angles listed.
   a. 45°.
   b. 210°.

14.6 A sine wave of alternating voltage has an instantaneous value of 45 V at an angle of 60°. Determine the peak value of the sine wave.

SECTION 14.4 Alternating Current

14.7 In Fig. 14-30, the sine wave of applied voltage has a peak or maximum value of 10 V, as shown. Calculate the instantaneous value of current for the phase angles listed.
   a. 30°.
   b. 300°.

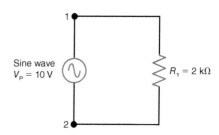

Figure 14-30

14.8 In Fig. 14-30, do electrons flow clockwise or counterclockwise in the circuit during
   a. the positive alternation?
   b. the negative alternation?

Note: During the positive alternation, terminal 1 is positive with respect to terminal 2.

SECTION 14.5 Voltage and Current Values for a Sine Wave

14.9 If the sine wave in Fig. 14-31 has a peak value of 15 V, then calculate
   a. the peak-to-peak value.
   b. the rms value.
   c. the average value.

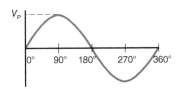

Figure 14-31

14.10 If the sine wave in Fig. 14-31 has an rms value of 40 V, then calculate
  a. the peak value.
  b. the peak-to-peak value.
  c. the average value.

SECTION 14.6 Frequency

14.11 What is the frequency, f, of a sine wave that completes
  a. 10 cycles per second?
  b. 500 cycles per second?
  c. 50,000 cycles per second?
  d. 2,000,000 cycles per second?

14.12 How many cycles per second (cps) do the following frequencies correspond to?
  a. 2 kHz.
  b. 15 MHz.
  c. 10 kHz.
  d. 5 GHz.

SECTION 14.7 Period

14.13 Calculate the period, T, for the following sine wave frequencies:
  a. 2 kHz.
  b. 4 kHz.
  c. 200 kHz.
  d. 2 MHz.

14.14 Calculate the frequency, f, of a sine wave whose period, T, is
  a. 5 m-s.
  b. 10 μs.
  c. 500 ns.
  d. 33.33 μs.

14.15 For a 5-kHz sine wave, how long does it take for
  a. ¼ cycle?
  b. ½ cycle?
  c. ¾ cycle?
  d. 1 full cycle?

SECTION 14.8 Wavelength

14.16 What is the velocity of an electromagnetic radio wave in
  a. miles per second (mi/s)?
  b. meters per second (m/s)?

14.17 What is the velocity in ft/s of a sound wave produced by mechanical vibrations?

14.18 What is the wavelength in meters of an electromag-netic radio wave whose frequency is 150 MHz?

14.19 What is the wavelength in ft of a sound wave whose frequency is
  a. 50 Hz?
  b. 4 kHz?

14.20 What is the frequency of an electromagnetic radio wave whose wavelength is
  a. 160 m?
  b. 17 m?

SECTION 14.9 Phase Angle

14.21 Describe the difference between a sine wave and a cosine wave.

14.22 Two voltage waveforms of the same amplitude, V_X and V_Y, are 45° out of phase with each other, with V_Y lagging V_X. Draw the phasors representing these voltage waveforms if
  a. V_X is used as the reference phasor.
  b. V_Y is used as the reference phasor.

SECTION 14.10 The Time Factor in Frequency and Phase

14.23 For two waveforms with a frequency of 1 kHz, how much time corresponds to a phase angle difference of
  a. 30°?
  b. 90°?

SECTION 14.11 Alternating Current Circuits with Resistance

14.24 In Fig. 14-32, solve for the following values: R_T, I, V_1, V_2, P_1, P_2, and P_T.

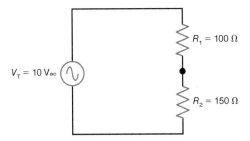

Figure 14-32

SECTION 14.12 Nonsinusoidal AC Waveforms

14.25 Determine the peak-to-peak voltage and frequency for the waveform in
  a. Fig. 14-33a.
  b. Fig. 14-33b.
  c. Fig. 14-33c.

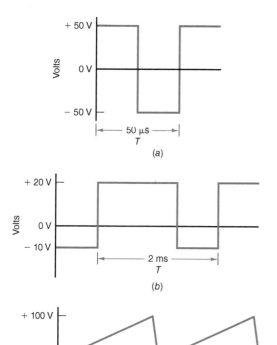

Figure 14-33

SECTION 14.13 The Frequency Domain and the Fourier Theory

14.26 State the Fourier theory.

14.27 What is the fifth harmonic of 2.5 GHz?

14.28 What is the fundamental frequency associated with a third harmonic of 76 MHz?

14.29 A nonsinusoidal waveform is made up of fundamental and harmonic
a. sine waves.
b. cosine waves.
c. either or both sine and cosine waves.

14.30 The speed of the rise and fall times of a pulse are determined by
a. low-frequency harmonics.
b. high-frequency harmonics.
c. pulse amplitude.
d. duty cycle.

14.31 Which nonsinusoidal waveform has only even harmonics?
a. Square waves.
b. Triangular waves.
c. Sawtooth waves.
d. Half-sine waves.

14.32 What instrument shows the frequency domain?
a. Spectrum analyzer.
b. Oscilloscope.
c. Signal generator.
d. Digital multimeter.

14.33 Square waves are made up of
a. even harmonics.
b. odd harmonics.
c. both odd and even harmonics.

14.34 Half sine waves are made up of
a. even harmonics.
b. odd harmonics.
c. both odd and even harmonics.

14.35 Rectangular waves (*not* square waves) are made up of
a. even harmonics.
b. odd harmonics.
c. both odd and even harmonics.

14.36 A dc square wave has a peak amplitude of 3 V. What is the average dc voltage? What is the amplitude of the third harmonic?

Chapter 15

Capacitance

Learning Outcomes

After studying this chapter, you should be able to:

> *Describe* how charge is stored in the dielectric of a capacitor.

> *Describe* how a capacitor charges and discharges.

> *Define* the farad unit of capacitance.

> *List* the physical factors affecting the capacitance of a capacitor.

> *List* several types of capacitors and the characteristics of each.

> *Explain* how an electrolytic capacitor is constructed.

> *Explain* how ultracapacitors can have such high values of capacitance.

> *Give* two examples of stray and distributed capacitance.

> *Calculate* the total capacitance of parallel-connected capacitors.

> *Calculate* the equivalent capacitance of series-connected capacitors.

Capacitance is the ability of a dielectric to hold or store an electric charge. The more charge stored for a given voltage, the higher the capacitance. The symbol for capacitance is C, and the unit is the farad (F).

A capacitor consists of an insulator (also called a *dielectric*) between two conductors. The conductors make it possible to apply voltage across the insulator. Different types of capacitors are manufactured for specific values of C. They are named according to the dielectric. Common types are air, ceramic, mica, paper, film, and electrolytic capacitors.

The most important property of a capacitor is its ability to block a steady dc voltage while passing ac signals. The higher the frequency, the less the opposition to ac voltage.

15.1 How Charge Is Stored in a Dielectric

Figure 15-1a shows the construction of a capacitor with two conducting plates separated by an insulator or dielectric. A battery is used to charge the capacitor. Electrons from the voltage source accumulate on the side of the capacitor connected to the negative terminal of V. The opposite side of the capacitor connected to the positive terminal of V loses electrons.

As a result, the excess of electrons produces a negative charge on one side of the capacitor, and the opposite side has a positive charge. As an example, if 6.25×10^{18} electrons are accumulated, the negative charge equals 1 coulomb (C). The charge on only one plate need be considered because the number of electrons accumulated on one plate is exactly the same as the number taken from the opposite plate.

What the voltage source does is simply redistribute some electrons from one side of the capacitor to the other. This process is called *charging* the capacitor. Charging continues until the potential difference across the capacitor is equal to the applied voltage. Without any series resistance, the charging is instantaneous. Practically, however, there is always some series resistance. This charging current is transient, or temporary; it flows only until the capacitor is charged to the applied voltage. Then there is no current in the circuit.

The result is a device for storing charge in the dielectric. Storage means that the charge remains even after the voltage source is disconnected. The measure of how much charge can be stored is the capacitance C. More charge stored for a given amount of applied voltage means more capacitance. Components made to provide a specified amount of capacitance are called *capacitors*.

Electrically, then, capacitance is the ability to store charge. A capacitor consists simply of two conductors separated by an insulator. For example, Fig. 15-1b shows a variable capacitor using air for the dielectric between the metal plates. There are many types with different dielectric materials, including paper, mica, and ceramics, but the schematic symbols shown in Fig. 15-1c apply to all capacitors.

Electric Field in the Dielectric

Any voltage has a field of electric lines of force between the opposite electric charges. The electric field corresponds to the magnetic lines of force of the magnetic field associated with electric current. What a capacitor does is concentrate the electric field in the dielectric between the plates. This concentration corresponds to a magnetic field concentrated in the turns of a coil. The only function of the capacitor plates and wire conductors is to connect the voltage source V across the dielectric. Then the electric field is concentrated in the capacitor, instead of being spread out in all directions.

Electrostatic Induction

The capacitor has opposite charges because of electrostatic induction by the electric field. Electrons that accumulate on the negative side of the capacitor provide electric lines of force that repel electrons from the opposite side. When this side loses electrons, it becomes positively charged. The opposite charges induced by an electric field correspond to the opposite poles induced in magnetic materials by a magnetic field.

15.2 Charging and Discharging a Capacitor

Charging and discharging are the two main functions of capacitors. Applied voltage puts charge in the capacitor. The accumulation of charge results in a buildup of potential difference across the capacitor plates. When the capacitor voltage equals the applied voltage, there is no more charging. The charge remains in the capacitor, with or without the applied voltage connected.

The capacitor discharges when a conducting path is provided across the plates, without any applied voltage. Actually, it is necessary only that the capacitor voltage be more than the applied voltage. Then the capacitor can serve as a voltage source, temporarily, to produce discharge current in the discharge path. The capacitor discharge continues until the capacitor voltage drops to zero or is equal to the applied voltage.

Applying the Charge

In Fig. 15-2a, the capacitor is neutral with no charge because it has not been connected to any source of applied voltage and there is no electrostatic field in the dielectric.

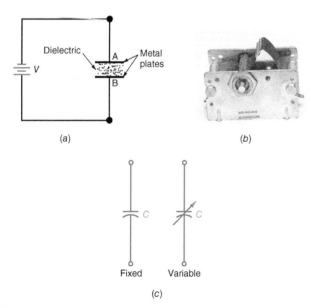

(a)

(b)

(c)

Figure 15-1 Capacitance stores the charge in the dielectric between two conductors. (a) Structure. (b) Air-dielectric variable capacitor. Length is 2 in. (c) Schematic symbols for fixed and variable capacitors.

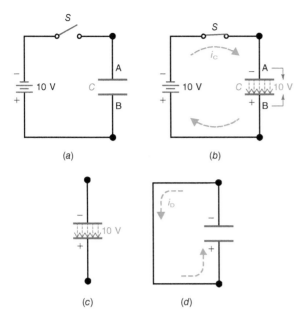

Figure 15-2 Storing electric charge in a capacitance. (*a*) Capacitor without any charge. (*b*) Battery charges capacitor to applied voltage of 10 V. (*c*) Stored charge remains in the capacitor, providing 10 V without the battery. (*d*) Discharging the capacitor.

Closing the switch in Fig. 15-2*b*, however, allows the negative battery terminal to repel free electrons in the conductor to plate A. At the same time, the positive terminal attracts free electrons from plate B. The side of the dielectric at plate A accumulates electrons because they cannot flow through the insulator, and plate B has an equal surplus of protons.

Remember that opposite charges have an associated potential difference, which is the voltage across the capacitor. The charging process continues until the capacitor voltage equals the battery voltage, which is 10 V in this example. Then no further charging is possible because the applied voltage cannot make free electrons flow in the conductors.

Note that the potential difference across the charged capacitor is 10 V between plates A and B. There is no potential difference from each plate to its battery terminal, however, which is why the capacitor stops charging.

Storing the Charge

The negative and positive charges on opposite plates have an associated electric field through the dielectric, as shown by the dotted lines in Fig. 15-2*b* and *c*. The direction of these electric lines of force is shown repelling electrons from plate B, making this side positive. The effect of electric lines of force through the dielectric results in storage of the charge. The electric field distorts the molecular structure so that the dielectric is no longer neutral. The dielectric is actually stressed by the invisible force of the electric field. As

evidence, the dielectric can be ruptured by a very intense field with high voltage across the capacitor.

The result of the electric field, then, is that the dielectric has charge supplied by the voltage source. Since the dielectric is an insulator that cannot conduct, the charge remains in the capacitor even after the voltage source is removed, as illustrated in Fig. 15-2*c*. You can now take this charged capacitor by itself out of the circuit, and it still has 10 V across the two terminals.

Discharging

The action of neutralizing the charge by connecting a conducting path across the dielectric is called *discharging* the capacitor. In Fig. 15-2*d*, the wire between plates A and B is a low-resistance path for discharge current. With the stored charge in the dielectric providing the potential difference, 10 V is available to produce discharge current. The negative plate repels electrons, which are attracted to the positive plate through the wire, until the positive and negative charges are neutralized. Then there is no net charge. The capacitor is completely discharged, the voltage across it equals zero, and there is no discharge current. Now the capacitor is in the same uncharged condition as in Fig. 15-2*a*. It can be charged again, however, by a source of applied voltage.

Nature of the Capacitance

A capacitor can store the amount of charge necessary to provide a potential difference equal to the charging voltage. If 100 V were applied in Fig. 15-2, the capacitor would charge to 100 V.

The capacitor charges to the applied voltage because it takes on more charge when the capacitor voltage is less. As soon as the capacitor voltage equals the applied voltage, no more charging current can flow. *Note that any charge or discharge current flows through the conducting wires to the plates but not through the dielectric.*

Charge and Discharge Currents

In Fig. 15-2*b*, i_C is in the opposite direction from i_D in Fig. 15-2*d*. In both cases, the current shown is electron flow. However, i_C is charging current to the capacitor and i_D is discharge current from the capacitor. The charge and discharge currents must always be in opposite directions. In Fig. 15-2*b*, the negative plate of C accumulates electrons from the voltage source. In Fig. 15-2*d*, the charged capacitor is a voltage source to produce electron flow around the discharge path.

More charge and discharge current result from a higher value of C for a given amount of voltage. Also, more V produces more charge and discharge current with a given amount of capacitance. However, the value of C does not change with the voltage because the amount of C depends on the physical construction of the capacitor.

15.3 The Farad Unit of Capacitance

With more charging voltage, the electric field is stronger and more charge is stored in the dielectric. The amount of charge Q stored in the capacitance is therefore proportional to the applied voltage. Also, a larger capacitance can store more charge. These relations are summarized by the formula

$$Q = CV \text{ coulombs} \qquad \textbf{(15-1)}$$

where Q is the charge stored in the dielectric in coulombs (C), V is the voltage across the plates of the capacitor, and C is the capacitance in farads.

The C is a physical constant, indicating the capacitance in terms of the amount of charge that can be stored for a given amount of charging voltage. When one coulomb is stored in the dielectric with a potential difference of one volt, the capacitance is one *farad*.

Practical capacitors have sizes in millionths of a farad, or smaller. The reason is that typical capacitors store charge of microcoulombs or less. Therefore, the common units are

$$1 \text{ microfarad} = 1 \ \mu\text{F} = 1 \times 10^{-6} \text{ F}$$
$$1 \text{ nanofarad} = 1 \text{ nF} = 1 \times 10^{-9} \text{ F}$$
$$1 \text{ picofarad} = 1 \text{ pF} = 1 \times 10^{-12} \text{ F}$$

EXAMPLE 15-1

How much charge is stored in a 2-μF capacitor connected across a 50-V supply?

Answer:

$$Q = CV = 2 \times 10^{-6} \times 50$$
$$= 100 \times 10^{-6} \text{ C}$$

EXAMPLE 15-2

How much charge is stored in a 40-μF capacitor connected across a 50-V supply?

Answer:

$$Q = CV = 40 \times 10^{-6} \times 50$$
$$= 2000 \times 10^{-6} \text{ C}$$

Note that the larger capacitor stores more charge for the same voltage, in accordance with the definition of capacitance as the ability to store charge.

The factors in $Q = CV$ can be inverted to

$$C = \frac{Q}{V} \qquad \textbf{(15-2)}$$

or

$$V = \frac{Q}{C} \qquad \textbf{(15-3)}$$

For all three formulas, the basic units are volts for V, coulombs for Q, and farads for C. Note that the formula $C = Q/V$ actually defines one farad of capacitance as one coulomb of charge stored for one volt of potential difference. The letter C (in italic type) is the symbol for capacitance. The same letter C (in roman type) is the abbreviation for the coulomb unit of charge. The difference between C and C will be made clearer in the examples that follow.

EXAMPLE 15-3

A constant current of 2 μA charges a capacitor for 20 s. How much charge is stored? Remember $I = Q/t$ or $Q = I \times t$.

Answer:

$$Q = I \times t$$
$$= 2 \times 10^{-6} \times 20$$
$$= 40 \times 10^{-6} \text{ or } 40 \ \mu\text{C}$$

EXAMPLE 15-4

The voltage across the charged capacitor in Example 15-3 is 20 V. Calculate C.

Answer:

$$C = \frac{Q}{V} = \frac{40 \times 10^{-6}}{20} = 2 \times 10^{-6}$$
$$= 2 \ \mu\text{F}$$

EXAMPLE 15-5

A constant current of 5 mA charges a 10-μF capacitor for 1 s. How much is the voltage across the capacitor?

Answer:

Find the stored charge first:

$$Q = I \times t = 5 \times 10^{-3} \times 1$$
$$= 5 \times 10^{-3} \text{ C or 5 mC}$$
$$V = \frac{Q}{C} = \frac{5 \times 10^{-3}}{10 \times 10^{-6}} = 0.5 \times 10^{3}$$
$$= 500 \text{ V}$$

Larger Plate Area Increases Capacitance

As illustrated in Fig. 15-3, when the area of each plate is doubled, the capacitance in Fig. 15-3b stores twice the charge of Fig. 15-3a. The potential difference in both cases is still 10 V. This voltage produces a given strength of electric field. A larger plate area, however, means that more of the dielectric surface can contact each plate, allowing more lines of force through the dielectric between the plates and less

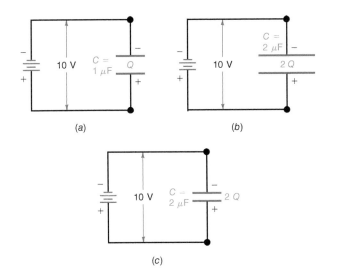

Figure 15-3 Increasing stored charge and capacitance by increasing the plate area and decreasing the distance between plates. (*a*) Capacitance of 1 μF. (*b*) A 2-μF capacitance with twice the plate area and the same distance. (*c*) A 2-μF capacitance with one-half the distance and the same plate area.

flux leakage outside the dielectric. Then the field can store more charge in the dielectric. The result of larger plate area is more charge stored for the same applied voltage, which means that the capacitance is larger.

Thinner Dielectric Increases Capacitance

As illustrated in Fig. 15-3c, when the distance between plates is reduced by one-half, the capacitance stores twice the charge of Fig. 15-3a. The potential difference is still 10 V, but its electric field has greater flux density in the thinner dielectric. Then the field between opposite plates can store more charge in the dielectric. With less distance between the plates, the stored charge is greater for the same applied voltage, which means that the capacitance is greater.

Dielectric Constant K_ϵ

The dielectric constant K_ϵ indicates the ability of an insulator to concentrate electric flux. Its numerical value is specified as the ratio of flux in the insulator compared with the flux in air or vacuum. The dielectric constant of air or vacuum is 1, since it is the reference.

Mica, for example, has an average dielectric constant of 6, which means that it can provide a density of electric flux six times as great as that of air or vacuum for the same applied voltage and equal size. Insulators generally have a dielectric constant K_ϵ greater than 1, as listed in Table 15-1. Higher values of K_ϵ allow greater values of capacitance.

Note that the aluminum oxide and tantalum oxide listed in Table 15-1 are used for the dielectric in electrolytic

Table 15-1 Dielectric Materials*

Material	Dielectric Constant K_ϵ	Dielectric Strength, V/mil[†]
Air or vacuum	1	20
Aluminum oxide	7	
Ceramics	80–1200	600–1250
Glass	8	335–2000
Mica	3–8	600–1500
Oil	2–5	275
Paper	2–6	1250
Plastic film	2–3	
Tantalum oxide	25	

* Exact values depend on the specific composition of different types.
† 1 mil equals one-thousandth of an inch or 0.001 in.

capacitors. Also, plastic film is often used instead of paper for the rolled-foil type of capacitor.

The dielectric constant for an insulator is actually its *relative permittivity*. The symbol ϵ_r, or K_ϵ, indicates the ability to concentrate electric flux. This factor corresponds to relative permeability, with the symbol μ_r or K_m, for magnetic flux. Both ϵ_r and μ_r are pure numbers without units as they are just ratios.*

These physical factors for a parallel-plate capacitor are summarized by the formula

$$C = K_\epsilon \times \frac{A}{d} \times 8.85 \times 10^{-12} \text{ F} \qquad \textbf{(15-4)}$$

where A is the area in square meters of either plate, d is the distance in meters between plates, K_ϵ is the dielectric constant, or relative permittivity, as listed in Table 15-1, and C is capacitance in farads. The constant factor 8.85×10^{-12} is the absolute permittivity of air or vacuum, in SI, since the farad is an SI unit.

EXAMPLE 15-6

Calculate C for two plates, each with an area 2 m², separated by 1 cm, or 10^{-2} m, with a dielectric of air.

Answer:
Substituting in Formula (15-4),

$$C = 1 \times \frac{2}{10^{-2}} \times 8.85 \times 10^{-12} \text{ F}$$
$$= 200 \times 8.85 \times 10^{-12}$$
$$= 1770 \times 10^{-12} \text{ F or 1770 pF}$$

* The absolute permittivity ϵ_0 is 8.854×10^{-12} F/m in SI units for electric flux in air or vacuum. This value corresponds to an absolute permeability μ_0 of $4\pi \times 10^{-7}$ H/m in SI units for magnetic flux in air or a vacuum.

This value means that the capacitor can store 1770×10^{-12} C of charge with 1 V. Note the relatively small capacitance, in picofarad units, with the extremely large plates of 2 m², which is really the size of a tabletop or a desktop.

If the dielectric used is paper with a dielectric constant of 6, then C will be six times greater. Also, if the spacing between plates is reduced by one-half to 0.5 cm, the capacitance will be doubled. Note that practical capacitors for electronic circuits are much smaller than this parallel-plate capacitor. They use a very thin dielectric with a high dielectric constant, and the plate area can be concentrated in a small space.

Dielectric Strength

Table 15-1 also lists breakdown voltage ratings for typical dielectrics. *Dielectric strength* is the ability of a dielectric to withstand a potential difference without arcing across the insulator. This voltage rating is important because rupture of the insulator provides a conducting path through the dielectric. Then it cannot store charge because the capacitor has been short-circuited. Since the breakdown voltage increases with greater thickness, capacitors with higher voltage ratings have more distance between plates. This increased distance reduces the capacitance, however, all other factors remaining the same.

15.4 Typical Capacitors

Commercial capacitors are generally classified according to the dielectric. Most common are air, mica, paper, plastic film, and ceramic capacitors, plus the electrolytic type. Electrolytic capacitors use a molecular-thin oxide film as the dielectric, resulting in large capacitance values in little space. These types are compared in Table 15-2 and discussed in the sections that follow.

Except for electrolytic capacitors, capacitors can be connected to a circuit without regard to polarity, since either side can be the more positive plate. Electrolytic capacitors are marked to indicate the side that must be connected to the positive or negative side of the circuit. *Note that the polarity of the charging source determines the polarity of the capacitor voltage.* Failure to observe the correct polarity can damage the dielectric and lead to the complete destruction of the capacitor.

Mica Capacitors

Thin mica sheets as the dielectric are stacked between tinfoil sections for the conducting plates to provide the required capacitance. Alternate strips of tinfoil are connected and brought out as one terminal for one set of plates, and the opposite terminal connects to the other set of interlaced plates. The construction is shown in Fig. 15-4a. The entire unit is generally in a molded Bakelite case. Mica capacitors are often used for small capacitance values of about 10 to 5000 pF; their length is ¾ in. or less with about ⅛-in. thickness. A typical mica capacitor is shown in Fig. 15-4b.

Paper Capacitors

In this construction shown in Fig. 15-5a, two rolls of tinfoil conductor separated by a paper dielectric are rolled into a compact cylinder. Each outside lead connects to its roll of tinfoil as a plate. The entire cylinder is generally placed in a cardboard container coated with wax or encased in plastic. Paper capacitors are often used for medium capacitance values of 0.001 to 1.0 μF, approximately. The size of a 0.05-μF capacitor is typically 1 in. long and ⅜-in. in diameter. A paper capacitor is shown in Fig. 15-5b. Paper capacitors are no longer used in new designs.

A black or a white band at one end of a paper capacitor indicates the lead connected to the outside foil. This lead should be used for the ground or low-potential side of the circuit to take advantage of shielding by the outside foil. There is no required polarity, however, since the capacitance is the same no matter which side is grounded. Also note that in the schematic symbol for C, the curved line usually indicates the low-potential side of the capacitor.

Table 15-2 Types of Capacitors			
Dielectric	**Construction**	**Capacitance**	**Breakdown, V**
Air	Meshed plates	10–400 pF	400 (0.02-in. air gap)
Ceramic	Tubular	0.5–1600 pF	500–20,000
	Disk	1 pF–1 μF	
Electrolytic	Aluminum	1–6800 μF	10–450
	Tantalum	0.047–330 μF	6–50
Mica	Stacked sheets	10–5000 pF	500–20,000
Paper	Rolled foil	0.001–1 μF	200–1600
Plastic film	Foil or metallized	100 pF–100 μF	50–600

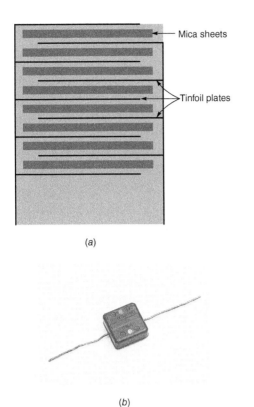

(a)

(b)

Figure 15-4 Mica capacitor. (a) Physical construction. (b) Example of a mica capacitor.

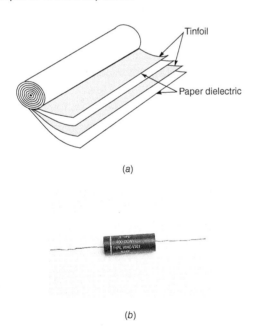

(a)

(b)

Figure 15-5 Paper capacitor. (a) Physical construction. (b) Example of a paper capacitor.

Film Capacitors

Film capacitors are constructed much like paper capacitors except that the paper dielectric is replaced with a plastic film such as polypropylene, polystyrene, polycarbonate, or polyethelene terepthalate (Mylar). There are two main types of film capacitors: the foil type and the metallized type. The foil type uses sheets of metal foil, such as aluminum or tin, for its conductive plates. The metallized type is constructed by depositing (spraying) a thin layer of metal, such as aluminum or zinc, on the plastic film. The sprayed-on metal serves as the plates of the capacitor. The advantage of the metallized type over the foil type is that the metallized type is much smaller for a given capacitance value and breakdown voltage rating. The reason is that the metallized type has much thinner plates because they are sprayed on. Another advantage of the metallized type is that it is self-healing. This means that if the dielectric is punctured because its breakdown voltage rating is exceeded, the capacitor is not damaged permanently. Instead, the capacitor heals itself. This is not true of the foil type.

Film capacitors are very temperature-stable and are therefore used frequently in circuits that require very stable capacitance values. Some examples are radio-frequency oscillators and timer circuits. Film capacitors are available with values ranging from about 100 pF to 100 μF. Figure 15-6 shows a typical film capacitor.

Ceramic Capacitors

The ceramic materials used in ceramic capacitors are made from earth fired under extreme heat. With titanium dioxide or one of several types of silicates, very high values of dielectric constant K_ϵ can be obtained. Most ceramic capacitors come in disk form, as shown in Fig. 15-7. In the disk form, silver is deposited on both sides of the ceramic dielectric to form the capacitor plates. Ceramic capacitors are available with values of 1 pF (or less) up to about 1 μF. The wide range of values is possible because the dielectric constant K_ϵ can be tailored to provide almost any desired value of capacitance.

Figure 15-6 Film capacitor.

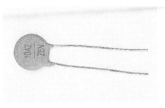

Figure 15-7 Ceramic disk capacitor.

Note that ceramic capacitors are also available in forms other than disks. Some ceramic capacitors are available with axial leads and use a color code similar to that of a resistor.

Surface-Mount Capacitors

Like resistors, capacitors are also available as surface-mounted components. Surface-mount capacitors are often called *chip capacitors*. Chip capacitors are constructed by placing a ceramic dielectric material between layers of conductive film which form the capacitor plates. The capacitance is determined by the dielectric constant K_ϵ and the physical area of the plates. Chip capacitors are available in many sizes. A common size is 0.125 in. long by 0.063 in. wide in various thicknesses. Another common size is 0.080 in. long by 0.050 in. wide in various thicknesses. Figure 15-8 shows two sizes of chip capacitors. Like chip resistors, chip capacitors have their end electrodes soldered directly to the copper traces of the printed circuit board. Chip capacitors are available with values ranging from a fraction of a picofarad up to several microfarads.

Variable Capacitors

Figure 15-1*b* shows a variable air capacitor. In this construction, the fixed metal plates connected together form the *stator*. The movable plates connected together on the shaft form the *rotor*. Capacitance is varied by rotating the shaft to make the rotor plates mesh with the stator plates. They do not touch, however, since air is the dielectric. Full mesh is

Figure 15-8 Chip capacitors.

maximum capacitance. Moving the rotor completely out of mesh provides minimum capacitance.

A common application is the tuning capacitor in radio receivers. When you tune to different stations, the capacitance varies as the rotor moves in or out of mesh. Combined with an inductance, the variable capacitance then tunes the receiver to a different resonant frequency for each station. Usually two or three capacitor sections are *ganged* on one common shaft.

Temperature Coefficient

Ceramic capacitors are often used for temperature compensation to increase or decrease capacitance with a rise in temperature. The temperature coefficient is given in parts per million (ppm) per degree Celsius, with a reference of 25°C. As an example, a negative 750-ppm unit is stated as N750. A positive temperature coefficient of the same value would be stated as P750. Units that do not change in capacitance are labeled NPO.

Capacitance Tolerance

Ceramic disk capacitors for general applications usually have a tolerance of ±20%. For closer tolerances, mica or film capacitors are used. These have tolerance values of ±2 to 20%. Silver-plated mica capacitors are available with a tolerance of ±1%.

The tolerance may be less on the minus side to make sure that there is enough capacitance, particularly with electrolytic capacitors, which have a wide tolerance. For instance, a 20-μF electrolytic with a tolerance of 10%, +50% may have a capacitance of 18 to 30 μF. However, the exact capacitance value is not critical in most applications of capacitors for filtering, ac coupling, and bypassing.

Voltage Rating of Capacitors

This rating specifies the maximum potential difference that can be applied across the plates without puncturing the dielectric. Usually the voltage rating is for temperatures up to about 60°C. Higher temperatures result in a lower voltage rating. Voltage ratings for general-purpose paper, mica, and ceramic capacitors are typically 200 to 500 V. Ceramic capacitors with ratings of 1 to 20 kV are also available.

Electrolytic capacitors are typically available in 15-, 35-, and 50-V ratings. For applications where a lower voltage rating is permissible, more capacitance can be obtained in a smaller size.

The potential difference across the capacitor depends on the applied voltage and is not necessarily equal to the voltage rating. A voltage rating higher than the potential difference applied across the capacitor provides a safety factor for long life in service. However, the actual capacitor voltage of electrolytic capacitors should be close to the rated voltage to produce the oxide film that provides the specified capacitance.

The voltage ratings are for dc voltage applied. The breakdown rating is lower for ac voltage because of the internal heat produced by continuous charge and discharge.

Capacitor Applications

In most electronic circuits, a capacitor has dc voltage applied, combined with a much smaller ac signal voltage. The usual function of the capacitor is to block the dc voltage but pass the ac signal voltage by means of the charge and discharge current. These applications include coupling, bypassing, and filtering of ac signals.

15.5 Electrolytic Capacitors

Electrolytic capacitors are commonly used for C values ranging from about 1 to 6800 μF because electrolytics provide the most capacitance in the smallest space with least cost.

Construction

Figure 15-9 shows the aluminum-foil type. The two aluminum electrodes are in an electrolyte of borax, phosphate, or carbonate. Between the two aluminum strips, absorbent gauze soaks up electrolyte to provide the required electrolysis that produces an oxide film. This type is considered a wet electrolytic, but it can be mounted in any position.

When dc voltage is applied to form the capacitance in manufacture, the electrolytic action accumulates a molecular-thin layer of aluminum oxide at the junction between the positive aluminum foil and the electrolyte. The oxide film is an insulator. As a result, capacitance is formed between the positive aluminum electrode and the electrolyte in the gauze separator. The negative aluminum electrode simply provides a connection to the electrolyte. Usually, the metal can itself is the negative terminal of the capacitor, as shown in Fig. 15-9c.

Because of the extremely thin dielectric film, very large C values can be obtained. The area is increased by using long strips of aluminum foil and gauze, which are rolled into a compact cylinder with very high capacitance. For example, an electrolytic capacitor the same size as a 0.1-μF paper capacitor, but rated at 10 V breakdown, may have 1000 μF of capacitance or more. Higher voltage ratings, up to 450 V, are available, with typical C values up to about 6800 μF. The very high C values usually have lower voltage ratings.

Polarity

Electrolytic capacitors are used in circuits that have a combination of dc voltage and ac voltage. The dc voltage maintains the required polarity across the electrolytic capacitor to form the oxide film. A common application is for electrolytic filter capacitors to eliminate the 60- or 120-Hz ac ripple in a dc power supply. Another use is for audio coupling capacitors in transistor amplifiers. In both applications, for filtering or coupling, electrolytics are needed for large C with a low-frequency ac component, whereas the circuit has a dc component for the required voltage polarity. Incidentally, the difference between filtering out an ac component or coupling it into a circuit is only a question of parallel or series connections. The filter capacitors for a power supply are typically 100 to 1000 μF. Audio capacitors are usually 10 to 47 μF.

If the electrolytic is connected in opposite polarity, the reversed electrolysis forms gas in the capacitor. It becomes hot and may explode. This is a possibility only with electrolytic capacitors.

Leakage Current

The disadvantage of electrolytics, in addition to the required polarization, is their relatively high leakage current compared with other capacitors, since the oxide film is not a perfect insulator. The problem with leakage current in a capacitor is that it allows part of the dc component to be coupled into the next circuit along with the ac component. In newer electrolytic capacitors, the leakage current is quite small.

Nonpolarized Electrolytics

Nonpolarized electrolytic capacitors are available for applications in circuits without any dc polarizing voltage, as in a 60-Hz ac power line. One application is the starting capacitor for ac motors. A nonpolarized electrolytic actually contains two capacitors, connected internally in series-opposing polarity.

Tantalum Capacitors

Another form of electrolytic capacitor, uses tantalum (Ta) instead of aluminum. Titanium (Ti) is also used. Typical tantalum capacitors are shown in Fig. 15-10. They feature

1. Larger C in a smaller size
2. Longer shelf life
3. Less leakage current

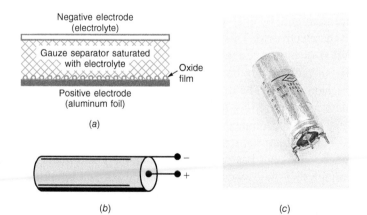

Figure 15-9 Construction of aluminum electrolytic capacitor. (a) Internal electrodes. (b) Foil rolled into cartridge. (c) Typical capacitor with multiple sections.

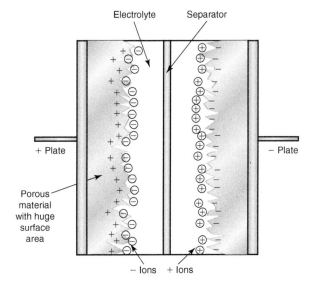

Figure 15-11　Construction of an ultracapacitor.

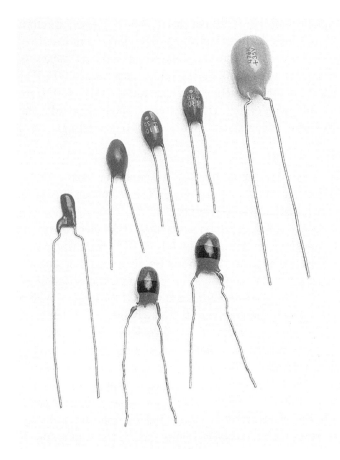

Figure 15-10　Tantalum capacitors.

However, tantalum electrolytics cost more than the aluminum type. Construction of tantalum capacitors include the wet-foil type and a solid chip or slug. The solid tantalum is processed in manufacture to have an oxide film as the dielectric. Referring back to Table 15-1, note that tantalum oxide has a dielectric constant of 25, compared with 7 for aluminum oxide.

15.6　Ultracapacitors

Ultracapacitors, or supercapacitors as they are also called, are special high-capacitance capacitors used for energy storage. Most capacitors used in electronics have smaller values in the microfarad and picofarad range. Electrolytic capacitors can have larger values up to 10,000 to 40,000 μF. Ultracapacitors are available with values as high as several hundred farads. With such high capacitance, ultracapacitors can store an enormous charge and can almost act as a battery.

Conventional capacitors store an electric charge on two plates separated by a dielectric. The capacitance is a function of the plate area, plate spacing, and the dielectric constant. Large plates spaced very close together produce the highest capacitance. However, there are physical limitations. These

limitations are nearly reached with electrolytic capacitors. Ultracapacitors use a different approach.

The general construction of an ultracapacitor is shown in Fig. 15-11. It is made up of two metal plates that are coated with a highly porous carbon material that provides a huge surface area. These plates are immersed in a liquid electrolyte material that is highly ionized with both positive and negative ions. A separator between the plates isolates the charges on each plate. This construction gives the ultracapacitor its alternate name, *electrochemical double-layer capacitor (EDLC)*.

When a voltage is applied to the capacitor plates, the positive plate attracts negative ions from the electrolyte. At the same time, the negative plate attracts positive ions from the electrolyte. This produces two separately charged surfaces on either side of the separator, which is the equivalent to two capacitors in series. The charges on each plate are separated by only a distance that is the size of the ions in the electrolyte. This is the same as having a plate separation of only a few angstroms. (*Note*: An angstrom is a distance of one ten-billionth of a meter, or 10^{-10} meter, or 0.1 nanometer.) The thin spacing combined with the large plate area offered by the porous plate coating produces a very large capacitance.

This unique construction achieves very high capacitance values but has one main limitation. The amount of voltage that the capacitor can withstand without breaking down is very small. A typical maximum working voltage value for a 1- to 100-F capacitor is only 2.7 V. This limitation can be overcome by connecting capacitors in series so that the externally applied voltage is spread across several capacitors. Ultracapacitor modules made up of multiple series capacitors are available with voltage ratings of up to about 150 V.

Ultracapacitors store energy for use later. They are often used in conjunction with a battery to provide short duration pulses of power or backup power. A common arrangement is for the ultracapacitor to be connected in parallel with a battery. The battery charges the ultracapacitor. Any load draws current from both the battery and the ultracapacitor. However, if the load suddenly increases, requiring extra current, that current will be supplied by the ultracapacitor rather than the battery. Batteries have peak pulse current limits, which are a function of the internal resistance of the battery. The demand for a high pulse of current is met by the ultracapacitor with its low internal resistance.

For example, in an automotive sound system, which is powered by the car's 12-V battery, high-amplitude bass notes will produce pulses of current from the battery. The battery alone cannot supply such high pulses of current; thus the bass amplitude becomes smaller or more washed out. However with a parallel ultracapacitor, a high-current pulse can be provided. Most high-power sound systems in automobiles have an ultracapacitor.

Ultracapacitors also used in consumer products that require a battery. It provides voltage to the circuits if the battery dies or is being replaced. Examples are products with memories that may be erased if power is lost or clocks that will lose their time setting. Other uses include backup power for uninterruptible power supplies and wind generators. Ultracapacitors are found in electric vehicles for short-term bursts of power and for storing energy from regenerative braking.

15.7 Stray and Distributed Capacitance

We normally think of capacitance as an electrical quantity that comes packaged as a component with two connecting leads. However, there are other forms of capacitance that exist and must be taken into consideration when analyzing or troubleshooting electronic circuits. Stray and distributed capacitance is present any place where two conductors separated by an insulating medium are present. The most common example is the capacitance between two wires in a cable. Refer to Fig. 15-12a. The wires are separated by an insulator so the combination forms a long capacitor distributed over the length of the cable. While the capacitance is generally small per increment of length, long sections of cable can have very large values of capacitance. It could be many picofarads or even low microfarad values. Most cable specifications actually designate the amount of capacitance per foot of the cable.

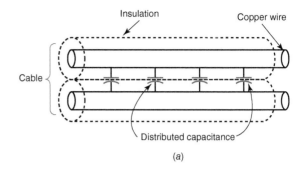

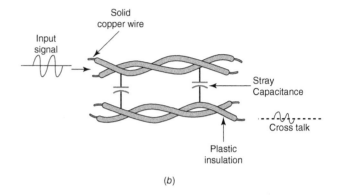

Figure 15-12 Distributed capacitance. (*a*) on a cable and (*b*) stray capacitance between two cables that produces cross talk.

Another example is the capacitance that exists between two copper patterns on a printed circuit board. There is also capacitance between two adjacent pins on a connector. Stray capacitance also exists between the pins on an integrated circuit or between the leads of a transistor or resistor. In general, stray capacitance exists everywhere in electronic circuits.

The important fact is to be aware that it exists. Just because you cannot see it, it is still present and doing what capacitors do: charge, discharge, and store energy. Because this capacitance is relatively small, for many applications it can be ignored. In low-frequency circuits, stray capacitance has minimal effects. However at high frequencies, stray capacitance does have a significant impact on circuit performance. You will learn more about this type of capacitance and its effects in ac circuits in the chapters to come. You will discover that it usually affects circuits negatively. For example, consider two cables running parallel and very close to one another as in Fig. 15-12b. The signal on one cable can be transferred to the other cable by the capacitance between them. The signals on each cable can interfere with one another. This is called *cross talk*, and it is a common problem associated with distributed capacitance. In addition, high capacitance can filter out high frequencies and greatly attenuate them.

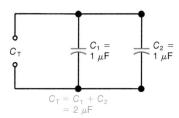

Figure 15-13   Capacitances in parallel.

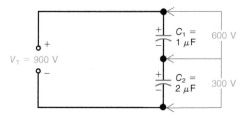

Figure 15-15   With series capacitors, the smaller C has more voltage for the same charge.

15.8 Parallel Capacitances

Connecting capacitances in parallel is equivalent to adding the plate areas. Therefore, the total capacitance is the sum of the individual capacitances. As illustrated in Fig. 15-13,

$$C_T = C_1 + C_2 + \cdots + \text{etc.} \qquad (15\text{-}5)$$

A 10-μF capacitor in parallel with a 5-μF capacitor, for example, provides a 15-μF capacitance for the parallel combination. The voltage is the same across the parallel capacitors. Note that adding parallel capacitances is opposite to inductances in parallel and resistances in parallel.

15.9 Series Capacitances

Connecting capacitances in series is equivalent to increasing the thickness of the dielectric. Therefore, the combined capacitance is less than the smallest individual value. As shown in Fig. 15-14, the combined equivalent capacitance is calculated by the reciprocal formula:

$$C_{EQ} = \frac{1}{\dfrac{1}{C_1} + \dfrac{1}{C_2} + \dfrac{1}{C_3} + \cdots + \text{etc.}} \qquad (15\text{-}6)$$

Any of the shortcut calculations for the reciprocal formula apply. For example, the combined capacitance of two equal capacitances of 10 μF in series is 5 μF.

Capacitors are used in series to provide a higher working voltage rating for the combination. For instance, each

of three equal capacitances in series has one-third the applied voltage. A common arrangement is having two capacitors in series. In this case, Formula (15-6) simplifies to

$$C_{EQ} = \frac{C_1 \times C_2}{C_1 + C_2}$$

Division of Voltage across Unequal Capacitances

In series, the voltage across each C is inversely proportional to its capacitance, as illustrated in Fig. 15-15. The smaller capacitance has the larger proportion of the applied voltage. The reason is that the series capacitances all have the same charge because they are in one current path. With equal charge, a smaller capacitance has a greater potential difference.

We can consider the amount of charge in the series capacitors in Fig. 15-15. Let the charging current be 600 μA flowing for 1 s. The charge Q equals $I \times t$ or 600 μC. Both C_1 and C_2 have Q equal to 600 μC because they are in the same series path for charging current.

Although the charge is the same in C_1 and C_2, they have different voltages because of different capacitance values. For each capacitor, $V = Q/C$. For the two capacitors in Fig. 15-15, then,

$$V_1 = \frac{Q}{C_1} = \frac{600 \ \mu C}{1 \ \mu F} = 600 \ V$$

$$V_2 = \frac{Q}{C_2} = \frac{600 \ \mu C}{2 \ \mu F} = 300 \ V$$

Charging Current for Series Capacitances

The charging current is the same in all parts of the series path, including the junction between C_1 and C_2, even though this point is separated from the source voltage by two insulators. At the junction, the current is the result of electrons repelled by the negative plate of C_2 and attracted by the positive plate of C_1. The amount of current in the circuit is determined by the equivalent capacitance of C_1 and C_2 in series. In Fig. 15-15, the equivalent capacitance is $\frac{2}{3} \ \mu F$.

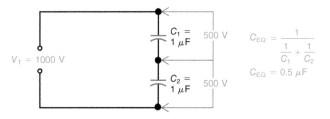

Figure 15-14   Capacitances in series.

Dynamic Random Access Memory

Most digital computers today use a type of memory called dynamic random access memory (DRAM). It is the main memory in personal computers (PCs), laptops, tablets, and cell phones. Programs and data are stored in the memory as binary numbers or codes. The binary system uses the digits 0 and 1 to represent numerical values, the alphabet, special symbols, and codes. The 0s and 1s are called *bits*. For example, the 8-bit number 11001101 is used to represent the decimal value 205. The code 01001010 represents the letter *J*. Bits are represented in electronic circuits by voltage levels. For example, a voltage of +2.5 volts could mean a binary 1, while a value of zero volts could represent a binary 0.

To store binary information, we need a circuit or device that can assume either of two states to store either a 0 or a 1. Such a circuit is called a *storage cell*. There have been a variety of storage cells invented and used over the years. One of the most common and widely used is the dynamic storage cell that has a capacitor.

Figure S15-1 shows a simplified dynamic storage cell. It consists of a capacitor and a switch. The switch is a transistor that can be turned off and on like any electrical switch. The capacitor stores a 0 if it is fully discharged or a 1 if it is charged.

To store a bit, the transistor is turned on by a signal on the address line. Then the bit voltage is applied to the data line and through the transistor switch to the capacitor. The capacitor is either charged or discharged according to the input. This operation is referred to as a write operation. The transistor switch is then turned off, isolating the capacitor that continues to hold the charge.

To access the cell, the transistor is turned on by the address line connecting the capacitor to the data line. The voltage stored on the capacitor is connected to a sense amplifier through a transistor switch where is it amplified and otherwise conditioned to appear at the memory output used by the computer. This is called a *read operation*.

Millions or billions of these cells are manufactured on a chip of silicon. They are arranged in a large array or matrix of rows and columns with digital circuitry that allows multiple numbers, words, or codes to be stored. See Fig. S15-2. These cells are organized into multibit words or numbers and each is assigned a unique number or address. By a computer addressing the memory, any word or number can be randomly accessed for either a read or write operation on its cells. Thus the name *random access memory (RAM)*. The time to access any cell is very short, on the order of a few nanoseconds.

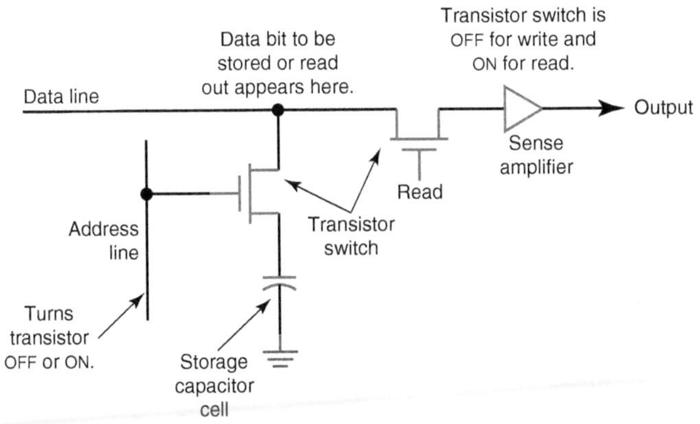

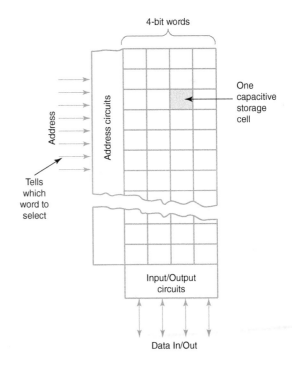

Figure S15-1   A dynamic RAM storage cell is made up of a transistor and a capacitor. The capacitor stores the bit while the transistors control the access to the capacitor.

Figure S15-2   The dynamic RAM cells are organized into a matrix of rows and columns where cells are grouped to store multibit words. Here 4-bit words are used. Each word is accessed for a read or write operation by giving the memory an address in binary form.

Since capacitance of each capacitor in a cell is very small and because of leakage resistance within the chip, the charge on each cell will eventually leak off and the data will be lost. However, in this type of RAM, the charge status of each cell is automatically restored periodically every few milliseconds, ensuring that the data are maintained. This refresh operation is what gives the memory the name "dynamic."

DRAM chips come in many sizes. Each can store millions, even billion, of bits. For example, one type of DRAM chip can store 512 Mb, or 512 million bits. The chip may be organized as 128 M of 4-bit words or numbers. Larger chips are available. Multiple chips are then connected together to form the main memory of most computers. DRAM chips are typically mounted on small printed circuit boards called *dual in-line memory modules (DIMM)* that plug into the motherboard of a PC. See Fig. S15-3. A common DRAM size in a computer is 2 to 8 GB. The letter *G* means giga or billion and *B* means byte. A *byte* is an 8-bit number or word.

Figure S15-3   A dual in-line memory module (DIMM) contains multiple DRAM chips. The DIMMs plug into the computer motherboard.

CHAPTER 15 REVIEW QUESTIONS

1. In general, a capacitor is a component that can
   a. pass a dc current.
   b. store an electric charge.
   c. act as a bar magnet.
   d. step up or step down an ac voltage.

2. The basic unit of capacitance is the
   a. farad.
   b. henry.
   c. tesla.
   d. ohm.

3. Which of the following factors affect the capacitance of a capacitor?
   a. the area, A, of the plates.
   b. the distance, d, between the plates.
   c. the type of dielectric used.
   d. all of the above.

4. How much charge in coulombs is stored by a 50-μF capacitor with 20 V across its plates?
   a. $Q = 100\ \mu C$.
   b. $Q = 2.5\ \mu C$.
   c. $Q = 1\ mC$.
   d. $Q = 1\ \mu C$.

5. A capacitor consists of
   a. two insulators separated by a conductor.
   b. a coil of wire wound on an iron core.

   c. two conductors separated by an insulator.
   d. none of the above.

6. A capacitance of 82,000 pF is the same as
   a. 0.082 μF.
   b. 82 μF.
   c. 82 nF.
   d. both *a* and *c*.

7. A 47-μF capacitor has a stored charge of 2.35 mC. What is the voltage across the capacitor plates?
   a. 50 V.
   b. 110 V approx.
   c. 5 V.
   d. 100 V.

8. Which of the following types of capacitors typically has the highest leakage current?
   a. plastic film.
   b. electrolytic.
   c. mica.
   d. air variable.

9. One of the main applications of a capacitor is to
   a. block ac and pass dc.
   b. block both dc and ac.
   c. block dc and pass ac.
   d. pass both dc and ac.

10. Which insulating material has the highest dielectric constant?
    a. Air.
    b. Glass.
    c. Plastic.
    d. Ceramic.

11. The equivalent capacitance, C_{EQ}, of a 10-μF and a 40-μF capacitor in series is
    a. 50 μF.
    b. 125 μF.
    c. 8 μF.
    d. 400 μF.

12. A 0.33-μF capacitor is in parallel with a 0.15-μF and a 220,000-pF capacitor. What is the total capacitance, C_T?
    a. 0.7 μF.
    b. 0.007 μF.
    c. 0.07 μF.
    d. 7 nF.

13. A 5-μF capacitor, C_1, and a 15-μF capacitor, C_2, are connected in series. If the charge stored in C_1 equals 90 μC, what is the voltage across the capacitor C_2?
    a. 18 V.
    b. 12 V.
    c. 9 V.
    d. 6 V.

14. The typical capacitance range of available ultracapacitors is approximately
    a. 1 to 400 pF.
    b. 1 to 400 μF.
    c. 1 to 400 nF.
    d. 1 to 400 F.

15. Capacitors are never coded in
    a. nanofarad units.
    b. microfarad units.
    c. picofarad units.
    d. femtofarad units.

16. Which type of capacitor could explode if the polarity of voltage across its plates is incorrect?
    a. air variable.
    b. mica.
    c. ceramic disk.
    d. aluminum electrolytic.

17. The voltage rating of a capacitor is not affected by
    a. the area of the plates.
    b. the distance between the plates.
    c. the type of dielectric used.
    d. both *b* and *c*.

18. The leakage resistance of a capacitor is typically represented as a(n)
    a. resistance in series with the capacitor plates.
    b. electric field between the capacitor plates.
    c. resistance in parallel with the capacitor plates.
    d. closed switch across the dielectric material.

19. A 2200-μF capacitor with a voltage rating of 35 V is most likely a(n)
    a. electrolytic capacitor.
    b. air variable capacitor.
    c. mica capacitor.
    d. paper capacitor.

20. A capacitor that can store 100 μC of charge with 10 V across its plates has a capacitance value of
    a. 0.01 μF.
    b. 10 μF.
    c. 10 nF.
    d. 100 mF.

21. Calculate the permissible capacitance range of a ceramic disk capacitor whose value is 0.068μF with a tolerance of +80 to –20%.
    a. 0.0544 μF to 0.1224 μF.
    b. 0.0136 μF to 0.0816 μF.
    c. 0.0136 μF to 0.1224 μF.
    d. 0.0544 pF to 0.1224 pF.

22. A two-wire cable has a capacitance of 30 pF per ft. What is the capacitance of a 15-ft piece of this cable?
    a. 2 pF
    b. 15 pF
    c. 30 pF
    d. 450 pF

23. The charge and discharge current of a capacitor flows
    a. through the dielectric.
    b. to and from the capacitor plates.
    c. through the dielectric only until the capacitor is fully charged.
    d. straight through the dielectric from one plate to the other.

24. Capacitance increases with
    a. larger plate area and greater distance between the plates.
    b. smaller plate area and greater distance between the plates.
    c. larger plate area and less distance between the plates.
    d. higher values of applied voltage.

25. Two 0.02-μF, 500-V capacitors in series have an equivalent capacitance and breakdown voltage rating of
    a. 0.04 μF, 1 kV.
    b. 0.01 μF, 250 V.
    c. 0.01 μF, 500 V.
    d. 0.01 μF, 1 kV.

SECTION 15.3 The Farad Unit of Capacitance

15.1 Calculate the amount of charge, Q, stored by a capacitor if
   a. $C = 10\ \mu F$ and $V = 5$ V.
   b. $C = 680$ pF and $V = 200$ V.

15.2 How much charge, Q, is stored by a $0.05\text{-}\mu F$ capacitor if the voltage across the plates equals
   a. 10 V?
   b. 500 V?

15.3 How much voltage exists across the plates of a $200\text{-}\mu F$ capacitor if a constant current of 5 mA charges it for
   a. 100 ms?
   b. 3 s?

15.4 Determine the voltage, V, across a capacitor if
   a. $Q = 2.5\ \mu C$ and $C = 0.01\ \mu F$.
   b. $Q = 49.5$ nC and $C = 330$ pF.

15.5 Determine the capacitance, C, of a capacitor if
   a. $Q = 15\ \mu C$ and $V = 1$ V.
   b. $Q = 15\ \mu C$ and $V = 30$ V.

15.6 List the physical factors that affect the capacitance, C, of a capacitor.

15.7 Calculate the capacitance, C, of a capacitor for each set of physical characteristics listed.
   a. $A = 0.1\ cm^2$, $d = 0.005$ cm, $K_\epsilon = 1$.
   b. $A = 0.05\ cm^2$, $d = 0.001$ cm, $K_\epsilon = 500$.

SECTION 15.6 Ultracapacitors

15.8 Give two other names for an ultracapacitor.

15.9 Ultracapacitors are polarized like electrolytic capacitors. (True or False)

15.10 The primary disadvantage of an ultracapacitor is
   a. low breakdown voltage.
   b. low capacitance.
   c. very high capacitance.
   d. high toxicity.

SECTION 15.8 Parallel Capacitances

15.11 A $5\text{-}\mu F$ and $15\text{-}\mu F$ capacitor are in parallel. How much is C_T?

15.12 A $0.1\text{-}\mu F$, $0.27\text{-}\mu F$, and $0.01\text{-}\mu F$ capacitor are in parallel. How much is C_T?

15.13 A 150-pF, 330-pF, and $0.001\text{-}\mu F$ capacitor are in parallel. How much is C_T?

15.14 In Fig. 15-16,
   a. how much voltage is across each individual capacitor?
   b. how much charge is stored by C_1?
   c. how much charge is stored by C_2?
   d. how much charge is stored by C_3?
   e. what is the total charge stored by all capacitors?
   f. how much is C_T?

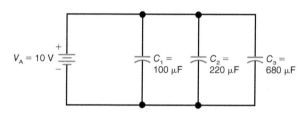

Figure 15-16

SECTION 15.9 Series Capacitances

15.15 A $0.1\text{-}\mu F$ and $0.4\text{-}\mu F$ capacitor are in series. How much is the equivalent capacitance, C_{EQ}?

15.16 A 1500-pF and $0.001\text{-}\mu F$ capacitor are in series. How much is the equivalent capacitance, C_{EQ}?

15.17 A $0.082\text{-}\mu F$, $0.047\text{-}\mu F$, and $0.012\ \mu F$ capacitor are in series. How much is the equivalent capacitance, C_{EQ}?

15.18 In Fig. 15-17, assume a charging current of 180 μA flows for 1 s. Solve for
   a. C_{EQ}.
   b. the charge stored by C_1, C_2, and C_3.
   c. the voltage across C_1, C_2, and C_3.
   d. the total charge stored by all capacitors.

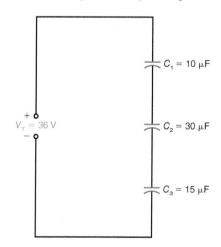

Figure 15-17

Chapter 16

Capacitive Reactance

Learning Outcomes

After studying this chapter, you should be able to:

> *Explain* how alternating current can flow in a capacitive circuit.

> *Calculate* the reactance of a capacitor when the frequency and capacitance are known.

> *Calculate* the total capacitive reactance of series-connected capacitors.

> *Calculate* the equivalent capacitive reactance of parallel-connected capacitors.

> *Explain* how Ohm's law can be applied to capacitive reactance.

> *State* the phase difference between the voltage across and current through a capacitor.

When a capacitor charges and discharges with a varying voltage applied, alternating current can flow. Although there cannot be any current through the dielectric of the capacitor, its charge and discharge produce alternating current in the circuit connected to the capacitor plates. The amount of I that results from the applied sine-wave V depends on the capacitor's capacitive reactance. The symbol for capacitive reactance is X_C, and its unit is the ohm. The X in X_C indicates reactance, whereas the subscript C specifies capacitive reactance.

The amount of X_C is a V/I ratio, but it can also be calculated as $X_C = 1/(2\pi fC)$ in terms of the value of the capacitance and the frequency of the varying V and I. With f and C in the units of the hertz and farad, X_C is in units of ohms. The reciprocal relation in $1/(2\pi fC)$ means that the ohms of X_C decrease for higher frequencies and with more C because more charge and discharge current results either with more capacitance or faster changes in the applied voltage.

16.1 Alternating Current in a Capacitive Circuit

The fact that current flows with ac voltage applied is demonstrated in Fig. 16-1, where the bulb lights in Fig. 16-1a and b because of the capacitor charge and discharge current. There is no current through the dielectric, which is an insulator. While the capacitor is being charged by increasing applied voltage, however, the charging current flows in one direction in the conductors to the plates. While the capacitor is discharging, when the applied voltage decreases, the discharge current flows in the reverse direction. With alternating voltage applied, the capacitor alternately charges and discharges.

First the capacitor is charged in one polarity, and then it discharges; next the capacitor is charged in the opposite polarity, and then it discharges again. The cycles of charge and discharge current provide alternating current in the circuit at the same frequency as the applied voltage. This is the current that lights the bulb.

In Fig. 16-1a, the 4-μF capacitor provides enough alternating current to light the bulb brightly. In Fig. 16-1b, the 1-μF capacitor has less charge and discharge current because of the smaller capacitance, and the light is not so bright. Therefore, the smaller capacitor has more opposition to alternating current as less current flows with the same applied voltage; that is, it has more reactance for less capacitance.

In Fig. 16-1c, the steady dc voltage will charge the capacitor to 120 V. Because the applied voltage does not change, though, the capacitor will just stay charged. Since the potential difference of 120 V across the charged capacitor is a voltage drop opposing the applied voltage, no current can flow. Therefore, the bulb cannot light. The bulb may flicker on for an instant because charging current flows when voltage is applied, but this current is only temporary until the capacitor is charged. Then the capacitor has the applied voltage of 120 V, but there is zero voltage across the bulb.

As a result, the capacitor is said to *block* direct current or voltage. In other words, after the capacitor has been charged by a steady dc voltage, there is no current in the dc circuit. All the applied dc voltage is across the charged capacitor with zero voltage across any series resistance.

In summary, then, this demonstration shows the following points:

1. Alternating current flows in a capacitive circuit with ac voltage applied.
2. A smaller capacitance allows less current, which means more X_C with more ohms of opposition.
3. Lower frequencies for the applied voltage result in less current and more X_C. With a steady dc voltage source, which corresponds to a frequency of zero, the opposition of the capacitor is infinite and there is no current. In this case, the capacitor is effectively an open circuit.

These effects have almost unlimited applications in practical circuits because X_C depends on frequency. A very common use of a capacitor is to provide little opposition for ac voltage but to block any dc voltage. Another example is to use X_C for less opposition to a high-frequency alternating current, compared with lower frequencies.

Capacitive Current

The reason that a capacitor allows current to flow in an ac circuit is the alternate charge and discharge. If we insert an ammeter in the circuit, as shown in Fig. 16-2, the ac meter will read the amount of charge and discharge current. In this example, I_C is 0.12 A. This current is the same in the voltage source, the connecting leads, and the plates of the capacitor.

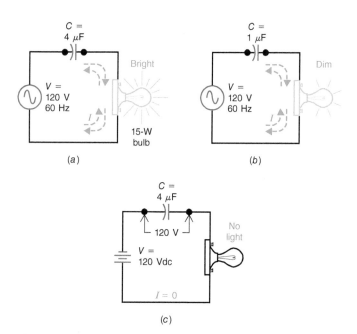

(a)

(b)

(c)

Figure 16-1 Current in a capacitive circuit. (a) The 4-μF capacitor allows enough current I to light the bulb brightly. (b) Less current with a smaller capacitor causes dim light. (c) The bulb cannot light with dc voltage applied because a capacitor blocks direct current.

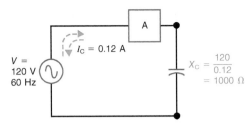

Figure 16-2 Capacitive reactance X_C is the ratio V_C/I_C.

In the figure:
$I_C = 0.12$ A
$V = 120$ V, 60 Hz
$X_C = \dfrac{120}{0.12} = 1000\ \Omega$

However, there is no current through the insulator between the plates of the capacitor.

Values for X_C

When we consider the ratio of V_C/I_C for the ohms of opposition to the sine-wave current, this value is $^{120}\!/_{0.12}$, which equals $1000\ \Omega$. This $1000\ \Omega$ is what we call X_C, to indicate how much current can be produced by sine-wave voltage applied to a capacitor. In terms of current, $X_C = V_C/I_C$. In terms of frequency and capacitance, $X_C = 1/(2\pi fC)$.

The X_C value depends on the amount of capacitance and the frequency of the applied voltage. If C in Fig. 16-2 were increased, it could take on more charge for more charging current and then produce more discharge current. Then X_C is less for more capacitance. Also, if the frequency in Fig. 16-2 were increased, the capacitor could charge and discharge faster to produce more current. This action also means that V_C/I_C would be less with more current for the same applied voltage. Therefore, X_C is less for higher frequencies. Reactance X_C can have almost any value from practically zero to almost infinite ohms.

16.2 The Amount of X_C Equals $1/(2\pi fC)$

The effects of frequency and capacitance are included in the formula for calculating ohms of reactance. The f is in hertz units and the C is in farads for X_C in ohms. As an example, we can calculate X_C for C of 2.65 μF and f of 60 Hz. Then

$$X_C = \frac{1}{2\pi fC} \qquad \text{(16-1)}$$

$$= \frac{1}{2\pi \times 60 \times 2.65 \times 10^{-6}} = \frac{1}{6.28 \times 159 \times 10^{-6}}$$

$$= 0.00100 \times 10^6$$

$$= 1000\ \Omega$$

Note the following factors in the formula $X_C = \dfrac{1}{2\pi fC}$.

1. The constant factor 2π is always $2 \times 3.14 = 6.28$. It indicates the circular motion from which a sine wave is derived. Therefore, the formula $X_C = \dfrac{1}{2\pi fC}$ applies only to sine-wave ac circuits. The 2π is actually 2π rad or $360°$ for a complete circle or cycle.

2. The frequency, f, is a time element. A higher frequency means that the voltage varies at a faster rate. A faster voltage change can produce more charge and discharge current for a given value of capacitance, C. The result is less X_C.

3. The capacitance, C, indicates the physical factors of the capacitor that determine how much charge and discharge current it can produce for a given change in voltage.

4. Capacitive reactance, X_C, is measured in ohms corresponding to the $\dfrac{V_C}{I_C}$ ratio for sine-wave ac circuits. The X_C value determines how much current C allows for a given value of applied voltage.

EXAMPLE 16-1

How much is X_C for (a) 0.1 μF of C at 1400 Hz? (b) 1 μF of C at the same frequency?

Answer:

a. $$X_C = \frac{1}{2\pi fC} = \frac{1}{6.28 \times 1400 \times 0.1 \times 10^{-6}}$$

$$= \frac{1}{6.28 \times 140 \times 10^{-6}} = 0.00114 \times 10^6$$

$$= 1140\ \Omega$$

b. At the same frequency, with ten times more C, X_C is one-tenth or $^{1140}\!/_{10}$, which equals 114 Ω.

EXAMPLE 16-2

How much is the X_C of a 47-pF value of C at (a) 1 MHz? (b) 10 MHz?

Answer:

a. $$X_C = \frac{1}{2\pi fC} = \frac{1}{6.28 \times 47 \times 10^{-12} \times 1 \times 10^6}$$

$$= \frac{1}{295.16 \times 10^{-6}} = 0.003388 \times 10^6$$

$$= 3388\ \Omega$$

b. At 10 times the frequency,

$$X_C = \frac{3388}{10} = 338\ \Omega.$$

Note that X_C in Example 16-2b is one-tenth the value in Example 16-2a because f is 10 times greater.

X_C Is Inversely Proportional to Capacitance

This statement means that X_C increases as capacitance is decreased. In Fig. 16-3, when C is reduced by a factor of $\frac{1}{10}$ from 1.0 to 0.1 μF, then X_C increases 10 times from 1000 to 10,000 Ω. Also, decreasing C by one-half from 0.2 to 0.1 μF doubles X_C from 5000 to 10,000 Ω.

This inverse relation between C and X_C is illustrated by the graph in Fig. 16-3. Note that values of X_C increase downward

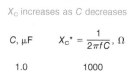

X_C increases as C decreases

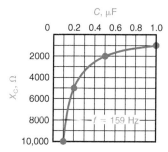

C, µF	$X_C* = \dfrac{1}{2\pi fC}$, Ω
1.0	1000
0.5	2000
0.2	5000
0.1	10,000

*For $f = 159$ Hz

Figure 16-3   A table of values and a graph to show that capacitive reactance X_C decreases with higher values of C. Frequency is constant at 159 Hz.

on the graph, indicating negative reactance that is opposite from inductive reactance. With C increasing to the right, the decreasing values of X_C approach the zero axis of the graph.

X_C Is Inversely Proportional to Frequency

Figure 16-4 illustrates the inverse relationship between X_C and f. With f increasing to the right in the graph from 0.1 to 1 MHz, the value of X_C for the 159-pF capacitor decreases from 10,000 to 1000 Ω as the X_C curve comes closer to the zero axis.

The graphs are nonlinear because of the inverse relation between X_C and f or C. At one end, the curves approach infinitely high reactance for zero capacitance or zero frequency. At the other end, the curves approach zero reactance for infinitely high capacitance or frequency.

Calculating C from Its Reactance

In some applications, it may be necessary to find the value of capacitance required for a desired amount of X_C. For this case, the reactance formula can be inverted to

$$C = \frac{1}{2\pi f X_C} \qquad \text{(16-2)}$$

The value of 6.28 for 2π is still used. The only change from Formula (16-1) is that the C and X_C values are inverted between denominator and numerator on the left and right sides of the equation.

X_C increases as f decreases

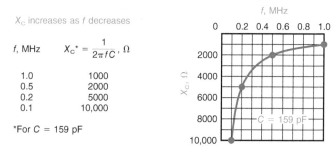

f, MHz	$X_C* = \dfrac{1}{2\pi fC}$, Ω
1.0	1000
0.5	2000
0.2	5000
0.1	10,000

*For $C = 159$ pF

Figure 16-4   A table of values and a graph to show that capacitive reactance X_C decreases with higher frequencies. C is constant at 159 pF.

EXAMPLE 16-3

What C is needed for X_C of 100 Ω at 3.4 MHz?

Answer:

$$C = \frac{1}{2\pi f X_C} = \frac{1}{6.28 \times 3.4 \times 10^6 \times 100}$$

$$= \frac{1}{628 \times 3.4 \times 10^6}$$

$$= 0.000468 \times 10^{-6}\,\text{F} = 0.000468\,\mu\text{F} \quad \text{or} \quad 468\,\text{pF}$$

A practical size for this capacitor would be 470 pF. The application is to have low reactance at the specified frequency of 3.4 MHz.

Calculating Frequency from the Reactance

Another use is to find the frequency at which a capacitor has a specified amount of X_C. Again, the reactance formula can be inverted to the form shown in Formula (16-3).

$$f = \frac{1}{2\pi C X_C} \qquad \text{(16-3)}$$

The following example illustrates the use of this formula.

EXAMPLE 16-4

At what frequency will a 10-µF capacitor have X_C equal to 100 Ω?

Answer:

$$f = \frac{1}{2\pi C X_C} = \frac{1}{6.28 \times 10 \times 10^{-6} \times 100}$$

$$= \frac{1}{6280 \times 10^{-6}}$$

$$= 0.000159 \times 10^6$$

$$= 159\,\text{Hz}$$

This application is a capacitor for low reactance at audio frequencies.

Summary of X_C Formulas

Formula (16-1) is the basic form for calculating X_C when f and C are known values. As another possibility, the value of X_C can be measured as V_C/I_C.

With X_C known, the value of C can be calculated for a specified f by Formula (16-2), or f can be calculated with a known value of C by using Formula (16-3).

16.3 Series or Parallel Capacitive Reactances

Because capacitive reactance is an opposition in ohms, series or parallel reactances are combined in the same way as resistances. As shown in Fig. 16-5a, series capacitive reactances are added arithmetically.

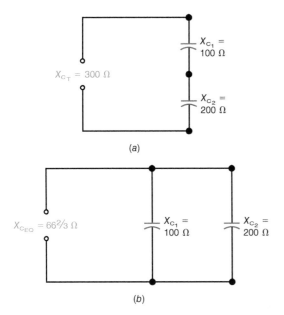

(a)

(b)

Figure 16-5 Reactances alone combine like resistances. (a) Addition of series reactances. (b) Two reactances in parallel equal their product divided by their sum.

Series capacitive reactance:

$$X_{C_T} = X_{C_1} + X_{C_2} + \cdots + \text{etc.} \qquad \textbf{(16-4)}$$

For parallel reactances, the combined reactance is calculated by the reciprocal formula, as shown in Fig. 16-5b. *Parallel capacitive reactance:*

$$X_{C_{EQ}} = \cfrac{1}{\cfrac{1}{X_{C_1}} + \cfrac{1}{X_{C_2}} + \cfrac{1}{X_{C_3}} + \cdots + \text{etc.}} \qquad \textbf{(16-5)}$$

In Fig. 16-5b, the parallel combination of 100 and 200 Ω is 66 $\frac{2}{3}$ Ω for $X_{C_{EQ}}$. The combined parallel reactance is less than the lowest branch reactance. Any shortcuts for combining parallel resistances also apply to parallel reactances.

Combining capacitive reactances is opposite to the way capacitances are combined. The two procedures are compatible, however, because capacitive reactance is inversely proportional to capacitance. The general case is that ohms of opposition add in series but combine by the reciprocal formula in parallel.

16.4 Ohm's Law Applied to X_C

The current in an ac circuit with X_C alone is equal to the applied voltage divided by the ohms of X_C. Three examples with X_C are illustrated in Fig. 16-6. In Fig. 16-6a, there is just one reactance of 100 Ω. The current I then is equal to V/X_C, or 100 V/100 Ω, which is 1 A.

For the series circuit in Fig. 16-6b, the total reactance, equal to the sum of the series reactances, is 300 Ω. Then the current is 100 V/300 Ω, which equals $\frac{1}{3}$ A. Furthermore,

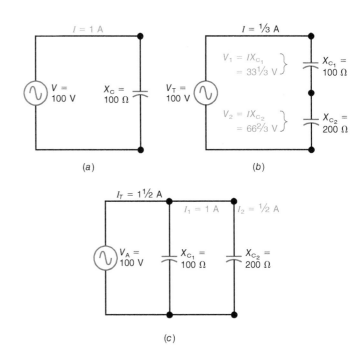

(a)

(b)

(c)

Figure 16-6 Example of circuit calculations with X_C. (a) With a single X_C, the $I = V/X_C$. (b) The sum of series voltage drops equals the applied voltage V_T. (c) The sum of parallel branch currents equals total line current I_T.

the voltage across each reactance is equal to its IX_C product. The sum of these series voltage drops equals the applied voltage.

For the parallel circuit in Fig. 16-6c, each parallel reactance has its individual branch current, equal to the applied voltage divided by the branch reactance. The applied voltage is the same across both reactances, since all are in parallel. In addition, the total line current of 1½ A is equal to the sum of the individual branch currents of 1 and ½ A each. Because the applied voltage is an rms value, all calculated currents and voltage drops in Fig. 16-6 are also rms values.

16.5 Sine-Wave Charge and Discharge Current

In Fig. 16-7, sine-wave voltage applied across a capacitor produces alternating charge and discharge current. The action is considered for each quarter-cycle. Note that the voltage v_C across the capacitor is the same as the applied voltage v_A at all times because they are in parallel. The values of current i, however, depend on the charge and discharge of C. When v_A is increasing, it charges C to keep v_C at the same voltage as v_A; when v_A is decreasing, C discharges to maintain v_C at the same voltage as v_A. When v_A is not changing, there is no charge or discharge current.

During the first quarter-cycle in Fig. 16-7a, v_A is positive and increasing, charging C in the polarity shown. The electron flow is from the negative terminal of the source voltage,

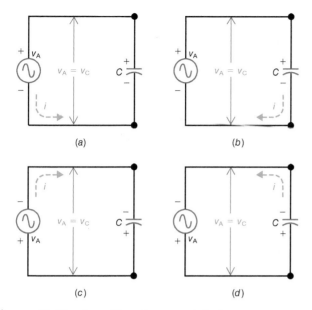

Figure 16-7 Capacitive charge and discharge currents. (a) Voltage v_A increases positive to charge C. (b) The C discharges as v_A decreases. (c) Voltage v_A increases negative to charge C in opposite polarity. (d) The C discharges as reversed v_A decreases.

producing charging current in the direction indicated by the arrow for i. Next, when the applied voltage decreases during the second quarter-cycle, v_C also decreases by discharging. The discharge current is from the negative plate of C through the source and back to the positive plate. Note that the direction of discharge current in Fig. 16-7b is opposite that of the charge current in Fig. 16-7a.

For the third quarter-cycle in Fig. 16-7c, the applied voltage v_A increases again but in the negative direction. Now C charges again but in reversed polarity. Here the charging current is in the direction opposite from the charge current in Fig. 16-7a but in the same direction as the discharge current in Fig. 16-7b. Finally, the negative applied voltage decreases during the final quarter-cycle in Fig. 16-7d. As a result, C discharges. This discharge current is opposite to the charge current in Fig. 16-7c but in the same direction as the charge current in Fig. 16-7a.

For the sine wave of applied voltage, therefore, the capacitor provides a cycle of alternating charge and discharge current. Notice that capacitive current flows for either charge or discharge, whenever the voltage changes, for either an increase or a decrease. Also, i and v have the same frequency.

Figure 16-8a shows a sine wave of voltage v_C across a 240-pF capacitance C. The capacitive current i_C depends on the rate of change of voltage, rather than on the absolute value of v. The rate of voltage change is greatest at the zero-crossing points and produces the peak amount of current. The rate of voltage change is least at the voltage peaks, both positive and negative, therefore producing zero current.

Figure 16-8b shows the actual capacitive current i_C.

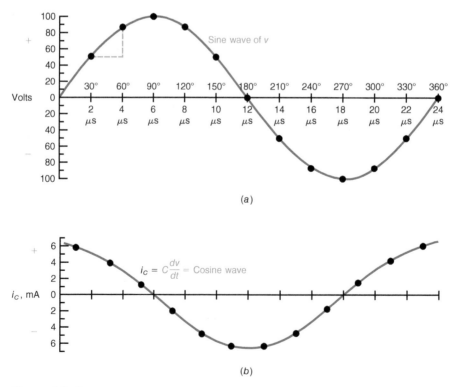

Figure 16-8 Waveshapes of capacitive circuits. (a) Waveshape of sine-wave voltage at top. (b) Changes in voltage cause current i_C charge and discharge waveshape.

90° Phase Angle

The i_C curve at the bottom of Fig. 16-8 has its zero values when the v_C curve at the top is at maximum. The maximum current values appear at the zero-voltage points. This comparison shows that the curves are 90° out of phase with one another. Since the voltage is a sine wave, the 90° difference makes i_C a cosine wave. More details of this 90° phase angle for capacitance are explained in the next chapter.

For each of the curves, the period T is 24 μs. Therefore, the frequency is $1/T$ or $1/24$, which equals 41.67 kHz. Each curve has the same frequency, although there is a 90° phase difference between i and v.

? CHAPTER 16 REVIEW QUESTIONS

1. The capacitive reactance, X_C, of a capacitor is
   a. inversely proportional to frequency.
   b. unaffected by frequency.
   c. directly proportional to frequency.
   d. directly proportional to capacitance.

2. The charge and discharge current of a capacitor flows
   a. through the dielectric.
   b. only when a dc voltage is applied.
   c. to and from the plates.
   d. both *a* and *b*.

3. For direct current (dc), a capacitor acts like a(n)
   a. closed switch.
   b. open.
   c. short.
   d. small resistance.

4. At the same frequency, a larger capacitance provides
   a. more charge and discharge current.
   b. less charge and discharge current.
   c. less capacitive reactance, X_C.
   d. both *a* and *c*.

5. How much is the capacitance, C, of a capacitor that draws 4.8 mA of current from a 12-Vac generator? The frequency of the ac generator is 636.6 Hz.
   a. 0.01 μF.
   b. 0.1 μF.
   c. 0.001 μF.
   d. 100 pF.

6. At what frequency does a 0.015-μF capacitor have an X_C value of 2 kΩ?
   a. 5.3 MHz.
   b. 5.3 Hz.
   c. 5.3 kHz.
   d. 106 kHz.

7. What is the capacitive reactance, X_C, of a 330-pF capacitor at a frequency of 1 MHz?
   a. 482 Ω.
   b. 48.2 Ω.
   c. 1 kΩ.
   d. 482 MΩ.

8. What is the instantaneous value of charging current, i_C, of a 10-μF capacitor if the voltage across the capacitor plates changes at the rate of 250 V per second?
   a. 250 μA.
   b. 2.5 A.
   c. 2.5 μA.
   d. 2.5 mA.

9. For a capacitor, the charge and discharge current, i_C,
   a. lags the capacitor voltage, v_C, by a phase angle of 90°.
   b. leads the capacitor voltage, v_C, by a phase angle of 90°.
   c. is in phase with the capacitor voltage, v_C.
   d. none of the above.

10. Two 1-kΩ X_C values in series have a total capacitive reactance of
    a. 1.414 kΩ.
    b. 500 Ω.
    c. 2 kΩ.
    d. 707 Ω.

11. Two 5-kΩ X_C values in parallel have an equivalent capacitive reactance of
    a. 7.07 kΩ.
    b. 2.5 kΩ.
    c. 10 kΩ.
    d. 3.53 kΩ.

12. For any capacitor,
    a. the stored charge increases with more capacitor voltage.
    b. the charge and discharge currents are in opposite directions.
    c. i_C leads v_C by 90°.
    d. all of the above.

13. The unit of capacitive reactance, X_C, is the
    a. ohm.
    b. farad.
    c. hertz.
    d. radian.

14. The main difference between resistance, R, and capacitive reactance, X_C, is that
  a. X_C is the same for both dc and ac, whereas R depends on frequency.
  b. R is the same for both dc and ac, whereas X_C depends on frequency.
  c. R is measured in ohms and X_C is measured in farads.
  d. none of the above.

15. A very common use for a capacitor is to
  a. block any dc voltage but provide very little opposition to an ac voltage.
  b. block both dc and ac voltages.
  c. pass both dc and ac voltages.
  d. none of the above.

CHAPTER 16 PROBLEMS

SECTION 16.1 Alternating Current in a Capacitive Circuit

16.1 With the switch, S_1, closed in Fig. 16-9, how much is
  a. the current, I, in the circuit?
  b. the dc voltage across the 12–V lamp?
  c. the dc voltage across the capacitor?

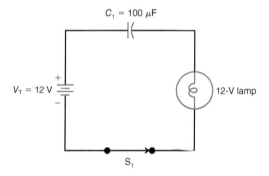

$C_1 = 100\ \mu F$

$V_T = 12\ V$

12-V lamp

S_1

Figure 16-9

16.2 In Fig. 16-9 explain why the bulb will light for just an instant when S_1 is initially closed.

16.3 In Fig. 16-10, the capacitor and the lightbulb draw 400 mA from the 120-Vac source. How much current flows
  a. to and from the terminals of the 120-Vac source?
  b. through the lightbulb?
  c. to and from the plates of the capacitor?
  d. through the connecting wires?
  e. through the dielectric of the capacitor?

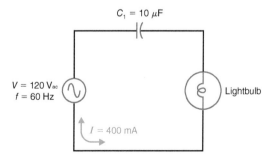

$C_1 = 10\ \mu F$

$V = 120\ V_{ac}$
$f = 60\ Hz$

Lightbulb

$I = 400\ mA$

Figure 16-10

16.4 In Fig. 16-11, calculate the capacitive reactance, X_C, for the following values of Vac and I?
  a. $Vac = 10\ V$ and $I = 20\ mA$.
  b. $Vac = 24\ V$ and $I = 8\ mA$.
  c. $Vac = 15\ V$ and $I = 300\ \mu A$.
  d. $Vac = 100\ V$ and $I = 50\ \mu A$.

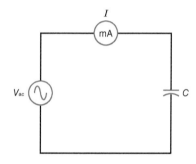

I

mA

V_{ac}

C

Figure 16-11

16.5 In Fig. 16-11, list three factors that can affect the amount of charge and discharge current flowing in the circuit.

SECTION 16.2 The Amount of X_C Equals $1/(2\pi fC)$

16.6 Calculate the capacitive reactance, X_C, of a 0.1-μF capacitor at the following frequencies:
  a. $f = 50\ Hz$.
  b. $f = 10\ kHz$.

16.7 Calculate the capacitive reactance, X_C, of a 10-pF capacitor at the following frequencies:
  a. $f = 60\ Hz$.
  b. $f = 2.4\ GHz$.

16.8 What value of capacitance will provide an X_C of 1 kΩ at the following frequencies?
  a. $f = 318.3\ Hz$.
  b. $f = 433\ MHz$.

16.9 At what frequency will a 0.047-μF capacitor provide an X_C value of
  a. 100 kΩ?
  b. 50 Ω?

16.10 How much is the capacitance of a capacitor that draws 2 mA of current from a 10-Vac generator whose frequency is 3.183 kHz?

16.11 At what frequency will a 820-pF capacitance have an X_C value of 250 Ω?

16.12 A 0.01-μF capacitor draws 50 mA of current when connected directly across a 50-Vac source. What is the value of current drawn by the capacitor when
  a. the frequency is doubled?
  b. the frequency is decreased by one-half?
  c. the capacitance is doubled to 0.02 μF?
  d. the capacitance is reduced by one–half to 0.005 μF?

16.13 A capacitor has an X_C value of 10 kΩ at a given frequency. What is the new value of X_C when the frequency is
  a. cut in half?
  b. doubled?
  c. quadrupled?
  d. increased by a factor of 10?

16.14 Calculate the capacitive reactance, X_C, for the following capacitance and frequency values:
  a. $C = 0.47\ \mu$F, $f = 1$ kHz.
  b. $C = 250$ pF, $f = 1$ MHz.

16.15 Determine the capacitance value for the following frequency and X_C values:
  a. $X_C = 1$ kΩ, $f = 3.183$ kHz.
  b. $X_C = 200\ \Omega$, $f = 63.66$ kHz.

16.16 Determine the frequency for the following capacitance and X_C values:
  a. $C = 0.05\ \mu$F, $X_C = 4$ kΩ.
  b. $C = 0.1\ \mu$F, $X_C = 1.591$ kΩ.

SECTION 16.3 Series or Parallel Capacitive Reactances

16.17 How much is the total capacitive reactance, X_{C_T}, for the following series capacitive reactances:
  a. $X_{C_1} = 1$ kΩ, $X_{C_2} = 1.5$ kΩ, $X_{C_3} = 2.5$ kΩ.
  b. $X_{C_1} = 500\ \Omega$, $X_{C_2} = 1$ kΩ, $X_{C_3} = 1.5$ kΩ.

16.18 What is the equivalent capacitive reactance, $X_{C_{EQ}}$, for the following parallel capacitive reactances:
  a. $X_{C_1} = 100\ \Omega$ and $X_{C_2} = 400\ \Omega$.
  b. $X_{C_1} = 1.2$ kΩ and $X_{C_2} = 1.8$ kΩ.

SECTION 16.4 Ohm's Law Applied to X_C

16.19 In Fig. 16-12, calculate the current, I.

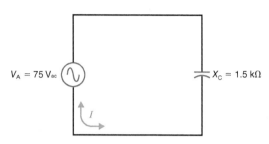

$V_A = 75$ Vac $X_C = 1.5$ kΩ

Figure 16-12

16.20 In Fig. 16-12, what happens to the current, I, when the frequency of the applied voltage
  a. decreases?
  b. increases?

16.21 In Fig. 16-13, solve for C_1, C_2, C_3, and C_T if the frequency of the applied voltage is 6.366 kHz.

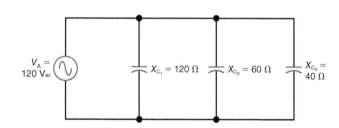

$V_A = 120$ Vac $X_{C_1} = 120\ \Omega$ $X_{C_2} = 60\ \Omega$ $X_{C_3} = 40\ \Omega$

Figure 16-13

Capacitive Circuits

This chapter analyzes circuits that combine capacitive reactance X_C and resistance R. The main questions are, How do we combine the ohms of opposition, how much current flows, and what is the phase angle? Although both X_C and R are measured in ohms, they have different characteristics. Specifically, X_C decreases with more C and higher frequencies for sine-wave ac voltage applied, whereas R is the same for dc and ac circuits. Furthermore, the phase angle for the voltage across X_C is at $-90°$ measured in the clockwise direction with i_C as the reference at $0°$.

In addition, the practical application of a coupling capacitor shows how a low value of X_C can be used to pass the desired ac signal variations while blocking the steady dc level of a fluctuating dc voltage. In a coupling circuit with C and R in series, the ac component is

Learning Outcomes

After studying this chapter, you should be able to:

> *Explain* why the current leads the voltage by 90° for a capacitor.

> *Define* the term *impedance*.

> *Calculate* the total impedance and phase angle of a series *RC* circuit.

> *Describe* the operation and application of an *RC* phase-shifter circuit.

> *Calculate* the total current, equivalent impedance, and phase angle of a parallel *RC* circuit.

> *Explain* how a capacitor can couple some ac frequencies but not others.

> *Calculate* the individual capacitor voltage drops for capacitors in series.

across R for the output voltage, but the dc component across C is not present across the output terminals.

Finally, the general case of capacitive charge and discharge current produced when the applied voltage changes is shown with nonsinusoidal voltage variations. In this case, we compare the waveshapes of v_C and i_C. Remember that the $-90°$ angle for an IX_C voltage applies only to sine waves.

17.1 Sine Wave v_C Lags i_C by $90°$

For a sine wave of applied voltage, a capacitor provides a cycle of alternating charge and discharge current, as shown in Fig. 17-1a. In Fig. 17-1b, the waveshape of this charge and discharge current i_C is compared with the voltage v_C.

Examining the v_C and i_C Waveforms

In Fig. 17-1b, note that the instantaneous value of i_C is zero when v_C is at its maximum value. At either its positive or its negative peak, v_C is not changing. For one instant at both peaks, therefore, the voltage must have a static value before changing its direction. Then v is not changing and C is not charging or discharging. The result is zero current at this time.

Also note that i_C is maximum when v_C is zero. When v_C crosses the zero axis, i_C has its maximum value because then the voltage is changing most rapidly.

Therefore, i_C and v_C are $90°$ out of phase, since the maximum value of one corresponds to the zero value of the other; i_C leads v_C because i_C has its maximum value a quarter-cycle before the time that v_C reaches its peak. The phasors in Fig. 17-1c show i_C leading v_C by the counterclockwise angle of $90°$. Here v_C is the horizontal phasor for the reference angle of $0°$. In Fig. 17-1d, however, the current i_C is the horizontal phasor for reference. Since i_C must be $90°$ leading, v_C is shown lagging by the clockwise angle of $-90°$. In series circuits, the current i_C is the reference, and then the voltage v_C can be considered to lag i_C by $90°$.

Why i_C Leads v_C by $90°$

The $90°$ phase angle results because i_C depends on the rate of change of v_C. In other words, i_C has the phase of dv/dt, not the phase of v. As shown previously in Fig. 16-8 for a sine wave of v_C, the capacitive charge and discharge current is a cosine wave. This $90°$ phase between v_C and i_C is true in any sine-wave ac circuit, whether C is in series or parallel and whether C is alone or combined with other components. We can always say that for any X_C, its current and voltage are $90°$ out of phase.

Capacitive Current Is the Same in a Series Circuit

The leading phase angle of capacitive current is only with respect to the voltage across the capacitor, which does not change the fact that the current is the same in all parts of a series circuit. In Fig. 17-1a, for instance, the current in the generator, the connecting wires, and both plates of the capacitor must be the same because they are all in the same path.

Capacitive Voltage Is the Same across Parallel Branches

In Fig. 17-1a, the voltage is the same across the generator and C because they are in parallel. There cannot be any lag or lead in time between these two parallel voltages. At any instant, whatever the voltage value is across the generator at that time, the voltage across C is the same. With respect to the series current, however, both v_A and v_C are $90°$ out of phase with i_C.

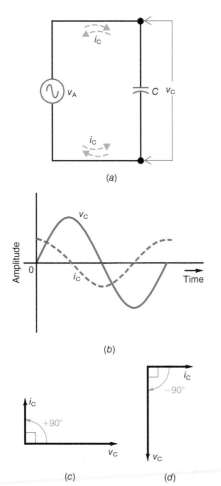

(a)

(b)

(c)   (d)

Figure 17-1 Capacitive current i_C leads v_C by $90°$. (a) Circuit with sine wave v_A across C. (b) Waveshapes of i_C $90°$ ahead of v_C. (c) Phasor diagram of i_C leading the horizontal reference v_C by a counterclockwise angle of $90°$. (d) Phasor diagram with i_C as the reference phasor to show v_C lagging i_C by an angle of $-90°$.

The Frequency Is the Same for v_C and i_C

Although v_C lags i_C by 90°, both waves have the same frequency. For example, if the frequency of the sine wave v_C in Fig. 17-1b is 100 Hz, this is also the frequency of i_C.

17.2 X_C and R in Series

When a capacitor and a resistor are connected in series, as in Fig. 17-2a, the current I is limited by both X_C and R. The current I is the same in both X_C and R since they are in series. However, each component has its own series voltage drop, equal to IR for the resistance and IX_C for the capacitive reactance.

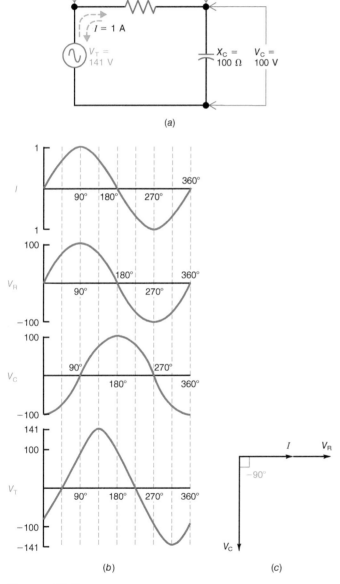

(a)

(b)

(c)

Figure 17-2 Circuit with X_C and R in series. (a) Schematic diagram. (b) Waveforms of current and voltages. (c) Phasor diagram.

Note the following points about a circuit that combines both X_C and R in series, like that in Fig. 17-2a.

1. The current is labeled I, rather than I_C, because I flows through all series components.

2. The voltage across X_C, labeled V_C, can be considered an IX_C voltage drop, just as we use V_R for an IR voltage drop.

3. The current I through X_C must lead V_C by 90° because this is the phase angle between the voltage and current for a capacitor.

4. The current I through R and its IR voltage drop are in phase. There is no reactance to sine-wave alternating current in any resistance. Therefore, I and IR have a phase angle of 0°.

It is important to note that the values of I and V may be in rms, peak, peak to peak, or instantaneous, as long as the same measure is applied to the entire circuit. Peak values will be used here for convenience in comparing waveforms.

Phase Comparisons

Note the following points about a circuit containing series resistance and reactance:

1. The voltage V_C is 90° out of phase with I.

2. However, V_R and I are in phase.

3. Since I is common to all circuit components and the voltage source, it is commonly used as the reference in phasor diagrams of series RC circuits.

4. If I is used as the reference, V_C is 90° out of phase with V_R.

Specifically, V_C lags V_R by 90° just as the voltage V_C lags the current I by 90°. The phase relationships between I, V_R, V_C, and V_T are shown by the waveforms in Fig. 17-2b. Figure 17-2c shows the phasors representing I, V_R, and V_C.

Combining V_R and V_C

As shown in Fig. 17-2b, when the voltage wave V_R is combined with the voltage wave V_C, the result is the voltage wave of the applied voltage V_T. The instantaneous values of the capacitor and resistor voltage drops are added together at each point along the horizontal time axis. The voltage drops, V_R and V_C, must add to equal the applied voltage V_T. The 100-V peak values for V_R and V_C total 141 V, however, instead of 200 V, because of the 90° phase difference.

Consider some instantaneous values in Fig. 17-2b, to see why the 100-V peak V_R and 100-V peak V_C cannot be added arithmetically. When V_R is at its maximum of 100 V, for instance, V_C is at zero. The total voltage V_T at this instant, then, is 100 V. Similarly, when V_C is at its maximum of 100 V, V_R is at zero and the total voltage V_T is again 100 V.

Actually, V_T reaches its maximum of 141 V when V_C and V_R are each at 70.7 V. When series voltage drops that are

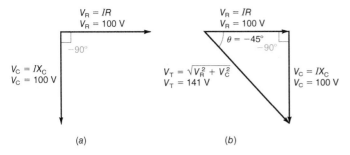

Figure 17-3 Addition of two voltages 90° out of phase. (a) Phasors for v_C and v_R are 90° out of phase. (b) Resultant of the two phasors is the hypotenuse of the right triangle for v_T.

out of phase are combined, therefore, they cannot be added without taking the phase difference into account.

Phasor Voltage Triangle

Instead of combining waveforms that are out of phase, as in Fig. 17-2*b*, we can add them more quickly by using their equivalent phasors, as shown in Fig. 17-3. The phasors in Fig. 17-3*a* show the 90° phase angle without any addition. The method in Fig. 17-3*b* is to add the tail of one phasor to the arrowhead of the other, using the angle required to show their relative phase. Note that voltages V_R and V_C are at right angles to each other because they are 90° out of phase. Note also that the phasor for V_C is downward at an angle of −90° from the phasor for V_R. Here V_R is used as the reference phasor because it has the same phase as the series current I, which is the same everywhere in the circuit. The phasor V_T, extending from the tail of the V_R phasor to the arrowhead of the V_C phasor, represents the applied voltage V_T, which is the phasor sum of V_R and V_C. Since V_R and V_C form a right angle, the resultant phasor V_T is the hypotenuse of a right triangle. The hypotenuse is the side opposite the 90° angle.

From the geometry of a right triangle, the Pythagorean theorem states that the hypotenuse is equal to the square root of the sum of the squares of the sides. For the voltage triangle in Fig. 17-3*b*, therefore, the resultant is

$$V_T = \sqrt{V_R^2 + V_C^2} \qquad (17\text{-}1)$$

where V_T is the phasor sum of the two voltages V_R and V_C 90° out of phase.

This formula is for V_R and V_C when they are in series, since they are 90° out of phase. All voltages must be expressed in the same units. When V_T is an rms value, V_R and V_C must also be rms values. For the voltage triangle in Fig. 17-3*b*,

$$V_T = \sqrt{100^2 + 100^2} = \sqrt{10{,}000 + 10{,}000}$$
$$= \sqrt{20{,}000}$$
$$= 141 \text{ V}$$

17.3 Impedance *Z* Triangle

A triangle of R and X_C in series corresponds to the voltage triangle, as shown in Fig. 17-4. It is similar to the voltage triangle in Fig. 17-3*b*, but the common factor I cancels because the series current I is the same in X_C and R. The resultant of the phasor addition of X_C and R is their total opposition in ohms, called *impedance*, with the symbol Z_T. The Z takes into account the 90° phase relation between R and X_C.

For the impedance triangle of a series circuit with capacitive reactance X_C and resistance R,

$$Z_T = \sqrt{R^2 + X_C^2} \qquad (17\text{-}2)$$

where R, X_C, and Z_T are all in ohms. For the phasor triangle in Fig. 17-4,

$$Z_T = \sqrt{100^2 + 100^2} = \sqrt{10{,}000 + 10{,}000}$$
$$= \sqrt{20{,}000}$$
$$= 141 \ \Omega$$

This is the total impedance Z_T in Fig. 17-2*a*.

Note that the applied voltage V_T of 141 V divided by the total impedance of 141 Ω results in 1 A of current in the series circuit. The IR voltage V_R is 1 A × 100 Ω or 100 V; the IX_C voltage is also 1 A × 100 Ω or 100 V. The series IR and IX_C voltage drops of 100 V each are added using phasors to equal the applied voltage V_T of 141 V. Finally, the applied voltage equals IZ_T or 1 A × 141 Ω, which is 141 V.

Summarizing the similar phasor triangles for voltage and ohms in a series *RC* circuit,

1. The phasor for R, IR, or V_R is used as a reference at 0°.
2. The phasor for X_C, IX_C, or V_C is at −90°.
3. The phasor for Z_T, IZ_T, or V_T has the phase angle θ of the complete circuit.

Phase Angle with Series X_C and R

The angle between the applied voltage V_T and the series current I is the phase angle of the circuit. Its symbol is θ (theta). In Fig. 17-3*b*, the phase angle between V_T and IR is −45°. Since IR and I have the same phase, the angle is also −45° between V_T and I.

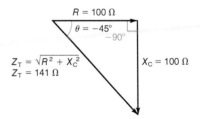

Figure 17-4 Addition of R and X_C 90° out of phase in a series *RC* circuit to find the total impedance Z_T.

In the corresponding impedance triangle in Fig. 17-4, the angle between Z_T and R is also equal to the phase angle. Therefore, the phase angle can be calculated from the impedance triangle of a series RC circuit by the formula

$$\tan \theta_Z = -\frac{X_C}{R} \qquad (17\text{-}3)$$

The tangent (tan) is a trigonometric function of an angle, equal to the ratio of the opposite side to the adjacent side of a triangle. In this impedance triangle, X_C is the opposite side and R is the adjacent side of the angle. We use the subscript Z for θ to show that θ_Z is found from the impedance triangle for a series circuit. To calculate this phase angle,

$$\tan \theta_Z = -\frac{X_C}{R} = -\frac{100}{100} = -1$$

The angle that has the tangent value of -1 is $-45°$ in this example. The numerical values of the trigonometric functions can be found by using a scientific calculator. The mathematical expression for this is

$$\theta = \tan^{-1}\left(\frac{X_C}{R}\right)$$

The term $\tan^{-1}$ is also called the inverse tangent, or arctangent (arctan). Note that the phase angle of $-45°$ is halfway between $0°$ and $-90°$ because R and X_C are equal.

EXAMPLE 17-1

If a 30-Ω R and a 40-Ω X_C are in series with 100 V applied, find the following: Z_T, I, V_R, V_C, and θ_Z. What is the phase angle between V_C and V_R with respect to I? Prove that the sum of the series voltage drops equals the applied voltage V_T.

Answer:

$$Z_T = \sqrt{R^2 + X_C^2} = \sqrt{30^2 + 40^2}$$
$$= \sqrt{900 + 1600}$$
$$= \sqrt{2500}$$
$$= 50\ \Omega$$
$$I = \frac{V_T}{Z_T} = \frac{100\ \text{V}}{50\ \Omega} = 2\ \text{A}$$
$$V_R = IR = 2\ \text{A} \times 30\ \Omega = 60\ \text{V}$$
$$V_C = IX_C = 2\ \text{A} \times 40\ \Omega = 80\ \text{V}$$
$$\tan \theta_Z = -\frac{X_C}{R} = -\frac{40}{30} = -1.333$$
$$\theta_Z = -53.1°$$

Therefore, V_T lags I by 53.1°. Furthermore, I and V_R are in phase, and V_C lags I by 90°. Finally,

$$V_T = \sqrt{V_R^2 + V_C^2} = \sqrt{60^2 + 80^2} = \sqrt{3600 + 6400}$$
$$= \sqrt{10,000}$$
$$= 100\ \text{V}$$

Note that the phasor sum of the voltage drops equals the applied voltage V_T.

Table 17-1 Series R and X_C Combinations

R, Ω	X_c, Ω	Z_T, Ω (Approx.)	Phase Angle θ_Z
1	10	$\sqrt{101} = 10$	$-84.3°$
10	10	$\sqrt{200} = 14$	$-45°$
10	1	$\sqrt{101} = 10$	$-5.7°$

Note: θ_Z is the phase angle of Z_T or V_T with respect to the reference phasor I in series circuits.

Series Combinations of X_C and R

In series, the higher the X_C compared with R, the more capacitive the circuit. There is more voltage drop across the capacitive reactance X_C, and the phase angle increases toward $-90°$. The series X_C always makes the series current I lead the applied voltage V_T. With all X_C and no R, the entire applied voltage V_T is across X_C and θ equals $-90°$.

Several combinations of X_C and R in series are listed in Table 17-1 with their resultant impedance values and phase angle. Note that a ratio of 10:1, or more, for X_C/R means that the circuit is practically all capacitive. The phase angle of $-84.3°$ is almost $-90°$, and the total impedance Z_T is approximately equal to X_C. The voltage drop across X_C in the series circuit is then practically equal to the applied voltage V_T with almost none across R.

At the opposite extreme, when R is 10 times more than X_C, the series circuit is mainly resistive. The phase angle of $-5.7°$ then means that the current is almost in phase with the applied voltage V_T; Z_T is approximately equal to R, and the voltage drop across R is practically equal to the applied voltage V_T with almost none across X_C. In summary, if the resistive voltage drop is greater than the capacitive voltage drop, the circuit is resistive. If the capacitive voltage drop is greater than the resistive voltage, the circuit is capacitive.

When X_C and R equal each other, the resultant impedance Z_T is 1.41 times either one. The phase angle then is $-45°$, halfway between $0°$ for resistance alone and $-90°$ for capacitive reactance alone.

17.4 RC Phase-Shifter Circuit

Figure 17-5 shows an application of X_C and R in series to provide a desired phase shift in the output V_R compared with the input V_T. The R can be varied up to 100 kΩ to change the phase angle. The C is 0.05 μF here for the 60-Hz ac power-line voltage, but a smaller C would be used for a higher frequency. The capacitor must have an appreciable value of reactance for the phase shift.

For the circuit in Fig. 17-5a, assume that R is set for 50 kΩ at its middle value. The reactance of the 0.05-μF capacitor

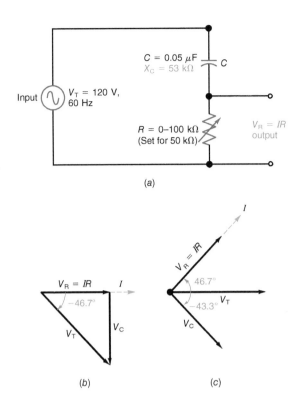

(a)

(b)   (c)

Figure 17-5 An *RC* phase-shifter circuit. (a) Schematic diagram. (b) Phasor triangle with *IR*, or V_R, as the horizontal reference. V_R leads V_T by 46.7° with *R* set at 50 kΩ. (c) Phasors shown with V_T as the horizontal reference.

at 60 Hz is approximately 53 kΩ. For these values of X_C and *R*, the phase angle of the circuit is −46.7°. This angle has a tangent of $-\frac{53}{50} = -1.06$.

The phasor triangle in Fig. 17-5b shows that *IR* or V_R is out of phase with V_T by the leading angle of 46.7°. Note that V_C is always 90° lagging V_R in a series circuit. The angle between V_C and V_T then becomes 90° − 46.7° = 43.3°.

This circuit provides a phase-shifted voltage V_R at the output with respect to the input. For this reason, the phasors are redrawn in Fig. 17-5c to show the voltages with the input V_T as the horizontal reference. The conclusion, then, is that the output voltage across *R* leads the input V_T by 46.7°, whereas V_C lags V_T by 43.3°.

Now let *R* be varied for a higher value at 90 kΩ, while X_C stays the same. The phase angle becomes −30.5°. This angle has a tangent of $-\frac{53}{90} = -0.59$. As a result, V_R leads V_T by 30.5°, and V_C lags V_T by 59.5°.

For the opposite case, let *R* be reduced to 10 kΩ. Then the phase angle becomes −79.3°. This angle has the tangent $-\frac{53}{10} = -5.3$. Then V_R leads V_T by 79.3° and V_C lags V_T by 10.7°. Notice that the phase angle between V_R and V_T becomes larger as the series circuit becomes more capacitive with less resistance.

A practical application for this circuit is providing a voltage of variable phase to set the conduction time of

semiconductors in power-control circuits. In this case, the output voltage is taken across the capacitor *C*. This provides a lagging phase angle with respect to the input voltage V_T. As *R* is varied from 0 Ω to 100 kΩ, the phase angle between V_C and V_T decreases from 0° to about −62°. If *R* were changed so that it varied from 0 to 1 MΩ, the phase angle between V_C and V_T would vary between 0° and −90° approximately. If the positions of the resistor and capacitor in Fig. 17-5a are reversed, the current and voltage drops will remain the same. However, the output is now taken from across the capacitor. For this reason, the output voltage will lag the input voltage. You can see this in Fig. 17-5c. The input is V_T, but the output is V_C rather than V_R. A negative angle means a lagging phase shift on the output.

17.5 X_C and *R* in Parallel

For parallel circuits with X_C and *R*, the 90° phase angle must be considered for each of the branch currents. Remember that any series circuit has different voltage drops but one common current. A parallel circuit has different branch currents but one common voltage.

In the parallel circuit in Fig. 17-6a, the applied voltage V_A is the same across X_C, *R*, and the generator, since they are all in parallel. There cannot be any phase difference between these voltages. Each branch, however, has its own individual current. For the resistive branch, $I_R = V_A/R$; for the capacitive branch, $I_C = V_A/X_C$.

The resistive branch current I_R is in phase with the generator voltage V_A. The capacitive branch current I_C leads V_A, however, because the charge and discharge current of a capacitor leads the capacitor voltage by 90°. The waveforms for V_A, I_R, I_C, and I_T in Fig. 17-6a are shown in Fig. 17-6b. The individual branch currents I_R and I_C must add to equal the total current I_T. The 10-A peak values for I_R and I_C total 14.14 A, however, instead of 20 A, because of the 90° phase difference.

Consider some instantaneous values in Fig. 17-6b to see why the 10-A peak for I_R and 10-A peak for I_C cannot be added arithmetically. When I_C is at its maximum of 10 A, for instance, I_R is at zero. The total for I_T at this instant then is 10 A. Similarly, when I_R is at its maximum of 10 A, I_C is at zero and the total current I_T at this instant is also 10 A.

Actually, I_T has its maximum of 14.14 A when I_R and I_C are each 7.07 A. When branch currents that are out of phase are combined, therefore, they cannot be added without taking the phase difference into account.

Figure 17-6c shows the phasors representing V_A, I_R, and I_C. Notice that I_C leads V_A and I_R by 90°. In this case, the applied voltage V_A is used as the reference phasor since it is the same across both branches.

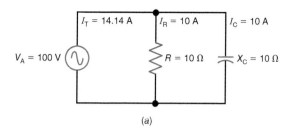

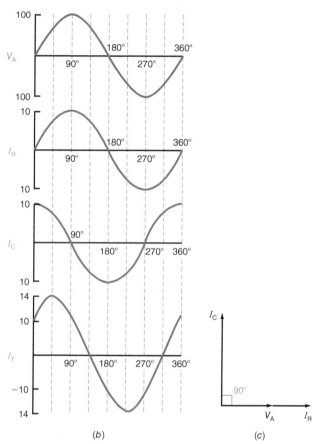

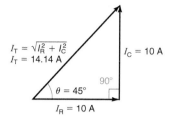

(a)

(b)

(c)

Figure 17-6 Capacitive reactance X_C and R in parallel. (a) Circuit. (b) Waveforms of the applied voltage V_A, branch currents I_R and I_C, and total current I_T. (c) Phasor diagram.

Phasor Current Triangle

Figure 17-7 shows the phasor current triangle for the parallel *RC* circuit in Fig. 17-6a. Note that the resistive branch current I_R is used as the reference phasor since V_A and I_R are in phase. The capacitive branch current I_C is drawn upward at an angle of $+90°$ since I_C leads V_A and thus I_R by $90°$. The sum of the I_R and I_C phasors is indicated by the phasor for I_T, which connects the tail of the I_R phasor to the tip of the I_C phasor. The I_T phasor is the hypotenuse of the right triangle. The phase angle between I_T and I_R represents the phase angle of the circuit. Peak values are shown here for convenience, but rms and peak-to-peak values could also be used.

Using the Pythagorean theorem, the total current I_T could be calculated by taking the square root of the sum of the

Figure 17-7 Phasor triangle of capacitive and resistive branch currents $90°$ out of phase in a parallel circuit to find the resultant I_T.

squares of the sides. For the current triangle in Fig. 17-7 therefore, the resultant I_T is

$$I_T = \sqrt{I_R^2 + I_C^2} \qquad \textbf{(17-4)}$$

For the values in Fig. 17-6,

$$I_T = \sqrt{10^2 + 10^2} = \sqrt{100 + 100}$$
$$= \sqrt{200}$$
$$= 14.14 \text{ A}$$

Impedance of X_C and R in Parallel

A practical approach to the problem of calculating the total or equivalent impedance of X_C and R in parallel is to calculate the total line current I_T and divide the applied voltage V_A by this value.

$$Z_{EQ} = \frac{V_A}{I_T} \qquad \textbf{(17-5)}$$

For the circuit in Fig. 17-6a, V_A is 100 V, and the total current I_T, obtained as the phasor sum of I_R and I_C, is 14.14 A. Therefore, we can calculate the equivalent impedance Z_{EQ} as

$$Z_{EQ} = \frac{V_A}{I_T} = \frac{100 \text{ V}}{14.14 \text{ A}}$$
$$= 7.07 \text{ } \Omega$$

This impedance, the combined opposition in ohms across the generator, is equal to the 10-Ω resistance in parallel with the 10-Ω X_C.

Note that the impedance Z_{EQ} for equal values of R and X_C in parallel is not one-half but instead equals 70.7% of either one. Still, the value of Z_{EQ} will always be less than the lowest ohm value in the parallel branches.

For the general case of calculating the Z_{EQ} of X_C and R in parallel, any number can be assumed for the applied voltage V_A because, in the calculations for Z_{EQ} in terms of the branch currents, the value of V_A cancels. A good value to assume for V_A is the value of either R or X_C, whichever is the larger number. This way, there are no fractions smaller than that in the calculation of the branch currents.

Table 17-2 Parallel Resistance and Capacitance Combinations*

R, Ω	X_c, Ω	I_R, A	I_C, A	I_T, A (Approx.)	Z_{EQ}, Ω (Approx.)	Phase Angle θ_I
1	10	10	1	$\sqrt{101} = 10$	1	5.7°
10	10	1	1	$\sqrt{2} = 1.4$	7.07	45°
10	1	1	10	$\sqrt{101} = 10$	1	84.3°

* V_A = 10 V. Note that θ_I is the phase angle of I_T with respect to the reference V_A in parallel circuits.

Phase Angle in Parallel Circuits

In Fig. 17-7, the phase angle θ is 45° because R and X_C are equal, resulting in equal branch currents. The phase angle is between the total current I_T and the generator voltage V_A. However, V_A and I_R are in phase. Therefore θ is also between I_T and I_R.

Using the tangent formula to find θ from the current triangle in Fig. 17-7 gives

$$\tan \theta_I = \frac{I_C}{I_R} \qquad (17\text{-}6)$$

The phase angle is positive because the I_C phasor is upward, leading V_A by 90°. This direction is opposite from the lagging phasor of series X_C. The effect of X_C is no different, however. Only the reference is changed for the phase angle.

Note that the phasor triangle of branch currents for parallel circuits gives θ_I as the angle of I_T with respect to the generator voltage V_A. This phase angle for I_T is labeled θ_I with respect to the applied voltage. For the phasor triangle of voltages in a series circuit, the phase angle for Z_T and V_T is labeled θ_Z with respect to the series current.

EXAMPLE 17-2

A 30-mA I_R is in parallel with another branch current of 40 mA for I_C. The applied voltage V_A is 72 V. Calculate I_T, Z_{EQ}, and θ_I.

Answer:

This problem can be calculated in mA units for I and kΩ for Z without powers of 10.

$$I_T = \sqrt{I_R^2 + I_C^2} = \sqrt{(30)^2 + (40)^2}$$
$$= \sqrt{900 + 1600} = \sqrt{2500}$$
$$= 50 \text{ mA}$$

$$Z_{EQ} = \frac{V_A}{I_T} = \frac{72 \text{ V}}{50 \text{ mA}}$$
$$= 1.44 \text{ k}\Omega$$

$$\tan \theta_I = \frac{I_C}{I_R} = \frac{40}{30} = 1.333$$
$$= \arctan (1.333)$$
$$\theta_I = 53.1°$$

Parallel Combinations of X_C and R

In Table 17-2, when X_C is 10 times R, the parallel circuit is practically resistive because there is little leading capacitive current in the main line. The small value of I_C results from the high reactance of shunt X_C. Then the total impedance of the parallel circuit is approximately equal to the resistance, since the high value of X_C in a parallel branch has little effect. The phase angle of 5.7° is practically 0° because almost all of the line current is resistive.

As X_C becomes smaller, it provides more leading capacitive current in the main line. When X_C is $\frac{1}{10}$ R, practically all of the line current is the I_C component. Then the parallel circuit is practically all capacitive with a total impedance practically equal to X_C. The phase angle of 84.3° is almost 90° because the line current is mostly capacitive. Note that these conditions are opposite to the case of X_C and R in series. With X_C and R equal, their branch currents are equal and the phase angle is 45°. In summary, the branch with the highest current determines the state of the circuit. A higher capacitive current means the circuit is mostly capacitive. A higher resistive current means that the circuit is resistive.

As additional comparisons between series and parallel RC circuits, remember that

1. The series voltage drops V_R and V_C have individual values that are 90° out of phase. Therefore, V_R and V_C are added by phasors to equal the applied voltage V_T. The negative phase angle $-\theta_Z$ is between V_T and the common series current I. More series X_C allows more V_C to make the circuit more capacitive with a larger negative phase angle for V_T with respect to I.

2. The parallel branch currents I_R and I_C have individual values that are 90° out of phase. Therefore, I_R and I_C are added by phasors to equal I_T, which is the main-line current. The positive phase angle θ_I is between the line current I_T and the common parallel voltage V_A. Less parallel X_C allows more I_C to make the circuit more capacitive with a larger positive phase angle for I_T with respect to V_A.

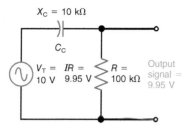

Figure 17-8 Series circuit for *RC* coupling. Small X_C compared with *R* allows practically all the applied voltage to be developed across *R* for the output, with little across *C*.

17.6 RF and AF Coupling Capacitors

In Fig. 17-8, C_C is used in the application of a coupling capacitor. Its low reactance allows developing practically all the ac signal voltage of the generator across *R*. Very little of the ac voltage is across C_C. Any dc in series with the ac signal is blocked.

The coupling capacitor is used for this application because it provides more reactance at lower frequencies, resulting in less ac voltage coupled across *R* and more across C_C. For dc voltage, all voltage is across *C* with none across *R*, since the capacitor blocks direct current. As a result, the output signal voltage across *R* includes the desired higher frequencies but not direct current or very low frequencies. This application of C_C, therefore, is called *ac coupling*.

The dividing line for C_C to be a coupling capacitor at a specific frequency can be taken as X_C one-tenth or less of the series *R*. Then the series *RC* circuit is primarily resistive. Practically all the voltage drop of the ac generator is across *R*, with little across *C*. In addition, the phase angle is almost 0°.

Typical values of a coupling capacitor for audio or radio frequencies can be calculated if we assume a series resistance of 16,000 Ω. Then X_C must be 1600 Ω or less. Typical values for C_C are listed in Table 17-3. At 100 Hz, a coupling capacitor must be 1 μF to provide 1600 Ω of reactance. Higher frequencies allow a smaller value of C_C for a

coupling capacitor having the same reactance. At 100 MHz in the VHF range, the required capacitance is only 1 pF.

Note that the C_C values are calculated for each frequency as a lower limit. At higher frequencies, the same size C_C will have less reactance than one-tenth of *R*, which improves coupling.

Choosing a Coupling Capacitor for a Circuit

As an example of using these calculations, suppose that we have the problem of determining C_C for an audio amplifier. This application also illustrates the relatively large capacitance needed with low series resistance. The *C* is to be a coupling capacitor for audio frequencies of 50 Hz and up with a series *R* of 4000 Ω. Then the required X_C is $^{4000}\!/_{10}$, or 400 Ω. To find *C* at 50 Hz,

$$C = \frac{1}{2\pi f X_C} = \frac{1}{6.28 \times 50 \times 400}$$

$$= \frac{1}{125,600} = 0.0000079$$

$$= 7.9 \times 10^{-6} \quad \text{or} \quad 7.9 \ \mu\text{F}$$

A 10-μF electrolytic capacitor would be a good choice for this application. The slightly higher capacitance value is better for coupling. The voltage rating should exceed the actual voltage across the capacitor in the circuit. Although electrolytic capacitors have a slight leakage current, they can be used for coupling capacitors in this application because of the low series resistance.

17.7 Capacitive Voltage Dividers

When capacitors are connected in series across a voltage source, the series capacitors serve as a voltage divider. Each capacitor has part of the applied voltage, and the sum of all the series voltage drops equals the source voltage.

The amount of voltage across each is inversely proportional to its capacitance. For instance, with 2 μF in series with 1 μF, the smaller capacitor has double the voltage of the larger capacitor. Assuming 120 V applied, one-third of this, or 40 V, is across the 2-μF capacitor, and two-thirds, or 80 V, is across the 1-μF capacitor.

The two series voltage drops of 40 and 80 V add to equal the applied voltage of 120 V. The phasor addition is the same as the arithmetic sum of the two voltages because they are in phase. When voltages are out of phase with each other, arithmetic addition is not possible and phasor addition becomes necessary.

AC Divider

With sine-wave alternating current, the voltage division between series capacitors can be calculated on the basis of reactance. In Fig. 17-9*a*, the total reactance is 120 Ω across the 120-V source. The current in the series circuit is 1 A. This

f	C_C	Remarks
100 Hz	1 μF	Low audio frequencies
1000 Hz	0.1 μF	Audio frequencies
10 kHz	0.01 μF	Audio frequencies
1000 kHz	100 pF	Radio frequencies
100 MHz	1 pF	Very high frequencies

Table 17-3 Coupling Capacitors with a Reactance of 1600 Ω*

* For an X_C one-tenth of a series *R* of 16,000 Ω.

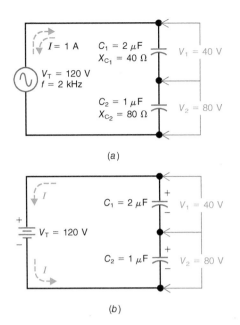

(a)

(b)

Figure 17-9 Series capacitors divide V_T inversely proportional to each C. The smaller C has more V. (a) An ac divider with more X_C for the smaller C. (b) A dc divider.

current is the same for X_{C_1} and X_{C_2} in series. Therefore, the IX_C voltage across C_1 is 40 V with 80 V across C_2.

The voltage division is proportional to the series reactances, as it is to series resistances. However, reactance is inversely proportional to capacitance. As a result, the smaller capacitance has more reactance and a greater part of the applied voltage.

DC Divider

In Fig. 17-9b, both C_1 and C_2 will be charged by the battery. The voltage across the series combination of C_1 and C_2 must equal V_T. When charging current flows, electrons repelled from the negative battery terminal accumulate on the negative plate of C_2, repelling electrons from its positive plate. These electrons flow through the conductor to the negative plate of C_1. As the positive battery terminal attracts electrons, the charging current from the positive plate of C_1 returns to the positive side of the dc source. Then C_1 and C_2 become charged in the polarity shown.

Since C_1 and C_2 are in the same series path for charging current, both have the same amount of charge. However, the

potential difference provided by the equal charges is inversely proportional to capacitance. The reason is that $Q = CV$, or $V = Q/C$. Therefore, the 1-μF capacitor has double the voltage of the 2-μF capacitor with the same charge in both.

If you measure across C_1 with a dc voltmeter, the meter reads 40 V. Across C_2, the dc voltage is 80 V. The measurement from the negative side of C_2 to the positive side of C_1 is the same as the applied battery voltage of 120 V.

If the meter is connected from the positive side of C_2 to the negative plate of C_1, however, the voltage is zero. These plates have the same potential because they are joined by a conductor of zero resistance.

The polarity marks at the junction between C_1 and C_2 indicate the voltage at this point with respect to the opposite plate of each capacitor. This junction is positive compared with the opposite plate of C_2 with a surplus of electrons. However, the same point is negative compared with the opposite plate of C_1, which has a deficiency of electrons.

In general, the following formula can be used for capacitances in series as a voltage divider:

$$V_C = \frac{C_{EQ}}{C} \times V_T \qquad \textbf{(17-7)}$$

The total equivalent capacitance, C_{EQ}, of two capacitors in series is

$$C_{EQ} = \frac{C_1 \times C_2}{C_1 + C_2}$$

The C_{EQ} is in the numerator, since it must be less than the smallest individual C with series capacitances. For the divider examples in Fig. 17-9a and b,

$$C_{EQ} = (2 \times 1)/(2 + 1) = 2/3 \ \mu F$$

$$V_1 = \frac{C_{EQ}}{C_1} \times 120 = \tfrac{2/3}{2} \times 120 = 40 \text{ V}$$

$$V_2 = \frac{C_{EQ}}{C_2} \times 120 = \tfrac{2/3}{1} \times 120 = 80 \text{ V}$$

This method applies to series capacitances as dividers for either dc or ac voltage, as long as there is no series resistance. Note that the case of capacitive dc dividers also applies to pulse circuits. Furthermore, bleeder resistors may be used across each of the capacitors to ensure more exact division.

❓ CHAPTER 17 REVIEW QUESTIONS

1. For a capacitor in a sine-wave ac circuit,
   a. V_C lags i_C by 90°.
   b. i_C leads V_C by 90°.
   c. i_C and V_C have the same frequency.
   d. all of the above.

2. In a series RC circuit,
   a. V_C leads V_R by 90°.
   b. V_C and I are in phase.
   c. V_C lags V_R by 90°.
   d. both b and c.

3. In a series RC circuit where $V_C = 15$ V and $V_R = 20$ V, how much is the total voltage, V_T?
   a. 35 V.
   b. 25 V.
   c. 625 V.
   d. 5 V.

4. A 10-Ω resistor is in parallel with a capacitive reactance of 10 Ω. The combined equivalent impedance, Z_{EQ}, of this combination is
   a. 7.07 Ω.
   b. 20 Ω.
   c. 14.14 Ω.
   d. 5 Ω.

5. In a parallel RC circuit,
   a. I_C lags I_R by 90°.
   b. I_R and I_C are in phase.
   c. I_C leads I_R by 90°.
   d. I_R leads I_C by 90°.

6. In a parallel RC circuit where $I_R = 8$ A and $I_C = 10$ A, how much is the total current, I_T?
   a. 2 A.
   b. 12.81 A.
   c. 18 A.
   d. 164 A.

7. In a series RC circuit where $R = X_C$, the phase angle, θ_Z, is
   a. +45°.
   b. −90°.
   c. 0°.
   d. −45°.

8. A 10-μF capacitor, C_1, and a 15-μF capacitor, C_2, are connected in series with a 12-Vdc source. How much voltage is across C_2?
   a. 4.8 V.
   b. 7.2 V.
   c. 12 V.
   d. 0 V.

9. The dividing line for a coupling capacitor at a specific frequency can be taken as
   a. X_C 10 or more times the series resistance.
   b. X_C equal to R.

c. X_C one–tenth or less the series resistance.
   d. none of the above.

10. A 100-Ω resistance is in series with a capacitive reactance of 75 Ω. The total impedance, Z_T, is
    a. 125 Ω.
    b. 25 Ω.
    c. 175 Ω.
    d. 15.625 kΩ.

11. In a series RC circuit,
    a. V_C and V_R are in phase.
    b. V_T and I are always in phase.
    c. V_R and I are in phase.
    d. V_R leads I by 90°.

12. In a parallel RC circuit,
    a. V_A and I_R are in phase.
    b. V_A and I_C are in phase.
    c. I_C and I_R are in phase.
    d. V_A and I_R are 90° out of phase.

13. When the frequency of the applied voltage increases in a parallel RC circuit,
    a. the phase angle, θ_I, increases.
    b. Z_{EQ} increases.
    c. Z_{EQ} decreases.
    d. both a and c.

14. When the frequency of the applied voltage increases in a series RC circuit,
    a. the phase angle, θ, becomes more negative.
    b. Z_T increases.
    c. Z_T decreases.
    d. both a and c.

15. Capacitive reactance, X_C,
    a. applies only to nonsinusoidal waveforms or dc.
    b. applies only to sine waves.
    c. applies to either sinusoidal or nonsinusoidal waveforms.
    d. is directly proportional to frequency.

CHAPTER 17 PROBLEMS

SECTION 17.1 Sine Wave v_C Lags i_C by 90°

17.1 In Fig. 17-10, what is the
   a. peak value of the capacitor voltage, V_C?
   b. peak value of the charge and discharge current, i_C?
   c. frequency of the charge and discharge current?
   d. phase relationship between V_C and i_C?

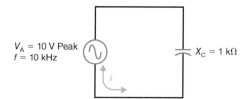

$V_A = 10$ V Peak
$f = 10$ kHz

$X_C = 1$ kΩ

Figure 17-10

17.2 In Fig. 17-10, draw the phasors representing V_C and i_C using
    a. V_C as the reference phasor.
    b. i_C as the reference phasor.

SECTION 17.2 X_C and R in Series

17.3 In Fig. 17-11, how much current, I, is flowing
    a. through the 30-Ω resistor, R?
    b. through the 40-Ω capacitive reactance, X_C?
    c. to and from the terminals of the applied voltage, V_T?

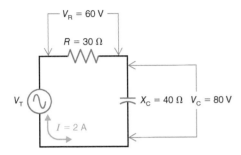

Figure 17-11

17.4 In Fig. 17-11, what is the phase relationship between
    a. I and V_R?
    b. I and V_C?
    c. V_C and V_R?

17.5 In Fig. 17-11, how much is the applied voltage, V_T?

17.6 Draw the phasor voltage triangle for the circuit in Fig. 17-11. (Use V_R as the reference phasor.)

17.7 In Fig. 17-12, solve for
    a. the resistor voltage, V_R.
    b. the capacitor voltage, V_C.
    c. the total voltage, V_T.

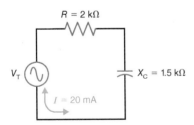

Figure 17-12

SECTION 17.3 Impedance Z Triangle

17.8 In Fig. 17-13, solve for Z_T, I, V_C, V_R, and θ_Z.

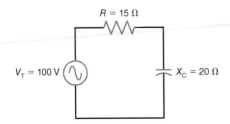

Figure 17-13

17.9 Draw the impedance triangle for the circuit in Fig. 17-13. (Use R as the reference phasor.)

17.10 In Fig. 17-14, what happens to each of the following quantities if the frequency of the applied voltage increases?
    a. X_C.
    b. Z_T.
    c. I.
    d. V_C.
    e. V_R.
    f. θ_Z.

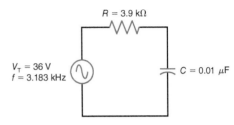

Figure 17-14

SECTION 17.4 RC Phase-Shifter Circuit

17.11 With R set to 50 kΩ in Fig. 17-15, solve for X_C, Z_T, I, V_R, V_C, and θ_Z.

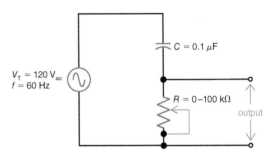

Figure 17-15

17.12 With R set to 50 kΩ in Fig. 17-15, what is the phase relationship between
    a. V_T and V_R?
    b. V_T and V_C?

17.13 Draw the phasors for V_R, V_C, and V_T in Fig. 17-15 with R set at 50 kΩ. Use V_T as the reference phasor.

SECTION 17.5 X_C and R in Parallel

17.14 In Fig. 17-16, how much voltage is across
    a. the 40-Ω resistor, R?
    b. the 30-Ω capacitive reactance, X_C?

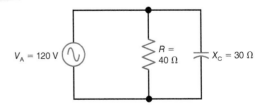

Figure 17-16

17.15 In Fig. 17-16, what is the phase relationship between
  a. V_A and I_R?
  b. V_A and I_C?
  c. I_C and I_R?

17.16 In Fig. 17-16, solve for I_R, I_C, I_T, Z_{EQ}, and θ_I.

17.17 Draw the phasor current triangle for the circuit in Fig. 17-16. (Use I_R as the reference phasor.)

17.18 In Fig. 17-17, solve for I_R, I_C, I_T, Z_{EQ}, and θ_I.

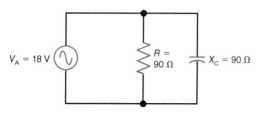

Figure 17-17

17.19 In Fig. 17-18, solve for X_C, I_R, I_C, I_T, Z_{EQ}, and θ_I.

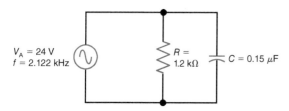

Figure 17-18

17.20 In Fig. 17-18, what happens to each of the following quantities if the frequency of the applied voltage increases?
  a. I_R.
  b. I_C.
  c. I_T.
  d. Z_{EQ}.
  e. θ_I.

SECTION 17.6 RF and AF Coupling Capacitors

17.21 In Fig. 17-19, calculate the minimum coupling capacitance, C_C, in series with the 1-kΩ resistance, R, if the frequency of the applied voltage is
  a. 1591 Hz.
  b. 15.91 kHz.

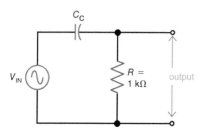

Figure 17-19

17.22 In Fig. 17-19, assume that $C_C = 0.047\ \mu F$ and $R = 1\ k\Omega$, as shown. For these values, what is the lowest frequency of the applied voltage that will provide an X_C of 100 Ω? At this frequency, what is the phase angle, θ_Z?

SECTION 17.7 Capacitive Voltage Dividers

17.23 In Fig. 17-20, calculate the following:
  a. X_{C_1}, X_{C_2}, X_{C_3}, X_{C_4}, and X_{C_T}.
  b. I.
  c. V_{C_1}, V_{C_2}, V_{C_3}, and V_{C_4}.

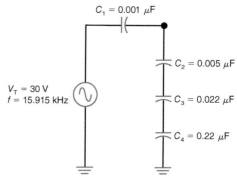

Figure 17-20

17.24 In Fig. 17-21, calculate V_{C_1}, V_{C_2}, and V_{C_3}.

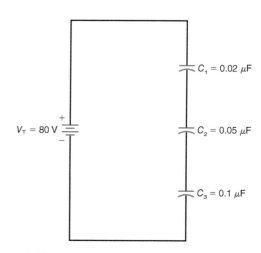

Figure 17-21

Inductance and Transformers

Learning Outcomes

After studying this chapter, you should be able to:

> *Explain* the concept of self-inductance.

> *Define* the henry unit of inductance.

> *Calculate* the inductance when the induced voltage and rate of current change are known.

> *List* the physical factors affecting the inductance of an inductor.

> *Describe* how a transformer works and *list* important transformer ratings.

> *Calculate* the currents, voltages, and impedances of a transformer circuit.

> *Identify* the different types of transformer cores.

> *Calculate* the total inductance of series and parallel connected inductors.

Inductance is the ability of a conductor to produce induced voltage when the current varies. Components manufactured to have a definite value of inductance are coils of wire, called *inductors*. The symbol for inductance is *L*, and the unit is the henry (H). Inductors are reactive components like capacitors. Changes in current through an inductor produces a self-induced voltage that opposes current changes. Inductors oppose current flow in ac circuits but do it in a different way than capacitors do. You will learn about inductors and their many forms in this chapter.

A special form of inductor is the transformer. It uses induction to produce useful effects like voltage transformation and impedance matching, as you will also see in this chapter.

18.1 Induction by Alternating Current

Induced voltage is the result of magnetic flux cutting across a conductor. This action can be produced by physical motion of either the magnetic field or the conductor. When the current in a conductor varies in amplitude, however, the variations of current and its associated magnetic field are equivalent to motion of the flux. As the current increases in value, the magnetic field expands outward from the conductor. When the current decreases, the field collapses into the conductor. As the field expands and collapses with changes of current, the flux is effectively in motion. Therefore, a varying current can produce induced voltage in the conductor without the need for motion of the conductor.

Figure 18-1 illustrates the changes in the magnetic field of a sine wave of alternating current. Since the alternating current varies in amplitude and reverses in direction, its magnetic field has the same variations. At point A, the current is zero and there is no flux. At B, the positive direction of current provides some field lines taken here in the counterclockwise direction. Point C has maximum current and maximum counterclockwise flux.

At D there is less flux than at C. Now the field is collapsing because of reduced current. At E, with zero current, there is no magnetic flux. The field can be considered as having collapsed into the wire.

The next half-cycle of current allows the field to expand and collapse again, but the directions are reversed. When the flux expands at points F and G, the field lines are clockwise, corresponding to current in the negative direction. From G to H and I, this clockwise field collapses into the wire.

The result of an expanding and collapsing field, then, is the same as that of a field in motion. This moving flux cuts across the conductor that is providing the current, producing induced voltage in the wire itself. Furthermore, any other conductor in the field, whether or not

carrying current, also is cut by the varying flux and has induced voltage.

It is important to note that induction by a varying current results from the change in current, not the current value itself. The current must change to provide motion of the flux. A steady direct current of 1000 A, as an example of a large current, cannot produce any induced voltage as long as the current value is constant. A current of 1 μA changing to 2 μA, however, does induce voltage. Also, the faster the current changes, the higher the induced voltage because when the flux moves at a higher speed, it can induce more voltage.

Since inductance is a measure of induced voltage, the amount of inductance has an important effect in any circuit in which the current changes. The inductance is an additional characteristic of a circuit beside its resistance. The characteristics of inductance are important in

1. *AC circuits.* Here the current is continuously changing and producing induced voltage. Lower frequencies of alternating current require more inductance to produce the same amount of induced voltage as a higher-frequency current. The current can have any waveform, as long as the amplitude is changing.

2. *DC circuits in which the current changes in value.* It is not necessary for the current to reverse direction. One example is a dc circuit turned on or off. When the direct current is changing between zero and its steady value, the inductance affects the circuit at the time of switching. This effect of a sudden change is called the circuit's *transient response.* A steady direct current that does not change in value is not affected by inductance, however, because there can be no induced voltage without a change in current.

18.2 Self-Inductance L

The ability of a conductor to induce voltage in itself when the current changes is its *self-inductance* or simply *inductance.* The symbol for inductance is *L*, for linkages of the magnetic flux, and its unit is the *henry* (H).

Definition of the Henry Unit

As illustrated in Fig. 18-2, one henry is the amount of inductance that allows one volt to be induced when the current changes at the rate of one ampere per second. The formula is

$$L = \frac{v_L}{di/dt} \qquad (18\text{-}1)$$

where v_L is in volts and di/dt is the current change in amperes per second.

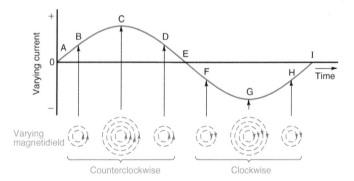

Figure 18-1 The magnetic field of an alternating current is effectively in motion as it expands and contracts with current variations.

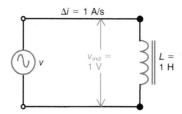

Figure 18-2 When a current change of 1 A/s induces 1 V across *L*, its inductance equals 1 H. Note that Δi is the equivalent expression of the factor *di/dt*.

The symbol *d* is used to indicate an infinitesimally small change in current with time. The factor *di/dt* for the current variation with respect to time specifies how fast the current's magnetic flux is cutting the conductor to produce v_L.

EXAMPLE 18-1

The current in an inductor changes from 12 to 16 A in 1 s. How much is the *di/dt* rate of current change in amperes per second?

Answer:

The *di* is the difference between 16 and 12, or 4 A in 1 s. Then

$$\frac{di}{dt} = 4 \text{ A/s}$$

EXAMPLE 18-2

The current in an inductor changes by 50 mA in 2 μs. How much is the *di/dt* rate of current change in amperes per second?

Answer:

$$\frac{di}{dt} = \frac{50 \times 10^{-3}}{2 \times 10^{-6}} = 25 \times 10^3$$
$$= 25,000 \text{ A/s}$$

EXAMPLE 18-3

How much is the inductance of a coil that induces 40 V when its current changes at the rate of 4 A/s?

Answer:

$$L = \frac{v_L}{di/dt} = \frac{40}{4}$$
$$= 10 \text{ H}$$

EXAMPLE 18-4

How much is the inductance of a coil that induces 1000 V when its current changes at the rate of 50 mA in 2 μs?

Answer:

For this example, the 1/*dt* factor in the denominator of Formula (18-1) can be inverted to the numerator.

$$L = \frac{v_L}{di/dt} = \frac{v_L \times dt}{di}$$
$$= \frac{1 \times 10^3 \times 2 \times 10^{-6}}{50 \times 10^{-3}}$$
$$= \frac{2 \times 10^{-3}}{50 \times 10^{-3}} = \frac{2}{50}$$
$$= 0.04 \text{ H or 40 mH}$$

Notice that the smaller inductance in Example 18-4 produces much more v_L than the inductance in Example 18-3. The very fast current change in Example 18-4 is equivalent to 25,000 A/s.

Inductance of Coils

In terms of physical construction, the inductance depends on how a coil is wound. Note the following factors.

1. A greater number of turns *N* increases *L* because more voltage can be induced. *L* increases in proportion to N^2. Double the number of turns in the same area with the same length will increase the inductance four times.

2. More area *A* enclosed by each turn increases *L*. This means that a coil with larger turns has more inductance. The *L* increases in direct proportion to *A* and as the square of the diameter of each turn.

3. The *L* increases with the permeability of the core. For an air core, μ_r is 1. With a magnetic core, L is increased by the μ_r factor because the magnetic flux is concentrated in the coil.

4. The *L* decreases with more length for the same number of turns because the magnetic field is less concentrated.

These physical characteristics of a coil are illustrated in Fig. 18-3. For a long coil, where the length is at least 10 times the diameter, the inductance can be calculated from the formula

$$L = \mu_r \times \frac{N^2 \times A}{l} \times 1.26 \times 10^{-6} \text{ H} \qquad \textbf{(18-2)}$$

where *L* is in henrys, *l* is in meters, and *A* is in square meters. The constant factor 1.26×10^{-6} is the absolute permeability of air or vacuum in SI units to calculate *L* in henrys.

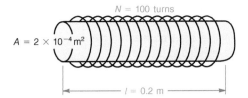

Figure 18-3 Physical factors for inductance L of a coil. See text for calculating L.

For the air-core coil in Fig. 18-3,

$$L = 1 \times \frac{10^4 \times 2 \times 10^{-4}}{0.2} \times 1.26 \times 10^{-6}$$

$$= 12.6 \times 10^{-6}\,\text{H} = 12.6\ \mu\text{H}$$

This value means that the coil can produce a self-induced voltage of 12.6 μV when its current changes at the rate of 1 A/s because $v_L = L(di/dt)$. Furthermore, if the coil has an iron core with $\mu_r = 100$, then L will be 100 times greater.

Typical Coil Inductance Values

Air-core coils for RF applications have L values in millihenrys (mH) and microhenrys (μH). A typical air-core RF inductor (called a *choke*) is shown with its schematic symbol in Fig. 18-4a. Note that

$$1\ \text{mH} = 1 \times 10^{-3}\,\text{H}$$
$$1\ \mu\text{H} = 1 \times 10^{-6}\,\text{H}$$
$$1\ \text{nH} = 1 \times 10^{-9}\,\text{H}$$

For example, an RF coil for the radio broadcast band of 535 to 1605 kHz may have an inductance L of 250 μH, or 0.250 mH. An inductor used in a cell phone at 850 MHz may have a value of 4 nH.

Iron-core inductors for the 60-Hz power line and for audio frequencies have inductance values of about 1 to 25 H. An iron-core choke is shown in Fig. 18-4b. The inductor in Fig. 18-4c has a unique core called a toroid. It uses a core of powdered iron or ferrite material. A typical toroidal inductor has an inductance in the 50- to 200-mH range.

18.3 Self-Induced Voltage v_L

The self-induced voltage across an inductance L produced by a change in current di/dt can be stated as

$$v_L = L\frac{di}{dt} \qquad \textbf{(18-3)}$$

where v_L is in volts, L is in henrys, and di/dt is in amperes per second. This formula is an inverted version of Formula (18-1), which defines inductance.

When the magnetic flux associated with the current varies the same as i, then Formula (18-3) gives the same results for calculating induced voltage. Remember also that

(a)

(b)

(c)

Figure 18-4 Typical inductors with symbols. (a) Air-core coil used as RF choke. Length is 2 in. (b) Iron-core coil used for 60 Hz. Height is 2 in. (c) Toroidal inductor.

the induced voltage across the coil is actually the result of inducing electrons to move in the conductor, so that there is also an induced current. In using Formula (18-3) to calculate v_L, multiply L by the di/dt factor.

EXAMPLE 18-5

How much is the self-induced voltage across a 4-H inductance produced by a current change of 12 A/s?

Answer:

$$v_L = L\frac{di}{dt} = 4 \times 12$$

$$= 48\ \text{V}$$

EXAMPLE 18-6

The current through a 200-mH L changes from 0 to 100 mA in 2 μs. How much is v_L?

Answer:

$$v_L = L\frac{di}{dt}$$

$$= 200 \times 10^{-3} \times \frac{100 \times 10^{-3}}{2 \times 10^{-6}}$$

$$= 10{,}000 \text{ V or } 10 \text{ kV}$$

Note the high voltage induced in the 200-mH inductance because of the fast change in current.

The induced voltage is an actual voltage that can be measured, although v_L is produced only while the current is changing. When di/dt is present for only a short time, v_L is in the form of a voltage pulse. For a sine-wave current, which is always changing, v_L is a sinusoidal voltage 90° out of phase with i_L.

18.4 How v_L Opposes a Change in Current

By Lenz's law, the induced voltage v_L must produce current with a magnetic field that opposes the change of current that induces v_L. The polarity of v_L, therefore, depends on the direction of the current variation di. When di increases, v_L has polarity that opposes the increase in current; when di decreases, v_L has opposite polarity to oppose the decrease in current.

In both cases, the change in current is opposed by the induced voltage. Otherwise, v_L could increase to an unlimited amount without the need to add any work. *Inductance, therefore, is the characteristic that opposes any change in current.* This is the reason that an induced voltage is often called a *counter emf* or *back emf.*

More details of applying Lenz's law to determine the polarity of v_L in a circuit are shown in Fig. 18-5. Note the directions carefully. In Fig. 18-5a, the electron flow is into the top of the coil. This current is increasing. By Lenz's law, v_L must have the polarity needed to oppose the increase. The induced voltage shown with the top side negative opposes the increase in current. The reason is that this polarity of v_L can produce current in the opposite direction, from minus to plus in the external circuit. Note that for this opposing current, v_L is the generator. This action tends to keep the current from increasing.

In Fig. 18-5b, the source is still producing electron flow into the top of the coil, but i is decreasing because the source voltage is decreasing. By Lenz's law, v_L must have the polarity needed to oppose the decrease in current. The induced

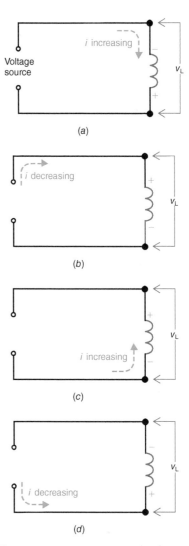

Figure 18-5 Determining the polarity of v_L that opposes the change in i. (a) The i is increasing, and v_L has the polarity that produces an opposing current. (b) The i is decreasing, and v_L produces an aiding current. (c) The i is increasing but is flowing in the opposite direction. (d) The same direction of i as in (c) but with decreasing values.

voltage shown with the top side positive now opposes the decrease. The reason is that this polarity of v_L can produce current in the same direction, tending to keep the current from decreasing.

In Fig. 18-5c, the voltage source reverses polarity to produce current in the opposite direction, with electron flow into the bottom of the coil. The current in this reversed direction is now increasing. The polarity of v_L must oppose the increase. As shown, now the bottom of the coil is made negative by v_L to produce current opposing the source current. Finally, in Fig. 18-5d, the reversed current is decreasing. This decrease is opposed by the polarity shown for v_L to keep the current flowing in the same direction as the source current.

Notice that the polarity of v_L reverses for either a reversal of direction for i or a reversal of change in di between increasing or decreasing values. When both the direction of the current and the direction of change are reversed, as in a comparison of Fig. 18-5a and d, the polarity of v_L remains unchanged.

Sometimes the formulas for induced voltage are written with minus signs to indicate that v_L opposes the change, as specified by Lenz's law. However, the negative sign is omitted here so that the actual polarity of the self-induced voltage can be determined in typical circuits.

In summary, Lenz's law states that the reaction v_L opposes its cause, which is the change in i. When i is increasing, v_L produces an opposing current. For the opposite case when i is decreasing, v_L produces an aiding current.

18.5 Mutual Inductance L_M

When the current in an inductor changes, the varying flux can cut across any other inductor nearby, producing induced voltage in both inductors. In Fig. 18-6, the coil L_1 is connected to a generator that produces varying current in the turns. The winding L_2 is not connected to L_1, but the turns are linked by the magnetic field. A varying current in L_1, therefore, induces voltage across L_1 and across L_2. If all flux of the current in L_1 links all turns of the coil L_2, each turn in L_2 will have the same amount of induced voltage as each turn in L_1. Furthermore, the induced voltage v_{L_2} can produce current in a load resistance connected across L_2.

When the induced voltage produces current in L_2, its varying magnetic field induces voltage in L_1. The two coils, L_1 and L_2, have mutual inductance, therefore, because current in one can induce voltage in the other.

The unit of mutual inductance is the henry, and the symbol is L_M. *Two coils have L_M of 1 H when a current change of 1 A/s in one coil induces 1 V in the other coil.*

The schematic symbol for two coils with mutual inductance is shown in Fig. 18-7a for an air core and in Fig. 18-7b for an iron core. Iron increases the mutual inductance, since it concentrates magnetic flux. Any magnetic lines that do not link the two coils result in *leakage flux.*

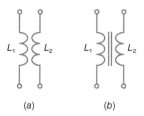

Figure 18-7 Schematic symbols for two coils with mutual inductance. (a) Air core. (b) Iron core.

Coefficient of Coupling

The fraction of total flux from one coil linking another coil is the coefficient of coupling k between the two coils. As examples, if all the flux of L_1 in Fig. 18-6 links L_2, then k equals 1, or unity coupling; if half the flux of one coil links the other, k equals 0.5. Specifically, the coefficient of coupling is

$$k = \frac{\text{flux linkages between } L_1 \text{ and } L_2}{\text{flux produced by } L_1}$$

There are no units for k, because it is a ratio of two values of magnetic flux. The value of k is generally stated as a decimal fraction, like 0.5, rather than as a percent.

The coefficient of coupling is increased by placing the coils close together, possibly with one wound on top of the other, by placing them parallel rather than perpendicular to each other, or by winding the coils on a common iron core. Several examples are shown in Fig. 18-8.

A high value of k, called *tight coupling,* allows the current in one coil to induce more voltage in the other coil. *Loose coupling,* with a low value of k, has the opposite effect. In the extreme case of zero coefficient of coupling, there is no

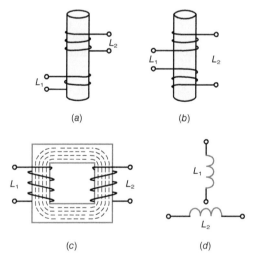

Figure 18-8 Examples of coupling between two coils linked by L_M. (a) L_1 or L_2 on paper or plastic form with air core; k is 0.1. (b) L_1 wound over L_2 for tighter coupling; k is 0.3. (c) L_1 and L_2 on the same iron core; k is 1. (d) Zero coupling between perpendicular air-core coils.

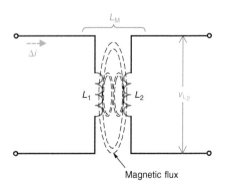

Magnetic flux

Figure 18-6 Mutual inductance L_M between L_1 and L_2 linked by magnetic flux.

mutual inductance. Two coils may be placed perpendicular to each other and far apart for essentially zero coupling to minimize interaction between the coils.

Air-core coils wound on one form have values of k equal to 0.05 to 0.3, approximately, corresponding to 5 to 30% linkage. Coils on a common iron core can be considered to have practically unity coupling, with k equal to 1. As shown in Fig. 18-8c, for both windings L_1 and L_2, practically all magnetic flux is in the common iron core. Mutual inductance is also called *mutual coupling*.

EXAMPLE 18-7

A coil L_1 produces 80 μWb of magnetic flux. Of this total flux, 60 μWb are linked with L_2. How much is k between L_1 and L_2?

Answer:

$$k = \frac{60 \ \mu Wb}{80 \ \mu Wb}$$
$$= 0.75$$

EXAMPLE 18-8

A 10-H inductance L_1 on an iron core produces 4 Wb of magnetic flux. Another coil L_2 is on the same core. How much is k between L_1 and L_2?

Answer:

Unity or 1. All coils on a common iron core have practically perfect coupling.

Calculating L_M

Mutual inductance increases with higher values for the primary and secondary inductances and tighter coupling:

$$L_M = k\sqrt{L_1 \times L_2} \qquad \textbf{(18-4)}$$

where L_1 and L_2 are the self-inductance values of the two coils, k is the coefficient of coupling, and L_M is the mutual inductance linking L_1 and L_2, in the same units as L_1 and L_2. The k factor is needed to indicate the flux linkages between the two coils.

As an example, suppose that $L_1 = 2$ H and $L_2 = 8$ H, with both coils on an iron core for unity coupling. Then the mutual inductance is

$$L_M = 1\sqrt{2 \times 8} = \sqrt{16} = 4 \text{ H}$$

The value of 4 H for L_M in this example means that when the current changes at the rate of 1 A/s in either coil, it will induce 4 V in the other coil.

EXAMPLE 18-9

Two 400-mH coils L_1 and L_2 have a coefficient of coupling k equal to 0.2. Calculate L_M.

Answer:

$$L_M = k\sqrt{L_1 \times L_2}$$
$$= 0.2\sqrt{400 \times 10^{-3} \times 400 \times 10^{-3}}$$
$$= 0.2 \times 400 \times 10^{-3}$$
$$= 80 \times 10^{-3} \text{ H or 80 mH}$$

EXAMPLE 18-10

If the two coils in Example 18-9 had a mutual inductance L_M of 40 mH, how much would k be?

Answer:

Formula (18-4) can be inverted to find k.

$$k = \frac{L_M}{\sqrt{L_1 \times L_2}}$$
$$= \frac{40 \times 10^{-3}}{\sqrt{400 \times 10^{-3} \times 400 \times 10^{-3}}}$$
$$= \frac{40 \times 10^{-3}}{400 \times 10^{-3}}$$
$$= 0.1$$

Notice that the same two coils have one-half the mutual inductance L_M because the coefficient of coupling k is 0.1 instead of 0.2.

18.6 Transformers

The transformer is an important application of mutual inductance. As shown in Fig. 18-9, a transformer has a primary winding inductance L_P connected to a voltage source that produces alternating current, and the secondary winding inductance L_S is connected across the load resistance R_L. The purpose of the transformer is to transfer power from the

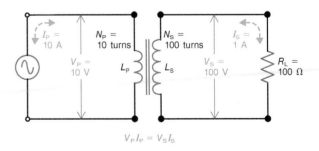

Figure 18-9 Iron-core transformer with a 1:10 turn ratio. Primary current I_P induces secondary voltage V_S, which produces current in secondary load R_L.

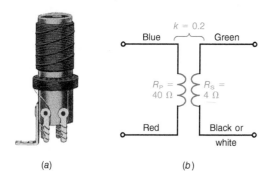

(a)   (b)

Figure 18-10   (a) Air-core RF transformer. Height is 2 in. (b) Color code and typical dc resistance of windings.

Figure 18-11   Iron-core power transformer.

primary, where the generator is connected, to the secondary, where the induced secondary voltage can produce current in the load resistance that is connected across L_S.

Although the primary and secondary are not physically connected to each other, power in the primary is coupled into the secondary by the magnetic field linking the two windings. The transformer is used to provide power for the load resistance R_L, instead of connecting R_L directly across the generator, whenever the load requires an ac voltage higher or lower than the generator voltage. By having more or fewer turns in L_S, compared with L_P, the transformer can step up or step down the generator voltage to provide the required amount of secondary voltage. Typical transformers are shown in Figs. 18-10 and 18-11. Note that a steady dc voltage cannot be stepped up or down by a transformer because a steady current cannot produce induced voltage.

Turns Ratio

The ratio of the number of turns in the primary to the number in the secondary is the turns ratio of the transformer:

$$\text{Turns ratio} = \frac{N_P}{N_S} \quad \textbf{(18-5)}$$

where N_P = number of turns in the primary and N_S = number of turns in the secondary. For example, 500 turns in the primary and 50 turns in the secondary provide a turns ratio of $^{500}/_{50}$, or 10:1, which is stated as "ten-to-one."

Voltage Ratio

With unity coupling between primary and secondary, the voltage induced in each turn of the secondary is the same as the self-induced voltage of each turn in the primary. Therefore, the voltage ratio is in the same proportion as the turns ratio:

$$\frac{V_P}{V_S} = \frac{N_P}{N_S} \quad \textbf{(18-6)}$$

When the secondary has more turns than the primary, the secondary voltage is higher than the primary voltage and the primary voltage is said to be stepped up. This principle is illustrated in Fig. 18-9 with a step-up ratio of $^{10}/_{100}$, or 1:10. When the secondary has fewer turns, the voltage is stepped down.

In either case, the ratio is in terms of the primary voltage, which may be stepped up or down in the secondary winding.

These calculations apply only to iron-core transformers with unity coupling. Air-core transformers for RF circuits (as shown in Fig. 18-10a) are generally tuned to resonance. In this case, the resonance factor is considered instead of the turns ratio.

EXAMPLE 18-11

A power transformer has 100 turns for N_P and 600 turns for N_S. What is the turns ratio? How much is the secondary voltage V_S if the primary voltage V_P is 120 V?

Answer:

The turns ratio is $^{100}/_{600}$, or 1:6. Therefore, V_P is stepped up by the factor 6, making V_S equal to 6 × 120, or 720 V.

EXAMPLE 18-12

A power transformer has 100 turns for N_P and 5 turns for N_S. What is the turns ratio? How much is the secondary voltage V_S with a primary voltage of 120 V?

Answer:

The turns ratio is $^{100}/_5$, or 20:1. The secondary voltage is stepped down by a factor of $^1/_{20}$, making V_S equal to $^{120}/_{20}$, or 6 V.

Secondary Current

By Ohm's law, the amount of secondary current equals the secondary voltage divided by the resistance in the secondary circuit. In Fig. 18-9, with a value of 100 Ω for R_L and negligible coil resistance assumed,

$$I_S = \frac{V_S}{R_L} = \frac{100 \text{ V}}{100 \text{ } \Omega} = 1 \text{ A}$$

Power in the Secondary

The power dissipated by R_L in the secondary is $I_S^2 \times R_L$ or $V_S \times I_S$, which equals 100 W in this example. The calculations are

$$P = I_S^2 \times R_L = 1 \times 100 = 100 \text{ W}$$
$$P = V_S \times I_S = 100 \times 1 = 100 \text{ W}$$

It is important to note that power used by the secondary load, such as R_L in Fig. 18-9, is supplied by the generator in the primary. How the load in the secondary draws power from the generator in the primary can be explained as follows.

With current in the secondary winding, its magnetic field opposes the varying flux of the primary current. The generator must then produce more primary current to maintain the self-induced voltage across L_P and the secondary voltage developed in L_S by mutual induction. If the secondary current doubles, for instance, because the load resistance is reduced by one-half, the primary current will also double in value to provide the required power for the secondary. Therefore, the effect of the secondary-load power on the generator is the same as though R_L were in the primary, except that the voltage for R_L in the secondary is stepped up or down by the turns ratio.

Current Ratio

With zero losses assumed for the transformer, the power in the secondary equals the power in the primary:

$$V_S I_S = V_P I_P \tag{18-7}$$

or

$$\frac{I_S}{I_P} = \frac{V_P}{V_S} \tag{18-8}$$

The current ratio is the inverse of the voltage ratio, that is, voltage step-up in the secondary means current step-down, and vice versa. The secondary does not generate power but takes it from the primary. Therefore, the current step-up or step-down is in terms of the secondary current I_S, which is determined by the load resistance across the secondary voltage. These points are illustrated by the following two examples.

EXAMPLE 18-13

A transformer with a 1:6 turns ratio has 720 V across 7200 Ω in the secondary. (a) How much is I_S? (b) Calculate the value of I_P.

Answer:

a.
$$I_S = \frac{V_S}{R_L} = \frac{720 \text{ V}}{7200 \text{ Ω}}$$
$$= 0.1 \text{ A}$$

b. With a turns ratio of 1:6, the current ratio is 6:1. Therefore,

$$I_P = 6 \times I_S = 6 \times 0.1$$
$$= 0.6 \text{ A}$$

EXAMPLE 18-14

A transformer with a 20:1 voltage step-down ratio has 6 V across 0.6 Ω in the secondary. (a) How much is I_S? (b) How much is I_P?

Answer:

a.
$$I_S = \frac{V_S}{R_L} = \frac{6 \text{ V}}{0.6 \text{ Ω}}$$
$$= 10 \text{ A}$$

b.
$$I_P = \frac{1}{20} \times I_S = \frac{1}{20} \times 10$$
$$= 0.5 \text{ A}$$

As an aid in these calculations, remember that the side with the higher voltage has the lower current. The primary and secondary V and I are in the same proportion as the number of turns in the primary and secondary.

Total Secondary Power Equals Primary Power

Figure 18-12 illustrates a power transformer with two secondary windings L_1 and L_2. There can be one, two, or more secondary windings with unity coupling to the primary as long as all the windings are on the same iron core. Each secondary winding has induced voltage in proportion to its turns ratio with the primary winding, which is connected across the 120 V source.

The secondary winding L_1 has a voltage step-up of 6:1, providing 720 V. The 7200-Ω load resistance R_1, across L_1, allows the 720 V to produce 0.1 A for I_1 in this secondary circuit. The power here is 720 V $\times$ 0.1 A = 72 W.

The other secondary winding L_2 provides voltage step-down with the ratio 20:1, resulting in 6 V across R_2. The 0.6-Ω load resistance in this circuit allows 10 A for I_2. Therefore, the power here is 6 V $\times$ 10 A, or 60 W. Since the windings have separate connections, each can have its individual values of voltage and current.

The total power used in the secondary circuits is supplied by the primary. In this example, the total secondary power is

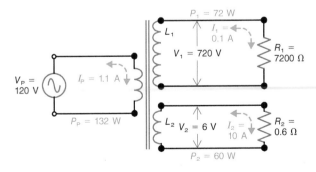

Figure 18-12 Total power used by two secondary loads R_1 and R_2 is equal to the power supplied by the source in the primary.

132 W, equal to 72 W for P_1 and 60 W for P_2. The power supplied by the 120-V source in the primary then is $72 + 60 = 132$ W.

The primary current I_P equals the primary power P_P divided by the primary voltage V_P. This is 132 W divided by 120 V, which equals 1.1 A for the primary current. The same value can be calculated as the sum of 0.6 A of primary current providing power for L_1 plus 0.5 A of primary current for L_2, resulting in the total of 1.1 A as the value of I_P.

This example shows how to analyze a loaded power transformer. The main idea is that the primary current depends on the secondary load. The calculations can be summarized as follows:

1. Calculate V_S from the turns ratio and V_P.
2. Use V_S to calculate I_S: $I_S = V_S/R_L$.
3. Use I_S to calculate P_S: $P_S = V_S \times I_S$.
4. Use P_S to find P_P: $P_P = P_S$.
5. Finally, I_P can be calculated: $I_P = P_P/V_P$.

With more than one secondary, calculate each I_S and P_S. Then add all P_S values for the total secondary power, which equals the primary power.

Autotransformers

As illustrated in Fig. 18-13, an autotransformer consists of one continuous coil with a tapped connection such as terminal 2 between the ends at terminals 1 and 3. In Fig. 18-13a, the autotransformer steps up the generator

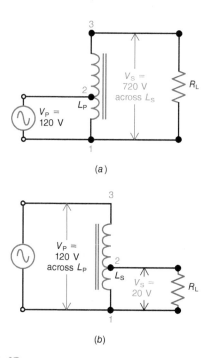

(a)

(b)

Figure 18-13 Autotransformer with tap at terminal 2 for 10 turns of the complete 60-turn winding. (a) V_P between terminals 1 and 2 stepped up across 1 and 3. (b) V_P between terminals 1 and 3 stepped down across 1 and 2.

voltage. Voltage V_P between 1 and 2 is connected across part of the total turns, and V_S is induced across all the turns. With six times the turns for the secondary voltage, V_S also is six times V_P.

In Fig. 18-13b, the autotransformer steps down the primary voltage connected across the entire coil. Then the secondary voltage is taken across less than the total turns.

The winding that connects to the voltage source to supply power is the primary, and the secondary is across the load resistance R_L. The turns ratio and voltage ratio apply the same way as in a conventional transformer having an isolated secondary winding.

Autotransformers are used often because they are compact and efficient and usually cost less since they have only one winding. Note that the autotransformer in Fig. 18-13 has only three leads, compared with four leads for the transformer in Fig. 18-9 with an isolated secondary.

Isolation of the Secondary

In a transformer with a separate winding for L_S, as in Fig. 18-9, the secondary load is not connected directly to the ac power line in the primary. This isolation is an advantage in reducing the chance of electric shock. With an autotransformer, as in Fig. 18-13, the secondary is not isolated. Another advantage of an isolated secondary is that any direct current in the primary is blocked from the secondary. Sometimes a transformer with a 1:1 turns ratio is used for isolation from the ac power line.

Transformer Efficiency

Efficiency is defined as the ratio of power out to power in. Stated as a formula,

$$\% \text{ efficiency} = \frac{P_{out}}{P_{in}} \times 100 \qquad \textbf{(18-9)}$$

For example, when the power out in watts equals one-half the power in, the efficiency is one-half, which equals 0.5×100, or 50%. In a transformer, power out is secondary power, and power in is primary power.

Assuming zero losses in the transformer, power out equals power in and the efficiency is 100%. Actual power transformers, however, have an efficiency slightly less than 100%. The efficiency is approximately 80 to 90% for transformers that have high power ratings. Transformers for higher power are more efficient because they require heavier wire, which has less resistance. In a transformer that is less than 100% efficient, the primary supplies more than the secondary power. The primary power that is lost is dissipated as heat in the transformer, resulting from I^2R in the conductors and certain losses in the core material. The R of the primary winding is generally about 10 Ω or less for power transformers.

18.7 Transformer Ratings

Like other components, transformers have voltage, current, and power ratings that must not be exceeded. Exceeding any of these ratings will usually destroy the transformer. What follows is a brief description of the most important transformer ratings.

Voltage Ratings

Manufacturers of transformers always specify the voltage rating of the primary and secondary windings. Under no circumstances should the primary voltage rating be exceeded. In many cases, the rated primary and secondary voltages are printed on the transformer. For example, consider the transformer shown in Fig. 18-14a. Its rated primary voltage is 120 V, and its secondary voltage is specified as 12.6–0–12.6, which indicates that the secondary is center-tapped. The notation 12.6–0–12.6 indicates that 12.6 V is available between the center tap connection and either outside secondary lead. The total secondary voltage available is 2 × 12.6 V or 25.2 V. In Fig. 18-14a, the black leads coming out of the top of the transformer provide connection to the primary winding. The two yellow leads coming out of the bottom of the transformer provide connection to the outer leads of the secondary winding. The bottom middle black lead connects to the center tap on the secondary winding.

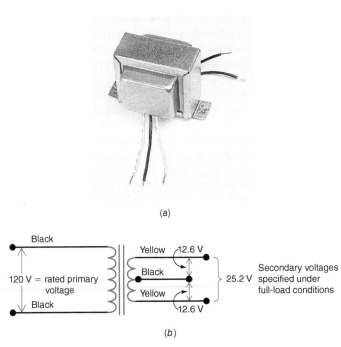

(a)

(b)

Figure 18-14 Transformer with primary and secondary voltage ratings. (a) Top black leads are primary leads. Yellow and black leads on bottom are secondary leads. (b) Schematic symbol.

Note that manufacturers may specify the secondary voltages of a transformer differently. For example, the secondary in Fig. 18-14a may be specified as 25.2 V CT, where CT indicates a center-tapped secondary. Another way to specify the secondary voltage in Fig. 18-14a would be 12.6 V each side of center.

Regardless of how the secondary voltage of a transformer is specified, the rated value is always specified under full-load conditions with the rated primary voltage applied. A transformer is considered fully loaded when the rated current is drawn from the secondary. When unloaded, the secondary voltage will measure a value that is approximately 5 to 10% higher than its rated value. Let's use the transformer in Fig. 18-14a as an example. It has a rated secondary current of 2 A. If 120 V is connected to the primary and no load is connected to the secondary, each half of the secondary will measure somewhere between 13.2 and 13.9 V approximately. However, with the rated current of 2 A drawn from the secondary, each half of the secondary will measure approximately 12.6 V.

Figure 18-14b shows the schematic diagram for the transformer in Fig. 18-14a. Notice that the colors of each lead are identified for clarity.

As you already know, transformers can have more than one secondary winding. They can also have more than one primary winding. The purpose is to allow using the transformer with more than one value of primary voltage. Figure 18-15 shows a transformer with two separate primaries and a single secondary. This transformer can be wired to work with a primary voltage of either 120 or 240 V. For either value of primary voltage, the secondary voltage is 24 V. Figure 18-15a shows the individual primary windings with phasing dots to identify those leads with the same instantaneous polarity. Figure 18-15b shows how to connect the primary windings to 240 V. Notice the connections of the leads with the phasing dots. With this connection, each half of the primary voltage is in the proper phase to provide a series-aiding connection of the induced voltages. Furthermore, the series connection of the primary windings provides a turns ratio N_P/N_S of 10:1, thus allowing a secondary voltage of 24 V. Figure 18-15c shows how to connect the primaries to 120 V. Again, notice the connection of the leads with the phasing dots. When the primary windings are in parallel, the total primary current I_p is divided evenly between the windings. The parallel connection also provides a turns ratio N_P/N_S of 5:1, thus allowing a secondary voltage of 24 V.

Figure 18-16 shows a transformer that can operate with a primary voltage of either 120 or 440 V. In this case, only one of the primary windings is used with a given primary voltage. For example, if 120 V is applied to the lower primary, the upper primary winding is not used. Conversely, if 440 V is applied to the upper primary, the lower primary winding is not used.

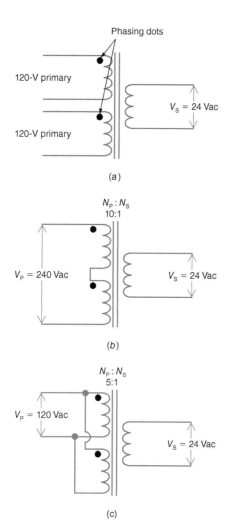

Figure 18-15 Transformer with multiple primary windings. (a) Phasing dots show primary leads with same instantaneous polarity. (b) Primary windings connected in series to work with a primary voltage of 240 V; $N_P/N_S = 10:1$. (c) Primary windings connected in parallel to work with a primary voltage of 120 V; $N_P/N_S = 5:1$.

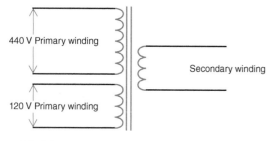

Figure 18-16 Transformer that has two primaries, which are used separately and never together.

Current Ratings

Manufacturers of transformers usually specify current ratings only for the secondary windings. The reason is quite simple. If the secondary current is not exceeded, there is no possible way the primary current can be exceeded. If the

secondary current exceeds its rated value, excessive I^2R losses will result in the secondary winding. This will cause the secondary, and perhaps the primary, to overheat, thus eventually destroying the transformer. The IR voltage drop across the secondary windings is the reason that the secondary voltage decreases as the load current increases.

EXAMPLE 18-15

In Fig. 18-14b, calculate the primary current I_P if the secondary current I_S equals its rated value of 2 A.

Answer:

Rearrange Formula (18-8) and solve for the primary current I_P.

$$I_P = \frac{V_S}{V_P} \times I_S$$

$$= \frac{25.2\ \text{V}}{120\ \text{V}} \times 2\ \text{A}$$

$$= 0.42\ \text{A} \qquad \text{or} \qquad 420\ \text{mA}$$

Power Ratings

The power rating of a transformer is the amount of power the transformer can deliver to a resistive load. The power rating is specified in volt-amperes (VA) rather than watts (W) because the power is not actually dissipated by the transformer. The product VA is called *apparent power*, since it is the power that is *apparently* used by the transformer. The unit of apparent power is VA because the watt unit is reserved for the dissipation of power in a resistance.

Assume that a power transformer whose primary and secondary voltage ratings are 120 and 25 V, respectively, has a power rating of 125 VA. What does this mean? It means that the product of the transformer's primary, or secondary, voltage and current must not exceed 125 VA. If it does, the transformer will overheat and be destroyed. The maximum allowable secondary current for this transformer can be calculated as

$$I_{S(max)} = \frac{125\ \text{VA}}{25\ \text{V}}$$

$$= 5\ \text{A}$$

The maximum allowable primary current can be calculated as

$$I_{P(max)} = \frac{125\ \text{VA}}{120\ \text{V}}$$

$$= 1.04\text{A}$$

With multiple secondary windings, the VA rating of each individual secondary may be given without any mention of the primary VA rating. In this case, the sum of all secondary VA ratings must be divided by the rated primary voltage to determine the maximum allowable primary current.

In summary, you will never overload a transformer or exceed any of its maximum ratings if you obey two fundamental rules:

1. Never apply more than the rated voltage to the primary.
2. Never draw more than the rated current from the secondary.

Frequency Ratings

All transformers have a frequency rating that must be adhered to. Typical frequency ratings for power transformers are 50, 60, and 400 Hz. A power transformer with a frequency rating of 400 Hz cannot be used at 50 or 60 Hz because it will overheat. However, many power transformers are designed to operate at either 50 or 60 Hz because many types of equipment may be sold in both Europe and the United States, where the power-line frequencies are 50 and 60 Hz, respectively. Power transformers with a 400-Hz rating are often used in aircraft because these transformers are much smaller and lighter than 50- or 60-Hz transformers having the same power rating. Modern power supplies operate at very high frequencies over the range from about 50 kHz to 4 MHz. Smaller transformers can be used, and the core is commonly a toroid.

18.8 Impedance Transformation

Transformers can be used to change or transform a secondary load impedance to a new value as seen by the primary. The secondary load impedance is said to be reflected back into the primary and is therefore called a *reflected impedance*. The reflected impedance of the secondary may be stepped up or down in accordance with the square of the transformer turns ratio.

By manipulating the relationships between the currents, voltages, and turns ratio in a transformer, an equation for the reflected impedance can be developed. This relationship is

$$Z_P = \left(\frac{N_P}{N_S}\right)^2 \times Z_S \qquad \textbf{(18-10)}$$

where Z_P = primary impedance and Z_S = secondary impedance (see Fig. 18-17). If the turns ratio N_P/N_S is greater than 1, Z_S will be stepped up in value. Conversely, if the turns ratio N_P/N_S is less than 1, Z_S will be stepped down in value. It should be noted that the term *impedance* is used rather loosely here, since the primary and secondary impedances may be purely resistive. In the discussions and examples that follow, Z_P and Z_S will be assumed to be purely resistive. The concept of reflected impedance has several practical applications in electronics.

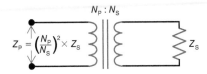

Figure 18-17 The secondary load impedance Z_S is reflected back into the primary as a new value that is proportional to the square of the turns ratio, N_P/N_S.

To find the required turns ratio when the impedance ratio is known, rearrange Formula (18-10) as follows:

$$\frac{N_P}{N_S} = \sqrt{\frac{Z_P}{Z_S}} \qquad \textbf{(18-11)}$$

EXAMPLE 18-16

Determine the primary impedance Z_P for the transformer circuit in Fig. 18-18.

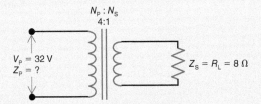

Figure 18-18 Circuit for Example 18-16.

Answer:

Use Formula (18-10). Since $Z_S = R_L$, we have

$$Z_P = \left(\frac{N_P}{N_S}\right)^2 \times R_L$$

$$= \left(\frac{4}{1}\right)^2 \times 8\ \Omega$$

$$= 16 \times 8\ \Omega$$

$$= 128\ \Omega$$

The value of 128 Ω obtained for Z_P using Formula (18-10) can be verified as follows.

$$V_S = \frac{N_S}{N_P} \times V_P$$

$$= \frac{1}{4} \times 32\ \text{V}$$

$$= 8\ \text{V}$$

$$I_S = \frac{V_S}{R_L}$$

$$= \frac{8\ \text{V}}{8\ \Omega}$$

$$= 1\ \text{A}$$

$$I_P = \frac{V_S}{V_P} \times I_S$$

$$= \frac{8\ \text{V}}{32\ \text{V}} \times 1\ \text{A}$$

$$= 0.25\ \text{A}$$

And finally,

$$Z_P = \frac{V_P}{I_P}$$

$$= \frac{32\ \text{V}}{0.25\ \text{A}}$$

$$= 128\ \Omega$$

EXAMPLE 18-17

In Fig. 18-19, calculate the turns ratio N_P/N_S that will produce a reflected primary impedance Z_P of (a) 75 Ω; (b) 600 Ω.

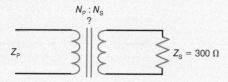

Figure 18-19 Circuit for Example 18-17.

Answer:

(a) Use Formula (18-11).

$$\frac{N_P}{N_S} = \sqrt{\frac{Z_P}{Z_S}}$$

$$= \sqrt{\frac{75\ \Omega}{300\ \Omega}}$$

$$= \sqrt{\frac{1}{4}}$$

$$= \frac{1}{2}$$

(b)
$$\frac{N_P}{N_S} = \sqrt{\frac{Z_P}{Z_S}}$$

$$= \sqrt{\frac{600\ \Omega}{300\ \Omega}}$$

$$= \sqrt{\frac{2}{1}}$$

$$= \frac{1.414}{1}$$

Impedance Matching for Maximum Power Transfer

Transformers are used when it is necessary to achieve maximum transfer of power from a generator to a load when the generator and load impedances are not the same. This application of a transformer is called *impedance matching*.

As an example, consider the amplifier and load in Fig. 18-20a. Notice that the internal resistance r_i of the amplifier is 200 Ω and the load R_L is 8 Ω. If the amplifier and load are connected directly, as in Fig. 18-20b, the load receives 1.85 W of power, which is calculated as

$$P_L = \left(\frac{V_G}{r_i + R_L}\right)^2 \times R_L$$

$$= \left(\frac{100\ \text{V}}{200\ \Omega + 8\ \Omega}\right)^2 \times 8\ \Omega$$

$$= 1.85\ \text{W}$$

To increase the power delivered to the load, a transformer can be used between the amplifier and load. This is shown

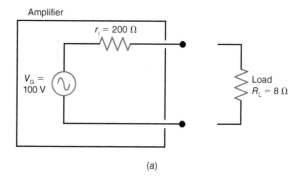

(a)

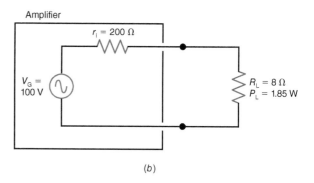

(b)

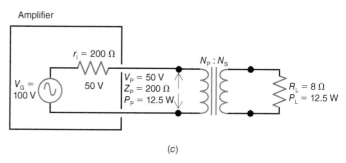

(c)

Figure 18-20 Transferring power from an amplifier to a load R_L. (a) Amplifier has $r_i = 200$ Ω and $R_L = 8$ Ω. (b) Connecting the amplifier directly to R_L. (c) Using a transformer to make the 8-Ω R_L appear like 200 Ω in the primary.

in Fig. 18-20c. We know that to transfer maximum power from the amplifier to the load, R_L must be transformed to a value equaling 200 Ω in the primary. With Z_P equaling r_i, maximum power will be delivered from the amplifier to the primary. Since the primary power P_P must equal the secondary power P_S, maximum power will also be delivered to the load R_L. In Fig. 18-20c, the turns ratio that provides a Z_P of 200 Ω can be calculated as

$$\frac{N_P}{N_S} = \sqrt{\frac{Z_P}{Z_S}}$$

$$= \sqrt{\frac{200\ \Omega}{8\ \Omega}}$$

$$= \frac{5}{1}$$

With r_i and Z_P equal, the power delivered to the primary can be calculated as

$$P_P = \left(\frac{V_G}{r_i + Z_P}\right)^2 \times Z_P$$

$$= \left(\frac{100 \text{ V}}{400 \text{ }\Omega}\right)^2 \times 200 \text{ }\Omega$$

$$= 12.5 \text{ W}$$

Since $P_P = P_S$, the load R_L also receives 12.5 W of power. As proof, calculate the secondary voltage.

$$V_S = \frac{N_S}{N_P} \times V_P$$

$$= \frac{1}{5} \times 50 \text{ V}$$

$$= 10 \text{ V}$$

(Notice that V_P is ½ V_G, since r_i and Z_P divide V_G evenly.) Next, calculate the load power P_L.

$$P_L = \frac{V_S^2}{R_L}$$

$$= \frac{10^2 \text{ V}}{8 \text{ }\Omega}$$

$$= 12.5 \text{ W}$$

Notice how the transformer has been used as an impedance matching device to obtain the maximum transfer of power from the amplifier to the load. Compare the power dissipated by R_L in Fig. 18-20b to that in Fig. 18-20c. There is a big difference between the load power of 1.85 W in Fig. 18-20b and the load power of 12.5 W in Fig. 18-20c.

18.9 Core Losses

The fact that the magnetic core can become warm, or even hot, shows that some of the energy supplied to the coil is used up in the core as heat. The two main effects are eddy-current losses and hysteresis losses.

Eddy Currents

In any inductance with an iron core, alternating current induces voltage in the core itself. Since it is a conductor, the iron core has current produced by the induced voltage. This current is called an *eddy current* because it flows in a circular path through the cross section of the core, as illustrated in Fig. 18-21.

The eddy currents represent wasted power dissipated as heat in the core. Note in Fig. 18-21 that the eddy-current flux opposes the coil flux, so that more current is required in the coil to maintain its magnetic field. The higher the frequency of the alternating current in the inductance, the greater the eddy-current loss.

Eddy currents can be induced in any conductor near a coil with alternating current, not only in its core. For instance,

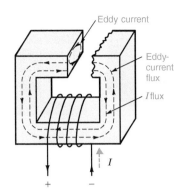

Figure 18-21 Cross-sectional view of iron core showing eddy currents.

a coil has eddy-current losses in a metal cover. In fact, the technique of induction heating is an application of heat resulting from induced eddy currents.

RF Shielding

The reason that a coil may have a metal cover, usually copper or aluminum, is to provide a shield against the varying flux of RF current. In this case, the shielding effect depends on using a good conductor for the eddy currents produced by the varying flux, rather than magnetic materials used for shielding against static magnetic flux.

The shield cover not only isolates the coil from external varying magnetic fields but also minimizes the effect of the coil's RF current for external circuits. The reason that the shield helps both ways is the same, as the induced eddy currents have a field that opposes the field that is inducing the current. Note that the clearance between the sides of the coil and the metal should be equal to or greater than the coil radius to minimize the effect of the shield in reducing the inductance.

Hysteresis Losses

Another loss factor present in magnetic cores is hysteresis, although hysteresis losses are not as great as eddy-current losses. The hysteresis losses result from the additional power needed to reverse the magnetic field in magnetic materials in the presence of alternating current. The greater the frequency, the more hysteresis losses.

Air-Core Coils

Note that air has practically no losses from eddy currents or hysteresis. However, the inductance for small coils with an air core is limited to low values in the microhenry or millihenry range.

18.10 Types of Cores

To minimize losses while maintaining high flux density, the core can be made of laminated steel layers insulated from each other. Insulated powdered-iron granules and

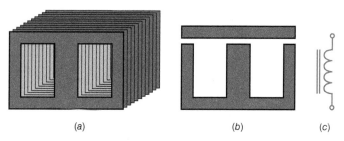

Figure 18-22 Laminated iron core. (*a*) Shell-type construction. (*b*) E- and I-shaped laminations. (*c*) Symbol for iron core.

Figure 18-23 Most toroid inductors use ferrite cores. Some also use powdered iron cores at radio frequencies.

ferrite materials can also be used. These core types are illustrated in Figs. 18-22 and 18-23. The purpose is to reduce the amount of eddy currents. The type of steel itself can help reduce hysteresis losses.

Laminated Core

Figure 18-22*a* shows a shell-type core formed with a group of individual laminations. Each laminated section is insulated by a very thin coating of iron oxide, silicon steel, or varnish. The insulating material increases the resistance in the cross section of the core to reduce the eddy currents but allows a low-reluctance path for high flux density around the core. Transformers for audio frequencies and 60-Hz power are generally made with a laminated iron core.

Powdered-Iron Core

Powdered iron is generally used to reduce eddy currents in the iron core of an inductance for radio frequencies. It consists of individual insulated granules pressed into one solid form. Some toroidal cores use powdered iron.

Ferrite Core

Ferrites are synthetic ceramic materials that are ferromagnetic. They provide high values of flux density, like iron, but have the advantage of being insulators. Therefore, a ferrite core can be

used for high frequencies with minimum eddy-current losses. Most toroidal cores use ferrite. See Fig. 18-23.

18.11 Variable Inductance

The inductance of a coil can be varied by one of the methods illustrated in Fig. 18-24. In Fig. 18-24*a*, more or fewer turns can be used by connection to one of the taps on the coil. Also, in Fig. 18-24*b*, a slider contacts the coil to vary the number of turns used. These methods are for large coils.

Figure 18-24*c* shows the schematic symbol for a coil with a slug of powdered iron or ferrite. The dotted lines indicate that the core is not solid iron. The arrow shows that the slug is variable. Usually, an arrow at the top means that the adjustment is at the top of the coil. An arrow at the bottom, pointing down, shows that the adjustment is at the bottom.

The symbol in Fig. 18-24*d* is a *variometer,* which is an arrangement for varying the position of one coil within the other. The total inductance of the series-aiding coils is minimum when they are perpendicular.

For any method of varying *L,* the coil with an arrow in Fig. 18-24*e* can be used. However, an adjustable slug is usually shown as in Fig. 18-24*c*.

A practical application of variable inductance is the *Variac.* The Variac is an autotransformer with a variable tap to change the turns ratio. The output voltage in the

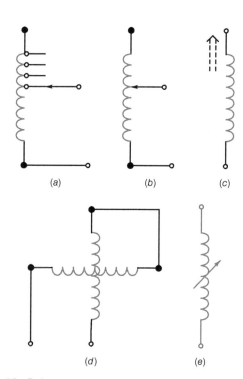

Figure 18-24 Methods of varying inductance. (*a*) Tapped coil. (*b*) Slider contact. (*c*) Adjustable slug. (*d*) Variometer. (*e*) Symbol for variable *L*.

secondary can be varied from 0 to approximately 140 V, with input from the 120-V, 60-Hz power line. One use is to test equipment with voltage above or below the normal line voltage.

The Variac is plugged into the power line, and the equipment to be tested is plugged into the Variac. Note that the power rating of the Variac should be equal to or more than the power used by the equipment being tested.

18.12 Inductances in Series or Parallel

As shown in Fig. 18-25, the total inductance of coils connected in series is the sum of the individual L values, as for series R. Since the series coils have the same current, the total induced voltage is a result of the total number of turns. Therefore, total series inductance is,

$$L_T = L_1 + L_2 + L_3 + \ldots + \text{etc.} \qquad \textbf{(18-12)}$$

where L_T is in the same units of inductance as L_1, L_2, and L_3. This formula assumes no mutual induction between the coils.

EXAMPLE 18-18

Inductance L_1 in Fig. 18-25 is 5 mH and L_2 is 10 mH. How much is L_T?

Answer:

$$L_T = 5 \text{ mH} + 10 \text{ mH} = 15 \text{ mH}.$$

With coils connected in parallel, the equivalent inductance is calculated from the reciprocal formula

$$L_{EQ} = \cfrac{1}{\cfrac{1}{L_1} + \cfrac{1}{L_2} + \cfrac{1}{L_3} + \ldots + \text{etc.}} \qquad \textbf{(18-13)}$$

Again, no mutual induction is assumed, as illustrated in Fig. 18-26.

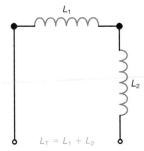

Figure 18-25 Inductances L_1 and L_2 in series without mutual coupling.

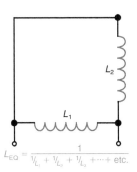

Figure 18-26 Inductances L_1 and L_2 in parallel without mutual coupling.

EXAMPLE 18-19

Inductances L_1 and L_2 in Fig. 18-26 are each 8 mH. How much is L_{EQ}?

Answer:

$$L_{EQ} = \frac{1}{\frac{1}{8} + \frac{1}{8}}$$

$$= 4 \text{ mH}$$

All shortcuts for calculating parallel R can be used with parallel L, since both are based on the reciprocal formula. In this example, L_{EQ} is $\frac{1}{2} \times 8 = 4$ mH.

Series Coils with L_M

This depends on the amount of mutual coupling and on whether the coils are connected series-aiding or series-opposing. *Series aiding* means that the common current produces the same direction of magnetic field for the two coils. The *series-opposing* connection results in opposite fields.

The coupling depends on the coil connections and direction of winding. Reversing either one reverses the field. Inductances L_1 and L_2 with the same direction of winding are connected series-aiding in Fig. 18-27a. However, they are series-opposing in Fig. 18-27b because L_1 is connected to the opposite end of L_2. To calculate the total inductance of two coils that are series-connected and have mutual inductance,

$$L_T = L_1 + L_2 \pm 2L_M \qquad \textbf{(18-14)}$$

The mutual inductance L_M is plus, increasing the total inductance, when the coils are series-aiding, or minus when they are series-opposing to reduce the total inductance.

Note the phasing dots above the coils in Fig. 18-27. Coils with phasing dots at the same end have the same direction of winding. When current enters the dotted ends for two coils, their fields are aiding and L_M has the same sense as L.

How to Measure L_M

Formula (18-14) provides a method of determining the mutual inductance between two coils L_1 and L_2 of known

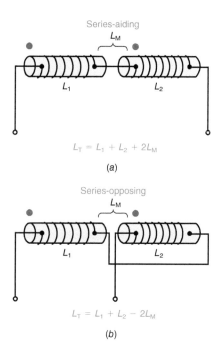

Series-aiding

L_M

L_1 L_2

$L_T = L_1 + L_2 + 2L_M$

(a)

Series-opposing

L_M

L_1 L_2

$L_T = L_1 + L_2 - 2L_M$

(b)

Figure 18-27 Inductances L_1 and L_2 in series but with mutual coupling L_M. (a) Aiding magnetic fields. (b) Opposing magnetic fields.

inductance. First, the total inductance is measured for the series-aiding connection. Let this be L_{T_a}. Then the connections to one coil are reversed to measure the total inductance for the series-opposing coils. Let this be L_{T_o}. Then

$$L_M = \frac{L_{T_a} - L_{T_o}}{4} \qquad \textbf{(18-15)}$$

When the mutual inductance is known, the coefficient of coupling k can be calculated from the fact that $L_M = k\sqrt{L_1 L_2}$.

EXAMPLE 18-20

Two series coils, each with an L of 250 μH, have a total inductance of 550 μH connected series-aiding and 450 μH series-opposing. (a) How much is the mutual inductance L_M between the two coils? (b) How much is the coupling coefficient k?

Answer:

a.

$$L_M = \frac{L_{T_a} - L_{T_o}}{4}$$

$$= \frac{550 - 450}{4} = \frac{100}{4}$$

$$= 25\,\mu\text{H}$$

b.

$$L_M = k\sqrt{L_1 L_2}, \quad \text{or}$$

$$k = \frac{L_M}{\sqrt{L_1 L_2}} = \frac{25}{\sqrt{250 \times 250}}$$

$$= \frac{25}{250} = \frac{1}{10}$$

$$= 0.1$$

Coils may also be in parallel with mutual coupling. However, the inverse relations with parallel connections and the question of aiding or opposing fields make this case complicated. Actually, it would hardly ever be used.

18.13 Stray Capacitive and Inductive Effects

Stray capacitive and inductive effects can occur in all circuits with all types of components. A capacitor has a small amount of inductance in the conductors. A coil has some capacitance between windings. A resistor has a small amount of inductance and capacitance. After all, physically a capacitance is simply an insulator between two conductors having a difference of potential. An inductance is basically a conductor carrying current.

Actually, though, these stray effects are usually quite small, compared with the concentrated or lumped values of capacitance and inductance. Typical values of stray capacitance may be 1 to 10 pF, whereas stray inductance is usually a fraction of 1 μH. For very high radio frequencies, however, when small values of L and C must be used, the stray effects become important. As another example, any wire cable has capacitance between the conductors.

A practical case of problems caused by stray L and C is a long cable used for RF signals. If the cable is rolled in a coil to save space, a serious change in the electrical characteristics of the line will take place. Specifically, for twin-lead or coaxial cable feeding the antenna input to a television receiver, the line should not be coiled because the added L or C can affect the signal. Any excess line should be cut off, leaving the little slack that may be needed. This precaution is not so important with audio cables.

Stray Circuit Capacitance

The wiring and components in a circuit have capacitance to the metal chassis. This stray capacitance C_S is typically 5 to 10 pF. To reduce C_S, the wiring should be short with the leads and components placed high off the chassis. Sometimes, for very high frequencies, stray capacitance is included as part of the circuit design. Then changing the placement of components or wiring affects the circuit operation. Such critical *lead dress* is usually specified in the manufacturer's service notes.

Stray Inductance

Although practical inductors are generally made as coils, all conductors have inductance. The amount of L is $v_L/(di/dt)$, as with any inductance producing induced voltage when the current changes. The inductance of any wiring not included in the conventional inductors can be considered stray inductance. In most cases, stray inductance is very small; typical values are less than 1 μH. For high radio frequencies, though, even a small L can have an appreciable inductive effect.

One source of stray inductance is connecting leads. A wire 0.04 in. in diameter and 4 in. long has an L of approximately 0.1 μH. At low frequencies, this inductance is negligible. However, consider the case of RF current, where i varies from 0- to 20-mA peak value, in the short time of 0.025 μs, for a quarter-cycle of a 10-MHz sine wave. Then v_L equals 80 mV, which is an appreciable inductive effect. This is one reason that connecting leads must be very short in RF circuits.

As another example, wire-wound resistors can have appreciable inductance when wound as a straight coil. This is why carbon resistors are preferred for minimum stray inductance in RF circuits. However, noninductive wire-wound resistors can also be used. These are wound so that adjacent turns have current in opposite directions and the magnetic fields oppose each other to cancel the inductance. Another application of this technique is twisting a pair of connecting leads to reduce the inductive effect.

Inductance of a Capacitor

Capacitors with coiled construction, particularly paper and electrolytic capacitors, have some internal inductance. The larger the capacitor, the greater its series inductance. Mica and ceramic capacitors have very little inductance, however, which is why they are generally used for radio frequencies.

For use above audio frequencies, the rolled-foil type of capacitor must have noninductive construction. This means that the start and finish of the foil winding must not be the terminals of the capacitor. Instead, the foil windings are offset. Then one terminal can contact all layers of one foil at one edge, and the opposite edge of the other foil contacts the second terminal. Most rolled-foil capacitors, including the paper and film types, are constructed this way.

Distributed Capacitance of a Coil

As illustrated in Fig. 18-28, a coil has distributed capacitance C_d between turns. Note that each turn is a conductor separated from the next turn by an insulator, which is the definition of capacitance. Furthermore, the potential of each turn is different from the next, providing part of the total voltage as a potential difference to charge C_d. The result then is the equivalent circuit shown for an RF coil. The L is the inductance and R_e its internal effective ac resistance in series with L, and the total distributed capacitance C_d for all turns is across the entire coil.

Special methods for minimum C_d include *space-wound* coils, where the turns are spaced far apart; the honeycomb

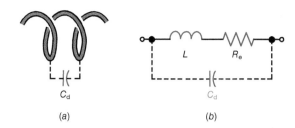

Figure 18-28 Equivalent circuit of an RF coil. (*a*) Distributed capacitance C_d between turns of wire. (*b*) Equivalent circuit.

or *universal* winding, with the turns crossing each other at right angles; and the *bank winding,* with separate sections called *pies*. These windings are for RF coils. In audio and power transformers, a grounded conductor shield, called a *Faraday screen,* is often placed between windings to reduce capacitive coupling.

Reactive Effects in Resistors

As illustrated by the high-frequency equivalent circuit in Fig. 18-29, a resistor can include a small amount of inductance and capacitance. The inductance of carbon-composition resistors is usually negligible. However, approximately 0.5 pF of capacitance across the ends may have an effect, particularly with large resistances used for high radio frequencies. Wire-wound resistors definitely have enough inductance to be evident at radio frequencies. However, special resistors are available with double windings in a noninductive method based on cancellation of opposing magnetic fields.

Capacitance of an Open Circuit

An open switch or a break in a conducting wire has capacitance C_O across the open. The reason is that the open consists of an insulator between two conductors. With a voltage source in the circuit, C_O charges to the applied voltage. Because of the small C_O, of the order of picofarads, the capacitance charges to the source voltage in a short time. This charging of C_O is the reason that an open series circuit has the applied voltage across the open terminals. After a momentary flow of charging current, C_O charges to the applied voltage and stores the charge needed to maintain this voltage.

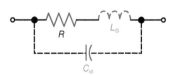

Figure 18-29 High-frequency equivalent circuit of a resistor.

AC Motors

There are two basic types of motors, dc and ac. You learned about the importance of motors and dc motor operation in the Chap. 11 sidebar. This sidebar introduces you to ac motors.

Generally speaking, ac motors are more widely used than dc motors. With ac power so common to home, business, and industry, it makes sense just to plug the motor into the existing ac source.

AC motors are also used where higher power is needed. While most dc motors have a power of less than several horsepower (hp), ac motors are available in power levels of hundreds, even thousands, of horsepower. Furthermore, an ac motor is usually smaller than a dc motor of the same horsepower. In addition, ac motors are typically more reliable and require less maintenance than a dc motor. The primary reason for this is that some ac motors do not have brushes and commutators, which are the most common failures.

One primary advantage of dc over ac motors is variable speed. It is easier to change the speed of a dc motor than the speed of an ac motor. Most ac motors are used in constant-speed applications. This advantage of dc over ac motors has mostly disappeared over the years as electronic drive circuits using high-power semiconductor devices like IGBTs (insulated gate bipolar transistors) make it possible to generate an ac voltage of any desired frequency. The circuits are called variable-frequency drives.

There are three basic types of ac motors: series or universal, induction, and synchronous. What follows is a brief look at each.

Universal Motor

The simplest and most widely used ac motor is the so-called universal motor. Also known as a *series motor,* it actually works with either dc or ac. Most ac universal motors are rated at less than 1 hp. Typical applications are appliances, fans, and drills. Figure S18-1 shows the basic universal motor configuration. The stator and rotor coils are in series. As the ac varies, it produces opposing magnetic fields in each of the coils producing the mechanical force that creates rotary motion. The universal motor has a commutator and set of brushes to keep the magnetic fields in the rotor and stator in opposition to sustain rotation.

Induction Motor

An induction motor works like a transformer. A magnetic field produced in the stator induces a voltage into a rotor winding. The resulting current flow produces a magnetic

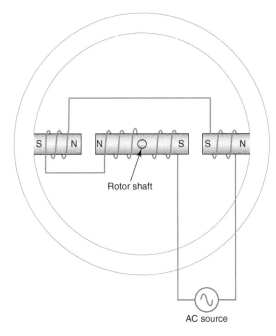

Figure S18-1   A universal ac or dc motor.

field that interacts with the magnetic field of the stator to produce rotation. The key to this operation is making the stator field rotate from pole to pole to produce the rotary motion. The magnetic field in the rotor essentially "chases" the stator field as the fields attract or repel one another. This rotary field generation is easiest to accomplish in a three-phase motor.

Figure S18-2*a* and *b* shows the stator poles and windings of a three-phase inductor motor. The windings are connected in a Y-configuration, as indicated in Fig. S18-2*c*. You can visualize how the fields rotate if you remember that the three-phase ac voltages applied to the windings are 120° out of phase with one another, as Fig. S18-2*d* illustrates.

The rotor of an induction motor is a group of windings arranged around a magnetic core. One common type is the squirrel-cage rotor shown in Fig. S18-3. The windings are the individual copper bars which, when shorted by the shorting rings, produce multiple shorted single loops. These loops act as secondary windings on a transformer. The magnetic fields of the stator induce voltages into the loops which, in turn, produce the magnetic fields that sustain rotation with the push and pull of the stator fields.

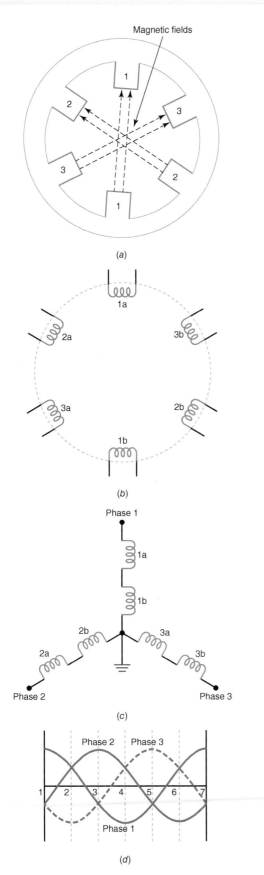

Figure S18-2 Stator of a three-phase induction motor.
(a) Poles. (b) coils. (c) Y-connected stator windings.
(d) applied ac voltages.

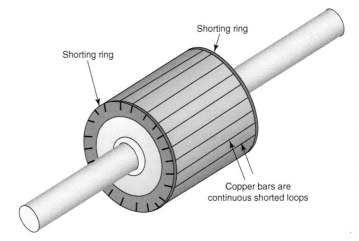

Figure S18-3 Induction motor rotor.

Single-Phase Induction Motor

A three-phase induction motor is overkill for many applications because of its power and the need for a three-phase source. But there are single-phase induction motors. In fact, the single-phase induction motor is the most commonly used appliance motor. It is used in air-conditioning and heating systems, washers, dryers, dishwashers, fans, and many other common devices. Without the three-phase voltages such motors are difficult to start. The three-phase voltages keep the fields rotating as well as the rotor in a three-phase induction motor. With a single fixed stator field that does not rotate, there is no rotation until there is relative motion between rotor and stator. What the rotor needs is a kick-start to get it moving, then the attraction and repulsion of the rotor and stator fields will sustain rotation. A manual start is impractical.

The solution to this problem is to introduce a starting winding and a phase shift, as shown in Fig. S18-4. The starting winding is placed at a 90° angle to the main stator winding and connected in series with a phase-shift capacitor and a starting switch. When the motor is turned on, the switch is closed. The ac is applied to both the starting winding and

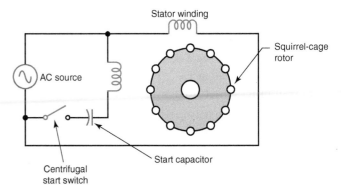

Figure S18-4 A single-phase induction motor with a starting capacitor and switch.

the stator winding in parallel. The capacitor introduces just enough phase shift to create a temporary rotational effect that gets the rotor going. Once the rotor is up to speed, the switch contacts are automatically opened by centrifugal force.

Synchronous Motor

A synchronous motor is similar to an induction motor in construction except that the rotor has a winding that is powered by a dc source. For that reason, the rotor has a commutator and brushes. Synchronous motors operate from three-phase ac, and the rotor is commonly a squirrel-cage design. When voltage is first applied, the dc is not connected. Instead the motor acts like a standard induction motor and begins to turn. Once the rotor is up to speed, a centrifugal switch turns the dc on. The strong dc field produces more torque for turning.

Synchronous motors produce the greatest amount of power of any type of ac motor. Some synchronous motors have ratings of up to several thousand hp. Another benefit of the synchronous motor is that its speed is not affected by connecting a load as it is with an induction motor. The speed of an induction motor slows as a load is applied. The speed of the induction motor rotor is always a bit less than the rotational speed of the stator fields in order for there to be relative motion between stator and rotor windings as required for magnetic induction to take place. The section to follow shows how the speed of an ac motor is determined.

AC Motor Speed

The speed, S, in revolutions per minute (rpm) of an ac motor is determined by the number of stator pole pairs, P, and the frequency, f, of the applied ac.

$$S = \frac{120f}{P}$$

For three pole pairs and 60-Hz ac the speed is

$$S = \frac{120 \times 60}{3} = 2400 \text{ rpm}$$

Most ac motors operate from the main ac power lines that supply 60-Hz voltage. Special electronic drive circuits are also used to produce variable-frequency ac so that the speed can be changed. Electric and hybrid cars use variable-speed induction motors.

❓ CHAPTER 18 REVIEW QUESTIONS

1. The unit of inductance is the
   a. henry.
   b. farad.
   c. ohm.
   d. volt-ampere.

2. The inductance, L, of an inductor is affected by
   a. number of turns.
   b. area enclosed by each turn.
   c. permeability of the core.
   d. all of the above.

3. A transformer cannot be used to
   a. step up or down an ac voltage.
   b. step up or down a dc voltage.
   c. match impedances.
   d. transfer power from primary to secondary.

4. The interaction between two inductors physically close together is called
   a. counter emf.
   b. self-inductance.
   c. mutual inductance.
   d. hysteresis.

5. If the secondary current in a step-down transformer increases, the primary current will
   a. not change.
   b. increase.
   c. decrease.
   d. drop a little.

6. Inductance can be defined as the characteristic that
   a. opposes a change in current.
   b. opposes a change in voltage.
   c. aids or enhances any change in current.
   d. stores electric charge.

7. If the number of turns in a coil is doubled in the same length and area, the inductance, L, will
   a. double.
   b. quadruple.
   c. stay the same.
   d. be cut in half.

8. An open coil has
   a. zero resistance and zero inductance.
   b. infinite inductance and zero resistance.
   c. normal inductance but infinite resistance.
   d. infinite resistance and zero inductance.

9. Two 10-H inductors are connected in series aiding and have a mutual inductance, L_M, of 0.75 H. The total inductance, L_T, of this combination is
   a. 18.5 H.
   b. 20.75 H.
   c. 21.5 H.
   d. 19.25 H.

10. How much is the self-induced voltage, V_L, across a 100-mH inductor produced by a current change of 50,000 A/s?
    a. 5 kV.
    b. 50 V.
    c. 5 MV.
    d. 500 kV.

11. The measured voltage across an unloaded secondary of a transformer is usually
    a. the same as the rated secondary voltage.
    b. 5 to 10% higher than the rated secondary voltage.
    c. 50% higher than the rated secondary voltage.
    d. 5 to 10% lower than the rated secondary voltage.

12. A laminated iron-core transformer has reduced eddy-current losses because
    a. the laminations are stacked vertically.
    b. more wire can be used with less dc resistance.
    c. the magnetic flux is in the air gap of the core.
    d. the laminations are insulated from each other.

13. How much is the inductance of a coil that induces 50 V when its current changes at the rate of 500 A/s?
    a. 100 mH.
    b. 1 H.
    c. 100 μH.
    d. 10 μH.

14. A 100-mH inductor is in parallel with a 150-mH and a 120-mH inductor. Assuming no mutual inductance between coils, how much is L_{EQ}?
    a. 400 mH.
    b. 370 mH.
    c. 40 mH.
    d. 80 mH.

15. A 400-μH coil is in series with a 1.2-mH coil without mutual inductance. How much is L_T?
    a. 401.2 μH.
    b. 300 μH.
    c. 160 μH.
    d. 1.6 mH.

16. A step-down transformer has a turns ratio, $\frac{N_P}{N_S}$, of 4:1. If the primary voltage, V_P, is 120 Vac, how much is the secondary voltage, V_S?
    a. 480 Vac.
    b. 120 Vac.
    c. 30 Vac.
    d. It cannot be determined.

17. If an iron-core transformer has a turns ratio, $\frac{N_P}{N_S}$, of 3:1 and $Z_S = 16\ \Omega$, how much is Z_P?
    a. 48 Ω.
    b. 144 Ω.
    c. 1.78 Ω.
    d. 288 Ω.

18. How much is the induced voltage, V_L, across a 5-H inductor carrying a steady dc current of 200 mA?
    a. 0 V.
    b. 1 V.
    c. 100 kV.
    d. 120 Vac.

19. The secondary current, I_S, in an iron-core transformer equals 1.8 A. If the turns ratio, $\frac{N_P}{N_S}$, equals 3:1, how much is the primary current, I_P?
    a. $I_P = 1.8$ A.
    b. $I_P = 600$ mA.
    c. $I_P = 5.4$ A.
    d. none of the above.

20. For a coil, the dc resistance, r_i, and inductance, L, are
    a. in parallel.
    b. infinite.
    c. the same thing.
    d. in series.

CHAPTER 18 PROBLEMS

SECTION 18.1 Induction by Alternating Current

18.1 Which can induce more voltage in a conductor, a steady dc current of 10 A or a small current change of 1 to 2 mA?

18.2 Examine the sine wave of alternating current in Fig. 18-1. Identify the points on the waveform (using the letters A–I) where the rate of current change, $\frac{di}{dt}$, is
    a. greatest.
    b. zero.

18.3 Which will induce more voltage across a conductor, a low-frequency alternating current or a high-frequency alternating current?

SECTION 18.2 Self-Inductance *L*

18.4 Convert the following current changes, $\frac{di}{dt}$, to amperes per second:
   a. 0 to 3 A in 2 s.
   b. 0 to 50 mA in 5 μs.

18.5 How much inductance, *L*, will be required to produce an induced voltage, V_L, of 15 V for each of the $\frac{di}{dt}$ values listed in Prob. 18-4?

18.6 How much is the inductance, *L*, of a coil that induces 75 V when the current changes at the rate of 2500 A/s?

18.7 Calculate the inductance, *L*, for the following long coils: (Note: 1 m = 100 cm and 1 m^2 = 10,000 cm^2.)
   a. air core, 20 turns, area 3.14 cm^2, length 25 cm.
   b. same coil as step a with ferrite core having a μ_r of 5000.
   c. iron core with μ_r of 2000, 100 turns, area 5 cm^2, length 10 cm.

18.8 What is another name for an RF inductor?

SECTION 18.3 Self-Induced Voltage V_L

18.9 How much is the self-induced voltage across a 5-H inductance produced by a current change of 100 to 200 mA in 1 ms?

18.10 How much is the self-induced voltage across a 33-mH inductance when the current changes at the rate of 1500 A/s?

18.11 Calculate the self-induced voltage across a 100-mH inductor for the following values of $\frac{di}{dt}$:
   a. 100 A/s.
   b. 1000 A/s.

SECTION 18.5 Mutual Inductance L_M

18.12 A coil, L_1, produces 40 μWb of magnetic flux. A coil, L_2, nearby, is linked with L_1 by 30 μWb of magnetic flux. What is the value of *k*?

18.13 Two inductors, L_1 and L_2, have a coefficient of coupling, *k*, equal to 0.5. L_1 = 100 mH and L_2 = 150 mH. Calculate L_M.

18.14 What is the assumed value of *k* for an iron-core transformer?

SECTION 18.6 Transformers

18.15 In Fig. 18-30, solve for
   a. the secondary voltage, V_S.
   b. the secondary current, I_S.
   c. the secondary power, P_{sec}.
   d. the primary power, P_{pri}.
   e. the primary current, I_P.

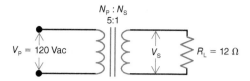

Figure 18-30

18.16 In Fig. 18-31, solve for
   a. the turns ratio $\frac{N_P}{N_S}$.
   b. the secondary current, I_S.
   c. the primary current, I_P.

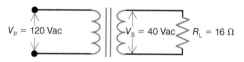

Figure 18-31

18.17 In Fig. 18-32, what turns ratio, $\frac{N_P}{N_S}$, is required to obtain a secondary voltage of
   a. 600 Vac?
   b. 24 Vac?

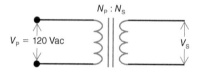

Figure 18-32

18.18 A transformer delivers 400 W to a load connected to its secondary. If the input power to the primary is 500 W, what is the efficiency of the transformer?

18.19 Explain the advantages of a transformer having an isolated secondary.

SECTION 18.7 Transformer Ratings

18.20 How is the power rating of a transformer specified?

18.21 To avoid overloading a transformer, what two rules should be observed?

18.22 What is the purpose of phasing dots on the schematic symbol of a transformer?

18.23 Refer to the transformer in Fig. 18-33. How much voltage would a DMM measure across the following secondary leads if the secondary current is 2 A?
   a. V_{AB}.
   b. V_{AC}.
   c. V_{BC}.

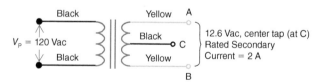

Figure 18-33

18.24 How much is the primary current, I_P, in Fig. 18-33 if the secondary current is 2 A?

SECTION 18.8 Impedance Transformation

18.25 In Fig. 18-34, calculate the primary impedance, Z_P, for a turns ratio $\dfrac{N_P}{N_S}$ of
  a. 2:1.
  b. 1:3.16.

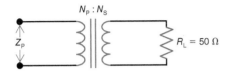

Figure 18-34

18.26 In Fig. 18-35, calculate the required turns ratio $\dfrac{N_P}{N_S}$ for
  a. $Z_P = 50\ \Omega$ and $R_L = 600\ \Omega$.
  b. $Z_P = 200\ \Omega$ and $R_L = 10\ \Omega$.

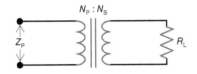

Figure 18-35

18.27 In Fig. 18-36, what turns ratio, $\dfrac{N_P}{N_S}$, will provide maximum transfer of power from the amplifier to the 4-Ω speaker?

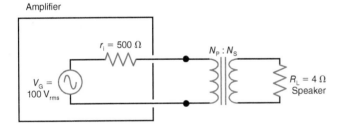

Figure 18-36

18.28 Using your answer from Prob. 18-27, calculate
  a. the primary impedance, Z_P.
  b. the power delivered to the 4-Ω speaker.
  c. the primary power.

SECTION 18.12 Inductances in Series or Parallel

18.29 Calculate the total inductance, L_T, for the following combinations of series inductors. Assume no mutual induction.
  a. $L_1 = 5$ mH and $L_2 = 15$ mH.
  b. $L_1 = 220\ \mu H$, $L_2 = 330\ \mu H$, and $L_3 = 450\ \mu H$.
  c. $L_1 = 1$ mH, $L_2 = 500\ \mu H$, $L_3 = 2.5$ mH, and $L_4 = 6$ mH.

18.30 Assuming that the inductor combinations listed in Prob. 18-29 are in parallel rather than series, calculate the equivalent inductance, L_{EQ}. Assume no mutual induction.

18.31 A 100-mH and 300-mH inductor are connected in series aiding and have a mutual inductance, L_M, of 130 mH. What is the total inductance, L_T?

18.32 A 20-mH and 40-mH inductor have a coefficient of coupling, k, of 0.4. Calculate L_T if the inductors are
  a. series-aiding.
  b. series-opposing.

Chapter 19

Inductive Reactance

When alternating current flows in an inductance L, the amount of current is much less than the dc resistance alone would allow. The reason is that the current variations induce a voltage across L that opposes the applied voltage. This additional opposition of an inductance to sine-wave alternating current is specified by the amount of its inductive reactance X_L. It is an opposition to current, measured in ohms. The X_L is the ohms of opposition, therefore, that an inductance L has for sine-wave current.

The amount of X_L equals $2\pi fL$ ohms, with f in hertz and L in henrys. Note that the opposition in ohms of X_L increases for higher frequencies and more inductance. The constant factor 2π indicates sine-wave variations.

Learning Outcomes

After studying this chapter, you should be able to:

> *Explain* how inductive reactance reduces the amount of alternating current.

> *Calculate* the reactance of an inductor when the frequency and inductance are known.

> *Calculate* the total reactance of series-connected inductors.

> *Calculate* the equivalent reactance of parallel-connected inductors.

> *Explain* how Ohm's law can be applied to inductive reactance.

> *Describe* the waveshape of induced voltage produced by sine-wave alternating current.

The requirements for X_L correspond to what is needed to produce induced voltage. There must be variations in current and its associated magnetic flux.

For a steady direct current without any changes in current, X_L is zero. However, with sine-wave alternating current, X_L is the best way to analyze the effect of L.

19.1 How X_L Reduces the Amount of I

Figure 19-1 illustrates the effect of X_L in reducing the alternating current for a lightbulb. The more ohms of X_L, the less current flows. When X_L reduces I to a very small value, the bulb cannot light.

In Fig. 19-1a, there is no inductance, and the ac voltage source produces a 2.4-A current to light the bulb with full brilliance. This 2.4-A I results from 120 V applied across the 50-Ω R of the bulb's filament.

In Fig. 19-1b, however, a coil is connected in series with the bulb. The coil has a dc resistance of only 1 Ω, which is negligible, but the reactance of the inductance is 1000 Ω. This 1000-Ω X_L is a measure of the coil's reaction to sine-wave current in producing a self-induced voltage that opposes the applied voltage and reduces the current. Now I is 120 V/1000 Ω, approximately, which equals 0.12 A. This I is not enough to light the bulb.

Although the dc resistance is only 1 Ω, the X_L of 1000 Ω for the coil limits the amount of alternating current to such a low value that the bulb cannot light. This X_L of 1000 Ω for a 60-Hz current can be obtained with an inductance L of approximately 2.65 H.

In Fig. 19-1c, the coil is also in series with the bulb, but the applied battery voltage produces a steady value of direct current. Without any current variations, the coil cannot induce any voltage and, therefore, it has no reactance. The amount of direct current, then, is practically the same as though the dc voltage source were connected directly across the bulb, and it lights with full brilliance. In this case, the coil is only a length of wire because there is no induced voltage without current variations. The dc resistance is the resistance of the wire in the coil.

In summary, we can draw the following conclusions:

1. An inductance can have appreciable X_L in ac circuits to reduce the amount of current. Furthermore, the higher the frequency of the alternating current, and the greater the inductance, the higher the X_L opposition.

2. There is no X_L for steady direct current. In this case, the coil is a resistance equal to the resistance of the wire.

These effects have almost unlimited applications in practical circuits. Consider how useful ohms of X_L can be for different kinds of current, compared with resistance, which always has the same ohms of opposition. One example is to use X_L where it is desired to have high ohms of opposition to alternating current but little opposition to direct current. Another example is to use X_L for more opposition to a high-frequency alternating current, compared with lower frequencies.

X_L Is an Inductive Effect

An inductance can have X_L to reduce the amount of alternating current because self-induced voltage is produced to oppose the applied voltage. In Fig. 19-2, V_L is the voltage across L, induced by the variations in sine-wave current produced by the applied voltage V_A.

The two voltages V_A and V_L are the same because they are in parallel. However, the current I_L is the amount that allows the self-induced voltage V_L to be equal to V_A. In this example, I is 0.12 A. This value of a 60-Hz current in the inductance produces a V_L of 120 V.

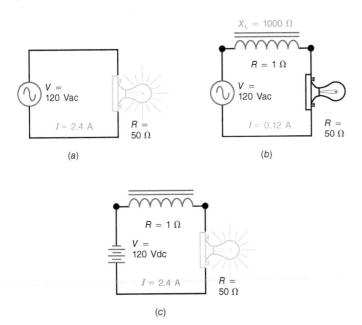

(a)

(b)

(c)

Figure 19-1 Illustrating the effect of inductive reactance X_L in reducing the amount of sine-wave alternating current. (a) Bulb lights with 2.4 A. (b) Inserting an X_L of 1000 Ω reduces I to 0.12 A, and the bulb cannot light. (c) With direct current, the coil has no inductive reactance, and the bulb lights.

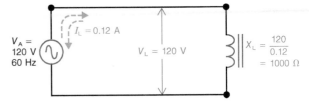

Figure 19-2 The inductive reactance X_L equals the V_L/I_L ratio in ohms.

The Reactance Is a *V/I* Ratio

The *V/I* ratio for the ohms of opposition to the sine-wave current is $^{120}\!/_{0.12}$, which equals 1000 Ω. This 1000 Ω is what we call X_L, to indicate how much current can be produced by sine-wave voltage across an inductance. The ohms of X_L can be almost any amount, but the 1000 Ω here is a typical example.

The Effect of *L* and *f* on X_L

The X_L value depends on the amount of inductance and on the frequency of the alternating current. If *L* in Fig. 19-2 were increased, it could induce the same 120 V for V_L with less current. Then the ratio of V_L/I_L would be greater, meaning more X_L for more inductance.

Also, if the frequency were increased in Fig. 19-2, the current variations would be faster with a higher frequency. Then the same *L* could produce the 120 V for V_L with less current. For this condition also, the V_L/I_L ratio would be greater because of the smaller current, indicating more X_L for a higher frequency.

19.2 $X_L = 2\pi f L$

The formula $X_L = 2\pi f L$ includes the effects of frequency and inductance for calculating the inductive reactance. The frequency is in hertz, and *L* is in henrys for an X_L in ohms. As an example, we can calculate X_L for an inductance of 2.65 H at the frequency of 60 Hz:

$$X_L = 2\pi f L \tag{19-1}$$
$$= 6.28 \times 60 \times 2.65$$
$$= 1000 \ \Omega$$

Note the following factors in the formula $X_L = 2\pi f L$.

1. The constant factor 2π is always $2 \times 3.14 = 6.28$. It indicates the circular motion from which a sine wave is derived. Therefore, this formula applies only to sine-wave ac circuits. The 2π is actually 2π rad or 360° for a complete circle or cycle.

2. The frequency *f* is a time element. Higher frequency means that the current varies at a faster rate. A faster current change can produce more self-induced voltage across a given inductance. The result is more X_L.

3. The inductance *L* indicates the physical factors of the coil that determine how much voltage it can induce for a given current change.

4. Inductive reactance X_L is in ohms, corresponding to a V_L/I_L ratio for sine-wave ac circuits, to determine how much current *L* allows for a given applied voltage.

Stating X_L as V_L/I_L and as $2\pi f L$ are two ways of specifying the same value of ohms. The $2\pi f L$ formula gives the effect of *L* and *f* on the X_L. The V_L/I_L ratio gives the result of $2\pi f L$ in reducing the amount of *I*.

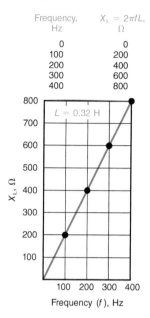

Figure 19-3 Graph of values to show linear increase of X_L for higher frequencies. The *L* is constant at 0.32 H.

The formula $2\pi f L$ shows that X_L is proportional to frequency. When *f* is doubled, for instance, X_L is doubled. This linear increase in inductive reactance with frequency is illustrated in Fig. 19-3.

The reactance formula also shows that X_L is proportional to the inductance. When the value of henrys for *L* is doubled, the ohms of X_L is also doubled. This linear increase of inductive reactance with inductance is illustrated in Fig. 19-4.

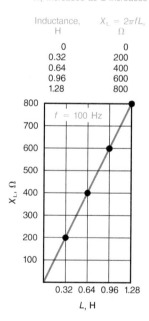

Figure 19-4 Graph of values to show linear increase of X_L for higher values of inductance *L*. The frequency is constant at 100 Hz.

EXAMPLE 19-1

How much is X_L of a 6-mH L at 41.67 kHz?

Answer:

$$X_L = 2\pi fL$$
$$= 6.28 \times 41.67 \times 10^3 \times 6 \times 10^{-3}$$
$$= 1570 \ \Omega$$

EXAMPLE 19-2

Calculate the X_L of (a) a 10-H L at 60 Hz and (b) a 5-H L at 60 Hz.

Answer:

a. For a 10-H L,

$$X_L = 2\pi fL = 6.28 \times 60 \times 10$$
$$= 3768 \ \Omega$$

b. For a 5-H L,

$$X_L = \tfrac{1}{2} \times 3768 = 1884 \ \Omega$$

EXAMPLE 19-3

Calculate the X_L of a 250-μH coil at (a) 1 MHz and (b) 10 MHz.

Answer:

a. At 1 MHz,

$$X_L = 2\pi fL = 6.28 \times 1 \times 10^6 \times 250 \times 10^{-6}$$
$$= 1570 \ \Omega$$

b. At 10 MHz,

$$X_L = 10 \times 1570 = 15,700 \ \Omega$$

The last two examples illustrate the fact that X_L is proportional to frequency and inductance. In Example 19-2b, X_L is one-half the value in Example 19-2a because the inductance is one-half. In Example 19-3b, the X_L is 10 times more than in Example 19-3a because the frequency is 10 times higher.

Finding L from X_L

Not only can X_L be calculated from f and L, but if any two factors are known, the third can be found. Very often X_L can be determined from voltage and current measurements. With the frequency known, L can be calculated as

$$L = \frac{X_L}{2\pi f} \qquad \textbf{(19-2)}$$

This formula has the factors inverted from Formula (19-1). Use the basic units with ohms for X_L and hertz for f to calculate L in henrys.

It should be noted that Formula (19-2) can also be stated as

$$L = \frac{1}{2\pi f} \times X_L$$

This form is easier to use with a calculator because $1/2\pi f$ can be found as a reciprocal value and then multiplied by X_L.

The following problems illustrate how to find X_L from V and I measurements and using X_L to determine L with Formula (19-2).

EXAMPLE 19-4

A coil with negligible resistance has 62.8 V across it with 0.01 A of current. How much is X_L?

Answer:

$$X_L = \frac{V_L}{I_L} = \frac{62.8 \ \text{V}}{0.01 \ \text{A}}$$
$$= 6280 \ \Omega$$

EXAMPLE 19-5

Calculate L of the coil in Example 19-4 when the frequency is 1000 Hz.

Answer:

$$L = \frac{X_L}{2\pi f} = \frac{6280}{6.28 \times 1000}$$
$$= 1 \ \text{H}$$

EXAMPLE 19-6

Calculate L of a coil that has 15,700 Ω of X_L at 12 MHz.

Answer:

$$L = \frac{X_L}{2\pi f} = \frac{1}{2\pi f} \times X_L$$
$$= \frac{1}{6.28 \times 12 \times 10^6} \times 15,700$$
$$= 0.0133 \times 10^{-6} \times 15,700$$
$$= 208.8 \times 10^{-6} \ \text{H} \quad \text{or} \quad 208.8 \ \mu\text{H}$$

Finding f from X_L

For a third version of the inductive reactance formula,

$$f = \frac{X_L}{2\pi L} \qquad \textbf{(19-3)}$$

Use the basic units of ohms fowr X_L and henrys for L to calculate the frequency in hertz.

Formula 19-3 can also be stated as

$$f = \frac{1}{2\pi L} \times X_L$$

This form is easier to use with a calculator. Find the reciprocal value and multiply by X_L, as explained before in Example 19-6.

EXAMPLE 19-7

At what frequency will an inductance of 1 H have a reactance of 1000 Ω?

Answer:

$$f = \frac{1}{2\pi L} \times X_L = \frac{1}{6.28 \times 1} \times 1000$$
$$= 0.159 \times 1000$$
$$= 159 \text{ Hz}$$

19.3 Series or Parallel Inductive Reactances

Since reactance is an opposition in ohms, the values of X_L in series or in parallel are combined the same way as ohms of resistance. With series reactances, the total is the sum of the individual values, as shown in Fig. 19-5a. For example, the series reactances of 100 and 200 Ω add to equal 300 Ω of X_L across both reactances. Therefore, in series,

$$X_{L_T} = X_{L_1} + X_{L_2} + X_{L_3} + \cdots + \text{etc.} \qquad \textbf{(19-4)}$$

The combined reactance of parallel reactances is calculated by the reciprocal formula. As shown in Fig. 19-5b, in parallel

$$X_{L_{EQ}} = \frac{1}{\dfrac{1}{X_{L_1}} + \dfrac{1}{X_{L_2}} + \dfrac{1}{X_{L_3}} + \cdots + \text{etc.}} \qquad \textbf{(19-5)}$$

The combined parallel reactance will be less than the lowest branch reactance. Any shortcuts for calculating parallel resistances also apply to parallel reactances. For instance, the combined reactance of two equal reactances in parallel is one-half either reactance.

19.4 Ohm's Law Applied to X_L

The amount of current in an ac circuit with only inductive reactance is equal to the applied voltage divided by X_L. Three examples are given in Fig. 19-6. No dc resistance is indicated, since it is assumed to be practically zero for the coils shown. In Fig. 19-6a, there is one reactance of 100 Ω. Then I equals V/X_L, or 100 V/100 Ω, which is 1 A.

In Fig. 19-6b, the total reactance is the sum of the two individual series reactances of 100 Ω each, for a total of 200 Ω. The current, calculated as V/X_{L_T}, then equals 100 V/200 Ω, which is 0.5 A. This current is the same in both series reactances. Therefore, the voltage across each reactance equals its IX_L product. This is 0.5 A × 100 Ω, or 50 V across each X_L.

In Fig. 19-6c, each parallel reactance has its individual branch current, equal to the applied voltage divided by the branch reactance. Then each branch current equals 100 V/100 Ω, which is 1 A. The voltage is the same across both reactances, equal to the generator voltage, since they are all in parallel.

The total line current of 2 A is the sum of the two individual 1-A branch currents. With the rms value for the applied voltage, all calculated values of currents and voltage drops in Fig. 19-6 are also rms values.

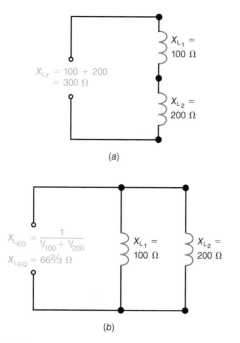

(a)

(b)

Figure 19-5 Combining ohms of X_L for inductive reactances. (a) X_{L_1} and X_{L_2} in series. (b) X_{L_1} and X_{L_2} in parallel.

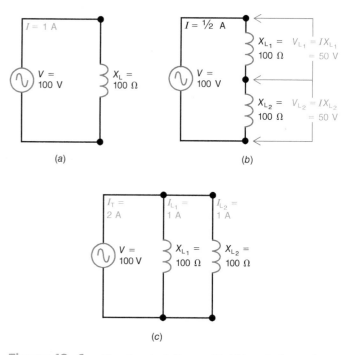

(a)

(b)

(c)

Figure 19-6 Circuit calculations with V, I, and ohms of reactance X_L. (a) One reactance. (b) Two series reactances. (c) Two parallel reactances.

19.5 Applications of X_L for Different Frequencies

The general use of inductance is to provide minimum reactance for relatively low frequencies but more for higher frequencies. In this way, the current in an ac circuit can be reduced for higher frequencies because of more X_L. There are many circuits in which voltages of different frequencies are applied to produce current with different frequencies. Then, the general effect of X_L is to allow the most current for direct current and low frequencies, with less current for higher frequencies, as X_L increases.

Compare this frequency factor for ohms of X_L with ohms of resistance. The X_L increases with frequency, but R has the same effect in limiting direct current or alternating current of any frequency.

19.6 Waveshape of v_L Induced by Sine-Wave Current

More details of inductive circuits can be analyzed by means of the waveshapes in Fig. 19-7, plotted from calculated values. The top curve shows a sine wave of current i_L flowing through a 6-mH inductance L. Since induced voltage

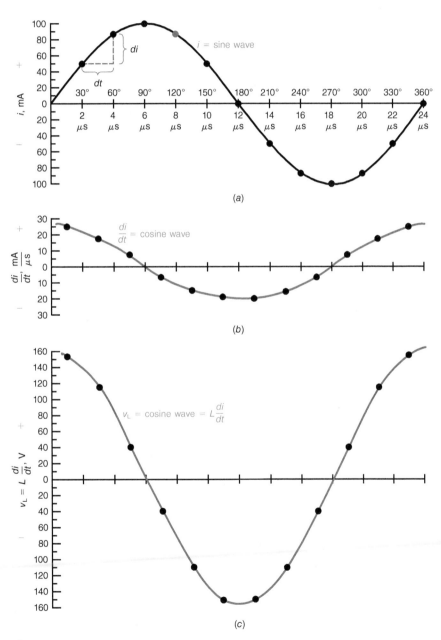

Figure 19-7 Waveshapes in inductive circuits. (a) Sine-wave current i. (b) changes in current with time di/dt. (c) induced voltage v_L.

depends on the rate of change of current rather than on the absolute value of i, the curve in Fig. 19-7b shows how much the current changes. In this curve, the di/dt values are plotted for the current changes every 30° of the cycle. The bottom curve shows the actual induced voltage v_L. This v_L curve is similar to the di/dt curve because v_L equals the constant factor L multiplied by di/dt. Note that di/dt indicates infinitely small changes in i and t.

90° Phase Angle

The v_L curve at the bottom of Fig. 19-7 has its zero values when the i_L curve at the top is at maximum. This comparison shows that the curves are 90° out of phase. The v_L is a cosine wave of voltage for the sine wave of current i_L.

The 90° phase difference results from the fact that v_L depends on the di/dt rate of change, rather than on i itself. More details of this 90° phase angle between v_L and i_L for inductance are explained in the next chapter.

Frequency

For each of the curves, the period T is 24 μs. Therefore, the frequency is $1/T$ or $\frac{1}{24}$ μs, which equals 41.67 kHz. Each curve has the same frequency.

Ohms of X_L

The ratio of v_L/i_L specifies the inductive reactance in ohms. For this comparison, we use the actual value of i_L, which has a peak value of 100 mA. The rate-of-change factor is included in the induced voltage v_L. Although the peak of v_L at 150 V is 90° before the peak of i_L at 100 mA, we can compare these two peak values. Then v_L/i_L is $\frac{150}{0.1}$, which equals 1500 Ω.

This X_L is only approximate because v_L cannot be determined exactly for the large dt changes every 30°. If we used smaller intervals of time, the peak v_L would be 157 V. Then

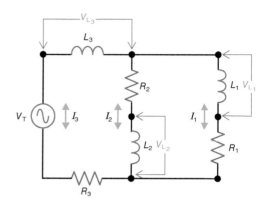

Figure 19-8 How a 90° phase angle for the V_L applies in a complex circuit with more than one inductance. The current I_1 lags V_{L_1} by 90°, I_2 lags V_{L_2} by 90°, and I_3 lags V_{L_3} by 90°.

X_L would be 1570 Ω, the same as $2\pi f L$ Ω with a 6-mH L and a frequency of 41.67 kHz. This is the same X_L problem as Example 19-1.

Application of the 90° Phase Angle in a Circuit

The phase angle of 90° between V_L and I will always apply for any L with sine-wave current. Remember, though, that the specific comparison is only between the induced voltage across any one coil and the current flowing in its turns. To emphasize this important principle, Fig. 19-8 shows an ac circuit with a few coils and resistors. The details of this complex circuit are not to be analyzed now. However, for each L in the circuit, the V_L is 90° out of phase with its I. The I lags V_L by 90°, or V_L leads I. For the three coils in Fig. 19-8,

> Current I_1 lags V_{L_1} by 90°.
> Current I_2 lags V_{L_2} by 90°.
> Current I_3 lags V_{L_3} by 90°.
> Note that I_3 is also I_T for the series-parallel circuit.

❓ CHAPTER 19 REVIEW QUESTIONS

1. The unit of inductive reactance, X_L, is the
   a. henry.
   b. ohm.
   c. farad.
   d. hertz.

2. The inductive reactance, X_L, of an inductor is
   a. inversely proportional to frequency.
   b. unaffected by frequency.
   c. directly proportional to frequency.
   d. inversely proportional to inductance.

3. For an inductor, the induced voltage, V_L,
   a. leads the inductor current, i_L, by 90°.
   b. lags the inductor current, i_L, by 90°.

   c. is in phase with the inductor current, i_L.
   d. none of the above.

4. For a steady dc current, the X_L of an inductor is
   a. infinite.
   b. extremely high.
   c. usually about 10 kΩ.
   d. 0 Ω.

5. What is the inductive reactance, X_L, of a 100-mH coil at a frequency of 3.183 kHz?
   a. 2 kΩ.
   b. 200 Ω.
   c. 1 MΩ.
   d. 4 Ω.

6. At what frequency does a 60-mH inductor have an X_L value of 1 kΩ?
  a. 377 Hz.
  b. 265 kHz.
  c. 2.65 kHz.
  d. 15.9 kHz.

7. What value of inductance will provide an X_L of 500 Ω at a frequency of 159.15 kHz?
  a. 5 H.
  b. 500 μH.
  c. 500 mH.
  d. 750 μH.

8. Two inductors, L_1 and L_2, are in series. If X_{L_1} = 4 kΩ and X_{L_2} = 2 kΩ, how much is X_{L_T}?
  a. 6 kΩ.
  b. 1.33 kΩ.

  c. 4.47 kΩ.
  d. 2 kΩ.

9. Two inductors, L_1 and L_2, are in parallel. If X_{L_1} = 1 kΩ and X_{L_2} = 1 kΩ, how much is $X_{L_{EQ}}$?
  a. 707 Ω.
  b. 2 kΩ.
  c. 1.414 kΩ.
  d. 500 Ω.

10. How much is the inductance of a coil that draws 25 mA of current from a 24-Vac source whose frequency is 1 kHz?
  a. 63.7 μH.
  b. 152.8 mH.
  c. 6.37 H.
  d. 15.28 mH.

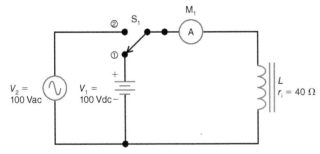
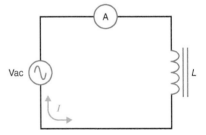

CHAPTER 19 PROBLEMS

SECTION 19.1 How X_L Reduces the Amount of I

19.1 How much is the inductive reactance, X_L, of a coil for a steady dc current?

19.2 List two factors that determine the amount of inductive reactance a coil will have.

19.3 In Fig. 19-9, how much dc current will be indicated by the ammeter, M_1, with S_1 in position 1?

Figure 19-9

19.4 In Fig. 19-9, how much inductive reactance, X_L, does the coil have with S_1 in position 1? Explain your answer.

19.5 In Fig. 19-9, the ammeter, M_1, reads an ac current of 25 mA with S_1 in position 2.
  a. Why is there less current in the circuit with S_1 in position 2 compared to position 1?
  b. How much is the inductive reactance, X_L, of the coil? (Ignore the effect of the coil resistance, r_i.)

19.6 In Fig. 19-10, how much is the inductive reactance, X_L, for each of the following values of Vac and I?
  a. Vac = 10 V and I = 2 mA.
  b. Vac = 50 V and I = 20 μA.
  c. Vac = 12 V and I = 15 mA.
  d. Vac = 6 V and I = 40 μA.
  e. Vac = 120 V and I = 400 mA.

Figure 19-10

SECTION 19.2 $X_L = 2\pi fL$

19.7 Calculate the inductive reactance, X_L, of a 100-mH inductor at the following frequencies:
  a. f = 60 Hz.
  b. f = 10 kHz.

19.8 Calculate the inductive reactance, X_L, of a 50-μH coil at the following frequencies:
  a. f = 60 Hz.
  b. f = 3.8 MHz.

19.9 What value of inductance, L, will provide an X_L value of 1 kΩ at the following frequencies?
a. $f = 1.591$ kHz.
b. $f = 5$ kHz.

19.10 At what frequency will a 30-mH inductor provide an X_L value of
a. 50 Ω?
b. 40 kΩ?

19.11 How much is the inductance of a coil that draws 15 mA from a 24-Vac source whose frequency is 1 kHz?

19.12 At what frequency will a stray inductance of 0.25 μH have an X_L value of 100 Ω?

19.13 A 25-mH coil draws 2 mA of current from a 10-Vac source. What is the value of current drawn by the inductor when
a. the frequency is doubled?
b. the inductance is reduced by one-half to 12.5 mH?

19.14 A coil has an inductive reactance, X_L, of 10 kΩ at a given frequency. What is the value of X_L when the frequency is
a. cut in half?
b. increased by a factor 10?

19.15 Calculate the inductive reactance, X_L, for the following inductance and frequency values:
a. $L = 7$ H, $f = 60$ Hz.
b. $L = 1$ mH, $f = 159.2$ kHz.

19.16 Determine the inductance value for the following frequency and X_L values:
a. $X_L = 2$ kΩ, $f = 5$ kHz.
b. $X_L = 10$ Ω, $f = 795.7$ kHz.

19.17 Determine the frequency for the following inductance and X_L values:
a. $L = 60$ μH, $X_L = 200$ Ω.
b. $L = 150$ mH, $X_L = 7.5$ kΩ.

SECTION 19.3 Series or Parallel Inductive Reactances

19.18 How much is the total inductive reactance, X_{L_T}, for the following series inductive reactances:
a. $X_{L_1} = 250$ Ω and $X_{L_2} = 1.5$ k Ω.
b. $X_{L_1} = 1.8$ kΩ, $X_{L_2} = 2.2$ kΩ and $X_{L_3} = 1$ kΩ.

19.19 What is the equivalent inductive reactance, $X_{L_{EQ}}$, for the following parallel inductive reactances?
a. $X_{L_1} = 1.2$ kΩ and $X_{L_2} = 1.8$ kΩ.
b. $X_{L_1} = 1.5$ kΩ and $X_{L_2} = 1$ kΩ.

SECTION 19.4 Ohm's Law Applied to X_L

19.20 In Fig. 19-11, calculate the current, I.

Figure 19-11

19.21 In Fig. 19-11, what happens to the current, I, when the frequency of the applied voltage
a. decreases?
b. increases?

19.22 In Fig. 19-12, solve for
a. X_{L_T}.
b. I.
c. V_{L_1}, V_{L_2}, and V_{L_3}.

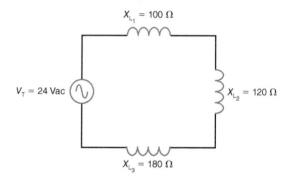

Figure 19-12

SECTION 19.5 Applications of X_L for Different Frequencies

19.23 Calculate the value of inductance, L, required to produce an X_L value of 500 Ω at the following frequencies:
a. $f = 250$ Hz.
b. $f = 7.957$ kHz.

SECTION 19.6 Waveshape of V_L Induced by Sine-Wave Current

19.24 For an inductor, what is the phase relationship between the induced voltage, V_L, and the inductor current, i_L? Explain your answer.

19.25 For a sine wave of alternating current flowing through an inductor, at what angles in the cycle will the induced voltage be
a. maximum?
b. zero?

Chapter 20

Inductive Circuits

‖‖‖

Learning Outcomes

After studying this chapter, you should be able to:

> *Explain* why the voltage leads the current by 90° for an inductor.

> *Calculate* the total impedance and phase angle of a series *RL* circuit.

> *Calculate* the total current, equivalent impedance, and phase angle of a parallel *RL* circuit.

> *Define* what is meant by the *Q* of a coil.

> *Explain* how an inductor can be used to pass some ac frequencies but block others.

> *Calculate* the real, apparent, and reactive power and power factor in an *RL* or *RC* circuit.

> *Express* the impedance of an *RL* or *RC* circuit as a complex number.

This chapter analyzes circuits that combine inductive reactance X_L and resistance R. The main questions are, How do we combine the ohms of opposition, how much current flows, and what is the phase angle? Although X_L and R are both measured in ohms, they have different characteristics. Specifically, X_L increases with more L and higher frequencies, when sine-wave ac voltage is applied, whereas R is the same for dc or ac circuits. Furthermore, the phase angle for the voltage across X_L is at 90° with respect to the current through L.

In addition, the practical application of using a coil as a choke to reduce the current for a specific frequency is explained here. For a circuit with L and R in series, the X_L can be high for an undesired ac signal frequency, whereas R is the same for either direct current or alternating current.

20.1 Sine Wave i_L Lags v_L by 90°

When sine-wave variations of current produce an induced voltage, the current lags its induced voltage by exactly 90°, as shown in Fig. 20-1. The inductive circuit in Fig. 20-1a has the current and voltage waveshapes shown in Fig. 20-1b. The phasors in Fig. 20-1c show the 90° phase angle between i_L and v_L. Therefore, we can say that i_L lags v_L by 90°, or v_L leads i_L by 90°.

This 90° phase relationship between i_L and v_L is true in any sine-wave ac circuit, whether L is in series or parallel and whether L is alone or combined with other components. We can always say that the voltage across any X_L is 90° out of phase with the current through it.

Why the Phase Angle Is 90°

The 90° phase angle results because v_L depends on the rate of change of i_L. As previously shown in Fig. 19-7 for a sine wave of i_L, the induced voltage is a cosine wave. In other words, v_L has the phase of di/dt, not the phase of i.

Why i_L Lags v_L

The 90° difference can be measured between any two points having the same value on the i_L and v_L waves. A convenient point is the positive peak value. Note that the i_L wave does not have its positive peak until 90° after the v_L wave. Therefore, i_L lags v_L by 90°. This 90° lag is in time. The time lag equals one quarter-cycle, which is one-quarter of the time for a complete cycle.

Inductive Current Is the Same in a Series Circuit

The time delay and resultant phase angle for the current in an inductance apply only with respect to the voltage across the inductance. This condition does not change the fact that the current is the same in all parts of a series circuit. In Fig. 20-1a, the current in the generator, the connecting wires, and L must be the same because they are in series. Whatever the current value is at any instant, it is the same in all series components. The time lag is between current and voltage.

Inductive Voltage Is the Same across Parallel Branches

In Fig. 20-1a, the voltage across the generator and the voltage across L are the same because they are in parallel. There cannot be any lag or lead in time between these two parallel voltages. Whatever the voltage value is across the generator at any instant, the voltage across L is the same. The parallel voltage v_A or v_L is 90° out of phase with the current.

The voltage across L in this circuit is determined by the applied voltage, since they must be the same. The inductive effect here is to make the current have values that produce $L(di/dt)$ equal to the parallel voltage.

The Frequency Is the Same for i_L and v_L

Although i_L lags v_L by 90°, both waves have the same frequency. The i_L wave reaches its peak values 90° later than the v_L wave, but the complete cycles of variations are repeated at the same rate. As an example, if the frequency of the sine wave v_L in Fig. 20-1b is 100 Hz, this is also the frequency for i_L.

20.2 X_L and R in Series

When a coil has series resistance, the current is limited by both X_L and R. This current I is the same in X_L and R, since they are in series. Each has its own series voltage drop, equal to IR for the resistance and IX_L for the reactance.

Note the following points about a circuit that combines series X_L and R, as in Fig. 20-2:

1. The current is labeled I, rather than I_L, because I flows through all series components.
2. The voltage across X_L, labeled V_L, can be considered an IX_L voltage drop, just as we use V_R for an IR voltage drop.
3. The current I through X_L must lag V_L by 90° because this is the phase angle between current through an inductance and its self-induced voltage.
4. The current I through R and its IR voltage drop are in phase. There is no reactance to sine-wave current in any resistance. Therefore, I and IR have a phase angle of 0°.

Resistance R can be either the internal resistance of the coil or an external series resistance. The I and V values may be rms, peak, or instantaneous, as long as the same measure is applied to all. Peak values are used here for convenience in comparing waveforms.

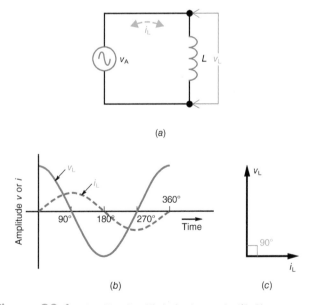

Figure 20-1 (a) Circuit with inductance L. (b) Sine wave of i_L lags v_L by 90°. (c) Phasor diagram.

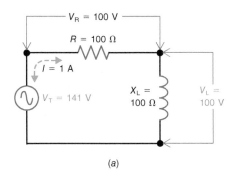

(a)

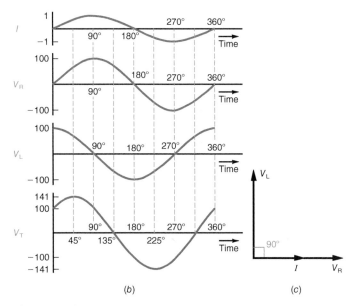

(b)   (c)

Figure 20-2 Inductive reactance X_L and resistance R in series. (a) Circuit. (b) Waveforms of current and voltage. (c) Phasor diagram.

Phase Comparisons

Note the following:

1. Voltage V_L is 90° out of phase with I.
2. However, V_R and I are in phase.
3. If I is used as the reference, V_L is 90° out of phase with V_R.

Specifically, V_R lags V_L by 90°, just as the current I lags V_L. These phase relations are shown by the waveforms in Fig. 20-2b and the phasors in Fig. 20-2c.

Combining V_R and V_L

As shown in Fig. 20-2b, when the V_R voltage wave is combined with the V_L voltage wave, the result is the voltage wave for the applied generator voltage V_T. The voltage drops must add to equal the applied voltage. The 100-V peak values for V_R and for V_L total 141 V, however, instead of 200 V, because of the 90° phase difference.

Consider some instantaneous values to see why the 100-V peak V_R and 100-V peak V_L cannot be added arithmetically.

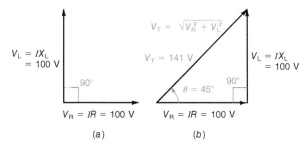

(a)   (b)

Figure 20-3 Addition of two voltages 90° out of phase. (a) Phasors for V_L and V_R are 90° out of phase. (b) Resultant of the two phasors is the hypotenuse of a right triangle for the value of V_T.

When V_R is at its maximum value of 100 V, for instance, V_L is at zero. The total for V_T then is 100 V. Similarly, when V_L is at its maximum value of 100 V, then V_R is zero and the total V_T is also 100 V.

Actually, V_T has its maximum value of 141 V when V_L and V_R are each 70.7 V. When series voltage drops that are out of phase are combined, therefore, they cannot be added without taking the phase difference into account.

Phasor Voltage Triangle

Instead of combining waveforms that are out of phase, we can add them more quickly by using their equivalent phasors, as shown in Fig. 20-3. The phasors in Fig. 20-3a show only the 90° angle without any addition. The method in Fig. 20-3b is to add the tail of one phasor to the arrowhead of the other, using the angle required to show their relative phase. Voltages V_R and V_L are at right angles because they are 90° out of phase. The sum of the phasors is a resultant phasor from the start of one to the end of the other. Since the V_R and V_L phasors form a right angle, the resultant phasor is the hypotenuse of a right triangle. The hypotenuse is the side opposite the 90° angle.

From the geometry of a right triangle, the Pythagorean theorem states that the hypotenuse is equal to the square root of the sum of the squares of the sides. For the voltage triangle in Fig. 20-3b, therefore, the resultant is

$$V_T = \sqrt{V_R^2 + V_L^2} \tag{20-1}$$

where V_T is the phasor sum of the two voltages V_R and V_L 90° out of phase.

This formula is for V_R and V_L when they are in series, since then they are 90° out of phase. All voltages must be in the same units. When V_T is an rms value, V_R and V_L are also rms values. For the example in Fig. 20-3,

$$V_T = \sqrt{100^2 + 100^2} = \sqrt{10,000 + 10,000}$$
$$= \sqrt{20,000}$$
$$= 141 \text{ V}$$

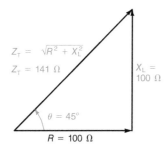

$Z_T = \sqrt{R^2 + X_L^2}$

$Z_T = 141\ \Omega$

$X_L = 100\ \Omega$

$\theta = 45°$

$R = 100\ \Omega$

Figure 20-4 Addition of R and X_L 90° out of phase in series circuit, to find the resultant impedance Z_T.

20.3 Impedance Z Triangle

A triangle of R and X_L in series corresponds to a voltage triangle, as shown in Fig. 20-4. It is similar to the voltage triangle in Fig. 20-3, but the common factor I cancels because the current is the same in X_L and R. The resultant of the phasor addition of R and X_L is their total opposition in ohms, called *impedance*, with the symbol Z_T.* The Z takes into account the 90° phase relation between R and X_L.

For the impedance triangle of a series circuit with reactance and resistance,

$$Z_T = \sqrt{R^2 + X_L^2} \tag{20-2}$$

where R, X_L, and Z_T are all in ohms. For the example in Fig. 20-4,

$$Z_T = \sqrt{100^2 + 100^2} = \sqrt{10{,}000 + 10{,}000}$$
$$= \sqrt{20{,}000}$$
$$= 141\ \Omega$$

Note that the applied voltage of 141 V divided by the total impedance of 141 Ω results in 1 A of current in the series circuit. The IR voltage is 1 × 100, or 100 V; the IX_L voltage is also 1 × 100, or 100 V. The total of the series IR and IX_L drops of 100 V each, added by phasors, equals the applied voltage of 141 V. Finally, the applied voltage equals IZ, or 1 × 141, which is 141 V.

Summarizing the similar phasor triangles for volts and ohms in a series circuit,

1. The phasor for R, IR, or V_R is used as a reference at 0°.
2. The phasor for X_L, IX_L, or V_L is at 90°.
3. The phasor for Z, IZ, or V_T has the phase angle θ of the complete circuit.

Phase Angle with Series X_L

The angle between the generator voltage and its current is the phase angle of the circuit. Its symbol is θ (theta). In Fig. 20-3, the phase angle between V_T and IR is 45°. Since

---

* Although Z_T is a passive component, we consider it a phasor here because it determines the phase angle of V and I.

IR and I have the same phase, the angle is also 45° between V_T and I.

In the corresponding impedance triangle in Fig. 20-4, the angle between Z_T and R is also equal to the phase angle. Therefore, the phase angle can be calculated from the impedance triangle of a series circuit by the formula

$$\tan \theta_Z = \frac{X_L}{R} \tag{20-3}$$

The tangent (tan) is a trigonometric function of any angle, equal to the ratio of the opposite side to the adjacent side of a right triangle. In this impedance triangle, X_L is the opposite side and R is the adjacent side of the angle. We use the subscript z for θ to show that θ_Z is found from the impedance triangle for a series circuit. To calculate this phase angle,

$$\tan \theta_Z = \frac{X_L}{R} = \frac{100}{100} = 1$$

The angle whose tangent is equal to 1 is 45°. Therefore, the phase angle is 45° in this example. The numerical values of the trigonometric functions can be found from a table or from a scientific calculator.

Note that the phase angle of 45° is halfway between 0° and 90° because R and X_L are equal.

EXAMPLE 20-1

If a 30-Ω R and a 40-Ω X_L are in series with 100 V applied, find the following: Z_T, I, V_R, V_L, and θ_Z. What is the phase angle between V_L and V_R with respect to I? Prove that the sum of the series voltage drops equals the applied voltage V_T.

Answer:

$$Z_T = \sqrt{R^2 + X_L^2} = \sqrt{900 + 1600}$$
$$= \sqrt{2500}$$
$$= 50\ \Omega$$
$$I = \frac{V_T}{Z_T} = \frac{100}{50} = 2\ A$$
$$V_R = IR = 2 \times 30 = 60\ V$$
$$V_L = IX_L = 2 \times 40 = 80\ V$$
$$\tan \theta_Z = \frac{X_L}{R} = \frac{40}{30} = \frac{4}{3} = 1.33$$
$$\theta_Z = 53.1°$$

Therefore, I lags V_T by 53.1°. Furthermore, I and V_R are in phase, and I lags V_L by 90°. Finally,

$$V_T = \sqrt{V_R^2 + V_L^2} = \sqrt{60^2 + 80^2} = \sqrt{3600 + 6400}$$
$$= \sqrt{10{,}000}$$
$$= 100\ V$$

Note that the phasor sum of the voltage drops equals the applied voltage.

Series Combinations of X_L and R

In a series circuit, the higher the value of X_L compared with R, the more inductive the circuit. This means that there is more voltage drop across the inductive reactance and the phase angle increases toward 90°. The series current lags the applied generator voltage. With all X_L and no R, the entire applied voltage is across X_L, and θ_Z equals 90°.

At the opposite extreme, when R is much larger than X_L, the series circuit is mainly resistive. The total impedance Z_T is approximately equal to R, and the voltage drop across R is practically equal to the applied voltage, with almost none across X_L.

When X_L and R equal each other, their resultant impedance Z_T is 1.41 times the value of either one. The phase angle then is 45°, halfway between 0° for resistance alone and 90° for inductive reactance alone.

20.4 X_L and R in Parallel

For parallel circuits with X_L and R, the 90° phase angle must be considered for each of the branch currents, instead of the voltage drops. Remember that any series circuit has different voltage drops but one common current. A parallel circuit has different branch currents but one common voltage.

In the parallel circuit in Fig. 20-5a, the applied voltage V_A is the same across X_L, R, and the generator, since they are all in parallel. There cannot be any phase difference between these voltages. Each branch, however, has its individual current. For the resistive branch, $I_R = V_A/R$; in the inductive branch, $I_L = V_A/X_L$.

The resistive branch current I_R is in phase with the generator voltage V_A. The inductive branch current I_L lags V_A, however, because the current in an inductance lags the voltage across it by 90°.

The total line current, therefore, consists of I_R and I_L, which are 90° out of phase with each other. The phasor sum of I_R and I_L equals the total line current I_T. These phase relations are shown by the waveforms in Fig. 20-5b, and the phasors in Fig. 20-5c. Either way, the phasor sum of 10 A for I_R and 10 A for I_L is equal to 14.14 A for I_T.

Both methods illustrate the general principle that quadrature components must be combined by phasor addition. The branch currents are added by phasors here because they are the factors that are 90° out of phase in a parallel circuit. This method is similar to combining voltage drops 90° out of phase in a series circuit.

Phasor Current Triangle

Note that the phasor diagram in Fig. 20-5c has the applied voltage V_A of the generator as the reference phasor because V_A is the same throughout the parallel circuit.

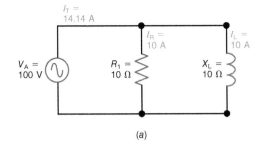

(a)

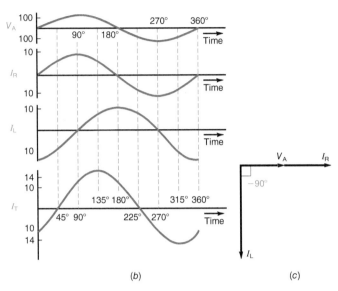

(b)   (c)

Figure 20-5 Inductive reactance X_L and R in parallel. (a) Circuit. (b) Waveforms of applied voltage and branch currents. (c) Phasor diagram.

The phasor for I_L is down, compared with up for an X_L phasor. Here the parallel branch current I_L lags the parallel voltage reference V_A. In a series circuit, the X_L voltage leads the series current reference I. For this reason, the I_L phasor is shown with a negative 90° angle. The −90° means that the current I_L lags the reference phasor V_A.

The phasor addition of the branch currents in a parallel circuit can be calculated by the phasor triangle for currents shown in Fig. 20-6. Peak values are used for convenience in

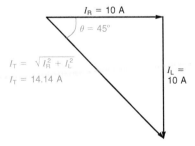

Figure 20-6 Phasor triangle of inductive and resistive branch currents 90° out of phase in a parallel circuit to find resultant I_T

this example, but when the applied voltage is an rms value, the calculated currents are also in rms values. To calculate the total line current,

$$I_T = \sqrt{I_R^2 + I_L^2} \qquad \text{(20-4)}$$

For the values in Fig. 20-6,

$$I_T = \sqrt{10^2 + 10^2} = \sqrt{100 + 100}$$

$$= \sqrt{200}$$

$$= 14.14 \text{ A}$$

Impedance of X_L and R in Parallel

A practical approach to the problem of calculating the total impedance of X_L and R in parallel is to calculate the total line current I_T and divide this value into the applied voltage V_A:

$$Z_{EQ} = \frac{V_A}{I_T} \qquad \text{(20-5)}$$

For example, in Fig. 20-5, V_A is 100 V and the resultant I_T, obtained as the phasor sum of the resistive and reactive branch currents, is equal to 14.14 A. Therefore, we calculate the impedance as

$$Z_{EQ} = \frac{V_A}{I_T} = \frac{100 \text{ V}}{14.14 \text{ A}}$$

$$= 7.07 \text{ }\Omega$$

This impedance is the combined opposition in ohms across the generator, equal to the resistance of 10 Ω in parallel with the reactance of 10 Ω.

Note that the impedance for equal values of R and X_L in parallel is not one-half but equals 70.7% of either one. Still, the combined value of ohms must be less than the lowest ohms value in the parallel branches.

For the general case of calculating the impedance of X_L and R in parallel, any number can be assumed for the applied voltage because the value of V_A cancels in the calculations for Z in terms of the branch currents. A good value to assume for V_A is the value of either R or X_L, whichever is the higher number. This way, there are no fractions smaller than 1 in the calculation of the branch currents. For X_L in parallel with R, Z_{EQ} can also be calculated as

$$Z_{EQ} = \frac{X_L R}{\sqrt{R^2 + X_L^2}}$$

EXAMPLE 20-2

What is the total Z of a 600-Ω R in parallel with a 300-Ω X_L? Assume 600 V for the applied voltage.

Answer:

$$I_R = \frac{600 \text{ V}}{600 \text{ }\Omega} = 1 \text{ A}$$

$$I_L = \frac{600 \text{ V}}{300 \text{ }\Omega} = 2 \text{ A}$$

$$I_T = \sqrt{I_R^2 + I_L^2}$$

$$= \sqrt{1 + 4} = \sqrt{5}$$

$$= 2.24 \text{ A}$$

Then, dividing the assumed value of 600 V for the applied voltage by the total line current gives

$$Z_{EQ} = \frac{V_A}{I_T} = \frac{600 \text{ V}}{2.24 \text{ A}}$$

$$= 268 \text{ }\Omega$$

The combined impedance of a 600-Ω R in parallel with a 300-Ω X_L is equal to 268 Ω, no matter how much the applied voltage is.

Phase Angle with Parallel X_L and R

In a parallel circuit, the phase angle is between the line current I_T and the common voltage V_A applied across all branches. However, the resistive branch current I_R has the same phase as V_A. Therefore, the phase of I_R can be substituted for the phase of V_A. This is shown in Fig. 20-5c. The triangle of currents is shown in Fig. 20-6. To find θ_I from the branch currents, use the tangent formula:

$$\tan \theta_I = -\frac{I_L}{I_R} \qquad \text{(20-6)}$$

We use the subscript I for θ to show that θ_I is found from the triangle of branch currents in a parallel circuit. In Fig. 20-6, θ_I is $-45°$ because I_L and I_R are equal. Then $\tan \theta_I = -1$.

The negative sign is used for this current ratio because I_L is lagging at $-90°$, compared with I_R. The phase angle of $-45°$ here means that I_T lags I_R and V_A by 45°.

Note that the phasor triangle of branch currents gives θ_I as the angle of I_T with respect to the generator voltage V_A. This phase angle for I_T is with respect to the applied voltage as the reference at 0°. For the phasor triangle of voltages in a series circuit, the phase angle θ_Z for Z_T and V_T is with respect to the series current as the reference phasor at 0°.

Parallel Combinations of X_L and R

When X_L is 10 times R, the parallel circuit is practically resistive because there is little inductive current in the line.

The small value of I_L results from the high X_L. The total impedance of the parallel circuit is approximately equal to the resistance, then, since the high value of X_L in a parallel branch has little effect.

As X_L becomes smaller, it provides more inductive current in the main line. When X_L is $\frac{1}{10}$ R, practically all of the line current is the I_L component. Then the parallel circuit is practically all inductive, with a total impedance practically equal to X_L. The phase angle of $-84.3°$ is almost $-90°$ because the line current is mostly inductive. Note that these conditions are opposite from those of X_L and R in series.

When X_L and R are equal, their branch currents are equal and the phase angle is $-45°$. All these phase angles are negative for parallel I_L and I_R.

20.5 Q of a Coil

The ability of a coil to produce self-induced voltage is indicated by X_L, since it includes the factors of frequency and inductance. However, a coil has internal resistance equal to the resistance of the wire in the coil. This internal r_i of the coil reduces the current, which means less ability to produce induced voltage. Combining these two factors of X_L and r_i, the *quality* or *merit* of a coil is indicated by

$$Q = \frac{X_L}{r_i} = \frac{2\pi f L}{r_i} \qquad \textbf{(20-7)}$$

As shown in Fig. 20-7, the internal r_i is in series with X_L.

As an example, a coil with X_L of 500 Ω and r_i of 5 Ω has a Q of $^{500}\!/_5 = 100$. The Q is a numerical value without any units, since the ohms cancel in the ratio of reactance to resistance. This Q of 100 means that the X_L of the coil is 100 times more than its r_i.

The Q of coils may range in value from less than 10 for a low-Q coil up to 1000 for a very high Q. Radio-frequency (RF) coils generally have Qs of about 30 to 300.

At low frequencies, r_i is just the dc resistance of the wire in the coil. However, for RF coils, the losses increase with higher frequencies and the effective r_i increases. The increased resistance results from eddy currents and other losses.

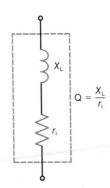

Figure 20-7 The Q of a coil depends on its inductive reactance X_L and resistance r_i.

$$Q = \frac{X_L}{r_i}$$

Because of these losses, the Q of a coil does not increase without limit as X_L increases for higher frequencies. Generally, Q can increase by a factor of about 2 for higher frequencies, within the range for which the coil is designed. The highest Q for RF coils generally results from an inductance value that provides an X_L of about 1000 Ω at the operating frequency.

More fundamentally, Q is defined as the ratio of reactive power in the inductance to the real power dissipated in the resistance. Then

$$Q = \frac{P_L}{P_{r_i}} = \frac{I^2 X_L}{I^2 r_i} = \frac{X_L}{r_i} = \frac{2\pi f L}{r_i}$$

which is the same as Formula (20-7).

Skin Effect

Radio-frequency current tends to flow at the surface of a conductor at very high frequencies, with little current in the solid core at the center. This skin effect results from the fact that current in the center of the wire encounters slightly more inductance because of the magnetic flux concentrated in the metal, compared with the edges, where part of the flux is in air. For this reason, conductors for VHF currents are often made of hollow tubing. The skin effect increases the effective resistance because a smaller cross-sectional area is used for the current path in the conductor.

AC Effective Resistance

When the power and current applied to a coil are measured for RF applied voltage, the $I^2 R$ loss corresponds to a much higher resistance than the dc resistance measured with an ohmmeter. This higher resistance is the ac effective resistance R_e. Although it is a result of high-frequency alternating current, R_e is not a reactance; R_e is a resistive component because it draws in-phase current from the ac voltage source.

The factors that make the R_e of a coil more than its dc resistance include skin effect, eddy currents, and hysteresis losses. Air-core coils have low losses but are limited to small values of inductance.

For a magnetic core in RF coils, a powdered-iron or ferrite slug is generally used. In a powdered-iron slug, the granules of iron are insulated from each other to reduce eddy currents. Ferrite materials have small eddy-current losses because they are insulators, although magnetic. A ferrite core is easily saturated. Therefore, its use must be limited to coils with low values of current.

As an example of the total effect of ac losses, assume that an air-core RF coil of 50-μH inductance has a dc resistance of 1 Ω measured with the battery in an ohmmeter. However, in an ac circuit with a 2-MHz current, the effective coil resistance R_e can increase to 12 Ω. The increased resistance reduces the Q of the coil.

Actually, the Q can be used to determine the effective ac resistance. Since Q is X_L/R_e, then R_e equals X_L/Q. For this 50-μH L at 2 MHz, its X_L, equal to $2\pi fL$, is 628 Ω. The Q of the coil can be measured on a Q meter, which operates on the principle of resonance. Let the measured Q be 50. Then $R_e = {}^{628}\!/\!_{50}$, equal to 12.6 Ω.

In general, the lower the internal resistance of a coil, the higher its Q.

EXAMPLE 20-3

An air-core coil has an X_L of 700 Ω and an R_e of 2 Ω. Calculate the value of Q for this coil.

Answer:

$$Q = \frac{X_L}{R_e} = \frac{700}{2}$$
$$= 350$$

EXAMPLE 20-4

A 200-μH coil has a Q of 40 at 0.5 MHz. Find R_e.

Answer:

$$R_e = \frac{X_L}{Q} = \frac{2\pi fL}{Q}$$
$$= \frac{2\pi \times 0.5 \times 10^6 \times 200 \times 10^{-6}}{40}$$
$$= \frac{628}{40}$$
$$= 15.7\ \Omega$$

The Q of a Capacitor

The quality Q of a capacitor in terms of minimum loss is often indicated by its power factor. The lower the numerical value of the power factor, the better the quality of the capacitor. Since the losses are in the dielectric, the power factor of the capacitor is essentially the power factor of the dielectric, independent of capacitance value or voltage rating. At radio frequencies, approximate values of power factor are 0.000 for air or vacuum, 0.0004 for mica, about 0.01 for paper, and 0.0001 to 0.03 for ceramics.

The reciprocal of the power factor can be considered the Q of the capacitor, similar to the idea of the Q of a coil. For instance, a power factor of 0.001 corresponds to a Q of 1000. A higher Q therefore means better quality for the capacitor. If the leakage resistance R_l is known, the Q can be calculated as $Q = 2\pi fR_lC$. Capacitors have Qs that are much higher than those of inductors. The Q of capacitors typically ranges into the thousands, depending on design.

20.6 AF and RF Chokes

Inductance has the useful characteristic of providing more ohms of reactance at higher frequencies. Resistance has the same opposition at all frequencies and for direct current. The skin effect for L at very high frequencies is not being considered here. These characteristics of L and R are applied to the circuit in Fig. 20-8 where X_L is much greater than R for the frequency of the ac source V_T. The result is that L has practically all the voltage drop in this series circuit with very little of the applied voltage across R.

The inductance L is used here as a *choke*. Therefore, a choke is an inductance in series with an external R to prevent the ac signal voltage from developing any appreciable output across R at the frequency of the source.

The dividing line in calculations for a choke can be taken as X_L 10 or more times the series R. Then the circuit is primarily inductive. Practically all the ac voltage drop is across L, with little across R. This case also results in θ of practically 90°, but the phase angle is not related to the action of X_L as a choke.

Figure 20-8b illustrates how a choke is used to prevent ac voltage in the input from developing voltage in the output for the next circuit. Note that the output here is V_R from point A to earth ground. Practically all ac input voltage is across X_L between points B and C. However, this voltage is not coupled out because neither B nor C is grounded.

The desired output across R could be direct current from the input side without any ac component. Then X_L has no effect on the steady dc component. Practically all dc voltage would be across R for the output, but the ac voltage would be just across X_L. The same idea applies to passing an AF signal through to R, while blocking an RF signal as IX_L across the choke because of more X_L at the higher frequency.

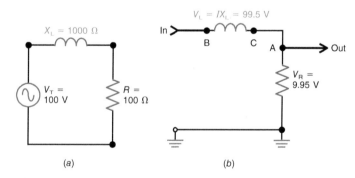

(a)

(b)

Figure 20-8 Coil used as a choke with X_L at least 10 × R. Note that R is an external resistor; V_L across L is practically all of the applied voltage with very little V_R. (a) Circuit with X_L and R in series. (b) Input and output voltages.

20.7 Power in AC Circuits

In any ac circuit with inductive or capacitive reactance, the current either lags or leads to applied voltage by some phase angle between 0 and 90°. With this condition, computing the power consumed by the circuit can be confusing. As it turns out, there are actually three different types of power involved in any reactive circuit: real power (P), reactive power (Q) and apparent power (S).

Real Power

Real power, or true power, is the power dissipated by the resistance in the circuit. You can always calculate it with the familiar expression

$$P = I^2R \qquad (20\text{-}8)$$

This is the power that generates the heat in the circuit.

Using the circuit in Fig. 20-9a, you can see that the current in the circuit is 2 A and the resistance is 100 Ω. The real power therefore is

$$P = I^2R = 4 \times 100 = 400 \text{ W}$$

You could also compute the real power with the expressions

$$P = \frac{V_R^2}{R}$$

$$P = V_R I$$

where V_R is the voltage drop across the resistor, not the applied voltage. Both these formulas will give you the same 400 W of real power.

The unit of real power is the watt, as in dc circuits.

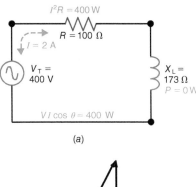

(a)

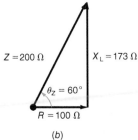

(b)

Figure 20-9 Real power, *P*, in a series circuit.
(a) Schematic diagram. (b) Impedance triangle with phase angle.

Reactive Power

Reactive power is the power that appears to be consumed by the reactive component in a circuit. In reality, reactive components do not consume any power. During part of an ac cycle, the component does consume power as it stores energy. In another part of the cycle, the reactive component gives that power back to the circuit. The net consumption is zero. However, any resistive part of the reactive component will dissipate power. Examples are the resistance of the wire in an inductor and the resistance of the plates and leads of a capacitor.

The reactive power Q is computed with the expression

$$Q = I^2X \qquad (20\text{-}9)$$

Further expressions for finding Q include

$$Q = \frac{V_X^2}{X}$$

$$Q = V_X I$$

where X is the capacitive or inductive reactance and V_X is the voltage across the reactance.

In Fig. 20-9a, the voltage across the inductor is

$$V_X = IX = 2 \times 173 = 364 \text{ V}$$

The reactive power then is

$$Q = \frac{V_X^2}{X} = \frac{364^2}{173} = 765.87 \text{ VAR}$$

The unit of reactive is the volt-ampere reactive (VAR). The watt is reserved only for real power.

Apparent Power

Apparent power is as the name implies is the power that is assumed to be consumed by the circuit. Apparent power is computed as

$$S = VI \qquad (20\text{-}10)$$

where V is the applied voltage V_T and I is the total circuit current. The apparent power in Fig. 20-9a is

$$S = 400 \times 2 = 800 \text{ VA}$$

The unit of apparent power is the volt-ampere (VA). Most ac power components that have some reactance like transformers, and motors are rated in VA rather than watts because VA gives a more accurate indication of what the actual current the total load will draw and use. And that, in turn, dictates the wire sizes that must be used for safety reasons.

Power Factor

The ideal condition of most power circuits is that there be no reactance so that the real power and the apparent power are the same. A reactive component in the circuit means that the generator capacity and the current-handling capability of any related power lines must be greater than if the load were purely resistive. Therefore, it is important to have some measure that indicates the relationship between the real and apparent power values. This measure is called the *power factor (PF)*. It is the ratio of the real to apparent power, or

$$PF = \frac{P}{S} \qquad (20\text{-}11)$$

It tells the percentage of real power in relationship to the apparent power. In the circuit of Fig. 20-9a, the power factor is

$$PF = \frac{P}{S} = \frac{400}{800} = 0.5$$

Therefore, 50% of the total apparent power is real power.

You can also express power factor as the ratio of the resistance in the circuit to the total impedance, or

$$PF = \frac{R}{Z}$$

In Fig. 20-9, the power factor is

$$PF = \frac{R}{Z} = \frac{100}{200} = 0.5$$

As it turns out, as you can see in Fig. 20-9b, the ratio of R to Z is the ratio of the adjacent side of the triangle to the hypotenuse. This is the same as the cosine of the phase angle. Therefore the power factor is

$$PF = R/Z = \cos\theta \qquad (20\text{-}12)$$

In the circuit of Fig. 20-9a, the phase angle is 60°.

$$PF = \cos\theta = \cos 60 = 0.5$$

You can compute the real power from the power factor and the apparent power:

$$P = S\cos\theta$$

In the example we have been using,

$$P = S\cos\theta = 80(0.5) = 400\text{ W}$$

Remember, the goal is a power factor of 1 where the real and apparent power are the same. That goal is rarely achieved, but the idea is to have the power factor as close to 1 as possible. A power factor of 0 would indicate no real power dissipation or only pure reactive power.

EXAMPLE 20-5

A series *RC* circuit contains a capacitor with a reactance of 70 Ω and a resistance of 90 Ω. The applied voltage is 240 V. What are the real, reactive, and apparent power values? What is the power factor? What is the phase angle?

Answer:

$$Z = \sqrt{R^2 + X^2} = \sqrt{90^2 + 70^2} = \sqrt{13{,}000} = 114\ \Omega$$
$$I = V/Z = 240/114 = 2.1\text{ A}$$
$$S = IV = 2.1(240) = 505.2\text{ VA}$$
$$Q = I^2X = 2.1^2 \times 70 = 308.7\text{ VAR}$$
$$P = I^2R = 2.1^2 \times 90 = 396.9\text{ W}$$
$$PF = R/Z = 90/114 = 0.79 \quad \text{or} \quad PF = P/S = 0.79$$
$$PF = \cos\theta = 0.79$$
$$\theta = \cos^{-1} 0.79 = \text{arcos } 0.79 = 37.86°$$

EXAMPLE 20-6

A parallel *RC* circuit has a resistance of 7 ohm and a capacitive reactance of 10 Ω. The applied voltage is 120 V. Calculate the real, reactive, and apparent power and the power factor.

Answer:

All of the formulas given earlier work for parallel circuits as well as series circuits.

$$P = V^2/R = 120^2/7 = 2057\text{ W} \quad \text{or} \quad 2.057\text{ kW}$$
$$Q = V^2/X = 120^2/10 = 1440\text{ VAR}$$
$$I_R = V/R = 120/7 = 17.14\text{ A}$$
$$I_C = V/X_C = 120/10 = 12\text{ A}$$

The total current is

$$I_T = \sqrt{I_R^2 + I_X^2} = \sqrt{17.14^2 + 12^2} = 20.93\text{ A}$$
$$Z = V/I_T = 120/20.93 = 5.73\ \Omega$$
$$S = VI_T = 120 \times 20.93 = 2511.6\text{ VA}$$
$$PF = P/S = 2057/2511.6 = 0.81$$

20.8 Complex Impedances

As you have seen in Formula (20-2), the impedance of a series *RL* or *RC* circuit is calculated with the expression

$$Z = \sqrt{R^2 + X^2}$$

This relationship is expressed in the form of an impedance triangle. As an example, assume we have a series *RL* circuit, as shown in Fig. 20-10a. With $R = 3\ \Omega$ and $X_L = 4\ \Omega$, the impedance is

$$Z = \sqrt{R^2 + X^2} = \sqrt{3^2 + 4^2} = 5\ \Omega$$

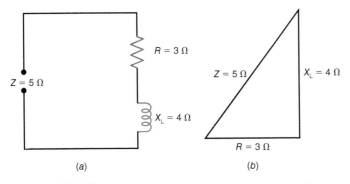

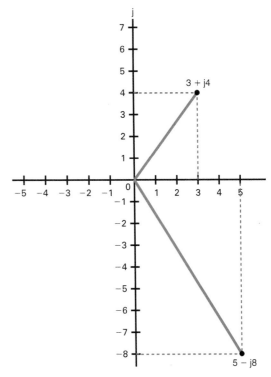

Figure 20-10 Impedance of (*a*) a curcuit expressed as (*b*) an impedance triangle.

Figure 20-11 Real and imaginary axis for plotting complex impedances.

This is illustrated in the impedance triangle shown in Fig.20-10*b*.

Another way to express this impedance is with a complex number such as

$$Z = R + jX = 3 + j4$$

This is called the *complex impedance*. That format is used in certain circuit analysis and calculation methods.

This complex number is derived by plotting the resistance and reactance on a set of rectangular coordinates, as shown in Fig. 20-11. There is a real axis and an imaginary axis. The horizontal real axis is used for plotting real numbers, either positive or negative. The real axis is used to plot the resistance. Since there is no negative resistance, only the positive axis to the right is used.

The vertical axis is called the j axis and is used to plot imaginary numbers. The imaginary numbers are the reactance values. The j factor simply indicates a 90° phase shift. (You do not multiply by j or 90.) It indicates that the reactance causes a 90° shift. An inductance produces a +90° shift, and a capacitance produces a −90° shift.

We plot the impedance $\underline{Z} = 3 +j4$ on the chart as shown in Fig. 20-11. Using the plotted points, we can then draw the related impedance triangle. This is the same impedance triangle you are familiar with, but it is on the set of axes.

A capacitive circuit is plotted in the same way. Assume a series *RC* circuit with $R = 5\ \Omega$ and $X_C = 8\ \Omega$. The complex impedance is $Z = 5 - j8$ and is plotted on the chart in Fig. 20-11.

Voltages and currents can also be expressed in complex form. Moreover, there are rules for adding, subtracting, multiplying, and dividing complex numbers.

You won't generally need to use these complex numbers, but it is helpful to know what they mean.

CHAPTER 20 SYSTEM SIDEBAR

Power Factor Correction

When sine-wave power is delivered to customers as the utility companies do, the desire is for a power factor of 1, where the actual real power consumed is equal to the apparent power. However, because there are so many inductive loads in the form of transformers and motors, the power factor is always less than 1. How much less is a matter of the actual number of inductive loads, which can vary widely from location to location as well as the time of day. The effect of this is to require the utility power generators to supply more power at a higher current level. This increases the losses in the system and makes it necessary for more power to be generated than actually used. This wastes energy. The solution to this problem is power factor correction.

Power factor correction is the process of adding a capacitor across the ac line to offset the effects of the inductive loads. See Fig. S20-1. Here the total inductive reactance of

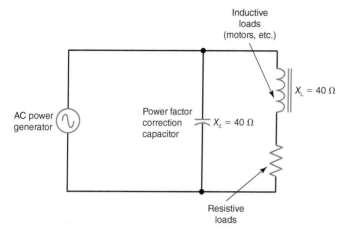

Figure S20-1 Power factor correction with a parallel capacitor.

the load is a hypothetical 40 Ω at 60 Hz. Assume that this is producing a lagging phase shift of 30°. This means that the power factor is

$$PF = \cos \theta = \cos 30° = 0.866$$

To correct for this phase shift, the idea is to connect a capacitor in parallel with the load that will offset or cancel the

inductive effects. What we need is a capacitor that provides a reactance of 40 Ω at 60 Hz.

Recall that the capacitive reactance is

$$X_C = \frac{1}{2\pi f C}$$

Knowing X_C and f, we can find C:

$$C = \frac{1}{2\pi f X_C} = \frac{1}{6.28(60)(40)} = 66.35 \times 10^{-6} \text{ or about } 66 \, \mu F$$

Placing the capacitor across the load will produce a leading phase shift that will offset the lagging phase shift of the inductive load. The capacitive current will almost completely cancel the inductive component of the load current. The resulting line current will be very low. The cancellation will not be complete because there still is a resistive part of the load. The power factor will be near 1.

All utility companies have a way to switch in power factor correction capacitors to help improve the efficiency of the power delivery. These capacitors may be located in substations or may actually be on power poles near transformers. By monitoring the phase angle of the power, the correct amount of capacitance can be switched in as needed. The inductive loading will vary over time so the process of monitoring and correcting is a continuous one. The resulting energy savings is enormous.

❓ CHAPTER 20 REVIEW QUESTIONS

1. Inductive reactance, X_L,
   a. applies only to nonsinusoidal waveforms or dc.
   b. applies only to sine waves.
   c. applies to either sinusoidal or nonsinusoidal waveforms.
   d. is inversely proportional to frequency.

2. For an inductor in a sine-wave ac circuit,
   a. V_L leads i_L by 90°.
   b. V_L lags i_L by 90°.
   c. V_L and i_L are in phase.
   d. none of the above.

3. In a series RL circuit,
   a. V_L lags V_R by 90°.
   b. V_L leads V_R by 90°.
   c. V_R and I are in phase.
   d. both b and c.

4. In a series RL circuit where $V_L = 9$ V and $V_R = 12$ V, how much is the total voltage, V_T?
   a. 21 V.
   b. 225 V.

   c. 15 V.
   d. 3 V.

5. A 50-Ω resistor is in parallel with an inductive reactance, X_L, of 50 Ω. The combined equivalent impedance, Z_{EQ} of this combination is
   a. 70.7 Ω.
   b. 100 Ω.
   c. 35.36 Ω.
   d. 25 Ω.

6. In a parallel RL circuit,
   a. I_L lags I_R by 90°.
   b. I_L leads I_R by 90°.
   c. I_L and I_R are in phase.
   d. I_R lags I_L by 90°.

7. In a parallel RL circuit, where $I_R = 1.2$ A and $I_L = 1.6$ A, how much is the total current, I_T?
   a. 2.8 A.
   b. 2 A.
   c. 4 A.
   d. 400 mA.

8. In a series *RL* circuit where $X_L = R$, the phase angle, θ_Z, is
   a. $-45°$.
   b. $0°$.
   c. $+90°$.
   d. $+45°$.

9. In a parallel *RL* circuit,
   a. V_A and I_L are in phase.
   b. I_L and I_R are in phase.
   c. V_A and I_R are in phase.
   d. V_A and I_R are 90° out of phase.

10. A 1-kΩ resistance is in series with an inductive reactance, X_L, of 2 kΩ. The total impedance, Z_T, is
    a. 2.24 kΩ.
    b. 3 kΩ.
    c. 1 kΩ.
    d. 5 MΩ.

11. When the frequency of the applied voltage decreases in a parallel *RL* circuit,
    a. the phase angle, θ_I, becomes less negative.
    b. Z_{EQ} increases.
    c. Z_{EQ} decreases.
    d. both *a* and *b*.

12. When the frequency of the applied voltage increases in a series *RL* circuit,
    a. θ_Z increases.
    b. Z_T decreases.
    c. Z_T increases.
    d. both *a* and *c*.

13. The dividing line for calculating the value of a choke inductance is to make
    a. X_L 10 or more times larger than the series *R*.
    b. X_L one-tenth or less than the series *R*.
    c. X_L equal to *R*.
    d. *R* 10 or more times larger than the series X_L.

14. The *Q* of a coil is affected by
    a. frequency.
    b. the resistance of the coil.

c. skin effect.
d. all of the above.

15. If the current through a 300-mH coil increases at the linear rate of 50 mA per 10 μs, how much is the induced voltage, V_L?
    a. 1.5 V.
    b. 1.5 kV.
    c. This is impossible to determine because X_L is unknown.
    d. This is impossible to determine because V_L also increases at a linear rate.

16. The unit for real power is
    a. watts.
    b. kilowatt-hours.
    c. VA.
    d. VAR.

17. Real power is the power dissipated by the
    a. complete circuit.
    b. capacitor.
    c. inductor.
    d. Resistance.

18. Ideal reactive components like inductors or capacitors dissipate heat.
    a. True.
    b. False.

19. Power factor is the ratio of the
    a. real power to reactive power.
    b. reactive power to real power.
    c. real power to apparent power.
    d. reactive power to apparent power.

20. The ideal power factor is
    a. 0.
    b. 1.
    c. less than 10.
    d. infinite.

 CHAPTER 20 PROBLEMS

SECTION 20.1 Sine Wave i_L Lags v_L by 90°

20.1 In Fig. 20-12, what is the
    a. peak value of the inductor voltage, v_L?
    b. peak value of the inductor current, i_L?
    c. frequency of the inductor current, i_L?
    d. phase relationship between v_L and i_L?

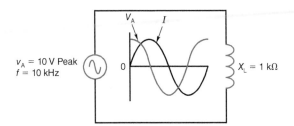

Figure 20-12

20.2 In Fig. 20-12, what is the value of the induced voltage, v_L, when i_L is at
  a. 0 mA?
  b. its positive peak of 10 mA?
  c. its negative peak of 10 mA?

20.3 In Fig. 20-12, draw the phasors representing v_L and i_L using
  a. i_L as the reference phasor.
  b. v_L as the reference phasor.

SECTION 20.2 X_L and R in Series

20.4 In Fig. 20-13, how much current, I, is flowing
  a. through the 15-Ω resistor, R?
  b. through the 20-Ω inductive reactance, X_L?
  c. to and from the terminals of the applied voltage, V_T?

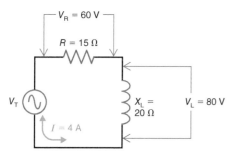

Figure 20-13

20.5 In Fig. 20-13, what is the phase relationship between
  a. I and V_R?
  b. I and V_L?
  c. V_L and V_R?

20.6 In Fig. 20-13, how much is the applied voltage, V_T?

20.7 Draw the phasor voltage triangle for the circuit in Fig. 20-13. (Use V_R as the reference phasor.)

20.8 In Fig. 20-14, solve for
  a. the resistor voltage, V_R.
  b. the inductor voltage, V_L.
  c. the total voltage, V_T.

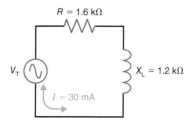

Figure 20-14

SECTION 20.3 Impedance Z Triangle

20.9 In Fig. 20-15, solve for Z_T, I, V_L, V_R, and θ_Z.

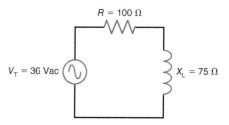

Figure 20-15

20.10 Draw the impedance triangle for the circuit in Fig. 20-15. (Use R as the reference phasor.)

20.11 In Fig. 20-16, what happens to each of the following quantities if the frequency of the applied voltage increases?
  a. X_L.
  b. Z_T.
  c. I.
  d. V_R.
  e. V_L.
  f. θ_Z.

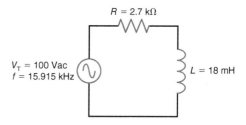

Figure 20-16

SECTION 20.4 X_L and R in Parallel

20.12 In Fig. 20-17, how much voltage is across
  a. the 30-Ω resistor, R?
  b. the 40-Ω inductive reactance, X_L?

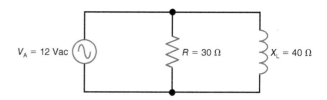

Figure 20-17

20.13 In Fig. 20-17, what is the phase relationship between
  a. V_A and I_R?
  b. V_A and I_L?
  c. I_L and I_R?

20.14 Draw the phasor current triangle for the circuit in Fig. 20-17. (Use I_R as the reference phasor.)

20.15 In Fig. 20-18, solve for I_R, I_L, I_T, Z_{EQ}, and θ_I.

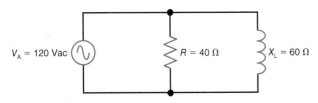

Figure 20-18

20.16 In Fig. 20-19, what happens to each of the following quantities if the frequency of the applied voltage increases?
a. I_R.
b. I_L.
c. I_T.
d. Z_{EQ}.
e. θ_I.

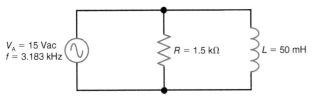

Figure 20-19

SECTION 20.5 *Q* of a Coil

20.17 For the inductor shown in Fig. 20-20, calculate the Q for the following frequencies:
a. $f = 500$ Hz.
b. $f = 10$ kHz.

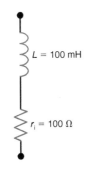

Figure 20-20

20.18 Why can't the Q of a coil increase without limit as the value of X_L increases for higher frequencies?

20.19 Calculate the ac effective resistance, R_e, of a 350-μH inductor whose Q equals 35 at 1.5 MHz.

SECTION 20.6 AF and RF Chokes

20.20 In Fig. 20.21, calculate the required value of the choke inductance, L, at the following frequencies:
a. $f = 500$ Hz.
b. $f = 1$ MHz.

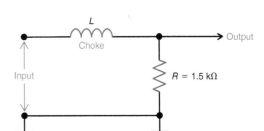

Figure 20-21

20.21 If $L = 50$ mH in Fig. 20-21, then what is the lowest frequency at which L will serve as a choke?

SECTION 20.7 Power in AC Circuits

20.22 In Fig. 20-13, calculate the real, reactive, and apparent power and power factor.

20.23 In Fig. 20-18, calculate the real, reactive, and apparent power and power factor.

SECTION 20.8 Complex Impedances

20.24 Express the impedance of the circuit in Fig. 20-4 as a complex number.

20.25 A series RC circuit has a resistance of 55 Ω and capacitive reactance of 90 Ω. What is the complex impedance?

Chapter 21

RC and *L/R* Time Constants

Many applications of inductance are for sine-wave ac circuits, but anytime the current changes, *L* has the effect of producing induced voltage. Examples of nonsinusoidal waveshapes include dc voltages that are switched on or off, square waves, sawtooth waves, and rectangular pulses. For capacitance, also, many applications are for sine waves, but whenever the voltage changes, *C* produces charge or discharge current.

With nonsinusoidal voltage and current, the effect of *L* or *C* is to produce a change in waveshape. This effect can be analyzed by means of the time constant for capacitive and inductive circuits. The time constant is the time for a change of 63.2% in the current through *L* or the voltage across *C*.

Actually, *RC* circuits are more common than *RL* circuits because capacitors are smaller and more economical and do not have strong magnetic fields.

Learning Outcomes

After studying this chapter, you should be able to:

> *Define* the term *time constant.*

> *Calculate* the time constant of a circuit containing resistance and inductance.

> *Explain* the effect of producing a high voltage when opening an *RL* circuit.

> *Calculate* the time constant of a circuit containing resistance and capacitance.

> *List* the criteria for proper differentiation and integration.

> *Explain* why a long time constant is required for an *RC* coupling circuit.

> *Use* the universal time constant graph to solve for voltage and current values in an *RC* or *RL* circuit that is charging or discharging.

21.1 Response of Resistance Alone

Figure 21-1a illustrates how an ordinary resistive circuit behaves. When the switch is closed, the battery supplies 10 V across the 10-Ω R and the resultant I is 1 A. The graph in Fig. 21-1b shows that I changes from 0 to 1 A instantly when the switch is closed. If the applied voltage is changed to 5 V, the current will change instantly to 0.5 A. If the switch is opened, I will immediately drop to zero.

Resistance has only opposition to current; there is no reaction to a change because R has no concentrated magnetic field to oppose a change in I, like inductance, and no electric field to store charge that opposes a change in V, like capacitance.

21.2 *L/R* Time Constant

Consider the circuit in Fig. 21-2, where L is in series with R. When S is closed, the current changes as I increases from zero. Eventually, I will reach the steady value of 1 A, equal to the battery voltage of 10 V divided by the circuit resistance of 10 Ω. While the current is building up from 0 to 1 A, however, I is changing and the inductance opposes the change. The action of the RL circuit during this time is its *transient response,* which means that a temporary condition exists only until the steady-state current of 1 A is reached. Similarly, when S is

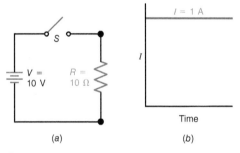

(a)   (b)

Figure 21-1   Response of circuit with R alone. When switch is closed, current I is 10 V/10 Ω = 1 A. (a) Circuit. (b) Graph of steady I.

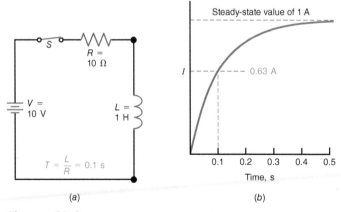

(a)   (b)

Figure 21-2   Transient response of circuit with R and inductance L. When the switch is closed, I rises from zero to the steady-state value of 1 A. (a) Circuit with time constant L/R of 1 H/10 Ω = 0.1 s. (b) Graph of I during five time constants. Compare with graph in Fig. 21-1b.

opened, the transient response of the RL circuit opposes the decay of current toward the steady-state value of zero.

The transient response is measured in terms of the ratio L/R, which is the time constant of an inductive circuit. To calculate the time constant,

$$T = \frac{L}{R} \tag{21-1}$$

where T is the time constant in seconds, L is the inductance in henrys, and R is the resistance in ohms. The resistance in series with L is either the coil resistance, an external resistance, or both in series. In Fig. 21-2,

$$T = \frac{L}{R} = \frac{1}{10} = 0.1 \text{ s}$$

Specifically, the time constant is a measure of how long it takes the current to change by 63.2%, or approximately 63%. In Fig. 21-2, the current increases from 0 to 0.63 A, which is 63% of the steady-state value, in a period of 0.1 s, which is one time constant. In a period of five time constants, the current is practically equal to its steady-state value of 1 A.

EXAMPLE 21-1

What is the time constant of a 20-H coil having 100 Ω of series resistance?

Answer:

$$T = \frac{L}{R} = \frac{20 \text{ H}}{100 \text{ }\Omega}$$
$$= 0.2 \text{ s}$$

EXAMPLE 21-2

An applied dc voltage of 10 V will produce a steady-state current of 100 mA in the 100-Ω coil of Example 21-1. How much is the current after 0.2 s? After 1 s?

Answer:

Since 0.2 s is one time constant, I is 63% of 100 mA, which equals 63 mA. After five time constants, or 1 s (0.2 s × 5), the current will reach its steady-state value of 100 mA and remain at this value as long as the applied voltage stays at 10 V.

EXAMPLE 21-3

If a 1-MΩ R is added in series with the coil of Example 21-1, how much will the time constant be for the higher resistance RL circuit?

Answer:

$$T = \frac{L}{R} = \frac{20 \text{ H}}{1,000,000 \text{ }\Omega}$$
$$= 20 \times 10^{-6} \text{ s}$$
$$= 20 \text{ }\mu\text{s}$$

The L/R time constant becomes longer with larger values of L. More series R, however, makes the time constant shorter. With more series resistance, the circuit is less inductive and more resistive.

21.3 High Voltage Produced by Opening an *RL* Circuit

When an inductive circuit is opened, the time constant for current decay becomes very short because L/R becomes smaller with the high resistance of the open circuit. Then the current drops toward zero much faster than the rise of current when the switch is closed. The result is a high value of self-induced voltage V_L across a coil whenever an RL circuit is opened. This high voltage can be much greater than the applied voltage.

There is no gain in energy, though, because the high-voltage peak exists only for the short time the current is decreasing at a very fast rate at the start of the decay. Then, as I decays at a slower rate, the value of V_L is reduced. After the current has dropped to zero, there is no voltage across L.

This effect can be demonstrated by a neon bulb connected across a coil, as shown in Fig. 21-3. The neon bulb requires 90 V for ionization, at which time it glows. The source here is only 8 V, but when the switch is opened, the self-induced voltage is high enough to light the bulb for an instant. The sharp voltage pulse or spike is more than 90 V just after the switch is opened, when I drops very fast at the start of the decay in current.

Note that the 100-Ω R_1 is the internal resistance of the 2-H coil. This resistance is in series with L whether S is closed or open. The 4-kΩ R_2 across the switch is in the circuit only when S is opened, to have a specific resistance across the open switch. Since R_2 is much more than R_1, the L/R time constant is much shorter with the switch open.

Closing the Circuit

In Fig. 21-3a, the switch is closed to allow current in L and to store energy in the magnetic field. Since R_2 is short-circuited by the switch, the 100-Ω R_1 is the only resistance.

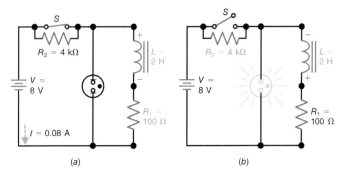

(a) (b)

Figure 21-3 Demonstration of high voltage produced by opening inductive circuit. (*a*) With switch closed, 8 V applied cannot light the 90-V neon bulb. (*b*) When the switch is opened, the short L/R time constant results in high V_L, which lights the bulb.

The steady-state I is $V/R_1 = 8/100 = 0.08$ A. This value of I is reached after five time constants.

One time constant is $L/R = 2/100 = 0.02$ s. Five time constants equal $5 \times 0.02 = 0.1$ s. Therefore, I is 0.08 A after 0.1 s, or 100 ms. The energy stored in the magnetic field is 64×10^{-4} J, equal to $\frac{1}{2}LI^2$.

Opening the Circuit

When the switch is opened in Fig. 21-3b, R_2 is in series with L, making the total resistance 4100 Ω, or approximately 4 kΩ. The result is a much shorter time constant for current decay. Then L/R is $2/4000$, or 0.5 ms. The current decays practically to zero in five time constants, or 2.5 ms.

This rapid drop in current results in a magnetic field collapsing at a fast rate, inducing a high voltage across L. The peak v_L in this example is 320 V. Then v_L serves as the voltage source for the bulb connected across the coil. As a result, the neon bulb becomes ionized, and it lights for an instant. One problem is arcing produced when an inductive circuit is opened. Arcing can destroy contact points on mechanical switches and completely ruin any series transistor switch.

Applications of Inductive Voltage Pulses

There are many uses for the high voltage generated by opening an inductive circuit. One example is the high voltage produced for the ignition system in an automobile. Here the circuit of the battery in series with a high-inductance spark coil is opened by the breaker points of the distributor to produce the high voltage needed for each spark plug. When an inductive circuit is opened very rapidly, voltages as high as 50,000 V can easily be produced.

21.4 *RC* Time Constant

The transient response of capacitive circuits is measured in terms of the product $R \times C$. To calculate the time constant,

$$T = R \times C \qquad \text{(21-2)}$$

where R is in ohms, C is in farads, and T is in seconds. In Fig. 21-4, for example, with an R of 3 MΩ and a C of 1 μF,

$$T = 3 \times 10^6 \times 1 \times 10^{-6}$$
$$= 3 \text{ s}$$

Note that the 10^6 for megohms and the 10^{-6} for microfarads cancel. Therefore, multiplying the units of M$\Omega \times \mu$F gives the RC product in seconds.

Common combinations of units for the RC time constant are

$$M\Omega \times \mu F = s$$
$$k\Omega \times \mu F = ms$$
$$M\Omega \times pF = \mu s$$

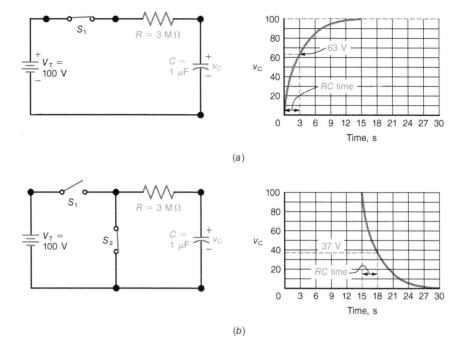

Figure 21-4 Details of how a capacitor charges and discharges in an *RC* circuit. (*a*) With S_1 closed, *C* charges through *R* to 63% of v_T in one *RC* time constant of 3 s and is almost completely charged in five time constants. (*b*) With S_1 opened to disconnect the battery and S_2 closed for *C* to discharge through *R*, V_C drops to 37% of its initial voltage in one time constant of 3 s and is almost completely discharged in five time constants.

The reason that the *RC* product is expressed in units of time can be illustrated as follows: $C = Q/V$. The charge Q is the product of $I \times T$. The factor V is IR. Therefore, *RC* is equivalent to $(R \times Q)/V$, or $(R \times IT)/IR$. Since I and R cancel, T remains to indicate the dimension of time.

The Time Constant Indicates the Rate of Charge or Discharge

RC specifies the time it takes *C* to charge to 63% of the charging voltage. Similarly, *RC* specifies the time it takes *C* to discharge 63% of the way down to the value equal to 37% of the initial voltage across *C* at the start of discharge.

In Fig. 21-4*a*, for example, the time constant on charge is 3 s. Therefore, in 3 s, *C* charges to 63% of the 100 V applied, reaching 63 V in *RC* time. After five time constants, which is 15 s here, *C* is almost completely charged to the full 100 V applied. If *C* discharges after being charged to 100 V, then *C* will discharge down to 36.8 V or approximately 37 V in 3 s. After five time constants, *C* discharges to zero.

A shorter time constant allows the capacitor to charge or discharge faster. If the *RC* product in Fig. 21-4 is 1 s, then *C* will charge to 63 V in 1 s instead of 3 s. Also, v_C will reach the full applied voltage of 100 V in 5 s instead of 15 s. Charging to the same voltage in less time means a faster charge.

On discharge, the shorter time constant will allow *C* to discharge from 100 to 37 V in 1 s instead of 3 s. Also, v_C will be down to zero in 5 s instead of 15 s.

For the opposite case, a longer time constant means slower charge or discharge of the capacitor. More *R* or *C* results in a longer time constant.

RC Applications

Several examples are given here to illustrate how the time constant can be applied to *RC* circuits.

EXAMPLE 21-4

What is the time constant of a 0.01-μF capacitor in series with a 1-MΩ resistance?

Answer:

$$T = R \times C = 1 \times 10^6 \times 0.01 \times 10^{-6}$$

$$= 0.01 \text{ s}$$

The time constant in Example 21-4 is for charging or discharging, assuming the series resistance is the same for charge or discharge.

EXAMPLE 21-5

With a dc voltage of 300 V applied, how much is the voltage across C in Example 21-4 after 0.01 s of charging? After 0.05 s? After 2 hours? After 2 days?

Answer:

Since 0.01 s is one time constant, the voltage across C then is 63% of 300 V, which equals 189 V. After five time constants, or 0.05 s, C will be charged practically to the applied voltage of 300 V. After 2 hours or 2 days, C will still be charged to 300 V if the applied voltage is still connected.

EXAMPLE 21-6

If the capacitor in Example 21-5 is allowed to charge to 300 V and then discharged, how much is the capacitor voltage 0.01 s after the start of discharge? The series resistance is the same on discharge as on charge.

Answer:

In one time constant, C discharges to 37% of its initial voltage, or 0.37×300 V, which equals 111 V.

EXAMPLE 21-7

Assume the capacitor in Example 21-5 is discharging after being charged to 200 V. How much will the voltage across C be 0.01 s after the beginning of discharge? The series resistance is the same on discharge as on charge.

Answer:

In one time constant, C discharges to 37% of its initial voltage, or 0.37×200, which equals 74 V.

Example 21-7 shows that the capacitor can charge or discharge from any voltage value. The rate at which it charges or discharges is determined by RC, counting from the time the charge or discharge starts.

EXAMPLE 21-8

If a 1-MΩ resistance is added in series with the capacitor and resistor in Example 21-4, how much will the time constant be?

Answer:

Now the series resistance is 2 MΩ. Therefore, RC is 2×0.01, or 0.02 s.

The RC time constant becomes longer with larger values of R and C. More capacitance means that the capacitor can store more charge. Therefore, it takes longer to store the charge needed to provide a potential difference equal to 63% of the applied voltage. More resistance reduces the charging current, requiring more time to charge the capacitor.

Note that the RC time constant only specifies a rate. The actual amount of voltage across C depends on the amount of applied voltage as well as on the RC time constant.

A capacitor takes on charge whenever its voltage is less than the applied voltage. The charging continues at the RC rate until the capacitor is completely charged, or the voltage is disconnected.

A capacitor discharges whenever its voltage is more than the applied voltage. The discharge continues at the RC rate until the capacitor is completely discharged, the capacitor voltage equals the applied voltage, or the load is disconnected.

To summarize these two important principles:

1. Capacitor C charges when the net charging voltage is more than v_C.
2. Capacitor C discharges when v_C is more than the net charging voltage.

The net charging voltage equals the difference between v_C and the applied voltage.

21.5 *RC* Charge and Discharge Curves

In Fig. 21-4, the rise is shown in the RC charge curve because the charging is fastest at the start and then tapers off as C takes on additional charge at a slower rate. As C charges, its potential difference increases. Then the difference in voltage between V_T and v_C is reduced. Less potential difference reduces the current that puts the charge in C. The more C charges, the more slowly it takes on additional charge.

Similarly, on discharge, C loses its charge at a declining rate. At the start of discharge, v_C has its highest value and can produce maximum discharge current. As the discharge continues, v_C goes down and there is less discharge current. The more C discharges, the more slowly it loses the remainder of its charge.

Charge and Discharge Current

There is often the question of how current can flow in a capacitive circuit with a battery as the dc source. The answer is that current flows anytime there is a change in voltage. When V_T is connected, the applied voltage changes from zero. Then charging current flows to charge C to the applied voltage. After v_C equals V_T, there is no net charging voltage and I is zero.

Similarly, C can produce discharge current anytime v_C is greater than V_T. When V_T is disconnected, v_C can discharge down to zero, producing discharge current in the direction opposite from the charging current. After v_C equals zero, there is no current.

Capacitance Opposes Voltage Changes across Itself

This ability of capacitance to oppose voltage changes across itself corresponds to the ability of inductance to oppose a change in current. When the applied voltage in an RC circuit increases, the voltage across the capacitance cannot increase until the charging current has stored enough charge in C. The increase in applied voltage is present across the resistance in series with C until the capacitor has charged to the higher applied voltage. When the applied voltage decreases, the voltage across the capacitor cannot go down immediately because the series resistance limits the discharge current.

The voltage across the capacitance in an RC circuit, therefore, cannot follow instantaneously the changes in applied voltage. As a result, the capacitance is able to oppose changes in voltage across itself. The instantaneous variations in V_T are present across the series resistance, however, since the series voltage drops must add to equal the applied voltage at all times.

21.6 High Current Produced by Short-Circuiting an RC Circuit

A capacitor can be charged slowly by a small charging current through a high resistance and then be discharged quickly through a low resistance to obtain a momentary surge, or pulse, of discharge current. This idea corresponds to the pulse of high voltage obtained by opening an inductive circuit.

The circuit in Fig. 21-5 illustrates the application of a battery-capacitor (BC) unit to fire a flashbulb for cameras.

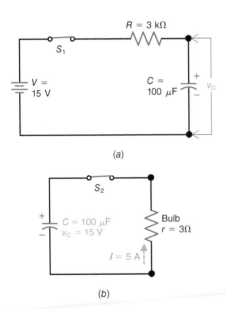

(a)

(b)

Figure 21-5 Demonstration of high current produced by discharging a charged capacitor through a low resistance. (a) When S_1 is closed, C charges to 15 V through 3 kΩ. (b) Without the battery, S_2 is closed to allow V_C to produce the peak discharge current of 5 A through the 3-Ω bulb. V_C in (b) is across the same C used in (a).

The flashbulb needs 5 A to ignite, but this is too much load current for the small 15-V battery, which has a rating of 30 mA for normal load current. Instead of using the bulb as a load for the battery, though, the 100-μF capacitor is charged by the battery through the 3-kΩ R in Fig. 21-5a, and then the capacitor is discharged through the bulb in Fig. 21-5b.

Charging the Capacitor

In Fig. 21-5a, S_1 is closed to charge C through the 3-kΩ R without the bulb. The time constant of the RC charging circuit is 0.3 s.

After five time constants, or 1.5 s, C is charged to the 15 V of the battery. The peak charging current, at the first instant of charge, is V/R or 15 V/3 kΩ, which equals 5 mA. This value is an easy load current for the battery.

Discharging the Capacitor

In Fig. 21-5b, v_C is 15 V without the battery. Now S_2 is closed, and C discharges through the 3-Ω resistance of the bulb. The time constant for discharge with the lower r of the bulb is $3 \times 100 \times 10^{-6}$, which equals 300 μs. At the first instant of discharge, when v_C is 15 V, the peak discharge current is $^{15}\!/_3$, which equals 5 A. This current is enough to fire the bulb.

Energy Stored in C

When the 100-μF C is charged to 15 V by the battery, the energy stored in the electric field is $CV^2/2$, which equals 0.01 J, approximately. This energy is available to maintain v_C at 15 V for an instant when the switch is closed. The result is the 5-A I through the 3-Ω r of the bulb at the start of the decay. Then v_C and i_C drop to zero in five time constants.

21.7 RC Waveshapes

The voltage and current waveshapes in the RC circuit in Fig. 21-6 show when a capacitor is allowed to charge through a resistance for RC time and then discharge through the same resistance for the same amount of time. Note that this particular case is not typical of practical RC circuits, but the waveshapes show some useful details about the voltage and current for charging and discharging. The RC time constant here equals 0.1 s to simplify the calculations.

Square Wave of Applied Voltage

The idea of closing S_1 to apply 100 V and then opening it to disconnect V_T at a regular rate corresponds to a square wave of applied voltage, as shown by the waveform in Fig. 21-6a. When S_1 is closed for charge, S_2 is open; when S_1 is open, S_2 is closed for discharge. Here the voltage is on for the RC time of 0.1 s and off for the same time of 0.1 s. The period of the square wave is 0.2 s, and f is 1/0.2 s, which equals 5 Hz for the frequency.

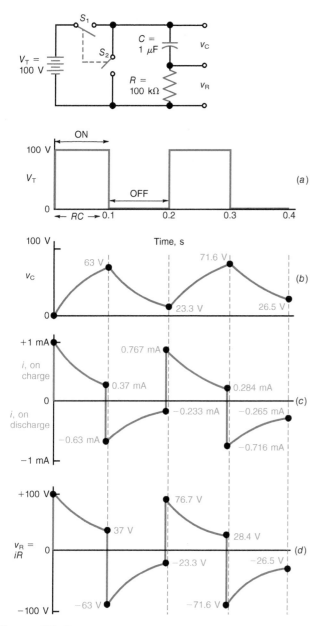

Figure 21-6 Waveshapes for the charge and discharge of an *RC* circuit in *RC* time. Circuit on top with S_1 and S_2 provides the square wave of applied voltage.

Capacitor Voltage v_C

As shown in Fig. 21-6*b*, the capacitor charges to 63 V, equal to 63% of the charging voltage, in the *RC* time of 0.1 s. Then the capacitor discharges because the applied V_T drops to zero. As a result, v_C drops to 37% of 63 V, or 23.3 V in *RC* time.

The next charge cycle begins with v_C at 23.3 V. The net charging voltage now is $100 - 23.3 = 76.7$ V. The capacitor voltage increases by 63% of 76.7 V, or 48.3 V. When 48.3 V is added to 23.3 V, v_C rises to 71.6 V. On discharge, after 0.3 s, v_C drops to 37% of 71.6 V, or to 26.5 V.

Charge and Discharge Current

As shown in Fig. 21-6*c*, the current *i* has its positive peak at the start of charge and its negative peak at the start of discharge. On charge, *i* is calculated as the net charging voltage, which is $(V_T - v_C)$ divided by *R*. On discharge, *i* always equals v_C/R.

At the start of charge, *i* is maximum because the net charging voltage is maximum before *C* charges. Similarly, the peak *i* for discharge occurs at the start, when v_C is maximum before *C* discharges.

Note that *i* is an ac waveform around the zero axis, since the charge and discharge currents are in opposite directions. We are arbitrarily taking the charging current as positive values for *i*.

Resistor Voltage v_R

This waveshape in Fig. 21-6*d* follows the waveshape of current because v_R is $i \times R$. Because of the opposite directions of charge and discharge current, the *iR* waveshape is an ac voltage.

Note that on charge, v_R must always be equal to $V_T - v_C$ because of the series circuit.

On discharge, v_R has the same values as v_C because they are in parallel, without V_T. Then S_2 is closed to connect *R* across *C*.

Why the i_C Waveshape Is Important

The v_C waveshape of capacitor voltage in Fig. 21-6 shows the charge and discharge directly, but the i_C waveshape is very interesting. First, the voltage waveshape across *R* is the same as the i_C waveshape. Also, whether *C* is charging or discharging, the i_C waveshape is the same except for the reversed polarity. We can see the i_C waveshape as the voltage across *R*. It generally is better to connect an oscilloscope for voltage waveshapes across *R*, especially with one side grounded.

Finally, we can tell what v_C is from the v_R waveshape. The reason is that at any instant, V_T must equal the sum of v_R and v_C. Therefore v_C is equal to $V_T - v_R$, when V_T is charging *C*. When *C* is discharging, there is no V_T. Then v_R is the same as v_C.

21.8 Long and Short Time Constants

Useful waveshapes can be obtained by using *RC* circuits with the required time constant. In practical applications, *RC* circuits are used more than *RL* circuits because almost any value of an *RC* time constant can be obtained easily. With coils, the internal series resistance cannot be short-circuited and the distributed capacitance often causes resonance effects.

Long *RC* Time

Whether an *RC* time constant is long or short depends on the pulse width of the applied voltage. We can arbitrarily define a long time constant as at least five times longer than the pulse width, in time, for the applied voltage. As a result, C takes on very little charge. The time constant is too long for v_C to rise appreciably before the applied voltage drops to zero and C must discharge. On discharge also, with a long time constant, C discharges very little before the applied voltage rises to make C charge again.

Short RC Time

A short time constant is defined as no more than one-fifth the pulse width, in time, for the applied voltage V_T. Then V_T is applied for a period of at least five time constants, allowing C to become completely charged. After C is charged, v_C remains at the value of V_T while the voltage is applied. When V_T drops to zero, C discharges completely in five time constants and remains at zero while there is no applied voltage. On the next cycle, C charges and discharges completely again.

Differentiation

The voltage across R in an RC circuit is called a *differentiated output* because v_R can change instantaneously. A short time constant is always used for differentiating circuits to provide sharp pulses of v_R.

Integration

The voltage across C is called an *integrated output* because it must accumulate over a period of time. A medium or long time constant is always used for integrating circuits.

21.9 Charge and Discharge with a Short RC Time Constant

Usually, the time constant is made much shorter or longer than a factor of 5 to obtain better waveshapes. In Fig. 21-7, RC is 0.1 ms. The frequency of the square wave is 25 Hz, with a period of 0.04 s, or 40 ms. One-half this period is the time when V_T is applied. Therefore, the applied voltage is on for 20 ms and off for 20 ms. The RC time constant of 0.1 ms is shorter than the pulse width of 20 ms by a factor of ½₀₀. Note that the time axis of all waveshapes is calibrated in seconds for the period of V_T, not in RC time constants.

Square Wave of V_T Is across C

The waveshape of v_C in Fig. 21-7b is the same as the square wave of applied voltage because the short time constant allows C to charge or discharge completely very soon after V_T is applied or removed. The charge or discharge time of five time constants is much less than the pulse width.

Sharp Pulses of i

The waveshape of i shows sharp peaks for the charge or discharge current. Each current peak is $V_T/R = 1$ mA, decaying to zero in five RC time constants. These pulses coincide with the leading and trailing edges of the square wave of V_T.

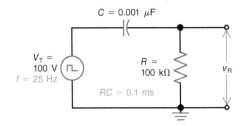

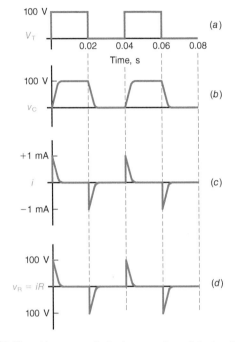

Figure 21-7 Charge and discharge of an RC circuit with a short time constant. Note that the waveshape of V_R in (d) has sharp voltage peaks for the leading and trailing edges of the square-wave applied voltage.

Actually, the pulses are much sharper than shown. They are not to scale horizontally to indicate the charge and discharge action. Also, v_C is actually a square wave, like the applied voltage, but with slightly rounded corners for the charge and discharge.

Sharp Pulses of v_R

The waveshape of voltage across the resistor follows the current waveshape because $v_R = iR$. Each current pulse of 1 mA across the 100-kΩ R results in a voltage pulse of 100 V.

More fundamentally, the peaks of v_R equal the applied voltage V_T before C charges. Then v_R drops to zero as v_C rises to the value of V_T.

On discharge, $v_R = v_C$, which is 100 V at the start of discharge. Then the pulse drops to zero in five time constants. The pulses of v_R in Fig. 21-7 are useful as timing pulses that match the edges of the square-wave applied voltage V_T. Either the positive or the negative pulses can be used.

The *RC* circuit in Fig. 21-7a is a good example of an *RC* differentiator. With the *RC* time constant much shorter than the pulse width of V_T, the voltage V_R follows instantaneously the changes in the applied voltage. Keep in mind that a differentiator must have a short time constant with respect to the pulse width of V_T to provide good differentiation. For best results, an *RC* differentiator should have a time constant which is one-tenth or less of the pulse width of V_T.

21.10 Long Time Constant for an *RC* Coupling Circuit

The *RC* circuit in Fig. 21-8 is the same as that in Fig. 21-7, but now the *RC* time constant is long because of the higher frequency of the applied voltage. Specifically, the *RC* time

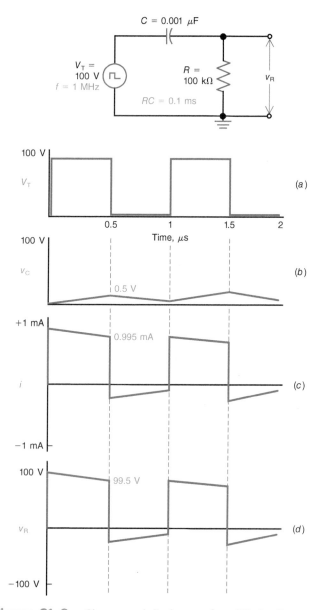

Figure 21-8 Charge and discharge of an *RC* circuit with a long time constant. Note that the waveshape of V_R in (*d*) has the same waveform as the applied voltage.

of 0.1 ms is 200 times longer than the 0.5-μs pulse width of V_T with a frequency of 1 MHz. Note that the time axis is calibrated in microseconds for the period of V_T, not in *RC* time constants.

Very Little of V_T Is across C

The waveshape of v_C in Fig. 21-8b shows very little voltage rise because of the long time constant. During the 0.5 μs when V_T is applied, *C* charges to only $\frac{1}{200}$ of the charging voltage. On discharge, also, v_C drops very little.

Square Wave of *i*

The waveshape of *i* stays close to the 1-mA peak at the start of charging. The reason is that v_C does not increase much, allowing V_T to maintain the charging current. On discharge, the reverse *i* for discharge current is very small because v_C is low.

Square Wave of V_T Is across R

The waveshape of v_R is the same square wave as *i* because $v_R = iR$. The waveshapes of *i* and v_R are essentially the same as the square-wave V_T applied. They are not shown to scale vertically to indicate the slight charge and discharge action.

Eventually, v_C will climb to the average dc value of 50 V, *i* will vary ± 0.5 mA above and below zero, and v_R will vary ± 50 V above and below zero. This application is an *RC* coupling circuit to block the average value of the varying dc voltage V_T as the capacitive voltage v_C, and v_R provides an ac voltage output having the same variations as V_T.

If the output is taken across *C* rather than *R* in Fig. 21-8a, the circuit is classified as an *RC* integrator. In Fig. 21-8b, it can be seen that *C* combines or integrates its original voltage with the new change in voltage. Eventually, however, the voltage across *C* will reach a steady-state value of 50 V after the input waveform has been applied for approximately five *RC* time constants. Keep in mind that an integrator must have a long time constant with respect to the pulse width of V_T to provide good integration. For best results, an *RC* integrator should have a time constant which is 10 or more times longer than the pulse width of V_T.

21.11 Advanced Time Constant Analysis

We can determine transient voltage and current values for any amount of time with the curves in Fig. 21-9. The rising curve *a* shows how v_C builds up as *C* charges in an *RC* circuit; the same curve applies to i_L, increasing in the inductance for an *RL* circuit. The decreasing curve *b* shows how v_C drops as *C* discharges or i_L decays in an inductance.

Note that the horizontal axis is in units of time constants rather than absolute time. Suppose that the time constant of

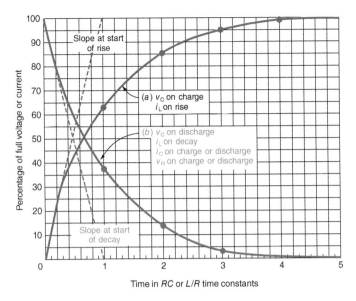

Figure 21-9 Universal time constant chart for *RC* and *RL* circuits. The rise or fall changes by 63% in one time constant.

an *RC* circuit is 5 μs. Therefore, one *RC* time unit = 5 μs, two *RC* units = 10 μs, three *RC* units = 15 μs, four *RC* units = 20 μs, and five *RC* units = 25 μs.

As an example, to find v_C after 10 μs of charging, we can take the value of curve *a* in Fig. 21-9 at two *RC*. This point is at 86% amplitude. Therefore, we can say that in this *RC* circuit with a time constant of 5 μs, v_C charges to 86% of the applied V_T after 10 μs. Similarly, some important values that can be read from the curve are listed in Table 21-1.

If we consider curve *a* in Fig. 21-9 as an *RC* charge curve, v_C adds 63% of the net charging voltage for each additional unit of one time constant, although it may not appear so. For instance, in the second interval of *RC* time, v_C adds 63% of the net charging voltage, which is 0.37 V_T. Then 0.63 × 0.37 equals 0.23, which is added to 0.63 to give 0.86, or 86%, as the total charge from the start.

Table 21-1 Time Constant Factors

Factor	Amplitude
0.2 time constant	20%
0.5 time constant	40%
0.7 time constant	50%
1 time constant	63%
2 time constants	86%
3 time constants	96%
4 time constants	98%
5 time constants	99%

Slope at *t* = 0

The curves in Fig. 21-9 can be considered approximately linear for the first 20% of change. In 0.1 time constant, for instance, the change in amplitude is 10%; in 0.2 time constant, the change is 20%. The dashed lines in Fig. 21-9 show that if this constant slope continued, the result would be 100% charge in one time constant. This does not happen, though, because the change is opposed by the energy stored in *L* and *C*. However, at the first instant of rise or decay, at *t* = 0, the change in v_C or i_L can be calculated from the dotted slope line.

Equation of the Decay Curve

The rising curve *a* in Fig. 21-9 may seem more interesting because it describes the buildup of v_C or i_L, but the decaying curve *b* is more useful. For *RC* circuits, curve *b* can be applied to

1. v_C on discharge
2. *i* and v_R on charge or discharge

If we use curve *b* for the voltage in *RC* circuits, the equation of this decay curve can be written as

$$v = V \times \epsilon^{-t/RC} \qquad \textbf{(21-3)}$$

where V is the voltage at the start of decay and *v* is the instantaneous voltage after the time *t*. Specifically, *v* can be v_R on charge and discharge or v_C only on discharge.

The constant ϵ is the base 2.718 for natural logarithms. The negative exponent $-t/RC$ indicates a declining exponential or logarithmic curve. The value of t/RC is the ratio of actual time of decline *t* to the *RC* time constant.

This equation can be converted to common logarithms for easier calculations. Since the natural base ϵ is 2.718, its logarithm to base 10 equals 0.434. Therefore, the equation becomes

$$v = \text{antilog}\left(\log V - 0.434 \times \frac{t}{RC}\right) \qquad \textbf{(21-4)}$$

Calculations for v_R

As an example, let us calculate v_R dropping from 100 V, after *RC* time. Then the factor t/RC is 1. Substituting these values,

$$v_R = \text{antilog}\,(\log 100 - 0.434 \times 1)$$
$$= \text{antilog}\,(2 - 0.434)$$
$$= \text{antilog}\,1.566$$
$$= 37\ \text{V}$$

All these logs are to base 10. Note that log 100 is taken first so that 0.434 can be subtracted from 2 before the antilog of the difference is found. The antilog of 1.566 is 37.

We can also use V_R to find V_C, which is $V_T - V_R$. Then 100 − 37 = 63 V for V_C. These answers agree with the fact that in one time constant, V_R drops 63% and V_C rises 63%.

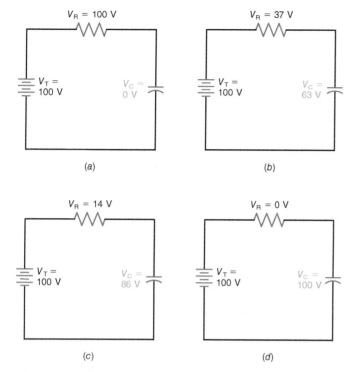

(a)   (b)

(c)   (d)

Figure 21-10 How v_C and v_R add to equal the applied voltage v_T of 100 V. (a) Zero time at the start of charging. (b) After one RC time constant. (c) After two RC time constants. (d) After five or more RC time constants.

Figure 21-10 illustrates how the voltages across R and C in series must add to equal the applied voltage V_T. The four examples with 100 V applied are

1. At time zero, at the start of charging, V_R is 100 V and V_C is 0 V. Then $100 + 0 = 100$ V.

2. After one time constant, V_R is 37 V and V_C is 63 V. Then $37 + 63 = 100$ V.

3. After two time constants, V_R is 14 V and V_C is 86 V. Then $14 + 86 = 100$ V.

4. After five time constants, V_R is 0 V and V_C is 100 V, approximately. Then $0 + 100 = 100$ V

It should be emphasized that Formulas (21-3) and (21-4) can be used to calculate any decaying value on curve b in Fig. 21-9. These applications for an RC circuit include V_R on charge or discharge, i on charge or discharge, and V_C only on discharge. For an RC circuit in which C is charging, Formula (21-5) can be used to calculate the capacitor voltage v_C at any point along curve a in Fig. 21-9:

$$v_C = V(1 - \epsilon^{-t/RC}) \tag{21-5}$$

In Formula (21-5), V represents the maximum voltage to which C can charge, whereas v_C is the instantaneous capacitor voltage after time t. Formula (21-5) is derived from the fact that v_C must equal $V_T - V_R$ while C is charging.

EXAMPLE 21-9

An RC circuit has a time constant of 3 s. The capacitor is charged to 40 V. Then C is discharged. After 6 s of discharge, how much is V_R?

Answer:

Note that 6 s is twice the RC time of 3 s. Then $t/RC = 2$.

$$V_R = \text{antilog } (\log 40 - 0.434 \times 2)$$
$$= \text{antilog } (1.602 - 0.868)$$
$$= \text{antilog } 0.734$$
$$= 5.42 \text{ V}$$

Note that in two RC time constants, the V_R is down to approximately 14% of its initial voltage, a drop of about 86%.

Calculations for t

Furthermore, Formula (21-4), can be transposed to find the time t for a specific voltage decay. Then

$$t = 2.3 \, RC \log \frac{V}{v} \tag{21-6}$$

where V is the higher voltage at the start and v is the lower voltage at the finish. The factor 2.3 is $\frac{1}{0.434}$.

As an example, let RC be 1 s. How long will it take for v_R to drop from 100 to 50 V? The required time for this decay is

$$t = 2.3 \times 1 \times \log \frac{100}{50} = 2.3 \times 1 \times \log 2$$
$$= 2.3 \times 1 \times 0.3$$
$$= 0.7 \text{ s} \quad \text{approximately}$$

This answer agrees with the fact that a drop of 50% takes 0.7 time constant. Formula (21-6) can also be used to calculate the time for any decay of v_C or v_R.

Formula (21-6) cannot be used for a rise in v_C. However, if you convert this rise to an equivalent drop in v_R, the calculated time is the same for both cases.

EXAMPLE 21-10

An RC circuit has an R of 10 kΩ and a C of 0.05 μF. The applied voltage for charging is 36 V. (a) Calculate the time constant. (b) How long will it take C to charge to 24 V?

Answer:

a. RC is 10 kΩ $\times$ 0.05 μF $= 0.5$ ms or 0.5×10^{-3} s.

b. The v_C rises to 24 V while v_R drops from 36 to 12 V. Then

$$t = 2.3 \, RC \log \frac{V}{v}$$
$$= 2.3 \times 0.5 \times 10^{-3} \times \log \frac{36}{12}$$
$$= 2.3 \times 0.5 \times 10^{-3} \times 0.477$$
$$= 0.549 \times 10^{-3} \text{ s} \quad \text{or} \quad 0.549 \text{ ms}$$

Data Rate vs. Bandwidth

The amount of information that can be transmitted over a signal path is limited by the bandwidth of that path. The signal path, or medium of communications, may just be a coaxial cable, some copper lines on a printed circuit board, a fiber-optic cable, or the free space of a wireless path. All those have a finite bandwidth either inherent or imposed by regulation. The greater the bandwidth, the greater the amount of information that can be sent.

While analog information like voice, music, and video is still transmitted over such communication paths, most information today is in the form of digital data. Communicated information is a serial stream of binary numbers or words that represent the information such as computer data, including digitized voice, music, or video. The data are a sequential train of voltage pulses representing the binary 0s and 1s as shown in Fig. S21-1.

Data Rate

The speed of the data refers to how many 0s and 1s, or bits, are transmitted per second, or bits per second (bps). The faster the better. Data rate or speed is simply the reciprocal of the time interval of each bit (t_b) or:

$$\text{bps} = \frac{1}{t_b}$$

For example, if the bit time is 100 ns, or 100×1^{-9}, the rate is

$$\text{bps} = 1/100 \times 10^{-9} = 10,000,000$$

This is the same as 10 megabits per second (Mbps).

The fastest data are alternating binary 0s and 1s as Fig. S21-1 shows. The frequency, f, of the data is the reciprocal of one binary 0 interval plus one binary 1 interval, creating one cycle. The period of one cycle, T, is twice the bit time, or

$$T = \frac{1}{2t_b}$$

The maximum frequency of the data then is

$$f = \frac{1}{T} = \frac{1}{2t_b}$$

For a 100-ns bit time, the maximum frequency is

$$f = \frac{1}{T} = \frac{1}{2t_b} = \frac{1}{2(100 \times 10^{-9})} = \frac{1}{200 \times 10^{-9}} = 5 \text{ MHz}$$

Bandwidth

Bandwidth (BW) is a segment of the frequency spectrum. All filters, amplifiers, and other circuits including cables have a finite bandwidth. For example a bandpass filter with a center frequency of 9 MHz may only pass frequencies in the range of 8,996,000 Hz to 9,004,000 Hz. See Fig. S21-2a. The bandwidth is the difference between these upper and lower cutoff frequencies, or

$$BW = 9,004,000 - 8,996,000 = 8000 \text{ Hz or 8 kHz}$$

This BW sets the upper limit on how fast digital data can pass through that circuit or path.

Another example is an audio amplifier that passes frequencies from dc to its upper cutoff frequency 20,000 Hz. Refer to Figure S21-2b. Its bandwidth is

$$BW = 20,000 - 0 = 20,000 \text{ Hz} = 20 \text{ kHz}$$

Relationship between Data Rate and Bandwidth

If you want to transmit a sine wave, then the bandwidth need only be that sufficient to pass a sine wave of the desired frequency. However, if the data are a series of square waves as it is when transmitting alternating 0s and 1s, then the transmission medium must also pass a sufficient number of harmonics to maintain the shape of the signal. Generally, if you can pass up to the fifth harmonic of the basic frequency of the data, then a reasonable waveform shape is retained to ensure reliable recovery the data. To amplify a 30-MHz

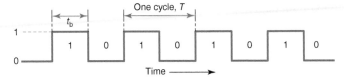

Figure S21-1 Binary or digital data transmitted as a streams of voltage pulses representing the binary 0s and 1s.

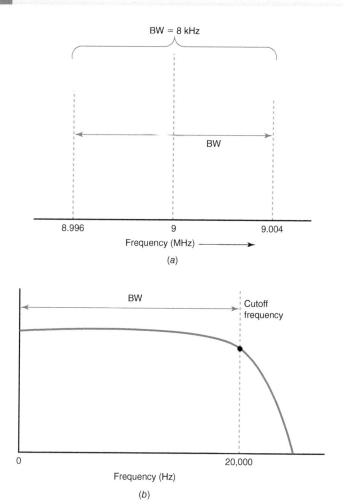

Figure S21-2   Examples of bandwidth. (a) Bandpass filter.
(b) Audio amplifier.

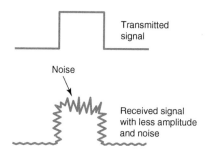

Figure S21-3   The transmission medium introduces
attenuation and noise to a transmitted signal.

square wave, you need to pass the third and fifth harmonics of 90 MHz and 150 MHz. The minimum bandwidth then is from 30 to 150 MHz or 120 MHz.

In terms of bits per second, the basic relationship is

$$C = 2B$$

where C is the channel capacity, another name for data rate, in bps, and B is the bandwidth in Hz.

For example, what is the channel capacity of the 8-kHz bandwidth of the 9-MHz channel mentioned earlier?

$$C = 2(8 \text{ kHz}) = 16 \text{ kbps}$$

This relationship assumes no errors and no noise or interference in the channel.

The Shannon-Hartley Law

There is no perfect communications medium. All media introduce *attenuation,* which is the loss of signal strength over the path. The received signal is always much smaller than the transmitted signal. Furthermore, noise gets added to the signal along the way. Figure S21-3 illustrates this.

Noise is any interference that conflicts with the main signal. It may be other signals picked up by inductive or capacitive coupling to other cables or equipment. Or it could be electrical interference from power lines, auto ignitions, fluorescent lights, motors, or other electronic equipment. In any case, the received signal is not only smaller at the receiver but also has noise added to it. The noise may be great enough to cause bit errors in the data. As a result, noise must be taken into account when determining the maximum possible data rate in a specific bandwidth. This is expressed as the Shannon-Hartley law:

$$C = B \log_2 (1 + S/N)$$

Converting the logarithm of the base 2 to standard common base 10 logarithm changes this to

$$C = 3.32B \log (1 + S/N)$$

where S is the signal power in watts and N is the noise power in watts.

As an example, assume the signal power is 500 μW and the noise power is 2 μW. Again assume a bandwidth of 8 kHz. The maximum data rate without error will be

$$C = 3.32(8 \text{ kHz}) \log (1 + 500/2) = 64 \text{ kbps}$$

This seems in conflict with the previously discussed relationship of $C = 28$. What is not explained in the Shannon-Hartley law specifically is that special multilevel modulation and data compression schemes may be used to achieve this maximum data rate.

The important takeaway information here is that the data rate is generally proportional to bandwidth. The higher the desired speed such as that needed to transmit digital video over a cell phone channel, the greater the amount of bandwidth needed. Noise, modulation, and other factors also play an important role.

1. What is the time constant of the circuit in Fig. 21-11 with S_1 closed?
   a. 250 μs.
   b. 31.6 μs.
   c. 50 μs.
   d. 5 ms.

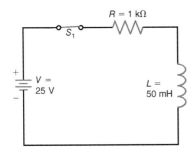

Figure 21-11

2. With S_1 closed in Fig. 21-11, what is the eventual steady-state value of current?
   a. 15.8 mA.
   b. 12.5 mA.
   c. 0 mA.
   d. 25 mA.

3. In Fig. 21-11, how long does it take the current, I, to reach its steady-state value after S_1 is closed?
   a. 50 μs.
   b. 250 μs.
   c. 500 μs.
   d. It cannot be determined.

4. In Fig. 21-11, how much is the resistor voltage at the very first instant ($t = 0$ s) S_1 is closed?
   a. 0 V.
   b. 25 V.
   c. 15.8 V.
   d. 9.2 V.

5. In Fig. 21-11, what is the value of the resistor voltage exactly one time constant after S_1 is closed?
   a. 15.8 V.
   b. 9.2 V.
   c. 6.32 V.
   d. 21.5 V.

6. If a 2-MΩ resistor is placed across the switch, S_1, in Fig. 21-11, how much is the peak inductor voltage, V_L, when S_1 is opened?
   a. 0 V.
   b. 25 V.
   c. 50 kV.
   d. It cannot be determined.

7. In Fig. 21-11, what is the value of the current 35 μs after S_1 is closed?
   a. Approximately 20 mA.
   b. Approximately 12.5 mA.
   c. 15.8 mA.
   d. 20 mA.

8. With S_1 closed in Fig. 21-11, the length of one time constant could be increased by
   a. decreasing L.
   b. decreasing R.
   c. increasing L.
   d. both b and c.

9. In Fig. 21-11, what is the value of the inductor voltage five time constants after S_1 is closed?
   a. 50 kV.
   b. 25 V.
   c. 0 V.
   d. 9.2 V.

10. In Fig. 21-11, how much is the resistor voltage exactly 100 μs after S_1 is closed?
    a. 12 V.
    b. 21.6 V.
    c. 3.4 V.
    d. 15.8 V.

11. In Fig. 21-12, what is the time constant of the circuit with S_1 in position 1?
    a. 2 s.
    b. 5 s.
    c. 10 s.
    d. 1 s.

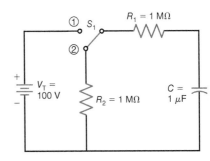

Figure 21-12

12. In Fig. 21-12, what is the time constant of the circuit with S_1 in position 2?
    a. 2 s.
    b. 5 s.
    c. 10 s.
    d. 1 s.

13. In Fig. 21-12, how long will it take for the voltage across C to reach 100 V after S_1 is placed in position 1?
  a. 1 s.
  b. 2 s.
  c. 10 s.
  d. 5 s.

14. In Fig. 21-12, how much voltage is across resistor, R_1, at the first instant the switch is moved from position 2 to position 1? (Assume that C was completely discharged with S_1 in position 2.)
  a. 100 V.
  b. 63.2 V.
  c. 0 V.
  d. 36.8 V.

15. In Fig. 21-12, assume that C is fully charged to 100 V with S_1 in position 1. How long will it take for C to discharge fully if S_1 is moved to position 2?
  a. 1 s.
  b. 5 s.
  c. 10 s.
  d. 2 s.

16. In Fig. 21-12, assume that C is completely discharged while in position 2. What is the voltage across C exactly 1s after S_1 is moved to position 1?
  a. 50 V.
  b. 63.2 V.
  c. 36.8 V.
  d. 100 V.

17. In Fig. 21-12, assume that C is completely discharged while in position 2. What is the voltage across R_1 exactly two time constants after S_1 is moved to position 1?
  a. 37 V.
  b. 13.5 V.
  c. 50 V.
  d. 86 V.

18. In Fig. 21-12, what is the steady-state value of current with S_1 in position 1?
  a. 100 μA.
  b. 50 μA.
  c. 1 A.
  d. 0 μA.

19. In Fig. 21-12, assume that C is fully charged to 100 V with S_1 in position 1. What is the value of the capacitor voltage 3s after S_1 is moved to position 2?
  a. 77.7 V.
  b. 0 V.

  c. 22.3 V.
  d. 36.8 V.

20. In Fig. 21-12, assume that C is charging with S_1 in position 1. At the instant the capacitor voltage reaches 75 V, S_1 is moved to position 2. What is the approximate value of the capacitor voltage 0.7 time constant after S_1 is moved to position 2?
  a. 75 V.
  b. 27.6 V.
  c. 50 V.
  d. 37.5 V.

21. For best results, an RC coupling circuit should have a
  a. short time constant.
  b. medium time constant.
  c. long time constant.
  d. zero time constant.

22. A differentiator is a circuit whose
  a. output combines its original voltage with the new change in voltage.
  b. output is always one-half of V_{in}.
  c. time constant is long with the output across C.
  d. output is proportional to the change in applied voltage.

23. An integrator is a circuit whose
  a. output combines its original voltage with the new change in voltage.
  b. output is always equal to V_{in}.
  c. output is proportional to the change in applied voltage.
  d. time constant is short with the output across R.

24. The time constant of an RL circuit is 47 μs. If $L = 4.7$ mH, calculate R.
  a. $R = 10$ kΩ.
  b. $R = 100$ Ω.
  c. $R = 10$ MΩ.
  d. $R = 1$ kΩ.

25. The time constant of an RC circuit is 330 μs. If $R = 1$ kΩ, calculate C.
  a. $C = 0.33$ μF.
  b. $C = 0.033$ μF.
  c. $C = 3.3$ μF.
  d. $C = 330$ pF.

SECTION 21.1 Response of Resistance Alone

21.1 In Fig. 21-13, how long does it take for the current, I, to reach its steady-state value after S_1 is closed?

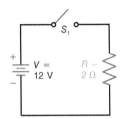

Figure 21-13

21.2 In Fig. 21-13, what is the current with S_1 closed?

SECTION 21.2 L/R Time Constant

21.3 In Fig. 21-14,
   a. what is the time constant of the circuit with S_1 closed?
   b. what is the eventual steady-state current with S_1 closed?
   c. what is the value of the circuit current at the first instant S_1 is closed? ($t = 0$ s)
   d. what is the value of the circuit current exactly one time constant after S_1 is closed?
   e. how long after S_1 is closed will it take before the circuit current reaches its steady-state value?

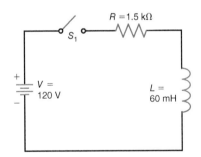

Figure 21-14

21.4 Calculate the time constant for an inductive circuit with the following values:
   a. $L = 500$ mH, $R = 2$ kΩ.
   b. $L = 250$ μH, $R = 50$ Ω.

21.5 List two ways to
   a. increase the time constant of an inductive circuit.
   b. decrease the time constant of an inductive circuit.

SECTION 21.3 High Voltage Produced by Opening an RL Circuit

21.6 Assume that the switch, S_1, in Fig. 21-14 has been closed for more than five L/R time constants. If a 1-MΩ resistor is placed across the terminals of the switch, calculate
   a. the approximate time constant of the circuit with S_1 open.
   b. the peak inductor voltage, V_L, when S_1 is opened.
   c. how long it takes for the current to decay to zero after S_1 is opened (approximately).

21.7 Without a resistor across S_1 in Fig. 21-14, is it possible to calculate the time constant of the circuit with the switch open? Also, what effect will probably occur inside the switch when it is opened?

SECTION 21.4 RC Time Constant

21.8 In Fig. 21-15, what is the time constant of the circuit with the switch, S_1, in position
   a. 1?
   b. 2?

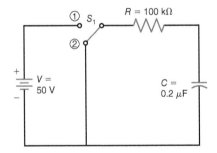

Figure 21-15

21.9 Assume that the capacitor in Fig. 21-15 is fully discharged with S_1 in position 2. How much is the capacitor voltage, V_C,
   a. exactly one time constant after S_1 is moved to position 1?
   b. five time constants after S_1 is moved to position 1?
   c. 1 week after S_1 is moved to position 1?

21.10 Assume that the capacitor in Fig. 21-15 is fully charged with S_1 in position 1. How much is the capacitor voltage, V_C,
   a. exactly one time constant after S_1 is moved to position 2?
   b. five time constants after S_1 is moved to position 2?
   c. 1 week after S_1 is moved to position 2?

21.11 Calculate the time constant of a capacitive circuit with the following values:
  a. $R = 1\text{M}\,\Omega$, $C = 1\,\mu\text{F}$.
  b. $R = 5\,\text{k}\Omega$, $C = 40\,\mu\text{F}$.

21.12 List two ways to
  a. increase the time constant of a capacitive circuit.
  b. decrease the time constant of a capacitive circuit.

SECTION 21.5 RC Charge and Discharge Curves

21.13 Assume that the capacitor in Fig. 21-15 is fully discharged with S_1 in position 2. What is
  a. the value of the charging current at the first instant S_1 is moved to position 1?
  b. the value of the charging current five time constants after S_1 is moved to position 1?
  c. the value of the resistor voltage exactly one time constant after S_1 is moved to position 1?
  d. the value of the charging current exactly one time constant after S_1 is moved to position 1?

21.14 Assume that the capacitor in Fig. 21-15 is fully charged to 50 V with S_1 in position 1. What is the value of the discharge current
  a. at the first instant S_1 is moved to position 2?
  b. exactly one time constant after S_1 is moved to position 2?
  c. five time constants after S_1 is moved to position 2?

SECTION 21.6 High Current Produced by Short-Circuiting an RC Circuit

21.15 In Fig. 21-16, what is the RC time constant with S_1 in position
  a. 1?
  b. 2?

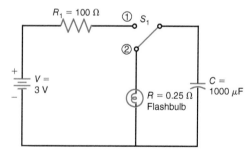

Figure 21-16

21.16 In Fig. 21-16, how long will it take the capacitor voltage to
  a. reach 3 V after S_1 is moved to position 1?
  b. discharge to 0 V after S_1 is moved to position 2?

SECTION 21.7 RC Waveshapes

21.17 For the circuit in Fig. 21-17,
  a. calculate the RC time constant.
  b. draw the capacitor voltage waveform and include voltage values at times t_0, t_1, t_2, t_3, and t_4.
  c. draw the resistor voltage waveform and include voltage values at times t_0, t_1, t_2, t_3, and t_4.
  d. draw the charge and discharge current waveform and include current values at times t_0, t_1, t_2, t_3, and t_4.

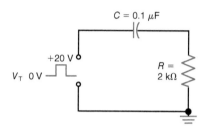

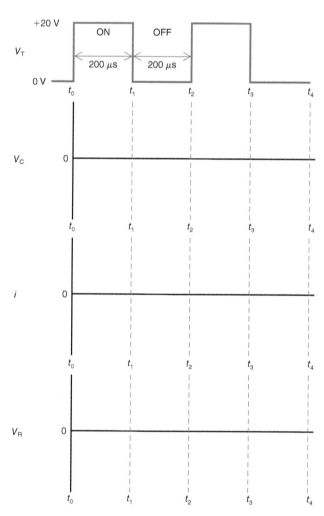

Figure 21-17

SECTION 21.8 Long and Short Time Constants

21.18 In Fig. 21-17, is the time constant of the circuit considered long or short with respect to the pulse width of the applied voltage, V_T, if the resistance, R, is
    a. increased to 10 kΩ?
    b. decreased to 400 Ω?

21.19 For an *RC* circuit used as a differentiator,
    a. across which component is the output taken?
    b. should the time constant be long or short with respect to the pulse width of the applied voltage?

21.20 For an *RC* circuit used as an integrator,
    a. across which component is the output taken?
    b. should the time constant be long or short with respect to the pulse width of the applied voltage?

SECTION 21.9 Charge and Discharge with a Short *RC* Time Constant

21.21 For the circuit in Fig. 21-18,
    a. calculate the *RC* time constant.
    b. draw the capacitor voltage waveform and include voltage values at times t_0, t_1, t_2, t_3, and t_4.

    c. draw the resistor voltage waveform and include voltage values at times t_0, t_1, t_2, t_3, and t_4.
    d. specify the ratio of the pulse width of the applied voltage to the *RC* time constant.

SECTION 21.10 Long Time Constant for an *RC* Coupling Circuit

21.22 Assume that the resistance, R, in Fig. 21-18 is increased to 100 kΩ but the frequency of the applied voltage, V_T, remains the same. Determine
    a. the new *RC* time constant of the circuit.
    b. the ratio of the pulse width of the applied voltage to the *RC* time constant.
    c. the approximate capacitor and resistor voltage waveforms, assuming that the input voltage has been applied for longer than five *RC* time constants.

SECTION 21.11 Advanced Time Constant Analysis

21.23 What is the time constant of the circuit in Fig. 21-19?

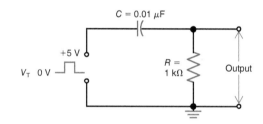

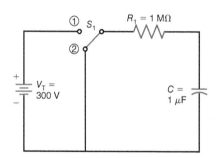

Figure 21-19

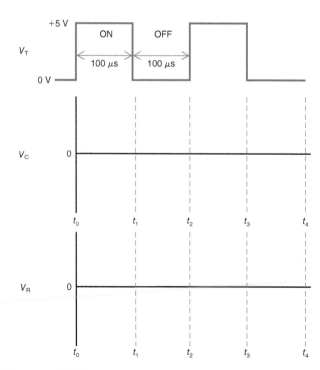

Figure 21-18

21.24 Assume that *C* in Fig. 21-19 is completely discharged with S_1 in position 2. If S_1 is moved to position 1, how much is the capacitor voltage at the following time intervals?
    a. $t = 0.7$ s.
    b. $t = 1.5$ s.
    c. $t = 3.5$ s.

21.25 Assume that *C* in Fig. 21-19 is fully charged with S_1 in position 1. If S_1 is moved to position 2, how much is the resistor voltage at the following time intervals?
    a. $t = 0.7$ s.
    b. $t = 2.5$ s.

21.26 What is the time constant of the circuit in Fig. 21-20?

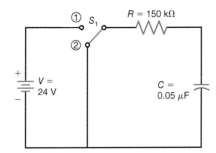

Figure 21-20

21.27 Assume that C in Fig. 21-20 is completely discharged with S_1 in position 2. If S_1 is moved back to position 1, how long will it take for the capacitor voltage to reach

a. 3 V?

b. 20 V?

21.28 Assume that C in Fig. 21-20 is completely discharged with S_1 in position 2. If S_1 is moved back to position 1, how much is the resistor voltage at the following time intervals?

a. $t = 4.5$ ms.

b. $t = 15$ ms.

21.29 Assume that C in Fig. 21-20 is fully charged with S_1 in position 1. If S_1 is moved to position 2, how long will it take the capacitor to discharge to

a. 4 V?

b. 18 V?

Chapter 22

Resonance

Learning Outcomes

After studying this chapter, you should be able to:

> *Define* the term *resonance*.

> *List* four characteristics of a series resonant circuit.

> *List* three characteristics of a parallel resonant circuit.

> *Calculate* the *Q* of a series or parallel resonant circuit.

> *Calculate* the equivalent impedance of a parallel resonant circuit.

> *Explain* what is meant by the *bandwidth* of a resonant circuit.

> *Calculate* the bandwidth of a series or parallel resonant circuit.

> *Explain* the effect of varying *L* or *C* in tuning an *LC* circuit.

> *Calculate* *L* or *C* for a resonant circuit.

This chapter explains how X_L and X_C can be combined to favor one particular frequency, the resonant frequency to which the *LC* circuit is tuned. The resonance effect occurs when the inductive and capacitive reactances are equal.

In radio-frequency (RF) circuits, the main application of resonance is for tuning to an ac signal of the desired frequency. Applications of resonance include tuning in communication receivers, transmitters, and electronic equipment in general.

Tuning by means of the resonant effect provides a practical application of selectivity. The resonant circuit can be operated to select a particular frequency for the output with many different frequencies at the input.

22.1 The Resonance Effect

Inductive reactance increases as the frequency is increased, but capacitive reactance decreases with higher frequencies. Because of these opposite characteristics, for any LC combination, there must be a frequency at which the X_L equals the X_C because one increases while the other decreases. This case of equal and opposite reactances is called *resonance,* and the ac circuit is then a *resonant circuit.*

Any LC circuit can be resonant. It all depends on the frequency. At the resonant frequency, an LC combination provides the resonance effect. Off the resonant frequency, either below or above, the LC combination is just another ac circuit.

The frequency at which the opposite reactances are equal is the *resonant frequency.* This frequency can be calculated as $f_r = 1/(2\pi\sqrt{LC})$, where L is the inductance in henrys, C is the capacitance in farads, and f_r is the resonant frequency in hertz that makes $X_L = X_C$.

In general, we can say that large values of L and C provide a relatively low resonant frequency. Smaller values of L and C allow higher values for f_r. The resonance effect is most useful for radio frequencies, where the required values of microhenrys for L and picofarads for C are easily obtained.

The most common application of resonance in RF circuits is called *tuning.* In this use, the LC circuit provides maximum voltage output at the resonant frequency, compared with the amount of output at any other frequency either below or above resonance. This idea is illustrated in Fig. 22-1, where the LC circuit resonant at 1000 kHz magnifies the effect of this particular frequency. The result is maximum output at 1000 kHz, compared with lower or higher frequencies. Another application of resonance is filtering. Filters are circuits that select one frequency or a band of frequencies and reject others. Resonant circuits are often used as filters or as parts of filters.

22.2 Series Resonance

When the frequency of the applied voltage is 1000 kHz in the series ac circuit in Fig. 22-2a, the reactance of the 239-μH inductance equals 1500 Ω. At the same frequency, the reactance of the 106-pF capacitance also is 1500 Ω. Therefore,

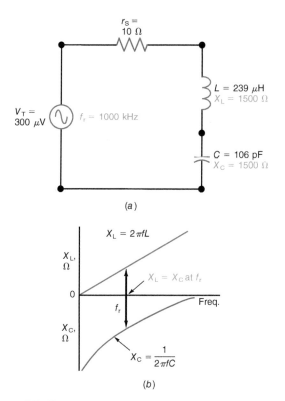

Figure 22-2 Series resonance. (*a*) Schematic diagram of series r_S, L, and C. (*b*) Graph to show that reactances X_C and X_L are equal and opposite at the resonant frequency f_r. Inductive reactance is shown up for X_L and capacitive reactance is down for $-X_C$.

this LC combination is resonant at 1000 kHz. This is f_r because the inductive reactance and capacitive reactance are equal at this frequency.

In a series ac circuit, inductive reactance leads by 90°, compared with the zero reference angle of the resistance, and capacitive reactance lags by 90°. Therefore, X_L and X_C are 180° out of phase. The opposite reactances cancel each other completely when they are equal.

Figure 22-2b shows X_L and X_C equal, resulting in a net reactance of zero ohms. The only opposition to current, then, is the coil resistance r_S, which limits how low the series resistance in the circuit can be. With zero reactance and just the low value of series resistance, the generator voltage produces the greatest amount of current in the series LC circuit at the resonant frequency. The series resistance should be as small as possible for a sharp increase in current at resonance.

Maximum Current at Series Resonance

The main characteristic of series resonance is the resonant rise of current to its maximum value of V_T/r_S at the resonant frequency. For the circuit in Fig. 22-2a, the maximum current at series resonance is 30 μA, equal to 300 μV/10 Ω. At any other frequency, either below or above the resonant frequency, there is less current in the circuit.

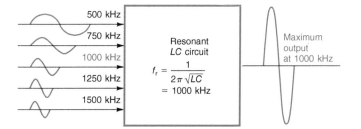

Figure 22-1 LC circuit resonant at f_r of 1000 kHz to provide maximum output at this frequency.

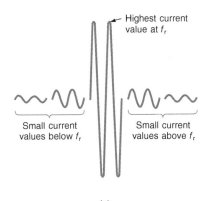

(a)

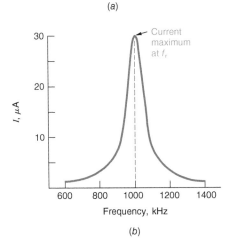

(b)

Figure 22-3 Graphs showing maximum current at resonance for the series circuit in Fig. 22-2. (a) Amplitudes of individual cycles. (b) Response curve to show the amount of I below and above resonance.

This resonant rise of current to 30 μA at 1000 kHz is shown in Fig. 22-3. In Fig. 22-3a, the amount of current is shown as the amplitude of individual cycles of the alternating current produced in the circuit by the ac generator voltage. Whether the amplitude of one ac cycle is considered in terms of peak, rms, or average value, the amount of current is greatest at the resonant frequency. In Fig. 22-3b, the current amplitudes are plotted on a graph for frequencies at and near the resonant frequency, producing a typical *response curve* for a series resonant circuit. The response curve in Fig. 22-3b can be considered an outline of the increasing and decreasing amplitudes of the individual cycles shown in Fig. 22-3a.

The response curve of the series resonant circuit shows that the current is small below resonance, rises to its maximum value at the resonant frequency, and then drops off to small values above resonance.

Below resonance, at 600 kHz, X_C is more than X_L and there is appreciable net reactance, which limits the current to a relatively low value. At the higher frequency of 800 kHz, X_C decreases and X_L increases, making the two reactances closer to the same value. The net reactance is then smaller, allowing more current.

At the resonant frequency, X_L and X_C are equal, the net reactance is zero, and the current has its maximum value equal to V_T/r_S.

Above resonance at 1200 and 1400 kHz, X_L is greater than X_C, providing net reactance that limits the current to values much smaller than at resonance. In summary,

1. Below the resonant frequency, X_L is small, but X_C has high values that limit the amount of current.
2. Above the resonant frequency, X_C is small, but X_L has high values that limit the amount of current.
3. At the resonant frequency, X_L equals X_C, and they cancel to allow maximum current.

Minimum Impedance at Series Resonance

Since reactances cancel at the resonant frequency, the impedance of the series circuit is minimum, equal to just the low value of series resistance. This minimum impedance at resonance is resistive, resulting in zero phase angle. At resonance, therefore, the resonant current is in phase with the generator voltage.

Resonant Rise in Voltage across Series *L* or *C*

The maximum current in a series *LC* circuit at resonance is useful because it produces maximum voltage across either X_L or X_C at the resonant frequency. As a result, the series resonant circuit can select one frequency by providing much more voltage output at the resonant frequency, compared with frequencies above and below resonance. Figure 22-4 illustrates the resonant rise in voltage across the capacitance in a series ac circuit. At the resonant frequency of 1000 kHz, the voltage across *C* rises to 45,000 μV, and the input voltage is only 300 μV.

The voltage across *C* is calculated as IX_C, and across *L* as IX_L. Below the resonant frequency, X_C has a higher value than at resonance, but the current is small. Similarly, above the resonant frequency, X_L is higher than at resonance, but the current has a low value because of inductive reactance. At resonance, although X_L and X_C cancel each other to allow maximum current, each reactance by itself has an

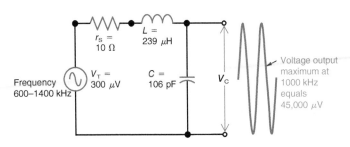

Figure 22-4 Series circuit selects frequency by producing maximum IX_C voltage output across *C* at resonance.

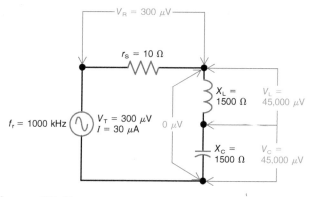

Figure 22-5 Voltage drops around series resonant circuit.

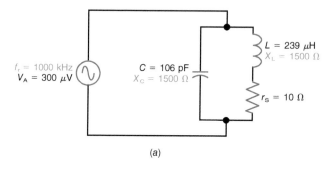

(a)

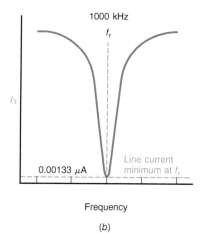

(b)

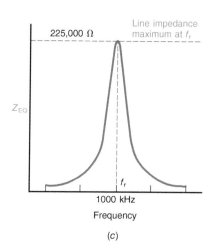

(c)

Figure 22-6 Parallel resonant circuit. (a) Schematic diagram of L and C in parallel branches. (b) Response curve of I_T shows that the line current dips to a minimum at f_r. (c) Response curve of Z_{EQ} shows that it rises to a maximum at f_r.

appreciable value. Since the current is the same in all parts of a series circuit, the maximum current at resonance produces maximum voltage IX_C across C and an equal IX_L voltage across L for the resonant frequency.

Although the voltage across X_C and X_L is reactive, it is an actual voltage that can be measured. In Fig. 22-5, the voltage drops around the series resonant circuit are 45,000 μV across C, 45,000 μV across L, and 300 μV across r_S. The voltage across the resistance is equal to and in phase with the generator voltage.

Across the series combination of both L and C, the voltage is zero because the two series voltage drops are equal and opposite. To use the resonant rise of voltage, therefore, the output must be connected across either L or C alone. We can consider the V_L and V_C voltages similar to the idea of two batteries connected in series opposition. Together, the resultant is zero for equal and opposite voltages, but each battery still has its own potential difference.

In summary, the main characteristics of a series resonant circuit are

1. The current I is maximum at the resonant frequency f_r.
2. The current I is in phase with the generator voltage, or the phase angle of the circuit is 0°.
3. The voltage is maximum across either L or C alone.
4. The impedance is minimum at f_r, equal only to the low r_S.

22.3 Parallel Resonance

When L and C are in parallel, as shown in Fig. 22-6, and X_L equals X_C, the reactive branch currents are equal and opposite at resonance. Then they cancel each other to produce minimum current in the main line. Since the line current is minimum, the impedance is maximum. These relations are based on r_S being very small compared with X_L at resonance. In this case, the branch currents are practically equal when X_L and X_C are equal.

Minimum Line Current at Parallel Resonance

With L and C the same as in the series circuit of Fig. 22-2, X_L and X_C have the same values at the same frequencies. Since L, C, and the generator are in parallel, the voltage applied across the branches equals the generator voltage of 300 μV. Therefore, each reactive branch current is calculated as 300 μV divided by the reactance of the branch.

At 600 kHz, the capacitive branch current equals 300 μV/2500 Ω, or 0.12 μA. The inductive branch current at this frequency is 300 μV/900 Ω, or 0.33 μA. Since this is a parallel ac circuit, the capacitive current leads by 90°, whereas the inductive current lags by 90°, compared with the reference angle of the generator voltage, which is applied across the parallel branches. Therefore, the opposite currents are 180° out of phase. The net current in the line, then, is the difference between 0.33 and 0.12, which equals 0.21 μA.

Following this procedure, the calculations show that as the frequency is increased toward resonance, the capacitive branch current increases because of the lower value of X_C and the inductive branch current decreases with higher values of X_L. As a result, there is less net line current as the two branch currents become more nearly equal.

At the resonant frequency of 1000 kHz, both reactances are 1500 Ω, and the reactive branch currents are both 0.20 μA, canceling each other completely.

Above the resonant frequency, there is more current in the capacitive branch than in the inductive branch, and the net line current increases above its minimum value at resonance.

The dip in I_T to its minimum value at f_r is shown by the graph in Fig. 22-6b. At parallel resonance, I_T is minimum and Z_{EQ} is maximum.

The in-phase current due to r_S in the inductive branch can be ignored off-resonance because it is so small compared with the reactive line current. At the resonant frequency when the reactive currents cancel, however, the resistive component is the entire line current. Its value at resonance equals 0.00133 μA in this example. This small resistive current is the minimum value of the line current at parallel resonance.

Maximum Line Impedance at Parallel Resonance

The minimum line current resulting from parallel resonance is useful because it corresponds to maximum impedance in the line across the generator. Therefore, an impedance that has a high value for just one frequency but a low impedance for other frequencies, either below or above resonance, can be obtained by using a parallel LC circuit resonant at the desired frequency. This is another method of selecting one frequency by resonance. The response curve in Fig. 22-6c shows how the impedance rises to a maximum for parallel resonance.

The main application of parallel resonance is the use of an LC tuned circuit as the load impedance Z_L in the output circuit of RF amplifiers. Because of the high impedance, then, the gain of the amplifier is maximum at f_r. The voltage gain of an amplifier is directly proportional to Z_L. The advantage of a resonant LC circuit is that Z is maximum

only for an ac signal at the resonant frequency. Also, L has practically no dc resistance, which means practically no dc voltage drop.

The total impedance of the parallel ac circuit is calculated as the generator voltage divided by the total line current. At 600 kHz, for example, Z_{EQ} equals 300 μV/0.21 μA, or 1400 Ω. At 800 kHz, the impedance is higher because there is less line current.

At the resonant frequency of 1000 kHz, the line current is at its minimum of 0.00133 μA. Then the impedance is maximum and is equal to 300 μV/0.00133 μA, or 225,000 Ω. At resonance, the parallel resonant circuit looks like a 225-kΩ resistor.

Above 1000 kHz, the line current increases, and the impedance decreases from its maximum.

How the line current can be very low even though the reactive branch currents are appreciable is illustrated in Fig. 22-7. In Fig. 22-7a, the resistive component of the total line current is shown as though it were a separate branch drawing an amount of resistive current from the generator in the main line equal to the current resulting from the coil resistance. Each reactive branch current has its value equal to the generator voltage divided by the reactance. Since they are equal and of opposite phase, however, in any part of

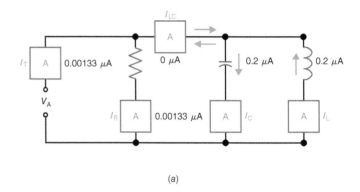

(a)

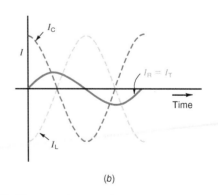

(b)

Figure 22-7 Distribution of currents in a parallel circuit at resonance. Resistive current shown as an equivalent branch for I_R. (a) Circuit with branch currents for R, L, and C. (b) Graph of equal and opposite reactive currents I_L and I_C.

the circuit where both reactive currents are present, the net amount of electron flow in one direction at any instant corresponds to zero current. The graph in Fig. 22-7b shows how equal and opposite currents for I_L and I_C cancel.

If a meter is inserted in series with the main line to indicate total line current I_T, it dips sharply to the minimum value of line current at the resonant frequency. With minimum current in the line, the impedance across the line is maximum at the resonant frequency. The maximum impedance at parallel resonance corresponds to a high value of resistance, without reactance, since the line current is then resistive with zero phase angle.

In summary, the main characteristics of a parallel resonant circuit are

1. The line current I_T is minimum at the resonant frequency.
2. The current I_T is in phase with the generator voltage V_A, or the phase angle of the circuit is 0°.
3. The impedance Z_{EQ}, equal to V_A/I_T, is maximum at f_r because of the minimum I_T.

The LC Tank Circuit

Note that the individual branch currents are appreciable at resonance, although I_T is minimum. At f_r, either the I_L or the I_C equals 0.2 μA. This current is greater than the I_C values below f_r or the I_L values above f_r.

The branch currents cancel in the main line because I_C is at 90° with respect to the source V_A while I_L is at −90°, making them opposite with respect to each other.

However, inside the LC circuit, I_L and I_C do not cancel because they are in separate branches. Then I_L and I_C provide a circulating current in the LC circuit, which equals 0.2 μA in this example. For this reason, a parallel resonant LC circuit is often called a *tank circuit*.

Because of the energy stored by L and C, the circulating tank current can provide full sine waves of current and voltage output when the input is only a pulse. The sine-wave output is always at the natural resonant frequency of the LC tank circuit. This ability of the LC circuit to supply complete sine waves is called the *flywheel effect*. Also, the process of producing sine waves after a pulse of energy has been applied is called *ringing* of the LC circuit.

Impedance at Resonance

As you have seen, a parallel LC circuit at resonance appears to the generator as a high value of resistance. That value can be calculated from the circuit values directly. The two formulas give approximately the same value:

$$Z_{EQ} = L/Cr_s \qquad \textbf{(22-1)}$$

$$Z_{EQ} = r_s(Q^2 + 1) \qquad \textbf{(22-2)}$$

Remember that $Q = X_L/r_s$.

Using the values given previously in Fig, 22-6, the equivalent resistance is

$$Z_{EQ} = \frac{L}{Cr_s} = \frac{239 \times 10^{-6}}{106 \times 10^{-12}} = 225{,}472 \ \Omega$$

Using the other formula we get

$$Q = X_L/r_s = 1500/10 = 150$$

$$Z_{EQ} = r_s(Q^2 + 1) = 10(150^2 + 1) = 225{,}010 \ \Omega$$

The slightly different values result from the formula derivations in which some approximations are used.

EXAMPLE 22-1

Calculate the equivalent resistance of a parallel resonant circuit with in inductance of 14 nH and a capacitance of 9 pF. The inductor resistance is 2 Ω. Assume X_L is 39.4 Ω.

Answer:

$$Z_{EQ} = \frac{L}{Cr_s} = \frac{14 \times 10^{-6}}{9 \times 10^{-12}} = 777.77 \ \Omega$$

$$Q = X_L/r_s = 39.4/2 = 19.7$$

$$Z_{EQ} = r_s(Q^2 + 1) = 2(19.7^2 + 1) = 779.77 \ \Omega$$

22.4 Resonant Frequency
$f_r = 1/(2\pi \sqrt{LC})$

The formula for the resonant frequency is derived from $X_L = X_C$. Using f_r to indicate the resonant frequency in the formulas for X_L and X_C,

$$2\pi f_r L = \frac{1}{2\pi f_r C}$$

Inverting the factor f_r gives

$$2\pi L(f_r)^2 = \frac{1}{2\pi C}$$

Inverting the factor $2\pi L$ gives

$$(f_r)^2 = \frac{1}{(2\pi)^2 LC}$$

The square root of both sides is then

$$f_r = \frac{1}{2\pi \sqrt{LC}} \qquad \textbf{(22-3)}$$

where L is in henrys, C is in farads, and the resonant frequency f_r is in hertz (Hz). For example, to find the resonant frequency of the LC combination in Fig. 22-2, the values of 239×10^{-6} and 106×10^{-12} are substituted for L and C. Then,

$$f_r = \frac{1}{2\pi\sqrt{LC}} = \frac{1}{2\pi\sqrt{239 \times 10^{-6} \times 106 \times 10^{-12}}}$$

$$= \frac{1}{6.28\sqrt{25{,}334 \times 10^{-18}}} = \frac{1}{6.28 \times 159.2 \times 10^{-9}}$$

$$= \frac{1}{1000 \times 10^{-9}}$$

$$= 1 \times 10^6 \text{ Hz} = 1 \text{ MHz} = 1000 \text{ kHz}$$

For any series or parallel LC circuit, the f_r equal to $1/(2\pi\sqrt{LC})$ is the resonant frequency that makes the inductive and capacitive reactances equal.

How the f_r Varies with L and C

It is important to note that higher values of L and C result in lower values of f_r. Either L or C, or both, can be varied. An LC circuit can be resonant at any frequency from a few hertz to many megahertz.

As examples, an LC combination with the relatively large values of an 8-H inductance and a 20-μF capacitance is resonant at the low audio frequency of 12.6 Hz. For a much higher frequency in the RF range, a small inductance of 2 μH will resonate with the small capacitance of 3 pF at an f_r of 64.9 MHz. These examples are solved in the next two problems for more practice with the resonant frequency formula. Such calculations are often used in practical applications of tuned circuits. Probably the most important feature of any LC combination is its resonant frequency, especially in RF circuits. The applications of resonance are mainly for radio frequencies.

EXAMPLE 22-2

Calculate the resonant frequency for an 8-H inductance and a 20-μF capacitance.

Answer:

$$f_r = \frac{1}{2\pi\sqrt{LC}}$$

$$= \frac{1}{2\pi\sqrt{8 \times 20 \times 10^{-6}}}$$

$$= \frac{1}{6.28\sqrt{160 \times 10^{-6}}}$$

$$= \frac{1}{6.28 \times 12.65 \times 10^{-3}}$$

$$= \frac{1}{79.44 \times 10^{-3}}$$

$$= 0.0126 \times 10^3$$

$$= 12.6 \text{ Hz} \qquad \text{(approx.)}$$

EXAMPLE 22-3

Calculate the resonant frequency for a 2-μH inductance and a 3-pF capacitance.

Answer:

$$f_r = \frac{1}{2\pi\sqrt{LC}}$$

$$= \frac{1}{2\pi\sqrt{2 \times 10^{-6} \times 3 \times 10^{-12}}}$$

$$= \frac{1}{6.28\sqrt{6 \times 10^{-18}}}$$

$$= \frac{1}{6.28 \times 2.45 \times 10^{-9}}$$

$$= \frac{1}{15.4 \times 10^{-9}} = 0.065 \times 10^9$$

$$= 65 \times 10^6 \text{ Hz} = 65 \text{ MHz}$$

Specifically, because of the square root in the denominator of Formula (22.3), the f_r decreases inversely as the square root of L or C. For instance, if L or C is quadrupled, the f_r is reduced by one-half. The ½ is equal to the square root of ¼.

As a numerical example, suppose that f_r is 6 MHz with particular values of L and C. If either L or C is made four times larger, then f_r will be reduced to 3 MHz.

Or, to take the opposite case of doubling the frequency from 6 MHz to 12 MHz, the following can be done:

1. Use one-fourth the L with the same C.
2. Use one-fourth the C with the same L.
3. Reduce both L and C by one-half.
4. Use any new combination of L and C whose product will be one-fourth the original product of L and C.

LC Product Determines f_r

There are any number of LC combinations that can be resonant at one frequency. With more L, then less C can be used for the same f_r. Or less L can be used with more C. When either L or C is increased by a factor of 10 or 2, the other is decreased by the same factor, resulting in a constant value for the LC product.

The reactance at resonance changes with different combinations of L and C, but X_L and X_C are equal to each other at 1000 kHz. This is the resonant frequency determined by the value of the LC product in $f_r = 1/(2\pi\sqrt{LC})$.

Measuring L or C by Resonance

Of the three factors L, C, and f_r in the resonant-frequency formula, any one can be calculated when the other two are known. The resonant frequency of the LC combination can be found experimentally by determining the frequency that produces the resonant response in an *LC* combination. With a known value of either *L* or *C*, and the resonant frequency determined, the third factor can be calculated. This method is commonly used for measuring inductance or capacitance. A test instrument for this purpose is the *Q* meter, which also measures the *Q* of a coil.

Calculating *C* from f_r

The *C* can be taken out of the square root sign or radical in the resonance formula, as follows:

$$f_r = \frac{1}{2\pi\sqrt{LC}}$$

Squaring both sides to eliminate the radical gives

$$f_r^2 = \frac{1}{(2\pi)^2 LC}$$

Inverting *C* and f_r^2 gives

$$C = \frac{1}{4\pi^2 f_r^2 C} \qquad \textbf{(22-4)}$$

where f_r is in hertz, *C* is in farads, and *L* is in henrys.

Calculating *L* from f_r

Similarly, the resonance formula can be transposed to find *L*. Then

$$L = \frac{1}{4\pi^2 f_r^2 C} \qquad \textbf{(22-5)}$$

With Formula (22-5), *L* is determined by f_r with a known value of *C*. Similarly, *C* is determined from Formula (22-4) by f_r with a known value of *L*.

EXAMPLE 22-4

What value of *C* resonates with a 239-μH *L* at 1000 kHz?

Answer:

$$C = \frac{1}{4\pi^2 f_r^2 L}$$

$$= \frac{1}{4\pi^2 (1000 \times 10^3)^2 239 \times 10^{-6}}$$

$$= \frac{1}{39.48 \times 1 \times 10^6 \times 239}$$

$$= \frac{1}{9435.75 \times 10^6}$$

$$= 0.000106 \times 10^{-6}\ \text{F} = 106\ \text{pF}$$

Note that 39.48 is a constant for $4\pi^2$.

EXAMPLE 22-5

What value of *L* resonates with a 106-pF *C* at 1000 kHz, equal to 1 MHz?

Answer:

$$L = \frac{1}{4\pi^2 f_r^2 C}$$

$$= \frac{1}{39.48 \times 1 \times 10^{12} \times 106 \times 10^{-12}}$$

$$= \frac{1}{4184.88}$$

$$= 0.000239\ \text{H} = 239\ \mu\text{H}$$

Note that 10^{12} and 10^{-12} in the denominator cancel each other. Also, 1×10^{12} is the square of 1×10^6, or 1 MHz.

The values in Examples 22-4 and 22-5 are from the *LC* circuit illustrated in Fig. 22-2 for series resonance and Fig. 22-6 for parallel resonance.

22.5 *Q* Magnification Factor of a Resonant Circuit

The quality, or *figure of merit,* of the resonant circuit, in sharpness of resonance, is indicated by the factor *Q*. In general, the higher the ratio of the reactance at resonance to the series resistance, the higher the *Q* and the sharper the resonance effect.

Q of Series Circuit

In a series resonant circuit, we can calculate *Q* from the following formula:

$$Q = \frac{X_L}{r_S} \qquad \textbf{(22-6)}$$

where *Q* is the figure of merit, X_L is the inductive reactance in ohms at the resonant frequency, and r_S is the resistance in ohms in series with X_L. For the series resonant circuit in Fig. 22-2,

$$Q = \frac{1500\ \Omega}{10\ \Omega} = 150$$

The *Q* is a numerical factor without any units, because it is a ratio of reactance to resistance and the ohms cancel. Since the series resistance limits the amount of current at resonance, the lower the resistance, the sharper the increase to maximum current at the resonant frequency, and the higher the *Q*. Also, a higher value of reactance at resonance allows the maximum current to produce higher voltage for the output.

The *Q* has the same value if it is calculated with X_C instead of X_L, since they are equal at resonance. However, the *Q* of the circuit is generally considered in terms of X_L, because usually the coil has the series resistance of the circuit. In this case, the Q of the coil and the Q of the series resonant circuit are the same. If extra resistance is added, the Q of the circuit will be less than the Q of the coil. The highest possible Q for the circuit is the Q of the coil.

The value of 150 can be considered a high Q. Typical values are 50 to 250, approximately. Less than 10 is a low Q; more than 300 is a very high Q.

Higher L/C Ratio Can Provide Higher Q

Different combinations of L and C can be resonant at the same frequency. However, the amount of reactance at resonance is different. More X_L can be obtained with a higher L and lower C for resonance, although X_L and X_C must be equal at the resonant frequency. Therefore, both X_L and X_C are higher with a higher L/C ratio for resonance.

More X_L can allow a higher Q if the ac resistance does not increase as much as the reactance. An approximate rule for typical RF coils is that maximum Q can be obtained when X_L is about 1000 Ω. In many cases, though, the minimum C is limited by stray capacitance in the circuit.

Q Rise in Voltage across Series L or C

The Q of the resonant circuit can be considered a magnification factor that determines how much the voltage across L or C is increased by the resonant rise of current in a series circuit. Specifically, the voltage output at series resonance is Q times the generator voltage:

$$V_L = V_C = Q \times V_{gen} \qquad \textbf{(22-7)}$$

In Fig. 22-4, for example, the generator voltage is 300 μV and Q is 150. The resonant rise of voltage across either L or C then equals 300 μV × 150, or 45,000 μV. Note that this is the same value calculated in Fig. 22-2 for V_C or V_L at resonance.

How to Measure Q in a Series Resonant Circuit

The fundamental nature of Q for a series resonant circuit is seen from the fact that the Q can be determined experimentally by measuring the Q rise in voltage across either L or C and comparing this voltage with the generator voltage. As a formula,

$$Q = \frac{V_{out}}{V_{in}} \qquad \textbf{(22-8)}$$

where V_{out} is the ac voltage measured across the coil or capacitor and V_{in} is the generator voltage.

Referring to Fig. 22-5, suppose that you measure with an ac voltmeter across L or C and this voltage equals 45,000 μV

at the resonant frequency. Also, measure the generator input of 300 μV. Then

$$Q = \frac{V_{out}}{V_{in}}$$

$$= \frac{45,000 \ \mu V}{300 \ \mu V}$$

$$= 150$$

This method is better than the X_L/r_S formula for determining Q because r_S is the ac resistance of the coil, which is not so easily measured. Remember that the coil's ac resistance can be more than double the dc resistance measured with an ohmmeter. In fact, measuring Q with Formula (25-8) makes it possible to calculate the ac resistance. These points are illustrated in the following examples.

EXAMPLE 22-6

A series circuit resonant at 0.4 MHz develops 100 mV across a 250-μH L with a 2-mV input. Calculate Q.

Answer:

$$Q = \frac{V_{out}}{V_{in}} = \frac{100 \ mV}{2 \ mV}$$

$$= 50$$

EXAMPLE 22-7

What is the ac resistance of the coil in the preceding example?

Answer:

The Q of the coil is 50. We need to know the reactance of this 250-μH coil at the frequency of 0.4 MHz. Then,

$$X_L = 2\pi f L = 6.28 \times 0.4 \times 10^6 \times 250 \times 10^{-6}$$

$$= 628 \ \Omega$$

$$\text{Also, } Q = \frac{X_L}{r_S} \quad \text{or} \quad r_S = \frac{X_L}{Q}$$

$$r_S = \frac{628 \ \Omega}{50}$$

$$= 12.56 \ \Omega$$

Q of Parallel Circuit

In a parallel resonant circuit where r_S is very small compared with X_L, the Q also equals X_L/r_S. Note that r_S is still the resistance of the coil in series with X_L (see Fig. 22-8). The Q of the coil determines the Q of the parallel circuit here because it is less than the Q of the capacitive branch. Capacitors used in tuned circuits generally have a very high

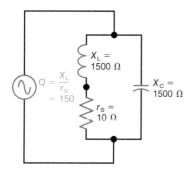

Figure 22-8 The Q of a parallel resonant circuit in terms of X_L and its series resistance r_S.

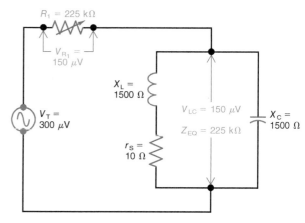

Figure 22-9 How to measure Z_{EQ} of a parallel resonant circuit. Adjust R_1 to make its V_R equal to V_{LC}. Then $Z_{EQ} = R_1$.

Q because of their low losses. In Fig. 22-8, the Q is 1500 Ω/ 10 Ω, or 150, the same as the series resonant circuit with the same values.

This example assumes that the generator resistance is very high and that there is no other resistance branch shunting the tuned circuit. Then the Q of the parallel resonant circuit is the same as the Q of the coil. Actually, shunt resistance can lower the Q of a parallel resonant circuit, as analyzed in Sec. 22-8.

Q Rise in Impedance across a Parallel Resonant Circuit

For parallel resonance, the Q magnification factor determines by how much the impedance across the parallel LC circuit is increased because of the minimum line current. Specifically, the impedance across the parallel resonant circuit is Q times the inductive reactance at the resonant frequency:

$$Z_{EQ} = Q \times X_L \qquad \textbf{(22-9)}$$

Referring back to the parallel resonant circuit in Fig. 22-6 as an example, X_L is 1500 Ω and Q is 150. The result is a rise of impedance to the maximum value of 150 × 1500 Ω, or 225,000 Ω, at the resonant frequency.

Since the line current equals V_A/Z_{EQ}, the minimum line current is 300 μV/225,000 Ω, which equals 0.00133 μA.

At f_r, the minimum line current is $1/Q$ of either branch current. In Fig. 22-7, I_L or I_C is 0.2 μA and Q is 150. Therefore, I_T is $^{0.2}/_{150}$, or 0.00133 μA, which is the same answer as V_A/Z_{EQ}. Or, stated another way, the circulating tank current is Q times the minimum I_T.

How to Measure Z_{EQ} of a Parallel Resonant Circuit

Formula (22-9) for Z_{EQ} is also useful in its inverted form as $Q = Z_{EQ}/X_L$. We can measure Z_{EQ} by the method illustrated in Fig. 22-9. Then Q can be calculated from the value of Z_{EQ} and the inductive reactance of the coil.

To measure Z_{EQ}, first tune the LC circuit to resonance. Then adjust R_1 in Fig. 22-9 to the resistance that makes

its ac voltage equal to the ac voltage across the tuned circuit. With equal voltages, the Z_{EQ} must have the same value as R_1.

For the example here, which corresponds to the parallel resonance shown in Figs. 22-6 and 22-8, Z_{EQ} is equal to 225,000 Ω. This high value is a result of parallel resonance. The X_L is 1500 Ω. Therefore, to determine Q, the calculations are

$$Q = \frac{Z_{EQ}}{X_L} = \frac{225,000}{1500} = 150$$

EXAMPLE 22-8

In Fig. 22-9, assume that with a 4-mVac input signal for V_T, the voltage across R_1 is 2 mV when R_1 is 225 kΩ. Determine Z_{EQ} and Q.

Answer:

Because they divide V_T equally, Z_{EQ} is 225 kΩ, the same as R_1. The amount of input voltage does not matter, as the voltage division determines the relative proportions between R_1 and Z_{EQ}. With 225 kΩ for Z_{EQ} and 1.5 kΩ for X_L, the Q is $^{225}/_{1.5}$, or $Q = 150$.

EXAMPLE 22-9

A parallel LC circuit tuned to 200 kHz with a 350-μH L has a measured Z_{EQ} of 17,600 Ω. Calculate Q.

Answer:

First, calculate X_L as $2\pi fL$ at f_r:

$$X_L = 2\pi \times 200 \times 10^3 \times 350 \times 10^{-6} = 440 \ \Omega$$

Then,

$$Q = \frac{Z_{EQ}}{X_L} = \frac{17,600}{440}$$
$$= 40$$

22.6 Bandwidth of a Resonant Circuit

When we say that an *LC* circuit is resonant at one frequency, this is true for the maximum resonance effect. However, other frequencies close to f_r also are effective. For series resonance, frequencies just below and above f_r produce increased current, but a little less than the value at resonance. Similarly, for parallel resonance, frequencies close to f_r can provide high impedance, although a little less than the maximum Z_{EQ}.

Therefore, any resonant frequency has an associated band of frequencies that provide resonance effects. How wide the band is depends on the *Q* of the resonant circuit. Actually, it is practically impossible to have an *LC* circuit with a resonant effect at only one frequency. The width of the resonant band of frequencies centered around f_r is called the *bandwidth* of the tuned circuit.

Measurement of Bandwidth

The group of frequencies with a response 70.7% of maximum, or more, is generally considered the bandwidth of the tuned circuit, as shown in Fig. 22-10b. The resonant response here is increasing current for the series circuit in

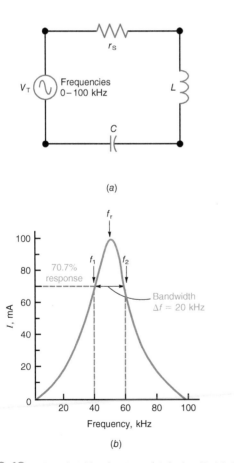

(a)

(b)

Figure 22-10 Bandwidth of a tuned *LC* circuit. (*a*) Series circuit with input of 0 to 100 kHz. (*b*) Response curve with bandwidth Δ*f* equal to 20 kHz between f_1 and f_2.

Fig. 22-10a. Therefore, the bandwidth is measured between the two frequencies f_1 and f_2 producing 70.7% of the maximum current at f_r.

For a parallel circuit, the resonant response is increasing impedance Z_{EQ}. Then the bandwidth is measured between the two frequencies allowing 70.7% of the maximum Z_{EQ} at f_r.

The bandwidth indicated on the response curve in Fig. 22-10b equals 20 kHz. This is the difference between f_2 at 60 kHz and f_1 at 40 kHz, both with 70.7% response.

Compared with the maximum current of 100 mA for f_r at 50 kHz, f_1 below resonance and f_2 above resonance each allows a rise to 70.7 mA. All frequencies in this band 20 kHz wide allow 70.7 mA, or more, as the resonant response in this example.

Bandwidth Equals f_r/Q

Sharp resonance with high *Q* means narrow bandwidth. The lower the *Q*, the broader the resonant response and the greater the bandwidth.

Also, the higher the resonant frequency, the greater the range of frequency values included in the bandwidth for a given sharpness of resonance. Therefore, the bandwidth of a resonant circuit depends on the factors f_r and *Q*. The formula is

$$f_2 - f_1 = \Delta f = \frac{f_r}{Q} \qquad \textbf{(22-10)}$$

where Δ*f* is the total bandwidth in the same units as the resonant frequency f_r. The bandwidth Δ*f* can also be abbreviated BW.

For example, a series circuit resonant at 800 kHz with a *Q* of 100 has a bandwidth of $^{800}/_{100}$, or 8 kHz. Then the *I* is 70.7% of maximum, or more, for all frequencies for a band 8 kHz wide. This frequency band is centered around 800 kHz, from 796 to 804 kHz.

With a parallel resonant circuit having a *Q* higher than 10, Formula (22-10) also can be used for calculating the bandwidth of frequencies that provide 70.7% or more of the maximum Z_{EQ}. However, the formula cannot be used for parallel resonant circuits with low *Q*, as the resonance curve then becomes unsymmetrical.

High *Q* Means Narrow Bandwidth

The effect for different values of *Q* is illustrated in Fig. 22-11. Note that a higher *Q* for the same resonant frequency results in less bandwidth. The slope is sharper for the sides or *skirts* of the response curve, in addition to its greater amplitude.

High *Q* is generally desirable for more output from the resonant circuit. However, it must have enough bandwidth to include the desired range of signal frequencies.

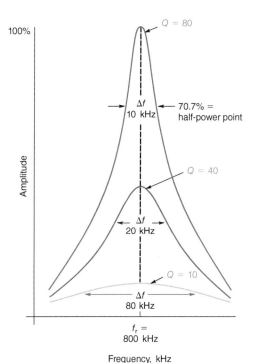

Figure 22-11 Higher Q provides a sharper resonant response. Amplitude is I for series resonance or Z_{EQ} for parallel resonance. Bandwidth at half-power frequencies is Δf.

The Edge Frequencies

Both f_1 and f_2 are separated from f_r by one-half of the total bandwidth. For the top curve in Fig. 22-11, as an example, with a Q of 80, Δf is ±5 kHz centered around 800 kHz for f_r. To determine the edge frequencies,

$$f_1 = f_r - \frac{\Delta f}{2} = 800 - 5 = 795 \text{ kHz}$$

$$f_2 = f_r + \frac{\Delta f}{2} = 800 + 5 = 805 \text{ kHz}$$

These examples assume that the resonance curve is symmetrical. This is true for a high-Q parallel resonant circuit and a series resonant circuit with any Q.

EXAMPLE 22-10

An LC circuit resonant at 2000 kHz has a Q of 100. Find the total bandwidth Δf and the edge frequencies f_1 and f_2.

Answer:

$$\Delta f = \frac{f_r}{Q} = \frac{2000 \text{ kHz}}{100} = 20 \text{ kHz}$$

$$f_1 = f_r - \frac{\Delta f}{2} = 2000 - 10 = 1990 \text{ kHz}$$

$$f_2 = f_r + \frac{\Delta f}{2} = 2000 + 10 = 2010 \text{ kHz}$$

EXAMPLE 22-11

Repeat Example 22-10 for an f_r equal to 6000 kHz and the same Q of 100.

Answer:

$$f = \frac{f_r}{Q} = \frac{6000 \text{ kHz}}{100} = 60 \text{ kHz}$$

$$f_1 = 6000 - 30 = 5970 \text{ kHz}$$

$$f_2 = 6000 + 30 = 6030 \text{ kHz}$$

Notice that Δf is three times as wide as Δf in Example 22-10 for the same Q because f_r is three times higher.

Half-Power Points

It is simply for convenience in calculations that the bandwidth is defined between the two frequencies having 70.7% response. At each of these frequencies, the net capacitive or inductive reactance equals the resistance. Then the total impedance of the series reactance and resistance is 1.4 times greater than R. With this much more impedance, the current is reduced to $\frac{1}{1.414}$, or 0.707, of its maximum value.

Furthermore, the relative current or voltage value of 70.7% corresponds to 50% in power, since power is I^2R or V^2/R and the square of 0.707 equals 0.50. Therefore, the bandwidth between frequencies having 70.7% response in current or voltage is also the bandwidth in terms of half-power points. Formula (22-10) is derived for Δf between the points with 70.7% response on the resonance curve.

Measuring Bandwidth to Calculate Q

The half-power frequencies f_1 and f_2 can be determined experimentally. For series resonance, find the two frequencies at which the current is 70.7% of maximum I, or for parallel resonance, find the two frequencies that make the impedance 70.7% of the maximum Z_{EQ}. The following method uses the circuit in Fig. 22-9 for measuring Z_{EQ}, but with different values to determine its bandwidth and Q:

1. Tune the circuit to resonance and determine its maximum Z_{EQ} at f_r. In this example, assume that Z_{EQ} is 10,000 Ω at the resonant frequency of 200 kHz.

2. Keep the same amount of input voltage, but change its frequency slightly below f_r to determine the frequency f_1 that results in a Z_1 equal to 70.7% of Z_{EQ}. The required value here is 0.707 × 10,000, or 7070 Ω, for Z_1 at f_1. Assume that this frequency f_1 is determined to be 195 kHz.

3. Similarly, find the frequency f_2 above f_r that results in the impedance Z_2 of 7070 Ω. Assume that f_2 is 205 kHz.

4. The total bandwidth between the half-power frequencies equals $f_2 - f_1$ or $205 - 195$. Then the value of $\Delta f = 10$ kHz.

5. Then $Q = f_r/\Delta f$ or 200 kHz$/10$ kHz $= 20$ for the calculated value of Q.

In this way, measuring the bandwidth makes it possible to determine Q. With Δf and f_r, Q can be determined for either parallel or series resonance.

22.7 Tuning

Tuning means obtaining resonance at different frequencies by varying either L or C. As illustrated in Fig. 22-12, the variable capacitance C can be adjusted to tune the series LC circuit to resonance at any one of the five different frequencies. Each of the voltages V_1 to V_5 indicates an ac input with a specific frequency. Which one is selected for maximum output is determined by the resonant frequency of the LC circuit.

When C is set to 424 pF, for example, the resonant frequency of the LC circuit is 500 kHz for f_{r_1}. The input voltage whose frequency is 500 kHz then produces a resonant rise of current that results in maximum output voltage across C. At other frequencies, such as 707 kHz, the

voltage output is less than the input. With C at 424 pF, therefore, the LC circuit tuned to 500 kHz selects this frequency by providing much more voltage output than other frequencies.

Suppose that we want maximum output for the ac input voltage that has the frequency of 707 kHz. Then C is set at 212 pF to make the LC circuit resonant at 707 kHz for f_{r_2}. Similarly, the tuned circuit can resonate at a different frequency for each input voltage. In this way, the LC circuit is tuned to select the desired frequency.

The variable capacitance C can be set to any capacitance value between 26.5 and 424 pF and can tune the 239-μH coil to resonance at any frequency in the range of 500 to 2000 kHz. Note that a parallel resonant circuit also can be tuned by varying C or L.

Radio Tuning Dial

Figure 22-13 illustrates a typical application of resonant circuits in tuning a receiver to the carrier frequency of a desired station in the AM broadcast band. The tuning is done by the air capacitor C, which can be varied from 360 pF with the plates completely in mesh to 40 pF out of mesh. The fixed plates form the *stator*, whereas the *rotor* has plates that move in and out.

Note that the lowest frequency F_L at 540 kHz is tuned with the highest C at 360 pF. Resonance at the highest frequency F_H at 1620 kHz results from the lowest C at 40 pF. Also note that the resonant circuit appears to be a parallel circuit. Because the RF input signal causes current to flow in its winding, it induces a voltage into the 239-μH inductor. That voltage appears in series with the inductor and capacitor. The output is taken from across the capacitor.

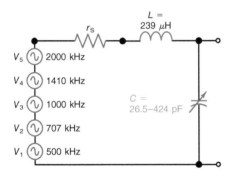

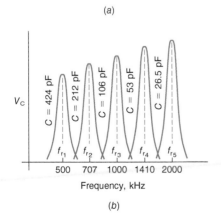

Figure 22-12 Tuning a series *LC* circuit. (*a*) Input voltages at different frequencies. (*b*) Relative response for each frequency when *C* is varied (not to scale).

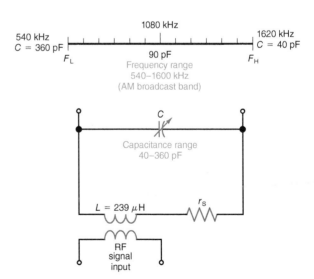

Figure 22-13 Application of tuning an *LC* circuit through the AM radio band.

The capacitance range of 40 to 360 pF tunes through the frequency range from 1620 kHz down to 540 kHz. Frequency F_L is one-third F_H because the maximum C is nine times the minimum C.

The same idea applies to tuning through the commercial FM broadcast band of 88 to 108 MHz with smaller values of L and C. Also, television receivers are tuned to a specific broadcast channel by resonance at the desired frequencies.

For electronic tuning, the C is varied by a *varactor*. This is a semiconductor diode that varies in capacitance when its voltage is changed.

22.8 Damping of Parallel Resonant Circuits

In Fig. 22-14a, the shunt R_P across L and C is a damping resistance because it lowers the Q of the tuned circuit. The R_P may represent the resistance of the external source driving the parallel resonant circuit, or R_P can be an actual resistor added for lower Q and greater bandwidth. Using the parallel R_P to reduce Q is better than increasing the series resistance

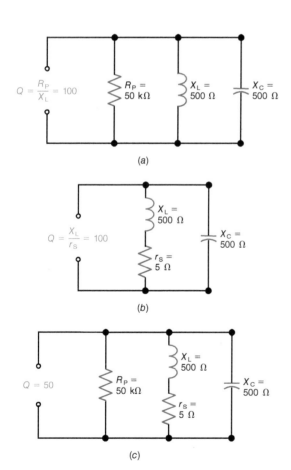

(a)

(b)

(c)

Figure 22-14   The Q of a parallel resonant circuit in terms of coil resistance r_s and parallel damping resistor R_P. See Formula (22-12) for calculating Q. (a) Parallel R_P but negligible r_s. (b) Series r_s but no R_P branch. (c) Both R_P and r_s.

r_s because the resonant response is more symmetrical with shunt damping.

The effect of varying the parallel R_P is opposite from that of the series r_s. A lower value of R_P lowers the Q and reduces the sharpness of resonance. Remember that less resistance in a parallel branch allows more current. This resistive branch current cannot be canceled at resonance by the reactive currents. Therefore, the resonant dip to minimum line current is less sharp with more resistive line current. Specifically, when Q is determined by parallel resistance

$$Q = \frac{R_P}{X_L} \tag{22-11}$$

This relationship with shunt R_P is the reciprocal of the Q formula with series r_s. Reducing R_P decreases Q, but reducing r_s increases Q. The damping can be done by series r_s, parallel R_P, or both.

Parallel R_P without r_s

In Fig. 22-14a, Q is determined only by the R_P, as no series r_s is shown. We can consider that r_s is zero or very small. Then the Q of the coil is infinite or high enough to be greater than the damped Q of the tuned circuit, by a factor of 10 or more. The Q of the damped resonant circuit here is $R_P/X_L = {}^{50,000}\!/_{500} = 100$.

Series r_s without R_P

In Fig. 22-14b, Q is determined only by the coil resistance r_s, as no shunt damping resistance is used. Then $Q = X_L/r_s = {}^{500}\!/_5 = 100$. This is the Q of the coil, which is also the Q of the parallel resonant circuit without shunt damping.

Conversion of r_s or R_P

For the circuits in both Fig. 22-14a and b, Q is 100 because the 50,000-Ω R_P is equivalent to the 5-Ω r_s as a damping resistance. One value can be converted to the other. Specifically,

$$r_s = \frac{X_L^2}{R_P}$$

or

$$R_P = \frac{X_L^2}{r_s}$$

In this example, r_s equals ${}^{250,000}\!/_{50,000} = 5$ Ω, or R_P is ${}^{250,000}\!/_5 = 50,000$ Ω.

Damping with Both r_s and R_P

Figure 22-14c shows the general case of damping where both r_s and R_P must be considered. Then the Q of the circuit can be calculated as

$$Q = \frac{X_L}{r_s + X_L^2/R_P} \tag{22-12}$$

Table 22-1 Comparison of Series and Parallel Resonance

Series Resonance	Parallel Resonance (high Q)
$f_r = \dfrac{1}{2\pi\sqrt{LC}}$	$f_r = \dfrac{1}{2\pi\sqrt{LC}}$
I maximum at f_r with θ of 0°	I_T minimum at f_r with θ of 0°
Impedance Z minimum at f_r	Impedance Z maximum at f_r
$Q = X_L/r_S$, or	$Q = X_L/r_S$, or
$Q = V_{out}/V_{in}$	$Q = Z_{max}/X_L$
Q rise in voltage $= Q \times V_{gen}$	Q rise in impedance $= Q \times V_L$
Bandwidth $\Delta f = f_r/Q$	Bandwidth $\Delta f = f_r/Q$
Circuit capacitive below f_r, but inductive above f_r	Circuit inductive below f_r, but capacitive above f_r
Needs low-resistance source for low r_S, high Q, and sharp tuning	Needs high-resistance source for high R_P, high Q, and sharp tuning
Source is inside LC circuit	Source is outside LC circuit

For the values in Fig. 22-14c,

$$Q = \frac{500}{5 + 250{,}000/50{,}000} = \frac{500}{5 + 5} = \frac{500}{10}$$
$$= 50$$

The Q is lower here compared with Fig. 22-14a or b because this circuit has both series and shunt damping.

Note that for an r_S of zero, Formula (22-12) can be inverted and simplified to $Q = R_P/X_L$. This is the same as Formula (22-11) for shunt damping alone.

For the opposite case where R_P is infinite, that is, an open circuit, Formula (22-12) reduces to X_L/r_S. This is the same as Formula (22-6) without shunt damping. For an overall comparison of series and parallel LC circuits, see Table 22-1.

❓ CHAPTER 22 REVIEW QUESTIONS

1. The resonant frequency of an LC circuit is the frequency where
   a. $X_L = 0\ \Omega$ and $X_C = 0\ \Omega$.
   b. $X_L = X_C$.
   c. X_L and r_S of the coil are equal.
   d. X_L and X_C are in phase.

2. The impedance of a series LC circuit at resonance is
   a. maximum.
   b. nearly infinite.
   c. minimum.
   d. both a and b.

3. The total line current, I_T, of a parallel LC circuit at resonance is
   a. minimum.
   b. maximum.
   c. equal to I_L and I_C.
   d. Q times larger than I_L or I_C.

4. The current at resonance in a series LC circuit is
   a. zero.
   b. minimum.

   c. different in each component.
   d. maximum.

5. The impedance of a parallel LC circuit at resonance is
   a. zero.
   b. maximum.
   c. minimum.
   d. equal to the r_S of the coil.

6. The phase angle of an LC circuit at resonance is
   a. 0°.
   b. +90°.
   c. 180°.
   d. −90°.

7. Below resonance, a series LC circuit appears
   a. inductive.
   b. resistive.
   c. capacitive.
   d. none of the above.

8. Above resonance, a parallel LC circuit appears
   a. inductive.
   b. resistive.

c. capacitive.

d. none of the above.

9. A parallel *LC* circuit has a resonant frequency of 3.75 MHz and a *Q* of 125. What is the bandwidth?

    a. 15 kHz.

    b. 30 kHz.

    c. 60 kHz.

    d. None of the above.

10. What is the resonant frequency of an *LC* circuit with the following values: $L = 100\ \mu H$ and $C = 63.3$ pF?

    a. $f_r = 1$ MHz.

    b. $f_r = 8$ MHz.

    c. $f_r = 2$ MHz.

    d. $f_r = 20$ MHz.

11. What value of capacitance is needed to provide a resonant frequency of 1 MHz if *L* equals $50\ \mu H$?

    a. 506.6 pF.

    b. 506.6 μF.

    c. 0.001 μF.

    d. 0.0016 μF.

12. When either *L* or *C* is increased, the resonant frequency of an *LC* circuit

    a. decreases.

    b. increases.

    c. doesn't change.

    d. cannot be determined.

13. A series *LC* circuit has a *Q* of 100 at resonance. If $V_{in} = 5$ mV$_{p-p}$, how much is the voltage across *C*?

    a. 50 μV_{p-p}.

    b. 5 mV$_{p-p}$.

    c. 50 mV$_{p-p}$.

    d. 500 mV$_{p-p}$.

14. In a low *Q* parallel resonant circuit, when $X_L = X_C$,

    a. $I_L = I_C$.

    b. I_L is less than I_C.

c. I_C is less than I_L.

d. I_L is more than I_C.

15. To double the resonant frequency of an *LC* circuit with a fixed value of *L*, the capacitance, *C*, must be

    a. doubled.

    b. quadrupled.

    c. reduced by one-half.

    d. reduced by one-quarter.

16. A higher *Q* for a resonant circuit provides a

    a. dampened response curve.

    b. wider bandwidth.

    c. narrower bandwidth.

    d. none of the above.

17. The current at the resonant frequency of a series *LC* circuit is 10 mA$_{p-p}$. What is the value of current at the half-power points?

    a. 7.07 mA$_{p-p}$.

    b. 14.14 mA$_{p-p}$.

    c. 5 mA$_{p-p}$.

    d. 10 mA$_{p-p}$.

18. The *Q* of a parallel resonant circuit can be lowered by

    a. placing a resistor in parallel with the tank.

    b. adding more resistance in series with the coil.

    c. decreasing the value of *L* or *C*.

    d. both *a* and *b*.

19. The ability of an *LC* circuit to supply complete sine waves when the input to the tank is only a pulse is called

    a. tuning.

    b. the flywheel effect.

    c. antiresonance.

    d. its *Q*.

20. What is the phase angle of the current with respect to the applied voltage in a resonant circuit?

    a. 0°.

    b. 90°.

    c. 180°.

    d. Depends on the frequency.

CHAPTER 22 PROBLEMS

SECTION 22.1 The Resonance Effect

22.1 Define what is meant by a resonant circuit.

22.2 What is the main application of resonance?

22.3 If an inductor in a resonant *LC* circuit has an X_L value of 1 kΩ, how much is X_C?

SECTION 22.2 Series Resonance

22.4 Figure 22-15 shows a series resonant circuit with the values of X_L and X_C at f_r. Calculate the

    a. net reactance, *X*.

    b. total impedance, Z_T.

    c. current, *I*.

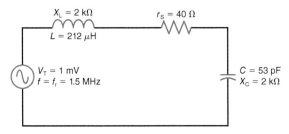

Figure 22-15

d. phase angle, θ.
e. voltage across L.
f. voltage across C.
g. voltage across r_S.

22.5 In Fig. 22-15, what happens to Z_T and I if the frequency of the applied voltage increases or decreases from the resonant frequency, f_r? Explain your answer.

22.6 In Fig. 22-15, why is the phase angle, θ, 0° at f_r?

SECTION 22.3 Parallel Resonance

22.7 Figure 22-16, shows a parallel resonant circuit with the same values for X_L, X_C, and r_S as in Fig. 22-15. With an applied voltage of 10 V and a total line current, I_T, of 100 μA, calculate the
a. inductive current, I_L (ignore r_S).
b. capacitive current, I_C.
c. net reactive branch current, I_X.
d. equivalent impedance, Z_{EQ}, of the tank circuit.

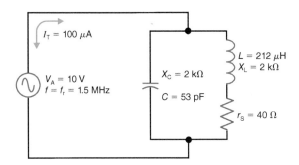

Figure 22-16

22.8 In Fig. 22-16, what happens to Z_{EQ} and I_T as the frequency of the applied voltage increases or decreases from the resonant frequency, f_r? Explain your answer.

SECTION 22.4 Resonant Frequency
$f_r = 1/(2\pi\sqrt{LC})$

22.9 Calculate the resonant frequency, f_r, of an LC circuit with the following values:
a. $L = 250\ \mu$H and $C = 633.25$ pF.
b. $L = 50\ \mu$H and $C = 20.26$ pF.

22.10 What value of inductance, L, must be connected in series with a 50-pF capacitance to obtain an f_r of 3.8 MHz?

22.11 What value of capacitance, C, must be connected in parallel with a 100-μH inductance to obtain an f_r of 1.9 MHz?

22.12 In Fig. 22-17, what is the range of resonant frequencies as C is varied from 40 to 400 pF?

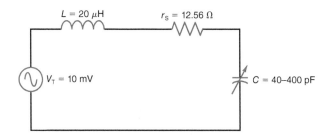

Figure 22-17

22.13 In Fig. 22-17, what value of C will provide an f_r of 2.5 MHz?

22.14 If C is set at 360 pF in Fig. 22-17, what is the resonant frequency, f_r? What value of C will double the resonant frequency?

SECTION 22.5 Q Magnification Factor of a Resonant Circuit

22.15 What is the Q of the series resonant circuit in Fig. 22-17 with C set at
a. 202.7 pF?
b. 50.67 pF?

22.16 A series resonant circuit has the following values: $L = 50\ \mu$H, $C = 506.6$ pF, $r_S = 3.14\ \Omega$, and $V_{in} = 10$ mV. Calculate the following:
a. f_r.
b. Q.
c. V_L and V_C.

22.17 What is the Q of a series resonant circuit if the output voltage across the capacitor is 15 $V_{p\text{-}p}$ with an input voltage of 50 m$V_{p\text{-}p}$?

22.18 Explain why the Q of a resonant circuit cannot increase without limit as X_L increases for higher frequencies.

SECTION 22.6 Bandwidth of a Resonant Circuit

22.19 With C set at 50.67 pF in Fig. 22-17, calculate the following:
a. the bandwidth, Δf.
b. the edge frequencies f_1 and f_2.
c. the current, I, at f_r, f_1, and f_2.

22.20 Does a higher Q correspond to a wider or narrower bandwidth?

22.21 In Fig. 22-18, calculate the following:
   a. f_r.
   b. X_L and X_C at f_r.
   c. Z_T at f_r.
   d. I at f_r.
   e. Q.
   f. V_L and V_C at f_r.
   g. θ_Z at f_r.
   h. Δf, f_1, and f_2.
   i. I at f_1 and f_2.

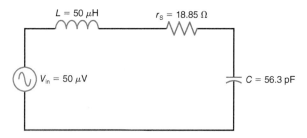

Figure 22-18

22.22 In Fig. 22-19, calculate the following:
   a. f_r.
   b. X_L and X_C at f_r.
   c. I_L and I_C at f_r.
   d. Q.
   e. Z_{EQ}.
   f. I_T.
   g. θ_I.
   h. Δf, f_1, and f_2.

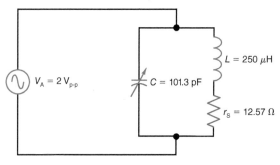

Figure 22-19

22.23 In Fig. 22-17, calculate the ratio of the highest to lowest resonant frequency when C is varied from its lowest to its highest value.

SECTION 22.8 Damping of Parallel Resonant Circuits

22.24 In Fig. 22-20, calculate the Q and bandwidth, Δf, if a 100-kΩ resistor is placed in parallel with the tank circuit.

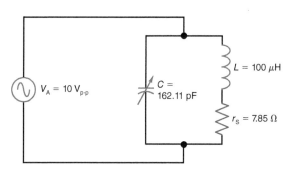

Figure 22-20

22.25 In Fig. 22-19, calculate the Q and bandwidth, Δf, if a 2-MΩ resistor is placed in parallel with the tank circuit.

22.26 In Fig. 22-20, convert the series resistance, r_S, to an equivalent parallel resistance, R_P.

22.27 Repeat Prob. 22.26 for Fig. 22-19.

Chapter 23

Filters

Learning Outcomes

After studying this chapter, you should be able to:

> *Name* and *define* the five basic types of filters.

> *List* the key specifications of a filter.

> *Determine* the attenuation of a filter is decibels (dB).

> *Plot* a filter response on log graph paper.

> *Calculate* the cutoff frequencies of *RC* and *LC* filters.

> *Explain* how crystal, ceramic, and SAW filters operate.

> *Describe* how a DSP filter works.

A filter is a frequency selective circuit. It passes some frequencies but rejects others. Filters are used to select one radio signal and block others. Filters are also used to reject noise and interference.

The key quality of a filter is its selectivity, the ability to discriminate between signals that are close in frequency. We say that a filter has good selectivity if its response curve is very steep and sharply defined rather than gradual.

Passive filters can be made with resistors and capacitors (*RC*) or with inductors and capacitors (*LC*). These *RC* and *LC* filters come in five basic forms: low pass, high pass, bandpass, bandstop, and all pass. There are many variations.

Other types are active filters made with *RC* networks and operational amplifiers (op amps). Crystal and ceramic filters are also available. A surface acoustic wave (SAW) filter is used at very high frequencies.

This chapter provides an overview of the most widely used passive filters and a brief look at some of the more specialized crystal and SAW filters.

23.1 Ideal Responses

The *frequency response of a filter* is the graph of its voltage gain versus frequency. There are five types of filters: *low pass, high pass, bandpass, bandstop,* and *all pass.* This section discusses the ideal frequency response of each. The next section describes the approximations for these ideal responses.

Low-Pass Filter

Figure 23-1 shows the ideal frequency response of a **low-pass filter.** It is sometimes called a *brick wall response* because the right edge of the rectangle looks like a brick wall. A low-pass filter passes all frequencies from zero to the cutoff frequency and blocks all frequencies above the cutoff frequency,

With a low-pass filter, the frequencies between zero and the cutoff frequency are called the **passband.** The frequencies above the cutoff frequency are called the **stopband.** The roll-off region between the passband and the stopband is called the **transition.** An ideal low-pass filter has zero *attenuation* (signal loss) in the passband, infinite attenuation in the stopband, and a vertical transition.

One more point: The ideal low-pass filter has zero phase shift for all frequencies in the passband. Zero phase shift is important when the input signal is nonsinusoidal. When a filter has zero phase shift, the shape of the nonsinusoidal signal is preserved as it passes through the ideal filter. For instance, if the input signal is a square wave, it has a fundamental frequency and harmonics. If the fundamental frequency and all significant harmonics (approximately the first 10) are inside the passband, the square wave will have approximately the same shape at the output.

High-Pass Filter

Figure 23-2 shows the ideal frequency response of a **high-pass filter.** A high-pass filter blocks all frequencies from zero up to the cutoff frequency and passes all frequencies above the cutoff frequency.

With a high-pass filter, the frequencies between zero and the cutoff frequency are the stopband. The frequencies above the cutoff frequency are the passband. An ideal high-pass filter has infinite attenuation in the stopband, zero attenuation in the passband, and a vertical transition.

Bandpass Filter

A **bandpass filter** is useful when you want to tune in a radio or television signal. It is also useful in telephone communications equipment for separating the different phone conversations that are being simultaneously transmitted over the same communication path.

Figure 23-3 shows the ideal frequency response of a bandpass filter. A brick wall response like this blocks all frequencies from zero up to the lower cutoff frequency. Then,

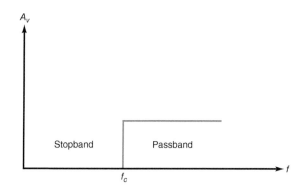

Figure 23-2   Ideal high-pass response.

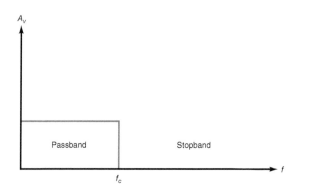

Figure 23-1   Ideal low-pass response.

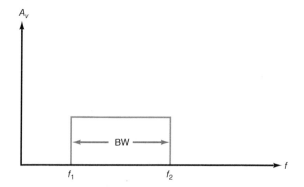

Figure 23-3   Ideal bandpass response.

it passes all the frequencies between the lower and upper cutoff frequencies. Finally, it blocks all frequencies above the upper cutoff frequency.

With a bandpass filter, the passband is all the frequencies between the lower and upper cutoff frequencies. The frequencies below the lower cutoff frequency and above the upper cutoff frequency are the stopband. An ideal bandpass filter has zero attenuation in the passband, infinite attenuation in the stopband, and two vertical transitions.

The *bandwidth (BW)* of a bandpass filter is the difference between its upper and lower 3-dB cutoff frequencies:

$$BW = f_2 - f_1 \qquad \textbf{(23-1)}$$

For instance, if the cutoff frequencies are 450 and 460 kHz, the bandwidth is

$$BW = 460\,\text{kHz} - 450\,\text{kHz} = 10\,\text{kHz}$$

As another example, if the cutoff frequencies are 300 and 3300 Hz, the bandwidth is

$$BW = 3300\,\text{Hz} - 300\,\text{Hz} = 3000\,\text{Hz}$$

The center frequency is symbolized by f_0 and is given by the **geometric average** of the two cutoff frequencies:

$$f_0 = \sqrt{f_1 f_2} \qquad \textbf{(23-2)}$$

For instance, telephone companies use a bandpass filter with cutoff frequencies of 300 and 3300 Hz to separate phone conversations. The center frequency of these filters is

$$f_0 = \sqrt{(300\,\text{Hz})(3300\,\text{Hz})} = 995\,\text{Hz}$$

To avoid interference between different phone conversations, the bandpass filters have responses that approach the brick wall response shown in Fig. 23-3.

The Q of a bandpass filter is defined as the center frequency divided by the bandwidth:

$$Q = \frac{f_0}{BW} \qquad \textbf{(23-3)}$$

For instance, if $f_0 = 200\,\text{kHz}$ and $BW = 40\,\text{kHz}$, then $Q = 5$.

When the Q is greater than 10, the center frequency can be approximated by the *arithmetic average* of the cutoff frequencies:

$$f_0 \cong \frac{f_1 + f_2}{2}$$

For instance, in a radio receiver the cutoff frequencies of the bandpass filter (IF stage) are 450 and 460 kHz. The center frequency is approximately

$$f_0 \cong \frac{450\,\text{kHz} + 460\,\text{kHz}}{2} = 455\,\text{kHz}$$

If Q is less than 1, the bandpass filter is called a **wideband filter.** If Q is greater than 1, the filter is called a **narrowband filter.** For example, a filter with cutoff frequencies

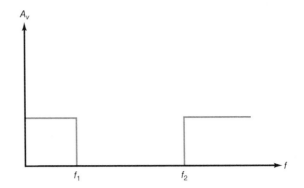

Figure 23-4   Ideal bandstop response.

of 95 and 105 kHz has a bandwidth of 10 kHz. This is a narrowband because Q is approximately 10. A filter with cutoff frequencies of 300 and 3300 Hz has a center frequency of approximately 1000 Hz and a bandwidth of 3000 Hz. This is wideband because Q is approximately 0.333.

Bandstop Filter

Figure 23-4 shows the ideal frequency response of a **bandstop filter.** This type of filter passes all frequencies from zero up to the lower cutoff frequency. Then, it blocks all the frequencies between the lower and upper cutoff frequencies. Finally, it passes all frequencies above the upper cutoff frequency.

With a bandstop filter, the stopband is all the frequencies between the lower and upper cutoff frequencies. The frequencies below the lower cutoff frequency and above the upper cutoff frequency are the passband. An ideal bandstop filter has infinite attenuation in the stopband, no attenuation in the passband, and two vertical transitions.

The definitions for bandwidth, narrowband, and center frequency are the same as before. In other words, with a bandstop filter, we use Formulas (23-l) through (23-3) to calculate BW, f_0, and Q. Incidentally, the bandstop filter is sometimes called a *notch filter* because it notches out or removes all frequencies in the stopband.

All-Pass Filter

Figure 23-5 shows the frequency response of an ideal **all-pass filter.** It has a pass-band and no stopband. Because of this, it passes all frequencies between zero and infinite frequency. It may seem rather unusual to call it a filter since it has zero attenuation for all frequencies. The reason it is called a filter is because of the effect it has on the *phase* of signals passing through it. The all-pass filter is useful when we want to produce a certain amount of phase shift for the signal being filtered without changing its amplitude.

The *phase response of a filter* is defined as the graph of phase shift versus frequency. As mentioned earlier, the ideal low-pass filter has a phase response of 0° at all frequencies.

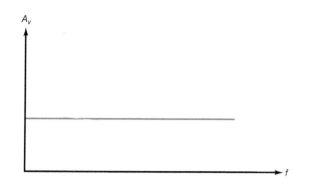

Figure 23-5 Ideal all-pass response.

Because of this, a nonsinusoidal input signal has the same shape after passing through an ideal low-pass filter, provided its fundamental frequency and all significant harmonics are in the passband.

The phase response of an all-pass filter is different from that of the ideal low-pass filter. With the all-pass filter, each distinct frequency can be shifted by a certain amount as it passes through the filter.

23.2 Filter Circuits

The most widely used types of filters are the low pass or high pass. A low-pass filter allows the lower-frequency components of the applied voltage to develop output voltage across the load resistance, whereas the higher-frequency components are attenuated, or reduced, in the output. A high-pass filter does the opposite, allowing the higher-frequency components of the applied voltage to develop voltage across the output load resistance.

An RC coupling circuit is an example of a high-pass filter because the ac component of the input voltage is developed across R while the dc voltage is blocked by the series capacitor. Furthermore, with higher frequencies in the ac component, more ac voltage is coupled. For the opposite case, a bypass capacitor is an example of a low-pass filter. The higher frequencies are bypassed, but the lower the frequency, the less the bypassing action. Then lower frequencies can develop output voltage across the shunt bypass capacitor.

To make the filtering more selective in terms of which frequencies are passed to produce output voltage across the load, filter circuits generally combine inductance and capacitance. Since inductive reactance increases with higher frequencies and capacitive reactance decreases, the two opposite effects improve the filtering action.

With combinations of L and C, filters are named to correspond to the circuit configuration. Most common types of filters are the L, T, and π. Any one of the three can function as either a low-pass filter or a high-pass filter.

The reactance X_L of either low-pass or high-pass filters with L and C increases with higher frequencies, while X_C decreases. The frequency characteristics of X_L and X_C cannot be changed. However, the circuit connections are opposite to reverse the filtering action.

The ability of any filter to reduce the amplitude of undesired frequencies is called the *attenuation* of the filter. The frequency at which the attenuation reduces the output to 70.7% is the *cutoff frequency*, usually designated f_c.

23.3 Low-Pass Filters

Figure 23-6 illustrates low-pass circuits from a single filter element with a shunt bypass capacitor in Fig. 23-6a or

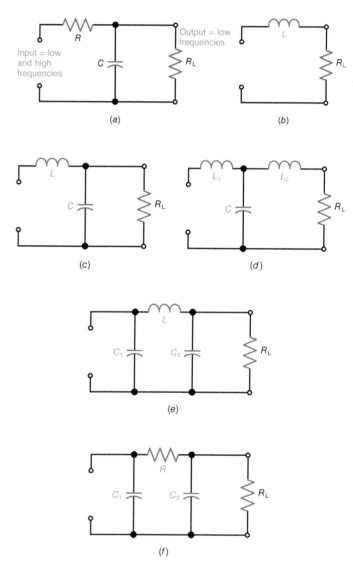

Figure 23-6 Low-pass filter circuits. (a) Bypass capacitor C in parallel with R_L. (b) Choke L in series with R_L. (c) Inverted-L type with choke and bypass capacitor. (d) The T type with two chokes and one bypass capacitor. (e) The π type with one choke and bypass capacitors at both ends. (f) The π type with a series resistor instead of a choke.

a series choke in *b*, to the more elaborate combinations of an inverted-L type filter in *c*, a T type in *d*, and a π type in *e* and *f*. With an applied input voltage having different frequency components, the low-pass filter action results in maximum low-frequency voltage across R_L, while most of the high-frequency voltage is developed across the series choke or resistance.

In Fig. 23-6*a*, the shunt capacitor *C* bypasses R_L at high frequencies. In Fig. 23-6*b*, the choke *L* acts as a voltage divider in series with R_L. Since *L* has maximum reactance for the highest frequencies, this component of the input voltage is developed across *L* with little across R_L. At lower frequencies, *L* has low reactance, and most of the input voltage can be developed across R_L.

In Fig. 23-6*c*, the use of both the series choke and the bypass capacitor improves the filtering by providing a sharper cutoff between the low frequencies that can develop voltage across R_L and the higher frequencies stopped from the load by producing maximum voltage across *L*. Similarly, the T-type circuit in Fig. 23-6*d* and the π-type circuits in *e* and *f* improve filtering.

Using the series resistance in Fig. 23-6*f* instead of a choke provides an economical π filter in less space.

Passband and Stopband

As illustrated in Fig. 23-7, a low-pass filter attenuates frequencies above the cutoff frequency f_c of 15 kHz in this example. Any component of the input voltage having a frequency lower than 15 kHz can produce output voltage across the load. These frequencies are in the *passband*. Frequencies of 15 kHz or more are in the *stopband*. The sharpness of filtering between the passband and the stopband depends on the type of circuit. In general, the more *L* and *C* components, the sharper the response of the filter. Therefore, π and T types are better filters than the L type and the bypass or choke alone.

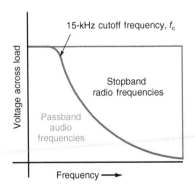

Figure 23-7 The response of a low-pass filter with cutoff at 15 kHz. The filter passes the audio signal but attenuates radio frequencies.

The response curve in Fig. 23-7 is illustrated for the application of a low-pass filter attenuating RF voltages while passing audio frequencies to the load. This is necessary when the input voltage has RF and AF components but only the audio voltage is desired for the AF circuits that follow the filter.

Circuit Variations

The choice between the T-type filter with a series input choke and the π type with a shunt input capacitor depends on the internal resistance of the generator supplying input voltage to the filter. A low-resistance generator needs the T filter so that the choke can provide high series impedance for the bypass capacitor. Otherwise, the bypass capacitor must have extremely large values to short-circuit the low-resistance generator at high frequencies.

The π filter is more suitable with a high-resistance generator when the input capacitor can be effective as a bypass. For the same reasons, the L filter can have the shunt bypass either in the input for a high-resistance generator or across the output for a low-resistance generator.

In all filter circuits, the series choke can be connected either in the high side of the line, as in Fig. 23-6, or in series in the opposite side of the line, without having any effect on the filtering action. Also, the series components can be connected in both sides of the line for a *balanced filter* circuit.

23.4 High-Pass Filters

As illustrated in Fig. 23-8, the high-pass filter passes to the load all frequencies higher than the cutoff frequency f_c, whereas lower frequencies cannot develop appreciable voltage across the load. The graph in Fig. 23-8*a* shows the response of a high-pass filter with a stopband of 0 to 50 Hz. Above the cutoff frequency of 50 Hz, the higher audio frequencies in the passband can produce AF voltage across the output load resistance.

The high-pass filtering action results from using C_C as a coupling capacitor in series with the load, as in Fig. 23-8*b*. The L, T, and π types use the inductance for a high-reactance choke across the line. In this way, the higher-frequency components of the input voltage can develop very little voltage across the series capacitance, allowing most of this voltage to be produced across R_L. The inductance across the line has higher reactance with increasing frequencies, allowing the shunt impedance to be no lower than the value of R_L.

For low frequencies, however, R_L is effectively short-circuited by the low inductive reactance across the line. Also, C_C has high reactance and develops most of the voltage at low frequencies, stopping these frequencies from developing voltage across the load.

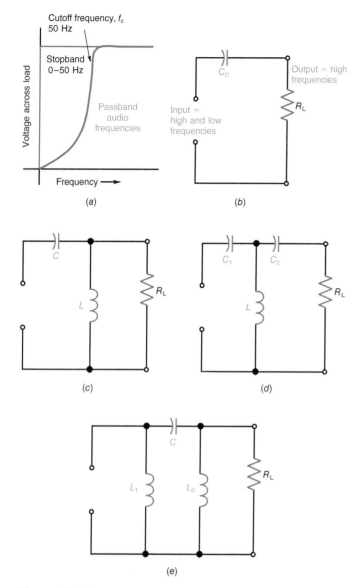

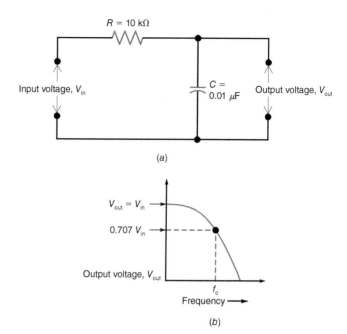

Figure 23-9 *RC* low-pass filter. (a) Circuit. (b) Graph of V_{out} versus frequency.

when $f = 0$ Hz (dc) and $f = \infty$ Hz. At $f = 0$ Hz, the capacitor C has infinite capacitive reactance X_C, calculated as

$$X_C = \frac{1}{2\pi fC}$$

$$= \frac{1}{2 \times \pi \times 0 \text{ Hz} \times 0.01 \ \mu\text{F}}$$

$$= \infty \ \Omega$$

Figure 23-10a shows the equivalent circuit for this condition. Notice that C appears as an open. Since all of the input voltage appears across the open in a series circuit, V_{out} must equal V_{in} when $f = 0$ Hz.

At the other extreme, consider the circuit when the frequency f is very high or infinitely high. Then $X_C = 0 \ \Omega$, calculated as

$$X_C = \frac{1}{2\pi fC}$$

$$= \frac{1}{2 \times \pi \times \infty \text{ Hz} \times 0.01 \ \mu\text{F}}$$

$$= 0 \ \Omega$$

Figure 23-10b shows the equivalent circuit for this condition. Notice that C appears as a short. Since the voltage across a short is zero, the output voltage for very high frequencies must be zero.

When the frequency of the input voltage is somewhere between zero and infinity, the output voltage can be determined by using Formula (23-4):

$$V_{out} = \frac{X_C}{Z_T} \times V_{in} \qquad \textbf{(23-4)}$$

Figure 23-8 High-pass filters. (a) The response curve for an audio frequency filter cutting off at 50 Hz. (b) An *RC* coupling circuit. (c) Inverted-L type. (d) The T type. (e) The π type.

23.5 Analyzing Filter Circuits

Any low-pass or high-pass filter can be thought of as a frequency-dependent voltage divider, since the amount of output voltage is a function of frequency. Special formulas can be used to calculate the output voltage for any frequency of the applied voltage. What follows is a more mathematical approach in analyzing the operation of the most basic low-pass and high-pass filter circuits.

RC Low-Pass Filter

Figure 23-9a shows a simple *RC* low-pass filter, and Fig. 23-9b shows how its output voltage V_{out} varies with frequency. Let's examine how the *RC* low-pass filter responds

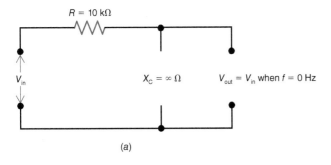

V_{in} $R = 10\ k\Omega$ $X_C = \infty\ \Omega$ $V_{out} = V_{in}$ when $f = 0$ Hz

(a)

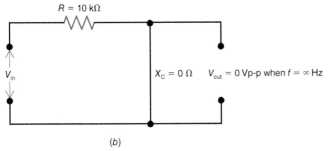

V_{in} $R = 10\ k\Omega$ $X_C = 0\ \Omega$ $V_{out} = 0$ Vp-p when $f = \infty$ Hz

(b)

Figure 23-10 *RC* low-pass equivalent circuits. (a) Equivalent circuit for $f = 0$ Hz. (b) Equivalent circuit for very high frequencies, or $f = \infty$ Hz.

where

$$Z_T = \sqrt{R^2 + X_C^2}$$

At very low frequencies, where X_C approaches infinity, V_{out} is approximately equal to V_{in}. This is true because the ratio X_C/Z_T approaches one as X_C and Z_T become approximately the same value. At very high frequencies, where X_C approaches zero, the ratio X_C/Z_T becomes very small, and V_{out} is approximately zero.

With respect to the input voltage V_{in}, the phase angle θ of the output voltage V_{out} can be calculated as

$$\theta = \arctan - \frac{R}{X_C} \qquad \textbf{(23-5)}$$

At very low frequencies, X_C is very large and θ is approximately $0°$. At very high frequencies, however, X_C is nearly zero and θ approaches $-90°$.

The frequency where $X_C = R$ is the *cutoff frequency*, designated f_c. At f_c, the series current I is at 70.7% of its maximum value because the total impedance Z_T is 1.41 times larger than the resistance of R. The formula for the cutoff frequency f_c of an *RC* low-pass filter is derived as follows. Because $X_C = R$ at f_c,

$$\frac{1}{2\pi f_c C} = R$$

Solving for f_c gives

$$f_c = \frac{1}{2\pi RC} \qquad \textbf{(23-6)}$$

The response curve in Fig. 23-9b shows that $V_{out} = 0.707 V_{in}$ at the cutoff frequency f_c.

EXAMPLE 23-1

In Fig. 23-9a, calculate (a) the cutoff frequency f_c; (b) V_{out} at f_c; (c) θ at f_c. (Assume $V_{in} = 10\ V_{p-p}$ for all frequencies.)

Answer:

a. To calculate f_c, use Formula (23-6):

$$f_c = \frac{1}{2\pi RC}$$

$$= \frac{1}{2 \times \pi \times 10\ k\Omega \times 0.01\ \mu F}$$

$$= 1.592\ kHz$$

b. To calculate V_{out} at f_c, use Formula (23-4). First, however, calculate X_C and Z_T at f_c:

$$X_C = \frac{1}{2\pi f_c C}$$

$$= \frac{1}{2 \times \pi \times 1.592\ kHz \times 0.01\ \mu F}$$

$$= 10\ k\Omega$$

$$Z_T = \sqrt{R^2 + X_C^2}$$

$$= \sqrt{10^2\ k\Omega + 10^2\ k\Omega}$$

$$= 14.14\ k\Omega$$

Next,

$$V_{out} = \frac{X_C}{Z_T} \times V_{in}$$

$$= \frac{10\ k\Omega}{14.14\ k\Omega} \times 10\ V_{p-p}$$

$$= 7.07\ V_{p-p}$$

c. To calculate θ, use Formula (23-5):

$$\theta = \arctan - \frac{R}{X_C}$$

$$= \arctan - \frac{10\ k\Omega}{10\ k\Omega}$$

$$= \arctan - 1$$

$$= -45°$$

The phase angle of $-45°$ tells us that V_{out} lags V_{in} by $45°$ at the cutoff frequency f_c.

RL Low-Pass Filter

Figure 23-11a shows a simple *RL* low-pass filter, and Fig. 23-11b shows how its output voltage V_{out} varies with frequency. The performance of an *RL* filter is similar to that of an *RC* low-pass filter. The *RL* version is not used as much because an inductor is larger, more expensive, and has higher resistive losses. Most low-pass filters are of the *RC* type.

The frequency at which $X_L = R$ is the cutoff frequency f_c. At f_c, the series current I is at 70.7% of its maximum value,

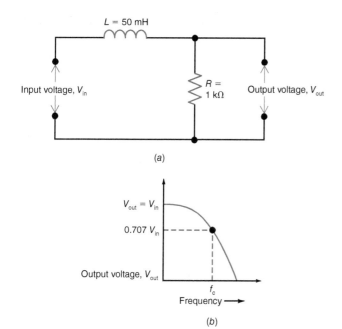

(a)

(b)

Figure 23-11 *RL* low-pass filter. (*a*) Circuit. (*b*) Graph of V_{out} versus frequency.

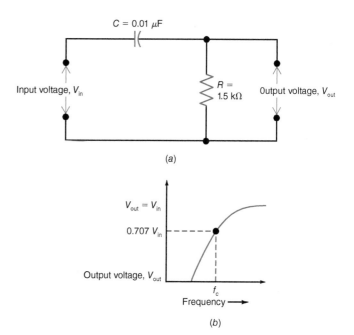

(a)

(b)

Figure 23-12 *RC* high-pass filter. (*a*) Circuit. (*b*) Graph of V_{out} versus frequency.

since $Z_T = 1.41R$ when $X_L = R$. The formula for the cutoff frequency of an *RL* low-pass filter is

$$f_c = \frac{R}{2\pi L} \qquad (23\text{-}7)$$

The response curve in Fig. 23-11*b* shows that $V_{out} = 0.707V_{in}$ at the cutoff frequency f_c.

RC High-Pass Filter

Figure 23-12*a* shows an *RC* high-pass filter. Notice that the output is taken across the resistor *R* rather than across the capacitor *C*. Figure 23-12*b* shows how the output voltage varies with frequency. To calculate the output voltage V_{out} at any frequency, use Formula (23-8):

$$V_{out} = \frac{R}{Z_T} \times V_{in} \qquad (23\text{-}8)$$

where

$$Z_T = \sqrt{R^2 + X_C^2}$$

At very low frequencies, the output voltage approaches zero because the ratio R/Z_T becomes very small as X_C and thus Z_T approach infinity. At very high frequencies, V_{out} is approximately equal to V_{in}, because the ratio R/Z_T approaches one as Z_T and R become approximately the same value.

The phase angle of V_{out} with respect to V_{in} for an *RC* high-pass filter can be calculated using Formula (23-9):

$$\theta = \arctan \frac{X_C}{R} \qquad (23\text{-}9)$$

At very low frequencies where X_C is very large, θ is approximately 90°. At very high frequencies where X_C approaches zero, θ is approximately 0°.

To calculate the cutoff frequency f_c for an *RC* high-pass filter, use Formula (23-6). Although this formula is used to calculate f_c for an *RC* low-pass filter, it can also be used to calculate f_c for an *RC* high-pass filter. The reason is that, in both circuits, $X_C = R$ at the cutoff frequency. In Fig. 23-12*b*, notice that $V_{out} = 0.707V_{in}$ at f_c.

RL High-Pass Filter

An *RL* high-pass filter is shown in Fig. 23-13*a*, and its response curve is shown in Fig. 23-13*b*. In Fig. 23-13*a*, notice that the output is taken across the inductor *L* rather than across the resistance *R*. Like an *RL* low-pass filter, an *RL* high-pass filter is not widely used because of the inductor disadvantages mentioned earlier.

EXAMPLE 23-2

Calculate the cutoff frequency for the *RC* high-pass filter in Fig. 23-12*a*.

Answer:
Use Formula (23-6):

$$f_c = \frac{1}{2\pi RC}$$

$$= \frac{1}{2 \times \pi \times 1.5 \text{ k}\Omega \times 0.01 \text{ }\mu\text{F}}$$

$$= 10.61 \text{ kHz}$$

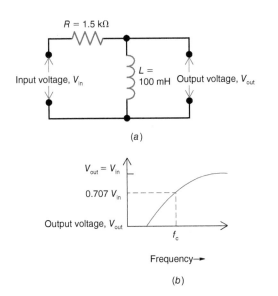

(a)

(b)

Figure 23-13 *RL* high-pass filter. (a) Circuit. (b) Graph of V_{out} versus frequency.

RC Bandstop Filter

A high-pass filter can also be combined with a low-pass filter when it is desired to block or severely attenuate a certain band of frequencies. Such a filter is called a *bandstop* or *notch filter*. Figure 23-14a shows an *RC* bandstop filter, and Fig. 23-14b shows how its output voltage varies with frequency. In Fig. 23-14a, the components identified as $2R_1$ and $2C_1$ constitute the low-pass filter section, and the components identified as R_1 and C_1 constitute the high-pass filter

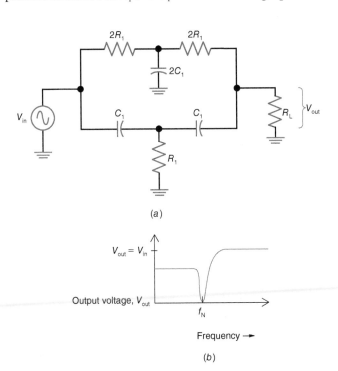

(a)

(b)

Figure 23-14 Notch filter. (a) Circuit. (b) Graph of V_{out} versus frequency.

section. Notice that the individual filters are in parallel. The frequency of maximum attenuation is called the *notch frequency*, identified as f_N in Fig. 23-14b. Notice that the maximum value of V_{out} below f_N is less than the maximum value of V_{out} above f_N. The reason for this is that the series resistances ($2R_1$) in the low-pass filter provide greater circuit losses than the series capacitors (C_1) in the high-pass filter.

To calculate the notch frequency f_N in Fig. 23-14a, use Formula (23-10):

$$f_N = \frac{1}{4\pi R_1 C_1} \qquad \textbf{(23-10)}$$

EXAMPLE 23-3

Calculate the notch frequency f_N in Fig. 23-14a if $R_1 = 1\ k\Omega$ and $C_1 = 0.01\ \mu F$. Also, calculate the required values for $2R_1$ and $2C_1$ in the low-pass filter.

Answer:
Use Formula (23-10):

$$f_N = \frac{1}{4\pi R_1 C_1}$$

$$= \frac{1}{4 \times \pi \times 1\ k\Omega \times 0.01\ \mu F}$$

$$= 7.96\ kHz$$

$$2R_1 = 2 \times 1\ k\Omega$$

$$= 2\ k\Omega$$

$$2C_1 = 2 \times 0.01\ \mu F$$

$$= 0.02\ \mu F$$

23.6 Decibels and Frequency Response Curves

In analyzing filters, the decibel (dB) unit is often used to describe the amount of attenuation offered by the filter. In basic terms, the *decibel* is a logarithmic expression that compares two power levels. Expressed mathematically,

$$N_{dB} = 10 \log \frac{P_{out}}{P_{in}} \qquad \textbf{(23-11)}$$

where

$$N_{dB} = \text{gain or loss in decibels}$$
$$P_{in} = \text{input power}$$
$$P_{out} = \text{output power}$$

If the ratio P_{out}/P_{in} is greater than one, the N_{dB} value is positive, indicating an increase in power from input to output. If the ratio P_{out}/P_{in} is less than one, the N_{dB} value is negative, indicating a loss or reduction in power from input to output. A reduction in power, corresponding to a negative N_{dB} value, is referred to as *attenuation*.

EXAMPLE 23-4

A certain amplifier has an input power of 1 W and an output power of 100 W. Calculate the dB power gain of the amplifier.

Answer:
Use Formula (23-11):

$$N_{dB} = 10 \log \frac{P_{out}}{P_{in}}$$

$$= 10 \log \frac{100 \text{ W}}{1 \text{ W}}$$

$$= 10 \times 2$$

$$= 20 \text{ dB}$$

EXAMPLE 23-5

The input power to a filter is 100 mW, and the output power is 5 mW. Calculate the attenuation, in decibels, offered by the filter.

Answer:

$$N_{dB} = 10 \log \frac{P_{out}}{P_{in}}$$

$$= 10 \log \frac{5 \text{ mW}}{100 \text{ mW}}$$

$$= 10 \times (-1.3)$$

$$= -13 \text{ dB}$$

The power gain or loss in decibels can also be computed from a voltage ratio if the measurements are made across equal resistances.

$$N_{dB} = 20 \log \frac{V_{out}}{V_{in}} \tag{23-12}$$

where

$$N_{dB} = \text{gain or loss in decibels}$$
$$V_{in} = \text{input voltage}$$
$$V_{out} = \text{output voltage}$$

The N_{dB} values of the passive filters discussed in this chapter can never be positive because V_{out} can never be greater than V_{in}.

Consider the *RC* low-pass filter in Fig. 23-15. The cutoff frequency f_c for this circuit is 1.592 kHz, as determined by Formula (23-6). Recall that the formula for V_{out} at any frequency is

$$V_{out} = \frac{X_C}{Z_T} \times V_{in}$$

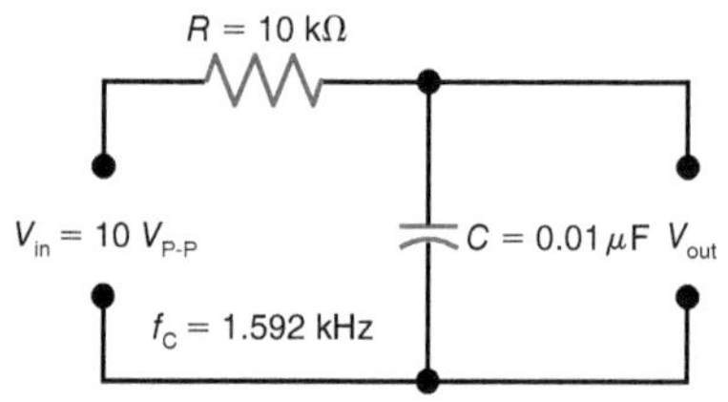

Figure 23-15 *RC* low-pass filter.

Dividing both sides of the equation by V_{in} gives

$$\frac{V_{out}}{V_{in}} = \frac{X_C}{Z_T}$$

Substituting X_C/Z_T for V_{out}/V_{in} in Formula (23-11) gives

$$N_{dB} = 20 \log \frac{X_C}{Z_T}$$

EXAMPLE 23-6

In Fig. 23-15, calculate the attenuation, in decibels, at the following frequencies: (a) 0 Hz; (b) 1.592 kHz; (c) 15.92 kHz. (Assume that $V_{in} = 10$ V$_{p-p}$ at all frequencies.)

Answer:

a. At 0 Hz, $V_{out} = V_{in} = 10$ V$_{p-p}$, since the capacitor *C* appears as an open. Therefore,

$$N_{dB} = 20 \log \frac{V_{out}}{V_{in}}$$

$$= 20 \log \frac{10 \text{ V}_{p-p}}{10 \text{ V}_{p-p}}$$

$$= 20 \log 1$$

$$= 20 \times 0$$

$$= 0 \text{ dB}$$

b. Since 1.592 kHz is the cutoff frequency f_c, V_{out} will be 0.707 × V_{in} or 7.07 V$_{p-p}$. Therefore,

$$N_{dB} = 20 \log \frac{V_{out}}{V_{in}}$$

$$= 20 \log \frac{7.07 \text{ V}_{p-p}}{10 \text{ V}_{p-p}}$$

$$= 20 \log 0.707$$

$$= 20 \times (-0.15)$$

$$= -3 \text{ dB}$$

c. To calculate N_{dB} at 15.92 kHz, X_C and Z_T must first be determined.

$$X_C = \frac{1}{2\pi f C}$$

$$= \frac{1}{2 \times \pi \times 15.92 \text{ kHz} \times 0.01 \text{ } \mu F}$$

$$= 1 \text{ k}\Omega$$

$$Z_T = \sqrt{R^2 + X_C^2}$$

$$= \sqrt{10^2 \text{ k}\Omega + 1^2 \text{ k}\Omega}$$

$$= 10.05 \text{ k}\Omega$$

Next,

$$N_{dB} = 20 \log \frac{X_C}{Z_T}$$

$$= 20 \log \frac{1 \text{ k}\Omega}{10.05 \text{ k}\Omega}$$

$$= 20 \log 0.0995$$

$$= 20(-1)$$

$$= -20 \text{ dB}$$

In Example 23-6, notice that N_{dB} is 0 dB at a frequency of 0 Hz, which is in the filter's passband. This may seem unusual, but the 0-dB value simply indicates that there is no attenuation at this frequency. For an ideal passive filter, $N_{dB} = 0$ dB in the passband. As another point of interest from Example 23-6, N_{dB} is -3 dB at the cutoff frequency of 1.592 kHz. Since $V_{out} = 0.707 \ V_{in}$ at f_c for any passive filter, N_{dB} is always -3 dB at the cutoff frequency of a passive filter.

Frequency Response Curves

The frequency response of a filter is typically shown by plotting its gain (or loss) versus frequency on logarithmic graph paper. The two types of logarithmic graph paper are log-log and semilog. On *semilog graph paper,* the divisions along one axis are spaced logarithmically, and the other axis has conventional linear spacing between divisions. On *log-log graph paper,* both axes have logarithmic spacing between divisions. Logarithmic spacing results in a scale that expands the display of smaller values and compresses the display of larger values. On logarithmic graph paper, a 2-to-1 range of frequencies is called an *octave,* and a 10-to-1 range of values is called a *decade.*

One advantage of logarithmic spacing is that a larger range of values can be shown in one plot without losing resolution in the smaller values. For example, if frequencies between 10 Hz and 100 kHz were plotted on 100 divisions of linear graph paper, each division would represent approximately 1000 Hz and it would be impossible to plot values in the decade between 10 Hz and 100 Hz. On the other hand, by using logarithmic graph paper, the decade between 10 Hz and 100 Hz would occupy the same space on the graph as the decade between 10 kHz and 100 kHz.

Log-log or semilog graph paper is specified by the number of decades it contains. Each decade is a *graph cycle.* For example, 2-cycle by 4-cycle log-log paper has two decades on one axis and four on the other. The number of cycles must be adequate for the range of data plotted. For example,

if the frequency response extends from 25 Hz to 40 kHz, 4 cycles are necessary to plot the frequencies corresponding to the decades 10 Hz to 100 Hz, 100 Hz to 1 kHz, 1 kHz to 10 kHz, and 10 kHz to 100 kHz. A typical sheet of log-log graph paper is shown in Fig. 23-16. Because there are three decades on the horizontal axis and five decades on the vertical axis, this graph paper is called 3-cycle by 5-cycle log-log paper. Notice that each octave corresponds to a 2-to-1 range in values and each decade corresponds to a 10-to-1 range in values. For clarity, several octaves and decades are shown in Fig. 23-16.

When semilog graph paper is used to plot a frequency response, the observed or calculated values of gain (or loss) must first be converted to decibels before plotting. On the other hand, since decibel voltage gain is a logarithmic function, the gain or loss values can be plotted on log-log paper without first converting to decibels.

RC Low-Pass Frequency Response Curve

Figure 23-17a shows an *RC* low-pass filter whose cutoff frequency f_c is 1.592 kHz as determined by Formula (23-6). Figure 23-17b shows its frequency response curve plotted on semilog graph paper. Notice there are 6 cycles on the horizontal axis, which spans a frequency range from 1 Hz to 1 MHz. Notice that the vertical axis specifies the N_{dB} loss, which is the amount of attenuation offered by the filter in decibels. Notice that $N_{dB} = -3$ dB at the cutoff frequency of 1.592 kHz. Above f_c, N_{dB} decreases at the rate of approximately 6 dB/octave, which is equivalent to a rate of 20 dB/decade.

EXAMPLE 23-7

From the graph in Fig. 23-17b, what is the attenuation in decibels at (a) 100 Hz; (b) 10 kHz; (c) 50 kHz?

Answer:

a. At $f = 100$ Hz, $N_{dB} = 0$ dB, as indicated by point A on the graph.
b. At $f = 10$ kHz, $N_{dB} = -16$ dB, as indicated by point B on the graph.
c. At $f = 50$ kHz, $N_{dB} = -30$ dB, as indicated by point C.

For filters such as the inverted-*L*, *T*, or π type, the response curve rolloff is much steeper beyond the cutoff frequency f_c. For example, a low-pass filter with a series inductor and a shunt capacitor has a rolloff rate of 12 dB/octave or 40 dB/decade above the cutoff frequency f_c. To increase the rate of rolloff, more inductors and capacitors must be used in the filter design. Filters are available whose rolloff rates exceed 36 dB/octave.

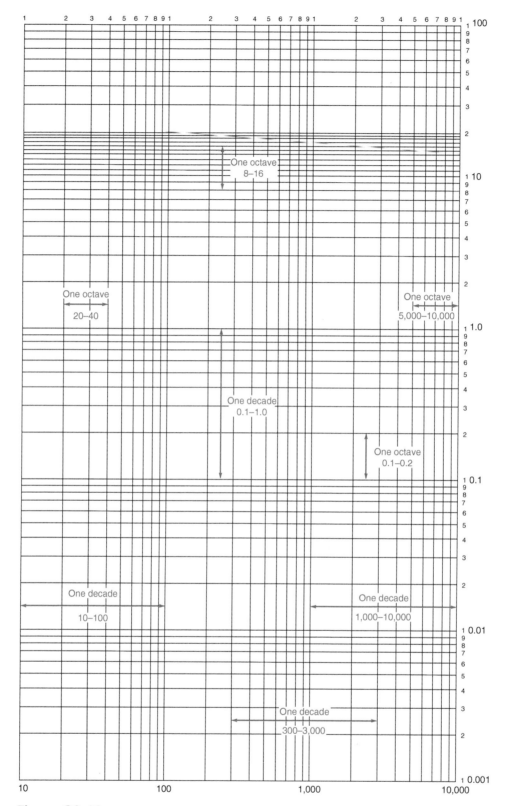

Figure 23-16   Log-log graph paper. Notice that each octave corresponds to a 2-to-1 range of values and each decade corresponds to a 10-to-1 range of values.

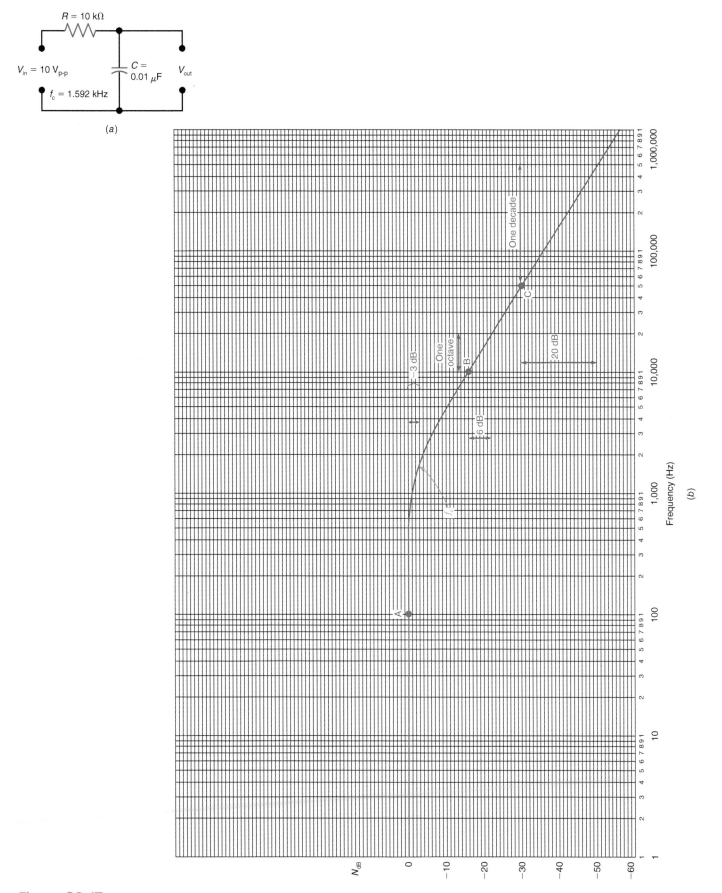

Figure 23-17 *RC* low-pass filter frequency response curve. (*a*) Circuit. (*b*) Frequency response curve.

23.7 Resonant Filters

Tuned circuits provide a convenient method of filtering a band of radio frequencies because relatively small values of L and C are necessary for resonance. A tuned circuit provides filtering action by means of its maximum response at the resonant frequency.

The width of the band of frequencies affected by resonance depends on the Q of the tuned circuit; a higher Q provides a narrower bandwidth. Because resonance is effective for a band of frequencies below and above f_r, resonant filters are called *bandstop* or *bandpass* filters. Series or parallel LC circuits can be used for either function, depending on the connections with respect to R_L. In the application of a bandstop filter to suppress certain frequencies, the LC circuit is often called a *wavetrap*.

Series Resonance Filters

A series resonant circuit has maximum current and minimum impedance at the resonant frequency. Connected in series with R_L, as in Fig. 23-18a, the series-tuned LC circuit allows frequencies at and near resonance to produce maximum output across R_L. Therefore, this is bandpass filtering.

When the series LC circuit is connected across R_L as in Fig. 23-18b, however, the resonant circuit provides a low-impedance shunt path that short-circuits R_L. Then there is minimum output. This action corresponds to a shunt bypass capacitor, but the resonant circuit is more selective, short-circuiting R_L just for frequencies at and near resonance. For the bandwidth of the tuned circuit, the series resonant circuit in shunt with R_L provides bandstop filtering.

The series resistor R_S in Fig. 23-18b is used to isolate the low resistance of the LC filter from the input source. At the resonant frequency, practically all of the input voltage is across R_S with little across R_L because the LC tuned circuit then has very low resistance due to series resonance.

Parallel Resonance Filters

A parallel resonant circuit has maximum impedance at the resonant frequency. Connected in series with R_L, as in Fig. 23-19a, the parallel-tuned LC circuit provides maximum impedance in series with R_L at and near the resonant frequency. Then these frequencies produce maximum voltage across the LC circuit but minimum output voltage across R_L. This is a bandstop filter, therefore, for the bandwidth of the tuned circuit.

The parallel LC circuit connected across R_L, however, as in Fig. 23-19b, provides a bandpass filter. At resonance, the high impedance of the parallel LC circuit allows R_L to develop its output voltage. Below resonance, R_L is short-circuited by the low reactance of L; above resonance, R_L is short-circuited by the low reactance of C. For frequencies at or near resonance, though, R_L is shunted by high impedance, resulting in maximum output voltage.

The series resistor R_S in Fig. 23-19b is used to improve the filtering effect. Note that the parallel LC combination and R_S divide the input voltage. At the resonant frequency,

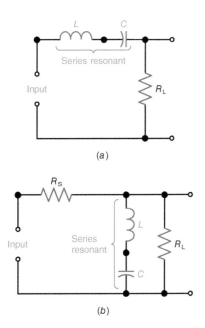

Figure 23-18 The filtering action of a series resonant circuit. (a) Bandpass filter when L and C are in series with R_L. (b) Bandstop filter when LC circuit is in shunt with R_L.

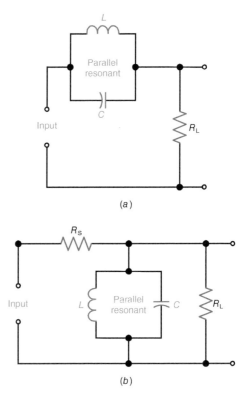

Figure 23-19 The filtering action of a parallel resonant circuit. (a) Bandstop filter when LC tank is in series with R_L. (b) Bandpass filter when LC tank is in shunt with R_L.

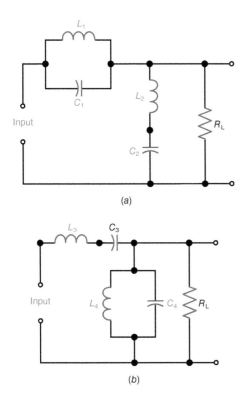

Figure 23-20 Inverted-L filter with resonant circuits. (*a*) Bandstop filtering action. (*b*) Bandpass filtering action.

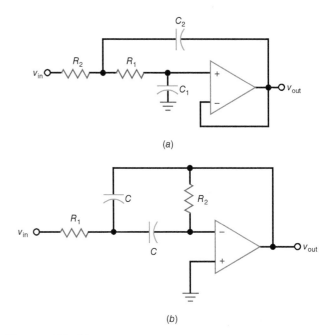

(*a*)

(*b*)

Figure 23-21 Active filters. (*a*) Second-order, low-pass filter. (*b*) Bandpass active filter.

though, the *LC* circuit has very high resistance for parallel resonance. Then most of the input voltage is across the *LC* circuit and R_L with little across R_S.

L-Type Resonant Filter

Series and parallel resonant circuits can be combined in *L*, *T*, or π sections for sharper discrimination of the frequencies to be filtered. Examples of an L-type filter are shown in Fig. 23-20.

The circuit in Fig. 23-20*a* is a bandstop filter. The reason is that the parallel resonant L_1C_1 circuit is in series with the load, whereas the series resonant L_2C_2 circuit is in shunt with R_L. There is a dual effect as a voltage divider across the input source voltage. The high resistance of L_1C_1 reduces voltage output to the load. Also, the low resistance of L_2C_2 reduces the output voltage.

For the opposite effect, the circuit in Fig. 23-20*b* is a bandpass filter. Now the series resonant L_3C_3 circuit is in series with the load. Here the low resistance of L_3C_3 allows more output for R_L at resonance. Also, the high resistance of L_4C_4 allows maximum output voltage.

23.8 Special Filter Types

RC and *LC* filters are widely used, but there are instances where they are less desirable because of their excessive attenuation, poor selectivity, or inability to work well at very

high frequencies. These disadvantages can be overcome with several special filters. These include active filters, crystal and ceramic filters, and surface acoustic wave filters.

Active Filters

Active filters are those that combine *RC* networks with operational amplifiers (op amps). The amplifiers provide gain to offset the normal loss of *RC* networks, and the feedback techniques provide improved selectivity. Any of the five basic types of filters can be implemented with active filter circuits.

Figure 23-21*a* shows a typical second-order, low-pass filter. Both *R* and C_1 form one of the *RC* sections, while *R* and C_2 form the other. It provides a rolloff rate of 40 dB per decade. The gain is usually set to one, but other op amps can be added to boost that as required. Greater selectivity can be achieved by cascading stages without the loss of gain. A bandpass active filter is shown in Fig. 23-21*b*. High-pass and notch filters can be formed in a similar way.

Most active filters are used at low frequencies to eliminate the need for large expensive inductors. They are also used at audio frequencies to set the frequency response of an amplifier or other circuit. Active filters are not widely used at the higher frequencies but with wide-band op amps they can function well into the 100-MHz RF range.

Crystal Filters

Crystal filters are thin slices of quartz that act as a resonant circuits. The crystal freely vibrates at a precise frequency when voltage is applied to it. Maximum vibration

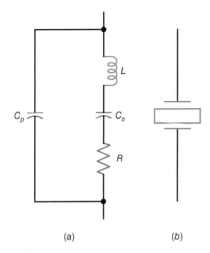

(a)   (b)

Figure 23-22 Equivalent circuit of a crystal filter and its schematic symbol.

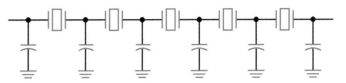

Figure 23-23 Crystal filter as a bandpass filter.

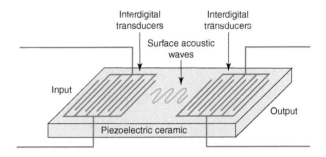

Figure 23-24 Basic construction of a surface acoustic wave (SAW) filter.

occurs at this resonant frequency. The equivalent circuit of a crystal and its schematic symbol are shown in Fig. 23-22. Depending on the frequency, the crystal may operate as a parallel resonant circuit or as a series resonant circuit. The Q of the crystal is very high, usually over 10,000 or more. As a result, the crystal can be used to build very selective filters.

Most of these filters are bandpass filters. An example is shown in Fig. 23-23. The response curve has extremely steep rolloffs, making the filter useful in separating closely spaced signals. Most crystal filters are used at radio frequencies from roughly 1 MHz to 100 MHz.

Special ceramic materials, such as lead titanate, can also be used like a crystal. Ceramic resonators are available to make filters. Ceramic filters are smaller, with a lower Q, but still have better selectivity than larger LC or crystal filters. Most are used at radio frequencies from 400 kHz to 50 MHz.

Surface Acoustic Wave Filters

A surface acoustic wave (SAW) filter is a special bandpass filter used primarily for radio-frequency selectivity. Figure 23-24 shows the basic construction. The base is a piezoelectric ceramic substrate such as lithium niobate. A pattern of interdigital fingers are made on the surface. The pattern on the left converts the signals into acoustic waves that travel across the filter surface. By controlling the shape, size, and spacing of the interdigital fingers, the response can be tailored to any frequency or desired shape. The interdigital pattern on the right in Fig. 23-24 converts the acoustic waves back into an electronic signal.

SAW filters are available with a frequency range of 10 MHz 4 GHz. The attenuation is very high and in the 10- to 35-dB range. As a result they are used with appropriate amplifiers. SAW filters are widely used in TV sets, cell phones, and many types of wireless equipment.

CHAPTER 23 SYSTEM SIDEBAR

DSP Filters

As you have seen in this chapter, filters are made up of components like resistors, capacitors, inductors, or op amps. Special filters are made of ceramic and crystal resonators and unique components, like SAW filters. All these filters are still widely in use. However, a newer and more complex filter has found many applications in modern electronic circuits and systems. This is the digital signal processing (DSP) filter.

DSP refers to a way to process analog signals with digital circuits. The analog signal to be processed, in this case, is first converted to digital form. Then a special processor or digital computer processes the equivalent digital data to perform the filtering function. Almost any type of filter, as described earlier, can be implemented with software to perform the filtering digitally. Once processed, the data are then converted back into analog form. The result is the desired filtering effect.

It may seem as though a DSP filter is overly complex and expensive. That was once true, but today, thanks to modern semiconductor technology, the circuitry and processors can be made small enough and cheap enough to make DSP filters practical. And they are easily integrated onto larger chips to create a complete system on a chip (SoC). Better still, the digital filters are usually more effective than equivalent analog filters. They are usually more selective with steeper rolloff and less attenuation, and with proper tailoring of the program, the phase and other characteristics can be controlled as desired.

Figure S23-1 shows a simplified block diagram of a DSP filter. The analog signal to be filtered is first digitized in a circuit called an analog-to-digital converter (ADC). This circuit produces a stream of binary numbers that represent samples of the analog signal at closely spaced time intervals. The binary numbers are stored in a data memory.

The analog signal data are then processed by the digital signal processor (DSP). The processing may be done by a special microcontroller designed for DSP, or it can be any microprocessor or controller with DSP capability.

Stored in the processor's memory is the program that does the processing. It is usually a special mathematical algorithm that performs any one of the normal filter functions like low pass, high pass, bandpass, or bandstop. The processed data are then stored back in data memory. Finally, the processed data are fed to a digital-to-analog converter (DAC) that translates the digital data back into an analog

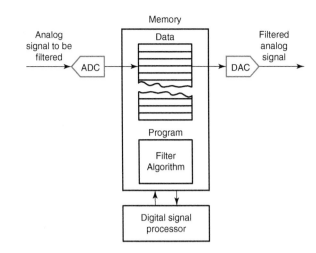

Figure S23-1 The concept of digital signal processing (DSP).

signal. The output is similar to that obtained with an analog filter. Typically, the filtering action is superior in some way. And the filter characteristics can be changed on the fly by using a new or modified processing algorithm.

DSP filters are invisible, since they are implemented inside a small processor or other chip. They are widely used in all forms of electronics, including TV sets, cell phones, MP3 players, military radios, and many other types of equipment.

The mathematics of DSP is well beyond the scope of this text, but you should know of its existence because it is not experimental but very widely used in many applications.

❓ CHAPTER 23 REVIEW QUESTIONS

1. A low-pass filter has a cutoff frequency of 20 kHz. Its bandwidth is
   a. 4 kHz.
   b. 10 kHz.
   c. 20 kHz.
   d. 40 kHz.

2. What kind of filter would you use to select one radio station from dozens or others around it in the same spectrum?
   a. High pass.
   b. Low pass.
   c. Bandpass.
   d. Bandstop.

3. What term describes a filter's ability to sharply discriminate one signal from another adjacent one?
   a. Selectivity.
   b. Sensitivity.

   c. Bandwidth.
   d. Discrimination.

4. In an RC low-pass filter, the output is taken across the
   a. resistor.
   b. inductor.
   c. capacitor.
   d. none of the above.

5. On logarithmic graph paper, a 10-to-1 range of frequencies is called a(n)
   a. octave.
   b. decibel (dB).
   c. harmonic.
   d. decade.

6. The cutoff frequency, f_c, of a filter is the frequency at which the output voltage is
   a. reduced to 50% of its maximum.
   b. reduced to 70.7% of its maximum.

c. practically zero.

d. exactly equal to the input voltage.

7. The decibel attenuation of a passive filter at the cutoff frequency is
   a. −3 dB.
   b. 0 dB.
   c. −20 dB.
   d. −6 dB.

8. To increase the cutoff frequency of an *RL* high-pass filter, you can
   a. decrease the value of *R*.
   b. decrease the value of *L*.
   c. increase the value of *R*.
   d. both *b* and *c*.

9. An *RC* low-pass filter uses a 2.2-kΩ *R* and a 0.01-μF *C*. What is its cutoff frequency?
   a. 3.5 MHz.
   b. 72.3 Hz.
   c. 7.23 kHz.
   d. 1.59 kHz.

10. For either an *RC* low-pass or high-pass filter,
    a. $X_c = 0\ \Omega$ at the cutoff frequency.
    b. $X_c = R$ at the cutoff frequency.
    c. X_c is infinite at the cutoff frequency.
    d. none of the above.

11. What type of filter would you use to get rid of a 156-kHz sine wave that is interfering with audio signals in the 20-Hz to 20-kHz range?
    a. Low pass.
    b. High pass.
    c. Bandpass.
    d. Bandstop.

12. A power-line filter used to reduce RF interference is an example of a
    a. low-pass filter.
    b. high-pass filter.
    c. notch filter.
    d. bandpass filter.

13. On logarithmic graph paper, a 2-to-1 range of frequencies is called a(n)
    a. decade.
    b. decibel (dB).
    c. harmonic.
    d. octave.

14. What is the decibel (dB) attenuation of a filter with a 100-mV input and a 1-mV output at a given frequency?
    a. −40 dB.
    b. −20 dB.
    c. −3 dB.
    d. 0 dB.

15. In an *RL* high-pass filter, the output is taken across the
    a. resistor.
    b. inductor.
    c. capacitor.
    d. none of the above.

16. An *RL* high-pass filter uses a 60-mH *L* and a 1-kΩ *R*. What is its cutoff frequency?
    a. 2.65 kHz.
    b. 256 kHz.
    c. 600 kHz.
    d. 32 kHz.

17. A T-type low-pass filter consists of
    a. series capacitors and a parallel inductor.
    b. series inductors and a bypass capacitor.
    c. series capacitors and a parallel resistor.
    d. none of the above.

18. A π-type high-pass filter consists of
    a. series inductors and parallel capacitors.
    b. series inductors and a parallel resistor.
    c. a series capacitor and parallel inductors.
    d. none of the above.

19. When examining the frequency response curve of an *RC* low-pass filter, it can be seen that the rate of roll-off well above the cutoff frequency is
    a. 6 dB/octave.
    b. 6 dB/decade.
    c. 20 dB/decade.
    d. both a and c.

20. For signal frequencies in the passband, an *RC* high-pass filter has a phase angle of approximately
    a. 45°.
    b. 0°.
    c. +90°.
    d. −90°.

21. The main component in an active filter besides the *RC* networks is a(n)
    a. transistor.
    b. op amp.
    c. inductor.
    d. transformer.

22. The equivalent circuit of a quartz crystal is a(n)
    a. *RC* network.
    b. mechanical vibrator.
    c. low-pass filter.
    d. *LC* resonant circuit.

23. Why is a crystal filter so much more selective than an *LC* or *RC* filter?
    a. Its high *Q*.
    b. Its small size.
    c. Its lower losses.
    d. Its power-handling capability.

24. Which of the following best describes a SAW filter?
　a. Low-frequency, low-pass filter.
　b. High-frequency bandpass filter.
　c. High-frequency, high-pass filter.
　d. A microwave notch filter.

25. Cascading filter circuits produce a larger filter with
　a. lower selectivity and higher attenuation.
　b. improved selectivity with gain.
　c. improved selectivity with greater attenuation.
　d. lower selectivity but higher gain.

CHAPTER 23 PROBLEMS

SECTION 23.1 Ideal Responses

23.1 A bandpass filter has an upper cutoff frequency of 10.8 MHz and a lower cutoff frequency 10.6 MHz. The bandwidth is
　a. 100 kHz.
　b. 200 kHz.
　c. 10.7 MHz.
　d. 21.4 MHz.

23.2 An all-pass filter has
　a. no lower cutoff frequency.
　b. no upper cutoff frequency.
　c. a constant phase shift.
　d. all of the above.

SECTION 23.2 Filter Circuits

23.3 What type of filter, low pass or high pass, uses
　a. series inductance and parallel capacitance?
　b. series capacitance and parallel inductance?

23.4 Suppose that a low-pass filter has a cutoff frequency of 1 kHz. If the input voltage for a signal at this frequency is 30 mV, how much is the output voltage?

SECTION 23.3 Low-Pass Filters

23.5 For a low-pass filter, define what is meant by the terms
　a. passband.
　b. stopband.

23.6 Assume that both the RC low-pass filter in Fig. 23-6a and the π-type filter in Fig. 23-6e have the same cutoff frequency, f_c. How do the filtering characteristics of these two filters differ?

SECTION 23.4 High-Pass Filters

23.7 Do the terms passband and stopband apply to high-pass filters?

23.8 In Fig. 23-8, does the T-type filter provide sharper filtering than the RC filter? If so, why?

SECTION 23.5 Analyzing Filter Circuits

23.9 Identify the filters in each of the following figures as either low-pass or high-pass:
　a. Fig. 23-25.
　b. Fig. 23-26.

c. Fig. 23-27.
d. Fig. 23-28.

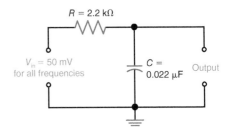

Figure 23-25

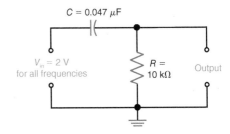

Figure 23-26

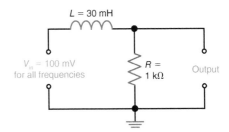

Figure 23-27

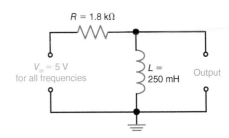

Figure 23-28

23.10 Calculate the cutoff frequency, f_c, for the filters in each of the following figures:
  a. Fig. 23-25.
  b. Fig. 23-26.
  c. Fig. 23-27.
  d. Fig. 23-28.

23.11 In Fig. 23-25, calculate the output voltage, V_{out}, and phase angle, θ, at the following frequencies:
  a. 50 Hz.
  b. 200 Hz.
  c. 1 kHz.
  d. f_c.
  e. 10 kHz.
  f. 20 kHz.
  g. 100 kHz.

23.12 In Fig. 23-26, calculate the output voltage, V_{out}, and phase angle, θ, at the following frequencies:
  a. 10 Hz.
  b. 50 Hz.
  c. 100 Hz.
  d. f_c.
  e. 1 kHz.
  f. 20 kHz.
  g. 500 kHz.

23.13 For the filters in Figs. 23-25 through 23-28, what is the ratio of V_{out}/V_{in} at the cutoff frequency?

23.14 Without regard to sign, what is the phase angle, θ, at the cutoff frequency for each of the filters in Figs. 23-25 through 23-28?

23.15 For a low-pass filter, what is the approximate phase angle, θ, for frequencies
  a. well below the cutoff frequency?
  b. well above the cutoff frequency?

23.16 Calculate the notch frequency, f_N, in Fig. 23-29.

23.17 Calculate the decibel (dB) power gain of an amplifier for the following values of P_{in} and P_{out}:
  a. $P_{in} = 1$ W, $P_{out} = 2$ W.
  b. $P_{in} = 1$ W, $P_{out} = 10$ W.
  c. $P_{in} = 50$ W, $P_{out} = 1$ kW.
  d. $P_{in} = 10$ W, $P_{out} = 400$ W.

23.18 Calculate the decibel (dB) attenuation of a filter for the following values of P_{in} and P_{out}:
  a. $P_{in} = 1$ W, $P_{out} = 500$ mW.
  b. $P_{in} = 100$ mW, $P_{out} = 10$ mW.
  c. $P_{in} = 5$ W, $P_{out} = 5$ μW.
  d. $P_{in} = 10$ W, $P_{out} = 100$ mW.

23.19 What is the rolloff rate of an RC low-pass filter for signal frequencies well beyond the cutoff frequency?

23.20 What determines the width of the band of frequencies that are allowed to pass through a resonant bandpass filter?

23.21 Identify the following configurations as either bandpass or bandstop filters:
  a. series LC circuit in series with R_L.
  b. parallel LC circuit in series with R_L.
  c. parallel LC circuit in parallel with R_L.
  d. series LC circuit in parallel with R_L.

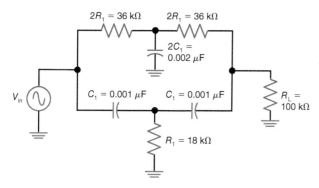

Figure 23-29

Chapter 24

AC Power

Learning Outcomes

After studying this chapter, you should be able to:

> *Explain* the operation of an ac generator.

> *Name* five different forms of energy used to generate electric power.

> *State* the function of a turbine.

> *Name* five alternative energy sources, and *explain* how the two most promising generate power.

> *Explain* why electric power is distributed at a high voltage level.

> *State* the main components at an electrical substation.

> *Name* the two ways that three-phase ac generators and transformers are connected.

Most of the energy used to power electronic equipment and systems comes from our ac electric grid. Electrical utility companies supply 60-Hz ac voltage to all of our homes, businesses, and industries. For this reason, it is important for every electronics technician and engineer to know something about the ac power grid and the related systems. This chapter introduces you to the ac power grid and provides details on common electric wiring and components. Three-phase electric power is also discussed.

> *Calculate* the output voltage of both Y and Δ three-phase generators and transformers.

> *State* the names and colors of the wires in a typical home electrica cable.

> *State* the voltages available in most homes and offices.

> *Explain* the purpose of a circuit breaker.

> *Explain* the purpose of the third ground wire in standard electric wiring.

24.1 Power Generation

The ac electric power system used throughout the world consists of four major components, as shown in Fig. 24-1. An energy source is used to generate the electric power which is then distributed to the end users. The external distribution system, also known as the grid, is part of the utility that generates the power. Once the electricity is delivered to the home or business, an internal distribution system or premises wiring is used to deliver the ac voltage to those devices requiring it. The final part of the system, of course, is the various loads that consist of the lights, appliances, and other equipment requiring electricity. This section discusses the power generation part of the system.

A utility power plant consists of three main components: an energy source, a turbine, and an electric generator. The energy source may be water, fossil fuels (such as coal, oil, and natural gas), nuclear, wind, or solar. While wind and solar power sources are increasing, today they still represent a small fraction of the power sources for the electrical utility business. In the United States, for example, nearly half of all energy is generated by burning coal because it is the least expensive and most abundant energy source.

Figure 24-2 shows the typical power generating process. The energy source could be coal, oil, or natural gas that is burned in a furnace. The furnace heats water. Or a nuclear reactor could also heat the water. The heated water inside a heat exchanger is used to heat another water source that is flowing through a series of water tubes. The water is heated until it is turned into steam. The steam is sent to a turbine that is used to rotate the generator.

A turbine is a machine that converts the motion of a fluid such as water or steam into rotational movement. The main element in a turbine is a set of rotor blades similar to that on a fan or a propeller. The high-pressure steam turns the rotor blades that, in turn, rotate a shaft. The shaft is mechanically

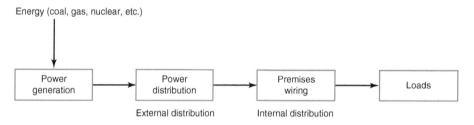

Figure 24-1 Simplified model of the electric power system.

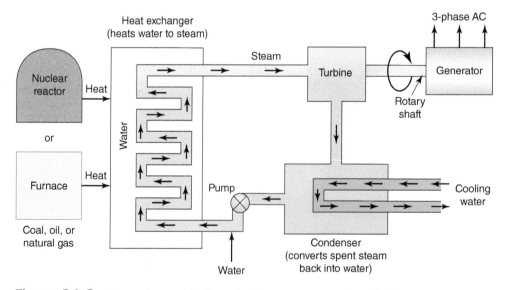

Figure 24-2 General concept of an electric power generating plant.

coupled to the electric generator, which produces the alternating current.

After turning the turbine rotor, the high-pressure steam goes into a condenser unit where it is cooled by circulating water. The steam is thus converted back into water, and the water is filtered and then again pumped into the heat exchanger where it is reheated.

As you can see, the energy in the form of heat is converted into mechanical motion by the turbine. The turbine in turn rotates the generator to produce the electric power.

In coal, oil, and natural gas plants the furnace and heat exchanger may be the same. In nuclear energy plants the reactor heats the heat exchanger. Natural gas is far more efficient and significantly cleaner burning than coal or oil. Nuclear energy is the cleanest of all energy whereby the nuclear fission process produces heat that generates steam for the turbine.

The process of wind and solar power generation is considerably different from that of a conventional power plant and will be discussed separately.

Electric Generators

In Chap. 14 you saw how a basic sine wave of voltage was generated by a simple generator circuit consisting of a coil rotating inside a magnetic field. Figure 24-3 shows the basic arrangement. As the coil rotates inside the magnetic field, a voltage is induced into it. This is transferred through the slip rings and brushes to the circuit to be powered. To produce the 60-Hz voltage commonly used in this country, the coil has to rotate 60 revolutions per second to produce the 60 cycles per second standard. This requires a rotational speed of 3600 revolutions per minute.

While the approach in Fig. 24-3 is useful in low-voltage generators, the technique cannot be used in the high-voltage generation process used by the utilities. Instead a

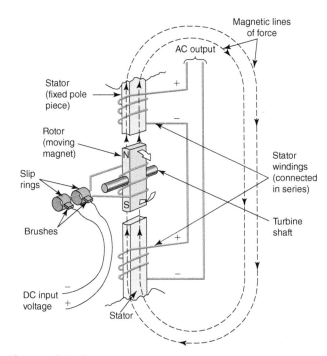

Figure 24-4   An alternative method of generating ac.

new arrangement is used whereby the coil is fixed and the magnetic field is rotated. The basic concept is illustrated in Fig. 24-4.

The generator in Fig. 24-4 uses a rotating electromagnet. Direct current is applied to the coil through a system of brushes and slip rings. The direct current flows through the coil, around the rotating core, and produces an electromagnet with north and south poles. A low voltage, approximately 250 V, is normally used to generate the magnetic field.

It is the turbine shaft that causes the electromagnet to rotate inside the fixed poles. These poles have coils, and the assembly is referred to as a *stator*. As the magnetic field rotates between the stator poles, alternating current is induced into the stator windings.

Recall that the amount of voltage induced into a coil is greatest when the maximum number of magnetic lines of force cut the turns of the coil. When the rotating magnet poles are even with the stator pole, the maximum magnetic field flows in the stator windings and induces the peak voltage. One complete rotation of the electromagnet produces one cycle of the sine wave.

A single set of stator windings produces a single sine wave. This is called a *single phase*. A utility generator produces three separate sine waves or phases of voltage simultaneously. Each phase of sine-wave voltage is shifted from the other by 120°, as shown in Fig. 24-5. To generate three separate sine waves of voltage requires three separate generators; however, the generators themselves can be combined

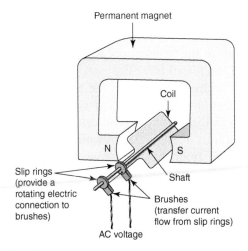

Figure 24-3   A basic ac generator.

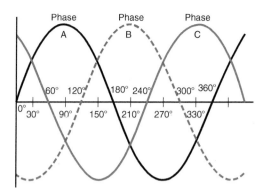

Figure 24-5 Three ac voltages or phases.

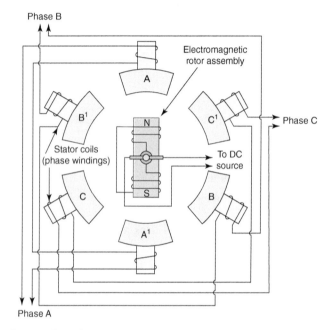

Figure 24-6 A basic 3-phase ac generator.

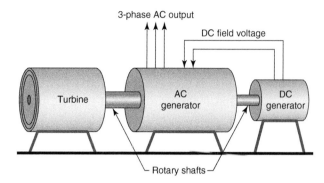

Figure 24-7 Physical arrangement of turbine, ac generator, and dc field generator.

into a common structure. A typical arrangement is shown in Fig. 24-6. It consists of three pairs of stator poles on which are wound fixed coils into which the voltage will be induced. Each north and south pair is mounted opposite each other, and the pairs are spaced 60° apart. A magnetic core is wrapped with the windings connected to the dc source, as described earlier. This produces the rotating magnetic field. As the turbine rotates the magnetic field, ac voltages are induced in each pair of stator poles. The coils on each opposite pair of poles are connected in series to form the A, B, and C output phases. Figure 24-7 shows the physical arrangement of the turbine, ac generator, and dc field generator. The turbine rotary shaft turns both generator shafts.

The voltage induced into each winding is typically very high. Most commercial power generators generally produce a voltage of 13,800 V. Some ac generators produce even higher voltages.

A critically important consideration in the construction and operation of the ac generator is the frequency produced. The standard power-line frequency in the United States and many other countries is 60 Hz. However, in Europe and other parts of the world the power-line frequency is 50 Hz. The frequency of the sine wave produced by the generator is dependent on the number of stator poles in the generator and the speed of rotation. This basic relationship is expressed as

$$f = \frac{N \times S}{120} \tag{24-1}$$

where f is the frequency in hertz, N is the number of poles per phase, and S is the speed of rotation in revolutions per minute (rpm). In the case of the generator in Fig. 24-6, there are two poles per phase. To compute the frequency of the generator in Fig. 24-6, assume that the rotating magnetic field turns at 3600 rpm. The frequency then is

$$f = \frac{N \times S}{120} = \frac{2(3600)}{120} = \frac{7200}{120} = 60 \text{ Hz}$$

Since the number of poles in a generator is fixed, it is strictly the speed of the rotation that determines the output frequency. Speed control is a very important part of the generating process. Usually some form of speed-control mechanism is built into the turbine to ensure a constant speed. This mechanism is called a *governor,* or *regulator,* and it provides automatic speed control. Speed control is excellent in most power plants as the frequency of ac generated typically has an error of less than 0.1%.

The three-phase output from the generator can be connected in two different ways. The two configurations are referred to as the Y (or wye) and delta (or Δ) connections. They are shown in Fig. 24-8, in which V_g is the generator output voltage from the series-connected stator windings and V_o is the output from the generator after the windings are connected.

The Y connection in Fig. 24-8*a* has all the phases connected to a common junction labeled X. This point is usually grounded for safety purposes. The three voltage outputs are taken between any two pairs of the wires.

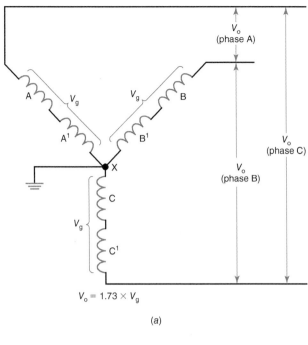

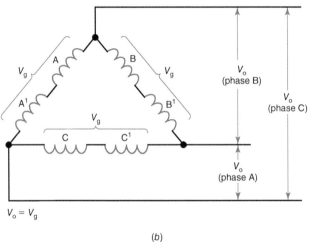

Figure 24-8 Three-phase generator stator winding connections. (*a*) Y or wye. (*b*) Delta or Δ.

The output voltage between any two output wires is 1.73 times the voltage produced by the stator coils, or

$$V_{out} = 1.73V_g \qquad \textbf{(24-2)}$$

The output voltage is basically the sum of two of the coil outputs connected in series. With the two voltages 120° out of phase with one another, the summation results in the 1.73 multiplication factor. For example, if the output voltage from the coils in series produces 8000 V, the output voltage between any pair of wires is

$$V_o = 1.73V_g = 1.73 \times 8000 = 13,840 \text{ V}$$

Another common arrangement for connecting the coils is the delta connection shown in Fig. 24-8*b*. Here each output

voltage is taken from across a coil so the output is just equal to the generator voltage.

Most power generators use the Y connection because it does provide a voltage step-up of 1.73 over the generated voltage and provides for a common ground.

Once the voltage is generated, it is ready for distribution. Distribution consists of transformers for stepping voltage up and down, as well as the cables over which the power is carried to the distribution system.

24.2 Power Distribution

The power distribution system consists of the wiring from the power generating station to the substations and the wiring from the substations to the ultimate source in homes, office buildings, and factories. The resulting network of wires is large and complex. Most wiring operates at very high voltages because of the efficiency of high-voltage transmission.

Power plants are usually located a considerable distance from the end user. The first step in getting power to the user is transmitting it over longer distances using high-voltage transmission lines.

The utility generator output of 13,800 V is typically stepped up to a higher voltage before being transmitted over long distances. A step-up transformer at the generating station typically increases the voltage to significantly higher levels depending on the distance of transmission. While voltages vary from one utility to another, typical voltage levels are 138 kV, 230 kV, 345 kV, 500 kV, and 765 kV. Some systems have even used voltages as high as 1 megavolt. A few systems even use high-voltage dc distribution. The higher voltages are used for transmitting the power over the longer distances which could be over 100 miles. Once the voltage reaches the various substations in selected areas, the voltage is stepped down to lower levels before it is distributed to the user. The purpose of the substation is to perform this voltage step-down process.

The reason why electric power is distributed in such high voltage is that it is more efficient to do so. That is, less power is lost in the wiring at higher voltages than it is for lower voltages. In any electric power distribution system, the distances involved are considerable, and the long distribution wires all have resistance. When loads are connected to consume the power, current flows in the distribution wiring. Because the wiring has resistance, the current flow causes voltage drops develop. The voltage drops subtract from the amount of voltage available at the load end of the line. Furthermore, current flow in the resistance of the lines generates heat, wasting energy.

One way to minimize the voltage drop of course is to simply increase the size of the wires. Increasing the wire diameter lowers its resistance and so lowers the voltage drop for a given current level. When large-diameter wires are used, the wire is heavier. The weight of the long wires requires heavy-duty

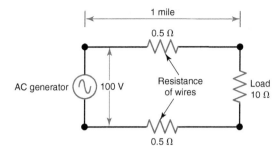

Figure 24-9 How power is lost in distribution wiring.

power poles to support that weight. In most cases, aluminum rather than copper conductors are used for the wires in order to reduce the amount of weight carried in the conductors. Large-diameter wires are used to ensure the lowest possible resistance. Yet resistance cannot be completely eliminated.

To illustrate the importance of using high-voltage transmission techniques, consider the basic example shown in Fig. 24-9. Here it is desired to distribute the 100 V generated by an ac generator to a 10-Ω load 1 mile away. Applying 100 V to a 10-Ω load will produce a load current of 10 A. This translates to a power of 1000 W.

Next, assume that each mile length of wire has a resistance of 0.5 Ω. These two cables are in effect two 0.5-Ω resistors connected in series with the 10-Ω load. The total additional resistance in the circuit then is 1 Ω. The total load on the generator, therefore, is the 10-Ω load plus this wiring resistance for a total of 11 Ω.

With a 100-V source, the current in the circuit is now 9.09 A instead of the desired 10 A. With this information, you can calculate the amount of voltage drop across each wire, or

$$V_{drop} = IR = 9.09(0.5) = 4.545 \text{ V}$$

The 4.545 V across each interconnecting wire represents a total voltage loss of 9.09 V. Subtracting this from the 100-V source leaves only 90.91 V across the load. Now the power consumed by the load is only

$$P = V^2/R = 90.91^2(10) = 8264.63/10 = 826.63 \text{ W}$$
$$\text{or approximately 827 W}$$

As you can see, with 827 W consumed by the load, the remaining 173 W is dissipated as heat in the conductors themselves. This is a significant waste of power. The solution to this problem is to step the voltage up before it is distributed.

High-Voltage Transmission

Refer to Fig. 24-10. Here the 100 V produced by the generator is stepped up by a transformer to 10,000 V. The 10,000 V is then connected to 1 mile of interconnecting cables. At the destination the voltage is stepped down by a transformer to the original 100 V, which is applied to the load.

Remembering how a transformer works, we can assume that the power output is approximately equal to the power input. While most transformers are not 100% efficient, they are nearly so, and for this example let's consider them to be 100% efficient. This means that the primary and secondary powers are the same. If the load consumes 1000 W, then the secondary of transformer T_2 must also have a power of 1000 W. Therefore, we can compute the amount of current flowing in the transmission line, which is

$$I = P/V = 1000/10,000 = 0.1 \text{ A}$$

With the voltage at such a high level but with the same power consumption, the amount of current flowing in the conductors is considerably less. With less current, less voltage will be dropped across the conductors. The voltage drop across each conductor, therefore, is

$$V_{drop} = IR = 0.1(0.5) = .05 \text{ V}$$

In this case only 0.05 V is dropped across each conductor for a total voltage loss of 0.1 volt. This tenth of a volt is a tiny fraction of the total 10,000 V being transmitted. In fact, it is so low that it is practically negligible. So while some loss still occurs at the high voltage, low current produces voltage drops that are a small percentage of the overall transmitted voltage. The power lost in the transmission process is, therefore, considerably less at the higher voltage levels. In this case

$$P = VI = 0.1(0.1) = 0.01 \text{ W} \quad \text{or} \quad 10 \text{ mW}$$

That is 10 mW out of a total of 1000 W. That is negligible.

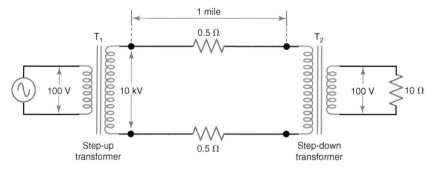

Figure 24-10 How stepping the voltage up for transmission reduces power loss.

The longer the distance that electric power must be transmitted, the higher the transmission-line voltage. That's why hundreds of thousands of voltage are often used in long-distance power distribution systems.

Substations

Once the high voltage has been distributed to the desired areas, it is ready for local distribution. The high voltage terminates in a facility called a substation. A *substation* is simply a facility that accepts the very high voltage input and steps it down to lower levels for further distribution in local wiring. Transformers at the substation reduce the high voltages down to various levels, depending on what the utility company uses. For example, one substation may step the voltage down to values of approximately 12,500 or 34,500 V. Additional transmission lines are used for secondary substations whose transformers further step the voltage down. Some systems step the voltage down to the 7000-V range or to the 2300- or 4100-V levels. The voltage level is determined by whether the distribution is for homes in neighborhoods or for industrial facilities.

Finally, the voltage reaches the neighborhood for final distribution. This is typically a 4100-V level that is applied to a transformer mounted on a pole or in some cases a steel housing on a concrete pad. The transformer steps the voltage down to the 120/240-V level typical of that distributed to homes and businesses.

The complete example of the voltage generating and distribution system is illustrated in Fig. 24-11. Note that three-phase voltages are transmitted. The 13,800 V produced by generator is stepped up in transformer T_1 to 138,000 V. At the first substation the 138 kV is stepped down to 34,500 V in T_2. At the next substation it is stepped down to 4,100 V in T_3 for further local distribution. The final transformers T_4, T_5, T_6, and T_7 step the voltage down to the 480-, 208-, and 120/240-V levels that supply homes, apartments, office buildings, and industrial plants.

Most substations also have a variety of other pieces of equipment. These include switches, fuses and circuit breakers, lightning arresters, and power factor correction capacitors. Large manually operated switches are usually installed at substations to physically connect and disconnect power from different parts of the system.

Fuses and circuit breakers are used to protect the system from shorts or excessively heavy loads. Excessive current drawn from the system will cause the breaker to trip, thereby disconnecting the power and protecting the transformers and other parts of the system.

A lightning arrester is a special component that protects the substation wiring and components from lightning strikes. The lightning arrester is an extremely high resistance or open-circuit device connected to the wiring. One side of the lightning arrester is connected to the power line, while the other side is connected to ground. Normally it is an open circuit and draws no current from the system. However, if an extremely high voltage from lightning hits the conductor, the lightning arrester will arc over and act as a low resistance or near short circuit. This directs the high-voltage lightning to the earth ground, thereby protecting the conductors, transformers, and other components in the subsystem.

Another important component is the power factor correction capacitors used in most substations. While a high percentage of the loads connected to the system are resistive, many motors and other inductive components like transformers are connected to the system. This will cause the loads to be somewhat inductive, which causes a phase shift that effectively reduces the efficiency of the system. Much of the power goes to power the inductive loads, and less is converted to the desired electric energy in the resistive loads.

To correct for the inductive loading, capacitors are connected across the line to offset the inductive reactance. Different size capacitors are available to switch across the line to correct for the inductive loading effects. The capacitive reactance offsets the inductive reactance, making the load appear as though it is purely resistive. Most substations have power factor correction capacitors that can be switched into the line as inductive load conditions are detected.

One final thought: There are literally hundreds of electric power generating plants and distribution systems throughout the country. These thousands of utility companies do not operate separately. Instead, most of them are connected together in one large power distribution system referred to as the power grid or power pool. The power grid is a huge network of cooperating utilities that come together to share electric power across a specific region. The United States and Canada, for example, are divided into nine regional power grid networks. The utility companies in each region are connected together to share power. The result is as if multiple power generating plants are all connected in parallel to provide voltage. The overall result is a higher current capacity. Various coordinating control stations within various regions measure the power demands and loads and then provide automatically switching in different systems to meet the needs of the various regions.

The grid itself is roughly divided into three large segments. There is an eastern segment that handles power in all the eastern states from the Atlantic Ocean to the Rocky Mountains. A western segment handles all the power from the Rockies to the Pacific Ocean. A third segment covers Texas.

The interconnection of utility systems provides an overall system reliability. A power failure in one system can usually be compensated for by power coming from a neighboring

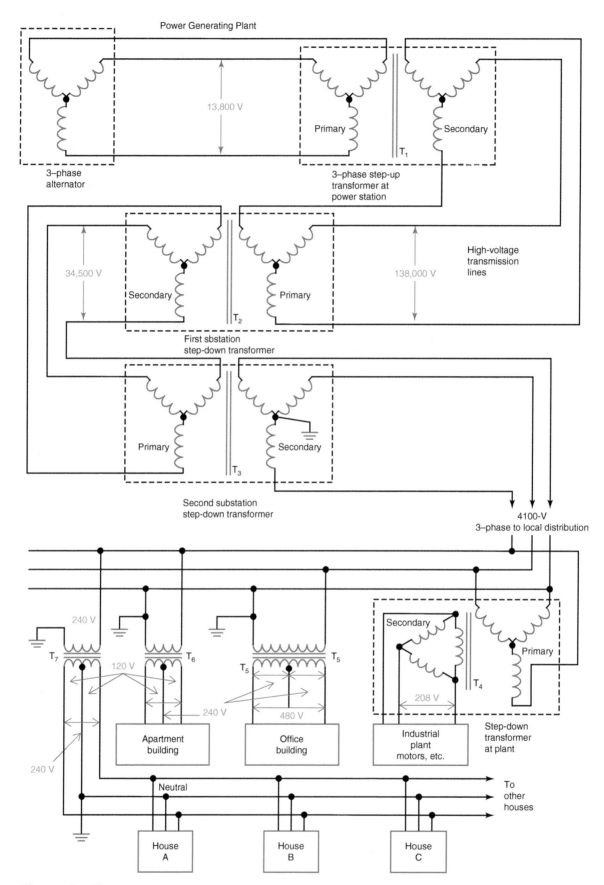

Figure 24-11 The complete power transmission system.

system. Computers at the various coordinating stations sense the power conditions and can automatically make the different switching connections required.

24.3 Home Electric Wiring

Most homes and small businesses use the standard 120/240-Vac power provided by the utility. This voltage is derived from the high voltage distributed by the utility to the various localities and neighborhoods. The power distribution in the form of a three-wire voltage connection terminates at the home or office in what we call the service entrance. The *service entrance* refers to all the wiring, components, and fittings that carry electricity from the utility's transformer to the home or office building. The main purpose of the service entrance is to measure the electric power consumed, protect the wiring from excessive current, and distribute the power to all the installed outlets, fixtures, and other equipment.

The service entrance includes the following:

1. The utility's wires from the transformer to the building
2. The service entrance wiring and associated conduit
3. A kilowatt-hour meter for measuring power consumption with the associated connections, enclosure, and wiring
4. The service entrance box or distribution panel containing circuit breakers and switches plus the connection points for distributing the wiring throughout the facility
5. An electric ground

All of these are discussed in more detail in this section.

Power Distribution

As you have seen earlier, the voltage from the utility's generating station is transmitted and distributed at very high voltage levels to minimize power loss. Transformers are used at some stations to reduce the voltage several times along the way to the consumer. The final voltage translation is done by a power transformer that is very near to the homes or offices being served. This transformer is mounted on a utility pole, as shown in Fig. 24-12. Two wires, with typical values being 4,100, 7,000 or 12,000 V, carry the final high voltage to the neighborhood and are connected to the step-down transformer, which reduces the voltage to the normal 120/240-V level. The transformer voltage is carried to the residence or buildings by three wires, two hot wires and a neutral, or ground, wire. The wires running from the utility transformer to the building are referred to as the service drop.

In older neighborhoods, the final distribution transformer is mounted on a nearby utility pole. In newer suburban neighborhoods, most power distribution is underground, and, therefore, the transformer is typically mounted in a steel enclosure on a concrete pad, as shown in Fig. 24-13.

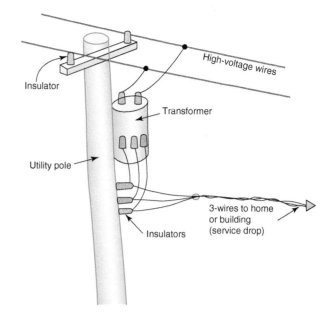

Figure 24-12 Step-down transformer on a poles for service drop distribution.

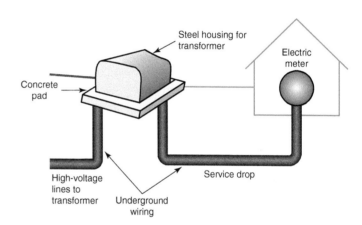

Figure 24-13 Underground wiring for the service drop.

For most modern homes, the service drop contains wiring that is capable of handling a current of up to 200 A. Larger homes and most businesses will typically have a higher-current capacity.

The connection to the utility's power transformer consists of three wires that come from a secondary winding that provides 240 V across the two outer connections. A center tap on the transformer, therefore, provides two 120-V connections. The first device encountered by this wiring is the kilowatt-hour meter, as shown in Fig. 24-14. All utilities require homes and offices to have a meter for measuring the amount of electric power consumed. This so-called kilowatt-hour meter typically has a heavy glass enclosure that houses four or five dials showing the amount of power used. The meter itself is an electromechanical device that converts the mechanical energy used into rotary motion to operate the

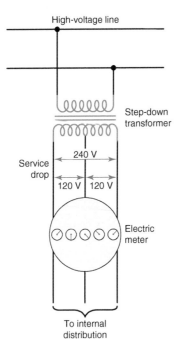

Figure 24-14 The service drop and the kilowatt-hour meter.

dials. The meter is almost always located outside the home or office so that the utility company can read the meter. For underground wiring source drops, the wiring comes into the meter from a conduit below the ground.

The electric meter itself measures the amount of power consumed in kilowatt-hours. A kilowatt, of course, is 1000 W. A *kilowatt-hour (kWh)* is equal to the total power used by a 1-kilowatt load over a 1-hour period. To calculate the number of kilowatt-hours for a device, simply multiply the power, in watts, consumed by the device by the number of hours the device is used. Divide the result by 1000. For example, an appliance draws 400 W of power during its normal use, and it is used for 5 hours, the total watt-hours consumed is $400 \times 5 = 2000$ Wh. Dividing by 1000 gives the final total of 2 kWh.

The kilowatt-hour meter usually consists of two sets of coils that operate the mechanism operating the dials. One set of coils is connected in series with the power line and monitors the voltage. The other coils produce a magnetic field that is proportional to the amount of current drawn.

Both the voltage and current coils are physically arranged so that the resulting magnetic field developed turns a small aluminum disk. The rotational speed of this disk is proportional to the amount of power being consumed. As the disk turns, it rotates a shaft connected to it. This shaft, in turn, is attached to a series of gears. The gears turn the indicating dials.

While most of the kilowatt-hour meters in use today are still of this variety, newer homes are beginning to receive so-called smart meters that use electronic circuits to measure the current and voltage. An internal microcontroller computes the amount of power and kilowatt-hours used. Most smart meters are designed so that there is some form of communications built in so that the power consumption figures can be communicated over the power line or wirelessly to an in-home energy monitoring device. Some meters also contain built-in wireless or communications connections to permit the utility to read the meter remotely. These smart meters become part of the overall larger utility power system generally known as the smart grid.

How to Read a Standard Electric Meter

Most electric meters have either four or five dials. Each dial is marked with the numbers 0 through 9, providing a total of 10 increments. Figure 24-15 shows an example. Notice that some of the dials have their numbers arranged clockwise, while others are incrementally counterclockwise. In all cases, it is the right-hand dial that rotates the fastest, and each succeeding dial to the left rotates at one-tenth the speed of the dial to the right. It takes one complete rotation of a dial to the right in order for the dial to the left to move one increment. Figure 24-15 is an example of how to read the meter.

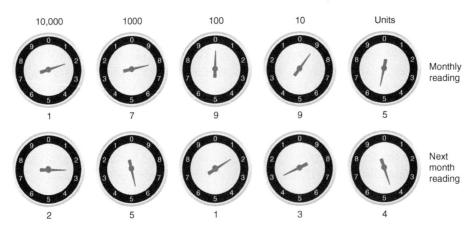

Figure 24-15 Reading a kilowatt-hour meter.

The upper dials indicate the reading at the end of a month, while the lower dials indicate the reading at the end of the next month. To read the dials, simply write down the numbers corresponding to the dial pointers from left to right. If a pointer is right on top of a number, write down that number. If the pointer is between the two numbers, write down the smaller number. If you can't tell if a pointer is directly on top of a number, then look at the dial directly to the right. If the pointer is a little bit past 0 or 1, then the previous pointer is on top of a number. If the pointer to the right is in the 8, 9, or 0 range, the pointer to be read is not yet on top of the number, so write down the next lower number.

To find the total power used, write down the reading for one day of the month. Then at the same time a month later, write down the new reading. Next, subtract the smaller reading from the larger. In this example, $25,134 - 17,995 = 7139$. This is the total number of kilowatt-hours used in one month. The utility will use this number to compute the electrical bill based on the price per kilowatt-hour. Assuming a rate of 10 cents per kilowatt-hour, the monthly bill would equal $7,139 \times 0.10 = \$713.90$.

The wiring from electric meter is then fed to the service entrance box. This enclosure is also called the breaker box or service panel. It is a rectangular steel enclosure that contains the circuit breakers and switches, as well as the bus bars and connectors for distributing the wiring to other parts of the building. The service entrance box is generally located in a garage, basement, or utility room. Some are mounted on the wall with service near the electric meter, while others are more remotely mounted and recessed into an existing wall.

The Service Panel

The main components of the service panel are

1. A master on/off switch
2. Voltage distribution bus bars
3. Overcurrent protection devices like circuit breakers
4. Connectors/buses for attaching the wires to the individual branch circuits
5. An earth ground

Figure 24-16 shows a basic schematic diagram of the wiring at the service entrance. The high-voltage, step-down transformer from the utility provides 240 V from the secondary winding. This secondary winding is center-tapped providing two 120-V circuits. The center tap and wire are known as the neutral, which is connected to earth ground. The other two sides of the transformer are known as hot wires. All three connections are attached to the bus bars. The bus bars then distribute the voltage to the various circuits. Note the main circuit breaker or switch. Then note that each of the individual circuits shown has its own circuit breaker with the peak current designated.

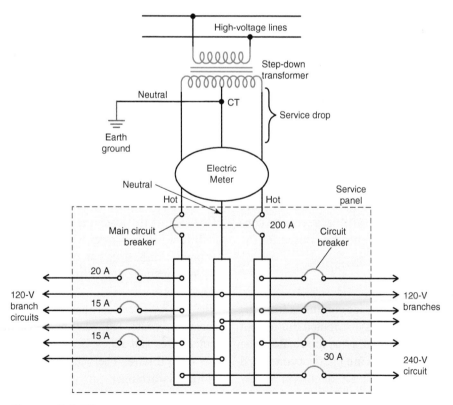

Figure 24-16   The service panel.

Some service panels contain a master switch that can be used to cut all the power to the building. In some installations, the master switch is in a separate enclosure near the kilowatt-hour meter. This switch is usually a double-pole device that will interrupt both the hot wires coming from the meter. In most modern installations, the master switch is also a circuit breaker. The circuit breaker is an overcurrent protection device that senses when a particular current level has been exceeded. Should the current level go over the rated value of the circuit breaker, the circuit breaker switch contacts open automatically and disconnect the power. This protects the wiring from overheating that could lead to a fire. A typical home may have a master circuit breaker switch with a rating of 200 A.

The bus bars are heavy rectangular copper rods that are insulated from the box itself. Screws or other methods of attachment are used to connect the circuit breakers and the circuit wiring to the bus bars. One of these bars is in each of the three main connections to the system.

The bus bars are used so that individual circuits or specific runs of cable can be developed. This divides the home or building into separate power areas. Each will get the voltage as required, but the current load is divided among the different circuits.

Each individual circuits that is developed from the bus bars will have its own circuit breaker for overcurrent detection. The circuit breakers typically have maximum current ratings of 15, 20, 25, or 30 A. There is one circuit breaker for each branch circuit to be used. The circuit breakers are grouped together inside the service panel box. The circuit breakers mount onto or connect to the bus bars. The circuit breakers also serve as switches so that you can conveniently turn off selected circuits for installation or maintenance purposes. This allows you to work on one circuit without disabling the others.

All service entrance boxes also contain a common ground bus bar. The ground is in every case connected to an earth ground at some point. It is usually done with a heavy copper wire that connects the neutral or ground crossbar to a copper water pipe, which is typically buried in the earth. In those instances if no copper water pipe is available, a separate copper ground rod driven into the earth nearby is used for the ground.

Branch Circuits

As you have seen, the power-line voltage coming in from the utility from the power transformer is distributed throughout the home or building by multiple cables that terminate at the circuit entrance box. All these cables are in effect connected in parallel. Each provides a separate path for current to flow through the various loads. These individual current paths are called *branch circuits.*

It would be possible to use one set of wires for every outlet, lighting fixture, or other load. However, this is not done in practice because an overload on a single circuit would cause all power to be lost. By using multiple parallel circuits, an overload on one shuts down only that circuit while allowing the power to continue to be provided to the other circuits. Power to the various outlets, fixtures, and appliances is divided into multiple paths for current flow. Each run is called a branch circuit. Each branch circuit is protected by its own circuit breaker.

Virtually all homes and businesses use three basic types of branch circuits: general purpose, appliance, and single purpose. The general-purpose branch circuit is used to supply 120 V to electric outlets and light fixtures. There may be as many as 10 or 20 individual general-purpose branch circuits for supplying outlets and overhead lights. Usually, 15-A circuit breakers are used on each of the general-purpose circuits.

An appliance branch circuit is designed to handle higher currents for operating small appliances. Appliance branch circuits generally terminate in the kitchen or laundry areas. These are usually set up for a maximum current of 20 A at 120 V. These circuits will handle such appliances as refrigerators, toasters, and dishwashers.

Single-purpose branch circuits are designed to handle a special appliance or device. It may be either a 120-V or 240-V circuit. Most single-purpose branch circuits supply 240 V to electric kitchen ranges, clothes dryers, or hot water heaters. Many central heating and air-conditioning units use a single-purpose branch circuit. They can handle 30 to 50 A of current, depending on the device.

An important part of connecting the various branch circuits to bus bars is to ensure that they are generally evenly divided between the two 120-V lines. The goal is to balance the branch currents on one of the 120-V outputs to be approximately equal to the loads on the other 120-V circuit.

The reason for attempting to balance the loads on the 120-V circuits is to ensure that the neutral wire never has to carry a large amount of current. If the loads on the two sides are perfectly equal, no current will flow in the neutral wire. However, if the loads are unbalanced, considerable current will flow in the neutral wire. If the neutral wire is not large enough, overheating can occur.

You can understand this concept better by considering the basic transformer connections from the utility. Referring to Fig. 24-17, you can see that the secondary of the power transformer is center-tapped. There is 240 V between the two outer wires and 120 V between each of the outer wires and the neutral wire. The neutral wire is connected to ground. With this arrangement, you have two 120-V sources, but they are 180° out of phase with one another. The reference point is the neutral wire or ground. When one 120-V line is

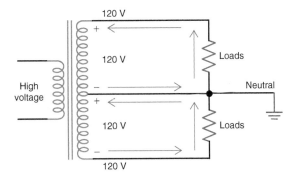

Figure 24-17   Current in the neutral wire.

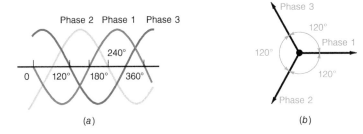

Figure 24-18   Three-phase alternating voltage or current with 120° between each phase. (a) Sine waves. (b) Phasor diagram.

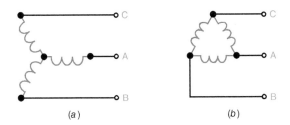

Figure 24-19   Types of connections for three-phase power. (a) Wye or Y. (b) Delta or Δ.

positive, the other is negative and vise versa. Since the two 120-V circuits are 180° out of phase with one another, the current in the neutral wire will be zero if the loads are equal. However, if one circuit has a greater load than the other, then the neutral wire will have some amount of current flowing.

Figure 24-17 shows the polarities of one half-cycle of the sine wave and the direction of electron flow in the wires and loads. Note the opposite directions in the neutral. If the loads are equal, the neutral currents will be equal and so will cancel one another. With unequal loads the two branch currents will cancel one another in the neutral but not completely.

Most home and building wire is designed to equally distribute the loads on the two 120-V circuits. Depending on the lights, appliances, and other devices that are being used, the loads will always be unequal, but in most cases, the neutral current will be fairly small, and, therefore, the neutral wire does not have to carry heavy currents.

24.4 Three-phase AC Power

As you saw earlier in Sec. 24-1, all ac generators produce three-phase output. While most homes and businesses simply use a single phase of either 120 V or 240 V, many industrial applications use three-phase power. The main advantage of three-phase ac voltage is that it is more efficient for power distribution. In addition, one of the most popular types of ac motors used in industry for high-power applications operate from three-phase voltage. The three-phase induction motors are also self-starting with three-phase voltage rather than requiring special starting circuitry on single-phase voltage. Finally, the ac ripple produced as a result of rectifying the ac is easier to filter. Rectification is the process of converting ac into dc. With single-phase ac, the rectification produces half-sine pulses that drop to zero at some point during the process. This is inefficient and more difficult to smooth into continuous dc voltage. With three-phase output, the ripple is smaller and continuous and simply easier to filter.

Figure 24-18a shows the three ac voltages being generated and distributed. They are spaced 120° from one another. Figure 24-18b shows the ac voltages represented as phasors.

The two basic generator output formats are the Y and delta connections described earlier. They are shown again here in Fig. 24-19. In the Y connection, all three coils are connected at one end, and the opposite ends are for the output terminals: A, B, and C. Note that any pair of terminals is across two coils in series. With each coil at 120 V, the output across any terminal is 1.73 × 120 or 208 V. The 208-V level is a standard three-phase ac voltage value and is widely used in industry.

In Fig. 24-19b the three windings are connected in the form of the Greek letter delta, or Δ. Any pair of output terminals is across one of the generator output windings. The output is then simply 120 V. However, the other coils are in a parallel branch. Therefore, the current capacity of the line is increased by a factor of 1.73.

Note in Fig. 24-20, the center point of the Y used as a fourth line distribution for the neutral wire in a three-phase

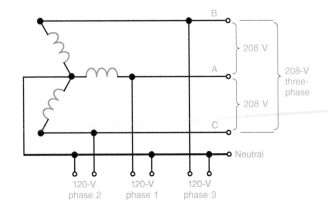

Figure 24-20   Y connections to a four-wire line with neutral.

power distribution system. With this arrangement, the voltage available is either 208-V three phase or 120-V single phase. Be sure to note that a three-phase voltage is 208 V and not the 240 V that is standard for single-phase distribution.

24.5 Wires, Cables, and Components

The wiring in virtually all homes, businesses, and industrial electrical systems is insulated solid copper wire. In Chap. 9, you learned about the basic characteristics of wire sizes and their basic characteristics. The wiring for ac electrical systems typically uses a wire no smaller than No. 14 and as large as type No. 2 or 1 AWG. The size of wiring is a function of the amount of current carried by the circuits used. The National Electrical Code (NEC) defines the amount of current different wire sizes can carry at some specified temperature. The idea is to prevent wiring with its insulation from overheating and causing a fire. Excessive current in small conductor wire can produce sufficient heat to cause electric insulation to burn.

One method for defining current ratings is called ampacity. *Ampacity* is the maximum current-carrying capability of popular wire sizes at a temperature of 86° Fahrenheit (30 Celsius).

As an example, consider the fact that most general-purpose electric circuits in a home have a maximum current rating of 15 A. This means a wire size of No. 14. A 20-A circuit requires a minimum wire size of No. 12. For a 40-A circuit, No. 8 or larger wire should be used.

Most wiring is in the form of cables rather than in individual wires. A cable, of course, is a collection of two or more wires combined in a common outer covering. The wires are generally insulated and color-coded. Standard electric cables come in a variety of sizes. Most of these cables contain three wires: two wires for the electric connection and a third ground wire.

The most commonly used cable for electric wiring is called *nonmetallic-sheathed cable*. It usually has three insulated wires contained in a common thermoplastic outer covering. Figure 24-21 shows the basic construction of this type of cable. The nonmetallic-sheathed cable is often called by its commercial brand name, Romex. These cables come in different types

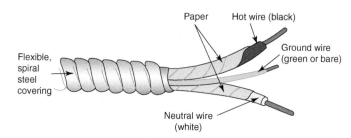

Figure 24-22 Armored cable.

for use in dry wiring situations or in wet conditions. Others are designed specifically for underground usage.

Note in Fig. 24-21 that the basic specifications of the cable are usually printed on it. This includes wire size, number of conductors, the presence of a ground wire, and voltage rating. The basic wire colors are white for the neutral wire, black for the hot wire, and green for the ground wire. In some cables the ground wire is not insulated and is just bare copper.

Another form of electric cable is known as *armored cable*. It is generally known in the trade as BX cable. It too has two standard insulated wires and a ground wire. The outer jacket is of flexible galvanized steel. The steel is wound in a spiral fashion to make it flexible. This type of cable is shown in Fig. 24-22. It is used in outdoor and hazardous environments to protect the wires.

There are many other different types of electric cables. Another common one is called lamp cord or zip cord. This is the common two-conductor cable you see connected to lamps and many other appliances. This cable uses stranded copper wire rather than solid. It comes in various sizes, but the most common are No. 16 and 18 wire, depending on the current-carrying requirements. The outer insulation is a rubber or thermoplastic. Stranded wires are more flexible than solid wires.

Some installations use low-voltage devices, such as doorbells and special lighting. Such devices typically operate at 24 V and carry only a small amount of current. In these cases, standard electric cable is not used. Instead a typical type of cable is preferred, known as twisted pair, which is shown in Fig. 24-23. Two solid copper wires with insulation are loosely twisted together to form the cable. Typical sizes are No. 16 through 24. The 120 V from one of the branches is usually fed to a small transformer that steps the voltage

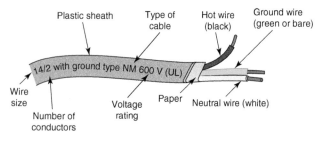

Figure 24-21 Standard nonmetallic-sheathed cable for electric wiring.

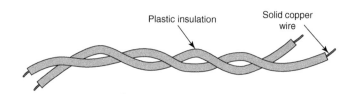

Figure 24-23 Twisted-pair cable.

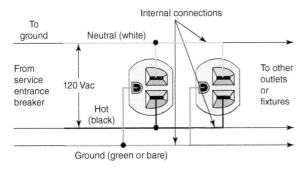

Figure 24-24 A common duplex outlet.

down to the 24-V level. The devices to be connected are then wired with the twisted-pair cable.

Twisted-pair cable is also used in telephone wiring, using size No. 24 or 26 wire. Special computer cables for networking also use twisted pair. A common form is four twisted pairs per cable. It is generally referred to as category 5, or CAT5. Multiple versions are available (CAT6, etc.).

Outlets and Fixtures

The most common electric connector is the duplex outlet shown in Fig. 24-24. The three wires from the service entrance box terminate at the outlet. The wires are attached to the outlet usually by screws, copper-colored screws for the hot wire and silver-colored screws for the neutral connection. A green screw is used to attach the ground wire. Note in the outlet itself that the large slot is the neutral connection. Some outlets are switched. That is, one of the outlets is designed to be enabled by a wall switch. In this case, the hot side of one outlet is opened and connected to the switch with a separate cable.

Another common connection point is a light fixture. A typical fixture is shown in Fig. 24-25. The power cable from the service box comes into the fixture box usually on the ceiling. Another cable connects the remote wall switch to

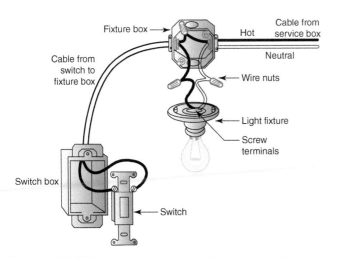

Figure 24-25 Wiring for lighting fixture and switch.

the fixture. Note that the actual connections are made with wire nuts. The two wires to be connected are tightly twisted together, and a wire nut is screwed on top of them. The wire nut ensures a tight copper-to-copper connection and at the same time insulates the connection.

Another important point is that the switch is connected in series with the hot wire and not the neutral. The neutral is never switched. Also notice that the ground wire is not shown. It is not often used in light fixtures. If it is, all ground wires from the switch cable and the service box cable will be tied together in the fixture box with a wire nut.

24.6 Electrical Safety

The voltages in standard home, business, and industrial wiring are considerably high and can be lethal. In working with electric wiring, components, and systems, it is essential that you use every safety measure possible to prevent shock.

Electric Shock

While you are working on electric circuits, there is often the possibility of receiving an electric shock by touching the "live" conductors when the power is on. The shock is a sudden involuntary contraction of the muscles, with a feeling of pain, caused by current through the body. If severe enough, the shock can be fatal. Safety first, therefore, should always be the rule.

The greatest shock hazard is from high-voltage circuits that can supply appreciable amounts of power. The resistance of the human body is also an important factor. If you hold a conducting wire in each hand, the resistance of the body across the conductors is about 10,000 to 50,000 Ω. Holding the conductors tighter lowers the resistance. If you hold only one conductor, your resistance is much higher. It follows that the higher the body resistance, the smaller the current that can flow through you.

A safety tip, therefore, is to work with only one of your hands if the power is ON. Place the other hand behind your back or in your pocket. Therefore, if a live circuit is touched with only one hand, the current will normally not flow directly through the heart. Also, keep yourself insulated from earth ground when working on power-line circuits, since one side of the power line is connected to earth ground. The final and best safety rule is to work on circuits with the power disconnected if at all possible and make resistance tests.

Note that it is current through the body, not through the circuit, which causes the electric shock. This is why high-voltage circuits are most important, since sufficient potential difference can produce a dangerous amount of current through the relatively high resistance of the body. For instance, 500 V across a body resistance of 25,000 Ω produces 0.02 A, or 20 mA, which can be fatal. As little as 1 mA through the body can cause an electric shock. The chart

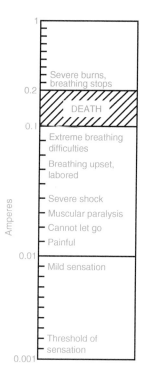

Figure 24-26 Physiological effects of electric current.

shown in Fig. 24-26 is a visual representation of the physiological effects of an electric current on the human body. As the chart shows, the threshold of sensation occurs when the current through the body is only slightly above 0.001 A or 1 mA. Slightly above 10 mA, the sensation of current through the body becomes painful, and the person can no longer let go or get free from the circuit. When the current through the body exceeds approximately 100 mA, the result is usually death.

In addition to high voltage, the other important consideration in how dangerous the shock can be is the amount of power the source can supply. A current of 0.02 A through 25,000 Ω means that the body resistance dissipates 10 W. If the source cannot supply 10 W, its output voltage drops with the excessive current load. Then the current is reduced to the amount corresponding to the amount of power the source can produce.

In summary, then, the greatest danger is from a source having an output of more than about 30 V with enough power to maintain the load current through the body when it is connected across the applied voltage. In general, components that can supply high power are physically big because of the need for dissipating heat.

Avoiding Electric Shock

With your life on the line or potential injury imminent, it is essential to avoid shock at any cost. Most electric shocks occur simply because individuals are working on live circuits. This means you are working with the wiring, components, and equipment to which the voltages are applied. Many who do this simply believe they are safe enough to get away with it, so to speak. However, the most prudent approach is to turn the power off first before you do the work. Many think it is time-consuming and inconvenient to go to the service entrance box and turn off the breaker for the circuit you are investigating.

The key point about electric shock is that it is the amount of current flowing that determines the degree of severity. And because of Ohm's law, we know that the level of the voltages will determine how much current flows. Generally speaking, any voltage below approximately 40 to 50 V will have little or no effect. That's why you cannot get shocked from a 12-V car battery or any of the lower-voltage batteries. Above the 40- to 50-V level, the currents will begin to be higher and fall into the range when some shock will occur.

The secret to avoiding shock is simply to have the proper attitude toward working on powered electric circuits. Shock is invariably caused by a combination of carelessness, laziness, and impatience.

Another precaution is to approach defective wiring, components or circuits with care. Defective parts, such as switches, light fixtures, outlets, or even appliances, can have unknown conditions that can lead to a shock. Again, the best approach is simply to disconnect the equipment from the ac power before you work on it. Perform your tests and repairs first before powering up again.

Finally, it is absolutely essential that ground connections in all electric wiring components be maintained. For some simple circuits, such as lamps and low-power equipment, electric connections use only the hot and neutral wires. For larger appliances and equipment, there is always a third ground wire. The ground wire is included specifically for safety purposes. Any attempt to eliminate this ground connection can produce an electric shock as you will see in the next section.

Grounds and Grounding

One of the most important safety features of electric wiring systems is the ground connection. A ground is the electric connection made between one side of the electric circuit and ground, or earth. Sometimes the ground connection is made to a large conductor, such as a metal frame or water pipe. Connecting one side of an electric circuit to ground provides a low-resistance path for current flow that may in many instances prevent shock for certain types of accidents and equipment failures.

The electric power to homes and businesses is supplied over three wires, as previously shown in Fig. 24-16. The three wires come from the step-down transformer provided by the utility company. The three wires from the power transformer terminate at the service entrance box. That box contains the circuit breakers and bus bars for wiring

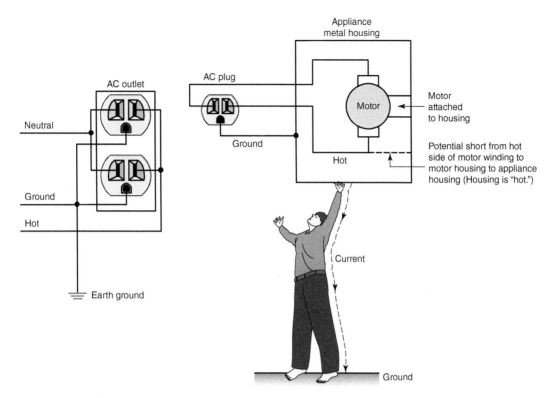

Figure 24-27 How grounds prevent shock.

distribution. In Fig. 24-16, you can see that the center tap (CT) wire is referred to as the neutral. This wire is usually white. The other two wires are referred to as the hot wires. These wires are usually black but could be red or blue. The voltage between either of the hot wires and the neutral wire is 120 V. The voltage between the two hot wires is 240 V. The higher voltage is used in heavy appliances, clothes dryers, air-conditioning and heating systems, and hot water heaters.

The neutral wire is connected to ground. This may be done at the utility company's transformer but also at the service entrance box in the facility. Note the special symbol used for ground. A large wire connected to the neutral bus bar is usually attached to a long copper rod or plate which is buried in the earth. This ground wire provides a path for current flow to the earth under some conditions. The ground exists primarily for safety purposes.

Obviously, if you touch two hot wires or one of the hot wires and the neutral, you will receive a shock. But you can also receive a shock by just touching one of the hot wires if you happen to be standing on the ground. Since the neutral wire is connected to ground, current can flow from the ground through the earth and through your shoes and socks. There may be sufficient resistance to prevent at least a minimal shock. The shock could be especially severe if the ground is damp and your shoes are wet or you happen to be standing in water. It is the third ground connection in the electric wiring that is designed to help eliminate or at least minimize the possibility of such a shock occurring.

Note in Fig. 24-27 that the ground connection on an electric outlet connects by a separate third wire back to the ground bus at the service entrance box. This is the same bus to which the neutral wire is connected. In effect then you actually have two ground wires running from the service entrance box to an outlet or other piece of equipment. The question becomes, How does this redundancy eliminate shock?

Figure 24-27 shows how the grounding arrangement can protect against shock. This arrangement is typical of large electric appliances containing motors, such as refrigerators, washing machines, dryers, and many electric tools including saws and drills. The motor winding is connected to the hot and neutral wires, with the neutral going to ground through the plug, outlet, and service box. The third ground connection is made to the external metal housing of the appliance. But assume for a minute that the ground connection at the appliance housing does not exist.

If an electric short occurs between the electric windings of the motor and its metal housing, the hot side of the line will actually come into contact through the housing of the appliance itself by way of its contact with the motor mounting equipment. If for some reason then you should touch the enclosure of the appliance while coming into contact with some other earth-grounded position, a shock will occur.

This problem is dealt with if the equipment has a third grounding wire. A third ground wire is connected to the metal housing of the motor and the appliance enclosure itself. This ground wire will be attached to the third prong in the electric outlet and will ultimately be connected back to the ground at the service entrance box. Now, if a short occurs between the motor windings and the metal housing of the appliance, the hot side will actually be connected back to ground. The ground wire, in effect, causes a short across the line that forces the circuit breaker to open, thereby protecting the equipment and eliminating the shock hazard.

Ground Fault Circuit Interrupters

A ground fault circuit interrupter (GFCI), or sometimes called just a ground fault interrupter (GFI), is an electronic circuit that is used to protect against shock and damage to electric equipment caused by a ground fault.

A ground fault is a type of short that occurs between one of the hot wires and ground. Should a load accidentally touch a hot wire while coming into contact with earth ground or something touching earth ground, then a ground fault occurs. This unintended load will cause current to flow which may or may not result in shock, depending on the load.

Most ground faults are not of the low-resistance short variety. Low-resistance shorts cause the circuit breaker to open. However, it is possible for a high-resistance short to occur. The high-resistance short will not trip the breaker but will cause current to flow between a hot wire and the earth ground. Figure 24-28 illustrates this concept. Such high-resistance shorts are sometimes referred to as *leakage resistance*. The current may be small, but in some instances it can cause serious shock or may cause equipment damage. If a ground fault interrupter is used in the circuit, such faults will be detected and the circuit will be automatically opened, eliminating the problem. Most such ground faults are caused by the presence of moisture or wetness in some form.

One example of a shock condition that may occur is if an individual is using a power tool such as an electric saw or electric drill/screwdriver. If the electric cord should become frayed or otherwise damaged, it could cause one of the hot wires to make contact with the metal housing of the tool. If this is the case, the person using the tool will come into contact with the hot wire. If that person is also at the same time standing on the ground, particularly moist or wet ground, then the connection between the hot side of the line and ground is complete and a shock will occur. A GFCI will detect this problem and automatically interrupt the power-line voltage.

Another example is that a hair dryer with a frayed cord can cause serious electric shock if the individual comes into contact with any exposed hot wire and ground. Ground is usually connected to water pipes, and should the individual touch the faucet or metal basin, then a shock may occur. Garage and outdoor electric outlets are always susceptible to moisture and ground faults. Any outdoor electric equipment, particularly pump motors associated with swimming pools, are subject to such dangerous conditions. The solution in all these cases is a ground fault circuit interrupter.

A GFCI is an electronic circuit that is used to sense high-resistance ground faults. A simplified diagram of one type of GFCI is shown in Fig. 24-29. AC power passes through the contacts of a double-pole relay. Normally the relay contacts are closed to complete the circuit. Note that both the hot and neutral wires pass through the center of the magnetic core. The hot and neutral wires act as the primary windings of a transformer. Another winding around the magnetic core acts as a secondary winding. The secondary winding is connected to a high-gain amplifier that operates a magnetic coil on a relay. The magnetic core is called a *differential transformer*.

Under normal conditions, the current in the hot wire is the same as in the ground or neutral wire, and, therefore, magnetic fields produced around the wires are equal and opposite and do not produce any magnetism in the core. As a result, no voltage is induced into the secondary winding. However if a ground fault occurs, the current in the hot wire is usually higher than the current in the neutral wire. This causes the magnetic fields to be unequal. The small difference created by the higher current in the hot wire causes a magnetic field to be induced into the magnetic core. The magnetic field induces a voltage into the secondary winding. The voltage is stepped up, amplified, and then used to operate the relay. When the relay is energized, the contacts on the hot and ground wires are opened, disconnecting the power.

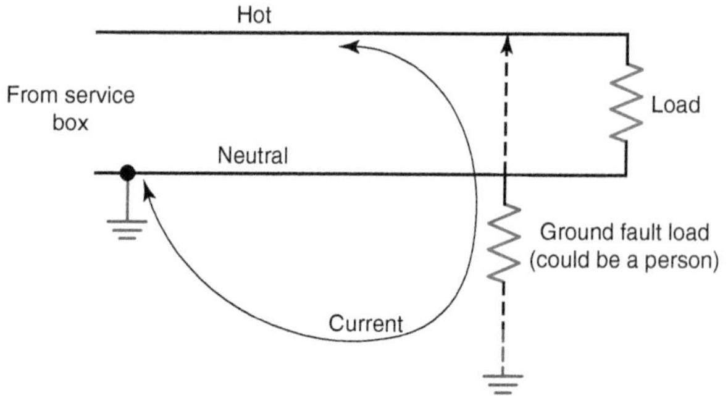

Figure 24-28 A ground fault.

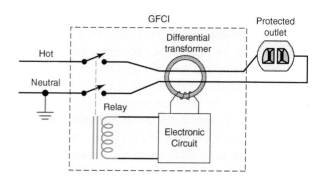

Figure 24-29 A ground fault circuit interrupter.

Figure 24-30 Ground-fault circuit interrupter (GFCI).

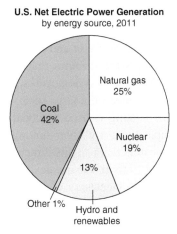

Figure 24-31 U.S. energy sources from 2011.
Source: U.S. Energy information Administration.

Most ground fault interrupters are designed to sense extremely low values of current. For example, GFCIs used in hospitals are the most sensitive for the typical maximum-sense current of 2 mA. GFCIs used in residences are usually set in the 5-mA range.

There are multiple types of GFCIs; however, most of them are built into the standard electric outlet, as shown in Fig. 24-30. The unit fits in a standard duplex receptacle and is usually accompanied by a push button that is used for resetting the unit if it is activated. Most GFCIs contain a built-in self-test and reset function. Pressing the test button produces a simulated ground fault that instantly triggers the unit to remove power. The purpose of the test button is to verify that the unit is operational. Some GFCIs also contain an indicator light to designate the status of the unit.

24.7 Alternative Energy

An estimated 86% of all power generated in the United States is produced by coal, natural gas, or nuclear. Refer to Fig. 24-31. Coal alone accounts for 42%.. The remainder of the power comes from alternative energy sources. *Alternative energy* refers to any sources other than coal, oil, gas, or nuclear and includes hydro, wind, solar, geothermal, bio fuels, and hydrogen. Geothermal is the use of hot water or steam from the interior of the earth to generate power. Bio fuels and ethanol are plant-derived fuels that can be burned like gasoline. Hydrogen refers to making hydrogen gas that can then be used to supply large fuel cells. Altogether, wind and solar only account for about 3% of the total energy generated despite their visibility and potential.

Over the past years there has been concern over the pollution effects of burning fossil fuel: coal, oil, and gas, which produce what is believed to be detrimental quantities of carbon dioxide (CO_2). There is concern that this produces global warming, and the consensus is that the use of fossil fuels should be decreased or eliminated completely if practical alternative energy sources can be found and developed. While there are many alternatives, only a few appear to be practical and affordable. Each has its advantages and disadvantages. Currently the primary disadvantage is that these alternative sources of energy are too inefficient and overly expensive. While the entire U.S. grid could theoretically be fully powered by wind, solar, or some combination, the price of electricity would be far greater than the average person, family, or business could afford.

Of the alternative energy sources, wind and solar are the most practical, and it is believed that ongoing research, development, and deployment will ultimately make them practical and affordable. Government subsidies and test deployments have made excellent progress, but it will be many years in the future before these two energy sources will substantially replace legacy sources.

Of all the sources, nuclear is the source with the greatest potential. It is nonpolluting and could easily handle all current and future demands. However, government regulations, politics, and a negative image has left nuclear energy's potential untapped. A cleaner alternative is natural gas. It is abundant and low in cost. Most new power plants are natural gas, and many coal plants are converting to gas. Natural gas appears to be the best combination of low cost and lower pollution for the years to come. However, most of the focus, attention, and funding is alternative energy, including wind and solar. Both are decades away from being practical and affordable, especially the latter.

Solar Power

The sun's energy is massive. Solar energy striking the earth could potentially power the entire world. The problem is converting that heat and light energy into electricity. There are two basic approaches. One technique uses shaped mirrors to focus the sun's rays on pipes through which water is flowing. The sun heats the water and converts it into steam. The steam then drives the turbines that operate the traditional generators. While this system works, it is complex and expensive and impractical on a large scale. This concept of using the sun for heating is used practically on a smaller scale. For example there are solar hot water heaters and swimming pool heaters that use this direct heating method.

A more practical method is to use devices that convert the sun's light energy into electricity directly. The device that does this is a semiconductor component called a solar cell or *photovoltaic (PV) cell*. This device is usually made of silicon and generates an output voltage of about 0.5 V of direct current (dc) when light strikes it. Figure 24-32 show the basic construction. The silicon is in two parts, *n*-type semiconductor material with extra electrons and *p*-type semiconductor material with extra positive charges. When light strikes the *n*-type side, electrons flow from it through a load and back to the *p*-type element. The current-supplying capacity is a function of the cell's size but is usually tens or hundreds of mA. Other materials have also been used, but silicon is still the most abundant and inexpensive.

To make the PV cell practical, many of them must be connected into series to get higher voltages and in parallel to achieve higher current-carrying capacity. Refer to Fig. 24-33. If each cell can provide 0.5 V at 100 mA, the array will produce 3 V with a maximum current capacity of 200 mA. Large arrays of cells are built to produce high voltage levels with high-current output potential. While PV cells can be connected to provide any desired voltage or current capacity, the most common panel voltage is 12 V or some multiple

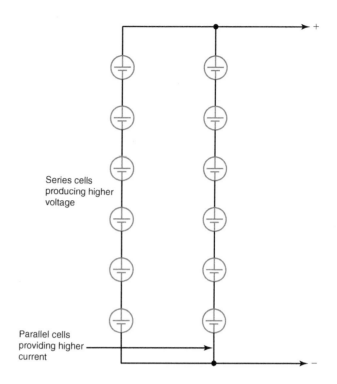

Figure 24-33 A solar array connects cells in series and parallel to provide higher voltage and current.

such as 24 or 48 V. Panels are rated by power capacity. Figure 24-34 shows a typical panel of multiple cells.

Solar panels are simple and generally affordable. But they do have disadvantages. First, they produce dc and not the ac required for most uses. Most solar electric systems therefore include a piece of equipment called an inverter. The *inverter* is a circuit that converts dc into ac. The 12 Vdc from the panels is converted to the standard 120-V, 60-Hz ac output. The output may be a pure sine wave or an approximated sine wave that works just as well in most applications.

Some inverters are designed to connect directly to standard house wiring. Their output is effectively in parallel with the power line. The inverter senses the phase and connects

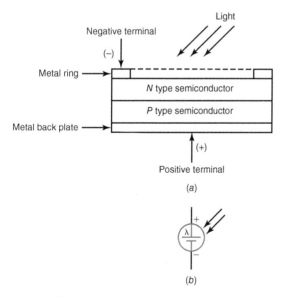

Figure 24-32 Solar cell. (*a*) Construction and (*b*) Schematic symbol.

Figure 24-34 A typical solar panel.

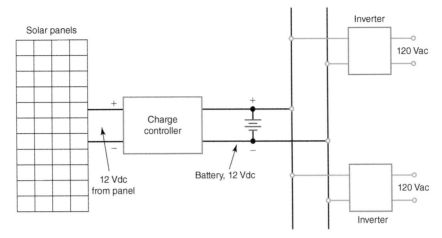

Figure 24-35 A complete home solar power system.

so that its voltage is in phase with the line phase. Such an inverter is called a grid-tied or grid-direct inverter. The inverter and/or the ac grid then supply the loads. If no load is present, the inverter just puts its voltage on the line and sends it back to the utility.

A key disadvantage of solar power is that the voltage disappears when the sun sets. There is no voltage at night. For that reason, some means must be provided to supply ac at night. If grid-tied inverters are used, they become inactive without sun power, so the power is supplied by the utility as usual. Another approach is to have the solar panels charge a large bank of batteries. Standard 12-V automobile batteries or some variation thereof are used for the energy storage. At night, the batteries supply power to the inverters. The batteries are then recharged during the day. Figure 24-35 shows the solar panel array supplying 12 Vdc to a charge controller, an electronic circuit that charges the batteries. The 12-V battery supplies power to one or more inverters. The inverters supply 120 Vac. Inverters come in many sizes, capable of a few hundred watts to larger units that can supply many kilowatts.

A third disadvantage of solar power is that very large panels are required to generate the kind of power used by the home or business. Daily averages vary, but typical average power consumption in a home is many kilowatts. This translates into huge panels that may occupy all or most of the roof space. The expense is rather high and typically runs tens of thousands of dollars for the average home. Such costs are generally out of reach of the average family despite the energy and cost savings of not using the existing utility power. Furthermore, in many areas, solar panels are still prohibited by homeowner's associations or cities because of aesthetics and appearances.

Overall, solar power is clean and readily available. But it is expensive. Cost of electric power is compared on a kilowatt-hour cost basis. Average utility prices are in the 10 to 20 cents per kilowatt-hour, but solar power costs almost 10 times that. Several utilities have built large fields of solar panels that can generate megawatts. The power is very expensive, but with subsidies from the utility and the government, power cost will decline. As efficiencies improve and costs decrease, solar power will be more affordable.

Wind Power

Wind power systems use the prevailing winds as an energy source. Large windmills and turbines use wind mechanical energy to turn generators that produce the electricity. Again, like the sun, the energy is free and readily available and very clean. However, the means for converting the energy is still complex, expensive, and even offensive to some.

Wind also has its disadvantages. First, the wind generators must be located where the prevailing winds are steady. In the United States, the area in the center of the country from the Texas panhandle to the Canadian border is best. Some coastal areas also have good wind locations. Even in the best areas, the wind does not blow continuously.

Another problem is that of appearances and aesthetics. Everyone likes the idea of a wind power source, but no one wants to live near one. That is the so-called not in my back yard, or NIMBY, effect. The towers are ugly, and they make annoying noises. Wind turbines kill millions of birds each year as well. Finally, since wind farms are built long distances from civilized areas, there is a need for a whole new structure of transmission lines to get the power to the utility and the end users. In many cases, these lines do not yet exist, and the funding is not available to build them. But despite these disadvantages, wind power has progressed thanks to government subsidies and the eagerness of utilities to build alternative energy sources.

Wind Turbine

The basic power generating unit is the wind turbine. See Fig. 24-36. A cylindrical steel tower anywhere from 200 to 300 ft tall contains a nacelle, or housing, that contains the

Figure 24-36   A wind tower with blades.

generator and all the control equipment. The turbine is powered by the wind with blades that can be from 60 to 120 ft long. Most have three blades. The wind blows, and the blades turn at a rate from 5 to 20 revolutions per minute (rpm). Wind speeds can be from 0 to over 100 miles per hour. Most turbines shut down if the wind is too great. The internal electric generator has a power capacity from 500 kW to over 2 MW.

Figure 24-37 shows what the turbine looks like inside. The shaft from the blades is connected to a gear box that steps up the speed from 5 to 20 rpm to some value in the 700- to 3600-rpm range. The gear box output shaft turns the electric generator. The generator may produce ac directly or dc which is converted to ac.

A major part of the wind turbine is the control unit. An anemometer measures wind speed and direction, and these inputs go to the control unit that generates control signals to a yaw motor and drive. This motor rotates the nacelle and blades to maximize facing the speed of the wind and the rotational speed and torque. Pitch motors in the blades rotate the blades to change pitch for maximum wind application. A braking system is also provided to stop rotation completely. The control unit can function alone or can be operated remotely via a communications link either wired or wireless.

The electric power is produced in several different ways. In some systems a standard ac generator is used. A speed governor system ensures that the shaft speed is accurate to product 60-Hz sine waves. In other systems, the turbine shaft turns a dc generator that does not require speed control. The dc output is then fed to inverters that provide the ac power to the grid. In some installations, battery backup is used. The dc generator charges the batteries that supply the inverters when the wind is not blowing.

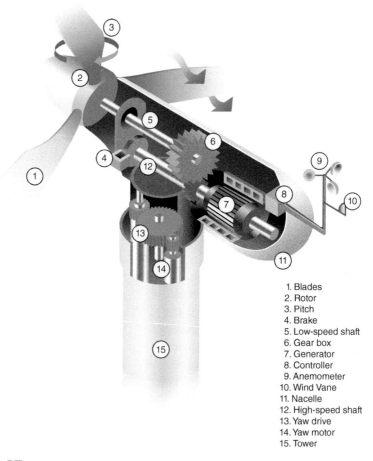

1. Blades
2. Rotor
3. Pitch
4. Brake
5. Low-speed shaft
6. Gear box
7. Generator
8. Controller
9. Anemometer
10. Wind Vane
11. Nacelle
12. High-speed shaft
13. Yaw drive
14. Yaw motor
15. Tower

Figure 24-37   Internal view of a wind turbine.

The Smart Grid

The *smart grid* is the term used to describe a fully revitalized and intelligent electric grid. The smart grid embodies a plan to modernize and update the aging electric grid with computers and communications networks. The goal is to improve the efficiency, reliability, and economics of the grid.

The electric grid is a huge one-way power system. It generates and distributes power to homes, offices, and manufacturing facilities. Once the power is delivered, the utility gets little or no feedback on the status of the power other than measuring and monitoring usage so that billing can be accomplished. To make the grid more efficient and reliable, the utility needs to more closely monitor and control the energy along the delivery path. The smart grid provides a way to increase the monitoring of generating plants, substations, and other points in the system to better understand the status of the grid. This means installing sensors in more locations and then providing a communications network to send the sensor data back to the utility where computers can diagnose, record, and display the grid status and, in some cases, take corrective control action if needed.

At the heart of the smart grid is a vast communications network to connect the grid back to computers at the utility. This network is only partially in place. It consists of both wired and wireless connections. Fiber-optic cable is used in some areas, while wireless links including the cellular network are used in others.

A key part of the smart grid is the smart meters that will eventually be installed in each home, office, and other facility. The currently used electric meters are not smart. They just monitor usage, and the data are sent to the utility. Some meters are still read by field personnel, while a few may be remotely read.

Smart meters have an internal microcontroller to monitor voltage and current used and compute power consumption. The data collected are then sent back to the utility. This is usually done by a wireless link to a neighborhood collection point that sends the data to the utility. In some cases power-line communication (PLC) is used. PLC converts the data with modulation and transmits them back over the power line. The data actually ride on the power line as the communication medium.

The smart meter also has another wireless capability built in to communicate with a home monitoring and control unit. Power usage is communicated to a device called an *energy dashboard* that shows the consumer the energy used, how and when. With this information, the consumer can

Figure S24-1   Smart energy home monitor control.

better monitor and control energy usage to save money. Figure S24-1 shows one such home monitoring device. It can be installed by the utility, service providers such as broadband service providers and home security companies who offer home management services, or the homeowner. One interesting and useful feature of such control devices is that they can also be used to monitor and control individual appliances. The appliance is plugged into a control unit that measures energy usage and can turn the appliance off or on typically via a home wireless network implemented in the control panel.

Most smart meters also provide the capability of allowing the utility to control the power to the home. The utility may turn the power off for short periods in peak usage periods to prevent overloads and brownouts. For example, during hot summers when all air conditioners are running at full capacity, the energy usage in some areas is more than the utility can supply. Cutting the power for short periods on many customers can prevent brownouts and even save money. Customers must first grant permission to the utility to do this.

The smart grid is an idea and a work in progress. It will be many years before it is fully implemented. And most homes still do not have smart meters. But even without a smart meter, a consumer can buy an energy monitoring unit to install at the service entrance box. With this unit, the consumer can conduct an energy audit and better manage energy usage to save money and energy.

Over the coming years, utility companies will slowly install smart meters and implement other improvements to their part of the grid. The cost is high, but the end result should be greater efficiency and better conservation of existing energy sources.

1. The most widely used energy source in the United States is
   a. oil.
   b. natural gas.
   c. coal.
   d. ethanol.

2. A turbine is a(n)
   a. electric generator that produces the ac voltage.
   b. type of furnace that burns coal, oil, or natural gas.
   c. heat exchanger that transfers heat energy to water flow.
   d. machine that converts liquid or gas flow into rotary motion.

3. Which energy source does *not* produce heat and steam?
   a. Hydro.
   b. Nuclear.
   c. Geothermal.
   d. Natural gas.

4. What rotates the electric generator?
   a. A motor.
   b. A turbine.
   c. A windmill.
   d. A water wheel.

5. Most ac generators use which configuration?
   a. Rotate a magnetic field inside the stator coils.
   b. Rotate a coil inside a fixed magnetic field.
   c. Both *a* and *b* are used equally.
   d. Both magnetic field and coil rotate in synchronism.

6. An electromagnetic generator produces what type of voltage?
   a. DC.
   b. Square wave.
   c. Sine wave.
   d. Triangular wave.

7. Most ac is generated is
   a. single phase.
   b. Two phase.
   c. polyphasic.
   d. three phase.

8. Most power generators produce a voltage in which range?
   a. 12 to 48 V.
   b. 120 to 240 V.
   c. 1 kV to 10 kV.
   d. 10 kV to 50 kV.

9. AC power is distributed by high voltage because it
   a. is more efficient.
   b. costs less.
   c. is safer.
   d. is a tradition.

10. What is the main component of a substation?
    a. Capacitor.
    b. Transformer.
    c. Lightning arrestor.
    d. Circuit breaker.

11. The longer the transmission line, the lower the ac voltage.
    a. True.
    b. False.

12. What range of voltage is used to distribute power to local neighborhoods?
    a. 120 to 240 V.
    b. 480 to 1200 V.
    c. 4 kV to 7 kV.
    d. 13 kV to 34 kV.

13. Which three-phase connection produces the highest output voltage for a given input?
    a. Y.
    b. Delta.
    c. Both *a* and *b* are the same.
    d. A combination of the two.

14. The three-phase sine waves are spaced by
    a. 45°.
    b. 90°.
    c. 120°.
    d. 180°.

15. A common three-phase ac voltage is
    a. 240 V.
    b. 208 V.
    c. 480 V.
    d. 4 kV.

16. The service drop to the building or house comes from a transformer that is
    a. mounted on a power pole.
    b. mounted on a concrete slab.
    c. either *a* or *b*.
    d. inside the building.

17. The service drop provides voltages of
    a. 120 V only.
    b. 208 V.
    c. 480 V.
    d. 120/240 V.

18. The service drop voltages are
    a. single phase.
    b. dual phase.
    c. three phase.
    d. delta-connected.

19. The first device encountered by the service drop voltages is the
    a. service entrance box.
    b. electric meter.
    c. master circuit breaker.
    d. distribution bus bars.

20. Each branch circuit from the bus bars has its own
    a. metering circuit.
    b. center tap.
    c. twisted-pair cable.
    d. circuit breaker.

21. Which wire connects to the earth?
    a. Hot.
    b. Neutral.
    c. Ground.
    d. Both *b* and *c*.

22. The wiring color scheme assigns which color to the hot wire?
    a. Red.
    b. White.
    c. Black.
    d. Green.

23. Why is three-phase power preferred over single phase?
    a. It is more efficient.
    b. It is better suited to large motors.
    c. It is best in rectification.
    d. All of the above.

24. What is the most common wire size in house wiring?
    a. No. 14, stranded.
    b. No. 14, solid.
    c. No. 12, stranded.
    d. No. 24, solid.

25. Electric shock is determined by
    a. voltage.
    b. resistance of the human.
    c. current level.
    d. all of the above.

26. Which voltage is most likely to provide a shock?
    a. 1.5-V AA cell.
    b. 12-V auto battery.
    c. 90-V power supply.
    d. Solar cell.

27. The wide slot on a standard duplex outlet is connected to which line?
    a. Neutral.
    b. Ground.

    c. Hot.
    d. Circuit breaker.

28. A device that helps ensure a tight electric connection between two copper wires is a
    a. screw terminal.
    b. wire nut.
    c. solder.
    d. weld.

29. The third or ground wire in an outlet or electric plug is provided for
    a. shock safety.
    b. backup redundancy.
    c. center-tap connections.
    d. 240-V connections.

30. A GFCI usually detects
    a. direct shorts.
    b. high-resistance leakage.
    c. presence of human contact.
    d. presence of a load.

31. Most electric energy in the United States is produced by
    a. nuclear.
    b. natural gas.
    c. oil.
    d. coal.

32. Which of the following is not an alternative energy source?
    a. Wind.
    b. Hydrogen.
    c. Wood.
    d. Solar.

33. What makes alternative energy sources less desirable than conventional methods?
    a. Politically correctness.
    b. High cost.
    c. Lack of national interest.
    d. Lagging technology.

34. The cleanest, most practical, and affordable energy source is
    a. nuclear.
    b. natural gas.
    c. ethanol.
    d. wind.

35. The output from a typical solar cell is about
    a. 0.5 V.
    b. 1.5 V.
    c. 3 V.
    d. 12 V.

36. A solar cell produces
    a. ac.
    b. dc.

37. The device that converts dc into ac is a(n)
a. power supply.
b. rectifier.
c. charge controller.
d. inverter.

38. Wind generators produce
a. ac directly.
b. dc.
c. either ac or dc.
d. ac and dc at the same time.

39. A key disadvantage of wind turbines is
a. no wind, no power.
b. large, ugly blight on landscape.
c. long-distance transmission lines needed.
d. all of the above.

40. What is the approximate percentage of all U.S. power generated by solar and wind?
a. 2%.
b. 10%.
c. 19%.
d. 35%.

CHAPTER 24 PROBLEMS

SECTION 24.1 Power Generation

24.1 In a four-pole generator producing a 50-Hz sine wave, what is the generator shaft speed?

24.2 How could a 50-Hz generator be used to produce 60 Hz?

SECTION 24.2 Power Distribution

24.3 A 120-V source is transmitted over a distance of 2 miles to a load of 3 Ω. If each wire has a resistance of 1.5 Ω, how much power is lost in the transmission lines? What is the efficiency?

24.4 To correct the efficiency described in Prob. 24-3, a step-up transformer with a ratio of 1:12 is used before the transmission line. A step-down transformer with a 12:1 turns ratio is used at the load. What is the power lost in the transmission lines if each has a resistance of 1.5 Ω?

SECTION 24.3 Home Electric Wiring

24.5 A big-screen TV set uses 750 W of power. It is on for 5 hours per day. What is the kilowatt-hour consumption?

24.6 An electric power meter reads 45,018 for the first month and 52,396 the following month. How much power was consumed? If the electric rate is 12 cents per kilowatt-hour, what is the monthly bill?

24.7 A 240-V hot water heater consumes 2500 W. Will a 15-A circuit breaker be adequate to operate and protect the unit?

SECTION 24.4 Three-Phase AC Power

24.8 A transformer secondary produces 240 V in each phase of a Y connection. What is the voltage from ground to each Y connection? What is the voltage between any two Y outputs?

24.9 A delta-to-delta transformer connection has a turns ratio of 1:1. If the input is 208 V, what is the output?

24.10 If a Y-connected output transformer has a voltage of 750 V between each of the Y outputs, what is the winding voltage?

SECTION 24.6 Electrical Safety

24.11 What level of current in a human body will cause a mild shock? Certain death?

24.12 Explain how the third or ground wire provides shock protection.

SECTION 24.7 Alternative Energy

24.13 A solar panel consists of 48 cells in series with a second 48-cell panel. Each cell can provide up to 75 mA. What is the output voltage and the maximum current capacity?

24.14 Name three disadvantages of solar power.

24.15 Name three disadvantages of wind power.

Chapter **25**

Cables and Transmission Lines

Learning Outcomes

After studying this chapter, you should be able to:

> *Define* the terms *cable* and *transmission line*.
> *Define* wavelength and *calculate* it given the frequency of operation.
> *State* the equivalent circuit of a transmission line.
> *Define* and *calculate* characteristic impedance of a transmission line.
> *State* the rule for properly terminating a transmission line.
> *Explain* why reflections occur on a transmission line.
> *Define* the terms *attenuation* and *delay* and *state* how they are measured in transmission lines.
> *List* the advantages and disadvantages of twisted pair line.
> *List* the advantages and disadvantages of coaxial line.
> *Name* the most commonly used connectors used with twisted pair and coaxial transmission lines.

Cables are the connecting media between pieces of equipment, from antenna to radio, from one printed circuit board to another, from test instrument to circuit, and over large networks of interconnected computers. They are also a source of trouble in many systems if they become defective or if they are applied incorrectly. And as you will find out in this chapter, many cables are just not pairs of interconnecting wires. Instead they are transmission lines, cables that behave like circuits. Knowledge of basic cables and transmission lines is essential for all technicians today.

25.1 Cables vs. Transmission Lines

Cables are two or more wires or other conductors packaged together in one unit. AC power-line cords are cables. The coaxial cable from your cable box to the TV is a cable. The wire connecting a microphone to an amplifier is a cable. The networking line from your PC to a router or network switch is a cable. Cables are everywhere. As a technician or engineer you will work with cables on a daily basis. They are literally the lifelines of electronics.

Cables can carry either analog or digital signals. Cables for analog signals are mostly for carrying audio like voice and music or video. Cables for digital carry binary signals for networks, the Internet, and even digital video. Most cables today handle more digital than analog signals. There are both long cables and short cables.

The actual determination of what is a "short" or "long" distance really depends on the frequencies of operation. The higher the frequency, the "longer" a cable looks to the transmitting and receiving circuits. Inside a computer, a bus operating at 100 MHz over a distance of about one foot is relatively slow, and few special precautions must be observed to design and use this bus. However, if you try to operate this bus with a microprocessor that runs at 2 GHz, suddenly the bus lines look like multiple "long" transmission lines, and special consideration must be given to the design of the bus if it is to work correctly, if at all.

On the other hand, slow-speed serial transmission at 9600 bps can take place over cables of 50 to 100 feet with no difficulty. But as the distance increases, even at this relatively low speed, the cable begins to look like a "long" transmission line. Special conditions must be observed to ensure proper operation.

For analog signals an audio signal with its low frequencies can tolerate a long cable without difficulty. For video signals in the 4- to 10-MHz range, any cable begins to look long. And the type of cable becomes an issue.

There are two basic kinds of media used in electronic systems: copper wire cables and fiber-optic cables. The most common types of wire cables are *coaxial* (or *coax*) and *twisted pair cable*. Fiber-optic cable is a thin glass or plastic cable that carries the digital data in the form of light pulses.

In this chapter, you will learn about transmission lines and the two most common types of wire media, twisted pair and coaxial cable.

25.2 Principles of Transmission Lines

A transmission line is a structure of two conductors used to carry electric energy from one place to another. It is a pair of wires that not only carry electrons but also guide the electric and magnetic fields set up by the applied voltage and the current flow.

Whenever wire conductors are used to carry signals at the lower ac frequencies, they are called cables. There are many such wire cables used in electronics. You are already familiar with the basic types, such as ac power cords, speaker cables, microphone cables, and telephone wire. Short cables, especially those carrying low-frequency ac, such as 60-Hz power or audio signals like voice and music in the 20-Hz to 20-kHz range, are essentially noncritical. They consist of two or more insulated copper conductors bundled in a common jacket and terminated on one or both ends by connectors. In most cases, we don't give much thought to cables as they are simply nothing more than copper conductors that make an electric connection between one point and another. The only critical factor in such cables is that the wire size be sufficiently large to carry the desired current without overheating. Electrons flow through them and carry power from one place to another. The cables themselves have very little or at least minimal effect on the signals themselves.

However, whenever the cables are required to carry high-frequency signals, the cables themselves begin to act more like complex electronic circuits than simply interconnecting wires. This is particularly true when the length of the cables begin to approach the size of one-tenth to one full wavelength at the operating frequency. When signal wavelength and cable lengths become the same order of magnitude, the cables themselves are more appropriately referred to as transmission lines. They have a significant effect on the signals being carried. These transmission lines must be analyzed, designed, and used as if they are circuits that "process" as well as transport the signal from one place to another. The goal is to minimize the effect of the transmission line on the signal so that the signal arrives at the receiver with essentially the same characteristics it has at the output of the transmitter.

The basic rule for determining whether a cable is a transmission line is that the line must have a length at least one-tenth wavelength or more at the frequency of operation.

Frequency and Wavelength

Figure 25-1 shows a sine-wave signal. It is the variation of voltage with respect to time. Many signals to be carried by a transmission line are analog signals like sine waves. Radio-frequency signals and telephone voice signals are good examples. Our main interest in this chapter is in digital signals, especially binary signals, which are rectangular waves of varying frequency and duty cycle. However, keep in mind that such signals, according to Fourier theory, are composed of a fundamental sine wave and harmonic sine wave added together. For that reason, we can understand pulse signals by using standard sine-wave analysis techniques.

The frequency of a sine wave (f) indicate how many complete cycles occur per second. *One cycle* is defined as one

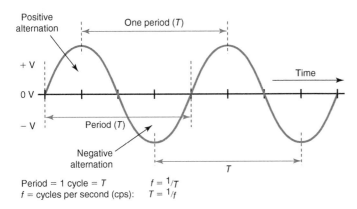

Figure 25-1 The relationship between frequency and period.

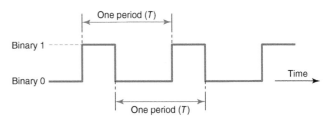

Figure 25-2 The period of a binary signal.

positive alternation and one negative alternation. The frequency is expressed in terms of cycles per second, or hertz (Hz).

Note also in Fig. 25-1 that the time for one cycle is referred to as the period, *T. One period* consists of one positive alternation and one negative alternation and has the time duration of *T*. You can also measure the period between every other zero-crossing point, but the zero crossings are hard to determine in practical measurements, so it is easier to measure the time between any two adjacent positive peaks or negative peaks as shown.

The relationship between frequency and period is the familiar reciprocal relationship indicated below.

$$T = \frac{1}{f} \tag{25-1}$$

$$f = \frac{1}{T} \tag{25-2}$$

If the frequency is 9600 Hz or 9.6 kHz, the period is

$$T = 1/9600 = 1/(9.6 \times 10^3) = 0.104 \times 10^{-3}$$
$$= 0.104 \text{ ms or } 104 \text{ } \mu s$$

If the period is 75 μs, the frequency is

$$f = 1/T = 1/(75 \times 10^{-6}) = 1.3333 \times 10^4 = 13,333 \text{ Hz}$$
$$= 13.333 \text{ kHz}$$

For rectangular waves like binary signals, the definitions are similar. One period is one "on" level or binary 1 interval plus one "off" level or binary 0. See Fig. 25-2. Frequency is the number of cycles per second as with a sine wave, but you will also hear the terms *pulse repetition rate (prr)* or *pulse repetition frequency (prf)* used instead of frequency. The relationship between *T* and prr is as described earlier.

$$\text{prr} = \frac{1}{T} \tag{25-3}$$

$$T = \frac{1}{\text{prr}} \tag{25-4}$$

Wavelength

Wavelength refers to the distance between the maximum amplitude points of the electric and magnetic fields produced by a signal in a cable. Wavelength is also the distance traveled by a signal in the time of one cycle or one period of the signal. Wavelength is illustrated in Fig. 25-3. It too is represented as a sine wave, but the horizontal axis is a physical distance rather than time. *Wavelength* is the distance between adjacent positive or adjacent negative peaks of the signal. When a sine-wave signal is applied to a transmission line, it produces patterns of electric and magnetic fields along the cable. Try to visual these patterns of maximum and minimum points along the line. The distance between the maximum amplitude peaks can actually be measured.

Wavelength is computed with the formula

$$\lambda = 300,000,000/f \tag{25-5}$$

In this expression, wavelength is represented by the lower-case Greek letter lambda. The quantity 300,000,000 is the speed of light which is also the speed of any radio or electric signal in free space. That speed is 300,000,000 meters (m) per second or about 186,400 miles per second. The frequency *f* is in cycles per second or hertz.

Consider the wavelength of an audio frequency signal, say 3 kHz or 3,000 Hz. Computing the wavelength we get:

$$\lambda = 300,000,000/3000 = 100,000 \text{ m}$$

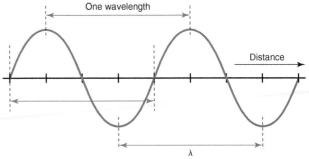

Wavelength = λ
$\lambda = 300,000,000/f$

Figure 25-3 Wavelength is the distance between peaks of the magnetic and electric fields produced by a signal.

Since there are 3.28 feet per meter, this distance represents 328,000 ft. Since there are 5280 ft per mile, this figure also represents a distance of

$$328,000/5280 = 62.12 \text{ miles}$$

As you can see, the wavelength of audio and other low-frequency signals is extremely great. Most of the cables that we use to carry signals at these low frequencies represent only a tiny fraction of one wavelength, and, therefore, the cables tend to have little or no real effect on the signal with the exception of resistive attenuation and capacitive loading. In practical length cables, which are rarely longer than several hundred feet, it is never possible to measure maximum magnetic or electric fields along the cable.

Now consider the wavelength of a higher-frequency signal. Assume a frequency of 100 MHz. This is a radio-frequency signal but is also a frequency that represents a common data transmission speed. The wavelength of a 100-MHz signal can be computed with Formula (25-5):

$$\lambda = 300,000,000/100,000,000 = 3 \text{ m}$$

When working with frequencies in the MHz range, the formula can be simplified to

$$\lambda = 300/f_{\text{MHz}} \qquad \textbf{(25-6)}$$

where f is in MHz.

$$\lambda = 300/100 = 3 \text{ m}$$

Three meters represents a distance of $3 \times 3.28 = 9.84$ ft. In this case, the wavelength is in the same ballpark as the length of cables used in practical applications. In fact, many of the cables in common use are actually multiple wavelengths long at the operating frequency. Such cables have a profound effect upon the signals that they carry. At high frequencies, these cables function like electronic circuits rather than simply conductors providing a connection from one place to another.

EXAMPLE 25-1

A signal has a frequency of 400 MHz. It is to be carried by a cable with a length of 15 ft. Is the cable a transmission line?

Answer:

$$\lambda = 300/f_{\text{MHz}} = 300/400 = 0.75 \text{ m}$$

The cable is 15 ft long. With 3.28 feet per meter, the length in meters is $15/3.28 = 4.573$ m.

If the wavelength of the signal is one-tenth or less of the cable length, the cable is a transmission line.

One-tenth of 4.573 m is 0.4573 m. The cable is definitely a transmission line.

Cables act like a low-pass filter passing dc and lower frequencies below some cutoff frequency that is a function of the cable characteristics and length. However, depending on the frequency of operation and how the cables are connected and terminated, they can also appear to be inductive or capacitive reactances or complex impedances. Thus the cables perform filtering effects and can highly distort nonsinusoidal signals. Under these conditions, the only proper way to analyze the effect of a cable on a signal is to use the techniques of analysis established for transmission lines.

Equivalent Circuit

The general equivalent circuit of a two-wire transmission line is shown in Fig. 25-4a. To a signal, the transmission line appears to be a low-pass filter made up of series inductances and shunt capacitances. Individual inductors, capacitors, and resistors are shown in the figure, but in reality, the capacitance, inductance, and resistance are evenly distributed along the line.

For example, the copper conductors making up the cable have dc resistance. The conductors also offer additional resistance at high frequencies caused by skin effect. *Skin effect* is a phenomenon that causes the electrons in a copper conductor to flow near the surface of the conductor rather than in the center of the conductor and thus reducing the effective cross-sectional area of the conductor and increasing its resistance. This effect becomes recognizable at several megahertz and gets progressively worse at higher frequencies. At VHF, UHF, and microwave frequencies, it is severely pronounced, and, in most cases, the current flow is by electrons on or right near the surface of the conductor. This effectively reduces the cross-sectional area of the conductor and, therefore, represents a significant increase in resistance.

As for inductance, we normally think of inductors as coils of wire. However, a straight piece of wire also has inductance. The inductance per foot is very low and may only be picohenrys (pH) or nanohenrys (nH). Just remember that the inductive reactance, however, is directly proportional to the frequency of operation.

$$X_{\text{L}} = 2\pi fL \qquad \textbf{(25-7)}$$

So while the inductance may be small, at high frequencies it produces significant reactance that can oppose high frequency signals.

The capacitance in the circuit, of course, is the result of having two conductors separated by an insulating medium. Capacitance is distributed continuously along the cable. This capacitance is rather significant and may only be a few picofarads (pF) per foot up to hundreds of picofarads for a long

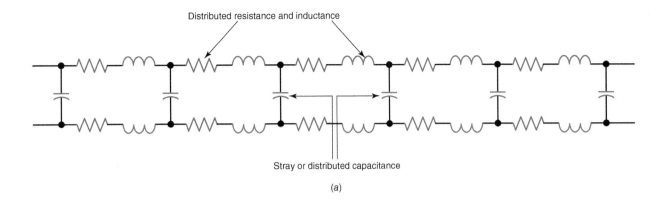

Distributed resistance and inductance

Stray or distributed capacitance

(a)

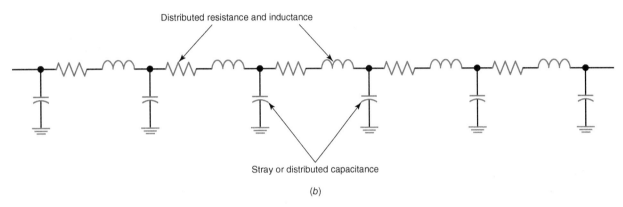

Distributed resistance and inductance

Stray or distributed capacitance

(b)

Figure 25-4 The equivalent electric circuit of a cable or transmission line is a low-pass filter. (a) Balanced. (b) Unbalanced.

cable. This has a dramatic effect on high frequency signals, especially pulse signals. Remember that the capacitive reactance decreases with frequency.

$$X_C = \frac{1}{2\pi f C} \qquad \textbf{(25-8)}$$

At higher frequencies, the capacitive reactance is low and, therefore, acts as a loading or shorting effect on the signal.

Although not illustrated in Fig. 25-4a, there is also a distributed shunt resistance or conductance along the line. It is caused by the resistance of any insulating material between the two conductors or any leakage resistance between the two. In most cases it is so high that it is negligible in determining the overall effect of the line on a signal being carried.

The equivalent circuit is an attempt to illustrate the distributed parameters. You can think of the cable being small low-pass filter segments made up of the inductors, resistors, and a capacitor. Many of these low-pass filter sections are then cascaded.

The equivalent circuit in Fig. 25-4a represents a balanced two-wire parallel transmission line. For unbalanced or single-ended lines, the equivalent circuit is best shown as in Fig. 25-4b. Again, the cable is made up of series

resistance and inductance and shunt capacitance, with one side of the cable being connected to ground. For the discussions to follow, assume that each LC segment represents one foot of cable.

25.3 Transmission Line Specifications and Characteristics

There are several important characteristics and specifications associated with transmission lines. These include characteristic impedance, attenuation, velocity factor, capacitance per foot, and delay. The specifications determine the type of transmission line to be used in different applications and also determine how the cable will affect the signal to be transmitted.

Characteristic Impedance

Perhaps the most important characteristic of a transmission line is its *characteristic impedance*. Also known as the *surge impedance,* it is the impedance that a generator driving the cable would see if the cable were infinitely long. The characteristic impedance represented by the designation Z_o is basically a complex impedance that is a function of the distributed inductance and capacitance of the cable.

The characteristic impedance of a transmission line is calculated using the expression

$$Z_o = \sqrt{\frac{L}{C}} \qquad \text{(25-9)}$$

The values for inductance (L) and capacitance (C) are the distributed values for any length of cable. A typical method of calculating the characteristic impedance is to determine the inductance and capacitance of the cable per foot and use these values in the expression.

Assume that the capacitance for one foot of cable is 28 pF and the total inductance per foot is 70 nH, the characteristics of the impedance of the cable is

$$Z_o = \sqrt{\frac{70 \times 10^{-9}}{28 \times 10^{-12}}} = \sqrt{2500} = 50 \ \Omega$$

The characteristic impedance of most cables and transmission lines is in the 50- to 600-Ω range. Coaxial cables typically have impedance values of 50 Ω, 75 Ω, and 93 Ω. By far the most common are 50-Ω coaxial cables.

A twisted pair transmission line typically has a characteristic impedance of something between 90 Ω and 200 Ω, with 100 to 150 Ω being the most common. The exact value depends considerably on the wire size and the tightness of the twist.

The importance of characteristic impedance is that it must match the internal impedance of the generator and the load. Figure 25-5 shows a generator with its output impedance (R_i or Z_g) driving a resistive load (R_L) through a transmission line. In order for this arrangement to work satisfactorily, the load impedance must match that of the characteristic impedance of the cable.

$$R_L = Z_o \qquad \text{(25-10)}$$

If the load is 150 Ω, then the characteristic impedance of the cable carrying the signal must be 150 Ω.

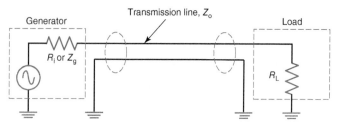

R_i = internal resistance of the source generator (or Z_g)
R_L = load resistance terminating the transmission line
Z_o = characteristic impedance of transmission line
$R_L = Z_o$
$R_L = R_i$ or $R_L = Z_g$

Figure 25-5 The load on the transmission line must equal the characteristic impedance of the line.

In addition, in order for maximum transfer of power to occur from the generator to the load, the generator output impedance and the load impedance should be equal. Recall that the maximum power transfer theorem states that maximum power will be dissipated in the load when the generator internal impedance and load impedances are equal.

If the load and generator impedances are not equal, maximum transfer of power will not take place. In addition, if the load resistance is not matched to the characteristic impedance of the transmission line, standing waves and reflections will occur that can seriously distort the signal as well as lower the amount of power delivered to the load. We discuss more about standing waves and reflections later.

Attenuation

All transmission lines attenuate the signal applied to them. The voltage appearing across the load at the end of the cable is less than the input voltage. The signal loss in the cable occurs as a result of the resistance, inductance, and capacitance of the cable. Looking back at Fig. 25-4b, you can see that the equivalent circuit appears to be a chain of voltage dividers made up of series resistance and inductance and shunt capacitance. The longer the cable, the greater the number of equivalent voltage divider circuits and the greater the signal loss.

Attenuation in a cable is usually expressed in decibels of loss per unit of length. Recall that a decibel is a number that gives a relative indication of the gain or loss in the circuit. The gain or loss of a circuit is expressed as the ratio of the output voltage to the input voltage or the output power to the input power. This ratio can be converted to decibels with the basic formulas given below.

$$\text{dB} = 20 \log \frac{V_o}{V_i} \qquad \text{(25-11)}$$

$$\text{dB} = 10 \log \frac{P_o}{P_i} \qquad \text{(25-12)}$$

Two examples of the use of these formulas to compute gain are given next.

The input voltage to an amplifier is 20 μV. The output voltage is 4 V. The gain of the amplifier in dB is

$$\text{dB} = 20 \log \frac{V_o}{V_i} = 20 \log \frac{4}{20 \times 10^{-6}}$$

$$= 20 \log 200{,}000 = 20(5.3)$$

$$= 106 \ \text{dB}$$

The input power to a filter is 80 mW. The output power is 500 μW.

The loss of the filter in dB is

$$\text{dB} = 10 \log \frac{P_o}{P_i} = 10 \log \frac{500 \times 10^{-6}}{80 \times 10^{-3}}$$

$$= 10 \log (6.25 \times 10^{-3}) = 10(-2.2)$$

$$= -22 \text{ dB}$$

Note in the last example that the power ratio is a fraction giving a negative logarithm. A negative dB value indicates a loss rather than a gain.

You can rearrange the dB formula so that if you know the dB gain or attenuation and either the input or output value, you can calculate the other.

$$\frac{P_o}{P_i} = \log^{-1} \frac{\text{dB}}{10} \qquad \textbf{(25-13)}$$

Alternative formulas include

$$\frac{P_o}{P_i} = \text{antilog } \frac{\text{dB}}{10} \qquad \text{or} \qquad 10^{\text{dB}/10}$$

For example, if a circuit or cable has an attenuation of -15 dB and the input to the cable is 500 mW, what is the output of the cable?

$$P_o/P_i = \log^{-1} (-15/10) = 0.0316$$

$$P_o/0.5 = 0.0316$$

$$P_o = 0.0316(0.5) = 0.0158 \text{ W or } 15.8 \text{ mW}$$

You can use the same dB formulas to compute the loss in a circuit or a cable. Since the output voltage or power will be lower than the input voltage or power, the ratio will be a fraction. Taking the logarithm of a fraction produces a negative value. A negative dB value indicates attenuation or loss.

EXAMPLE 25-2

a. The input to an amplifier is 30 W. The output is 100 W. What is the gain?
b. A cable has an input 4 W and an output 1 W. What is the loss?
c. The output of a cable is measured to be 24 mW. The cable loss is -8 dB. What is the input power?

Answer:

a. From Formula (25-12),

$$\text{dB} = 10 \log \frac{100}{30} = 10(0.523)$$

$$= 5.23$$

b. From Formula (25-12),

$$\text{dB} = 10 \log \frac{1}{4} = 10(-0.602)$$

$$= -6.02$$

c. From Formula (25-13),

$$P_o/P_i = \log^{-1} (-8/10) = 6.3$$

$$24/P_i = 6.3$$

$$P_i = 24(6.3) = 151.4 \text{ mW}$$

The attenuation in a cable is generally stated as the dB loss per 100 ft. This is a power dB loss rather than voltage loss. For example, one type of coaxial cable has an attenuation of 6 dB per 100 ft. The dB attenuation per foot then is $6/100 = 0.06$ dB/ft. The important thing to remember about this specification is that the attenuation value is only valid at a single frequency. The attenuation rate will be stated at a frequency of 100 MHz, 400 MHz, 1 GHz, or as determined by the manufacturer. The higher the frequency, the greater the attenuation of that cable for a given length.

Assume that a cable has an attenuation of 4.8 dB/100 ft at 400 MHz. What is the attenuation of 265 ft of this cable at that frequency?

The attenuation per foot is $4.8/100 = 0.048$ dB/ft

Attenuation $= 0.048(265) = 12.72$ dB

EXAMPLE 25-3

A coaxial cable has an attenuation of 3.5 dB per 100 ft at 100 MHz.

a. What is the attenuation of 60 ft of cable?
b. How is attenuation affected if the frequency is raised to 300 MHz?

Answer:

a. Attenuation per foot is $3.5/100 = 0.035$ dB/ft, so attenuation for 60 ft is $0.35 \times 60 = 2.1$ dB.
b. The same cable used at 300 MHz would have significantly higher attenuation.

Capacitance per Foot

Capacitance per foot is another typical cable specification. It is a measure of the amount of capacitance across the cable conductors for each foot of length. The capacitance can be anything from approximately 5 pF for a loosely twisted pair of wires to over 30 pF per foot for certain types of coaxial. The capacitance affects the upper cutoff frequency or bandwidth of the cable. The capacitance will also cause significant signal distortion, especially of digital signals.

In general, the lower the capacitance per foot, the better the cable, because it will produce less attenuation and signal distortion and will have a wider bandwidth.

Velocity Factor

The *velocity factor* is a fraction that indicates the speed of the signal in the cable. Light and radio waves travel at the speed of light, which is 300,000,000 meters per second. This translates to a speed of approximately 186,400 miles per second. These speed figures are for the velocity in free space, meaning in air or a vacuum.

Whenever electric signals are applied to a cable, the inductance and capacitance of the cable slow the signal down. The velocity factor indicates the percentage of the speed of light at which the signal will travel in the cable.

Velocity factors of coaxial and twisted pair cable range from approximately 0.5 to 0.95. Typical coaxial cable velocity factors are 0.66 and 0.8. Twisted pair cable has a higher velocity factor of 0.8 to 0.95. You will also see velocity factors expressed as a percentage, such as 66%, 78%, or 80%. The velocity factor is usually quoted by the manufacturer in the cable specifications.

25.4 How a Transmission Line Works

You have already learned that the equivalent circuit of any type of transmission line is a series of low-pass filter sections. Refer to Fig. 25-6. For our discussion here, assume that each series inductance is the combined inductance of both conductors of the transmission line, one foot in length, and that each capacitor represents the capacitance between the conductors for each foot of line. The resistance in the circuit represents the resistance of both conductors for each foot of line.

Now let's consider the operation of this line with a dc voltage pulse. The battery and switch will be used to apply a dc step voltage to the line so that we may analyze its result. Only three sections of line have been shown for simplicity, and the final section is terminated in a resistance equal to the characteristic resistance of the transmission line.

Keep in mind during this analysis that an inductor opposes changes in current, while the capacitor opposes changes in voltage. In the equivalent circuit of the transmission line, the capacitors will charge to the applied dc voltage through the inductive and resistive sections.

When switch SW is closed, the battery voltage will be applied to the line. Inductor L_1 opposes any current change, while capacitor C_1 initially has zero volts across it. Eventually, capacitor C_1 begins to charge to the applied voltage through R_1 and L_1. Remember that it takes a finite amount of time for the capacitor to charge through the inductance and resistance.

When C_1 is nearly fully charged, C_2 will begin charging to the voltage across C_1 through R_2 and L_2. Again, it takes a finite amount of time for C_2 to charge. As the voltage across C_2 becomes fully charged, capacitor C_3 begins to charge through R_3 and L_3. Depending on the number of LC sections in the line, the charging action will continue until all the capacitors are charged. It is at that time that the dc applied voltage appears across the resistive load R_L. The dc voltage at the end of the line will be less than the applied battery voltage by an amount dependent upon the dc resistance of the conductors in the line, which is usually low compared to the terminating load resistance. Therefore, most of the battery voltage will appear across the load.

The amount of time that it takes for the dc voltage to appear across the load is dependent on the values of inductance and capacitance on the line. This delay time can be computed with the expression

$$t_d = \sqrt{LC} \qquad \text{(25-14)}$$

where L is the inductance per foot in henrys, C is the capacitance per foot in farads, and t is the delay time in seconds. A typical way to determine the delay time is to determine the inductance and capacitance values for a given length of line, such as one foot. For example, if a transmission line has a capacitance of 31.5 pF per foot and an inductance of 96.8 nH per foot, the amount of delay introduced by one foot of this line is

$$t_d = \sqrt{LC} = \sqrt{96.8 \times 10^{-9} \times 31.5 \times 10^{-12}}$$
$$= \sqrt{3.05 \times 10^{-18}} = 1.746 \times 10^{-9}$$
$$= 1.746 \text{ ns/ft}$$

Given the amount of delay per foot, you can calculate the total delay of a transmission line by simply multiplying the delay per foot by the number of feet in the total length. For a 15-ft length of the line given above, the total propagation delay (t_p) is

$$t_p = 1.746 \times 15 = 26.2 \text{ ns}$$

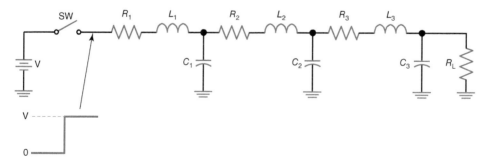

Figure 25-6 The equivalent circuit of a transmission line driven by a dc step pulse.

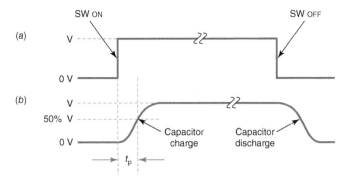

Figure 25-7 (a) Input and (b) Output pulses of the transmission line showing propagation delay t_p.

This is a significant amount of time in electronics. The delay is the same order of magnitude of many modern circuits.

Figure 25-7 shows this delay from the application of the dc voltage to the time it appears fully across the load. Note that the voltage across the load is a nonlinear rise because of the charging action of the capacitors. The propagation delay is measured at the 50% amplitude point, which is the standard measuring point for pulse times.

If the switch SW in Fig. 25-6 is now opened to terminate the pulse, the voltage applied to the line is removed. All the capacitors in the line are charged to the battery voltage V. They will now discharge into the load resistance. The voltage will be maintained across the load for a short period of time. Eventually all the capacitors will be discharged, and the voltage across the resistive load will be zero. See Fig. 25-7. The pulse at the load has been rounded, delayed, and attenuated but still usable.

Now assume a continuous square wave is applied to the line. The pulses are simply dc voltage levels that switch off and on at a relatively high rate of speed. The capacitance in the line will charge to the applied voltage and then discharge when the pulse is removed. The longer the transmission line, the greater amount of inductance and capacitance and the longer it takes for the line to charge and discharge. Thus the rise and fall times of the pulse across the load increases with the length of the line. Furthermore, the greater the length of the line, the greater the amount of resistance and the lower the amplitude of the pulse at the end of the line.

Figure 25-8 shows the effect on the signal for different lengths of line. The longer the line, the lower the amplitude of the signal across the load, the greater the rise and fall times, and the longer the signal delay. The line greatly distorts the square wave into a more rounded wave. Furthermore the wave amplitude is also attenuated.

One way to understand the effect of the transmission line on the square wave input signal is to recognize that the transmission line itself is a low-pass filter. The response of a typical low-pass filter is illustrated in Fig. 25-9. Signals below the cutoff frequency are passed unattenuated. Above the cutoff frequency, the higher frequencies are gradually "rolled off"; that is, they are gradually attenuated by a greater amount. In a transmission line, the cutoff frequency is determined by the inductance and capacitance per foot and the length of the line. The greater the length, the lower the cutoff frequency.

The higher-frequency harmonics in the square wave are being filtered out by this low-pass effect. This leaves only the lower harmonics and the fundamental. If the line is long enough or the frequency of the square wave high enough, virtually all the harmonics will be filtered out, leaving only a low-amplitude fundamental sine wave at the output.

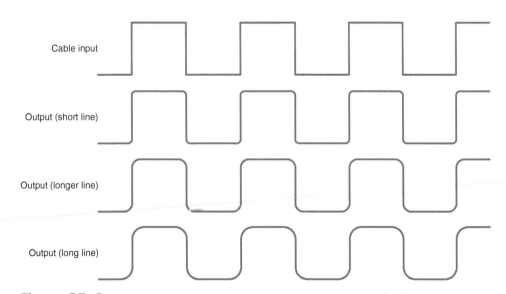

Figure 25-8 Input and output signals of transmission lines of different lengths.

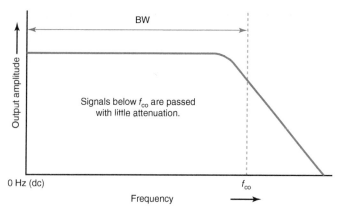

BW = bandwidth of line
BW = f_{co} signals above the cutoff frequency are attenuated.

Figure 25-9 The low-pass filter response of a transmission line.

Improperly Terminated Transmission Line

A transmission line terminated in a load resistance equal to the characteristic impedance of the line works perfectly well, and its performance is predictable. However, if the load on the transmission line is not equal to the characteristic impedance of the line, some unusual effects develop. These effects are particularly pronounced if the load is open or shorted. An open, shorted, or otherwise mismatched line causes a signal to be developed at the load end of the transmission line that is reflected back toward the source. Such reflections are undesirable because they change the shape of the signal and thus introduce distortion and data errors. Furthermore, the reflected signal could damage the generator.

If the load is equal to the line impedance, all is well. The signal from the generator is propagated down the line and is absorbed by the load. Pulses applied to the line appear at the load somewhat delayed by the line, attenuated in amplitude, and rounded by the filtering effect. Overall, the load signal is very usable. For an analog signal like a sine wave, the load signal is again delayed and attenuated but more than acceptable. Power is delivered to the load as desired.

Now consider the effect of an open line or no load. Pulses travel down the line charging the capacitance. As the pulse arrives at the end of the line, its energy is not absorbed since there is no load. What happens is that the signal turns around and starts its way back down the line toward the generator. We say that the signal is reflected. The result is that a pulse similar to the originally transmitted pulse appears back at the generator a time later equal to twice the propagation delay (t_p) of the line. See Fig. 25-10a. Now imagine a serial string of pulses representing binary data applied to the line with its mix of binary 0 and 1 voltage levels. The reflected

signal will add to the forward signal creating a highly distorted and totally incoherent pulse stream. Massive data errors will occur.

A similar thing happens if the line is shorted at the end. The energy reaching the shorted load is reflected back as before. If a single pulse is transmitted, the reflection will be a similar but attenuated pulse that is inverted in polarity, as Fig. 25-10b shows. Again, if a binary pulse train were applied with a shorted load, no data reach the end of the line but massive reflections add to the forward signal to create a real mishmash of a signal back at the generator.

If the cable does have a load but its impedance is not equal to the line impedance, there will be some reflected signal but not as large. A minimal mismatch will generally not affect the signal as much, and operation will be close to normal. The detrimental effects will be in proportion to the degree of mismatch.

A similar thing happens with analog signals like sine waves. If the load is not equal to the line impedance, reflections will occur and the signal will be modified. Normally,

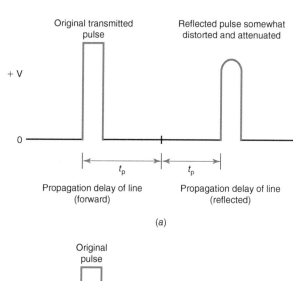

(a)

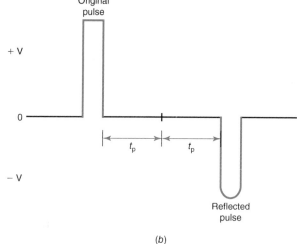

(b)

Figure 25-10 Pulse transmitted on a line showing reflected pulse at the generator. (a) Open line. (b) Shorted line.

if the load is matched to the line, then the voltage and current along the line are constant. In other words, if you measured the voltage and current anywhere along the line, they would be the same from generator to load. This is what is called a *flat line*.

Now if the load does not match the line impedance, the sine wave travels down the line and does not get fully absorbed by the load. Power is lost because some of the signal is reflected back toward the generator. The reflected sine wave adds to the forward signal creating a new sine wave of a different phase and amplitude. The result is that the voltage and current, as well as the accompanying electric and magnetic fields along the line, are no longer constant but vary in patterns of half-sine pulses. These are called *standing waves*. The distribution of voltage and current along the line depends on the length of the line and the degree of mismatch. The goal is to have no standing waves by eliminating the reflected signal. In this way, most of the power reaches the load as desired.

Figure 25-11 shows the standing wave effect. In Fig. 25-11a, a shorted line produces the voltage and current distribution shown. Note that the voltage and current actually go to zero every $\lambda/2$. With a short at the end of the line, the voltage is zero but the current is at its peak.

The effect of an open line is shown in Fig. 25-11b. The standing wave pattern is similar, but the voltage is at its peak at the load end of the line while current is zero.

Fig. 25-11c shows a mismatched line. The load R_L does not match the characteristic impedance Z_o of the line. The standing waves are similar but not as pronounced and the voltage and current do not go to zero. The degree of mismatch is usually expressed as the ratio of the load resistance to the line impedance. This is called the standing wave ratio (SWR).

$$SWR = R_L/Z_o \qquad (25\text{-}15)$$

Or if Z_o is greater than R_L,

$$SWR = Z_o/R_L$$

When the load and line impedances match, the SWR is 1. This is the ideal condition. As the mismatch and SWR become greater, the power lost at the load becomes significant because of the reflection. For example, an SWR of 2; 1 produces a loss of 10% power at the load.

In summary, the proper operation of a transmission line using pulse signals occurs when the transmission line is terminated in a resistance equal to the characteristic impedance of the line. In addition, the source impedance should also be equal to the load impedance to ensure maximum power transfer. Any line mismatch will produce reflections which, if severe enough, will alter the data of a binary signal.

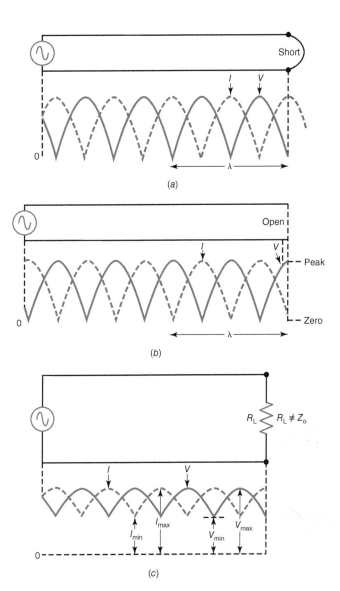

Figure 25-11 Standing waves on a transmission line. (a) Shorted. (b) Open. (c) Mismatched.

A similar thing occurs with analog signals if the transmission line is improperly terminated. If the signal is a sine wave and the transmission line is open, shorted, or mismatched, reflections occur. This sets up a pattern of standing waves along the line. The signal at the load is severely attenuated.

25.5 Twisted Pair Cables

The two most popular types of transmission lines are coaxial and twisted pair. Coaxial is the primary cable used in radio-frequency (RF) work for connecting transmitters and receivers to antennas. Coaxial is also the preferred cable for analog video. It is also sometimes used in data communications applications. Twisted pair is the primary type of wiring used in the telephone system. It is also used in most types of data communications systems, especially networking.

Cable Requirements for Binary Communications

If cables are to carry high-speed digital data, they must meet specific requirements. First, they must certainly have the bandwidth or frequency response suitable for passing the range of frequencies of interest. Coaxial cable is useful at frequencies up to approximately 10 GHz or more if they are kept short. Twisted pair can operate at frequencies over 100 MHz. Both types of cables are widely used to carry high-speed digital data, but only RF coaxial cable is used in radio applications.

Another requirement is that the cable have minimum attenuation. All cables exhibit resistive and other losses and, therefore, will attenuate any signal applied to them. While attenuation is inherent in any cable, it must be minimized to prevent excessive reduction in signal strength that occurs over long sections of cable.

Finally, a primary requirement is that the cable not radiate energy. Whenever voltage is applied across a cable, current flows in the conductors. The voltage causes the buildup of an electric field between the conductors. Furthermore, current flowing in the conductors produces a magnetic field. At very high frequencies, these electric and magnetic fields combine to form an electromagnetic wave which can radiate. An electromagnetic wave is a radio signal that can propagate over long distances if the conditions are correct. Radiation from a cable means loss of energy and possible interference to other adjacent cables or equipment.

A simple cable made of two parallel conductors, such as an ac power cord, speaker cable, or certain types of telephone wire, will radiate. The electric and magnetic fields around the cables appear something like that shown in Fig. 25-12. Note the polarity of the conductors which will reverse periodically if the conductors are carrying ac. The electric field is shown as solid lines, while the magnetic field is shown as dashed lines. Note the magnetic fields aid one another, that is, are in the same direction, between the conductors. This varying magnetic field around the conductors can cause a voltage to be induced into any nearby conductor. Capacitance between one cable and another adjacent cable can cause a signal to be transferred. This is called *cross talk,* and it is an undesirable characteristic of a cable. It must be minimized.

Both twisted pair and coaxial meet these technical requirements. Both have a wideband frequency response and thus are capable of handling high-frequency signals. Both are made so as to minimize attenuation. And the construction of both coaxial and twisted pair is such that radiation is greatly minimized or completely eliminated. Radiation is almost completely eliminated in a coaxial cable by its unique construction, while it is greatly minimized in twisted pair by twisting the two conductors together. The act of twisting

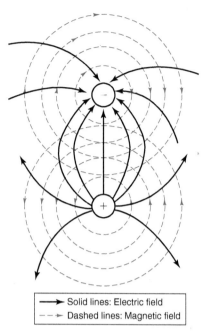

Figure 25-12 Electric and magnetic fields around parallel current-carrying conductors can radiate.

the wires around one another produces cancellation of the magnetic fields, thus minimizing the radiation to surrounding cables and other objects.

Twisted Pair Cable Construction

Twisted pair cable is made up of two insulated copper wires twisted loosely together. Figure 25-13a shows a common twisted pair cable. The wire is solid copper usually in the 22- to 28-gage range. Each wire is insulated with a plastic covering. The two wires are loosely twisted together not only to hold them together physically but also to help minimize radiation as indicated earlier. There are anywhere from three twists per foot to eight twists per foot depending on the type of twisted pair. Such a cable is usually referred to as *unshielded twisted pair (UTP).*

A popular variation of the twisted pair is the *shielded twisted pair (STP)* cable shown in Fig. 25-13b. In this case, a conductive shield is placed around the twisted pair. This may be a thin continuous foil attached to a wire or a fine-braided mesh of copper wires. The shield is usually grounded, and its purpose is to eliminate radiation from the twisted pair and the pickup of stray signals from other cables. Most shielded twisted pair is covered with an outer insulated jacket.

Twisted pair cables are used individually, meaning a single twisted pair by itself. However, most twisted pairs usually come bundled two or more to a cable inside a common outer jacket, as shown in Fig. 25-13c. There are many variations. Cables with two and four pairs per cable are available.

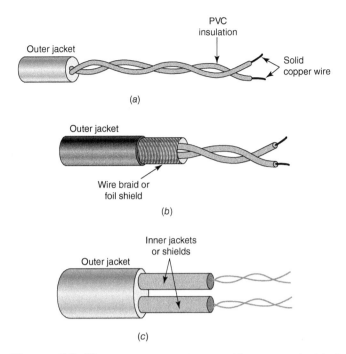

(a)

(b)

(c)

Figure 25-13 Types twisted pair cables. (*a*) Unshielded twisted pair (UTP). (*b*) Shielded twisted pair (STP). (*c*) Multiple twisted pairs per cable.

Each twisted pair may be contained within its own insulated jacket or with its own individual shield. In all cases, the individual pairs are contained within a larger outer jacket. The wires are color-coded so that they can be easily identified at each end of the cable.

Types of Twisted Pair Cable

There are many varieties of twisted pair cable in use. The oldest and the most common is that used in the telephone network. Telephones are usually connected to the telephone central office by a twisted pair referred to as the local or subscriber loop. Twisted pair cable is also widely used in private telephone systems like those in big companies and other large organizations. And it is used in most local area networks.

National and international standards have been developed for twisted pair cable to provide consistency in design and application. These standards have been set and agreed on by several organizations, including the American National Standards Institute (ANSI), the Electronic Industries Association (EIA), the Telecommunications Industries Association (TIA), and the Institute of Electrical and Electronic Engineers (IEEE). The most widely followed standard is the joint EIA/TIA document 568, known as the Commercial Building Wiring Standard. This standard defines several basic categories of twisted pair wiring.

You will see UTP designated as CAT *n* or Cat *n*, where *n* is the category number.

Category 1: Standard unshielded twisted pair used for voice grade telephone applications. Most telephone cable installed before the early 1980s was this type. It can support baseband digital transmissions at rates up to about 1 Mbps.

Category 2: For data communications applications at data rates up to 4 Mbps. It is typically used in unshielded form with four pairs per cable. It is not a widely used type of cable today.

Category 3: For data rates up to 16 Mbps and once widely used in early Ethernet and other local area networks. It has now been replaced with CAT 5 wiring in most applications.

Category 4: Unshielded twisted pair, suitable for data rates up to 20 Mbps. It is no longer widely used.

Category 5: By far the most popular and widely used type of UTP. It contains four pairs, and each supports data rates up to and over 100 Mbps. A more recent version, CAT 5e, is available to support rates up to 1 Gbps.

Category 6: For data rates up to about 10 Gbps. Version CAT 6a can support higher rates.

Category 7: Shielded cable that is sometimes referred to as Class F cable. It is for ultrahigh-speed applications. It uses a special connector unlike that used on the other classes of UTP cables.

Most of these cables come with four twisted pairs. The wire size is either No. 23 or 24. A color code is used to distinguish the different wires. The EIA/TIA 568 color codes are as follows:

Pair 1:	Solid blue or blue-white stripe
Pair 2:	Solid orange or orange-white stripe
Pair 3:	Solid green or green-white stripe
Pair 4:	Solid brown or brown-white stripe

All these cables can be purchased in two variations, standard and plenum. Plenum cable means that the insulation is made of plastic materials that do not produce toxic fumes or excessive smoke should they burn in a fire. Plenum cables are designed for use in long cable runs inside walls, ceilings, and vertical plenum chambers in tall buildings. The local fire codes may require plenum cable. Plenum cable has identical specifications and performance as standard cable, but it is more expensive.

Twisted Pair Specifications

Despite the simplicity of twisted pair and the similarities from one cable to another, twisted pair cables have a wide range of specifications. The most important specifications when you are designing or installing systems or replacing wiring are discussed below.

Wire Size Solid copper wire is used in twisted pair. The size varies from the larger No. 22 to the smaller No. 28. The most widely used size is No. 24. There are definite trade-offs to selecting a wire size. The smaller wire sizes give the cable less capacitance per foot. This makes the bandwidth of the cable greater so higher data rates can be accommodated. However, a smaller cable has a much higher dc ohmic resistance. Therefore its attenuation of the signal is greater, especially over longer distances at the higher data rates. The final choice is usually a compromise.

Capacitance per Foot Capacitance per foot is another important specification as it determines the bandwidth of the cable and sets the upper data rate for a given length. The lower the capacitance per foot (C/ft), the greater the bandwidth and higher the data speed the cable can support.

The amount of capacitance per foot is dependent on the wire size, the thickness and type of insulation, and the number of twists per foot. Larger wire sizes and thinner insulation produce higher capacitance figures. The greater the number of twists per foot, the higher the capacitance. The best twisted pair cable has only two or three twists per foot to keep capacitance low.

The amount of capacitance per foot varies from about 10 pF to 20 pF per foot. The popular CAT 5 twisted pair has 14 to 15 pF/ft.

Characteristic Impedance Twisted pair is a transmission line so it has a definite characteristic impedance (Z_o). The value runs between about 90 and 200 Ω. Common values are 100, 110, and 150 Ω. CAT 5 cable has a Z_o of 100 Ω. Remember that all twisted pair cable must be terminated with a resistance value equal to the characteristic impedance to minimize reflections.

Speed Some cables are rated by the upper data rate they can handle. Even the lowly CAT 1 voice grade telephone cable can handle rates up to about 1 Mbps. CAT 5 cable can pass data at a rate of 100 Mbps and beyond.

Just keep in mind that the speed a cable can accommodate depends not only on the capacitance per foot but also on the actual length of the cable. At greater lengths the bandwidth becomes more limited. This true for any type of cable.

Bandwidth Bandwidth is the low-pass filter response usually for 100 m of cable. For CAT 5e cable the bandwidth is 100 MHz. Bandwidth is related to speed, but the relationship of the two depends on the format of the data being transmitted.

Insertion Loss Insertion loss is the same as attenuation. It is usually given in dB per 100 m at a specific frequency. For CAT 5e the attenuation is 24 dB per 100 m.

Near-End Cross Talk (NEXT) Cross talk, as indicated earlier, is the transfer of a signal from one cable to a nearby cable. With four pairs in one cable, there will be some signal transfer by capacitive or inductive coupling from one pair to the others. The twisted wires do help cancel or minimize this effect, but it is not complete. Near-end cross talk (NEXT) is the transfer of one strong signal at the transmission end of one cable to the receiving end of an adjacent cable carrying a weak signal. It is measured in dB.

Far End Cross Talk (FEXT) Far-end cross talk (FEXT) is similar to NEXT but is the signal transferred from a signal entering the cable to an adjacent cable with an outgoing signal. The measure is dB.

Propagation Delay Propagation delay is the amount of delay in nanoseconds for 100 m of cable or some similar measure.

Twisted Pair Connectors

All cables, including twisted pair, use terminating connectors at their ends to provide a convenient and reliable way to attach them to the equipment and to one another. The most widely used connectors are the modular type similar to those used in on most modern telephones. The smallest is the RJ-11 modular connector, as shown in Fig. 25-14a. It has four connections and can accommodate two pair.

By far the most popular twisted pair connector is the RJ-45, shown in Fig. 25-14b. It is similar in shape and construction to the RJ-11, but it is larger. There are two kinds of modular connectors, plugs, and jacks. The plugs are male, like those shown in Fig. 25-14. These are mounted on the ends of the cable. The plugs connect to the female jacks that are mounted in the equipment.

The RJ-45 connectors can handle eight connections or four pairs. The connections to the RJ-45 vary, depending on the application and the manufacturer of the equipment or system. Figure 25-15 shows some of the ways that the pairs can be wired. Connections are shown for both the jacks and plugs. The T568A and T568B connections conform to the EIA/TIA standards.

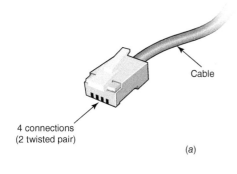

4 connections
(2 twisted pair)

(a)

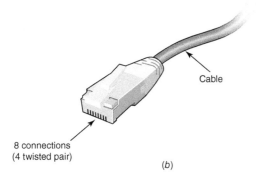

Cable

8 connections
(4 twisted pair)

(b)

Figure 25-14 Modular connectors used with twisted pair cable. (*a*) RJ-11. (*b*) RJ-45.

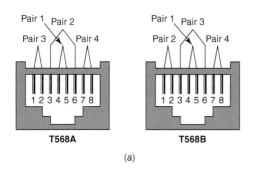

Pair 1 Pair 2
Pair 3   Pair 4
1 2 3 4 5 6 7 8
T568A

Pair 1 Pair 3
Pair 2   Pair 4
1 2 3 4 5 6 7 8
T568B

(a)

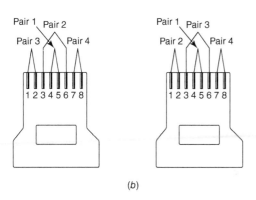

Pair 1 Pair 2
Pair 3   Pair 4
1 2 3 4 5 6 7 8

Pair 1 Pair 3
Pair 2   Pair 4
1 2 3 4 5 6 7 8

(b)

Figure 25-15 Connections on some types of RJ-45 UTP connectors. (*a*) Jacks (female). (*b*) Plugs (male).

25.6 Coaxial Cable

Coaxial Cable Construction

Coaxial, or coax, cable is named for the fact that the two conductors in the cable share a common central axis. The construction of the typical coaxial cable is shown in Fig. 25-16*a*. One conductor is a copper wire, while the other is a copper shield. The central copper wire is usually solid but may also be stranded. It is covered by an insulator, usually solid or foam plastic. The insulator is usually called the dielectric. Polyethylene is the most common insulation. Other insulators such as Teflon are also used. Some large cables are constructed so the dielectric is air or some other gas such as nitrogen. The dielectric is usually covered with a continuous shield braid. The braid is made of many fine copper wires arranged in a crossing pattern, as shown in the illustration. This shield continues for the length of the cable. Larger cables use a solid pipe or conduit for the outer conductor. An outer plastic insulating jacket covers the shield.

A special variation of coaxial cable is the so-called twinaxial cable, shown in Fig. 25-16*b*. It has the same basic construction as standard coaxial but has two inner conductors instead of one.

The primary benefit of coaxial cable over twisted pair is that the shield totally contains the signal. Any magnetic field

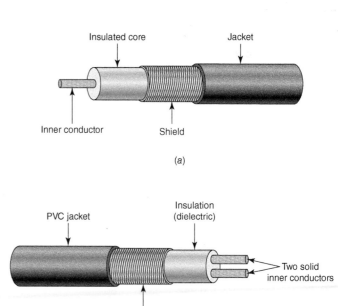

(a)

(b)

Figure 25-16 Physical construction of (*a*) coaxial cable and (*b*) twinaxial cable.

is also fully contained, so no signal is radiated. As a result, the coaxial cable cannot radiate magnetic fields that can be picked up by adjacent cables or other circuits.

Coaxial cable is preferred over twisted pair in those situations where radiation cannot tolerated. It is also the cable of choice in noisy environments, since coaxial cable is less immune to the pickup noise and interfering signals.

On the other hand, coaxial is more expensive. It also has greater attenuation per foot than twisted pair. Yet despite these limitations, it very useful and probably at least as popular as twisted pair. The choice between twisted pair and coaxial depends on the specific application where all the advantages and disadvantages of each type of cable are compared and contrasted and trade-offs weighed.

Types of Coaxial Cable

Coaxial cable comes in a variety of sizes and characteristics. Popular cable diameters are just slightly less than one-quarter inch and one-half inch although many other sizes are available. Coaxial is flexible like most cable but is also available in large sizes where the shield and the conductors themselves are rigid like pipe or tubing. Larger and rigid coaxial cables are used primarily for carrying signals to and from high-frequency antennas. Cell phone sites are a good example. The smaller sizes are widely used wireless applications.

There are literally dozens of types of coaxial cable manufactured for RF and other applications. However, only a few types are widely used in data communications. Listed in Table 25-1 are the most popular types and their major characteristics. The cable designation has the letters *RG* followed by a number and sometimes includes a slash and the letter *U*. This numbering designation was initiated many years ago by the military and is still widely used. Most manufacturers use it although many manufacturers have their own numbering system. A four-digit number is common.

The main point to keep in mind is that there are many variations of the basic types given in Table 25-1. The variations are usually indicated by an *A* or *B* at the end of the designation. The primary difference is the use of a different type of dielectric or conductor size that usually lowers the capacitance and the attenuation per foot and changes the velocity factor.

The cable designation is usually printed on the side of the cable. Always check to see what type it is, and by all means get a manufacturer's catalog or specification sheet so you know exactly what you are dealing with.

As for applications, the RG-8 and RG-58 types are used in radio installations connecting transmitters and receivers to antennas. RG-6A is used in cable TV and consumer electronic equipment.

Coaxial Specifications

The specifications for coaxial cable are essentially the same as those for twisted pair. However, let's discuss them again in the context of coaxial.

Characteristic Impedance Characteristic impedance (Z_o) of coaxial is dependent on the physical dimensions of the cable, which establishes the inductance and capacitance per foot. The characteristic impedance of coaxial is lower than that of twisted pair. Values in the 50- and 75-Ω range are the most common.

Capacitance per Foot Because of its construction, the capacitance per foot of coaxial is much higher than that of twisted pair. This capacitance depends on the dimensions of the cable, but is greatly influenced by the type of dielectric material used as the inner insulator. Values typically run from 20 to 30 pF per foot, although a few types have less capacitance. Lower capacitance per foot is preferred as it produces less loading on the driving circuit and has less attenuation at the higher frequencies.

Velocity Factor Velocity factor (VF) is the ratio of the velocity (speed) of a signal in the cable (V_p) to the speed of light (V_c) in free space. As you saw earlier, it takes time for a signal to travel down a cable as the capacitance of the cable charges. The speed of the signal in a cable is less than that of the same signal in free space. For example, the speed of light in free space of a vacuum is usually rounded off to the value 300,000,000 m per second. The speed of the signal in the cable may be 210,000,000 m/s. This produces a velocity factor of

$$\text{VF} = V_p/V_c = 210,000,000/300,000,000 = 0.70$$

Table 25-1 Popular Types of Coaxial Cables

Designation	Z_o	OD	C/ft	VF (%)	Attenuation (dB)
RG-8	52	0.405	29.5	66	2.5
RG-6A	75	0.332	20.6	66	2.5
RG-58	53.5	0.195	28.5	66	4.5
RG-59	73	0.242	21	66	3.6
RG-62	93	0.242	13.5	86	2.7
Twinaxial	100	0.33	15.5	–	–
CAT 5 UTP	100	–	14	.85	6.7

Note: The terms include the characteristic impedance (Z_o); the outside diameter (OD) in inches; the capacitance per foot (C/ft) in pF; the velocity factor (VF) given as a percentage; and the attenuation, which is usually given in dB per 100 ft at some given frequency, 100 MHz here. The specifications of popular CAT 5 UTP is given in contrast.

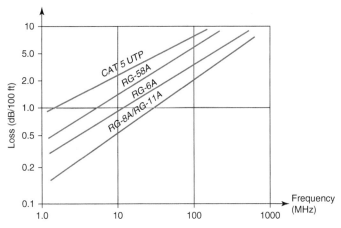

Figure 25-17 Attenuation vs. frequency for popular coaxial cable types.

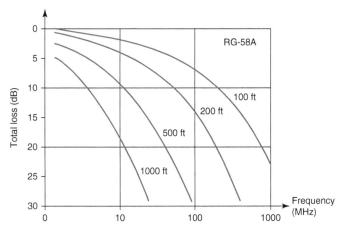

Figure 25-18 Attenuation vs. length of RG-58A coaxial cable.

This is normally expressed as a percentage, or 70%.

You can calculate the speed of propagation in a cable with the above expression if you know the velocity factor of the type of cable. For instance, the speed of propagation in a RG-58 coaxial cable with a VF of 66%, or 0.66, is

$$V_p = 300,000,000 \times 0.66 = 198,000,000 \text{ m/s}$$

Velocity is usually not a specification quoted for twisted pair. However, it does apply. The velocity factor is much higher in twisted pair with values running from 0.8 to 0.95, depending on the type.

Attenuation The attenuation or loss in coaxial cable is high but typically less than that of twisted pair. But it is still useful even over very long distances. Cable runs of many miles are routinely used in cable TV systems. Many large LANs cover distances of hundreds and even thousands of feet with coaxial. However, in all cases where the distances are great, booster amplifiers and repeater units are used to regenerate a signal.

As in any other transmission line, attenuation varies directly with distance logarithmically with frequency. Figure 25-17 shows the attenuation with frequency for several popular types of coaxial cables. A plot for CAT 5 unshielded pair is show for comparison. The horizontal frequency scale is logarithmic. Figure 25-18 shows the attenuation versus frequency for different lengths of RG-58A cable. Notice the low-pass response curve and how attenuation greatly increases with distance and frequency.

Speed Coaxial cable can handle frequencies above 1 GHz, and it is regularly used in RF applications at these frequencies. Although coaxial can be used with data speeds up to about 150 Mbps, it rarely is. Twisted pair is preferred for the newer 1-Gbps and 10-Gbps LANs.

Coaxial Connectors

Like twisted pair, coaxial must be terminated in some type of connector to make it convenient to connect and disconnect cables. The connector also maintains a constant impedance through the connection. There are many different types of coaxial connectors, and the most widely used are discussed below.

A widely used one for video and cable TV is the F-connector, shown in Fig. 25-19a. You will find it in TV sets, cable boxes, and DVD players. It uses the solid center conductor as the connector pin. The cable is usually RG-6A.

One of the oldest and most common coaxial connectors for audio and video is the RCA phono connector shown in Fig. 25-19b. It was originally developed to connect phonographs to audio amplifiers but is still widely used for connecting video equipment. Audio and video cables are all 75-Ω types and use this connector. The connector is also used for some RF applications to frequencies as high as 150 MHz. Common cable types are RG-58 and RG-59.

Another popular connector for the smaller coaxial types, like RG-58 and RG-59 coaxial, is the BNC connector shown in Fig. 25-19c. Its most common use is connecting test equipment like an oscilloscope and its probe or function generator to other equipment. This male BNC connector is usually attached to each end of the cable. It has a metallic outer structure that attaches to the coaxial shield braid and a center pin to which the coaxial center conductor is connected. It mates to a female jack that is mounted in the equipment to be connected. The outer housing of the plug has a camlike slot that links with a pin on the jack for fast easy connect and disconnect.

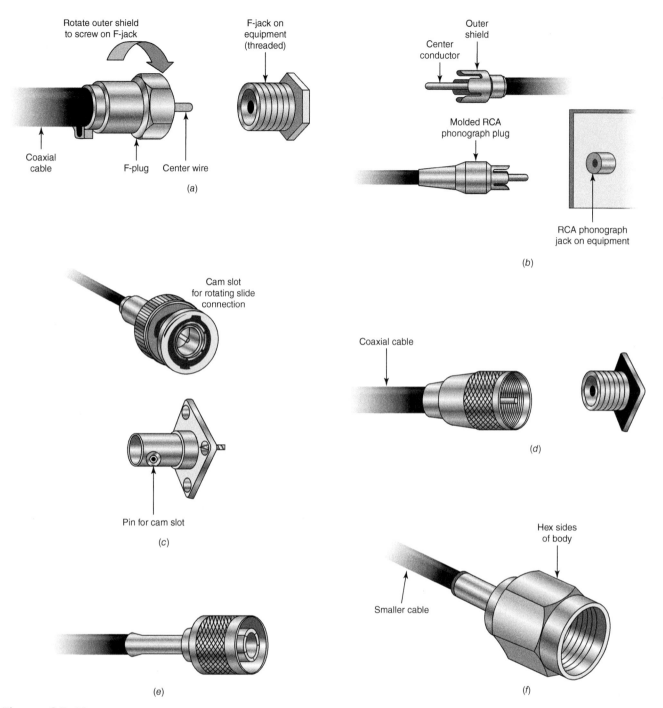

Figure 25-19 Common coaxial connectors. (*a*) F-connector. (*b*) RCA connectors. (*c*) BNC connectors. (*d*) UHF connectors. (*e*) N-type connector. (*f*) SMA connector.

A popular coaxial connector for radio and other RF connections is the UHF connector, shown in Fig. 25-19*d*. The male connector is also known as a PL-259, while the matching socket is known as an SO-239. Both attach with screw threads. They are used with 50-Ω coaxial cable like RG-8 and RG-11 or equivalents.

Another common connector for the larger coaxial cables like RG-8 and RG-11 is the N-type connector shown in

Fig. 25-19*e*. It is considerably larger than the UHF and BNC connectors and uses screw threads for attachment to the female connector.

A smaller connector for high-frequency RF applications is the SMA connector, shown in Fig. 25-19*f*. It is primarily used with smaller 50-Ω cables in test instruments and some radio equipment.

Fiber-Optic Cable and Its Application

While most cable uses good conductors like copper wires, fiber-optic cable uses a good insulator like glass or plastic. The signals to be carried by the cable are first converted into light and then applied to the cable. The light received at the other end of the cable is converted back into an electric signal. Both analog and digital signals can be transmitted this way.

Structure of a Fiber-Optic Cable

Figure S25-1 shows the basic structure of a fiber-optic cable. The light is actually carried by a thin glass or plastic core. Glass has less loss or attenuation, but it is more fragile and expensive. It is also best for very long runs of cable. Plastic is larger and has more attenuation but is less expensive. It is used for shorter runs. Most fiber cables are very thin. Typical size is a core diameter of 62.5 µm (micrometer or micron) where a µm is one-millionth of a meter. The outer cladding, or covering, is 125 µm. Even smaller sizes are used.

The benefits of using fiber include the ability to carry higher-speed digital signals and higher-frequency analog signals over a longer distance than with copper cables. Fiber cables easily carry pulses at a rate up to several hundred billion per second. Such signals are not subject to electric noise or interference caused by inductive or capacitive coupling.

Fiber cables are also more secure for sending sensitive information since they cannot be easily tapped or otherwise compromised.

How Fiber-Optic Cable Works

Most fiber carries binary data in the form of pulses. The stream of binary 1s and 0s are applied to a driver circuit that operates a light-emitting diode (LED) or laser. See Fig. S25-2. The LED or laser produces infrared (IR) light. Infrared is less distorted and attenuated by the cable than other frequencies of light. Typical IR wavelengths are in the 800- to 1600-nm range.

The digital pulses turn the IR light off and on at a rapid rate. Pulses can occur from 10 million pulses per second to over 100 billion pulses per second. Common pulse rates are 2.5, 10, 40, and 100 Gbps.

At the receiving end of the cable, a light-sensitive diode detects the pulses and sends them to an amplifier to regenerate the original pulses. These are then used by a computer.

Fiber Applications

A common fiber application is in cable TV systems. A modern cable system is shown in Fig. S25-3. The main source of the video and other services is the cable office called the head end. It collects the video from TV antennas, satellites, and the Internet and processes the data for transmission over cable. Most cable TV systems also offer high-speed Internet service. The TV and Internet data from the cable company are put on a fiber-optic trunk cable that is then routed by underground paths or on poles to various neighborhood connections points. Distribution amplifiers along the way keep the signal amplitude up to overcome cable attenuation. From there, the signals go by coaxial cable to homes or other facilities. Such a system is called a hybrid fiber coaxial (HFC) system. A few cable TV systems with Internet connectivity use only fiber from the cable office directly to the home.

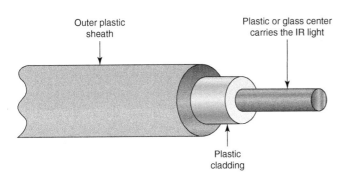

Outer plastic sheath

Plastic or glass center carries the IR light

Plastic cladding

Figure S25-1   Structure of a fiber-optic cable.

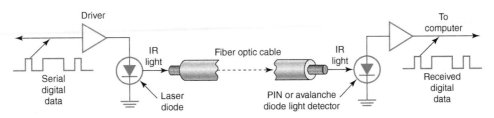

Driver

To computer

IR light

Fiber optic cable

IR light

Serial digital data

Laser diode

PIN or avalanche diode light detector

Received digital data

Figure S25-2   How a fiber-optic cable works.

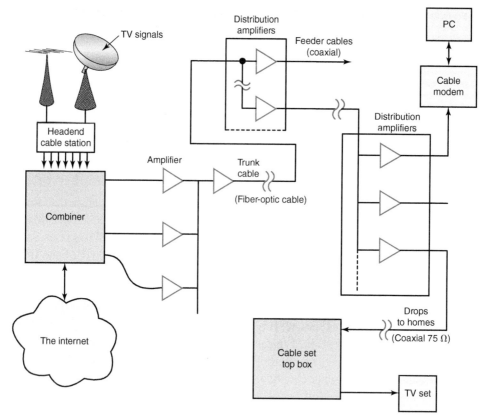

Figure S25-3   Modern fiber-optic cable system.

Fiber-optic cable also forms the core or backbone of the Internet. All Internet data accesses and transfers go by fiber cable, whether local or worldwide.

Another common application is in computer networks where servers, routers, switches, and other networking equipment are used to interconnect computers.

1. All cables are transmission lines.
   a. True.
   b. False.
   c. Depends on the cable length.
   d. Depends on signal frequency.
   e. Both *c* and *d*.

2. Cables at certain frequencies act like a
   a. high-pass filter.
   b. low-pass filter.
   c. bandpass filter.
   d. phase shifter.

3. Cables are more likely to be transmission lines at
   a. high frequencies.
   b. low frequencies.

4. Wavelength is the
   a. distance between amplitude peaks of a sine wave.
   b. the distance traveled by a wave in one cycle.
   c. length of one cycle.
   d. all of the above.

5. Wavelength is shorter at the higher frequencies.
   a. True.
   b. False.

6. The wavelength at a frequency of 1500 kHz is
   a. 2 meters.
   b. 50 meters.
   c. 200 meters.
   d. 2 km.

7. The frequency associated with a wavelength of 70 cm is
   a. 70 MHz.
   b. 428 MHz.
   c. 450 MHz.
   d. 915 MHz.

8. A cable is a transmission line if the cable length is
   a. one wavelength.
   b. ten times one wavelength.
   c. at least one-tenth wavelength.
   d. any length over a wavelength.

9. The most identifying specification of a transmission line is the
   a. attenuation.
   b. characteristic impedance.
   c. capacitance per foot.
   d. velocity factor.

10. What factor determine the characteristic impedance of a cable?
    a. Capacitance per foot.
    b. Inductance per foot.
    c. Both *a* and *c*.
    d. Length.

11. How is cable attenuation specified?
    a. Decibels.
    b. Voltage loss per foot.
    c. Power lost per foot.
    d. Resistance per foot.

12. The speed of a signal in a cable is
    a. the same as the speed of light.
    b. depends on the frequency.
    c. higher than the speed of light.
    d. less than the speed of light.

13. The effect of a mismatched transmission line is
    a. power loss.
    b. high standing waves.
    c. reflections.
    d. all of the above.

14. To avoid reflections in a transmission lines, the load impedance should be
    a. higher than the line impedance.
    b. lower than the line impedance.
    c. same as the line impedance.
    d. Twice the line impedance.

15. The capacitance per foot of a twisted pair line is dependent on
    a. twists per foot.
    b. wire size.
    c. type of insulation.
    d. all of the above.

16. The most widely used type of twisted pair line is
    a. CAT 3.
    b. CAT 5.
    c. CAT 7.
    d. the shielded type.

17. Cross talk is a common problem with twisted pair cable.
    a. True.
    b. False.

18. A common network cable connector used with twisted pair is the
    a. RJ-11.
    b. F-connector.
    c. RJ-45.
    d. N-type connector.

19. Twisted pair cable has higher attenuation per foot than coaxial.
    a. True.
    b. False.

20. Coaxial is preferred over twisted pair because it
    a. has lower attenuation.
    b. better shielding.
    c. higher characteristic impedance.

21. The approximate characteristic impedance of twisted pair line is
    a. 50 Ω.
    b. 75 Ω.
    c. 100 Ω.
    d. 300 Ω.

22. The most common coaxial line characteristic impedance is
    a. 50 Ω.
    b. 75 Ω.
    c. 100 Ω.
    d. 300 Ω.

23. Which coaxial connector is widely used in TV?
    a. N-type connector.
    b. UHF connector.
    c. SMA connector.
    d. F-connector.

24. Standing waves are
    a. patterns of varying electric and magnetic waves on a line.
    b. reflections.
    c. constant voltage and current along the line.
    d. signals that are not moving.

25. The most ideal SWR of a transmission line is
    a. 0.
    b. 1.
    c. 50.
    d. infinite.

SECTION 25.2 Principles of Transmission Lines

25.1 What is the period of a 100-MHz signal?

25.2 A sine wave has a period of 42 ns. What is its frequency?

25.3 What is the wavelength of a signal at 162 MHz?

25.4 A signal has a wavelength of 20 m. What is the frequency?

25.5 How far does a 150-MHz signal travel in one period? What is that period?

25.6 A signal of 433 MHz uses a coaxial cable of one foot. Is it a transmission line?

25.7 At what frequency does a 15-ft cable become a transmission line?

SECTION 25.3 Transmission Line Specifications and Characteristics

25.8 A transmission line has a capacitance per foot of 22 pF and an inductance of 0.6 nH per foot. What is the characteristic impedance?

25.9 What is the time delay produced by the line in Prob. 25-8?

25.10 A 50-Ω transmission line has a capacitance per foot of 32 pF. What is the inductance per foot?

25.11 A transmission line has an input power of 100 W and an output of 70 W. What is the attenuation in dB? If the line is 50 ft long, what is the line attenuation per foot?

25.12 A transmission line has an input of 15 W. The line length is 180 ft with an attenuation of 5 dB per 100 ft. What is the output at the load?

25.13 A transmission line has an attenuation of 2.6 dB per 100 ft. What is the attenuation for 230 ft?

25.14 A cell phone tower is 300 ft high with a transmission line length between transmitter and antenna of 350 ft. The power into the line is 40 W. How much power is radiated by the antenna if the line attenuation is 0.03 dB per foot?

SECTION 25.4 How a Transmission Line Works

25.15 A transmission line has a capacitance per foot of 28.5 pF. The inductance per foot is 87 nH per foot. What is the characteristic impedance of the line? What is the signal delay per foot?

25.16 A 300-m transmission line has a delay of 2.4 pF per foot. What is the total delay?

25.17 A twisted pair line has an impedance of 120 Ω. What is the desired load resistance?

25.18 A coaxial transmission line of 50 Ω has a load of 72 Ω. What is the SWR?

25.19 What SWR would be produced by a 75-Ω coaxial cable with an open load?

25.20 A transmission line has an impedance of 50 Ω and a delay of 1.2 ns per foot. It is driven by a 700-MHz sine wave. What is the amount of phase shift that the signal sees from input to output?

Chapter 26

AC Testing and Troubleshooting

Learning Outcomes

After studying this chapter, you should be able to:

> *Explain* the basic operation of both analog and digital oscilloscopes.

> *Display* an analog signal on an oscilloscope.

> *Measure* signal amplitude, period, frequency, and phase on the oscilloscope.

> *Set* and *adjust* a function generator.

> *Explain* the concept of virtual instrumentation.

> *Use* the scope and function generator to troubleshoot analog circuits.

Troubleshooting and testing circuits that process both analog and digital ac signals involve the same basic procedures outlined in Chapter 13. You may wish to review that chapter briefly before beginning this one. The key ac test instrument is the oscilloscope, which is introduced in this chapter. Basic measurements are outlined. The chapter then covers the function generator, an instrument that generates sine and rectangular waves over a wide frequency range. Then some basic troubleshooting procedures that work with many types of circuits and equipment are outlined.

26.1 Introduction to the Oscilloscope

The cathode-ray oscilloscope or "scope," as it is commonly known, is one of the most versatile test instruments in electronics. Oscilloscopes have the ability to measure the time, frequency, and voltage level of a signal; view rapidly changing waveforms; and determine if an output signal is distorted. The technician must therefore be able to operate this instrument and understand how and where it is used.

Oscilloscopes can be classified as either analog or digital. Both types are shown in Fig. 26-1. Analog oscilloscopes directly apply the voltage being measured to an electron beam moving across the oscilloscope screen. This voltage deflects the beam up, down, and across, thus tracing the waveform on the screen. Digital oscilloscopes sample the input waveform and then use an analog-to-digital converter (ADC) to change the voltage being measured into digital information. The digital information is then used to reconstruct the waveform to be displayed on the screen.

A digital or analog oscilloscope may be used for many of the same applications. Each type of oscilloscope possesses

(a)

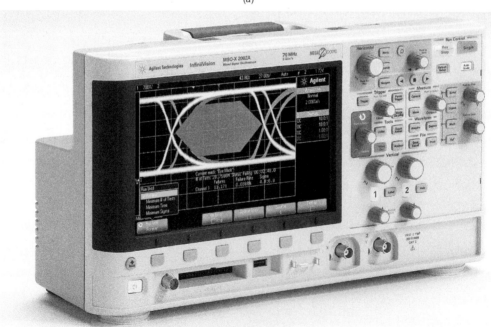

(b)

Figure 26-1 Oscilloscope types. (a) Analog oscilloscope. (b) Digital oscilloscope.

unique characteristics and capabilities. The analog oscilloscope can display high-frequency varying signals in "real time," whereas a digital oscilloscope allows you to capture and store information which can be accessed at a later time or be interfaced to a computer. While analog scopes are still found in some labs, most scopes in use today are digital.

26.2 Analog Oscilloscopes

An analog oscilloscope displays the instantaneous amplitude of an ac voltage waveform versus time on the screen of a cathode-ray tube (CRT). Basically, the oscilloscope is a graph-displaying device. It has the ability to show how signals change over time. As shown in Fig. 26-2, the vertical axis (y) represents voltage and the horizontal axis (x) represents time. The z axis or intensity is sometimes used in special measurement applications. Inside the cathode-ray tube is an electron gun assembly, vertical and horizontal deflection plates, and a phosphorous screen. The electron gun emits a high-velocity, low-inertia beam of electrons that strike the chemical coating on the inside face of the CRT, causing it to emit light. The brightness (called intensity) can be varied by a control located on the oscilloscope front panel. The motion of the beam over the CRT screen is controlled by the deflection voltages generated in the oscilloscope's circuits outside of the CRT and the deflection plates inside the CRT to which the deflection voltages are applied.

Figure 26-3 is an elementary block diagram of an analog oscilloscope. The block diagram is composed of a CRT and four system blocks. These blocks include the display system, vertical system, horizontal system, and trigger system. The CRT provides the screen on which waveforms of electric signals are viewed. These signals are applied to the vertical input system. Depending on how the Volts/Div control is set, the vertical attenuator—a variable voltage divider—reduces the input signal voltage to the desired signal level for the

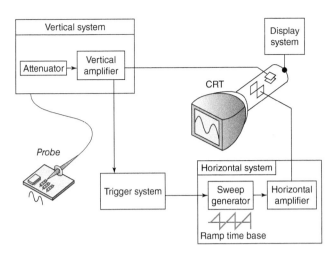

Figure 26-3 Analog oscilloscope block diagram.

vertical amplifier. This is necessary because the oscilloscope must handle a wide range of signal-voltage amplitudes. The vertical amplifier then processes the input signal to produce the required voltage levels for the vertical deflection plates. The signal voltage applied to the vertical deflection plates causes the electron beam of the CRT to be deflected vertically. The resulting up-and-down movement of the beam on the screen, called the trace, is significant in that *the extent of vertical deflection is directly proportional to the amplitude of the signal voltage applied to the vertical, or V, input.* A portion of the input signal, from the vertical amplifier, travels to the trigger system to start or trigger a horizontal sweep. The trigger system determines *when* and *if* the sweep generator will be activated. With the proper LEVEL and SLOPE control adjustment, the sweep will begin at the same trigger point each time. This will produce a stable display as shown in Fig. 26-4. The sweep generator produces a linear time-based deflection voltage. The resulting time-based signal is amplified by the horizontal amplifier and applied to the CRT's horizontal deflection plates. This makes it possible for the oscilloscope to graph a time-varying voltage. The sweep generator may be triggered from sources other than the vertical amplifier. External trigger input signals or internal 60-Hz (line) sources may be selected.

The display system includes the controls and circuits necessary to view the CRT signal with optimum clarity and position. Typical controls include intensity, focus, and trace rotation along with positioning controls.

Dual-Trace Oscilloscopes

Most oscilloscopes have the ability to measure two input signals at the same time. These dual-trace oscilloscopes have two separate vertical amplifiers and an electronic switching circuit. It is then possible to observe two time-related waveforms simultaneously at different points in an electric circuit.

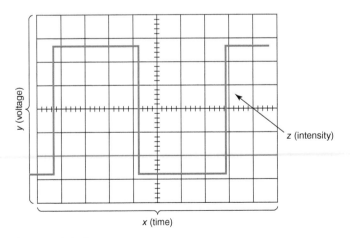

Figure 26-2 The *x, y,* and *z* components of a displayed waveform.

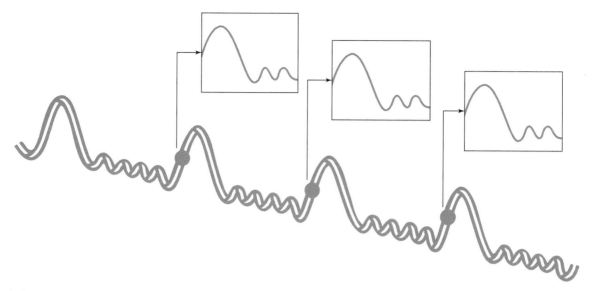

Figure 26-4 Triggering produces a stable display because the same trigger point starts the sweep each time. The SLOPE and LEVEL controls define the trigger points on the trigger signal. The waveform on the screen is all those sweeps overlaid in what appears to be a single picture.

Operating Controls of a Triggered Oscilloscope

The type, location, and function of the front panel controls of an analog oscilloscope differ from manufacturer to manufacturer and from model to model. The descriptions that follow apply to the broadest range of general-use analog scope models.

Intensity This control sets the level of brightness or intensity of the light trace on the CRT. Rotation in a clockwise (CW) direction increases the brightness. Too high an intensity can damage the phosphorous coating on the inside of the CRT screen.

Focus This control is adjusted in conjunction with the intensity control to give the sharpest trace on the screen. There is interaction between these two controls, so adjustment of one may require readjustment of the other.

Astigmatism This is another beam-focusing control found on older oscilloscopes that operates in conjunction with the focus control for the sharpest trace. The astigmatism control is sometimes a screwdriver adjustment rather than a manual control.

Horizontal and Vertical Positioning or Centering These are trace-positioning controls. They are adjusted so that the trace is positioned or centered both vertically and horizontally on the screen. In front of the CRT screen is a faceplate called the *graticule,* on which is etched a grid of horizontal and vertical lines. Calibration markings are sometimes placed on the center vertical and horizontal lines on this faceplate. This is shown in Fig. 26-5.

Volts/Div This control attenuates the vertical input signal waveform that is to be viewed on the screen. This is

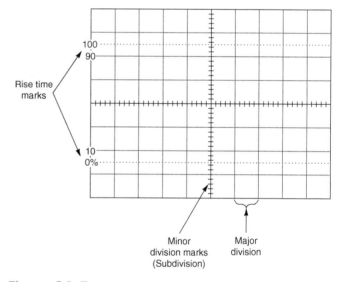

Figure 26-5 An oscilloscope graticule.

frequently a click-stop control that provides step adjustment of vertical sensitivity. A separate Volts/Div control is available for each channel of a dual-trace scope. Some scopes mark this control Volts/cm.

Variable In some scopes this is a concentric control in the center of the Volts/Div control. In other scopes this is a separately located control. In either case, the functions are similar. The variable control works with the Volts/Div control to provide a more sensitive control of the vertical height of the waveform on the screen. The variable control also has a calibrated position (CAL) either at the extreme counterclockwise or clockwise position. In the CAL position the Volts/Div control is calibrated at some set value—for example, 5 mV/div, 10 mV/div, or 2 V/div. This allows the scope to be used for peak-to-peak voltage measurements of the vertical input signal. Dual-trace scopes have a separate variable control for each channel.

Input Coupling AC-GND-DC Switches This three-position switch selects the method of coupling the input signal into the vertical system.

> *AC*—The input signal is capacitively coupled to the vertical amplifier. The dc component of the input signal is blocked.
>
> *GND*—The vertical amplifier's input is grounded to provide a zero volt (ground) reference point. It does not ground the input signal.
>
> *DC*—This direct-coupled input position allows all signals (ac, dc, or ac-dc combinations) to be applied directly to the vertical system's input.

Vertical Mode Switches These switches select the mode of operation for the vertical amplifier system.

> *CH1*—Selects only the Channel 1 input signal for display.
>
> *CH2*—Selects only the Channel 2 input signal for display.
>
> *Both*—Selects both Channel 1 and Channel 2 input signals for display. When in this position, ALT, CHOP, or ADD operations are enabled.
>
> *ALT*—Alternatively displays Channel 1 and Channel 2 input signals. Each input is completely traced before the next input is traced. Effectively used at sweep speeds of 0.2 ms per division or faster.
>
> *CHOP*—During the sweep the display switches between Channel 1 and Channel 2 input signals. The switching rate is approximately at 500 kHz. This is useful for viewing two waveforms at slow sweep speeds of 0.5 ms per division or slower.
>
> *ADD*—This mode algebraically sums the Channel 1 and Channel 2 input signals.
>
> *INVERT*—This switch inverts Channel 2 (or Channel 1 on some scopes) to enable a differential measurement when in the ADD mode.

Time/Div This is usually two concentric controls that affect the timing of the horizontal sweep or time-base generator. The outer control is a click-stop switch that provides step selection of the sweep rate. The center control provides a more sensitive adjustment of the sweep rate on a continuous basis. In its extreme clockwise position, usually marked CAL, the sweep rate is calibrated. Each step of the outer control is therefore equal to an exact time unit per scale division. Thus,

the time it takes the trace to move horizontally across one division of the screen graticule is known. Dual-trace scopes generally have one Time/Div control. Some scopes mark this control Time/cm.

X-Y Switch When this switch is engaged, one channel of the dual-trace scope becomes the horizontal, or *x*, input, while the other channel becomes the vertical, or *y*, input. In this condition the trigger source is disabled. On some scopes, this setting occurs when the Time/Div control is fully counterclockwise.

Triggering Controls The typical dual-trace scope has a number of controls associated with the selection of the triggering source, the method by which it is coupled, the level at which the sweep is triggered, and the selection of the slope at which triggering takes place:

1. *Level Control.* This is a rotary control which determines the point on the triggering waveform where the sweep is triggered. When no triggering signal is present, no trace will appear on the screen. Associated with the level control is an Auto switch, which is often an integral part of the level rotary control or may be a separate push button. In the Auto position the rotary control is disengaged and automatic triggering takes place. In this case a sweep is always generated, and therefore a trace will appear on the screen even in the absence of a triggering signal. When a triggering signal is present, the normal triggering process takes over.

2. *Coupling.* This control is used to select the manner in which the triggering is coupled to the signal. The types of coupling and the way they are labeled vary from one manufacturer and model to another. For example, ac coupling usually indicates the use of capacitive coupling that blocks dc; line coupling indicates the 50- or 60-Hz line voltage is the trigger. If the oscilloscope was designed for television testing, the coupling control might be marked for triggering by the horizontal or vertical sync pulses.

3. *Source.* The trigger signal may be external or internal. As already noted, the line voltage may also be used as the triggering signal.

4. *Slope.* This control determines whether triggering of the sweep occurs at the positive-going or negative-going portion of the triggering signal. The switch itself is usually labeled positive or negative, or simply + or −.

26.3 Oscilloscope Probes

Oscilloscope probes are the test leads used for connecting the vertical input signal to the oscilloscope. There are three types: a direct lead that is just a shielded cable, the low-capacitance

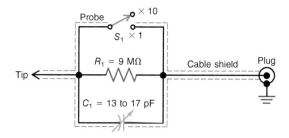

Figure 26-6 Circuit for low-capacitance probe (LCP) for an oscilloscope.

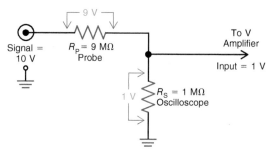

Figure 26-7 Voltage division of 1:10 with a low-capacitance probe.

probe (LCP) with a series-isolating resistor, and a demodulator probe. Figure 26-6 shows a circuit for an LCP for an oscilloscope. The LCP usually has a switch to short out the isolating resistor so that the same probe can be used either as a direct lead or with low capacitance. (See S_1 in Fig. 26-6.)

Direct Probe

The direct probe is just a shielded wire without any isolating resistor. A shielded cable is necessary to prevent any pickup of interfering signals, especially with the high resistance at the vertical input terminals of the oscilloscope. The higher the resistance, the more voltage that can be developed by induction. Any interfering signals in the test lead produce distortion of the trace pattern. The main sources of interference are 60-Hz magnetic fields from the power line and stray RF signals.

The direct probe as a shielded lead has relatively high capacitance. A typical value is 90 pF for 3 ft (0.9 m) of 50-Ω coaxial cable. Also, the vertical input terminals of the oscilloscope have a shunt capacitance of about 40 pF. The total C then is $90 + 40 = 130$ pF. This much capacitance can have a big effect on the circuit being tested. For example, it could detune a resonant circuit. Also, nonsinusoidal waveshapes are distorted. Therefore, the direct probe can be used only when the added C has little or no effect. These applications include voltages for the 60-Hz power line or sine-wave audio signals in a circuit with a relatively low resistance of several kilohms or less. The advantage of the direct probe is that it does not divide down the amount of input signal, since there is no series-isolating resistance.

Low-Capacitance Probe (LCP)

Refer to the diagram in Fig. 26-6. The 9-MΩ resistor in the probe isolates the capacitance of the cable and the oscilloscope from the circuit connected to the probe tip. With an LCP, the input capacitance of the probe is only about 10 pF. The LCP must be used for oscilloscope measurements when

1. The signal frequency is above audio frequencies.
2. The circuit being tested has R higher than about 50 kΩ.

3. The waveshape is nonsinusoidal, especially with square waves and sharp pulses.

Without the LCP, the observed waveform can be distorted. The reason is that too much capacitance changes the circuit while it is being tested.

The 1:10 Voltage Division of the LCP

Refer to the voltage divider circuit in Fig. 26-7. The 9-MΩ of R_P is a series resistor in the probe. Also, R_S of 1 MΩ is a typical value for the shunt resistance at the vertical terminals of the oscilloscope. Then $R_T = 9 + 1 = 10$ MΩ. The voltage across R_S for the scope equals R_S/R_T or 1/10 of the input voltage. For the example in Fig. 26-7 with 10 V at the tip of the LCP, 1 V is applied to the oscilloscope.

Remember, when using the LCP, multiply by 10 for the actual signal amplitude. As an example, for a trace pattern on the screen that measures 2.4 V, the actual signal input at the probe is 24 V. For this reason, the LCP is generally called the "× 10" probe. Check to see whether or not the switch on the probe is in the direct or LCP position. Even though the scope trace is reduced by the factor of 1/10, it is preferable to use the LCP for almost all oscilloscope measurements to minimize distortion of the waveshapes.

Trimmer Capacitor of the LCP

Referring back to Fig. 26-6, note that the LCP has an internal variable capacitor C_1 across the isolating resistor R_1. The purpose of C_1 is to compensate the LCP for high frequencies. Its time constant with R_1 should equal the RC time constant of the circuit at the vertical input terminals of the oscilloscope. When necessary, C_1 is adjusted for minimum tilt on a square-wave signal.

26.4 Digital Oscilloscopes

Digital oscilloscopes have replaced analog oscilloscopes in most electronic industries and educational facilities. In addition to being able to make the traditional voltage, time, and phase measurements, digital scopes can also store a measured waveform for later viewing. Digital scopes are also much smaller and weigh less than their analog counterparts.

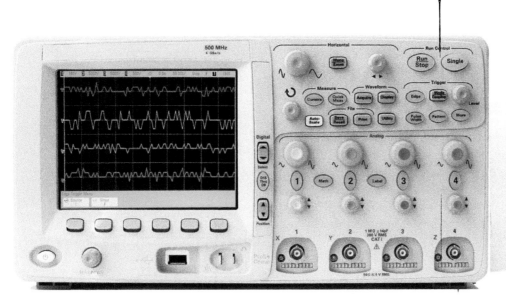

Figure 26-8 Four-channel digital oscilloscope.

These two advantages alone have prompted many schools and industries to make the switch from analog to digital scopes.

Like any piece of test equipment there is a learning curve involved before you will be totally comfortable operating a digital oscilloscope. The biggest challenge facing you will be familiarizing yourself with the vast number of menus and submenus in which a digital scope uses to access its features and functions. But it's not too bad once you sit down and start with some simple and straightforward measurements. It's always best if you can obtain the operating manual and educational materials for the digital scope you are learning to use. Keep these materials nearby so you can refer to them when you need help in making a measurement. This is not an uncommon practice, even for very experienced users of digital oscilloscopes.

Figure 26-8 shows a modern TDS-224 (four-channel) digital oscilloscope. You will often see digital scopes referred to as digital storage oscilloscopes (DSO) because they store the digitized analog input signal in memory. Another common term is mixed signal oscilloscope (MSO). An MSO can display analog and digital signals simultaneously. The digital signals are applied to separate inputs. A typical MSO can display 8 or 16 digital signals along with two or four channels of analog signals. Figure 26-9 shows an MSO with an analog signal being displayed along with eight channels of digital signals.

How a Digital Oscilloscope Works

Figure 26-10 shows a simplified block diagram of a DSO. The input signal to be displayed is applied to an attenuator that selects the attenuation level based on the amplitude of the signal. This attenuator is controlled by the

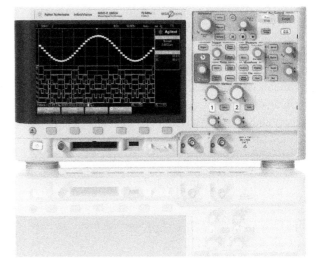

Figure 26-9 A mixed signal oscilloscope (MSO) that can display analog signals and multiple binary signals simultaneously.

(Courtesy Agilent Technologies.)

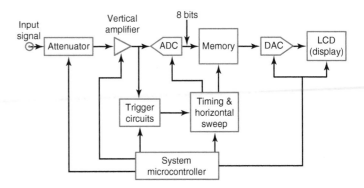

Figure 26-10 Simplified block diagram of a DSO.

front panel controls. The correctly sized signal is then applied to the vertical amplifier that boosts the signal level to a range making it compatible with the ADC. The ADC digitizes the signal into a stream of 8-bit binary numbers proportional to multiple samples of the input signal. The digital samples are stored in a memory. The stored signals are then sent to a digital-to-analog converter (DAC) that recovers the original input signal for the liquid-crystal display (LCD). The timing circuitry controls the digitizing process, as well as the storage and retrieval rate of the data from memory, and provides the horizontal sweep timing for the display. The trigger circuits ensure a stable display.

DSO Controls and Operation

The controls on a DSO vary from model to model, but the functions are mostly similar. Here is a brief description of the controls of the DSO shown in Fig. 26-8. Just remember to have for reference the instruction/operation manual for the scope you are using.

Vertical Controls (See Fig. 26-11)

Ch. 1, 2, 3, 4 and Cursor 1 and 2 Position Positions the waveform vertically. When cursors are turned on and the cursor menu is displayed, these knobs position the cursors. (*Note:* Cursors are horizontal or vertical lines that can be moved up and down or left and right to make either voltage or time measurements.)

Ch. 1, Ch. 2, Ch. 3, and Ch. 4 Menu Displays the channel input menu selections, and toggles the channel display on and off.

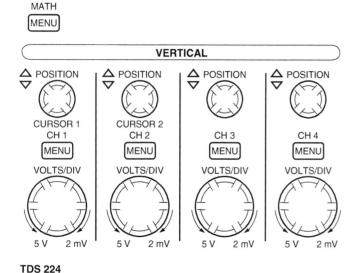

TDS 224

Figure 26-11

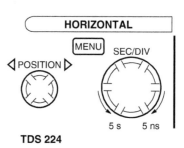

TDS 224

Figure 26-12

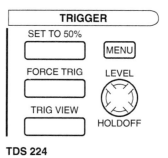

TDS 224

Figure 26-13

Volts/Div (Ch. 1, Ch. 2, Ch. 3, and Ch. 4) Selects calibrated scale factors also referred to as Volts/Div. settings.

Math Menu Displays waveform math operations menu, and can also be used to toggle the math waveform on and off.

Horizontal Controls (See Fig. 26-12)

Position Adjusts the horizontal position of all channels and math waveforms. The resolution of this control varies with the time base.

Horizontal Menu Displays the horizontal menu.

Sec/Div Selects the horizontal time/div setting (scale factor).

Trigger Controls (See Fig. 26-13)

Level and Holdoff This control has a dual purpose. As an edge trigger level control, it sets the amplitude level the signal must cross to cause an acquisition. As a holdoff control, it sets the amount of time before another trigger event can be accepted. (*Note:* The term *acquisition* refers to the process of sampling signals from input channels, digitizing the samples, processing the results into data points, and assembling the data points into a waveform record. The waveform record is stored in memory.)

Trigger Menu Displays the trigger menu.

Set Level to 50% The trigger level is set to the vertical midpoint between the peaks of the trigger signal.

Force Trigger Starts an acquisition regardless of an adequate trigger signal. This button has no effect if the acquisition is already stopped.

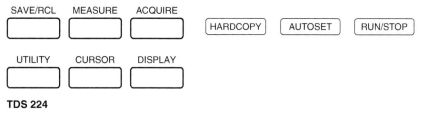

TDS 224

Figure 26-14

Trigger View Displays the trigger waveform in place of the channel waveform while the TRIGGER VIEW button is held down. You can use this to see how the trigger settings affect the trigger signal, such as trigger coupling.

Menu and Control Buttons (See Fig. 26-14)

Save/Recall Displays the save/recall menu for setups and waveforms.

Measure Displays the automated measurements menu.

Acquire Displays the acquisition menu.

Display Displays the display menu.

Cursor Displays the cursor menu. Vertical position controls adjust cursor position while displaying the cursor menu and the cursors are turned on. Cursors remain displayed (unless turned off) after leaving the cursor menu but are not adjustable.

Utility Displays the utility menus.

Autoset Automatically sets the scopes controls to produce a usable display of the input signal.

Hardcopy Starts print operations.

Run/Stop Starts and stops waveform acquisition.

26.5 Voltage and Time Measurements

In general, an oscilloscope is normally used to make two basic measurements: amplitude and time. After making these two measurements, other values can be determined. Figure 26-15 shows the screen of a typical oscilloscope.

As mentioned earlier, the vertical or *y* axis represents values of voltage amplitude whereas the horizontal or *x* axis represents values of time. The Volts/Div control on the oscilloscope determines the amount of voltage needed at the scope input to deflect the electron beam one division vertically on the *y* axis. The Seconds/Div control on the oscilloscope determines the time it takes for the scanning electron beam to scan one horizontal division. In Fig. 26-15, note that there are eight vertical divisions and ten horizontal divisions.

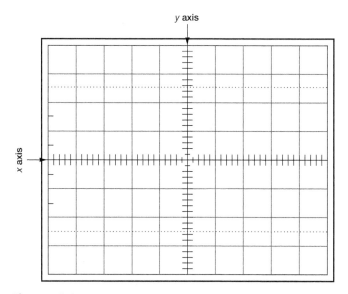

Figure 26-15 Oscilloscope screen (graticule).

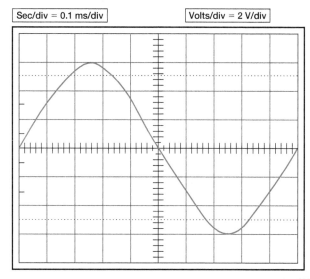

Figure 26-16 Determining V_{P-P}, *I*, and *t* from the sine wave displayed on the scope graticule.

Refer to the sine wave being displayed on the oscilloscope graticule in Fig. 26-16. To calculate the peak-to-peak value of the waveform, simply count the number of vertical divisions occupied by the waveform and then multiply this number by the Volts/Div setting. Expressed as a formula,

$$V_{\text{P-P}} = \text{\# vertical divisions} \times \frac{\text{volts}}{\text{division}} \text{ setting} \quad \textbf{(26-1)}$$

In Fig. 26-16, the sine wave occupies 6 vertical divisions. Since the volts/div setting equals 2 V/division, the peak-to-peak calculations are as follows.

$$V_{\text{P-P}} = 6 \text{ vertical divisions} \times \frac{2 \text{ V}}{\text{division}} = 12 \text{ V}_{\text{P-P}}$$

To calculate the period, T, of the waveform, all you do is count the number of horizontal divisions occupied by one cycle. Then, simply multiply the number of horizontal divisions by the sec/div setting. Expressed as a formula,

$$T = \text{\# horizontal divisions} \times \frac{\text{sec}}{\text{division}} \text{ setting} \quad \textbf{(26-2)}$$

In Fig. 26-16, one cycle of the sine wave occupies exactly 10 horizontal divisions. Since the sec/div setting is set to 0.1 ms/div, the calculations for T are as follows:

$$T = 10 \text{ horizontal divisions} \times \frac{0.1 \text{ ms}}{\text{div}} = 1 \text{ ms}$$

With the period, T, known, the frequency, f, can be found as follows:

$$f = \frac{1}{T}$$
$$= \frac{1}{1 \text{ ms}}$$
$$= 1 \text{ kHz}$$

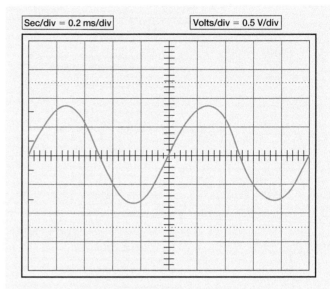

Sec/div = 0.2 ms/div   Volts/div = 0.5 V/div

Figure 26-17 Determining $V_{\text{P-P}}$, T, and f from the sine wave displayed on the scope graticule.

EXAMPLE 26-1

In Fig. 26-17, determine the peak-to-peak voltage, the period, T, and the frequency, f, of the displayed waveform.

Answer:
Careful study of the scope's graticule reveals that the height of the waveform occupies 3.4 vertical divisions. With the volts/div setting at 0.5 V/div, the peak-to-peak voltage is calculated as follows:

$$V_{\text{P-P}} = 3.4 \text{ vertical divisions} \times \frac{0.5 \text{ V}}{\text{div}} = 1.7 \text{ V}_{\text{P-P}}$$

To find the period, T, of the displayed waveform, count the number of horizontal divisions occupied by just one cycle. By viewing the scope's graticule, we see that one cycle occupies 5 horizontal divisions. Since the Sec/Div control is set to 0.2 ms/div, the period, T, is calculated as

$$T = 5 \text{ horizontal divisions} \times \frac{0.2 \text{ ms}}{\text{div}} = 1 \text{ ms}$$

To calculate the frequency, f take the reciprocal of the period, T.

$$f = \frac{1}{T} = \frac{1}{1 \text{ ms}} = 1 \text{ kHz}$$

EXAMPLE 26-2

In Fig. 26-18, determine the pulse time, t_p, pulse repetition time, prt, and the peak value, V_{pk}, of the displayed waveform. Also, calculate the waveform's % duty cycle and the pulse repetition frequency, prf.

Answer:
To find the pulse time, t_p, count the number of horizontal divisions occupied by just the pulse. In Fig. 26-18, the pulse occupies exactly 4 horizontal divisions. With the Sec/Div control set to 1 μs/div, the pulse time, t_p, is calculated as

$$t_p = 4 \text{ horizontal divisions} \times \frac{1 \text{ }\mu\text{s}}{\text{div}} = 4 \text{ }\mu\text{s}$$

The pulse repetition time, prt, is found by counting the number of horizontal divisions occupied by one cycle of the waveform. Since one cycle occupies 10 horizontal divisions, the pulse repetition time, prt, is calculated as follows:

$$\text{prt} = 10 \text{ horizontal divisions} \times \frac{1 \text{ }\mu\text{s}}{\text{Div}} = 10 \text{ }\mu\text{s}$$

With t_p and prt known, the % duty cycle is calculated as follows:

$$\% \text{ duty cycle} = \frac{t_p}{\text{prt}} \times 100$$
$$= \frac{4 \text{ }\mu\text{s}}{10 \text{ }\mu\text{s}} \times 100$$
$$= 40\%$$

The pulse repetition frequency, prf, is calculated by taking the reciprocal of prt.

$$\text{prf} = \frac{1}{\text{prt}}$$
$$= \frac{1}{10 \text{ }\mu\text{s}}$$
$$= 100 \text{ kHz}$$

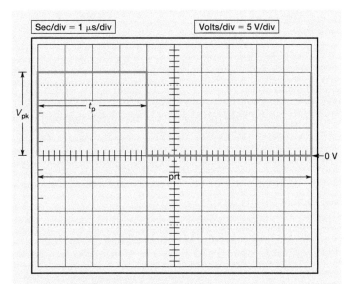

Note: t_p = the length of time the pulse exists.
prt = pulse repetition time (period)
prf = pulse repetition frequency
% duty cycle = the percentage of prt
for which the pulse exists.

$$\% \text{ duty cycle} = \frac{t_p}{prt} \times 100$$

Figure 26-18 Determining V_{pk}, t_p, prt, prf, and % duty cycle from the rectangular wave displayed on the scope graticule.

The peak value of the waveform is based on the fact that the baseline value of the waveform is 0 V as shown. The positive peak of the waveform is shown to be three vertical divisions above zero. Since the volts/div setting of the scope is 5 V/div, the peak value of the waveform is

$$V_{pk} = 3 \text{ vertical divisions} \times \frac{5 \text{ V}}{\text{div}} = 15 \text{ V}$$

Notice that the waveform shown in Fig. 26-18 is entirely positive because the waveform's pulse makes a positive excursion from the zero-volt reference.

Phase Measurement

Phase measurements can be made with a dual-trace oscilloscope when the signals are of the same frequency. To make this measurement, the following procedure can be used:

1. Preset the scope's controls and obtain a baseline trace (the same for both channels). Set the Trigger Source to whichever input is chosen to be the reference input. Channel 1 is often used as the reference, but Channel 2 as well as External Trigger or Line could be used.

2. Set both Vertical Input Coupling switches to the same position, depending on the type of input.

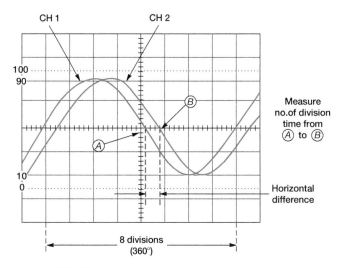

Figure 26-19 Oscilloscope phase-shift measurement.

3. Set the Vertical MODE to Both; then select either ALT or CHOP, depending on the input frequency.

4. Although not necessary, set both Volts/Div and both Variable controls so that both traces are approximately the same height.

5. Adjust the TRIGGER LEVEL to obtain a stable display. Typically set so that the beginning of the reference trace begins at approximately zero volts.

6. Set the Time/Div switch to display about one full cycle of the reference waveform.

7. Use the Position controls, Time/Div switch, and Variable time control so that the reference signal occupies exactly 8 horizontal divisions. The entire cycle of this waveform represents 360°, and each division of the graticule now represents 45° of the cycle.

8. Measure the horizontal difference between corresponding points of each waveform on the horizontal graticule line as shown in Fig. 26-19.

9. Calculate the phase shift by using the formula:

Phase Shift = (no. of horizontal difference divisions) × (no. of degrees per division)

As an example, Fig. 26-19 displays a difference of 0.6 division at 45° per division. The phase shift = (0.6 div) × (45°/div) = 27°.

Oscilloscope Specifications

While an oscilloscope is a versatile instrument capable of being used for many electronic measurements, you must be sure the scope specifications match the type of measurements you need to make. Specifically, you must take into consideration the frequency and time characteristics of the signals being measured to be sure the scope can display and measure them accurately. Following are main specifications to be considered.

Bandwidth Bandwidth refers to the frequency range of the signals to be displayed. For example, the digital scope in Fig. 26-1b has a basic bandwidth of 70 MHz. That means it can display signals with frequencies up to 70 MHz. This number is the upper cutoff (3 dB down) frequency of the vertical amplifier. Scopes with greater bandwidth are also available. For the scope in Fig. 26-1b, models with bandwidths of 100 and 200 MHz are available. Other scope models can be obtained with bandwidths of 1 GHz and above.

The bandwidth also defines the limit on how fast a pulse waveform displays. The bandwidth essentially sets the fastest rise and fall times of pulses to be displayed. The minimum rise time (t_f) that can be displayed is computed with the expression

$$t_f = \frac{0.35}{\text{BW}} \qquad (26\text{-}3)$$

where t_f is the shortest rise time possible in microseconds and BW is the scope bandwidth in MHz.

The 70-MHz bandwidth will let you display rise times as short as

$$t_f = 0.35/70 = 0.005 \ \mu s \text{ or } 5 \text{ ns}$$

Remember that to display digital pulses, the scope amplifier must be able to pass the harmonics and the fundamental frequency of the signal. The greater the number of harmonics passed, the greater the fidelity of the displayed signal. As a general rule of thumb, the scope must pass up to at least the fifth harmonic for a reasonable representation of the signal. For instance, if the basic pulse frequency is 26 MHz, then the scope needs to display up to the fifth harmonic, or $5 \times 26 = 130$ MHz. Consequently, the scope bandwidth should be greater than 130 MHz. The 70-MHz bandwidth scope can faithfully display pulses up to a rate of $70/5 = 14$ MHz.

Sample Rate The speed with which the ADC samples or digitizes the input signal is the sample rate. An ADC takes a measurement at a specific time and rate and converts that measurement into a binary number. The result is a stepped approximation of the analog input. The greater the number of samples taken per second, the more faithful the digital reproduction. As a general rule, a signal should be sampled at least 10 times or more than the frequency of the signal. An 8-MHz signal then needs to be sampled at least 10×8 MHz, or 80 samples per second (S/s). The Agilent scope shown in Fig. 26-1b has a minimum sample rate of 1 GS/s.

Memory Size of Depth The memory size of depth is the actual number of samples the memory can hold. Once the memory is full, a new samples replace the previous ones. This process is continuous. The more samples stored, the better the display, and the refresh rate on the screen can be increased. The Agilent scope in Fig. 26-1b has a basic memory size of 100 kilobytes (kB) or 100 kilopoints (kpts). (A byte is 8 bits.) Some DSOs can have their memory size increased to many megasamples or megapoints (Mpts). The screen refresh rate is 50 kpts per second.

Input Voltage Range The measurement resolution states the range of signal amplitudes that can be displayed. An example is a range from 1 mV per division to 5 V per division. Smaller signals (<1 mV) need to be preamplified before display, and large signals (>50 V) need to be attenuated before display.

Options When selecting a scope, take a look at the various options. For example, the Agilent scope of Fig. 26-1b has the option of a 20-MHz function generator and/or digital voltmeter (DVM) built in. Other options are greater sample memory size and software options for specific measurements. The mixed signal option is to add the digital signal measurement capability to the scope.

26.6 Function Generators

A function generator is a signal generator. The term *function* refers to the mathematical expression for the signal it is generating, such as a sine wave. Most function generators generate sine waves, rectangular/square waves, and triangular waves. The function generator is usable in a wide range of testing applications and can cover a wide frequency range continuously from a fraction of a Hz to 10 MHz or even higher. The signal amplitude is also variable from a few millivolts to many volts.

Most of the newer generators are usually referred to as *arbitrary waveform generators (AWG)*. Because of the unique way that function generators produce their signals digitally, it is possible to program the generator yourself to produce waves of almost any shape, either analog or digital or pulse. Most modern generators also provide amplitude modulation (AM) and frequency modulation (FM) of a sine-wave carrier for testing radio and wireless circuits.

Figure 26-20 shows the basic idea behind the operation of a function generator. A digital version of a sine wave is stored in memory. It consists of a sequence of binary numbers that are precise values of sine-wave amplitude samples spaced at close intervals. These digital values are fed to a DAC that converts them into the actual analog values of the binary numbers. A high-speed clock circuit steps the memory and the DAC. The rate of feeding the samples to

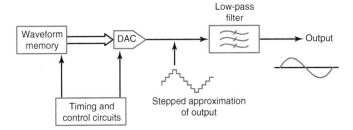

Figure 26-20 The concept of a modern function generator.

the DAC determines the actual frequency of the output sine wave. The DAC output is a stepped approximation of the desired sine wave. A low-pass filter is then used to smooth out the steps into a continuous sine wave.

The function generator has stored digital patterns for sine, square, and triangular waves as well as AM and FM signals. However, an arbitrary wave generator has extra room in memory that you can use to store any special waves you want to create for test purposes. Some generators have auxiliary software that lets you program a desired custom waveform.

Function generators come in a wide range of sizes and shapes. Most have the appearance of the unit shown in Fig. 26-21. A large knob on the right adjusts frequency. A keyboard lets you enter a desired frequency directly. Push buttons select the type of waveform and adjust output signal amplitude. This generator also has a small LCD display that shows the shape, frequency, and amplitude of the signal selected for the output. Lower-cost generators do not have this screen, so you will need to observe the generator output on an oscilloscope. The generator output is on the lower right. It uses a BNC connector and has a 50-Ω output impedance. A coaxial cable connects the output signal to the circuit or equipment being tested.

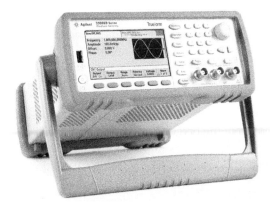

Figure 26-21 A modern arbitrary waveform generator. (Courtesy Agilent Technologies.)

Most function generators are easy to use once you become familiar with them. This takes some practice and experience. Because of the ease of use, it is best to go ahead and start using the generator. You should also have the instruction manual available for reference. The procedure is to connect the generator output to an oscilloscope. Then select a waveform type, set the frequency and output amplitude, and make the measurements on an oscilloscope. Repeat this with a square wave and any other signal of interest, just to get the feel of the instrument.

Some special features to try out are the swept frequency output and the AM and FM modulations to see what they look like. The sweep frequency essentially linearly increases the output frequency continuously over time. This is useful for testing filters and plotting their output response. The AM and FM functions may have provision for different modulating signals, such as another sine wave. Using binary pulses for modulation will allow you to create amplitude shift keying (ASK) and frequency shift keying (FSK), both of which are widely used in wireless equipment.

26.7 Virtual Instruments

Virtual instruments (VIs) are digital instruments implemented on a personal computer, either a desktop or laptop. Digital instruments are capable of measuring and generating both analog and digital signals. The most common virtual instruments are the digital multimeter, oscilloscope, and function generator.

Virtual instruments work like the previously described digital scope and function generator. For the DMM, the voltage to be measured is connected to an ADC to produce a digital version. The binary representation of the input is stored in memory and then analyzed by software. The software is stored on the host PC. The software also implements an output screen on the PC video monitor that will show the numerical value of the input signal. With virtual instruments, the computer screen is used for simulating the front panel of a real instrument.

A virtual oscilloscope works the same way. The signal to be measured is digitized by the ADC, and a binary version is stored in memory. The values are measured and analyzed by the software. Then the binary values representing the signal are sent to a DAC to re-create the input analog signal. It is then sent to the PC where the waveform is displayed on a graticule as it would be on an oscilloscope.

A function generator works like the one described earlier. Sine waves and other signals are stored in the PC software and then fed to DAC that reproduces the desired output. The whole idea of virtual instruments is to let the PC and its special software create the instrument front panel, displays, and controls. VIs also use the power of the PC and the software to analyze and measure the signal captured by the ADC. You operate the VI with mouse and keyboard and see the results on the PC screen.

Virtual instruments have not replaced standard benchtop instruments, but they have become popular because they are smaller and less expensive than traditional instruments. Many laboratories already have computers, so it is a low-cost option to add a VI, creating a complete working laboratory for electronic testing and experimentation. The cost of some units has dropped so low that it allows students, professors, engineers, or technicians to afford a testing lab if they already own a PC, as most do. Two good examples of low-cost VIs are National Instruments (NI) myDAQ and the Digilent Discovery.

Figure 26-22 shows the myDAQ made NI. It contains all the circuitry for implementing a digital multimeter (DMM), an oscilloscope, a function generator, and multiple dc power supplies. The whole unit is housed in a 3.5- by 5.8-in. package. It attaches to the PC or laptop with a standard USB 2.0 cable. The software implements the instruments. As you can see in the figure, the DMM inputs are on the front of the unit and use standard banana jacks and test leads. Other connections to the power supplies, scope, function generator, and multiple digital inputs and outputs are made with screw terminals on the right side of the unit. The PC uses NI's popular LabVIEW software for taking measurements and displaying the controls on screen. The unit is designed to be used with a standard breadboarding socket where you can build circuits for testing and experimenting.

Figure 26-23 The Digilent Discovery is a complete set of virtual instruments that works with a host computer. (Courtesy Digilent Inc.)

The specifications for the myDAQ are as follows:

DMM: Standard dc and ac voltage measurements up to 60 Vdc and 20-V rms maximum, dc and ac current to 1 A, and resistance. A diode test mode is provided.

Power Supplies: +5 Vdc, 100 mA max, ±15 V, 32 mA max.

Analog Inputs (oscilloscope): 400-kHz bandwidth, ±10 V max, sample rate 200 kHz max.

Analog Outputs (function generator): 200-kHz sample rate maximum, ±10 V max.

Digital I/O: 8 lines of either input or output, +5-V or 3.3-V logic levels.

Analog I/O: 3.5-mm stereo audio jack, ±2 V max.

A similar VI unit is the Digilent Discovery, shown in Fig. 26-23. It houses essentially the same components as the myDAQ except the DMM. It uses software called Waveforms to implement the virtual instruments on a PC. Its specifications are as follows:

Power Supplies: ±4.5–5 V with 50 mA max.

Oscilloscope: 2 channels with 5-MHz bandwidth and 50 megasamples/s, ±20 V max.

Function Generator: 2-outputs, up to 5 MHz, ±4 V max; AM, FM, and sweep.

Digital I/O: Up to 16 input or output channels to be user configurable.

External wires from the unit connect to plug-in pins on the accompanying breadboarding socket.

There are numerous other similar virtual instrument products on the market. Their specifications do not match standard benchtop instruments, but they are very capable and affordable. They can easily replace college and university lab equipment for basic courses and are great for students, hobbyists, experimenters, and do-it-yourselfers (DIYs).

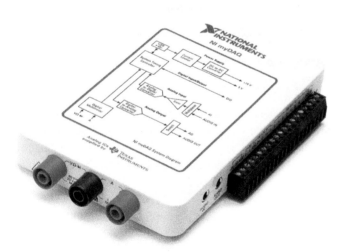

Figure 26-22 The National Instruments myDAQ is a complete DMM, scope, function generator, and power supply designed to be used with a personal computer. (Courtesy National Instruments.)

26.8 AC Testing and Troubleshooting Examples

Here are two examples of how to test and troubleshoot typical passive electronic circuits. One is an attenuator circuit, and the other is a low-pass filter. Standard test equipment is used.

T-Pad Attenuator

An *attenuator* is a circuit designed to introduce a specific amount of loss in a circuit. These are used in audio, video, and RF wireless testing. There are two common attenuator types, the T-pad and the π, both shown in Fig. 26-24. These devices are designed for a specific amount of attenuation in decibels for a defined load impedance. The attenuation is power attenuation, rather than voltage attenuation, so the appropriate formula is

$$dB = 10 \log \frac{P_o}{P_i} \qquad \textbf{(26-4)}$$

where P_o and P_i are the power output and power input, respectively.

Figure 26-25 shows a T-pad for a 50-Ω load impedance and a 12-dB attenuation. The attenuator and its load look like a 50-Ω resistance to the generator with its 50-Ω output impedance. This means that the load and source are matched for maximum power transfer. Assume that the attenuator was used and found to be defective. Your job is to test it and find the problem. Following are two possible approaches you can try.

The quickest method is simply to test the attenuator with an ohmmeter. If you measure between the input and output connections (with the load disconnected), you should measure the sum of R_1 and R_3, or 60 Ω. A resistance measurement

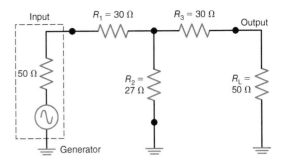

Figure 26-25 A 12-dB, T-pad attenuator.

between the input and ground should produce a reading of 57 Ω (30 + 27). That should be the same value between the output and ground. If you get these readings, the attenuator is good.

If you do not get these readings, then either a resistor is defective or the connections are open or shorted. For example if R_1 is open, you will read an open (infinite R) between input and output and between input and ground. Similar readings will occur if R_3 is open. If R_2 is open, you will read 60 Ω from input to output but have an open reading from input or output to ground. Verify all these possibilities for yourself by tracing through the schematic diagram.

As a final test of the attenuator, apply a signal of a known value, measure the output, and compute the actual attenuation.

Assume that you connect a function generator to the attenuator. Its output impedance is 50 Ω. You set the sine-wave voltage at the attenuator input to 2 V (rms). Measure this with an oscilloscope. Be sure to use a frequency that is near the standard operating conditions of the attenuator.

The power input to the attenuator is

$$P = \frac{V^2}{R} = \frac{2^2}{50} = \frac{4}{50} = 0.08 \text{ W or } 80 \text{ mW}$$

With 12 dB of attenuation you can calculate the expected output power from Formula (26-4),

$$dB = 10 \log \frac{P_o}{P_i}$$

Be sure to use a negative sign on the attenuation value. Then rearranging the formula to solve for the output power:

$$\frac{dB}{10} = \log^{-1} \frac{P_o}{P_i}$$

$$\frac{-12}{10} = -1.2 = \log^{-1} \frac{P_o}{80}$$

$$10^{-1.2} = \frac{P_o}{80}$$

$$0.063 = \frac{P_o}{80}$$

$$p_o = 0.063 \times 80 = 5 \text{ mW}$$

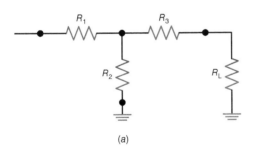

(a)

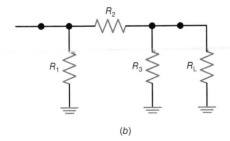

(b)

Figure 26-24 Attenuator types. (a) T-pad. (b) π.

Since we are measuring output with an oscilloscope, we need to find the voltage across 50 Ω that produces 5 mW.

$$P = \frac{V^2}{R}$$

$$V = \sqrt{PR} = \sqrt{0.005 \times 50} = \sqrt{0.25} = 0.5 \text{ V}$$

This should be the rms value of voltage, not the peak-to-peak value.

If you get near this value, the attenuator is working properly.

Power Supply

One of the most common electronic circuits is a power supply. Practically all electronic devices have one, and it is often the point of failure. Power supplies convert the standard 120-Vac, 60-Hz power-line voltage to one or more fixed dc voltages that power the electronic circuits. A power supply is an excellent example to use in troubleshooting. While power supplies will be covered in detail in a later chapter, there is no reason you cannot learn how they work and how to troubleshoot one in this chapter to illustrate troubleshooting procedures.

A basic ac to dc power supply is shown in Fig. 26-26. The input is the 120 Vac from the wall outlet. It then encounters an ON-OFF switch SW_1 and a protective fuse F_1. If the switch is on and the fuse is good, 120 V is applied through the line filter to the primary of the power transformer T_1. The line filter is a low-pass filter that keeps high-frequency noise out of the power supply and keeps the pulses and ripple voltage from the rectifiers out of the power line. This filter is usually a single unit sealed in a metal can, and generally the internal components cannot be accessed or repaired.

The transformer typically steps the line voltage down to some lower level, such as 16 V. In this circuit assume a total secondary voltage of 16 V rms. The center tap splits that into two 8-V windings, one for each of the rectifier diodes.

The diodes perform rectification, which is the conversion of ac into pulsating dc. Figure 26-27 shows the process. The top waveform shows the ac voltage across one half of T_1. It is 8 V rms, or 8 × 1.414 = 11.3 V peak.

The waveform at Fig. 26-27b shows the output of the rectifiers but without filter C_3. The diodes produce half-sine pulses for each half of the input sine wave. The peak voltage is +10.5 V because the voltage drop across each diode subtracts about 0.7 V.

The purpose of C_3 is to filter or smooth the pulses into a constant dc average voltage. Thus C_3 charges to the peak value of the dc pulses, or 10.5 volts. When the pulses go to zero, the capacitor discharges into the load and so loses some of its charge. If C_3 is large enough, the discharge is small. Still, what you see across C_3 is a small ac voltage called *ripple*. It occurs at twice the line frequency, or 120 Hz. The ripple is actually riding on top of the average dc output. It should only be several millivolts peak to peak. Making C_3 larger will make the ripple smaller.

The filtered dc with ripple is applied to the integrated circuit (IC) regulator. This is a complex electronic circuit in a three-lead package that maintains the output voltage at a desired level. Any variation in load or input voltage is

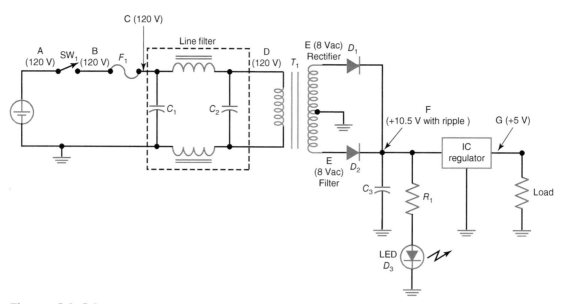

Figure 26-26 Troubleshooting a power supply.

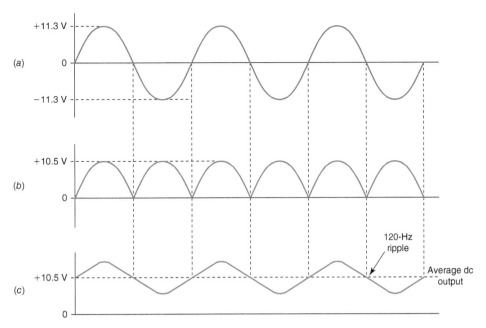

Figure 26-27 The rectification and filtering process showing the ripple voltage. (a) Voltage across half of T_1 secondary. (b) Rectifier output without filter C_3. (c) DC output with ripple.

automatically compensated for. It also reduces the size of the ripple. The desired output is +5 V.

Now assume that the power supply fails. This is usually noted by the absence of the +5 V at the output. Troubleshooting will quickly find the problem. Both a DMM and oscilloscope are useful in finding the problem. The basic process is signal tracing from input to output or vice versa.

Refer back to Fig. 26-26. To begin, turn the power supply on. If the LED does not light, then there is no dc at point F. Start by making sure you have ac at the input. Is the power supply plugged in, and is the ac output getting power? Did a breaker blow? Check the voltage at point A with respect to ground with a DMM or the oscilloscope. If voltage is present, go to point B. (Voltages at the lettered points are made with respect to ground.) If the switch is good, you will get 120 Vac at point B. If not, check the switch for continuity with an ohmmeter (with the power off). If the switch is good, measure the voltage at point C. You will get 120 V here if the fuse is okay. If not, check the fuse. Take it out if you can, and check it for continuity with an ohmmeter. Replace if necessary.

Next, check for voltage at point D. Again you should measure 120 V. If not, the filter is bad. Perhaps one of the inductors is open or a capacitor shorted. Take it out, and check with an ohmmeter to see if the inductors are good. Replace the filter if bad.

If all is well up to this point, T_1 will be getting the 120 V. The voltage at points E should be 8 V rms. Be sure to check both sides of the secondary winding at points E with respect to ground. You can verify this with the oscilloscope as well.

If secondary voltage is present, look at point F. You should measure about 10.5 Vdc. If you are using an oscilloscope, you will see some 120-Hz ripple, which should only be a few millivolts. LED D_3 should be on at this time. If you see the full rectified pulses, then C_3 is defective. It is open. Replace C_3. Also, C_3 is almost always an electrolytic, so be sure to observe the polarity connections if you replace it.

Other problems may be open or shorted diodes. An open diode will not produce the dc pulses. A shorted diode will put ac on C_3. Since it is usually an electrolytic and polarity-sensitive, it will be destroyed.

Next, look at the output. It should be +5 V. If not, the IC is bad. It cannot be repaired. Replace it.

The opposite approach is to start at the output and work backward. You should see +5 V at the output. If not check for the output at point F. It should be +10.5 V. If not, keep working your way back toward the input looking for the place where you do not see a desired voltage. That is the point of failure.

1. An oscilloscope displays signals in the
   a. time domain.
   b. frequency domain.

2. The horizontal display in a scope is
   a. frequency.
   b. voltage.
   c. time.
   d. current.

3. Most scopes in use today are
   a. analog.
   b. digital.
   c. equally analog and digital.
   d. mixed signal.

4. The display on an analog scope is a(n)
   a. LCD panel.
   b. LED.
   c. fluorescent.
   d. cathode-ray tube.

5. The display on a digital scope is a(n)
   a. LCD panel.
   b. LED.
   c. fluorescent.
   d. cathode-ray tube.

6. The vertical controls on a scope control the signal
   a. frequency.
   b. amplitude.
   c. timing.
   d. triggering.

7. The horizontal controls on a scope control the
   a. amplitude.
   b. frequency.
   c. sweep rate.
   d. triggering.

8. The main purpose of the scope trigger is to
   a. measure amplitude accurately.
   b. create a stable display.
   c. measure period accurately.
   d. selectively display parts of a signal.

9. A MSO displays
   a. only analog signals.
   b. only digital signals.

c. one signal at a time.
d. analog and digital signals simultaneously.

10. An oscilloscope can measure a dc voltage.
    a. True.
    b. False.

11. Which of the following is *not* a common function generator output wave?
    a. Sine.
    b. Triangle.
    c. Trapezoid.
    d. Square.

12. The waveforms produced by a function generator come from a(n)
    a. DAC fed by a digital memory.
    b. oscillator.
    c. phase-locked loop.
    d. amplifier.

13. Which of the following in *not* usually adjustable on a function generator?
    a. Amplitude.
    b. Phase.
    c. Frequency.
    d. Sweep frequency.

14. A virtual instrument
    a. is an imaginary piece of equipment.
    b. a common benchtop instrument.
    c. has not front panel.
    d. uses an ADC and software on a PC.

15. Which of the following is typically *not* part of a virtual instrument?
    a. Oscilloscope.
    b. Digital multimeter.
    c. Spectrum analyzer.
    d. Function generator.

16. The most common approach to troubleshooting involves
    a. signal tracing.
    b. dc measurement only.
    c. ac measurement only.
    d. scientific and logical guessing.

SECTION 26.5 Voltage and Time Measurements

26.1 In Fig. 26-28, what is the peak-to-peak voltage of the waveform and its period and frequency?

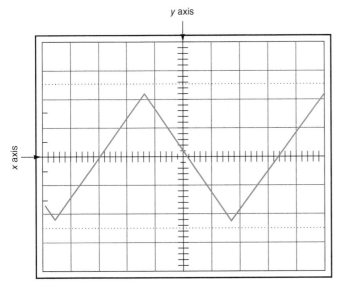

Figure 26-28 Vertical: 0.5 V/div; horizontal: 5 μs/div.

26.2 What is the rms value of the sine wave shown in Fig. 26-29? Its frequency?

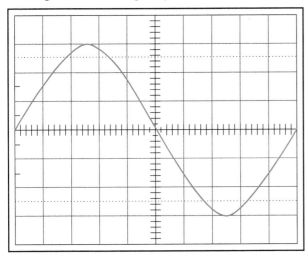

Figure 26-29 Vertical: 2 mV/div; horizontal: 2 ms/div.

26.3 In Fig. 26-30, what is the width of the positive pulse width and the period? Calculate the duty cycle.

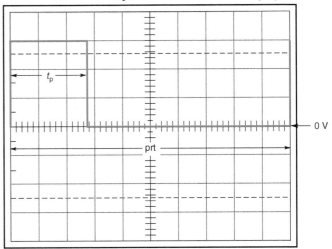

Figure 26-30 Vertical: 1 V/div; horizontal: 10 ns/div.

26.4 Determine the rise time of the pulse in Fig. 26-31.

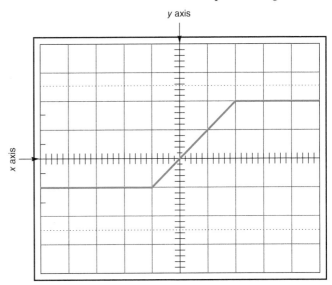

Figure 26-31 Vertical: 2 V/div; horizontal: 5 ns/div.

26.5 What is the amount of phase shift between the two sine waves in Fig. 26-32?

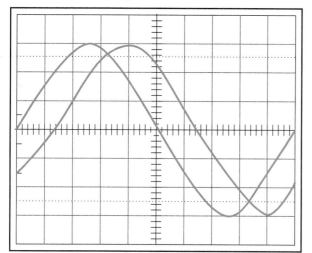

Figure 26-32 Vertical: 200 mV/div; horizontal: 2 μs/div.

26.6 An oscilloscope has a vertical bandwidth of 50 MHz. What is minimum pulse rise time that can be displayed?

26.7 If you wish to display a pulse rise time of 3 ns, what is the minimum bandwidth of the vertical amplifier?

26.8 If you wish to display at least the fifth harmonic of an 12-MHz square wave, what is the minimum bandwidth required?

SECTION 26.8 AC Testing and Troubleshooting Examples

26.9 A stereo system consists of a CD player and amplifier. The amplifier drives two speakers connected to the amplifier with speaker cables of about 20 ft long.

One of the speakers is not working. See Fig. 26-33. Explain how you would troubleshoot this problem and what test instruments you would use.

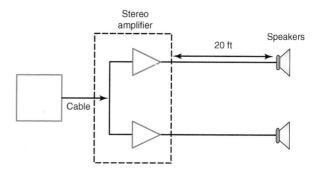

Figure 26-33 Circuit for Prob. 26-9.

26.10 A music PA system consists of a microphone, mixer preamplifier, a power amplifier, and several speakers. Refer to Fig. 26-34. The system is not working. Explain how you would troubleshoot it and what test instruments you would use.

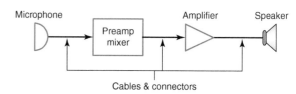

Figure 26-34 Circuit for Prob. 26-10.

Chapter 27

A Systems Overview of Electronics

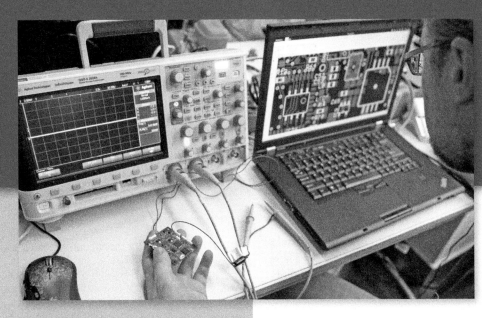

Learning Outcomes

After studying this chapter, you should be able to:

> *Define* electronic systems.

> *State* the hierarchy of electronic systems.

> *Name* the most common types of linear circuits and *state* their key specifications.

> *Name* the most common types of digital circuits and *state* their key specifications.

> *Identify* linear and digital circuits in a system block diagram and *follow* signal flow.

Before you begin your study of the electronic components and circuits in the coming chapters, it is helpful for you to know something about the circuits themselves first and how they are put together to form a piece of equipment or a system. With this context, you will better understand the circuit applications. Since most circuits are in integrated circuit form, this overview will have more meaning when it comes to understanding how circuits are used.

27.1 Introduction to Systems

A *system* is any collection of components or devices that work together to form a complete entity that performs some useful function. Systems can be large or small. A large electronic system example is the Federal Aviation Agency's air traffic control system with all its many radars, navigational aids, control towers, computers, radio communications, and networks. A small system example is an Apple iPod. It stores digitized music from Apple's iTunes website in flash memory. The music can then be heard by converting it into analog signals for amplification and then played through headphones. A small internal computer called an embedded controller controls the selection of the songs and manages the memory and display. Most electronic products are in the form of systems. There are small systems or equipment and large systems that are combined to make a larger system.

A Typical System

This concept is illustrated in Fig. 27-1. Electronic components are combined to make circuits; then circuits are combined to make modules, usually on printed circuit boards (PCBs). In the past, most circuits were made with individual or discrete components. Today, integrated circuits (ICs) are more common. Many modules are a combination of ICs and discrete components. The PCBs are then packaged to make subassemblies that may consist of two or more PCBs. These are then interconnected to make a piece of electronic equipment. That piece of equipment forms a small system that has some useful function. In some cases, the equipment is considered as a subsystem that may be interconnected to other equipment to form a larger system. For instance, a personal computer may be part of a larger automated test system.

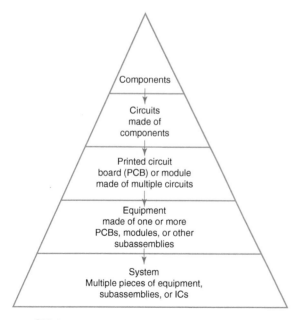

Figure 27-1 Hierarchy of electronic systems.

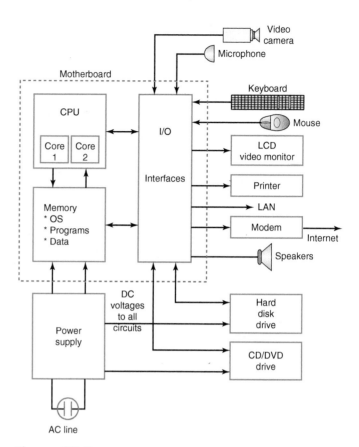

Figure 27-2 A personal computer as a system.

A good example of a small system is a personal computer (PC). See Fig. 27-2. A microprocessor known as a central processing unit (CPU) like an Intel Pentium or AMD Athlon is combined with memory ICs and input/output (I/O) circuits on a printed circuit board called the motherboard. Note that PCs usually have a dual-core processor, meaning two CPUs inside one microprocessor chip. The motherboard is then surrounded by other modules or subsystems. The power supply contains a printed circuit board with components that converts the ac line voltage into multiple dc voltages to power all the other circuits. The disk drives are made up of the mechanical subassemblies with their motors and other circuitry. All these are combined into a single package that may be a desktop or a laptop computer. The computer is then connected to an external LCD video display, keyboard, mouse, speakers, printer, modem, and/or local area network (LAN). The result is a computer system.

The hierarchy of electronics is the same today as it has always been; however, significant changes have occurred over the years to modify it. For example, circuits were initially made up of individual discrete electronic components like resistors, capacitors, transistors, and transformers. Today, most electronic circuits are integrated circuits where all components are formed on a tiny chip of silicon or other semiconductor material. These ICs are interconnected to form larger circuits and subassemblies. In some cases, external discrete components like capacitors, inductors, or transistors are used but to a far lesser extent than previously. In fact, the trend is to put more components and circuits together on a single silicon chip forming a system on a chip (SoC). Today, SoCs can have upward of hundreds of millions and even billions of components on a single chip.

How Integrated Circuits Have Changed Electronics

The impact of semiconductor technology has been to make electronic circuits and equipment ever smaller and lighter than before. More usefulness and functionality can be incorporated at virtually no extra cost. Electronic equipment is also easier to manufacturer and is lower in cost overall. By using modern ICs, electronic equipment is more reliable and so does not fail as often or rarely needs repair. One key factor in making electronic equipment with ICs is that the equipment cannot easily be tested or repaired. Most circuitry within an IC is inaccessible, so it cannot be analyzed, tested, or otherwise inspected to determine its functionality. So most electronic equipment is frequently not repaired. It is simply discarded and replaced.

Integrated circuits have in fact made a major change in how electronic circuits and equipment are designed and built. Electronic circuits are designed by engineers who use complex computer software that simulates the circuits and then provides the physical layout for the semiconductor manufacturers. This is called computer-aided design, and the process is referred to as *electronic design automation (EDA)*.

Engineers who design the equipment often do not need to design the circuits they need as these are commonly available in integrated circuit form. The design process is largely one of assembling the various ICs, interconnecting them, and making sure that they perform as required. In a few cases, discrete component circuit design is required. High-voltage and high-power circuits typically use discrete components because ICs do not withstand high voltages or power levels and the heat they generate.

A major part of all design is programming the small single-chip microcomputer called an *embedded controller* or *microcontroller*. This device controls, monitors, and manages virtually every electronic product built today. Programming the device and interfacing to all the other chips constitute well over 50% of most electronic design today.

How the Work of Technicians Has Changed

The impact on technicians and their work has been major. For example, a great deal of electronic design in the past has been for an engineer to design the circuit on paper and give it to a tech to build a breadboard prototype to test. That process typically went through multiple stages of design and testing until a workable circuit was obtained. Today, the design process is largely on the computer with multiple simulations fine-tuning the design instead of multiple iterations of prototype builds. While prototypes are still built and tested, these are more typically done by an engineer. As a result, there is significantly less need for engineering technicians. While they have not disappeared from industry, this is no longer a good job objective for most technicians.

Today the main work of technicians is to troubleshoot, test, measure, and repair, but even that work has changed drastically. In the past, most faulty electronic equipment was repaired by troubleshooting down to the component level. That bad resistor or other part was identified and replaced. Today, that is less possible, as defects inside ICs cannot be identified and replacements may not be possible. Of course, it is possible to find the bad IC and replace it, and that is sometimes done.

The main problem is finding the trouble. That usually takes time and often very expensive test equipment. The cost of that time and equipment is frequently way beyond what it costs to simply discard the defective unit and replace it completely with a new unit. A new unit is commonly less expensive than the time to troubleshoot and repair the old unit. This has made electronic products almost throwaway devices in many cases. Only very large and expensive equipment like large-screen HDTV sets, medical equipment (MRI machines), and military gear is actually repaired. And that repair is often simply the replacement of a printed circuit board, module, or other subassembly, rather than repairing the defective one. The economics of repair has drastically changed how technicians work. Luckily, the high reliability of electronic equipment means fewer repairs and more replacements.

Today, technicians work at the systems level. Technicians work with the end equipment and its subassemblies. They install, test, troubleshoot, service and maintain, and operate the equipment. They do not design or analyze it or otherwise engage in the work generally done by an engineer. As a result, a technician commonly works at a higher level than before. Rather than dealing with individual components of a circuit, the tech more often works with the integrated circuits or with PC boards and modules. A good example is servicing personal computers. Motherboards are not repaired, but memory modules are replaced. A defective power supply is replaced and not repaired. That is also true of disk drives. The tech disconnects and interconnects these subassemblies, tests them, and replaces them if they fail.

The modern tech does not often get involved with component level repairs unless it is with equipment that uses them. Some examples where considerable discrete component circuitry is still used are power supplies, power amplifiers, and special high-voltage, high-power equipment that is not implemented with ICs. Otherwise, techs do not commonly need to know circuit details, analysis procedures, or the design process.

Technicians essentially work at the block diagram and signal flow level. While schematic diagrams are still used, these are often only drawings of how integrated circuits and other devices are interconnected. The work at this level involves the inputs and outputs of the ICs and the PC boards or modules. For this reason, knowledge of circuit details is less valuable than it once was. With ICs, circuit details are typically just not available. The circuitry inside most integrated circuits today is designated as intellectual property (IP) that is protected by patents, copyrights, or simply company confidential methods. The important information is the operation of the IC and its specifications. And how to test and measure the circuits.

This chapter introduces you to the more common classes and types of electronic circuits at the block diagram level. You will learn about amplifiers, oscillators, filters, digital circuits, and others that are commonly used in modern equipment. Later chapters will deal with the individual components and circuits.

27.2 How Circuits and Systems Work

All electronic circuits and systems work essentially as portrayed in Fig. 27-3. Electronic signals are applied as inputs to a circuit or system where they are processed in some way. The signals are converted or transformed in some way or used to generate new signals. In a circuit or system, some output is sent back to the processing circuits to enhance or change the process based on the inputs. This signal is called *feedback*. The new output signals become the desired end result. Alternatively, the outputs may become the inputs to other circuits or equipment. The model in Fig. 27-3 can be used for a single simple circuit or for a more complex subassembly of multiple circuits or for a complete piece of equipment.

A systems example is shown in Fig. 27-4. A satellite TV signal from an antenna is amplified at the antenna and converted to a lower frequency and transmitted on coaxial cable to the satellite receiver box. The received video and audio signals are sent to the TV set where multiple circuits process the audio and video signals into the outputs that drive the large high-definition video display and the speakers.

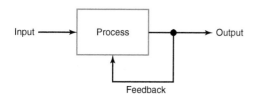

Figure 27-3 How electronic circuits and systems work.

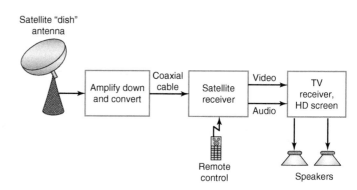

Figure 27-4 A satellite TV receiver system.

Remember that there are two basic types of signals, analog and digital. Analog signals like a sine wave vary smoothly and continuously over time. Voice, music, video, and radio signals are examples. Digital signals are pulses or OFF-ON signals. Digital voltages are usually binary in that they have two discrete values. These signals represent numbers and codes in the binary system.

The two types of signals require two different types of electronic circuits, linear and digital. Linear circuits process analog signals. They are called linear because their input-output response is a straight line or proportional variation. Digital circuits process binary signals. Digital circuits can also be called pulse circuits as they work with OFF-ON pulses rather than analog signals. Some circuits actually combine analog and digital functions and can be referred to as hybrid circuits. The various types of circuits in each category are summarized and defined in the sections to follow.

27.3 Linear, or Analog, Circuits

Linear circuits process analog signals. The most common types are amplifiers, oscillators, mixers, modulators and demodulators, and phase-locked loops (PLLs). Then there are categories within each type. Here is how each circuit type breaks down.

Amplifiers

Amplifiers are circuits that have gain. They take a small signal at the input and produce a larger signal at the output. Gain (A) is ratio of the output to the input.

$$A = \text{output/input} \qquad \textbf{(27-1)}$$

The inputs and outputs may be voltage, current, or power, with voltage and power being the most common. The response is linear, as shown in Fig. 27-5. This is the plot of an amplifier with a gain of 20. If the input is 100 mV the output will be 2000 mV, or 2 V.

Amplifiers are usually categorized as being small-signal or large-signal amplifiers. Small-signal amplifiers typically work with small signals less than about 1 V. Large-signal

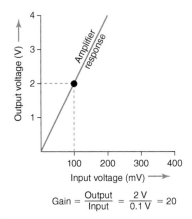

$$\text{Gain} = \frac{\text{Output}}{\text{Input}} = \frac{2\text{ V}}{0.1\text{ V}} = 20$$

Figure 27-5 Amplifiers have a linear relationship between input and output.

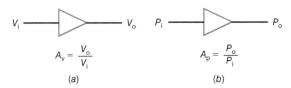

$$A_v = \frac{V_o}{V_i}$$

(a)

$$A_p = \frac{P_o}{P_i}$$

(b)

Figure 27-6 Schematic symbols for an amplifier. (a) Voltage amplifier. (b) Power amplifier.

amplifiers are usually power amplifiers. There are current amplifiers, although they are not widely used.

The symbol used to represent an amplifier is a triangle, as shown in Fig. 27-6. The inputs and output are labeled. Note the difference between the voltage and power amplifier expressions for gain (A).

$$A_v = V_o/V_i \qquad \textbf{(27-2)}$$

$$A_p = P_o/P_i \qquad \textbf{(27-3)}$$

Because most amplifiers are in IC form these days, this triangle is usually all that you will see on a schematic diagram. Your primary concerns are the levels of input and output signals.

While gain is often given as the ratio of output to input, it is also common to express gain in decibels (dB). Decibel expression uses logarithms to compress large and small numbers, which makes gain calculations simpler. To calculate the gain in dB, you use the formulas below:

$$\text{Voltage gain in DB} = 20 \log \frac{V_o}{V_i} \qquad \textbf{(27-4)}$$

$$\text{Power gain in DB} = 20 \log \frac{P_o}{P_i} \qquad \textbf{(27-5)}$$

Use the standard common (base 10) log functions on your calculator. More details will be given on dB calculations in a later chapter.

Another way to categorize amplifiers is by frequency response. Frequency response is, of course, the measure of the frequency range over which the amplifier provides its

gain. Some amplifiers have a narrow range; others have a wide range. Most IC amplifiers are generally optimized for a particular frequency range. Typical categorizations are the following:

- *DC amplifiers:* Amplify direct current. They usually have a restricted upper-frequency response.
- *Audio amplifiers:* Designed for audio signals usually in the 20-Hz to 20-kHz range. They boost signals levels of voice and music.
- *RF amplifiers:* Amplify radio-frequency signals, which could range from a few kHz to many GHz. Most are optimized for specific ranges.
- *Video amplifiers:* Optimized for video signals in the dc to several hundred MHz range.
- *Microwave amplifiers:* Amplify only signals typically above 1 GHz.

Each type could also be a voltage amplifier or a power amplifier.

Another categorization for amplifiers has to do with their input and output configurations. The two main types are singled-ended and differential. Single-ended amplifiers have their input and outputs referenced to a common ground, as shown in Fig. 27-7a. These are often referred to as unbalanced amplifiers.

A differential amplifier has two inputs, as shown in Fig. 27-7b. The output is a function of the difference between the two input signals and the gain (A). For example the output of a differential voltage amplifier is expressed as

$$V_o = A(V_B - V_A) \qquad \textbf{(27-6)}$$

Each input is referenced to ground, but a single input could be applied between inputs A and B with no ground reference. The output is single-ended; however, a differential output is also possible. Differential amplifiers have what we call balanced inputs and/or outputs.

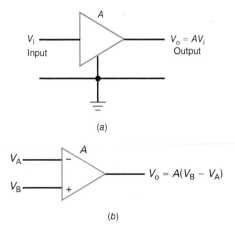

Figure 27-7 Amplifier configurations. (a) Single-ended (unbalanced). (b) Differential (balanced).

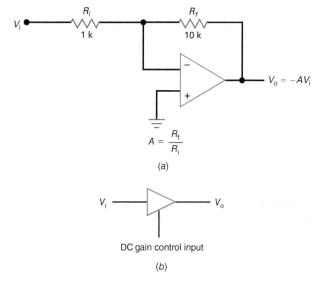

$A = \dfrac{R_f}{R_i}$

(a)

V_i ───▷─── V_o

DC gain control input

(b)

Figure 27-8 Amplifier examples. (a) Op-amp inverter. (b) Programmable gain amplifier (PGA).

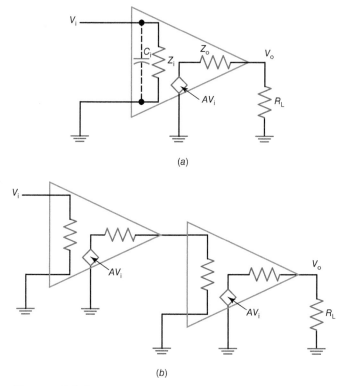

(a)

(b)

Figure 27-9 (a) Input and output impedances of an amplifier. (b) How one amplifier "loads" another.

There are many other classifications. For example, a popular type of amplifier is the *op amp,* or operational amplifier. It is a general-purpose dc differential amplifier with a very high gain and wide frequency response. The op amp was originally developed as the main computing element in analog computers. It can be configured to perform almost any basic math operation, thus the name. Connecting different resistors, capacitors, diodes, or other components to the inputs and in the feedback path configures the amplifier to perform a specific function. Figure 27-8a shows an op amp connected as an amplifier with a gain of 10. The resistor values set the gain. It also inverts the input 180°, or reverses the polarity, as indicated by the negative sign in the gain expression.

Another category is programmable gain amplifier (PGA). See Fig. 27-8b. This amplifier has a special input that can be used to adjust the gain of the amplifier in increments or, in some cases, continuously over a given range. Typically the gain of the amplifier is adjusted by varying a dc input voltage.

The main amplifier specifications are gain, frequency response, dc operating voltage range, voltage or power input and output ranges, input impedance, and output impedance. Most amplifiers are designed to operate with a specific dc power supply. Some supply levels are as low as 1.5 V but most are greater, usually from 3 to 15 V. Power amplifiers often operate with higher voltages, from about 12 V up to about 48 V. It is this dc operating voltage that sets the output voltage swing or range. The output generally cannot exceed the dc operating voltage, as you will see later. Input voltage range is also usually specified.

Input and output impedances refer to the fact that all practical amplifiers have a specific input resistance (R_i) to a driving signal and an internal output resistance (R_o). See Fig. 27-9. Input resistance is more commonly called the input impedance (Z_i). This impedance is mostly resistive and can range from a few hundred ohms to many megohms. Generally, the higher the better. Some amplifiers also specify an input shunt capacitance (C_i)

Amplifiers generate an output signal that is an enlarged version of the smaller input. That signal appears to an external load as coming from a generator. The generator is illustrated as a diamond shape. It generates an output voltage of AV_i that is in series with the output impedance, Z_o. The output impedance appears in series with the load, R_L. You can think of the amplifier as a signal generator with a series internal resistance called the output impedance. When amplifiers are connected to one another or cascaded, the input impedance of one becomes the load for the driving amplifier, as Fig. 27-9b shows. The key fact to recognize here is that the input and output impedances form a voltage divider that, in turn, actually reduces the amount of voltage gain that can be obtained.

When using power amplifiers, the input and output impedances must be matched to provide maximum power transfer from one stage to another. For example, for maximum output to a 50-Ω antenna, the output impedance of transmitter power amplifier should be 50 Ω.

Oscillators and Frequency Synthesizers

An oscillator is a signal-generating circuit. Oscillators can generate either sine waves or rectangular waves like pulses or binary square waves. A pulse output oscillator is often

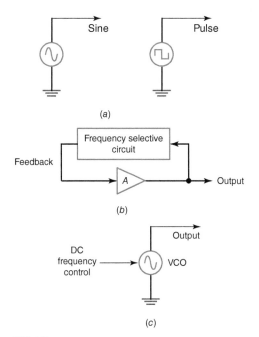

Figure 27-10 (a) Oscillator schematic symbols. (b) An amplifier with positive feedback creates an oscillator. (c) VCO output frequency is controlled by a dc input.

called a clock, since its pulses occur at accurate intervals of time. Some oscillators can also produce triangular or sawtooth output signals. The basic symbols for an oscillator in a schematic diagram are shown in Fig. 27-10a.

Many oscillators are actually amplifiers that have positive feedback. If some of the output signal from an amplifier is fed back to its input with the correct phase and amplitude, the amplifier becomes a self-sustaining signal generator. See Fig. 27-10b. A circuit in the feedback path is used to select the frequency of oscillation. That frequency selective circuit can be an *RC* phase shifter, an *LC* series or parallel resonant circuit, or a quartz crystal.

Oscillators can have a fixed frequency or can be made variable. A variable-frequency oscillator (VFO) lets you tune the oscillator to a desired value for the application. The frequency variation may be achieved by varying a resistance, a capacitance, or an inductance. Quartz crystal oscillators are fixed to a single stable frequency as determined by the crystal.

Another type of oscillator is the *voltage-controlled oscillator (VCO)*. This type of oscillator can have its frequency varied by a dc input voltage, as Fig. 27-10c shows. For example, increasing the dc input voltage may increase output frequency.

In many applications, a simple oscillator is replaced by a signal source called a *frequency synthesizer*. A frequency synthesizer is an oscillator, usually a VCO, that is controlled by other circuits to provide a precise output frequency in increments. The synthesizer typically receives a digital input code that causes the synthesizer to be set to a specific frequency. The stability of the synthesizer is determined by a fixed-frequency crystal oscillator.

One common type of frequency synthesizer is based on a circuit called a phase-locked loop (PLL). This circuit is shown in Fig. 27-11. The output comes from the VCO whose frequency is controlled by a dc voltage derived from the phase detector and low-pass filter. The phase detector compares the VCO output to that of a precise quartz crystal reference oscillator. If the two are the same, the circuit is said to be locked to the desired frequency. If there is a frequency difference, a phase shift will be detected and the dc output from the phase detector and low-pass filter will change, forcing the VCO back to the desired frequency.

The frequency divider is a digital circuit that divides an input frequency by some integer *n*. Note that *n* can be changed by an external digital input. By changing the division ratio of the frequency divider circuit in the feedback path from the VCO to phase detector, the output frequency can be varied in steps. In Fig. 27-11, as *n* is varied in integer steps, the output frequency varies in 1-MHz steps. The PLL is useful not only as a signal source but also as a filter, frequency multiplier, or demodulator. Other types of frequency synthesizers are also available.

Power Supply Circuits

A power supply is a set of electronic circuits that supply power to other circuits or equipment. In portable equipment like a laptop, cell phone, digital camera, or MP3 player, the supply is a battery. However, the most common power supply is one that converts standard ac line voltage of 120 V, 60 Hz into one or more dc voltages. Some of the circuits

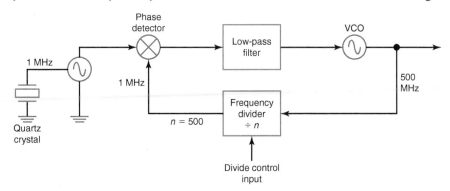

Figure 27-11   A phase-locked loop (PLL) frequency synthesizer.

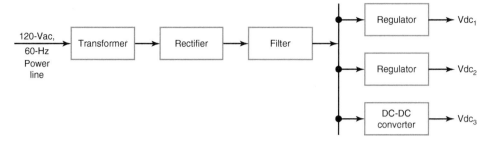

Figure 27-12 A common ac-to-dc power supply.

that make up a power supply are rectifiers, filters, regulators, dc-dc converters, inverters, and power management circuits.

Figure 27-12 is a block diagram of a power supply. It is really a subsystem of any piece of equipment. A transformer at the input steps up or steps down the ac input voltage to the desired level. The rectifier is a circuit made up of diodes that converts ac into pulses of dc. A filter is used to smooth those pulses into a more constant, continuous dc voltage. The filter is a low-pass filter, usually made up of a single large capacitor or an inductor and capacitor.

A regulator is a circuit that takes a varying dc input and converts it into a steady, constant voltage of a specific value. Most ICs require a precise dc operation voltage that does not vary. Changes in the ac input voltage or load resistance are automatically compensated for by the regulator to maintain its design voltage output.

A dc-dc converter is a circuit that takes in one value of dc voltage and outputs another dc value. A good analogy of a dc-dc converter is a transformer. By using different turns ratios, a transformer can step up or step down an ac voltage as desired. The dc-dc converter provides this for dc.

An inverter is a dc-to-ac converter. It takes a dc input and generates an ac output, usually a sine wave. For example, common inverters convert 12 Vdc from a car battery into 120-V, 60-Hz ac. Inverters are used to provide battery backup for equipment that needs continuous power even in a power failure. These devices are called uninterruptible power supplies.

Power management circuits that do just that manage power. Some typical functions are battery charging, overvoltage detection and protection, undervoltage detection and protection, temperature measurement, and power sequencing of multiple dc voltages.

27.4 Digital Circuits

Digital circuits are circuits that operate on binary voltages. Binary signals are pulses that switch rapidly from one voltage level to another. As an example, some digital circuits switch between 0 V or near ground and +3.3 V. Switching speeds are in the nanosecond and picosecond range at speeds exceeding 1 Gb/s. The binary pulses represent the 0s and 1s of the binary number system.

Binary numbers make up the base 2 numbering system that represents quantities with just two digits or bits, 0 and 1. For instance the decimal number 98 in binary is 1100010. Binary data represent any decimal quantity, as well as special codes that designate letters and symbols or other information. Digital circuits process the binary data. You will learn about digital technology in another course if you have not already done so. But since so much of electronic equipment is digital, the basic digital circuits are described here briefly. This basic introduction is necessary since all systems use digital methods in some way.

Gates

The most basic circuits for processing digital data are gates and flip-flops. Gates are circuits that make decisions based on their binary inputs, and their unique operational characteristics then generate an appropriate binary output condition. The basic gate functions are the AND, OR, and NOT. The symbols for circuits that perform these functions are given in Fig. 27-13. Note beside each gate symbol is a simple "truth" table that explains what each gate produces at its output given the input conditions. The AND gate produces a binary 1 output when all its inputs are 1. The OR gate simply produces a 1 output anytime either or both inputs are 1s. The NOT circuit has only one input and is called an inverter, as it takes any input and produces the opposite value. These circuits are the basic building blocks of all other digital circuits.

Other variations are the NAND and NOR gates in Fig. 27-13. The NAND means NOT AND and is equivalent to inverting the output of an AND gate. The circle or bubble on the gate symbol output means inversion. The NOR is similar, meaning a NOT OR and is an inverted OR gate. A special gate called the exclusive OR, or XOR, is also shown. It produces a 1 output only when the inputs are different. The XOR gate is useful for many functions, such as comparison and addition.

Flip-Flops

A *flip-flop (FF)* is a binary storage element. A common FF is the latch shown in Fig. 27-14a. It has two inputs—set (*S*) and reset (*R*)—and two outputs—*Q* and *Q**, where *Q** is *Q* NOT, or the opposite or complement of *Q*. The FF has two stable states, which we label 0 and 1. If the flip-flop is put into the

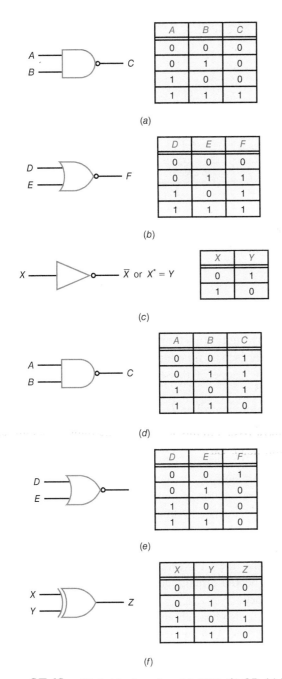

(a)

A	B	C
0	0	0
0	1	0
1	0	0
1	1	1

(b)

D	E	F
0	0	0
0	1	1
1	0	1
1	1	1

(c)

$\overline{X}$ or $X^* = Y$

X	Y
0	1
1	0

(d)

A	B	C
0	0	1
0	1	1
1	0	1
1	1	0

(e)

D	E	F
0	0	1
0	1	0
1	0	0
1	1	0

(f)

X	Y	Z
0	0	0
0	1	1
1	0	1
1	1	0

Figure 27-13 Digital logic gates. (a) AND. (b) OR. (c) NOT. (d) NAND. (e) NOR. (f) XOR.

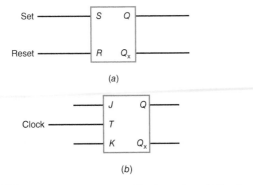

Figure 27-14 Flip-flops. (a) Latch flip-flop. (b) J-K flip-flop.

binary 1 state, it is said to be set and storing a binary 1. If the FF is put into the 0 state, it is said to be reset and storing a binary 0. The FF stores one bit of data as long as the circuit receives power. Multiple FFs can be used to store a multibit binary number or code. This circuit is called a *storage register.*

FFs also have the ability to toggle between the two states. One such FF is the J-K FF shown in Fig. 27-14b. Each time the toggle input, T, switches from 0 to 1 with the input from a clock signal, the output, Q, changes state. This characteristic makes the FF able to count and to perform frequency division. Multiple FFs can also be combined to produce a shift register that can shift bits to the right or to the left.

Gates and FFs are the key building blocks of digital circuits and can be used to produce any digital function, including computing.

Functional Circuits

Functional circuits are digital circuits made from gates and/ or FFs. The most common ones are listed below:

- *Encoder:* Converts one binary input code into another.
- *Decoder:* Made up of AND gates, a decoder recognizes the presence of specific codes at the input.
- *Multiplexer:* Selects one of multiple inputs to appear as the single output.
- *Demultiplexer:* Selects one of multiple outputs to which the single input is connected.
- *Adder:* Adds and subtracts binary numbers.
- *Counter:* Multiple FFs make up a circuit that counts up (or down) in the binary number format.
- *Storage register:* Multiple FFs store a binary number. Multiple FFs can be put together to form storage registers that can store a multibit number. Eight FFs are used as an 8-bit register.
- *Shift register:* Multiple FFs store a binary number and can shift the bits right or left to move the data into or out of the register.

Memory Circuits

Memory circuits are extremely important to all digital applications, especially computers. Memory circuits store multiple binary numbers or words. The simplest memory is a storage register or multiple storage registers put together to store multiple words. The basic storage element is the FF. Another common storage element is a capacitor. A capacitor can be discharged, which represents a binary 0. When it is charged, the capacitor is storing a binary 1. A type of transistor known as a metal-oxide semiconductor field-effect transistor (MOSFET) can also be made to store a 0 or 1 by keeping the transistor conducting or nonconducting.

There are many different types of memory, static, dynamic, RAM, ROM, and flash. Static memory is made up of many FF

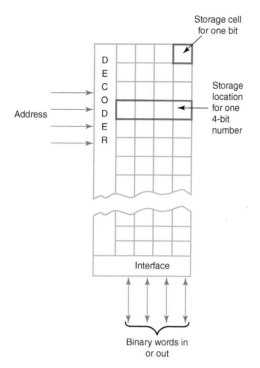

Figure 27-15   Basic concept of a RAM.

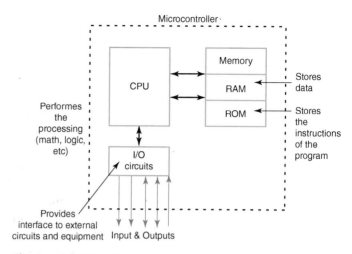

Figure 27-16   The major components of a microcontroller.

storage elements organized in such a way to store thousands or even millions of bits or words. A dynamic memory stores data in capacitors organized again to store millions of bits.

A RAM is *random access memory*. This can be either a static or dynamic type that can access any one of millions of locations for data by just giving it an address. RAMs can be used to read or write data. Writing data means storing data into the memory. Reading data means accessing it for use elsewhere. RAM is a volatile memory, meaning that if you remove the power from the circuits, all data are lost. Figure 27-15 shows the concept of a RAM. A decoder is used to accept a binary input number called the address. The *address* is a unique code that defines one storage location that will be written into or have data accessed and read out.

A ROM is read-only memory. As the name implies, a ROM has data permanently or semipermanently stored in it. The only operation is to read data or access it. Data are retained even if the power is turned off, so ROMs are nonvolatile. The organization of a ROM is similar to that in Fig. 27-15.

A special type of ROM is the electrically programmable ROM which we now call *flash memory*. You can store data in it and update or change it like a RAM, but flash memory also retains any data even if power is removed.

Microcontrollers

A microcontroller is a complete computer on a chip. Every electronic product has at least one microcontroller inside. This miniature computer can be programmed to perform almost any digital function, including even the functions of gates, FFs, or any functional circuit. Microcontrollers are beyond the scope of this book, but a little background is helpful.

The microcontroller is made up of three main parts: the central processing unit (CPU), memory, and the input/output (I/O) circuits. Refer to Fig. 27-16. The CPU does the actual data processing. The CPU is programmable with a set of commands or instructions that tell the CPU what to do. A program is formed when multiple instructions are put together in sequence to do a desired operation. The instructions are represented by codes that are stored in a RAM or ROM, more usually a ROM. The instructions make up the program that the CPU executes to perform the desired operation. The I/O circuits get data into and out of the CPU and memory.

You will also hear about microcontrollers called embedded controllers. And there are other variations, such as a microcontroller unit (MCU) and microcomputer (μP). A CPU by itself is called a microprocessor. Some computers are made up of a microprocessor and separate memory and I/O circuits. Microcontrollers have CPU, memory, and I/O plus other circuits on a single silicon wafer.

Field-Programmable Gate Array

A field-programmable gate array (FPGA) is a SoC made up of hundreds or thousands of gates, flip-flops, and memory circuits. These can be interconnected to one another on the chip by a programming process. As a result, you can use the FPGA to create almost any digital function, including a complete computer. The programming dedicates the FPGA to a specific function. Most digital equipment today is implemented with FPGAs or embedded controllers rather than individual gates and flip-flops.

27.5 Interface or Mixed Signal Circuits

Interface circuits are circuits that can be linear, digital, or some hybrid or mixed signal combination. These connect one circuit or piece of equipment to another. Most involve both analog and digital functions. The most common interface circuit is the comparator illustrated in Fig. 27-17. It has two inputs, V_A and V_B, for analog signals and a single digital output. The comparator looks at the two input voltages and develops a binary output that says that A is greater than B or B is greater than A. The output is zero if the two voltages are equal.

Another widely used interface circuit is the analog-to-digital converter (ADC), shown in Fig. 27-18a. This circuit looks at an analog input signal and generates a binary output proportional to the input voltage. Entire analog signals may be digitized by simply taking multiple sequential samples of the varying analog signal and storing the binary words representing each sample in a memory, as illustrated in Fig. 27-18b and c, respectively. Each sample is a voltage that

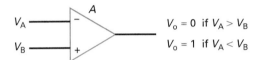

$$V_o = 0 \text{ if } V_A > V_B$$
$$V_o = 1 \text{ if } V_A < V_B$$

Figure 27-17 A comparator.

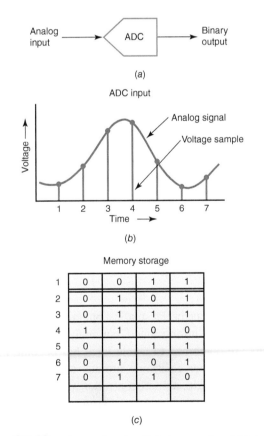

Figure 27-18 An analog-to-digital converter. (a) Interface circuit. (b) Voltage sampling. (c) ADC output stored in RAM.

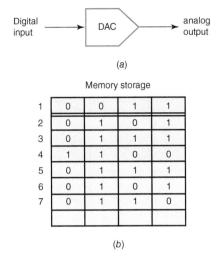

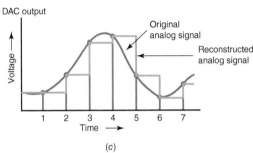

Figure 27-19 A digital-to-analog converter. (a) Interface circuit. (b) Data in memory sent to DAC. (c) Reconstructing the analog signal.

is converted into a binary number. These samples can then be used to reconstruct the analog signal.

The digital-to-analog converter (DAC) presented in Fig. 27-19a takes a binary input and generates a proportional dc or analog output voltage. Given a sequence of previously stored binary numbers, the DAC reconstruct the analog signal represented by the binary values. See Fig. 27-19b and c. The DAC output is a stepped approximation of the original analog signal. Using a large number of samples at high speeds, a digitized and reconstructed analog signal is very close in appearance to the original.

27.6 System Examples

Just to be sure you understand the system concept, here are a few examples of both linear and digital systems. As you will see, you can fully understand how a product or system works by tracing the signals through a block diagram. Most circuits are in IC form, so individual circuit operations do not need to be known except by their function and, in some cases, specifications.

Recording Studio

Figure 27-20 shows a block diagram of an audio recording studio for making music CDs, voice podcasts, or other audio materials. The inputs come from multiple sources, including

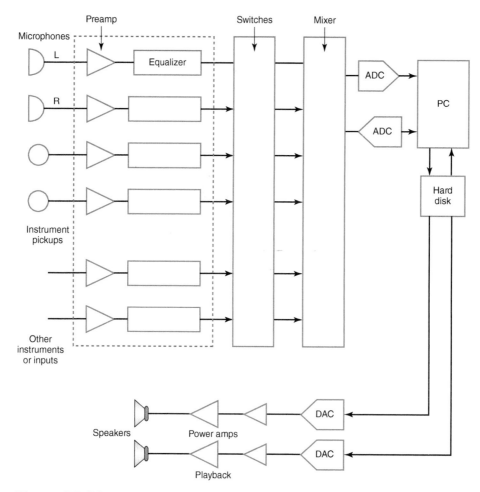

Figure 27-20 A generic block diagram of a recording studio.

microphones and electrical pickups from guitars, piano keyboards, and other instruments. Audio signals may also come from electronic sources, such as previously recorded material on CDs, tape, or computer memory.

The microphone and pickup signals are weak, so they must first be amplified. These preamplifiers are usually part of a console containing level (volume) and tone controls called equalizers. The amplified and frequency-corrected signals are sent to a switching matrix that selects which signals are to be included in the final signal. The switch output goes to a mixer where the multiple inputs are combined at the right levels to create a composite of all voice and musical instrument inputs. If stereo is being developed, then there are separate left (L) and right (R) channels for audio that are mixed separately.

The final audio signals are amplified and sent to the recorder. In the past, music was recorded on discs where a needle cut a varying grove in the surface storing the audio. Later, magnetic tape was used to capture and store the music. Today, most recording is digital, by which the analog audio signals to be recorded are digitized in analog-to-digital converters and then sent to a personal computer where they are

stored in digital memory circuits and on a hard disk drive. From there, the digitized music may be manipulated by a computer to create the format for a compact disc (CD), or the music may be compressed for use in an MP3 player or iPod.

All recording studios also have playback equipment that consists of amplifiers and speakers so the recording artists can hear what they record. The stored signals are passed through digital-to-analog converters to re-create the analog audio, which is then amplified in power amplifiers and sent to one or more speakers.

Cell Phone

A cell phone is one of the smallest but most complex systems we use. We take it for granted, but many different electronic circuits are combined to create what has become the most popular consumer electronic product available. It consists of a two-way radio, a microcontroller with memory, special LCD displays and input keyboards, and a battery power supply. Figure 27-21 shows a simplified block diagram.

The main radio is made up of a receiver (RX) and a transmitter (TX). The PA is the transmitter power amplifier. The

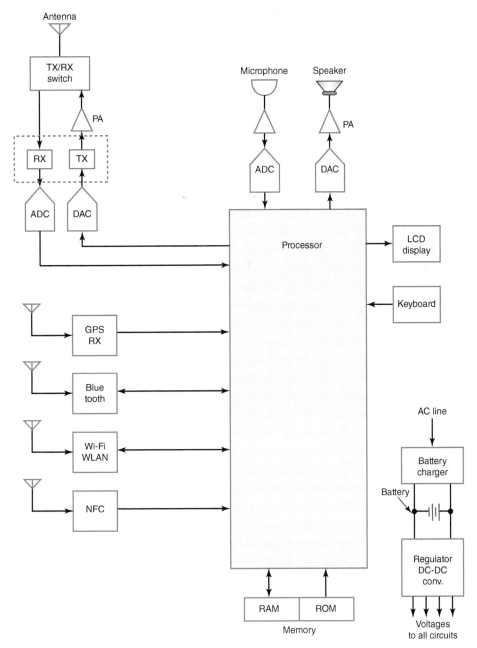

Figure 27-21 General block diagram of a cell phone.

TX/RX switch lets the transmitter and receiver share the same antenna.

For phone calls, the microphone accepts the input voice, which is amplified, then digitized by an ADC, and translated into a binary code. The binary voice modulates the transmitter via a DAC to generate the radio signal that is then transmitted to the nearest cell site or base station. The base station sends the voice signal to the desired destination via the cellular company's network. The voice signal being received from the person being called is also in digital form, and it comes through the network to the base station, where it is transmitted to the cell phone, picked up by the receiver, digitized in an ADC, and sent to the processor. The voice signal is demodulated, then amplified, and applied to the headphone or speaker via a power amplifier.

The microcontroller takes care of dialing, storing numbers, operating the display, operating the touch screen if one is used, and other functions, including the management of all the other circuits in the phone.

The larger, more complex cell phones called smartphones, like the Apple iPhone, have additional circuitry, including multiple radios such as Bluetooth for the wireless headset, an MP3 player, a global positioning system (GPS) navigation receiver, and a Wi-Fi wireless local area network (LAN) transceiver for connecting to the Internet through hot spots. The near field communication (NFC) section is a small short-range transceiver that lets you use your cell phone as a credit card. The smartphone is a truly complex system in a handheld format. Note the battery, its charger, and the multiple regulators and dc-dc converters that power the circuits.

Keep in mind one important thing about these systems. They are all made up of integrated circuits that contain the individual amplifiers, microcontrollers, and other circuits. You generally do not have access to any of these circuits. Very few discrete components are used, and these are mostly capacitors for filtering, quartz crystals for frequency settings, and occasionally a resistor. High-power circuits like audio power amplifiers will typically use discrete transistors and related resistors and capacitors. Your personal involvement with all circuits is with the inputs and outputs to the device itself and the ICs. While you will never have to design or analyze any of the circuits you cannot see, it is useful to know how they operate to process the inputs and produce new outputs. In the coming chapters you will learn about transistors, diodes, and all the basic linear circuits and some digital circuits that make up these complex ICs. Other system examples will be given throughout the remaining chapters as examples of the different circuits being discussed.

DVD Player

Figure 27-22 shows the system diagram of a DVD player. The DVD contains an invisible spiral of digital data representing the video and audio of a movie or other entertainment. That data are accessed by shining a tiny infrared (IR) laser beam on the underside of the spinning disc. The laser beam is reflected by the data and picked up by a photo sensor or detector. The sensor output is a series of digital pulses representing the video and audio. The photo detector signal is amplified and applied to processing circuits that separate the data into video and audio streams. Then DACs are used to convert the digital data into the analog video and audio signals that can be presented on a TV set.

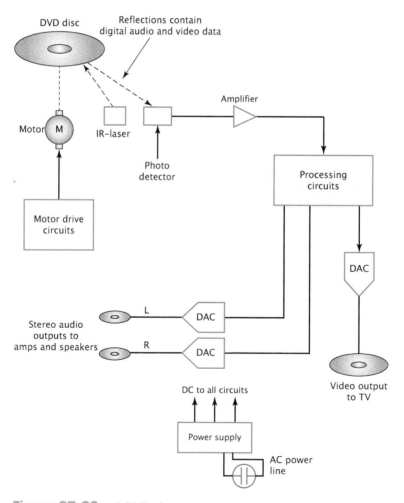

Figure 27-22 A DVD player.

1. The most accurate description of a system is
   a. any collection of multiple electronic components.
   b. a group of electronic components, circuits, or devices that collectively perform a useful function.
   c. a piece of electronic equipment.
   d. large complex interconnections of electronic equipment.

2. Most electronic circuits
   a. are in integrated circuit form.
   b. are collections of discrete components wired together.
   c. contain at least one transistor.
   d. are designed by a technician.

3. Which of the following is *not* involved in troubleshooting and repair of modern electronic equipment?
   a. Finding and replacing the bad component.
   b. Replacing a bad IC.
   c. Replacing a defective board or module.
   d. Discarding the bad device and replacing it with a new one.

4. The primary work of a technician is
   a. designing electronic circuits.
   b. analyzing electronic circuits.
   c. troubleshooting, testing, and repairing.
   d. breadboarding prototypes.

5. Modern technicians work mostly with
   a. schematic diagrams.
   b. block diagrams.

6. An MP3 music player can be considered as a system.
   a. True.
   b. False.

7. Which of the following is *not* a linear circuit?
   a. Amplifier.
   b. Oscillator.
   c. Power supply.
   d. Flash memory.

8. The most common linear circuit is the
   a. amplifier.
   b. oscillator.
   c. rectifier.
   d. regulator.

9. Another name for a large-signal amplifier is
   a. op amp.
   b. RF amplifier.
   c. dc amplifier.
   d. power amplifier.

10. Most amplifiers are classified by
    a. power level.
    b. voltage level.
    c. frequency range.
    d. input impedance.

11. The name of a general-purpose dc differential amplifier that can be easily configured is the
    a. power amplifier.
    b. operational amplifier.
    c. microwave amplifier.
    d. small signal amplifier.

12. A circuit that generates a sine-wave or rectangular-wave signal is called a(n)
    a. oscillator.
    b. amplifier.
    c. rectifier.
    d. filter.

13. Which of the following is *not* a method of setting the frequency of an oscillator?
    a. Resistor-capacitor combination.
    b. Capacitor-inductor combination.
    c. Quartz crystal.
    d. Transformer

14. A circuit that generates a precise frequency that can be changed in increments is called a(n)
    a. amplifier.
    b. oscillator.
    c. frequency synthesizer.
    d. VCO.

15. The power supply circuit that converts ac into dc is the
    a. regulator.
    b. transformer.
    c. filter.
    d. rectifier.

16. A power supply circuit that maintains a constant dc output voltage is called a
    a. rectifier.
    b. regulator.
    c. filter.
    d. dc-dc converter.

17. A power supply circuit that converts dc into ac is known as a(n)
    a. rectifier.
    b. dc-dc converter.
    c. inverter.
    d. filter.

18. Digital circuits process which type of signals?
    a. Analog.
    b. Binary.

19. The basic building blocks of all digital circuits are
    a. resistors and capacitors.
    b. counters and registers.
    c. gates and flip-flops.
    d. diodes and transistors.

20. A circuit that generates a binary 1 only when all inputs are binary 1 is the
    a. AND gate.
    b. OR gate.
    c. inverter.
    d. NAND gate.

21. A circuit that store one bit of information is called a(n)
    a. AND gate.
    b. exclusive OR gate.
    c. inverter.
    d. flip-flop.

22. A circuit that stores one complete binary number is called a
    a. NAND gate.
    b. multiplexer.
    c. decoder.
    d. register.

23. A circuit that digitizes an analog signal is called a(n)
    a. analog-to-digital converter.
    b. digital-to-analog converter.
    c. comparator.
    d. counter.

24. A complete digital computer on a chip is called a(n)
    a. microcomputer.
    b. microcontroller.
    c. embedded controller.
    d. all of the above.

25. Which of the following is *not* a part of a microcontroller?
    a. Memory.
    b. Power supply.
    c. CPU.
    d. Input/output circuits.

26. Which electronic component can be used as a memory circuit?
    a. Capacitor.
    b. Resistor.
    c. Inductor.
    d. Transformer.

27. Which type of memory loses its stored data if power is removed?
    a. ROM.
    b. Flash.

c. RAM.
d. Nonvolatile memory.

28. A circuit that selects one digital input from memory and routes it to an output is a
    a. differential amplifier.
    b. multiplexer.
    c. decoder.
    d. demultiplexer.

29. What electronic circuit is found in almost every electronic product?
    a. Microcontroller.
    b. FPGA.
    c. PLL.
    d. DAC.

30. An FPGA is a digital circuit that
    a. is the same as a microcontroller.
    b. incorporates a PLL frequency synthesizer.
    c. contains multiple functional digital circuits.
    d. can be programmed to implement any digital function.

31. The circuit used to digitize an analog signal is a(n)
    a. analog-to-digital converter.
    b. digital-to-analog converter.
    c. comparator.
    d. a hybrid circuit.

32. The output is an ADC is
    a. an analog signal.
    b. a sequence of binary numbers.
    c. either high or low.
    d. a single binary number.

33. The circuit that reconstructs an analog signal from binary words stored in a memory is a(n)
    a. analog-to-digital converter.
    b. digital-to-analog converter.
    c. comparator.
    d. a hybrid circuit.

34. The output of a comparator is
    a. an analog signal.
    b. a sequence of binary numbers.
    c. either high or low.
    d. a single binary number.

35. Which of the following would *not* be considered as an electronic system?
    a. Garage door opener.
    b. TV set.
    c. Filter.
    d. Traffic light controller.

SECTION 27.3 Linear, or Analog, Circuits

27.1 An amplifier has an input of 40 mV and an output of 2 V. What is the gain?

27.2 An amplifier has an output power of 50 W and a gain of 800. What is the input power?

27.3 Express the voltage gain of 300 in dB.

27.4 Express the power gain of 75 in dB.

SECTION 27.6 System Examples

27.6 Draw a basic block diagram representing any system of your choice.

Chapter 28

Semiconductors

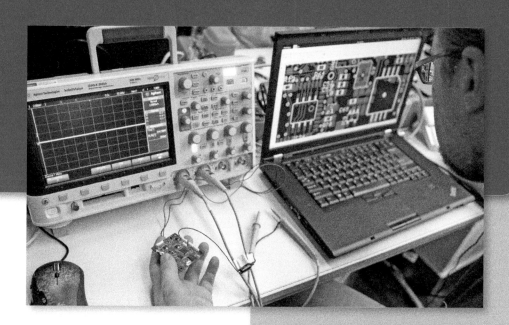

To understand how diodes, transistors, and integrated circuits work, you first have to study semiconductors: materials that are neither conductors nor insulators. Semiconductors contain some free electrons, but what makes them unusual is the presence of holes. In this chapter, you will learn about semiconductors, holes, and related topics.

Learning Outcomes

After studying this chapter, you should be able to:

> *Recognize*, at the atomic level, the characteristics of good conductors and semiconductors.

> *Describe* the structure of a silicon crystal.

> *List* the two types of carriers and *name* the type of impurity that causes each to be a majority carrier.

> *Explain* the conditions that exist at the *pn* junction of an unbiased diode, a forward-biased diode, and a reverse-biased diode.

> *Describe* the types of breakdown current caused by excessive reverse voltage across a diode.

28.1 Conductors

Copper is a good conductor. The reason is clear when we look at its atomic structure (Fig. 28-1). The nucleus of the atom contains 29 protons (positive charges). When a copper atom has a neutral charge, 29 electrons (negative charges) circle the nucleus like planets around the sun. The electrons travel in distinct *orbits* (also called *shells*). There are 2 electrons in the first orbit, 8 electrons in the second, 18 in the third, and 1 in the outer orbit.

Stable Orbits

The positive nucleus of Fig. 28-1 attracts the planetary electrons. The reason why these electrons are not pulled into the nucleus is the centrifugal (outward) force created by their circular motion. This centrifugal force is exactly equal to the inward pull of the nucleus, so that the orbit is stable. The idea is similar to a satellite that orbits the earth. At the right speed and height, a satellite can remain in a stable orbit above the earth.

The larger the orbit of an electron, the smaller the attraction of the nucleus. In a larger orbit, an electron travels more slowly, producing less centrifugal force. The outermost electron in Fig. 28-1 travels very slowly and feels almost no attraction to the nucleus.

The Core

In electronics, all that matters is the outer orbit. It is called the *valence orbit*. This orbit controls the electrical properties of the atom. To emphasize the importance of the valence orbit, we define the *core* of an atom as the nucleus and all the inner orbits. For a copper atom, the core is the nucleus (+29) and the first three orbits (−28).

The core of a copper atom has a net charge of +1 because it contains 29 protons and 28 inner electrons. Figure 28-2 can help in visualizing the core and the valence orbit. The valence electron is in a large orbit around a core and has a net charge of only +1. Because of this, the inward pull felt by the valence electron is very small.

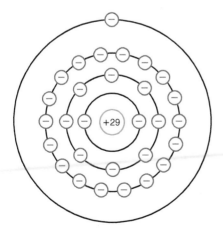

Figure 28-1 Copper atom.

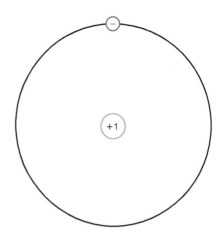

Figure 28-2 Core diagram of copper atom.

Free Electron

Since the attraction between the core and the valence electron is very weak, an outside force can easily dislodge this electron from the copper atom. This is why we often call the valence electron a **free electron.** This is also why copper is a good conductor. The slightest voltage causes the free electrons to flow from one atom to the next. The best conductors are silver, copper, and gold. All have a core diagram like Fig. 28-2.

EXAMPLE 28-1

Suppose an outside force removes the valence electron of Fig. 28-2 from a copper atom. What is the net charge of the copper atom? What is the net charge if an outside electron moves into the valence orbit of Fig. 28-2?

Answer:

When the valence electron leaves, the net charge of the atom becomes +1. Whenever an atom loses one of its electrons, it becomes positively charged. We call a positively charged atom a *positive ion*.

When an outside electron moves into the valence orbit of Fig. 28-2, the net charge of the atom becomes −1. Whenever an atom has an extra electron in its valence orbit, we call the negatively charged atom a *negative ion*.

28.2 Semiconductors

The best conductors (silver, copper, and gold) have one valence electron, whereas the best insulators have eight valence electrons. A **semiconductor** is an element with electrical properties between those of a conductor and those of an insulator. As you might expect, the best semiconductors have four valence electrons.

Germanium

Germanium is an example of a semiconductor. It has four electrons in the valence orbit. Many years ago, germanium was the only material suitable for making semiconductor devices. But these germanium devices had a fatal flaw (their excessive reverse current, discussed in a later section) that engineers could not overcome. Eventually, another semiconductor named **silicon** became practical and made germanium obsolete in most electronic applications.

Silicon

Next to oxygen, silicon is the most abundant element on the earth. But there were certain refining problems that prevented the use of silicon in the early days of semiconductors. Once these problems were solved, the advantages of silicon (discussed later) immediately made it the semiconductor of choice. Without it, modern electronics, communications, and computers would be impossible.

An isolated silicon atom has 14 protons and 14 electrons. As shown in Fig. 28-3a, the first orbit contains 2 electrons and the second orbit contains 8 electrons. The 4 remaining electrons are in the valence orbit. In Fig. 28-3a, the core has a net charge of +4 because it contains 14 protons in the nucleus and 10 electrons in the first two orbits.

Figure 28-3b shows the core diagram of a silicon atom. The 4 valence electrons tell us that silicon is a semiconductor.

EXAMPLE 28-2

What is the net charge of the silicon atom in Fig. 28-3b if it loses one of its valence electrons? If it gains an extra electron in the valence orbit?

Answer:

If it loses an electron, it becomes a positive ion with a charge of +1. If it gains an extra electron, it becomes a negative ion with a charge of −1.

28.3 Silicon Crystals

When silicon atoms combine to form a solid, they arrange themselves into an orderly pattern called a *crystal*. Each silicon atom shares its electrons with four neighboring atoms in such a way as to have eight electrons in its valence orbit. For instance, Fig. 28-4a shows a central atom with four neighbors. The shaded circles represent the silicon cores. Although the central atom originally had four electrons in its valence orbit, it now has eight.

Covalent Bonds

Each neighboring atom shares an electron with the central atom. In this way, the central atom has four additional electrons, giving it a total of eight electrons in the valence orbit. The electrons no longer belong to any single atom. Each central atom and its neighbors share the electrons. The same idea is true for all the other silicon atoms. In other words, every atom inside a silicon crystal has four neighbors.

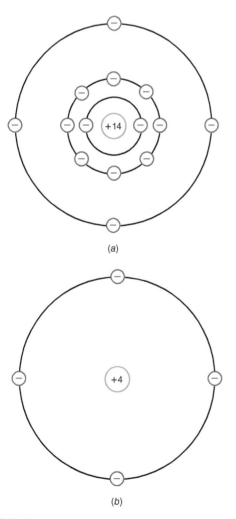

(a)

(b)

Figure 28-3   (a) Silicon atom. (b) Core diagram.

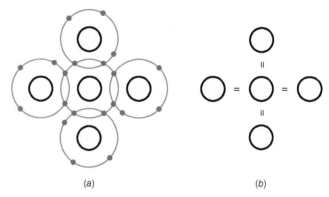

(a)

(b)

Figure 28-4   (a) Atom in crystal has four neighbors. (b) Covalent bonds.

In Fig. 28-4a, each core has a charge of +4. Look at the central core and the one to its right. These two cores attract the pair of electrons between them with equal and opposite force. This pulling in opposite directions is what holds the silicon atoms together. The idea is similar to tug-of-war teams pulling on a rope. As long as both teams pull with equal and opposite force, they remain bonded together.

Since each shared electron in Fig. 28-4a is being pulled in opposite directions, the electron becomes a bond between the opposite cores. We call this type of chemical bond a **covalent bond.** Figure 28-4b is a simpler way to show the concept of the covalent bonds. In a silicon crystal, there are billions of silicon atoms, each with eight valence electrons. These valence electrons are the covalent bonds that hold the crystal together—that give it solidity.

Valence Saturation

Each atom in a silicon crystal has eight electrons in its valence orbit. These eight electrons produce a chemical stability that results in a solid piece of silicon material. No one is quite sure why the outer orbit of all elements has a predisposition toward having eight electrons. When eight electrons do not exist naturally in an element, there seems to be a tendency for the element to combine and share electrons with other atoms so as to have eight electrons in the outer orbit.

There are advanced equations in physics that partially explain why eight electrons produce chemical stability in different materials, but no one knows the reason why the number eight is so special. It is one of those laws like the law of gravity, Coulomb's law, and other laws that we observe but cannot fully explain.

When the valence orbit has eight electrons, it is *saturated* because no more electrons can fit into this orbit. Stated as a law:

$$\text{Valence saturation: } n = 8 \qquad \textbf{(28-1)}$$

In words, *the valence orbit can hold no more than eight electrons.* Furthermore, the eight valence electrons are called *bound electrons* because they are tightly held by the atoms. Because of these bound electrons, a silicon crystal is almost a perfect insulator at room temperature, approximately 25°C.

The Hole

The **ambient temperature** is the temperature of the surrounding air. When the ambient temperature is above absolute zero (−273°C), the heat energy in this air causes the atoms in a silicon crystal to vibrate. The higher the ambient temperature, the stronger the mechanical vibrations become. When you pick up a warm object, the warmth you feel is the effect of the vibrating atoms.

In a silicon crystal, the vibrations of the atoms can occasionally dislodge an electron from the valence orbit. When

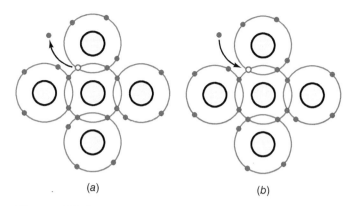

Figure 28-5 (a) Thermal energy produces electron and hole. (b) Recombination of free electron and hole.

this happens, the released electron gains enough energy to go into a larger orbit, as shown in Fig. 28-5a. In this larger orbit, the electron is a free electron.

But that's not all. The departure of the electron creates a vacancy in the valence orbit called a **hole** (see Fig. 28-5a). This hole behaves like a positive charge because the loss of the electron produces a positive ion. The hole will attract and capture any electron in the immediate vicinity. The existence of holes is the critical difference between conductors and semiconductors. Holes enable semiconductors to do all kinds of things that are impossible with conductors.

At room temperature, thermal energy produces only a few holes and free electrons. To increase the number of holes and free electrons, it is necessary to *dope* the crystal. More is said about this in a later section.

Recombination and Lifetime

In a pure silicon crystal, **thermal** (heat) **energy** creates an equal number of free electrons and holes. The free electrons move randomly throughout the crystal. Occasionally, a free electron will approach a hole, feel its attraction, and fall into it. **Recombination** is the merging of a free electron and a hole (see Fig. 28-5b).

The amount of time between the creation and disappearance of a free electron is called the *lifetime*. It varies from a few nanoseconds to several microseconds, depending on how perfect the crystal is and other factors.

Main Ideas

At any instant, the following is taking place inside a silicon crystal:

1. Some free electrons and holes are being created by thermal energy.
2. Other free electrons and holes are recombining.
3. Some free electrons and holes exist temporarily, awaiting recombination.

EXAMPLE 28-3

If a pure silicon crystal has 1 million free electrons inside it, how many holes does it have? What happens to the number of free electrons and holes if the ambient temperature increases?

Answer:

Look at Fig. 28-5a. When heat energy creates a free electron, it automatically creates a hole at the same time. Therefore, a pure silicon crystal always has the same number of holes and free electrons. If there are 1 million free electrons, there are 1 million holes.

A higher temperature increases the vibrations at the atomic level, which means that more free electrons and holes are created. But no matter what the temperature is, a pure silicon crystal has the same number of free electrons and holes.

Crystal Lattice Structure

The covalent bond shown in Fig. 28-4 is only two dimensional to simplify the discussion of bonding. Of course, the covalent bonds are actually three dimensional. The bonding process usually produces an arrangement such as that shown in Fig. 28-6a. This is called a unit cell. It represents five atoms that arrange themselves within an imaginary cubic structure. Then this unit cell is repeated again and again in

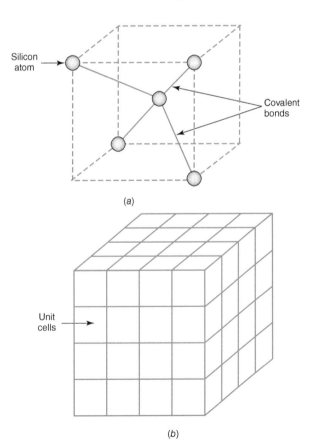

(a)

(b)

Figure 28-6 Bonding process. (a) Unit cell. (b) Unit cells arrange themselves into a crystalline structure.

a three-dimensional way to form a crystal lattice structure as Fig. 28-6b shows. All semiconductor materials have a similar crystal lattice arrangement.

Compound Semiconductors

While any semiconductor material such as silicon, germanium or carbon will form covalent bonds among the atoms, a semiconductor material can also be formed with two or more materials. Such semiconductor materials are known as *compound semiconductors.* An example of a compound semiconductor with elements having four valence electrons is silicon germanium (SiGe). Another less common combination is silicon carbon (SiC) and gallium nitride (GaN). The most common compound semiconductor materials are formed between an element with three valence electrons and another with five valence electrons. These are elements from columns III (13) and V (15) of the periodic table and are often referred to as III-V compounds. When such elements are combined, the result is a total of eight valence electrons producing valence saturation. Some common elements with three valence electrons are boron (B), aluminum (Al), gallium (Ga), and indium (In). Elements with five valence electrons are nitrogen (N), phosphorus (P), arsenic (As), antimony (Sb), and bismuth (Bi).

Some of the most common compound semiconductors are silicon germanium, gallium arsenide (GaAs), and indium phosphide (InP). These combinations produce a semiconductor in which current flows faster, making them useful at very high microwave and even light frequencies. Compounds of three or four of these materials (e.g., AlGaSb or AlGaPAs) are used to form light-emitting diodes (LEDs) and lasers.

28.4 Intrinsic Semiconductors

An **intrinsic semiconductor** is a pure semiconductor. A silicon crystal is an intrinsic semiconductor if every atom in the crystal is a silicon atom. At room temperature, a silicon crystal acts like an insulator because it has only a few free electrons and holes produced by thermal energy.

Flow of Free Electrons

Figure 28-7 shows part of a silicon crystal between charged metallic plates. Assume that thermal energy has produced a free electron and a hole. The free electron is in a large orbit at the right end of the crystal. Because of the negatively charged plate, the free electron is repelled to the left. This free electron can move from one large orbit to the next until it reaches the positive plate.

Flow of Holes

Notice the hole at the left of Fig. 28-7. This hole attracts the valence electron at point A. This causes the valence electron to move into the hole.

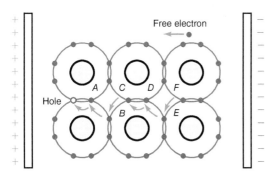

Figure 28-7   Hole flow through a semiconductor.

When the valence electron at point *A* moves to the left, it creates a new hole at point *A*. The effect is the same as moving the original hole to the right. The new hole at point *A* can then attract and capture another valence electron. In this way, valence electrons can travel along the path shown by the arrows. This means the hole can move the opposite way, along path *A-B-C-D-E-F*, acting the same as a positive charge.

28.5 Two Types of Flow

Figure 28-8 shows an intrinsic semiconductor. It has the same number of free electrons and holes. This is because *thermal energy produces free electrons and holes in pairs.* The applied voltage will force the free electrons to flow left and the holes to flow right. When the free electrons arrive at the left end of the crystal, they enter the external wire and flow to the positive battery terminal.

On the other hand, the free electrons at the negative battery terminal will flow to the right end of the crystal. At this point, they enter the crystal and recombine with holes that arrive at the right end of the crystal. In this way, a steady flow of free electrons and holes occurs inside the semiconductor. Note that there is no hole flow outside the semiconductor.

In Fig. 28-8, *the free electrons and holes move in opposite directions.* From now on, we will visualize the current in a semiconductor as the combined effect of the two types of flow: the flow of free electrons in one direction and the flow of holes in the other direction. Free electrons and holes are often called *carriers* because they carry a charge from one place to another.

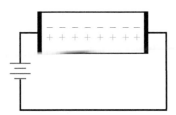

Figure 28-8   Intrinsic semiconductor has equal number of free electrons and holes.

28.6 Doping a Semiconductor

One way to increase conductivity of a semiconductor is by **doping.** This means adding impurity atoms to an intrinsic crystal to alter its electrical conductivity. A doped semiconductor is called an **extrinsic semiconductor.**

Increasing the Free Electrons

How does a manufacturer dope a silicon crystal? The first step is to melt a pure silicon crystal. This breaks the covalent bonds and changes the silicon from a solid to a liquid. To increase the number of free electrons, *pentavalent atoms* are added to the molten silicon. Pentavalent atoms have five electrons in the valence orbit. Examples of pentavalent atoms include arsenic, antimony, and phosphorus. Because these materials *will donate an extra electron* to the silicon crystal, they are often referred to as *donor impurities.*

Figure 28-9a shows how the doped silicon crystal appears after it cools down and re-forms its solid crystal structure. A pentavalent atom is in the center, surrounded by four silicon atoms. As before, the neighboring atoms share an electron with the central atom. But this time, there is an extra electron left over. Remember that each pentavalent atom has five valence electrons. Since only eight electrons can fit into the valence orbit, the extra electron remains in a larger orbit. In other words, it is a free electron.

Each pentavalent or donor atom in a silicon crystal produces one free electron. This is how a manufacturer controls the conductivity of a doped semiconductor. The more impurity that is added, the greater the conductivity. In this way, a semiconductor may be lightly or heavily doped. A lightly doped semiconductor has a high resistance, whereas a heavily doped semiconductor has a low resistance.

Increasing the Number of Holes

How can we dope a pure silicon crystal to get an excess of holes? By using a *trivalent impurity,* one whose atoms have

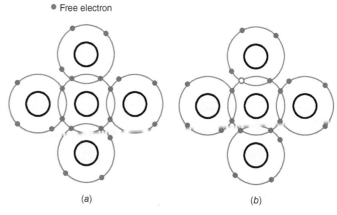

• Free electron

(a)   (b)

Figure 28-9   (a) Doping to get more free electrons. (b) Doping to get more holes.

only three valence electrons. Examples include aluminum, boron, and gallium.

Figure 28-9b shows a trivalent atom in the center. It is surrounded by four silicon atoms, each sharing one of its valence electrons. Since the trivalent atom originally had only three valence electrons and each neighbor shares one electron, only seven electrons are in the valence orbit. This means that a hole exists in the valence orbit of each trivalent atom. A trivalent atom is also called an *acceptor atom* because each hole it contributes can accept a free electron during recombination.

28.7 Two Types of Extrinsic Semiconductors

A semiconductor can be doped to have an excess of free electrons or an excess of holes. Because of this, there are two types of doped semiconductors.

N-Type Semiconductor

Silicon that has been doped with a pentavalent impurity is called an ***n*-type semiconductor,** where the *n* stands for negative. Figure 28-10 shows an *n*-type semiconductor. Since the free electrons outnumber the holes in an *n*-type semiconductor, the free electrons are called the **majority carriers** and the holes are called the **minority carriers.**

Because of the applied voltage, the *free electrons move to the left* and the *holes move to the right.* When a hole arrives at the right end of the crystal, one of the free electrons from the external circuit enters the semiconductor and recombines with the hole.

The free electrons shown in Fig. 28-10 flow to the left end of the crystal, where they enter the wire and flow on to the positive terminal of the battery.

P-Type Semiconductor

Silicon that has been doped with a trivalent impurity is called a ***p*-type semiconductor,** where the *p* stands for positive. Figure 28-11 shows a *p*-type semiconductor. Since holes outnumber free electrons, the holes are referred to as

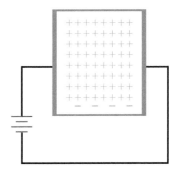

Figure 28-11   *P*-type semiconductor has many holes.

the majority carriers and the free electrons are known as the minority carriers.

Because of the applied voltage, the *free electrons move to the left* and the *holes move to the right.* In Fig. 28-11, the holes arriving at the right end of the crystal will recombine with free electrons from the external circuit.

There is also a flow of minority carriers in Fig. 28-11. The free electrons inside the semiconductor flow from right to left. Because there are so few minority carriers, they have almost no effect in this circuit.

28.8 The Unbiased Diode

By itself, a piece of *n*-type semiconductor is about as useful as a carbon resistor; the same can be said for a *p*-type semiconductor. But when a manufacturer dopes a crystal so that one-half of it is *p*-type and the other half is *n*-type, something new comes into existence.

The border between *p*-type and *n*-type is called the ***pn junction.*** The *pn* junction has led to all kinds of inventions including diodes, transistors, and integrated circuits. Understanding the *pn* junction enables you to understand all kinds of semiconductor devices.

The Unbiased Diode

As discussed in the preceding section, each trivalent atom in a doped silicon crystal produces one hole. For this reason, we can visualize a piece of *p*-type semiconductor as shown on the left side of Fig. 28-12. Each circled minus sign is the trivalent atom, and each plus sign is the hole in its valence orbit.

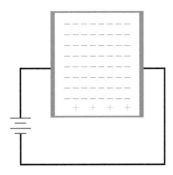

Figure 28-10   *N*-type semiconductor has many free electrons.

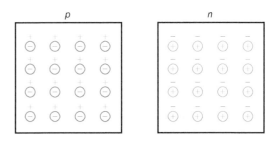

Figure 28-12   Two types of semiconductor.

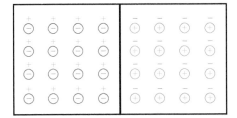

Figure 28-13   The *pn* junction.

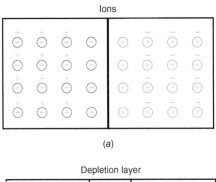

(a)

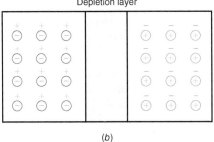

(b)

Figure 28-14   (a) Creation of ions at junction. (b) Depletion layer.

Similarly, we can visualize the pentavalent atoms and free electrons of an *n*-type semiconductor as shown on the right side of Fig. 28-12. Each circled plus sign represents a pentavalent atom, and each minus sign is the free electron it contributes to the semiconductor. Notice that each piece of semiconductor material is *electrically neutral because the number of pluses and minuses is equal.*

A manufacturer can produce a single crystal with *p*-type material on one side and *n*-type on the other side, as shown in Fig. 28-13. The junction is the border where the *p*-type and the *n*-type regions meet, and **junction diode** is another name for a *pn* crystal. The word **diode** is a contraction of two electrodes, where *di* stands for "two."

The Depletion Layer

Because of their repulsion for each other, the free electrons on the *n* side of Fig. 28-13 tend to diffuse (spread) in all directions. Some of the free electrons diffuse across the junction. When a free electron enters the *p* region, it becomes a minority carrier. With so many holes around it, this minority carrier has a short lifetime. Soon after entering the *p* region, the free electron recombines with a hole. When this happens, the *hole disappears* and the *free electron becomes a valence electron.*

Each time an electron diffuses across a junction, it creates a pair of ions. When an electron leaves the *n* side, it leaves behind a pentavalent atom that is short one negative charge; this pentavalent atom becomes a positive ion. After the migrating electron falls into a hole on the *p* side, it makes a negative ion out of the trivalent atom that captures it.

Figure 28-14*a* shows these ions on each side of the junction. The circled plus signs are the positive ions, and the circled minus signs are the negative ions. The ions are fixed in the crystal structure because of covalent bonding, and they cannot move around like free electrons and holes.

Each pair of positive and negative ions at the junction is called a *dipole*. The creation of a dipole means that one free electron and one hole have been taken out of circulation. As the number of dipoles builds up, the region near the junction is emptied of carriers. We call this charge-empty region the **depletion layer** (see Fig. 28-14*b*).

Barrier Potential

Each dipole has an electric field between the positive and negative ions. Therefore, if additional free electrons enter the depletion layer, the electric field tries to push these electrons back into the *n* region. The strength of the electric field increases with each crossing electron until equilibrium is reached. To a first approximation, this means that the electric field eventually stops the diffusion of electrons across the junction.

In Fig. 28-14*a*, the electric field between the ions is equivalent to a difference of potential called the **barrier potential.** At 25°C, the barrier potential equals approximately 0.3 V for germanium diodes and 0.7 V for silicon diodes.

28.9 Forward Bias

Figure 28-15 shows a dc source across a diode. The negative source terminal is connected to the *n*-type material, and the positive terminal is connected to the *p*-type material. This connection produces what is called **forward bias.**

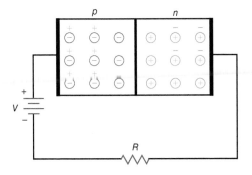

Figure 28-15   Forward bias.

Flow of Free Electrons

In Fig. 28-15, the battery pushes holes and free electrons toward the junction. If the battery voltage is less than the barrier potential, the free electrons do not have enough energy to get through the depletion layer. When they enter the depletion layer, the ions will push them back into the *n* region. Because of this, there is no current through the diode.

When the dc voltage source is greater than the barrier potential, the battery again pushes holes and free electrons toward the junction. This time, the free electrons have enough energy to pass through the depletion layer and re-combine with the holes. If you visualize all the holes in the *p* region moving to the right and all the free electrons moving to the left, you will have the basic idea. Somewhere in the vicinity of the junction, these opposite charges re-combine. Since free electrons continuously enter the right end of the diode and holes are being continuously created at the left end, there is a continuous current through the diode.

The Flow of One Electron

Let us follow a single electron through the entire circuit. After the free electron leaves the negative terminal of the battery, it enters the right end of the diode. It travels through the *n* region until it reaches the junction. When the battery voltage is greater than 0.7 V, the free electron has enough energy to get across the depletion layer. Soon after the free electron has entered the *p* region, it recombines with a hole.

In other words, the free electron becomes a valence electron. As a valence electron, it continues to travel to the left, passing from one hole to the next until it reaches the left end of the diode. When it leaves the left end of the diode, a new hole appears and the process begins again. Since there are billions of electrons taking the same journey, we get a continuous current through the diode. A series resistor is used to limit the amount of forward current.

What to Remember

Current flows easily in a forward-biased diode. As long as the applied voltage is greater than the barrier potential, there will be a large continuous current in the circuit. In other words, if the source voltage is greater than 0.7 V, a silicon diode allows a continuous current in the forward direction.

28.10 Reverse Bias

Turn the dc source around and you get Fig. 28-16. This time, the negative battery terminal is connected to the *p* side, and the positive battery terminal to the *n* side. This connection produces what is called **reverse bias.**

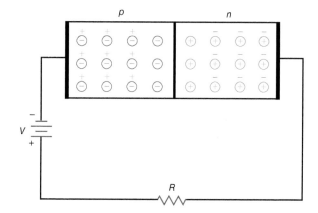

Figure 28-16   Reverse bias.

Depletion Layer Widens

The negative battery terminal attracts the holes, and the positive battery terminal attracts the free electrons. Because of this, holes and free electrons flow away from the junction. Therefore, the depletion layer gets wider.

How wide does the depletion layer get in Fig. 28-17*a*? When the holes and electrons move away from the junction, the newly created ions increase the difference of potential across the depletion layer. The wider the depletion layer, the greater the difference of potential. The depletion layer stops growing when its difference of potential equals the applied reverse voltage. When this happens, electrons and holes stop moving away from the junction.

Sometimes the depletion layer is shown as a shaded region like that of Fig. 28-17*b*. The width of this shaded region is proportional to the reverse voltage. *As the reverse voltage increases, the depletion layer gets wider.*

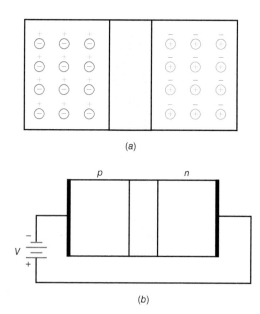

Figure 28-17   (*a*) Depletion layer. (*b*) Increasing reverse bias widens depletion layer.

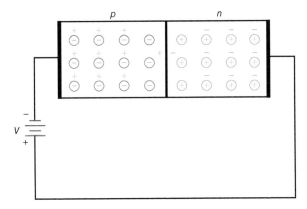

Figure 28-18 Thermal production of free electron and hole in depletion layer produces reverse minority saturation current.

Minority-Carrier Current

Is there any current after the depletion layer stabilizes? Yes. A small current exists with reverse bias. Recall that thermal energy continuously creates pairs of free electrons and holes. This means that a few minority carriers exist on both sides of the junction. Most of these recombine with the majority carriers. But those inside the depletion layer may exist long enough to get across the junction. When this happens, a small current flows in the external circuit.

Figure 28-18 illustrates the idea. Assume that thermal energy has created a free electron and hole near the junction. The depletion layer pushes the free electron to the right, forcing one electron to leave the right end of the crystal. The hole in the depletion layer is pushed to the left. This extra hole on the *p* side lets one electron enter the left end of the crystal and fall into a hole. Since thermal energy is continuously producing electron-hole pairs inside the depletion layer, a small continuous current flows in the external circuit.

The reverse current caused by the thermally produced minority carriers is called the **saturation current.** In equations, the saturation current is symbolized by I_S. The name *saturation* means that we cannot get more minority-carrier current than is produced by the thermal energy. In other words, *increasing the reverse voltage will not increase the number of thermally created minority carriers.*

Surface-Leakage Current

Besides the thermally produced minority-carrier current, does any other current exist in a reverse-biased diode? Yes. A small current flows on the surface of the crystal. Known as the **surface-leakage current,** it is caused by surface impurities and imperfections in the crystal structure.

The reverse current in a diode consists of a minority-carrier current and a surface-leakage current. In most applications, the reverse current in a silicon diode is so small that you don't even notice it. The main idea to remember is this: *Current is approximately zero in a reverse-biased silicon diode.*

28.11 Breakdown

Diodes have maximum voltage ratings. There is a limit to how much reverse voltage a diode can withstand before it is destroyed. If you continue increasing the reverse voltage, you will eventually reach the **breakdown voltage** of the diode. For many diodes, breakdown voltage is at least 50 V. The breakdown voltage is shown on the *data sheet* for the diode. We discuss data sheets in Chap. 29.

Once the breakdown voltage is reached, a large number of the minority carriers suddenly appears in the depletion layer and the diode conducts heavily.

Where do the carriers come from? They are produced by the **avalanche effect** (see Fig. 28-19), which occurs at higher reverse voltages. Here is what happens. As usual, there is a small reverse minority-carrier current. When the reverse voltage increases, it forces the minority carriers to move more quickly. These minority carriers collide with the atoms of the crystal. When these minority carriers have enough energy, they can knock valence electrons loose, producing free electrons. These new minority carriers then join the existing minority carriers to collide with other atoms. The process is geometric, because one free electron liberates one valence electron to get two free electrons. These two free electrons then free two more electrons to get four free electrons. The process continues until the reverse current becomes huge.

Figure 28-20 shows a magnified view of the depletion layer. The reverse bias forces the free electron to move to the right. As it moves, the electron gains speed. The larger the reverse bias, the faster the electron moves. If the high-speed electron has enough energy, it can bump the valence electron of the first atom into a larger orbit. This results in

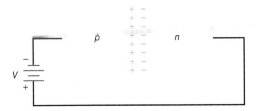

Figure 28-19 Avalanche produces many free electrons and holes in depletion layer.

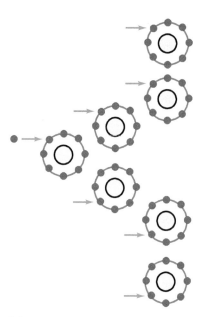

Figure 28-20 The process of avalanche is a geometric progression: 1, 2, 4, 8, . . .

two free electrons. Both of these then accelerate and go on to dislodge two more electrons. In this way, the number of minority carriers may become quite large and the diode can conduct heavily.

The breakdown voltage of a diode depends on how heavily doped the diode is. With rectifier diodes (the most common type), the breakdown voltage is usually greater than 50 V. Table 28-1 illustrates the difference between a forward- and reverse-biased diode.

28.12 Advances in Semiconductors

Two major advances in semiconductors are further affecting electronics, MEMS and nanotechnology. *MEMS* is an acronym for microelectromechanical systems. MEMS are mechanical devices made using semiconductor technology. The techniques for making transistors and integrated circuits are now being used to build tiny mechanical machines. Dimensions are in the less than 1-mm range or in the 10- to 100-micron (10^{-6} meter) range. These devices are made up of gears, levers, ratchets, and other moving structures. Today, such process techniques can make switches, relays, mirrors, and vibrating components. Miniature mirror structures are what make up the digital light-processing (DLP) devices that implement some large HDTV screens. Vibrating elements can also now be made precise enough to replace quartz crystals as electronic frequency-setting components. MEMS are being used more and more to create sensors and actuators for medical and scientific applications. A key feature is that any related electronic circuitry can also be created simultaneously on the same silicon wafer.

Nanotechnology is a new science based on the idea that you can now build any device by manipulating molecules and atoms individually. Techniques are now being developed to literally assemble a structure an atom or molecule at a time. The devices could be transistors and other electronic components or mechanical devices. Nanotechnology is a developing technology still primarily in the research phase. Few actual nanotechnology devices or structures exist, and these are experimental items.

Both these technologies bear watching as they affect electronics.

Table 28-1 Diode Bias Summary

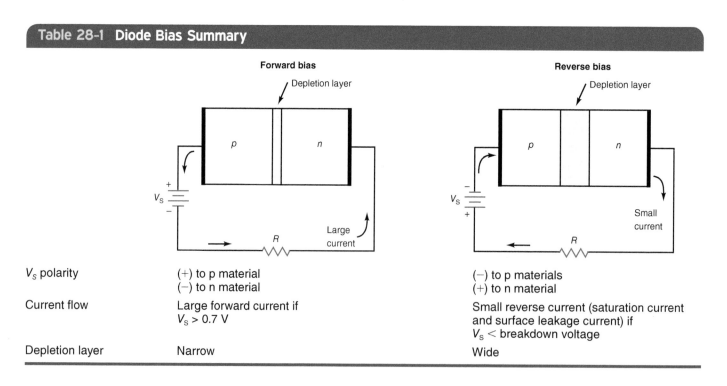

	Forward bias	Reverse bias
V_S polarity	(+) to p material (−) to n material	(−) to p materials (+) to n material
Current flow	Large forward current if $V_S > 0.7$ V	Small reverse current (saturation current and surface leakage current) if $V_S <$ breakdown voltage
Depletion layer	Narrow	Wide

1. The nucleus of a copper atom contains how many protons?
   a. 1
   b. 4
   c. 18
   d. 29

2. The net charge of a neutral copper atom is
   a. 0.
   b. +1.
   c. −1.
   d. +4.

3. Assume the valence electron is removed from a copper atom. The net charge of the atom becomes
   a. 0.
   b. +1.
   c. −1.
   d. +4.

4. The valence electron of a copper atom experiences what kind of attraction toward the nucleus?
   a. None.
   b. Weak.
   c. Strong.
   d. Impossible to say.

5. How many valence electrons does a silicon atom have?
   a. 0
   b. 1
   c. 2
   d. 4

6. Which is the most widely used semiconductor?
   a. Copper.
   b. Germanium.
   c. Silicon.
   d. None of the above.

7. How many protons does the nucleus of a silicon atom contain?
   a. 4
   b. 14
   c. 29
   d. 32

8. Silicon atoms combine into an orderly pattern called a
   a. covalent bond.
   b. crystal.
   c. semiconductor.
   d. valence orbit.

9. An intrinsic semiconductor has some holes in it at room temperature. What causes these holes?
   a. Doping.
   b. Free electrons.
   c. Thermal energy.
   d. Valence electrons.

10. When an electron is moved to a higher orbit level, its energy level with respect to the nucleus
   a. increases
   b. decreases
   c. remains the same
   d. depends on the type of atom

11. The merging of a free electron and a hole is called
   a. covalent bonding.
   b. lifetime.
   c. recombination.
   d. thermal energy.

12. At room temperature an intrinsic silicon crystal acts approximately like
   a. a battery.
   b. a conductor.
   c. an insulator.
   d. a piece of copper wire.

13. The amount of time between the creation of a hole and its disappearance is called
   a. doping.
   b. lifetime.
   c. recombination.
   d. valence.

14. The valence electron of a conductor can also be called a
   a. bound electron.
   b. free electron.
   c. nucleus.
   d. proton.

15. A conductor has how many types of flow?
   a. 1
   b. 2
   c. 3
   d. 4

16. A semiconductor has how many types of flow?
   a. 1
   b. 2
   c. 3
   d. 4

17. When a voltage is applied to a semiconductor, holes will flow
    a. away from the negative potential.
    b. toward the positive potential.
    c. in the external circuit.
    d. none of the above.

18. For semiconductor material, its valence orbit is saturated when it contains
    a. 1 electron.
    b. equal (+) and (−) ions.
    c. 4 electrons.
    d. 8 electrons.

19. In an intrinsic semiconductor, the number of holes
    a. equals the number of free electrons.
    b. is greater than the number of free electrons.
    c. is less than the number of free electrons.
    d. none of the above.

20. Absolute zero temperature equals
    a. −273°C.
    b. 0°C.
    c. 25°C.
    d. 50°C.

21. At absolute zero temperature an intrinsic semiconductor has
    a. a few free electrons.
    b. many holes.
    c. many free electrons.
    d. no holes or free electrons.

22. At room temperature an intrinsic semiconductor has
    a. a few free electrons and holes.
    b. many holes.
    c. many free electrons.
    d. no holes.

23. The number of free electrons and holes in an intrinsic semiconductor decreases when the temperature
    a. decreases.
    b. increases.
    c. stays the same.
    d. none of the above.

24. The flow of valence electrons to the right means that holes are flowing to the
    a. left.
    b. right.
    c. either way.
    d. none of the above.

25. Holes act like
    a. atoms.
    b. crystals.

    c. negative charges.
    d. positive charges.

26. Trivalent atoms have how many valence electrons?
    a. 1
    b. 3
    c. 4
    d. 5

27. An acceptor atom has how many valence electrons?
    a. 1
    b. 3
    c. 4
    d. 5

28. If you wanted to produce an *n*-type semiconductor, which of these would you use?
    a. Acceptor atoms.
    b. Donor atoms.
    c. Pentavalent impurity.
    d. Silicon.

29. Electrons are the minority carriers in which type of semiconductor?
    a. Extrinsic.
    b. Intrinsic.
    c. *N*-type.
    d. *P*-type.

30. How many free electrons does a *p*-type semiconductor contain?
    a. Many.
    b. None.
    c. Only those produced by thermal energy.
    d. Same number as holes.

31. Silver is the best conductor. How many valence electrons do you think it has?
    a. 1
    b. 4
    c. 18
    d. 29

32. Suppose an intrinsic semiconductor has 1 billion free electrons at room temperature. If the temperature drops to 0°C, how many holes are there?
    a. Fewer than 1 billion.
    b. 1 billion.
    c. More than 1 billion.
    d. Impossible to say.

33. An external voltage source is applied to a *p*-type semiconductor. If the left end of the crystal is positive, which way do the majority carriers flow?
    a. Left.
    b. Right.

c. Neither.

d. Impossible to say.

34. Which of the following doesn't fit in the group?
    a. Conductor.
    b. Semiconductor.
    c. Four valence electrons.
    d. Crystal structure.

35. Which of the following is approximately equal to room temperature?
    a. 0°C.
    b. 25°C.
    c. 50°C.
    d. 75°C.

36. How many electrons are there in the valence orbit of a silicon atom within a crystal?
    a. 1
    b. 4
    c. 8
    d. 14

37. Negative ions are atoms that have
    a. gained a proton.
    b. lost a proton.
    c. gained an electron.
    d. lost an electron.

38. Which of the following describes an *n*-type semiconductor?
    a. Neutral.
    b. Positively charged.
    c. Negatively charged.
    d. Has many holes.

39. A *p*-type semiconductor contains holes and
    a. positive ions.
    b. negative ions.
    c. pentavalent atoms.
    d. donor atoms.

40. Which of the following describes a *p*-type semiconductor?
    a. Neutral.
    b. Positively charged.
    c. Negatively charged.
    d. Has many free electrons.

41. What is the correct name for a large three-dimensional structure of semiconductor atoms?
    a. Crystal lattice structure.
    b. Atomic matrix.
    c. Covalent bonds.
    d. Interlinked atoms.

42. What causes the depletion layer?
    a. Doping.
    b. Recombination.

c. Barrier potential.

d. Ions.

43. What is the barrier potential of a silicon diode at room temperature?
    a. 0.3 V.
    b. 0.7 V.
    c. 1 V.
    d. 2 mV per degree Celsius.

44. Which of the following is *not* a common compound semiconductor?
    a. SiGe
    b. GaAs
    c. AlSb
    d. InP

45. In a silicon diode the reverse current is usually
    a. very small.
    b. very large.
    c. zero.
    d. in the breakdown region.

46. While maintaining a constant temperature, a silicon diode has its reverse-bias voltage increased. The diode's saturation current will
    a. increase.
    b. decrease.
    c. remain the same.
    d. equal its surface-leakage current.

47. The voltage where avalanche occurs is called the
    a. barrier potential.
    b. depletion layer.
    c. knee voltage.
    d. breakdown voltage.

48. Compound semiconductors have the following advantage over silicon.
    a. Smaller.
    b. Electrons flow faster for high-frequency operation.
    c. lower-power consumption.
    d. Lower leakage current.

49. When the reverse voltage decreases from 10 to 5 V, the depletion layer
    a. becomes smaller.
    b. becomes larger.
    c. is unaffected.
    d. breaks down.

50. A reverse voltage of 10 V is across a diode. What is the voltage across the depletion layer?
    a. 0 V.
    b. 0.7 V.
    c. 10 V.
    d. None of the above.

51. MEMS refers to the process of building which devices on a silicon chip?
- **a.** Mechanical devices.
- **b.** Electric/electronic circuits.
- **c.** Both *a* and *b*.
- **d.** Nano devices.

52. Nanotechnology refers to process of building miniature structures with
- **a.** silicon.
- **b.** compound semiconductors.
- **c.** MEMS.
- **d.** atoms.

 CHAPTER 28 PROBLEMS

28.1 What is the net charge of a copper atom if it gains two electrons?

28.2 What is the net charge of a silicon atom if it gains three valence electrons?

28.3 Classify each of the following as conductor or semiconductor:
- **a.** Germanium
- **b.** Silver
- **c.** Silicon
- **d.** Gold

28.4 If a pure silicon crystal has 500,000 holes inside it, how many free electrons does it have?

28.5 A diode is forward-biased. If the current is 5 mA through the *n* side, what is the current through each of the following?
- **a.** *P* side.
- **b.** External connecting wires.
- **c.** Junction.

Chapter 29

Diode Types and Operation

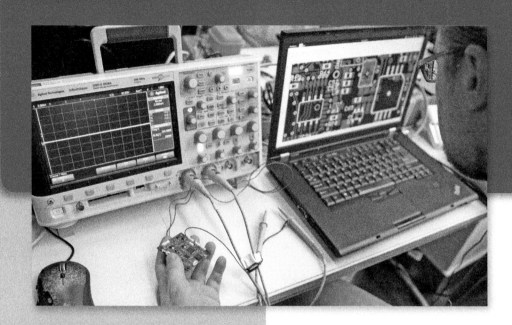

Learning Outcomes

After studying this chapter, you should be able to:

> *Draw* a diode symbol and *label* the anode and cathode.

> *Describe* the ideal diode.

> *List* four basic characteristics of diodes shown on a data sheet.

> *Show* how the zener diode is used and *calculate* various values related to its operation.

> *List* several optoelectronic devices and *describe* how each works.

> *Recall* two advantages Schottky diodes have over common diodes.

> *Explain* how a varactor works.

> *State* a primary use of the varistor.

This chapter continues your study of diodes. After discussing the diode curve, we look at approximations of a diode. We need approximations to simplify and speed up troubleshooting and circuit analysis.

Rectifier diodes are the most common type of diode. They are used in power supplies to convert ac voltage to dc voltage. You will learn about rectifiers in Chap. 30. The chapter continues with zener diodes, which are important because they are the key to voltage regulation. The chapter also covers optoelectronic diodes, Schottky diodes, varactors, and PIN diodes.

29.1 Basic Diode Characteristics

An ordinary resistor is a **linear device** because the graph of its current versus voltage is a straight line. A diode is different. It is a **nonlinear device** because the graph of its current versus voltage is not a straight line. The reason is the barrier potential. When the diode voltage is less than the barrier potential, the diode current is small. When the diode voltage exceeds the barrier potential, the diode current increases rapidly.

The Schematic Symbol and Case Styles

Figure 29-1a shows the schematic symbol of a diode. The *p* side is called the **anode,** and the *n* side the **cathode.** The diode symbol looks like an arrow that points from the *p* side to the *n* side, from the anode to the cathode. Figure 29-1b shows some of the many typical diode case styles. Many, but not all, diodes have the cathode lead (K) identified by a colored band.

Basic Diode Circuit

Figure 29-1c shows a diode circuit. In this circuit, the diode is forward biased. How do we know? Because the positive battery terminal drives the *p* side through a resistor, and the negative battery terminal is connected to the *n* side. With this connection, the circuit is trying to push holes and free electrons toward the junction.

In more complicated circuits, it may be difficult to decide whether the diode is forward-biased. Here is a guideline. Ask yourself this question, Is the external circuit pushing current in the *easy direction* of flow? If the answer is yes, the diode is forward-biased.

What is the easy direction of flow? If you use conventional current, the easy direction is the same direction as the diode arrow. If you prefer electron flow, the easy direction is the other way.

When the diode is part of a complicated circuit, we also can use Thevenin's theorem to determine whether it is forward-biased. For instance, assume that we have reduced a complicated circuit with Thevenin's theorem to get Fig. 29-1c. We would know that the diode is forward-biased.

The Forward Region

Figure 29-1c is a circuit that you can set up in the laboratory. After you connect this circuit, you can measure the diode current and voltage. You can also reverse the polarity of the dc source and measure diode current and voltage for reverse bias. If you plot the diode current versus the diode voltage, you will get a graph that looks like Fig. 29-2.

This is a visual summary of the ideas discussed in the preceding chapter. For instance, when the diode is forward-biased, there is no significant current until the diode voltage is greater than the barrier potential. On the other hand, when the diode is reverse-biased, there is almost no reverse current until the diode voltage reaches the breakdown voltage. Then avalanche produces a large reverse current, destroying the diode.

Knee Voltage

In the forward region, the voltage at which the current starts to increase rapidly is called the **knee voltage** of the diode. The knee voltage equals the barrier potential. Analysis of diode circuits usually comes down to determining whether the diode voltage is more or less than the knee voltage. If it's more, the diode conducts easily. If it's less, the diode conducts poorly. We define the knee voltage of a silicon diode as

$$V_K \approx 0.7 \text{ V} \qquad \text{(29-1)}$$

(*Note:* The symbol $\approx$ means "approximately equal to.")

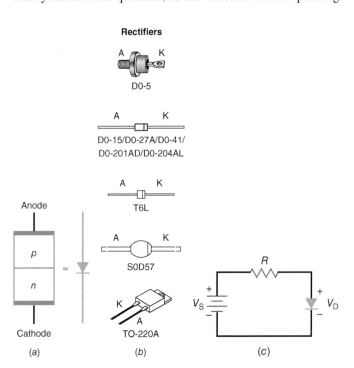

Figure 29-1 Diode. (*a*) Schematic symbol. (*b*) Diode case styles. (*c*) Forward bias.

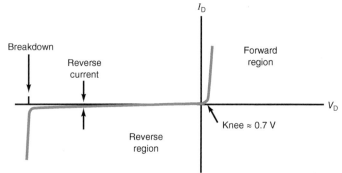

Figure 29-2 Diode curve.

Schottky diodes that you will learn about later in this chapter have a knee voltage of about 0.4 V. The knee voltage of a germanium diode is approximately 0.3 V. This lower knee voltage is an advantage and accounts for the use of a germanium and Schottky diodes in small-signal applications.

Bulk Resistance

Above the knee voltage, the diode current increases rapidly. This means that small increases in the diode voltage cause large increases in diode current. After the barrier potential is overcome, all that impedes the current is the **ohmic resistance** of the p and n regions. In other words, if the p and n regions were two separate pieces of semiconductor, each would have a resistance that you could measure with an ohmmeter, the same as an ordinary resistor.

The sum of the ohmic resistances is called the **bulk resistance** of the diode. It is defined as

$$R_B = R_P + R_N \qquad \text{(29-2)}$$

The bulk resistance depends on the size of the p and n regions, and how heavily doped they are. Often, the bulk resistance is less than 1 Ω.

Maximum DC Forward Current

If the current in a diode is too large, the excessive heat can destroy the diode. For this reason, a manufacturer's data sheet specifies the maximum current a diode can safely handle without shortening its life or degrading its characteristics.

The **maximum forward current** is one of the maximum ratings given on a data sheet. This current may be listed as I_{max}, $I_{F(max)}$, I_O, etc., depending on the manufacturer. For instance, a 1N4001 has a maximum forward current rating of 1 A. This means that it can safely handle a continuous forward current of 1 A.

Power Dissipation

You can calculate the power dissipation of a diode the same way as you do for a resistor. It equals the product of diode voltage and current. As a formula,

$$P_D = V_D I_D \qquad \text{(29-3)}$$

The **power rating** is the maximum power the diode can safely dissipate without shortening its life or degrading its properties. In symbols, the definition is

$$P_{max} = V_{max} I_{max} \qquad \text{(29-4)}$$

where V_{max} is the voltage corresponding to I_{max}. For instance, if a diode has a maximum voltage and current of 1 V and 2 A, its power rating is 2 W.

EXAMPLE 29-1

Is the diode of Fig. 29-3a forward-biased or reverse-biased?

Answer:

The voltage across R_2 is positive; therefore, the circuit is trying to push current in the easy direction of flow. If this is not clear, visualize the Thevenin circuit facing the diode as shown in Fig. 29-3b. In this series circuit, you can see that the dc source is trying to push current in the easy direction of flow. Therefore, the diode is forward-biased.

Whenever in doubt, reduce the circuit to a series circuit. Then it will be clear whether the dc source is trying to push current in the easy direction or not.

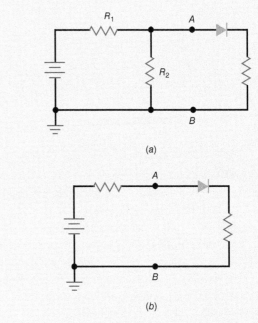

(a)

(b)

Figure 29-3

EXAMPLE 29-2

A diode has a power rating of 5 W. If the diode voltage is 1.2 V and the diode current is 1.75 A, what is the power dissipation? Will the diode be destroyed?

Answer:

$$P_D = (1.2 \text{ V})(1.75 \text{ A}) = 2.1 \text{ W}$$

This is less than the power rating, so the diode will not be destroyed.

29.2 The Ideal Diode

Figure 29-4 shows a detailed graph of the forward region of a diode. Here you see the diode current I_D versus diode voltage V_D. Notice how the current is approximately zero until the diode voltage approaches the barrier potential. Somewhere

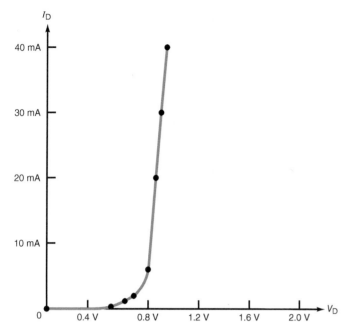

Figure 29-4 Graph of forward current.

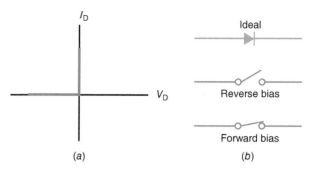

Figure 29-5 (a) Ideal diode curve. (b) Ideal diode acts like a switch.

in the vicinity of 0.6 to 0.7 V, the diode current increases. When the diode voltage is greater than 0.8 V, the diode current is significant and the graph is almost linear.

Depending on how a diode is doped and its physical size, it may differ from other diodes in its maximum forward current, power rating, and other characteristics. If we need an exact solution, we would have to use the graph of the particular diode. Although the exact current and voltage points will differ from one diode to the next, the graph of any diode is similar to Fig. 29-4. All silicon diodes have a knee voltage of approximately 0.7 V.

Most of the time, we do not need an exact solution. This is why we can and should use approximations for a diode. We will begin with the simplest approximation, called an **ideal diode.** In the most basic terms, what does a diode do? It conducts well in the forward direction and poorly in the reverse direction. Ideally, a diode acts like a perfect conductor (zero resistance) when forward-biased and like a perfect insulator (infinite resistance) when reverse biased.

Figure 29-5a shows the current-voltage graph of an ideal diode. It echoes what we just said: zero resistance when forward-biased and infinite resistance when reverse-biased. It is impossible to build such a device, but this is what manufacturers would produce if they could.

Is there any device that acts like an ideal diode? Yes. An ordinary switch has zero resistance when closed and infinite resistance when open. Therefore, an ideal diode acts like a switch that closes when forward-biased and opens when reverse-biased. Figure 29-5b summarizes the switch idea.

EXAMPLE 29-3

Use the ideal diode to calculate the load voltage and load current in Fig. 29-6a.

Answer:

Since the diode is forward-biased, it is equivalent to a closed switch. Visualize the diode as a closed switch. Then you can see that all the source voltage appears across the load resistor:

$$V_L = 10 \text{ V}$$

With Ohm's law, the load current is

$$I_L = \frac{10 \text{ V}}{1 \text{ k}\Omega} = 10 \text{ mA}$$

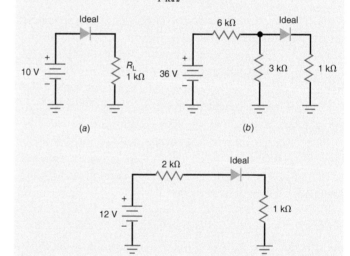

Figure 29-6

EXAMPLE 29-4

Calculate the load voltage and load current in Fig. 29-6b using an ideal diode.

Answer:

One way to solve this problem is to thevenize the circuit to the left of the diode. Looking from the diode back toward the source, we see a voltage divider with 6 kΩ and 3 kΩ. The Thevenin voltage is 12 V, and the Thevenin resistance is 2 kΩ. Figure 29-6c shows the Thevenin circuit driving the diode.

Now that we have a series circuit, we can see that the diode is forward-biased. Visualize the diode as a closed switch. Then the remaining calculations are

$$I_L = \frac{12\ V}{3\ k\Omega} = 4\ mA$$

and

$$V_L = (4\ mA)(1\ k\Omega) = 4\ V$$

You don't have to use Thevenin's theorem. You can analyze Fig. 29-6b by visualizing the diode as a closed switch. Then you have 3 kΩ in parallel with 1 kΩ, equivalent to 750 Ω. Using Ohm's law, you can calculate a voltage drop of 32 V across the 6 kΩ. The rest of the analysis produces the same load voltage and load current.

29.3 The Second and Third Approximations

The ideal approximation is all right in most troubleshooting situations. But we are not always troubleshooting. Sometimes, we want a more accurate value for load current and load voltage. This is where the *second approximation* comes in.

Figure 29-7a shows the graph of current versus voltage for the second approximation. The graph says that no current exists until 0.7 V appears across the diode. At this point, the diode turns on. Thereafter, only 0.7 V can appear across the diode, no matter what the current.

Figure 29-7b shows the equivalent circuit for the second approximation of a silicon diode. We think of the diode as a switch in series with a barrier potential of 0.7 V. If the Thevenin voltage facing the diode is greater than 0.7 V, the switch will close. When conducting, then the diode voltage is 0.7 V for any forward current.

On the other hand, if the Thevenin voltage is less than 0.7 V, the switch will open. In this case, there is no current through the diode.

EXAMPLE 29-5

Use the second approximation to calculate the load voltage, load current, and diode power in Fig. 29-8.

Answer:
Since the diode is forward-biased, it is equivalent to a battery of 0.7 V. This means that the load voltage equals the source voltage minus the diode drop:

$$V_L = 10\ V - 0.7\ V = 9.3\ V$$

With Ohm's law, the load current is

$$I_L = \frac{93\ V}{1\ k\Omega} = 9.3\ mA$$

The diode power is

$$P_D = (0.7\ V)(9.3\ mA) = 6.51\ mW$$

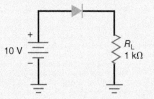

Figure 29-8

The *third approximation* of a diode includes the bulk resistance, R_B. Figure 29-9a shows the effect that R_B has on the diode curve. After the silicon diode turns on, the voltage increases linearly with an increase in current. The greater the current, the larger the diode voltage because of the voltage drop across the bulk resistance.

The equivalent circuit for the third approximation is a switch in series with a barrier potential of 0.7 V and a resistance of R_B (see Fig. 29-9b). When the diode voltage is larger than 0.7 V, the diode conducts. During conduction, the total voltage across the diode is

$$V_D = 0.7\ V + I_D R_B \qquad (29\text{-}5)$$

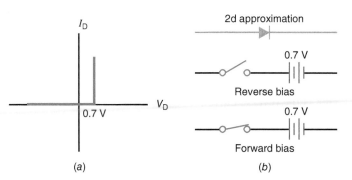

Figure 29-7 (a) Diode curve for second approximation. (b) Equivalent circuit for second approximation.

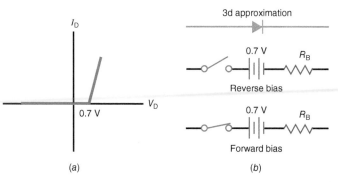

Figure 29-9 (a) Diode curve for third approximation. (b) Equivalent circuit for third approximation.

Often, the bulk resistance is less than 1 Ω, and we can safely ignore it in our calculations.

EXAMPLE 29-6

The 1N4001 of Fig. 29-10a has a bulk resistance of 0.23 Ω. What is the load voltage, load current, and diode power?

Answer:

Replacing the diode by its third approximation, we get Fig. 29-10b. The bulk resistance is small enough to ignore because it is less than 1/100 of the load resistance. The load voltage, load current, and diode power are 9.3 V, 9.3 mA, and 6.51 mW, respectively.

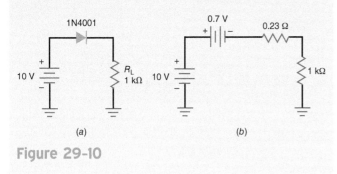

(a)   (b)

Figure 29-10

29.4 Key Diode Specifications

Following are the most important specifications to consider when replacing or selecting a diode. Such specifications are given in the manufacturer's data sheet.

Reverse Breakdown Voltage

The 1N4001 is a rectifier diode used in power supplies (circuits that convert ac voltage to dc voltage). The data sheet has an entry under "Absolute Maximum Ratings":

	Symbol	1N4001
Peak Repetitive Reverse Voltage	V_{RRM}	50 V

The breakdown voltage for this diode is 50 V. This breakdown occurs because the diode goes into avalanche when a huge number of carriers suddenly appear in the depletion layer. With a rectifier diode like the 1N4001, breakdown is usually destructive.

Reverse breakdown voltage may also be designated *PIV*, *PRV*, or *BV*.

Maximum Forward Current

Another specification of interest is average rectified forward current, which looks like this on the data sheet:

	Symbol	Value
Average Rectified Forward Current @ $T_A = 75°C$	$I_{F(AV)}$	1 A

This tells us that the 1N4001 can handle up to 1 A in the forward direction when used as a rectifier. You will learn more about average rectified forward current in the next chapter. For now, all you need to know is that 1 A is the level of forward current when the diode burns out because of excessive power dissipation. The average current may also be designated as I_o.

Forward Voltage Drop

Forward voltage drop is another key specification.

Characteristic and Conditions	Symbol	Maximum Value
Forward Voltage Drop (i_F) = 1.0 A, T_A = 25°C	V_F	1.1 V

The typical 1N4001 has a forward voltage drop of 0.93 V when the current is 1 A and the junction temperature is 25°C. If you test thousands of 1N4001s, you will find that a few will have as much as 1.1 V across them when the current is 1 A.

Maximum Reverse Current

Another specification worth discussing is maximum reverse current.

Characteristic and Conditions	Symbol	Typical Value	Maximum Value
Reverse Current	I_R		
T_A = 25°C		0.05 μA	10 μA
T_A = 100°C		1.0 μA	50 μA

This is the reverse current at the maximum reverse dc rated voltage (50 V for a 1N4001). At 25°C, the typical 1N4001 has a maximum reverse current of 5.0 μA. But notice how it increases to 500 μA at 100°C. Remember that this reverse current includes thermally produced saturation current and surface-leakage current. You can see from these numbers that temperature is important.

29.5 DC Resistance of a Diode

If you take the ratio of total diode voltage to total diode current, you get the *dc resistance* of the diode. In the forward direction, this dc resistance is symbolized by R_F; in the reverse direction, it is designated R_R.

Forward Resistance

Because the diode is a nonlinear device, its dc resistance varies with the current through it. For example, here are some pairs of forward current and voltage for a 1N914: 10 mA at 0.65 V, 30 mA at 0.75 V, and 50 mA at 0.85 V. At the first point, the dc resistance is

$$R_F = \frac{0.65\ V}{10\ mA} = 65\ \Omega$$

At the second point,

$$R_F = \frac{0.75\ V}{30\ mA} = 25\ \Omega$$

And at the third point,

$$R_F = \frac{0.85\ mV}{50\ mA} = 17\ \Omega$$

Notice how the dc resistance decreases as the current increases. In any case, the forward resistance is low compared to the reverse resistance.

Reverse Resistance

Similarly, here are two sets of reverse current and voltage for a 1N914: 25 nA at 20 V; 5 μA at 75 V. At the first point, the dc resistance is

$$R_R = \frac{20\ V}{25\ nA} = 800\ M\Omega$$

At the second point,

$$R_R = \frac{75\ V}{5\ \mu A} = 15\ M\Omega$$

Notice how the dc resistance decreases as we approach the breakdown voltage (75 V).

DC Resistance versus Bulk Resistance

The dc resistance of a diode is different from the bulk resistance. The dc resistance of a diode equals the bulk resistance *plus* the effect of the barrier potential. In other words, the dc resistance of a diode is its total resistance, whereas the bulk resistance is the resistance of only the *p* and *n* regions. For this reason, the dc resistance of a diode is always greater than bulk resistance.

29.6 Surface-Mount Diodes

Surface-mount (SM) diodes can be found anywhere there is a need for diode applications. SM diodes are small, efficient, and relatively easy to test, remove, and replace on the circuit board. Although there are a number of SM package styles, two basic styles dominate the industry: SM (surface mount) and SOT (small outline transistor).

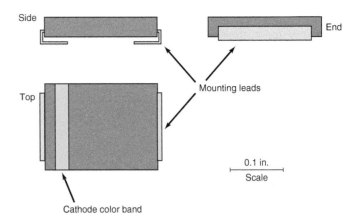

Figure 29-11 The two-terminal SM-style package, used for SM diodes.

The SM package has two L-bend leads and a colored band on one end of the body to indicate the cathode lead. Figure 29-11 shows a typical set of dimensions. The length and width of the SM package are related to the current rating of the device. The larger the surface area, the higher the current rating. So an SM diode rated at 1 A might have a surface area given by 0.181 by 0.115 in. The 3 A version, on the other hand, might measure 0.260 by 0.236 in. The thickness tends to remain at about 0.103 in for all current ratings.

Increasing the surface area of an SM-style diode increases its ability to dissipate heat. Also, the corresponding increase in the width of the mounting terminals increases the thermal conductance to a virtual heat sink made up of the solder joints, mounting lands, and the circuit board itself.

29.7 The Zener Diode

Small-signal and rectifier diodes are never intentionally operated in the breakdown region because this may damage them. A **zener diode** is different; it is a silicon diode that the manufacturer has optimized for operation in the breakdown region. The zener diode is the backbone of voltage regulators, circuits that hold the load voltage almost constant despite large changes in line voltage and load resistance.

I-V Graph

Figure 29-12*a* shows the schematic symbol of a zener diode; Fig. 29-12*b* is an alternative symbol. In either symbol, the lines resemble a *z*, which stands for "zener." By varying the doping level of silicon diodes, a manufacturer can produce zener diodes with breakdown voltages from about 2 to over 1000 V. These diodes can operate in any of three regions: forward, leakage, and breakdown.

Figure 29-12*c* shows the *I-V* graph of a zener diode. In the forward region, it starts conducting around 0.7 V, just like an ordinary silicon diode. In the **leakage region** (between zero and breakdown), it has only a small reverse current. In a zener diode, the breakdown has a very sharp knee, followed

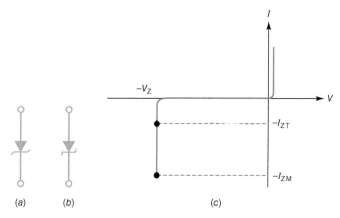

Figure 29-12 Zener diode. (a) Schematic symbol.
(b) Alternative symbol. (c) Graph of current versus voltage.

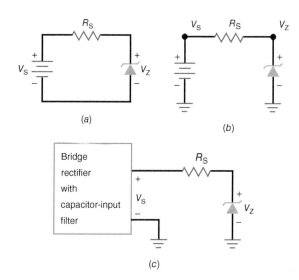

Figure 29-13 Zener regulator. (a) Basic circuit. (b) Same circuit with grounds. (c) Power supply drives regulator.

by an almost vertical increase in current. Note that the voltage is almost constant, approximately equal to V_Z over most of the breakdown region. Data sheets usually specify the value of V_Z at a particular test current I_{ZT}.

Figure 29-12c also shows the maximum reverse current I_{ZM}. As long as the reverse current is less than I_{ZM}, the diode is operating within its safe range. If the current is greater than I_{ZM}, the diode will be destroyed. To prevent excessive reverse current, a *current-limiting resistor* must be used (discussed later).

Zener Resistance

In the third approximation of a silicon diode, the forward voltage across a diode equals the knee voltage plus the additional voltage across the bulk resistance.

Similarly, in the breakdown region, the reverse voltage across a diode equals the breakdown voltage plus the additional voltage across the bulk resistance. In the reverse region, the bulk resistance is referred to as the **zener resistance.**

In Fig. 29-12c, the zener resistance means that an increase in reverse current produces a slight increase in reverse voltage. The increase in voltage is very small, typically only a few tenths of a volt.

Zener Regulator

A zener diode is sometimes called a *voltage-regulator diode* because it maintains a constant output voltage even though the current through it changes. For normal operation, you have to reverse-bias the zener diode, as shown in Fig. 29-13a. Furthermore, to get breakdown operation, the source voltage V_S must be greater than the zener breakdown voltage, V_Z. A series resistor R_S is always used to limit the zener current to less than its maximum current rating. Otherwise, the zener diode will burn out like any device with too much power dissipation.

Figure 29-13b shows an alternative way to draw the circuit with grounds. Whenever a circuit has grounds, you can measure voltages with respect to ground.

For instance, suppose you want to know the voltage across the series resistor of Fig. 29-13b. Here is the one way to find it when you have a built-up circuit. First, measure the voltage from the left end of R_S to ground. Second, measure the voltage from the right end of R_S to ground. Third, subtract the two voltages to get the voltage across R_S. If you have a floating VOM or DMM, you can connect directly across the series resistor.

Figure 29-13c shows the output of a power supply connected to a series resistor and a zener diode. This circuit is used when you want a dc output voltage that is less than the output of the power supply. A circuit like this is called a *zener voltage regulator,* or simply a **zener regulator.**

Ohm's Law Applies

In Fig. 29-13, the voltage across the series or current-limiting resistor equals the difference between the source voltage and the zener voltage. Therefore, the current through the resistor is

$$I_S = \frac{V_S - V_Z}{R_S} \qquad \textbf{(29-6)}$$

Once you have the value of series current, you also have the value of zener current. This is because Fig. 29-13 is a series circuit. Note that I_S must be less than I_{ZM}.

EXAMPLE 29-6

Suppose the zener diode of Fig. 29-14 has a breakdown voltage of 10 V. What are the minimum and maximum zener currents?

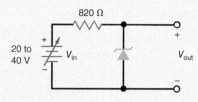

Figure 29-14 Example.

Answer:

The applied voltage may vary from 20 to 40 V. The zener voltage is 10 V; therefore, the output voltage is 40 V for any source voltage between 20 and 40 V.

The minimum current occurs when the source voltage is minimum. Visualize 20 V on the left end of the resistor and 10 V on the right end. Then you can see that the voltage across the resistor is 20 V − 10 V, or 10 V. The rest is Ohm's law:

$$I_S = \frac{10\ V}{820\ \Omega} = 12.2\ mA$$

The maximum current occurs when the source voltage is 40 V. In this case, the voltage across the resistor is 30 V, which gives a current of

$$I_S = \frac{30\ V}{820\ \Omega} = 36.6\ mA$$

In a voltage regulator like Fig. 29-14, the output voltage is held constant at 10 V, despite the change in source voltage from 20 to 40 V. The larger source voltage produces more zener current, but the output voltage holds rock-solid at 10 V. (If the zener resistance is included, the output voltage increases slightly when the source voltage increases.)

29.8 The Loaded Zener Regulator

Figure 29-15a shows a *loaded* zener regulator, and Fig. 29-15b shows the same circuit with grounds. The zener diode operates in the breakdown region and holds the load voltage constant. Even if the source voltage changes or the load resistance varies, the load voltage will remain fixed and equal to the zener voltage.

Breakdown Operation

How can you tell whether the zener diode of Fig. 29-15 is operating in the breakdown region? Because of the voltage divider, the Thevenin voltage facing the diode is

$$V_{TH} = \frac{R_L}{R_S + R_L} V_S \qquad \textbf{(29-7)}$$

This is the voltage that exists when the zener diode is disconnected from the circuit. This Thevenin voltage has to be greater than the zener voltage; otherwise, breakdown cannot occur.

Series Current

Unless otherwise indicated, in all subsequent discussions we assume that the zener diode is operating in the breakdown region. In Fig. 29-15, the current through the series resistor is given by

$$I_S = \frac{V_S - V_Z}{R_S} \qquad \textbf{(29-8)}$$

This is Ohm's law applied to the current-limiting resistor. It is the same whether or not there is a load resistor. In

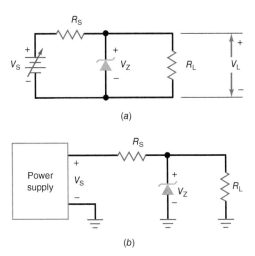

(a)

(b)

Figure 29-15 Loaded zener regulator. (*a*) Basic circuit. (*b*) Practical circuit.

other words, if you disconnect the load resistor, the current through the series resistor still equals the voltage across the resistor divided by the resistance.

Load Current

Ideally, the load voltage equals the zener voltage because the load resistor is in parallel with the zener diode. As an equation,

$$V_L = V_Z \qquad \textbf{(29-9)}$$

This allows us to use Ohm's law to calculate the load current:

$$I_L = \frac{V_L}{R_L} \qquad \textbf{(29-10)}$$

Zener Current

With Kirchhoff's current law,

$$I_S = I_Z + I_L$$

The zener diode and the load resistor are in parallel. The sum of their currents has to equal the total current, which is the same as the current through the series resistor.

We can rearrange the foregoing equation to get this important formula:

$$I_Z = I_S - I_L \qquad \textbf{(29-11)}$$

This tells you that the zener current no longer equals the series current, as it does in an unloaded zener regulator. Because of the load resistor, the zener current now equals the series current minus the load current.

EXAMPLE 29-7

Is the zener diode of Fig. 29-16a operating in the breakdown region?

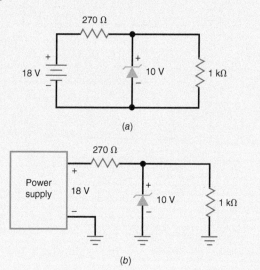

(a)

(b)

Figure 29-16 Example.

Answer:

$$V_{TH} = \frac{1\text{ k}\Omega}{270\ \Omega + 1\text{ k}\Omega}\ (18\text{ V}) = 14.2\text{ V}$$

Since this Thevenin voltage is greater than the zener voltage, the zener diode is operating in the breakdown region.

EXAMPLE 29-8

What does the zener current equal in Fig. 29-16b?

Answer:

You are given the voltage on both ends of the series resistor. Subtract the voltages, and you can see that 8 V is across the series resistor. Then Ohm's law gives

$$I_S = \frac{8\text{ V}}{270\ \Omega} = 29.6\text{ mA}$$

Since the load voltage is 10 V, the load current is

$$I_L = \frac{10\text{ V}}{1\text{ k}\Omega} = 10\text{ mA}$$

The zener current is the difference between the two currents:

$$I_Z = 29.6\text{ mA} - 10\text{ mA} = 19.6\text{ mA}$$

EXAMPLE 29-9

What does the circuit of Fig. 29-17 do?

Answer:

In most applications, zener diodes are used in voltage regulators where they remain in the breakdown region. But there are

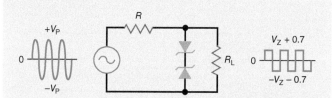

Figure 29-17 Zener diodes used for waveshaping.

exceptions. Sometimes zener diodes are used in waveshaping circuits like Fig. 29-17.

Notice the back-to-back connection of two zener diodes. On the positive half-cycle, the upper diode conducts and the lower diode breaks down. Therefore, the output is clipped as shown. The clipping level equals the zener voltage (broken-down diode) plus 0.7 V (forward-biased diode).

On the negative half-cycle, the action is reversed. The lower diode conducts, and the upper diode breaks down. In this way, the output is almost a square wave. The larger the input sine wave, the better looking the output square wave.

EXAMPLE 29-10

Briefly describe the circuit action for each of the circuits in Fig. 29-18.

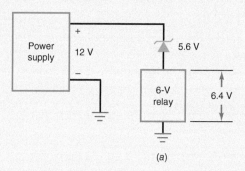

(a)

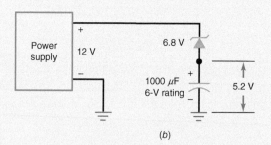

(b)

Figure 29-18 Zener applications. (a) Using a 6-V relay in a 12-V system. (b) Using a 6-V capacitor in a 12-V system.

Answer:

If you try to connect a 6-V relay to a 12-V system, you will probably damage the relay. It is necessary to drop some of the voltage. Figure 29-18a shows one way to do this. By connecting a 5.6-V zener diode in series with the relay, only 6.4 V appears across the relay, which is usually within the tolerance of the relay's voltage rating.

Large electrolytic capacitors often have small voltage ratings. For instance, an electrolytic capacitor of 1000 μF may have a voltage rating of only 6 V. This means that the maximum voltage across the capacitor should be less than 6 V. Figure 29-18b shows a workaround solution in which a 6-V electrolytic capacitor is used with a 12-V power supply. Again, the idea is to use a zener diode to drop some of the voltage. In this case, the zener diode drops 6.8 V, leaving only 5.2 V across the capacitor. This way, the electrolytic capacitor can filter the power supply and still remain with its voltage rating.

29.9 Zener Specifications

Following are the main zener specifications you should consider in replacing or selecting a zener diode.

Maximum Power

The power dissipation of a zener diode equals the product of its voltage and current:

$$P_Z = V_Z I_Z \qquad \text{(29-12)}$$

For instance, if $V_Z = 12$ V and $I_Z = 10$ mA, then

$$P_Z = (12 \text{ V})(10 \text{ mA}) = 120 \text{ mW}$$

As long as P_Z is less than the power rating, the zener diode can operate in the breakdown region without being destroyed. Commercially available zener diodes have power ratings from ¼ to more than 50 W.

For example, the data sheet for the 1N4728A series lists a maximum power rating of 1 W. A safe design includes a safety factor to keep the power dissipation well below this 1-W maximum.

Maximum Current

Data sheets often include the *maximum current* a zener diode can handle without exceeding its power rating. If this value is not listed, the maximum current can be found as follows:

$$I_{ZM} = \frac{P_{ZM}}{V_Z} \qquad \text{(29-13)}$$

where

$$I_{ZM} = \text{maximum-rated zener current}$$

$$P_{ZM} = \text{power rating}$$

$$V_Z = \text{zener voltage}$$

For example, the 1N4742A has a zener voltage of 12 V and a 1-W power rating. Therefore, it has a maximum current rating of

$$I_{ZM} = \frac{1 \text{ W}}{12 \text{ V}} = 83.3 \text{ mA}$$

If you satisfy the current rating, you automatically satisfy the power rating. For instance, if you keep the maximum zener current less than 83.3 mA, you are also keeping the maximum power dissipation less than 1 W.

Tolerance

Most zener diodes will have a suffix A, B, C, or D to identify the zener voltage tolerance. Because these suffix markings are not always consistent, be sure to identify any special notes included on the zener's data sheet that indicate that specific tolerance. For instance, the data sheet for the 1N4728A series shows its tolerance to equal ±5%. A suffix of C generally indicates ±2%, D ±1%, and no suffix ±20%.

Zener Resistance

The zener resistance (also called *zener impedance*) may be designated R_{ZT} or Z_{ZT}. For instance, the 1N4740A has a zener resistance of 7 Ω measured at a test current of 25 mA. As long as the zener current is beyond the knee of the curve, you can use 7 Ω as the approximate value of the zener resistance. But note how the zener resistance increases at the knee of the curve (700 Ω). The point is this: Operation should be at or near the test current, if at all possible. Then you know that the zener resistance is relatively small.

The data sheet contains a lot of additional information, but it is primarily aimed at designers. If you do get involved in design work, then you have to read the data sheet carefully, including the notes that specify how quantities were measured.

29.10 Optoelectronic Devices

Optoelectronics is the technology that combines optics and electronics. This field includes many devices based on the action of a *pn* junction. Examples of optoelectronic devices are **light-emitting diodes (LEDs),** photodiodes, optocouplers, and laser diodes. Our discussion begins with the LED.

Light-Emitting Diode

Figure 29-19a shows a source connected to a resistor and an LED. The outward arrows symbolize the radiated light. In a forward-biased LED, free electrons cross the junction and fall into holes. As these electrons fall from a higher to a lower energy level, they radiate energy. In ordinary diodes, this energy is radiated in the form of heat. But in an LED, the energy is radiated as light. LEDs made from different elements have the ability to radiate energy across a wide wavelength spectrum. LEDs have replaced incandescent lamps in many applications because of their low voltage, long life, and fast ON-OFF switching.

By using elements like gallium, arsenic, and phosphorus, a manufacturer can produce LEDs that radiate red, green, yellow, blue, orange, or infrared (invisible). LEDs that produce visible radiation are useful with instruments, calculators, and so on. The infrared LED finds applications in burglar alarm systems, remote controls, CD players, and other devices requiring invisible radiation.

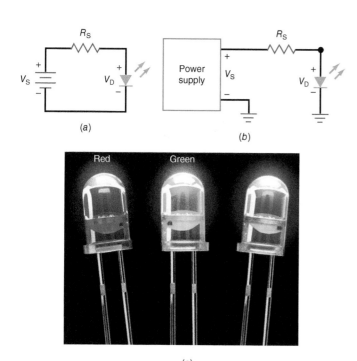

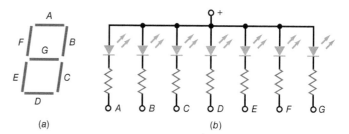

Figure 29-20 Seven-segment indicator. (a) Physical layout of segments. (b) Schematic diagram.

Figure 29-19 LED indicator. (a) Basic circuit. (b) Practical circuit. (c) Typical LEDs.

LED Voltage and Current

The resistor of Fig. 29-19b is the usual current-limiting resistor that prevents the current from exceeding the maximum current rating of the diode. Since the resistor has a node voltage of V_S on the left and a node voltage of V_D on the right, the voltage across the resistor is the difference between the two voltages. With Ohm's law, the series current is

$$I_S = \frac{V_S - V_D}{R_S} \tag{29-14}$$

For most commercially available LEDs, the typical voltage drop is from 1.5 to 2.5 V for currents between 10 and 50 mA. The exact voltage drop depends on the LED current, color, tolerance, and so on. Unless otherwise specified, we will use a nominal drop of 2 V when troubleshooting or analyzing the LED circuits in this book. Figure 29-19c shows typical LEDs.

LED Brightness

The brightness of an LED depends on the current. When V_S is much greater than V_D in Formula (29-14), the brightness of the LED is approximately constant. For instance, a TIL222 is a green LED with a forward voltage of between 1.8 (minimum) and 3 V (maximum), for a current of 25 mA. If a circuit like Fig. 29-19b is mass-produced using a TIL222, the brightness of the LED will be almost constant if V_S is much greater than V_D. If V_S is only slightly more than V_D, the LED brightness will vary noticeably from one circuit to the next.

Breakdown Voltage

LEDs have very low breakdown voltages, typically between 3 and 5 V. Because of this, they are easily destroyed if reverse-biased with too much voltage. When troubleshooting an LED circuit in which the LED will not light, check the polarity of the LED connection to make sure that it is forward-biased.

An LED is often used to indicate the presence of power-line voltage into equipment. In this case, a rectifier diode may be used in parallel with the LED to prevent reverse-bias destruction of the LED.

Seven-Segment Display

Figure 29-20a shows a **seven-segment display.** It contains seven rectangular LEDs (A through G). Each LED is called a *segment* because it forms part of the character being displayed. Figure 29-20b is a schematic diagram of the seven-segment display. External series resistors are included to limit the currents to safe levels. By grounding one or more resistors, we can form any digit from 0 through 9. For instance, by grounding A, B, and C, we get a 7. Grounding A, B, C, D, and G produces a 3.

A seven-segment display can also display capital letters A, C, E, and F, plus lowercase letters b and d. Microprocessor trainers often use seven-segment displays that show all digits from 0 through 9, plus A, b, C, d, E, and F.

The seven-segment indicator of Fig. 29-20b is referred to as the **common-anode** type because all anodes are connected together. Also available is the **common-cathode** type, in which all cathodes are connected together.

Photodiode

As previously discussed, one component of reverse current in a diode is the flow of minority carriers. These carriers exist because thermal energy keeps dislodging valence electrons from their orbits, producing free electrons and holes in the process. The lifetime of the minority carriers is short, but while they exist, they can contribute to the reverse current.

When light energy bombards a *pn* junction, it can dislodge valence electrons. The more light striking the junction, the larger the reverse current in a diode. A **photodiode** has

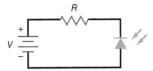

Figure 29-21 Incoming light increases reverse current in photodiode.

been optimized for its sensitivity to light. In this diode, a window lets light pass through the package to the junction. The incoming light produces free electrons and holes. The stronger the light, the greater the number of minority carriers and the larger the reverse current.

Figure 29-21 shows the schematic symbol of a photodiode. The arrows represent the incoming light. Especially important, the source and the series resistor reverse-bias the photodiode. As the light becomes brighter, the reverse current increases. With typical photodiodes, the reverse current is in the tens of microamperes.

Optocoupler

An optocoupler (also called an *optoisolator*) combines an LED and a photodiode in a single package. Figure 29-22 shows an optocoupler. It has an LED on the input side and a photodiode on the output side. The left source voltage and the series resistor set up a current through the LED. Then the light from the LED hits the photodiode, and this sets up a reverse current in the output circuit. This reverse current produces a voltage across the output resistor. The output voltage then equals the output supply voltage minus the voltage across the resistor.

When the input voltage is varying, the amount of light is fluctuating. This means that the output voltage is varying in step with the input voltage. This is why the combination of an LED and a photodiode is called an **optocoupler.** The device can couple an input signal to the output circuit. Other types of optocouplers use phototransistors, photothyristors, and other photo devices in their output circuit side.

The key advantage of an optocoupler is the electrical isolation between the input and output circuits. With an optocoupler, the only contact between the input and the output is a beam of light. Because of this, it is possible to have an insulation resistance between the two circuits in the thousands of megohms. Isolation like this is useful in high-voltage

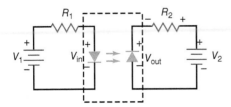

Figure 29-22 Optocoupler combines an LED and a photodiode.

applications in which the potentials of the two circuits may differ by several thousand volts.

Laser Diode

In an LED, free electrons radiate light when falling from higher-energy levels to lower ones. The free electrons fall randomly and continuously, resulting in light waves that have every phase between 0 and 360°. Light that has many different phases is called *noncoherent light.* An LED produces noncoherent light.

A **laser diode** is different. It produces a *coherent light.* This means that all the light waves are *in phase with each other.* The basic idea of a laser diode is to use a mirrored resonant chamber that reinforces the emission of light waves at a single frequency of the same phase. Because of the resonance, a laser diode produces a narrow beam of light that is very intense, focused, and pure.

Laser diodes are also known as *semiconductor lasers.* These diodes can produce visible light (red, green, or blue) and invisible light (infrared). Laser diodes are used in a large variety of applications. They are used in telecommunications, data communications, broadband access, industrial, aerospace, test and measurement, and medical and defense industries. They are also used in laser printers and consumer products requiring large-capacity optical disc systems, such as compact disc (CD) and digital video disc (DVD) players. In broadband communication, they are used with fiber-optic cables to increase the speed of the Internet.

A *fiber-optic cable* is analogous to a stranded wire cable, except that the strands are thin flexible fibers of glass or plastic that transmit light beams instead of free electrons. The advantage is that much more information can be sent through a fiber-optic cable than through a copper cable.

New applications are being found as the lasing wavelength is pushed lower into the visible spectrum with visible laser diodes (VLDs). Also, near-infrared diodes are being used in machine vision systems, sensors, and security systems.

EXAMPLE 29-11

Figure 29-23a shows a voltage-polarity tester. It can be used to test a dc voltage of unknown polarity. When the dc voltage is positive, the green LED lights up. When the dc voltage is negative, the red LED lights up. What is the approximate LED current if the dc input voltage is 50 V and the series resistance is 2.2 kΩ?

Answer:

We will use a forward voltage of approximately 2 V for either LED. With Formula (29-14),

$$I_S = \frac{50\ V - 2\ V}{2.2\ k\Omega} = 21.8\ mA$$

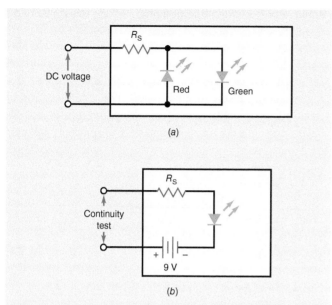

Figure 29-23 (a) Polarity indicator. (b) Continuity tester.

EXAMPLE 29-12

Figure 29-23b is a continuity tester. After you turn off all the power in a circuit under test, you can use this circuit to check for the continuity of cables, connectors, and switches. How much LED current is there if the series resistance is 470 Ω?

Answer:

When the input terminals are shorted (continuity), the internal 9-V battery produces an LED current of

$$I_S = \frac{9\text{ V} - 2\text{ V}}{470\ \Omega} = 14.9\text{ mA}$$

29.11 The Schottky Diode

As frequency increases, the action of small-signal rectifier diodes begins to deteriorate. They are no longer able to switch off fast enough to produce a well-defined half-wave signal. The solution to this problem is the *Schottky diode.* Before describing this special-purpose diode, let us look at the problem that arises with ordinary small-signal diodes.

Charge Storage

When a *pn* junction diode is forward-biased, electrons and holes freely flow across the junction. If the junction is suddenly reverse-biased, it takes a finite period of time for the electrons and holes to clear out of the depletion region. The charges are temporarily stored there.

When you try to switch a diode from ON to OFF, the stored charges will flow in the reverse direction for a while. This limits the effectiveness of the diode as a rectifier at high frequencies.

Reverse Recovery Time

The time it takes to turn off a forward-biased diode is called the *reverse recovery time, t_{rr}.* The conditions for measuring t_{rr} vary from one manufacturer to the next. As a guide, t_{rr} is the time it takes for the reverse current to drop to 10% of the forward current.

For instance, the 1N4148 has a t_{rr} of 4 ns. If this diode has a forward current of 10 mA and it is suddenly reverse-biased, it will take approximately 4 ns for the reverse current to decrease to 1 mA. Reverse recovery time is so short in small-signal diodes that you don't even notice its effect at frequencies below 10 MHz or so. It's only when you get well above 10 MHz that you have to take t_{rr} into account.

Poor Rectification at High Frequencies

What effect does reverse recovery time have on rectification? Take a look at the half-wave rectifier shown in Fig. 29-24a. At low frequencies the output is a half-wave rectified signal. As the frequency increases well into megahertz, however, the output signal begins to deviate from the half-wave shape, as shown in Fig. 29-24b. Some reverse conduction (called *tails*) is noticeable near the beginning of the reverse half-cycle.

The problem is that the reverse recovery time has become a significant part of the period, allowing conduction during the early part of the negative half-cycle. For instance, if $t_{rr} = 4$ ns and the period is 50 ns, the early part of the reverse half-cycle will have tails similar to those shown in Fig. 29-24b. As the frequency continues to increase, the rectifier becomes useless.

Eliminating Charge Storage

The solution to the problem of tails is a special-purpose device called a **Schottky diode.** This kind of diode uses a metal such as gold, silver, or platinum on one side of the junction and doped silicon (typically *n*-type) on the other side. See Fig. 29-25. Because of the metal on one side of the junction, the Schottky diode has no depletion layer. The lack

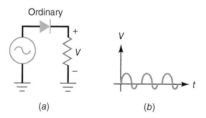

Figure 29-24 Stored charges degrade rectifier behavior at high frequencies. (a) Rectifier circuit with ordinary small-signal diode. (b) Tails appear on negative half-cycles at higher frequencies.

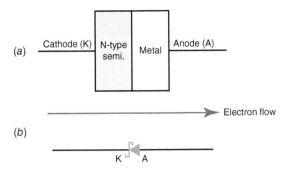

(a) Cathode (K) | N-type semi. | Metal | Anode (A)

(b)

Electron flow

K   A

Figure 29-25 The Schottky diode is a metal-semiconductor junction. Note (*a*) direction of electron flow and (*b*) schematic symbol.

of a depletion layer means that there are *no stored charges at the junction*. The Schottky diode is sometimes called a *hot-carrier diode*.

High-Speed Turnoff

The lack of charge storage means that the Schottky diode can switch off faster than an ordinary diode can. In fact, a Schottky diode can easily rectify frequencies above 500 MHz. When it is used in a circuit like Fig. 29-26*a*, the Schottky diode produces a perfect half-wave signal like Fig. 29-26*b* even at frequencies above 300 MHz.

Figure 29-26*a* shows the schematic symbol of a Schottky diode. Notice the cathode side. The lines look like a rectangular *S*, which stands for *Schottky*.

Low-Junction Voltage

Another major advantage of the Schottky diode is its low conduction voltage drop. It is usually in the 0.2- to 0.4-V range. This low barrier potential makes the Schottky diode ideal for small-signal applications. It also results in less power consumption in power supplies.

Applications

The most important application of Schottky diodes is in high-speed switching power supplies. Because it has no charge storage, the Schottky diode is also widely used in

digital integrated circuits to achieve operating speeds up to 200 MHz.

A final point: Since a Schottky diode has a barrier potential of only 0.25 V, you may occasionally see it used in low-voltage bridge rectifiers because you subtract only 0.25 V instead of the usual 0.7 V for each diode when using the second approximation. In a low-voltage supply, this lower diode voltage drop is an advantage.

29.12 The Varactor

The **varactor** (also called the *voltage-variable capacitance, varicap, epicap,* and *tuning diode*) is widely used in television receivers, FM receivers, and other communications equipment because it can be used for electronic tuning.

Basic Idea

In Fig. 29-27*a*, the depletion layer is between the *p* region and the *n* region. The *p* and *n* regions are like the plates of a capacitor, and the depletion layer is like the dielectric. When a diode is reverse-biased, the width of the depletion layer increases with the reverse voltage. Since the depletion layer gets wider with more reverse voltage, the capacitance becomes smaller. It's as though you moved apart the plates of a capacitor. The key idea is that capacitance is controlled by reverse voltage.

Equivalent Circuit and Symbol

Figure 29-27*b* shows the ac equivalent circuit for a reverse-biased diode. In other words, as far as an ac signal is concerned, the varactor acts the same as a variable capacitance. Figure 29-27*c* shows the schematic symbol for a varactor. The inclusion of a capacitor in series with the diode is a reminder that a varactor is a device that has been optimized for its variable-capacitance properties.

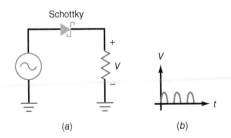

Figure 29-26 Schottky diodes eliminate tails at high frequencies. (*a*) Circuit with Schottky diode. (*b*) Half-wave signal at 300 MHz.

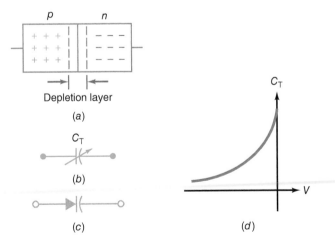

Figure 29-27 Varactor. (*a*) Doped regions are like capacitor plates separated by a dielectric. (*b*) AC equivalent circuit. (*c*) Schematic symbol. (*d*) Graph of capacitance versus reverse voltage.

	C_t, Diode Capacitance $V_R = 3.0$ Vdc, $f = 1.0$ MHz pF			Q, Figure of Merit $V_R = 3.0$ Vdc $f = 50$ MHz	C_R, Capacitance Ratio C_3/C_{25} $f = 1.0$ MHz[1]	
Device	**Min**	**Nom**	**Max**	**Min**	**Min**	**Max**
MMBV109LT1, MV209	26	29	32	200	5.0	6.5

1. C_R is the ratio of C_t measured at 3 Vdc divided by C_t measured at 25 Vdc.

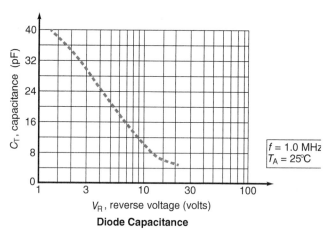

Figure 29-28   MV209 partial data sheet.

Capacitance Decreases at Higher Reverse Voltages

Figure 29-27d shows how the capacitance varies with reverse voltage. This graph shows that the capacitance gets smaller when the reverse voltage gets larger. The really important idea here is that reverse dc voltage controls capacitance.

How is a varactor used? It is connected in parallel with an inductor to form a parallel resonant circuit. This circuit has only one frequency at which maximum impedance occurs. This frequency is called the *resonant frequency*. If the dc reverse voltage to the varactor is changed, the resonant frequency is also changed. This is the principle behind electronic tuning of a radio station, a TV channel, and so on.

Varactor Characteristics

Because the capacitance is voltage-controlled, varactors have replaced mechanically tuned capacitors in many applications such as television receivers and automobile radios. Data sheets for varactors list a reference value of capacitance measured at a specific reverse voltage, typically −3 V

to −4 V. Figure 29-28 shows a partial data sheet for a MV209 varactor diode. It lists a reference capacitance C_t of 29 pF at −3 V.

In addition to providing the reference value of capacitance, data sheets normally list a capacitance ratio, C_R, or tuning range associated with a voltage range. For example, along with the reference value of 29 pF, the data sheet of a MV209 shows a minimum capacitance ratio of 5:1 for a voltage range of −3 V to −25 V. This means that the capacitance, or tuning range, decreases from 29 to 6 pF when the voltage varies from −3 V to −25 V.

EXAMPLE 29-13

What does the circuit of Fig. 29-29a do?

Answer:

In Fig. 29-29a, the transistor circuit feeds a fixed number of milliamperes into the resonant LC tank circuit. A negative dc voltage reverse-biases the varactor. By varying this dc control voltage, we can vary the resonant frequency of the LC circuit.

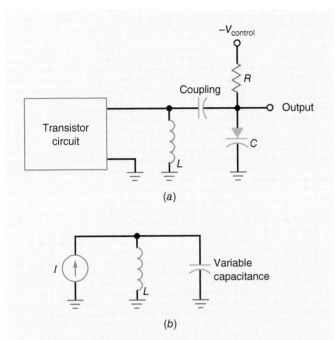

Figure 29-29 Varactors can tune resonant circuits. (a) Transistor (current source) drives tuned LC tank. (b) AC equivalent circuit.

As far as the ac signal is concerned, we can use the equivalent circuit shown in Fig. 29-29b. The coupling capacitor acts like a short circuit. An ac current source drives a resonant *LC* tank circuit. The varactor acts like variable capacitance, which means that we can change the resonant frequency by changing the dc control voltage. This is the basic idea behind the tuning of radio and television receivers and other wireless devices.

29.13 Other Diodes

Besides the special-purpose diodes discussed so far, there are others you should know about. Because they are so specialized, only a brief description follows.

Varistors and Clippers

Lightning, power-line faults, and transients can pollute the ac line voltage by superimposing dips and spikes on the normal 120 V rms. *Dips* are severe voltage drops lasting microseconds or less. *Spikes* are very brief overvoltages up to 2000 V or more. In some equipment, filters are used between the power line and the primary of the transformer to eliminate the problems caused by ac line transients.

One of the devices used for line filtering is the **varistor** (also called a *metal-oxide varistor,* or *MOV,* or *transient*

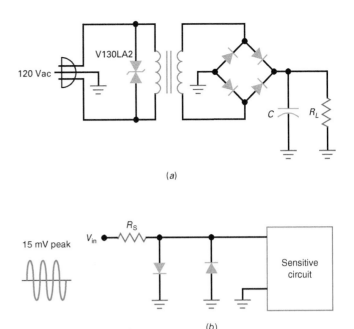

Figure 29-30 (a) A varistor protects the primary winding of a transformer in a basic power supply. (b) Diode clamp protecting a sensitive circuit.

suppressor). This semiconductor device is like two back-to-back zener diodes with a high breakdown voltage in both directions. Varistors are commercially available with breakdown voltages from 10 to 1000 V. They can handle peak transient currents in the hundreds or thousands of amperes.

For instance, a V130LA2 is a varistor with a breakdown voltage of 184 V (equivalent to 130 V rms) and a peak current rating of 400 A. Connect one of these across the primary winding as shown in Fig. 29-30a, and you don't have to worry about spikes. The varistor will clip all spikes at the 184-V level and protect your power supply.

The same concept can be used to protect other sensitive circuits. For example, two standard silicon diodes can be connected in parallel, as shown in Fig. 29-30b, to form what is sometimes called a clamp or clipper. Any ac signal that exceeds the forward voltage of the diode, approximately 0.7 V, will cause the diodes to conduct, one for each polarity. This clamps the input voltage at ± 0.7 V and clips off any spikes over that value. Such an arrangement is widely used to protect the input circuits of integrated circuit amplifiers. Most ac signals to be amplified are less than ± 0.7 V and so never cause the diodes to conduct. Any noise or undesired high-voltage spike is suppressed by the diodes. Other diodes can also be used in a similar way. Back-to-back zener diodes work well, as do Schottky diodes.

PIN Diodes

A **PIN diode** is a semiconductor device that operates as a variable resistor at RF and microwave frequencies. Figure 29-31a shows its construction. It consists of an intrinsic (pure) semiconductor material sandwiched between p-type and n-type materials. Figure 29-31b shows the schematic symbol for the PIN diode.

When the diode is forward biased, it acts like a current-controlled resistance. Figure 29-31c shows how the PIN diode's series resistance, R_S, decreases as its forward current increases. When reverse-biased, the PIN diode acts like a fixed capacitor. The PIN diode is widely used in RF and microwave modulator circuits. When properly biased the PIN diode can be used as a switch for microwave frequencies (>1 GHz).

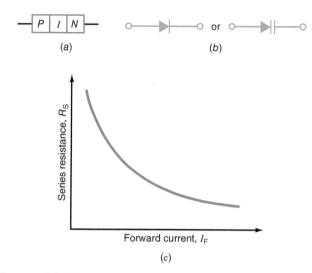

Figure 29-31 PIN diode. (*a*) Construction. (*b*) Schematic symbol. (*c*) Series resistance.

CHAPTER 29 REVIEW QUESTIONS

1. When the graph of current versus voltage is a straight line, the device is referred to as
   a. active.
   b. linear.
   c. nonlinear.
   d. passive.

2. What kind of device is a resistor?
   a. Unilateral.
   b. Linear.
   c. Nonlinear.
   d. Bipolar.

3. What kind of a device is a diode?
   a. Bilateral.
   b. Linear.
   c. Nonlinear.
   d. Unipolar.

4. How is a nonconducting diode biased?
   a. Forward.
   b. Inverse.
   c. Poorly.
   d. Reverse.

5. When the diode current is large, the bias is
   a. forward.
   b. inverse.
   c. poor.
   d. reverse.

6. The knee voltage of a diode is approximately equal to the
   a. applied voltage.
   b. barrier potential.
   c. breakdown voltage.
   d. forward voltage.

7. The reverse current consists of minority-carrier current and
   a. avalanche current.
   b. forward current.
   c. surface-leakage current.
   d. zener current.

8. How much voltage is there across the second approximation of a silicon diode when it is forward-biased?
   a. 0 V.
   b. 0.3 V.
   c. 0.7 V.
   d. 1 V.

9. How much current is there through the second approximation of a silicon diode when it is reverse-biased?
   a. 0 mA.
   b. 1 mA.
   c. 300 mA.
   d. None of the above.

10. How much forward diode voltage is there with the ideal-diode approximation?
   a. 0 V.
   b. 0.7 V.
   c. More than 0.7 V.
   d. 1 V.

11. What is true about the breakdown voltage in a zener diode?
   a. It decreases when current increases.
   b. It destroys the diode.
   c. It equals the current times the resistance.
   d. It is approximately constant.

12. Which of these is the best description of a zener diode?
   a. It is a rectifier diode.
   b. It is a constant-voltage device.
   c. It is a constant-current device.
   d. It works in the forward region.

13. A zener diode
   a. is a battery.
   b. has a constant voltage in the breakdown region.
   c. has a barrier potential of 1 V.
   d. is forward-biased.

14. The voltage across the zener resistance is usually
   a. small.
   b. large.
   c. measured in volts.
   d. subtracted from the breakdown voltage.

15. If the series resistance increases in an unloaded zener regulator, the zener current
   a. decreases.
   b. stays the same.
   c. increases.
   d. equals the voltage divided by the resistance.

16. The load voltage is approximately constant when a zener diode is
   a. forward-biased.
   b. reverse-biased.
   c. operating in the breakdown region.
   d. unbiased.

17. In a loaded zener regulator, which is the largest current?
   a. Series current.
   b. Zener current.
   c. Load current.
   d. None of these.

18. If the load resistance increases in a zener regulator, the zener current
   a. decreases.
   b. stays the same.
   c. increases.
   d. equals the source voltage divided by the series resistance.

19. If the load resistance decreases in a zener regulator, the series current
   a. decreases.
   b. stays the same.
   c. increases.
   d. equals the source voltage divided by the series resistance.

20. When the source voltage increases in a zener regulator, which of these currents remains approximately constant?
   a. Series current.
   b. Zener current.
   c. Load current.
   d. Total current.

21. If the zener diode in a zener regulator is connected with the wrong polarity, the load voltage will be closest to
   a. 0.7 V.
   b. 10 V.
   c. 14 V.
   d. 18 V.

22. At high frequencies, ordinary diodes don't work properly because of
   a. forward bias.
   b. reverse bias.
   c. breakdown.
   d. charge storage.

23. The capacitance of a varactor diode increases when the reverse voltage across it
   a. decreases.
   b. increases.
   c. breaks down.
   d. stores charges.

24. Breakdown does not destroy a zener diode, provided the zener current is less than the
   a. breakdown voltage.
   b. zener test current.
   c. maximum zener current rating.
   d. barrier potential.

25. As compared to a silicon rectifier diode, an LED has a
   a. lower forward voltage and lower breakdown voltage.
   b. lower forward voltage and higher breakdown voltage.
   c. higher forward voltage and lower breakdown voltage.
   d. higher forward voltage and higher breakdown voltage.

26. To display the digit 0 in a seven-segment indicator,
   a. *C* must be OFF.
   b. *G* must be OFF.
   c. *F* must be ON.
   d. all segments must be lighted.

27. A photodiode is normally
   a. forward-biased.
   b. reverse-biased.
   c. neither forward- nor reverse-biased.
   d. emitting light.

28. When the light decreases, the reverse minority-carrier current in a photodiode
   a. decreases.
   b. increases.
   c. is unaffected.
   d. reverses direction.

29. The device associated with voltage-controlled capacitance is a
   a. light-emitting diode.
   b. photodiode.
   c. varactor diode.
   d. zener diode.

30. If the depletion layer width decreases, the capacitance
   a. decreases.
   b. stays the same.
   c. increases.
   d. is variable.

31. When the reverse voltage decreases, the capacitance
   a. decreases.
   b. stays the same.
   c. increases.
   d. has more bandwidth.

32. The varactor is usually
   a. forward-biased.
   b. reverse-biased.
   c. unbiased.
   d. operated in the breakdown region.

33. The device to use for rectifying a weak ac signal is a
   a. zener diode.
   b. light-emitting diode.
   c. varistor.
   d. Schottky diode.

34. To isolate an output circuit from an input circuit, which is the device to use?
   a. Varactor.
   b. Optocoupler.
   c. Seven-segment indicator.
   d. PIN diode.

35. The diode with a forward voltage drop of approximately 0.25 V is the
   a. zener diode.
   b. Schottky diode.
   c. silicon diode.
   d. varactor.

36. For typical operation, you need to use reverse bias with
   a. a zener diode.
   b. a photodiode.
   c. a varactor.
   d. all of the above.

37. As the forward current through a PIN diode decreases, its resistance
   a. increases.
   b. decreases.
   c. remains constant.
   d. cannot be determined.

 CHAPTER 29 PROBLEMS

SECTION 29.1 Basic Diode Characteristics

29.1 A diode is in series with 220 Ω. If the voltage across the resistor is 6 V, what is the current through the diode?

29.2 A diode has a voltage of 0.7 V and a current of 100 mA. What is the diode power?

29.3 Two diodes are in series. The first diode has a voltage of 0.75 V and the second has a voltage of 0.8 V. If the current through the first diode is 400 mA, what is the current through the second diode?

SECTION 29.2 The Ideal Diode

29.4 In Fig. 29-32*a*, calculate the load current, load voltage, load power, diode power, and total power.

29.5 If the resistor is doubled in Fig. 29-32*a*, what is the load current?

29.6 In Fig. 29-32*b*, calculate the load current, load voltage, load power, diode power, and total power.

29.7 If the diode polarity is reversed in Fig. 29-32*b*, what is the diode current? The diode voltage?

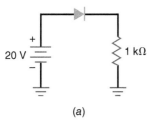

(a)

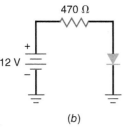

(b)

Figure 29-32

SECTION 29.3 The Second and Third Approximations

29.8 In Fig. 29-32a, calculate the load current, load voltage, load power, diode power, and total power.

29.9 In Fig. 29-32b, calculate the load current, load voltage, load power, diode power, and total power.

29.10 If the diode polarity is reversed in Fig. 29-32b, what is the diode current? The diode voltage?

SECTION 29.7 The Zener Diode

29.11 An unloaded zener regulator has a source voltage of 24 V, a series resistance of 470 Ω, and a zener voltage of 15 V. What is the zener current?

29.12 If the source voltage in Prob. 29-11 varies from 24 to 40 V, what is the maximum zener current?

29.13 If the series resistor of Prob. 29-11 has a tolerance of ±5%, what is the maximum zener current?

SECTION 29.8 The Loaded Zener Regulator

29.14 If the zener diode is disconnected in Fig. 29-33, what is the load voltage?

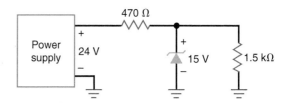

Figure 29-33

29.15 Calculate all three currents in Fig. 29-33.

29.16 Suppose the supply voltage of Fig. 29-33 can vary from 24 to 40 V. What is the maximum zener current?

SECTION 29.9 Zener Specifications

29.17 A zener diode has a voltage of 10 V and a current of 20 mA. What is the power dissipation?

29.18 What is the power dissipation in the resistors and zener diode of Fig. 29-33?

SECTION 29.10 Optoelectronic Devices

29.19 What is the current through the LED of Fig. 29-34?

29.20 If the supply voltage of Fig. 29-34 increases to 40 V, what is the LED current?

29.21 If the resistor is decreased to 1 kΩ, what is the LED current in Fig. 29-34?

29.22 The resistor of Fig. 29-34 is decreased until the LED current equals 13 mA. What is the value of the resistance?

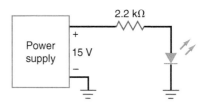

Figure 29-34

Chapter 30

Power Supply Circuits

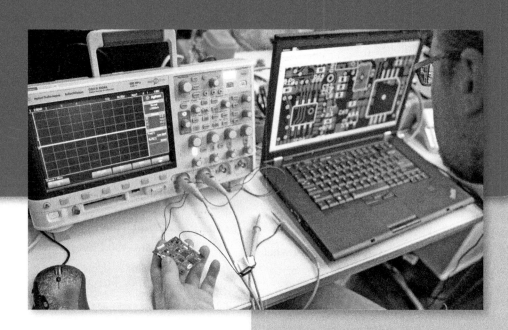

Most electronic systems, like HDTVs, DVD/CD players, and computers, need a dc voltage to work properly. Since the power-line voltage is alternating, the first thing we need to do is to convert the ac line voltage to dc voltage. The section of the electronic system that produces this dc voltage is called the power supply. Within the power supply are circuits that allow current to flow in only one direction. These circuits are called rectifiers. This chapter discusses rectifier circuits, filters, regulators, and voltage multipliers.

Learning Outcomes

After studying this chapter, you should be able to:

> *Draw* a diagram of a half-wave rectifier and *explain* how it works.

> *Describe* the role of the input transformer in power supplies.

> *Draw* a diagram of a full-wave rectifier and *explain* how it works.

> *Draw* a diagram of a bridge rectifier and *explain* how it works.

> *Analyze* a capacitor input filter and its surge current.

> *List* three important specifications found on a rectifier data sheet.

> *Describe* the action of voltage multipliers.

> *Define* regulation and *show* how it is achieved with a zener diode.

30.1 The Power Supply Subsystem

All electronic products, equipment, and systems have a power supply. It is the main subsystem that makes the rest of the circuits and equipment work. A *power supply* is a collection of components and circuits that supply one or more dc voltages to the circuits. In modern portable and mobile equipment, the power supply is a battery that is accompanied by a battery charger and other circuits that condition the battery voltage prior to use. Another common power supply is the ac operated power supply. It usually gets its input from the common 60-Hz ac power line. In some modern power supplies, the ac is internally generated by an oscillator to produce a high-frequency ac signal that is a superior and more efficient source of power.

AC Power Supply

Figure 30-1 shows a general block diagram of a common ac operated power supply. The main power source is the ac power line with its 120-V, 60-Hz sine wave. That input voltage is first passed through a low-pass filter to prevent high-frequency spikes and noise on the power line from entering the power supply and to prevent any high-frequency signals generated by the power supply from getting onto the ac line. Next the filtered ac is supplied to a transformer to step up or step down the voltage to a level appropriate to the equipment to be powered. In most cases, the 120 Vac is too high for modern integrated circuits, so a step-down transformer is used to lower the voltage.

Next, a rectifier circuit, converts the sine wave into dc pulses at a 60-Hz rate. Rectifiers are diodes, used individually or in various combinations to produce the dc pulses. In some applications transistors are used as rectifiers. In this chapter you will learn about the most common diode rectifier circuits.

Next, the dc pulses are passed to a filter. This is a low-pass filter that is used to smooth the pulses into a constant dc voltage. The most common filter is a large capacitor, but inductors, or inductors in combination with capacitors, are used to create the constant dc output. The dc voltage is commonly placed on an easily accessed distribution bus that can be tapped into by various loads that need the power.

The output of the filter is a dc voltage that is suitable to be used for some applications. However, in most applications the dc voltage is further processed with a regulator circuit. A *regulator* is a circuit that generates a specific dc output voltage and provides a way to maintain that voltage at a constant level despite input ac voltage variations or dc load changes. Regulators are necessary as most ICs and other circuit must have an operating voltage at a very specific level such as 1.8, 2.5, 3.3, or 5 V or other level, and the voltage must be maintained with in <1% of that value. Most regulators sense any output variations from the set value and proceed to make internal adjustments to correct the change. A regulator may be as simple as a zener diode, but in most cases it is an integrated circuit designed for the job.

Finally, many power supplies today must generate multiple dc voltages. These are created from the main filtered dc supply with multiple regulators and, in some cases, dc-dc converters. A *dc-dc converter* is a circuit that converts one dc input level to another dc level. AC signals are easily stepped up or down with a transformer, but you cannot do that with dc. The dc-dc converter provides a way to produce a needed dc value when only another dc voltage is available. DC-DC converters are available in IC form or in modules that can be used like an IC.

DC Power Supply

Another common power supply is the one used in portable equipment, such as laptop computers, cell phones, and digital cameras. Figure 30-2 is a block diagram of such a power supply. The main power source is a battery. It can be a pair of alkaline AA cells, a lithium-ion battery, or a 12-V car battery. No rectifiers are needed since the power is already in dc form. However, that dc voltage is then conditioned by regulators and dc-dc converters as needed.

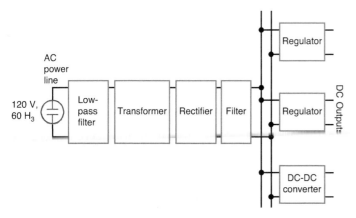

Figure 30-1 Generic block diagram of an ac-to-dc power supply.

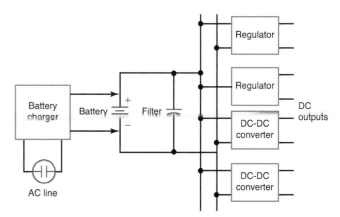

Figure 30-2 A typical battery power supply.

A key circuit in this power supply is the battery charger. It is basically a simple ac-to-dc power supply with an output voltage suitable for recharging the battery when necessary. Finally, most dc battery supplies have a power management capability, usually on the basis of multiple circuits in an IC that monitor voltage levels, battery condition, temperature, and other conditions. It can do things like sense overvoltage or overcurrent conditions or excessive temperature and turn the power off. It can also turn the regulators or dc-dc converters on or off in sequence as may be required by the circuits to be powered.

This chapter focuses on the basic ac-to-dc power supply and its circuits. You will learn more about regulators, dc-dc converters, and power management in a later chapter.

30.2 The Half-Wave Rectifier

Figure 30-3a shows a **half-wave rectifier** circuit. The ac source produces a sinusoidal voltage. Assuming an ideal diode, the positive half-cycle of source voltage will forward-bias the diode. Since the switch is closed, as shown in Fig. 30-3b, the positive half-cycle of source voltage will appear across the load resistor. On the negative half-cycle, the diode is reverse-biased. In this case, the ideal diode will appear as an open switch, as shown in Fig. 30-3c, and no voltage appears across the load resistor.

Ideal Waveforms

Figure 30-4a shows a graphical representation of the input voltage waveform. It is a sine wave with an instantaneous value of v_{in} and a peak value of $V_{p(in)}$. A pure sinusoid like this has an average value of zero over one cycle because each instantaneous voltage has an equal and opposite voltage half a cycle later. If you measure this voltage with a dc voltmeter, you will get a reading of zero because a dc voltmeter indicates the average value.

In the half-wave rectifier of Fig. 30-4b, the diode is conducting during the positive half-cycle but is nonconducting during the negative half-cycle. Because of this, the circuit clips off the negative half-cycle, as shown in Fig. 30-4c. We call a waveform like this a *half-wave signal*. This half-wave voltage produces a **unidirectional load current.** This

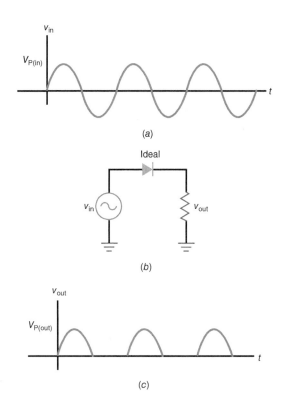

Figure 30-4 (a) Input to half-wave rectifier. (b) Circuit. (c) Output of half-wave rectifier.

means that it flows in only one direction. If the diode were reversed, the output pulses would be negative.

A half-wave signal like the one in Fig. 30-4c is a pulsating dc voltage that increases to a maximum, decreases to zero, and then remains at zero during the negative half-cycle. This is not the kind of dc voltage we need for electronics equipment. What we need is a constant voltage, the same as you get from a battery. To get this kind of voltage, we need to **filter** the half-wave signal (discussed later in this chapter).

When you are troubleshooting, you can use the ideal diode to analyze a half-wave rectifier. It's useful to remember that the peak output voltage equals the peak input voltage:

$$\text{Ideal half-wave: } V_{p(out)} = V_{p(in)} \quad \textbf{(30-1)}$$

DC Value of Half-Wave Signal

The **dc value of a signal** is the same as the average value. If you measure a signal with a dc voltmeter, the reading will equal the average value. The formula is

$$\text{Half-wave: } V_{dc} = \frac{V_p}{\pi} \quad \textbf{(30-2)}$$

Since $1/\pi \approx 0.318$, you may see Formula (30-2) written as

$$V_{dc} \approx 0.318 V_p$$

When the equation is written in this form, you can see that the dc or average value equals 31.8% of the peak value. For instance, if the peak voltage of the half-wave signal is 100 V, the dc voltage or average value is 31.8 V.

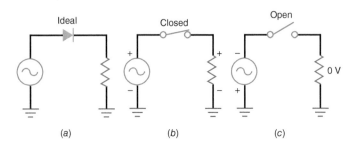

Figure 30-3 (a) Ideal half-wave rectifier. (b) On positive half-cycle. (c) On negative half-cycle.

Output Frequency

The output frequency is the same as the input frequency. This makes sense when you compare Fig. 30-4c with Fig. 30-4a. Each cycle of input voltage produces one cycle of output voltage. Therefore, we can write

$$\text{Half-wave: } f_{out} = f_{in} \qquad (30\text{-}3)$$

We will use this derivation later with filters.

Second Approximation

We don't get a perfect half-wave voltage across the load resistor. Because of the barrier potential, the diode does not turn on until the ac source voltage reaches approximately 0.7 V. When the peak source voltage is much greater than 0.7 V, the load voltage will resemble a half-wave signal. For instance, if the peak source voltage is 100 V, the load voltage will be very close to a perfect half-wave voltage. If the peak source voltage is only 5 V, the load voltage will have a peak of only 4.3 V. When you need to get a better answer, use this derivation:

$$\text{2d half-wave: } V_{p(out)} = V_{p(in)} - 0.7 \text{ V} \qquad (30\text{-}4)$$

30.3 The Transformer

Power companies in the United States supply a nominal line voltage of 120 V rms and a frequency of 60 Hz. The actual voltage coming out of a power outlet may vary from 105 to 125 V rms, depending on the time of day, the locality, and other factors. Line voltage is too high for most of the circuits used in electronics equipment. This is why a transformer is commonly used in the power supply section of almost all electronics equipment. The transformer steps the line voltage down to safer and lower levels that are more suitable for use with diodes, transistors, and other semiconductor devices.

Basic Idea

Earlier courses discussed the transformer in detail. All we need in this chapter is a brief review. Figure 30-5 shows a transformer. Here you see line voltage applied to the primary winding of a transformer. Usually, the power plug has a third prong to ground the equipment. Because of the turns ratio N_1/N_2, the secondary voltage is stepped down when N_1 is greater than N_2.

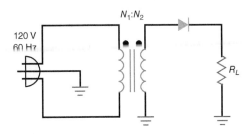

Figure 30-5 Half-wave rectifier with transformer.

Phasing Dots

Recall the meaning of the phasing dots shown at the upper ends of the windings. Dotted ends have the same instantaneous phase. In other words, when a positive half-cycle appears across the primary, a positive half-cycle appears across the secondary. If the secondary dot were on the ground end, the secondary voltage would be 180° out of phase with the primary voltage.

On the positive half-cycle of primary voltage, the secondary winding has a positive half sine wave across it and the diode is forward-biased. On the negative half-cycle of primary voltage, the secondary winding has a negative half-cycle and the diode is reverse-biased. Assuming an ideal diode, we will get a half-wave load voltage.

Turns Ratio

Recall from your earlier course work the following derivation:

$$V_2 = \frac{V_1}{N_1/N_2} \qquad (30\text{-}5)$$

This says that the secondary voltage equals the primary voltage divided by the turns ratio. Sometimes you will see this equivalent form:

$$V_2 = \frac{N_2}{N_1} V_1$$

This says that the secondary voltage equals the inverse turns ratio times the primary voltage.

You can use either formula for rms, peak values, and instantaneous voltages. Most of the time, we will use Formula (30-5) with rms values because ac source voltages are almost always specified as rms values.

The terms *step up* and *step down* are also encountered when dealing with transformers. These terms always relate the secondary voltage to the primary voltage. This means that a step-up transformer will produce a secondary voltage that is larger than the primary, and a step-down transformer will produce a secondary voltage that is smaller than the primary.

EXAMPLE 30-1

What are the peak load voltage and dc load voltage in Fig. 30-6?

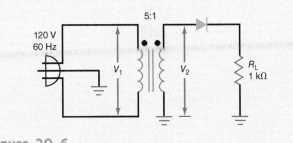

Figure 30-6

Answer:

The transformer has a turns ratio of 5 : 1. This means that the rms secondary voltage is one-fifth of the primary voltage:

$$V_2 = \frac{120 \text{ V}}{5} = 24 \text{ V}$$

and the peak secondary voltage is

$$V_p = \frac{24 \text{ V}}{0.707} = 34 \text{ V}$$

With an ideal diode, the peak load voltage is

$$V_{p(out)} = 34 \text{ V}$$

The dc load voltage is

$$V_{dc} = \frac{V_p}{\pi} = \frac{34 \text{ V}}{\pi} = 10.8 \text{ V}$$

With the second approximation, the peak load voltage is

$$V_{p(out)} = 34 \text{ V} - 0.7 \text{ V} = 33.3 \text{ V}$$

and the dc load voltage is

$$V_{dc} = \frac{V_p}{\pi} = \frac{33.3 \text{ V}}{\pi} = 10.6 \text{ V}$$

30.4 The Full-Wave Rectifier

Figure 30-7a shows a **full-wave rectifier** circuit. Notice the grounded center tap on the secondary winding. The full-wave rectifier is equivalent to two half-wave rectifiers. Because of the center tap, each of these rectifiers has an input voltage equal to half the secondary voltage. Diode D_1 conducts on the positive half-cycle, and diode D_2 conducts on the negative half-cycle. As a result, the rectified load current flows during both half-cycle. The full-wave rectifier acts the same as two back-to-back half-wave rectifiers.

Figure 30-7b shows the equivalent circuit for the positive half-cycle. As you see, D_1 is forward-biased. This produces a positive load voltage as indicated by the plus-minus polarity across the load resistor. Figure 30-7c shows the equivalent circuit for the negative half-cycle. This time, D_2 is forward-biased. As you can see, this also produces a positive load voltage.

During both half-cycle, the load voltage has the same polarity and the load current is in the same direction. The circuit is called a *full-wave rectifier* because it has changed the ac input voltage to the pulsating dc output voltage shown in Fig. 30-7d.

DC or Average Value

Since the full-wave signal has twice as many positive cycles as the half-wave signal, the dc or average value is twice as much, given by

$$\text{Full wave: } V_{dc} = \frac{2V_p}{\pi} \qquad \textbf{(30-6)}$$

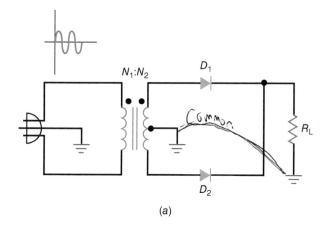

(a)

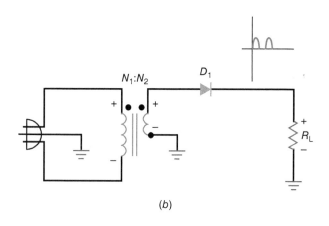

(b)

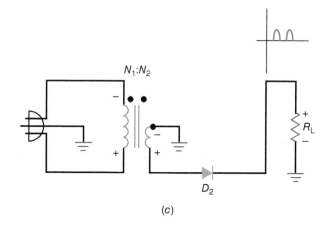

(c)

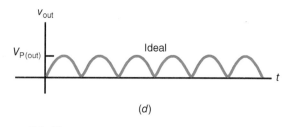

(d)

Figure 30-7 (a) Full-wave rectifier. (b) Equivalent circuit for positive half-cycle. (c) Equivalent circuit for negative half-cycle. (d) Full-wave output.

Since $2/\pi = 0.636$, you may see Formula (30-6) written as

$$V_{dc} \approx 0.636 V_p$$

In this form, you can see that the dc or average value equals 63.6% of the peak value. For instance, if the peak voltage of the full-wave signal is 100 V, the dc voltage or average value is 63.6 V.

Output Frequency

With a half-wave rectifier, the output frequency equals the input frequency. But with a full-wave rectifier, something unusual happens to the output frequency. The ac line voltage has a frequency of 60 Hz. Therefore, the input period equals

$$T_{in} = \frac{1}{f} = \frac{1}{60\ Hz} = 16.7\ ms$$

Because of the full-wave rectification, the period of the full-wave signal is half the input period:

$$T_{out} = 0.5(16.7\ ms) = 8.33\ ms$$

When we calculate the output frequency, we get

$$f_{out} = \frac{1}{T_{out}} = \frac{1}{8.33\ ms} = 120\ Hz$$

The frequency of the full-wave signal is double the input frequency. This makes sense. A full-wave output has twice as many cycles as the sine-wave input has. The full-wave rectifier inverts each negative half-cycle, so that we get double the number of positive half-cycle. The effect is to double the frequency. As a derivation,

$$\text{Full wave: } f_{out} = 2f_{in} \qquad \textbf{(30-7)}$$

Second Approximation

Since the full-wave rectifier is like two back-to-back half-wave rectifiers, we can use the second approximation given earlier. The idea is to subtract 0.7 V from the ideal peak output voltage. The following example will illustrate the idea.

EXAMPLE 30-2

Figure 30-7a shows a full-wave rectifier. Assume a 120-V, 60-Hz input; a turns ratio, $N_1 : N_2$, of 10 : 1; 1N4001 diodes; and a 1-kΩ load. Calculate the peak input and output voltages.

Answer:

The peak primary voltage is

$$V_{p(1)} = \frac{V_{rms}}{0.707} = \frac{120\ V}{0.707} = 170\ V$$

Because of the 10 : 1 step-down transformer, the peak secondary voltage is

$$V_{p(2)} = \frac{V_{p(1)}}{N_1/N_2} = \frac{170\ V}{10} = 17\ V$$

The full-wave rectifier acts like two back-to-back half-wave rectifiers. Because of the center tap, the input voltage to each half-wave rectifier is only half the secondary voltage

$$V_{p(in)} = 0.5(17\ V) = 8.5\ V$$

Ideally, the output voltage is

$$V_{p(out)} = 8.5\ V$$

Using the second approximation,

$$V_{p(out)} = 8.5\ V - 0.7\ V = 7.8\ V$$

EXAMPLE 30-3

If one of the diodes in Fig. 30-7 were open, what would happen to the different voltages?

Answer:

If one of the diodes is open, the circuit reverts to a half-wave rectifier. In this case, half the secondary voltage is still 8.5 V, but the load voltage will be a half-wave signal rather than a full-wave signal. This half-wave voltage will still have a peak of 8.5 V (ideally) or 7.8 V (second approximation).

30.5 The Bridge Rectifier

Figure 30-8a shows a **bridge rectifier** circuit. The bridge rectifier is similar to a full-wave rectifier because it produces a full-wave output voltage. Diodes D_1 and D_2 conduct on the positive half-cycle, and D_3 and D_4 conduct on the negative half-cycle. As a result, the rectified load current flows during both half-cycle.

Figure 30-8b shows the equivalent circuit for the positive half-cycle. As you can see, D_1 and D_2 are forward-biased. This produces a positive load voltage as indicated by the plus-minus polarity across the load resistor. As a memory aid, visualize D_2 shorted. Then the circuit that remains is a half-wave rectifier, which we are already familiar with.

Figure 30-8c shows the equivalent circuit for the negative half-cycle. This time, D_3 and D_4 are forward-biased. This also produces a positive load voltage. If you visualize D_3 shorted, the circuit looks like a half-wave rectifier. So the bridge rectifier acts like two back-to-back half-wave rectifiers.

During both half-cycle, the load voltage has the same polarity and the load current is in the same direction. The circuit has changed the ac input voltage to the pulsating dc output voltage shown in Fig. 30-8d. Note the advantage of this type of full-wave rectification over the center-tapped version in the previous section: *The entire secondary voltage can be used.*

Fig. 30-8e shows bridge rectifier packages that contain all four diodes.

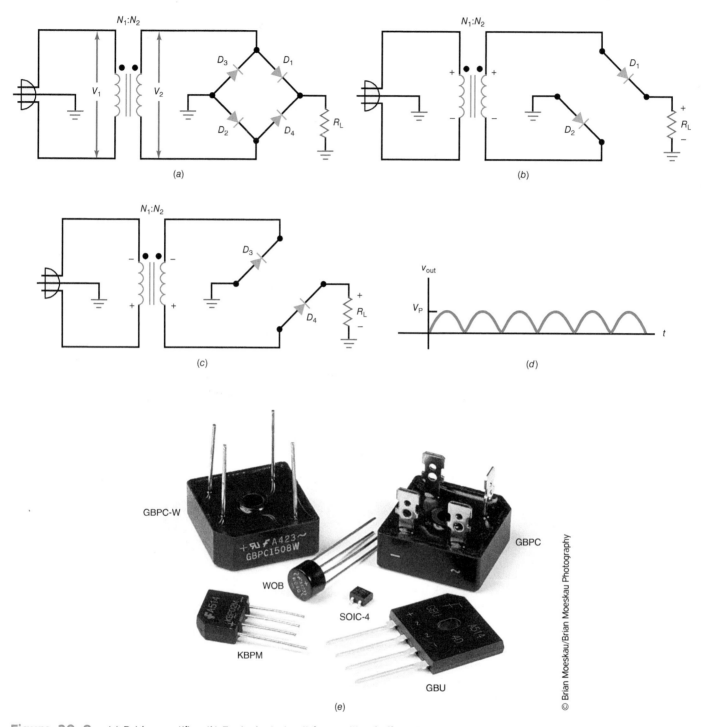

Figure 30-8 (a) Bridge rectifier. (b) Equivalent circuit for positive half-cycle. (c) Equivalent circuit for negative half-cycle. (d) Full-wave output. (e) Bridge rectifier packages.

Average Value and Output Frequency

Because a bridge rectifier produces a full-wave output, the equations for average value and output frequency are the same as given for a full-wave rectifier:

$$V_{dc} = \frac{2V_p}{\pi}$$

and

$$f_{out} = 2f_{in}$$

The average value is 63.6% of the peak value, and the output frequency is 120 Hz, given a line frequency of 60 Hz.

One advantage of a bridge rectifier is that all the secondary voltage is used as the input to the rectifier. Given

Table 30-1 Unfiltered Rectifiers*

	Half-wave	Full-wave	Bridge
Number of diodes	1	2	4
Rectifier input	$V_{p(2)}$	$0.5V_{p(2)}$	$V_{p(2)}$
Peak output (ideal)	$V_{p(2)}$	$0.5V_{p(2)}$	$V_{p(2)}$
Peak output (2d)	$V_{p(2)} - 0.7$ V	$0.5V_{p(2)} - 0.7$ V	$V_{p(2)} - 1.4$ V
DC output	$V_{p(out)}/\pi$	$2V_{p(out)}/\pi$	$2V_{p(out)}/\pi$
Ripple frequency	f_{in}	$2f_{in}$	$2f_{in}$

*$V_{p(2)}$ = peak secondary voltage; $V_{p(out)}$ = peak output voltage.

the same transformer, we get twice as much peak voltage and twice as much dc voltage with a bridge rectifier as with a full-wave rectifier. Doubling the dc output voltage compensates for having to use two extra diodes. As a rule, you will see *the bridge rectifier used a lot more than the full-wave rectifier.*

Second Approximation and Other Losses

Since the bridge rectifier has two diodes in the conducting path, the peak output voltage is given by

$$\text{2d bridge: } V_{p(out)} = V_{p(in)} - 1.4 \text{ V} \qquad \textbf{(30-8)}$$

As you can see, we have to subtract two diode drops from the peak to get a more accurate value of peak load voltage. Table 30-1 compares the three rectifiers and their properties.

EXAMPLE 30-4

Calculate the peak input and output voltages in Fig. 30-8. Then, compare the theoretical values to the measured values.
Notice the circuit uses a bridge rectifier package.

Answer:

The peak primary and secondary voltages are the same as in Example 30-2:

$$V_{p(1)} = 170 \text{ V}$$
$$V_{p(2)} = 17 \text{ V}$$

With a bridge rectifier, all of the secondary voltage is used as the input to the rectifier. Ideally, the peak output voltage is

$$V_{p(out)} = 17 \text{ V}$$

To a second approximation,

$$V_{p(out)} = 17 \text{ V} - 1.4 \text{ V} = 15.6 \text{ V}$$

30.6 Power Supply Filtering

The purpose of the filter in a power supply is to take the dc pulses produced by the rectifier and smooth them into a constant dc voltage. There are two ways to visualize the filtering process, one in the frequency domain and the other in the time domain.

In frequency domain analysis you view the rectifier output pulses as nonsinusoidal signals. You should immediately recognize that these pulses are made up of a fundamental sine wave and an infinite number of harmonics, as described by Fourier theory. Half- and full-wave rectified sine waves are made up of a fundamental sine wave and even harmonic sine waves. With the 60-Hz pulses of the half-wave rectifier, this means a 60-Hz sine wave and harmonics of 120, 240, 360, and 480 Hz and other even-numbered sine waves. A full-wave rectifier produces 120-Hz pulses with its even harmonics of 240, 480, and 600 Hz and others.

To get rid of these sine waves, a low-pass filter is the obvious choice. Selecting a cutoff frequency well below base pulse frequency of 60 Hz or 120 Hz, the attenuation will be high enough to eliminate most of the harmonics.

Figure 30-9 has a frequency domain plot of the output of a full-wave rectifier showing the fundamental and harmonic components. Overlaid with that is a low-pass filter response curve showing how the filter will attenuate the harmonics and fundamental. The result is a near constant dc voltage. Note that because of the gradual slope of the filter, not all the harmonics are completely filtered out. Therefore, some residual signal, called *ripple*, will still be present. The ripple is usually low enough not to be a problem.

A typical low-pass filter can be made from *RC* and *LC* circuits. A large shunt capacitor at the output of the rectifier is the most common filter. It forms a low-pass filter with the series resistance of the transformer secondary winding and the diode resistances. An *LC* low-pass filter can also be formed with a series inductor (choke) and the shunt capacitor.

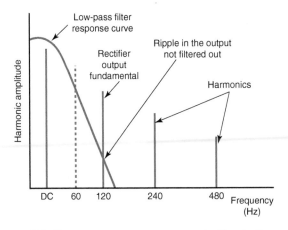

Figure 30-9 Viewing the power supply filter in the frequency domain: a full-wave rectifier with 60-Hz input.

By selecting the L and C values to produce a low cutoff frequency, a near constant dc output will result.

A more common way to analyze the filter is with traditional time domain analysis. This makes use of RC and LR time constants. By considering the capacitor charge-discharge curve, the time constant, and the pulse frequency, values can be selected to remove the ripple. Time domain analysis is used in the following section about filters.

The Capacitor-Input Filter

The most widely used filter is a large capacitor. It is called a capacitor-input filter. The **capacitor-input filter** produces a dc output voltage equal to the peak value of the rectified voltage.

Basic Idea

Figure 30-10a shows an ac source, a diode, and a capacitor. The key to understanding a capacitor-input filter is understanding what this simple circuit does during the first quarter-cycle.

Initially, the capacitor is uncharged. During the first quarter-cycle of Fig. 30-10b, the diode is forward-biased. Since it ideally acts like a closed switch, the capacitor charges, and its voltage equals the source voltage at each instant of the first quarter-cycle. The charging continues until the input reaches its maximum value. At this point, the capacitor voltage equals V_p.

After the input voltage reaches the peak, it starts to decrease. As soon as the input voltage is less than V_p, the diode turns off. In this case, it acts like the open switch of Fig. 30-10c. During the remaining cycles, the capacitor stays fully charged and the diode remains open. This is why the output voltage of Fig. 30-10b is constant and equal to V_p.

Ideally, all that the capacitor-input filter does is charge the capacitor to the peak voltage during the first quarter-cycle. This peak voltage is constant, the perfect dc voltage we need

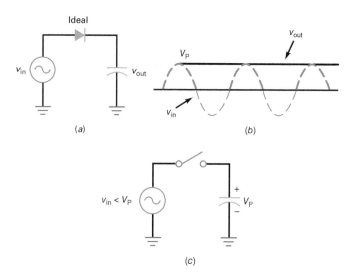

Figure 30-10 (a) Unloaded capacitor-input filter. (b) Output is pure dc voltage. (c) Capacitor remains charged when diode is off.

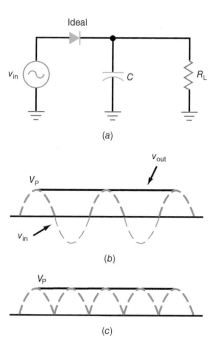

Figure 30-11 (a) Loaded capacitor-input filter. (b) Output is direct current with small ripple. (c) Full-wave output has less ripple.

for electronics equipment. There's only one problem: There is no load resistor.

Effect of Load Resistor

For the capacitor-input filter to be useful, we need to connect a load resistor across the capacitor, as shown in Fig. 30-11a. As long as the R_LC time constant is much greater than the period, the capacitor remains almost fully charged and the load voltage is approximately V_p. The only deviation from a perfect dc voltage is the small ripple seen in Fig. 30-11b. The smaller the peak-to-peak value of this ripple, the more closely the output approaches a perfect dc voltage.

Between peaks, the diode is off and the capacitor discharges through the load resistor. In other words, the capacitor supplies the load current. Since the capacitor discharges only slightly between peaks, the peak-to-peak ripple is small. When the next peak arrives, the diode conducts briefly and recharges the capacitor to the peak value. A key question is, What size should the capacitor be for proper operation?

Full-Wave Filtering

If we connect a full-wave or bridge rectifier to a capacitor-input filter, the peak-to-peak ripple is cut in half. Figure 30-11c shows why. When a full-wave voltage is applied to the RC circuit, the capacitor discharges for only half as long. Therefore, the peak-to-peak ripple is half the size it would be with a half-wave rectifier.

Figure 30-12 shows the details for the capacitor charge and discharge process. The capacitor charges to the peak of the sine-wave input less any diode drops. Then as the dc pulse decreases, the diode turns off and the capacitor discharges

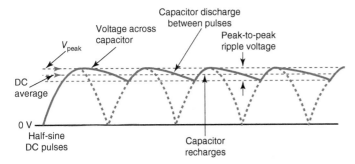

Figure 30-12 Capacitor charge and discharge process.

through the load. On the next dc pulse, the diode turns on, the capacitor recharges to the peak, and the process is repeated. Note that the charging and discharging produces a sawtooth-like waveform and appears to be riding on the average dc. This ac voltage component is called ripple and, of course, is undesired. It can be minimized by using a lighter load (higher load resistance) and/or a larger capacitor. This increases the time constant and reduces the peak-to-peak ripple voltage.

The Ripple Formula

Here is a derivation we will use to estimate the peak-to-peak ripple out of any capacitor-input filter:

$$V_R = \frac{I}{fC} \quad \textbf{(30-9)}$$

where V_R = peak-to-peak ripple voltage
 I = dc load current
 f = ripple frequency
 C = capacitance

This is an approximation, not an exact derivation. We can use this formula to estimate the peak-to-peak ripple. When a more accurate answer is needed, one solution is to use a computer with a circuit simulator like Multisim.

For instance, if the dc load current is 10 mA and the capacitance is 200 μF, the ripple with a bridge rectifier and a capacitor-input filter is

$$V_R = \frac{10 \text{ mA}}{(120 \text{ Hz})(200 \ \mu\text{F})} = 0.417 \text{ V p-p}$$

In using this derivation, remember two things. First, the ripple is in peak-to-peak (p-p) voltage. This is useful because you normally measure ripple voltage with an oscilloscope. Second, the formula works with half-wave or full-wave voltages. Use 60 Hz for half-wave, and 120 Hz for full-wave.

You should use an oscilloscope for ripple measurements if one is available. If not, you can use an ac voltmeter, although there will be a significant error in the measurement. Most ac voltmeters are calibrated to read the rms value of a sine wave. Since the ripple is not a sine wave, you may get a measurement error of as much as 25%, depending on the design of the ac voltmeter.

Exact DC Load Voltage

It is difficult to calculate the exact dc load voltage in a bridge rectifier with a capacitor-input filter. To begin with, we have the two diode drops that are subtracted from the peak voltage. Besides the diode drops, an additional voltage drop occurs, as follows: The diodes conduct heavily when recharging the capacitor because they are on for only a short time during each cycle. This brief but large current has to flow through the transformer windings and the bulk resistance of the diodes. In our examples, we will calculate either the ideal output or the output with the second approximation of a diode, remembering that the actual dc voltage is slightly lower.

EXAMPLE 30-5

What is the dc load voltage and ripple in Fig. 30-13?

Answer:
The rms secondary voltage is

$$V_2 = \frac{120 \text{ V}}{5} = 24 \text{ V}$$

The peak secondary voltage is

$$V_p = \frac{24 \text{ V}}{0.707} = 34 \text{ V}$$

Assuming an ideal diode and small ripple, the dc load voltage is

$$V_L = 34 \text{ V}$$

To calculate the ripple, we first need to get the dc load current

$$I_L = \frac{V_L}{R_L} = \frac{34 \text{ V}}{5 \text{ k}\Omega} = 6.8 \text{ mA}$$

Now we can use Formula (30-9) to get

$$V_R = \frac{6.8 \text{ mA}}{(60 \text{ Hz})(100 \ \mu\text{F})} = 1.13 \text{ Vpp} \approx 1.1 \text{ Vp-p}$$

We rounded the ripple to two significant digits because it is an approximation and cannot be accurately measured with an oscilloscope with greater precision.

Here is how to improve the answer slightly: There is about 0.7 V across a silicon diode when it is conducting. Therefore, the peak voltage across the load will be closer to 33.3 V than to 34 V. The ripple also lowers the dc voltage slightly. So the actual dc load voltage will be closer to 33 V than to 34 V. But these are minor deviations. Ideal answers are usually adequate for trouble-shooting and preliminary analysis.

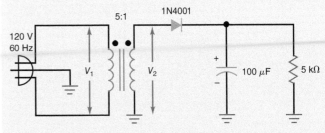

Figure 30-13 Half-wave rectifier and capacitor-input filter.

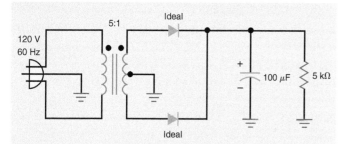

Figure 30-14   Full-wave rectifier and capacitor-input filter.

A final point about the circuit: The plus sign on the filter capacitor indicates a **polarized capacitor**, one whose plus side must be connected to the positive rectifier output. In Fig. 30-14, the plus sign on the capacitor case is correctly connected to the positive output voltage. You must look carefully at the capacitor case when you are building or troubleshooting a circuit to find out whether it is polarized or not.

Power supplies often use polarized electrolytic capacitors because this type can provide high values of capacitance in small packages. As discussed in earlier courses, *electrolytic capacitors must be connected with the correct polarity* to produce the oxide film. If an electrolytic capacitor is connected in opposite polarity, *it becomes hot and may explode.*

EXAMPLE 30-6

What is the dc load voltage and ripple in Fig. 30-14?

Answer:
Since the transformer is 5:1 step down like the preceding example, the peak secondary voltage is still 34 V. Half this voltage is the input to each half-wave section. Assuming an ideal diode and small ripple, the dc load voltage is

$$V_{\rm L} = 17 \text{ V}$$

The dc load current is

$$I_{\rm L} = \frac{17 \text{ V}}{5 {\rm k}\Omega} = 3.4 \text{ mA}$$

Now, Formula (30-9) gives

$$V_{\rm R} = \frac{3.4 \text{ mA}}{(120 \text{ Hz})(100 \ \mu\text{F})} = 0.283 \text{ Vp-p} \approx 0.28 \text{ Vp-p}$$

Because of the 0.7 V across the conducting diode, the actual dc load voltage will be closer to 16 V than to 17 V.

EXAMPLE 30-7

What is the dc load voltage and ripple in Fig. 30-15? Compare the answers with those in the two preceding examples.

Answer:
Since the transformer is 5:1 step down as in the preceding example, the peak secondary voltage is still 34 V. Assuming an ideal diode and small ripple, the dc load voltage is

$$V_{\rm L} = 34 \text{ V}$$

The dc load current is

$$I_{\rm L} = \frac{34 \text{ V}}{5 {\rm k}\Omega} = 6.8 \text{ mA}$$

Now, Formula (30-9) gives

$$V_{\rm R} = \frac{6.8 \text{ mA}}{(120 \text{ Hz})(100 \ \mu\text{F})} = 0.566 \text{ Vp-p} \approx 0.57 \text{ Vp-p}$$

Because of the 1.4 V across two conducting diodes and the ripple, the actual dc load voltage will be closer to 32 V than to 34 V.

We have calculated the dc load voltage and ripple for the three different rectifiers. Here are the results:

Half-wave: 34 V and 1.13 V

Full wave: 17 V and 0.288 V

Bridge: 34 V and 0.566 V

For a given transformer, the bridge rectifier is better than the half-wave rectifier because it has less ripple, and it's better than the full-wave rectifier because it produces twice as much output voltage. Of the three, *the bridge rectifier has emerged as the most popular.*

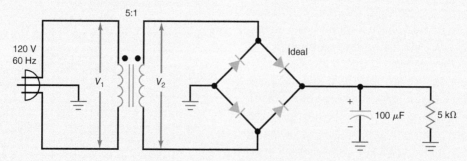

Figure 30-15   Bridge rectifier and capacitor-input filter.

The Choke-Input Filter

While the capacitor filter is the commonly used, a choke or inductor is also sometimes used with a capacitor to improve filtering and lessen the ripple.

Basic Idea

Look at Fig. 30-16a. This type of filter is called a **choke-input filter.** The ac source produces a current in the inductor, capacitor, and resistor. The ac current in each component depends on the inductive reactance, capacitive reactance, and the resistance. The inductor has a reactance given by

$$X_L = 2\pi f L$$

The capacitor has a reactance given by

$$X_C = \frac{1}{2\pi f C}$$

As you learned in previous courses, the choke (or inductor) has the primary characteristic of opposing a change in current. Because of this, a choke-input filter ideally reduces the ac current in the load resistor to zero. To a second approximation, it reduces the ac load current to a very small value. Let us find out why.

The first requirement of a well-designed choke-input filter is to have X_C at the input frequency be much smaller than R_L. When this condition is satisfied, we can ignore the load resistance and use the equivalent circuit of Fig. 30-16b. The second requirement of a well-designed choke-input filter is to have X_L be much greater than X_C at the input frequency. When this condition is satisfied, the ac output voltage approaches zero. On the other hand, since the choke approximates a short circuit at 0 Hz and the capacitor approximates an open at 0 Hz, the dc current can be passed to the load resistance with minimum loss.

In Fig. 30-16b, the circuit acts like a reactive voltage divider. When X_L is much greater than X_C, almost all the ac voltage is dropped across the choke. In this case, the ac output voltage equals

$$V_{out} \approx \frac{X_C}{X_L} V_{in} \qquad \textbf{(30-10)}$$

For instance, if $X_L = 10 \text{ k}\Omega$, $X_C = 100 \text{ }\Omega$, and $V_{in} = 15 \text{ V}$, the ac output voltage is

$$V_{out} \approx \frac{100 \text{ }\Omega}{10 \text{ k}\Omega} 15 \text{ V} = 0.15 \text{ V}$$

In this example, the choke-input filter reduces the ac voltage by a factor of 100.

Filtering the Output of a Rectifier

Figure 30-17a shows a choke-input filter between a rectifier and a load. The rectifier can be a half-wave, full-wave, or bridge type. What effect does the choke-input filter have on the load voltage?

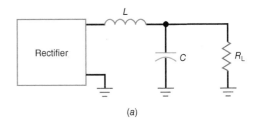

(a)

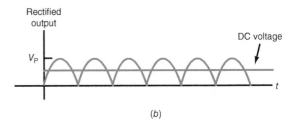

(b)

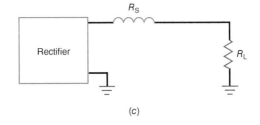

(c)

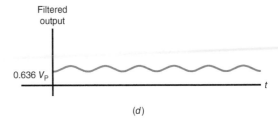

(d)

Figure 30-17   (a) Rectifier with choke-input filter. (b) Rectifier output has dc and ac components. (c) DC equivalent circuit. (d) Filter output is direct current with small ripple.

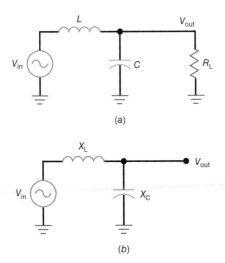

(a)

(b)

Figure 30-16   (a) Choke-input filter. (b) AC equivalent circuit.

The rectifier output has two different components: a dc voltage (the average value) and an ac voltage (the fluctuating part), as shown in Fig. 30-17b. Each of these voltages acts like a separate source. As far as the ac voltage is concerned, X_L is much greater than X_C, and this results in very little ac voltage across the load resistor. Even though the ac component is not a pure sine wave, Formula (30-10) is still a close approximation for the ac load voltage.

The circuit acts like Fig. 30-17c as far as dc voltage is concerned. At 0 Hz, the inductive reactance is zero and the capacitive reactance is infinite. Only the series resistance of the inductor windings remains. Making R_S much smaller than R_L causes most of the dc component to appear across the load resistor.

That's how a choke-input filter works: Almost all the dc component is passed on to the load resistor, and almost all the ac component is blocked. In this way, we get an almost perfect dc voltage, one that is almost constant, like the voltage out of a battery. Figure 30-17d shows the filtered output for a full-wave signal. The only deviation from a perfect dc voltage is the small ac load voltage shown in Fig. 30-17d. This small ac load voltage is called ripple. With an oscilloscope, we can measure its peak-to-peak value.

Main Disadvantage

A *power supply* is the circuit inside electronics equipment that converts the ac input voltage to an almost perfect dc output voltage. It includes a rectifier and a filter. With low-voltage, high-current power supplies and a line frequency of 60 Hz, large inductances have to be used to get enough reactance for adequate filtering. But large inductors have large winding resistances, which create a serious design problem with large load currents. In other words, too much dc voltage is dropped across the choke resistance. Furthermore, bulky inductors are not suitable for modern semiconductor circuits, where the emphasis is on lightweight designs.

Switching Regulators

One important application does exist for the choke-input filter. A **switching regulator** is a newer form of power supply. The frequency used in a switching regulator is much higher than 60 Hz. Typically, the frequency being filtered is above 100 kHz. At this much higher frequency, we can use much smaller inductors to design efficient choke-input filters. We will discuss the details in a later chapter.

30.7 Peak Inverse Voltage and Surge Current

The **peak inverse voltage (PIV)** is the maximum voltage across the nonconducting diode of a rectifier. *This voltage must be less than the breakdown voltage of the diode; otherwise, the diode will be destroyed.* The peak inverse voltage

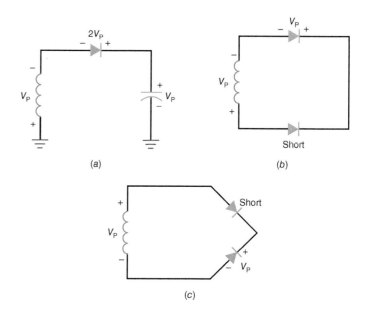

Figure 30-18 (a) Peak inverse voltage in half-wave rectifier. (b) Peak inverse voltage in full-wave rectifier. (c) Peak inverse voltage in bridge rectifier.

depends on the type of rectifier and filter. The worst case occurs with the capacitor-input filter.

Manufacturers use many different symbols to indicate the maximum reverse voltage rating of a diode. Sometimes, these symbols indicate different conditions of measurement. Some of the data sheet symbols for the maximum reverse voltage rating are PIV, PRV, V_B, V_{BR}, V_R, V_{RRM}, V_{RWM}, and $V_{R(max)}$.

Half-Wave Rectifier with Capacitor-Input Filter

Figure 30-18a shows the critical part of a half-wave rectifier. This is the part of the circuit that determines how much reverse voltage is across the diode. The rest of the circuit has no effect and is omitted for the sake of clarity. In the worse case, the peak secondary voltage is on the negative peak and the capacitor is fully charged with a voltage of V_p. Apply Kirchhoff's voltage law, and you can see right away that the peak inverse voltage across the nonconducting diode is

$$\text{PIV} = 2V_p \qquad \textbf{(30-11)}$$

For instance, if the peak secondary voltage is 15 V, the peak inverse voltage is 30 V. As long as the breakdown voltage of the diode is greater than this, the diode will not be damaged.

Full-Wave Rectifier with Capacitor-Input Filter

Figure 30-18b shows the essential part of a full-wave rectifier needed to calculate the peak inverse voltage. Again, the secondary voltage is at the negative peak. In this case, the lower diode acts like a short (closed switch) and the upper diode is open. Kirchhoff's law implies

$$\text{PIV} = V_p \qquad \textbf{(30-12)}$$

Bridge Rectifier with Capacitor-Input Filter

Figure 30-18c shows part of a bridge rectifier. This is all you need to calculate the peak inverse voltage. Since the upper diode is shorted and the lower one is open, the peak inverse voltage across the lower diode is

$$PIV = V_p \qquad \text{(30-13)}$$

Another advantage of the bridge rectifier is that it has the lowest peak inverse voltage for a given load voltage. To produce the same load voltage, the full-wave rectifier would need twice as much secondary voltage.

30.8 Other Power Supply Topics

You have a basic idea of how power supply circuits work. In the preceding sections, you have seen how an ac input voltage is rectified and filtered to get a dc voltage. There are a few additional ideas you need to know about.

Commercial Transformers

The use of turns ratios with transformers applies only to ideal transformers. Iron-core transformers are different. In other words, the transformers you buy from a parts supplier are not ideal because the windings have resistance, which produces power losses. Furthermore, the laminated core has eddy currents, which produce additional power losses. Because of these unwanted power losses, the turns ratio is only an approximation. In fact, the data sheets for transformers rarely list the turns ratio. Usually, all you get is the secondary voltage at a rated current.

For instance, Fig. 30-19a shows an F-25X, an industrial transformer whose data sheet gives only the following specifications: For a primary voltage of 115 Vac, the secondary voltage is 12.6 Vac when the secondary current is 1.5 A. If the secondary current is less than 1.5 A in Fig. 30-19a, the secondary voltage will be more than 12.6 Vac because of lower power losses in the windings and laminated core.

If it is necessary to know the primary current, you can estimate the turns ratio of a real transformer by using this definition:

$$\frac{N_1}{N_2} = \frac{V_1}{V_2} \qquad \text{(30-14)}$$

For instance, the F25X has $V_1 = 115$ V and $V_2 = 12.6$ V. The turns ratio at the rated load current of 1.5 A is

$$\frac{N_1}{N_2} = \frac{115}{12.6} = 9.13$$

This is an approximation because the calculated turns ratio decreases when the load current decreases. Other common transformer secondary voltages with a 115-Vac primary are 6.3, 8, 9, 16, 18, 24, 25.5, 28, and 36 V. Some windings are center-tapped. Transformers are also available in the popular wall adapter format that plugs directly into an ac outlet.

Calculating Fuse Current

When troubleshooting, you may need to calculate the primary current to determine whether a fuse is adequate or not. The easiest way to do this with a real transformer is to assume that the input power equals the output power: $P_{in} = P_{out}$. For instance, Fig. 30-19b shows a fused transformer driving a filtered rectifier. Is the 0.1-A fuse adequate?

Here is how to estimate the primary current when troubleshooting. The output power equals the dc load power,

$$P_{out} = VI = (15 \text{ V})(1.2 \text{ A}) = 18 \text{ W}$$

Ignore the power losses in the rectifier and the transformer. Since the input power must equal the output power,

$$P_{in} = 18 \text{ W}$$

Since $P_{in} = V_1 I_1$, we can solve for the primary current:

$$I_1 = \frac{18 \text{ W}}{115 \text{ V}} = 0.156 \text{ A}$$

This is only an estimate because we ignored the power losses in the transformer and rectifier. The actual primary current will be higher by about 5 to 20% because of these additional losses. In any case, the fuse is inadequate. It should be at least 0.25 A.

Slow-Blow Fuses

Assume that a capacitor-input filter is used in Fig. 30-19b. If an ordinary 0.25-A fuse is used in Fig. 30-19b, it will blow out when you turn the power on. The reason is the surge current, described earlier. Most power supplies use a slow-blow

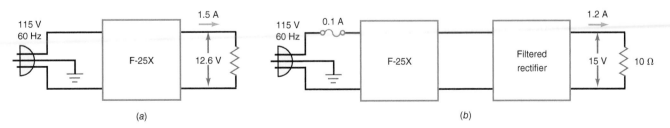

(a)   *(b)*

Figure 30-19   (a) Rating on real transformer. (b) Calculating fuse current.

fuse, one that can temporarily withstand overloads in current. For instance, a 0.25-A slow-blow fuse can withstand

2 A for 0.1 s

1.5 A for 1 s

1 A for 2 s

and so on. With a slow-blow fuse, the circuit has time to charge the capacitor. Then the primary current drops down to its normal level with the fuse still intact.

Regulators

As indicated earlier in the chapter, most power supplies today use a regulator after the filter. The regulator keeps the output voltage constant at a desired level and provides the added bonus of reducing the ripple.

The simplest form of regulator is the zener diode discussed in Chap. 29. The regulator output voltage will be equal to one of the standard zener diode voltage values. Since the zener has a low internal resistance, it forms a voltage divider with the series input resistor. For that reason, it reduces the ripple by the amount of the voltage divider ratio.

In most applications, the regulator is an integrated circuit (IC) that automatically adjusts the output voltage if input ac line voltage or load changes occur. Any variation is corrected, keeping the output voltage constant and usually within a fraction of a percent of the design value. Since ripple is a variation, the regulator greatly diminishes the ripple at the output as well.

You will learn more about regulators in a later chapter.

30.9 Troubleshooting

Almost every piece of electronics equipment has a power supply, typically a rectifier driving a capacitor-input filter followed by a voltage regulator (discussed later). This power supply produces the dc voltages needed by transistors and other devices. If a piece of electronics equipment is not working properly, start your troubleshooting with the power supply. More often than not, *equipment failure is caused by troubles in the power supply.*

Procedure

Assume that you are troubleshooting the circuit of Fig. 30-20. You can start by measuring the dc load voltage.

It should be approximately the same as the peak secondary voltage. If not, there are two possible courses of action.

First, if there is no load voltage, you can use a floating VOM or DMM to measure the secondary voltage (ac range). The reading is the rms voltage across the secondary winding. Convert this to peak value. You can estimate the peak value by adding 40% to the rms value. If this is normal, the diodes may be defective. If there is no secondary voltage, either the fuse is blown or the transformer is defective.

Second, if there is dc load voltage, but it is lower than it should be, look at the dc load voltage with an oscilloscope and measure the ripple. A peak-to-peak ripple around 10% of the ideal load voltage is reasonable. The ripple may be somewhat more or less than this, depending on the design. Furthermore, the ripple frequency should be 120 Hz for a full-wave or bridge rectifier. If the ripple is 60 Hz, one of the diodes may be open.

Common Troubles

Here are the most common troubles that arise in bridge rectifiers with capacitor-input filters:

1. If the fuse is open, there will be no voltages anywhere in the circuit.
2. If the filter capacitor is open, the dc load voltage will be low because the output will be an unfiltered full-wave signal.
3. If one of the diodes is open, the dc load voltage will be low because there will be only half-wave rectification. Also, the ripple frequency will be 60 Hz instead of 120 Hz. If all diodes are open, there will be no output.
4. If the load is shorted, the fuse will be blown. Possibly, one or more diodes may be ruined or the transformer may be damaged.
5. Sometimes the filter capacitor becomes leaky with age, and this reduces the dc load voltage.
6. Occasionally, shorted windings in the transformer reduce the dc output voltage. In this case, the transformer often feels very warm to the touch.
7. Besides these troubles, you can have solder bridges, cold-solder joints, bad connections, and so on.

Table 30-2 lists these troubles and their symptoms.

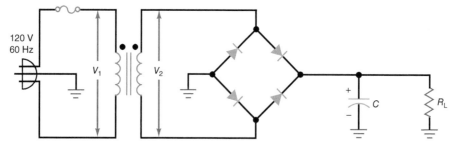

Figure 30-20 Troubleshooting.

Table 30-2 Typical Troubles for Capacitor-Input Filtered Bridge Rectifier

	V_1	V_2	$V_{L(dc)}$	V_R	f_{ripple}	Scope on Output
Fuse blown	Zero	Zero	Zero	Zero	Zero	No output
Capacitor open	OK	OK	Low	High	120 Hz	Full-wave signal
One diode open	OK	OK	Low	High	60 Hz	Half-wave ripple
All diodes open	OK	OK	Zero	Zero	Zero	No output
Load shorted	Zero	Zero	Zero	Zero	Zero	No output
Leaky capacitor	OK	OK	Low	High	120 Hz	Low output
Shorted windings	OK	Low	Low	OK	120 Hz	Low output

EXAMPLE 30-8

When the circuit of Fig. 30-21 is working normally, it has an rms secondary voltage of 12.7 V, a load voltage of 18 V, and a peak-to-peak ripple of 318 mV. If the filter capacitor is open, what happens to the dc load voltage?

Answer:

With an open filter capacitor, the circuit reverts to a bridge rectifier with no filter capacitor. Because there is no filtering, an oscilloscope across the load will display a full-wave signal with a peak value of 18 V. The average value is 63.6% of 18 V, which is 11.4 V.

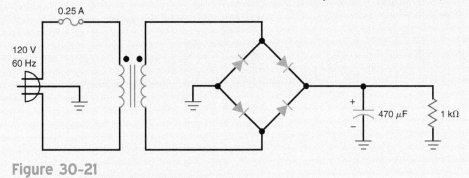

Figure 30-21

EXAMPLE 30-9

Suppose the load resistor of Fig. 30-21 is shorted. Describe the symptoms.

Answer:
A short across the load resistor will increase the current to a very high value. This will blow out the fuse. Furthermore, it is possible that one or more diodes will be destroyed before the fuse blows. Often, when one diode shorts, it will cause the other rectifier diodes to also short. Because of the blown fuse, all voltages will measure zero. When you check the fuse visually or with an ohmmeter, you will see that it is open.

With the power off, you should check the diodes with an ohmmeter to see whether any of them have been destroyed. You should also measure the load resistance with an ohmmeter. If it measures zero or very low, you have more troubles to locate.

The trouble could be a solder bridge across the load resistor, incorrect wiring, or any number of possibilities. Fuses do occasionally blow out without a permanent short across the load. But the point is this: *When you get a blown fuse, check the diodes for possible damage and the load resistance for a possible short.*

A troubleshooting exercise at the end of the chapter has eight different troubles, including open diodes, filter capacitors, shorted loads, blown fuses, and open grounds.

30.10 Voltage Multipliers

A *voltage multiplier* is a type of rectifier or filter circuit that is used to generate dc voltages greater than the peak ac input.

Voltage Doubler

Figure 30-22a is a *voltage doubler*. On the negative half-cycle of the secondary voltage, D_1 conducts and charges C_1 to the peak value with the polarity shown. On the positive half-cycle of the secondary voltage, the peak secondary voltage adds in series with the C_1 capacitor voltage to turn on D_2 and charge C_2 to a value almost double the peak sine voltage less the diode drops. The capacitors are usually large electrolytic types, and the ripple is relatively high.

Why bother using a voltage doubler when you can change the turns ratio to get more output voltage? The answer is that you don't need to use a voltage doubler at lower voltages. The only time you run into a problem is when you are trying to produce very high dc output voltages.

For instance, line voltage is 120 V rms, or 170 V peak. If you are trying to produce 3400 Vdc, you will need to use a 1:20 step-up transformer. Here is where the problem

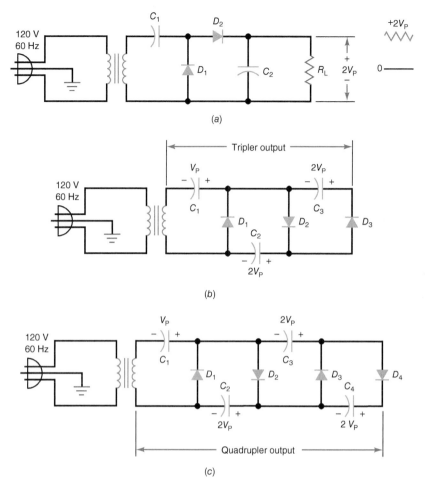

Figure 30-22 Voltage multipliers with floating loads. (*a*) Doubler. (*b*) Tripler. (*c*) Quadrupler.

comes in. Very high secondary voltages can be obtained only with bulky and expensive transformers. It is simpler to use a voltage doubler and a smaller transformer.

Voltage Tripler

By connecting another section, we get the *voltage tripler* of Fig. 30-22*b*. The first two sections act like a doubler. At the peak of the negative half-cycle, D_3 is forward-biased. This charges C_3 to $2V_p$ with the polarity shown in Fig. 30-22*b*. The tripler output appears across C_1 and C_3. The load resistance can be connected across the tripler output. As long as the time constant is long, the output equals approximately $3V_p$.

Voltage Quadrupler

Figure 30-22*c* is a *voltage quadrupler* with four sections in *cascade* (one after another). The first three sections are a tripler, and the fourth makes the overall circuit a quadrupler. The first capacitor charges to V_p. All others charge to $2V_p$. The quadrupler output is across the series connection of C_2 and C_4. We can connect a load resistance across the quadrupler output to get an output of $4V_p$.

Theoretically, we can add sections indefinitely, but the ripple gets much worse with each new section. Increased ripple is another reason why **voltage multipliers** (doublers, triplers, and quadruplers) are not used in low-voltage power supplies. As stated earlier, voltage multipliers are almost always used to produce high voltages, well into the hundreds or thousands of volts. Voltage multipliers are the natural choice for high-voltage and low-current devices.

Variations

All the voltage multipliers shown in Fig. 30-22 use load resistances that are *floating*. This means that neither end of the load is grounded. Figure 30-23*a*, *b*, and *c* shows variations of the voltage multipliers. Figure 30-23*a* merely adds grounds to Fig. 30-22*a*. On the other hand, Fig. 30-23*b* and *c* are redesigns of the tripler (Fig. 30-22*b*) and quadrupler (Fig. 30-22*c*).

Full-Wave Voltage Doubler

Figure 30-23*d* shows a full-wave voltage doubler. On the positive half-cycle of the source, the upper capacitor charges

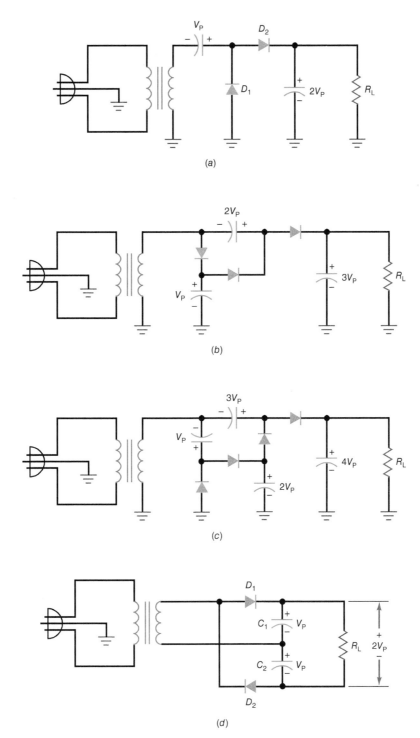

Figure 30-23 Voltage multipliers with grounded loads, except full-wave doubler. (a) Doubler. (b) Tripler. (c) Quadrupler. (d) Full-wave doubler.

to the peak voltage with the polarity shown. On the next half-cycle, the lower capacitor charges to the peak voltage with the indicated polarity. For a light load, the final output voltage is approximately $2V_p$.

The voltage multipliers discussed earlier are half-wave designs; that is, the output ripple frequency is 60 Hz. On the other hand, the circuit of Fig. 30-23d is called a *full-wave voltage doubler* because one of the output capacitors is being charged during each half-cycle. Because of this, the output ripple is 120 Hz. This ripple frequency is an advantage because it is easier to filter. Another advantage of the full-wave doubler is that the PIV rating of the diodes need only be greater than V_p.

1. Which of the following is not a major circuit in an ac power supply?
   a. Rectifier.
   b. Filter.
   c. Battery charger.
   d. Regulator.

2. The output of most power supplies is
   a. ac.
   b. dc.

3. The outputs of most power supplies are regulated.
   a. True.
   b. False.

4. In a battery-operated power supply, how are different dc voltages obtained?
   a. DC-DC converter.
   b. Regulators.
   c. Zener diodes.
   d. All of the above.

5. If $N_1/N_2 = 4$, and the primary voltage is 120 V, what is the secondary voltage?
   a. 0 V.
   b. 30 V.
   c. 60 V.
   d. 480 V.

6. In a step-down transformer, which is larger?
   a. Primary voltage.
   b. Secondary voltage.
   c. Neither.
   d. No answer possible.

7. A transformer has a turns ratio of 2:1. What is the peak secondary voltage if 115 V rms is applied to the primary winding?
   a. 57.5 V.
   b. 81.3 V.
   c. 230 V.
   d. 325 V.

8. With a half-wave rectified voltage across the load resistor, load current flows for what part of a cycle?
   a. 0°.
   b. 90°.
   c. 180°.
   d. 360°.

9. Suppose line voltage may be as low as 105 V rms or as high as 125 rms in a half-wave rectifier. With a 5:1

step-down transformer, the minimum peak load voltage is closest to
   a. 21 V.
   b. 25 V.
   c. 29.7 V.
   d. 35.4 V.

10. The voltage out of a bridge rectifier is a
    a. half-wave signal.
    b. full-wave signal.
    c. bridge-rectified signal.
    d. sine wave.

11. If the line voltage is 115 V rms, a turns ratio of 5:1 means the rms secondary voltage is closest to
    a. 15 V.
    b. 23 V.
    c. 30 V.
    d. 35 V.

12. What is the peak load voltage in a full-wave rectifier if the secondary voltage is 20 V rms?
    a. 0 V.
    b. 0.7 V.
    c. 14.1 V.
    d. 28.3 V.

13. We want a peak load voltage of 40 V out of a bridge rectifier. What is the approximate rms value of secondary voltage?
    a. 0 V.
    b. 14.4 V.
    c. 28.3 V.
    d. 56.6 V.

14. With a full-wave rectified voltage across the load resistor, load current flows for what part of a cycle?
    a. 0°.
    b. 90°.
    c. 180°.
    d. 360°.

15. What is the peak load voltage out of a bridge rectifier for a secondary voltage of 12.6 V rms? (Use second approximation.)
    a. 7.5 V.
    b. 16.4 V.
    c. 17.8 V.
    d. 19.2 V.

16. If line frequency is 60 Hz, the output frequency of a half-wave rectifier is
    a. 30 Hz.
    b. 60 Hz.
    c. 120 Hz.
    d. 240 Hz.

17. If line frequency is 60 Hz, the output frequency of a bridge rectifier is
    a. 30 Hz.
    b. 60 Hz.
    c. 120 Hz.
    d. 240 Hz.

18. With the same secondary voltage and filter, which has the most ripple?
    a. Half-wave rectifier.
    b. Full-wave rectifier.
    c. Bridge rectifier.
    d. Impossible to say.

19. With the same secondary voltage and filter, which produces the least load voltage?
    a. Half-wave rectifier.
    b. Full-wave rectifier.
    c. Bridge rectifier.
    d. Impossible to say.

20. The purpose of a filter on the 60-Hz input to the power supply is to
    a. filter the rectifier output.
    b. keep noise and spikes from getting into the power supply.
    c. keep harmonics generated in the power supply from getting on the ac line.
    d. both b and c.

21. The kind of filter on the rectifier output is
    a. low pass.
    b. high pass.
    c. bandpass.
    d. Bandstop.

22. The output of a bridge rectifier connected to a 600-kHz ac source will have frequency of
    a. 120 Hz.
    b. 600 kHz.
    c. 1.2 MHz.
    d. 2.4 MHz.

23. Most rectifiers produce which kind of harmonics?
    a. None.
    b. Even.
    c. Odd.
    d. Both odd and even.

24. The most common power supply filter is a(n)
    a. Capacitor.
    b. Inductor.

25. If the filtered load current is 10 mA, which of the following has a diode current of 10 mA?
    a. Half-wave rectifier.
    b. Full-wave rectifier.
    c. Bridge rectifier.
    d. Impossible to say.

26. If the load current is 5 mA and the filter capacitance is 1000 μF, what is the peak-to-peak ripple out of a bridge rectifier?
    a. 21.3 pV.
    b. 56.3 nV.
    c. 21.3 mV.
    d. 41.7 mV.

27. The diodes in a bridge rectifier each have a maximum dc current rating of 2 A. This means the dc load current can have a maximum value of
    a. 1 A.
    b. 2 A.
    c. 4 A.
    d. 8 A.

28. What is the PIV across each diode of a bridge rectifier with a secondary voltage of 20 V rms?
    a. 14.1 V.
    b. 20 V.
    c. 28.3 V.
    d. 34 V.

29. If the secondary voltage increases in a bridge rectifier with a capacitor-input filter, the load voltage will
    a. decrease.
    b. stay the same.
    c. increase.
    d. none of these.

30. If the filter capacitance is increased, the ripple will
    a. decrease.
    b. stay the same.
    c. increase.
    d. none of these.

31. Voltage multipliers are circuits best used to produce
    a. low voltage and low current.
    b. low voltage and high current.
    c. high voltage and low current.
    d. high voltage and high current.

SECTION 30.2 The Half-Wave Rectifier

30.1 What is the peak output voltage in Fig. 30-24a using the second approximation of a diode? The average value? The dc value? Sketch the output waveform.

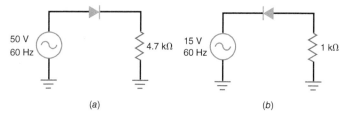

Figure 30-24

30.2 Repeat the preceding problem for Fig. 30-24b.

SECTION 30.3 The Transformer

30.3 If a transformer has a turns ratio of 1:12, what is the rms secondary voltage? The peak secondary voltage? Assume a primary voltage 120 V rms.

30.4 Calculate the peak output voltage and the dc output voltage in Fig. 30-25 using the second approximation.

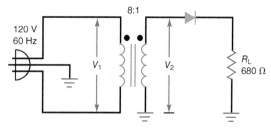

Figure 30-25

SECTION 30.4 The Full-Wave Rectifier

30.5 A center-tapped transformer with 120-V input has a turns ratio of 4:1. What is the rms voltage across the upper half of the secondary winding? The peak voltage? What is the rms voltage across the lower half of the secondary winding?

30.6 What is the peak output voltage in Fig. 30-26 if the diodes are the second approximation? The average value? The dc value? Sketch the output waveform.

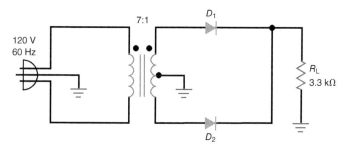

Figure 30-26

SECTION 30.5 The Bridge Rectifier

30.7 In Fig. 30-27, what is the peak output voltage if the diodes are ideal? The average value? The dc value? Sketch the output waveform.

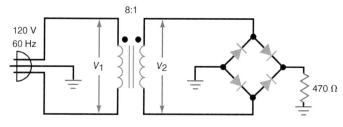

Figure 30-27

30.8 If the line voltage in Fig. 30-27 varies from 105 to 125 V rms, what is the minimum dc output voltage? The maximum?

SECTION 30.6 Power Supply Filtering

30.9 In Fig. 30-28, calculate the dc output voltage and ripple.

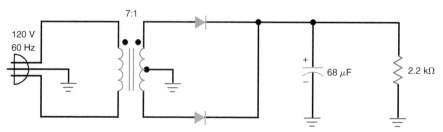

Figure 30-28

30.10 What happens to the ripple in Fig. 30-28 if the capacitance value is reduced to half?

30.11 In Fig. 30-28, what happens to the ripple if the resistance is reduced to 500 Ω?

30.12 What is the dc output voltage in Fig. 30-29? The ripple? Sketch the output waveform.

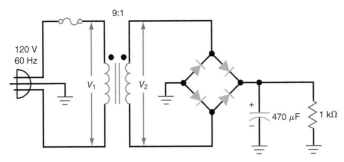

Figure 30-29

30.13 If the line voltage decreases to 105 V in Fig. 30-29, what is the dc output voltage?

30.14 A full-wave signal with a peak of 14 V is the input to a choke-input filter. If $X_L = 2$ kΩ and $X_C = 50$ Ω, what is the approximate peak-to-peak ripple across the capacitor?

SECTION 30.7 Peak Inverse Voltage and Surge Current

30.15 What is the peak inverse voltage in Fig. 30-29?

SECTION 30.8 Other Power Supply Topics

30.16 An F-25X replaces the transformer of Fig. 30-29. What is the approximate peak voltage across the secondary winding? The approximate dc output voltage? Is the transformer being operated at its rated output current? Will the dc output voltage be higher or lower than normal?

30.17 What is the primary current in Fig. 30-29?

SECTION 30.9 Troubleshooting

30.18 If the filter capacitor in Fig. 30-29 is open, what is the dc output voltage?

30.19 If only one diode in Fig. 30-29 is open, what is the dc output voltage?

30.20 If somebody builds the circuit of Fig. 30-29 with the electrolytic capacitor reversed, what kind of trouble is likely to happen?

30.21 If the load resistance of Fig. 30-29 opens, what changes will occur in the output voltage?

SECTION 30.10 Voltage Multipliers

30.22 Calculate the dc output voltage in Fig. 30-30.

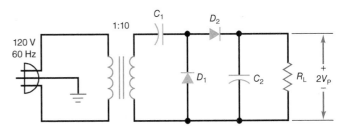

Figure 30-30

Chapter 31

Introduction to Transistors

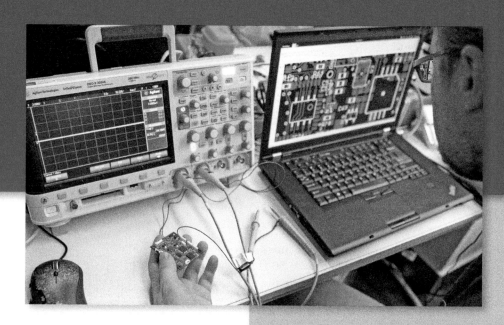

Just as the vacuum tube brought forth the field of electronics and started a major technical revolution, so did the transistor usher in a major revolutionary cycle in electronics. This small solid-state device was orders of magnitude smaller than a vacuum tube and consumed orders of magnitude less power. Transistors changed electronic equipment from large, heavy, and hot to small, light, and cooler. And it made portable battery-operated equipment practical. Today the transistor is at the heart of all integrated circuits, those complete circuits and systems made on a single chip of silicon. This makes the transistor a key part of electronics worth knowing about.

For decades the transistor was a very visible component in electronic equipment. But today, transistors have mostly disappeared into the integrated circuits where we

Learning Outcomes

After studying this chapter, you should be able to:

> *State* the basic structure of a generic transistor.

> *Explain* the concept of a transistor.

> *Explain* the operation of a transistor as a switch.

> *Explain* the operation of a transistor as an amplifier.

> *Name* the two basic types of transistors in common usage and *identify* their elements and schematic symbols.

cannot see them, access them, or otherwise manipulate them. While individual discrete transistors are still used in many designs, for the most part better than 99% of all transistors are inside ICs. Even though these devices cannot be accessed, it is important to know what they are and how they work. Detailed discussions of how to design and analyze transistors circuits are no longer essential, but the concepts of the transistor and its applications form the core of most electronics curricula.

31.1 The Concept of a Transistor

A transistor is a three-terminal semiconductor device in which the current flowing between two of the terminals is controlled by an input signal on a third terminal. With this basic device, the functions of switching and amplification can be performed. All electronic circuits and applications rely on one or both of these basic operations.

Figure 31-1*a* shows a generic diagram of a transistor in a basic electric circuit. The device is designed so that current flows freely through the device from terminal 1 to terminal 3. The amount of current flowing is determined by a voltage or current input on terminal 2. An external dc voltage source (power supply) +V and a current limiting resistor, R_x, set the initial amount of current that can flow.

In most schematic diagrams, the power supply voltage is not shown as a battery as it is here. Instead, it is removed, and we just assume that the power supply voltage is applied between +V and ground, as Fig. 31-1*b* shows. Note that the input to the transistor is between terminal 2 and ground, while the output is between terminal 3 and ground.

Now, varying the voltage or current on terminal 2, causes the current in the device to increase or decrease. With this arrangement, the transistor can be used to turn current off

This chapter introduces you to the concept of a transistor and shows how a transistor works to implement the two basic electronic functions, switching and amplification, that form the basis of all other electronic circuits. The most common transistors like the metal-oxide semiconductor field-effect transistor (MOSFET) and bipolar junction transistor (BJT) are introduced. Each is covered in more detail in later chapters.

or on, as with a switch, or to vary the current in such a way that a small input variation on terminal 2 produces a larger variation in current between terminals 1 and 3.

The ability of a transistor to perform these functions lies with its inherent characteristic called gain. *Gain* refers to the ability of a device to have a larger output variation than the corresponding input variation. A small variation in input produces a much larger change in the output. It is this gain characteristic that allows the transistor to perform its basic operations as an amplifier or a switch.

31.2 The Transistor as a Switch

Of the two basic functions performed by a transistor, amplification and switching, switching is by far the common application. Transistor switches are at the heart of all digital circuits, microcomputers, memory devices, and other pulse-type electronics.

Transistor Switch Operation

Figure 31-2 shows the transistor set up as a switch. Assume that the transistor is a device in which the voltage on terminal 2 causes the current between terminals 1 and 3 to vary. With no input voltage or zero voltage applied to terminal 2, no current will flow in the transistor (Fig. 31-2*a*). The output then is just +3 V, as seen through the resistor. It is as if the transistor was just an open or OFF switch (Fig. 31-2*b*). Another way to look at this is as if the resistance between terminals 1 and 3 was very high, many megohms to infinity. If a load is connected to the output, it will form a voltage divider with the resistor. Assume no load at this time.

Now, assume that we apply a voltage of 1 V to terminal 2. This causes the transistor to conduct heavily (Fig. 31-2*c*). In this conductive state, the resistance between terminals 1 and 3 is very low. It could be a few ohms down to a fraction of an ohm approaching zero ohms, or what we could call a dead short. The transistor is acting as a closed or on switch. The output voltage is near zero volts (Fig. 31-2*d*).

Actually, the transistor resistance between terminals 1 and 3 will never be zero. Instead, it will be some small value that

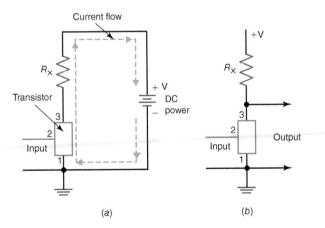

(a)　　　　　　　　　(b)

Figure 31-1　Basic circuit for most transistor applications.

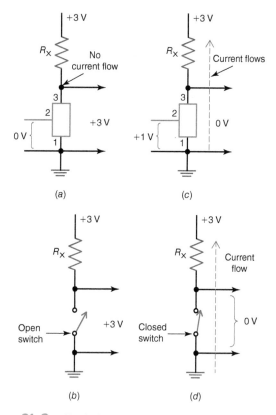

Figure 31-2 Basic transistor switch operation.
(a) Nonconducting transistor. (b) Equivalent to an open switch.
(c) Conducting transistor. (d) Equivalent to a closed switch.

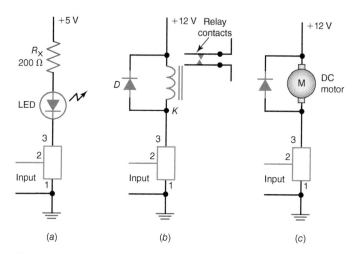

Figure 31-3 Transistor switch applications. (a) LED driver.
(b) Relay driver. (c) DC motor driver.

forms a voltage divider with the resistor. Since the resistor value is usually much higher than the transistor on resistance, the output will be just a fraction of the 3-V supply voltage. A few millivolts may be typical but the actual value depends on the transistor characteristics. If we assume R_x is 1000 Ω and the transistor on resistance is 1 Ω, the output voltage will be

$$V_o = 3 \times \frac{1}{1000 + 1} = 0.002997 \text{ V} \quad \text{or} \quad \text{about 3 mV}$$

Transistor Switch Applications

The transistor switch makes a good driver switch for devices such as lightbulbs, LEDs, relays, and motors. Figure 31-3 shows some of these applications. In Fig. 31-3a, a small control voltage on the transistor terminal 2 can turn the LED off and on. With zero volts applied to terminal 2, the transistor does not conduct, so no current can flow through the LED and it is off. However, applying 1 V to terminal 2 causes the transistor to turn on and become a low resistance, or closed switch. Current then flows in the LED, turning it on. A lightbulb could be used in place of the LED. The resistor in series with the LED sets the current level and the brightness of the LED. If the transistor has zero ON resistance, the current in the LED with the values shown will be

$$I = 5/200 = 0.025 \text{ A} \quad \text{or} \quad 25 \text{ mA}$$

This same principle applies to the relay driver circuit in Fig. 31-3b. Recall that a relay is a magnetic switch. Applying a voltage to the relay coil produces a magnetic field that closes a set of switch contacts. Removing the voltage causes the magnetic field to disappear, so the contacts spring open.

In most cases, you can apply the input voltage directly to the relay coil to produce the desired switching. But more often the case is that only a very small voltage is available. What if only 1 V is available to operate a 12-V relay coil? In this case, you can use the relay driver circuit in Fig. 31-3b. With no or zero voltage on terminal 2, the transistor is off, so the relay is off. Applying 1 V to terminal 2 turns the transistor on, making its resistance very low. Now the full 12 V is applied to the relay coil, turning it on.

One important detail to note in Fig. 31-3b: The diode, *D*, is needed across the relay coil to prevent the transistor from being destroyed when the relay is turned off. Remember that when current is flowing through a coil, a large magnetic field appears around the coil. The field is steady. But if you suddenly turn off the current by switching off the transistor, the magnetic field collapses, and as it does, it induces a huge voltage, hundreds or even thousands of volts, across the coil. This short duration induced pulse of voltage can destroy almost any transistor. By putting the diode across the coil, the diode will conduct when the high voltage is induced and current will flow through it momentarily until the magnetic field disappears. This clamps the voltage on terminal 3 of the transistor to 12 V plus about 0.7 V if a silicon diode is used. The transistor is protected.

The operation of the circuit in Fig. 31-3c is similar. Here the device in series with the transistor is a dc motor. Turning the transistor on with a small input voltage causes the motor to turn on. With zero input to the transistor, the transistor is off, so no current flows in the motor. Notice that a diode is used across the motor as well since the motor winding is an inductor.

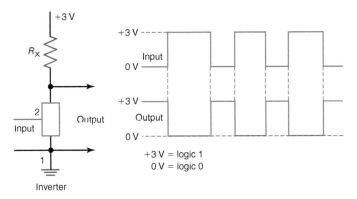

Figure 31-4   A transistor invertor.

Digital Logic

Another major use of the transistor is to perform basic digital logic operations. You will learn more about these in later courses but for now just know that there are three basic logic operations used in all digital equipment and computers. These are inversion, AND, and OR.

The transistor switch makes a great inverter, as Fig. 31-4 shows. Here a digital signal that switches from 0 to +3 V and back periodically is applied to the input at terminal 2. When the input is zero, the transistor is off, so the output as seen through the resistor is +3 V. When the input switches to +3 V, the transistor turns on and becomes a very low resistance. Therefore the output is very low, near zero volts. As you can see from the input and output waveforms, when the input is zero, the output is +3 V. When the input is +3 V, the output is zero. Therefore the transistor "inverts" the input or produces the opposite voltage level. This circuit is therefore referred to as an inverter.

Digital circuits have two voltage states and are called binary. We usually call these states either a logic 0 or logic 1. It is common for a zero-volt level to represent a logic 0, while a + voltage level represents a logic 1. We say that the inverter changes a logic 0 to 1 and logic 1 to 0. The two states are said to be complementary. Logic 0 is the complement to logic 1 and vice versa.

Another logic operation is called the AND function. It is implemented with something called an AND gate. It has two or more inputs and one output. The output is a logic 1 only if both inputs are logic 1. Otherwise, the output is logic 0 for all other conditions. This logic operation can be implemented by connecting two transistors in series as shown in Fig. 31-5a. If one or both transistors is off, the series current path is broken, so no current can flow. With no current across the resistor, the output is 0. Now if both inputs have +3 V applied to them, both transistors turn on and the path is completed between the +3 volt input and the output. If the transistor on resistances are low, +3 V appears across the resistor. Therefore, the logic AND condition is fulfilled. The truth table in Fig. 31-5b shows all possible combinations of inputs and output.

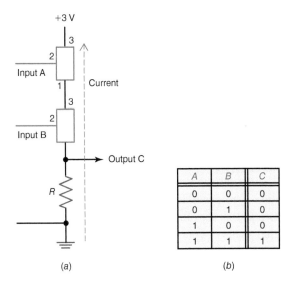

Figure 31-5   An AND gate. (a) Circuit. (b) Truth table.

The logical OR function performed by an OR gate is that the output will be logic 1 of one or both inputs are logic 1. The circuit in Fig. 31-6a implements this operation. With both inputs zero, both transistors are off, so there is no current in the resistor and the output is zero. Now, if a logic 1 or +3 V is applied to either transistor's terminal 2, it will conduct, connecting the +3-V supply to the output resistor, R, and producing a +3-V output. The same is true if either or both transistors are turned on. The truth table (Fig. 31-6b) sums up the complete operation of the circuit.

With inverters, AND gates, and OR gates, you can make virtually any other digital logic circuit. These occur most commonly in integrated circuit form.

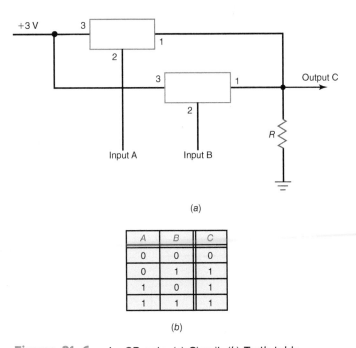

Figure 31-6   An OR gate. (a) Circuit. (b) Truth table.

31.3 The Transistor as an Amplifier

Amplification is inherent to most electronic operations. Small signals must be made larger to process them or to allow them to generate the proper level of output signal. For example, amplification is necessary to operate a speaker, an electromagnetic device that converts mechanical vibrations of a speaker cone into sound waves. High power is needed. In a radio, the wireless signal is very small at the antenna, and it must usually be amplified so that it can be processed to reproduce the transmitted information. The key to amplification is to accurately reproduce the input signal to be amplified so that the amplification does not distort or change the signal shape or content.

Transistor Amplifier Operation

The same basic circuit used for switching can be used for amplification, as shown in Fig. 31-7. Here instead of just turning the transistor off or on, it will be made to conduct continuously, and then the input signal to be amplified will be applied to terminal 2 to vary its current. The inherent gain of the transistor will cause a larger variation in output current and voltage that is applied to the input. If the output response is a linear or straight line or proportional to the input variation, no distortion will occur. We call this the linear operation of a transistor. A linear response implies amplification.

A key part of making the transistor amplify is biasing it properly. *Bias* refers to an external voltage usually applied to terminal 2 of the transistor along with the signal to be amplified. Bias can be a separate dc voltage source, but usually it is derived from the main transistor dc supply voltage. This is the arrangement in Fig. 31-7, for which a voltage divider made of R_1 and R_2 supplies a bias voltage to terminal 2. What this does is to set the current flowing between terminals 1 and 3. The current flows continuously as long as the bias is applied.

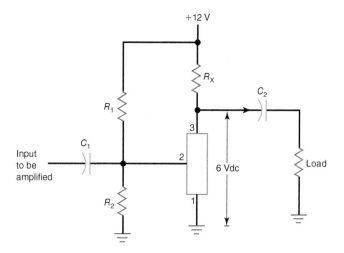

Figure 31-7   A basic transistor amplifier.

The bias sets the current in the transistor so that the output voltage is centered in the middle of the transistor's linear operating range. This is the point at which the dc output voltage is approximately half the dc supply voltage. In this example, with a 12-V supply, the output voltage with bias is 6 V.

Now, the signal to be amplified is applied to terminal 2 as well. In this case, the input is applied through a capacitor, C_1, so that the dc bias voltage does not affect or damage the source of the input signal. The input signal here is a sine wave. When the sine wave goes positive, it causes the current in the transistor to increase. This, in turn, causes a larger current variation between terminals 1 and 3. Look closely at the circuit, and you will see that the transistor and resistor R_X form a voltage divider. If the current goes up, the voltage across the resistor goes up. This will force the output voltage across the transistor to go down. Why? Because the circuit obeys Kirchhoff's voltage law, which, as you recall, says that the voltage across the components in a series circuit must always add up to the supply voltage value. So if the current goes up, the voltage across the resistor must increase. That means that the voltage between terminals 1 and 3 must go down. It is equivalent to saying that the resistance between terminals 1 and 3 goes down as the current goes up.

Now, if the input voltage goes down, the transistor current will decrease. The voltage across the resistor R_x goes down, so the output voltage across the transistor goes up. That is equivalent to the transistor resistance going up.

One way for you to visualize what is going on in the transistor is to assume its resistance between terminals 1 and 3 is being varied by the input signal. While that is not entirely correct for some types of transistors, the analogy is useful for understanding how the circuit amplifies.

In any case, remember that even a very small variation in input voltage will cause a larger current variation and even larger output voltage variation because the transistor has gain. If the characteristics of the transistor are linear, the output will be a larger variation of the input voltage. Figure 31-8 shows the input and output voltages. The input is ac, but the larger output voltage is a dc voltage with the same sine variations. It is as if the sine wave were riding on a 6-Vdc level. We can get rid of that dc by just passing the output to a load resistor through a capacitor, C_2, which blocks the dc and lets the ac pass, as Fig. 31-8 shows. Another thing to note is that the output is 180° out of phase with the input. We call this *phase inversion*.

The main thing to point out here is that the transistor does not really take a small signal and make it larger. Instead, it simply uses a small input signal to control the current flow to produce a larger variation with the same shape in the output. It is the gain of the transistor that makes this

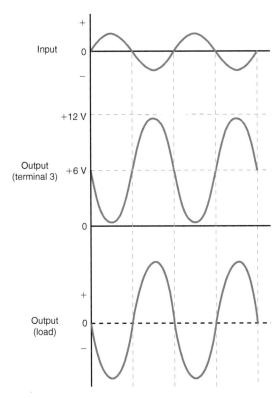

Figure 31-8 The input and output of a transistor amplifier.

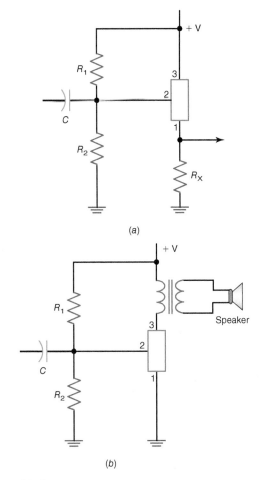

(a)

(b)

Figure 31-9 Other common amplifier circuits. (a) Follower. (b) Power amplifier.

happen. Millivolt- or microvolt-level inputs can produce outputs of many output volts, depending on the transistor characteristics.

Transistor Amplifier Applications

The basic circuit shown in Fig. 31-7 is the most common amplifier configuration. But there are other arrangements. The circuit in Fig. 31-9a is an example. Here the resistor is placed in series with terminal 1 rather than terminal 3. This creates a circuit called a *follower*. The transistor still has gain, but the output voltage is essentially the same as the input voltage for essentially no gain. We call that a gain of 1, or *unity gain*. However, the circuit amplifies power rather than voltage.

Figure 31-9b shows one other variation. Here the circuit is the same as that in Fig. 31-7, but a transformer primary winding replaces the resistor. The secondary of the transformer connects to a speaker. The impedance levels between the transistor circuit and the speaker are so great that a transformer is needed to ensure maximum power transfer. So this too is a power amplifier. The operation is the same as that described earlier but with a transformer stepping the output voltage down and the power up.

31.4 Types of Transistors

The two basic types of transistors are *bipolar junction transistors (BJTs)* and *field-effect transistors (FETs)*. They are made with different segments of both *n*- and *p*-type semiconductor material. And there are two types of FETs: junction FETs and insulated gate FETs called metal-oxide semiconductor (MOS) FETs, or *MOSFETs*. Figure 31-10 illustrates the various types. You will learn each of these in greater detail in the coming chapters. Here is a preview.

Field-Effect Transistors

The earliest transistors developed were of the FET type. The first FET was invented by Julius Lilienfeld, a professor at the University of Leipzig in 1925. Later in 1928, Lilienfeld patented the first MOSFET. In 1934, Oskar Heil, a physicist at the University of Gottingen in Germany, patented another type of FET. Strangely, none of these devices ever got built simply because pure semiconductor material was just not available or practical.

In the late 1940s, a team at the famous research company Bell Labs created a point contact transistor that was never

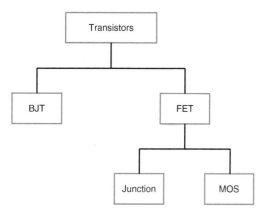

Figure 31-10 Types of transistors.

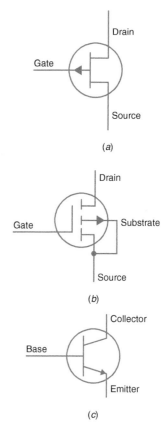

Figure 31-11 Schematic symbols of transistors. (a) JFET. (b) MOSFET. (c) BJT.

practical. But that work spurred on William Shockley to create the first BJT in the early 1950s. The BJT was easier to make and more forgiving of inferior semiconductor materials. For this reason, BJTs became the first commercial transistors of practical value. The *junction FET (JFET)* came about next, followed by the first practical MOSFETs in the early 1960s.

Most early transistor equipment was made with BJTs. Today, while BJTs are still widely used, most transistors making up integrated circuits and even discrete component circuits are MOSFETs. They can be made smaller than BJTs, so you can fit more on a chip of any given size. ICs with over a billion MOSFETs are commonplace today.

The JFET schematic diagram is shown in Fig. 31-11a. Notice that it has three elements, the source, the drain, and the gate. The source and drain are equivalent to terminals 1 and 3 on the generic transistor discussed earlier. Current flows from source to drain through the FET. A voltage applied to the gate, which is equivalent to terminal 2, controls the current from source to drain. The JFET is used mostly as an amplifier and rarely as a switch.

The schematic of a MOSFET is shown in Fig. 31-11b. There are several versions, but this one is the most common. It too has a source, drain, and gate. The fourth connection shown here is the substrate or semiconductor base on which the transistor is made. The voltage on the gate controls the

current from source to drain. The MOSFET is used mostly as a switch but can be biased as a linear amplifier as well. Most ICs (over 90%) are made with MOSFETs.

The schematic symbol for a BJT is shown in Fig. 31.11c. Like the FET it has three terminals. These are the emitter and collector, which correspond to terminals 1 and 3 in the generic transistor discussed earlier. The base is equivalent to terminal 2. Current flows from emitter to collector, while the signal on the base controls the amount of current flowing. In the chapters to come you will learn more about how each transistor works, its characteristics, and the most common switch and amplifier circuit configurations.

❓ CHAPTER 31 REVIEW QUESTIONS

1. A transistor
   a. controls current flow.
   b. creates current flow.
   c. allows current flow in two directions.
   d. measures current flow.

2. In Fig. 31-1, the main path for current flow is between terminals
   a. 1 and 2.
   b. 2 and 3.
   c. 1 and 3.
   d. All of the above.

3. The current in a transistor is larger than the input controlling it.
   a. True.
   b. False.

4. The main use of a transistor is
   a. switching.
   b. amplification.
   c. both *a* and *b*.
   d. none of these.

·5. What characteristic of a transistor makes it useful?
   a. Low voltage operation.
   b. Gain.
   c. Low power consumption.
   d. High-frequency operation.

6. When a transistor is cut off,
   a. full current flows.
   b. no current flows.

7. The approximate resistance of a heavily conducting transistor is
   a. zero.
   b. near zero.
   c. $100\ \Omega$ to $10\ k\Omega$.
   d. infinity.

8. When used as an amplifier, a transistor
   a. is cut off.
   b. fully conducting.
   c. switches off and on.
   d. conducts partially all the time.

9. What is the name of the external voltage applied to make an amplifier conduct?
   a. Supply voltage.
   b. Input voltage.
   c. Bias voltage.
   d. Output voltage.

10. Which of the following is the most true about an amplifier?
    a. The transistor generates the extra voltage so the output is larger than the input.
    b. The small input controls the larger dc supply voltage to produce an equivalent output.

11. If a transistor amplifier has a +15-Vdc supply, what is the output voltage with no input signal applied?
    a. Zero.
    b. 6 V.
    c. 7.5 V.
    d. 15 V.

12. The phase relationship between the inputs and outputs to an amplifier are
    a. in phase.
    b. 90° out of phase.
    c. 180° out of phase.
    d. all of the above.

13. A transistor switching inverter has a +5-Vdc supply. When the input is +5 V and the transistor is fully conducting, the output is
    a. Zero.
    b. 2.5 V.
    c. 5 V.
    d. 7.5 V.

14. A transistor AND gate has a logic 1 output when
    a. both inputs are 0.
    b. both inputs are 1.
    c. when the inputs are 0 and 1.
    d. when the inputs are 1 and 0.

15. The only time the output of an OR gate is zero is when
    a. both inputs are 0.
    b. both inputs are 1.
    c. when the inputs are 0 and 1.
    d. when the inputs are 1 and 0.

16. Which of the following is not a widely used transistor type?
    a. BJT.
    b. JFET.
    c. MOSFET.
    d. IGBT.

17. Which is the most widely used transistor?
    a. BJT.
    b. JFET.
    c. MOSFET.
    d. IGBT.

18. Most transistors are inside ICs.
    a. True.
    b. False.

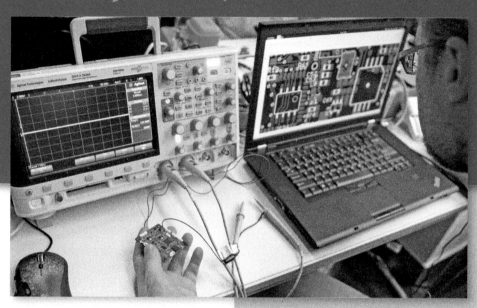

Chapter **32**

Field-Effect Transistors, Amplifiers, and Switches

The very first transistor was a field-effect transistor (FET) that was conceived by Julius Lilienfeld in 1925 and later by Oskar Heil in 1934. No practical devices were ever made to test the concept as suitable pure semiconductor material was not available. However, patents were filed. In 1947, Walter Brattain and John Bardeen at the Bell Labs invented the point contact transistor. Their colleague William Shockley invented the bipolar junction transistor (BJT) in 1951. The BJT became the first practical commercial transistor and went on to revolutionize electronics as it quickly replaced the vacuum tube in virtually every application.

Practical FETs called junction FETs (JFETs) came along in 1952. The MOSFET was invented in 1960, and the first practical devices became available in the early 1960s. Over the years the MOSFET has evolved into

Learning Outcomes

After studying this chapter, you should be able to:

> *Describe* the basic construction of a JFET.
> *Show* common JFET biasing circuits.
> *Calculate* the proportional pinchoff voltage and *determine* which region a JFET in operation in.
> *Determine* transconductance and *use* it to calculate gain in JFET amplifiers.
> *Explain* the characteristics and operation of both depletion-mode and enhancement-mode MOSFETs.
> *Describe* how E-MOSFETs are used as switches.
> *Draw* a diagram of a CMOS switching circuit and explain its operation.
> *Name* and *describe* several power FET applications.
> *Explain* the operation of several types of D-MOSFET and E-MOSFET amplifier circuits.

the most important device in electronics. Because it can be easily manufactured in integrated circuit form, it has become the most widely used type of transistor. It is, in fact, the dominant type of transistor in use today as it makes up over 90% of all electronic circuits.

While instruction in transistors traditionally begins with BJTs, in this book that honor is given to the MOSFET, the transistor most likely to be encountered in everyday use. This chapter covers the MOSFET as well as JFETs. BJTs are covered in a later chapter.

32.1 Introduction to FETs

As Chap. 31 points out, there are two major classes of transistors, field-effect transistors and bipolar junction transistors. In this chapter, we will concentrate on the field-effect transistors. These are the most widely used types of transistor today both in integrated circuits and in discrete form. The chapter will cover how they work and how they are used. First, there are two basic types of field-effect transistors, the junction FET and the metal-oxide semiconductor FET as shown in Fig. 32-1. The MOSFET is by far the most widely used.

Note in Fig. 32-1 that there are two types of MOSFETs. These are the enhancement-mode type, which is the most common, and the depletion-mode type. Note that there are two variations of each type of FET, n-type and p-type. These designations refer to dominant current carriers used, either holes for p-type or electrons for n-type.

All FETs have three basic elements: the source, the gate and the drain. The device is designed to permit current flow from the source through the device to the drain. The amount of current flowing is determined by the control element known as the gate. Like all transistors, the FETs are used in two basic ways, as amplifiers or linear devices or as switches. While the application as a switch predominates, FETs also make good linear circuit devices.

32.2 Junction FET Basics

Figure 32-2a shows a piece of n-type semiconductor. The lower end is called the **source,** and the upper end is called the **drain.** The supply voltage V_{DD} forces free electrons to flow from the source to the drain. To produce a JFET, a manufacturer diffuses two areas of p-type semiconductor into the n-type semiconductor, as shown in Fig. 32-2b. These p regions are connected internally to get a single external **gate** *lead.*

Field Effect

Figure 32-3 shows the normal biasing voltages for a JFET. The drain supply voltage is positive, and the gate supply voltage is negative. The term **field effect** is related to the depletion layers around each p region. These depletion layers exist because free electrons diffuse from the n regions into the p regions. The recombination of free electrons and holes creates the depletion layers shown by the colored areas.

Reverse Bias of Gate

In Fig. 32-3, the p-type gate and the n-type source form the gate-source diode. With a JFET, we always *reverse-bias* the gate-source diode. Because of reverse bias, the gate current I_G is approximately zero, which is equivalent to saying that the JFET has an almost infinite input resistance.

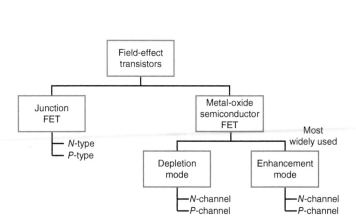

Figure 32-1 Types of field-effect transistors (FETs).

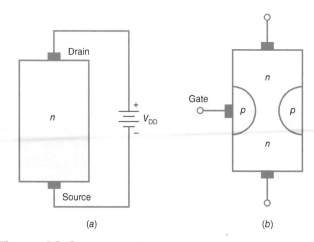

Figure 32-2 (a) Part of JFET. (b) Single-gate JFET.

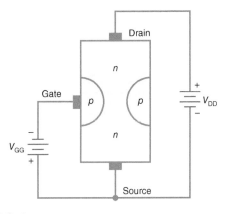

Figure 32-3 Normal biasing of JFET.

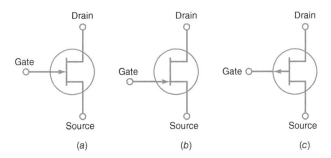

Figure 32-4 (a) Schematic symbol. (b) Offset-gate symbol. (c) P-channel symbol.

Keep in mind that the reverse-biased, gate-source junction also acts as a low-value capacitor. This capacitance has little or no effect at low frequencies but can greatly impact high-frequency signals.

A typical JFET has an input resistance in the hundreds of megohms. This is the big advantage that a JFET has over a bipolar transistor. It is the reason that JFETs excel in applications in which a high input impedance is required. One of the most important applications of the JFET is the *source follower,* a circuit like the emitter follower, except that the input impedance is in the hundreds of megohms for lower frequencies.

Gate Voltage Controls Drain Current

In Fig. 32-3, electrons flowing from the source to the drain must pass through the narrow **channel** between the depletion layers. When the gate voltage becomes more negative, the depletion layers expand and the conducting channel becomes narrower. The more negative the gate voltage, the smaller the current between the source and the drain.

The JFET is a **voltage-controlled device** because an input voltage controls an output current. In a JFET, the gate-to-source voltage V_{GS} determines how much current flows between the source and the drain. When V_{GS} is zero, maximum drain current flows through the JFET. This is why a JFET is referred to as a normally on device. On the other hand, if V_{GS} is negative enough, the depletion layers touch and the drain current is cut off.

Schematic Symbol

The JFET of Fig. 32-3 is an *n-channel JFET* because the channel between the source and the drain is an *n*-type semiconductor. Figure 32-4*a* shows the schematic symbol for an *n*-channel JFET.

Figure 32-4*b* shows an alternative symbol for an *n*-channel JFET.

There is also a *p*-channel JFET. The schematic symbol for a *p*-channel JFET, shown in Fig. 32-4*c*, is similar to that for the *n*-channel JFET, except that the gate arrow points in the opposite direction. The action of a *p*-channel JFET is complementary; that is, all voltages and currents are reversed. To reverse-bias a *p*-channel JFET, the gate is made positive in respect to the source. Therefore, V_{GS} is made positive.

32.3 Drain Curves

Figure 32-5*a* shows a JFET with normal biasing voltages. In this circuit, the gate-source voltage V_{GS} equals the gate supply voltage V_{GG}, and the drain-source voltage V_{DS} equals the drain supply voltage V_{DD}.

Maximum Drain Current

If we short the gate to the source, as shown in Fig. 32-5*b*, we will get maximum drain current because $V_{GS} = 0$. Figure 32-5*c* shows the graph of drain current I_D versus drain-source voltage V_{DS} for this shorted-gate condition. Notice how the drain current increases rapidly and then becomes almost horizontal when V_{DS} is greater than V_P.

Why does the drain current become almost constant? When V_{DS} increases, the depletion layers expand. When $V_{DS} = V_P$, the depletion layers are almost touching. The narrow conducting channel therefore pinches off or prevents a further increase in current. This is why the current has an upper limit of I_{DSS}.

The active region of a JFET is between V_P and $V_{DS(max)}$. The minimum voltage V_P is called the **pinch-off voltage,** and the maximum voltage $V_{DS(max)}$ is the *breakdown voltage.* Between pinch-off and breakdown, the JFET acts like a current source of approximately I_{DSS} when $V_{GS} = 0$.

I_{DSS} stands for the current drain to source with a shorted gate. This is the maximum drain current a JFET can produce. The data sheet of any JFET lists the value of I_{DSS}. This is one of the most important JFET quantities, and you should always look for it first because it is the upper limit on the JFET current.

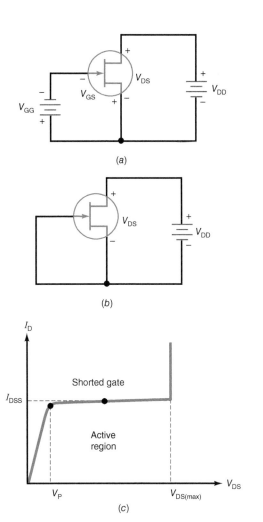

Figure 32-5 (a) Normal bias. (b) Zero gate voltage. (c) Shorted-gate drain current.

The Ohmic Region

In Fig. 32-6, the pinch-off voltage separates two major operating regions of the JFET. The almost-horizontal region is the active region. The almost-vertical part of the drain curve below pinch-off is called the **ohmic region.**

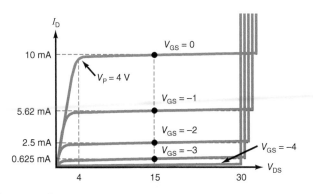

Figure 32-6 Drain curves.

When operated in the ohmic region, a JFET is equivalent to a resistor with a value of approximately

$$R_{DS} = \frac{V_P}{I_{DSS}} \tag{32-1}$$

where R_{DS} is called the *ohmic resistance of the JFET*. In Fig. 32-6, $V_P = 4$ V and $I_{DSS} = 10$ mA. Therefore, the ohmic resistance is

$$R_{DS} = \frac{4 \text{ V}}{10 \text{ mA}} = 400 \text{ }\Omega$$

If the JFET is operating anywhere in the ohmic region, it has an ohmic resistance of 400 Ω.

Gate Cutoff Voltage

Figure 32-6 shows the drain curves for a JFET with an I_{DSS} of 10 mA. The top curve is always for $V_{GS} = 0$, the shorted-gate condition. In this example, the pinch-off voltage is 4 V and the breakdown voltage is 30 V. The next curve down is for $V_{GS} = -1$ V, the next for $V_{GS} = -2$ V, and so on. As you can see, the more negative the gate-source voltage, the smaller the drain current.

The bottom curve is important. Notice that a V_{GS} of -4 V reduces the drain current to almost zero. This voltage is called the **gate-source cutoff voltage** and is symbolized by $V_{GS(off)}$ on data sheets. At this cutoff voltage the depletion layers touch. In effect, the conducting channel disappears. This is why the drain current is approximately zero.

In Fig. 32-6, notice that

$$V_{GS(off)} = -4 \text{ V} \quad \text{and} \quad V_P = 4 \text{ V}$$

This is not a coincidence. The two voltages always have the same magnitude because they are the values where the depletion layers touch or almost touch. Data sheets may list either quantity, and you are expected to know that the other has the same magnitude. As an equation:

$$V_{GS(off)} = -V_P \tag{32-2}$$

32.4 The Transconductance Curve

The **transconductance curve** of a JFET is a graph of I_D versus V_{GS}. By reading the values of I_D and V_{GS} of each drain curve in Fig. 32-6, we can plot the curve of Fig. 32-7a. Notice that the curve is nonlinear because the current increases faster when V_{GS} approaches zero.

Any JFET has a transconductance curve like Fig. 32-7b. The endpoints on the curve are $V_{GS(off)}$ and I_{DSS}. The equation for this graph is

$$I_D = I_{DSS}\left(1 - \frac{V_{GS}}{V_{GS(off)}}\right)^2 \tag{32-3}$$

Because of the squared quantity in this equation, JFETs are often called *square-law devices*. The squaring of the quantity produces the nonlinear curve of Fig. 32-7b.

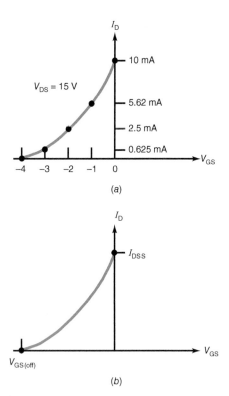

(a)

(b)

Figure 32-7   Transconductance curve.

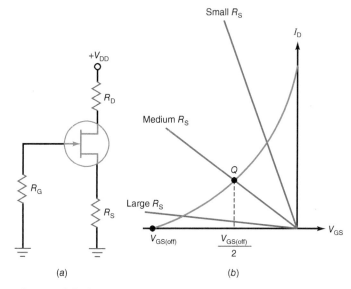

(a)

(b)

Figure 32-8   Self-bias.

32.5 Biasing the JFET

Bias refers to the external dc voltages applied to a transistor to establish an operating point in one of the operating regions of the device. The JFET can be biased in the ohmic or in the active region. When biased in the ohmic region, the JFET is equivalent to a resistance. When biased in the active region, the JFET is equivalent to a current source. The ohmic region is rarely used, so we will focus on biasing for the active region.

JFET amplifiers need to have a fixed operating point called the Q point in the active region. The most common bias method is self-bias.

Figure 32-8a shows **self-bias.** Since drain current flows through the source resistor R_S, a voltage exists between the source and ground, given by

$$V_S = I_D R_S \qquad \text{(32-4)}$$

Since V_G is zero,

$$V_{GS} = -I_D R_S \qquad \text{(32-5)}$$

This says that the gate-source voltage equals the negative of the voltage across the source resistor. Basically, the circuit creates its own bias by using the voltage developed across R_S to reverse-bias the gate.

Figure 32-8b shows the effect of different source resistors. There is a medium value of R_S at which the gate-source

voltage is half the cutoff voltage. An approximation for this medium resistance is

$$R_S \approx R_{DS} \qquad \text{(32-6)}$$

This equation says that the source resistance should equal the ohmic resistance of the JFET. When this condition is satisfied, the V_{GS} is roughly half the cutoff voltage and the drain current is roughly one-quarter of I_{DSS}. This is the desired Q point.

Voltage Divider Bias

Figure 32-9a shows **voltage divider bias.** The voltage divider produces a gate voltage that is a fraction of the supply voltage. By subtracting the gate-source voltage, we get the voltage across the source resistor:

$$V_S = V_G - V_{GS} \qquad \text{(32-7)}$$

Since V_{GS} is a negative, the source voltage will be slightly larger than the gate voltage. When you divide this source voltage by the source resistance, you get the drain current:

$$I_D = \frac{V_G - V_{GS}}{R_S} \approx \frac{V_G}{R_S} \qquad \text{(32-8)}$$

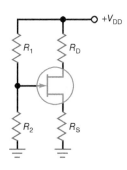

Figure 32-9   Voltage divider bias.

When the gate voltage is large, it can swamp out the variations in V_{GS} from one JFET to the next. Ideally, the drain current equals the gate voltage divided by the source resistance. As a result, the drain current is almost constant for any JFET. Voltage divider bias is no longer widely used in JFET circuits. However, it is used in this chapter to illustrate basic JFET amplifier circuits.

32.6 Transconductance

To analyze JFET amplifiers, we need to discuss **transconductance,** designated g_m and defined as

$$g_m = \frac{i_d}{v_{gs}} \qquad \textbf{(32-9)}$$

This says that transconductance equals the ac drain current divided by the ac gate-source voltage. Transconductance tells us how effective the gate-source voltage is in controlling the drain current. The higher the transconductance, the more control the gate voltage has over the drain current.

For instance, if $i_d = 0.2$ mA p-p when $v_{gs} = 0.1$ V p-p, then

$$g_m = \frac{0.2 \text{ mA}}{0.1 \text{ V}} = 2(10^{-3}) \text{ mho} = 2000 \ \mu\text{mho}$$

On the other hand, if $i_d = 1$ mA p-p when $v_{gs} = 0.1$ V p-p, then

$$g_m = \frac{1 \text{ mA}}{0.1 \text{ V}} = 10,000 \ \mu\text{mho}$$

In the second case, the higher transconductance means that the gate is more effective in controlling the drain current.

Siemen

The unit *mho* is the ratio of current to voltage. An equivalent and modern unit for the mho is the *siemen (S),* so the foregoing answers can be written as 2000 μS and 10,000 μS. On data sheets, either quantity (mho or siemen) may be used. Data sheets may also use the symbol g_{fs} instead of g_m. As an example, the data sheet of a 2N5451 lists a g_{fs} of 2000 μS for a drain current of 1 mA. This is identical to saying that the 2N5451 has a g_m of 2000 μmho for a drain current of 1 mA.

Slope of Transconductance Curve

Figure 32-10a brings out the meaning of g_m in terms of the transconductance curve. Between points A and B, a change in V_{GS} produces a change in I_D. The change in I_D divided by the change in V_{GS} is the value of g_m between A and B. If we select another pair of points farther up the curve at C and D, we get a bigger change in I_D for the same change in V_{GS}. Therefore, g_m has a larger value higher up the curve. Stated another way, g_m is the slope of the transconductance

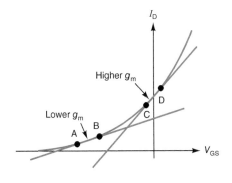

(a)

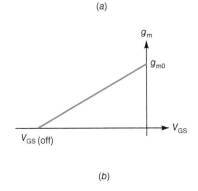

(b)

Figure 32-10 (a) Transconductance. (b) Variation of g_m.

curve. The steeper the curve is at the Q point, the higher the transconductance.

Transconductance and Gate-Source Cutoff Voltage

The quantity $V_{GS(off)}$ is difficult to measure accurately. On the other hand, I_{DSS} and g_{m0} are easy to measure with high accuracy. For this reason, $V_{GS(off)}$ is often calculated with the following equation:

$$V_{GS(off)} = \frac{-2I_{DSS}}{g_{m0}} \qquad \textbf{(32-10)}$$

In this equation, g_{m0} is the value of transconductance when $V_{GS} = 0$. Typically, a manufacturer will use the foregoing equation to calculate the value of $V_{GS(off)}$ for use on data sheets.

The quantity g_{m0} is the maximum value of g_m for a JFET because it occurs when $V_{GS} = 0$. When V_{GS} becomes negative, g_m decreases. Here is the equation for calculating g_m for any value of V_{GS}:

$$g_m = g_{m0}\left(1 - \frac{V_{GS}}{V_{GS(off)}}\right) \qquad \textbf{(32-11)}$$

Notice that g_m decreases linearly when V_{GS} becomes more negative, as shown in Fig. 32-10b. Changing the value of g_m is useful in *automatic gain control,* which is discussed later.

EXAMPLE 32-1

A 2N5457 has I_{DSS} = 5 mA and g_{m0} = 5000 μS. What is the value of $V_{GS(off)}$? What does g_m equal when V_{GS} = −1 V?

Answer:

With Formula (32-10),

$$V_{GS(off)} = \frac{-2(5\ mA)}{5000\ \mu S} = -2\ V$$

Next, use Formula (32-11) to get

$$g_m = (5000\ \mu S)\left(1 - \frac{1\ V}{2\ V}\right) = 2500\ \mu S$$

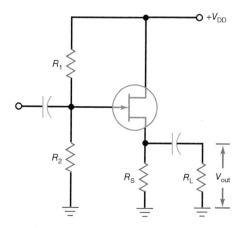

Figure 32-12 Source follower.

32.7 JFET Amplifiers

Figure 32-11 shows a **common-source (CS) amplifier.** The coupling and bypass capacitors are ac shorts. Because of this, the signal is coupled directly into the gate. Since the source is bypassed to ground, all the ac input voltage appears between the gate and the source. This produces an ac drain current. Since the ac drain current flows through the drain resistor, we get an amplified and inverted ac output voltage. This output signal is then coupled to the load resistor.

Voltage Gain of CS Amplifier

The voltage gain is

$$A_v = \frac{v_{out}}{v_{in}} = g_m r_d \qquad \textbf{(32-12)}$$

This says that the voltage gain of a CS amplifier equals the transconductance times the ac drain resistance. The ac drain resistance r_d is defined as

$$r_d = R_D \parallel R_L$$

Source Follower

Figure 32-12 shows a **source follower.** The input signal drives the gate, and the output signal is coupled from the source to the load resistor. The source follower has a voltage gain less than 1. The main advantage of the source follower is its very high input resistance. Often, you will see a source follower used at the front end of a system, followed by additional stages of voltage gain.

In Fig. 32-12, the ac source resistance is defined as

$$r_s = R_S \parallel R_L$$

It is possible to derive this equation for the voltage gain of a source follower:

$$A_v = \frac{g_m r_s}{1 + g_m r_s} \qquad \textbf{(32-13)}$$

Because the denominator is always greater than the numerator, the voltage gain is always less than 1.

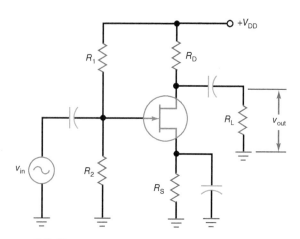

Figure 32-11 CS amplifier.

EXAMPLE 32-2

If g_m = 5000 μS in Fig. 32-13, what is the output voltage?

Answer:

The ac drain resistance is

$$r_d = 3.6\ k\Omega \parallel 10\ k\Omega = 2.65\ k\Omega$$

The voltage gain is

$$A_v = (5000\ \mu S)(2.65\ k\Omega) = 13.3$$

The output voltage is

$$v_{out} = 13.3(1\ mV\ p\text{-}p) = 13.3\ mV\ p\text{-}p$$

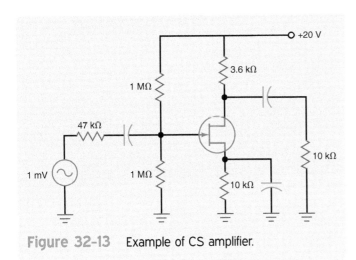

Figure 32-13 Example of CS amplifier.

32.8 Special Types of JFETs

Several special types of JFETs are used for very high frequency signal amplification. These include the MESFET, the HEMT, and the pHEMT.

MESFET

The *metal-semiconductor junction field-effect transistor (MESFET)* is different from a standard JFET. Instead of being made of silicon, the MESFET is normally made of gallium arsenide (GaAs) or some other compound semiconductor that offers a shorter transit time for electrons through the material than silicon does. As a result, GaAs transistors can operate as amplifiers at frequencies up to about 50 GHz. The MESFET is sometimes called a GASFET.

Figure 32-14 shows the general cross section of a MESFET. The *n*-channel is made on a *p* substrate. The source and drain metal connections are on either end. Electrons flow from source to drain as in any JFET. Instead of a *p*-type gate diffused into the channel, a metal gate is placed directly on top of the *n*-channel, forming a metal-semiconductor diode.

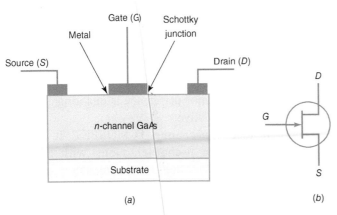

Figure 32-14 (a) Cross section of a MESFET and (b) its schematic symbol.

This type of diode is called a hot carrier or *Schottky diode*. It works like any other diode but has a lower barrier voltage in the 0.2- to 0.4-V range, rather than the 0.7-V level of a silicon diode. This gate diode is reversed-biased as in other JFETs. The operation is otherwise similar.

HEMT and pHEMT

A *high electron mobility transistor (HEMT)* is similar to a MESFET, but instead of the metal-semiconductor junction, the gate junction uses different semiconductor materials for the gate and the channel. This is called a *heterojunction*. A typical combination is GaAs for the channel and aluminum gallium arsenide (AlGaAs) for the gate. A HEMT made of gallium nitride (GaN) also makes a good power transistor, providing several watts of power into the low GHz range.

A special form of HEMT is the pseudomorphic HEMT, or *pHEMT*, that uses additional layers of different semiconductor materials including compounds of indium (In). The layers are optimized to further speed electron transit, pushing their ability to amplify well into the millimeter wave range (30 to 100 GHz).

Today, the JFET is the least used type of FET. It will occasionally be found in discrete form in older equipment or in specialized high-frequency military equipment. Most JFETs are used as input stages to radio receivers, such as amplifiers or mixers, and to certain types of integrated circuit amplifiers because of their high input impedance and very low noise generation.

32.9 The Depletion-Mode MOSFET

Figure 32-15*a* shows a **depletion-mode MOSFET** with *n*-type source (*S*), drain (*D*), and channel diffused into a *p*-type substrate. Electrons flowing from source to drain must pass through the narrow channel between the gate and the *p* substrate.

A thin layer of silicon dioxide (SiO_2) is deposited on the top of the channel. Silicon dioxide is the same as glass, which is an insulator. In a MOSFET, the gate is metallic or highly conductive silicon. Because the metallic gate is insulated from the channel, negligible gate current flows even when the gate voltage is positive.

Figure 32-15*a* shows a depletion-mode MOSFET with a negative gate voltage. The V_{DD} supply forces free electrons to flow from source to drain. These electrons flow through the narrow channel on the top of the *p* substrate. As with a JFET, the gate voltage controls the width of the channel. The more negative the gate voltage, the smaller the drain current. When the gate voltage is negative enough, the drain current is cut off. Therefore, the operation of a depletion-mode MOSFET is similar to that of a JFET when V_{GS} is negative.

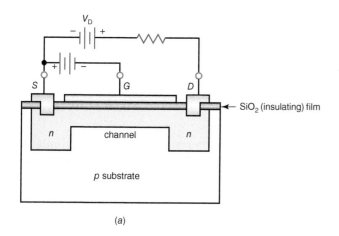

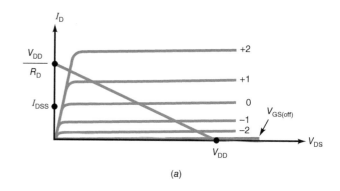

(a)

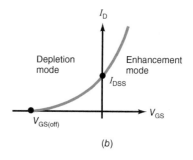

(b)

Figure 32-16 An *n*-channel, depletion-mode MOSFETs. (*a*) Drain curves. (*b*) Transconductance curve.

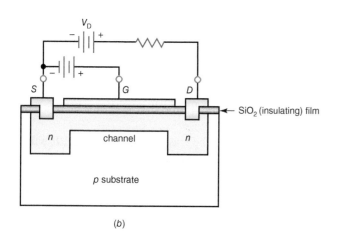

Figure 32-15 (*a*) D-MOSFET with negative gate. (*b*) D-MOSFET with positive gate.

Since the gate is insulated, we can also use a positive input voltage, as shown in Fig. 32-15*b*. The positive gate voltage increases the number of free electrons flowing through the channel. The more positive the gate voltage, the greater the conduction from source to drain.

32.10 D-MOSFET Curves

Figure 32-16*a* shows the set of drain curves for a typical *n*-channel, depletion-mode MOSFET. Notice that the curves above $V_{GS} = 0$ are positive and the curves below $V_{GS} = 0$ are negative. As with a JFET, the bottom curve is for $V_{GS} = V_{GS(off)}$ and the drain current will be approximately zero. As shown, when $V_{GS} = 0$ V, the drain current will equal I_{DSS}. This demonstrates that the depletion-mode MOSFET, or D-MOSFET, is a *normally on* device. When V_{GS} is made negative, the drain current will be reduced. In contrast to an *n*-channel JFET, the *n*-channel D-MOSFET can have V_{GS} made positive and still function properly. This is because there is no *pn* junction to become forward-biased. When V_{GS}

becomes positive, I_D will increase following the square-law equation

$$I_D = I_{DSS}\left(1 - \frac{V_{GS}}{V_{GS(off)}}\right)^2 \qquad \textbf{(32-14)}$$

When V_{GS} is negative, the D-MOSFET is operating in the depletion mode. When V_{GS} is positive, the D-MOSFET is operating in the enhancement mode. Like the JFET, the D-MOSFET curves display an ohmic region, a current-source region, and a cutoff region.

Figure 32-16*b* is the transconductance curve for a D-MOSFET. Again, I_{DSS} is the drain current with the gate shorted to the source. Here I_{DSS} is no longer the maximum possible drain current. The parabolic transconductance curve follows the same square-law relation that exists with a JFET. As a result, the analysis of a depletion-mode MOSFET is almost identical to that of a JFET circuit. The major difference is enabling V_{GS} to be either negative or positive.

There is also a *p*-channel D-MOSFET. It consists of a drain-to-source *p*-channel, along with a *n*-type substrate. Once again, the gate is insulated from the channel. The action of a *p*-channel MOSFET is complementary to the *n*-channel MOSFET. The schematic symbols for both *n*-channel and *p*-channel D-MOSFETs are shown in Fig. 32-17.

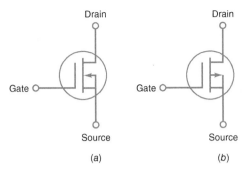

Figure 32-17 D-MOSFET schematic symbols.
(a) N-channel. (b) P-channel.

32.11 Depletion-Mode MOSFET Amplifiers

A depletion-mode MOSFET is unique because it can operate with a positive or a negative gate voltage. Because of this, we can set its Q point at $V_{GS} = 0$ V, as shown in Fig. 32-18a. When the input signal goes positive, it increases I_D above I_{DSS}. When the input signal goes negative, it decreases I_D below I_{DSS}. Because there is no pn junction to forward-bias, the input resistance of the MOSFET remains very high. Being able to use zero V_{GS} allows us to build the very simple bias circuit of Fig. 32-18b. Because I_G is zero, $V_{GS} = 0$ V and $I_D = I_{DSS}$. The drain voltage is

$$V_{DS} = V_{DD} - I_{DSS}\, R_D \qquad \textbf{(32-15)}$$

Due to the fact that a D-MOSFET is a normally on device, it is also possible to use self-bias by adding a source resistor. The operation becomes the same as a self-biased JFET circuit.

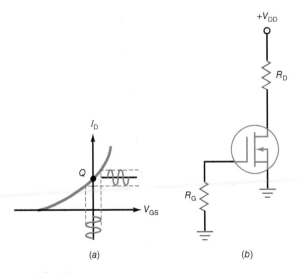

(a) (b)

Figure 32-18 Zero bias.

EXAMPLE 32-3

The D-MOSFET amplifier shown in Fig. 32-19 has $V_{GS(off)} = -2$ V, $I_{DSS} = 4$ mA, and $g_{mo} = 2000$ μS. What is the circuit's output voltage?

Answer:

With the source grounded, $V_{GS} = 0$ V and $I_D = 4$ mA.

$$V_{DS} = 15 \text{ V} - (4 \text{ mA})(2 \text{ k}\Omega) = 7 \text{ V}$$

Since $V_{GS} = 0$ V, $gm = g_{mo} = 2000$ μS.
The amplifier's voltage gain is found by

$$A_V = g_m r_d$$

The ac drain resistance is equal to

$$r_d = R_D \,\|\, R_L = 2 \text{ K} \,\|\, 10 \text{ K} = 1.76 \text{ k}\Omega$$

and A_V is

$$A_V = (2000 \ \mu S)(1.67 \text{ k}\Omega) = 3.34$$

Therefore,

$$V_{out} = (V_{in})(A_V) = (20 \text{ mV})(3.34) = 66.8 \text{ mV}$$

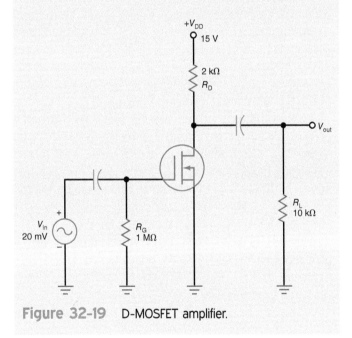

Figure 32-19 D-MOSFET amplifier.

As shown by Example 32-3, the D-MOSFET has a relatively low voltage gain. One of the major advantages of this device is its extremely high input resistance. This allows us to use this device when circuit loading could be a problem. Also, MOSFETs have excellent low-noise properties. This is a definite advantage for any stage near the front end of a system where the signal is weak. This is very common in many types of electronic communications circuits.

32.12 The Enhancement-Mode MOSFET

The enhancement-mode MOSFET, or E-MOSFET is the most widely used type of FET and transistor. Virtually all digital circuits use E-MOSFETs, and there is a growing number of linear circuits that use it too.

The Basic Idea

Figure 32-20 shows an E-MOSFET. Two-type regions are diffused into a p substrate. As you can see, there no longer is an n channel between the source and the drain. How does an E-MOSFET work? Figure 32-20 also shows normal biasing polarities. When the gate voltage is zero, the current between source and drain is zero. For this reason, an E-MOSFET is *normally off* when the gate voltage is zero.

The only way to get current is with a positive gate voltage. When the gate is positive, it attracts free electrons into the p region. The free electrons recombine with the holes next to the silicon dioxide. When the gate voltage is positive enough, all the holes touching the silicon dioxide are filled and free electrons begin to flow from the source to the drain. The effect is the same as creating a thin layer of n-type material next to the silicon dioxide. This thin conducting layer is called the *n-type inversion layer*. When it exists, free electrons can flow easily from the source to the drain. See Fig. 32-21.

The minimum V_{GS} that creates the n-type inversion layer is called the **threshold voltage,** symbolized $V_{GS(th)}$. When V_{GS} is less than $V_{GS(th)}$, the drain current is zero. When V_{GS} is greater than $V_{GS(th)}$, an n-type inversion layer connects the source to the drain and the drain current can flow. Typical values of $V_{GS(th)}$ for small-signal devices are from 1 to 3 V.

The JFET is referred to as a *depletion-mode device* because its conductivity depends on the action of depletion layers. The E-MOSFET is classified as an

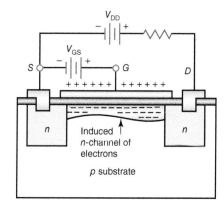

Figure 32-21 Conduction between source and drain in an E-MOSFET.

enhancement-mode device because a gate voltage greater than the threshold voltage enhances its conductivity. With zero gate voltage, a JFET is *on,* whereas an E-MOSFET is *off.* Therefore, the E-MOSFET is considered to be a normally off device.

Drain Curves

A small-signal E-MOSFET has a power rating of 1 W or less. Figure 32-22a shows a set of drain curves for a typical small-signal E-MOSFET. The lowest curve is the $V_{GS(th)}$ curve. When V_{GS} is less than $V_{GS(th)}$, the drain current is approximately zero. When V_{GS} is greater than $V_{GS(th)}$, the device turns on and the drain current is controlled by the gate voltage.

The almost-vertical part of the graph is the ohmic region, and the almost-horizontal parts are the active region. When biased in the ohmic region, the E-MOSFET is equivalent to a resistor. When biased in the active region, it is equivalent to a current source. Although the E-MOSFET can operate in the active region, the main use is the ohmic region.

Figure 32-22b shows a typical transconductance curve. There is no drain current until $V_{GS} = V_{GS(th)}$. The drain current then increases rapidly until it reaches the saturation current $I_{D(sat)}$. Beyond this point, the device is biased in the ohmic region. Therefore, I_D cannot increase, even though V_{GS} increases. To ensure hard saturation, a gate voltage of $V_{GS(on)}$ well above $V_{GS(th)}$ is used, as shown in Fig. 32-22b.

Schematic Symbol

When $V_{GS} = 0$, the E-MOSFET is off because there is no conducting channel between source and drain. The schematic symbol of Fig. 32-23a has a broken channel line to indicate this normally off condition. As you know, a gate

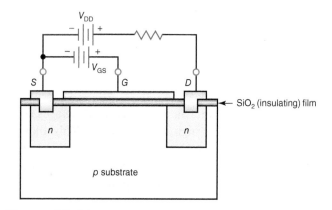

Figure 32-20 Enhancement-mode MOSFET.

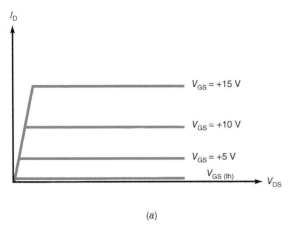

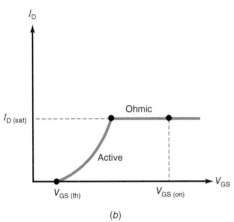

Figure 32-22 EMOS graphs. (a) Drain curves. (b) Transconductance curve.

voltage greater than the threshold voltage creates an *n*-type inversion layer that connects the source to the drain. The arrow points to this inversion layer, which acts like an *n*-channel when the device is conducting.

There is also a *p*-channel E-MOSFET. The schematic symbol is similar, except that the arrow points outward, as shown in Fig. 32-23*b*.

The circle around the MOSFET schematic symbols in Fig. 32-23 indicates a discrete component device. If the circle is omitted, the device is part of an integrated circuit. Other simplified schematic symbols for E-MOSFETs, shown in Fig. 32-24, are more widely used.

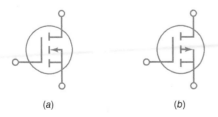

Figure 32-23 EMOS schematic symbols. (a) *N*-channel device. (b) *P*-channel device.

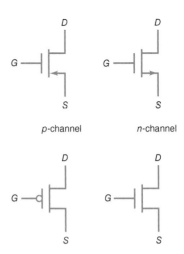

Figure 32-24 Simplified E-MOSFET schematic symbols used in integrated circuit devices.

Maximum Gate-Source Voltage

MOSFETs have a thin layer of silicon dioxide, an insulator that prevents gate current for positive as well as negative gate voltages. This insulating layer is kept as thin as possible to give the gate more control over the drain current. Because the insulating layer is so thin, it is easily destroyed by excessive gate-source voltage.

For instance, a 2N7000 has a $V_{GS(max)}$ rating of ± 20 V. If the gate-source voltage becomes more positive than $+20$ V or more negative than -20 V, the thin insulating layer will be destroyed.

Aside from directly applying an excessive V_{GS}, you can destroy the thin insulating layer in more subtle ways. If you remove or insert a MOSFET into a circuit while the power is on, transient voltages caused by inductive kickback may exceed the $V_{GS(max)}$ rating. Even picking up a MOSFET may deposit enough static charge to exceed the $V_{GS(max)}$ rating. This is the reason why MOSFETs are often shipped with a wire ring around the leads, or wrapped in tin foil, or inserted into conductive foam.

Some MOSFETs are protected by a built-in zener diode in parallel with the gate and the source. The zener voltage is less than the $V_{GS(max)}$ rating. Therefore, the zener diode breaks down before any damage to the thin insulating layer occurs. The disadvantage of these internal zener diodes is that they reduce the MOSFET's high input resistance. The trade-off is worth it in some applications because expensive MOSFETs are easily destroyed without zener protection.

In conclusion, MOSFET devices are delicate and can be easily destroyed. You have to handle them carefully. Furthermore, you should never connect or disconnect them while the power is on. Finally, before you pick up a MOSFET device, you should ground your body by touching the chassis of the equipment you are working on.

32.13 The Ohmic Region

Although the E-MOSFET can be biased in the active region, it is more commonly used as a switching device. The typical input voltage is either low or high. Low voltage is 0 V, and high voltage is $V_{GS(on)}$, a value specified on data sheets.

Drain Source on Resistance

When an E-MOSFET is biased in the ohmic region, it is equivalent to a resistance of $R_{DS(on)}$. Almost all data sheets will list the value of this resistance at a specific drain current and gate-source voltage.

Figure 32-25 illustrates the idea. There is a Q_{test} point in the ohmic region of the $V_{GS} = V_{GS(on)}$ curve. The manufacturer measures $I_{D(on)}$ and $V_{DS(on)}$ at this Q_{test} point. From this, the manufacturer calculates the value of $R_{DS(on)}$ using this definition:

$$R_{DS(on)} = \frac{V_{DS(on)}}{I_{D(on)}} \qquad \textbf{(32-16)}$$

For instance, at the test point on one device has $V_{DS(on)} = 1$ V and $I_{D(on)} = 100$ mA. With Formula (32-15),

$$R_{DS(on)} = \frac{1 \text{ V}}{100 \text{ mA}} = 10 \text{ }\Omega$$

Typical ON resistance values range from about 10 Ω to less than 1 Ω for newer devices.

An E-MOSFET is biased in the ohmic region when this condition is satisfied:

$$I_{D(sat)} \; I_{D(on)} \qquad \text{when} \qquad V_{GS} = V_{GS(on)} \qquad \textbf{(32-17)}$$

Formula (32-17) is important. It tells us whether an E-MOSFET is operating in the active region or the ohmic region. Given an EMOS circuit, we can calculate the $I_{D(sat)}$. If $I_{D(sat)}$ is less than $I_{D(on)}$ when $V_{GS} = V_{GS(on)}$, we will know that the device is biased in the ohmic region and is equivalent to a small resistance.

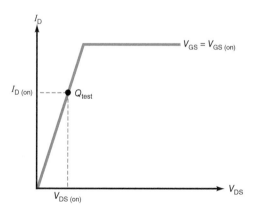

Figure 32-25   Measuring $R_{DS(on)}$.

EXAMPLE 32-4

What is the output voltage in Fig. 32-26a?

Answer:

For the 2N7000, the most important values are

$$V_{GS(on)} = 4.5 \text{ V}$$
$$I_{D(on)} = 75 \text{ mA}$$
$$R_{DS(on)} = 6 \text{ }\Omega$$

Since the input voltage swings from 0 to 4.5 V, the 2N7000 is being switched on and off.

The drain saturation current in Fig. 32-26a is

$$I_{D(sat)} = \frac{20 \text{ V}}{1 \text{ k}\Omega} = 20 \text{ mA}$$

Since 20 mA is less than 75 mA, the value of $I_{D(on)}$, the 2N7000 is biased in the ohmic region when the gate voltage is high.

Figure 32-26b is the equivalent circuit for a high-input gate voltage. Since the E-MOSFET has a resistance of 6 Ω, the output voltage is

$$V_{out} = \frac{6 \text{ }\Omega}{1 \text{ k}\Omega + 6 \text{ }\Omega} (20 \text{ V}) = 0.12 \text{ V}$$

On the other hand, when V_{GS} is low, the E-MOSFET is open (Fig. 32-26c), and the output voltage is pulled up to the supply voltage,

$$V_{out} = 20 \text{ V}$$

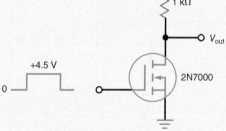

(a)

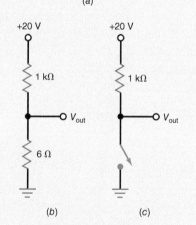

(b)   (c)

Figure 32-26   Switching between cutoff and saturation.

EXAMPLE 32-5

What is the LED current in Fig. 32-27?

Answer:

When V_{GS} is low, the LED is off. When V_{GS} is high, the action is similar to that in the preceding example because the 2N7000 goes into hard saturation.

If we allow 2 V for the LED drop and assume V_{DS} is zero,

$$I_D = \frac{20\ V - 2\ V}{1\ k\Omega} = 18\ mA$$

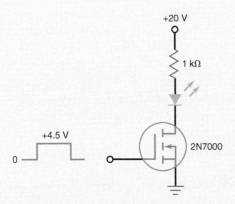

Figure 32-27 Turning an LED on and off.

EXAMPLE 32-6

What does the circuit of Fig. 32-28a do if a coil current of 30 mA or more closes the relay contacts?

Answer:

The E-MOSFET is being used to turn a relay on and off. Since the relay coil has a resistance of 500 Ω, the saturation current is

$$I_{D(sat)} = \frac{24\ V}{500\ \Omega} = 48\ mA$$

Assume this is less than the $I_{D(on)}$ of the MOSFET that has an ON resistance of only 10 Ω.

Figure 32-28b shows the equivalent circuit for high V_{GS}. The current through the relay coil is approximately 48 mA, more than enough to close the relay. When the relay is closed, the contact circuit looks like Fig. 32-28c. Therefore, the final load current is 8 A (120 V divided by 15 Ω).

In Figure 32-28a, an input voltage of only +2.5 V and almost zero input current control a load voltage of 120 Vac and a load current of 8 A. A circuit like this is useful with remote control. The input voltage could be a signal that has been transmitted a long distance through copper wire, fiber-optic cable, or outer space.

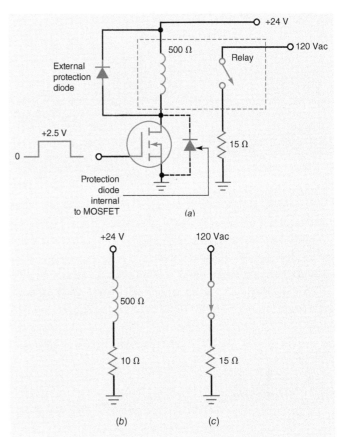

Figure 32-28 Low-input current signal controls large-output current.

Note: The diode across the relay coil is used to protect the transistor from the large voltage spike generated by the coil when the drain current cuts off. The rapidly dropping current creates a huge voltage spike that can damage the transistor. The polarity of this induced voltage is such that it causes the diode to conduct, momentarily clamping the coil voltage to 0.7 V. Some E-MOSFETs have an internal protection diode from source to drain.

32.14 Digital Switching

Why has the E-MOSFET revolutionized the computer industry? Because of its threshold voltage, it is ideal for use as a switching device. When the gate voltage is well above the threshold voltage, the device switches from cutoff to saturation. This OFF-ON action is the key to building computers and other digital devices. A typical computer uses millions of E-MOSFETs as OFF-ON switches to process data.

Analog, Digital, and Switching Circuits

The word **analog** means "continuous," like a sine wave. When we speak of an analog signal, we are talking about signals that continuously change in voltage like the one in Fig. 32-29a. The signal does not have to be sinusoidal. As long as there are no sudden jumps between two distinct voltage levels, the signal is referred to as an *analog signal.*

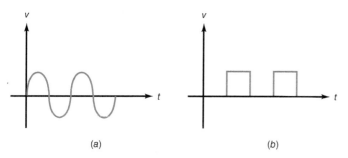

(a)   (b)

Figure 32-29 (a) Analog signal. (b) Digital signal.

The word **digital** refers to a discontinuous signal. This means that the signal jumps between two distinct voltage levels like the waveform of Fig. 32-29b. Digital signals like these are the kind of signals inside computers. These signals are computer codes that represent numbers, letters, and other symbols.

The word *switching* is a broader word than *digital*. Switching circuits include digital circuits as a subset. In other words, switching circuits can also refer to circuits that turn on motors, lamps, heaters, and other heavy-current devices.

Passive-Load Switching

Figure 32-30 shows an E-MOSFET with a passive load. The word *passive* refers to ordinary resistors like R_D. In this circuit, v_{in} is either low or high. When v_{in} is low, the MOSFET is cut off, and v_{out} equals the supply voltage V_{DD}. When v_{in} is high, the MOSFET saturates and v_{out} drops to a low value. For the circuit to work properly, the drain saturation current $I_{D(sat)}$ has to be less than $I_{D(on)}$ when the input voltage is equal to or greater than $V_{GS(on)}$. This is equivalent to saying that the resistance in the ohmic region has to be much smaller than the passive drain resistance. In symbols,

$$R_{DS(on)} \ll R_D$$

A circuit like Fig. 32-30 is the simplest computer circuit that can be built. It is called an *inverter* because the output voltage is the opposite of the input voltage. When the input voltage is low, the output voltage is high. When the input voltage is high, the output voltage is low. Great accuracy is not necessary when analyzing switching circuits. All that matters is that the input and output voltages can be easily recognized as low or high.

Active-Load Switching

Integrated circuits (ICs) consist of thousands to hundreds of millions of microscopically small transistors, mostly MOSFETs. The earliest integrated circuits used passive-load resistors like the one of Fig. 32-30. But a passive-load resistance presents a major problem. It is physically much larger than a MOSFET. Because of this, integrated circuits with passive-load resistors were too big until somebody invented **active-load resistors.** This greatly reduced the size of integrated circuits and led to the personal computers that we have today.

The key idea was to get rid of passive-load resistors. Figure 32-31a shows the invention: *active-load switching.* The lower MOSFET acts like a switch, but the upper MOSFET acts like large resistance. Notice that the upper MOSFET has its gate connected to its drain. Because of this, it becomes a *two-terminal device* with an active resistance that is a function of the V_{DS} and I_D values, as well as the doping and physical geometry of the transistor.

For the circuit to work properly, the R_D of the upper MOSFET has to be large compared to the $R_{DS(on)}$ of the lower MOSFET. For instance, if the upper MOSFET acts like an R_D of 5 kΩ and the lower one like an $R_{DS(on)}$ of 667 Ω, as shown in Fig. 32-31b, then the output voltage will be low.

Exact values don't matter with digital switching circuits as long as the voltages can be easily distinguished as low or high. Therefore, the exact value of R_D does not matter. It can be 5, 6.25, or 7.2 kΩ. Any of these values is large enough to produce a low output voltage in Fig. 32-31b.

Conclusion

Active-load resistors are necessary with digital ICs because a small physical size is important with digital ICs. The designer makes sure that the R_D of upper MOSFET is large compared to the $R_{D(on)}$ of the lower MOSFET. When you see a circuit like Fig. 32-31a, all you have to remember is the basic idea: The circuit acts like a resistance of R_D in series with a switch. As a result, the output voltage is either high or low.

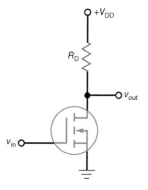

Figure 32-30 Passive load.

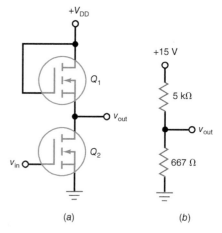

(a)   (b)

Figure 32-31 (a) Active load. (b) Equivalent circuit.

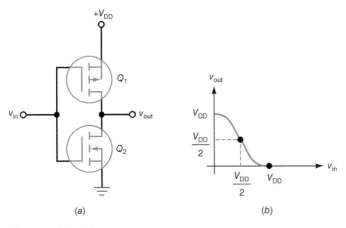

Figure 32-32 CMOS inverter. (a) Circuit. (b) Input-output graph.

32.15 CMOS

With active-load switching, the current drain with a low output is approximately equal to $I_{D(sat)}$. This may create a problem with battery-operated equipment. One way to reduce the current drain of a digital circuit is with **complementary MOS (CMOS).** In this approach, the IC designer combines *n*-channel and *p*-channel MOSFETs.

Figure 32-32*a* shows the idea. Here Q_1 is a *p*-channel MOSFET and Q_2 is an *n*-channel MOSFET. These two devices are complementary; that is, they have equal and opposite values of $V_{GS(th)}$, $V_{GS(on)}$, $I_{D(on)}$, and so on. The circuit is similar to a class B amplifier because one MOSFET conducts while the other is off.

Basic Action

When a CMOS circuit like Fig. 32-32*a* is used in a switching application, the input voltage is either high ($+V_{DD}$) or low (0 V). When the input voltage is high, Q_1 is off and Q_2 is on. In this case, the shorted Q_2 pulls the output voltage down to ground. On the other hand, when the input voltage is low, Q_1 is on and Q_2 is off. Now, the shorted Q_1 pulls the output voltage up to $+V_{DD}$. Since the output voltage is inverted, the circuit is called a *CMOS inverter.*

Figure 32-32*b* shows how the output voltage varies with the input voltage. When the input voltage is zero, the output voltage is high. When the input voltage is high, the output voltage is low. Between the two extremes, there is a crossover point where the input voltage equals $V_{DD}/2$. At this point, both MOSFETs have equal resistances and the output voltage equals $V_{DD}/2$.

Power Consumption

The main advantage of CMOS is its extremely low power consumption. Because both MOSFETs are in series in Fig. 32-32*a*, the quiescent current drain is determined by the nonconducting device. Since its resistance is in the megohms, the *quiescent* (idling) power consumption approaches zero.

The power consumption increases when the input signal switches from low to high, and vice versa. The reason is this:

At the midway point in a transition from low to high, or vice versa, both MOSFETs are on. This means that the drain current temporarily increases. Since the transition is very rapid, only a brief pulse of current occurs. The product of the drain supply voltage and the brief pulse of current means that the average *dynamic* power consumption is greater than the quiescent power consumption. In other words, a CMOS device dissipates more average power when it has transitions than when it is quiescent. Power consumption is proportional to switching frequency.

Since the pulses of current are very short, however, the average power dissipation is very low even when CMOS devices are switching states. In fact, the average power consumption is so small that CMOS circuits are often used for battery-powered applications such as cell phones, MP3 players, calculators, digital watches, and hearing aids.

32.16 Power FETs

In earlier discussions, we emphasized small-signal E-MOSFETs, that is, low-power MOSFETs. Although some discrete low-power E-MOSFETs are commercially available, the major use of low-power EMOS is with digital integrated circuits.

High-power EMOS is different. With high-power EMOS, the E-MOSFET is a discrete device widely used in applications that control motors, lamps, disk drives, printers, power supplies, and so on. In these applications, the E-MOSFET is called a **power FET.**

Discrete Devices

Manufacturers are producing different devices such as VMOS, TMOS, hexFET, trench MOSFET, and waveFET. All these power FETs use different channel geometries to increase their maximum ratings. These devices have current ratings from 1 A to more than 200 A, and power ratings from 1 W to more than 500 W.

Figure 32-33 shows the structure of a **vertical MOS (VMOS)** device. It has two sources at the top, which are usually connected, and the substrate acts like the drain. When V_{GS} is greater than $V_{GS(th)}$, free electronics flow vertically downward from the two sources to the drain. Because the conducting channel is much wider along both sides of the V groove, the current can be much larger. This enables the VMOS device to act as a power FET.

Another type of power MOSFET is the laterally diffused MOS (LDMOS) FET, shown in Fig. 32-33*b*. It is usually an *n*-channel enhancement-mode type with large elements to handle the high power and heat. The device is built on a highly doped *p*-type substrate. Above this is a lightly doped epitaxial layer to help accommodate the higher voltages these devices are normally subjected to. The source is connected to the substrate. The *n*-type areas are diffused or implanted into the substrate to form the source and drain. A drain extension (DEX) is used to aid the high-voltage tolerance. The gate is a polysilicon conductor on top of the silicon dioxide insulating layer. A shield

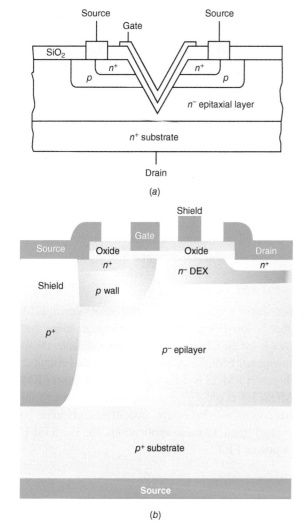

Figure 32-33   MOS structures. (a) VMOS structure.
(b) LDMOS.

connected to the source is used to reduce the drain to gate feedback capacitance to extend the high-frequency operating range.

LDMOS FETs are used primarily as RF power amplifiers in radio transmitters for cellular base stations. Other RF uses are in radar and other high-power transmitters for broadcast and two-way radios. With drain voltages as high as 50 Vdc, output power levels to 600 W are achievable to frequencies up to about 5 GHz.

Discrete power MOSFETs can handle currents of a few amperes to several hundred amperes. Power ratings range from less than 100 W to over 1 kW. Typical ON resistances are less than 1 Ω and as low as thousandths of an ohm. Power MOSFETs are widely used in switch-mode power supply circuits like regulators and dc-dc converters, as you will see in a later chapter. Other applications include switching in lighting, heating, and motor control. Power MOSFETs can switch many amperes of current in just tens of nanoseconds.

A newer form of discrete power FET is one made from gallium nitride (GaN). Those semiconductor materials permit the FET to operate at higher voltages up to 100 V and

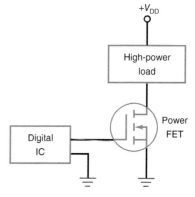

Figure 32-34   Power FET is the interface between low-power digital IC and high-power load.

several amperes. The GaN FETs handle heat better than silicon transistors, and GaN FETs can also operate at higher frequencies up to 10 GHz and above. They also switch faster than silicon power FETs. The GaN FETs are of the depletion-mode pHEMT type. They are used in RF power amplifiers for radar, satellite, and cellular base stations, as well as in very high speed switching power supplies.

Power FET as an Interface

Digital ICs are low-power devices because they can supply only small load currents. If we want to use the output of a digital IC to drive a high-current load, we can use a power FET as an **interface** (a device B that allows device A to communicate with or control device C).

Figure 32-34 shows how a digital IC can control a high-power load. The output of the digital IC drives the gate of the power FET. When the digital output is high, the power FET is like a closed switch. When the digital output is low, the power FET is like an open switch. Interfacing digital ICs (small-signal EMOS and CMOS) to high-power loads is one of the important applications of power FETs.

A common motor control circuit is shown in Fig. 32-35, which has simplified schematic symbols. MOSFETs M_1 and

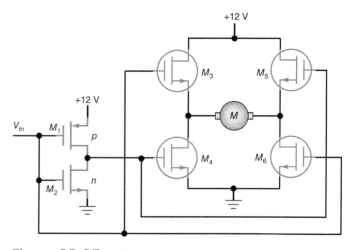

Figure 32-35   An H-bridge motor controller.

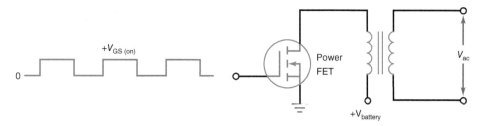

Figure 32-36 A rudimentary dc-to-ac converter.

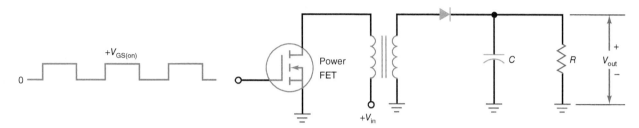

Figure 32-37 A rudimentary dc-to-dc converter.

M_2 form a CMOS inverter that is usually a part of an integrated circuit. The remaining M_3 to M_6 are power MOSFETs that act like switches and control the direction of rotation of a dc motor. The circuit is referred to as an *H-bridge.*

When v_{in} is low or zero, M_3 and M_6 are cut off. The output of M_1 and M_2 is high, so M_4 and M_5 are switched on. Electrons flow through the motor from left to right, turning the motor in one direction. If v_{in} is high, the reverse condition occurs: M_3 and M_6 turn on while M_4 and M_5 are off. Electrons flow from right to left in the motor, reversing its direction of rotation.

DC-to-AC Converters

When there is a sudden power failure, computers will stop operating and valuable data may be lost. One solution is to use an **uninterruptible power supply (UPS).** A UPS contains a battery and a dc-to-ac converter. The basic idea is this: When there is a power failure, the battery voltage is converted to an ac voltage to drive the computer.

Figure 32-36 shows a **dc-to-ac converter,** also known as a power inverter. When the power fails, other circuits (op amps, discussed later) are activated and generate a square wave to drive the gate. The square-wave input switches the power FET on and off. Since a square wave will appear across the transformer windings, the secondary winding can supply the ac voltage needed to keep the computer running. A commercial UPS is more complicated than this, but the basic idea of converting dc to ac is the same.

DC-to-DC Converters

Figure 32-37 is a **dc-to-dc** (or dc-dc) **converter,** a circuit that converts an input dc voltage to an output dc voltage that is either higher or lower. The power FET switches on and off, producing a square wave across the secondary winding. The half-wave rectifier diode and capacitor-input filter then produce the dc output voltage V_{out}. By using different turns ratios, we can get a dc output voltage that is higher or lower than the input voltage V_{in}. The dc-to-dc converter is one of the important sections of a switching or switch-mode power supply. This application will be examined in a later chapter.

Another Switching Example

During the day, the photodiode of Fig. 32-38 is conducting heavily, and the gate voltage is below the MOSFET gate threshold. At night, the photodiode is off, and the gate voltage rises to +10 V. Therefore, the circuit turns the MOSFET and lamp on automatically at night.

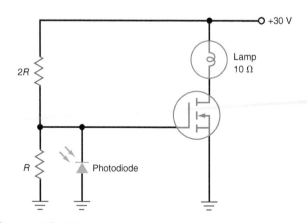

Figure 32-38 Automatic light control.

32.17 E-MOSFET Amplifiers

The E-MOSFET is used primarily as a switch. Applications do exist for this device to be used as an amplifier, however. Discrete component MOSFETs are used as front-end, high-frequency RF amplifiers in communications equipment, and power E-MOSFETs are used in class AB power amplifiers. Several types of integrated circuit E-MOSFET amplifiers are also available.

With E-MOSFETs, V_{GS} has to be greater than $V_{GS(th)}$ for drain current to flow. This eliminates self-bias, current-source bias, and zero bias because all these favor depletion-mode operation. This leaves gate bias and voltage divider bias. Both these biasing arrangements will work with E-MOSFETs because they can achieve enhancement-mode operation.

Discrete E-MOSFET amplifiers are rare, but when used, they can be biased in two ways. Figure 32-39 shows a biasing method for E-MOSFETs called **drain-feedback bias.** When

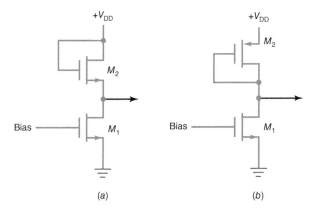

Figure 32-41 Using a MOSFET as a drain resistor. (a) N-type. (b) P-type.

the MOSFET is conducting, it has a drain current of $I_{D(on)}$ and a drain voltage of $V_{DS(on)}$. Because there is virtually no gate current, $V_{GS} = V_{DS(on)}$. As with collector-feedback, drain-feedback bias tends to compensate for changes in FET characteristics. For example, if $I_{D(on)}$ tries to increase for some reason, $V_{DS(on)}$ decreases. This reduces V_{GS} and partially offsets the original increase in $I_{D(on)}$.

Voltage divider bias can also be used with an E-MOSFET. In Figure 32-40, the 1-MΩ and 350-kΩ resistors set the gate voltage above $V_{GS(th)}$. Otherwise, the circuit works like a JFET or D-MOSFET amplifier.

When biasing E-MOSFETs for amplifiers in integrated circuits, resistors are not used because they take up too much space. Instead, active-drain resistors are created with another MOSFET, as shown in Fig. 32-41a, in which simplified schematic symbols are used. Transistor M_2 is the drain resistor. An E-MOSFET acts like a resistor when its gate is connected to its drain. In Fig. 32-41b, a p-type MOSFET is used as a drain resistor by turning it upside-down so that it conducts. Both these circuits are common-source, inverting linear amplifiers.

One way to obtain the gate bias voltage without resistors is to use a MOSFET voltage divider as shown in Fig. 32-42.

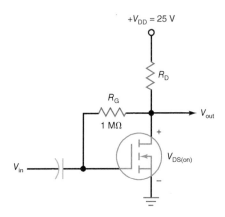

Figure 32-39 Drain-feedback biasing method.

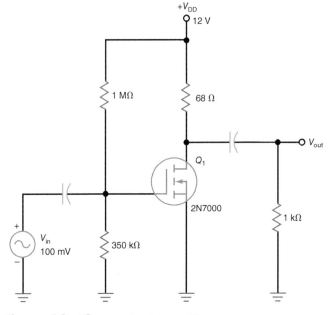

Figure 32-40 E-MOSFET amplifier.

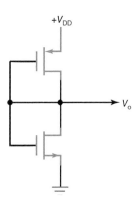

Figure 32-42 Using biased MOSFETs as a resistive voltage divider.

It uses two MOSFETs connected in series. Both are biased to act as resistors. The drain currents are equal, but the characteristics of each transistor are controlled during manufacturing to get the desired voltage division ratio.

32.18 Differential Amplifiers

A *differential amplifier* is one with two inputs and two outputs. Refer to Fig. 32-43a. The output is the difference between the two inputs multiplied by the gain, A. Inputs v_1 and v_2 produce outputs v_{o1} and v_{o2}. All signals are referenced to ground. The output V_{out} is taken between the two output lines, meaning that it is not referenced to ground. It is the difference between the two voltages:

$$V_{out} = v_{o1} - v_{o2} \qquad \textbf{(32-18)}$$

A more common arrangement is shown in Fig. 32-43b. Here only one output is referenced to ground. In this case the output is

$$V_{out} = A(v_2 - v_1) \qquad \textbf{(32-19)}$$

Figure 32-44 shows a simplified JFET differential amplifier. It is essentially two JFET amplifiers sharing a common-source resistor for bias. When only one output is used, a drain resistor can be eliminated and the output taken from the other.

The amplifier can also be used with one input while the other is grounded or set to zero. By selecting which input and output to be used, the output may be in phase with the input or inverted (180° out of phase).

There are several important reasons for using a differential amplifier. Using two inputs and the differential characteristic of the circuit, common noise on both inputs is cancelled out. Second, differential amplifiers can be more easily used in direct or dc-coupled circuits, thereby eliminating the need for coupling and other types of capacitors. This is beneficial in integrated circuits where capacitors are not used because they take up too much area on the chip.

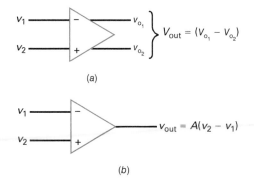

(a)

(b)

Figure 32-43 Differential amplifier configurations.
(a) Dual or differential output. (b) Single-ended or ground-referenced output.

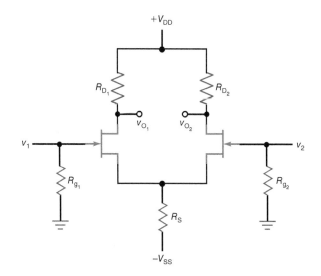

Figure 32-44 A JFET differential amplifier.

Discrete JFETs are not widely used in differential amplifiers. However, they are widely used as the input stages of IC operational amplifiers. JFETs appear only in the input stages, while the remainder of the op amp is usually bipolar transistors. The primary appeal of a JFET op amp is its extremely high input impedance, which can reach $10^{12}\ \Omega$.

MOSFETs are also used to create differential amplifiers. The primary application is in op amps where the entire amplifier is made of enhancement-mode MOSFETs. Figure 32-45 shows a typical MOSFET differential amplifier as it appears in most IC amplifiers. Note the use of the M designation for each E-MOSFET. Also, Q is used to designate bipolar transistors in a schematic diagram. The

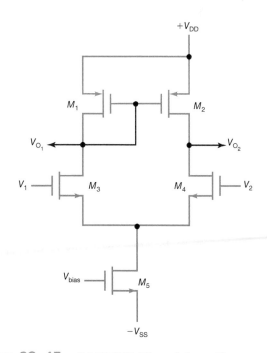

Figure 32-45 E-MOSFET differential amplifier.

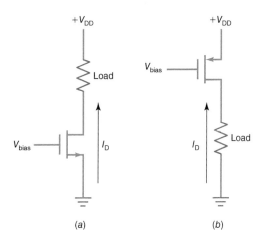

(a)　　　　(b)

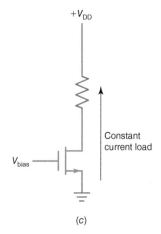

(c)

Figure 32-46 (a) Current source. (b) P-type current source. (c) Current sink.

differential pair is made up of M_3 and M_4, whereas M_1 and M_2 are active loads made from a current mirror. A current source, M_5, supplies a constant current to the differential transistors. The inputs V_1 and V_2 see an input impedance of over 10^{12} Ω.

Current Sources and Sinks

A current source is a transistor that supplies a constant current to a load. Figure 32-46a shows a MOSFET current source. Its gate bias is set by an n-type MOSFET voltage divider like that described earlier in this chapter. A p-type current source is shown in Fig. 32-46b.

A current sink, shown in Fig. 32-46c, is the reverse of a current source. It acts as a load that always draws a constant current. The bias voltage comes from a MOSFET voltage divider.

Current Mirror

A current mirror is a circuit that is used to supply bias to other MOSFETs. One type is shown in Fig. 32-47, in which M_1 is a current source with M_2 as an active-load resistance.

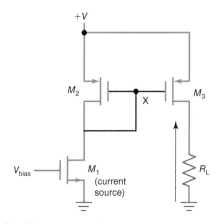

Figure 32-47 A MOSFET current mirror.

The voltage, a point X in the circuit, is used to bias M_3. If M_3 has the same characteristics as M_2, then the current in M_3 is the same as that in M_1. The current in M_3 is called the "mirrored" current. By connecting other MOSFET gates at point X in the circuit, additional mirrored currents can be supplied to other circuits.

The current mirror is commonly used in place of drain resistors in MOSFET differential amplifiers, as Fig. 32-45 has shown earlier.

Cascode Amplifier

Another popular integrated MOSFET amplifier circuit is the *cascode amplifier,* shown in Fig. 32-48. Here M_3 is the main amplifying transistor, while M_1 is a current source used as a resistive load. Note that M_2 isolates M_1 from M_3. Bias comes from MOSFET voltage dividers.

The benefit of the cascode circuit is that the gain can be very large because the current source M_3 has a very high

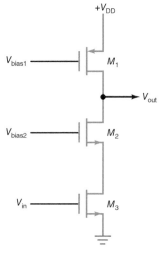

Figure 32-48 MOSFET cascode amplifier.

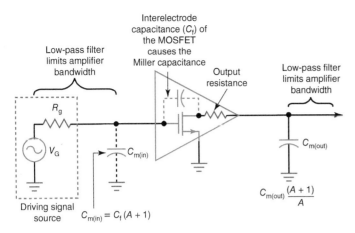

Figure 32-49 How Miller capacitance limits the upper-frequency response of a MOSFET amplifier.

effective resistance and a high drain resistance produces a high gain.

A second benefit is that the cascode connection allows the circuit to operate at higher frequencies. The presence of M_2 helps reduce the Miller effect. *Miller effect* causes the gate-drain capacitance that naturally exists in all FETs to reduce the high-frequency response of the amplifier. This interelectrode, or feedback capacitance, C_f, is multiplied by the gain of the amplifier.

Refer to Fig. 32-49. This Miller effect essentially introduces shunt Miller capacitances, $C_{m(in)}$ and $C_{m(out)}$, that appear across the input and output of an amplifier. Both $C_{m(in)}$ and the generator resistance, R_g, form a low-pass filter at the input. Moreover, $C_{m(out)}$ and output resistance of the amplifier also form a low-pass filter. Both these filters cause the higher frequencies to be rolled off, thereby reducing the upper frequencies that the amplifier can pass. The cascode connection minimizes this problem, allowing MOSFET amplifiers to operate well into the upper MHz range and lower GHz frequencies.

32.19 MOSFET Circuits and Static Electricity

MOSFETs are very sensitive to external voltages, especially static electricity. You will hear this referred to as *electrostatic discharge (ESD)*. The very thin silicon dioxide insulator between the gate and substrate is very easily penetrated, causing the gate to short to the substrate if the voltage on the gate exceeds some small value.

Static electricity voltages can be as high as several hundred or even thousands of volts. You have probably experienced a mild shock yourself on a dry winter day after walking on a carpet and then touching a metal item.

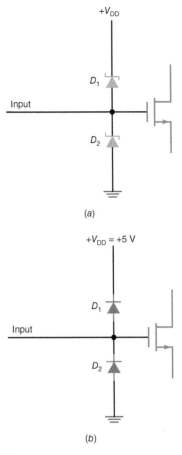

Figure 32-50 Protecting MOSFETs from static electricity. (*a*) Using zener diodes. (*b*) Using standard silicon diodes.

In discrete MOSFETs, this breakdown voltage may be 10 to 30 V. In IC MOSFET-based circuits, the penetration voltage may be only a few volts. If the voltage rating is exceeded, the device will be destroyed.

To overcome this problem, most discrete MOSFETs and the inputs to IC MOSFETs are protected by built-in surge protection diodes. See Fig. 32-50*a*. With two back-to-back zener diodes, the input is protected for either positive or negative voltages between the source and gate. If the input is higher than the zener breakdown voltage, one of the zener diodes will conduct and clamp the input to the zener value, thereby saving the device.

Another diode protection method is shown in Fig. 32-50*b*. Two standard silicon diodes are connected as voltage clamps. If a negative input voltage occurs, D_2 conducts and clamps the input to a maximum of 0.7 V. If the input is positive and more than the +5-V supply voltage, D_1 conducts, clamping the input to +5 V and saving the transistor.

ESD is always a problem when handling MOSFETs. Remember to ground yourself to a metal object before touching or handling a MOSFET on an IC. Some MOSFETs are packaged in conductive material to keep the leads shorted together to prevent accidental damage.

1. A JFET
   a. is a voltage-controlled device.
   b. is a current-controlled device.
   c. has a low input resistance.
   d. has a very large voltage gain.

2. A field-effect transistor uses
   a. both free electrons and holes.
   b. only free electrons.
   c. only holes.
   d. either one or the other, but not both.

3. The input impedance of a JFET
   a. approaches zero.
   b. approaches one.
   c. approaches infinity.
   d. is impossible to predict.

4. The gate controls
   a. the width of the channel.
   b. the drain current.
   c. the gate voltage.
   d. all the above.

5. The gate-source diode of a JFET should be
   a. forward-biased.
   b. reverse-biased.
   c. either forward- or reverse-biased.
   d. none of the above.

6. The pinch-off voltage has the same magnitude as the
   a. gate voltage.
   b. drain-source voltage.
   c. gate-source voltage.
   d. gate-source cutoff voltage.

7. When the drain saturation current is less than I_{DSS}, a JFET acts like a
   a. bipolar junction transistor.
   b. current source.
   c. resistor.
   d. battery.

8. R_{DS} equals pinch-off voltage divided by
   a. drain current.
   b. gate current.
   c. ideal drain current.
   d. drain current for zero gate voltage.

9. The transconductance curve is
   a. linear.
   b. similar to the graph of a resistor.
   c. nonlinear.
   d. like a single drain curve.

10. The transconductance increases when the drain current approaches
    a. 0.
    b. $I_{D(sat)}$.
    c. I_{DSS}.
    d. I_S.

11. A CS amplifier has a voltage gain of
    a. $g_m r_d$.
    b. $g_m r_s$.
    c. $g_m r_s/(1 + g_m r_s)$.
    d. $g_m r_d/(1 + g_m r_d)$.

12. A source follower has a voltage gain of
    a. $g_m r_d$.
    b. $g_m r_s$.
    c. $g_m r_s/(1 + g_m r_s)$.
    d. $g_m r_d/(1 + g_m r_d)$.

13. When the input signal is large, a source follower has
    a. a voltage gain of less than 1.
    b. some distortion.
    c. a high input resistance.
    d. all of these.

14. When a JFET is cut off, the depletion layers are
    a. far apart.
    b. close together.
    c. touching.
    d. conducting.

15. When the gate voltage becomes more negative in an n-channel JFET, the channel between the depletion layers
    a. shrinks.
    b. expands.
    c. conducts.
    d. stops conducting.

16. Self-bias produces
    a. positive feedback.
    b. negative feedback.
    c. forward feedback.
    d. reverse feedback.

17. To get a negative gate-source voltage in a self-biased JFET circuit, you must have a
    a. voltage divider.
    b. source resistor.
    c. ground.
    d. negative gate supply voltage.

18. Transconductance is measured in
    a. ohms.
    b. amperes.

c. volts.

d. mhos or siemens.

19. Transconductance indicates how effectively the input voltage controls the
    a. voltage gain.
    b. input resistance.
    c. supply voltage.
    d. output current.

20. A type of JFET used as a microwave amplifier is the
    a. MOSFET.
    b. MESFET.
    c. BJT.
    d. SCR.

21. A D-MOSFET can operate in the
    a. depletion mode only.
    b. enhancement mode only.
    c. depletion mode or enhancement mode.
    d. low-impedance mode.

22. When an n-channel D-MOSFET has $I_D > I_{DSS}$, it
    a. will be destroyed.
    b. is operating in the depletion mode.
    c. is forward-biased.
    d. is operating in the enhancement mode.

23. The voltage gain of a D-MOSFET amplifier is dependent on
    a. R_D.
    b. R_L.
    c. g_m.
    d. All of the above.

24. Which of the following devices revolutionized the computer industry?
    a. JFET.
    b. D-MOSFET.
    c. E-MOSFET.
    d. Power FET.

25. The voltage that turns on an EMOS device is the
    a. gate-source cutoff voltage.
    b. pinch-off voltage.
    c. threshold voltage.
    d. knee voltage.

26. An ordinary resistor is an example of
    a. a three-terminal device.
    b. an active load.
    c. a passive load.
    d. a switching device.

27. An E-MOSFET with its gate connected to its drain is an example of
    a. a three-terminal device.
    b. an active load.

c. a passive load.

d. a switching device.

28. An E-MOSFET that operates at cutoff or in the ohmic region is an example of
    a. a current source.
    b. an active load.
    c. a passive load.
    d. a switching device.

29. VMOS devices generally
    a. switch off faster than BJTs.
    b. carry low values of current.
    c. have a negative temperature coefficient.
    d. are used as CMOS inverters.

30. A D-MOSFET is considered to be a
    a. normally off device.
    b. normally on device.
    c. current-controlled device.
    d. high-power switch.

31. CMOS stands for
    a. common MOS.
    b. active-load switching.
    c. p-channel and n-channel devices.
    d. complementary MOS.

32. With active-load switching, the upper E-MOSFET is a
    a. two-terminal device.
    b. three-terminal device.
    c. switch.
    d. small resistance.

33. CMOS devices use
    a. bipolar transistors.
    b. complementary E-MOSFETs.
    c. class A operation.
    d. DMOS devices.

34. The main advantage of CMOS is its
    a. high power rating.
    b. small-signal operation.
    c. switching capability.
    d. low power consumption.

35. Power FETs are
    a. integrated circuits.
    b. small-signal devices.
    c. used mostly with analog signals.
    d. used to switch large currents.

36. When the internal temperature increases in a power FET, the
    a. threshold voltage increases.
    b. gate current decreases.
    c. drain current decreases.
    d. saturation current increases.

37. Most small-signal E-MOSFETs are found in
  a. heavy-current applications.
  b. discrete circuits.
  c. disk drives.
  d. integrated circuits.

38. Most power FETS are
  a. used in high-current applications.
  b. digital computers.
  c. RF stages.
  d. integrated circuits.

39. With CMOS, the upper MOSFET is
  a. a passive load.
  b. an active load.
  c. nonconducting.
  d. complementary.

40. The high output of a CMOS inverter is
  a. $V_{DD}/2$.
  b. V_{GS}.

  c. V_{DS}.
  d. V_{DD}.

41. A type of power FET that has superior high power, voltage, and frequency performance is made of
  a. silicon.
  b. GaAs.
  c. SiGe.
  d. GaN.

42. One key advantage of a differential amplifier is that it
  a. has higher gain.
  b. cancels noise common to both inputs.
  c. has higher input impedance.
  d. uses less power.

43. Transistors replace resistors in most biasing circuits in integrated circuits.
  a. True.
  b. False.

CHAPTER 32 PROBLEMS

SECTION 32.2 Junction FET Basics

32.1 A 2N5458 has a gate current of 1 nA when the reverse voltage is -15 V. What is the input resistance of the gate?

SECTION 32.3 Drain Curves

32.2 A JFET has $I_{DSS} = 20$ mA and $V_P = 4$ V. What is the maximum drain current? The gate-source cutoff voltage? The value of R_{DS}?

32.3 A 2JFET has $I_{DSS} = 16$ mA and $V_{GS(off)} = -2$ V. What is the pinch-off voltage for this JFET? What is the drain-source resistance R_{DS}?

SECTION 32.4 The Transconductance Curve

32.4 A 2N5462 has $I_{DSS} = 16$ mA and $V_{GS(off)} = -6$ V. What are the gate voltage and drain current at the half cutoff point?

32.5 A 2N5670 has $I_{DSS} = 10$ mA and $V_{GS(off)} = -4$ V. What are the gate voltage and drain current at the half cutoff point?

32.6 If a 2N5486 has $I_{DSS} = 14$ mA and $V_{GS(off)} = -4$ V, what is the drain current when $V_{GS} = -1$ V? When $V_{GS} = -3$ V?

SECTION 32.5 Biasing the JFET

32.7 What is the ideal drain voltage in Fig. 32-51?

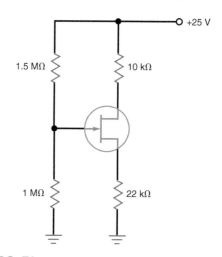

Figure 32-51

32.8 In Fig. 32-52a, the drain current is 1.5 mA. What does V_{GS} equal? What does V_{DS} equal?

32.9 The voltage across the 1 kΩ of Fig. 32-52a is 1.5 V. What is the voltage between the drain and ground?

32.10 In Fig. 32-52a, find V_{GS} and I_D using the transconductance curve of Fig. 32-52b.

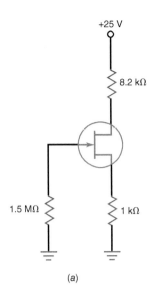

(a)

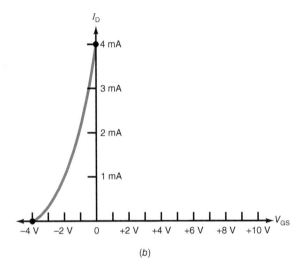

(b)

Figure 32-52

32.11 Change R_S in Fig. 32-52a from 1 kΩ to 2 kΩ. Use the curve of Fig. 32-52b to find V_{GS}, I_D, and V_{DS}.

SECTION 32.6 Transconductance

32.12 A 2N4416 has I_{DSS} = 10 mA and g_{m0} = 4000 μS. What is its gate-source cutoff voltage? What is the value of g_m for V_{GS} = −1 V?

32.13 The JFET of Fig. 32-53 has g_{m0} = 6000 μS. If I_{DSS} = 12 mA, what is the approximate value of I_D for V_{GS} of −2 V? Find the g_m for this I_D.

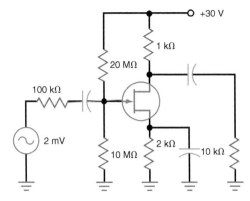

Figure 32-53

SECTION 32.7 JFET Amplifiers

32.14 If g_m = 3000 μS in Fig. 32-53, what is the ac output voltage?

32.15 If the source follower of Fig. 32-54 has g_m = 2000 μS, what is the ac output voltage?

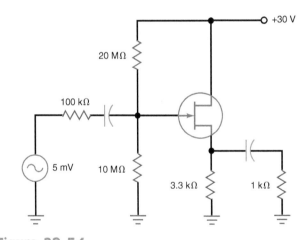

Figure 32-54

SECTION 32.10 D-MOSFET Curves

32.16 An *n*-channel D-MOSFET has the specifications $V_{GS(off)}$ = −2 V and I_{DSS} = 4 mA. Given V_{GS} values of −0.5 V, −1.0 V, −1.5 V, +0.5 V, +1.0 V, and +1.5 V, determine I_D in the depletion mode only.

32.17 Given the same values as in the previous problem, calculate I_D for the enhancement mode only.

SECTION 32.11 Depletion-Mode MOSFET Amplifiers

32.18 The D-MOSFET in Fig. 32-55 has $V_{GS(off)} = -3$ V and $I_{DSS} = 12$ mA. Determine the circuit's drain current and V_{DS} values.

32.19 In Fig. 32-55, what are the values of r_d, A_v, and V_{out} using a g_{mo} of 4000 μS?

32.20 What is the approximate input impedance of Fig. 32-55?

SECTION 32.12 The Ohmic Region

32.21 Calculate $R_{DS(on)}$ for each of these E-MOSFET values:

a. $V_{DS(on)} = 0.1$ V and $I_{D(on)} = 10$ mA

b. $V_{DS(on)} = 0.75$ V and $I_{D(on)} = 100$ mA

32.22 What is the approximate drain current in each of the circuits in Fig. 32-56?

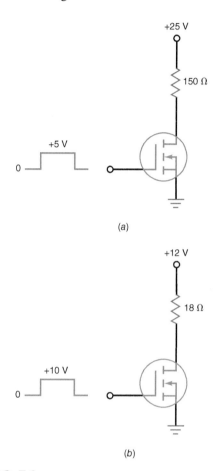

(a)

(b)

Figure 32-56

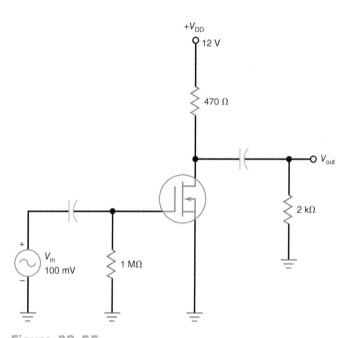

Figure 32-55

Chapter **33**

Bipolar Junction Transistors

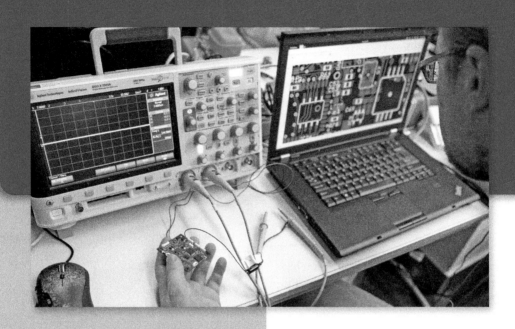

Learning Outcomes

After studying this chapter, you should be able to:

> *Describe* the relationships among the base, emitter, and collector currents of a bipolar junction transistor.

> *Draw* a diagram of the common emitter (CE) circuit and *label* each terminal, voltage, and resistance.

> *Name* the three regions of operation on a bipolar junction transistor.

> *Calculate* the respective CE transistor current and voltage values in a typical circuit

> *Explain* the operation of typical common emitter circuits.

> *Calculate* input and output impeqances and gain of typical amplifier circuits.

> *Explain* the operation and benefits of a differential amplifier.

> *Explain* the operation of the BJT as a switch.

In 1951, William Schockley invented the first junction transistor, a semiconductor device that can amplify (enlarge) electronic signals, such as radio and television signals. The transistor has led to many other semiconductor inventions, including the integrated circuit (IC).

This chapter introduces the bipolar junction transistor (BJT). The word *bipolar* is an abbreviation for "two polarities." This chapter explores how a BJT works and how it can be used as an amplifier and as a switch.

33.1 The Unbiased Transistor

A transistor has three doped regions, as shown in Fig. 33-1. The bottom region is called the **emitter,** the middle region is the **base,** and the top region is the **collector.** In an actual transistor, the base region is much thinner as compared to the collector and emitter regions. The transistor of Fig. 33-1 is an *npn device* because there is a *p* region between two *n* regions. Recall that the majority carriers are free electrons in *n*-type material and holes in *p*-type material.

Transistors are also manufactured as *pnp* devices. A *pnp* transistor has an *n* region between two *p* regions. Our discussions will focus on the *npn* transistor.

Emitter and Collector Diodes

The transistor of Fig. 33-1 has two junctions: one between the emitter and the base, and another between the collector and the base. Because of this, a transistor is like two back-to-back diodes. The lower diode is called the *emitter-base diode,* or simply the **emitter diode.** The upper diode is called the *collector-base diode,* or the **collector diode.**

Before and After Diffusion

Figure 33-1 shows the transistor regions before diffusion has occurred. Free electrons in the *n* region diffuse across the junction and recombine with the holes in the *p* region. Visualize the free electrons in each *n* region crossing the junction and recombining with holes.

The result is two depletion layers, as shown in Fig. 33-2. For each of these depletion layers, the barrier potential is approximately 0.7 V at 25°C for a silicon transistor.

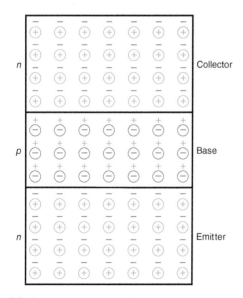

Figure 33-1   Structure of a bipolar transistor.

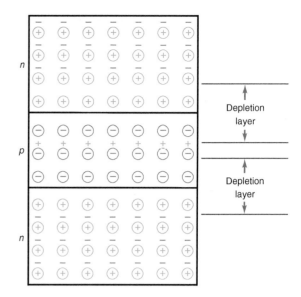

Figure 33-2   Depletion layers.

33.2 The Biased Transistor

An unbiased transistor is like two back-to-back diodes. Each diode has a barrier potential of approximately 0.7 V. When you connect external voltage sources to the transistor, you will get currents through the different parts of the transistor.

Emitter Electrons

Figure 33-3 shows a biased transistor. The minus signs represent free electrons. The heavily doped emitter has the following job: to emit or inject its free electrons into the base. The lightly doped base also has a well-defined purpose: to pass emitter-injected electrons on to the collector. The collector is so named because it collects or gathers most of the electrons from the base.

Figure 33-3 is the usual way to bias a transistor. The left source V_{BB} of Fig. 33-3 forward-biases the emitter diode, and the right source V_{CC} reverse-biases the collector diode.

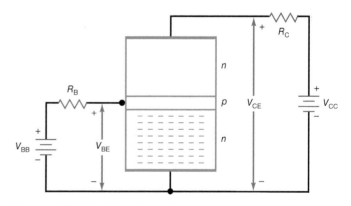

Figure 33-3   Biased transistor.

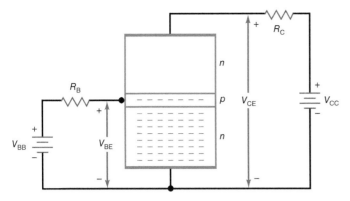

Figure 33-4 Emitter injects free electrons into base.

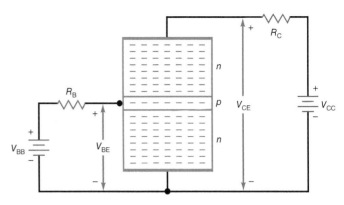

Figure 33-5 Free electrons from base flow into collector.

Base Electrons

At the instant that forward bias is applied to the emitter diode of Fig. 33-3, the electrons in the emitter have not yet entered the base region. If V_{BB} is greater than the emitter-base barrier potential, emitter electrons will enter the base region, as shown in Fig. 33-4. Theoretically, these free electrons can flow in either of two directions. First, they can flow to the left and out of the base, passing through R_B on the way to the positive source terminal. Second, the free electrons can flow into the collector.

Which way will the free electrons go? Most will continue on to the collector. Why? Two reasons: The base is *lightly doped* and *very thin*. The light doping means that the free electrons have a long lifetime in the base region. The very thin base means that the free electrons have only a short distance to go to reach the collector. For these two reasons, almost all the emitter-injected electrons pass through the base to the collector.

Only a few free electrons will recombine with holes in the lightly doped base of Fig. 33-4. Then, as valence electrons, they will flow through the base resistor to the positive side of the V_{BB} supply.

Collector Electrons

Almost all the free electrons go into the collector, as shown in Fig. 33-5. Once they are in the collector, they feel the attraction of the V_{CC} source voltage. Because of this, the free electrons flow through the collector and through R_C until they reach the positive terminal of the collector supply voltage.

Here's a summary of what's going on: In Fig. 33-5, V_{BB} forward-biases the emitter diode, forcing the free electrons in the emitter to enter the base. The thin and lightly doped base gives almost all these electrons enough time to diffuse into the collector. These electrons flow through the collector, through R_C, and into the positive terminal of the V_{CC} voltage source.

33.3 Transistor Currents

Figure 33-6a and b shows the schematic symbol for an *npn* transistor. If you prefer conventional flow, use Fig. 33-6a. If you prefer electron flow, use Fig. 33-6b. In Fig. 33-6, there are three different currents in a transistor: emitter current I_E, base current I_B, and collector current I_C.

How the Currents Compare

Because the emitter is the source of the electrons, it has the largest current. Since most of the emitter electrons flow to the collector, the collector current is almost as large as the emitter current. The base current is very small by comparison, *often less than 1% of the collector current.*

Relation of Currents

Recall Kirchhoff's current law. It says that the sum of all currents into a point or junction equals the sum of all currents out of the point or junction. When applied to a transistor, Kirchhoff's current law gives us this important relationship:

$$I_E = I_C + I_B \tag{33-1}$$

This says that the emitter current is the sum of the collector current and the base current. Since the base current is so

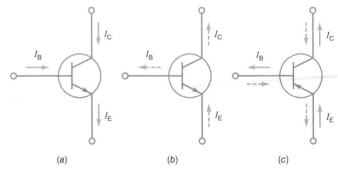

Figure 33-6 Three transistor currents: (*a*) conventional flow, (*b*) electron flow, and (*c*) *pnp* currents.

small, the collector current approximately equals the emitter current,

$$I_C \approx I_E$$

and the base current is much smaller than the collector current,

$$I_B \ll I_C$$

(*Note:* $\ll$ means *much smaller than*.)

Figure 33-6c shows the schematic symbol for a *pnp* transistor and its currents. Notice that the current directions are opposite that of the *npn*. Again notice that Formula (33-1) holds true for the *pnp* transistor currents.

Beta

The **dc beta** (symbolized β_{dc}) of a transistor is defined as the ratio of the dc collector current to the dc base current:

$$\beta_{dc} = \frac{I_C}{I_B} \qquad \text{(33-2)}$$

The dc beta is also known as the **current gain** because a small base current controls a much larger collector current.

The current gain is a major advantage of a transistor and has led to all kinds of applications. For low-power transistors (under 1 W), the current gain is typically 100 to 300. High-power transistors (over 1 W) usually have current gains of 20 to 100.

Two Derivations

Formula (33-2) may be rearranged into two equivalent forms. First, when you know the value of β_{dc} and I_B, you can calculate the collector current with this derivation:

$$I_C = \beta_{dc} I_B \qquad \text{(33-3)}$$

Second, when you have the value of β_{dc} and I_C, you can calculate the base current with this derivation:

$$I_B = \frac{I_C}{\beta_{dc}} \qquad \text{(33-4)}$$

EXAMPLE 33-1

A transistor has a collector current of 10 mA and a base current of 40 μA. What is the current gain of the transistor?

Answer:
Divide the collector current by the base current to get

$$\beta_{dc} = \frac{10 \text{ mA}}{40 \text{ } \mu\text{A}} = 250$$

EXAMPLE 33-2

A transistor has a current gain of 175. If the base current is 0.1 mA, what is the collector current?

Answer:
Multiply the current gain by the base current to get

$$I_C = 175(0.1 \text{ mA}) = 17.5 \text{ mA}$$

EXAMPLE 33-3

A transistor has a collector current of 2 mA. If the current gain is 135, what is the base current?

Answer:
Divide the collector current by the current gain to get

$$I_B = \frac{2 \text{ mA}}{135} = 14.8 \text{ } \mu\text{A}$$

33.4 The CE Connection

There are three useful ways to connect a transistor: with a common emitter (CE), a common collector (CC), or a common base (CB). In this chapter, we will focus on the CE connection because it is the most widely used.

Common Emitter

In Fig. 33-7a, the common or ground side of each voltage source is connected to the emitter. Because of this, the

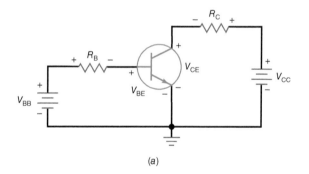

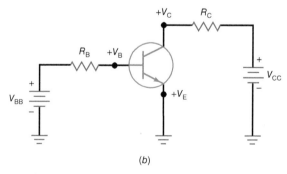

Figure 33-7 CE connection. (*a*) Basic circuit. (*b*) Circuit with grounds.

circuit is called a **common emitter (CE)** connection. The circuit has two loops. The left loop is the base loop, and the right loop is the collector loop.

In the base loop, the V_{BB} source forward-biases the emitter diode with R_B as a current-limiting resistance. By changing V_{BB} or R_B, we can change the base current. Changing the base current will change the collector current. In other words, *the base current controls the collector current*. This is important. It means that a small current (base) controls a large current (collector).

In the collector loop, a source voltage V_{CC} reverse-biases the collector diode through R_C. The supply voltage V_{CC} must reverse-bias the collector diode as shown, or else the transistor won't work properly. Stated another way, the collector must be positive in Fig. 33-7a to collect most of the free electrons injected into the base.

In Fig. 33-7a, the flow of base current in the left loop produces a voltage across the base resistor R_B with the polarity shown. Similarly, the flow of collector current in the right loop produces a voltage across the collector resistor R_C with the polarity shown.

Double Subscripts

Double-subscript notation is used with transistor circuits. When the subscripts are the same, the voltage represents a source (V_{BB} and V_{CC}). When the subscripts are different, the voltage is between the two points (V_{BE} and V_{CE}).

For instance, the subscripts of V_{BB} are the same, which means that V_{BB} is the base voltage source. Similarly, V_{CC} is the collector voltage source. On the other hand, V_{BE} is the voltage between points B and E, between the base and the emitter. Likewise, V_{CE} is the voltage between points C and E, between the collector and the emitter.

Single Subscripts

Single subscripts are used for node voltages, that is, voltages between the subscripted point and ground. For instance, if we redraw Fig. 33-7a with grounds, we get Fig. 33-7b. Voltage V_B is the voltage between the base and ground, voltage V_C is the voltage between the collector and ground, and voltage V_E is the voltage between the emitter and ground. (In this circuit, V_E is zero.)

You can calculate a double-subscript voltage of different subscripts by subtracting its single-subscript voltages. Here are three examples:

$$V_{CE} = V_C - V_E$$
$$V_{CB} = V_C - V_B$$
$$V_{BE} = V_B - V_E$$

This is how you could calculate the double-subscript voltages for any transistor circuit: Since V_E is zero in this CE connection (Fig. 33-7b), the voltages simplify to

$$V_{CE} = V_C$$
$$V_{CB} = V_C - V_B$$
$$V_{BE} = V_B$$

33.5 The Base Curve

What do you think the graph of I_B versus V_{BE} looks like? It looks like the graph of an ordinary diode as shown in Fig. 33-8a. And why not? This is a forward-biased emitter diode, so we would expect to see the usual diode graph of current versus voltage. What this means is that we can use any of the diode approximations discussed earlier.

Applying Ohm's law to the base resistor of Fig. 33-7b gives this derivation:

$$I_B = \frac{V_{BB} - V_{BE}}{R_B} \qquad \textbf{(33-5)}$$

If you use an ideal diode, $V_{BE} = 0$. A good approximation is $V_{BE} = 0.7$ V.

Most of the time, you will find this approximation to be the best compromise between the speed of using the ideal diode and accuracy of higher approximations. All you need to remember for the approximation is that V_{BE} is 0.7 V, as shown in Fig. 33-8a.

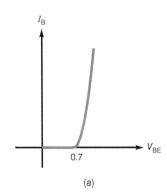

(a)

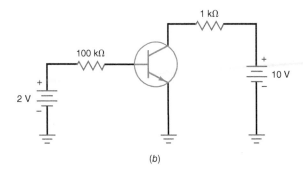

(b)

Figure 33-8 (a) Diode curve. (b) Example.

EXAMPLE 33-4

Use the approximation to calculate the base current in Fig. 33-8b. What is the voltage across the base resistor? The collector current if $\beta_{dc} = 200$?

Answer:

The base source voltage of 2 V forward-biases the emitter diode through a current-limiting resistance of 100 kΩ. Since the emitter diode has 0.7 V across it, the voltage across the base resistor is

$$V_{BB} - V_{BE} = 2\,V - 0.7\,V = 1.3\,V$$

The current through the base resistor is

$$I_B = \frac{V_{BB} - V_{BE}}{R_B} = \frac{1.3\,V}{100\,k\Omega} = 13\,\mu A$$

With a current gain of 200, the collector current is

$$I_C = \beta_{dc} I_B = (200)(13\,\mu A) = 2.6\,mA$$

33.6 Collector Curves

Now let's look at the collector loop. We can vary V_{BB} and V_{CC} in Fig. 33-9a to produce different transistor voltages and currents. By measuring I_C and V_{CE}, we can get data for a graph of I_C versus V_{CE}.

For instance, suppose we change V_{BB} as needed to get $I_B = 10\,\mu A$. With this fixed value of base current, we can now vary V_{CC} and measure I_C and V_{CE}. Plotting the data gives the graph shown in Fig. 33-9b. (*Note:* This graph is for a 2N3904, a widely used low-power transistor. With other transistors, the numbers may vary but the shape of the curve will be similar.)

When V_{CE} is zero, the collector diode is not reverse-biased. This is why the graph shows a collector current of zero when V_{CE} is zero. When V_{CE} increases from zero, the collector current rises sharply in Fig. 33-9b. When V_{CE} is a few tenths of a volt, the collector current becomes *almost constant* and equal to 1 mA.

The constant-current region in Fig. 33-9b is related to our earlier discussions of transistor action. After the collector diode becomes reverse-biased, it is gathering all the electrons that reach its depletion layer. Further increases in V_{CE} cannot increase the collector current. Why? Because the collector can collect only those free electrons that the emitter injects into the base. The number of these injected electrons depends only on the base circuit, not on the collector circuit. This is why Fig. 33-9b shows a constant collector current between a V_{CE} of less than 1 V to a V_{CE} of more than 40 V.

If V_{CE} is greater than 40 V, the collector diode breaks down and normal transistor action is lost. The transistor is not intended to operate in the breakdown region. For this reason, one of the maximum ratings to look for on a transistor data sheet is the collector-emitter breakdown voltage $V_{CE(max)}$. If the transistor breaks down, it will be destroyed.

Collector Voltage and Power

Kirchhoff's voltage law says that the sum of voltages around a loop or closed path is equal to zero. When applied to the collector circuit of Fig. 33-9a, Kirchhoff's voltage law gives us this derivation:

$$V_{CE} = V_{CC} - I_C R_C \qquad \text{(33-6)}$$

This says that the collector-emitter voltage equals the collector supply voltage minus the voltage across the collector resistor.

In Fig. 33-9a, the transistor has a power dissipation of approximately

$$P_D = V_{CE} I_C \qquad \text{(33-7)}$$

This says that the transistor power equals the collector-emitter voltage times the collector current. This power dissipation causes the junction temperature of the collector diode to increase. The higher the power, the higher the junction temperature.

Transistors will burn out when the junction temperature is between 150 and 200°C. One of the most important pieces of information on a data sheet is the maximum power

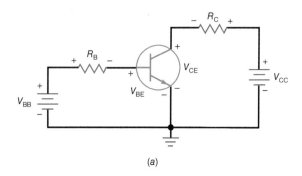

(a)

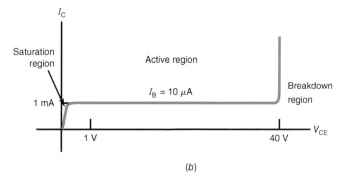

(b)

Figure 33-9 (a) Basic transistor circuit. (b) Collector curve.

rating $P_{D(max)}$. The power dissipation given by Formula (33-7) must be less than $P_{D(max)}$. Otherwise, the transistor will be destroyed.

Regions of Operation

The curve of Fig. 33-9*b* has different regions where the action of a transistor changes. First, there is the region in the middle where V_{CE} is between 1 and 40 V. This represents the normal operation of a transistor. In this region, the emitter diode is forward-biased, and the collector diode is reverse-biased. Furthermore, the collector is gathering almost all the electrons that the emitter has sent into the base. This is why changes in collector voltage have no effect on the collector current. This region is called the active region. Graphically, the active region is the horizontal part of the curve. In other words, the collector current is *constant* in this region. This is the region where amplification occur.

Another region of operation is the **breakdown region.** The transistor should never operate in this region because it will be destroyed.

Third, there is the early rising part of the curve, where V_{CE} is between 0 V and a few tenths of a volt. This sloping part of the curve is called the **saturation region.** In this region, the collector diode has insufficient positive voltage to collect all the free electrons injected into the base. In this region, the base current I_B is larger than normal and the current gain β_{dc} is smaller than normal. In this region, the transistor is used as a switch.

More Curves

If we measure I_B and V_{CE} for $I_B = 20\ \mu A$, we can plot the second curve of Fig. 33-10. The curve is similar to the first curve, except that the collector current is 2 mA in the active region. Again, the collector current is constant in the active region.

When we plot several curves for different base currents, we get a set of collector curves like those in Fig. 33-10. Another way to get this set of curves is with a *curve tracer* (a test instrument that can display I_C versus V_{CE} for a transistor). In the active region of Fig. 33-10, each collector current is 100 times greater than the corresponding base current. For instance, the top curve has a collector current of 7 mA and a base current of 70 μA. This gives a current gain of

$$\beta_{dc} = \frac{I_C}{I_B} = \frac{7\ mA}{70\ \mu A} = 100$$

If you check any other curve, you get the same result: a current gain of 100.

With other transistors, the current gain may be different from 100, but the shape of the curves will be similar. All transistors have an active region, a saturation region, and a breakdown region. The active region is the most important because amplification (enlargement) of signals is possible in the active region.

Cutoff Region

Figure 33-10 has an unexpected curve, the one on the bottom. This represents a fourth possible region of operation. Notice that the base current is zero, but there still is a small collector current. On a curve tracer, this current is usually so small that you cannot see it. We have exaggerated the bottom curve by drawing it larger than usual. This bottom curve is called the **cutoff region** of the transistor, and the small collector current is called the *collector cutoff current.*

Why does the collector cutoff current exist? Because the collector diode has reverse minority-carrier current and surface-leakage current. In a well-designed circuit, the collector cutoff current is small enough to ignore. For instance, a 2N3904 has a collector cutoff current of 50 nA. If the actual collector current is 1 mA, ignoring a collector cutoff current of 50 nA produces a calculation error of less than 5%.

Recap

A transistor has four distinct operating regions: *active, cutoff, saturation,* and *breakdown.* Transistors operate in the active region when they are used to amplify weak signals. Sometimes, the active region is called the *linear region* because changes in the input signal produce proportional changes in the output signal. The saturation and cutoff regions are useful in circuits referred to as **switching circuits.**

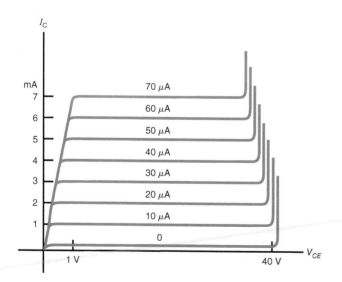

Figure 33-10   Set of collector curves.

EXAMPLE 33-5

The transistor of Fig. 33-11a has $\beta_{dc} = 300$. Calculate I_B, I_C, V_{CE}, and P_D.

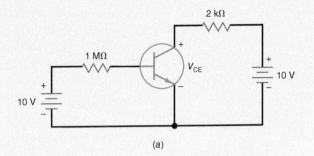

(a)

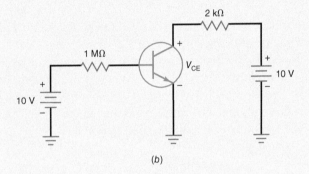

(b)

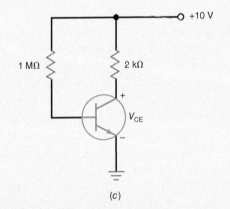

(c)

Figure 33-11 Transistor circuit. (a) Basic schematic diagram. (b) Circuit with grounds. (c) Simplified schematic diagram.

Answer:

Figure 33-11b shows the same circuit with grounds. The base current equals

$$I_B = \frac{V_{BB} - V_{BE}}{R_B} = \frac{10\ V - 0.7\ V}{1\ M\Omega} = 9.3\ \mu A$$

The collector current is

$$I_C = \beta_{dc}I_B = (300)(9.3\ \mu A) = 2.79\ mA$$

and the collector-emitter voltage is

$$V_{CE} = V_{CC} - I_C R_C = 10\ V - (2.79\ mA)(2\ k\Omega) = 4.42\ V$$

The collector power dissipation is

$$P_D = V_{CE}I_C = (4.42\ V)(2.79\ mA) = 12.3\ mW$$

Incidentally, when both the base and the collector supply voltages are equal, as in Fig. 33-11b, you usually see the circuit drawn in the simpler form of Fig. 33-11c.

33.7 Reading Data Sheets

Although you rarely need to read a transistor data sheet, it is good to know what to look for. Here are a few important factors.

Small-signal transistors can dissipate less than a watt; **power transistors** can dissipate more than a watt. When you look at a data sheet for either type of transistor, you should start with the maximum ratings because these are the limits on the transistor currents, voltages, and other quantities.

Breakdown Ratings

In the data sheet for a 2N3904 transistor, the following maximum ratings are given:

$$\begin{array}{ll} V_{CEO} & 40\ V \\ V_{CBO} & 60\ V \\ V_{EBO} & 6\ V \end{array}$$

These voltage ratings are reverse breakdown voltages, and V_{CEO} is the voltage between the collector and the emitter with the base open. The second rating is V_{CBO}, which stands for the voltage from collector to base with the emitter open. Likewise, V_{EBO} is the maximum reverse voltage from emitter to base with the collector open. As usual, a conservative design never allows voltages to get even close to the foregoing maximum ratings. If you recall, even getting close to maximum ratings can shorten the lifetime of some devices.

Maximum Current and Power

Also on the data sheet are these values:

$$\begin{array}{ll} I_C & 200\ mA \\ P_D & 625\ mW \end{array}$$

Here I_C is the maximum dc collector current rating. This means that a 2N3904 can handle up to 200 mA of direct current, provided the power rating is not exceeded. The next rating, P_D, is the maximum power rating of the device. This power rating depends on whether any attempt is being made to keep the transistor cool. If the transistor is not being fan-cooled and does not have a heat sink (discussed below), its case temperature T_C will be much higher than the ambient temperature T_A.

In most applications, a small-signal transistor like the 2N3904 is not fan-cooled and it does not have a heat sink. In this case, the 2N3904 has a power rating of 625 mW when the ambient temperature T_A is 25°C.

The case temperature T_C is the temperature of the transistor package or housing. In most applications, the case temperature will be higher than 25°C because the internal heat of the transistor increases the case temperature.

The only way to keep the case temperature at 25°C when the ambient temperature is 25°C is by fan cooling or by using a large heat sink. If fan cooling or a large heat sink is used, it is possible to reduce the temperature of the transistor case to 25°C. For this condition, the power rating can be increased to 1.5 W.

Heat Sinks

One way to increase the power rating of a transistor is to get rid of the internal heat faster. This is the purpose of a **heat sink** (a mass of metal). If we increase the surface area of the transistor case, we allow the heat to escape more easily into the surrounding air. For instance, Fig. 33-12*a* shows one type of heat sink. When this is pushed onto the transistor

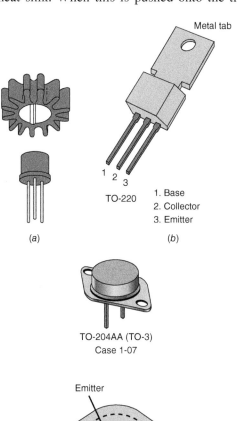

1. Base
2. Collector
3. Emitter

TO-220

(a)   (b)

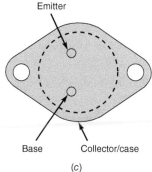

TO-204AA (TO-3)
Case 1-07

Emitter

Base   Collector/case

(c)

Figure 33-12 (*a*) Push-on heat sink. (*b*) Power-tab transistor. (*c*) Power transistor with collector connected to case.

case, heat radiates more quickly because of the increased surface area of the fins.

Figure 33-12*b* shows another approach. This is the outline of a power-tab transistor. A metal tab provides a path out of the transistor for heat. This metal tab can be fastened to the chassis of electronic equipment. Because the chassis is a massive heat sink, heat can easily escape from the transistor to the chassis.

Large power transistors like Fig. 33-12*c* have the collector connected to the case to let heat escape as easily as possible. The transistor case is then fastened to the chassis. To prevent the collector from shorting to chassis ground, a thin insulating washer and heat-conducting compound is used between the transistor case and the chassis. The important idea here is that heat can leave the transistor more rapidly, which means that the transistor has a higher power rating at the same ambient temperature. Sometimes, the transistor is fastened to a large heat sink with fins; this is even more efficient in removing heat from the transistor.

No matter what kind of heat sink is used, the purpose is to lower the case temperature because this will lower the internal or junction temperature of the transistor. The data sheet includes other quantities called **thermal resistances.** These allow a designer to work out the case temperature for different heat sinks.

Current Gain

In another system of analysis called **h parameters**, h_{FE} rather than β_{dc} is defined as the symbol for current gain. The two quantities are equal:

$$\beta_{dc} = h_{FE} \qquad (33\text{-}8)$$

Remember this relation because data sheets use the symbol h_{FE} for the current gain.

In the section labeled "On Characteristics," the data sheet of a 2N3904 lists the values of h_{FE} as follows:

I_C, mA	Min. h_{FE}	Max. h_{FE}
0.1	40	—
1	70	—
10	100	300
50	60	—
100	30	—

The 2N3904 works best when the collector current is in the vicinity of 10 mA. At this level of current, the minimum current gain is 100 and the maximum current gain is 300. What does this mean? It means that if you mass-produce a circuit using 2N3904s and a collector current of 10 mA, some of the transistors will have a current gain as low as 100, and some will have a current gain as high as 300. Most of the transistors will have a current gain in the middle of this range.

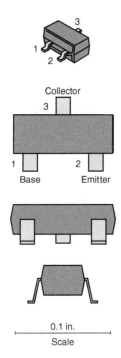

Figure 33-13 The SOT-23 package is suitable for SM transistors with power ratings less than 1 W.

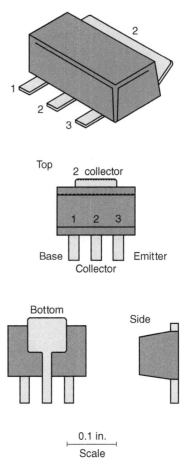

Figure 33-14 The SOT-223 package is designed to dissipate the heat generated by transistors operating in the 1-W range.

33.8 Surface-Mount Transistors

Surface-mount transistors are usually found in a simple three-terminal, gull-wing package. The SOT-23 package is the smaller of the two and is used for transistors rated in the milliwatt range. The SOT-223 is the larger package and is used when the power rating is about 1 W.

Figure 33-13 shows a typical SOT-23 package. Viewed from the top, the terminals are numbered in a counterclockwise direction, with terminal 3 the lone terminal on one side. The terminal assignments are fairly well standardized for bipolar transistors: 1 is the base, 2 is the emitter, and 3 is the collector. (The usual terminal assignments for FETs: 1 is the drain, 2 is the source, and 3 is the gate.)

The SOT-223 package in Fig. 33-14 is designed to dissipate the heat generated by transistors operating in the 1-W range. This package has a larger surface area than the SOT-23; this increases its ability to dissipate heat. Some of the heat is dissipated from the top surface, and much is carried away by the contact between the device and the circuit board below. The special feature of the SOT-223 case, however, is the extra collector tab that extends from the side opposite the main terminals. The bottom view in Fig. 33-14 shows that the two collector terminals are electrically identical.

The standard terminal assignments are different for the SOT-23 and SOT-223 packages. The three terminals located on one edge are numbered in sequence, from left to right as viewed from the top. Terminal 1 is the base, 2 is the collector

(electrically identical to the large tab at the opposite edge), and 3 is the emitter.

The SOT-23 packages are too small to have any standard part identification codes printed on them. Usually the only way to determine the standard identifier is by noting the part number printed on the circuit board and then consulting the parts list for the circuit. SOT-223 packages are large enough to have identification codes printed on them, but these codes are rarely standard transistor identification codes. The typical procedure for learning more about a transistor in an SOT-223 package is the same as for the smaller SOT-23 configurations.

Occasionally a circuit uses SOIC packages that house multiple transistors. The SOIC package resembles the tiny dual in-line package commonly used for ICs and the older feed-through circuit board technology. The terminals on the SOIC, however, have the gull-wing shape required for SM technology.

33.9 Basic Biasing

Biasing a transistor is done by applying one or more dc voltages to the device to set the initial or quiescent operating conditions. This is called the Q point. For a bipolar

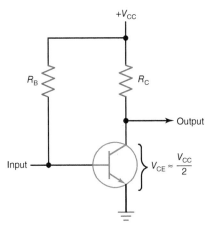

Figure 33-15 Basic base bias.

transistor used as an amplifier, this means setting the base current to a value that puts the Q point into roughly the center of the active region of operation. Then as the small signal input varies the base current, the collector current will vary in direct proportion without going into cutoff or saturation.

The simplest bias arrangement is shown in Fig. 33-15. Note that both the base and collector receive their bias voltage from a single supply, V_{CC}. The Q point is set from the currents that make the collector output, V_{CE}, about one-half of V_{CC}. Both R_C and R_B are chosen accordingly.

Assume that we want to establish a collector current of 10 mA in a 2N3904 with a 12-V supply. At this collector current, the h_{FE} or β_{dc} is approximately 100. This means that we need a base current of

$$I_B = I_C/h_{FE} = 10/100 = 0.1 \text{ mA} \quad \text{or} \quad 100 \, \mu\text{A}$$

With a base supply of 12 V, we will need a base resistor of

$$R_B = (12 - 0.7)/(100 \, \mu\text{A}) = 11.3/(100 \, \mu\text{A}) = 113 \text{ k}\Omega$$

A standard resistor value of 110 or 120 kΩ could be used.

Having 10 mA flowing in R_C with V_{CE} approximately one-half of V_{CC}, or 6 V, leaves 6 V across R_C. The value of R_C is

$$R_C = 6/(10 \text{ mA}) = 600 \, \Omega$$

Now when an input to be amplified is applied to the base, I_C will vary above and below 10 mA, producing an output voltage variation at the collector. The collector output, V_{CE}, can vary roughly from V_{CC} to zero.

This bias method is simple and effective, but it has a major disadvantage. If β_{dc} is anything other than 100, the Q point will change. This will affect the output voltage variation and could introduce clipping distortion. Note that β_{dc} variations are very common.

The β_{dc} varies with individual transistors of the same type due to manufacturing differences. Also, β_{dc} increases with temperature. Therefore, if you have to replace a

transistor even if it is the same part number, you will get a new operating point and potentially poor or unacceptable performance.

In Fig. 33-15, we set the base current to 100 μA. If we replace the transistor with one having an h_{FE} of 50, the collector will be

$$I_C = 50(100 \, \mu\text{A}) = 5 \text{ mA}$$

The output voltage will be

$$V_{CE} = 12 - (5 \text{ mA})(600) = 12 - (3 \text{ A}) = 9 \text{ V}$$

If h_{FE} is 150, the collector current will be

$$I_C = 150(100 \, \mu\text{A}) = 15 \text{ mA}$$

With this condition, the output voltage will be

$$V_{CE} = 12 - (15 \text{ mA})(600) = 12 - (9 \text{ A}) = 3 \text{ V}$$

This is an enormous range that will affect the amplifier performance. Obviously, another biasing method is needed.

33.10 Emitter Bias

Emitter bias solves the current gain sensitivity problem. Figure 33-16 shows **emitter bias.** As you can see, the bias resistor has been moved from the base circuit to the emitter circuit. The Q point of this new circuit is now rock solid. When the current gain changes from 50 to 150, the Q point shows almost no movement.

Basic Idea

The base supply voltage, V_{BB}, is now applied directly to the base. The emitter is no longer grounded. Now the emitter is above the ground and has a voltage given by

$$V_E = V_{BB} - V_{BE} \tag{33-9}$$

Finding the Q Point

Let us analyze the emitter-biased circuit of Fig. 33-17. The base supply voltage is only 5 V, so the voltage between the base and ground is 5 V. From now on, we refer to this

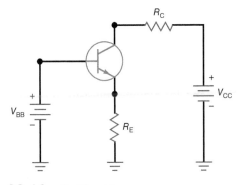

Figure 33-16 Emitter bias.

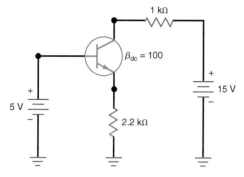

Figure 33-17 Finding the Q point.

base-to-ground voltage as the *base voltage*, or V_B. The voltage across the base-emitter terminals is 0.7 V. We refer to this voltage as the *base-emitter voltage*, or V_{BE}.

The voltage between the emitter and ground is called the *emitter voltage*. It equals

$$V_E = 5\ V - 0.7\ V = 4.3\ V$$

This voltage is across the emitter resistance, so we can use Ohm's law to find the emitter current:

$$I_E = \frac{4.3\ V}{2.2\ k\Omega} = 1.95\ mA$$

This means that the collector current is 1.95 mA to a close approximation. When this collector current flows through the collector resistor, it produces a voltage drop of 1.95 V. Subtracting this from the collector supply voltage gives the voltage between the collector and ground:

$$V_C = 15\ V - (1.95\ mA)(1\ k\Omega) = 13.1\ V$$

From now on, we will refer to this collector-to-ground voltage as the *collector voltage*.

This is the voltage a troubleshooter would measure when testing a transistor circuit. One lead of the voltmeter would be connected to the collector, and the other lead would be connected to ground. If you want the collector-emitter voltage, you have to subtract the emitter voltage from the collector voltage as follows:

$$V_{CE} = 13.1\ V - 4.3\ V = 8.8\ V$$

So the emitter-biased circuit of Fig. 33-17 has a Q point with these coordinates: $I_C = 1.95$ mA and $V_{CE} = 8.8$ V.

The collector-emitter voltage is the voltage used for drawing load lines and for reading transistor data sheets. As a formula,

$$V_{CE} = V_C - V_E \qquad \textbf{(33-10)}$$

Here is why emitter bias excels. The Q point of an emitter-biased circuit is immune to changes in current gain.

The proof lies in the process used to analyze the circuit. Here are the steps we used earlier:

1. Get the emitter voltage.
2. Calculate the emitter current.
3. Find the collector voltage.
4. Subtract the emitter from the collector voltage to get V_{CE}.

At no time do we need to use the current gain in the foregoing process. Since we don't use it to find the emitter current, collector current, and so on, the exact value of current gain no longer matters.

By moving the resistor from the base to the emitter circuit, we force the base-to-ground voltage to equal the base supply voltage. Before, almost all this supply voltage was across the base resistor, setting up a *fixed base current*. Now, all this supply voltage minus 0.7 V is across the emitter resistor, setting up a *fixed emitter current*.

33.11 Voltage Divider Bias

Figure 33-18a shows the most widely used biasing circuit. Notice that the base circuit contains a voltage divider (R_1 and R_2). Because of this, the circuit is called **voltage divider bias (VDB).**

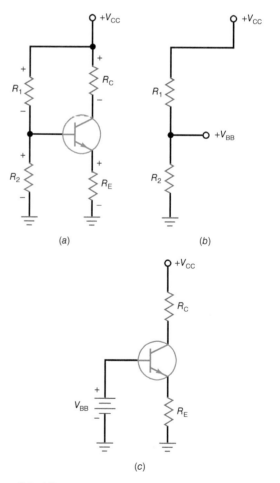

Figure 33-18 Voltage divider bias. (*a*) Circuit. (*b*) Voltage divider. (*c*) Simplified circuit.

Simplified Analysis

For troubleshooting and preliminary analysis, use the following method. In any well-designed VDB circuit, the base current is much smaller than the current through the voltage divider. Since the base current has a negligible effect on the voltage divider, we can mentally open the connection between the voltage divider and the base to get the equivalent circuit of Fig. 33-18b. In this circuit, the output of the voltage divider is

$$V_{BB} = \frac{R_2}{R_1 + R_2} V_{CC}$$

Ideally, this is the base supply voltage as shown in Fig. 33-18c.

As you can see, voltage divider bias is really emitter bias in disguise. In other words, Fig. 33-18c is an equivalent circuit for Fig. 33-18a. This is why VDB sets up a fixed value of emitter current, resulting in a solid Q point that is independent of the current gain.

There is an error in this simplified approach, but the crucial point is this: In any well-designed circuit, the error in using Fig. 33-18c is very small. In other words, a designer deliberately chooses circuit values so that Fig. 33-18a acts like Fig. 33-18c.

Conclusion

After you calculate V_{BB}, the rest of the analysis is the same as discussed earlier for emitter bias. Here is a summary of the equations you can use to analyze VDB:

$$V_{BB} = \frac{R_2}{R_1 + R_2} V_{CC}$$

$$V_E = V_{BB} - V_{BE}$$

$$I_E = \frac{V_E}{R_E}$$

$$I_C \approx I_E$$

$$V_C = V_{CC} - I_C R_C$$

$$V_{CE} = V_C - V_E$$

These equations are based on Ohm's and Kirchhoff's laws. Here are the steps in the analysis:

1. Calculate the base voltage V_{BB} out of the voltage divider.
2. Subtract 0.7 V to get the emitter voltage (use 0.3 V for germanium).
3. Divide by the emitter resistance to get the emitter current.
4. Assume that the collector current is approximately equal to the emitter current.
5. Calculate the collector-to-ground voltage by subtracting the voltage across the collector resistor from the collector supply voltage.

6. Calculate the collector-emitter voltage by subtracting the emitter voltage from the collector voltage.

Since these six steps are logical, they should be easy to remember. After you analyze a few VDB circuits, the process becomes automatic.

EXAMPLE 33-6

What is the collector-emitter voltage in Fig. 33-19?

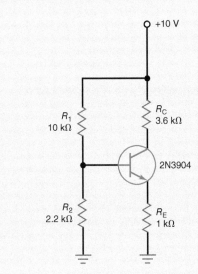

Figure 33-19   Example.

Answer:

The voltage divider produces an unloaded output voltage of

$$V_{BB} = \frac{2.2\ k\Omega}{10\ k\Omega + 2.2\ k\Omega} 10\ V = 1.8\ V$$

Subtract 0.7 V from this to get

$$V_E = 1.8\ V - 0.7\ V = 1.1\ V$$

The emitter current is

$$I_E = \frac{1.1\ V}{1\ k\Omega} = 1.1\ mA$$

Since the collector current almost equals the emitter current, we can calculate the collector-to-ground voltage like this

$$V_C = 10\ V - (1.1\ mA)(3.6\ k\Omega) = 6.04\ V$$

The collector-emitter voltage is

$$V_{CE} = 6.04 - 1.1\ V = 4.94\ V$$

Here is an important point: The calculations in this preliminary analysis do not depend on changes in the transistor, the collector current, or the temperature. This is why the Q point of this circuit is stable, almost rock solid.

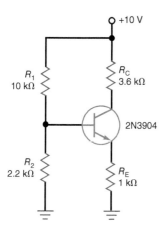

Figure 33-20 Troubleshooting.

33.12 Troubleshooting Transistor Circuits

Let us discuss troubleshooting voltage divider bias because this biasing method is the most widely used. Figure 33-20 shows the VDB circuit analyzed earlier.

The most common problems are no supply voltage, a damaged transistor (open or shorted), or an open resistor or connection. Shorted resistors are extremely rare as are other shorts.

A quick test is to measure V_{CC}. If correct, check V_C which should be about half of V_{CC}. This indicates an operating circuit.

If V_C is equal to V_{CC}, the transistor is not conducting. This may be an open transistor or R_E or lack of bias. If V_C equals V_E, the transistor is shorted or saturated. Another measure is V_{BE}. If it is about 0.7 V, the transistor is conducting.

33.13 *PNP* Transistors

To this point, we have concentrated on bias circuits using *npn* transistors. Many circuits also use *pnp* transistors. This type of transistor is often used when the electronics equipment has a negative power supply. Also, *pnp* transistors are used as complements to *npn* transistors when dual (positive and negative) power supplies are available.

Figure 33-21 shows the structure of a *pnp* transistor along with its schematic symbol. Because the doped regions are

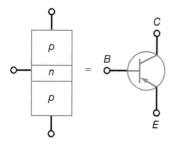

Figure 33-21 *PNP* transistor.

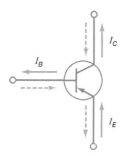

Figure 33-22 *PNP* currents.

of the opposite type, we have to turn our thinking around. Specifically, holes are the majority carriers in the emitter instead of free electrons.

Basic Ideas

Briefly, here is what happens at the atomic level: The emitter injects holes into the base. The majority of these holes flow onto the collector. For this reason, the collector current is almost equal to the emitter current.

Figure 33-22 shows the three transistor currents. Solid arrows represent conventional current, and dashed arrows represent electron flow.

Negative Supply

Figure 33-23a shows voltage divider bias with a *pnp* transistor and a negative supply voltage of −10 V. The 2N3906 is the complement of the 2N3904; that is, its characteristics have the same absolute values as those of the 2N3904, but all currents and voltage polarities are reversed.

The point is this: Whenever you have a circuit with *npn* transistors, you can often use the same circuit with a negative power supply and *pnp* transistors.

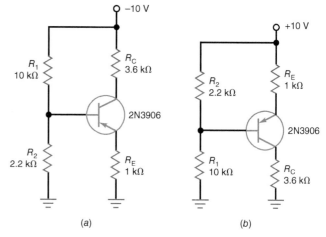

Figure 33-23 *PNP* circuit. (a) Negative supply. (b) Positive supply.

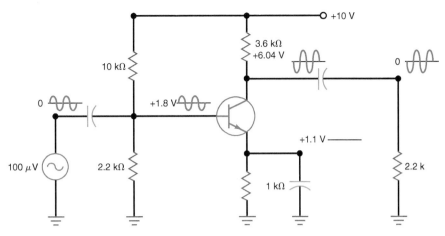

Figure 33-24 VDB amplifier with waveforms.

Positive Power Supply

Positive power supplies are used more often in transistor circuits than negative power supplies. Because of this, you often see *pnp* transistors drawn upside-down, as shown in Fig. 33-23*b*. Here is how the circuit works: The voltage across R_2 is applied to the emitter diode in series with the emitter resistor. This sets up the emitter current. The collector current flows through R_C, producing a collector-to-ground voltage.

33.14 Typical Bipolar Amplifiers

While most modern amplifiers are integrated circuits, you will still occasionally see discrete component transistor amplifiers. The two most common forms are the common emitter amplifier and the emitter follower. In this section, these two basic circuits are analyzed.

Common Emitter Amplifier

A complete amplifier circuit is shown in Fig. 33-24. This is the same amplifier discussed in a previous section. The dc bias conditions are

$$V_B = 1.8 \text{ V}$$
$$V_E = 1.1 \text{ V}$$
$$V_C = 6.04 \text{ V}$$
$$I_C = 1.1 \text{ mA}$$

Note the signal source delivers 100 μV. This signal could come from another amplifier, an oscillator, a microphone, or some other voltage source. The coupling capacitor has a very low reactance at this signal frequency, so it acts as a near short circuit. Therefore, all the 100 μV is applied to the base. The coupling capacitor blocks the dc from the voltage divider from getting back to the signal source. Another coupling capacitor is used to connect the collector output to the load while blocking V_C from the load.

The capacitor across the emitter resistor is called a *bypass capacitor*. It has a very low reactance at the operating frequency, so it effectively shorts the 1-kΩ resistor for ac connecting the emitter to ground.

Without this bypass capacitor, the ac variation across the 1-kΩ emitter resistor becomes negative feedback that subtracts from the input signal, thereby effectively lowering the amplifier gain. The emitter bypass capacitor shorts this feedback to ground, increasing the amplifier gain.

Notice the voltage waveforms in Fig. 33-24. The ac source voltage is a small sinusoidal voltage with an average value of zero. The base voltage is an ac voltage superimposed on a dc voltage of +1.8 V. The collector voltage is an amplified and inverted ac voltage superimposed on the dc collector voltage of +6.04 V. The load voltage is the same as the collector voltage, except that it has an average value of zero. Notice also the voltage on the emitter. It is a pure dc voltage of +1.1 V. There is no ac emitter voltage because emitter is at ac ground, a direct result of using a bypass capacitor.

AC Beta

The current gain in all discussions up to this point has been *dc current gain*. This was defined as

$$\beta_{dc} = \frac{I_C}{I_B} \qquad \textbf{(33-11)}$$

The currents in this formula are the currents at the Q point in Fig. 33-25. Because of the curvature in the graph of I_C versus I_B, the dc current gain depends on the location of the Q point.

Definition

The **ac current gain** is different. It is defined as

$$\beta = \frac{i_c}{i_b} \qquad \textbf{(33-12)}$$

In words, the ac current gain equals the ac collector current divided by the ac base current. In Fig. 33-25, the ac signal uses only a small part of the graph on both sides of the

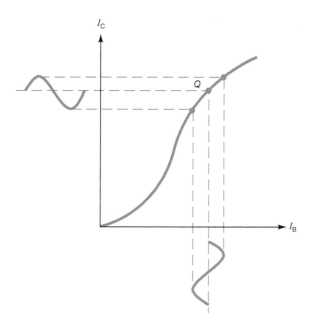

Figure 33-25 AC current gain equals ratio of changes.

Q point. Because of this, the value of the ac current gain is different from the dc current gain, which uses almost all the graph.

Graphically, b equals the slope of the curve at the Q point in Fig. 33-25. If we were to bias the transistor to a different Q point, the slope of the curve would change, which means that b would change. In other words, the value of b depends on the amount of dc collector current.

On data sheets, b_{dc} is listed as h_{FE} and b is shown as h_{fe}. Notice that capital subscripts are used with dc current gain, and lowercase subscripts with ac current gain. The two current gains are comparable in value, not differing by a large amount. For this reason, if you have the value of one, you can use the same value for the other in preliminary analysis.

Notation

To keep dc quantities distinct from ac quantities, it is standard practice to use capital letters and subscripts for dc quantities. For instance, we have been using:

I_E, I_C, and I_B for the dc currents
V_E, V_C, and V_B for the dc voltages
V_{BE}, V_{CE}, and V_{CB} for the dc voltages between terminals

For ac quantities, we will use lowercase letters and subscripts as follows:

i_e, i_c, and i_b for the ac currents
v_e, v_c, and v_b for the ac voltages
v_{be}, v_{ce}, and v_{cb} for the ac voltages between terminals

Also worth mentioning is the use of capital R for dc resistances and lowercase r for ac resistances.

AC Emitter Resistance

The ac emitter resistance is the resistance that an ac input signal sees between the base and emitter. This value is important as it determines the gain of the amplifier.

The following formula is used to calculate the ac emitter resistance:

$$r_e' = \frac{25 \text{ mV}}{I_E} \qquad \textbf{(33-13)}$$

This says that the ac resistance of the emitter diode equals 25 mV divided by the dc emitter current.

This formula is remarkable because of its simplicity and the fact that it applies to all transistor types. It is widely used in industry to calculate a preliminary value for the ac resistance of the emitter diode.

EXAMPLE 33-7

In Fig. 33-24, what does r_e' equal?

Answer:

We analyzed this VDB amplifier earlier and calculated a dc emitter current of 1.1 mA. The ac resistance of the emitter diode is

$$r_e' = \frac{25 \text{ mV}}{1.1 \text{ mA}} = 22.7 \ \Omega$$

Voltage Gain

The voltage gain (A) of a common emitter amplifier is the ratio of the ac output voltage to the ac input voltage, or

$$A_V = \frac{v_{out}}{v_{in}}$$

This gain can be calculated from the circuit values:

$$A_V = \frac{(R_C \parallel R_L)}{r_e'} \qquad \textbf{(33-14)}$$

In Fig. 33-24, the total ac load resistance seen by the collector is the parallel combination of R_C and R_L. This total resistance is called the **ac collector resistance,** symbolized r_c. As a definition,

$$r_c = R_C \parallel R_L \qquad \textbf{(33-15)}$$

Now, we can rewrite Formula (33-14) as

$$A_V = \frac{r_c}{r_e'} \qquad \textbf{(33-16)}$$

The voltage gain equals the ac collector resistance divided by the ac resistance of the emitter diode.

EXAMPLE 33-8

What is the voltage gain in Fig. 33-24? The output voltage across the load resistor?

Answer:
The ac collector resistance is

$$r_c = R_C \| R_L = (3.6 \text{ k}\Omega \| 2.2 \text{ k}\Omega) = 1.37 \text{ k}\Omega$$

The dc emitter current is approximately

$$I_E = \frac{1.8 \text{ V} - 0.7 \text{ V}}{1 \text{ k}\Omega} = 1.1 \text{ mA}$$

The ac resistance of the emitter diode is

$$r_e' = \frac{25 \text{ mV}}{1.1 \text{ mA}} = 22.7 \text{ }\Omega$$

The voltage gain is

$$A_V = \frac{r_c}{r_e'} = \frac{1.37 \text{ k}\Omega}{22.7 \text{ }\Omega} = 60.35$$

The output voltage is

$$v_{out} = A_V v_{in} = (60.35)(100) \text{ }\mu\text{V} = 6.035 \text{ mV}$$

If we remove the emitter bypass capacitor, the resulting ac feedback greatly reduces the gain. It becomes

$$A_V = \frac{r_c}{(r_e + R_E)}$$

where R_E is the emitter resistor value, or 1 kΩ. Recall that r_e is 22.7 Ω. Since r_e is much less than R_E, you can ignore it in some cases and just write the gain as

$$A_V = \frac{r_c}{R_E}$$

EXAMPLE 33-9

What is the voltage gain of the amplifier in Fig. 33-24 if the emitter bypass capacitor is removed?

Answer:
With r_c = 1.37 kΩ and R_E = 1 kΩ, the gain is

$$A_V = \frac{r_c}{R_E} = 1.37 \text{ k}\Omega/1 \text{ }\Omega\text{k} = 1.37$$

The Loading Effect of Input Impedance

Input impedance is the resistance seen by a driving source or generator. In Fig. 33-26a, an ac voltage source v_g has an internal resistance of R_G. (The subscript g stands for "generator," a synonym for *source*.) When the ac generator is not stiff, some of the ac source voltage is dropped across its internal resistance. As a result, the ac voltage between the base and ground is less than ideal.

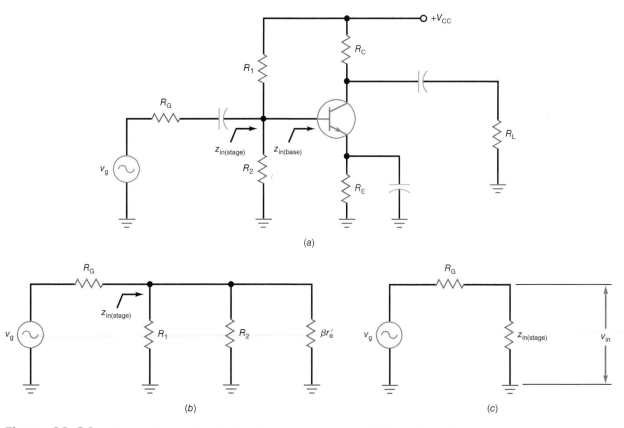

(a)

(b)

(c)

Figure 33-26 CE amplifier. (a) Circuit. (b) AC equivalent circuit. (c) Effect of input impedance.

The ac generator has to drive the input impedance of the stage $z_{\text{in(stage)}}$. This input impedance includes the effects of the biasing resistors R_1 and R_2, in parallel with the input impedance of the base $z_{\text{in(base)}}$. Figure 33-26b illustrates the idea. The input impedance of the stage equals

$$z_{\text{in(stage)}} = R_1 \| R_2 \| \beta r_e'$$

When the generator is not stiff meaning that R_G, is not much less than $z_{\text{in(stage)}}$, the ac input voltage, v_{in}, of Fig. 33-26c is less than v_g. With the voltage divider theorem, we can write

$$v_{\text{in}} = \frac{z_{\text{in(stage)}}}{R_G + z_{\text{in(stage)}}} v_g \qquad \textbf{(33-17)}$$

This equation is valid for any amplifier. After you calculate or estimate the input impedance of the stage, you can determine what the input voltage is. *Note:* The generator is stiff when R_G is less than $0.01z_{\text{in(stage)}}$.

EXAMPLE 33-10

In Fig. 33-27, the ac generator has an internal resistance of 600 Ω. What is the output voltage in Fig. 33-27 if $\beta = 300$?

Answer:
Here are two quantities calculated in earlier examples: $r_e' = 22.7\ \Omega$ and $A_V = 60.35$. We use these values in solving the problem.
When $\beta = 300$, the input impedance of the base is

$$z_{\text{in(base)}} = (300)(22.7\ \Omega) = 6.8\ \text{k}\Omega$$

The input impedance of the stage is

$$z_{\text{in(stage)}} = 10\ \text{k}\Omega \| 2.2\ \text{k}\Omega \| 6.8\ \text{k}\Omega = 1.42\ \text{k}\Omega$$

With Formula (33-17), we can calculate the input voltage:

$$v_{\text{in}} = \frac{1.42\ \text{k}\Omega}{600\ \Omega + 1.42\ \text{k}\Omega}\,2\text{mV} = 1.41\ \text{mV}$$

This is the ac voltage that appears at the base of the transistor, equivalent to the ac voltage across the emitter diode. The amplified output voltage equals

$$v_{\text{out}} = A_V v_{\text{in}} = (60.35)(1.41\ \text{mV}) = 85\ \text{mV}$$

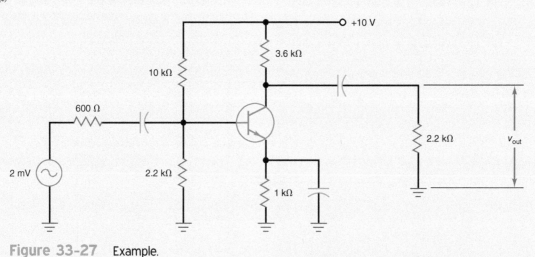

Figure 33-27 Example.

33.15 Emitter Follower

The **emitter follower** is also called a **common collector (CC) amplifier.** The input signal is coupled to the base, and the output signal is taken from the emitter.

Basic Idea

Figure 33-28a shows an emitter follower. Because the collector is at ac ground, the circuit is a CC amplifier. The input voltage is coupled to the base. This sets up an ac emitter current and produces an ac voltage across the emitter resistor. This ac voltage is then coupled to the load resistor.

Figure 33-28b shows the total voltage between the base and ground. It has a dc component and an ac component. As you can see, the ac input voltage rides on the quiescent base voltage, V_{BQ}. Similarly, Fig. 33-28c shows the total voltage between the emitter and ground. This time, the ac input voltage is centered on a quiescent emitter voltage, V_{EQ}.

The ac emitter voltage is coupled to the load resistor. This output voltage is shown in Fig. 33-28d, a pure ac voltage. This output voltage is in phase and is approximately equal to the input voltage. The reason the circuit is called an *emitter follower* is because the output voltage follows the input voltage.

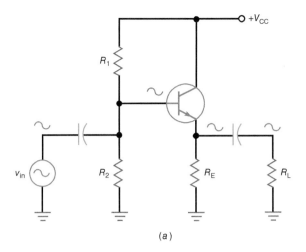

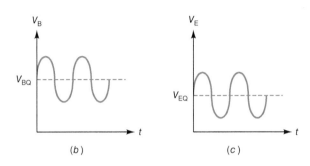

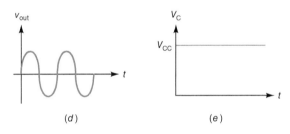

Figure 33-28 Emitter follower and waveforms.

Since there is no collector resistor, the total voltage between the collector and ground equals the supply voltage. If you look at the collector voltage with an oscilloscope, you will see a constant dc voltage like Fig. 33-28e. There is no ac signal on the collector because it is an ac ground point.

Negative Feedback

The emitter follower uses negative feedback. The ac signal developed across R_E effectively appears in series with the input voltage and is out of phase with it, making the effective input voltage lower. But with the emitter follower, the negative feedback is massive because the feedback resistance equals all the emitter resistance. As a result, the voltage gain is ultrastable, the distortion is almost nonexistent, and the input impedance of the base is very high.

The trade-off is the voltage gain, which has a maximum value of 1.

AC Emitter Resistance

In Fig. 33-28a, the ac signal coming out of the emitter sees R_E in parallel with R_L. Let us define the ac emitter resistance as follows:

$$r_e = R_E \parallel R_L \qquad \text{(33-18)}$$

This is the external ac emitter resistance, which is different from the internal ac emitter resistance r'_e.

Voltage Gain

The voltage gain of the emitter follower is

$$A_V = \frac{r_e}{r_e + r'_e} \qquad \text{(33-19)}$$

Usually, a designer makes r_e much greater than r'_e, so that the voltage gain equals 1 (approximately).

Why is an emitter follower called an *amplifier* if its voltage gain is only 1? Because it has a current gain of *b*. The stages near the end of a system need to produce more current because the final load is usually a low impedance. The emitter follower can produce the large output currents needed by low-impedance loads. In short, although it is not a voltage amplifier, the emitter follower is a current or power amplifier.

Input Impedance of the Base

The input impedance of the base is

$$z_{\text{in(base)}} = \beta(r_e + r'_e) \qquad \text{(33-20)}$$

Since r_e is usually much greater than r'_e, the input impedance is approximately βr_e.

The step up in impedance is the major advantage of an emitter follower. Small load resistances that would overload a CE amplifier can be used with an emitter follower because it steps up the impedance and prevents overloading.

Input Impedance of the Stage

When the ac source is not stiff, some of the ac signal will be lost across the internal resistance. If you want to calculate the effect of the internal resistance, you will need to use the input impedance of the stage, given by

$$z_{\text{in(stage)}} = R_1 \parallel R_2 \parallel \beta(r_e + r'_e) \qquad \text{(33-21)}$$

With the input impedance and the source resistance, you can use the voltage divider to calculate the input voltage reaching the base. The calculations are the same as shown in earlier examples.

EXAMPLE 33-11

What is the voltage gain of the emitter follower in Fig. 33-29? If $\beta = 150$, what is the ac load voltage?

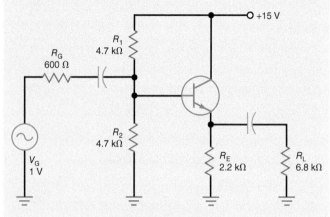

Figure 33-29 Example.

Answer:
The dc base voltage is half the supply voltage:

$$V_B = 7.5 \text{ V}$$

The dc emitter current is

$$I_E = \frac{6.8 \text{ V}}{2.2 \text{ k}\Omega} = 3.09 \text{ mA}$$

and the ac resistance of the emitter diode is

$$r_e' = \frac{25 \text{ mV}}{3.09 \text{ mA}} = 8.09 \text{ }\Omega$$

The external ac emitter resistance is

$$r_e = 2.2 \text{ k}\Omega \,\|\, 6.8 \text{ k}\Omega = 1.66 \text{ k}\Omega$$

The voltage gain equals

$$A_V = \frac{1.66 \text{ k}\Omega}{1.66 \text{ k}\Omega + 8.09 \text{ }\Omega} = 0.995$$

The input impedance of the base is

$$z_{in(base)} = 150(1.66 \text{ k}\Omega + 8.09 \text{ }\Omega) = 250 \text{ k}\Omega$$

This is much larger than the biasing resistors. Therefore, to a close approximation, the input impedance of the emitter follower is

$$z_{in(stage)} = 4.7 \text{ k}\Omega \,\|\, 4.7 \text{ k}\Omega = 2.35 \text{ k}\Omega$$

The ac input voltage is

$$v_{in} = \frac{2.35 \text{ k}\Omega}{600 \text{ }\Omega + 2.35 \text{ k}\Omega} 1 \text{ V} = 0.797 \text{ V}$$

The ac output voltage is

$$v_{out} = 0.995(0.797 \text{ V}) = 0.793 \text{ V}$$

Output Impedance

The output impedance of an amplifier is the same as its Thevenin impedance. One of the advantages of an emitter follower is its low output impedance.

Maximum power transfer occurs when the load impedance is *matched* (made equal) to the source (Thevenin) impedance. Sometimes, when maximum load power is wanted, a designer can match the load impedance to the output impedance of an emitter follower. For instance, the low impedance of a speaker can be matched to the output impedance of an emitter follower to deliver maximum power to the speaker.

Figure 33-30*a* shows an ac generator driving an amplifier. If the source is not stiff, some of the ac voltage is dropped across the internal resistance R_G. In this case, we need to analyze the voltage divider shown in Fig. 33-30*b* to get the input voltage v_{in}.

A similar idea can be used with the output side of the amplifier. In Fig. 33-30*c*, we can apply the Thevenin theorem at the load terminals. Looking back into the amplifier, we see an output impedance z_{out}. In the Thevenin

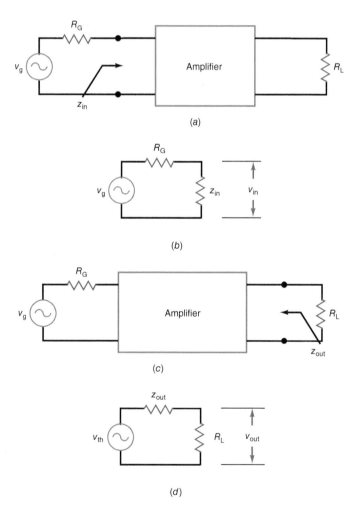

Figure 33-30 Input and output impedances.

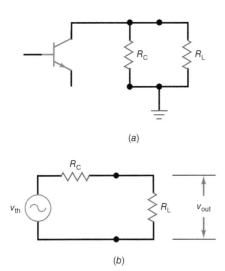

(a)

(b)

Figure 33-31 Output impedance of CE stage.

equivalent circuit, this output impedance forms a voltage divider with the load resistance, as shown in Fig. 33-30d. If z_{out} is much smaller than R_L, the output source is stiff and v_{out} equals v_{th}.

Figure 33-31a shows the ac equivalent circuit for the output side of the common emitter (CE) discussed earlier. When we apply Thevenin's theorem, we get Fig. 33-31b. In other words, the output impedance facing the load resistance is R_C, and R_C is effectively the output impedance z_{out} of the CE amplifier. Since the voltage gain of a CE amplifier depends on R_C, a designer cannot make R_C too small without losing voltage gain. Stated another way, it is very difficult to get a small output impedance with a CE amplifier. Because of this, CE amplifiers are not suited to driving small load resistances.

The output impedance z_{out} of an emitter follower is much smaller than you can get with a CE amplifier. It equals

$$z_{out} = R_E \parallel \left(r_e' + \frac{R_G \parallel R_1 \parallel R_2}{\beta} \right) \qquad \textbf{(33-22)}$$

In some designs the biasing resistances and the ac resistance of the emitter diode become negligible. In this case, the output impedance of an emitter follower can be approximated by

$$z_{out} = \frac{R_G}{\beta} \qquad \textbf{(33-23)}$$

This brings out the key idea of an emitter follower: It steps the impedance of the ac source down by a factor of b. As a result, the emitter follower allows us to build stiff ac sources. Instead of using a stiff ac source that maximizes the load voltage, a designer may prefer to maximize the load power. In this case, instead of designing for

$$z_{out} \ll R_L \qquad \text{(stiff voltage source)}$$

the designer will select values to get

$$z_{out} = R_L \qquad \text{(maximum power transfer)}$$

In this way, the emitter follower can deliver maximum power to a low-impedance load such as a stereo speaker. By basically removing the effect of R_L on the output voltage, the circuit is acting as a buffer between the input and output.

Formula (33-23) is an ideal formula. You can use it to get an approximate value for the output impedance of an emitter follower. With discrete circuits, the equation usually gives only an estimate of the output impedance. Nevertheless, it is adequate for troubleshooting and preliminary analysis. When necessary, you can use Formula (33-22) to get an accurate value for the output impedance.

EXAMPLE 33-12

Estimate the output impedance of the emitter follower of Fig. 33-29a.

Answer:
Ideally, the output impedance equals the generator resistance divided by the current gain of the transistor:

$$z_{out} = \frac{600 \ \Omega}{150} = 4 \ \Omega$$

Cascading CE and CC

To illustrate the buffering action of a CC amplifier, suppose we have a load resistance of 270 Ω. If we try to couple the output of a CE amplifier directly into this load resistance, we may overload the amplifier. One way to avoid this overload is by using an emitter follower between the CE amplifier and the load resistance. The signal can be coupled capacitively (this means through coupling capacitors), or it may be **direct-coupled** as shown in Fig. 33-32.

As you can see, the base of the second transistor is connected directly to the collector of the first transistor. Because of this, the dc collector voltage of the first transistor is used to bias the second transistor. If the dc current gain of the second transistor is 100, the dc resistance looking into the base of the second transistor is $R_{in} = 100(270 \ \Omega) = 27 \ k\Omega$.

Because 27 kΩ is large compared to the 3.6 kΩ, the dc collector voltage of the first stage is only slightly disturbed.

In Fig. 33-32, the amplified voltage out of the first stage drives the emitter follower and appears across the final load resistance of 270 Ω. Without the emitter follower, the 270 Ω would overload the first stage. But with the emitter follower, its impedance effect is increased by a factor of b. Instead of appearing like 270 Ω, it now looks like 27 kΩ in both the dc and the ac equivalent circuits.

This demonstrates how an emitter follower can act as a **buffer** between a high-output impedance and a low-resistance load.

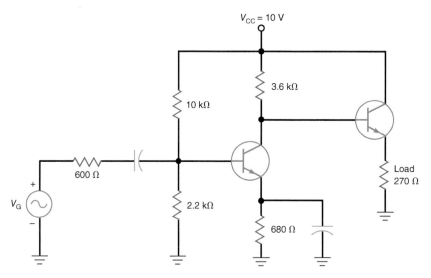

Figure 33-32   Direct-coupled output stage.

Darlington Connections

A **Darlington connection** is a connection of two transistors whose overall current gain equals the product of the individual current gains. Since its current gain is much higher, a Darlington connection can have a very high input impedance and can produce very large output currents. Darlington connections are often used with voltage regulators, power amplifiers, and high-current switching applications.

Darlington Pair

Figure 33-33a shows a **Darlington pair.** Since the emitter current of Q_1 is the base current for Q_2, the Darlington pair has an overall current gain of

$$\beta = \beta_1\beta_2 \qquad \textbf{(33-24)}$$

For instance, if each transistor has a current gain of 200, the overall current gain is

$$\beta = (200)(200) = 40,000$$

Semiconductor manufacturers can put a Darlington pair inside a single case like Fig. 33-33b. This device, known as a **Darlington transistor,** acts like a single transistor

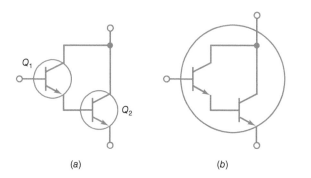

(a)   (b)

Figure 33-33   (a) Darlington pair. (b) Darlington transistor.

with a very high current gain. For instance, the 2N6725 is a Darlington transistor with a current gain of 25,000 at 200 mA. As another example, the TIP102 is a power Darlington with a current gain of 1000 at 3 A.

The analysis of a circuit using a Darlington transistor is almost identical to the emitter follower analysis. With the Darlington transistor, since there are two transistors, there are two V_{BE} drops. The base current of Q_2 is the same as the emitter current of Q_1. Also, the input impedance at the base of Q_1 can be found by $z_{in(base)} \cong \beta_1\beta_2r_e$ or stated as

$$z_{in(base)} \cong \beta_{re} \qquad \textbf{(33-25)}$$

33.16 The Differential Amplifier

Transistors, diodes, and resistors are the only practical components in typical ICs. Capacitors may also be used, but they are small, usually less than 50 pF. For this reason, IC designers cannot use coupling and bypass capacitors the way a discrete circuit designer can. Instead, the IC designer has to use direct coupling between stages and also needs to eliminate the emitter bypass capacitor without losing too much voltage gain.

The **differential amplifier (diff amp)** has two inputs and either one or two outputs. The output is the difference between the two input voltages multiplied by the amplifier gain (A). They are widely used in integrated circuits because they permit dc or direct coupling of stages and eliminate the need for an emitter bypass capacitor. For this and other reasons, the diff amp is used as the input stage of almost every IC op amp.

Differential Input and Output

Figure 33-34 shows a diff amp. It is two CE stages in parallel with a common emitter resistor. Although it has two input voltages (v_1 and v_2) and two collector voltages (v_{cl} and

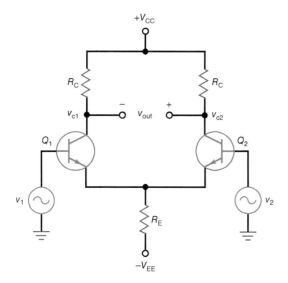

Figure 33-34 Differential input and differential output.

v_{c2}), the overall circuit is considered to be one stage. Because there are no coupling or bypass capacitors, there is no lower cutoff frequency.

The ac output voltage v_{out} is defined as the voltage between the collectors with the polarity shown in Fig. 33-34:

$$v_{out} = v_{c2} - v_{c1} \qquad (33\text{-}26)$$

This voltage is called a **differential output** because it combines the two ac collector voltages into one voltage that equals the difference of the collector voltages. *Note:* We will use lowercase letters for v_{out}, v_{c1}, and v_{c2} because they are ac voltages that include zero hertz (0 Hz) as a special case.

Ideally, the circuit has identical transistors and equal collector resistors. With perfect symmetry, v_{out} is zero when the two input voltages are equal. When v_1 is greater than v_2, the output voltage has the polarity shown in Fig. 33-34. When v_2 is greater than v_1, the output voltage is inverted and has the opposite polarity.

The diff amp of Fig. 33-34 has two separate inputs. Input v_1 is called the **noninverting input** because v_{out} is in phase with v_1. On the other hand, v_2 is called the **inverting input** because v_{out} is 180° out of phase with v_2. In some applications, only the noninverting input is used and the inverting input is grounded. In other applications, only the inverting input is active and the noninverting input is grounded.

When both the noninverting and inverting input voltages are present, the total input is called a **differential input** because the output voltage equals the voltage gain times the difference of the two input voltages. The equation for the output voltage is

$$v_{out} = A_v(v_1 - v_2) \qquad (33\text{-}27)$$

where A_v is the voltage gain. We will show the equation for voltage gain in the subsection "AC Analysis of a Diff App."

Single-Ended Output

A differential output like that of Fig. 33-34 requires a floating load because neither end of the load can be grounded. This is inconvenient in many applications since loads are often single-ended; that is, one end is grounded.

Figure 33-35a shows a widely used form of the diff amp. This has many applications because it can drive single-ended loads like CE stages, emitter followers, and other circuits. As you can see, the ac output signal is taken from the collector on the right side. The collector resistor on the left has been removed because it serves no useful purpose.

Because the input is differential, the ac output voltage is still given by $A_v(v_1 - v_2)$. With a single-ended output, however, the voltage gain is half as much as with a differential output. We get half as much voltage gain with a single-ended output because the output is coming from only one of the collectors.

Incidentally, Fig. 33-35b shows the block-diagram symbol for a diff amp with a differential input and a single-ended output. The same symbol is used for an op amp. The plus sign (+) represents the noninverting input, and the minus sign (−) is the inverting input.

Input Configurations

Often, only one of the inputs is active and the other is grounded. In Fig. 33-35, if the base input to Q_1 is grounded,

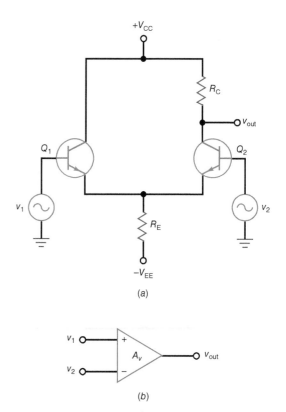

(a)

(b)

Figure 33-35 (a) Differential input and single-ended output. (b) Block diagram symbol.

the output is an inverted version of v_2. If the input to Q_2 is grounded, the output is a non-inverted version of v_1.

Biasing a Diff Amp

Figure 33-36 shows the dc equivalent circuit for a diff amp. Throughout this discussion, we will assume identical transistors and equal collector resistors. Also, both bases are grounded in this preliminary analysis.

The base bias comes from a second emitter power supply V_{EE}. Ignoring V_{BE}, most of V_{EE} appears across the emitter resistor which sets the emitter current.

A diff amp is sometimes called a *long-tail pair* because the two transistors share a common resistor R_E. The current through this common resistor is called the **tail current.** If we ignore the V_{BE} drops across the emitter diodes of Fig. 33-36, then the top of the emitter resistor is ideally a dc ground point. In this case, all of V_{EE} appears across R_E and the tail current is

$$I_T = \frac{V_{EE}}{R_E} \qquad \textbf{(33-28)}$$

This equation is fine for troubleshooting and preliminary analysis because it quickly gets to the point, which is that almost all the emitter supply voltage appears across the emitter resistor.

When the two halves of Fig. 33-36 are perfectly matched, the tail current will split equally. Therefore, each transistor has an emitter current of

$$I_E = \frac{I_T}{2} \qquad \textbf{(33-29)}$$

The dc voltage on either collector is given by this familiar equation:

$$V_C = V_{CC} - I_C R_C \qquad \textbf{(33-30)}$$

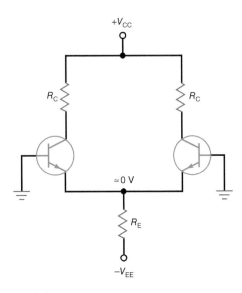

Figure 33-36 Ideal dc analysis.

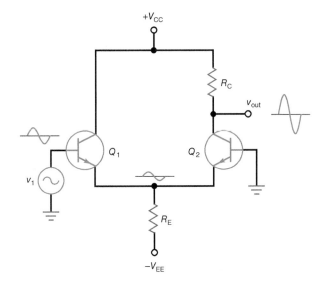

Figure 33-37 Noninverting input and single-ended output.

AC Anlysis of a Diff Amp

Figure 33-37 shows a noninverting input and single-ended output. With a large R_E, the tail current is almost constant when a small ac signal is present. Because of this, the two halves of a diff amp respond in a complementary manner to the noninverting input. In other words, an increase in the emitter current of Q_1 produces a decrease in the emitter current of Q_2. Conversely, a decrease in the emitter current of Q_1 produces an increase in the emitter current of Q_2.

In Fig. 33 37, the left transistor Q_1 acts like an emitter follower that produces an ac voltage across the emitter resistor. This ac voltage is half of the input voltage v_1. On the positive half-cycle of input voltage, the Q_1 emitter current increases, the Q_2 emitter current decreases, and the Q_2 collector voltage increases. Similarly, on the negative half-cycle of input voltage, the Q_1 emitter current decreases, the Q_2 emitter current increases, and the Q_2 collector voltage decreases. This is why the amplified output sine wave is in phase with the noninverting input.

Single-Ended Output Gain

Without showing the derivation, the gain of a differential amplifier with a single-ended output is

$$A_v = \frac{R_C}{2r_e'} \qquad \textbf{(33-31)}$$

Remember that r_e' is

$$r_e' = \frac{0.025}{I_E}.$$

In Fig. 33-37, a quiescent dc voltage V_C exists at the output terminal. This voltage is not part of the ac signal. The ac voltage v_{out} is any change from the quiescent voltage. In an op amp, the quiescent dc voltage is removed in a later stage because it is unimportant.

Differential-Output Gain

The differential output voltage is twice the single-ended version since there are two collector resistors:

$$A_v = \frac{R_C}{r_e'} \qquad \textbf{(33-32)}$$

This is easy to remember because it is the same as the voltage gain for a CE stage.

Inverting-Input Configurations

For an inverting input and single-ended output, the ac analysis is almost identical to the noninverting analysis. In this circuit, the inverting input v_2 produces an amplified and inverted ac voltage at the final output. The r_e' of each transistor is still part of a voltage divider in the ac equivalent circuit. This is why the ac voltage across R_E is half of the inverting input voltage. If a differential output is used, the voltage gain is twice as much, as previously discussed.

Differential-Input Configurations

The differential-input configurations have both inputs active at the same time.

The output voltage for a noninverting input is $A_v(v_1)$, and the output voltage for an inverting input is

$$v_{out} = -A_v(v_2)$$

By combining the two results, we get the equation for a differential input:

$$v_{out} = A_v(v_1 - v_2)$$

Input Impedance

In a CE stage, the input impedance of the base is

$$z_{in} = \beta r_e'$$

In a diff amp, the input impedance of either base is twice as high:

$$z_{in} = 2\beta r_e' \qquad \textbf{(33-33)}$$

The input impedance of a diff amp is twice as high because there are two ac emitter resistances r_e' in the ac equivalent circuit instead of one. Formula (33-33) is valid for all configurations because any ac input signal sees two ac emitter resistances in the path between the base and ground.

Common-Mode Gain

Figure 33-38a shows a differential input and single-ended output. The same input voltage, $v_{in(CM)}$ is being applied to each base. This voltage is called a **common-mode signal.** If the diff amp is perfectly symmetrical, there is no ac output voltage with a common-mode input signal because $v_1 = v_2$. When a diff amp is not perfectly symmetrical, there will be a small ac output voltage.

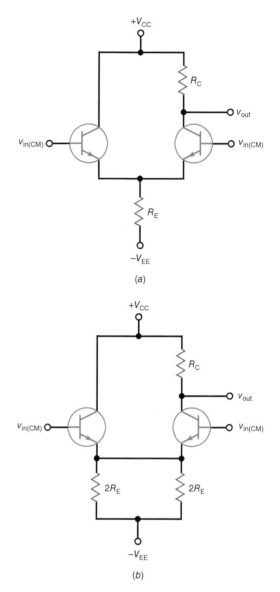

Figure 33-38 (a) Common-mode input signal. (b) Equivalent circuit.

In Fig. 33-38a, equal voltages are applied to the noninverting and inverting inputs. Nobody would deliberately use a diff amp this way because the output voltage is ideally zero. The reason for discussing this type of input is because most static, interference, and other kinds of undesirable pickup are common-mode signals.

If the diff amp is operating in an environment with a lot of electromagnetic interference, each base acts like a small antenna that picks up an unwanted signal voltage. One of the reasons the diff amp is so popular is because it discriminates against these common-mode signals. In other words, a diff amp does not amplify common-mode signals.

Here is an easy way to find the voltage gain for a common-mode signal: We can redraw the circuit, as shown in

Fig. 33-38b. Since equal voltages $v_{in(CM)}$ drive both inputs simultaneously, there is almost no current through the wire between the emitters. Therefore, we can remove the connecting wire.

The voltage gain is approximately

$$A_v(CM) = \frac{R_C}{2R_E} \qquad \textbf{(33-34)}$$

With typical values of R_C and R_E, the common-mode voltage gain is usually less than 1.

Common-Mode Rejection Ratio

The **common-mode rejection ratio (CMRR)** is defined as the voltage gain divided by common-mode voltage gain.

$$CMRR = \frac{A_v}{A_{v(CM)}} \qquad \textbf{(33-35)}$$

For instance, if $A_v = 200$ and $A_{v(CM)} = 0.5$, CMRR = 400.

The higher the CMRR, the better. A high CMRR means that the diff amp is amplifying the wanted signal and discriminating against the common-mode signal.

Data sheets usually specify CMRR in decibels, using the following formula for the decibel conversion:

$$CMRR_{dB} = 20 \log CMRR \qquad \textbf{(33-36)}$$

As an example, if CMRR = 400,

$$CMRR_{dB} = 20 \log 400 = 52 \text{ dB}$$

EXAMPLE 33-13

A differential amplifier has, $A_v = 150$, $A_{v(CM)} = 0.5$, and $v_{in} = 1$ mV. If the base leads are picking up a common-mode signal of 1 mV, what is the output voltage?

Answer:
The input has two components, the desired signal and a common-mode signal. Both are equal in amplitude. The desired component is amplified to get an output of

$$v_{out1} = 150(1 \text{ mV}) = 150 \text{ mV}$$

The common-mode signal is attenuated to get an output of

$$v_{out2} = 0.5(1 \text{ mV}) = 0.5 \text{ mV}$$

The total output is the sum of these two components:

$$v_{out} = v_{out1} + v_{out2}$$

The output contains both components, but the desired component is 300 times greater than the unwanted component.

This example shows why the diff amp is useful as the input stage of an op amp. It attenuates the common-mode signal. This is a distinct advantage over the ordinary CE amplifier, which amplifies a stray pickup signal the same way it amplifies the desired signal.

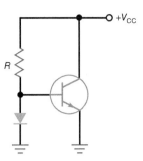

Figure 33-39 The current mirror.

The Current Mirror

With ICs, there is a way to increase the voltage gain and CMRR of a diff amp. Figure 33-39 shows a **compensating diode** in parallel with the emitter diode of a transistor. The current through the resistor is given by

$$I_R = \frac{V_{CC} - V_{BE}}{R} \qquad \textbf{(33-37)}$$

If the compensating diode and the emitter diode have identical current-voltage curves, the collector current will equal the current through the resistor:

$$I_C = I_R \qquad \textbf{(33-38)}$$

A circuit like Fig. 33-39 is called a **current mirror** because the collector current is a mirror image of the resistor current. With ICs, it is relatively easy to match the characteristics of the compensating diode and the emitter diode because both components are on the same chip. Current mirrors are used as current sources and active loads in IC op amps.

Current Mirror Sources the Tail Current

With a single-ended output, the voltage gain of a diff amp is $R_C/2r_e'$ and the common-mode voltage gain is $R_C/2R_E$. The ratio of the two gains gives

$$CMRR = \frac{R_E}{r_e'}$$

The larger we can make R_E, the greater the CMRR.

One way to get a high equivalent R_E is to use a current mirror to produce the tail current, as shown in Fig. 32-40. The current through the compensating diode is

$$I_R = \frac{V_{CC} + V_{EE} - V_{BE}}{R} \qquad \textbf{(33-39)}$$

Because of the current mirror, the tail current has the same value. Since Q_4 acts like a current source, it has a very high output impedance. As a result, the equivalent R_E of the diff amp is in hundreds of megohms and the CMRR is dramatically improved.

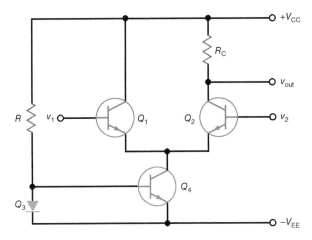

Figure 33-40 Current mirror sources the tail current.

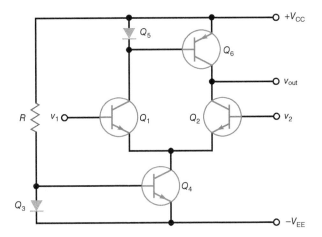

Figure 33-41 Current mirror is an active load.

Active Load

The voltage gain of a single-ended diff amp is $R_C/2r_e'$. The larger we can make R_C, the greater the voltage gain. Figure 33-41 shows a current mirror used as an **active load resistor.** Since Q_6 is a *pnp* current source, Q_2 sees an equivalent R_C that is hundreds of megohms. As a result, the voltage gain is much higher with an active load than with an ordinary resistor. Active loading like this is used in most *IC* op amps.

33.17 The Transistor Switch

There are two basic kinds of transistor circuits: **amplifying** and **switching.** With amplifying circuits, the Q point must remain in the active region under all operating conditions. If it does not, the output signal will be distorted on the peak where saturation or cutoff occurs. With switching circuits, the Q point usually switches between saturation and cutoff. Switching circuits are usually called digital or pulse circuits.

Base bias is useful in *digital circuits* because these circuits are usually designed to operate at saturation and cutoff. Because of this, they have either low output voltage or high

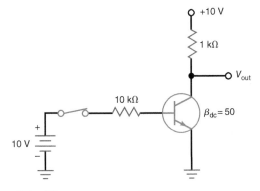

Figure 33-42 Hard saturation.

output voltage. In other words, none of the Q points between saturation and cutoff are used. For this reason, variations in the Q point don't matter, because the transistor remains in saturation or cutoff when the current gain changes.

Here is an example of using a base-biased circuit to switch between saturation and cutoff. Figure 33-42 shows an example of a transistor in hard saturation. Therefore, the output voltage is approximately 0 V.

When the switch opens, the base current drops to zero. Because of this, the collector current drops to zero. With no current through the 1 kΩ, all the collector supply voltage will appear across the collector-emitter terminals. Therefore, the output voltage rises to +10 V.

The circuit can have only two output voltages: 0 or +10 V. This is how you can recognize a digital circuit. It has only two output levels: low or high. The exact values of the two output voltages are not important. All that matters is that you can distinguish the voltages as low or high.

Another name often used is **two-state circuits or binary circuits,** referring to the low and high outputs.

The key to understanding switching circuits is to be able to recognize when saturation occurs. There are several methods. Keep in mind that in saturation the V_{CE} is only a few tenths of a volt or, for all practical purposes, zero. This puts the full V_{CE} across R_C defining the maximum possible I_C value.

Saturation-Current Method

Figure 33-42 shows a base-biased circuit. Start by calculating the saturation current:

$$I_{C(sat)} = \frac{10 \text{ V}}{1 \text{ k}\Omega} = 10 \text{ mA}$$

The base current is about 1 mA. Assuming a current gain of 50 as shown, the collector current is

$$I_C = 50(1 \text{ mA}) = 50 \text{ mA}$$

The answer is impossible because the collector current cannot be greater than the saturation current. Therefore, the transistor cannot be operating in the active region; it must be operating in the saturation region.

Hard Saturation

A designer who wants a transistor to operate in the saturation region under all conditions often selects a base resistance that produces a current gain of 10. This is called **hard saturation,** because there is more than enough base current to saturate the transistor. For example, a base resistance of 10 kΩ in Fig. 33-42 will produce a current gain of

$$\beta_{dc} = \frac{10 \text{ mA}}{1 \text{ mA}} = 10$$

For the transistor of Fig. 33-42, it takes only

$$I_B = \frac{10 \text{ mA}}{50} = 0.2 \text{ mA}$$

to saturate the transistor. Therefore, a base current of 1 mA will drive the transistor deep into saturation.

Recognizing Hard Saturation at a Glance

Here is how you can quickly tell whether a transistor is in hard saturation. Often, the base supply voltage and the collector supply voltage are equal: $V_{BB} = V_{CC}$. When this is the case, a designer will use the 10:1 rule, which says to make the base resistance approximately 10 times as large as the collector resistance.

Figure 33-42 was designed by using the 10:1 rule. Therefore, whenever you see a circuit with a 10:1 ratio (R_B to R_C), you can expect it to be saturated.

Another way to recognize saturation is to measure the transistor junction voltages to determine their value and polarity. In saturation, both the emitter and collector junctions are forward-biased. Figure 33-43 shows typical values. With $V_{CE} = 0.1$ V and $V_{BE} = 0.7$ V, V_{CB} can only be 0.6 V with the polarity shown.

33.18 Driver Circuits

A *driver circuit* is one used to operate some other device, like a light, relay, solenoid, or motor. Usually the driver is a transistor switch that is used to turn the device off or on. The key point of a driver is that a very small input current can control a much larger output current. While most drivers are switches, linear circuit drivers are also used.

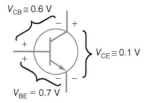

Figure 33-43   In a saturated bipolar transistor, both junctions are biased.

Base-Biased LED Driver

A good example of a driver is an LED driver. The base current is zero in Fig. 33-44a, which means that the transistor is at cutoff. When the switch of Fig. 33-44a closes, the transistor goes into hard saturation. Visualize a short between the collector-emitter terminals. Then the collector supply voltage (15 V) appears across the series connection of the 1.5 kΩ and the LED. If we ignore the voltage drop across the LED, the collector current is ideally 10 mA. But if we allow 2 V across the LED, then there is 13 V across the 1.5 kΩ, and the collector current is 13 V divided by 1.5 kΩ, or 8.67 mA.

This LED driver is designed for hard saturation, where the current gain doesn't matter. If you want to change the LED current in this circuit, you can change either the collector resistance or the collector supply voltage. The base resistance is made 10 times larger than the collector resistance because we want hard saturation when the switch is closed.

Emitter-Biased LED Driver

The emitter current is zero in Fig. 33-44b, which means that the transistor is at cutoff. When the switch of Fig. 33-44b closes, the transistor goes into the active region. Ideally, the emitter voltage is 15 V. This means that we get an emitter current of 10 mA. The transistor acts as a

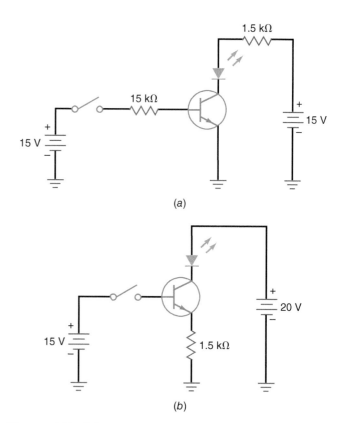

Figure 33-44   (a) Base-biased. (b) Emitter-biased.

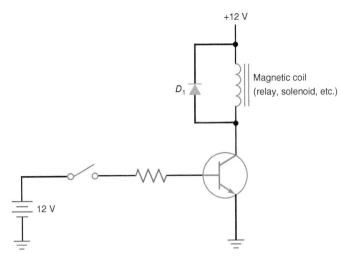

Figure 33-45 A driver for inductive devices with protective diode.

current source where the base voltage and emitter resistor set the emitter and the collector currents. Here the LED voltage drop has no effect. It doesn't matter whether the exact LED voltage is 1.8, 2, or 2.5 V. This is an advantage of the emitter-biased design over the base-biased design. The LED current is independent of the LED voltage. Another advantage is that the circuit doesn't require a collector resistor.

The emitter-biased circuit of Fig. 33-44b operates in the active region when the switch is closed. To change the LED current, you can change the base supply voltage or the emitter resistance. For instance, if you vary the base supply voltage, the LED current varies in direct proportion.

Another common driver is one for a magnetic relay or solenoid. Figure 33-45 shows an example. The coil in the collector is usually rated for a specific voltage, such as 12 V. When the transistor turns on, the transistor connects the coil to ground. The coil resistance sets the collector current, which produces a magnetic field in the coil. The magnetic field activates mechanical relay switching contacts or a magnetic rod on a solenoid that creates linear motion.

An important part of this circuit is diode D_1 across the relay coil. When the transistor turns off, the magnetic field in the coil drops suddenly. This causes a voltage spike of several hundred volts to be induced across the coil. This pulse can damage the transistor. When the spike occurs, the diode becomes forward-biased, eliminating the spike and clamping the collector voltage to a safe 0.7-V level.

33.19 Special Bipolar Transistors

Two examples of special bipolar transistors are covered in this section: heterojunction bipolar transistors and phototransistors.

Heterojunction Bipolar Transistor

Most bipolar transistors are made with silicon. Different areas of the silicon are doped from the p- and n-type emitter, base, and collector. One primary limitation of the silicon BJT is that its upper-operating frequency range is just beyond 1 GHz. This limits the BJT to amplifying signals below this range. To amplify microwave signals beyond 1 GHz, special bipolar transistors are used. These are made with compound semiconductors like silicon germanium (SiGe), gallium arsenide (GaAs), or indium phosphide (InP). These transistors amplify well into the lower microwave frequencies to about 20 GHz.

An even better BJT is the *heterojunction bipolar transistor (HBT)*. It uses different semiconductor materials for the emitter and base. The most common combination is gallium arsenide (GaAs) for the emitter and aluminum gallium arsenide (AlGaAs) for the base. Such a device can amplify signals up to roughly 200 GHz. The frequency range from 30 to 300 GHz is known as millimeter waveband. It is now regularly used by radar, satellites, and data/cellular backhaul applications.

Phototransistors and Other Optoelectronic Devices

A transistor with an open base has a small collector current consisting of thermally produced minority carriers and surface leakage. By exposing the collector junction to light, a manufacturer can produce a **phototransistor**, a device that has more sensitivity to light than a photodiode.

Basic Idea of Phototransistors

Figure 33-46a shows a transistor with an open base. As mentioned earlier, a small collector current exists in this circuit. Ignore the surface-leakage component, and concentrate on the thermally produced carriers in the collector diode. Visualize the reverse current produced by these carriers as an ideal current source in parallel with the collector-base junction of an ideal transistor (Fig. 33-46b).

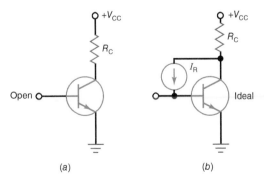

Figure 33-46 (a) Transistor with open base. (b) Equivalent circuit.

Because the base lead is open, all the reverse current is forced into the base of the transistor. The resulting collector current is

$$I_{CEO} = \beta_{dc} I_R$$

where I_R is the reverse minority-carrier current. This says that the collector current is higher than the original reverse current by a factor of β_{dc}.

The collector diode is sensitive to light as well as heat. In a phototransistor, light passes through a window and strikes the collector-base junction. As the light increases, I_R increases, and so does I_{CEO}.

Phototransistor versus Photodiode

The main difference between a phototransistor and a photodiode is the current gain b_{dc}. The same amount of light striking both devices produces b_{dc} times more current in a phototransistor than in a photodiode. The increased sensitivity of a phototransistor is a big advantage over that of a photodiode.

Figure 33-47a shows the schematic symbol of a phototransistor. Notice the open base. This is the usual way to operate a phototransistor. You can control the sensitivity with a variable base return resistor (Fig. 33-47b), but the base is usually left open to get maximum sensitivity to light.

The price paid for increased sensitivity is reduced speed. A phototransistor is more sensitive than a photodiode, but it cannot turn on and off as fast. A photodiode has typical output currents in microamperes and can switch on and off in nanoseconds. The phototransistor has typical output currents in milliamperes but switches on and off in microseconds. A typical phototransistor is shown in Fig. 33-47c.

Optocoupler

Figure 33-48a shows an LED driving a phototransistor. This is a much more sensitive optocoupler than an LED-photodiode combination. The idea is straightforward. Any changes in V_S produce changes in the LED current, which changes the current through the phototransistor. In turn, this produces a changing voltage across the collector-emitter terminals. Therefore, a signal voltage is coupled from the input circuit to the output circuit.

Again, the big advantage of an optocoupler is the electrical isolation between the input and output circuits. Stated another way, the common for the input circuit is different from the common for the output circuit. Because of this, no conductive path exists between the two circuits. This means that you can ground one of the circuits and float the other. For instance, the input circuit can be grounded to the chassis of the equipment, while the common of the output side is ungrounded. Figure 33-48b shows a typical optocoupler IC.

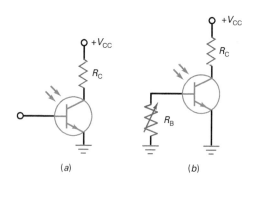

Figure 33-47 Phototransistor. (a) Open base gives maximum sensitivity. (b) Variable base resistor changes sensitivity. (c) Typical phototransistor.

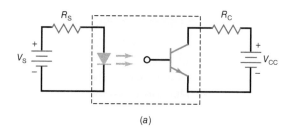

Figure 33-48 (a) Optocoupler with LED and phototransistor. (b) Optocoupler IC.

1. A bipolar transistor has how many *pn* junctions?
   a. 1
   b. 2
   c. 3
   d. 4

2. What is one important thing transistors do?
   a. Amplify weak signals.
   b. Rectify line voltage.
   c. Step down voltage.
   d. Emit light.

3. Who invented the first junction transistor?
   a. Bell.
   b. Faraday.
   c. Marconi.
   d. Schockley.

4. In an *npn* transistor, the majority carriers in the emitter are
   a. free electrons.
   b. holes.
   c. neither *a* nor *b*.
   d. both *a* and *b*.

5. The barrier potential across each silicon depletion layer is
   a. 0 V.
   b. 0.3 V.
   c. 0.7 V.
   d. 1 V.

6. The emitter diode is usually
   a. forward-biased.
   b. reverse-biased.
   c. nonconducting.
   d. operating in the breakdown region.

7. For normal linear operation of the transistor, the collector diode has to be
   a. forward-biased.
   b. reverse-biased.
   c. nonconducting.
   d. operating in the breakdown region.

8. Most of the electrons in the base of an *npn* transistor flow
   a. out of the base lead.
   b. into the collector.
   c. into the emitter.
   d. into the base supply.

9. Most of the electrons that flow through the base will
   a. flow into the collector.
   b. flow out of the base lead.

c. recombine with base holes.
d. recombine with collector holes.

10. The beta of a transistor is the ratio of the
    a. collector current to emitter current.
    b. collector current to base current.
    c. base current to collector current.
    d. emitter current to collector current.

11. Increasing the collector supply voltage will increase
    a. base current.
    b. collector current.
    c. emitter current.
    d. none of the above.

12. The fact that there are many free electrons in a transistor emitter region means the emitter is
    a. lightly doped.
    b. heavily doped.
    c. undoped.
    d. none of the above.

13. In a normally biased *npn* transistor, the electrons in the emitter have enough energy to overcome the barrier potential of the
    a. base-emitter junction.
    b. base-collector junction.
    c. collector-base junction.
    d. recombination path.

14. In a *pnp* transistor, the major carriers in the emitter are
    a. free electrons.
    b. holes.
    c. neither.
    d. both.

15. What is the most important fact about the collector current?
    a. It is measured in milliamperes.
    b. It equals the base current divided by the current gain.
    c. It is small.
    d. It approximately equals the emitter current.

16. If the current gain is 100 and the collector current is 10 mA, the base current is
    a. 10 μA.
    b. 100 μA.
    c. 1 A.
    d. 10 A.

17. The collector-emitter voltage is usually.
    a. less than the collector supply voltage.
    b. equal to the collector supply voltage.

c. more than the collector supply voltage.

d. cannot answer.

18. The power dissipated by a transistor approximately equals the collector current times

a. base-emitter voltage.

b. collector-emitter voltage.

c. base supply voltage.

d. 0.7 V.

19. A transistor acts like a diode and a

a. voltage source.

b. current source.

c. resistance.

d. power supply.

20. If the base current is 100 mA and the current gain is 30, the emitter current is

a. 3.33 mA.

b. 3 A.

c. 3.1 A.

d. 10 A.

21. The base-emitter voltage of an ideal transistor is

a. 0 V.

b. 0.3 V.

c. 0.7 V.

d. 1 V.

22. In the active region, the collector current is not changed significantly by

a. base supply voltage.

b. base current.

c. current gain.

d. collector resistance.

23. If the base resistor is open, what is the collector current?

a. 0 mA.

b. 1 mA.

c. 2 mA.

d. 10 mA.

24. When comparing the power dissipation of a 2N3904 transistor to the PZT3904 surface-mount version, the 2N3904

a. can handle less power.

b. can handle more power.

c. can handle the same power.

d. is not rated.

25. The current gain of a transistor is defined as the ratio of the collector current to the

a. base current.

b. emitter current.

c. supply current.

d. collector current.

26. When the collector current increases, what does the current gain do?

a. Decreases.

b. Stays the same.

c. Increases.

d. Any of the above.

27. When the base resistor increases, the collector voltage will probably

a. decrease.

b. stay the same.

c. increase.

d. do all of the above.

28. If the base resistor is very small, the transistor will operate in the

a. cutoff region.

b. active region.

c. saturation region.

d. breakdown region.

29. Ignoring the bulk resistance of the collector diode, the collector-emitter saturation voltage is

a. 0 V.

b. A few tenths of a volt.

c. 1 V.

d. supply voltage.

30. If the base supply voltage is disconnected, the collector-emitter voltage will equal

a. 0 V.

b. 6 V.

c. 10.5 V.

d. collector supply voltage.

31. If the base resistor has zero resistance, the transistor will probably be

a. saturated.

b. in cutoff.

c. destroyed.

d. none of the above.

32. The collector current is 1.5 mA. If the current gain is 50, the base current is

a. 3 μA.

b. 30 μA.

c. 150 μA.

d. 3 mA.

33. The base current is 50 mA. If the current gain is 100, the collector current is closest in value to

a. 50 μA.

b. 500 μA.

c. 2 mA.

d. 5 mA.

34. When there is no base current in a transistor switch, the output voltage from the transistor is
a. low.
b. high.
c. unchanged.
d. unknown.

35. If the current gain is unknown in an emitter-biased circuit, you cannot calculate the
a. emitter voltage.
b. emitter current.
c. collector current.
d. base current.

36. If the emitter resistor is open, the collector voltage is
a. low.
b. high.
c. unchanged.
d. unknown.

37. If the collector resistor is open, the collector voltage is
a. low.
b. high.
c. unchanged.
d. unknown.

38. When the current gain increases from 50 to 300 in an emitter-biased circuit, the collector current
a. remains almost the same.
b. decreases by a factor of 6.
c. increases by a factor of 6.
d. is zero.

39. If the emitter resistance increases, the collector voltage
a. decreases.
b. stays the same.
c. increases.
d. breaks down the transistor.

40. What DMM polarity connection is needed on an *npn* transistor's base to get a 0.7-V reading?
a. Positive.
b. Negative.
c. Either positive or negative.
d. Unknown.

41. The major advantage of a phototransistor as compared to a photodiode is its
a. response to higher frequencies.
b. ac operation.
c. increased sensitivity.
d. durability.

42. For the emitter bias, the voltage across the emitter resistor is the same as the voltage between the emitter and the
a. base.
b. collector.

c. emitter.
d. ground.

43. For emitter bias, the voltage at the emitter is 0.7 V less than the
a. base voltage.
b. emitter voltage.
c. collector voltage.
d. ground voltage.

44. VDB is noted for its
a. unstable collector voltage.
b. varying emitter current.
c. large base current.
d. stable Q point.

45. VDB normally operates in the
a. active region.
b. cutoff region.
c. saturation region.
d. breakdown region.

46. The collector voltage of a VDB circuit is not sensitive to changes in the
a. supply voltage.
b. emitter resistance.
c. current gain.
d. collector resistance.

47. If the emitter resistance decreases in a VDB circuit, the collector voltage
a. decreases.
b. stays the same.
c. increases.
d. doubles.

48. The Q point of a VDB circuit is
a. hypersensitive to changes in current gain.
b. somewhat sensitive to changes in current gain.
c. almost totally insensitive to changes in current gain.
d. greatly affected by temperature changes.

49. The current gain of a *pnp* transistor is
a. the negative of the *npn* current gain.
b. the collector current divided by the emitter current.
c. near zero.
d. the ratio of collector current to base current.

50. Which is the largest current in a *pnp* transistor?
a. Base current.
b. Emitter current.
c. Collector current.
d. None of these.

51. In a well-designed VDB circuit, the base current is
a. much larger than the voltage divider current.
b. equal to the emitter current.
c. much smaller than the voltage divider current.
d. equal to the collector current.

52. In a VDB circuit, the base input resistance R_{in} is
   a. equal to $\beta_{dc} R_E$.
   b. normally smaller than R_{TH}.
   c. equal to $\beta_{dc} R_C$.
   d. independent of β_{dc}.

53. A coupling capacitor is
   a. a dc short.
   b. an ac open.
   c. a dc open and an ac short.
   d. a dc short and an ac open.

54. The capacitor that produces an ac ground is called a(n)
   a. bypass capacitor.
   b. coupling capacitor.
   c. dc open.
   d. ac open.

55. AC emitter resistance equals 25 mV divided by the
   a. quiescent base current.
   b. dc emitter current.
   c. ac emitter current.
   d. change in collector current.

56. The output voltage of a CE amplifier is
   a. amplified.
   b. inverted.
   c. 180° out of phase with the input.
   d. all of the above.

57. The emitter of a CE amplifier has no ac voltage because of the
   a. dc voltage on it.
   b. bypass capacitor.
   c. coupling capacitor.
   d. load resistor.

58. The voltage across the load resistor of a capacitor-coupled CE amplifier is
   a. dc and ac.
   b. dc only.
   c. ac only.
   d. Neither dc nor ac.

59. The voltage gain equals the output voltage divided by the
   a. input voltage.
   b. ac emitter resistance.
   c. ac collector resistance.
   d. generator voltage.

60. The input impedance of the base decreases when
   a. β increases.
   b. supply voltage increases.
   c. β decreases.
   d. ac collector resistance increases.

61. Voltage gain is directly proportional to
   a. β.
   b. r_e.
   c. dc collector voltage.
   d. ac collector resistance.

62. If the emitter-bypass capacitor opens, the ac output voltage will
   a. decrease.
   b. increase.
   c. remain the same.
   d. equal zero.

63. An emitter follower has a voltage gain that is
   a. much less than one.
   b. approximately equal to one.
   c. greater than one.
   d. zero.

64. The input impedance of the base of an emitter follower is usually
   a. low.
   b. high.
   c. shorted to ground.
   d. open.

65. The output voltage of an emitter follower is across the
   a. emitter diode.
   b. dc collector resistor.
   c. load resistor.
   d. emitter diode and external ac emitter resistance.

66. If $\beta = 200$ and $r_e = 150\ \Omega$, the input impedance of the base is
   a. 30 kΩ.
   b. 600 Ω.
   c. 3 kΩ.
   d. 5 kΩ.

67. The input voltage to an emitter follower is usually
   a. less than the generator voltage.
   b. equal to the generator voltage.
   c. greater than the generator voltage.
   d. equal to the supply voltage.

68. The output voltage of an emitter follower is approximately
   a. 0 V.
   b. V_G.
   c. v_{in}.
   d. V_{CC}.

69. The output voltage of an emitter follower is
   a. in phase with v_{in}.
   b. much greater than v_{in}.
   c. 180° out of phase.
   d. generally much less than v_{in}.

70. For maximum power transfer, a CC amplifier is designed so
a. $R_G \ll z_{in}$.
b. $z_{out} \gg R_L$.
c. $z_{out} \ll R_L$.
d. $z_{out} = R_L$.

71. If a CE stage is directly coupled to an emitter follower,
a. low and high frequencies will be passed.
b. only high frequencies will be passed.
c. high-frequency signals will be blocked.
d. low-frequency signals will be blocked.

72. A Darlington transistor has
a. a very low input impedance.
b. three transistors.
c. a very high current gain.
d. one V_{BE} drop.

73. The amplifier configuration that produces a 180° phase shift is the
a. CB.
b. CC.
c. CE.
d. All of the above.

74. If the generator voltage is 5 mV in an emitter follower, the output voltage across the load is closest to
a. 5 mV.
b. 150 mV.
c. 0.25 V.
d. 0.5 V.

75. Usually, the distortion in an emitter follower is
a. very low.
b. very high.
c. large.
d. not acceptable.

76. If a CE stage is direct coupled to an emitter follower, how many coupling capacitors are there between the two stages?
a. 0
b. 1
c. 2
d. 3

77. A Darlington transistor has a b of 8000. If $R_E = 1\ k\Omega$ and $R_L = 100\ \Omega$, the input impedance of the base is closest to
a. 8 kΩ.
b. 80 kΩ.
c. 800 kΩ.
d. 8 MΩ.

78. The tail current of a diff amp is
a. half of either collector current.
b. equal to either collector current.
c. two times either collector current.
d. equal to the difference in base currents.

79. The voltage gain of a diff amp with an unloaded differential output is equal to R_C divided by
a. r_e'.
b. $r_e'/2$.
c. $2r_e'$.
d. R_E.

80. The input impedance of a diff amp equals r_e' times
a. 0.
b. R_C.
c. R_E.
d. 2β.

81. A dc signal has a frequency of
a. 0 Hz.
b. 60 Hz.
c. 0 to more than 1 MHz.
d. 1 MHz.

82. When the two input terminals of a diff amp are grounded,
a. the base currents are equal.
b. the collector currents are equal.
c. an output error voltage usually exists.
d. the ac output voltage is zero.

83. A common-mode signal is applied to
a. the noninverting input.
b. the inverting input.
c. both inputs.
d. the top of the tail resistor.

84. The common-mode voltage gain is
a. smaller than the voltage gain.
b. equal to the voltage gain.
c. greater than the voltage gain.
d. none of the above.

85. The input stage of an op amp is usually a
a. diff amp.
b. class B push-pull amplifier.
c. CE amplifier.
d. swamped amplifier.

86. The common-mode rejection ratio is
a. very low.
b. often expressed in decibels.
c. equal to the voltage gain.
d. equal to the common-mode voltage gain.

87. The typical input stage of an op amp has a
a. single-ended input and single-ended output.
b. single-ended input and differential output.
c. differential input and single-ended output.
d. differential input and differential output.

88. Active loads, current source, and current mirrors are used to eliminate what components in an IC?
a. Capacitors.
b. Resistors.

c. Diodes.

d. Inductors.

89. What is the common practice for getting higher gain in an IC differential amplifier?

a. Use larger collector or drain resistors.

b. Use a current source.

c. Use a current mirror.

d. Use higher-gain transistors.

90. What replaces the emitter resistor in an IC differential amplifier?

a. Current source.

b. Current mirror.

c. MOSFET.

d. Zener diode.

91. Cascode amplifiers are popular for which frequency range?

a. DC.

b. Audio.

c. Medium.

d. Very high.

92. A driver circuit can operate in the linear region.

a. True.

b. False.

93. The collector-emitter voltage in a saturated transistor driver when on is

a. V_{CC}.

b. ground.

c. <0.5 V.

d. I_V to V_{CC}.

94. A diode is needed in a magnetic device driver to

a. bias the transistor.

b. control current in the inductive load.

c. speed up switching.

d. protect against voltage spikes.

95. A transistor has a base current of 0.5 mA, a current gain of 150. The collector resistance is 2 kΩ and the collector supply is +12 V. The transistor

a. is open.

b. is operating in the linear region.

c. is saturated.

d. has insufficient information to tell.

96. Which is true about a saturated transistor?

a. EB forward-biased; CB reverse-biased.

b. Both EB and CB forward-biased.

c. EB reverse-biased; CB forward-biased.

d. Both EB and CB reverse-biased.

CHAPTER 33 PROBLEMS

SECTION 33.3 Transistor Currents

33.1 A transistor has an emitter current of 10 mA and a collector current of 9.95 mA. What is the base current?

33.2 The collector current is 10 mA, and the base current is 0.1 mA. What is the current gain?

33.3 A transistor has a current gain of 150 and a base current of 30 *m*A. What is the collector current?

33.4 If the collector current is 100 mA and the current gain is 65, what is the emitter current?

SECTION 33.5 The Base Curve

33.5 What is the base current in Fig. 33-49?

SECTION 33.6 Collector Curves

33.6 If a transistor has a collector current of 100 mA and a collector-emitter voltage of 3.5 V, what is its power dissipation?

SECTION 33.7 Reading Data Sheets

33.7 A transistor has a power rating of 1 W. If the collector-emitter voltage is 10 V and the collector current is 120 mA, what happens to the transistor?

SECTION 33.9 Basic Biasing

33.8 In Fig. 33-50, what is V_{CE}? Is this correct biasing for this circuit?

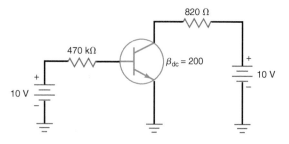

Figure 33-49

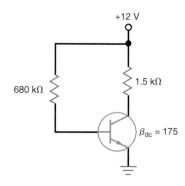

Figure 33-50

SECTION 33.10 Emitter Bias

33.9 What is the collector voltage in Fig. 33-51*a*? The emitter voltage?

33.10 What is the collector voltage in Fig. 33-51*b* if $V_{BB} = 2$ V?

33.11 If the base supply voltage is 2 V in Fig. 33-51*c*, what is the current through the LED?

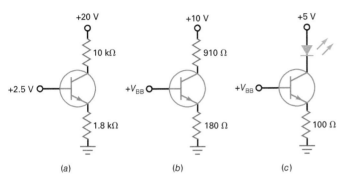

(a)   (b)   (c)

Figure 33-51

SECTION 33.11 Voltage Divider Bias

33.12 What is the emitter voltage in Fig. 33-52? The collector voltage?

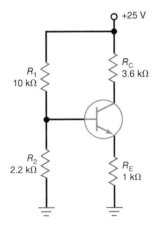

Figure 33-52

SECTION 33.12 Troubleshooting Transistor Circuits

33.13 What is the approximate value of the collector voltage in Fig. 33-53 for each of these troubles?
- **a.** R_1 open
- **b.** R_2 open
- **c.** R_E open
- **d.** R_C open
- **e.** Collector-emitter open

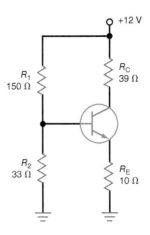

Figure 33-53

SECTION 33.13 *PNP* Transistors

33.14 What is the collector voltage in Fig. 33-54?

33.15 What is the collector-emitter voltage in Fig. 33-54?

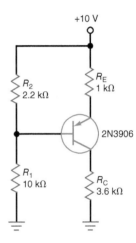

Figure 33-54

SECTION 33.14 Typical Bipolar Amplifiers

33.16 What is the ac resistance of the emitter diode in Fig. 33-55?

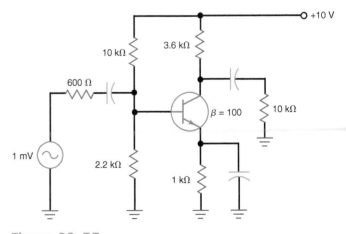

Figure 33-55

33.17 What is the output voltage of Fig. 33-55?

33.18 What is the gain of the amplifier in Fig. 33-55 if the emitter bypass capacitor is removed?

SECTION 33.15 Emitter Follower

33.19 In Fig. 33-56, what is the input impedance of the base if $\beta = 200$? The input impedance of the stage?

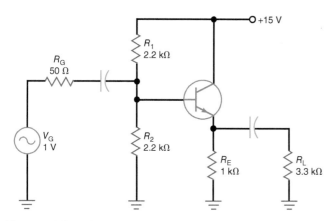

Figure 33-56

33.20 If $\beta = 150$ in Fig. 33-56, what is the ac input voltage to the emitter follower?

33.21 What is the voltage gain in Fig. 33-56? If $\beta = 175$, what is the ac load voltage?

33.22 If the Darlington pair of Fig. 33-57 has an overall current gain of 5000, what is the input impedance of the Q_1 base? What is the output voltage?

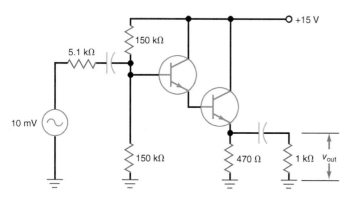

Figure 33-57

SECTION 33.16 The Differential Amplifier

33.23 What are the ideal currents and voltages in Fig. 33-58?

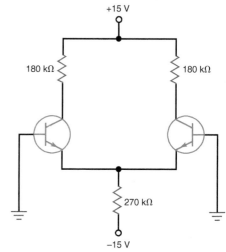

Figure 33-58

33.24 What are the ideal currents and voltages in Fig. 33-59?

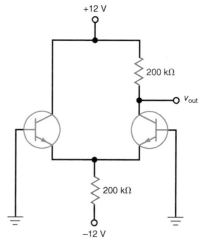

Figure 33-59

33.25 In Fig. 33-60, what is the ac output voltage? If $\beta = 275$, what is the input impedance of the diff amp?

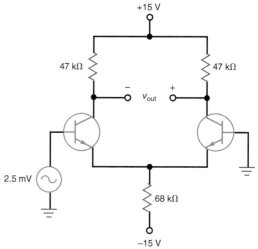

Figure 33-60

33.26 What is the common-mode voltage gain of Fig. 33-61? If a common-mode voltage of 20 μV exists on both bases, what is the common-mode output voltage?

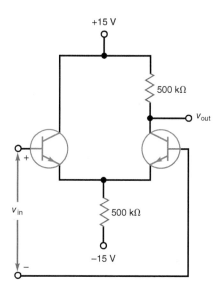

Figure 33-61

33.27 In Fig. 33-61, v_{in} = 2 mV and $v_{in(CM)}$ = 5 mV. What is the ac output voltage?

33.28 A 741C is an op amp with A_v = 100,000 and a minimum $CMRR_{dB}$ = 70 dB. What is the common-mode voltage gain? If a desired and common-mode signal each has a value of 5 μV, what is the output voltage?

SECTION 33.17 The Transistor Switch

33.29 In Fig. 33-62, use the circuit values shown unless otherwise indicated. Determine whether the transistor is saturated for each of these changes:
a. R_B = 51 kΩ and h_{FE} = 100
b. V_{BB} = 10 Ω and h_{FE} = 500
c. R_C = 10 kΩ and h_{FE} = 100
d. V_{CC} = 10 Ω and h_{FE} = 100

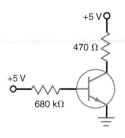

Figure 33-62

33.30 The 680 kΩ in Fig. 33-62 is replaced by 4.7 kΩ and a series switch. Assuming an ideal transistor, what is the collector voltage if the switch is open? What is the collector voltage if the switch is closed?

SECTION 33.18 Driver Circuits

33.31 If the base supply voltage is 2 V in Fig. 33-63, what is the current through the LED?

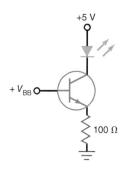

Figure 33-63

33.32 What is the maximum possible value of current through the 2 kΩ of Fig. 33-64?

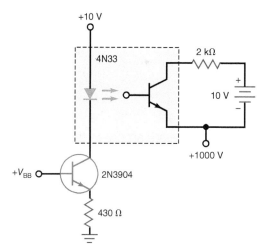

Figure 33-64

Chapter 34

Amplifier Fundamentals

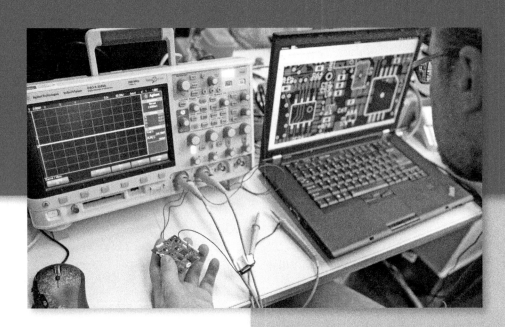

Most amplifiers are defined by the range of frequencies they cover. There are amplifiers that amplify signals from dc to well into the gigahertz region. The typical amplifier today is in integrated circuit form. All of them use similar ways to define gain and bandwidth. This chapter explains the gain versus the frequency response fundamentals.

Learning Outcomes

After studying this chapter, you should be able to:

> *List* five common amplifier types.

> *Calculate* decibel power gain and decibel voltage gain and *state* the implications of the impedance-matched condition.

> *Explain* how input and output impedances affect overall gain.

> *Sketch* Bode plots for both magnitude and phase.

> *Describe* the rise time–bandwidth relationship.

34.1 Types of Amplifiers

There are two basic ways to categorize amplifiers: by the amplitudes of the signals they handle and by the frequencies they amplify. Signal levels are designated small signals and large signals. Small signals are generally less than 1 or 2 V. These signals are more often in the millivolt and microvolt range. Amplifiers for such signals are voltage amplifiers.

Large signals are those several volts and higher. These amplifiers usually provide power amplification to heavy loads.

Frequency coverage defines a wide range of different amplifiers optimized for various frequency ranges. The most common frequency ranges are discussed below.

DC Amplifiers

DC amplifiers boost the signal levels of small dc voltages, usually from sensors or transducers that convert physical characteristics like temperature or pressure into electrical variables. Slowly varying signals are the norm, so dc amplifiers have a restricted upper-frequency limit, typically less than several kilohertz.

Audio Amplifiers

Audio amplifiers are optimized for audio signals like voice or music in the 20-Hz to 20-kHz range. Some amplifiers restrict the upper range to 4 kHz, limiting the application to voice frequencies like phones. Others extend the upper-frequency range beyond 20 kHz to 30 kHz to ensure passage of music harmonics. Small-signal amplifiers, often called *preamplifiers (preamps),* are usually of the low-noise type. Such amplifiers contribute very low noise, so they can amplify small signals from microphones, magnetic tape heads, CD optical pickups, MP3 players, and other audio devices. Audio power amplifiers boost signal levels to drive speakers and headphones. Power levels range from several hundred milliwatts to about 100 W. Both small-signal and power amplifiers are available in IC form.

Video Amplifiers

Video amplifiers are small-signal amplifiers designed for amplifying video signals, including TV. Their range extends from dc up to about several hundred megahertz, and they are found in digital HDTV sets, LCD monitors, camcorders, and other video gear. They are available in IC form.

RF Amplifiers

RF amplifiers are mostly small-signal amplifiers for radio signals, but RF power amplifiers are also available. Frequency coverage is typically above 1 MHz up to well over 10 GHz. The IC RF amplifiers typically only cover a limited range in a desired band. For example, there are amplifiers with a range of up to 50 MHz, from 30 to 300 MHz, and amplifiers that cover the lower-gigahertz range. The frequency range is often deliberately limited by external tuned circuits or filters.

There are also special RF amplifier categories. For example, the low-noise amplifier (LNA) is used at the front end of most radio receivers to amplify the small wireless signals from an antenna. Intermediate-frequency (IF) amplifiers amplify signals at common receiver circuit frequencies like 455 kHz, 9 MHz, 45 MHz, and 140 MHz. Most of these amplifiers are manufactured in IC form.

Microwave Amplifiers

Microwave amplifiers amplify signals at frequencies above 1 GHz. The upper range can extend to well over 100 GHz.

Op Amps

Operational amplifiers (op amps) are a class of general-purpose amplifiers that can be used to implement many specific types of amplifiers mentioned in this section. These high-gain dc IC amplifiers can be customized to many different applications by using external resistors, capacitors, diodes, and transistors. Two entire chapters are devoted to these popular amplifiers later in this book.

34.2 Frequency Response of an Amplifier

The **frequency response** of an amplifier is the graph of its gain versus the frequency. In this section, we discuss the frequency response of ac and dc amplifiers. Earlier, we discussed a transistor amplifiers with coupling and bypass capacitors. These are examples of *ac amplifiers,* ones designed to amplify ac signals. It is also possible to design a *dc amplifier,* one that can amplify dc signals as well as *ac* signals.

Response of an AC Amplifier

Figure 34-1a shows the *frequency response* of an ac amplifier. In the middle range of frequencies, the voltage gain is maximum. This middle range of frequencies is where the amplifier is normally operated. At low frequencies, the voltage gain decreases because the coupling and bypass capacitors no longer act like short circuits. Instead, their capacitive reactances are large enough to drop some of the ac signal voltage. The result is a loss of voltage gain as we approach zero hertz.

At high frequencies, the voltage gain decreases for other reasons. To begin with, a transistor has **internal capacitances** across its junctions, as shown in Fig. 34-1b. These capacitances provide bypass paths for the ac signal. As the frequency increases, the capacitive reactances become low enough to prevent normal transistor action. The result is a

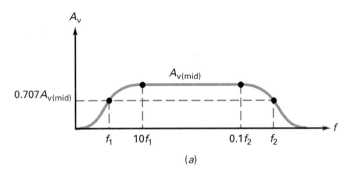

(a)

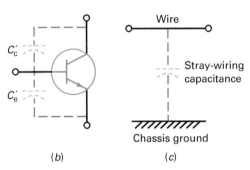

(b)   (c)

Figure 34-1 (a) Frequency response of ac amplifier. (b) Internal capacitance of transistor. (c) Connecting wire forms capacitance with chassis.

loss of voltage gain. The upper frequency response is caused by low-pass filter effects.

Stray-wiring capacitance is another reason for a loss of voltage gain at high frequencies. Figure 34-1c illustrates the idea. Any connecting wire in a transistor circuit acts like one plate of a capacitor, and the chassis ground acts like the other plate. The closely spaced copper patterns on a printed circuit board (PCB) form distributed low-pass filters. The stray-wiring capacitance that exists between this wire and ground is unwanted. At higher frequencies, its low capacitive reactance prevents the ac current from reaching the load resistor. This is equivalent to saying that the voltage gain drops off.

Cutoff Frequencies

The frequencies at which the voltage gain equals 0.707 of its maximum value are called the **cutoff frequencies.** In Fig. 34-1a, f_1 is the lower cutoff frequency and f_2 is the upper cutoff frequency. The cutoff frequencies are also referred to as the **half-power frequencies** because the load power is half of its maximum value at these frequencies.

Why is the output power half of maximum at the cutoff frequencies? When the voltage gain is 0.707 of the maximum value, the output voltage is 0.707 of the maximum value. Recall that power equals the square of voltage divided by resistance. When you square 0.707, you get 0.5. This is why the load power is half of its maximum value at the cutoff frequencies.

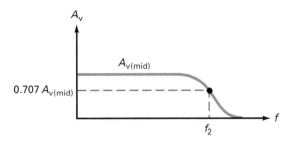

Figure 33-2 Frequency response of dc amplifier.

Midband

We define the **midband of an amplifier** as the band of frequencies between $10f_1$ and $0.1f_2$. In the midband, the voltage gain of the amplifier is approximately maximum, designated by $A_{v(mid)}$. Three important characteristics of any ac amplifier are its $A_{v(mid)}$, f_1, and f_2. Given these values, we know how much voltage gain there is in the midband and where the voltage gain is down to $0.707A_{v(mid)}$.

Response of a DC Amplifier

It is possibles to use direct coupling between amplifier stages. This allows the circuit to amplify all the way down to zero hertz. This type of amplifier is called a **dc amplifier.**

Figure 34-2 shows the frequency response of a dc amplifier. Since there is no lower cutoff frequency, the two important characteristics of a dc amplifier are $A_{v(mid)}$ and f_2. Given these two values on a data sheet, we have the voltage gain of the amplifier in the midband and its upper cutoff frequency.

The dc amplifier is more widely used than the ac amplifier because most amplifiers are now being designed with op amps instead of with discrete transistors. An *op amp* is a dc amplifier that has high voltage gain, high input impedance, and low output impedance. A wide variety of op amps is commercially available as integrated circuits (ICs). Most dc amplifiers are designed with one dominant capacitance that produces the cutoff frequency f_2.

EXAMPLE 34-1

Figure 34-3a shows an ac amplifier with a midband voltage gain of 200. If the cutoff frequencies are f_1 = 20 Hz and f_2 = 20 kHz, what does the frequency response look like?

Answer:

In the midband, the voltage gain is 200. At either cutoff frequency, it equals

$$A_v = 0.707(200) = 141$$

Figure 34-3b shows the frequency response.

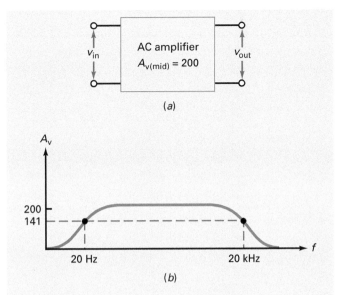

Figure 34-3 AC amplifier and its frequency response.

EXAMPLE 34-2

Figure 34-4a shows a 741C, an op amp with a midband voltage gain of 100,000. If f_2 = 10 Hz, what does the frequency response look like?

Answer:

At the cutoff frequency of 10 Hz, the voltage gain is 0.707 of its midband value:

$$A_v = 0.707(100,000) = 70,700$$

Figure 34-4b shows the frequency response. Notice that the voltage gain is 100,000 at a frequency of zero hertz. As the input frequency approaches 10 Hz, the voltage gain decreases until it equals approximately 70% of maximum.

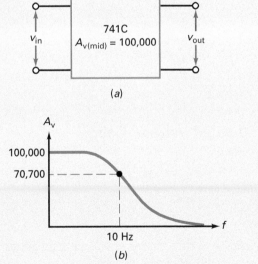

Figure 34-4 The 741C and its frequency response.

Bandwidth

Another name for frequency response is bandwidth (BW). *Bandwidth* is defined as the frequency range over which a circuit or component operates. It is calculated as the difference between the upper and lower cutoff frequencies.

$$\text{BW} = f_2 - f_1 \qquad \text{(34-1)}$$

An amplifier with f_2 = 2.7 GHz and f_1 = 400 MHz has a bandwidth of

$$\text{BW} = 2700 - 400 = 2300 \text{ MHz}$$

A dc amplifier with f_2 = 3 MHz has a bandwidth of

$$\text{BW} = 3 - 0 = 3 \text{ MHz}$$

34.3 Decibel Power Gain

We are about to discuss **decibels**, a useful method for describing frequency response. But before we do, we need to review some ideas from basic mathematics.

Review of Logarithms

Suppose we are given this equation:

$$x = 10^y \qquad \text{(34-2)}$$

This equation can be solved for y in terms of x to get

$$y = \log_{10} x$$

This says that y is the logarithm (or exponent) of 10 that gives x. Usually, the 10 is omitted, and the equation is written as

$$y = \log x \qquad \text{(34-3)}$$

With a calculator that has the common log function, you can quickly find the y value for any x value. For instance, here is how to calculate the value of y for x = 10, 100, and 1000:

$$y = \log 10 = 1$$
$$y = \log 100 = 2$$
$$y = \log 1000 = 3$$

As you can see, each time x increases by a factor of 10, y increases by 1.

You can also calculate y values, given decimal values of x. For instance, here are the values of y for x = 0.1, 0.01, and 0.001:

$$y = \log 0.1 = -1$$
$$y = \log 0.01 = -2$$
$$y = \log 0.001 = -3$$

Each time x decreases by a factor of 10, y decreases by 1.

Definition of $A_{p(dB)}$

Power gain A_p is defined as the output power divided by the input power:

$$A_p = \frac{p_{out}}{p_{in}}$$

Decibel power gain is defined as

$$A_{p(dB)} = 10 \log A_p \qquad (34\text{-}4)$$

Since A_p is the ratio of output power to input power, A_p has no units or dimensions. When you take the logarithm of A_p, you get a quantity that has no units or dimensions. But to make sure that $A_{p(dB)}$ is never confused with A_p, we attach the unit *decibel* (abbreviated *dB*) to all answers for $A_{p(dB)}$.

For instance, if an amplifier has a power gain of 100, it has a decibel power gain of

$$A_{p(dB)} = 10 \log 100 = 20 \text{ dB}$$

As another example, if $A_p = 100,000,000$, then

$$A_{p(dB)} = 10 \log 100,000,000 = 80 \text{ dB}$$

In both these examples, the log equals the number of zeros: 100 has two zeros, and 100,000,000 has eight zeros. You can use the zero count to find the logarithm whenever the number is a multiple of 10. Then, you can multiply by 10 to get the decibel answer. For instance, a power gain of 1000 has three zeros; multiply by 10 to get 30 dB. A power gain of 100,000 has five zeros; multiply by 10 to get 50 dB. This shortcut is useful for finding decibel equivalents and checking answers.

Decibel power gain is often used on data sheets to specify the power gain of devices. One reason for using decibel power gain is that logarithms compress numbers. For instance, if an amplifier has a power gain that varies from 100 to 100,000,000, the decibel power gain varies from 20 to 80 dB. As you can see, decibel power gain is a more compact notation than ordinary power gain.

Two Useful Properties

Decibel power gain has two useful properties:

1. Each time the ordinary power gain increases (decreases) by a factor of 2, the decibel power gain increases (decreases) by 3 dB.

2. Each time the ordinary power gain increases (decreases) by a factor of 10, the decibel power gain increases (decreases) by 10 dB.

Table 34-1 shows these properties in compact form. The following examples will demonstrate these properties.

Table 34-1 Properties of Power Gain

Factor	Decibel, dB
×2	+3
×0.5	−3
×10	+10
×0.1	−10

EXAMPLE 34-3

Calculate the decibel power gain for the following values: $A_p = 1$, 2, 4, and 8.

Answer:

With a calculator, we get the following answers:

$$A_{p(dB)} = 10 \log 1 = 0 \text{ dB}$$
$$A_{p(dB)} = 10 \log 2 = 3 \text{ dB}$$
$$A_{p(dB)} = 10 \log 4 = 6 \text{ dB}$$
$$A_{p(dB)} = 10 \log 8 = 9 \text{ dB}$$

Each time A_p increases by a factor of 2, the decibel power gain increases by 3 dB. This property is always true. Whenever you double the power gain, the decibel power gain increases by 3 dB.

EXAMPLE 34-4

Calculate the decibel power gain for each of these values: $A_p = 1$, 0.5, 0.25, and 0.125.

Answer:

$$A_{p(dB)} = 10 \log 1 = 0 \text{ dB}$$
$$A_{p(dB)} = 10 \log 0.5 = -3 \text{ dB}$$
$$A_{p(dB)} = 10 \log 0.25 = -6 \text{ dB}$$
$$A_{p(dB)} = 10 \log 0.125 = -9 \text{ dB}$$

Each time A_p decreases by a factor of 2, the decibel power gain decreases by 3 dB.

When the output power is less than the input power, we no longer have gain. Instead we have attenuation or loss. This is indicated by a negative logarithm and dB value. Many circuits have losses instead of gains. *RC* and *LC* filters are the best example. Voltage dividers offer attenuation. There are also special attenuator circuits for adding a desired amount loss to a circuit.

34.4 Decibel Voltage Gain

Voltage measurements are more common than power measurements. For this reason, decibels are even more useful with voltage gain.

Definition

As defined in earlier chapters, voltage gain is the output voltage divided by the input voltage:

$$A_v = \frac{v_{out}}{v_{in}}$$

Decibel voltage gain is defined as

$$A_{v(dB)} = 20 \log A_v \qquad (34\text{-}5)$$

The reason for using 20 instead of 10 in this definition is because power is proportional to the square of voltage. As will be discussed in the next section, this definition produces an important derivation for impedance-matched systems.

If an amplifier has a voltage gain of 100,000, it has a decibel voltage gain of

$$A_{v(dB)} = 20 \log 100,000 = 100 \text{ dB}$$

We can use a shortcut whenever the number is a multiple of 10. Count the number of zeros and multiply by 20 to get the decibel equivalent. In the foregoing calculation, count five zeros and multiply by 20 to get the decibel voltage gain of 100 dB.

As another example, if an amplifier has a voltage gain that varies from 100 to 100,000,000, then its decibel voltage gain varies from 40 to 160 dB.

Basic Rules for Voltage Gain

Here are the useful properties for decibel voltage gain:

1. Each time the voltage gain increases (decreases) by a factor of 2, the decibel voltage gain increases (decreases) by 6 dB.
2. Each time the voltage gain increases (decreases) by a factor of 10, the decibel voltage gain increases (decreases) by 20 dB.

Table 34-2 summarizes these properties.

Cascaded Stages

In Fig. 34-5, the total voltage gain of the two-stage amplifier is the product of the individual voltage gains:

$$A_v = (A_{v_1})(A_{v_2}) \qquad (34\text{-}6)$$

Table 34-2 Properties of Voltage Gain

Factor	Decibel, dB
×2	+6
×0.5	−6
×10	+20
×0.1	−20

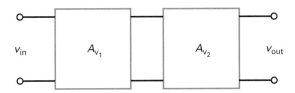

Figure 34-5 Two stages of voltage gain.

For instance, if the first stage has a voltage gain of 100 and the second stage has a voltage gain of 50, the total voltage gain is

$$A_v = (100)(50) = 5000$$

Something unusual happens in Formula (34-6) when we use the decibel voltage gain instead of the ordinary voltage gain:

$$A_{v(dB)} = 20 \log A_v = 20 \log (A_{v_1})(A_{v_2})$$
$$= 20 \log A_{v_1} + 20 \log A_{v_2}$$

This can be written as

$$A_{v(dB)} = A_{v_1(dB)} + A_{v_2(dB)} \qquad (34\text{-}7)$$

This equation says that the total decibel voltage gain of two cascaded stages equals the sum of the individual decibel voltage gains. The same idea applies to any number of stages. This additive property of decibel gain is one reason for its popularity.

EXAMPLE 34-5

What is the total voltage gain in Fig. 34-6a? Express this in decibels. Next, calculate the decibel voltage gain of each stage and the total decibel voltage gain using Formula (34-7).

Answer:

With Formula. (34-6), the total voltage gain is

$$A_v = (100)(200) = 20,000$$

In decibels, this is

$$A_{v(dB)} = 20 \log 20,000 = 86 \text{ dB}$$

You can use a calculator to get 86 dB, or you can use the following shortcut: The number 20,000 is the same as 2 times 10,000. The number 10,000 has four zeros, which means that the decibel

equivalent is 80 dB. Because of the factor of 2, the final answer is 6 dB higher, or 86 dB.

Next, we can calculate the decibel voltage gain of each stage as follows:

$$A_{v1(dB)} = 20 \log 100 = 40 \text{ dB}$$
$$A_{v2(dB)} = 20 \log 200 = 46 \text{ dB}$$

Figure 34-6b shows these decibel voltage gains. With Formula (34-7), the total decibel voltage gain is

$$A_{v(dB)} = 40 \text{ dB} + 46 \text{ dB} = 86 \text{ dB}$$

As you can see, adding the decibel voltage gain of each stage gives us the same answer calculated earlier.

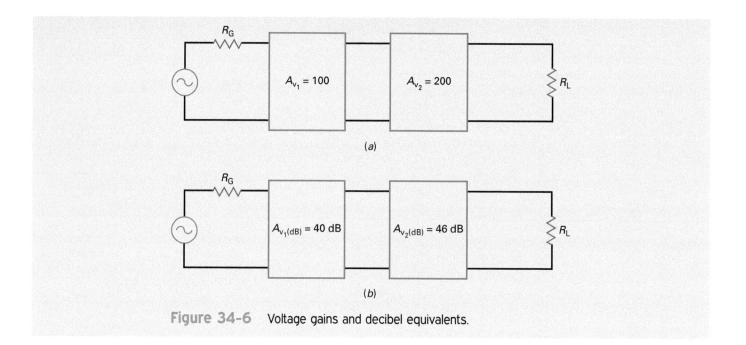

Figure 34-6   Voltage gains and decibel equivalents.

34.5 Impedance Matching

Figure 34-7a shows an amplifier stage with a generator resistance of R_G, an input resistance of R_{in}, an output resistance of R_{out}, and a load resistance of R_L. Up to now, most of our discussions have used different impedances.

In many communication systems (microwave, television, and telephone), all impedances are matched; that is, $R_G = R_{in} = R_{out} = R_L$. Figure 34-7b illustrates the idea. As indicated, all impedances equal R. The impedance R is 50 Ω in microwave systems, 75 Ω (coaxial cable) or 300 Ω (twin-lead) in television systems, and 600 Ω in telephone systems.

Impedance matching is used in these systems because it produces maximum power transfer.

In an impedance-matched system, the power gain equals the square of the voltage gain.

$$A_p = A_v^2 \tag{34-8}$$

In terms of decibels,

$$A_{p(dB)} = 10 \log A_p = 10 \log A_v^2 = 20 \log A_v$$

or

$$A_{p(dB)} = A_{v(dB)} \tag{34-9}$$

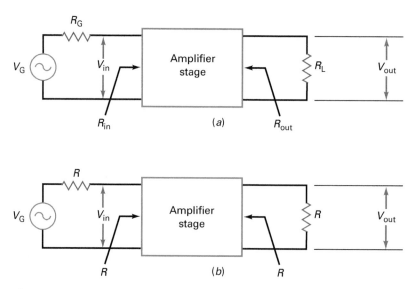

Figure 34-7   Impedance matching.

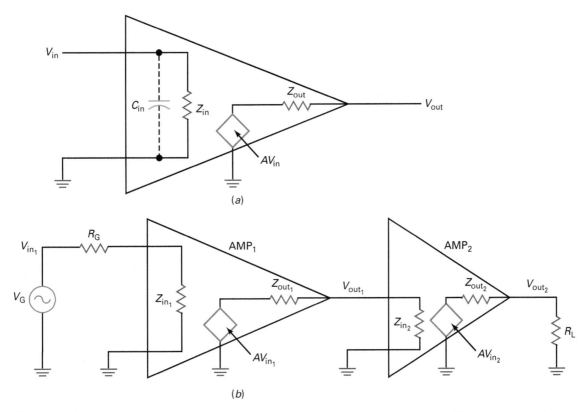

Figure 34-8   (a) Input and output impedances of an amplifier. (b) How one amplifier "loads" another.

This says that the decibel power gain equals the decibel voltage gain. Formula (34-9) is true for any impedance-matched system. If a data sheet states that the gain of a system is 40 dB, then both decibel power gain and voltage gain equal 40 dB.

In cascaded voltage amplifiers, there is no matching of input and output impedances as the goal is maximum voltage amplification rather than maximum power transfer to a load. The output impedance, Z_{out}, of one amplifier drives the input impedance, Z_{in}, of the next amplifier as shown in Fig. 34-8. These two impedances, which are primarily resistive, form a voltage divider that offsets the voltage gain in the amplifiers.

EXAMPLE 34-6

The gain, A_N, and impedance, Z, values for amplifiers AMP$_1$ and AMP$_2$ in Fig. 34-8 are

$$V_G = 50 \text{ mV}$$
$$R_G = 600 \ \Omega$$
$$Z_{in_1} = 1200 \ \Omega$$
$$A_1 = 15$$
$$Z_{out_1} = 4 \text{ k}\Omega$$
$$Z_{in_2} = 20 \text{ k}\Omega$$
$$A_2 = 4$$
$$Z_{out_2} = 100 \ \Omega$$
$$R_L = 200 \ \Omega$$

The actual input voltage, v_{in_1}, to AMP$_1$ is

$$v_{in_1} = V_G \left(\frac{Z_{in_1}}{Z_{in_1} + R_G} \right)$$

$$= 50 \text{ mV} \left(\frac{1200}{1200 + 600} \right) = 33.33 \text{ mV}$$

Amplifier AMP$_1$ amplifies this by 15 to produce a V_{out_1} of

$$V_{out_1} = A_1(v_{in_1}) = 15(33.33 \text{ mV}) = 500 \text{ mV}$$

The input voltage to AMP$_2$ is

$$V_{in_2} = V_{out_1} \left(\frac{Z_{in_2}}{Z_{in_2} + Z_{out_1}} \right)$$

$$= 500 \text{ mV} \left(\frac{20 \text{ k}\Omega}{20 \text{ k}\Omega + 4 \text{ k}\Omega} \right) = 416.7 \text{ mV}$$

Using AMP$_2$ amplifies this to

$$V_{out_2} = A_2(V_{in_2})$$
$$= 4(416.7 \text{ mV}) = 1666.7 \text{ mV or } 1.67 \text{ V}$$

The voltage across the load is

$$V_{RL} = V_{out_2}\left(\frac{R_L}{R_L + Z_{out_2}}\right)$$
$$= 1.67 \text{ V}\left(\frac{200}{200 + 100}\right) = 1.11 \text{ V}$$

Note: Without the voltage divider effect, the total gain would be

$$A_T = A_1A_2 = 15(4) = 60$$

With 50 mV in, the voltage on the load would be

$$V_{RL} = A_T V_G = 60(50 \text{ mV}) = 3 \text{ V}$$

As you can see, this load effect of one amplifier by another has a significant effect. That is why most voltage amplifiers are designed with high input impedance and low output impedance to minimize this effect.

Converting Decibel to Ordinary Gain

When a data sheet specifies the decibel power gain or voltage gain, you can convert the decibel gain to ordinary gain with the following equations:

$$A_p = \text{antilog} \frac{A_{p(dB)}}{10} \quad \textbf{(34-10)}$$

and

$$A_v = \text{antilog} \frac{A_{v(dB)}}{20} \quad \textbf{(34-11)}$$

The antilog is the inverse logarithm. These conversions are easily done on a scientific calculator that has a log function and an inverse key.

EXAMPLE 34-7

Figure 34-9 shows impedance-matched stages with $R = 50\ \Omega$. What is the total decibel gain? What is the total power gain? The total voltage gain?

Answer:

The total decibel voltage gain is

$$A_{v(dB)} = 23 \text{ dB} + 36 \text{ dB} + 31 \text{ dB} = 90 \text{ dB}$$

The total decibel power gain also equals 90 dB because the stages are impedance-matched.

With Formula (34-10), the total power gain is

$$A_p = \text{antilog} \frac{90 \text{ dB}}{10} = 1{,}000{,}000{,}000$$

and the total voltage gain is

$$A_v = \text{antilog} \frac{90 \text{ dB}}{20} = 31{,}623$$

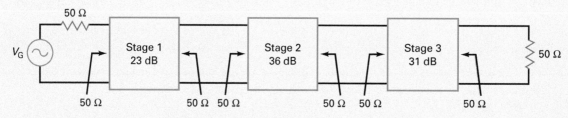

Figure 34-9 Impedance matching in a 50-Ω system.

EXAMPLE 34-8

In the preceding example, what is the ordinary voltage gain of each stage?

Answer:

The first stage has a voltage gain of

$$A_{v1} = \text{antilog} \frac{23 \text{ dB}}{20} = 14.1$$

The second stage has a voltage gain of

$$A_{v2} = \text{antilog} \frac{36 \text{ dB}}{20} = 63.1$$

The third stage has a voltage gain of

$$A_{v3} = \text{antilog} \frac{31 \text{ dB}}{20} = 35.5$$

34.6 Decibels above a Reference

In this section, we will discuss two more ways to use decibels. Besides applying decibels to power and voltage gains, we can use *decibels above a reference*. The reference levels used in this section are the milliwatt and the volt.

The Milliwatt Reference

Decibels are sometimes used to indicate the power level above 1 mW. In this case, the label *dBm* is used instead of dB. The *m* at the end of dBm reminds us of the milliwatt reference. The dBm equation is

$$P_{dBm} = 10 \log \frac{P}{1 \text{ mW}} \qquad \text{(34-12)}$$

where P_{dBm} is the power expressed in dBm. For instance, if the power is 2 W, then

$$P_{dBm} = 10 \log \frac{2\text{W}}{1 \text{ mW}} = 10 \log 2000 = 33 \text{ dBm}$$

Using dBm is a way of comparing the power to 1 mW. If a data sheet says that the output of a power amplifier is 33 dBm, it is saying that the output power is 2 W. Table 34-3 shows some dBm values.

You can convert any dBm value to its equivalent power by using this equation:

$$P = \text{antilog} \frac{P_{dBm}}{10} \qquad \text{(34-13)}$$

where P is the power in milliwatts.

The Volt Reference

Decibels can also be used to indicate the voltage level above 1 V. In this case, the label *dBV* is used. The dBV equation is

$$V_{dBV} = 20 \log \frac{V}{1 \text{ V}}$$

Since the denominator equals 1, we can simplify the equation to

$$V_{dBV} = 20 \log V \qquad \text{(34-14)}$$

Table 34-3 Power in dBm

Power	P_{dBm}
1 μW	−30
10 μW	−20
100 μW	−10
1 mW	0
10 mW	10
100 mW	20
1 W	30

Table 34-4 Voltage in dBV

Voltage	V_{dBV}
10 μV	−100
100 μV	−80
1 mV	−60
10 mV	−40
100 mV	−20
1 V	0
10 V	+20
100 V	+40

where V is dimensionless. For instance, if the voltage is 25 V, then

$$V_{dBV} = 20 \log 25 = 28 \text{ dBV}$$

Using dBV is a way of comparing the voltage to 1 V. If a data sheet says that the output of a voltage amplifier is 28 dBV, it is saying that the output voltage is 25 V. If the output level or sensitivity of a microphone is specified as −40 dBV, its output voltage is 10 mV. Table 34-4 shows some dBV values.

You can convert any dBV value to its equivalent voltage using this equation:

$$V = \text{antilog} \frac{V_{dBV}}{20} \qquad \text{(34-15)}$$

where V is the voltage in volts.

EXAMPLE 34-9

A data sheet says that the output of an amplifier is 24 dBm. What is the output power?

Answer:
With a calculator and Formula (34-13),

$$P = \text{antilog} \frac{24 \text{ dBm}}{10} = 251 \text{ mW}$$

EXAMPLE 34-10

If a data sheet says that the output of an amplifier is −34 dBV, what is the output voltage?

Answer:
With Formula (34-14),

$$V = \text{antilog} \frac{-34 \text{ dBV}}{20} = 20 \text{ mV}$$

34.7 Bode Plots

Figure 34-10 shows the frequency response of an ac amplifier. Although it contains some information such as the midband voltage gain and the cutoff frequencies, it is an incomplete picture of the amplifier's behavior. This is where the **Bode plot** comes in. Because this type of graph uses decibels, it can give us more information about the amplifier's response outside the midband.

Octaves

The middle C on a piano has a frequency of 256 Hz. The next-higher C is an octave higher, and it has a frequency of 512 Hz. The next-higher C has a frequency of 1024 Hz, and so on. In music, the word *octave* refers to a doubling of the frequency. Every time you go up one octave, you have doubled the frequency.

In electronics, an octave has a similar meaning for ratios like f_1/f and f/f_2. For instance, if $f_1 = 100$ Hz and $f = 50$ Hz, the f_1/f ratio is

$$\frac{f_1}{f} = \frac{100 \text{ Hz}}{50 \text{ Hz}} = 2$$

We can describe this by saying that f is one octave below f_1. As another example, suppose $f = 400$ kHz and $f_2 = 200$ kHz. Then

$$\frac{f_1}{f_2} = \frac{400 \text{ kHz}}{200 \text{ kHz}} = 2$$

This means that f is one octave above f_2.

Decades

A *decade* has a similar meaning for ratios like f_1/f and f/f_2, except that a factor of 10 is used instead of a factor of 2. For instance, if $f_1 = 500$ Hz and $f = 50$ Hz, the f_1/f ratio is

$$\frac{f_1}{f} = \frac{500 \text{ Hz}}{50 \text{ Hz}} = 10$$

We can describe this by saying that f is one decade below f_1. As another example, suppose $f = 2$ MHz and $f_2 = 200$ kHz. Then

$$\frac{f}{f_2} = \frac{2 \text{ MHz}}{200 \text{ kHz}} = 10$$

This means that f is one decade above f_2.

Linear and Logarithmic Scales

Ordinary graph paper has a *linear scale* on both axes. This means that the spaces between the numbers are the same for all numbers, as shown in Fig. 34-11a. With a linear scale, you start at 0 and proceed in uniform steps toward higher numbers. All the graphs discussed up to now have used linear scales.

Sometimes we may prefer to use a **logarithmic scale** because it compresses very large values and allows us to see over many decades. Figure 34-11b shows a logarithmic scale. Notice that the numbering begins with 1. The space between 1 and 2 is much larger than the space between 9 and 10. By compressing the scale logarithmically as shown here, we can take advantage of certain properties of logarithms and decibels.

Both ordinary graph paper and semilogarithmic paper are available. Semilogarithmic graph paper has a linear scale on the vertical axis and a logarithmic scale on the horizontal axis. People use semilogarithmic paper when they want to graph a quantity like voltage gain over many decades of frequency. The main advantage of using logarithmic spacing is that a larger range of values can be shown in one plot without losing resolution in the smaller values.

Graph of Decibel Voltage Gain

Figure 34-12a shows the frequency response of a typical ac amplifier. The graph is similar to Fig. 34-10, but this time we are looking at the decibel voltage gain versus frequency as it would appear on semilogarithmic paper. A graph like this is called a *Bode plot*. The vertical axis uses a linear scale, and the horizontal axis uses a logarithmic scale.

As shown, the decibel voltage gain is maximum in the midband. At each cutoff frequency, the decibel voltage gain is down slightly from the maximum value. Below f_1, the decibel voltage gain decreases 20 dB per decade. Above f_2, the decibel voltage gain decreases 20 dB per decade. Decreases of 20 dB per decade occur in an amplifier where there is one dominant capacitor producing the

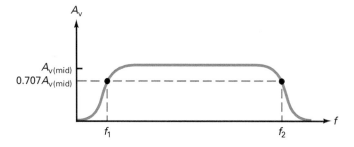

Figure 34-10 Frequency response of an ac amplifier.

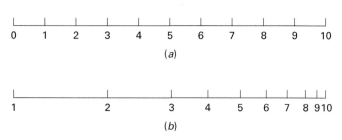

Figure 34-11 Linear and logarithmic scales.

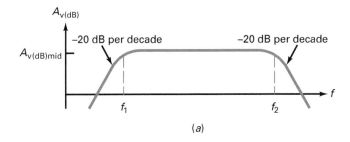

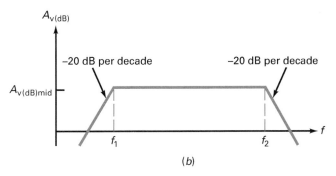

Figure 34-12   (a) Bode plot. (b) Ideal Bode plot.

lower cutoff frequency and one dominant bypass capacitor producing the upper cutoff frequency, as discussed in Sec. 34-2.

At the cutoff frequencies, f_1 and f_2, the voltage gain is 0.707 of the midband value. In terms of decibels,

$$A_{v(dB)} = 20 \log 0.707 = -3 \text{ dB}$$

We can describe the frequency response of Fig. 34-12a in this way: In the midband, the voltage gain is maximum. Between the midband and each cutoff frequency, the voltage gain gradually decreases until it is down 3 dB at the cutoff frequency. Then, the voltage gain rolls off (decreases) at a rate of 20 dB per decade.

Note: The 20 dB per decade rolloff rate is the same as the 6 dB per octave rate.

Ideal Bode Plot

Figure 34-12b shows the frequency response in *ideal* form. Many people prefer using the ideal Bode plot because it is easy to draw and gives approximately the same information. Anyone looking at this ideal graph knows that the decibel voltage gain is down 3 dB at the cutoff frequencies. The ideal Bode plot contains all the original information when this correction of 3 dB is mentally included.

Ideal Bode plots are approximations that allow us to draw the frequency response of an amplifier quickly and easily. They let us concentrate on the main issues rather than being caught in the details of exact calculations. For instance, an ideal Bode plot like Fig. 34-13 gives us a quick visual summary of an amplifier's frequency response. We can see the midband voltage gain (40 dB), the cutoff frequencies (1 kHz and 100 kHz), and rolloff rate (20 dB per decade). Also notice that the voltage gain equals 0 dB (unity or 1) at $f = 10$ Hz and $f = 10$ MHz. Ideal graphs like these are very popular in industry.

Incidentally, many technicians and engineers use the term *corner frequency* instead of *cutoff frequency*. This is because the ideal Bode plot has a sharp corner at each cutoff frequency. Another term often used is *break frequency*. This is because the graph breaks at each cutoff frequency and then decreases at a rate of 20 dB per decade.

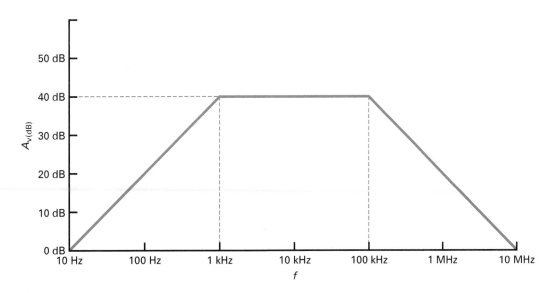

Figure 34-13   Ideal Bode plot of an ac amplifier.

EXAMPLE 34-11

The data sheet for a 741C op amp gives a midband voltage gain of 100,000, a cutoff frequency of 10 Hz, and rolloff rate of 20 dB per decade. Draw the ideal Bode plot. What is the ordinary voltage gain at 1 MHz?

Answer:

As mentioned in Sec. 34-2, op amps are dc amplifiers, so they have only an upper cutoff frequency. For a 741C, $f_2 = 10$ Hz. The midband voltage gain in decibels is

$$A_{v(dB)} = 20 \log 100{,}000 = 100 \text{ dB}$$

The ideal Bode plot has a midband voltage gain of 100 dB up to 10 Hz. Then it decreases 20 dB per decade.

Figure 34-14 shows the ideal Bode plot. After breaking at 10 Hz, the response rolls off 20 dB per decade until it equals 0 dB at 1 MHz. The ordinary voltage is unity (1) at this frequency. Data sheets often list the **unity-gain frequency** (symbolized f_{unity}) because it immediately tells you the frequency limitation of the op amp. The device can provide voltage gain up to unity-gain frequency but not beyond it.

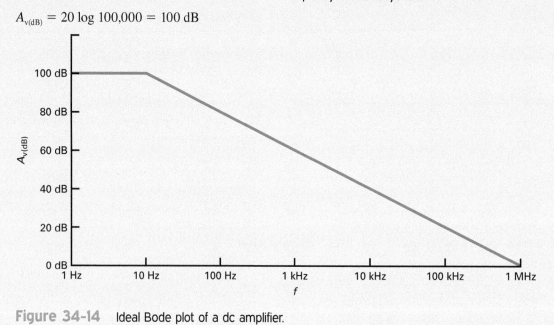

Figure 34-14 Ideal Bode plot of a dc amplifier.

34.8 More Bode Plots

Ideal Bode plots are useful approximations for preliminary analysis. But sometimes, we need more accurate answers. For instance, amplifiers also have a phase response that becomes a factor in their performance.

Lag Circuit

Most op amps include an *RC* lag circuit that rolls off the voltage gain at a rate of 20 dB per decade. This prevents *oscillations*, unwanted signals that can appear under certain conditions. Later chapters will explain oscillations and how the internal lag circuit of an op amp prevents these unwanted signals.

Figure 34-15 shows a circuit with bypass capacitor. The resistor, *R*, represents the Thevenized resistance facing the capacitor. This circuit is often called a **lag circuit** because the output voltage lags the input voltage at higher frequencies. Stated another way: If the input voltage has a phase angle of 0°, the output voltage has a phase angle between 0° and −90°.

At low frequencies, the capacitive reactance approaches infinity, and the output voltage equals the input voltage. As

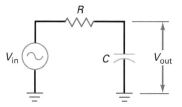

Figure 34-15 An *RC* bypass circuit.

the frequency increases, the capacitive reactance decreases, which decreases the output voltage. Recall from basic courses in electricity that the output voltage for this circuit is

$$V_{out} = \frac{X_C}{\sqrt{R^2 + X_C^2}} V_{in}$$

If we rearrange the foregoing equation, the voltage gain of Fig. 34-15 is

$$A_v = \frac{X_C}{\sqrt{R^2 + X_C^2}} \qquad \textbf{(34-16)}$$

Because the circuit has only passive devices, the voltage gain is always less than or equal to 1.

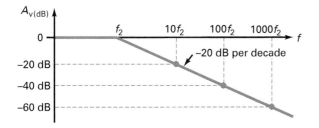

Figure 34-16   Ideal Bode plot of a lag circuit.

Table 34-5 Response of Lag Circuit	
f/f_2	ϕ
0.1	−5.71°
1	−45°
10	−84.3°
100	−89.4°
1000	−89.9°

The cutoff frequency of a lag circuit is where the voltage gain is 0.707. The equation for cutoff frequency is

$$f_2 = \frac{1}{2\pi RC} \qquad (34\text{-}17)$$

At this frequency, $X_C = R$ and the voltage gain is 0.707.

Figure 34-16 shows the ideal Bode plot of a lag circuit. In the midband, the decibel voltage gain is 0 dB. The response breaks at f_2 and then rolls off at a rate of 20 dB per decade.

6 dB per Octave

Above the cutoff frequency, the decibel voltage gain of a lag circuit decreases 20 dB per decade. This is equivalent to 6 dB per octave.

In other words, you can describe the frequency response of a lag circuit above the cutoff frequency in either of two ways: You can say that the decibel voltage gain decreases at a rate of 20 dB per decade, or you can say that it decreases at a rate of 6 dB per octave.

Phase Angle

The charging and discharging of a capacitor produce a lag in the output voltage of an RC bypass circuit. In other words, the output voltage will lag the input voltage by a phase angle ϕ. Figure 34-17 shows how ϕ varies with frequency. At zero hertz, the phase angle is 0°. As the frequency increases, the phase angle of the output voltage changes gradually from 0 to −90°. At very high frequencies, $\phi = -90°$.

When necessary, we can calculate the phase angle with this equation from basic courses:

$$\phi = -\arctan \frac{R}{X_C} \qquad (34\text{-}18)$$

By substituting $X_C = 1/2\pi fC$ into Formula (34-18) and rearranging, we can derive this equation:

$$\phi = -\arctan \frac{f}{f_2} \qquad (34\text{-}19)$$

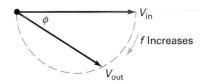

Figure 34-17   Phasor diagram of lag circuit.

With a calculator that has the tangent function and an inverse key, we can easily calculate the phase angle for any value of f/f_2. Table 34-5 shows a few values for ϕ. For example, when $f/f_2 = 0.1$, 1, and 10, the phase angles are

$$\phi = -\arctan 0.1 = -5.71°$$
$$\phi = -\arctan 1 = -45°$$
$$\phi = -\arctan 10 = -84.3°$$

Bode Plot of Phase Angle

Figure 34-18 shows how the phase angle of a lag circuit varies with the frequency. At very low frequencies, the phase angle is zero. When $f = 0.1f_2$, the phase angle is approximately −6°. When $f = f_2$, the phase angle equals −45°. When $f = 10f_2$, the phase angle is approximately −84°. Further increases in frequency produce little change because the limiting value is −90°. As you can see, the phase angle of a lag circuit is between 0 and −90°.

A graph like Fig. 34-18a is a Bode plot of the phase angle. Knowing that the phase angle is −6° at $0.1f_2$ and 84° at $10f_2$ is of little value except to indicate how close the phase angle is to its limiting value. The ideal Bode plot of Fig. 34-18b

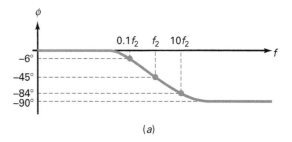

(a)

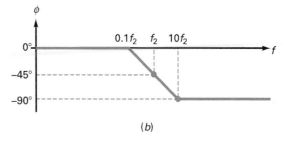

(b)

Figure 34-18   Bode plots of phase angle.

is more useful for preliminary analysis. This is the one to remember because it emphasizes these ideas:

1. When $f = 0.1f_2$, the phase angle is approximately zero.
2. When $f = f_2$, the phase angle is $-45°$.
3. When $f = 10f_2$, the phase angle is approximately $-90°$.

Another way to summarize the Bode plot of the phase angle is this: At the cutoff frequency, the phase angle equals $-45°$. A decade below the cutoff frequency, the phase angle is approximately $0°$. A decade above the cutoff frequency, the phase angle is approximately $-90°$.

EXAMPLE 34-12

Draw the ideal Bode plot for the lag circuit of Fig. 34-19a.

Answer:

With Formula (34-17), we can calculate the cutoff frequency:

$$f_2 = \frac{1}{2\pi(5\ \text{k}\Omega)(100\ \text{pF})} = 318\ \text{kHz}$$

Figure 34-19b shows the ideal Bode plot. The voltage gain is 0 dB at low frequencies. The frequency response breaks at 318 kHz and then rolls off at a rate of 20 dB/decade.

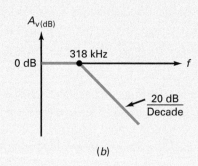

(a)

(b)

Figure 34-19 A lag circuit and its Bode plot.

EXAMPLE 34-13

In Fig. 34-20a, the dc amplifier stage has a midband voltage gain of 100. If the Thevenin resistance facing the bypass capacitor is 2 kΩ, what is the ideal Bode plot? Ignore all capacitances inside the amplifier stage.

Answer:

The Thevenin resistance and the bypass capacitor are a lag circuit with a cutoff frequency of

$$f_2 = \frac{1}{2\pi(2\ \text{k}\Omega)(500\ \text{pF})} = 159\ \text{kHz}$$

The amplifier has a midband voltage gain of 100, which is equivalent to 40 dB.

Figure 34-20b shows the ideal Bode plot. The decibel voltage gain is 40 dB from zero to the cutoff frequency of 159 kHz. The response then rolls off at a rate of 20 dB per decade until it reaches an f_{unity} of 15.9 MHz.

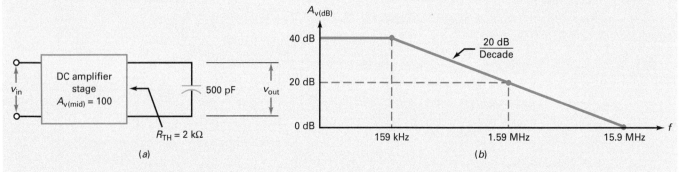

(a)

(b)

Figure 34-20 (a) DC amplifier and bypass capacitor. (b) Ideal Bode plot.

34.9 Rise Time–Bandwidth Relationship

Sine-wave testing of an amplifier means that we use a sinusoidal input voltage and measure the sinusoidal output voltage. To find the upper cutoff frequency, we have to vary the input frequency until the voltage gain drops 3 dB from the midband value. Sine-wave testing is one approach. But there is a faster and simpler way to test an amplifier by using a square wave instead of a sine wave.

Rise Time

The capacitor is initially uncharged in Fig. 34-21a. If we close the switch, the capacitor voltage will rise exponentially toward the supply voltage, V. The **rise time** T_R is the time it takes the capacitor voltage to go from $0.1V$ (called the *10% point*) to $0.9V$ (called the *90% point*). If it takes 10 μs for the exponential waveform to go from the 10% point to the 90% point, the waveform has a rise time of

$$T_R = 10 \ \mu s$$

Instead of using a switch to apply the sudden step in voltage, we can use a square-wave generator. For instance, Fig. 34-21b shows the leading edge of a square wave driving the same RC circuit as before. The rise time is still the time it takes for the voltage to go from the 10% point to the 90% point.

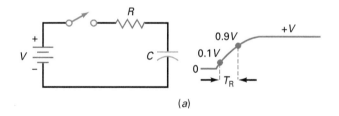

(a)

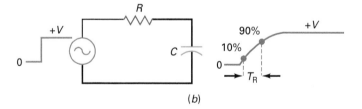

(b)

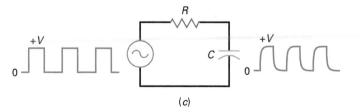

(c)

Figure 34-21 (a) Risetime. (b) Voltage step produces output exponential. (c) Square-wave testing.

Figure 34-21c shows how several cycles will look. Although the input voltage changes almost instantly from one voltage level to another, the output voltage takes much longer to make its transitions because of the bypass capacitor. The output voltage cannot suddenly step, because the capacitor has to charge and discharge through the resistance.

Relationship between T_R and RC

By analyzing the exponential charge of a capacitor, it is possible to derive this equation for the rise time:

$$T_R = 2.2RC \qquad \textbf{(34-20)}$$

This says that the rise time is slightly more than two RC time constants. For instance, if R equals 10 kΩ and C is 50 pF, then

$$RC = (10 \ k\Omega)(50 \ pF) = 0.5 \ \mu s$$

The rise time of the output waveform equals

$$T_R = 2.2RC = 2.2(0.5 \ \mu s) = 1.1 \ \mu s$$

Data sheets often specify the rise time because it is useful to know the response to a voltage step when analyzing switching circuits.

An Important Relationship

As mentioned earlier, a dc amplifier typically has one dominant lag circuit that rolls off the voltage gain at a rate of 20 dB per decade until f_{unity} is reached. The cutoff frequency of this lag circuit is given by

$$f_2 = \frac{1}{2\pi RC}$$

which can be solved for RC to get

$$RC = \frac{1}{2\pi f_2}$$

When we substitute this into Formula (34-20) and simplify, we get this widely used equation:

$$f_2 = \frac{0.35}{T_R} \qquad \textbf{(34-21)}$$

This is an important result because it converts rise time to cutoff frequency. It means that we can test an amplifier with a square wave to find the cutoff frequency. Since square-wave testing is much faster than sine-wave testing, many engineers and technicians use Formula (34-21) to find the upper cutoff frequency of an amplifier.

Formula (34-21) is called the *rise time–bandwidth relationship*. In a dc amplifier, the word *bandwidth* refers to all the frequencies from zero up to the cutoff frequency. Often, bandwidth is used as a synonym for *cutoff frequency*. If the data sheet for a dc amplifier gives a bandwidth of 100 kHz, it means that the upper cutoff frequency equals 100 kHz.

EXAMPLE 34-14

What is the upper cutoff frequency for the circuit shown in Fig. 34-22a?

Answer:

In Fig. 34-22a, the rise time is 1 μs. With Formula (34-21),

$$f_2 = \frac{0.35}{1\ \mu s} = 350\ \text{kHz}$$

Therefore, the circuit of Fig. 34-22a has an upper cutoff frequency of 350 kHz. An equivalent statement is that the circuit has a bandwidth of 350 kHz.

Figure 34-22b illustrates the meaning of sine-wave testing. If we change the input voltage from a square wave to a sine wave, we will get a sine-wave output. By increasing the input frequency, we can eventually find the cutoff frequency of 350 kHz. In other words, we would get the same result with sine-wave testing, except that it is slower than square-wave testing.

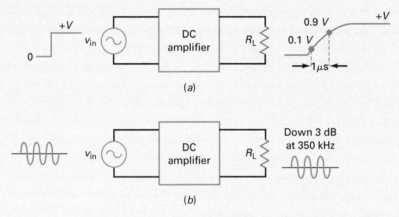

Figure 34-22 Risetime and cutoff frequency are related.

34.10 Frequency Effects of Surface-Mount Circuits

Stray capacitance and inductance become serious considerations for discrete and IC devices that are operating above 100 kHz. With conventional feed-through components, there are three sources of stray effects:

1. The geometry and internal structure of the device.

2. The printed circuit layout, including the orientation of the devices and the conductive tracks.

3. The external leads on the device.

Using SM components virtually eliminates item 3 from the list, thus increasing the amount of control design engineers have over stray effects among components on a circuit board. The result is that surface-mount devices typically have much higher frequency cutoffs.

❓ CHAPTER 34 REVIEW QUESTIONS

1. Frequency response is a graph of voltage gain versus
 a. frequency.
 b. power gain.
 c. input voltage.
 d. output voltage.

2. At low frequencies, the coupling capacitors produce a decrease in
 a. input resistance.
 b. voltage gain.
 c. generator resistance.
 d. generator voltage.

3. The stray-wiring capacitance has an effect on the
 a. lower cutoff frequency.
 b. midband voltage gain.
 c. upper cutoff frequency.
 d. input resistance.

4. At the lower or upper cutoff frequency, the voltage gain is
 a. $0.35A_{v(mid)}$.
 b. $0.5A_{v(mid)}$.
 c. $0.707A_{v(mid)}$.
 d. $0.995A_{v(mid)}$.

5. If the power gain doubles, the decibel power gain increases by
   a. a factor of 2.
   b. 3 dB.
   c. 6 dB.
   d. 10 dB.

6. If the voltage gain doubles, the decibel voltage gain increases by
   a. a factor of 2.
   b. 3 dB.
   c. 6 dB.
   d. 10 dB.

7. If the voltage gain is 10, the decibel voltage gain is
   a. 6 dB.
   b. 20 dB.
   c. 40 dB.
   d. 60 dB.

8. If the voltage gain is 100, the decibel voltage gain is
   a. 6 dB.
   b. 20 dB.
   c. 40 dB.
   d. 60 dB.

9. If the voltage gain is 2000, the decibel voltage gain is
   a. 40 dB.
   b. 46 dB.
   c. 66 dB.
   d. 86 dB.

10. Two stages have decibel voltage gains of 20 and 40 dB. The total ordinary voltage gain is
    a. 1.
    b. 10.
    c. 100.
    d. 1000.

11. Two stages have voltage gains of 100 and 200. The total decibel voltage gain is
    a. 46 dB.
    b. 66 dB.
    c. 86 dB.
    d. 106 dB.

12. One frequency is 8 times another frequency. How many octaves apart are the two frequencies?
    a. 1.
    b. 2.
    c. 3.
    d. 4.

13. If $f = 1$ MHz, and $f_2 = 10$ Hz, the ratio f/f_2 represents how many decades?
    a. 2.
    b. 3.
    c. 4.
    d. 5.

14. Semilogarithmic paper means that
    a. one axis is linear, and the other is logarithmic.
    b. one axis is linear, and the other is semilogarithmic.
    c. both axes are semilogarithmic.
    d. neither axis is linear.

15. If you want to improve the high-frequency response of an amplifier, which of these approaches would you try?
    a. Decrease the coupling capacitances.
    b. Increase the emitter bypass capacitance.
    c. Shorten leads as much as possible.
    d. Increase the generator resistance.

16. The rollout rate for a single RC section or op amp is
    a. 6 dB per octave.
    b. 6 dB per decade.
    c. 20 dB per octave.
    d. 20 dB per decade.
    e. a and c.
    f. b and d.

17. An amplifier with an upper cutoff frequency of 20 kHz is typically what kind of amplifier?
    a. DC.
    b. Audio.
    c. Video.
    d. RF.

18. A negative dB value indicates a(n)
    a. gain of 0.
    b. gain of 1.
    c. infinite gain.
    d. attenuation.

19. Stray-wiring capacitance produces effects similar to a
    a. low-pass filter.
    b. high-pass filter.
    c. bandpass filter.
    d. leading phase shifter.

20. A circuit with a loss has a gain of
    a. 0.
    b. 1.
    c. <1.
    d. infinity.

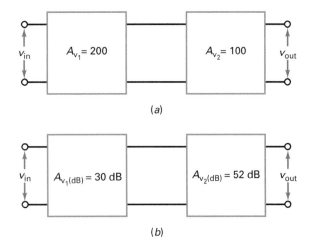
SECTION 34.2 Frequency Response of an Amplifier

34.1 An amplifier has a midband voltage gain of 1000. If cutoff frequencies are $f_1 = 100$ Hz and $f_2 = 100$ kHz, what does the frequency response look like?

34.2 Suppose an op amp has a midband voltage gain of 500,000. If the upper cutoff frequency is 15 Hz, what does the frequency response look like?

34.3 A dc amplifier has a midband voltage gain of 300. If the upper cutoff frequency is 10 kHz, what is the voltage gain at the cutoff frequency?

SECTION 34.3 Decibel Power Gain

34.4 Calculate the decibel power gain for $A_p = 5$, 10, 20, and 40.

34.5 Calculate the decibel power gain for $A_p = 0.4$, 0.2, 0.1, and 0.05.

34.6 Calculate the decibel power gain for $A_p = 2$, 20, 200, and 2000.

34.7 Calculate the decibel power gain for $A_p = 0.4$, 0.04, and 0.004.

SECTION 34.4 Decibel Voltage Gain

34.8 What is the total voltage gain in Fig. 34-23a? Convert the answer to decibels.

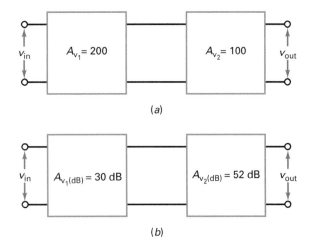

(a)

(b)

Figure 34-23

34.9 Convert each stage gain in Fig. 34-23a to decibels.

34.10 What is the total decibel voltage gain in Fig. 34-23b? Convert this to ordinary voltage gain.

34.11 What is the ordinary voltage gain of each stage in Fig. 34-24b?

34.12 What is the decibel voltage gain of an amplifier if it has an ordinary voltage gain of 100,000?

34.13 The data sheet of an LM380, an audio power amplifier, gives a decibel voltage gain of 34 dB. Convert this to ordinary voltage gain.

34.14 A two-stage amplifier has these stage gains: $A_{v1} = 25.8$ and $A_{v2} = 117$. What is the decibel voltage gain of each stage? The total decibel voltage gain?

SECTION 34.5 Impedance Matching

34.15 If Fig. 34-24 is an impedance-matched system, what is the total decibel voltage gain? The decibel voltage gain of each stage?

34.16 If the stages of Fig. 34-24 are impedance-matched, what is the load voltage? The load power?

34.17 A generator with an output voltage of 2 mV has a generator output resistance of 75 Ω. It drives an amplifier with an input impedance of 200 Ω. The amplifier gain is 30, the output impedance is 300 Ω, and the amplifier load is 600 Ω. What is the output voltage?

SECTION 34.6 Decibels above a Reference

34.18 If the output power of a preamplifier is 20 dBm, how much power is this in milliwatts?

34.19 How much output voltage does a microphone have when its output is −45 dBV?

34.20 Convert the following powers to dBm: 25 mW, 93.5 mW, and 4.87 W.

34.21 Convert the following voltages to dBV: 1 μV, 34.8 mV, 12.9 V, and 345 V.

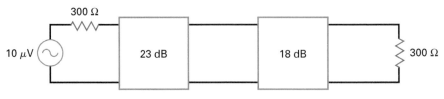

Figure 34-24

SECTION 34.7 Bode Plots

34.22 The LF351 is an op amp with a voltage gain of 316,000, a cutoff frequency of 40 Hz, and a roll-off rate of 20 dB per decade. Draw the ideal Bode plot.

SECTION 34.8 More Bode Plots

34.23 Draw the ideal Bode plot for the lag circuit of Fig. 34-25a.

34.24 Draw the ideal Bode plot for the lag circuit of Fig. 34-25b.

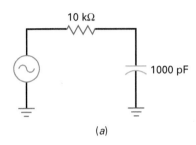

(a)

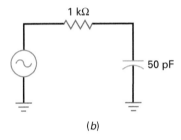

(b)

Figure 34-25

34.25 What is the ideal Bode plot for the stage of Fig. 34-26?

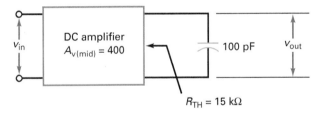

Figure 34-26

SECTION 34.9 Rise Time–Bandwidth Relationship

34.26 An amplifier has the step response shown in Fig. 34-27a. What is its upper cutoff frequency?

34.27 What is the bandwidth of an amplifier if the rise time is 0.25 μs?

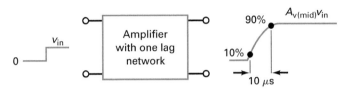

Figure 34-27

34.28 The upper cutoff frequency of an amplifier is 100 kHz. If it is square-wave-tested, what would the rise time of the amplifier output be?

34.29 In Fig. 34-28, what is the decibel voltage gain when $f = 100$ kHz?

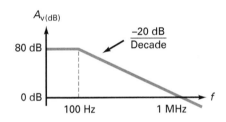

Figure 34-28

34.30 The amplifier of Fig. 34-27 has a midband voltage gain of 100. If the input voltage is a step of 20 mV, what is the output voltage at the 10% point? The 90% point?

34.31 You have two data sheets for amplifiers. The first shows a cutoff frequency of 1 MHz. The second gives a rise time of 1 μs. Which amplifier has the greater bandwidth?

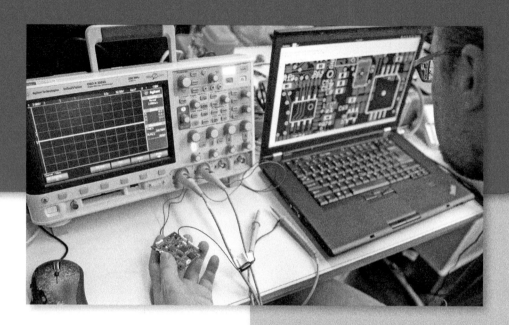

Chapter 35

Operational Amplifiers

Op amps are the most widely used linear IC. Their versatility makes them usable in a wide range of applications. Some op amps are optimized for their bandwidth, others for low input offsets, others for low noise, and so on. This is why the variety of commercially available op amps is so large. You can find an op amp for almost any analog application.

Op amps are some of the most basic active components in analog systems. For instance, by connecting two external resistors, we can adjust the voltage gain and bandwidth of an op amp to our exact requirements. Furthermore, with other external components, we can build waveform converters, oscillators, active filters, and other interesting circuits.

Learning Outcomes

After studying this chapter, you should be able to:

> *List* the characteristics of ideal op amps and 741 op amps.

> *Define* slew rate and *use* it to find the power bandwidth of an op amp.

> *Analyze* an op-amp inverting amplifier.

> *Analyze* an op-amp noninverting amplifier.

> *Explain* how summing amplifiers and voltage followers work.

> *Explain* the operation and characteristics of differential amplifiers and instrumentation amplifiers.

> *Explain* circuit showing how an op amp can be operated from a single power supply.

35.1 Introduction to Op Amps

Figure 35-1 shows a block diagram of an op amp. The input stage is a diff amp, followed by more stages of gain, and a class B push-pull emitter follower. Because a diff amp is the first stage, it determines the input characteristics of the op amp. In most op amps the output is single-ended, as shown. With positive and negative supplies, the single-ended output is designed to have a quiescent value of zero. This way, zero input voltage ideally results in zero output voltage.

Not all op amps are designed like Fig. 35-1. For instance, some do not use a class B push-pull output, and others may have a double-ended output. Also, op amps are not as simple as Fig. 35-1 suggests. The internal design of a monolithic op amp is very complicated, using dozens of transistors as current mirrors, active loads, and other innovations that are not possible in discrete designs. For our needs, Fig. 35-1 captures two important features that apply to typical op amps: the differential input and the single-ended output.

Figure 35-2a is the schematic symbol of an op amp. It has noninverting and inverting inputs and a single-ended output. Ideally, this symbol means that the amplifier has infinite voltage gain, infinite input impedance, and zero output impedance. The ideal op amp represents a perfect voltage amplifier that we can visualize, as shown in Fig. 35-2b, where R_{in} is infinite and R_{out} is zero.

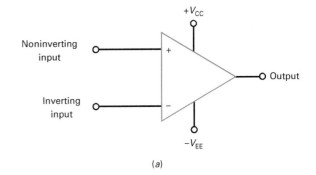

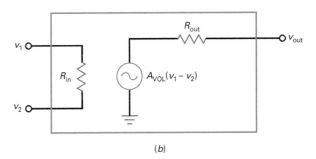

Figure 35-2 (a) Schematic symbol for op amp. (b) Equivalent circuit of op amp.

Table 35-1 summarizes the characteristics of an ideal op amp. The ideal op amp has infinite voltage gain, infinite unity-gain frequency, infinite input impedance, and infinite common-mode rejection ratio (CMRR). It also has zero output resistance, zero bias current, and zero offsets. This is what manufacturers would build if they could. What they actually can build approaches these ideal values.

For instance, the LM741C of Table 35-1 is a standard op amp, a classic that has been available since the 1960s. Its characteristics are the minimum of what to expect from a monolithic op amp. The LM741C has a voltage gain of 100,000, a unity-gain frequency of 1 MHz, an input impedance of 2 MΩ, and so on. Because the voltage gain is so high, the input offsets can easily saturate the op amp. This

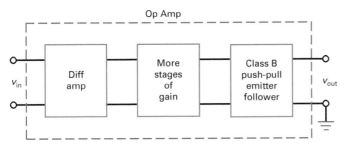

Figure 35-1 Block diagram of an op amp.

Table 35-1 Typical Op-Amp Characteristics

Quantity	Symbol	Ideal	LM741C	TL081A
Open-loop voltage gain	A_{VOL}	Infinite	100,000	200,000
Unity-gain frequency	f_{unity}	Infinite	1 MHz	3 MHz
Input resistance	R_{in}	Infinite	2 MΩ	$10^{12}\ \Omega$
Output resistance	R_{out}	Zero	75 Ω	100 Ω
Input bias current	$I_{in(bias)}$	Zero	80 nA	30 pA
Input offset current	$I_{in(off)}$	Zero	20 nA	5 pA
Input offset voltage	$V_{in(off)}$	Zero	2 mV	3 mV
Common-mode rejection ratio	CMRR	Infinite	90 dB	86 dB

is why practical circuits need external components between the input and output of an op amp to stabilize the voltage gain. For instance, in many applications negative feedback is used to adjust the overall voltage gain to a much lower value in exchange for stable linear operation.

When no feedback path (or loop) is used, the voltage gain is maximum and is called the **open-loop voltage gain,** designated A_{VOL}. In Table 35-1, notice that the A_{VOL} of the LM741C is 100,000. Although not infinite, this open-loop voltage gain is very high. For instance, an input as small as 10 μV produces an output of 1 V. Because the open-loop voltage gain is very high, we can use heavy negative feedback to improve the overall performance of a circuit.

The 741C has a unity-gain frequency of 1 MHz. This means that we can get usable voltage gain almost as high as 1 MHz. The 741C has an input resistance of 2 MΩ, an output resistance of 75 Ω, an input bias current of 80 nA, an input offset current of 20 nA, an input offset voltage of 2 mV, and a CMRR of 90 dB.

When higher input resistance is needed, a **BIFET op amp** can be used. This type of op amp incorporates JFETs and bipolar transistors on the same chip. The JFETs are used in the input stage to get smaller input bias and offset currents; the bipolar transistors are used in the later stages to get more voltage gain.

The TL081A is an example of a BIFET op amp. As shown in Table 35-1, the input bias current is only 30 pA, and the input resistance if 10^{12} Ω. The TL081A has a voltage gain of 200,000 and a unity-gain frequency of 3 MHz. With this device, we can get voltage gain up to 13 MHz.

While bipolar op amps are still used, they have been mostly replaced by the newer BIFET bipolar and CMOS op amps. Both offer the super high input impedances of FETs that are ideal for op amps. The newer FET op amps also tend to have lower bias currents and offset voltages and currents leading to fewer errors and simpler circuits. Over the years, transistor sizes have decreased, providing higher operating frequencies and wider gain bandwidth (GBW). Op amps with unity GBW of hundreds of MHz are readily available. Furthermore, CMOS op amps consume far less power.

There are literally tens of thousands of different types of IC op amps available for just about any application. New op amps are categorized and optimized for one or two major features, such as precision, wide bandwidth, low power, low noise, minimum error sources, slew rate, output voltage levels, or power output. Otherwise, all op amps operate the same, so knowing one lets you understand all of them.

35.2 The 741 Op Amp

In 1965 Fairchild Semiconductor introduced the μA709, the first widely used monolithic op amp. Although successful, this first-generation op amp had many disadvantages. These led to an improved op amp known as the μA741. Because it is inexpensive and easy to use, the μA741 has been an enormous success. Other 741 designs have appeared from various manufacturers. All these monolithic op amps are equivalent to the μA741 because they have the same specifications on their data sheets. For convenience, most people drop the prefixes and refer to this widely used op amp simply as the 741.

An Industry Standard

The 741 has become an industry standard. Despite its age, it is an inexpensive op amp for learning and is used here as an example. Once you understand the 741, you can branch out to other op amps.

Incidentally, the 741 has different versions with two and four similar op amps in one IC. All have an open-loop voltage gain of 100,000, an input impedance of 2 MΩ, and an output impedance of 75 Ω. Figure 35-3a shows the most popular, dual in-line package (DIP) and its pin-outs. The schematic symbol is shown in Fig. 35-3b.

Power Connections

Most op amps are designed to be operated with two power supplies, one positive ($+V_{CC}$) and the other negative ($-V_{EE}$), with respect to ground. These supplies connect to pins 7 and 4, respectively, as shown in Fig. 35-4. The dual supplies allow the output to swing both positive and negative above and below ground. Note that there is no ground pin on the IC. The ground is formed at the junction of the two power

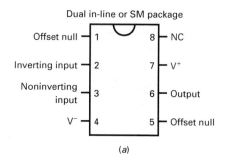

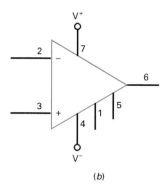

Figure 35-3 The 741 (a) dual in-line package and (b) schematic symbol.

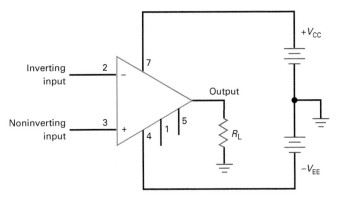

Figure 35-4 Power supply connections.

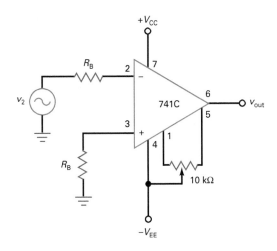

Figure 35-5 Compensation and nulling used with 741C.

supplies. All inputs and the output are referenced to this ground. Note the load R_L connected to the output. You will learn later in the chapter how to connect the inputs and the offset null connections on pins 1 and 5.

What follows next is a brief discussion of some of the key specifications of the 741 op amp.

Bias and Offsets

The differential input amplifier of an op amp is not perfect. This means that the transistors and related circuitry are not ideally balanced, matched, and equal. As a result, there are bias currents and voltages that are slightly different. These bias and offset currents and voltages represent errors that will produce an output from the op amp even with both inputs at zero. The very high gain of the amplifier multiplies these small errors into outputs that add to or subtract from the actual desired output, making the output incorrect for the given input.

Such errors arise mainly from transistor differences. Slightly different gains, bias currents, and emitter-base voltage drop differences are the most troublesome.

As discussed in Chap. 34, a diff amp has input bias and offsets that produce an output error when there is no input signal. In many applications, the output error is small enough to ignore. But when the output error cannot be ignored, a designer can reduce it by using equal base resistors. This eliminates the problem of bias current, but not the offset current or offset voltage.

This is why it is best to eliminate output error by using the **nulling circuit** given on the data sheet. This recommended nulling circuit works with the internal circuitry to eliminate the output error and also to minimize *thermal drift,* a slow change in output voltage caused by the effect of changing temperature on op-amp parameters. Sometimes, the data sheet of an op amp does not include a nulling circuit. In this case, we have to apply a small input voltage to null the output. We will discuss this method later.

Figure 35-5 shows the nulling method suggested on the data sheet of a 741C. The ac source driving the inverting

input has a Thevenin resistance of R_B. To neutralize the effect of input bias current (80 nA) flowing through this source resistance, a discrete resistor of equal value is added to the noninverting input, as shown.

To eliminate the effect of an input offset current of 20 nA and an input offset voltage of 2 mV, the data sheet of a 741C recommends using a 10-kΩ potentiometer between pins 1 and 5. By adjusting this potentiometer with no input signal, we can null or zero the output voltage.

Bias and offset errors are less of a problem in the newer op amps. Most newer op amps feature JFET or MOSFET differential input amplifiers that are less susceptible to these errors. In many applications, the errors are small enough to ignore. In those cases where they do matter, just remember there are ways to correct the problem with external resistances and/or voltages. Always use the recommended error-correcting procedures suggested by the IC manufacturer.

Common-Mode Rejection Ratio

For a 741C, CMRR is 90 dB at low frequencies. Given equal signals, one a desired signal and the other a common-mode signal, the desired signal will be 90 dB larger at the output than the common-mode signal. In ordinary numbers, this means that the desired signal will be approximately 30,000 times larger than the common-mode signal. At higher frequencies, reactive effects degrade CMRR, as shown in Fig. 35-6a. Notice that CMRR is approximately 75 dB at 1 kHz, 56 dB at 10 kHz, and so on.

Maximum Peak-to-Peak Output

The MPP value of an amplifier is the maximum peak-to-peak output that the amplifier can produce. Since the quiescent output of an op amp is ideally zero, the ac output voltage can swing positively or negatively. For load resistances that

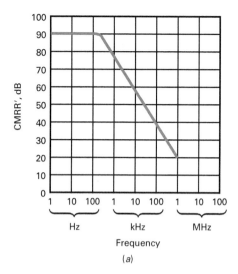

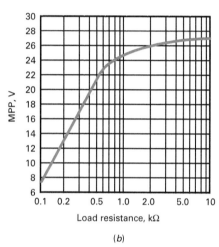

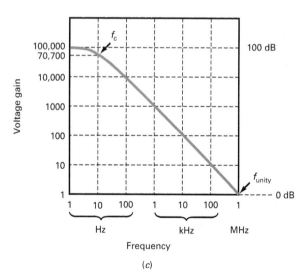

Figure 35-6 Typical 741C graphs for CMRR, MPP, and A_{VOL}.

are much larger than R_{out}, the output voltage can swing almost to the supply voltages. For instance, if $V_{CC} = +15$ V and $V_{EE} = -15$ V, the MPP value with a load resistance of 10 kΩ is ideally 30 V.

With a nonideal op amp, the output cannot swing all the way to the value of the supply voltages because there are small voltage drops in the final stage of the op amp. Furthermore, when the load resistance is not large compared to R_{out}, some of the amplified voltage is dropped across R_{out}, which means that the final output voltage is smaller.

Figure 36-6b shows MPP versus load resistance for a 741C with supply voltages of +15 V and −15 V. Notice that MPP is approximately 27 V for an R_L of 10 kΩ. This means that the output saturates positively at +13.5 V and negatively at −13.5 V. When the load resistance decreases, MPP decreases as shown. For instance, if the load resistance is only 275 Ω, MPP decreases to 16 V, which means that the output saturates positively at +8 V and negatively at −8 V.

Short-Circuit Current

In some applications, an op amp may drive a load resistance of approximately zero. In this case, you need to know the value of the **short-circuit output current.** The data sheet of a 741C lists a short-circuit output current of 25 mA. This is the maximum output current the op amp can produce. If you are using small load resistors (less than 75 Ω), don't expect to get a large output voltage because the voltage cannot be greater than the 25 mA times the load resistance.

Frequency Response

Figure 35-6c shows the small-signal frequency response of a 741C. In the midband, the voltage gain is 100,000. The 741C has a cutoff frequency f_c of 10 Hz. As indicated, the voltage gain is 70,700 (down 3 dB) at 10 Hz. Above the cutoff frequency, the voltage gain decreases at a rate of 20 dB per decade (first-order response).

The unity-gain frequency is the frequency at which the voltage gain equals 1. In Fig. 35-6c, f_{unity} is 1 MHz. Data sheets usually specify the value of f_{unity} because it represents the upper limit on the useful gain of an op amp. For instance, the data sheet of a 741C lists an f_{unity} of 1 MHz. This means that the 741C can amplify signals up to 1 MHz. Beyond 1 MHz, the voltage gain is less than 1 and the 741C is useless. If a designer needs a higher f_{unity}, better op amps are available. For instance, the LM318 has an f_{unity} of 15 MHz, which means that it can produce usable voltage gain all the way to 15 MHz.

Slew Rate

The compensating capacitor inside a 741C performs a very important function: It prevents oscillations that would interfere with the desired signal. But there is a disadvantage. The compensating capacitor needs to be charged and discharged.

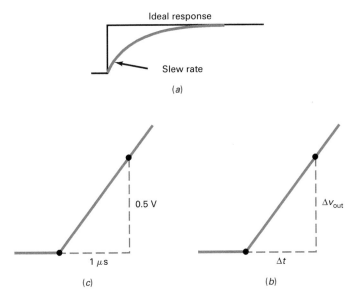

(a)

0.5 V

1 μs

(c)

Δv_{out}

Δt

(b)

Figure 35-7 (a) Ideal and actual responses to an input step voltage. (b) Illustrating definition of slew rate. (c) Slew rate equals 0.5 V/μs.

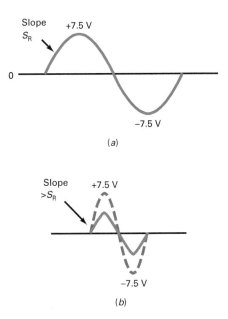

Figure 35-8 (a) Initial slope of a sine wave. (b) Distortion occurs if initial slope exceeds slew rate.

This creates a speed limit on how fast the output of the op amp can change.

Here is the basic idea: Suppose the input voltage to an op amp is a positive **voltage step,** a sudden transition in voltage from one dc level to a higher dc level. If the op amp were perfect, we would get the ideal response shown in Fig. 35-7a. Instead, the output is the positive exponential waveform shown. This occurs because the compensating capacitor must be charged before the output voltage can change to the higher level.

In Fig. 35-7a, the initial slope of the exponential waveform is called the **slew rate,** symbolized S_R. The definition of slew rate is

$$S_R = \frac{\Delta v_{out}}{\Delta t} \qquad \textbf{(35-1)}$$

where the Greek letter Δ (delta) stands for "the change in." In words, the equation says that slew rate equals the change in output voltage divided by the change in time.

Figure 35-7b illustrates the meaning of slew rate. The initial slope equals the vertical change divided by the horizontal change between two points on the early part of the exponential wave. For instance, if the exponential wave increases 0.5 V during the first microsecond, as shown in Fig. 35-7c, the slew rate is

$$S_R = \frac{0.5 \text{ V}}{1 \text{ μs}} = 0.5 \text{ V/μs}$$

The slew rate represents the fastest response that an op amp can have. For instance, the slew rate of a 741C is 0.5 V/μs. This means that the output of a 741C can change no

faster than 0.5 V in a microsecond. In other words, if a 741C is driven by a large step in input voltage, we do not get a sudden step in output voltage. Instead, we get an exponential output wave. The initial part of this output waveform will look like Fig. 35-7c.

We can also get slew-rate limiting with a sinusoidal signal. Here is how it occurs: In Fig. 35-8a, the op amp can produce the output sine wave shown only if the initial slope of the sine wave is less than the slew rate. For instance, if the output sine wave has an initial slope of 0.1 V/μs, a 741C can produce this sine wave with no trouble at all because its slew rate is 0.5 V/μs. On the other hand, if the sine wave has an initial slope of 1 V/μs, the output is smaller than it should be and it looks triangular instead of sinusoidal, as shown in Fig. 35-8b.

The data sheet of an op amp always specifies the slew rate because this quantity limits the large-signal response of an op amp. If the output sine wave is very small or the frequency is very low, slew rate is no problem. But when the signal is large and the frequency is high, slew rate will distort the output signal.

With calculus, it is possible to derive this equation:

$$S_S = 2\pi f V_p$$

where S_S is the initial slope of the sine wave, f is its frequency, and V_p is its peak value. To avoid slew-rate distortion of a sine wave, S_S has to be less than or equal to S_R. When the two are equal, we are at the limit, on the verge of slew-rate distortion. In this case:

$$S_R = S_S = 2\pi f V_p$$

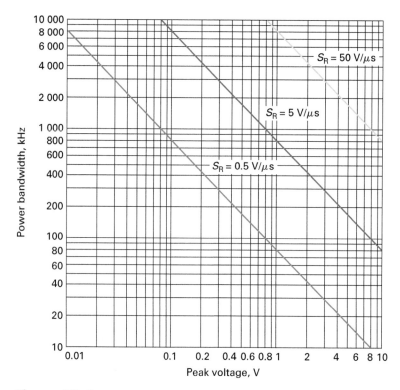

Figure 35-9 Graph of power bandwidth versus peak voltage.

Solving for f gives:

$$f_{max} = \frac{S_R}{2\pi V_P} \qquad (35\text{-}2)$$

where f_{max} is the highest frequency that can be amplified without slew-rate distortion. Given the slew rate of an op amp and the peak output voltage desired, we can use Formula (35-2) to calculate the maximum undistorted frequency. Above this frequency, we will see slew-rate distortion on an oscilloscope.

The frequency f_{max} is sometimes called the **power bandwidth** or *large-signal bandwidth* of the op amp. Figure 35-9 is a graph of Formula (35-2) for three slew rates. Since the bottom graph is for a slew rate of 0.5 V/μs, it is useful with a 741C. Since the top graph is for a slew rate of 50 V/μs, it is useful with an LM318 (it has a minimum slew rate of 50 V/μs).

For instance, suppose we are using a 741C. To get an undistorted output peak voltage of 8 V, the frequency can be no higher than 10 kHz (see Fig. 35-9). One way to increase the f_{max} is to accept less output voltage. By trading off peak value for frequency, we can improve the power bandwidth. As an example, if our application can accept a peak output voltage of 1 V, f_{max} increases to 80 kHz.

There are two bandwidths to consider when analyzing the operation of an op-amp circuit: the small-signal bandwidth determined by the first-order response of the op amp and the large-signal or power bandwidth determined by the slew rate. More will be said about these two bandwidths later.

EXAMPLE 35-1

How much inverting input voltage does it take to drive the 741C of Fig. 35-10a into negative saturation?

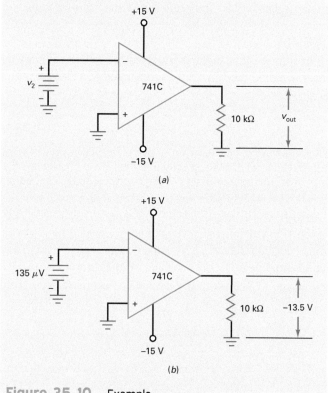

Figure 35-10 Example.

Answer:

Figure 35-10b shows that MPP equals 27 V for a load resistance of 10 kΩ, which translates into an output of −13.5 V for negative saturation. Since the 741C has an open-loop voltage gain of 100,000, the required input voltage is:

$$v_2 = \frac{13.5 \text{ V}}{100,000} = 135 \ \mu\text{V}$$

Figure 35-10b summarizes the answer. As you can see, an inverting input of 135 μV produces negative saturation, an output voltage of −13.5 V.

EXAMPLE 35-2

What is the open-loop voltage gain of a 741C when the input frequency is 1 kHz? 10 kHz? 100 kHz?

Answer:

In Fig. 35-6c, the voltage gain is 1000 for 1 kHz, 100 for 10 kHz, and 10 for 100 kHz. As you can see, the voltage gain decreases by a factor of 10 each time the frequency increases by a factor of 10.

EXAMPLE 35-3

The input voltage to an op amp is a large voltage step. The output is an exponential waveform that changes to 0.25 V in 0.1 μs. What is the slew rate of the op amp?

Answer:

With Formula (35-1),

$$S_R = \frac{0.25 \text{ V}}{0.1 \ \mu\text{s}} = 2.5 \text{ V}/\mu\text{s}$$

EXAMPLE 35-4

An op amp has a slew rate of 15 V/μs. What is the power bandwidth for a peak output voltage of 10 V?

Answer:

With Formula (35-2),

$$f_{\max} = \frac{S_R}{2\pi V_P} = \frac{15 \text{ V}/\mu\text{s}}{2\pi (10 \text{ V})} = 239 \text{ kHz}$$

35.3 The Inverting Amplifier

The **inverting amplifier** is the most basic op-amp circuit. It uses negative feedback to stabilize the overall voltage gain. The reason we need to stabilize the overall voltage gain is because A_{VOL} is too high and unstable to be of any use without some form of feedback. For instance, the 741C has a minimum A_{VOL} of 20,000 and a maximum A_{VOL} of more than 200,000. An unpredictable voltage gain of this magnitude and variation is useless without feedback.

Inverting Negative Feedback

Figure 35-11 shows an inverting amplifier. To keep the drawing simple, the power supply voltages are not shown. In other words, we are looking at the ac equivalent circuit. An input voltage v_{in} drives the inverting input through resistor R_1. This results in an inverting input voltage of v_2. The input voltage is amplified by the open-loop voltage gain to produce an inverted output voltage. The output voltage is fed back to the input through feedback resistor R_f. This results in negative feedback because the output is 180° out of phase with the input. In other words, any changes in v_2 produced by the input voltage are opposed by the output signal.

Here is how the negative feedback stabilizes the overall voltage gain: If the open-loop voltage gain A_{VOL} increases for any reason, the output voltage will increase and feed back more voltage to the inverting input. This opposing feedback voltage reduces v_2. Therefore, even though A_{VOL} has increased, v_2 has decreased, and the final output increases much less than it would without the negative feedback. The overall result is a very slight increase in output voltage, so small that it is hardly noticeable.

Virtual Ground

When we connect a piece of wire between some point in a circuit and ground, the voltage of the point becomes zero. Furthermore, the wire provides a path for current to flow to ground. A *mechanical ground* (a wire between a point and ground) is ground to both voltage and current.

A **virtual ground** is different. This type of ground is a widely used shortcut for analyzing an inverting amplifier. With a virtual ground, the analysis of an inverting amplifier and related circuits becomes incredibly easy.

The concept of a virtual ground is based on an ideal op amp. When an op amp is ideal, it has infinite open-loop

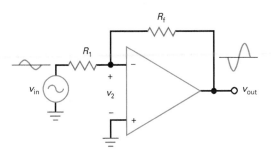

Figure 35-11 The inverting amplifier.

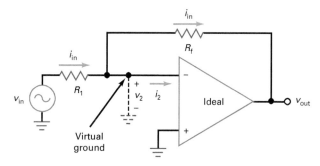

Figure 35-12 The concept of virtual ground: shorted to voltage and open to current.

voltage gain and infinite input resistance. Because of this, we can deduce the following ideal properties for the inverting amplifier of Fig. 35-12:

1. Since R_{in} is infinite, i_2 is zero.
2. Since A_{VOL} is infinite, v_2 is zero.

Since i_2 is zero in Fig. 35-12, the current through R_f must equal the input current through R_1, as shown. Furthermore, since v_2 is zero, the virtual ground shown in Fig. 35-12 means that the inverting input acts like a ground for voltage but an open for current!

Virtual ground is very unusual. It is like half of a ground because it is a short for voltage but an open for current. To remind us of this half-ground quality, Fig. 35-12 uses a dashed line between the inverting input and ground. The dashed line means that no current can flow to ground. Although virtual ground is an ideal approximation, it gives very accurate answers when used with heavy negative feedback.

Voltage Gain

In Fig. 35-13, visualize a virtual ground on the inverting input. Then, the right end of R_1 is a voltage ground, so we can write

$$v_{in} = i_{in}R_1$$

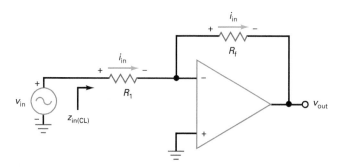

Figure 35-13 Inverting amplifier has same current through both resistors.

Similarly, the left end of R_f is a voltage ground, so the magnitude of output voltage is

$$v_{out} = -i_{in}R_f$$

Divide v_{out} by v_{in} to get the voltage gain

$$A_{v(CL)} = \frac{-R_f}{R_1} \quad \textbf{(35-3)}$$

where $A_{v(CL)}$ is the closed-loop voltage gain. This is called the **closed-loop voltage gain** because it is the voltage when there is a feedback path between the output and the input. Because of the negative feedback, the closed-loop voltage gain $A_{v(CL)}$ is always smaller than the open-loop voltage gain A_{VOL}.

Look at how simple and elegant Formula (35-3) is. The closed-loop voltage gain equals the ratio of the feedback resistance to the input resistance. For instance, if $R_1 = 1$ kΩ and $R_f = 50$ kΩ, the closed-loop voltage gain is 50. Because of the heavy negative feedback, this closed-loop voltage gain is very stable. If A_{VOL} varies because of temperature change, supply voltage variations, or op-amp replacement, $A_{v(CL)}$ will still be very close to 50. The negative sign in the voltage gain equation indicates a 180° phase shift.

Input Impedance

In some applications, a designer may want a specific input impedance. This is one of the advantages of an inverting amplifier; it is easy to set up a desired input impedance. Here is why: Since the right end of R_1 is virtually grounded, the closed-loop input impedance is

$$z_{in(CL)} = R_1 \quad \textbf{(35-4)}$$

This is the impedance looking into the left end of R_1, as shown in Fig. 35-13. For instance, if an input impedance of 2 kΩ and a closed-loop voltage gain of 50 is needed, a designer can use $R_1 = 2$ kΩ and $R_f = 100$ kΩ.

Bandwidth

The **open-loop bandwidth** or cutoff frequency of an op amp is very low because of the internal compensating capacitor. For a 741C,

$$f_{2(OL)} = 10 \text{ Hz}$$

At this frequency, the open-loop voltage gain breaks and rolls off in a first-order response.

When negative feedback is used, the overall bandwidth increases. Here is the reason: When the input frequency is greater than $f_{2(OL)}$, A_{VOL} decreases 20 dB per decade. When v_{out} tries to decrease, less opposing voltage is fed back to the inverting input. Therefore, v_2 increases and

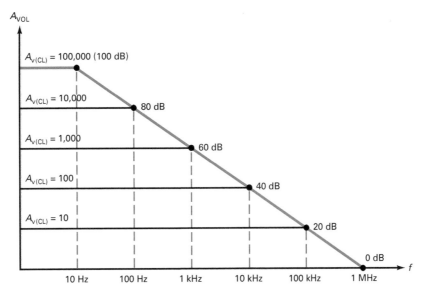

Figure 35-14 Lower voltage gain produces more bandwidth.

compensates for the decrease in A_{VOL}. Because of this, $A_{v(CL)}$ breaks at a higher frequency than $f_{2(OL)}$. The greater the negative feedback, the higher the closed-loop cutoff frequency. Stated another way: The smaller $A_{v(CL)}$ is, the higher $f_{2(CL)}$ is.

Figure 35-14 illustrates how the closed-loop bandwidth increases with negative feedback. As you can see, the heavier the negative feedback (smaller $A_{v(CL)}$), the greater the closed-loop bandwidth. Here is the equation for closed-loop bandwidth:

$$f_{2(CL)} = \frac{f_{unity}}{A_{v(CL)} + 1} \qquad \text{(inverting amplifier only)}$$

In most applications, $A_{v(CL)}$ is greater than 10 and the equation simplifies to

$$f_{2(CL)} = \frac{f_{unity}}{A_{v(CL)}} \qquad \text{(noninverting)} \qquad \textbf{(35-5)}$$

For instance, when $A_{v(CL)}$ is 10,

$$f_{2(CL)} = \frac{1\,\text{MHz}}{10} = 100\,\text{kHz}$$

which agrees with Fig. 35-14. If $A_{v(CL)}$ is 100,

$$f_{2(CL)} = \frac{1\,\text{MHz}}{100} = 10\,\text{kHz}$$

which also agrees.

Formula (35-5) can be rearranged into

$$f_{unity} = A_{v(CL)} f_{2(CL)} \qquad \textbf{(35-6)}$$

Notice that the unity-gain frequency equals the product of gain and bandwidth. For this reason, many data sheets refer to the unity-gain frequency as the **gain-bandwidth (GBW) product.**

Bias and Offsets

Negative feedback reduces the output error caused by input bias current, input offset current, and input offset voltage.

When $A_{v(CL)}$ is small, the total output error may be small enough to ignore. If not, resistor compensation and offset nulling will be necessary.

In an inverting amplifier, R_{B2} is the Thevenin resistance seen when looking back from the inverting input toward the source. This resistance is given by

$$R_{B2} = R_1 \parallel R_f \qquad \textbf{(35-7)}$$

If it is necessary to compensate for input bias current, an equal resistance R_{B1} should be connected to the noninverting input. This resistance has no effect on the virtual-ground approximation because no ac signal current flows through it.

EXAMPLE 35-5

Figure 35-15a is an ac equivalent circuit, so we can ignore the output error caused by input bias and offsets. What are closed-loop voltage gain and bandwidth? What is the output voltage at 1 kHz? At 1 MHz?

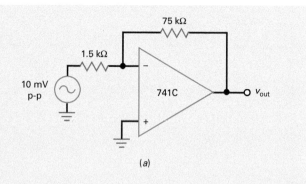

(a)

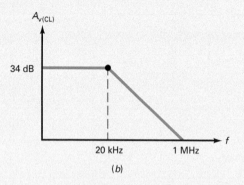

(b)

Figure 35-15 Example.

Answer:
With Formula (35-3), the closed-loop voltage gain is

$$A_{v(CL)} = \frac{-75\ k\Omega}{1.5\ k\Omega} = -50$$

With Formula (35-5), the closed-loop bandwidth is

$$f_{2(CL)} = \frac{1\ MHz}{50} = 20\ kHz$$

Figure 35-15b shows the ideal Bode plot of the closed-loop voltage gain. The decibel equivalent of 50 is 34 dB. (Shortcut: 50 is half of 100, or down 6 dB from 40 dB.)
The output voltage at 1 kHz is

$$v_{out} = (-50)(10\ mV\ p\text{-}p) = -500\ mV\ p\text{-}p$$

Since 1 MHz is the unity-gain frequency, the output voltage at 1 MHz is

$$v_{out} = -10\ mV\ p\text{-}p$$

Again, the minus (−) output value indicates a 180° phase shift between the input and output.

35.4 The Noninverting Amplifier

The **noninverting amplifier** is another basic op-amp circuit. It uses negative feedback to stabilize the overall voltage gain. With this type of amplifier, the negative feedback also increases the input impedance and decreases the output impedance.

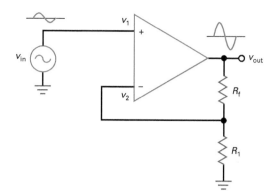

Figure 35-16 The noninverting amplifier.

Basic Circuit

Figure 35-16 shows the ac equivalent circuit of a noninverting amplifier. An input voltage v_{in} drives the noninverting input. This input voltage is amplified to produce the in-phase output voltage shown. Part of output voltage is fed back to the input through a voltage divider. The voltage across R_1 is the feedback voltage applied to the inverting input. This feedback voltage is almost equal to the input voltage. Because of the high open-loop voltage gain, the difference between v_1 and v_2 is very small. Since the feedback voltage opposes the input voltage, we have negative feedback.

Here is how the negative feedback stabilizes the overall voltage gain: If the open-loop voltage gain A_{VOL} increases for any reason, the output voltage will increase and feed back more voltage to the inverting input. This opposing feedback voltage reduces the net input voltage $v_1 - v_2$. Therefore, even though A_{VOL} increases, $v_1 - v_2$ decreases, and the final output increases much less than it would without the negative feedback. The overall result is only a very slight increase in output voltage.

Virtual Short

When we connect a piece of wire between two points in a circuit, the voltage of both points with respect to ground is equal. Furthermore, the wire provides a path for current to flow between the two points. A *mechanical short* (a wire between two points) is a short for both voltage and current.

A **virtual short** is different. This type of short can be used for analyzing noninverting amplifiers. With a virtual short, we can quickly and easily analyze noninverting amplifiers and related circuits.

The virtual short uses these two properties of an ideal op amp:

1. Since R_{in} is infinite, both input currents are zero.
2. Since A_{VOL} is infinite, $v_1 - v_2$ is zero.

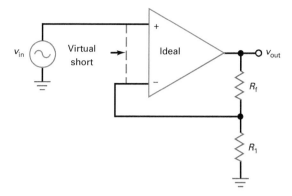

Figure 35-17 A virtual short exists between the two op-amp inputs.

Figure 35-17 shows a virtual short between the input terminals of the op amp. The virtual short is a short for voltage but an open for current. As a reminder, the dashed line means that no current can flow through it. Although the virtual short is an ideal approximation, it gives very accurate answers when used with heavy negative feedback.

Here is how we will use the virtual short: Whenever we analyze a noninverting amplifier or a similar circuit, we can visualize a virtual short between the input terminals of the op amp. As long as the op amp is operating in the linear region (not positively or negatively saturated), the open-loop voltage gain approaches infinity and a virtual short exists between the two input terminals.

One more point: Because of the virtual short, the inverting input voltage follows the noninverting input voltage. If the noninverting input voltage increases or decreases, the inverting input voltage immediately increases or decreases to the same value. This follow-the-leader action is called **bootstrapping** (as in "pulling yourself up by your bootstraps"). The noninverting input pulls the inverting input up or down to an equal value. Described another way, the inverting input is bootstrapped to the noninverting input.

Voltage Gain

In Fig. 35-18, visualize a virtual short between the input terminals of the op amp. Then, the virtual short means that the input voltage appears across R_1, as shown. So we can write

$$v_{in} = i_1 R_1$$

Since no current can flow through a virtual short, the same i_1 current must flow through R_f, which means that the output voltage is given by

$$v_{out} = i_1(R_f + R_1)$$

Divide v_{out} by v_{in} to get the voltage gain

$$A_{v(CL)} = \frac{R_f + R_1}{R_1}$$

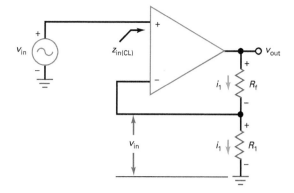

Figure 35-18 Input voltage appears across R_1 and same current flows through resistors.

or

$$A_{v(CL)} = \frac{R_f}{R_1} + 1 \qquad \textbf{(35-8)}$$

This is easy to remember because it is the same as the equation for an inverting amplifier, except that we add 1 to the ratio of resistances. Also note that the output is in phase with the input. Therefore, no (−) sign is used in the voltage gain equation.

Other Quantities

The closed-loop input impedance approaches infinity. Later, we will mathematically analyze the effect of negative feedback and will show that negative feedback increases the input impedance. Since the open-loop input impedance is already very high (2 MΩ for a 741C), the closed-loop input impedance will be even higher.

The effect of negative feedback on bandwidth is the same as with an inverting amplifier:

$$f_{2(CL)} = \frac{f_{unity}}{A_{v(CL)}}$$

Again, we can trade off voltage gain for bandwidth. The smaller the closed-loop voltage gain, the greater the bandwidth.

The input error voltages caused by input bias current, input offset current, and input offset voltage are analyzed the same way as with an inverting amplifier. After calculating each input error, we can multiply by the closed-loop voltage gain to get the total output error.

R_{B2} is the Thevenin resistance seen when looking from the inverting input toward the voltage divider. This resistance is the same as for an inverting amplifier:

$$R_{B2} = R_1 \parallel R_f$$

If it is necessary to compensate for input bias current, an equal resistance R_{B1} should be connected to the noninverting input. This resistance has no effect on the virtual-short approximation because no ac signal current flows through it.

EXAMPLE 35-6

In Fig. 35-19a, what is closed-loop voltage gain and bandwidth? What is the output voltage at 250 kHz?

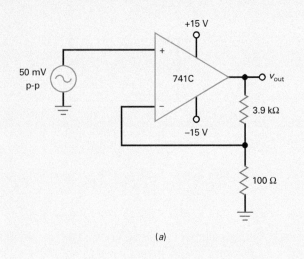

(a)

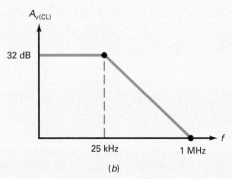

(b)

Figure 35-19 Example.

Answer:
With Formula (35-8),

$$A_{v(CL)} = \frac{3.9 \text{ k}\Omega}{100\Omega} + 1 = 40$$

Dividing the unity-gain frequency by the closed-loop voltage gain gives

$$f_{2(CL)} = \frac{1 \text{ MHz}}{40} = 25 \text{ kHz}$$

Figure 35-19b shows the ideal Bode plot of closed-loop voltage gain. The decibel equivalent of 40 is 32 dB. (Shortcut: $40 = 10 \times 2 \times 2$ or 20 dB + 6 dB + 6 dB = 32 dB.) Since the $A_{v(CL)}$ breaks at 25 kHz, it is down 20 dB at 250 kHz. This means that $A_{v(CL)} = 12$ dB at 250 kHz, which is equivalent to an ordinary voltage gain of 4. Therefore, the output voltage at 250 kHz is

$$v_{out} = 4 \text{ (50 mV p-p)} = 200 \text{ mV p-p}$$

35.5 Common Op-Amp Applications

Op-amp applications are so broad and varied that it is impossible to discuss them comprehensively in this chapter. Besides, we need to understand negative feedback better before looking at some of the more advanced applications. For now, let us take a look at two practical circuits.

The Summing Amplifier

Whenever we need to combine two or more analog signals into a single output, the **summing amplifier** of Fig. 35-20a is a natural choice a good example is an audio mixer for microphones and musical instruments. For simplicity, the circuit shows only two inputs, but we can have as many

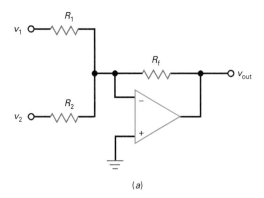

(a)

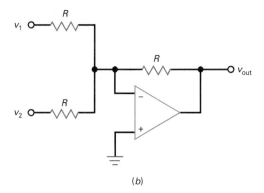

(b)

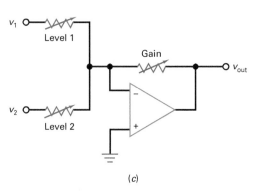

(c)

Figure 35-20 Summing amplifier.

inputs as needed for the application. A circuit like this amplifies each input signal. The gain for each *channel* or input is given by the ratio of the feedback resistance to the appropriate input resistance. For instance, the closed-loop voltage gains of Fig. 35-20*a* are

$$A_{v1(CL)} = \frac{-R_f}{R_1} \quad \text{and} \quad A_{v2(CL)} = \frac{-R_f}{R_2}$$

The summing circuit combines all the amplified input signals into a single output, given by

$$v_{out} = A_{v1(CL)}v1 + A_{v2(CL)}v2 \qquad \textbf{(35-9)}$$

It is easy to prove Formula (35-9). Since the inverting input is a virtual ground, the total input current is

$$i_{in} = i_1 + i_2 = \frac{v_1}{R_1} + \frac{v_2}{R_2}$$

Because of the virtual ground, all this current flows through the feedback resistor, producing an output voltage with a magnitude of

$$v_{out} = (i_1 + i_2)R_f = -\left(\frac{R_f}{R_1}v_1 + \frac{R_f}{R_2}v2\right)$$

Here you see that each input voltage is multiplied by its channel gain and added to produce the total output. The same result applies to any number of inputs.

In some applications, all resistances are equal, as shown in Fig. 35-20*b*. In this case, each channel has a closed-loop voltage gain of unity (1) and the output is given by:

$$v_{out} = -(v_1 + v_2 + \ldots + v_n)$$

This is a convenient way of combining input signals and maintaining their relative sizes. The combined output signal can then be processed by more circuits.

Figure 35-20*c* is a **mixer,** a convenient way to combine audio signals in a high-fidelity audio system. The adjustable resistors allow us to set the level of each input, and the gain control allows us to adjust the combined output volume. By decreasing *level 1,* we can make the v_1 signal louder at the output. By decreasing *level 2,* we can make the v_2 signal louder. By increasing *gain,* we can make both signals louder.

Voltage Follower

The **voltage follower** is the equivalent of an emitter follower, except that it works much better.

Figure 35-21 shows the ac equivalent circuit for a voltage follower. Although it appears deceptively simple, the circuit is very close to ideal because the negative feedback is maximum. As you can see, the feedback resistance is zero. Therefore, all the output voltage is fed back to the inverting input. Because of the virtual short

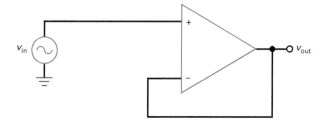

Figure 35-21 Voltage follower has unity gain and maximum bandwidth.

between the op-amp inputs, the output voltage equals the input voltage

$$v_{out} = v_{in}$$

which means that the closed-loop voltage gain is

$$A_{v(CL)} = 1 \qquad \textbf{(35-10)}$$

We can get the same result by calculating the closed-loop voltage gain with Formula (35-9). Since $R_f = 0$ and $R_1 = \infty$,

$$A_{v(CL)} = \frac{R_f}{R_f} + 1 = 1$$

Therefore, the voltage follower is a perfect follower circuit because it produces an output voltage that is exactly equal to the input voltage (or close enough to satisfy almost any application).

Furthermore, the maximum negative feedback produces a closed-loop input impedance that is much higher than the open-loop input impedance (2 MΩ for a 741C). Also, a maximum negative feedback produces a closed-loop output impedance that is much lower than the open-loop output impedance (75 Ω for a 741C). Therefore, we have an almost perfect method for converting a high-impedance source to a low-impedance source.

The crucial point to understand is this: The voltage follower is the ideal interface to use between a high-impedance source and a low-impedance load. Basically, it transforms the high-impedance voltage source into a low-impedance voltage source. You will see the voltage follower used a great deal in practice.

Since $A_{v(CL)} = 1$ in a voltage follower, the closed-loop bandwidth is maximum and equal to

$$f_{2(CL)} = f_{unity} \qquad \textbf{(35-11)}$$

Another advantage is the low output offset error because the input errors are not amplified. Since $A_{v(CL)} = 1$, the total output error voltage equals the worst-case sum of the input errors.

Finally, any amplifier that produces a voltage across a smaller load resistor that is the same as the equal input voltage across a higher impedance is really a power amplifier.

EXAMPLE 35-7

Three audio signals drive the summing amplifier of Fig. 35-22. What is the ac output voltage?

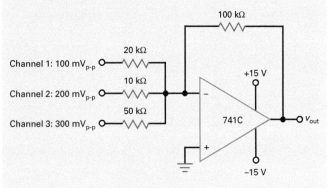

Figure 35-22 Example.

Answer:
The channels have closed-loop voltage gains of

$$A_{v1(CL)} = \frac{-100k\Omega}{20k\Omega} = -5$$

$$A_{v2(CL)} = \frac{-100k\Omega}{10k\Omega} = -10$$

$$A_{v3(CL)} = \frac{-100k\Omega}{50k\Omega} = -2$$

The output voltage is

$$v_{out} = (-5)(100 \text{ mVp-p}) + (-10)(200 \text{ mVp-p}) \\ + (-2)(300 \text{ mVp-p}) = -3.1 \text{ Vp-p}$$

Again, the negative sign indicates a 180° phase shift.

EXAMPLE 35-8

An ac voltage source of 10 mVp-p with an internal resistance of 100 kΩ drives a voltage follower with a load resistance of 1 Ω. What is the output voltage? The bandwidth?

Answer:
The closed-loop voltage gain is unity. Therefore,

$$v_{out} = 10 \text{ mVp-p}$$

and the bandwidth is

$$f_{2(CL)} = 1 \text{ MHz}$$

This example echoes the idea discussed earlier. The voltage follower is an easy way to transform a high-impedance source into a low-impedance source. It does what the emitter follower does, only far better.

Table 35-2 shows the basic op-amp circuits we have discussed to this point.

35.6 Op Amps as Surface-Mount Devices

Operational amplifiers and similar kinds of analog circuits are more often available in surface-mount (SM) packages as well as in the more tradition dual in-line IC forms. Because the pin-out for most op amp tends to be relatively simple, the small-outline package (SOP), also called the small-outline integrated circuit (SOIC), is the preferred SM style. Virtually all op amps, including the older 741 are available in the SOIC package. See Fig. 35.23.

35.7 Differential Amplifiers

This section discuss as how to build a **differential amplifier** using an op amp. One of the most important characteristics of a differential amplifier is its CMRR because the typical input signal is a small differential voltage and a large common-mode voltage.

Basic Differential Amplifier

Figure 35-24 shows an op amp connected as a differential amplifier. The resistor R_1' has the same nominal value as R_1 but differs slightly in value because of tolerances. For instance, if the resistors are 1 kΩ ±1%, R_1 may be as high as 1010 Ω and R_1' may be as low as 990 Ω, and vice versa. Similarly, R_2 and R_2' are nominally equal but may differ slightly because of tolerances.

In Fig. 35-24, the desired input voltage v_{in} is called the **differential input voltage** to distinguish it from the common-mode input voltage, $v_{in(CM)}$. A circuit like Fig. 35-24 amplifies the differential input voltage v_{in} to get an output voltage of v_{out}. It can be shown that

$$v_{out} = A_v v_{in}$$

where

$$A_v = \frac{-R_2}{R_1} \qquad (35\text{-}12)$$

This voltage gain is called the **differential voltage gain** to distinguish it from the common-mode voltage gain, $A_{v(CM)}$. By using precision resistors, we can build a differential amplifier with a precise voltage gain.

A differential amplifier is often used in applications in which the differential input signal v_{in} is a small dc voltage (millivolts) and the common-mode input signal is a large dc voltage (volts). As a result, the CMRR of the circuit becomes a critical parameter. For instance, if the differential input signal is 7.5 mV and the common-mode signal is 7.5 V, the differential input signal is 60 dB less than the common mode input signal. Unless the circuit has a very high CMRR, the common-mode output signal will be objectionably large.

Table 35-2 Basic Op-Amp Configurations

Inverting amp

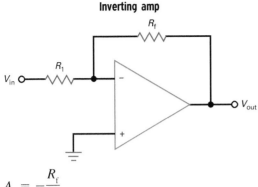

$$A_v = -\frac{R_f}{R_1}$$

Noninverting amp

$$A_v = \frac{R_f}{R_1} + 1$$

Summing amp

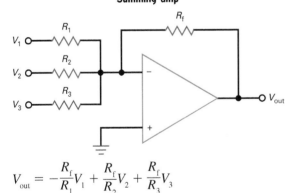

$$V_{out} = -\frac{R_f}{R_1}V_1 + \frac{R_f}{R_2}V_2 + \frac{R_f}{R_3}V_3$$

Voltage follower

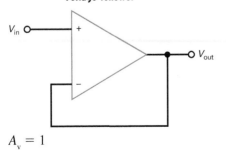

$$A_v = 1$$

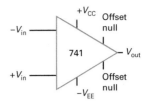

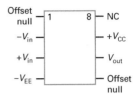

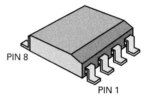

Figure 35-23 The SM version of the LM741 op amp.

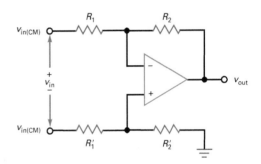

Figure 35-24 Differential amplifier.

CMRR of the Differential Amp

In Fig. 35-24, two factors determine the overall CMRR of the circuit. First, there is the CMRR of the op amp itself. For best results, use a precision op amp with a very high CMRR (>100 dB). Second, the external resistors must be closely matched for best CMRR. Use precision resistors with tolerances as small as possible (< 1%).

Buffered Inputs

The source resistances driving the differential amplifier of Fig. 35-24 effectively become part of R_1 and R_1', which changes the voltage gain and may degrade the CMRR. This is a very serious disadvantage. The solution is to increase the input impedance of the circuit.

Figure 35-25 shows one way to do it. The first stage (the preamp) consists of two voltage followers that **buffer** (isolate) the inputs. This can increase the input impedance to well over 100 MΩ. The voltage gain of the first stage is unity for both the differential and the common-mode input signal.

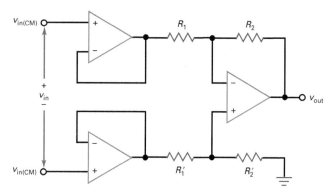

Figure 35-25 Differential input with buffered inputs.

Therefore, the second stage (the differential amplifier) still has to provide all the CMRR for the circuit.

Wheatstone Bridge

The differential input signal is usually a small dc voltage. The reason it is small is because it is the output of a Wheatstone bridge like that in Fig. 35-26a. A Wheatstone bridge is balanced when the ratio of resistances on the left side equals the ratio of resistances on the right side:

$$\frac{R_1}{R_2} = \frac{R_3}{R_4} \qquad \textbf{(35-13)}$$

When this condition is satisfied, the voltage across R_2 equals the voltage across R_4 and the output voltage of the bridge is zero.

The Wheatstone bridge can detect small changes in one of the resistors. For instance, suppose we have a bridge with three resistors of 1 kΩ and a fourth resistor of 1010 Ω, as shown in Fig. 35-26b. The voltage across R_2 is

$$v_2 = \frac{1\ k\Omega}{2\ k\Omega}(15\ V) = 7.5\ V$$

and the voltage across R_4 is approximately

$$v_4 = \frac{1010\ \Omega}{2010\ \Omega}(15\ V) = 7.537\ V$$

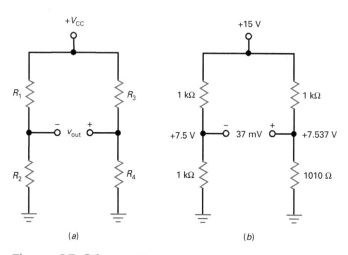

Figure 35-26 (a) Wheatstone bridge. (b) Slightly unbalanced bridge.

The output voltage of the bridge is approximately

$$v_{out} = v_4 - v_2 = 7.537\ V - 7.5\ V = 37\ mV$$

Transducers

Resistance R_4 may be an **input transducer,** a device that converts a nonelectrical quantity into an electrical quantity. For instance, a photoresistor converts a change in light intensity into a change in resistance, a **thermistor** converts a change in temperature into a change in resistance, and a strain gauge converts pressure into a resistance change.

There is also the **output transducer,** a device that converts an electrical quantity into a nonelectrical quantity. For instance, an LED converts current into light and a loudspeaker converts ac voltage into sound waves.

A wide variety of transducers are commercially available for quantities such as temperature, sound, light, humidity, velocity, acceleration, force, radioactivity, strain, and pressure, to mention a few. These transducers can be used with a Wheatstone bridge to measure nonelectrical quantities. Because the output of a Wheatstone bridge is a small dc voltage with a large common-mode voltage, we need to use dc amplifiers that have very high CMRRs.

A Typical Application

Figure 35-27 shows a typical application. Three of the bridge resistors have a value of

$$R = 1\ k\Omega$$

The transducer has a resistance of

$$R + \Delta R = 1010\ \Omega$$

The common-mode signal is

$$v_{in(CM)} = 0.5\,V_{CC} = 0.5(15\ V) = 7.5\ V$$

This is the voltage across each of the lower bridge resistors when $\Delta R = 0$.

When a bridge transducer is acted on by an outside quantity such as light, temperature, or pressure, its resistance will change. Figure 35-27 shows a transducer resistance of 1010 Ω, which implies that $\Delta R = 10\ \Omega$. Without showing the derivations, the input voltage is 37.5 mV.

Since the differential amplifier has a voltage gain of −100, the differential output voltage is

$$v_{out} = -100(37.5\ mV) = -3.75\ V$$

Again, without showing the math, using 0.1% tolerance resistors, the magnitude of CMRR is 25,000, which is equivalent to 88 dB.

That gives you the basic idea of how a differential amplifier is used with a Wheatstone bridge. A circuit like Fig. 35-27 is adequate for some applications but can be improved, as will be discussed in the following section.

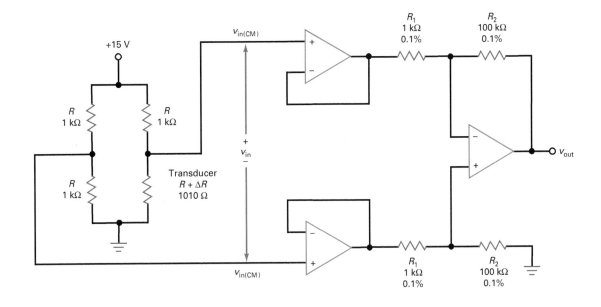

$$A_v = \frac{-R_2}{R_1}$$

$$v_{in} = \frac{\Delta R}{4R} V_{CC}$$

Figure 35-27 Bridge with transducer drives instrumentation amplifier.

35.8 Instrumentation Amplifiers

An **instrumentation amplifier** is a differential amplifier optimized for its dc performance. An instrumentation amplifier has a large voltage gain, a high CMRR, low input offsets, low temperature drift, and high input impedance.

Basic Instrumentation Amplifier

Figure 35-28 shows the classic design used for most instrumentation amplifiers. The output op amp is a differential amplifier with the voltage gain of unity. The resistors used in this output stage are usually matched to within ±0.1% or better. This means that the CMRR of the output stage is at least 54 dB.

Precision resistors are commercially available from less than 1 Ω to more than 10 MΩ, with tolerances of ±0.01 to ±1%. If we use matched resistors that are within ±0.01% of each other, the CMRR of the output stage can be as high as 74 dB. Also, temperature drift of precision resistors can be as low as 1 ppm/°C.

The first stage consists of two input op amps that act like a preamplifier. The design of the first stage is extremely clever. What makes it so ingenious is the action of point A, the junction between the two R_1 resistors. Point A acts like a virtual ground for a differential input signal and like a floating point for the common-mode signal. Because of this action, the differential signal is amplified but the common-mode signal is not.

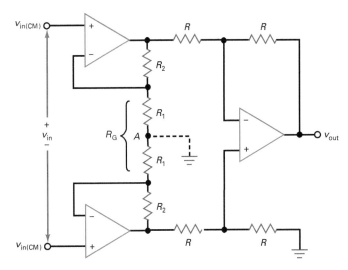

Figure 35-28 Standard three op-amp instrumentation amplifier.

To have high CMRR and low offsets, precision op amps must be used when building the instrumentation amplifier of Fig. 35-28.

A final point about Fig. 35-28: Since point A is a virtual ground rather than a mechanical ground, the R_1 resistors in the first stage do not have to be separate resistors. We can use a single resistor R_G that equals $2R_1$ without changing the operation of the first stage. The only difference is that the differential voltage gain is written as

$$A_v = \frac{2R_2}{R_G} + 1 \qquad (35\text{-}14)$$

The factor of 2 appears because $R_G = 2R_1$.

Integrated Instrumentation Amplifiers

The classic design of Fig. 35-28 can be integrated on a chip with all the components shown in Fig. 35-28, except R_G. This external resistance is used to control the voltage gain of the instrumentation amplifier. For instance, the AD620 is a monolithic instrumentation amplifier. The data sheet gives this equation for its voltage gain:

$$A_v = \frac{49.4 \text{ k}\Omega}{R_G} + 1 \qquad (35\text{-}15)$$

The quantity 49.4 kΩ is the sum of the two R_2 resistors. The IC manufacturer uses **laser trimming** to get a precise value of 49.4 kΩ. The word *trim* refers to a fine adjustment rather than a coarse adjustment. Laser trimming means burning off resistor areas on a semiconductor chip with a laser to get an extremely precise value of resistance.

Figure 35-29 shows the AD620 with an R_G of 499 Ω. This is a precision resistor with a tolerance of $\pm0.1\%$. The voltage gain is

$$A_v = \frac{49.4 \text{ k}\Omega}{499} + 1 = 100$$

The *pin-out* (pin numbers) of the AD620 is similar to that of a 741C since pins 2 and 3 are for the input signals, pins 4 and 7 are for the supply voltages, and pin 6 is the output. Pin 5 is shown grounded, the usual case for the AD620. But this pin does not have to be grounded. If necessary for interfacing with another circuit, we can offset the output signal by applying a dc voltage to pin 5.

In summary, monolithic instrumentation amplifiers typically have a voltage gain between 1 and 1000 that can be set with one external resistor, a CMRR greater than 100 dB, an input impedance greater than 100 MΩ, an input offset voltage less than 0.1 mV, a drift of less than 0.5 μV/°C, and other outstanding parameters.

EXAMPLE 35-9

In Fig. 35-28, $R_1 = 1$ kΩ, $R_2 = 100$ kΩ, and $R = 10$ kΩ. What is the differential voltage gain of the instrumentation amplifier?

Answer:
The voltage gain of the preamp is

$$A_v = \frac{100 \text{ k}\Omega}{1 \text{ k}\Omega} + 1 = 101$$

Since the voltage gain of the second stage is −1, the voltage gain of the instrumentation amplifier is −101.

35.9 Single-Supply Operation

Using dual supplies is the typical way to power op amps. But this is not necessary or even desirable in some applications. This section discusses the inverting and noninverting amplifiers running off a single positive supply.

Inverting Amplifier

Figure 35-30 shows a single-supply inverting voltage amplifier that can be used with ac signals. Capacitors C_1 and C_2 pass the ac but block any dc components of the signal. The usual negative V_{EE} supply (pin 4) is grounded, and a voltage divider applies half the V_{CC} supply to the noninverting input. Because the two inputs are virtually shorted, the inverting input has a quiescent voltage of approximately $+0.5V_{CC}$.

In the dc equivalent circuit, all capacitors are open and the circuit is a voltage follower that produces a dc output voltage of $+0.5V_{CC}$. Input offsets are minimized because the voltage gain is unity.

In the ac equivalent circuit, all capacitors are shorted and the circuit is an inverting amplifier with a voltage gain of $-R_2/R_1$.

A bypass capacitor C_3 is used on the noninverting input, as shown in Fig. 35-30. This reduces the power supply ripple

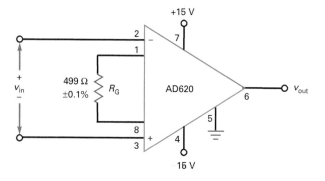

Figure 35-29 A monolithic instrumentation amplifier.

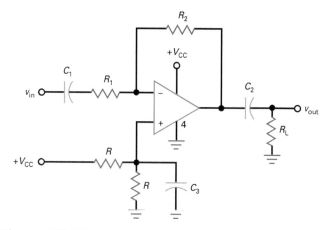

Figure 35-30 Single-supply inverting amplifier.

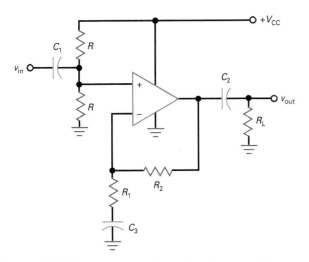

Figure 35-31 Single-supply noninverting amplifier.

and noise appearing at the noninverting input. To be effective, the cutoff frequency of this bypass circuit should be much lower than the ripple frequency out of the power supply.

Noninverting Amplifier

In Fig. 35-31, only a positive supply is being used. To get maximum output swing, you need to bias the noninverting input at half the supply voltage, which is conveniently done with an equal-resistor voltage divider. This produces a dc input of $+0.5V_{CC}$ at the noninverting input. Because of the negative feedback, the inverting input is bootstrapped to the same value.

In the dc equivalent circuit, all capacitors are open and the circuit has a voltage gain of unity, which minimizes the output offset voltage. The dc output voltage of the op amp is $+0.5V_{CC}$, but this is blocked from the final load by the output coupling capacitor.

In the ac equivalent circuit, all capacitors are shorted. When an ac signal drives the circuit, an amplified output signal appears across R_L. If a rail-to-rail op amp is used, the maximum peak-to-peak unclipped output is V_{CC}.

Single-Supply Op Amps

Although we can use ordinary op amps with a single supply, as shown in Figs. 35-30 and 35-31, there are some op amps that are optimized for single-supply operation. For instance, the LM324 is a quad op amp that eliminates the need for dual supplies. It contains four internally compensated op amps in a single package, each with an open-loop voltage gain of 100 dB, input biasing current of 45 nA, input offset current of 5 nA, and input offset voltage of 2 mV. It runs off a single positive supply voltage that can have any value between 3 and 32 V. Because of this, the LM324 is convenient to use as an interface with digital circuits that run off a single positive supply of +5 V.

CHAPTER 35 SYSTEM SIDEBAR

Balanced and Unbalanced Signal Wiring in Electronic Systems

Electronic systems are made up of multiple parts that must be connected together. On a small scale, ICs must be connected together on a printed circuit board (PCB). Modules or subassemblies must be interconnected inside an instrument or computer. Then on a larger scale, multiple pieces of equipment need to be connected to one another. Cables and connectors are used for many of these connections. On a PCB, no connectors may be involved but thin copper lines on the board form the connections. In all cases, there are two different ways that signal voltage levels are carried over a cable or a pair of wires, single-ended and balanced or differential. Here is a brief introduction to these methods that apply to both binary digital and analog signals.

Single-Ended Transmission

Single-ended transmission lines use two wires, one of them connected to ground and the other to carry the data voltage referenced to ground. This kind of line is also called an *unbalanced line* since one side is grounded. Figure S35-1 shows a simple data communications application with a data source, the transmitter (TX), and the data destination, or the receiver (RX). The two are connected to one another with a pair of wires. One of the wires is connected to ground. Recall that in any piece of electronic equipment there is a common reference point called ground. All voltages in the circuitry are referenced to ground; that is, voltages at any point in the circuit are measured with respect to ground. In most circuits, the ground is the negative side of the dc power

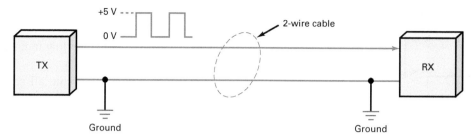

Figure S35-1 Single-ended serial data transmission.

supply but, in some cases, may be the positive side of the power supply, depending on the design.

If you wish to transmit more than one signal, analog or digital, over the same cable, all you have to do is add one additional wire for each signal you want to send. The signal on each wire is also referenced to the single ground wire. The reason why this system is called *single-ended* is that only one wire is required for each signal, which is referenced to the single ground wire or connection. For example, a five-wire cable would be needed to transmit four independent signals. The fifth wire is the common ground wire.

This method of transmission works fine, but it does sometimes suffer from a problem that is caused by an undesired signal referred to as common-mode voltage. Common-mode voltage, V_{cm}, is the voltage difference that exists between the transmitter ground and the receiver ground. We normally think of ground as being at zero volts potential, and that is why we used it as the reference for all other signals. And ground is indeed at zero potential inside any given piece of equipment. The problem is that the ground in one piece of equipment may actually be at a different voltage level than the ground in another piece of equipment at a distance.

To explain this problem, you must understand that the ultimate ground or reference point is really the earth. All electrical systems use the ground or earth as the common reference point. This starts with the electrical utility that distributes the ac power to you. One of the wires used to distribute power to your home, office, factory, or other facility is connected to ground or earth. It is called the neutral. One side of every ac power cord is connected to the neutral. In newer ac plugs, this is the prong or blade that is the wider of the two. This ground is usually isolated from the electric circuits in the equipment by the power supply, which usually incorporates a power transformer that provides isolation of the earth ground or neutral from the circuitry ground. But this may not be the case. In some equipment, the neutral is connected to the metal cabinet of the equipment, which may, in turn, make connection with the circuitry ground at some point in the equipment.

Because of long distribution lines and the differences and inefficiencies of earth grounding, there can easily be a potential difference between the grounds or neutral points from one piece of equipment to another. If you could use an ac voltmeter to measure the voltage between the grounded sides of the various outlets in a room, house, or building, you will actually see a voltage difference between them. While both grounds should theoretically be at the same potential, they are not. This problem becomes worse as the distance between the two pieces of equipment increases. There could be several volts difference between the ground at the transmitter, which may be a computer, and the ground at the receiver, which may be a printer at the other end of a 50-ft cable. In bad cases, the voltage difference could be tens of volts or more.

The problem with this ground potential difference, which we call the common-mode voltage, is that it adds to or subtracts from the transmitted logic signal so that the receiver sees a signal level that is different from the agreed-on standard voltage levels established during the design. The transmitter sends the correct levels, but the receiver sees these levels plus the common-mode voltage so it may or may not recognize them. Thus transmission errors occur. In fact, in some cases, the voltage differences may so great that they could damage the receiver circuits.

Figure S35-2 shows the common-mode voltage and how it appears in relationship to the transmitter and receiver. You would think that if any voltage existed between the two grounds that the ground wire itself would short out this voltage, but such is not the case. What the receiver sees is the signal voltage level transmitter, V_s, plus the common-mode voltage, V_{cm}, or $V_s + V_{cm}$. This is with respect to the transmitter ground, which we use as the main reference here since the transmitter is the origin of the signal.

The big question here is, What is the nature of the common-mode voltage? Is it dc, ac, or what? It is usually a 60-Hz, sine-wave ac accompanied by various types of noise and "garbage." It may be voltage spikes caused by transients on the ac line or noise picked up by capacitive or magnetic induction on the ac power lines. It may be noise generated by digital signals that occur on the power supply power bus or ground. Such signals, when added to the binary signal to be transmitted, create a messy, noisy signal that can be easily misinterpreted by the receiver.

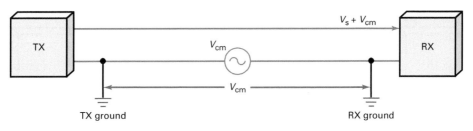

V_s = signal voltage
V_{cm} = common-mode voltage

Figure S35-2 Single-ended transmission showing common-mode voltage, V_{CM}, problem.

Assume the transmitter outputs digital logic levels of 0 and +5 V. If the common-mode voltage is a +3-V spike with respect to the TX ground, then the input to the receiver will be a pulse that will switch between 0 + 3 = +3 V and +5 + 3 = +8 V. If the common-mode voltage is a 12-V peak-to-peak sine wave, the receiver input will vary between a peak positive voltage between +6 V and 6 + 5 = +11 V and a negative peak that could be between −1 and −6 V. In this case, the receiver is not likely to recognize the data signal at all, resulting in complete loss of information.

Despite this problem, single-ended operation is common. In practice, there will always be some common-mode voltage. In most cases it is a relatively low value compared to the signal levels commonly used. So while it does shift these levels a bit, the tolerance of the receiver input circuits is such that the signals are still fully recognizable. In fact, the selection of the signal levels in single-ended systems fully takes into consideration the fact that some common-mode voltage will exist. Most systems handle the difference with minimum difficulty. In extreme cases, special techniques must be used.

The solution to the common-mode voltage problem is to use some kind of isolation device that transfers the desired signals but ignores any common-mode voltage. Figure S35-3a shows how a transformer is used for isolation. Only one transformer at the receiver is used in this illustration, although, in many cases, a transformer is used at both the transmitter and receiver ends. The signal from the transmitter is applied to the primary winding of the transformer. The signal creates a magnetic field that transfers the signal to the secondary winding by induction. The voltage induced into the secondary winding is the desired signal, which is connected to the receiver input. The secondary winding, being a separate coil of wire, is electrically insulated from the primary winding. The signal is transferred from one winding to the other by magnetic induction. This physical isolation allows the receiver ground to be separate from the transmitter ground, thus eliminating the common-mode voltage problem. The key to the success of this solution is in the transformer, which

is usually a special unit with a wide bandwidth designed for the type signals being transmitted.

For low-speed digital transmission, an optoisolator, as shown in Fig. S35-3b, can be used. An optoisolator is a semiconductor device that contains an LED and a phototransistor. If the LED is turned on, the transistor conducts as a closed switch. If the LED is off, the transistor is off and an open circuit. The optoisolator is applied at the receiver end of the cable. As the binary pulses turn off and on, they turn the LED off and on, thereby operating the transistor that reproduces the transmitted signal at the output. There is no common-mode voltage transference with this arrangement. The optoisolator does not work for analog signals, and only low-speed digital signals (<1 Mbps) can be transmitted.

Balanced Transmission

The best solution to the common-mode voltage problem is to use what are called *balanced transmission lines,* or *differential lines.* This system uses two conductors for transmission, but neither of them is grounded. Both the lines have voltages on them that are referenced to a ground, but they are opposites of one another. In digital terms, this means that the voltages on each line are complementary logic levels. If one line is binary 1, the other is binary 0, or vice versa. In analog signal terms, it means that the two signals are 180° out of phase with one another, or one signal is the inverted version of the other.

Figure S35-4 shows a balanced transmission line and the circuits normally used to process the signals. The inverters are used to create complementary binary signals that are placed on the transmission line. Remember: Each signal on each wire comes from a circuit whose output is referenced to a ground. What is being transmitted is the difference between the voltages on the two lines. The received signal is applied to a differential amplifier that converts the balanced or differential signal back into a single-ended signal like the original data. Since no ground wire is used with the transmission line, there is no common-mode voltage. Thus one sure way to eliminate the common-mode voltage problem is to use a balanced transmission line.

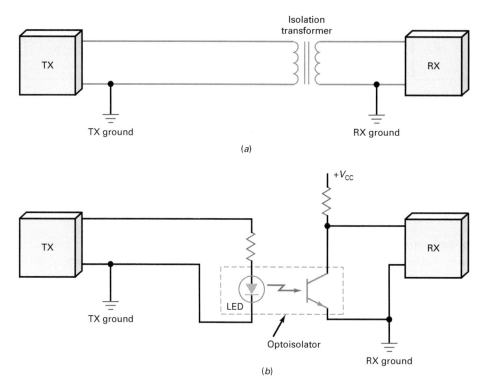

Figure S35-3 (*a*) How a transformer is used to separate and isolate the TX and RX grounds. (*b*) Using an optoisolator to eliminate the common-mode voltage problem in a single-ended system.

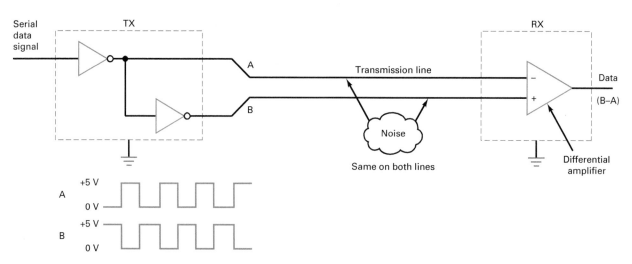

Figure S35-4 Balanced, or differential, signal transmission.

Besides eliminating the common-mode problem, balanced transmission lines also help minimize noise pickup on the transmission line. If the transmission line is run near other signal-carrying lines, the magnetic and electric fields produced by the adjacent lines can induce undesired voltages into the transmission line. This noise can cause transmission errors. The noise pickup problem is especially bad for single-ended lines.

On the other hand, the balanced line eliminates the noise problem. What happens is that the same voltage is induced into both the transmissions lines since they are both above ground. Remember that a differential amplifier generates the difference between the two input signals, which means that the noise signal on one line is subtracted from the same noise on the other line. The result is that the noise is canceled out. This advantage of the balanced method is just as important as the elimination of the common-mode problem. That is why most long-distance transmission lines are of the balanced type.

1. Which of the following is *not* a characteristic of an ideal op amp?
   a. Infinite gain.
   b. Infinite input impedance.
   c. Infinite output impedance.
   d. Zero output impedance.

2. The ground pin on a 741 op amp is
   a. pin 3.
   b. pin 4.
   c. pin 5.
   d. none of the above.

3. At the unity-gain frequency, the open-loop voltage gain is
   a. 1.
   b. $A_{v(mid)}$.
   c. Zero.
   d. Very large.

4. The cutoff frequency of an op amp equals the unity-gain frequency divided by
   a. the cutoff frequency.
   b. closed-loop voltage gain.
   c. unity.
   d. common-mode voltage gain.

5. If the cutoff frequency is 20 Hz and the midband open-loop voltage gain is 1,000,000, the unity-gain frequency is
   a. 20 Hz.
   b. 1 MHz.
   c. 2 MHz.
   d. 20 MHz.

6. If the unity-gain frequency is 5 MHz and the midband open-loop voltage gain is 100,000, the cutoff frequency is
   a. 50 Hz.
   b. 1 MHz.
   c. 1.5 MHz.
   d. 15 MHz.

7. The initial slope of a sine wave is directly proportional to
   a. slew rate.
   b. frequency.
   c. voltage gain.
   d. capacitance.

8. When the initial slope of a sine wave is greater than the slew rate,
   a. distortion occurs.
   b. linear operation occurs.

   c. voltage gain is maximum.
   d. the op amp works best.

9. The power bandwidth increases when
   a. frequency decreases.
   b. peak value decreases.
   c. initial slope decreases.
   d. voltage gain increases.

10. A 741C contains
    a. BJTs.
    b. FETs.
    c. MOSFETs.
    d. all of the above.

11. A 741C amplifier does not work with
    a. dc.
    b. ac.
    c. both *a* and *b*.
    d. digital.

12. The input impedance of a BIFET op amp is
    a. low.
    b. medium.
    c. high.
    d. extremely high.

13. An TL081 is a
    a. diff amp.
    b. source follower.
    c. Bipolar op amp.
    d. BIFET op amp.

14. If the two supply voltages are ±12 V, the MPP value of an op amp is closest to
    a. 0 V.
    b. +12 V.
    c. −12 V.
    d. 24 V.

15. The open-loop cutoff frequency of a 741C is controlled by
    a. a coupling capacitor.
    b. the output short circuit current.
    c. the power bandwidth.
    d. a compensating capacitor.

16. The 741C has a unity-gain frequency of
    a. 10 Hz.
    b. 20 kHz.
    c. 1 MHz.
    d. 15 MHz.

17. The unity-gain frequency equals the product of closed-loop voltage gain and the
    a. compensating capacitance.
    b. tail current.

c. closed-loop cutoff frequency.
d. load resistance.

18. If f_{unity} is 10 MHz and midband open-loop voltage gain is 200,000, then the open-loop cutoff frequency of the op amp is
   a. 10 Hz.
   b. 20 Hz.
   c. 50 Hz.
   d. 100 Hz.

19. The initial slope of a sine wave increases when
   a. frequency decreases.
   b. peak value increases.
   c. C_c increases.
   d. slew rate decreases.

20. If the frequency of the input signal is greater than the power bandwidth,
   a. slew-rate distortion occurs.
   b. a normal output signal occurs.
   c. output offset voltage increases.
   d. distortion may occur.

21. An op amp has an open input. The output voltage will be
   a. zero.
   b. slightly different from zero.
   c. maximum positive or negative.
   d. an amplified sine wave.

22. An op amp has a voltage gain of 200,000. If the output voltage is 1 V, the input voltage is
   a. 2 μV.
   b. 5 μV.
   c. 10 mV.
   d. 1 V.

23. A 741C has supply voltages of ±15 V. If the load resistance is large, the MPP value is approximately
   a. 0 V.
   b. +15 V.
   c. 27 V.
   d. 30 V.

24. Above the cutoff frequency, the voltage gain of a 741C decreases approximately
   a. 10 dB per decade.
   b. 20 dB per octave.
   c. 10 dB per octave.
   d. 20 dB per decade.

25. The voltage gain of an op amp is unity at the
   a. cutoff frequency.
   b. unity-gain frequency.
   c. generator frequency.
   d. power bandwidth.

26. When slew-rate distortion of a sine wave occurs, the output
   a. is larger.
   b. appears triangular.
   c. is normal.
   d. has no offset.

27. A 741C has
   a. a voltage gain of 100,000.
   b. an input impedance of 2 MΩ.
   c. an output impedance of 75 Ω.
   d. all of the above.

28. The closed-loop voltage gain of an inverting amplifier equals
   a. the ratio of the input resistance to the feedback resistance.
   b. the open-loop voltage gain.
   c. the feedback resistance divided by the input resistance.
   d. the input resistance.

29. The noninverting amplifier has a
   a. large closed-loop voltage gain.
   b. small open-loop voltage gain.
   c. large closed-loop input impedance.
   d. large closed-loop output impedance.

30. The voltage follower has a
   a. closed-loop voltage gain of unity.
   b. small open-loop voltage gain.
   c. closed-loop bandwidth of zero.
   d. large closed-loop output impedance.

31. A summing amplifier can have
   a. no more than two input signals.
   b. two or more input signals.
   c. a closed-loop input impedance of infinity.
   d. a small open-loop voltage gain.

32. An instrumentation amplifier has a high
   a. output impedance.
   b. power gain.
   c. CMRR.
   d. supply voltage.

33. In a differential amplifier, the CMRR is limited mostly by the
   a. CMRR of the op amp.
   b. gain-bandwidth product.
   c. supply voltages.
   d. tolerance of the resistors.

34. The input signal for an instrumentation amplifier usually comes from
   a. an inverting amplifier.
   b. a resistor.
   c. a differential amplifier.
   d. a Wheatstone bridge.

35. In the classic three op-amp instrumentation amplifier, the differential voltage gain is usually produced by the
 a. first stage.
 b. second stage.
 c. mismatched resistors.
 d. output op amp.

36. Another name for a summable amplifier is
 a. follower.
 b. mixer.
 c. instrumentation amp.
 d. modulator.

37. An input transducer converts
 a. voltage to current.
 b. current to voltage.
 c. an electrical quantity to a nonelectrical quantity.
 d. a nonelectrical quantity to an electrical quantity.

38. A thermistor converts
 a. light to resistance.
 b. temperature to resistance.
 c. voltage to sound.
 d. current to voltage.

39. When we trim a resistor, we are
 a. making a fine adjustment.
 b. reducing its value.
 c. increasing its value.
 d. making a coarse adjustment.

40. If an op amp has only a positive supply voltage, its output cannot
 a. be negative.
 b. be zero.
 c. equal the supply voltage.
 d. be ac-coupled.

CHAPTER 35 PROBLEMS

SECTION 35.2 The 741 Op Amp

35.1 Assume that negative saturation occurs at 1 V less than the supply voltage with an 741C. How much inverting input voltage does it take to drive the op amp of Fig. 35-32 into negative saturation?

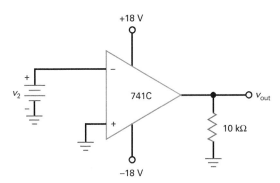

Figure 35-32

35.2 What is the common-mode rejection ratio of an TL081A at low frequencies? Convert this decibel value to an ordinary number. (See Table 35-1.)

35.3 What is the open-loop voltage gain of an 741A when the input frequency is 1 kHz? 10 kHz? 100 kHz? (Assume a first-order response, that is, 20 dB per decade rolloff.)

35.4 The input voltage to an op amp is a large voltage step. The output is an exponential waveform that changes 2.0 V in 0.4 μs. What is the slew rate of the op amp?

35.5 An op amp has a slew rate of 70 V/μs. What is the power bandwidth for a peak output voltage of 7 V?

SECTION 35.3 The Inverting Amplifier

35.6 What are closed-loop voltage gain and bandwidth in Fig. 35-33? What is the output voltage at 1 kHz? At 10 MHz? Draw the ideal Bode plot of closed-loop voltage gain.

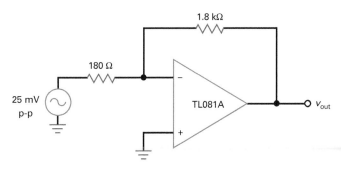

Figure 35-33

SECTION 35.4 The Noninverting Amplifier

35.7 In Fig. 35-34, what are the closed-loop voltage gain and bandwidth? The ac output voltage at 100 kHz?

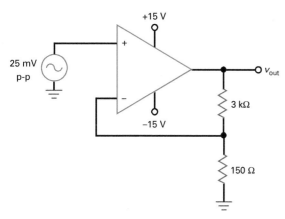

Figure 35-34

SECTION 35.5 Common Op-Amp Applications

35.8 In Fig. 35-35a, what is the ac output voltage?

35.9 What is the output voltage in Fig. 35-35b? The bandwidth?

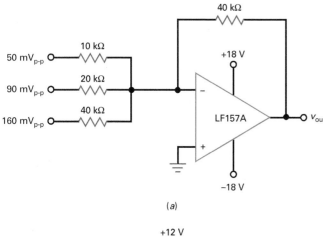

(a)

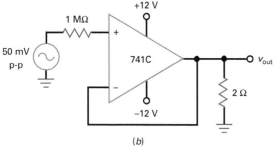

(b)

Figure 35-35

SECTION 35.7 Differential Amplifiers

35.10 The differential amplifier of Fig. 35-24 has $R_1 =$ 1.5 kΩ and $R_2 = 30$ kΩ. What is the differential voltage gain?

35.11 In Fig. 35-25, $R_1 = 1$ kΩ and $R_2 = 20$ kΩ. What is the differential voltage gain?

35.12 In the Wheatstone bridge of Fig. 35-26, $R_1 = 10$ kΩ, $R_2 = 20$ kΩ, $R_3 = 20$ kΩ, and $R_4 = 10$ kΩ. Is the bridge balanced?

35.13 In the typical application of Fig. 35-27, transducer resistance changes to 985 Ω. What is the final output voltage?

SECTION 35.8 Instrumentation Amplifiers

35.14 In the instrumentation amplifier of Fig. 35-28, $R_1 =$ 1 kΩ and $R_2 = 99$ kΩ. What is the output voltage if $v_{in} = 2$ mV?

35.15 The value of R_G is changed to 1008 Ω in Fig. 35-29. What is the differential output voltage if the differential input voltage is 20 mV?

Op-Amp Applications

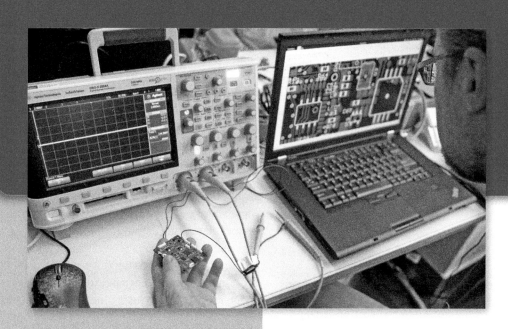

Learning Outcomes

After studying this chapter, you should be able to:

> *Explain* how a comparator works and *describe* the importance of a reference point.

> *Discuss* comparators that have positive feedback and *calculate* the trip points and hysteresis for these circuits.

> *Identify* and *discuss* waveform conversion circuits.

> *Identify* and *discuss* waveform generation circuits.

> *Describe* how several active diode circuits work.

> *Explain* the operation of an integrator.

> *Recognize* the different types of active filters.

The previous chapter presented the basic op-amp circuits and their most common uses. This chapter continues with additional coverage of common op-amp applications. Because of the versatility of the op amp, there are hundreds if not thousands of ways to use it. The applications presented here represent some of the most popular, including comparators, integrators, waveform generators, rectifiers, and active filters.

36.1 Comparators with Zero Reference

Often we want to compare one voltage with another to see which is larger. In this situation, a **comparator** may be the perfect solution. A comparator is similar to an op amp because it has two input voltages (noninverting and inverting) and one output voltage. It differs from a linear op-amp circuit because it has a two-state output, either a low or a high voltage. Because of this, comparators are often used to interface with analog and digital circuits.

Basic Idea

The simplest way to build a comparator is to connect an op amp without feedback resistors, as shown in Fig. 36-1a. Because of the high open-loop voltage gain, a positive input voltage produces positive saturation, and a negative input voltage produces negative saturation.

The comparator of Fig. 36-1a is called a **zero-crossing detector** because the output voltage ideally switches from low to high or vice versa whenever the input voltage crosses zero. Figure 36-1b shows the input-output response of a zero-crossing detector. The minimum input voltage that produces saturation is

$$v_{in(min)} = \frac{\pm V_{sat}}{A_{VOL}} \quad \text{(36-1)}$$

If $V_{sat} = 14$ V, the output swing of the comparator is from approximately -14 to $+14$ V. If the open-loop voltage gain is 100,000, the input voltage needed to produce saturation is

$$v_{in(min)} = \frac{\pm 14 \text{ V}}{100,000} = \pm 0.14 \text{ mV}$$

This means that an input voltage more positive than $+0.014$ mV drives the comparator into positive saturation, and an input voltage more negative than -0.014 mV drives it into negative saturation.

Input voltages used with comparators are usually much greater than ± 0.014 mV. This is why the output voltage is a two-state output, either $+V_{sat}$ or $-V_{sat}$. By looking at the output voltage, we can instantly tell whether the input voltage is greater than or less than zero.

Inverting Comparator

Sometimes, we may prefer to use an inverting comparator like Fig. 36-2. The noninverting input is grounded. The input signal drives the inverting input of the comparator. In this case, a slightly positive input voltage produces a maximum negative output. On the other hand, a slightly negative input voltage produces a maximum positive output.

Diode Clamps

Diodes are often used to protect sensitive circuits. Figure 36-2 is a practical example. Here we see two diode clamps protecting the comparator against excessively large input voltages. For instance, one IC comparator has an absolute maximum input rating of ± 15 V. If the input voltage exceeds these limits, the device will be destroyed.

With some comparators, the maximum input voltage rating may be as little as ± 5 V, whereas with others it may be more than ± 30 V. In any case, we can protect a comparator against destructively large input voltages by using the diode clamps shown in Fig. 36-2. These diodes have no effect on the operation of the circuit as long as the magnitude of the input voltage is less than 0.7 V. When the magnitude of the input voltage is greater than 0.7 V, one of the diodes will turn on and clamp the magnitude of the inverting input voltage to approximately 0.7 V.

Some ICs are optimized for use as comparators. These IC comparators often have diode clamps built into their input stages. When using one of these comparators, we have to add an external resistor in series with the input terminal. This series resistor will limit the internal diode currents to a safe level.

Converting Sine Waves to Square Waves

The **trip point** (also called the **threshold** or *reference*) of a comparator is the input voltage that causes the output voltage to switch states (from low to high or from high to low). In the noninverting and inverting comparators discussed earlier, the trip point is zero because this is the value of input voltage where the output switches states. Since a zero-crossing detector has a *two-state output,* any periodic input signal that crosses zero threshold will produce a rectangular output waveform.

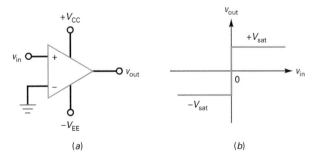

Figure 36-1 (a) Comparator. (b) Input-output response.

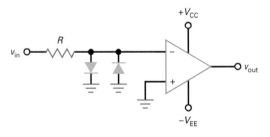

Figure 36-2 Inverting comparator with clamping diodes.

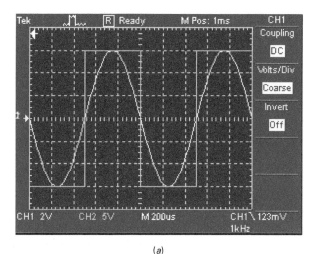

(a)

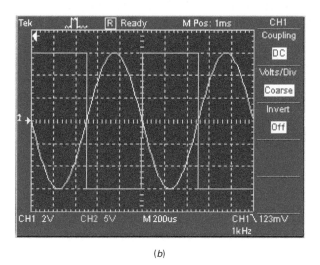

(b)

Figure 36-3 Comparator converts sine waves to square waves. (a) Noninverting. (b) Inverting.

For instance, if a sine wave is the input to a noninverting comparator with a threshold of 0 V, the output will be the square wave shown in Fig. 36-3a. As we can see, the output of a zero-crossing detector switches states each time the input voltage crosses the zero threshold.

Figure 36-3b shows the input sine wave and the output square wave for an inverting comparator with a threshold of 0 V. With this zero-crossing detector, the output square wave is 180° out of phase with the input sine wave.

Linear Region

Figure 36-4 shows a zero-crossing detector. If this comparator had an infinite open-loop gain, the transition between negative and positive saturation would be vertical.

In reality, the transition is not vertical. It takes approximately ±100 μV to get positive or negative saturation. This is typical for a comparator. The narrow input region between approximately −100 and +100 μV is called the *linear region of the comparator*. During a zero crossing, a changing input

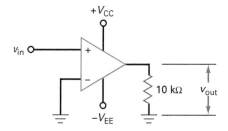

Figure 36-4 Narrow linear region of typical comparator.

signal usually passes through the linear region so quickly that we see only a sudden jump between negative and positive saturation, or vice versa.

Interfacing Analog and Digital Circuits

Comparators usually interface at their outputs with digital circuits such as CMOS, E-MOSFET, or TTL (stands for *transistor-transistor logic,* a family of digital circuits).

Figure 36-5a shows how a zero-crossing detector can interface with an E-MOSFET circuit. Whenever the input voltage is greater than zero, the output of the comparator is high. This turns on the power FET and produces a large load current.

Figure 36-5b shows a zero-crossing detector interfacing with a CMOS inverter. The idea is basically the same. A comparator input greater than zero produces a high input to the CMOS inverter.

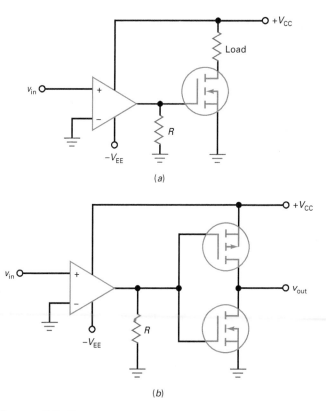

(a)

(b)

Figure 36-5 Comparator interfaces with (a) power FET and (b) CMOS.

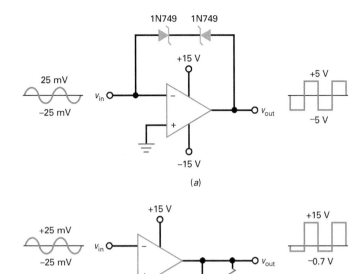

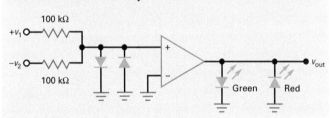

Figure 36-6 Bounded outputs. (a) Zener diodes. (b) Rectifier diode.

Most E-MOSFET devices can handle input voltages greater than ±15 V, and most CMOS devices can handle input voltages up to ±15 V. Therefore, we can interface the output of a typical comparator without any level shifting or clamping. TTL logic, on the other hand, operates with lower input voltages. Because of this, interfacing a comparator with TTL requires a different approach (to be discussed in the next section).

Bounded Output

The output swing of a zero-crossing detector may be too large in some applications. If so, we can *bound the output* by using back-to-back zener diodes, as shown in Fig. 36-6a. In this circuit, the inverting comparator has a bounded output because

one of the diodes will be conducting in the forward direction and the other will be operating in the breakdown region.

For instance, a 1N749 has a zener voltage of 4.3 V. Therefore, the voltage across the two diodes will be approximately ±5 V. If the input voltage is a sine wave with a peak value of 25 mV, then the output voltage will be an inverted square wave with a peak voltage of 5 V.

Figure 36-6b shows another example of a bounded output. This time, the output diode will clip off the negative half-cycles of the output voltage. Given an input sine wave with a peak of 25 mV, the output is bounded between −0.7 and +15 V as shown.

A third approach to bounding the output is to connect zener diodes across the output. For instance, if we connect the back-to-back zener diodes of Fig. 36-6a across the output, the output will be bounded at ±5 V.

EXAMPLE 36-1

What does the circuit of Fig. 36-7 do?

Figure 36-7 Comparing voltages of different polarities.

Answer:
This circuit compares two voltages of opposite polarity to determine which is greater. If the magnitude of v_1 is greater than the magnitude of v_2, the noninverting input is positive, the comparator output is positive, and the green LED is on. On the other hand, if the magnitude of v_1 is less than the magnitude of v_2, the noninverting input is negative, the comparator output is negative, and the red LED is on.

EXAMPLE 36-2

What does the circuit of Fig. 36-8 do?

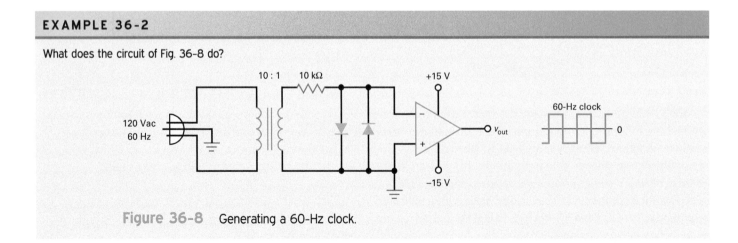

Figure 36-8 Generating a 60-Hz clock.

Answer:

This is one way to create a *60-Hz clock*, a square-wave signal used as the basic timing mechanism for inexpensive digital clocks. The transformer steps the line voltage down to 12 Vac. The diode clamps then binds the input to ±0.7 V. The inverting comparator produces an output square wave with a frequency of 60 Hz. The output signal is called a *clock* because its frequency can be used to get seconds, minutes, and hours.

A digital circuit called a *frequency divider* can divide the 60 Hz by 60 to get a square wave with a period of 1 s. Another divide-by-60 circuit can divide this signal to get a square wave with a period of 1 min. A final divide-by-60 circuit produces a square wave with a period of 1 hr. Using the three square waves (1 s, 1 min, 1 hr) with other digital circuits and seven-segment LED indicators, we can display the time of day numerically. The 60-Hz ac power line maintains a frequency precision of better than 0.1% so timing accuracy is very good.

36.2 Comparators with Nonzero References

In some applications a threshold voltage different from zero may be preferred. By biasing either input, we can change the threshold voltage as needed.

Moving the Trip Point

In Fig. 36-9a, a voltage divider produces the following reference voltage for the inverting input:

$$v_{ref} = \frac{R_2}{R_1 + R_2} V_{CC} \qquad (36\text{-}2)$$

When v_{in} is greater than v_{ref}, the differential input voltage is positive and the output voltage is high. When v_{in} is less than v_{ref}, the differential input voltage is negative and the output voltage is low.

A bypass capacitor is typically used on the inverting input, as shown in Fig. 36-9a. This reduces the amount of power supply ripple and other noise appearing at the inverting input. To be effective, the cutoff frequency of this bypass circuit should be much lower than the ripple frequency of the power supply. The cutoff frequency is given by

$$f_c = \frac{1}{2\pi(R_1 \parallel R_2)C_{BY}} \qquad (36\text{-}3)$$

Figure 36-9b shows the **transfer characteristic** (input-output response). The trip point is now equal to v_{ref}. When v_{in} is greater than v_{ref}, the output of the comparator goes into positive saturation. When v_{in} is less than v_{ref}, the output goes into negative saturation.

A comparator like this is sometimes called a *limit detector* because a positive output indicates that the input voltage exceeds a specific limit. With different values of R_1 and R_2, we can set the limit anywhere between 0 and V_{CC}. If a negative limit is preferred, connect $-V_{EE}$ to the voltage divider, as shown in Fig. 36-9c. Now a negative reference voltage is applied to the inverting input. When v_{in} is more positive than v_{ref}, the differential input voltage is positive and the output is high, as shown in Fig. 36-9d. When v_{in} is more negative than v_{ref}, the output is low.

Single-Supply Comparator

A typical op amp like the 741C can run on a single positive supply by grounding the $-V_{EE}$ pin, as shown in Fig. 36-10a. The output voltage has only one polarity, either a low or a

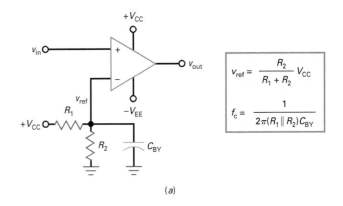

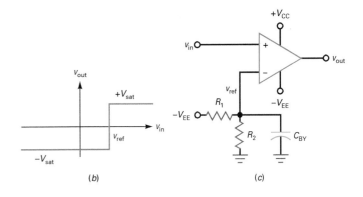

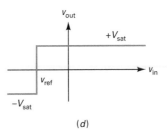

Figure 36-9 (a) Positive threshold. (b) Positive input-output response. (c) Negative threshold. (d) Negative input-output response.

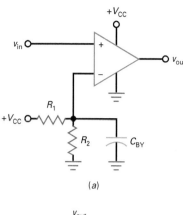

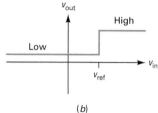

(a)

(b)

Figure 36-10 (a) Single-supply comparator. (b) Input-output response.

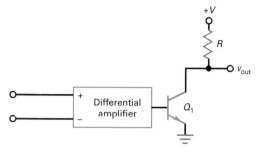

Figure 36-11 Using pullup resistor with open-collector output stage of a comparator.

high positive voltage. For instance, with V_{CC} equal to +15 V, the output swing is from approximately +1.5 V (low state) to around +13.5 V (high state).

When v_{in} is greater than v_{ref}, the output is high, as shown in Fig. 36-10b. When v_{in} is less than v_{ref}, the output is low. In either case, the output has a positive polarity. For many digital applications, this kind of positive output is preferred.

IC Comparators

An op amp like a 741C can be used as a comparator, but it has speed limitations because of its slew rate. With a 741C, the output can change no faster than 0.5 V/μs. Because of this, a 741C takes more than 50 μs to switch output states with supplies of ±15 V. One solution to the slew-rate problem is to use a faster op amp. For example, with a slew rate of 70 V/μs, the comparator can switch from $-V_{sat}$ to $+V_{sat}$ in approximately 0.3 μs.

Open-Collector Devices

Most op amps have an output stage that can be described as an *active-pullup stage* because it contains two devices in a class B push-pull connection. With the active pullup, the upper device turns on and pulls the output up to the high output state. On the other hand, an open-collector output stage of Fig. 36-11 needs external components to be connected to it.

For the output stage to work properly, the user has to connect the open collector to an external resistor and

supply voltage, as shown in Fig. 36-11. The resistor is called a **pullup resistor** because it pulls the output voltage up to the supply voltage when Q_1 is cut off. When Q_1 is saturated, the output voltage is low. Since the output stage is a transistor switch, the comparator produces a two-state output.

The main limitation on the switching speed is the amount of capacitance across Q_1. This output capacitance is the sum of the internal collector capacitance and the external stray wiring capacitance.

The output time constant is the product of the pullup resistance and the output capacitance. For this reason, the smaller the pullup resistance, the faster the output voltage can change. Typically, R is from a couple of hundred to a couple of thousand ohms.

Examples of IC comparators are the LM311, LM339, and NE529. They all have an open-collector output stage, which means that you have to connect the output pin to a pullup resistor and a positive supply voltage. Because of their high slew rates, these IC comparators can switch output states in a microsecond or less.

The LM339 is a *quad comparator*—four comparators in a single IC package. It can run off a single supply or off dual supplies. Because it is inexpensive and easy to use, the LM339 is a popular comparator for general-purpose applications.

Not all IC comparators have an open-collector output stage. Some have an active-collector output stage. The active pullup produces faster switching. These high-speed IC comparators require dual supplies.

Driving TTL

The LM339 is an open-collector device. Figure 36-12a shows how an LM339 can be connected to interface with TTL devices. A positive supply of +15 V is used for the comparator, but the open collector of the LM339 is connected to a supply of +5 V through a pullup resistor of 1 kΩ. Because of this, the output swings between 0 and +5 V, as shown in Fig. 36-12b. This output signal is ideal for TTL devices because they are designed to work with supplies of +5 V.

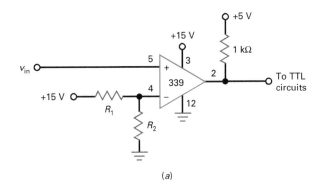

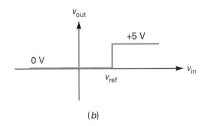

Figure 36-12 (a) LM339 comparator. (b) Input-output response.

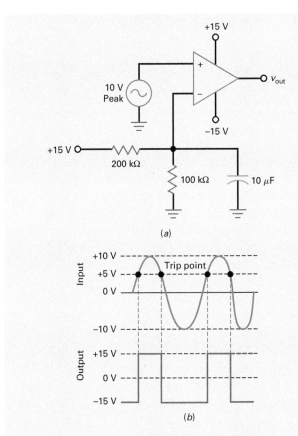

Figure 36-13 (a) Calculating duty cycle. (b) Output waveform.

EXAMPLE 36-3

In Fig. 36-13a, the input voltage is a sine wave with a peak value of 10 V. What is the trip point of the circuit? What is the cutoff frequency of the bypass circuit? What does the output waveform look like?

Answer:
Since +15 V is applied to a 3:1 voltage divider, the reference voltage is

$$v_{ref} = +5 \text{ V}$$

This is the trip point of the comparator. When the sine wave crosses through this level, the output voltage switches states.

With Formula (36-3), the cutoff frequency of the bypass circuit is

$$f_c = \frac{1}{2\pi(200 \text{ k}\Omega \parallel 100 \text{ k}\Omega)(10 \text{ }\mu\text{F})}$$

$$= 0.239 \text{ Hz}$$

This low cutoff frequency means that any 60-Hz ripple on the reference supply voltage will be heavily attenuated.

Figure 36-13b shows the input sine wave. It has a peak value of 10 V. The rectangular output has a peak value of approximately 15 V. Notice how the output voltage switches states when the input sine wave crosses the trip point of +5 V.

EXAMPLE 36-4

What is the duty cycle of the output waveform in Fig. 36-13b?

Answer:
Recall that the *duty cycle* is defined as the pulse width divided by the period. Duty cycle equals the conduction angle divided by 360°.

In Fig. 36-13b, the sine wave has a peak value of 10 V. Therefore, the input voltage is given by

$$v_{in} = 10 \sin \theta$$

The rectangular output switches states when the input voltage crosses +5 V. At this point, the foregoing equation becomes

$$5 = 10 \sin \theta$$

Now, we can solve for the angle θ where switching occurs:

$$\sin \theta = 0.5$$

or

$$\theta = \arcsin 0.5 = 30° \text{ and } 150°$$

The first solution, $\theta = 30°$, is where the output switches from low to high. The second solution, $\theta = 150°$, is where the output switches from high to low. The duty cycle is

$$D = \frac{\text{conduction angle}}{360°} = \frac{150° - 30°}{360°} = 0.333$$

The duty cycle in Fig. 36-13b can be expressed as 33.3%.

Op-Amp Applications   **671**

36.3 Comparators with Hysteresis

If the input to a comparator contains a large amount of noise, the output will be erratic when v_{in} is near the trip point. One way to reduce the effect of noise is by using a comparator with positive feedback. The positive feedback produces two separate trip points that prevent a noisy input from producing false transitions.

Noise

Noise is any kind of unwanted signal that is not derived from or harmonically related to the input signal. Electric motors, neon signs, power lines, car ignitions, lightning, and so on, produce electromagnetic fields that can induce noise voltages into electronic circuits. Power supply ripple is also classified as noise since it is not related to the input signal. By using regulated power supplies and shielding, we usually can reduce the ripple and induced noise to an acceptable level.

Thermal noise, on the other hand, is caused by the random motion of free electrons inside a resistor (see Fig. 36-14a). The energy for this electron motion comes from the thermal energy of the surrounding air. The higher the ambient temperature, the more active the electrons.

The motion of billions of free electrons inside a resistor is pure chaos. At some instants, more electrons move up than down, producing a small negative voltage across the resistor. At other instants, more electrons move down than up, producing a positive voltage. If this type of noise were amplified and viewed on an oscilloscope, it would resemble Fig. 36-14b. Like any voltage, noise has an rms or effective value. As an approximation, the highest noise peaks are about four times the rms value.

The randomness of the electron motion inside a resistor produces a distribution of noise at virtually all frequencies. The rms value of this noise increases with temperature, bandwidth, and resistance. For our purposes, we need to be aware of how noise may affect the output of a comparator.

Noise Triggering

The high open-loop gain of a comparator means that an input of only 100 μV may be enough to switch the output from one state to another. If the input contains noise with a peak of 100 μV or more, the comparator will detect the zero crossings produced by the noise.

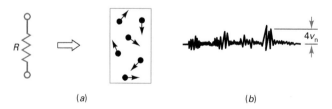

(a)   (b)

Figure 36-14 Thermal noise. (a) Random electron motion in resistor. (b) Noise on oscilloscope.

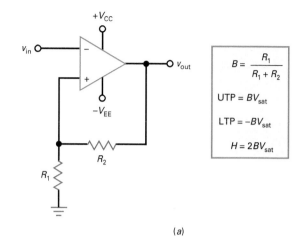

(a)

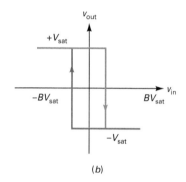

(b)

Figure 36-15 (a) Inverting Schmitt trigger. (b) Input-output response has hysteresis.

When the noise peaks are large enough, they produce unwanted changes in the comparator output. When an input signal is present, the noise is superimposed on the input signal and produces erratic triggering.

Schmitt Trigger

The standard solution for a noisy input is to use a comparator like the one shown in Fig. 36-15a. The input voltage is applied to the inverting input. Because the feedback voltage is aiding the input voltage, the feedback is *positive*. A comparator using positive feedback like this is usually called a **Schmitt trigger.**

When the comparator is positively saturated, a positive voltage is fed back to the noninverting input. This positive feedback voltage holds the output in the high state. Similarly, when the output voltage is negatively saturated, a negative voltage is fed back to the noninverting input, holding the output in the low state. In either case, the positive feedback reinforces the existing output state.

The feedback fraction is

$$B = \frac{R_1}{R_1 + R_2} \qquad \textbf{(36-4)}$$

When the output is positively saturated, the reference voltage applied to the noninverting input is

$$v_{\text{ref}} = +BV_{\text{sat}} \tag{36-5a}$$

When the output is negatively saturated, the reference voltage is

$$v_{\text{ref}} = -BV_{\text{sat}} \tag{36-5b}$$

The output voltage will remain in a given state until the input voltage exceeds the reference voltage for that state. For instance, if the output is positively saturated, the reference voltage is $+BV_{\text{sat}}$. The input voltage must be increased to slightly more than $+BV_{\text{sat}}$ to switch the output voltage from positive to negative, as shown in Fig. 36-15b. Once the output is in the negative state, it will remain there indefinitely until the input voltage becomes more negative than $-BV_{\text{sat}}$. Then, the output switches from negative to positive (Fig. 36-15b).

Hysteresis

The unusual response of Fig. 36-15b has a useful property called **hysteresis.** To understand this concept, put your finger on the upper end of the graph where it says $+V_{\text{sat}}$. Assume that this is the current value of output voltage. Move your finger to the right along the horizontal line. Along this horizontal line, the input voltage is changing but the output voltage is still equal to $+V_{\text{sat}}$. When you reach the upper right corner, v_{in} equals $+BV_{\text{sat}}$. When v_{in} increases to slightly more than $+BV_{\text{sat}}$, the output voltage goes into the transition region between the high and the low states.

If you move your finger down along the vertical line, you will simulate the transition of the output voltage from high to low. When your finger is on the lower horizontal line, the output voltage is negatively saturated and equal to $-V_{\text{sat}}$.

To switch back to the high output state, move your finger until it reaches the lower left corner. At this point, v_{in} equals $-BV_{\text{sat}}$. When v_{in} becomes slightly more negative than $-BV_{\text{sat}}$, the output voltage goes into the transition from low to high. If you move your finger up along the vertical line, you will simulate the switching of the output voltage from low to high.

In Fig. 36-15b, the trip points are defined as the two input voltages where the output voltage changes states. The *upper trip point (UTP)* has the value

$$\text{UTP} = BV_{\text{sat}} \tag{36-6}$$

and the *lower trip point (LTP)* has the value

$$\text{LTP} = -BV_{\text{sat}} \tag{36-7}$$

The difference between these trip points is defined as the hysteresis (also called the *deadband*):

$$H = \text{UTP} - \text{LTP} \tag{36-8}$$

With Formulas (36-6) and (36-7), this becomes

$$H = BV_{\text{sat}} - (-BV_{\text{sat}})$$

which equals

$$H = 2BV_{\text{sat}} \tag{36-9}$$

Positive feedback causes the hysteresis of Fig. 36-15b. If there were no positive feedback, B would equal zero and the hysteresis would disappear, because both trip points would equal zero.

Hysteresis is desirable in a Schmitt trigger because it prevents noise from causing false triggering. If the peak-to-peak noise voltage is less than the hysteresis, the noise cannot produce false triggering. For instance, if UTP = +1 V and LTP = −1 V, then $H = 2$ V. In this case, the Schmitt trigger is immune to false triggering as long as the peak-to-peak noise voltage is less than 2 V.

EXAMPLE 36-5

If $V_{\text{sat}} = 13.5$ V, what are the trip points and hysteresis in Fig. 36-16?

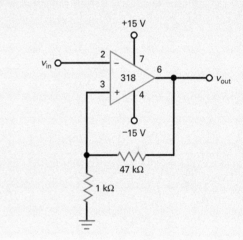

Figure 36-16 Example.

Answer:
With Formula (36-4), the feedback fraction is

$$B = \frac{1\ \text{k}\Omega}{48\ \text{k}\Omega} = 0.0208$$

With Formulas (36-6) and (36-7), the trip points are

$$\text{UTP} = 0.0208(13.5\ \text{V}) = 0.281\ \text{V}$$
$$\text{LTP} = -0.0208(13.5\ \text{V}) = -0.281\ \text{V}$$

With Formula (36-9), the hysteresis is

$$H = 2(0.0208\ \text{V})(13.5\ \text{V}) = 0.562\ \text{V}$$

This means that the Schmitt trigger of Fig. 36-16 can withstand a peak-to-peak noise voltage up to 0.562 V without false triggering.

36.4 The Integrator

An **integrator** is a circuit that performs a mathematical operation called *integration*. The most popular application of an integrator is in producing a *ramp* of output voltage, which is a linearly increasing or decreasing voltage.

Basic Circuit

Figure 36-17a is an op-amp integrator. As you can see, the feedback component is a capacitor instead of a resistor. The usual input to an integrator is a rectangular pulse like the one shown in Fig. 36-17b. The width of this pulse is equal to T. When the pulse is low, $v_{in} = 0$. When the pulse is high, $v_{in} = V_{in}$. Visualize this pulse applied to the left end of R. Because of the virtual ground on the inverting input, a high input voltage produces an input current of

$$I_{in} = \frac{V_{in}}{R}$$

All this input current goes into the capacitor. As a result, the capacitor charges and its voltage increases with the polarity shown in Fig. 36-17a. The virtual ground implies that the output voltage equals the voltage across the capacitor. For a positive input voltage, the output voltage will increase negatively, as shown in Fig. 36-17c.

Since a constant current is flowing into the capacitor, the charge Q increases linearly with time. This means that the capacitor voltage increases linearly, which is equivalent to a negative ramp of output voltage, as shown in Fig. 36-17c. At the end of the pulse period in Fig. 36-17b, the input voltage returns to zero and the capacitor charging stops. Because the capacitor retains its charge, the output voltage remains constant at a negative voltage of $-V$. The magnitude of this voltage is given by

$$V = \frac{T}{RC} V_{in} \qquad (36\text{-}10)$$

Note: If the integrator is allowed to continuing charging beyond the input pulse time, the output will hit the negative saturation level.

For the integrator to work properly, the closed-loop time constant should be much greater than the width of the input pulse (at least 10 times greater). As a formula,

$$\tau > 10T \qquad (36\text{-}11)$$

In the typical op-amp integrator, the closed-loop time constant is extremely long, so this condition is easily satisfied.

Eliminating Output Offset

The circuit of Fig. 36-17a needs a slight modification to make it practical. Because a capacitor is open to dc signals, there is no negative feedback at zero frequency. Without negative feedback, the circuit treats any input offset voltage as a valid input voltage. The result is that the capacitor charges and the output goes into positive or negative saturation, where it stays indefinitely.

One way to reduce the effect of input offset voltage is to decrease the voltage gain at zero frequency by inserting a resistor in parallel with the capacitor, as shown in Fig. 36-18. This resistor should be at least 10 times larger than the input resistor. If the added resistance equals $10R$, the closed-loop voltage gain is 10 and the output offset voltage is reduced to an acceptable level. When a valid input voltage is present, the additional resistor has almost no effect on the charging of a capacitor, so the output voltage is still almost a perfect ramp.

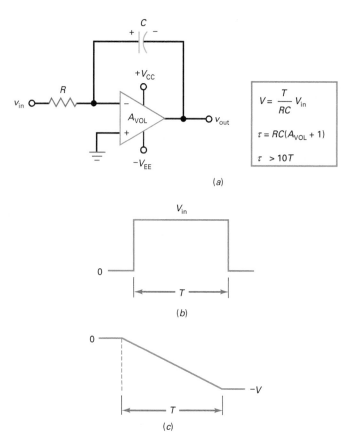

Figure 36-17 (a) Integrator. (b) Typical input pulse. (c) Output ramp.

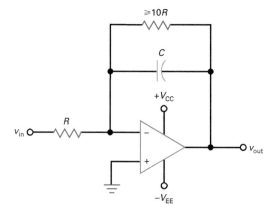

Figure 36-18 Resistor across capacitor reduces output offset voltage.

36.5 Waveform Conversion

With op amps we can convert sine waves to rectangular waves, rectangular waves to triangular waves, and so on. This section is about some basic circuits that convert an input waveform to an output waveform of a different shape.

Sine to Rectangular

Figure 36-19a shows a Schmitt trigger, and Fig. 36-19b is the graph of output voltage versus input voltage. When the input signal is *periodic* (repeating cycles), the Schmitt trigger produces a rectangular output, as shown. This assumes that the input signal is large enough to pass through both trip points of Fig. 36-19c. When the input voltage exceeds UTP on the upward swing of the positive half-cycle, the output voltage switches to $-V_{sat}$. One half cycle later, the input voltage becomes more negative than LTP, and the output switches back to $+V_{sat}$.

A Schmitt trigger always produces a rectangular output, regardless of the shape of the input signal. In other words, the input voltage does not have to be sinusoidal. As long as the waveform is periodic and has an amplitude large enough to pass through the trip points, we get a rectangular output from the Schmitt trigger. This rectangular wave has the same frequency as the input signal.

As an example, Fig. 36-19d shows a Schmitt trigger with trip points of approximately UTP = +0.1 V and LTP = −0.1 V. If the input voltage is repetitive and has a peak-to-peak value greater than 0.2 V, the output voltage is a rectangular wave with a peak-to-peak value of approximately $2V_{sat}$.

Rectangular to Triangular

In Fig. 36-20a, a rectangular wave is the input to an integrator. Since the input voltage has a dc or average value of zero, the dc or average value of the output is also zero. As shown in Fig. 36-20b, the ramp is decreasing during the positive half-cycle of input voltage and increasing during the negative half-cycle. Therefore, the output is a triangular wave with the same frequency as the input. It can be shown that the triangular output waveform has a peak-to-peak value of

$$V_{out(p-p)} = \frac{T}{2RC} V_p \qquad (36\text{-}12)$$

where T is the period of the signal. An equivalent expression in terms of frequency is

$$V_{out(p-p)} = \frac{V_p}{2fRC} \qquad (36\text{-}13)$$

where V_p is the peak input voltage and f is the input frequency.

Triangle to Pulse

Figure 36-21a shows a circuit that converts a triangular input to a rectangular output. By varying R_2, we can change the width of the output pulses, which is equivalent

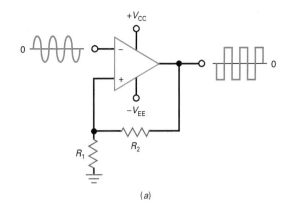

(a)

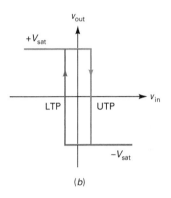

(b)

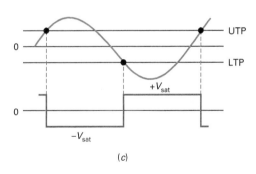

(c)

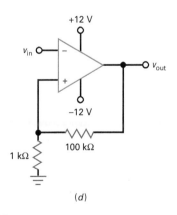

(d)

Figure 36-19 Schmitt trigger always produces rectangular output.

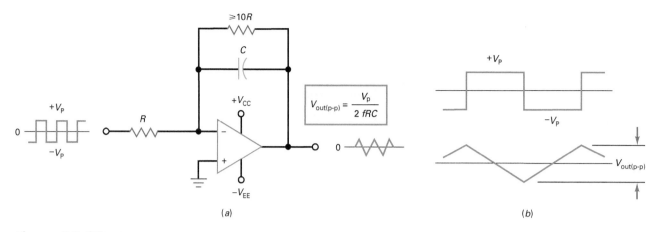

Figure 36-20 (a) Square wave into integrator produces triangular output. (b) Input and output waveforms.

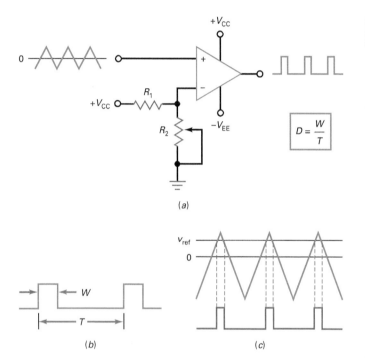

Figure 36-21 Triangular input to limit detector produces rectangular output.

to varying the duty cycle. In Fig. 36-21b, W represents the width of the pulse and T is the period. As previously discussed, the duty cycle D is the width of the pulse divided by the period.

In some applications, we want to vary the duty cycle. The adjustable limit detector of Fig. 36-21a is ideal for this purpose. With this circuit, we can move the trip point from zero to a positive level. When the triangular input voltage exceeds the trip point, the output is high, as shown in Fig. 36-21c. Since v_{ref} is adjustable, we can vary the width of the output pulse, which is equivalent to changing the duty cycle. With a circuit like this, we can vary the duty cycle from approximately 0 to 50%.

EXAMPLE 36-6

What is the output voltage in Fig. 36-22 if the input frequency is 1 kHz?

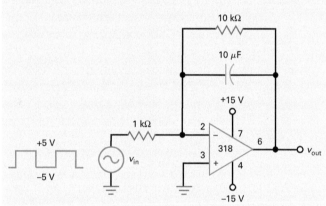

Figure 36-22 Example.

Answer:
With Formula (36-13), the output is a triangular wave with a peak-to-peak voltage of

$$V_{out(p-p)} = \frac{5V}{2(1 \text{ kHz})(1 \text{ k}\Omega)(10 \text{ } \mu\text{F})} = 0.25 \text{ V p-p}$$

EXAMPLE 36-7

A triangular input drives the circuit of Fig. 36-23a. The variable resistance has a maximum value of 10 kΩ. If the triangular input has a frequency of 1 kHz, what is the duty cycle when the wiper is at the middle of its range?

Answer:
When the wiper is at the middle of its range, it has a resistance of 5 kΩ. This means that the reference voltage is

$$v_{ref} = \frac{5 \text{ k}\Omega}{15 \text{ k}\Omega} 15 \text{ V} = 5 \text{ V}$$

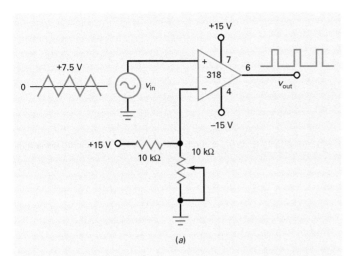

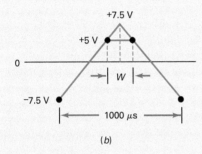

Figure 36-23 Example.

The period of the signal is

$$T = \frac{1}{1\ \mathrm{kHz}} = 1000\ \mu s$$

Figure 36-23*b* shows this value. It takes 500 *ms* for the input voltage to increase from −7.5 to +7.5 V because this is half of the cycle. The trip point of the comparator is +5 V. This means that the output pulse has a width of *W*, as shown in Fig. 36-23*b*.

Because of the geometry of Fig. 36-23*b*, we can set up a proportion between voltage and time as follows:

$$\frac{W/2}{500\ \mu s} = \frac{7.5\ \mathrm{V} - 5\ \mathrm{V}}{15\ \mathrm{V}}$$

Solving for *W* gives

$$W = 167\ \mu S$$

The duty cycle is

$$D = \frac{167\ \mu s}{1000\ \mu s} = 0.167$$

In Fig. 36-23*a*, moving the wiper down will increase the reference voltage and decrease the output duty cycle. Moving the wiper up will decrease the reference voltage and increase the output duty cycle. For all values given in Fig. 36-23*a*, the duty cycle can vary from 0 to 50%.

36.6 Waveform Generation

With positive feedback, we can build **oscillators,** circuits that generate or create an output signal with no external input signal. This section discusses some op-amp circuits that can generate nonsinusoidal signals.

Relaxation Oscillator

In Fig. 36-24*a*, there is no input signal. Nevertheless, the circuit produces a rectangular output signal. This output is a square wave that swings between $-V_{sat}$ and $+V_{sat}$. How is this possible? Assume that the output of Fig. 36-24*a* is in positive saturation. Because of feedback resistor *R*, the capacitor will charge exponentially toward $+V_{sat}$, as shown in Fig. 36-24*b*. But the capacitor voltage never reaches $+V_{sat}$ because the voltage crosses the UTP. When this happens, the output square wave switches to $-V_{sat}$.

With the output now in negative saturation, the capacitor discharges, as shown in Fig. 36-24*b*. When the capacitor voltage crosses through zero, the capacitor starts charging negatively toward $-V_{sat}$. When the capacitor voltage crosses the LTP, the output square wave switches back to $+V_{sat}$. The cycle then repeats.

Because of the continuous charging and discharging of the capacitor, the output is a rectangular wave with a duty

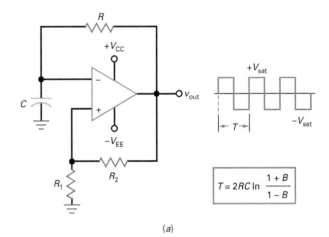

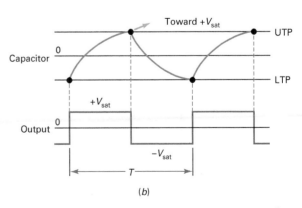

Figure 36-24 (*a*) Relaxation oscillator. (*b*) Capacitor charging and output waveform.

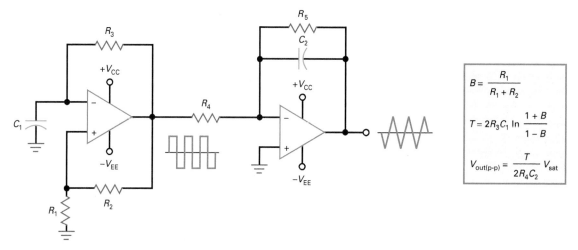

Figure 36-25 Relaxation oscillator drives integrator to produce triangular output.

cycle of 50%. By analyzing the exponential charge and discharge of the capacitor, we can derive this formula for the period of the rectangular output,

$$T = 2RC \ln \frac{1 + B}{1 - B} \qquad \textbf{(36-14)}$$

where B is the feedback fraction given by

$$B = \frac{R_1}{R_1 + R_2}$$

Formula (36-14), uses the *natural logarithm,* which is a logarithm to base e. A scientific calculator or table of natural logarithms must be used with this equation.

Figure 36-24a is called a **relaxation oscillator,** defined as a circuit that generates an output signal whose frequency

depends on the charging of a capacitor. If we increase the RC time constant, it takes longer for the capacitor voltage to reach the trip points. Therefore, the frequency is lower. By making R adjustable, we can get a 50:1 tuning range.

Generating Triangular Waves

By cascading a relaxation oscillator and an integrator, we get a circuit that produces the triangular output shown in Fig. 36-25. The rectangular wave out of the relaxation oscillator drives the integrator, which produces a triangular output waveform. The rectangular wave swings between $+V_{sat}$ and $-V_{sat}$. You can calculate its period with Formula (36-14). The triangular wave has the same period and frequency. You can calculate its peak-to-peak value with Formula (36-12).

EXAMPLE 36-8

What is the frequency of the output signal in Fig. 36-26?

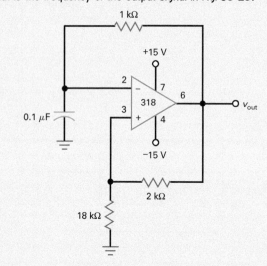

Figure 36-26 Example.

Answer:
The feedback fraction is

$$B = \frac{18 \text{ k}\Omega}{20 \text{ k}\Omega} = 0.9$$

With Formula (36-14),

$$T = 2RC \ln \frac{1 + B}{1 - B} = 2(1 \text{ k}\Omega)(0.1 \text{ } \mu\text{F})$$

$$\ln \frac{1 + 0.9}{1 - 0.9} = 589 \text{ } \mu\text{s}$$

The frequency is

$$f = \frac{1}{589 \text{ } \mu\text{s}} = 1.7 \text{ kHz}$$

The square-wave output voltage has a frequency of 1.7 kHz and a peak-to-peak value of $2V_{sat}$, approximately 27 V for the circuit of Fig. 36-26.

EXAMPLE 36-9

The relaxation oscillator of Example 36-8 is used in Fig. 36-25 to drive the integrator. Assume that the peak voltage out of the relaxation oscillator is 13.5 V. If the integrator has $R_4 = 10$ kΩ and $C_2 = 10$ μF, what is the peak-to-peak value of the triangular output wave?

Answer:
With the equations shown in Fig. 36-25, we can analyze the circuit. In Example 36-8, we calculated a feedback fraction of 0.9 and a period of 589 μs. Now, we can calculate the peak-to-peak value of the triangular output:

$$V_{out(p\text{-}p)} = \frac{589 \ \mu s}{2(10 \ k\Omega)(10 \ \mu F)} (13.5 \ \text{V}) = 39.8 \ \text{mV p-p}$$

The circuit generates a square wave with a peak-to-peak value of approximately 27 V and a triangular wave with a peak-to-peak value of 39.8 mV.

36.7 Active Diode Circuits

Op amps can enhance the performance of diode circuits. For one thing, an op amp with negative feedback reduces the effect of the knee voltage, allowing us to rectify, peak-detect, clip, and clamp low-level signals (those with amplitudes less than the knee voltage). And because of their buffering action, op amps can eliminate the effects of the source and load on diode circuits.

Figure 36-27 is an **active half-wave rectifier.** When the input signal goes positive, the output goes positive and turns on the diode. The circuit then acts like a voltage follower, and the positive half-cycle appears across the load resistor. When the input goes negative, the op-amp output goes negative and turns off the diode. Since the diode is open, no voltage appears across the load resistor. The final output is almost a perfect half-wave signal.

There are two distinct *modes* or regions of operation. First, when the input voltage is positive, the diode is conducting and the operation is linear. In this case, the output voltage is fed back to the input, and we have negative feedback. Second, when the input voltage is negative, the diode is nonconducting and the feedback path is open. In this case, the op-amp output is isolated from the load resistor.

The high open-loop voltage gain of the op amp almost eliminates the effect of the knee voltage. For instance, if the knee voltage is 0.7 V and A_{VOL} is 100,000, the input voltage that just turns on the diode is 7 μV.

The closed-loop knee voltage is given by

$$V_{K(CL)} = \frac{V_K}{A_{VOL}}$$

where $V_K = 0.7$ V for a silicon diode. Because the closed-loop knee voltage is so small, the active half-wave rectifier may be used with low-level signals in the microvolt region.

36.8 Active Filters

As you saw in an earlier chapter, filters are commonly made from passive components, with RC and LC filters being the most common. In some applications, active filters have replaced passive RC or LC filters. Active filters commonly use op amps with RC feedback and input components to produce common low-pass, high-pass, bandpass, and notch filters.

There are several advantages to active filters. First, they can provide gain rather than the inherent attenuation of passive filters. Second, active filters can eliminate bulky and costly inductors, especially in low-frequency designs. Furthermore, by cascading active filters, exceptional selectivity can be achieved without loss.

The disadvantages of active filters are that they require dc power supplies to power the op amps. And the upper frequency is limited by the bandwidth of the op amp and the practical values of resistors and capacitors. In general, active filters are used primarily in the audio frequency range and other applications with frequencies below 10 MHz.

36.9 First-Order Stages

First-order or 1-pole active-filter stages have only one capacitor. Because of this, they can produce only a low-pass or a high-pass response. Bandpass and bandstop filters can be implemented only when n is greater than 1.

Low-Pass Stage

Figure 36-28a shows the simplest way to build a first-order low-pass active filter. It is nothing more than an RC lag circuit and a voltage follower. The voltage gain is

$$A_v = 1$$

The 3-dB cutoff frequency is given by

$$f_c = \frac{1}{2\pi R_1 C_1} \qquad \textbf{(36-15)}$$

When the frequency increases above the cutoff frequency, the capacitive reactance decreases and reduces the noninverting input voltage. Since the R_1C_1 lag circuit is outside the

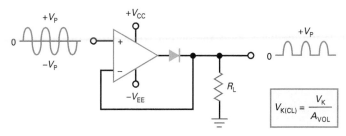

Figure 36-27 Active half-wave rectifier.

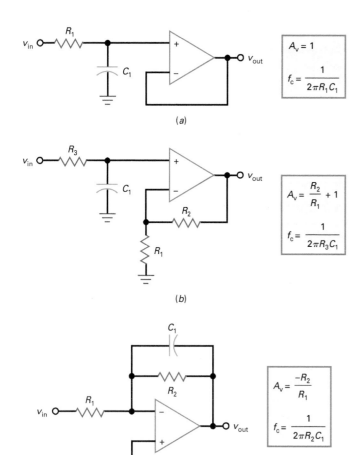

Figure 36-28 First-order low-pass stages.
(a) Noninverting unity gain. (b) Noninverting with voltage gain. (c) Inverting with voltage gain.

feedback loop, the output voltage rolls off. As the frequency approaches infinity, the capacitor becomes a short and there is zero input voltage.

Figure 36-28b shows another noninverting first-order low-pass filter. Although it has two additional resistors, it has the advantage of voltage gain. The voltage gain well below the cutoff frequency is given by

$$A_v = \frac{R_2}{R_1} + 1 \qquad (36\text{-}16)$$

The cutoff frequency is given by

$$f_c = \frac{1}{2\pi R_3 C_1} \qquad (36\text{-}17)$$

Above the cutoff frequency, the lag circuit reduces the noninverting input voltage. Since the $R_3 C_1$ lag circuit is outside the feedback loop, the output voltage rolls off at a rate of 20 dB per decade.

Figure 36-28c shows an inverting first-order low-pass filter and its equations. At low frequencies, the capacitor

appears to be open and the circuit acts like an inverting amplifier with a voltage gain of

$$A_v = \frac{-R_2}{R_1} \qquad (36\text{-}18)$$

As the frequency increases, the capacitive reactance decreases and reduces the impedance of the feedback branch. This implies less voltage gain. As the frequency approaches infinity, the capacitor becomes a short and there is no voltage gain. As shown in Fig. 36-28c, the cutoff frequency is given by

$$f_c = \frac{1}{2\pi R_2 C_1} \qquad (36\text{-}19)$$

A final point about all first-order stages. They can implement only a Butterworth response. The reason is that a first-order stage has no resonant frequency. Therefore, it cannot produce the peaking that produces a rippled passband. This means that all first-order stages are maximally flat in the passband and monotonic in the stopband, and they roll off at a rate of 20 dB per decade.

High-Pass Stage

Figure 36-29a shows the simplest way to build a first-order high-pass active filter. The voltage gain is

$$A_v = 1$$

The 3-dB cutoff frequency is given by

$$f_c = \frac{1}{2\pi R_1 C_1} \qquad (36\text{-}20)$$

When the frequency decreases below the cutoff frequency, the capacitive reactance increases and reduces the noninverting input voltage. Since the $R_1 C_1$ circuit is outside the feedback loop, the output voltage rolls off. As the frequency approaches zero, the capacitor becomes an open and there is zero input voltage.

Figure 36-29b shows another noninverting first-order high-pass filter. The voltage gain well above the cutoff frequency is given by

$$A_v = \frac{R_2}{R_1} + 1 \qquad (36\text{-}21)$$

The 3-dB cutoff frequency is given by

$$f_c = \frac{1}{2\pi R_3 C_1} \qquad (36\text{-}22)$$

Well below the cutoff frequency, the RC circuit reduces the noninverting input voltage. Since the $R_3 C_1$ lag circuit is outside the feedback loop, the output voltage rolls off at a rate of 20 dB per decade.

Figure 36-29c shows another first-order high-pass filter and its equations. At high frequencies, the circuit acts like an inverting amplifier with a voltage gain of

$$A_v = \frac{-X_{C2}}{X_{C1}} = \frac{-C_1}{C_2} \qquad (36\text{-}23)$$

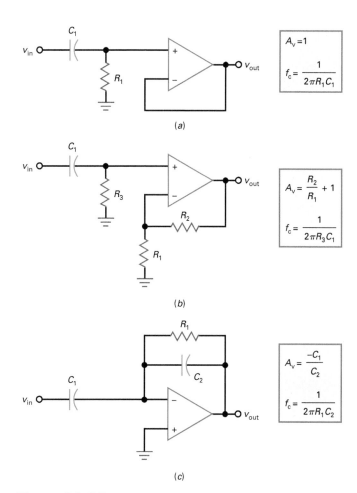

Figure 36-29 First-order high-pass stages.
(a) Noninverting unity gain. (b) Noninverting with voltage gain.
(c) Inverting with voltage gain.

As the frequency decreases, the capacitive reactances increase and eventually reduce the input signal and the feedback. This implies less voltage gain. As the frequency approaches zero, the capacitors become open and there is no input signal. As shown in Fig. 36-29c, the 3-dB cutoff frequency is given by

$$f_c = \frac{1}{2\pi R_1 C_2} \qquad (36\text{-}24)$$

EXAMPLE 36-10

What is the voltage gain in Fig. 36-30a? What is the cutoff frequency? What is the frequency response?

Answer:
This is a noninverting first-order low-pass filter. With Formulas (36-16) and (36-17), the voltage gain and cutoff frequencies are

$$A_v = \frac{39 \text{ k}\Omega}{1 \text{ k}\Omega} + 1 = 40$$

$$f_c = \frac{1}{2\pi(12 \text{ k}\Omega)(680 \text{ pF})} = 19.5 \text{ kHz}$$

Figure 36-30b shows the frequency response. The voltage gain is 32 dB in the passband. The response breaks at 19.5 kHz and then rolls off at a rate of 20 dB per decade.

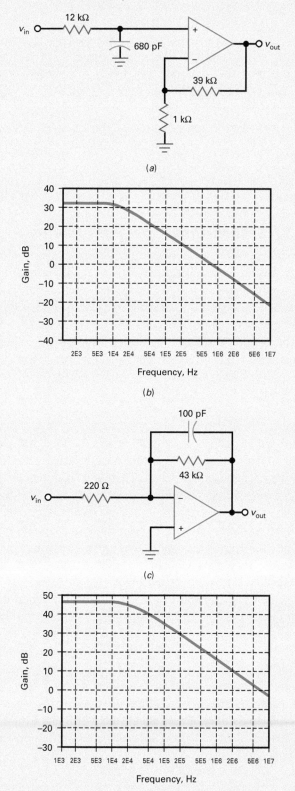

Figure 36-30 Example.

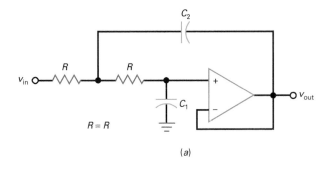

(a)

EXAMPLE 36-11

What is the voltage gain in Fig. 36-30c? What is the cutoff frequency? What is the frequency response?

Answer:

This is an inverting first-order low-pass filter. With Formulas (36-18) and (36-19), the voltage gain and cutoff frequencies are

$$A_v = \frac{-43\ \text{k}\Omega}{220\ \text{k}\Omega} = -195$$

$$f_c = \frac{1}{2\pi(43\ \text{k}\Omega)(100\ \text{pF})} = 37\ \text{kHz}$$

Figure 36-30d shows the frequency response. The voltage gain is 45.8 dB in the passband. The response breaks at 37 kHz and then rolls off at a rate of 20 dB per decade.

36.10 Second-Order Low-Pass Filters

Second-order or 2-pole stages are the most common because they are easy to build and analyze. Higher-order filters are usually made by cascading second-order stages. Each second-order stage has a resonant frequency and a Q to determine how much peaking occurs.

This section discusses the **Sallen-Key low-pass filters** (named after the inventors). These filters can implement three of the basic approximations: Butterworth, Chebyshev, and Bessel.

Circuit Implementation

Figure 36-31a shows a Sallen-Key second-order low-pass filter. Notice that the two resistors have the same value, but the two capacitors are different. There is a lag circuit on the noninverting input, but this time there is a feedback path through a second capacitor C_2. At low frequencies, both capacitors appear to be open and the circuit has a unity gain because the op amp is connected as a voltage follower.

As the frequency increases, the impedance of C_1 decreases and the noninverting input voltage decreases. At the same time, capacitor C_2 is feeding back a signal that is in phase with the input signal. Since the feedback signal adds to the source signal, the feedback is *positive*. As a result, the decrease in the noninverting input voltage caused by C_1 is not as large as it would be without the positive feedback.

Figure 36-31b shows another variation of the second-order low-pass filter. It uses equal values of R and C. It also has gain as determined by the usual noninverting op-amp formula:

$$A_v = \frac{R_2}{R_1} + 1 \qquad \textbf{(36-25)}$$

These second-order low-pass filters are more complex to analyze and design. Using the component values,

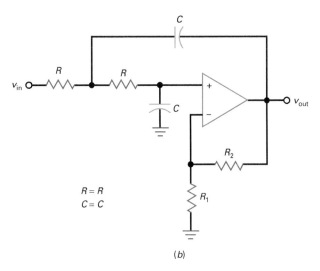

(b)

Figure 36-31 Second-order low-pass filter. (a) Unity gain. (b) With gain.

the pole frequency (f_p) and Q are calculated. Then using these values plus some special coefficients (K_x), the actual Bode corner frequencies and 3-dB cutoff frequencies are calculated. Different values of K are used for the different frequency responses, such as Butterworth or Bessel. The approximate rolloff rate is 40 dB per decade.

36.11 High-Pass Filters

Figure 36-32a shows the Sallen-Key unity-gain high-pass filters. Notice that the positions of resistors and capacitors have been reversed from these in the low-pass filters.

36.12 Higher-Order Filters and Filter Design

The standard approach in building higher-order filters is to cascade first- and second-order stages. When the order is even, we need to cascade only second-order stages. When the order is odd, we need to cascade second-order stages and a single first-order stage. For instance, if we want to build a sixth-order filter, we can cascade three second-order stages. If we want to build a fifth-order

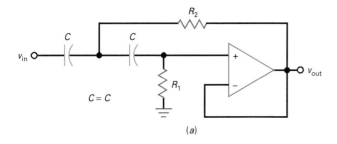

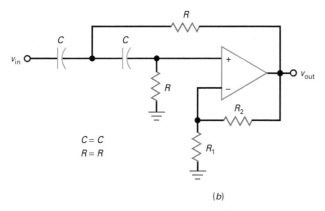

Figure 36-32 Second-order high-pass stages. (a) Unity gain. (b) Voltage gain greater than unity.

filter, we can cascade two second-order stages and one first-order stage.

Filter design is beyond the scope of this book. But that brings us to an important point: All serious filter design is done on computers because the calculations are too difficult and time-consuming to attempt by hand. An active-filter computer program stores all the equations, tables, and circuits needed to implement the five approximations (Butterworth, Chebyshev, inverse Chebyshev, elliptic, and Bessel). The circuits used to build filters range from a simple one op-amp stage to complex five op-amp stages.

36.13 Bandpass Filters

A bandpass filter has a center frequency and a bandwidth. Recall the basic equations for a bandpass response:

$$BW = f_2 - f_1$$

$$f_0 = \sqrt{f_1 f_2}$$

$$Q = \frac{f_0}{BW}$$

When Q is less than 1, the filter has a wideband response. In this case, a bandpass filter is usually built by cascading a low-pass stage with a high-pass stage. When Q is greater than 1, the filter has a narrowband response and a different approach is used.

Wideband Filters

Suppose we want to build a bandpass filter with a lower cut-off frequency of 300 Hz and an upper cutoff frequency of 3.3 kHz. The center frequency of the filter is

$$f_0 = \sqrt{f_1 f_2} = \sqrt{(300 \text{ Hz})(3.3 \text{ kHz})} = 995 \text{ Hz}$$

The bandwidth is

$$BW = f_2 - f_1 = 3.3 \text{ kHz} - 300 \text{ Hz} = 3 \text{ kHz}$$

The Q is

$$Q = \frac{f_0}{BW} = \frac{995 \text{ Hz}}{3 \text{ kHz}} = 0.332$$

Since Q is less than 1, we can use cascaded low-pass and high-pass stages, as shown in Fig. 36-33. The high-pass filter has a cutoff frequency of 300 Hz, and the low-pass filter has a cutoff frequency of 3.3 kHz. When the two decibel responses are added, we get a bandpass response with cutoff frequencies of 300 Hz and 3.3 kHz.

When Q is greater than 1, the cutoff frequencies are much closer than shown in Fig. 36-33. Because of this, the sum of the passband attenuations is greater than 3 dB at the cutoff

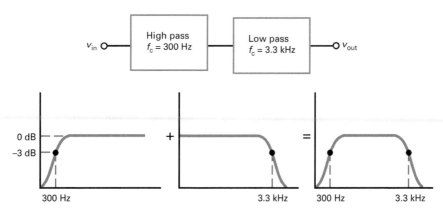

Figure 36-33 Wideband filter uses cascade of low-pass and high-pass stages.

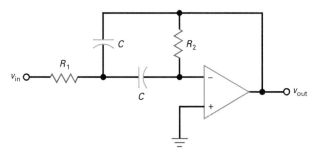

Figure 36-34 Multiple-feedback bandpass stage.

frequencies. This is why we use another approach for narrowband filters.

Narrowband Filters

When Q is greater than 1, we can use the **multiple-feedback (MFB)** filter shown in Fig. 36-34. First, notice that the input signal goes to the inverting input rather than the noninverting input. Second, notice that the circuit has two feedback paths, one through a capacitor and another through a resistor.

At low frequencies the capacitors appear to be open. Therefore, the input signal cannot reach the op amp, and the output is zero. At high frequencies the capacitors appear to be shorted. In this case, the voltage gain is zero because the feedback capacitor has zero impedance. Between the low and high extremes in frequency, there is a band of frequencies where the circuit acts like an inverting amplifier.

The voltage gain at the center frequency is given by

$$A_v = \frac{-R_2}{2R_1} \tag{36-26}$$

This is almost identical to the voltage gain of an inverting amplifier, except for the factor of 2 in the denominator. The Q of the circuit is given by

$$Q = 0.5 \sqrt{\frac{R_2}{R_1}} \tag{36-27}$$

which is equivalent to

$$Q = 0.707 \sqrt{-A_v} \tag{36-28}$$

For instance, if $A_v = -100$,

$$Q = 0.707 \sqrt{100} = 7.07$$

Formula (36-28) tells us that the greater the voltage gain, the higher the Q.

The center frequency is given by

$$f_0 = \frac{1}{2\pi \sqrt{R_1 R_2 C_1 C_2}} \tag{36-29}$$

Since $C_1 = C_2$ in Fig. 36-34, the equation simplifies to

$$f_0 = \frac{1}{2\pi C \sqrt{R_1 R_2}} \tag{36-30}$$

EXAMPLE 36-12

Calculate the gain, center frequency, Q, and bandwidth of the bandpass filter in Fig. 36.35.

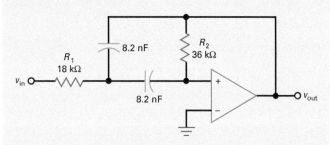

Figure 36-35 Example.

Answer

$$A = \frac{-R_2}{2R_1} = \frac{-36 \text{ k}\Omega}{2(18 \text{ k}\Omega)} = -1$$

$$f_0 = \frac{1}{2\pi C \sqrt{R_1 R_2}} = \frac{1}{2\pi (8.2 \text{ nF}) \sqrt{(18 \text{ k}\Omega)(36 \text{ k}\Omega)}}$$

$$= 762.85 \text{ Hz}$$

$$Q = 0.5 \sqrt{\frac{R_z}{R_1}} = 0.5 \sqrt{\frac{36 \text{ k}\Omega}{18 \text{ k}\Omega}} = 0.707$$

$$\text{BW} = \frac{f_0}{Q} = \frac{762.85}{0.707} = 1079 \text{ Hz}$$

36.14 Bandstop Filters

There are many circuit implementations for bandstop filters. They use from one to four op amps in each second-order stage. In many applications, a bandstop filter needs to block only a single frequency. For instance, the ac power lines may induce a hum of 60 Hz in sensitive circuits; this may interfere with a desired signal. In this case, we can use a bandstop filter to notch out the unwanted hum signal.

Figure 36-36 shows a **Sallen-Key second-order notch filter** and its analysis equations. At low frequencies all

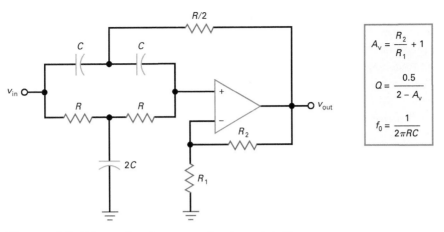

Figure 36-36 Sallen-Key second-order notch filter.

capacitors are open. As a result, all the input signal reaches the noninverting input. The circuit has a passband voltage gain of

$$A_v = \frac{R_2}{R_1} + 1 \qquad (36\text{-}31)$$

At very high frequencies, the capacitors are shorted. Again, all the input signal reaches the noninverting input.

Between the low and high extremes in frequency, there is a center frequency given by

$$f_0 = \frac{1}{2\pi RC} \qquad (36\text{-}32)$$

At this frequency, the feedback signal returns with the correct amplitude and phase to attenuate the signal on the noninverting input. Because of this, the output voltage drops to a very low value.

The Q of the circuit is given by

$$f = \frac{0.5}{2 - A_v} \qquad (36\text{-}33)$$

The voltage gain of a Sallen-Key notch filter must be less than 2 to avoid oscillations. Because of the tolerance of the R_1 and R_2 resistors, the circuit Q should be much less than 10. At higher Qs, the tolerance of these resistors may produce a voltage gain greater than 2, which would produce oscillations.

EXAMPLE 36-13

What are the voltage gain, center frequency, and Q for the band-stop filter shown in Fig. 36-36 if $R = 22\ \text{k}\Omega$, $C = 120\ \text{nF}$, $R_1 = 13\ \text{k}\Omega$, and $R_2 = 10\ \text{k}\Omega$?

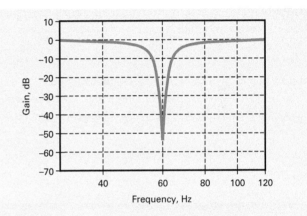

Figure 36-37 Second-order notch filter at 60 Hz.

Answer:
With Formulas (36-32) and (36-33),

$$A_v = \frac{10\ \text{k}\Omega}{13\ \text{k}\Omega} + 1 = 1.77$$

$$f_0 = \frac{1}{2\pi(22\ \text{k}\Omega)\,(120\ \text{nF})} = 60.3\ \text{Hz}$$

$$Q = \frac{0.5}{2 - A_v} = \frac{0.5}{2 - 1.77} = 2.17$$

Figure 36-37 shows the response. Notice how sharp the notch is for a second-order filter.

36.15 The All-Pass Filter

A more descriptive name for an all-pass filter is a *phase filter* because the filter shifts the phase of the output signal without changing the magnitude. Another descriptive title would be the *time-delay filter,* since time delay is related to a phase shift.

The all-pass filter has a constant voltage gain for all frequencies. This type of filter is useful when we want to produce a certain amount of phase shift for a signal without changing the amplitude.

Figure 36-38*a* shows a *first-order all-pass lag filter.* It is first order because it has only one capacitor. This is the phase shifter, which shifts the phase of the output signal between 0 and −180°. The center frequency of an all-pass filter is where the phase shift is half of maximum. For a first-order lag filter, the center frequency has a phase shift of −90°.

Figure 36-38*b* shows a *first-order all-pass lead filter.* In this case, the circuit shifts the phase of the output signal between 180 and 0°. This means that the output signal can lead the input signal by up to +180°. For a first-order lead filter, the phase shift is +90° at the center frequency.

Second-order all-pass filters are possible and have a wider phase-shift range and a faster phase response. A second-order all-pass filter has at least one op amp, two capacitors, and several resistors that can shift the phase between 0 and ±360°. Furthermore, it is possible to adjust the Q of a second-order all-pass filter to change the shape of the phase response between 0 and ±360°. The center frequency of a second-order filter is where the phase shift equals ±180°.

With a second-order all-pass filter, we can set the center frequency and Q of the circuit. By increasing the Q, we can get a faster phase response. The higher Q does not change the center frequency, but the phase change is steeper near the center frequency.

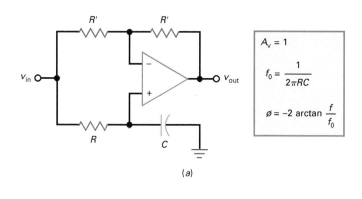

$$A_v = 1$$

$$f_0 = \frac{1}{2\pi RC}$$

$$\phi = -2 \arctan \frac{f}{f_0}$$

(a)

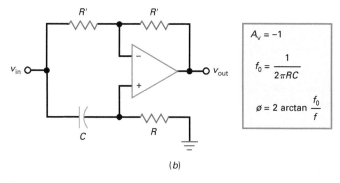

$$A_v = -1$$

$$f_0 = \frac{1}{2\pi RC}$$

$$\phi = 2 \arctan \frac{f_0}{f}$$

(b)

Figure 36-38   First-order all-pass stages. (*a*) Lagging output phase. (*b*) Leading output phase.

CHAPTER 36 SYSTEM SIDEBAR

Video Surveillance

One of the most pervasive electronic systems in the United States and around the world is video surveillance. It is everywhere. Most of these systems are not hidden, but we generally do not notice or pay attention to them. Video surveillance is used in stores to minimize shoplifting or to capture other events that are unwanted. It is used in companies and factories primarily for security. A major use is in gambling casinos. Additional surveillance is used in parking lots and outdoor storage areas. Even some high-end homes have video security. Considerable work goes into designing, building, and installing these systems. You will most likely encounter one in your future.

Analog Systems

Video surveillance systems have been around as long as there has been TV. Many older systems still exist, which use analog cameras and technology. Many of the simpler systems were black and white only, but today most analog systems use color. A typical system is shown in Fig. S36-1. It

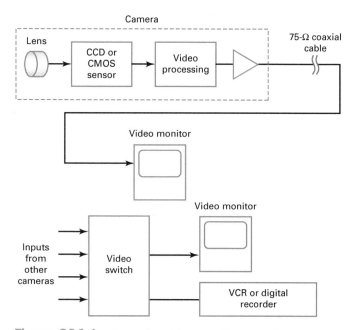

Figure S36-1   An analog video surveillance system.

consists of one or more analog color TV cameras mounted in areas that are to be observed.

The analog video outputs are connected back to a central monitoring office by way of 75-Ω coaxial cable such as RG-6/U, RG-59/U, or the equivalent. Maximum cable length is about 300 feet. Each camera is usually connected to a video monitor. In some systems, multiple cameras are connected to a video switch that selects which camera to display on the monitor. Many systems also have recording capability. Older systems used standard VHS VCRs. Newer systems use digital recording to a computer hard drive.

Most analog systems use some form of the original U.S. analog TV standard called the National Television Standards Committee (NTSC), established back in the 1950s and 1960s. It uses 525 scan lines (480 visible) interlaced and a 30-frame per second (fps) raster update rate. The greater the number of scan lines, the higher the resolution of the system, meaning the greater the amount of picture detail you can see.

Digital Systems

Most new video surveillance systems are digital. See Fig. S36-2. Digital surveillance systems use newer higher-resolution digital cameras that have charge-coupled devices (CCDs) or CMOS sensors to pick up the information from the scene being observed. The pixels convert light to voltage, which is then read out as the pixels are rapidly scanned, creating an analog signal. That analog video signal is then digitized in an analog-to-digital converter (ADC). This creates millions of words of binary data that are stored in memory. This SDRAM is inside the camera itself.

Video creates so much binary data that it rapidly overwhelms most digital memory or storage systems. For that reason, the digital video is subjected to a special process called compression. Digital compression takes the video data and processes them in a special circuit that uses algorithms to greatly reduce the number of bits and words needed to represent the picture information. The reverse process, called decompression, is carried out before the video is displayed. Most compression uses a standard called H.264 or MPEG-4. The compression takes place in a special video processing circuit, usually an FPGA.

Next, the compressed data are formatted for transmission over a local area network (LAN), usually the popular Ethernet LAN. This is done in an embedded processer in the camera. Then the data are transmitted back to a monitoring site by way of a standard Ethernet cable. This is a four twisted-pair cable, usually designated CAT5. The data rate is usually 10 Mbps or 100 Mbps, although newer systems use 1-Gbps Ethernet. The absolute maximum range is 100 meters. DC power for the camera is usually supplied over the Ethernet cable by the Ethernet switch that receives the camera data.

Many new systems eliminate the cable and use some form of wireless commuications. The popular 2.4-GHz Wi-Fi wireless method is used in some systems. Special wireless devices are also used in more secure systems or those covering longer distances. Even cell phone systems can be used in some installations.

Many new digital cameras are actually Internet-enabled, meaning that they have an IP (Internet protocol) address. With this you can actually access the camera over the Internet from anywhere to see what is going on at that location.

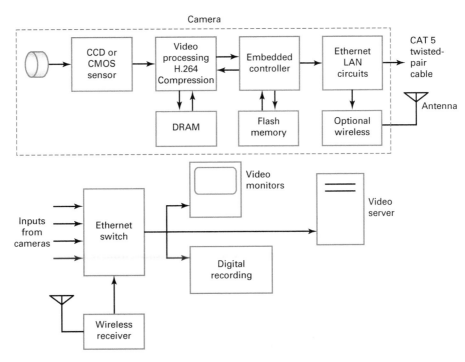

Figure S36-2 Modern digital video surveillance system.

At the monitoring station, the digital data are recovered usually in an Ethernet switch that determines which camera output will be selected for display or recording. It is then decompressed and connected to video monitors. The data are also stored digitally in compressed form on hard disk drives. These may be in a dedicated video unit on in a special server computer designated for video storage. Some digital systems can also accept inputs from analog cameras. The analog video is digitized and compressed and either stored or displayed.

Digital systems can have much higher resolution. There are multiple video standards used. Lower-resolution systems use the common intermediate format (CIF), which has a resolution of 352- by 288-pixel format at 30 fps. For higher resolution, formats similar to standard digital TV with resolutions of 480 lines (standard) or 720 lines and even 1080 lines for high definition.

The most advanced surveillance systems use special software called *video analytics*. The software uses artificial intelligence and machine vision algorithms to process the video scenes. It can look for motion, recognize objects, and do facial identification. In these advanced systems, little or no human monitoring is needed.

❓ CHAPTER 36 REVIEW QUESTIONS

1. To detect when the input is greater than a particular value, use a
   a. comparator.
   b. clamper.
   c. limiter.
   d. relaxation oscillator.

2. The voltage out of a Schmitt trigger is
   a. a low voltage.
   b. a high voltage.
   c. either a low or a high voltage.
   d. a sine wave.

3. Hysteresis prevents false triggering associated with
   a. a sinusoidal input.
   b. noise voltages.
   c. stray capacitances.
   d. trip points.

4. If the input is a rectangular pulse, the output of an integrator is a
   a. sine wave.
   b. square wave.
   c. ramp.
   d. rectangular pulse.

5. When a large sine wave drives a Schmitt trigger, the output is a
   a. rectangular wave.
   b. triangular wave.
   c. rectified sine wave.
   d. series of ramps.

6. If pulse width decreases and the period stays the same, the duty cycle
   a. decreases.
   b. stays the same.
   c. increases.
   d. is zero.

7. The output of a relaxation oscillator is a
   a. sine wave.
   b. square wave.
   c. ramp.
   d. spike.

8. If $A_{VOL} = 100,000$, the closed-loop knee voltage of a silicon diode is
   a. 1 μV.
   b. 3.5 μV.
   c. 7 μV.
   d. 14 μV.

9. The input to a positive limiter is a triangular wave with a peak-to-peak value of 8 V and an average value of 0. If the reference level is 2 V, the output has a peak-to-peak value of
   a. 0 V.
   b. 2 V.
   c. 6 V.
   d. 8 V.

10. A comparator with a trip point of zero is sometimes called a
    a. threshold detector.
    b. zero-crossing detector.
    c. positive limit detector.
    d. half-wave detector.

11. To work properly, many IC comparators need an external
    a. compensating capacitor.
    b. pullup resistor.
    c. bypass circuit.
    d. output stage.

12. A Schmitt trigger uses
    a. positive feedback.
    b. negative feedback.
    c. compensating capacitors.
    d. pullup resistors.

13. A Schmitt trigger
    a. is a zero-crossing detector.
    b. has two trip points.
    c. produces triangular output waves.
    d. is designed to trigger on noise voltage.

14. A relaxation oscillator depends on the charging of a capacitor through a
    a. resistor.
    b. inductor.
    c. capacitor.
    d. noninverting input.

15. A ramp of voltage
    a. always increases.
    b. is a rectangular pulse.
    c. increases or decreases at a linear rate.
    d. is produced by hysteresis.

16. The op-amp integrator uses
    a. capacitors.
    b. the Miller effect.
    c. sinusoidal inputs.
    d. hysteresis.

17. The trip point of a comparator is the input voltage that causes
    a. the circuit to oscillate.
    b. peak detection of the input signal.
    c. the output to switch states.
    d. clamping to occur.

18. In an op-amp integrator, the current through the input resistor flows into the
    a. inverting input.
    b. noninverting input.
    c. bypass capacitor.
    d. feedback capacitor.

19. An active half-wave rectifier has a knee voltage of
    a. V_K.
    b. 0.7 V.
    c. More than 0.7 V.
    d. Much less than 0.7 V.

20. The center frequency of a bandpass filter is always equal to the
    a. bandwidth.
    b. geometric average of the cutoff frequencies.
    c. bandwidth divided by Q.
    d. 3-dB frequency.

21. The Q of a narrowband filter is always
    a. small.
    b. equal to BW divided by f_0.
    c. less than 1.
    d. greater than 1.

22. A bandstop filter is sometimes called a
    a. snubber.
    b. phase shifter.
    c. notch filter.
    d. time-delay circuit.

23. The all-pass filter has
    a. no passband.
    b. one stopband.
    c. the same gain at all frequencies.
    d. a fast rolloff above cutoff.

24. A first-order active-filter stage has
    a. one capacitor.
    b. two op amps.
    c. three resistors.
    d. a high Q.

25. A first-order stage cannot have a
    a. Butterworth response.
    b. Chebyshev response.
    c. maximally flat passband.
    d. rolloff rate of 20 dB per decade.

26. To build a tenth-order filter, we should cascade
    a. 10 first-order stages.
    b. 5 second-order stages.
    c. 3 third-order stages.
    d. 2 fourth-order stages.

27. If bandwidth increases,
    a. the center frequency decreases.
    b. Q decreases.
    c. the rolloff rate increases.
    d. ripples appear in the stopband.

28. The all-pass filter is used when
    a. high rolloff rates are needed.
    b. phase shift is important.
    c. a maximally flat passband is needed.
    d. a rippled stopband is important.

29. A second-order all-pass filter can vary the output phase from
    a. 90 to −90°.
    b. 0 to −180°.
    c. 0 to −360°.
    d. 0 to −720°.

30. The all-pass filter is sometimes called a
    a. Tow-Thomas filter.
    b. delay equalizer.
    c. KHN filter.
    d. state-variable filter.

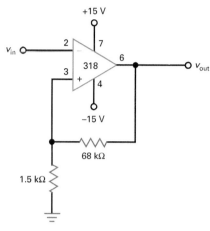
SECTION. 36.1 Comparators with Zero Reference

36.1 In Fig. 36-1a, the comparator has an open-loop voltage gain of 106 dB. What is the input voltage that produces positive saturation if the supply voltages are ±20 V?

36.2 If the input voltage is 50 V in Fig. 36-2a, what is the approximation current through the left clamping diode if $R = 10$ kΩ?

36.3 The dual supplies of Fig. 36-6b are reduced to ±12 V, and the diode is reversed. What is the output voltage?

SECTION 36.2 Comparators with Nonzero References

36.4 In Fig. 36-9a, the dual supply voltages are ±15 V. If $R_1 = 47$ kΩ and $R_2 = 12$ kΩ, what is the reference voltage? If the bypass capacitance is 0.5 μF, what is the cutoff frequency?

36.5 In Fig. 36-10, $V_{CC} = 9$ V, $R_1 = 22$ kΩ, and $R_2 = 4.7$ kΩ. What is the output duty cycle if the input is a sine wave with a peak of 7.5 V?

36.6 In Fig. 36-39, what is the output duty cycle if the input is a sine wave with a peak of 5 V?

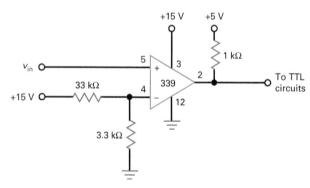

Figure 36-39

SECTION 36.3 Comparators with Hysteresis

36.7 In Fig. 36-15a, $R_1 = 2.2$ kΩ, and $R_2 = 18$ kΩ. If $V_{sat} = 14$ V, what are the trip points? What is the hysteresis?

36.8 If $V_{sat} = 13.5$ V in Fig. 36-40, what are the trip points and hysteresis?

Figure 36-40

36.9 What are the trip points and hysteresis if, $V_{sat} = 14$ V in Fig. 36-41?

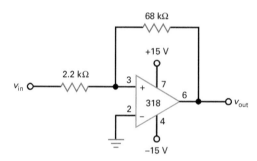

Figure 36-41

SECTION 36.4 The Integrator

36.10 What is the capacitor charging current in Fig. 36-42 when the input pulse is high?

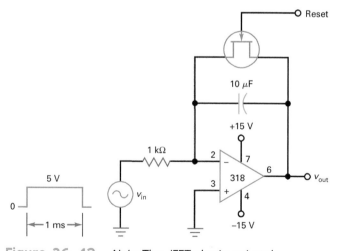

Figure 36-42 *Note:* The JFET shorts out and discharges the capacitor, resetting the integrator when a positive voltage is applied to the gate.

36.11 In Fig. 36-42, the output voltage is reset just before the pulse begins. What is the output voltage at the end of the pulse?

SECTION 36.5 Waveform Conversion

36.12 What is the output voltage in Fig. 36-43?

36.13 If the capacitance is changed to 0.068 μF in Fig. 36-43, what is the output voltage?

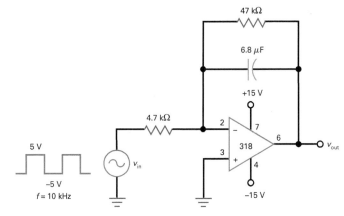

Figure 36-43

36.14 What is the duty cycle in Fig. 36-44 when the wiper is one-half of the way from the top?

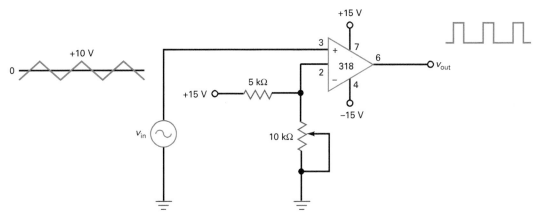

Figure 36-44

SECTION 36.6 Waveform Generation

36.15 What is the frequency of the output signal in Fig. 36-45?

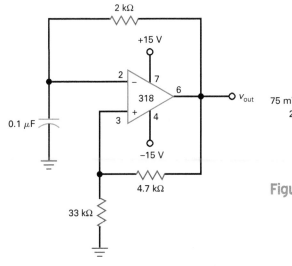

Figure 36-45

SECTION 36.7 Active Diode Circuits

36.16 In Fig. 36-27, the input sine wave has a peak of 100 mV. What is the output voltage?

36.17 What is the output voltage in Fig. 36-46?

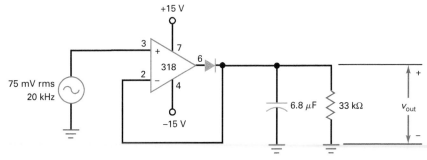

Figure 36-46

36.18 Suppose the diode of Fig. 36-46 is reversed. What is the output voltage?

36.19 The input voltage of Fig. 36-46 is changed from 75 mV rms to 150 mV p-p. What is the output voltage?

SECTION 36.9 First-Order Stages

36.20 In Fig. 36-28a, $R_1 = 15$ kΩ and $C_1 = 270$ nF. What is the cutoff frequency?

36.21 In Fig. 36-28b, $R_1 = 7.5$ kΩ, $R_2 = 33$ kΩ, $R_3 = 20$ kΩ, and $C_1 = 680$ pF. What is the cutoff frequency? What is the voltage gain in the passband?

36.22 In Fig. 36-28c, $R_1 = 2.2$ kΩ, $R_2 = 47$ kΩ, and $C_1 = 330$ pF. What is the cutoff frequency? What is the voltage gain in the passband?

SECTION 36.13 Bandpass Filters

36.23 In Fig. 36-34, $R_1 = 2$ kΩ, $R_2 = 56$ kΩ, and $C = 270$ pF. What are the voltage gain, Q, and center frequency?

SECTION 36.14 Bandstop Filters

36.24 What are the voltage gain, center frequency, and Q for the bandstop filter shown in Fig. 36-36 if $R = 56$ kΩ, $C = 180$ nF, $R_1 = 20$ kΩ, and $R_2 = 10$ kΩ? What is the bandwidth?

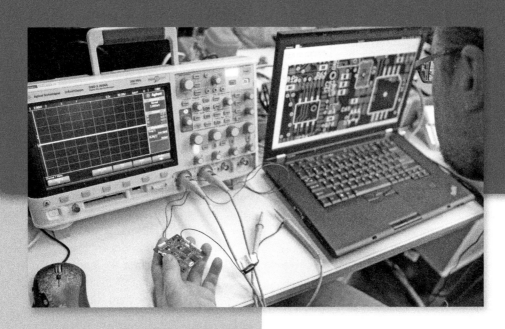

Chapter 37

Power Amplifiers

Learning Outcomes

After studying this chapter, you should be able to:

> *Describe* the characteristics of amplifiers, including classes of operation and frequency ranges.

> *Draw* a schematic of class B/AB push-pull amplifier and *explain* its operation.

> *Explain* the operation and application of class C, D, E, and F amplifiers.

> *Determine* the efficiency of transistor power amplifiers.

> *Discuss* the factors that limit the power rating of a transistor and what can be done to improve the power rating.

Power amplifiers are circuits that amplify power rather than voltage. They are used where the load demands power to perform its function. A good example is an audio amplifier that drives a speaker. A speaker is an electromechanical device that converts audio signals into the mechanical vibrations that result in sound. Speakers require a substantial amount of power. Amplifiers that drive motors also need high power. Radio-frequency transmitters must deliver power to drive an antenna to achieve wireless communications.

This chapter covers common power amplifier circuits, how they work, the most common types, and their specifications.

37.1 Amplifier Terms

There are different ways to describe amplifiers. For instance, we can describe them by their class of operation or by their frequency range.

Classes of Operation

Class A operation of an amplifier means that the transistor operates in the active region at all times. This implies that collector or drain current flows for 360° of the ac cycle, as shown in Fig. 37-1a. With a class A amplifier, the designer usually tries to locate the Q point somewhere near the middle of the load line. This way, the signal can swing over the maximum possible range without saturating or cutting off the transistor, which would distort the signal.

Class B operation is different. It means that collector or drain current flows for only half the cycle (180°), as shown in Fig. 37-1b. To have this kind of operation, a designer locates the Q point at cutoff. Then, only the positive half of ac base voltage can produce collector or drain current. This reduces the wasted heat in power transistors.

Class C operation means that collector or drain current flows for less than 180° of the ac cycle, as shown in Fig. 37-1c. With class C operation, only part of the positive half cycle of ac base voltage produces collector or drain current. As a result, we get brief pulses of collector or drain current like those of Fig. 37-1c.

Ranges of Frequency

Another way to describe amplifiers is by stating their frequency range. For instance, an **audio amplifier** refers to an amplifier that operates in the range of 20 Hz to 20 kHz. On the other hand, a **radio-frequency (RF) amplifier** is

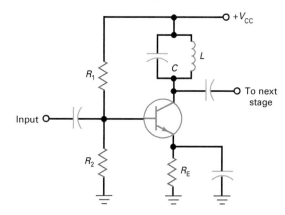

Figure 37-2   Tuned RF amplifiers.

one that amplifies frequencies above 20 kHz, usually much higher. For instance, the RF amplifiers in AM radios amplify frequencies between 535 and 1605 kHz, and the RF amplifiers in FM radios amplify frequencies between 88 and 108 MHz. Microwave amplifiers amplify signals in the 1-GHz to 50-GHz range.

Amplifiers are also classified as **narrowband** or **wideband.** A narrowband amplifier works over a small frequency range like 450 to 460 kHz. A wideband amplifier operates over a large frequency range like 0 to 1 MHz or several GHz.

Narrowband amplifiers are usually **tuned RF amplifiers,** which means that their ac load is a high-Q resonant tank tuned to a radio signal. Wideband amplifiers are usually untuned; that is, their ac load is resistive.

Figure 37-2a is an example of a tuned RF amplifier. The LC tank is resonant at some frequency. If the tank has a high Q, the bandwidth is narrow. The output is capacitively coupled to the next stage.

Signal Levels

We have already described *small-signal operation,* in which the peak-to-peak swing in collector current is less than 10% of quiescent collector current. In **large-signal operation,** a peak-to-peak signal uses all or most of the load line. In a stereo system, the small signal from a radio tuner, tape player, or compact disc player is used as the input to a **preamp,** an amplifier that produces a larger output suitable for driving tone and volume controls. The signal is then used as the input to a **power amplifier,** which produces output power ranging from a few hundred milliwatts up to hundreds of watts.

37.2 Class A Operation

Almost any voltage amplifier is also a power amplifier. It is just not optimized for power applications. The amplifier of Fig. 37-3a is a class A amplifier as long as the output signal is not clipped. With this kind of amplifier, collector current flows throughout the cycle. Stated another way, no clipping of the output signal occurs at any time during the cycle. Now,

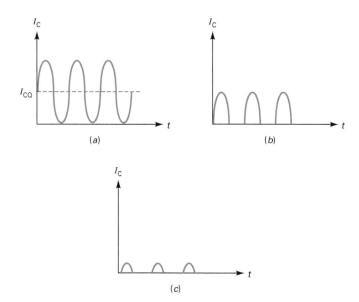

Figure 37-1   Collector current. (a) Class A. (b) Class B. (c) Class C.

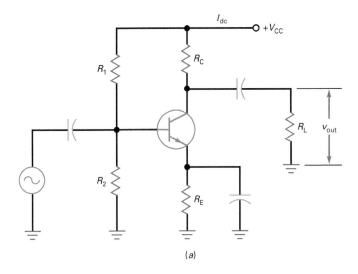

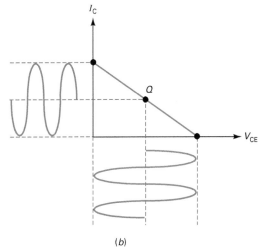

Figure 37-3 Class A amplifier.

we discuss a few equations that are useful in the analysis of class A amplifiers.

Power Gain

Besides voltage gain, any amplifier has a **power gain,** defined as

$$A_p = \frac{p_{out}}{p_{in}} \tag{37-1}$$

In words, the power gain equals the ac output power divided by the ac input power.

For instance, if the amplifier of Fig. 37-3a has an output power of 10 mW and an input power of 10 μW, it has a power gain of

$$A_p = \frac{10\ \text{mW}}{10\ \mu\text{W}} = 1000$$

Output Power

If we measure the output voltage of Fig. 37-3a in rms volts, the output power is given by

$$p_{out} = \frac{v_{rms}^2}{R_L} \tag{37-2}$$

Usually, we measure the output voltage in peak-to-peak volts with an oscilloscope. In this case, a more convenient equation to use for output power is

$$p_{out} = \frac{v_{out}^2}{8R_L} \tag{37-3}$$

The factor of 8 in the denominator occurs because $v_{p\text{-}p} = 2\sqrt{2}\ v_{rms}$. When you square $2\sqrt{2}$, you get 8.

The maximum output power occurs when the amplifier is producing the maximum peak-to-peak output voltage, as shown in Fig. 37-3b. In this case, $v_{p\text{-}p}$ equals the maximum peak-to-peak output voltage and the maximum output power is

$$p_{out(max)} = \frac{MPP^2}{8R_L} \tag{37-4}$$

Transistor Power Dissipation

When no signal drives the amplifier of Fig. 37-3a, the quiescent power dissipation is

$$P_{DQ} = V_{CEQ}I_{CQ} \tag{37-5}$$

This makes sense. It says that the quiescent power dissipation equals the dc voltage times the dc current.

When a signal is present, the power dissipation of a transistor decreases because the transistor converts some of the quiescent power to signal power. For this reason, the quiescent power dissipation is the worst case. Therefore, the power rating of a transistor in a class A amplifier must be greater than P_{DQ}; otherwise, the transistor will be destroyed.

Current Drain

As shown in Fig. 37-3a, the dc voltage source has to supply a dc current I_{dc} to the amplifier. This dc current has two components: the biasing current through the voltage divider and the collector current through the transistor. The dc current is called the **current drain** of the stage. If you have a multistage amplifier, you have to add the individual current drains to get the total current drain.

Efficiency

The dc power supplied to an amplifier by the dc source is

$$P_{dc} = V_{CC}I_{dc} \tag{37-6}$$

To compare the design of power amplifiers, we can use the **efficiency,** defined by

$$\eta = \frac{p_{out}}{P_{dc}} \times 100\% \tag{37-7}$$

This equation says that the efficiency equals the ac output power divided by the dc input power.

The efficiency of any amplifier is between 0 and 100%. Efficiency gives us a way to compare two different designs because it indicates how well an amplifier converts the dc input power to ac output power. The higher the efficiency, the better the amplifier is at converting dc power to ac power. This is important in battery-operated equipment because high efficiency means that the batteries last longer.

Since all resistors except the load resistor waste power, the efficiency is less than 100% in a class A amplifier. In fact, it can be shown that the maximum efficiency of a class A amplifier with a dc collector resistance and a separate load resistance is 25%.

In some applications, the low efficiency of class A is acceptable. For instance, the small-signal stages near the front of a system usually work fine with low efficiency because the dc input power is small. In fact, if the final stage of a system needs to deliver only a few hundred milliwatts, the current drain on the power supply may still be low enough to accept. But when the final stage needs to deliver watts of power, the current drain usually becomes too large with class A operation.

EXAMPLE 37-1

If the peak-to-peak output voltage is 18 V and the input impedance of the circuit is 37.4 Ω, what is the power gain in Fig. 37-4?

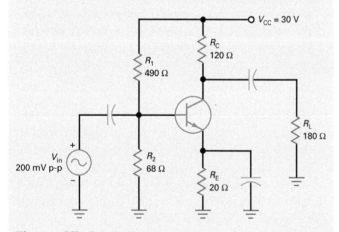

Figure 37-4 Example.

Answer:
The ac input power is

$$P_{in} = \frac{(200 \text{ mV})^2}{8 (37.4)} = 133.7 \ \mu\text{W}$$

The ac output power is

$$P_{out} = \frac{(18 \text{ V})^2}{8 (180 \ \Omega)} = 225 \text{ mW}$$

The power gain is

$$A_p = \frac{225 \text{ mW}}{133.7 \ \mu\text{W}} = 1683$$

EXAMPLE 37-2

What is the transistor power dissipation and efficiency of Fig. 37-4?

Answer:
The dc emitter current is

$$I_E = \frac{3 \text{ V}}{20 \ \Omega} = 150 \text{ mA}$$

The dc collector voltage is

$$V_C = 30 \text{ V} - (150 \text{ mA})(120 \ \Omega) = 12 \text{ V}$$

and the dc collector-emitter voltage is

$$V_{CEQ} = 12 \text{ V} - 3 \text{ V} = 9 \text{ V}$$

The transistor power dissipation is

$$P_{DQ} = V_{CEQ} I_{CQ} = (9 \text{ V})(150 \text{ mA}) = 1.35 \text{ W}$$

To find the stage efficiency,

$$I_{bias} = \frac{30 \text{ V}}{490 \ \Omega + 68 \ \Omega} = 53.8 \text{ mA}$$

$$I_{dc} = I_{bias} + I_{CQ} = 53.8 \text{ mA} + 150 \text{ mA} = 203.8 \text{ mA}$$

The dc input power to the stage is

$$P_{dc} = V_{CC} I_{dc} = (30 \text{ V})(203.8 \text{ mA}) = 6.11 \text{ W}$$

Since the output power (found in Example 37-1) is 225 mW, the efficiency of the stage is

$$\eta = \frac{225 \text{ mW}}{6.11 \text{ W}} \times 100\% = 3.68\%$$

An Improved Power Amplifier

The basic common emitter amplifier as shown in Fig. 37-4 can be made much more efficient by placing an inductor in series with the collector lead. The concept was illustrated in Fig. 37-2. The inductor replaces the normal collector resistor. This inductance has a higher reactance at the operating frequency as well as a very low dc resistance. The inductor is generally referred to as a radio-frequency choke (RFC), but it may also be the primary winding of a transformer. The inductor with its energy-storing capabilities provides an ac output voltage across the load that will be almost two times the dc supply voltage.

As the ac input signal to the amplifier varies the base current, the collector current will also vary in step. As the collector current varies, a self-induced voltage will be developed across the inductor. The self-induced voltage appears in series with the dc supply voltage. This results in an output voltage that will have a peak-to-peak value equal to two times the supply voltage. For example, if the dc supply voltage is 12 V, the peak-to-peak voltage across the load can be as much as 24 V. It produces a considerably greater output power across the load than before, and the efficiency is

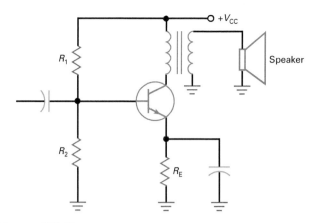

Figure 37-5   Class A power amplifier.

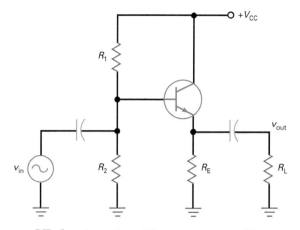

Figure 37-6   An emitter-follower power amplifier.

considerably improved. The efficiency can approach a maximum value of 50%.

Figure 37-5 shows an example of this in a class A power amplifier. The transistor is biased as usual with the voltage divider and emitter resistor. The primary winding of a transformer is connected in series with the collector. The secondary winding is connected to a speaker. The speaker has a very low impedance, typically in the 4- to 8-Ω range. The output impedance of the amplifier itself will be considerably more, generally in the range of several hundred ohms.

The purpose of the transformer, of course, is to match the impedance of the speaker to the higher impedance of the class A amplifier. Matching impedances provides maximum power transfer from the amplifier to the speaker. Remember the basic impedance matching relationship for which the speaker load resistance appears $(N_p/N_s)^2$ times larger at the collector.

A class A amplifier with the transformer impedance matching will have an impedance that approaches the maximum theoretical limit of 50%. Such a circuit is used when the desired output power is no more than several watts. Such amplifiers are no longer widely used because there are smaller, more efficient integrated circuit amplifiers. Circuits with no inductance in series with the collector are still occasionally found in some radio transmitters as a driving circuit for a higher-power amplifier. The configuration is similar to that shown in Fig. 37-2.

Emitter-Follower Power Amplifier

An emitter follower also makes a good class A power amplifier. Figure 37-6 shows the familiar emitter-follower circuit. It is generally biased so that approximately half the supply voltage is developed across the emitter resistor and the remaining across the emitter collector of the transistor. Remembering that the voltage gain of an emitter follower is approximately 1, you may wonder how such an amplifier can produce a power gain. The answer to this question lies in the fact that the amplifier has vastly different input and output impedances. With a high input impedance and a low output impedance, power gain is achieved.

Assume that the input impedance, Z_{in}, to the emitter follower is 500 Ω. Also assume an output impedance of 16 Ω. A 16-Ω load is connected to the emitter output capacitor to provide impedance matching and maximum power transfer.

Computing the input and output power levels using a 2-V rms signal, you can see below that the emitter follower delivers more power to the load than consumed by the input impedance.

$$P_{in} = \frac{V_{in}^2}{Z_{in}} = \frac{2^2}{500} = \frac{4}{500} = 0.008 \text{ W} \quad \text{or} \quad 8 \text{ mW}$$

$$P_{out} = \frac{V_{out}^2}{16} = \frac{4}{16} = 0.25 \text{ W} \quad \text{or} \quad 250 \text{ mW}$$

In this case, the power gain is $250/8 = 31.25$ or 15 dB.

Keep in mind that the circuit in Fig. 37-6 can also be implemented with an E-MOSFET to provide the same result. An emitter-follower or source-follower circuit such as this will generally be used in applications where power levels are no more than a few watts.

High-Power Class A Amplifiers

When very high power is needed and good linearity, special high-voltage, high-power transistors can be used. Rarely are high-power class A amplifiers used in audio amplifiers as there are other better circuits available, as you will see later in this chapter. However, for some high-power RF applications, a single-transistor, high-power amplifier is suitable. It would be used for an application where good linearity is required to pass all the frequency components of a modulated RF carrier.

A good example of such a high-power class A, RF amplifier is shown in Fig. 37-7. The circuit uses a single n-channel E-MOSFET. It is biased into conduction with the circuit components R_1, R_2, R_3, and D_1. A voltage of 12 V

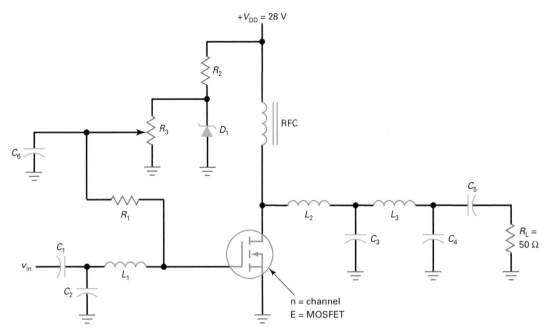

Figure 37-7 A high-power class A, E-MOSFET RF amplifier.

is developed across the zener diode D_1. This voltage is applied to a potentiometer, R_3, and the resulting dc voltage is applied through R_1 to the gate of the E-MOSFET. Adjusting R_3 optimizes the bias for maximum linearity. Capacitor C_6 provides bypassing to ensure a pure dc voltage at the gate. Once the biased voltage has been adjusted so that the amplifier operates in the linear region, it is ready to accept signals.

Capacitors C_1 and C_2 and inductor L_1 are used as an impedance-matching network to accept the input signal and transfer maximum power to the E-MOSFET gate. The remaining components, L_2, L_3, and C_3 through C_5, also provide impedance matching to the output load R_L, usually an antenna. These components match the relatively high output impedance of the E-MOSFET amplifier to a 50-Ω load. The most widely used input and output impedance for high-frequency radio transmitters is 50 Ω. Note that an inductor is used in the drain circuit of the amplifier to double the output voltage swing across the load. Using modern high-frequency, high-power E-MOSFETs, output powers of several hundred watts are possible at frequencies well into the several hundred megahertz range.

Keep one important thing in mind regarding such high-power class A amplifiers: Even while operating at the maximum theoretical efficiency of 50%, the amplifying transistor itself is dissipating approximately the same amount of power as that delivered to the load. An amplifier delivering 150 W of power to the 50-Ω load will also be dissipating 150 W as heat. Therefore, good high-power transistors must be used, and typically they will require large heat sink and good air circulation to remove the excess heat.

37.3 Class B Operation

Class A is the common way to run a transistor in linear circuits because it leads to the simplest and most stable biasing circuits. But class A is not the most efficient way to operate a transistor. In some applications, like battery-powered systems, current drain and stage efficiency become important considerations in the design. This section introduces the basic idea of class B operation.

Push-Pull Circuit

Figure 37-8 shows a basic class B amplifier. When a transistor operates as class B, it clips off half a cycle. To avoid the resulting distortion, we can use two transistors in a push-pull arrangement like that of Fig. 37-8. **Push-pull** means that one transistor conducts for half a cycle while the other is off, and vice versa.

Here is how the circuit works: On the positive half-cycle of input voltage, the secondary winding of T_1 has voltages v_1 and v_2, as shown. Therefore, the upper transistor conducts and the lower one cuts off. The collector current through Q_1 flows through the upper half of the output primary winding. This produces an amplified and inverted voltage, which is transformer-coupled to the loudspeaker.

On the next half-cycle of input voltage, the polarities reverse. Now, the lower transistor turns on and the upper transistor turns off. The lower transistor amplifies the signal, and the alternate half-cycle appears across the loudspeaker.

Since each transistor amplifies one-half of the input cycle, the loudspeaker receives a complete cycle of the amplified signal.

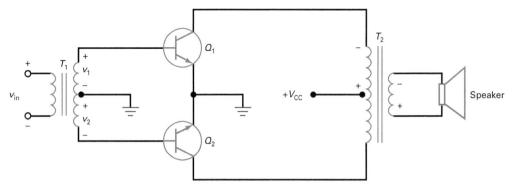

Figure 37-8   Class B push-pull amplifier.

Advantages and Disadvantages

Since there is no bias in Fig. 37-8, each transistor is at cutoff when there is no input signal, an advantage because there is no current drain when the signal is zero. With zero current, the power dissipation is also zero.

Another advantage is improved efficiency where there is no input signal. The maximum efficiency of a class B push-pull amplifier is 78.5%, so a class B push-pull power amplifier is more commonly used for an output stage than a class A power amplifier.

The main disadvantage of the amplifier shown in Fig. 37-8 is the use of transformers. Audio transformers are bulky and expensive. Although widely used at one time, a transformer-coupled amplifier like Fig. 37-8 is no longer used in audio applications. Newer audio designs have eliminated the need for transformers in most applications.

However, the class B circuit is still a viable option at radio frequencies as smaller, more efficient transformers are common. Most class B push-pull RF amplifiers use MOSFETs instead of BJTs.

37.4 Class B Push-Pull Emitter Follower

Class B operation means that the collector current flows for only 180° of the ac cycle. For this to occur, there is no dc bias. The advantage of class B amplifiers is lower current drain and higher stage efficiency.

Push-Pull Circuit

Figure 37-9 shows one way to make a transformerless class B push-pull amplifier using emitter followers. Here, we have an *npn* emitter followers and a *pnp* emitter follower connected in a push-pull arrangement. Because of the *npn* and *pnp* transistors, this circuit is usually referred to as a complementary symmetry amplifier.

The bias is so that the voltage to the emitter diode of each transistor between 0.6 and 0.7 V on the verge of conduction. The collecter current with no signal is about zero.

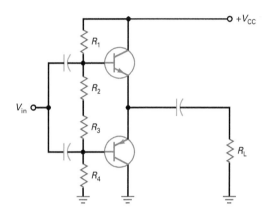

Figure 37-9   Class B push-pull emitter follower.

Because the biasing resistors are equal, each emitter diode is biased with the same value of voltage. As a result, half the supply voltage is dropped across each transistor's collector-emitter terminals. The dc output is one-half the supply voltage.

Overall Action

On the positive half-cycle of input voltage, the upper transistor of Fig. 37-9 conducts and the lower one cuts off. The upper transistor acts like an ordinary emitter follower, so that the output voltage approximately equals the input voltage.

On the negative half-cycle of input voltage, the upper transistor cuts off and the lower transistor conducts. The lower transistor acts like an ordinary emitter follower and produces a load voltage approximately equal to the input voltage. The upper transistor handles the positive half-cycle of input voltage, and the lower transistor takes care of the negative half-cycle. During either half-cycle, the source sees a relatively high input impedance looking into either base.

Crossover Distortion

Figure 37-10*a* shows the ac equivalent circuit of a class B push-pull emitter follower. Suppose that no bias is applied to the emitter diodes. Then, the incoming ac voltage has to rise to about 0.7 V to overcome the barrier potential of the

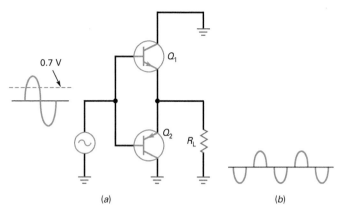

Figure 37-10 (a) AC equivalent circuit. (b) Crossover distortion.

emitter diodes. Because of this, no current flows through Q_1 when the signal is less than 0.7 V.

The action is similar on the other half-cycle. No current flows through Q_2 until the ac input voltage is more negative than −0.7 V. For this reason, if no bias is applied to the emitter diodes, the output of a class B push-pull emitter follower looks like Fig. 37-10b.

Because of clipping between half-cycles, the output is distorted. Since the clipping occurs between the time one transistor cuts off and the other one comes on, we call it **crossover distortion.** This distortion produces unwanted harmonics in the output. To eliminate crossover distortion, we need to apply a slight forward bias to each emitter diode.

Class AB

If we apply a slight forward bias to the transistors, the conduction angle will be slightly greater than 180° because the transistor will conduct for a bit more than half a cycle. Strictly speaking, we no longer have class B operation. Because of this, the operation is sometimes referred to as **class AB,** defined as a conduction angle between 180° and 360°. But it is barely class AB. For this reason, many people still refer to the circuit as a *class B push-pull amplifier* because the operation is class B to a close approximation.

Diode Bias

The most common way to ac have class AB bias is with diodes, as shown in Fig. 37-11. The idea is to use **compensating diodes** to produce the bias voltage for the emitter diodes. This arrangement eliminates the problem of thermal runaway that could occur at high temperatures. For this scheme to work, the diode characteristics must match the V_{BE} characteristics of the transistors. Then, any increase in temperature reduces the bias voltage developed by the compensating diodes by just the right amount.

For instance, assume that a bias voltage of 0.65 V sets up 2 mA of collector current. If the temperature rises 30°C, the voltage across each compensating diode drops 60 mV. Since

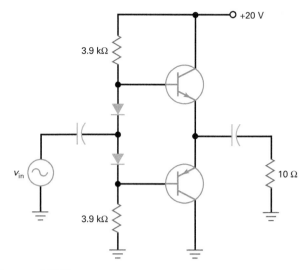

Figure 37-11 Diode bias in a class AB amplifier.

the required V_{BE} also decreases by 60 mV, the collector current remains fixed at 2 mA.

For diode bias to be immune to changes in temperature, the diode curves must match the V_{BE} curves over a wide temperature range. This is not easily done with discrete circuits because of the tolerance of components. But diode bias is easy to implement with integrated circuits because the diodes and transistors are on the same chip, which means that they have almost identical curves.

CE Driver

A complementary symmetry power amplifier normally needs a preamplifier called a driver for the output transistors to achieve full output.

The stage that precedes the output stage is called a **driver.** Rather than capacitively couple into the output push-pull stage, we can use the direct-coupled common emitter (CE) driver shown in Fig. 37-12. Transistor Q_1 is a current source

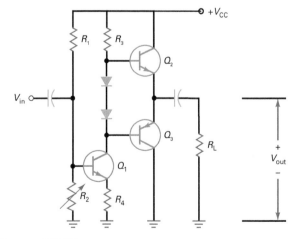

Figure 37-12 Direct-coupled CE driver for the complementary symmetry amplifier.

that sets up the dc biasing current through the diodes. By adjusting R_2, we can control the dc emitter current through R_4. This means that Q_1 sources the biasing current through the compensating diodes.

The amplified and inverted ac signal at the Q_1 collector drives the bases of Q_2 and Q_3. On the positive half-cycle, Q_2 conducts and Q_3 cuts off. On the negative half-cycle, Q_2 cuts off and Q_3 conducts. Because the output coupling capacitor is an ac short, the ac signal is coupled to the load resistance.

The driver stage is an amplifier whose amplified and inverted output drives both bases of the output transistors with the same signal. Often, the input impedance of the output transistors is very high, and we can approximate the voltage gain of the driver by

$$A_V = \frac{R_3}{R_4}$$

Two-Stage Negative Feedback

Figure 37-13 is another example of using a CE stage to drive a class B/AB push-pull emitter follower. The input signal is amplified and inverted by the Q_1 driver. The push-pull stage then provides the current gain needed to drive the low-impedance loudspeaker. Notice that the driver has its emitter connected to ground. As a result, this driver has more voltage gain than the driver of Fig. 37-12.

The resistance R_2 does two useful things: First, since it is connected to a dc voltage of $+V_{CC}/2$, this resistance provides the dc bias for Q_1. Second, R_2 produces negative feedback for the ac signal. Here's why: A positive-going signal on the base of Q_1 produces a negative-going signal on the Q_1 collector. The output of the emitter follower is therefore

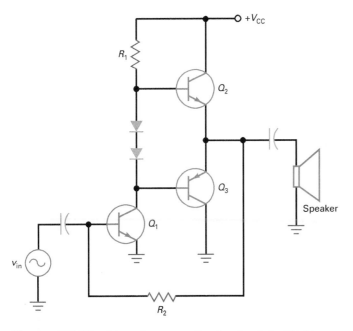

Figure 37-13   Two-stage negative feedback to CE driver.

negative-going. When fed back through R_2 to the Q_1 base, this returning signal opposes the original input signal. This is negative feedback, which stabilizes the bias and the voltage gain of the overall amplifier.

Most modern complementary symmetry power amplifiers are in integrated circuits.

Integrated Circuit Amplifiers

Most power amplifiers whose power level is below 100 W or so are available as integrated circuits. Discrete component power amplifiers are used only when very high voltages and very high powers are required. In audio applications, IC power amplifiers are used as drivers for external class AB complementary symmetry amplifiers made with discrete transistors. For most audio and RF products, IC power amplifiers are the most common. This section shows several IC power amplifiers.

Low-Voltage, Low-Power IC Amplifier

One of the most popular and widely used IC power amplifiers is the LM386. It typically operates from dc supply voltages in the 4- to 18-V range. Power output depends on the dc supply voltage and is generally in the 250-mW to 1-W range. The amplifier is designed to drive an 8-Ω speaker.

The schematic diagram of the amplifier is shown in Fig. 37-14, along with its block diagram equivalent. Obviously the circuit is not accessible, but the IC pins provide access to the power and ground and various inputs and outputs, as you can see. The transistors on the left side of the circuit make up a differential amplifier with inverting (−) and noninverting (+) inputs on pins 2 and 3. This amplifier drives the output stage. This is actually an op-amp power amplifier.

The output stage is not the usual complementary symmetry arrangement with *npn* and *pnp* transistors. Instead, two *npn* transistors are connected in series to form an arrangement generally known as a *totem pole amplifier*. Its operation is similar to that of the complementary symmetry amplifier discussed earlier. When the upper output transistor is conducting heavily, the lower *npn* transistor is conducting less and vice versa. This results in a very low output impedance for driving a speaker. Note the two diodes used for temperature compensation biasing of the output stage.

Figure 37-15 shows the typical connection of the amplifier circuit. The input is applied across a 10-kΩ potentiometer for volume control. The inputs at pins 2 and 3 are connected to form a noninverting op amp. With no external connections, the gain of the amplifier is set to 20. However, by connecting an external resistor capacitor combination between pins 1 and 8, the gain can be increased to a value of 200. The output of the amplifier at pin 5 is capacitively coupled to the 8-Ω speaker. The series resistor-capacitor combination at the output of the amplifier is used to shape the output

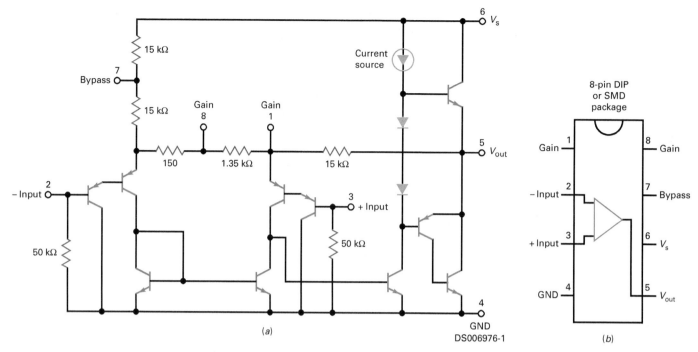

Figure 37-14 The popular LM386 audio power amplifier IC. (*a*) Schematic. (*b*) IC pin-out.

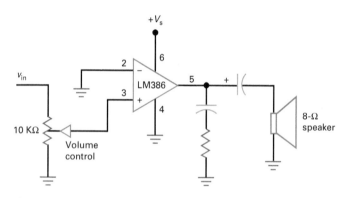

Figure 37-15 A simple audio power amplifier using the LM386 IC.

frequency response. With a gain of 20, the amplifier has an ac frequency response of several hundred kilohertz, well beyond most audio applications.

Bridged Amplifier

The output power of an amplifier is directly related to the supply voltage value. As more power is required, higher supply voltages are required. Nevertheless, power transistors do have their limitations. Furthermore, a trend over time has been to reduce the supply voltages on most ICs. Therefore, it has become more difficult to obtain the power required. This is particularly true of portable and mobile devices, such as cell phones, MP3 audio players like iPods, and similar devices. All these units operate from very low battery voltages in the 3- to 6-V range.

One technique that is widely used is to boost amplifier output power with low voltage is to use what is called a *bridged amplifier*. An example of such a circuit is shown in Fig. 37-16. Here two power op amps operated from relatively low voltage are used to operate the speaker. Note the input wiring connections to the op-amp power amplifiers.

The input signal is connected to the noninverting input of the upper amplifier and the inverting input of the lower amplifier. This means that if the power amplifiers are the same, the outputs of the two amplifiers will be equal and 180° out of phase.

Assume that each amplifier has an output voltage swing of ±5 V. When the upper amplifier is supplying +5 V, the

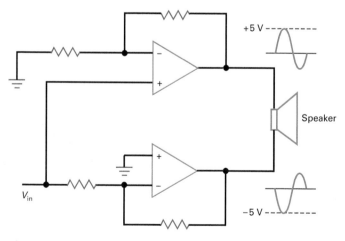

Figure 37-16 A bridged amplifier.

lower amplifier is providing −5 V to the other side of the speaker. Consequently, the speaker itself will have 10 V peak to peak across it. Since power is directly proportional to the square of the voltage, this means that the bridged amplifier can normally produce four times the output power with the same dc supply voltage as a single amplifier. This arrangement is widely used in portable battery-operated equipment. The LM4951 is an example of a single-IC bridged amplifier. It contains two power amplifiers and the related circuit. The typical supply voltage range is 2.7 to 9 V. With a 7.5-V supply, the typical power supplied to an 8-Ω speaker is 1.8 W.

Even higher IC amplifiers are available. For example, the LM1875 can deliver 20 W of power into a 4- or 8-Ω speaker load using ±25-V supplies. Boosting the supply voltages to ±30 V will provide over 30 W of power. It can deliver up to 68 W of continuous power using ±28-V supplies. And, of course, even higher powers can be obtained by using these devices in a bridged arrangement as described earlier.

Again, remember that with high-power amplifiers, with a maximum efficiency of about 50%, the amplifier itself will be dissipating almost as much power as that delivered to the load under maximum-power conditions. If 50 W is delivered to the load, the amplifier itself must dissipate 50 W of power in the form of heat. That's why all these higher-power IC amplifiers are designed to use large metal heat sinks to help dissipate the heat.

Audio Amplifier Specifications

Since it is not necessary to design a power amplifier or analyze its circuits, the key to using the amplifier lies in its specifications. Here are the main specifications to consider.

- Output power: Maximum possible power into the load.
- Output impedance: Usually low, less than 50 Ω.
- Input impedance: Usually high, more than several thousand Ω.
- Supply voltage range: As low as 3 V but can be as high as 36 V.
- Gain: Usually very high in the 20- to 50-dB range. Can be less with added feedback.
- Frequency response: Most amplifiers are direct-coupled to handle signal frequencies down to dc. Upper cutoff frequency is typically above 100 kHz to several megahertz.
- Total harmonic distortion (THD): THD is a measure of the linearity of the amplifiers; it indicates the amount of harmonic content in the output at a certain power level compared to the output power. Expressed as a percentage, the lower the better.

37.5 Class C Operation

With class B, we need to use a push-pull arrangement. That's why almost all class B amplifiers are push-pull amplifiers. With class C, we need to use a resonant circuit for the load. This is why almost all class C amplifiers are tuned amplifiers.

Resonant Frequency

With class C operation, the collector current flows for less than half a cycle. A parallel resonant circuit can filter the pulses of collector current and produce a pure sine wave of output voltage. The main application for class C is with tuned RF amplifiers. The maximum efficiency of a tuned class C amplifier is 100%.

Figure 37-17a shows a tuned RF amplifier. The ac input voltage drives the base, and an amplified output voltage appears at the collector. The amplified and inverted signal is then capacitively coupled to the load resistance. Because of the parallel resonant circuit, the output voltage is maximum at the resonant frequency, given by

$$f_r = \frac{1}{2\pi\sqrt{LC}} \tag{37-8}$$

On either side of the resonant frequency f_r, the voltage gain drops off as shown in Fig. 37-17b. For this reason, a tuned

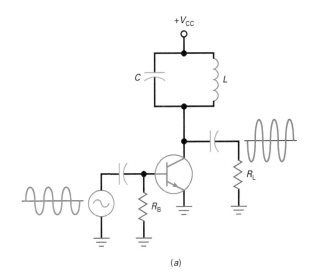

(a)

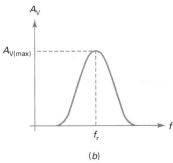

(b)

Figure 37-17 (a) Tuned class C amplifier. (b) Voltage gain versus frequency.

class C amplifier is always intended to amplify a narrow band of frequencies. This makes it ideal for amplifying radio signals because each station or channel is assigned a narrow band of frequencies on both sides of a center frequency.

In operation, the input signal drives the emitter diode, and the amplified current pulses drive the resonant tank circuit.

Only the positive peaks of the input signal can turn on the emitter diode. For this reason, the collector current flows in brief pulses.

Filtering the Harmonics

Previous chapters discussed the concept of harmonics. The basic idea is this: Nonsinusoidal waveforms like the collector current pulses are rich in **harmonics,** multiples of the input frequency. In other words, the pulses are equivalent to a group of sine waves with frequencies of $f, 2f, 3f, \ldots, nf$.

The resonant tank circuit of Fig. 37-17 has a high impedance only at the fundamental frequency f. This produces a large voltage gain at the fundamental frequency. On the other hand, the tank circuit has a very low impedance to the higher harmonics, producing very little voltage gain. This is why the voltage across the resonant tank looks almost like a pure sine wave. Since all higher harmonics are filtered, only the fundamental frequency appears across the tank circuit.

Class C Operation

Refer to Fig. 37-18. The circuit has a resonant frequency of

$$f_r = \frac{1}{2\pi \sqrt{(2 \, \mu H)(470 \, pF)}} = 5.19 \text{ MHz}$$

If the input signal has this frequency, the tuned class C circuit will amplify the input signal.

In Fig. 37-18, the input signal has a peak-to-peak value of 10 V. When the input signal goes to $+5$ V, the emitter-base junction is forward-biased so that the transistor conducts. The 0.01 μF capacitor charges up to 5 V less the forward base-emitter drop of 0.7 V, or 4.3 V. When the input signal goes in the negative direction, the -5-V peak of the input adds to the 4.3-V charge on the capacitor, putting the bias of -9.3 V on the base. As you can see then, the voltage at the base varies from $+0.7$ to -9.3 V. We say that the signal is negatively clamped at the base of the transistor. The average base voltage is -4.3 V. In any case, the transistor only conducts for a short period of time when the base goes positive. The collector current is a series of pulses that flow for less than 180° of the input signal.

The pulses of the collector current cause the parallel resonant circuit in the collector lead to oscillate at the resonant frequency, thereby providing a complete sine wave at the collector and across the load.

The collector signal is inverted because of the CE connection. The dc or average voltage of the collector waveform is $+15$ V, the supply voltage. Therefore, the peak-to-peak collector voltage is 30 V. This voltage is capacitively coupled to the load resistance. The final output voltage has a positive peak of $+15$ V and a negative peak of -15 V.

Bandwidth

A tuned class C amplifier is usually a narrowband amplifier. The input signal in a class C circuit is amplified to get large output power with an efficiency approaching 100 percent. The **bandwidth (BW)** of a resonant circuit is defined as

$$BW = f_2 - f_1 \tag{37-9}$$

where f_1 = lower half-power frequency
f_2 = upper half-power frequency

The half-power frequencies are identical to the frequencies at which the voltage gain equals 0.707 times the maximum gain, as shown in Fig. 37-19. The smaller BW is, the narrower the bandwidth of the amplifier.

With Formula (37-9), it is possible to derive this new relation for bandwidth,

$$BW = \frac{f_r}{Q} \tag{37-10}$$

where Q is the quality factor of the circuit. Formula (37-10) says that the bandwidth is inversely proportional to Q. The higher the Q of the circuit, the smaller the bandwidth.

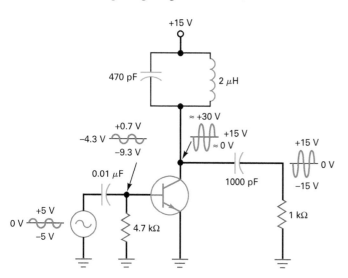

Figure 37-18 An example of a class C amplifier.

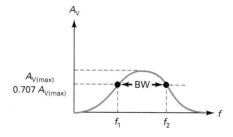

Figure 37-19 Bandwidth.

Class C amplifiers almost always have a circuit Q that is greater than 10. This means that the bandwidth is less than 10% of the resonant frequency. For this reason, class C amplifiers are narrowband amplifiers. The output of a narrowband amplifier is a large sinusoidal voltage at resonance with a rapid drop-off above and below resonance.

Stage Efficiency

The dc collector current depends on the conduction angle. For a conduction angle of 180° (a half-wave signal), the average or dc collector current is $I_{C(sat)}/\pi$. For smaller conduction angles, the dc collector current is less than this. The dc collector current is the only current drain in a class C amplifier because it has no biasing resistors.

In a class C amplifier, most of the dc input power is converted into ac load power because the transistor and coil losses are small. For this reason, a class C amplifier has high stage efficiency.

The optimum stage efficiency varies with conduction angle. When the angle is 180°, the stage efficiency is 78.5%, the theoretical maximum for a class B amplifier. When the conduction angle decreases, the stage efficiency increases. As indicated, class C has a maximum efficiency of 100%, approached at very small conduction angles.

Inhancement-mode MOSFETs are also widely used as class C amplifiers. Power levels of several hundred watts to over 1 kW can be achieved with bipolar and MOSFET class C amplifiers.

37.6 Switching Amplifiers

In the quest for more efficient amplifiers, the switching amplifier has emerged as the clear winner over more traditional class A and AB amplifiers. Switching amplifiers use transistor switches, mainly MOSFETs, to provide linear amplification for audio amplifications. And switching amplifiers are also used for high-efficiency RF amplification. Switching amplifiers are designated class D for audio and classes E and F for RF. In all cases, efficiencies of 70% or greater are commonplace. Some IC switching amplifiers have efficiencies of up to 95%. These are ideal for battery-powered devices as they not only reduce the heat produced but also permit smaller circuits.

Class D Amplifier

The class B or class AB amplifier has been the main choice of many designers for audio amplifiers. This linear amplifier configuration has been able to provide the necessary conventional performance and cost requirements. Now, products such as LCD TVs, auto sound systems, and laptop PCs are driving the necessity for greater power output while maintaining or reducing the form-factor, without increasing costs. Portable powered devices, such as iPod music players, cell phones, and notebook PCs, are demanding higher circuit efficiencies. Due to very high efficiency and low heat dissipation, the class D amplifier is now mostly replacing the class AB amplifiers in many applications.

Instead of being biased for linear operation, a **class D amplifier** uses output transistors operated as switches. This enables each transistor to either be in a cutoff or saturated mode. When in cutoff, its current is zero. When it is saturated, the voltage across it is low. In each mode, its power dissipation is very low. This concept increases the circuit efficiency, therefore, requires less power from the power supply and enables the use of smaller heat sinks for the amplifier.

A basic class D amplifier is shown in Fig. 37-20. The amplifier consists of a comparator op amp driving two MOSFETs

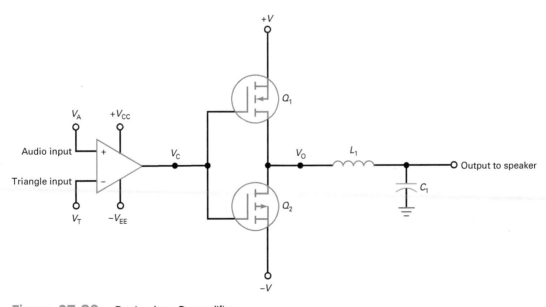

Figure 37-20 Basic class D amplifier.

operating as switches. The comparator has two input signals: one signal is the audio signal V_A, and the other input is a triangle wave V_T with a much higher frequency. The voltage value out of the comparator, V_C, will be approximately at either $+V_{CC}$ or $-V_{EE}$. When $V_A > V_T$, $V_C = +V_{CC}$. When $V_A < V_T$, $V_C = -V_{EE}$.

The comparator's positive or negative output voltage drives two complementary common-source MOSFETs. When V_C is positive, Q_1 is switched on and Q_2 is off. When V_C is negative, Q_2 is switched on and Q_1 is off. The output voltage of each transistor will be slightly less than their +V and −V supply values. Both L_1 and C_1 act as a low-pass filter. When their values are properly chosen, this filter passes the average value of the switching transistors' output to the speaker. If the audio input signal V_A were zero, V_O would be a symmetrical square wave with an average value of zero volts.

To illustrate the operation of this circuit, examine Fig. 37-21. A 1-kHz sine wave is applied to the input at V_A, and a 20-kHz triangle wave is applied to input V_T. In practice, the triangle-wave input frequency would be many times higher than in this illustration. A frequency of 300 kHz to 3 MHz is used. The frequency should be as high as possible compared to the cutoff frequency, f_c, of L_1C_1 for minimum output distortion. Also, note that the maximum voltage of V_A is at approximately 70% of V_T.

The resulting output V_O of the switching transistors is a **pulse-width-modulated (PWM)** waveform. With PWM, the pulse frequency stays the same but the duty cycle is varied. The duty cycle of the waveform produces an output whose average value follows the audio input signal. This is shown in Fig. 37-22.

More sophisticated class D amplifiers use a MOSFET H-bridge circuit configuration for the switching devices

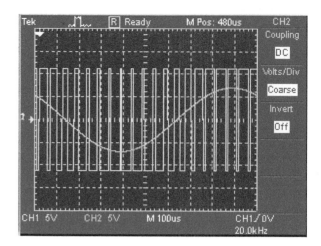

Figure 37-22 Output waveform following the input.

and incorporate active low-pass filters. Resulting efficiencies can reach upwards from 85–90%, even at lower power levels. This exceeds the efficiency of the class AB amplifier, whose efficiency reaches a theoretical maximum of 78% at high-output levels and is much less efficient at lower power levels.

Class D amplifiers are widely available in IC form. Power levels to several watts are common. Dual-amplifier ICs are used in stereo amplifier products like MP3 players. Discrete power MOSFETs are used to achive higher power levels to several hundred watts for outs sound systems.

Class E and F Amplifiers

Class E and F switching amplifiers are used in RF applications. Either bipolar transistors or MOSFETs can be used, with MOSFETs being the preferred type in most modern applications. Class E and F amplifiers are similar in operation to a class C amplifier. Class C amplifiers work with truncated half-sine pulses, but class E and F amplifiers work with square-wave pulses. All these circuits use a tuned circuit in the output to filter out harmonics and produce a sine wave into the load.

Figure 37-23a shows a MOSFET class E amplifier. The input is a square wave that turns the transistor off and on at the operating frequency. The square-wave output is sent to a series resonant circuit filter and the load. The resonant circuit is an effective filter for the many harmonics produced. Capacitor C_1 is used to assist the transistor in switching off and on faster. It is during the switching period that most of the heat is generated.

A class F RF power amplifier is shown in Fig. 37-23b. Again, the basic circuit is a MOSFET switch. A parallel resonant circuit across the load filters out the harmonics to

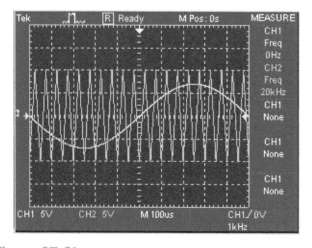

Figure 37-21 Input waveforms.

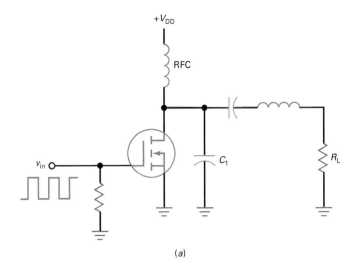

(a)

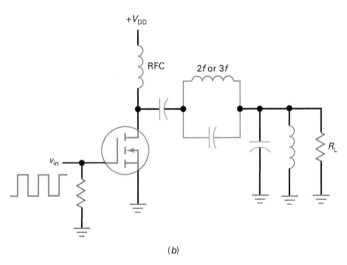

(b)

Figure 37-23 Class E and F switching amplifiers. (a) Class E. (b) Class F.

provide a sine-wave output. The other parallel resonant circuit, made up of C_1 and L_1, is set to resonate at two or three times the operating frequency. This harmonic resonator helps improve switching times to boost efficiency.

Both class E and F amplifiers can achieve efficiencies of 70 to 90% at frequencies to several hundred megahertz. Power levels can be as high as several hundred watts with the appropriate transistors.

37.7 Transistor Power Rating

The temperature at the collector junction or drain places a limit on the allowable power dissipation, P_D. Depending on the transistor type, a temperature in the range of 150 to 200°C will destroy the transistor. Data sheets specify this maximum temperature as $T_{J(max)}$.

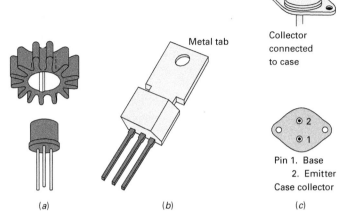

Figure 37-24 (a) Push-on heat sink. (b) Power-tab transistor. (c) Power transistor with collector connected to case.

Ambient Temperature

The heat produced at the junction passes through the transistor case (metal or plastic housing) and radiates to the surrounding air. The temperature of this air, known as the *ambient temperature,* is around 25°C, but it can get much higher on hot days. Also, the ambient temperature may be much higher inside a piece of electronic equipment.

Heat Sinks

One way to increase the power rating of a transistor is to get rid of the heat faster. This is why heat sinks are used. If we increase the surface area of the transistor case, we allow the heat to escape more easily into the surrounding air. Look at Fig. 37-24a. When this type of heat sink is pushed onto the transistor case, heat radiates more quickly because of the increased surface area of the fins.

Figure 37-24b shows the power-tab transistor. The metal tab provides a path out of the transistor for heat. This metal tab can be fastened to the chassis of electronics equipment. Because the chassis is a massive heat sink, heat can easily escape from the transistor to the chassis.

Large power transistors like Fig. 37-24c have the collector connected directly to the case to let heat escape as easily as possible. The transistor case is then fastened to the chassis. To prevent the collector from shorting to the chassis ground, a thin insulating washer and a thermal conductive paste are used between the transistor case and the chassis. The important idea here is that heat can leave the transistor more rapidly, which means that the transistor has a higher power rating at the same ambient temperature.

Class G and Class H Amplifiers

Class G and H amplifiers are special versions of class AB power amplifiers that are designed for greater efficiency. *Efficiency* (η) is the ratio of the output power of the amplifier to the total dc power used to power the amplifier:

$$\eta = P_o/P_{dc}$$

Recall that a class A power amplifier has a maximum efficiency of 25% with a resistive load. This occurs when the amplifier is supplying its maximum power. At lower output levels, the efficiency is considerably less. At the lower, less than maximum, settings, little power is used, so efficiency is not a major consideration. However, the power used is a function of the dc supply voltage.

A class B amplifier is more efficient with a maximum efficiency of 78.5% at full power. Adding minimal forward bias makes the class B amplifier a class AB amplifier, so that it is more linear. Most high-power audio and low-power, battery-operated amplifiers operate class AB for best linearity and minimal distortion. Their efficiency is less than the 78.5% but still above the 60% level at maximum power. At lower power levels, the efficiency is significantly less, so that the amplifier transistors dissipate and waste a considerable amount of power. Furthermore, large heat sinks and ventilation are required to get rid of the heat and provide reliable operation. The solution to this problem is to use a class G or class H amplifier.

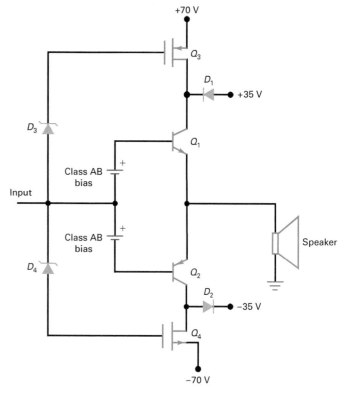

Figure S37-1 A class G amplifier.

Class G Amplifier

Figure S37-1 shows the concept of a class G amplifier. It uses the familiar complementary symmetry push-pull design. It is a class AB amplifier; however, the complex forward-biasing arrangement is not shown. Note that the bipolar transistors Q_1 and Q_2 form the class AB amplifier. The MOSFETs Q_3 and Q_4 are switches. Also note that there are two sets of dc supply voltages, ±70 V and ±35 V.

The basic idea is that at low signal levels, the amplifier only operates from the lower supply voltages. The MOSFET switches are off at this time. The lower supply voltages minimize the power dissipation while fully supplying the desired output power. However, at higher input levels, the MOSFETs switch on and connect the higher supply voltages to supply the higher output power demanded of the higher input. Diodes D_1 and D_2 keep the two sets of supply voltages isolated from one another.

The control of the MOSFETs is determined by the input signal level. When a desired higher-level input is reached, zener diodes D_3 and D_4 conduct, causing the MOSFETs to turn on. The MOSFET control can also come from the output rather than directly from the input or some separate control circuit driven by the input.

The improved efficiency means that smaller or no heat sinks are required on the output power transistors. Many professional high-power audio power amplifiers use the class G method, making their size smaller because of the reduced need for larger heat sinks or other cooling methods.

In battery-powered audio devices like MP3 players, class G headphone amplifiers are used to ensure the sound quality of a linear class AB amplifier while improving efficiency to minimize heat dissipation in a portable device and also greatly extending battery life. Class G amplifier ICs are widely available.

Class H Amplifiers

A class H amplifier extends the concept of class G by making the power supplies continuously variable. By making the supply voltages tracking only a few volts above the peak outputs, efficiency is maximized at all signal levels. To do this, the input signal needs to modulate the power supplies, making them continuously variable. This arrangement is shown in the block diagram of Fig. S37-2.

Note: Class G and H amplifiers are more efficient that standard class AB amplifiers but not as efficient as a class D switching amplifier. However, the compromise is a good one, and the filtering needs and noise of the class D amplifier are eliminated.

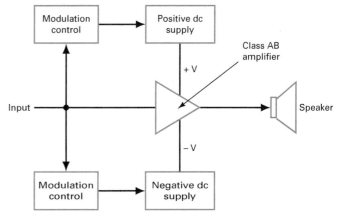

Figure S37-2 A class H amplifier.

? CHAPTER 37 REVIEW QUESTIONS

1. For class B operation, the collector or drain current flows for
  a. the whole cycle.
  b. half the cycle.
  c. less than half a cycle.
  d. less than a quarter of a cycle.

2. Power amplifiers *not* used in which of the following?
  a. Small-signal amplification.
  b. Speaker drivers.
  c. Motor drivers.
  d. Transmitter amplifiers.

3. An audio amplifier operates in the frequency range of
  a. 0 to 20 Hz.
  b. 20 Hz to 2 kHz.
  c. 20 to 20 kHz.
  d. Above 20 kHz.

4. A tuned RF amplifier is
  a. narrowband.
  b. wideband.
  c. direct-coupled.
  d. a dc amplifier.

5. The first stage of a preamp is
  a. a tuned RF stage.
  b. large signal.
  c. small signal.
  d. a dc amplifier.

6. An inductor in the collector or drain lead of a class A amplifier is used to
  a. reduce distortion.
  b. double output power.

  c. improve efficiency.
  d. reduce power consumption.

7. An emitter or source follower is a power amplifier because it
  a. has a high voltage gain.
  b. can sustain higher dc supply voltages.
  c. has a high input impedance and low output impedance.
  d. uses transistors with higher gains.

8. The most common name for a class AB amplifier with no transformers is
  a. emitter-follower amplifier.
  b. class B amplifier.
  c. diode-biased amplifier.
  d. complementary symmetry amplifier.

9. Push-pull is almost always used with
  a. class A.
  b. class B.
  c. class C.
  d. all of the above.

10. One advantage of a class B push-pull amplifier is
  a. no quiescent current drain.
  b. maximum efficiency of 78.5%.
  c. greater efficiency than class A.
  d. all of the above.

11. Class C amplifiers are almost always
  a. transformer-coupled between stages.
  b. operated at audio frequencies.
  c. tuned RF amplifiers.
  d. wideband.

12. The input signal of a class C amplifier
    a. is negatively clamped at the base.
    b. is amplified and inverted.
    c. produces brief pulses of collector current.
    d. all of the above.

13. The collector current of a class C amplifier
    a. is an amplified version of the input voltage.
    b. has harmonics.
    c. is negatively clamped.
    d. flows for half a cycle.

14. The bandwidth of a class C amplifier decreases when the
    a. resonant frequency increases.
    b. Q increases.
    c. X_L decreases.
    d. load resistance decreases.

15. The transistor dissipation in a class C amplifier decreases when the
    a. resonant frequency increases.
    b. coil Q increases.
    c. load resistance decreases.
    d. capacitance increases.

16. The power rating of a transistor can be increased by
    a. raising the temperature.
    b. using a heat sink.
    c. using a derating curve.
    d. operating with no input signal.

17. Which class of amplifier produces the least signal distortion?
    a. A.
    b. B.
    c. C.
    d. D.

18. The maximum efficiency of a class AB amplifier is
    a. 25%.
    b. 50%.
    c. 78.5%.
    d. Over 90%.

19. The quiescent collector current is the same as the
    a. dc collector current.
    b. ac collector current.
    c. total collector current.
    d. voltage divider current.

20. The main advantage of a bridged amplifier is
    a. greater efficiency.
    b. improved frequency response.
    c. higher output power with lower supply voltage.
    d. no heat sinks are needed.

21. Most modern power amplifiers use
    a. MOSFETs.
    b. bipolar transistors.

22. In a class A amplifier, the collector current flows for
    a. less than half the cycle.
    b. half the cycle.
    c. less than the whole cycle.
    d. the entire cycle.

23. With class A, the output signal should be
    a. unclipped.
    b. clipped on positive voltage peak.
    c. clipped on negative voltage peak.
    d. clipped on negative current peak.

24. Most audio power amplifiers are in IC form.
    a. True.
    b. False.

25. The power gain of an amplifier is often expressed in dB. What is the dB power gain of an amplifier with a power gain of 400?
    a. 10 dB.
    b. 26 dB.
    c. 52 dB.
    d. 400 dB.

26. The power gain of an amplifier
    a. is the same as the voltage gain.
    b. is smaller than the voltage gain.
    c. equals output power divided by input power.
    d. equals load power.

27. Heat sinks reduce the
    a. transistor power.
    b. ambient temperature.
    c. junction temperature.
    d. collector current.

28. When the ambient temperature increases, the maximum transistor power rating
    a. decreases.
    b. increases.
    c. remains the same.
    d. none of the above.

29. If the load power is 300 mW and the dc power is 1.5 W, the efficiency is
    a. 0%.
    b. 2%.
    c. 3%.
    d. 20%.

30. Switching amplifiers are used for
    a. audio.
    b. RF.
    c. both *a* and *b*.
    d. none of the above.

31. Class D amplifiers operated by
    a. clipping off all amplitude variations of the input.
    b. converting the input signal to a PWM signal.

c. using a tuned circuit to restore the signal.

d. filtering out the harmonics.

32. Class E and F amplifiers amplify

    a. audio.

    b. RF.

    c. Both *a* and *b*.

    d. none of the above

33. Most IC power amplifiers must have

    a. higher-operating voltages.

    b. higher gain.

    c. cooling fans.

    d. heat sinks.

34. The maximum efficiency of a class B push-pull amplifier is

    a. 25%.

    b. 50%.

    c. 78.5%.

    d. 100%.

35. A small quiescent current is necessary with a class AB push-pull amplifier to avoid

    a. crossover distortion.

    b. destroying the compensating diodes.

    c. excessive current drain.

    d. loading the driver stage.

Chapter 38

Oscillators and Frequency Synthesizers

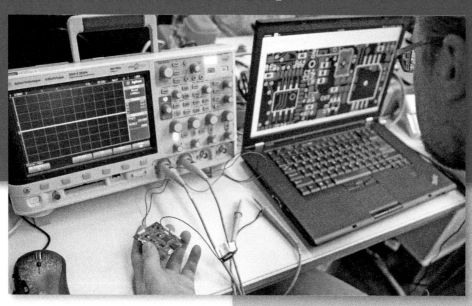

Oscillators are signal sources. They are circuits that generate the sine waves, square waves, and other signals processed by electronic equipment. Virtually all electronic equipment contains at least one oscillator, and many devices have two or more. Oscillators produce signals from a fraction of a hertz to well over 100 GHz and everything in between. Low-frequency oscillators generate signals at frequencies less than 1 MHz using resistor-capacitor (*RC*) networks to determine their operating frequency. Above about 1 MHz, oscillators use either inductor-capacitor (*LC*) circuits or quartz crystals to set the frequency. In this chapter you will learn about the basic *RC* and *LC* oscillator types. Also introduced in the chapter is the phase-locked loop (PLL), which has an oscillator at its heart. The PLL is widely used to create frequency synthesizers that produce precise frequency signals for use in communications equipment.

Learning Outcomes

After studying this chapter, you should be able to:

> *Explain* loop gain and phase and how they relate to sinusoidal oscillators.

> *Describe* the operation of a Wien-bridge *RC* sinusoidal oscillator.

> *Describe* the operation of several *LC* sinusoidal oscillators.

> *Explain* how crystal-controlled oscillators work.

> *Discuss* the 555 timer IC, its modes of operation, and how it is used as an oscillator.

> *Explain* the operation of phase-locked loops.

> *Name* two types of frequency synthesizers and *explain* how they work.

38.1 Theory of Sinusoidal Oscillation

To build a sinusoidal oscillator, we use an amplifier with positive feedback. The idea is to use the feedback signal in place of the input signal. If the feedback signal is large enough and has the correct phase, there will be an output signal even though there is no external input signal.

Loop Gain and Phase

Figure 38-1a shows an ac voltage source driving the input terminals of an amplifier. The amplified output voltage is

$$v_{out} = A_v(v_{in})$$

This voltage drives a feedback circuit that is usually a resonant circuit. Because of this, we get maximum feedback at one frequency. In Fig. 38-1a, the feedback voltage returning to point x is given by

$$v_f = A_v B(v_{in})$$

where B is the feedback fraction.

If the phase shift through the amplifier and feedback circuit is equivalent to $0°$, $A_v B(v_{in})$ is in phase with v_{in}.

Suppose we connect point x to point y and simultaneously remove voltage source v_{in}. Then the feedback voltage $A_v B(v_{in})$ drives the input of the amplifier, as shown in Fig. 38-1b.

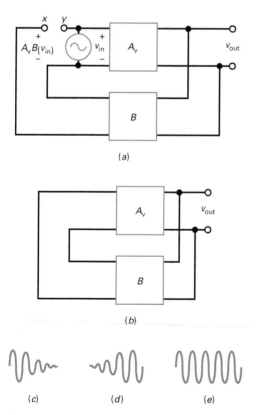

(a)

(b)

(c)   (d)   (e)

Figure 38-1 (a) Feedback voltage returns to point x. (b) Connecting points x and y. (c) Oscillations die out. (d) Oscillations increase. (e) Oscillations are fixed in amplitude.

What happens to the output voltage? If $A_v B$ is less than 1, $A_v B(v_{in})$ is less than v_{in} and the output signal will die out, as shown in Fig. 38-1c. However, if $A_v B$ is greater than 1, $A_v B(v_{in})$ is greater than v_{in} and the output voltage builds up (Fig. 38-1d). If $A_v B$ equals 1, then $A_v B(v_{in})$ equals v_{in} and the output voltage is a steady sine wave like the one in Fig. 38-1e. In this case, the circuit supplies its own input signal.

In any oscillator the loop gain $A_v B$ is greater than 1 when the power is first turned on. A small starting voltage is applied to the input terminals, and the output voltage builds up, as shown in Fig. 38-1d. After the output voltage reaches a certain level, $A_v B$ automatically decreases to 1, and the peak-to-peak output becomes constant (Fig. 38-1e).

Starting Voltage Is Thermal Noise

Where does the starting voltage come from? Every resistor contains some free electrons. Because of the ambient temperature, these free electrons move randomly in different directions and generate a noise voltage across the resistor. The motion is so random that it contains frequencies to over 1000 GHz. You can think of each resistor as a small ac voltage source producing all frequencies.

In Fig. 38-1b, here is what happens: When you first turn on the power, the only signals in the system are the noise voltages generated by the resistors. These noise voltages are amplified and appear at the output terminals. The amplified noise, which contains all frequencies, drives the resonant feedback circuit. By deliberate design, we can make the loop gain greater than 1 and the loop phase shift equal to $0°$ at the resonant frequency. Above and below the resonant frequency, the phase shift is different from $0°$. As a result, oscillations will build up only at the resonant frequency of the feedback circuit.

$A_v B$ Decreases to Unity

There are two ways in which $A_v B$ can decrease to 1. Either A_v can decrease or B can decrease. In some oscillators, the signal is allowed to build up until clipping occurs because of saturation and cutoff. This is equivalent to reducing voltage gain A_v. In other oscillators, the signal builds up and causes B to decrease before clipping occurs. In either case, the product $A_v B$ decreases until it equals 1.

Here are the key ideas behind any feedback oscillator:

1. Initially, loop gain $A_v B$ is greater than 1 at the frequency where the loop phase shift is $0°$.

2. After the desired output level is reached, $A_v B$ must decrease to 1 by reducing either A_v or B.

38.2 The Wien-Bridge Oscillator

The **Wien-bridge oscillator** is the standard oscillator circuit for low to moderate frequencies, in the range of 5 Hz to about 1 MHz. It is almost always used in commercial

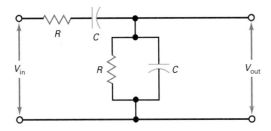

Figure 38-2 Lead-lag circuit.

audio generators and is usually preferred for other low-frequency applications. An *RC* network is used for the feedback circuit.

Lead-Lag Circuit

The Wien-bridge oscillator uses a resonant feedback circuit called a **lead-lag circuit** (Fig. 38-2). At very low frequencies, the series capacitor appears open to the input signal, and there is no output signal. At very high frequencies, the shunt capacitor looks shorted, and there is no output. In between these extremes, the output voltage reaches a maximum value (see Fig. 38-3a). The frequency where the output is maximum is the **resonant frequency, f_r**. At this frequency, the feedback fraction B reaches a maximum value of $\frac{1}{3}$.

Figure 38-3b shows the phase angle of the output voltage versus input voltage. At very low frequencies, the phase angle is positive (leading). At very high frequencies, the phase angle is negative (lagging). At the resonant frequency, the phase shift is 0°. Figure 38-3c shows the phasor diagram of the input and output voltages. The tip of the phasor can lie anywhere on the dashed circle. Because of this, the phase angle may vary from +90° to −90°.

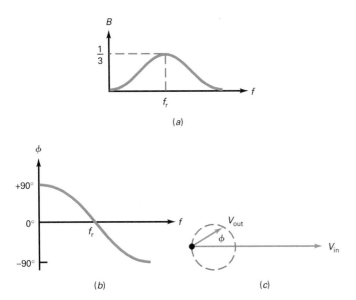

Figure 38-3 (a) Voltage gain. (b) Phase response. (c) Phasor diagram.

The lead-lag circuit of Fig. 38-2 acts like a resonant circuit. At the resonant frequency, f_r, the feedback fraction B reaches a maximum value of $\frac{1}{3}$, and the phase angle equals 0°. Above and below the resonant frequency, the feedback fraction is less than $\frac{1}{3}$, and the phase angle no longer equals 0°.

Formula for Resonant Frequency

The resonant frequency of the circuit in Fig. 38-2 is

$$f_r = \frac{1}{2\pi RC} \tag{38-1}$$

At this frequency the output of this circuit is $\frac{1}{3}$ the input.

How It Works

Figure 38-4a shows a Wien-bridge oscillator. It uses positive and negative feedback because there are two paths for feedback. There is a path for positive feedback from the output through the lead-lag circuit to the noninverting input. There is also a path for negative feedback from the output through the voltage divider to the inverting input. Note in figure 38-4 that the *RC* network forms one side of a bridge circuit, while the $2R' = R'$ voltage divider makes up the other side of the bridge. The bridge is balanced at the resonant frequency.

When the circuit is initially turned on, there is more positive feedback than negative feedback. This allows the oscillations to build up, as previously described. After the output signal reaches a desired level, the negative feedback becomes large enough to reduce loop gain A_vB to 1.

Here is why A_vB decreases to 1: At power-up, the tungsten lamp has a low resistance, and the negative feedback is small. For this reason, the loop gain is greater than 1, and the oscillations can build up at the resonant frequency. As the oscillations build up, the tungsten lamp heats slightly and its resistance increases. In most circuits, the current through the lamp is not enough to make the lamp glow, but it is enough to increase the resistance.

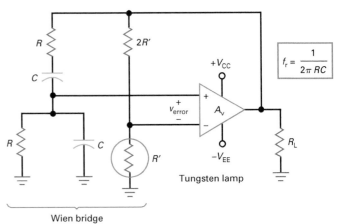

Figure 38-4 Wien-bridge oscillator.

At some high output level, the tungsten lamp has a resistance of exactly R'. At this point, the closed-loop voltage gain from the noninverting input to the output decreases to

$$A_{v(CL)} = \frac{2R'}{R'} + 1 = 3$$

Since the lead-lag circuit has a B of $\frac{1}{3}$, the loop gain is

$$A_{v(CL)}B = 3(\frac{1}{3}) = 1$$

When the power is first turned on, the resistance of the tungsten lamp is less than R'. As a result, the closed-loop voltage gain from the noninverting input to the output is greater than 3 and $A_{v(CL)}B$ is greater than 1.

As the oscillations build up, the peak-to-peak output becomes large enough to increase the resistance of the tungsten lamp. When its resistance equals R', the loop gain $A_{v(CL)}B$ is exactly equal to 1. At this point, the oscillations become stable, and the output voltage has a constant peak-to-peak value.

EXAMPLE 38-1

Calculate the minimum and maximum frequencies in Fig. 38-5. The two variable resistors are *ganged*, which means that they change together and have the same value for any wiper position.

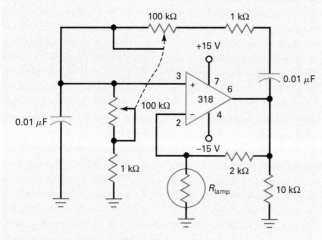

Figure 38-5 Example.

Answer:
With Formula (38-1), the minimum frequency of oscillation is

$$f_r = \frac{1}{2\pi(101\ k\Omega)(0.01\ \mu F)} = 158\ Hz$$

The maximum frequency of oscillation is

$$f_r = \frac{1}{2\pi(1\ k\Omega)(0.01\ \mu F)} = 15.9\ kHz$$

38.3 The Colpitts Oscillator

Although it is superb at low frequencies, the Wien-bridge oscillator is not suited to high frequencies (well above 1 MHz). Above 1 MHz, most oscillators use LC circuits or quartz crystals.

LC Oscillators

One way to produce high-frequency oscillations is with an LC oscillator, a circuit that can be used for frequencies between 1 MHz and over 1 GHz. This frequency range is beyond the f_{unity} of most op amps. This is why a bipolar junction transistor or an FET is typically used for the amplifier. With an amplifier and LC tank circuit, we can feed back a signal with the right amplitude and phase to sustain oscillations.

CE Connection

The most popular LC oscillator is called a Colpitts oscillator. Figure 38-6 shows a **Colpitts oscillator.** The voltage divider bias sets up a quiescent operating point. The RF choke has a very high inductive reactance, so it appears open to the ac signal. The circuit has a low-frequency voltage gain of r_c/r'_e, where r_c is the ac collector resistance. Because the RF choke appears open to the ac signal, the ac collector resistance is primarily the ac resistance of the resonant tank circuit. This ac resistance has a maximum value at resonance.

You will encounter many variations of the Colpitts oscillator. One way to recognize a Colpitts oscillator is by the capacitive voltage divider formed by C_1 and C_2. It produces the feedback voltage necessary for oscillations. In other kinds of oscillators, the feedback voltage is produced by transformers, inductive voltage dividers, and so on.

AC Equivalent Circuit

Figure 38-7 is a simplified ac equivalent circuit for the Colpitts oscillator. The circulating or loop current in the tank flows through C_1 in series with C_2. Notice that v_{out} equals the ac voltage across C_1. Also, the feedback voltage v_f appears across C_2. This feedback voltage drives the base and

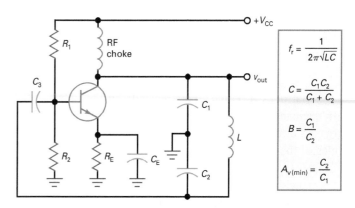

Figure 38-6 Colpitts oscillator.

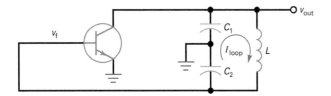

Figure 38-7 Equivalent circuit of Colpitts oscillator.

sustains the oscillations developed across the tank circuit, provided there is enough voltage gain at the oscillation frequency. Since the emitter is at ac ground, the circuit is a CE connection.

Resonant Frequency

Most LC oscillators use tank circuits with a Q greater than 10. Because of this, we can calculate the approximate resonant frequency as

$$f_r = \frac{1}{2\pi\sqrt{LC}} \qquad (38\text{-}2)$$

This is accurate to better than 1% when Q is greater than 10.

The capacitance to use in Formula (38-2) is the equivalent capacitance through which the circulating current passes. In the Colpitts tank circuit of Fig. 38-7, the circulating current flows through C_1 in series with C_2. Therefore, the equivalent capacitance is

$$C = \frac{C_1 C_2}{C_1 + C_2} \qquad (38\text{-}3)$$

For instance, if C_1 and C_2 are 100 pF each, you would use 50 pF in Formula (38-2).

Starting Condition

The required starting condition for any oscillator is $A_v B > 1$ at the resonant frequency of the tank circuit. This is equivalent to $A_v > 1/B$. In Fig. 38-7, the output voltage appears across C_1 and the feedback voltage appears across C_2. The feedback fraction in this type of oscillator is given by

$$B = \frac{C_1}{C_2} \qquad (38\text{-}4)$$

For the oscillator to start, the minimum voltage gain is

$$A_{v(min)} = \frac{C_2}{C_1} \qquad (38\text{-}5)$$

What does A_v equal? This depends on the upper cutoff frequencies of the amplifier. There are base and collector bypass circuits in a bipolar amplifier. If the cutoff frequencies of these bypass circuits are greater than the oscillation frequency, A_v is approximately equal to r_c/r'_e. If the cutoff frequencies are lower than the oscillation frequency, the voltage gain is less than r_c/r'_e and there is additional phase shift through the amplifier.

EXAMPLE 38-2

What is the frequency of oscillation in Fig. 38-6? $C_1 = 0.001\ \mu F$, $C_2 = 0.01\ \mu F$, and $L = 15\ \mu H$. What is the feedback fraction? How much voltage gain does the circuit need to start oscillating?

Answer:
This is a Colpitts oscillator using the CE connection of a transistor. With Formula (38-3), the equivalent capacitance is

$$C = \frac{(0.001\ \mu F)(0.01\ \mu F)}{0.001\ \mu F + 0.01\ \mu F} = 909\ pF$$

The inductance is 15 μH. With Formula (38-2), the frequency of oscillation is

$$f_r = \frac{1}{2\pi\sqrt{(15\ \mu H)(909\ pF)}} = 1.36\ MHz$$

With Formula (38-4), the feedback fraction is

$$B = \frac{0.001\ \mu F}{0.01\ \mu F} = 0.1$$

To start oscillating, the circuit needs a minimum voltage gain of

$$A_{v(min)} = \frac{0.01\ \mu F}{0.001\ \mu F} = 10$$

38.4 Other *LC* Oscillators

The Colpitts oscillator is the most widely used LC oscillator. The capacitive voltage divider in the resonant circuit is a convenient way to develop the feedback voltage. But other kinds of oscillators can also be used.

Hartley Oscillator

Figure 38-8 is an example of the **Hartley oscillator.** When the LC tank is resonant, the circulating current flows through L_1 in series with L_2. The equivalent L to use in Formula (38-2) is

$$L = L_1 + L_2 \qquad (38\text{-}6)$$

In a Hartley oscillator, the feedback voltage is developed by the inductive voltage divider, L_1 and L_2. Since the output

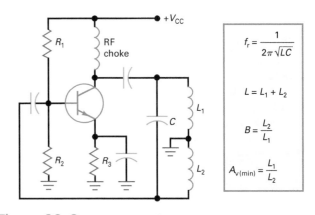

$$f_r = \frac{1}{2\pi\sqrt{LC}}$$

$$L = L_1 + L_2$$

$$B = \frac{L_2}{L_1}$$

$$A_{v(min)} = \frac{L_1}{L_2}$$

Figure 38-8 Hartley oscillator.

voltage appears across L_1 and the feedback voltage appears across L_2, the feedback fraction is

$$B = \frac{L_2}{L_1} \qquad \text{(38-7)}$$

As usual, this ignores the loading effects of the base. For oscillations to start, the voltage gain must be greater than $1/B$.

Often a Hartley oscillator uses a single tapped inductor instead of two separate inductors. Another variation sends the feedback signal to the emitter instead of to the base. Also, you may see an FET used instead of a bipolar junction transistor. The output signal can be either capacitively coupled or link-coupled.

Clapp Oscillator

The **Clapp oscillator** of Fig. 38-9 is a refinement of the Colpitts oscillator. The capacitive voltage divider produces the feedback signal as before. An additional capacitor C_3 is in series with the inductor. Since the circulating tank current flows through C_1, C_2, and C_3 in series, the equivalent capacitance used to calculate the resonant frequency is

$$C = \frac{1}{1/C_1 + 1/C_2 + 1/C_3} \qquad \text{(38-8)}$$

In a Clapp oscillator, C_3 is much smaller than C_1 and C_2. As a result, C is approximately equal to C_3, and the resonant frequency is given by

$$f_r \cong \frac{1}{2\pi\sqrt{LC_3}} \qquad \text{(38-9)}$$

Why is this important? Because C_1 and C_2 are shunted by transistor and stray capacitances. These extra capacitances alter the values of C_1 and C_2 slightly. In a Colpitts oscillator, the resonant frequency therefore depends on the transistor and stray capacitances. But in a Clapp oscillator, the transistor and stray capacitances have no effect on C_3, so the oscillation frequency is more stable and accurate. This is why you occasionally see the Clapp oscillator used.

Integrated *LC* Oscillators

While you will occasionally see discrete transistor oscillators like that described in the previous sections, be aware that most oscillators today are in integrated circuit form. They are integrated on a chip along with amplifiers and other circuits making up a larger IC.

An example of an *LC* oscillator commonly used in radio-frequency ICs is shown in Fig. 38-10*a*. It uses cross-coupled *n*-channel MOSFETs. While M_1 conducts, M_2 is cut off and

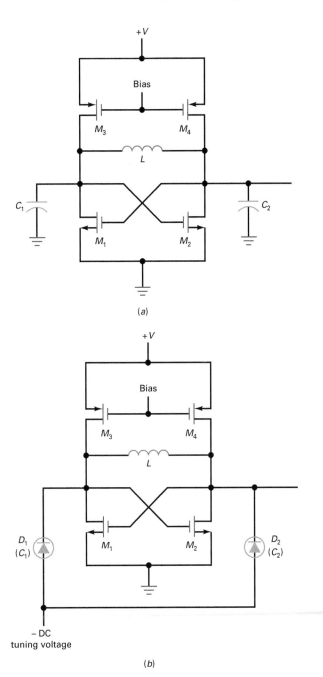

(a)

(b)

Figure 38-10 (a) A typical integrated circuit *LC* oscillator. (b) An integrated *LC* voltage-controlled oscillator whose frequency can be varied by changing the dc reverse bias on the varactor diodes.

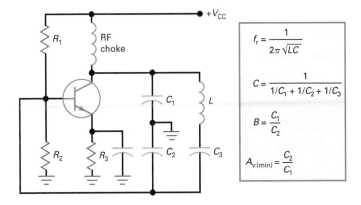

Figure 38-9 Clapp oscillator.

$$f_r = \frac{1}{2\pi\sqrt{LC}}$$

$$C = \frac{1}{1/C_1 + 1/C_2 + 1/C_3}$$

$$B = \frac{C_1}{C_2}$$

$$A_{v(min)} = \frac{C_2}{C_1}$$

vice versa. Both M_3 and M_4 are current sources. Inductor L and capacitors C_1 and C_2 set the frequency of oscillation, f.

$$f = \frac{1}{2\pi\sqrt{LC_t}} \qquad \text{(38-10)}$$

$$C_t = \frac{C_1 C_2}{C_1 + C_2} \qquad \text{(38-11)}$$

The inductor is a spiral pattern of aluminum or other conductive material integrated with the transistors and capacitors on a silicon chip.

There are numerous variations of this circuit with different combinations of inductors and capacitors. One of the most popular is that in Fig. 38-10b. Here the two capacitors have been replaced with varactor diodes D_1 and D_2. These reverse-biased diodes act as capacitors whose values can be changed by varying the amount of reverse dc bias voltage. This allows the frequency to be adjusted with an external dc voltage. This combination creates a voltage-controlled oscillator (VCO). VCOs are widely used for tuning and in phase-locked loops, to be discussed later in this chapter. They are also used to produce frequency modulation (FM) where a voice or data signal varies the frequency of a fixed radio carrier frequency.

38.5 Quartz Crystals

When the frequency of oscillation needs to be accurate and stable, a crystal oscillator is the natural choice. Radio transmitters and receivers require precise and stable frequencies. Microcontrollers and communications and networking equipment use crystal oscillators because they provide an accurate clock frequency.

Piezoelectric Effect

Some crystals found in nature exhibit the **piezoelectric effect.** When you apply an ac voltage across them, they vibrate at the frequency of the applied voltage. Conversely, if you mechanically force them to vibrate, they generate an ac voltage of the same frequency. The main substances that produce the piezoelectric effect are quartz, Rochelle salts, and tourmaline.

Rochelle salts have the greatest piezoelectric activity. For a given ac voltage, they vibrate more than quartz or tourmaline. Mechanically, they are the weakest because they break easily. Rochelle salts have been used to make microphones, headsets, and loudspeakers. Tourmaline shows the least piezoelectric activity but is the strongest of the three. It is also the most expensive. It is occasionally used at very high frequencies.

Quartz is a compromise between the piezoelectric activity of Rochelle salts and the strength of tourmaline. Because it is inexpensive and readily available in nature, quartz is widely used for RF oscillators and filters.

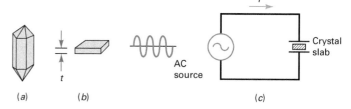

Figure 38-11 (a) Natural quartz crystal. (b) Slab. (c) Input current is maximum at resonance.

Crystal Slab

The natural shape of a quartz crystal is a hexagonal prism with pyramids at the ends (see Fig. 38-11a). To get a usable crystal out of this, a manufacturer slices a rectangular slab out of the natural crystal. Figure 38-11b shows this slab with thickness t. The number of slabs we can get from a natural crystal depends on the size of the slabs and the angle of the cut.

For use in electronic circuits, the slab must be mounted between two metal plates, as shown in Fig. 38-11c. In this circuit the amount of crystal vibration depends on the frequency of the applied voltage. By changing the frequency, we can find resonant frequencies at which the crystal vibrations reach a maximum. Since the energy for the vibrations must be supplied by the ac source, the ac current is maximum at each resonant frequency.

Fundamental Frequency and Overtones

Most of the time, the crystal is cut and mounted to vibrate best at one of its resonant frequencies, usually the **fundamental frequency,** or lowest frequency. Higher resonant frequencies, called *overtones,* are almost exact multiples of the fundamental frequency. As an example, a crystal with a fundamental frequency of 1 MHz has a first overtone of approximately 2 MHz, a second overtone of approximately 3 MHz, and so on.

Since the fundamental frequency is inversely proportional to the thickness, there is a limit to the highest fundamental frequency. The thinner the crystal, the more fragile it becomes and the more likely it is to break when vibrating.

Quartz crystals work well up to 100 MHz on the fundamental frequency. To reach higher frequencies, we can use a crystal that vibrates on overtones. In this way, we can reach frequencies up to 500 MHz.

AC Equivalent Circuit

What does the crystal look like to an ac source? When the crystal of Fig. 38-12a is not vibrating, it is equivalent to a capacitance C_m because it has two metal plates separated by a dielectric. The capacitance C_m is known as the **mounting capacitance.**

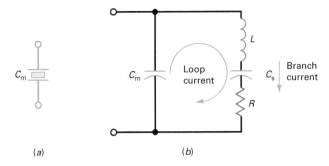

Figure 38-12 (a) Mounting capacitance. (b) AC equivalent circuit of vibrating crystal.

When a crystal is vibrating, it acts like a tuned circuit. Figure 38-12b shows the ac equivalent circuit of a crystal vibrating at its fundamental frequency. Typical values are L in henrys, C_s in fractions of a picofarad, R in hundreds of ohms, and C_m in picofarads. For instance, a crystal can have values such as $L = 3$ H, $C_s = 0.05$ pF, $R = 2$ kΩ, and $C_m = 10$ pF.

Crystals have an incredibly high Q. For the values just given, Q is almost 4000. The Q of a crystal can easily be over 10,000. The extremely high Q of a crystal means that crystal oscillators have a very stable frequency.

By comparison, the L and C values of a Colpitts oscillator have large tolerances, which means that the frequency is less precise.

Series and Parallel Resonance

The *series resonant frequency* f_s of a crystal is the resonant frequency of the *LCR* branch in Fig. 38-12b. At this frequency, the *branch current* reaches a maximum value because L resonates with C_s. The formula for this resonant frequency is

$$f_s = \frac{1}{2\pi \sqrt{LC_s}} \tag{38-12}$$

The *parallel resonant frequency* f_p of the crystal is the frequency at which the circulating or loop current of Fig. 38-12b reaches a maximum value. Since this loop current must flow through the series combination of C_s and C_m, the equivalent parallel capacitance is

$$C_p = \frac{C_m C_s}{C_m + C_s} \tag{38-13}$$

and the parallel resonant frequency is

$$f_p = \frac{1}{2\pi \sqrt{LC_p}} \tag{38-14}$$

In any crystal, C_s is much smaller than C_m. Because of this, f_p is only slightly greater than f_s. When we use a crystal in a circuit, the additional circuit capacitances appear in shunt with C_m. Because of this, the oscillation frequency will lie between f_s and f_p.

Crystal Stability

The frequency of any oscillator tends to change slightly with time. This *drift* is produced by temperature, aging, and other causes. In a crystal oscillator, the frequency drift is very small, typically less than 1 part in 10^6 per day. Stability like this is important in radio equipment because they use quartz-crystal oscillators as the basic timing device.

By putting a crystal oscillator in a temperature-controlled oven, we can get a frequency drift of less than 1 part in 10^{10} per day. A clock with this drift will take 300 years to gain or lose 1 s. Stability like this is needed in frequency and time standards.

38.6 Crystal Oscillators

Virtually all modern electronic products contain a crystal oscillator, and many like cell phones contain several. Frequency precision and stability are important for ensuring that timing operations in computers and other products are accurate and that frequency spectrum assignments in wireless equipment are met. Only a crystal oscillator can provide the kind of precision and stability that has been mandated for most equipment.

While some discrete component crystal oscillators are still found in mainly older equipment, most new ones are integrated circuits packaged with the crystal in separate housing. The result is crystal oscillator modules that are treated as a separate component. You will hear such oscillators called clocks because they maintain a repetitive, precise timing signal.

Crystal Oscillator Circuits

Figure 38-13 shows several popular discrete crystal oscillator circuits. The circuit in Fig. 38-13a is a Colpitts with an emitter-follower design for which the feedback from the emitter goes directly to the capacitor network. A key benefit of this circuit is its low output impedance. The circuit works well up to about 30 MHz. Note that external capacitors may be connected in parallel or series with the crystal to permit minor frequency adjustments. This process is called crystal or frequency "pulling."

Figure 38-13b shows another Colpitts circuit, but here the crystal is connected in series with a varactor diode. The diode's capacitance increases the frequency of oscillation by a small amount. By varying the diode reverse bias, the frequency of oscillation can be change or pulled over a narrow range. The end result is a voltage-controlled oscillator (VCO) or a frequency modulator.

Figure 38-13c is a FET Clapp oscillator. The intention is to improve the frequency stability by reducing the effect of stray capacitances. Figure 38-13d is a circuit called a *Pierce oscillator.* Its main advantage is simplicity.

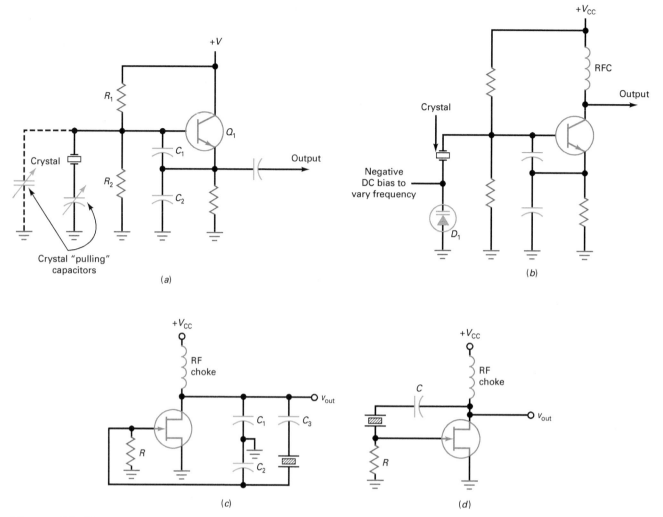

Figure 38-13 Crystal oscillators. (*a*) Emitter-follower Colpitts. (*b*) VCO Colpitts with a varactor. (*c*) Clapp. (*d*) Pierce.

Crystal Oscillator Specifications

Outlined in the section are the key specifications you need to understand in working with crystal oscillators.

Frequency of Operation Crystal oscillators generally operate in the frequency range of approximately 1 to 70 MHz. Special lower-frequency crystals like the popular 32.768 kHz digital watch crystal are also available. The upper-frequency range is limited by the physical thinness of the crystal. That limit has gone from about 30 MHz in past years to about 200 MHz as new manufacturing techniques have been developed. The operating frequency is usually stated at a temperature of 25°C. Higher-frequency oscillators can be obtained by using overtone crystals that take the output to over 200 MHz.

Frequency Accuracy You will also see frequency accuracy called *frequency tolerance,* which is a measure of how close the crystal frequency is to the desired value as determined

by the application. It is often expressed as a percentage deviation from the specified frequency or in parts per million (ppm). For example, an accuracy of ±100 ppm of a 10-MHz crystal oscillator means that the actual frequency could deviate from 10 MHz by ±1000 Hz. To compute the frequency deviation from the ppm value, simply form a fraction of the ppm value with 1 million and multiply by the crystal frequency. An accuracy of 100 ppm translates to a frequency deviation of

$$\frac{100}{1,000,000} \times 10,000,000 = 1000 \text{ Hz}$$

Frequency accuracy may also be expressed as a percentage using the deviation value. For example, a deviation of 1000 Hz would be $1000/10,000,000 = 0.0001 = 10^{-4}$ or 0.01%.

Typical oscillator accuracies range from 1 to 1000 ppm. These are stated at an initial temperature of 25°C.

EXAMPLE 38-3

What is the maximum frequency deviation of a 26-MHz crystal oscillator with an accuracy of 50 ppm? Express the accuracy as a percentage.

Answer:

Maximum deviation $= (50/1{,}000{,}000)26{,}000{,}000 = 1300$ Hz

% accuracy $= 50/26{,}000{,}000 = 1.923 \times 10^{-6}$ or 0.0001923%

Frequency Stability Frequency stability is a measure of how much the frequency deviates from the desired value over a specific temperature range. Common ranges are 0 to 70°C and −40 to +85°C. The stability is also stated in ppm and varies widely depending on the oscillator type from 10 to 1000 ppm.

Output Crystal oscillators are available with different types of output signals. Most are pulse or logic levels, but sine wave and clipped sine outputs are also available. The maximum load is also specified as a fan-out number or as a capacitance in pF.

Operating Voltage Most crystal oscillators operate from 5 Vdc. But newer designs are available to operate from 1.8, 2.5, and 3.3 V.

Phase Noise A critical specification at very high frequencies and in applications requiring exceptional stability is phase noise. This is the rapid, short-term random variation in output frequency. It produces a type of phase or frequency modulation and is also called *jitter*. Phase noise is measured in the frequency domain with a spectrum analyzer and is usually stated in terms of decibels relative to the carrier per hertz (dBc/Hz). A sine-wave output from an oscillator with no phase noise is called the carrier. The phase noise produces sidebands, signals above and below the carrier. The amount of phase noise is expressed as the ratio of the sideband power amplitude (P_s) to the carrier power amplitude (P_c) in decibel form.

$$\text{Phase noise in dBc} = 10 \log P_s/P_c$$

Phase noise is measured at increments from the carrier of 10 kHz or 100 kHz, although other frequency increments down to 10 or 100 Hz are also used. Typical phase noise values are in the −80 dBc to −160 dBc, depending on the frequency increment from the carrier.

Pullability Pullability is a measure of the amount of frequency variation that can be achieved by applying an external control voltage to a voltage-controlled crystal oscillator. It represents the maximum deviation possible. The pullability is usually expressed in ppm. The level of the control voltage is also given, and sometimes a linearity value in percent is provided. Typical dc control voltage values are in the 0- to 5-V range. The linearity of the frequency variation with the control voltage may be an issue.

Packaging There are a large number of different types of crystal oscillator packages. In the past, the metal can packages were the most common, but today these have been overtaken by the newer surface-mount packages. The metal packages commonly have standard DIP through-hole pins. Common surface-mount package sizes are 2.5 by 3.2 mm and 5 by 7 mm. The trend has been to make the packages thinner, as demanded by the cell phone manufacturers.

Types of Crystal Oscillators

There are four basic types of prepackaged crystal oscillators available: crystal oscillators (XO), voltage-controlled crystal oscillators (VCXO), temperature-compensated crystal oscillators (TCXO), and oven-controlled crystal oscillators (OCXO).

XO A crystal oscillator is a quartz crystal packaged with its oscillator circuitry. The frequency is usually fixed. XOs are commonly have a digital output and are used as clock oscillators for microprocessors or other digital circuits. The term *clock oscillator* is appropriate since the oscillator provides accurate timing pulses to the processor and related digital circuits. Typical accuracies range from 10 to several hundred ppm with aging rates from ±1 to ±5 ppm per year.

TCXO A temperature-compensated crystal oscillator is one that incorporates circuitry to compensate for the frequency variations that accompany temperature variations. This results in a far more precise and stable output frequency that is demanded by many applications. Cell phones and two-way radios are common examples. The simplest form of temperature compensation uses a thermistor temperature sensor in a circuit that operates a varactor (voltage-variable capacitor) in a feedback circuit to keep the crystal frequency more constant. Oscillator frequencies are available from about 1 to 60 MHz. Typical stability specifications range from ±0.2 to ±2.5 ppm with aging rates of ±0.5 to ±2 ppm per year. Figure 38-14 shows a packaged TCXO that is typical of those used today.

VCXO A voltage-controlled crystal oscillator is an XO but optimized for external frequency control by way of a dc input. The dc input varies an internal varactor pulling capacitor to provide a narrow range of output frequency adjustment. VCXOs are designed primarily for use in a narrowband phase-locked loop (PLL). Commonly available frequencies range from 1 to 60 MHz. A typical pullability range is from ±10 to ±2000 ppm.

Figure 38-14 A TCXO in a 3.2- by 2.5-mm package.

Source: Courtesy Fox Electronics, Inc.

OCXO An oven-compensated crystal oscillator puts the crystal and sometimes the whole oscillator circuit in a small oven. A dc heating element in a feedback loop keeps the temperature virtually constant, giving a very precise and stable output frequency. It is the best choice for critical applications, such as cellular base stations, telecom, communications networks, and GPS. However, the trade-off is the higher power (generally 1 to 3 W) it draws. The typical stability figure is $\pm 1 \times 10^{-8}$. A typical aging rate is from about ± 0.2 ppm per year to $\pm 2.0 \times 10^{-8}$ per year. Even improved accuracy and stability figures can be obtained with an OCXO that encloses one OCXO inside a second oven.

38.7 Square-Wave, Pulse, and Function Generators

Over the years, hundreds of different discrete component circuits have been developed to generate square waves and other rectangular pulse trains. The predominance of digital circuits in electronics today creates a need for circuits that can serve not only as a clock for microcontroller-based equipment but also for special needs in pulse generation and timing operations. Today, most square waves are generated by integrated circuits, either special circuits or the previously described crystal clock oscillators. This section takes a look at several very popular signal generation ICs.

One of the oldest and most popular integrated circuits is the 555. It is still made by multiple IC manufacturers and is available in both bipolar and CMOS forms. Its main function is to generate pulses and square waves, but because of its versatility, it has found hundreds of other uses. The architecture of the 555 is an excellent example of how basic circuits can be applied to meet a variety of electronic needs.

The 555 can run in either of two modes: **monostable** (one stable state) or **astable** (no stable states). In the monostable mode, it can produce accurate time delays from microseconds to hours. In the astable mode, it can produce rectangular waves with a variable duty cycle.

Monostable Operation Figure 38-15a illustrates monostable operation. Initially, the 555 timer has a low output voltage at which it can remain indefinitely. When the 555 timer receives a *trigger* at point A in time, the output voltage switches from low to high, as shown. The output remains high for a while and then returns to the low state after a time delay of W. The output will remain in the low state until another trigger arrives.

A **multivibrator** is a two-state circuit that has zero, one, or two stable output states. When the 555 timer is used in the monostable mode, it is sometimes called a *monostable multivibrator* because it has only one stable state. It is stable in the low state until it receives a trigger, which causes the output to temporarily change to the high state. The high state, however, is not stable because the output returns to the low state when the pulse ends.

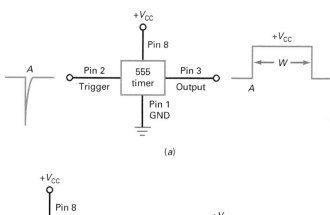

(a)

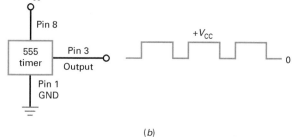

(b)

Figure 38-15 (a) The 555 timer used in monostable (one-shot) mode. (b) The 555 timer used in astable (free-running) mode.

When operating in the monostable mode, the 555 timer is often referred to as a *one shot* because it produces only one output pulse for each input trigger. The duration of this output pulse can be precisely controlled with an external resistor and capacitor.

The 555 timer is an 8-pin IC. Figure 38-15a shows four of the pins. Pin 1 is connected to ground, and pin 8 is connected to the positive supply voltage. The 555 timer will work with any supply voltage between +4.5 and +18 V. The trigger goes into pin 2, and the output comes from pin 3. The other pins, which are not shown here, are connected to external components that determine the pulse width of the output.

Astable Operation The 555 timer can also be connected to run as an *astable multivibrator*. When used in this way, the 555 timer has no stable states, which means that it cannot remain indefinitely in either state. Stated another way, it oscillates when operated in the astable mode and it produces a rectangular output signal.

Figure 38-15b shows the 555 timer used in the astable mode. As you can see, the output is a series of rectangular pulses. Since no input trigger is needed to get an output, the 555 timer operating in the astable mode is sometimes called a *free-running multivibrator*, clock oscillator, or square-wave generator.

Functional Block Diagram Figure 38-16 shows a functional diagram of the 555 timer. This diagram captures all the key ideas we need for our discussion of the 555 timer.

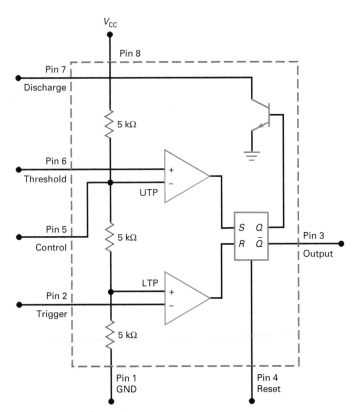

Figure 38-16 Simplified functional block diagram of a 555 timer.

As shown in Fig. 38-16, the 555 timer contains a voltage divider, two comparators, an *RS* flip-flop, and an *npn* transistor. Since the voltage divider has equal resistors, the top comparator has a trip point of

$$UTP = \frac{2V_{CC}}{3} \qquad (38\text{-}15)$$

The lower comparator has a trip point of

$$LTP = \frac{V_{CC}}{3} \qquad (38\text{-}16)$$

In Fig. 38-16, pin 6 is connected to the upper comparator. The voltage on pin 6 is called the *threshold*. This voltage comes from external components not shown. When the *threshold voltage* is greater than the UTP, the upper comparator has a high output.

Pin 2 is connected to the lower comparator. The voltage on pin 2 is called the *trigger*. This is the trigger voltage that is used for the monostable operation of the 555 timer. When the timer is inactive, the trigger voltage is high. When the trigger voltage falls to less than the LTP, the lower comparator produces a high output.

Pin 4 may be used to reset the output voltage to zero. Pin 5 may be used to control the output frequency when the 555 timer is used in the astable mode. In many applications, these two pins are made inactive as follows: Pin 4 is connected to $+V_{CC}$, and pin 5 is bypassed to ground through a capacitor. Later, we discuss how pins 4 and 5 are used in some advanced circuits.

RS Flip-Flop Before we can understand how a 555 timer works with external components, we need to discuss the action of the block that contains, *S*, *R*, *Q*, and $\overline{Q}$. This block is called an *RS flip-flop*, a circuit that has two stable states, set and reset.

The *RS* flip-flop (FF) has two outputs, *Q* and $\overline{Q}$. These are two-state outputs, either low or high voltages. Furthermore, the two outputs are always in opposite states. When *Q* is low, $\overline{Q}$ is high. When *Q* is high, $\overline{Q}$ is low. For this reason, $\overline{Q}$ is called the *complement of Q*. The overbar on $\overline{Q}$ is used to indicate that it is the complement of *Q*.

We can control the output states with the *S* and *R* inputs. If we apply a positive voltage to the *S* input, the FF will be set, *Q* will be high, and $\overline{Q}$ will be low. The high *S* input can then be removed, because the saturated left transistor will keep the right transistor in cutoff.

Similarly, we can apply a positive voltage to the *R* input. This will reset the FF with *Q* low and $\overline{Q}$ high. After this transition has occurred, the high *R* input can be removed because it is no longer needed.

Since the circuit is stable in either of two states, it is sometimes called a **bistable multivibrator.** A bistable multivibrator latches in either of two states. A high *S* input forces *Q* into the high state, and a high *R* input forces *Q* to return to

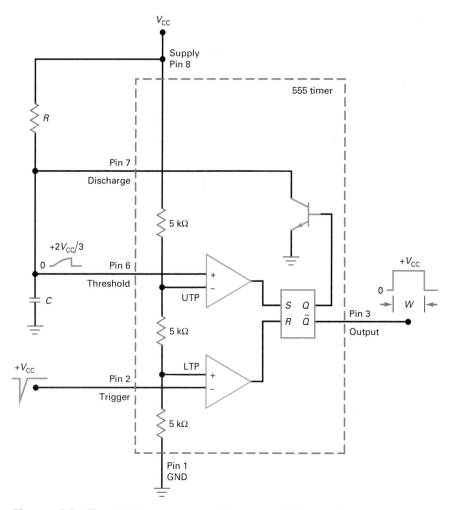

Figure 38-17 555 timer connected for monostable operation.

the low state. The output Q remains in a given state until it is triggered into the opposite state.

Incidentally, the S input is sometimes called the *set input* because it sets the Q output to high. The R input is called the *reset input* because it resets the Q output to low.

Monostable Operation Figure 38-17 shows the 555 timer connected for monostable operation. The circuit has an external resistor R and a capacitor C. The voltage across the capacitor is used for the threshold voltage to pin 6. When the trigger arrives at pin 2, the circuit produces a rectangular output pulse from pin 3.

Initially, the Q output of the RS flip-flop is high. This saturates the transistor and clamps the capacitor voltage at ground. The circuit will remain in this state until a trigger arrives. Because of the voltage divider, the trip points are the same as previously discussed: UTP = $2V_{CC}/3$ and LTP = $V_{CC}/3$.

When the trigger input falls to slightly less than $V_{CC}/3$, the lower comparator resets the flip-flop. Since Q has changed to low, the transistor goes into cutoff, allowing the capacitor to charge. At this time, $\overline{Q}$ has changed to high. The capacitor now charges exponentially as shown. When the capacitor

voltage is slightly greater than $2V_{CC}/3$, the upper comparator sets the flip-flop. The high Q turns on the transistor, which discharges the capacitor almost instantly. At the same instant, $\overline{Q}$ returns to the low state and the output pulse ends. Note that $\overline{Q}$ remains low until another input trigger arrives.

The complementary output $\overline{Q}$ comes out of pin 3. The width of the rectangular pulse depends on how long it takes to charge the capacitor through resistance R. The longer the time constant, the longer it takes for the capacitor voltage to reach $2V_{CC}/3$. In one time constant, the capacitor can charge to 63.2% of V_{CC}. Since $2V_{CC}/3$ is equivalent to 66.7% of V_{CC}, it takes slightly more than one time constant for the capacitor voltage to reach $2V_{CC}/3$. By solving the exponential charging equation, it is possible to derive this formula for the pulse width

$$W = 1.1RC \qquad \text{(38-17)}$$

Figure 38-18 shows the schematic diagram for the monostable 555 circuit as it usually appears. Only the pins and external components are shown. Notice that pin 4 (reset) is connected to $+V_{CC}$. As discussed earlier, this prevents pin 4 from having any effect on the circuit. In some applications, pin 4 may be temporarily grounded to suspend the operation.

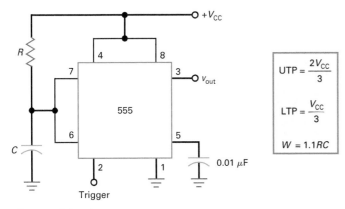

Figure 38-18 Monostable timer circuit.

$$UTP = \frac{2V_{CC}}{3}$$

$$LTP = \frac{V_{CC}}{3}$$

$$W = 1.1RC$$

When pin 4 is taken high, the operation resumes. A later discussion describes this type of reset in more detail.

Pin 5 (control) is a special input that can be used to change the UTP, which changes the width of the pulse. Later, we will discuss *pulse-width modulation,* in which an external voltage is applied to pin 5 to change the pulse width. For now, we will bypass pin 5 to ground as shown. By ac-grounding pin 5, we prevent stray electromagnetic noise from interfering with the operation of the 555 timer.

In summary, the monostable 555 timer produces a single pulse whose width is determined by the external R and C used in Fig. 38-18. The pulse begins with the leading edge of the input trigger. A one-shot operation like this has a number of applications in digital and switching circuits.

EXAMPLE 38-4

In Fig. 38-18, $V_{CC} = 12$ V, $R = 33$ kΩ, and $C = 0.47$ μF. What is the minimum trigger voltage that produces an output pulse? What is the maximum capacitor voltage? What is the width of the output pulse?

Answer:
As shown in Fig. 38-17, the lower comparator has a trip point of LTP. Therefore, the input trigger on pin 2 has to fall from $+V_{CC}$ to slightly less than LTP. With the equations shown in Fig. 38-18,

$$LTP = \frac{12 \text{ V}}{3} = 4 \text{ V}$$

After a trigger arrives, the capacitor charges from 0 V to a maximum of UTP, which is

$$UTP = \frac{2(12 \text{ V})}{3} = 8 \text{ V}$$

The pulse width of the one-shot output is

$$W = 1.1(33 \text{ k}\Omega)(0.47 \text{ } \mu\text{F}) = 17.1 \text{ ms}$$

This means that the falling edge of the output pulse occurs 17.1 ms after the trigger arrives. You can think of this 17.1 ms as a time delay, because the falling edge of the output pulse can be used to trigger some other circuit.

EXAMPLE 38-5

What is the pulse width in Fig. 38-18 if $R = 10$ MΩ and $C = 470$ μF?

Answer:

$$W = 1.1(10 \text{ M}\Omega)(470 \text{ } \mu\text{F}) = 5170 \text{ s} = 86.2 \text{ min} = 1.44 \text{ hr}$$

Here we have a pulse width of more than an hour. The falling edge of the pulse occurs after a time delay of 1.44 hr.

Astable Operation Figure 38-19 shows the 555 timer connected for astable operation. The trip points are the same as for monostable operation:

$$UTP = \frac{2V_{CC}}{3}$$

$$LTP = \frac{V_{CC}}{3}$$

When Q is low, the transistor is cut off and the capacitor is charging through a total resistance of

$$R = R_1 + R_2$$

Because of this, the charging time constant is $(R_1 + R_2)C$. As the capacitor charges, the threshold voltage (pin 6) increases.

Eventually, the threshold voltage exceeds $+2V_{CC}/3$. Then, the upper comparator sets the flip-flop. With Q high, the transistor saturates and grounds pin 7. The capacitor now discharges through R_2. Therefore, the discharging time constant is R_2C. When the capacitor voltage drops to slightly less than $V_{CC}/3$, the lower comparator resets the flip-flop.

Figure 38-20 shows the waveforms. The timing capacitor has exponentially rising and falling voltages between UTP and LTP. The output is a rectangular wave that swings between 0 and V_{CC}. Since the charging time constant is longer than the discharging time constant, the output is nonsymmetrical. Depending on resistances R_1 and R_2, the duty cycle is between 50 and 100%.

By analyzing the equations for charging and discharging, we can derive the following formulas. The pulse width is given by

$$W = 0.693(R_1 + R_2)C \tag{38-18}$$

The period of the output equals

$$T = 0.693(R_1 + 2R_2)C \tag{38-19}$$

The reciprocal of the period is the frequency

$$f = \frac{1.44}{(R_1 + 2R_2)C} \tag{38-20}$$

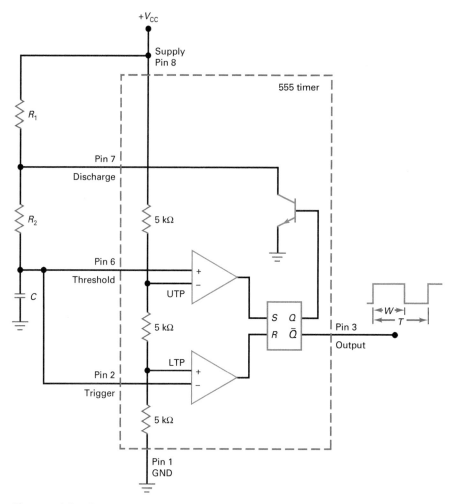

Figure 38-19   555 timer connected for astable operation.

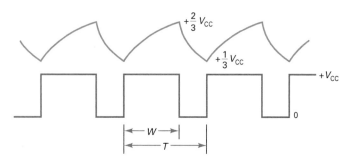

Figure 38-20   Capacitor and output waveforms for astable operation.

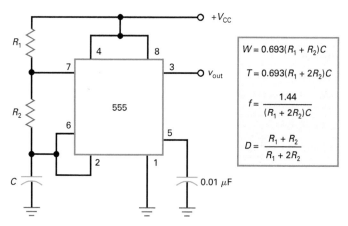

Figure 38-21   Astable multivibrator.

$$W = 0.693(R_1 + R_2)C$$

$$T = 0.693(R_1 + 2R_2)C$$

$$f = \frac{1.44}{(R_1 + 2R_2)C}$$

$$D = \frac{R_1 + R_2}{R_1 + 2R_2}$$

Dividing the pulse width by the period gives the duty cycle

$$D = \frac{R_1 + R_2}{R_1 + 2R_2} \qquad (38\text{-}21)$$

When R_1 is much smaller than R_2, the duty cycle approaches 50%. Conversely, when R_1 is much greater than R_2, the duty cycle approaches 100%.

Figure 38-21 shows the astable 555 timer as it usually appears on a schematic diagram. Again notice how pin 4 (reset) is tied to the supply voltage and how pin 5 (control) is bypassed to ground through a 0.01-μF capacitor.

The circuit of Fig. 38-21 can be modified to enable the duty cycle to become less than 50%. By placing a diode

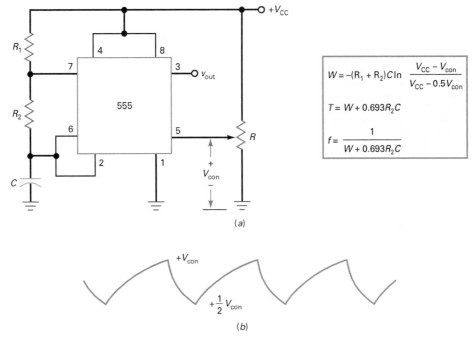

$$W = -(R_1 + R_2)C\ln\frac{V_{CC} - V_{con}}{V_{CC} - 0.5V_{con}}$$

$$T = W + 0.693R_2C$$

$$f = \frac{1}{W + 0.693R_2C}$$

(a)

(b)

Figure 38-22 (a) Voltage-controlled oscillator. (b) Capacitor voltage waveform.

in parallel with R_2 (anode connected to pin 7), the capacitor will effectively charge through R_1 and the diode. The capacitor will discharge through R_2. Therefore, the duty cycle becomes

$$D = \frac{R_1}{R_1 + R_2}$$ (38-22)

VCO Operation Figure 38-22a shows a **voltage-controlled oscillator (VCO),** another application for a 555 timer. The circuit is sometimes called a **voltage-to-frequency converter** because an input voltage can change the output frequency.

Here is how the circuit works: Recall that pin 5 connects to the inverting input of the upper comparator (Fig. 38-16). Normally, pin 5 is bypassed to ground through a capacitor, so that UTP equals $+2V_{CC}/3$. In Fig. 38-22a, however, the voltage from a stiff potentiometer overrides the internal voltage. In other words, UTP equals V_{con}. By adjusting the potentiometer, we can change UTP to a value between 0 and V_{CC}.

Figure 38-22b shows the voltage waveform across the timing capacitor. Notice that the waveform has a minimum value of $+V_{con}/2$ and a maximum value of $+V_{con}$. If we increase V_{con}, it takes the capacitor longer to charge and discharge. Therefore, the frequency decreases. As a result, we can change the frequency of the circuit by varying the control voltage. Incidentally, the control voltage may come from a potentiometer, as shown, or it may be the output of a transistor circuit, an op amp, or some other device.

Figure 38-22 shows the equations for calculating the frequency of the VCO.

EXAMPLE 38-6

The 555 timer of Fig. 38-21 has $R_1 = 75$ kΩ, $R_2 = 30$ kΩ, and $C = 47$ nF. What is frequency of the output signal? What is the duty cycle?

Answer:
With the equations shown in Fig. 38-21,

$$f = \frac{1.44}{(75\text{ k}\Omega + 60\text{ k}\Omega)(47\text{ nF})} = 227\text{ Hz}$$

$$D = \frac{75\text{ k}\Omega + 30\text{ k}\Omega}{75\text{ k}\Omega + 60\text{ k}\Omega} = 0.778$$

This is equivalent to 77.8%.

Other 555 Circuits The output stage of a 555 timer can *source* 200 mA. This means that a high output can produce up to 200 mA of load current (sourcing). Because of this, the 555 timer can drive relatively heavy loads such as relays, lamps, and loudspeakers. The output stage of a 555 timer can also *sink* 200 mA. This means that a low output can allow up to 200 mA to flow to ground (sinking). In this section, we discuss some applications for a 555 timer.

The 555 is also a good generator of audio tones. Figure 38-23 shows how to use an astable 555 timer as an audible ALARM. Normally, the ALARM switch is closed, which pulls pin 4 down to ground. In this case, the 555 timer is inactive and there is no output. When the ALARM switch is opened, however, the circuit will generate a rectangular output whose frequency is

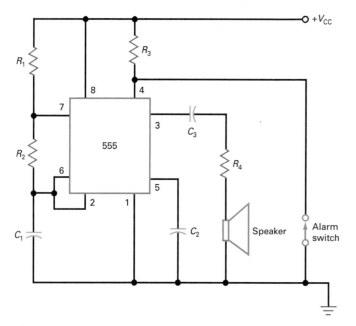

Figure 38-23 Astable 555 circuit used for siren or alarm.

Another potential application is a two-tone siren. By inserting a telegraph key in series with the pin 1 ground lead, a Morse code practice oscillator is produced.

Figure 38-24 shows a circuit used for **pulse-width modulation (PWM).** PWM produces a constant-frequency pulse train whose duty cycle is valued by some modulating signal. PWM is widely used in data communications, motor speed control, and switch-mode power supplies.

To generate PWM, the 555 timer is connected in the monostable mode. The values of R, C, UTP, and V_{CC} determine the width of the output pulse as follows:

$$W = -RC \ln\left(1 - \frac{\text{UTP}}{V_{CC}}\right) \tag{38-23}$$

A low-frequency signal called a **modulating signal** is capacitively coupled into pin 5. This modulating signal is voice or computer data. Since pin 5 controls the value of UTP, v_{mod} is being added to the quiescent UTP. Therefore, the instantaneous UTP is given by

$$\text{UTP} = \frac{2V_{CC}}{3} + v_{\text{mod}} \tag{38-24}$$

For instance, if $V_{CC} = 12$ V and the modulating signal has a peak value of 1 V,

$$\text{UTP}_{\text{max}} = 8 \text{ V} + 1 \text{ V} = 9 \text{ V}$$

$$\text{UTP}_{\text{min}} = 8 \text{ V} - 1 \text{ V} = 7 \text{ V}$$

This means that the instantaneous UTP varies sinusoidally between 7 and 9 V.

A train of triggers called the *clock* is the input to pin 2. This clock signal could be produced by another 555 timer in

determined by R_1, R_2, and C_1. An example of an ALARM switch is a magnetically operated reed switch commonly used in home alarms.

The output from pin 3 drives a loudspeaker through a resistance of R_4. The size of this resistance depends on the supply voltage and the impedance of the loudspeaker. The impedance of the branch with R_4 and the speaker should limit the output current to 200 mA or less because this is the maximum current a 555 timer can source.

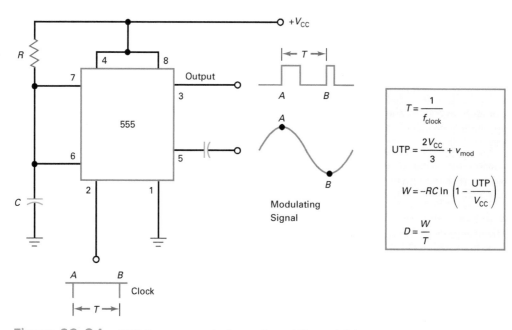

Figure 38-24 555 timer connected as pulse-width modulator.

astable mode. Each trigger produces an output pulse. Since the period of the triggers is T, the output will be a series of rectangular pulses with a period of T. The modulating signal has no effect on the period T, but it does change the width of each output pulse. At point A, the positive peak of the modulating signal, the output pulse is wide as shown. At point B, the negative peak of the modulating signal, the output pulse is narrow.

PWM is used in communications. It allows a low-frequency modulating signal (voice or data) to change the pulse width of a high-frequency signal called the **carrier.** The modulated carrier can be transmitted over copper wire, over fiber-optic cable, or through space to a receiver. The receiver recovers the modulating signal to drive a speaker (voice) or a computer (data).

EXAMPLE 38-7

A pulse-width modulator like Fig. 38-24 has V_{cc} = 12 V, R = 9.1 kΩ, and C = 0.01 μF. The clock has a frequency of 2.5 kHz. If a modulating signal has a peak value of 2 V, what is the period of the output pulses? What is the quiescent pulse width? What are the minimum and maximum pulse widths? What are the minimum and maximum duty cycles?

Answer:
The period of the output pulses equals the period of the clock:

$$T = \frac{1}{2.5 \text{ kHz}} = 400 \ \mu s$$

The quiescent pulse width is

$$W = 1.1RC = 1.1(9.1 \text{ k}\Omega)(0.01 \ \mu F) = 100 \ \mu s$$

With Formula (38-24), calculate the minimum and maximum UTP:

$$\text{UTP}_{min} = 8 \text{ V} - 2 \text{ V} = 6 \text{ V}$$

$$\text{UTP}_{max} = 8 \text{ V} + 2 \text{ V} = 10 \text{ V}$$

Now, calculate the minimum and maximum pulse widths with Formula (38-23),

$$W_{min} = -(9.1 \text{ k}\Omega)(0.01 \ \mu F) \ln\left(1 - \frac{6 \text{ V}}{12 \text{ V}}\right) = 63.1 \ \mu s$$

$$W_{max} = -(9.1 \text{ k}\Omega)(0.01 \ \mu F) \ln\left(1 - \frac{10 \text{ V}}{12 \text{ V}}\right) = 163 \ \mu s$$

The minimum and maximum duty cycles are

$$D_{min} = \frac{63.1 \ \mu s}{400 \ \mu s} = 0.158$$

$$D_{max} = \frac{163 \ \mu s}{400 \ \mu s} = 0.408$$

Function Generator ICs

Special function generator ICs have been developed that combine many of the individual circuit capabilities we have been discussing. These ICs are able to provide waveform generation including sine, square, triangle, ramp, and pulse signals. The output waveforms can be made to vary in amplitude and frequency by changing the values of external resistors and capacitors or by applying an external voltage. This external voltage enables the IC to perform useful applications such as voltage-to-frequency (V/F) conversion, AM and FM signal generation, voltage-controlled oscillation (VCO), and frequency-shift keying (FSK).

The XR-2206 An example of a special function generator IC is the XR-2206. This monolithic IC can provide externally controlled frequencies from 0.01 Hz to more than 1.00 MHz. A circuit diagram of this IC is shown in Fig. 38-25. The diagram shows four main functional blocks, which include a VCO, an analog multiplier and sine shaper, a unity gain buffer amplifier, and a set of current switches.

The output frequency of the VCO is proportional to an input current, which is determined by a set of external timing resistors. These resistors connect to pins 7 and 8, respectively, and ground. Because there are two timing pins, two discrete output frequencies can be obtained. A high or low input signal on pin 9 controls the current switches. The current switches then select which one of the timing resistors will be used. If the input signal on pin 9 alternately changes from high to low, the output frequency of the VCO will be shifted from one frequency to another. This action is referred to as **frequency-shift keying (FSK)** and is used in electronic communication applications.

The output of the VCO drives the multiplier and sine-shaper block, along with an output switching transistor. The output switching transistor is driven to cutoff and saturation, which provides a square-wave output signal at pin 11. The output of the multiplier and sine-shaper block is connected to a unity gain buffer amplifier, which determines the IC's output current capability and its output impedance. The output at pin 2 can be either a sine wave or a triangle wave.

Figure 38-25 shows the external circuit connections and components for generating sine waves or triangle waves. The frequency of oscillation f_0 is determined by the timing resistor R, connected at either pin 7 or pin 8, and the external capacitor C, connected across pins 5 and 6. The value of oscillation is found by

$$f_0 = \frac{1}{RC} \qquad \textbf{(38-25)}$$

Even though R can be up to 2 MΩ, maximum temperature stability occurs when 4 kΩ < R < 200 kΩ. Also, the recommended value of C should be from 1000 pF to 100 μF.

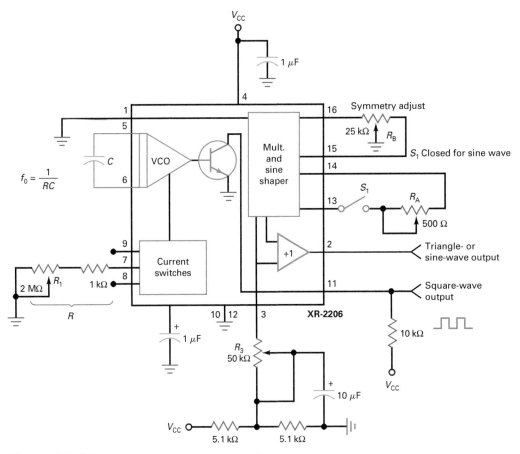

Figure 38-25   Sine-wave generation circuit.

In Fig. 38-25, when the switch S_1 is closed, the output at pin 2 will be a sine wave. The potentiometer R_1 at pin 7 provides the desired frequency tuning. Adjustable resistors R_A and R_B enable the output waveform to be modified for proper waveform symmetry and distortion levels. When S_1 is open, the output at pin 2 changes from a sine wave to a triangle wave. Resistor R_3, connected at pin 3, controls the amplitude of the output waveform. The output amplitude is directly proportional to the value of R_3. Notice that the value of the triangle waveform is approximately double the output of a sine waveform for a given R_3 setting.

Figure 38-26 shows the external connections of the circuit used to create sawtooth (ramp) and pulse outputs. Notice that the square-wave output at pin 11 is shorted to the FSK terminal at pin 9. This allows the circuit to automatically frequency-shift between two separate frequencies. This frequency shift occurs when the output at pin 11 changes from a high-level output to a low-level output or from a low-level output to a high-level output. The output frequency is found by

$$f = \frac{2}{C}\left[\frac{1}{R_1 + R_2}\right] \qquad \text{(38-26)}$$

and the circuit's duty cycle is found by

$$D = \frac{R_1}{R_1 + R_2} \qquad \text{(38-27)}$$

If operated with a single positive supply voltage, the supply can range from 10 V to 26 V. If a split- or dual-supply voltage is used, notice how the values range from ± 5 V to ± 13 V. The triangle- and sine-wave output has an output impedance value of 600 Ω. This makes the XR-2206 function generator IC well suited for many electronic communications applications.

Microcontroller Clocks and Timers

ICs like the 555 and 2206 rely on RC networks to set the frequency and duty cycle. For some applications this arrangement works fine. But because of the tolerances of the capacitors and resistors and their temperature sensitivity, the precision of the frequency or time is not satisfactory for some critical applications. Standard resistors and capacitors have tolerances in the 5 to 10% range. For capacitors some tolerances are even greater. While precision can be improved by using components with 1 or 2% tolerances or better, the cost of doing so is much higher. For these applications crystal control is preferred. ICs like the 555 and 2206

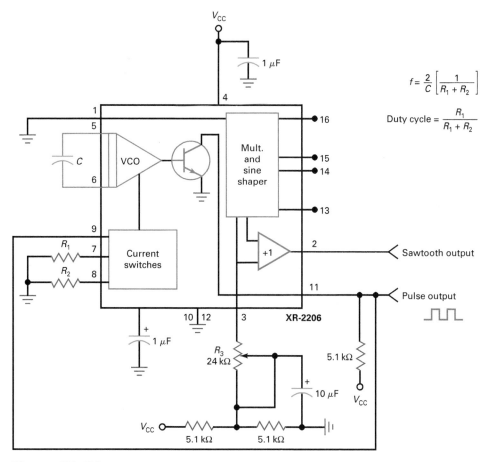

$$f = \frac{2}{C}\left[\frac{1}{R_1 + R_2}\right]$$

$$\text{Duty cycle} = \frac{R_1}{R_1 + R_2}$$

Figure 38-26   Pulse and ramp generation.

do not work with crystals so some other approach is needed. The most common solution is to use a very low cost single-chip microcontroller.

A microcontroller is a single-chip computer that can be programmed to perform almost any digital, ON-OFF operation. It consists of a processor that performs math and logic (AND, OR, etc.) operations, one or more memories to store programs and data, and input/output (I/O) circuits that accept inputs and produce outputs. Most microcontrollers also have built-in timers or counters that make it easy to implement oscillators.

Figure 39-27 shows a simplified block diagram of a microcontroller. Note the random access memory (RAM) for storing and retrieving data and the read-only memory (ROM) for storing the program. This controller is programmed to flash an external LED on for 1 s and off for 0.5 s.

Most microcontrollers also contain an internal clock oscillator for operating all the digital circuits inside. The frequency of this clock is usually set with an external RC network for noncritical functions or with any external crystal or ceramic resonator for more accurate outputs. When crystal control is desired, the crystal is connected to the OSC$_1$ and OSC$_2$ terminals along with two capacitors C_1 and C_2. The oscillator is an internal amplifier-inverter biased into

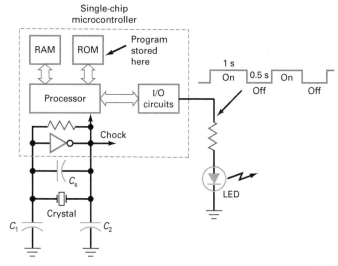

Figure 38-27   A microcontroller used as a timer.

the linear region by an internal resistor. The crystal and the capacitors C_1 and C_2 provide the feedback with the correct 180° phase shift for oscillation. The frequency of oscillation is determined by the crystal as well as the series combination of C_1 and C_2 in parallel with any stray capacitance (C_s)

on the PCB. With crystal precision, the microcontroller can be programmed to generated square waves or ON-OFF signals with any frequency or duty cycle.

The program in the ROM to generate an ON-OFF signal looks something like the list below. Each step is implemented by one or more of the microcontroller instructions.

1. Set I/O output high.
2. Count clock pulses or instruction increments for the desired duration (1 s).
3. Set I/O line low.
4. Count clock pulses or instruction increments for the desired duration (0.5 s).
5. Repeat.

The resolution of the pulse ON and OFF times is determined by the clock frequency and, in most cases, by the time of execution of the instructions used in the program. Resolutions of microseconds are easy to obtain. Nanosecond resolutions are possible with faster processors and clocks.

Today, a simple single-chip microcontroller can be purchased for less than $1 in quantity, so it can compete favorably with other low-cost *RC* timers. Its disadvantage is that it has to be programmed. But that may be an advantage, as the program can usually be changed if necessary. Most industrial timers, toys, and alarm systems are implemented this way.

38.8 The Phase-Locked Loop

A **phase-locked loop (PLL)** is a closed-loop feedback electronic control circuit. It contains a phase detector, a dc amplifier, a low-pass filter, and a voltage-controlled oscillator (VCO). When a PLL has an input signal with a frequency of f_{in}, its VCO will produce an output frequency that equals f_{in}.

Phase Detector

Figure 38-28*a* shows a **phase detector,** the first stage in a PLL. This circuit produces an output voltage proportional to the phase difference between two input signals. For instance, Fig. 38-28*b* shows two input signals with a phase difference of $\Delta\phi$. The phase detector responds to this phase difference by producing a dc output voltage, which is proportional to $\Delta\phi$, as shown in Fig. 38-28*c*.

When v_1 leads v_2, as shown in Fig. 38-28*b*, $\Delta\phi$ is positive. If v_1 were to lag v_2, $\Delta\phi$ would be negative. The typical phase detector produces a linear response between $-90°$ and $+90°$, as shown in Fig. 38-28*c*. As we can see, the output of the phase detector is zero when $\Delta\phi = 0°$. When $\Delta\phi$ is between $0°$ and $90°$, the output is a positive voltage. When $\Delta\phi$ is between $0°$ and $-90°$, the output is a negative voltage. The key idea here is that the phase detector produces an output voltage that is directly proportional to the phase difference between its two input signals.

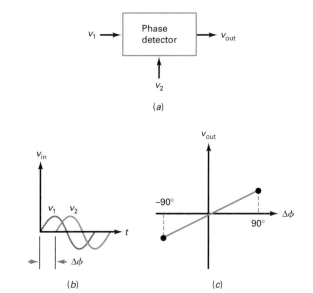

(a)

(b)   (c)

Figure 38-28 (*a*) Phase detector has two input signals and one output signal. (*b*) Equal-frequency sine waves with phase difference. (*c*) Output of phase detector is directly proportional to phase difference.

The VCO

In Fig. 38-29*a*, the input voltage v_{in} to the VCO determines the output frequency f_{out}. A typical VCO can be varied over a 10:1 range of frequency. Furthermore, the variation is linear as shown in Fig. 38-29*b*. When the input voltage to the VCO is zero, the VCO is free-running at a quiescent frequency f_0. When the input voltage is positive, the VCO frequency is greater than f_0. If the input voltage is negative, the VCO frequency is less than f_0.

Block Diagram of a PLL

Figure 38-30 is a block diagram of a PLL. The phase detector produces a dc voltage that is proportional to the phase difference of its two input signals. The output voltage of the phase detector is usually small. This is why the second stage is a dc amplifier. The amplified phase difference is filtered in a low-pass filter before being applied to the VCO. Notice that the VCO output is being fed back to the phase detector.

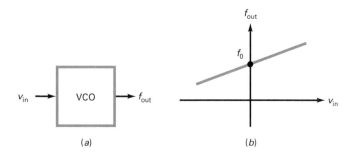

(a)   (b)

Figure 38-29 (*a*) Input voltage controls output frequency of VCO. (*b*) Output frequency is directly proportional to input voltage.

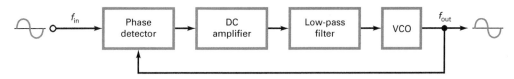

Figure 38-30   Block diagram of a phase-locked loop.

Input Frequency Equals Free-Running Frequency

To understand PLL action, let us start with the case of input frequency equal to f_0, the free-running frequency of the VCO. In this case, the two input signals to the phase detector have the same frequency and phase. Because of this, the phase difference $\Delta\phi$ is 0° and the output of the phase detector is zero. As a result, the input voltage to the VCO is zero, which means that the VCO is free-running with a frequency of f_0. As long as the frequency and phase of the input signal remain the same, the input voltage to the VCO will be zero.

Input Frequency Differs from Free-Running Frequency

Let us assume that the input and free-running VCO frequencies are each 10 kHz. Now, suppose the input frequency increases to 11 kHz. This increase will appear to be an increase in phase because v_1 leads v_2 at the end of the first cycle, as shown in Fig. 38-31a. Since input signal leads the VCO signal, $\Delta\phi$ is positive. In this case, the phase detector of Fig. 38-30 produces a positive output voltage. After being amplified and filtered, this positive voltage increases the VCO frequency.

The VCO frequency will increase until it equals 11 kHz, the frequency of the input signal. When the VCO frequency equals the input frequency, the VCO is *locked on* to the input signal. Even though each of the two input signals to the phase detector has a frequency of 11 kHz, the signals have a different phase, as shown in Fig. 38-31b. This positive phase difference produces the voltage needed to keep the VCO frequency slightly above its free-running frequency.

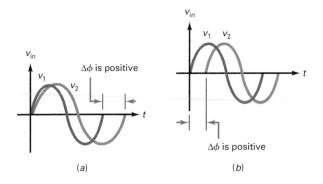

Figure 38-31   (a) An increase in the frequency of v_1 produces a phase difference. (b) A phase difference exists after the VCO frequency increases.

If the input frequency increases further, the VCO frequency also increases as needed to maintain the lock. For instance, if the input frequency increases to 12 kHz, the VCO frequency increases to 12 kHz. The phase difference between the two input signals will increase as needed to produce the correct control voltage for the VCO.

Lock Range

The **lock range** of a PLL is the range of input frequencies over which the VCO can remain locked onto and track the input frequency. It is related to the maximum phase difference that can be detected. In our discussion, we are assuming that the phase detector can produce an output voltage for $\Delta\phi$ between −90° and +90°. At these limits, the phase detector produces a maximum output voltage, either negative or positive.

If the input frequency is too low or too high, the phase difference is outside the range of −90° and +90°. Therefore, the phase detector cannot produce the additional voltage needed for the VCO to remain locked on. At these limits, therefore, the PLL loses its lock on the input signal.

The lock range is usually specified as a percentage of the VCO frequency. For instance, if the VCO frequency is 10 kHz and the lock range is ±20%, the PLL can remain locked on any input frequency between 8 and 12 kHz.

Capture Range

The capture range is different. Assume that the input frequency is outside the lock range. Then, the VCO is free-running at 10 kHz. Now, assume that the input frequency changes toward the VCO frequency. At some point, the PLL will be able to lock onto the input frequency. The range of input frequencies within which the PLL can reestablish the lock is called the **capture range.**

The capture range is specified as a percentage of the free-running frequency. If $f_0 = 10$ kHz and the capture range is ±5%, the PLL can lock on to an input frequency between 9.5 and 10.5 kHz. Typically, the capture range is less than the lock range because the capture range depends on the cutoff frequency of the low-pass filter. The lower the cutoff frequency, the smaller the capture range.

The cutoff frequency of the low-pass filter is kept low to prevent high-frequency components like noise or other unwanted signals from reaching the VCO. The lower the cutoff frequency of the filter, the cleaner the signal driving the VCO. Therefore, a designer has to trade off capture range against low-pass bandwidth to get a clean signal for the VCO.

Applications

A PLL can be used in two fundamentally different ways. First, it can be used to lock onto the input signal. The output frequency then equals the input frequency. This has the advantage of cleaning up a noisy input signal because the low-pass filter will remove high-frequency noise and other components. The PLL acts as a bandpass filter because of the narrow capture range and the noise-reducing effects of the low-pass filter. Since the output signal comes from the VCO, the final output is stable and almost noise-free.

Second, a PLL can be used as an FM demodulator. The theory of **frequency modulation (FM)** is covered in communication courses, so we discuss only the basic idea. The *LC* oscillator of Fig. 38-32*a* has a variable capacitance. If a modulating signal controls this capacitance, the oscillator output will be *frequency-modulated,* as shown in Fig. 38-32*b*. Notice how the frequency of this FM wave varies from a minimum to a maximum, corresponding to the minimum and maximum peaks of the modulating signal.

If the FM signal is the input to a PLL, the VCO frequency will lock onto the FM signal. Since the VCO frequency varies, $\Delta\phi$ follows the variations in the modulating signal. Therefore, the output of the phase detector will be a low-frequency signal that is a replica of the original modulating signal. When used in this way, the PLL is being used as an *FM demodulator,* a circuit that recovers the modulating signal from the FM wave.

PLLs are available as monolithic ICs. For instance, the NE565 is a PLL that contains a phase detector, a VCO, and a dc amplifier. The user connects external components like a timing resistor and capacitor to set the free-running frequency of the VCO. Another external capacitor sets the cutoff frequency of the low-pass filter. The NE565 can be used for FM demodulation, frequency synthesis, telemetry receivers, modems, tone decoding, and so forth. Most PLLs are fully integrated into other ICs.

Figure 38-32 (*a*) Variable capacitance changes resonant frequency of *LC* oscillator. (*b*) Sine wave has been frequency-modulated.

38.9 Frequency Synthesizers

A *frequency synthesizer* is a signal-generating circuit whose output frequency is varied in increments. The output may be a sine wave or square wave, depending on how the synthesizer is used. It has the frequency precision and stability of a crystal oscillator, but its output frequency is changed in steps rather than varied continuously.

A frequency synthesizer is used in virtually all wireless equipment for tuning. Most transmitters and receivers have their frequency of operation set by a synthesizer. This includes all radio and TV sets; two-way marine, aircraft, public service, and other radios; and all cell phones. Frequency synthesizers are also at the heart of all electronic test signal generators. These are often called *function generators.* Arbitrary waveform generators (AWG) with synthesis techniques can generate a signal of any shape with or without modulation.

There are two basic types of synthesizers, one based on the phase-locked loop and the other based on digital techniques. A brief introduction to each is given here.

PLL Frequency Synthesizers

Figure 38-33 shows a PLL synthesizer. Notice the alternative symbols used to represent oscillators and the phase detector. Oscillators are commonly represented by a circle with a sine or square wave inside to designate the type of oscillator.

The input to the PLL is a crystal oscillator whose frequency determines the step-frequency increment. Because the PLL will be locked to a crystal, the VCO output will have the precision and stability of this crystal oscillator.

The key component added to the PLL to create a synthesizer is a variable-frequency divider. A frequency divider, as the name implies, divides its input frequency by some integer value *n*. If *n* is 10, then a 30-kHz input will produce a 3-kHz output. Frequency dividers are usually made with

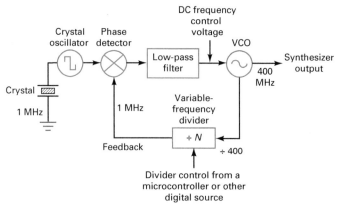

Figure 38-33 A PLL frequency synthesizer.

digital flip-flops and can be designed so that the frequency division factor can be varied in increments. Some form of digital input signal is given to change the division factor.

Note in Fig. 38-33 that the frequency divider output is connected to the second input of the phase detector. Since the two inputs to the phase detector must be equal to achieve a locked condition, the VCO output must be a factor of n higher than the crystal input frequency. For example, with a 1-MHz crystal oscillator input and a frequency division factor of 400, the VCO output frequency will be 400 MHz. The 400-MHz output is reduced by the divider to match the 1-MHz crystal oscillator input.

Now assume that the frequency division factor is changed to 399. The 400-MHz signal from the VCO is divided by 399, giving an output of 1.0025 MHz. The phase detector senses the difference from the 1-MHz input and produces a dc voltage that will rapidly change the VCO output to 399 MHz, bringing the PLL into lock. An important specification in frequency synthesizers is how fast the VCO can change and the PLL can achieve lock, the faster the better. In most PLL synthesizers this is several microseconds. The time span is determined by the low-pass filter characteristics.

As you can see, by changing the frequency division factor, the VCO output can be varied in 1-MHz steps. For some applications, a 1-MHz step size may be too big. In that case, another crystal frequency can be used. Alternatively, another frequency divider can be placed between the 1-MHz oscillator and the phase detector. For example, if a ÷100 divider is used, the phase detector input will be 1 MHz/100 = 10 kHz. The VCO output and the frequency divider output will then be changed to 10-kHz increments as the main frequency divider factor is varied. The frequency divider factor must be increased to $n = 40,000$ to maintain a 400-MHz output, but the changes provide for the 10-kHz incremental frequency change.

An important concept to take away from this PLL synthesizer discussion is that the VCO output has a frequency that is n times the phase detector input. In the circuit of Fig. 38-33, the input is 1 MHz from the crystal oscillator and the VCO output is 400 MHz. The synthesizer is acting as a ×400 frequency multiplier. This concept is widely used to create crystal oscillators at frequencies beyond their natural capabilities. For example, a packaged crystal or a microelectromechanical systems (MEMS) oscillator like those discussed earlier can use a PLL multiplier to create a 915-MHz output signal from a 18.3-MHz crystal by using a fixed division factor of 50. The 915-MHz output has the same frequency accuracy and stability as the 18.3-MHz crystal oscillator.

PLL synthesizers are available in integrated circuit form. However, most are buried inside other ICs. Wireless receivers and transmitters typically have the PLL synthesizer integrated into the other circuitry on a single chip.

EXAMPLE 38-8

A PLL synthesizer uses a 1-MHz crystal followed by a ÷40 divider. The feedback divider has division factor of 123. What are the VCO output frequency and the frequency step increment?

Answer:
The input to the phase detector is

$$1,000,000 \div 40 = 25,000 \text{ Hz or } 25 \text{ kHz.}$$

This is the step increment.
   The VCO output is

$$25 \text{ kHz} \times 123 = 3075 \text{ kHz or } 3.075 \text{ MHz}$$

Direct Digital Synthesis

While most frequency synthesizers still have some form of the PLL circuit, just discussed, a newer synthesis technique is now widely used in communications equipment as well as test instruments. Known as *direct digital synthesis (DDS),* it has been made feasible by large-scale ICs available from Analog Devices Inc. that contains all the necessary circuitry.

The basic concept is shown in Fig. 38-34. Assume that the desired output is a sine wave. The digital version of a sine (or cosine) wave is stored in an IC read-only memory (ROM) or a semiconductor random access memory (RAM). Each memory location or address contains one amplitude increment of the sine wave in binary format. A binary counter supplies sequential addresses to the memory to access the binary numbers. When the binary numbers are accessed and sequentially sent to a digital-to-analog converter (DAC), a proportional analog output amplitude is produced for each binary input. The output of the DAC is a stepped approximation of the sine wave. A low-pass filter is used to smooth the signal into a more continuous analog sine wave.

Figure 38-35 shows the process of converting binary numbers in a memory into approximately half a sine wave. In this example, it is assumed that there is a binary sine value for every 15° increment stored in memory. As each increment number is applied to the DAC, a proportional analog voltage is generated. The DAC output voltage stays constant

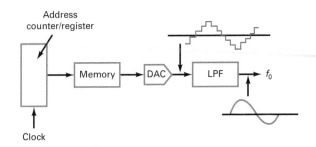

Figure 38-34 Concept of a direct digital synthesizer.

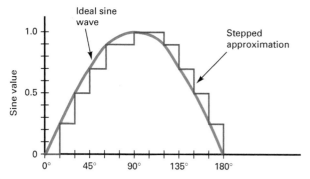

Figure 38-35 A stepped approximation of a sine wave.

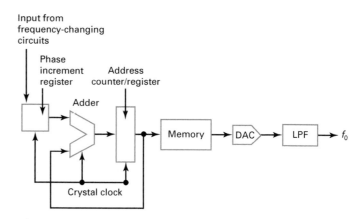

Figure 38-36 A complete DDS synthesizer.

until the next increment is received by the DAC. The result is a stepped approximation of a sine wave, as shown.

The number of bits used to represent the binary sine values determines the resolution of the sine wave and the DAC output voltage increment size. For example, if 8-bits are used for each number, there will be $2^8 = 256$ voltage levels. That will give better resolution and a stepped output closer to a true sine wave. Some synthesizers can use 10-, 12-, or 14-bit numbers and DACs to produce a very fine-grained sine output.

The method shown in Fig. 38-34 works fine, but the output frequency, f_o, depends on the clock frequency, f_c, driving the address counter and the number of sine samples, M, in memory, or $f_o = f_c/M$. With 120 sine samples and a clock frequency of 80 MHz, the output sine wave would have a frequency of

$$f_o = f_c/M = 80 \text{ MHz}/120 = 666.67 \text{ kHz}$$

This means that the clock frequency must be very high to produce even a low-frequency output. And you must change the clock frequency to change the output frequency, which does not make it easy to get crystal accuracy or stability.

A DDS uses this same waveform generation method but replaces the address counter with a sophisticated phase accumulator circuit, as shown in Fig. 38-36. A clock signal from a crystal-based oscillator clocks the adder and the address register that feeds the address to the DAC. The address register output is added to a phase increment register value

to get the next ROM address value. By changing the phase register value, which comes from a microcontroller or other circuit, the phase increment of the output is changed. The address for the waveform memory is derived from an initial value in the address register based on memory size plus a phase increment value supplied by the external frequency selection circuitry. The output frequency is changed by varying the phase register value. This permits a constant clock frequency from a crystal-stabilized source clock to achieve the desired crystal precision and stability.

DDS ICs are available with clock input values to 1 GHz and with phase increments of 24, 32, or 48 bits. This gives 2^n phase resolution for 360°. DAC update rates are also available to 1 GHz with output bit resolutions of 10, 12, and 14 bits. This allows DDS synthesizers to produce outputs in the hundreds of megahertz range with very high frequency and phase increment precision. And DDS ICs are available with AM, FM, and PM modulation as well as two channels for producing in-phase, I, and quadrature, Q, signals with precise, high 90° separation.

DDS synthesizers offer the benefits of high frequency and phase resolution as well as their ability to change frequency in nanoseconds vs. microseconds or more typical of PLL synthesizers.

CHAPTER 38 SYSTEM SIDEBAR

The Concept of Modulation

Modulation is the process of having one signal modify another for the purpose of communicating information over a medium. In some instances, information such as audio signals, video, or digital data can be transmitted from one place to another by just applying it to a cable. Audio voltages of

voice and music can be transmitted over speaker cables, surveillance video signals can be sent over coaxial cables, and computer data can be applied to twisted-pair network cables. However, such information cannot be transmitted directly by radio.

Wireless communication requires that the information or data modulate another signal called the *carrier*. The carrier is at a higher radio frequency that can be transmitted more easily over the air via antennas. The carrier signal sets the frequency of operation. Modulation varies the carrier at the transmitter in a unique way such that the receiver can recover it.

There are many different forms of modulation, but all are a variation of the three basic modulation types: amplitude modulation (AM), frequency modulation (FM), and phase modulation (PM). Following is a brief description of each.

Amplitude Modulation

Figure S38-1 shows AM. A sine wave information signal is varying the amplitude of a higher-frequency, sine-wave carrier. The information signal could be voice, music, or video. The carrier frequency remains constant.

Figure S38-2 shows binary digital data being transmitted by AM. This is called amplitude-shift keying (ASK). The carrier is shifted between two different amplitude levels. The binary data signal can also turn the carrier off and on, as the binary 0s and 1s occur, creating a form of ASK called ON-OFF keying (OOK).

Frequency Modulation

FM is illustrated in Fig. S38-3. A sine-wave information signal varies the frequency of the carrier while

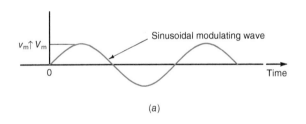

(a)

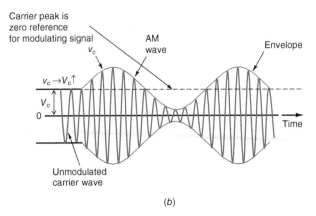

(b)

Figure S38-1 Amplitude modulation. (a) The modulating or information signal. (b) The modulated carrier.

Source: From Frenzel, *Principles of Electronic Communication Systems,* 3d ed.

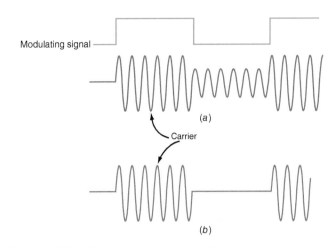

Figure S38-2 Amplitude modulation of a sine-wave carrier by a binary or rectangular wave is called amplitude-shift keying. (a) ASK. (b) OOK.

Source: From Frenzel, *Principles of Electronic Communications Systems,* 3d ed.

the carrier amplitude remains constant. The greater the amplitude of the modulating signal, the greater the frequency deviation.

The binary digital data transmission with FM is called frequency-shift keying (FSK). As shown in Fig. S38-4, the digital data shifts the carrier frequency between two fixed frequency values.

Phase Modulation

Phase modulation produces a result similar to FM when the modulating signal is analog like a sine wave, audio, or video. The modulating signal shifts the instantaneous phase of the carrier and that, in turn, produces frequency modulation.

Phase modulation using digital data results in what is called binary phase-shift keying (BPSK). Figure S38-5 shows BPSK. Note the 180° phase shift that occurs when there is a binary 0 to 1 or 1 to 0 transition.

Sidebands

When modulation occurs, the process produces new signals called sidebands. The sidebands are located above and below the carrier frequency by an amount equal to the modulating frequency value. Figure S38-6a shows the sidebands produced by AM in a frequency domain plot. The carrier frequency is f_c and the modulating signal frequency is f_m. The upper sideband (USB) is $f_c + f_m$. The lower sideband (LSB) is $f_c - f_m$. Note in Figure S38-6b that a nonsinusoidal or distorted wave contains harmonics that produce sidebands as well. The modulated carrier with its sidebands occupies a finite amount of bandwidth in the radio-frequency spectrum.

FM and PM produce multiple sidebands, as shown in Fig. S38-7. The number and amplitude of the sidebands depend on

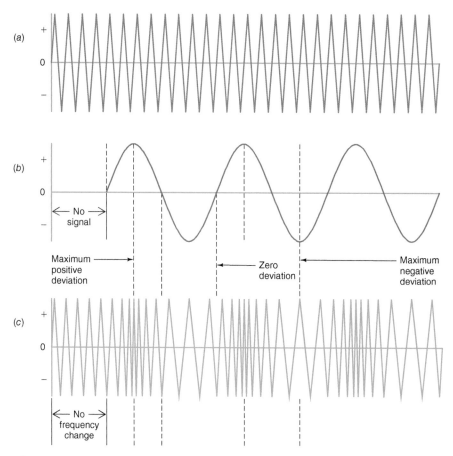

Figure S38-3 FM and PM signals. The carrier is drawn as a triangular wave for simplicity but is usually a sine wave. (*a*) Carrier. (*b*) Modulating signal. (*c*) FM signal.

Source: From Frenzel, *Principles of Electronic Communications Systems,* 3d ed.

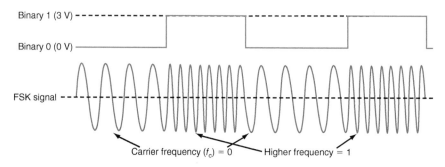

Figure S38-4 Frequency modulating of a carrier with binary data produces FSK.

Source: From Frenzel, *Principles of Electronic Communications Systems,* 3d ed.

the frequency of the modulating signal and other factors. Again, the total FM signal takes up a large segment of the spectrum.

There are dozens of different forms of modulation, but all are variations or combinations of the basic AM, FM, and PM techniques illustrated. Most of these modulation schemes are designed for the efficient transmission of digital data. For example, quadrature amplitude modulation (QAM) is a combination amplitude and phase modulation. It allows higher data rates in the same or less bandwidth.

Spread spectrum is a method by which binary data are combined with a special "chipping" code to spread the signal over a wider bandwidth. Orthogonal frequency division

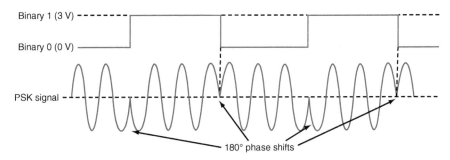

Figure S38-5 Phase modulation of a carrier by binary data produces PSK.

Source: From Frenzel, *Principles of Electronic Communications Systems,* 3d ed.

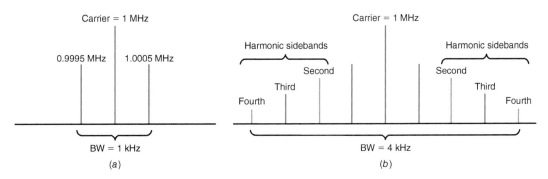

Figure S38-6 AM signal bandwidth. (*a*) Sine wave of 500 Hz modulating a 1-MHz carrier. (*b*) Distorted 500-Hz sine wave with significant second, third, and fourth harmonics.

Source: From Frenzel, *Principles of Electronic Communications Systems,* 3d ed.

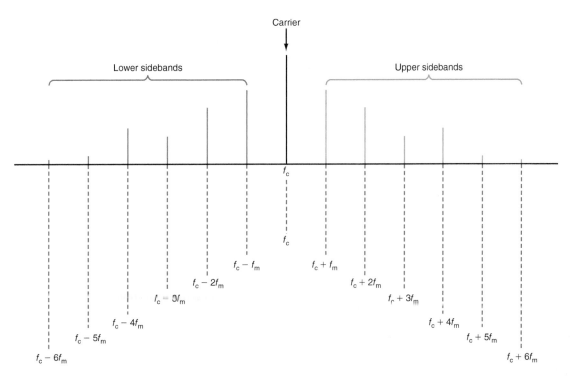

Figure S38-7 Frequency spectrum of an FM signal. Note that the carrier and sideband amplitudes shown are just examples. The amplitudes depend on the nature of the modulating signal.

Source: From Frenzel, *Principles of Electronic Communications Systems,* 3d ed.

multiplexing (OFDM) is a method that divides a serial binary data stream into multiple slower streams. Each of the slower streams modulates separate multiple adjacent carriers using BPSK or QAM, as shown in Fig. S38-8. All these specials forms of modulation are designed to enable the highest data speeds in limited bandwidth assignments and to allow multiple users to share the same bandwidth concurrently.

Modulation Applications

Here is a summary of the modulation methods used for the most popular wireless applications.

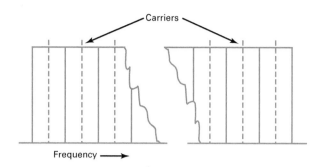

Figure S38-8 Orthogonal frequency division multiplexing (OFDM).

Application	Modulation	Comments
AM broadcast radio	AM	10-kHz bandwidth
FM broadcast radio	FM	200-kHz bandwidth
TV broadcast	8VSB	A combination of AM and PM with a partially suppressed lower sideband
Marine radio	FM	VHF frequency band
Aircraft radio	AM	VHF frequency band
Public radio (police, fire)	FM, FSK, PSK	VHF frequency band
Wireless LAN (Wi-Fi)	BPSK, QAM, OFDM	2.4-GHz and 5-GHz bands
HD radio (broadcast)	OFDM	Simultaneous transmission with AM and FM.
Computer modems	QAM and/or OFDM	Cable TV and DSLmodems
Satellites	BPSK, QAM	Microwave bands
Cell phones	GMSK, CDMA, OFDM, with QPSK, 16QAM, 64QAM	2G-GMSK, 3G-DCMA, 4G-OFDM

CHAPTER 38 REVIEW QUESTIONS

1. An oscillator always needs an amplifier with
   a. positive feedback.
   b. negative feedback.
   c. both types of feedback.
   d. an LC tank circuit.

2. The voltage that starts an oscillator is caused by
   a. ripple from the power supply.
   b. noise voltage in resistors.
   c. the input signal from a generator.
   d. positive feedback.

3. The Wien-bridge oscillator is useful
   a. at low frequencies.
   b. at high frequencies.

   c. with LC tank circuits.
   d. at small input signals.

4. A Wien-bridge oscillator uses
   a. positive feedback.
   b. negative feedback.
   c. both types of feedback.
   d. an LC tank circuit.

5. Initially, the loop gain of a Wien bridge is
   a. 0.
   b. 1.
   c. low.
   d. high.

6. A Wien bridge is sometimes called a
   a. notch filter.
   b. twin-T oscillator.
   c. phase shifter.
   d. Wheatstone bridge.

7. To vary the frequency of a Wien bridge, you can vary
   a. one resistor.
   b. two resistors.
   c. three resistors.
   d. one capacitor.

8. For oscillations to start in a circuit, the loop gain must be greater than 1 when the phase shift around the loop is
   a. 90°.
   b. 180°.
   c. 270°.
   d. 360°.

9. The most widely used LC oscillator is the
   a. Armstrong.
   b. Clapp.
   c. Colpitts.
   d. Hartley.

10. Heavy feedback in an LC oscillator
    a. prevents the circuit from starting.
    b. causes saturation and cutoff.
    c. produces maximum output voltage.
    d. means that B is small.

11. When Q decreases in a Colpitts oscillator, the frequency of oscillation
    a. decreases.
    b. remains the same.
    c. increases.
    d. becomes erratic.

12. Link coupling refers to
    a. capacitive coupling.
    b. transformer coupling.
    c. resistive coupling.
    d. power coupling.

13. The Hartley oscillator uses
    a. negative feedback.
    b. two inductors.
    c. a tungsten lamp.
    d. a tickler coil.

14. To vary the frequency of an LC oscillator, you can vary
    a. one resistor.
    b. two resistors.
    c. three resistors.
    d. one capacitor.

15. Of the following oscillators, the one with the most stable frequency is the
    a. Armstrong.
    b. Clapp.
    c. Colpitts.
    d. Hartley.

16. The material that has the piezoelectric effect is
    a. quartz.
    b. rochelle salts.
    c. tourmaline.
    d. all the above.

17. Crystals have a very
    a. low Q.
    b. high Q.
    c. small inductance.
    d. large resistance.

18. The series and parallel resonant frequencies of a crystal are
    a. very close together.
    b. very far apart.
    c. equal.
    d. low frequencies.

19. The kind of oscillator found in an electronic wristwatch is the
    a. Armstrong.
    b. Clapp.
    c. Colpitts.
    d. quartz crystal.

20. A monostable 555 timer has the following number of stable states:
    a. 0.
    b. 1.
    c. 2.
    d. 3.

21. An astable 555 timer has the following number of stable states:
    a. 0.
    b. 1.
    c. 2.
    d. 3.

22. The pulse width from a one-shot multivibrator increases when the
    a. supply voltage increases.
    b. timing resistor decreases.
    c. UTP decreases.
    d. timing capacitance increases.

23. The output waveform of a 555 timer is
    a. sinusoidal.
    b. triangular.

c. rectangular.
d. elliptical.

24. The quantity that remains constant in a pulse-width modulator is
   a. pulse width.
   b. period.
   c. duty cycle.
   d. space.

25. When a PPL is locked on the input frequency, the VCO frequency
   a. is less than f_0.
   b. is greater than f_0.
   c. equals f_0.
   d. equals f_{in}.

26. The bandwidth of the low-pass filter in a PPL determines the
   a. capture range.
   b. lock range.
   c. free-running frequency.
   d. phase difference.

27. FSK is a method of controlling the output
   a. functions.
   b. amplitude.
   c. frequency.
   d. phase.

28. Most oscillators in use today are
   a. discrete component types.
   b. integrated circuit types.

29. What component is used to create a VCO from an *LC* oscillator?
   a. Capacitor.
   b. Inductor.
   c. Resistor.
   d. Varactor.

30. Crystal oscillators are known and used primarily for their
   a. frequency stability.
   b. reliability.
   c. small size.
   d. low power consumption.

31. Which crystal oscillator delivers the most precise and constant frequency?
   a. VCXO.
   b. TCXO.
   c. OCXO.
   d. XO.

32. A crystal oscillator used to time digital circuits is usually called a
   a. microcontroller.
   b. clock.

c. timer.
d. synchronizer.

33. A microcontroller can generate rectangular pulses of any frequency and duty cycle.
   a. True.
   b. False.

34. A phase-locked loop can be used as a
   a. frequency divider.
   b. FM modulator.
   c. bandpass filter.
   d. phase shifter.

35. The main difference between a PLL and a PLL synthesizer is a
   a. different type of phase detector.
   b. no VCO.
   c. sine-wave output.
   d. a frequency divider in the feedback path.

36. How is the frequency of a PLL synthesizer changed?
   a. Change the feedback division factor.
   b. Vary the VCO frequency with a dc voltage.
   c. Change the input oscillator frequency.
   d. Vary the phase detector sensitivity.

37. Most modern radios and other wireless gear are tuned with a
   a. Colpitts oscillator.
   b. crystal oscillator.
   c. frequency synthesizer.
   d. variable capacitor.

38. What circuit in a DDS actually produces the analog output signal?
   a. Low-pass filter.
   b. DAC.
   c. ROM.
   d. Address register.

39. A DDS can potentially generate any shape output signal.
   a. True.
   b. False.

40. A PLL synthesizer makes a good
   a. adder.
   b. accumulator register.
   c. frequency multiplier.
   d. frequency divider.

SECTION 38.2 The Wien–Bridge Oscillator

38.1 What is the frequency of oscillation of a Wien-bridge oscillator if $C = 200$ pF and $R = 22$ kΩ?

SECTION 38.3 The Colpitts Oscillator

38.2 What is the approximate frequency of oscillation in Fig. 38-37? The value of B? For the oscillator to start, what is the minimum value of A_v?

38.3 What can you do to the inductance of Fig. 38-37 to double the frequency of oscillation?

SECTION 38.4 Other LC Oscillators

38.4 If 47 pF is connected in series with the 10 μH of Fig. 38-37, the circuit becomes a Clapp oscillator. What is the frequency of oscillation?

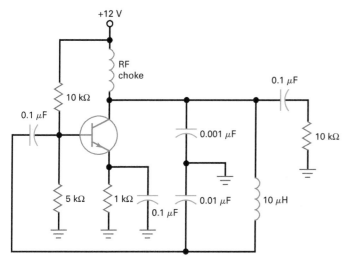

Figure 38-37

SECTION 38.5 Quartz Crystals

38.5 A crystal has a fundamental frequency of 5 MHz. What is the approximate value of the first overtone frequency? The second overtone? The third?

38.6 A crystal has a thickness of t. If you reduce t by 1%, what happens to the frequency?

SECTION 38.6 Crystal Oscillators

38.7 What is the maximum frequency deviation and percent accuracy of a crystal oscillator with a frequency of 40 MHz and a stability of 500 ppm?

SECTION 38.7 Square-Wave, Pulse, and Function Generators

38.8 A 555 timer is connected for monostable operation. If $R = 10$ kΩ and $C = 0.047$ μF, what is the width of the output pulse?

38.9 An astable 555 timer has $R_1 = 10$ kΩ, $R_2 = 2$ kΩ, and $C = 0.0022$ μF. What is the frequency?

38.10 The 555 timer of Fig. 38-21 has $R_1 = 20$ kΩ, $R_2 = 10$ kΩ, and $C = 0.047$ μF. What is frequency of the output signal? What is the duty cycle?

38.11 A pulse-width modulator like the one in Fig. 38-24 has $V_{CC} = 10$ V, $R = 5.1$ kΩ, and $C = 1$ nF. The clock has a frequency of 10 kHz. If a modulating signal has a peak value of 1.5 V, what is the period of the output pulses? What is the quiescent pulse width? What are the minimum and maximum pulse widths? What are the minimum and maximum duty cycles?

38.12 Select a value of L in Fig. 38-37 to get an oscillation frequency of 2.5 MHz.

SECTION 38.9 Frequency Synthesizers

38.13 A PLL synthesizer has a crystal oscillator input from a 4-MHz crystal oscillator followed by a ÷40 frequency divider. What is the frequency step increment?

38.14 A PLL frequency synthesizer has a crystal input of 10 MHz. The frequency divider is set to a division factor of 240. What is the VCO output frequency?

Chapter **39**

Regulated Power Supplies and Power Conversion

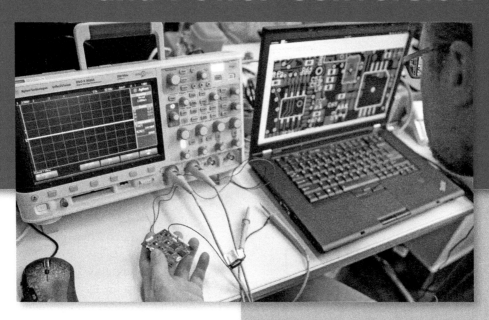

All electronic circuits and equipment require one or more dc voltages for proper operation. These voltages may come from a battery in portable equipment or from an ac-to-dc power supply like that previously discussed in Chap. 30. This chapter continues the power supply discussion with additional circuits and techniques, including voltage regulation and switch-mode power circuits like switching regulators, dc-dc converters and dc-to-ac inverters.

Learning Outcomes

After studying this chapter, you should be able to:

> *Define* and *calculate* regulation.

> *Describe* how series regulators work.

> *Explain* the operation and characteristics of IC voltage regulators.

> *Describe* the three basic topologies of switching regulators.

> *Explain* how dc-dc converters work.

> *State* the benefits of switch-mode power suppliers.

> *Explain* the operation and application of inverters.

39.1 Power Supply Review

Modern electronic power supplies can be categorized in the following ways:

Battery: The primary voltage source is a battery but multiple regulator and dc-dc converter circuits convert the main battery voltage into several different needed dc voltages.

Battery charger: A circuit designed to recharge a secondary battery. It may be another battery or an ac operated power supply with charge control circuits designed for battery charging.

AC-to-dc power supply: A basic supply with a transformer, rectifier, and filter as discussed in Chap. 30. It may also include one or more regulators and dc-dc converters to provide the needed voltages.

Inverter: A dc-to-ac power supply. The dc voltage, generally from a battery, is converted into an ac voltage, usually a 60-Hz sine wave or equivalent.

Power Supply Trends

The older-style "linear" ac-to-dc power supplies introduced in Chap. 30 are still used, but several trends have also necessitated major changes in power supplies. First, the dc operating voltage values are lower. In older equipment, dc voltages of 15, 12, and 5 V were the most common. Today, as a way to cut power consumption and reduce heat, lower voltages are used. The lower voltages are necessary since the smaller geometries of the newer ICs cannot withstand higher voltages. While you will still see the common 15, 12, and 5 V used, most newer ICs use voltages of 3.3, 2.5, 1.8, 1.5, and 0.9 V.

Second, these voltages must be very precise and stable. That means that the voltages must be within a few percent of that specified for the various ICs and that the voltage remains constant over a very narrow range. Third, the "green" movement means that efficiency is more essential than ever. This has resulted in a major change in power supply circuits to those using switching circuits rather than more traditional linear circuits. Today most power supplies are of the "switch-mode" type for which regulators, dc-dc converters, and inverters use switching transistors and switching methods that produce efficiencies in the 70 to 95% range compared to the 30 to 60% range for linear supplies. The result is now not only lower power consumption and less heat but also smaller, lighter power supplies.

39.2 Supply Characteristics

The quality of a power supply depends on its load regulation, line regulation, and output resistance. In this section, we will look at these characteristics because they are often used on data sheets to specify power supplies.

Load Regulation

Figure 39-1 shows a traditional ac-to-dc linear power supply, a bridge rectifier with a capacitor-input filter. Changing the load resistance will change the load voltage. If we reduce the load resistance, we get more ripple and additional voltage

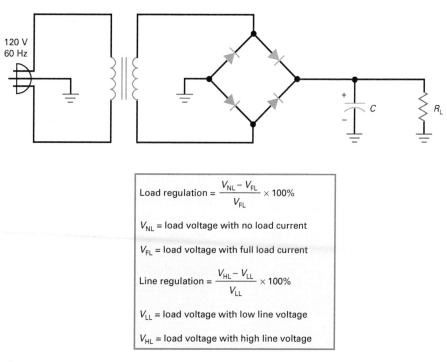

$$\text{Load regulation} = \frac{V_{NL} - V_{FL}}{V_{FL}} \times 100\%$$

V_{NL} = load voltage with no load current

V_{FL} = load voltage with full load current

$$\text{Line regulation} = \frac{V_{HL} - V_{LL}}{V_{LL}} \times 100\%$$

V_{LL} = load voltage with low line voltage

V_{HL} = load voltage with high line voltage

Figure 39-1   Power supply with capacitor-input filter.

drop across the transformer windings and diodes. Because of this, an increase in load current always decreases the load voltage.

Load regulation indicates how much the load voltage changes when the load current changes. The definition for load regulation is

$$\text{Load regulation} = \frac{V_{NL} - V_{FL}}{V_{FL}} \times 100\% \qquad \textbf{(39-1)}$$

where V_{NL} = load voltage with no load current

$\qquad V_{FL}$ = load voltage with full load current

With this definition, V_{NL} occurs when the load current is zero, and V_{FL} occurs when the load current is the maximum value for the design.

For instance, suppose that the power supply of Fig. 39-1 has these values:

$$V_{NL} = 10.6 \text{ V for } I_L = 0$$

$$V_{FL} = 9.25 \text{ V for } I_L = 1 \text{ A}$$

Then, Formula (39-1) gives

$$\text{Load regulation} = \frac{10.6 \text{ V} - 9.25 \text{ V}}{9.25 \text{ V}} \times 100\% = 14.6\%$$

The smaller the load regulation, the better the power supply. For instance, a well-regulated power supply can have a load regulation of less than 1%. This means that the load voltage varies less than 1% over the full range of load current.

Line Regulation

In Fig. 39-1, the input line voltage has a nominal value of 120 V. The actual voltage coming out of a power outlet may vary from 105 to 125 V rms, depending on the time of day, the locality, and other factors. Since the secondary voltage is directly proportional to the line voltage, the load voltage in Fig. 39-1 will change when line voltage changes.

Another way to specify the quality of a power supply is by its **line regulation,** defined as

$$\text{Line regulation} = \frac{V_{HL} - V_{LL}}{V_{LL}} \times 100\% \qquad \textbf{(39-2)}$$

where V_{HL} = load voltage with high line

$\qquad V_{LL}$ = load voltage with low line

For instance, suppose that the power supply of Fig. 39-1 has these measured values:

$$V_{LL} = 9.2 \text{ V for line voltage} = 105 \text{ V rms}$$

$$V_{HL} = 11.2 \text{ V for line voltage} = 125 \text{ V rms}$$

Then, Formula (39-2) gives

$$\text{Line regulation} = \frac{11.2 \text{ V} - 9.2 \text{ V}}{9.2 \text{ V}} \times 100\% = 21.7\%$$

As with load regulation, the smaller the line regulation, the better the power supply. For example, a well-regulated power supply can have a line regulation of less than 0.1%. This means that the load voltage varies less than 0.1% when the line voltage varies from 105 to 125 V rms.

Output Resistance

The Thevenin or output resistance of a power supply determines the load regulation. If a power supply has a low output resistance, its load regulation will also be low. Here is one way to calculate the output resistance:

$$R_{TH} = \frac{V_{NL} - V_{FL}}{I_{FL}} \qquad \textbf{(39-3)}$$

For example, here are the values given earlier for Fig. 39-1:

$$V_{NL} = 10.6 \text{ V for } I_L = 0$$

$$V_{FL} = 9.25 \text{ V for } I_L = 1 \text{ A}$$

For this power supply, the output resistance is

$$R_{TH} = \frac{10.6 \text{ V} - 9.25 \text{ V}}{1 \text{ A}} = 1.35 \text{ } \Omega$$

Figure 39-2 shows a graph of load voltage versus load current. As you can see, the load voltage decreases when the load current increases. The change in load voltage ($V_{NL} - V_{FL}$) divided by the change in current (I_{FL}) equals the output resistance of the power supply. The output resistance is related to the slope of this graph. The more horizontal the graph, the lower the output resistance.

In Fig. 39-2, the maximum load current I_{FL} occurs when the load resistance is minimum. Because of this, an equivalent expression for load regulation is

$$\text{Load regulation} = \frac{R_{TH}}{R_{L(min)}} \times 100\% \qquad \textbf{(39-4)}$$

For example, if a power supply has an output resistance of 1.5 Ω and the minimum load resistance is 10 Ω, it has a load regulation of

$$\text{Load regulation} = \frac{1.5 \text{ } \Omega}{10 \text{ } \Omega} \times 100\% = 15\%$$

39.3 Basic Regulators

The line regulation and load regulation of an unregulated power supply are too high for most applications. By using a voltage regulator between the power supply and the load, we can significantly improve the line and load regulation. A linear voltage regulator uses a device operating in the linear region to hold the load voltage constant. The simplest regulators are based on the zener diode.

Zener Regulator

The simplest **shunt regulator** is the zener-diode circuit of Fig. 39-3. The zener diode operates in the breakdown region,

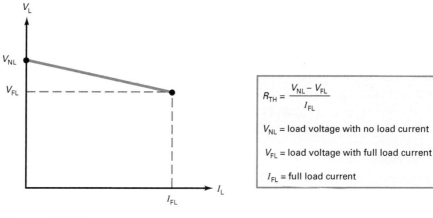

Figure 39-2 Graph of load voltage versus load current.

$$R_{TH} = \frac{V_{NL} - V_{FL}}{I_{FL}}$$

V_{NL} = load voltage with no load current

V_{FL} = load voltage with full load current

I_{FL} = full load current

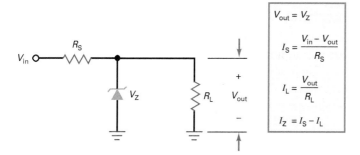

$V_{out} = V_Z$

$$I_S = \frac{V_{in} - V_{out}}{R_S}$$

$$I_L = \frac{V_{out}}{R_L}$$

$$I_Z = I_S - I_L$$

Figure 39-3 Zener regulator is a shunt regulator.

producing an output voltage equal to the zener voltage. When the load current changes, the zener current increases or decreases to keep the current through R_S constant. With any shunt regulator, a change in load current is complemented by an opposing change in shunt current. If load current increases by 1 mA, the shunt current decreases by 1 mA. Conversely, if the load current decreases by 1 mA, the shunt current increases by 1 mA.

As shown in Fig. 39-3, the equation for the current through the series resistor is

$$I_S = \frac{V_{in} - V_{out}}{R_S}$$

This series current equals the *input current* to the regulator. When the input voltage is constant, the input current is almost constant when the load current changes. A change in load current has almost no effect on the input current.

A final point: In Fig. 39-3, the maximum load current with regulation occurs when the zener current is almost zero. Therefore, the maximum load current in Fig. 39-3 equals the input current. The maximum load current with a regulated output voltage is equal to the input current.

The Zener Follower

The primary disadvantage of the basic zener regulator is its limited current- and power-handling capability. Typical zener power ratings are 400, 500, and 1000 mW. This limits the load power to roughly 5 to 10 times those values. For example, using a 12-V, 1-W zener diode, the maximum zener current is $1/12 = 0.0833$ A. Assuming a typical zener current of 20% of load current, the maximum load current would be $0.0833/0.2 = 0.4166$ or just over 400 mA. To overcome this limitation, a common practice is to use an emitter follower between the zener and the load, as shown in Fig. 39-4. The load is the emitter resistance.

The zener diode operates in the breakdown region, producing a base voltage equal to the zener voltage. The transistor is connected as an emitter follower. Therefore, the load voltage equals

$$V_{out} = V_Z - V_{BE} \tag{39-5}$$

If the line voltage or load current changes, the zener voltage and base-emitter voltage will change only slightly. Because of this, the output voltage shows only small changes for large changes in line voltage or load current.

The emitter follower is a series regulator because the load current passes through the transistor. The load current approximately equals the input current because the current through R_S is usually small enough to ignore. The transistor of a series regulator is called a **pass transistor** because all the load current passes through it.

By using a power transistor, loads of many amperes can be regulated. The primary concern is that for the transistor

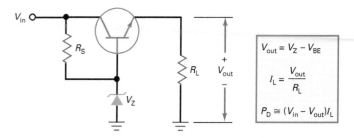

$V_{out} = V_Z - V_{BE}$

$$I_L = \frac{V_{out}}{R_L}$$

$$P_D \cong (V_{in} - V_{out})I_L$$

Figure 39-4 Zener follower is a series regulator.

to remain effective in controlling the output, it must continue to operate in the linear region. For that reason, V_{CE} must remain above about 2 or 3 V. This means that the unregulated input voltage must be higher than the desired output by that amount. Some power will be dissipated in the transistor.

While the emitter-follower regulator is occasionally used, its best regulation is still only a few percent. Newer designs call for improved regulation as well as better efficiency.

39.4 Series Regulators

Improved series regulators use negative feedback in a closed-loop control circuit to greatly improve regulation. A basic series regulator is shown in Fig. 39-5.

Headroom, Power Dissipation, and Efficiency

In Fig. 39-5, the **headroom voltage** is defined as the difference between the input and output voltage:

$$\text{Headroom voltage} = V_{in} - V_{out} \qquad \textbf{(39-6)}$$

The current through the pass transistor of Fig. 39-5 equals

$$I_C = I_L + I_2$$

where I_2 is the current through R_2. To keep the efficiency high, a designer will make I_2 much smaller than the full-load value of I_L. Therefore, we can ignore I_2 at larger load currents and write

$$I_C \cong I_L$$

At high load currents, the power dissipation in the pass transistor is given by the product of headroom voltage and load current:

$$P_D \cong (V_{in} - V_{out})I_L \qquad \textbf{(39-7)}$$

The power dissipation in the pass transistor is very large in some series regulators. In this case, a large heat sink is used. Sometimes, a fan is needed to remove the excess heat inside enclosed equipment.

At full load current, most of the regulator power dissipation is in the pass transistor. Since the current in the pass transistor is approximately equal to the load current, the efficiency is given by

$$\text{Efficiency} \cong \frac{V_{out}}{V_{in}} \times 100\% \qquad \textbf{(39-8)}$$

With this approximation, the best efficiency occurs when the output voltage is almost as large as the input voltage. It implies that the smaller the headroom voltage, the better the efficiency.

To improve the operation of the series regulator, a Darlington connection is often used for the pass transistor. This allows us to use a low-power transistor to drive a power transistor. The Darlington connection allows us to use larger values of R_1 to R_4 to improve the efficiency.

Current Limiting

The series regulator of Fig. 39-5 has no *short-circuit protection*. If we accidentally short the load terminals, the load current will try to approach infinity, which destroys the pass transistor. It may also destroy one or more diodes in the unregulated power supply that is driving the series regulator. To protect against an accidental short across the load, series regulators usually include some form of **current limiting.**

Figure 39-6 shows one way to limit the load current to safe values. In the figure, R_4 is a small resistor called a **current-sensing resistor.** For our discussion, we will use an R_4 of 1 Ω. Since the load current has to pass through R_4, the current-sensing resistor produces the base-emitter voltage for Q_1.

When the load current is less than 600 mA, the voltage across R_4 is less than 0.6 V. In this case, Q_1 is cut off and the regulator works as previously described. When the load current is between 600 and 700 mA, the voltage across R_4 is between 0.6 and 0.7 V. This turns on Q_1. The collector current of Q_1 flows through R_5. This decreases the base voltage to Q_2, which reduces the load voltage and the load current.

When the load is shorted, Q_1 conducts heavily and brings the base voltage of Q_2 down to approximately 1.4 V (two base-emitter voltage drops above ground). The current through the pass transistor is typically limited to 700 mA.

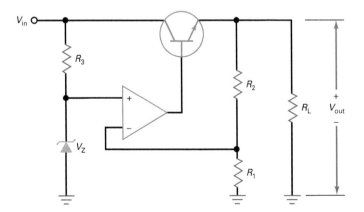

Figure 39-5   Series regulator with large negative feedback.

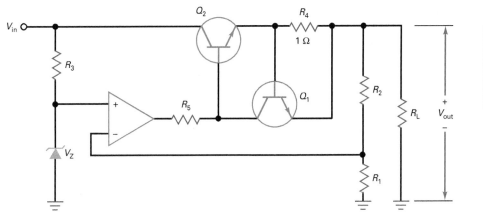

Figure 39-6 Series regulator with current limiting.

$$V_{out} = \frac{R_1 + R_2}{R_1} V_Z$$

$$I_{SL} = \frac{V_{BE}}{R_4}$$

It may be slightly more or less than this, depending on the characteristics of the two transistors.

Incidentally, resistor R_5 is added to the circuit because the output impedance of the op amp is very low (75 Ω is typical). Without R_5, the current-sensing transistor does not have enough voltage gain to produce sensitive current limiting. Typical values of R_5 are from a few hundred to a few thousand ohms.

Current limiting is a big improvement because it will protect the pass transistor and rectifier diodes in case the load terminals are accidentally shorted. But it has the disadvantage of a large power dissipation in the pass transistor when the load terminals are shorted. With a short across the load, almost all the input voltage appears across the pass transistor.

To avoid excessive power dissipation in the pass transistor under shorted-load conditions, some regulator designs incorporate **foldback current limiting.**

When the load current is high enough, further decreases in load resistance cause the current to fold back (decrease). As a result, the shorted-load current is much smaller than it would be without the foldback limiting.

EXAMPLE 39-1

What is the approximate output voltage in Fig. 39-7? Why is a Darlington transistor used?

Answer:

With the equations of Fig. 39-6,

$$V_{out} = \frac{2.7 \text{ k}\Omega + 2.2 \text{ k}\Omega}{2.7 \text{ k}\Omega}(5.6 \text{ V}) = 10.2 \text{ V}$$

The load current is

$$I_L = \frac{10.2 \text{ V}}{4 \Omega} = 2.55 \text{ A}$$

If an ordinary transistor with a current gain of 100 were used for the pass transistor, the required base current would be

$$I_B = \frac{2.55 \text{ A}}{100} = 25.5 \text{ mA}$$

This is too much output current for a typical op amp. If a Darlington transistor is used, the base current of the pass transistor is reduced to a much lower value. For instance, a Darlington transistor with a current gain of 1000 would require a base current of only 2.55 mA.

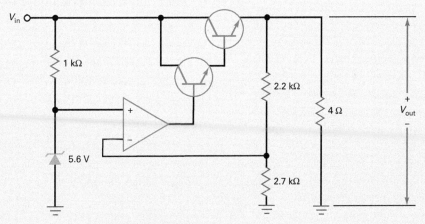

Figure 39-7 Series regulator with Darlington transistor.

EXAMPLE 39-2

When the series regulator of Fig. 39-7 is built and tested, the following values are measured: V_{NL} = 10.16 V, V_{FL} = 10.15 V, V_{HL} = 10.16 V, and V_{LL} = 10.07 V. What is the load regulation? What is the line regulation?

Answer:

$$\text{Load regulation} = \frac{10.16\text{ V} - 10.15\text{ V}}{10.15\text{ V}} \times 100\% = 0.0985\%$$

$$\text{Line regulation} = \frac{10.16\text{ V} - 10.07\text{ V}}{10.07\text{ V}} \times 100\% = 0.894\%$$

This example shows how effective negative feedback is in reducing the effects of line and load changes. In both cases, the change in the regulated output voltage is less than 1%.

39.5 Linear IC Regulators

Most power supply regulators are integrated circuits. Some IC regulators use external resistors can set the current limiting, the output voltage, and so on. By far, the most widely used IC regulators are those with only three pins: one for the unregulated input voltage, one for the regulated output voltage, and one for ground.

Available in plastic or metal packages, the three-terminal regulators have become extremely popular because they are inexpensive and easy to use. Aside from two optional bypass capacitors, three-terminal IC voltage regulators require no external components.

Basic Types of IC Regulators

Most IC voltage regulators have one of these types of output voltage: fixed positive, fixed negative, or adjustable. IC regulators with fixed positive or negative outputs are factory-trimmed to get different fixed voltages with magnitudes from about 5 to 24 V. IC regulators with an adjustable output can vary the regulated output voltage from less than 2 to more than 40 V.

IC regulators are also classified as standard, low power, and low dropout. Standard IC regulators are designed for straightforward and noncritical applications. With heat sinks, a standard IC regulator can have a load current of more than 1 A.

If load currents up to 100 mA are adequate, *low-power IC regulators* are available in TO-92 packages, the same size used for small-signal transistors like the 2N3904. Since these regulators do not require heat sinking, they are convenient and easy to use.

The **dropout voltage** of an IC regulator is defined as the minimum headroom voltage needed for regulation. For instance, standard IC regulators have a dropout voltage of 2 to 3 V. This means that the input voltage has to be at least 2 to 3 V greater than the regulated output voltage for the chip to regulate to specifications. In applications in which 2 to 3 V of headroom is not available, *low-dropout IC regulators* can be used. These regulators have typical dropout voltages of 0.15 V for a load current of 100 mA and 0.7 V for a load current of 1 A.

On-Card Regulation vs. Single-Point Regulation

With *single-point regulation,* we need to build a power supply with a large voltage regulator and then distribute the regulated voltage to all the different *cards* (printed-circuit boards) in the system. This creates problems. To begin with, the single regulator has to provide a large load current equal to the sum of all the card currents. Second, noise or other **electromagnetic interference (EMI)** can be induced on the connecting wires between the regulated power supply and the cards.

Because IC regulators are inexpensive, electronic systems that have many cards often use *on-card regulation.* This means that each card has its own three-terminal regulator to supply the voltage used by the components on that card. By using on-card regulation, we can deliver an unregulated voltage from a power supply to each card and have a local IC regulator take care of regulating the voltage for its card. This eliminates the problems of the large load current and noise pickup associated with single-point regulation.

Load and Line Regulation Redefined

Up to now, we have used the original definitions for load and line regulation. Manufacturers of fixed IC regulators prefer to specify the change in load voltage for a range of load and line conditions. Here are definitions for load and line regulation used on the data sheets of fixed regulators:

$$\text{Load regulation} = \Delta V_{out} \text{ for a range of load current}$$

$$\text{Line regulation} = \Delta V_{out} \text{ for a range of input voltage}$$

For instance, the LM7815 is an IC regulator that produces a fixed positive output voltage of 15 V. The data sheet lists the typical load and line regulation as follows:

$$\text{Load regulation} = 12\text{ mV for } I_L = 5\text{ mA to 1.5 A}$$

$$\text{Line regulation} = 4\text{ mV for } V_{in} = 17.5\text{ V to 30 V}$$

The load regulation will depend on the conditions of measurement. The foregoing load regulation is for T_J = 25°C and V_{in} = 23 V. Similarly, the foregoing line regulation is for T_J = 25°C and I_L = 500 mA. In each case, the junction temperature of the device is 25°C.

The LM7800 Series

The LM78XX series (where XX = 05, 06, 08, 10, 12, 15, 18, or 24) is typical of the three-terminal voltage regulators. The 7805 produces an output of +5 V, the 7806 produces +6 V, the 7808 produces +8 V, and so on, up to the 7824, which produces an output of +24 V.

Figure 39-8 shows the functional block diagram for the 78XX series. A built-in reference voltage V_{ref} drives the

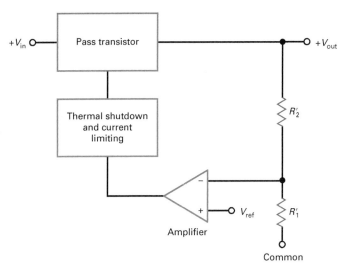

Figure 39-8 Functional block diagram of three-terminal IC regulator.

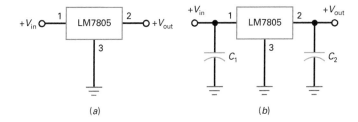

Figure 39-9 (a) Using a 7805 for voltage regulation. (b) Input capacitor prevents oscillations and output capacitor improves frequency response.

noninverting input of an amplifier. The voltage regulation is the similar to our earlier discussion. A voltage divider consisting of R_1' and R_2' samples the output voltage and returns a feedback voltage to the inverting input of a high-gain amplifier. The output voltage is given by

$$V_{out} = \frac{R_1' + R_2'}{R_1'} V_{ref}$$

In this equation, the reference voltage is equivalent to the zener voltage in our earlier discussions. The primes attached to R_1' and R_2' indicate that these resistors are inside the IC itself, rather than being external resistors. These resistors are factory-trimmed to get the different output voltages (5 to 24 V) in the 78XX series. The tolerance of the output voltage is ±4%.

The LM78XX includes a pass transistor that can handle 1 A of load current, provided that adequate heat sinking is used. Also included are thermal shutdown and current limiting. **Thermal shutdown** means that the chip will shut itself off when the internal temperature becomes too high, around 175°C. This is a precaution against excessive power dissipation, which depends on the ambient temperature, type of heat sinking, and other variables. Because of thermal shutdown and current limiting, devices in the 78XX series are almost indestructible.

Fixed Regulator

Figure 39-9a shows an LM7805 connected as a fixed voltage regulator. Pin 1 is the input, pin 2 is the output, and pin 3 is ground. The LM7805 has an output voltage of +5 V and a maximum load current over 1 A. The typical load regulation is 10 mV for a load current between 5 mA and 1.5 A. The typical line regulation is 3 mV for an input voltage of 7 to 25 V. It also has a ripple rejection of 80 dB, which means that it will reduce the input ripple by a factor of

10,000. With an output resistance of approximately 0.01 Ω, the LM7805 is a very stiff voltage source to all loads within its current rating.

When an IC is more than 6 in. from the filter capacitor of the unregulated power supply, the inductance of the connecting wire may produce oscillations inside the IC. This is why manufacturers recommend using a bypass capacitor C_1 on pin 1 (Fig. 39-9b). To improve the transient response of the regulated output voltage, a bypass capacitor C_2 is sometimes used on pin 2. Typical values for either bypass capacitor are from 0.1 to 1 μF. The data sheet of the 78XX series suggests 0.22 μF for the input capacitor and 0.1 μF for the output capacitor.

Any regulator in the 78XX series has a dropout voltage of 2 to 3 V, depending on the output voltage. This means that the input voltage must be at least 2 to 3 V greater than the output voltage. Otherwise, the chip stops regulating. Also, there is a maximum input voltage because of excessive power dissipation. For instance, the LM7805 will regulate over an input range of approximately 8 to 20 V. The data sheet for the 78XX series gives the minimum and maximum input voltages for the other preset output voltages.

The LM79XX Series

The LM79XX series is a group of negative voltage regulators with preset voltages of −5, −6, −8, −10, −12, −15, −18, or −24 V. For instance, an LM7905 produces a regulated output voltage of −5 V. At the other extreme, an LM7924 produces an output of −24 V. With the LM79XX series, the load-current capability is over 1 A with adequate heat sinking. The LM79XX series is similar to the 78XX series and includes current limiting, thermal shutdown, and excellent ripple rejection.

Regulated Dual Supplies

By combining an LM78XX and an LM79XX, as shown in Fig. 39-10, we can regulate the output of a dual supply. The LM78XX regulates the positive output, and the LM79XX handles the negative output. The input capacitors prevent oscillations, and the output capacitors improve transient response. The manufacturer's data sheet recommends the

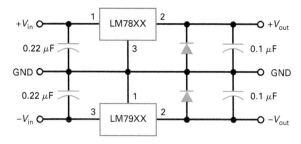

Figure 39-10 Using the LM78XX and LM79XX for dual outputs.

addition of two diodes to ensure that both regulators can turn on under all operating conditions.

An alternative solution for dual supplies is to use a dual-tracking regulator. This is an IC that contains a positive and a negative regulator in a single IC package. When adjustable, this type of IC can vary the dual supplies with a single variable resistor.

Adjustable Regulators

A number of IC regulators (LM317, LM337, LM338, and LM350) are adjustable. These have maximum load currents from 1.5 to 5 A. For instance, the LM317 is a three-terminal positive voltage regulator that can supply 1.5 A of load current over an adjustable output range of 1.25 to 37 V. The ripple rejection is 80 dB. This means that the input ripple is 10,000 smaller at the output of the IC regulator.

Again, manufacturers redefine the load and line regulation to suit the characteristics of the IC regulator. Here are definitions for load and line regulation used on the data sheets of adjustable regulators:

Load regulation = percent change in V_{out} for a range in load current

Line regulation = percent change in V_{out} per volt of input change

For instance, the data sheet of an LM317 lists these typical load and line regulations:

Load regulation = 0.3% for I_L = 10 mA to 1.5 A

Line regulation = 0.02% per volt

Since the output voltage is adjustable between 1.25 and 37 V, it makes sense to specify the load regulation as a percent. For instance, if the regulated voltage is adjusted to 10 V, the foregoing load regulation means that the output voltage will remain within 0.3% of 10 V (or 30 mV) when the load current changes from 10 mA to 1.5 A.

The line regulation is 0.02% per volt. This means that the output voltage changes only 0.02% for each volt of input change. If the regulated output is set at 10 V and the input voltage increases by 3 V, the output voltage will increase by 0.06%, equivalent to 60 mV.

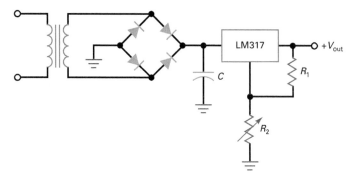

Figure 39-11 Using an LM317 to regulate output voltage.

Figure 39-11 shows an unregulated supply driving an LM317 circuit. The data sheet of an LM317 gives this formula for output voltage:

$$V_{out} = \frac{R_1 + R_2}{R_1} V_{ref} + I_{ADJ}R_2 \qquad \textbf{(39-9)}$$

In this equation, V_{ref} has a value of 1.25 V and I_{ADJ} has a typical value of 50 μA. In Fig. 39-11, I_{ADJ} is the current flowing through the middle pin (the one between the input and the output pins). Because this current can change with temperature, load current, and other factors, a designer usually makes the first term in Formula (39-9) much greater than the second. This is why we can use the following equation for all preliminary analyses of an LM317:

$$V_{out} \cong \frac{R_1 + R_2}{R_1} (1.25 \text{ V}) \qquad \textbf{(39-10)}$$

Ripple Rejection

The ripple rejection of an IC voltage regulator is high, from about 65 to 80 dB. This is a tremendous advantage because it means that we do not have to use bulky *LC* filters in the power supply to minimize the ripple. All we need is a capacitor-input filter that reduces the peak-to-peak ripple to about 10% of the unregulated voltage out of the power supply.

For instance, the LM7805 has a typical ripple rejection of 80 dB. If a bridge rectifier and a capacitor-input filter produce an unregulated output voltage of 10 V with a peak-to-peak ripple of 1 V, we can use an LM7805 to produce a regulated output voltage of 5 V with a peak-to-peak ripple of only 0.1 mV. Eliminating bulky *LC* filters in an unregulated power supply is a bonus that comes with IC voltage regulators.

Low-Dropout Regulators

Most (50–90%) of the lost power and heat generated in a linear regulator occurs in the series pass transistor. To minimize this loss, the overhead voltage across the transistor should be reduced to as low a value as possible. This is a minimum of about 3 V for most regulators. Yet even at such low voltages, the power loss can be considerable. One way to overcome this problem is to use a low-dropout (LDO) IC regulator.

An LDO regulator uses a *pnp* or *p*-channel MOSFET pass transistor instead of the *npn* transistor connected as an emitter follower. A simplified diagram of a LDO regulator using a MOSFET is shown in Fig. 39-12. With this arrangement, the overhead can be as low as several hundred millivolts. A typical range is 0.1 to 1.0 V over which regulation can be maintained. LDO regulators are used where low power consumption is a key design criterion, as in battery-powered electronic products. They have high efficiency and a fast response time to feedback.

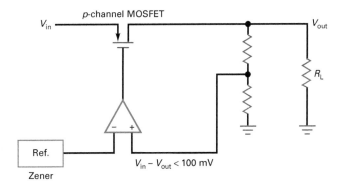

Figure 39-12   A basic LDO regulator circuit.

EXAMPLE 39-3

What is the load current in Fig. 39-13? What is the output ripple?

Answer:
The LM7812 produces a regulated output voltage of +12 V. Therefore, the load current is

$$I_L = \frac{12 \text{ V}}{100 \text{ }\Omega} = 120 \text{ mA}$$

We can calculate the peak-to-peak input ripple with the equation given in Chap. 30:

$$V_R = \frac{I_L}{fC} = \frac{120 \text{ mA}}{(120 \text{ Hz})(1000 \text{ }\mu\text{F})} = 1 \text{ V}$$

The typical ripple rejection for the LM7812 is 72 dB. With a scientific calculator, the exact ripple rejection is

$$RR = \text{antilog}\,\frac{72 \text{ dB}}{20} = 3981$$

The peak-to-peak output ripple is approximately

$$V_R = \frac{1 \text{ V}}{4000} = 0.25 \text{ mV}$$

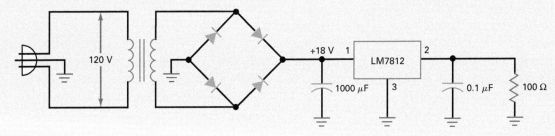

Figure 39-13   Example.

EXAMPLE 39-4

If $R_1 = 2$ kΩ and $R_2 = 22$ kΩ in Fig. 39-11, what is the output voltage? If R_2 is increased to 46 kΩ, what is the output voltage?

Answer:
With Formula (39-10),

$$V_{out} = \frac{2 \text{ k}\Omega + 22 \text{ k}\Omega}{2 \text{ k}\Omega}(1.25 \text{ V}) = 15 \text{ V}$$

When R_2 is increased to 46 kΩ, the output voltage increases to

$$V_{out} = \frac{2 \text{ k}\Omega + 46 \text{ k}\Omega}{2 \text{ k}\Omega}(1.25 \text{ V}) = 30 \text{ V}$$

EXAMPLE 39-5

The LM7805 can regulate to specifications with an input voltage between 7.5 and 20 V. What is the maximum efficiency? What is the minimum efficiency?

Answer:
The LM7805 produces an output of 5 V. With Formula (39-8), the maximum efficiency is

$$\text{Efficiency} \cong \frac{V_{out}}{V_{in}} \times 100\% = \frac{5\text{V}}{7.5 \text{ V}} \times 100\% = 67\%$$

This high efficiency is possible only because the headroom voltage is approaching the dropout voltage.

On the other hand, the minimum efficiency occurs when the input voltage is maximum. For this condition, the headroom voltage is maximum and the power dissipation in the pass transistor is maximum. The minimum efficiency is

$$\text{Efficiency} \cong \frac{5\text{ V}}{20\text{ V}} \times 100\% = 25\%$$

Since the unregulated input voltage is usually somewhere between the extremes in input voltage, the efficiency we can expect with an LM7805 is in the range of 40 to 50%.

39.6 Introduction to Switch-Mode Power Supplies

A switch-mode power supply (SMPS) ia a power supply that uses switching techniques, rather than linear methods, to provide regulation and other power supply functions. Switches are much more efficient than linear amplifier techniques and offer a wide range of other benefits as well. As a result, over the years, most power supply designs have abandoned linear techniques in favor of switching methods. High-speed, high-power transistors such as MOSFETs have made this possible. Today over 90% of all power supplies use switch-mode techniques.

The Efficient Switch

With a linear amplifier, the transistors must be biased so that some control can occur over a given range. This means that emitter-collector or source-drain voltages must be in the 2- to 3-V range at a minimum. With such a voltage drop, considerable power is still dissipated, especially at higher current levels. The result is lower efficiency and more heat to get rid of with heat sinks or fans.

With a switch to control current flow, greater efficiency is inherent. When a switch is open, zero current flows and zero power is dissipated. When the switch is on or closed, its resistance is very low, usually only a fraction of an ohm. So while current does flow, the switch voltage drop is also very low as is the power dissipated.

Both bipolar junction transistors and MOSFETs have been created to meet the switching requirements in SMPS. Their ON resistance is very low, a few ohms for BJTs and less than an ohm for most MOSFETs. Better still, both types of transistors switch very fast. It is during the ON-to-OFF and OFF-to-ON transitions that the transistor must pass through the linear region. Some power is dissipated during these transitions. If the switching time is very fast, power dissipation is minimal. Today, most SMPS use power E-MOSFETs for power generation and control.

Pulse-Width Modulation

SMPS voltages are set and controlled by pulse-width modulation (PWM). PWM is the technique of generating a rectangular wave with a fixed frequency and a varying duty cycle. By varying the duty cycle, the average output voltage may be precisely controlled.

Refer to Fig. 39-14a. The dc output of an unregulated power supply is connected to a load through a series switch. A clock oscillator operates the switch causing it to turn off and on at a fast rate. The average dc voltage across the load is directly related to the duty cycle and the dc input voltage:

$$V_{avg} = V_{dc}(\text{DC}) \tag{39-11}$$

where V_{avg} is that dc voltage across the load measured with a dc multimeter, V_{dc} is the dc input from the power supply, and DC is the duty cycle expressed as a fraction.

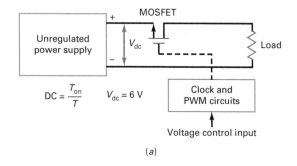

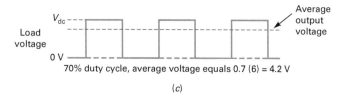

(a)

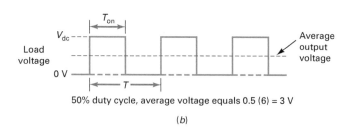

(b)

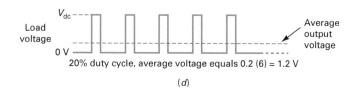

(c)

(d)

Figure 39-14 How PWM varies the duty cycle.

If the power supply voltage is 6 V and the duty cycle is 50%, as in Fig. 39-14b, the average load voltage will be

$$V_{avg} = 6(0.5) = 3 \text{ V}$$

Usually an *RC* low-pass filter is used between the switch and the load to smooth out the pulses into the constant dc average voltage. This is similar to the removal or reduction of the ripple in a conventional power supply filter.

Now note how the duty cycle has increased to 70% in Fig. 39-14c, giving an average output of 6(0.7) = 4.2 V. Decreasing the duty cycle to 20% produces an average output of 6(0.2) = 1.2 V, as shown in Fig. 39-14d. The output can literally be varied continuously from 0 to 6 V by changing the duty cycle.

The key to this variation is the PWM circuit that generates the pulses and controls the duty cycle with an external input. A variety of methods are used in modern ICs. Such PWM techniques are used in most SMPS circuits. Switching frequencies are usually in the 50-kHz to 3-MHz range for modern SMPS ICs.

Generating PWM

Figure 39-15 illustrates one common way that PWM is generated. An error voltage is developed by comparing the dc output of the power supply to a reference voltage like that from a zener diode in a differential amplifier. The difference is a dc error voltage that is used to control the duty cycle. The error voltage is sent to a switching comparator along with a triangular wave generated by an internal oscillator. The frequency of the triangular wave is in the 50-kHz to 3-MHz range. The output of the comparator is a pulse train at the oscillator frequency but one whose duty cycle varies depending on the amplitude of the error voltage. Figure 39-16 shows the process. As the error voltage increases, indicating an increased deviation of the power supply output from its assigned value, the duty cycle decreases. The pulses are then applied to the gate of a power MOSFET or base of a BJT connected as a series switch. This technique is used in both switching regulators and dc-dc converters.

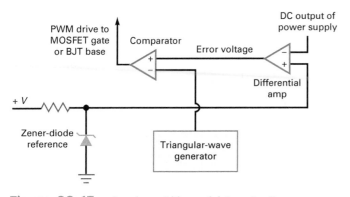

Figure 39-15 A pulse-width modulator circuit.

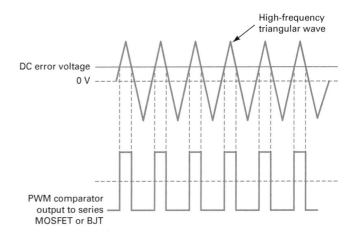

Figure 39-16 Generating PWM by comparing an error voltage to a reference and then using the error voltage to control duty cycle.

Types of SMPS

Switching techniques are used in the three main types of SMPS circuits: regulators, dc-dc converters, and inverters. SMPS regulators use a series MOSFET, and the duty cycle is varied to maintain a constant output with line and load variations. The output is sensed and compared to a zener reference to create a control signal that varies the duty cycle in the PWM circuit.

A dc-dc converter is a circuit that takes one dc input value and changes it to another dc value. The process is one of switching a constant dc input into voltage pulses that can be stepped up or down with a transformer. The ac output pulses are then rectified and filtered back into the new value of dc. Again PWM is the main method of controlling the final dc output.

An inverter is a dc-to-ac power supply. It takes a dc input from a battery or solar panel and converts it to an ac output, typically the 60-Hz, 120-V power-line voltage. Again, switching methods are use to "chop" the input dc into pulses that can be stepped up or down in a transformer. The switched output may be used or filtered into a sine wave as desired. PWM is use to control the output voltage in some designs.

Keep in mind that this same PWM technique is the basis for class D switching amplifiers discussed in Chap. 37.

Advantages and Disadvantages of SMPS

The main advantage of SMPS is high efficiency. Standard linear power supplies rarely have an efficiency greater than 50%. SMPS can achieve efficiencies in the 70 to 95% range. The result is lower overall power consumption, the conservation of energy, and much less heat.

Another major advantage of SMPS is lower cost and smaller size. SMPS usually operate at frequencies above

50 kHz and as high as 3 MHz. As a result, most components like transformers and filter capacitors are considerably smaller and lighter. Less cooling is also required, usually eliminating heat sinks. Overall cost is lower as a result.

The primary disadvantages of SMPS are the switching transients and noise generated. Fast switching produces current spikes and the generation of harmonics that can interfere with other circuits. High-frequency transients can produce electromagnetic interference (EMI) that will interfere with radio signals. However, this disadvantage has been overcome by good filtering and shielding as well as by proper grounding methods.

39.7 Switching Regulators

A *switching regulator* provides voltage regulation, typically by pulse-width modulation controlling the ON-OFF time of the transistor. By changing the duty cycle, a switching regulator can hold the output voltage constant under varying line and load conditions.

The Pass Transistor

In a series regulator the power dissipation of the pass transistor approximately equals the headroom voltage times the load current:

$$P_D = (V_{in} - V_{out})I_L$$

If the headroom voltage equals the output voltage, the efficiency is approximately 50%. For instance, if 10 V is the input to a 7805, the load voltage is 5 V and the efficiency is around 50%.

Three-terminal series regulators are very popular because they are easy to use and fill most of our needs when the load power is less than about 10 W. When the load power is 10 W and the efficiency is 50%, the power dissipation of the pass transistor is also 10 W. This represents a lot of wasted power, as well as heat created inside the equipment. Around load powers of 10 W, heat sinks get very bulky and the temperature of enclosed equipment may rise to objectionable levels.

Switching the Pass Transistor On and Off

The ultimate solution to the problem of low efficiency and high equipment temperature is the switching regulator. With this type of regulator, the pass transistor is switched between cutoff and saturation. When the transistor is cut off, the power dissipation is virtually zero. When the transistor is fully conducting, the power dissipation is still very low because $V_{CE(sat)}$ or $V_{DS(on)}$ is much less than the headroom voltage in a series regulator. As mentioned earlier, switching regulators can have efficiencies from about 75 to more than 95%. Because of the high efficiency and small size, switching regulators have become the primary regulator type.

Topologies

Topology is a term often used in switching-regulator literature. It is the design technique or fundamental layout of a circuit. Many topologies have evolved for switching regulators because some are better suited to an application than others.

Table 39-1 shows the topologies used for switching regulators. The first three are the most basic. They use the fewest number of parts and can deliver load power up to about 150 W. Because their complexity is low, they are widely used, especially with IC switching regulators.

When transformer isolation is preferred, the flyback and the half-forward topologies can be used for load power up to 150 W. When the load power is from 150 to 2000 W, the push-pull, half-bridge, and full-bridge topologies are used. Since the last three topologies use more components, the circuit complexity is high.

Buck Regulator

Figure 39-17a shows a **buck regulator,** the most basic topology for switching regulators. A buck regulator always steps the voltage down. The switch is usually a power MOSFET, but bipolar function transistors are still widely used. A rectangular signal out of the pulse-width modulator closes and opens the switch. A comparator controls the duty cycle of the pulses.

Table 39-1 Switching-Regulator Topologies

Topology	Step	Choke	Transformer	Diodes	Transistors	Power, W	Complexity
Buck	Down	Yes	No	1	1	0–150	Low
Boost	Up	Yes	No	1	1	0–150	Low
Buck-boost	Both	Yes	No	1	1	0–150	Low
Flyback	Both	No	Yes	1	1	0–150	Medium
Half-forward	Both	Yes	Yes	1	1	0–150	Medium
Push-pull	Both	Yes	Yes	2	2	100–1000	High
Half-bridge	Both	Yes	Yes	4	2	100–500	High
Full-bridge	Both	Yes	Yes	4	4	400–2000	Very high

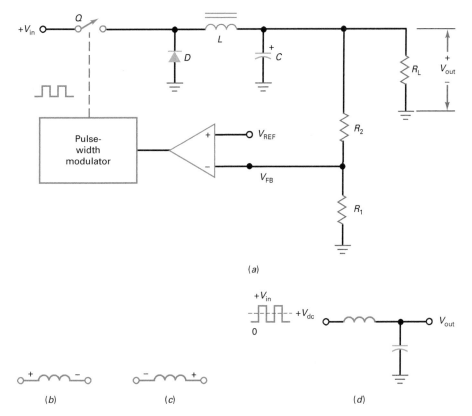

Figure 39-17 (a) Buck regulator. (b) Polarity with closed switch. (c) Polarity with open switch. (d) Choke-input filter passes dc value to output.

When the pulse is high, the switch is closed. This reverse-biases the diode, so that all the input current flows through the inductor. This current creates a magnetic field around the inductor.

The current through the inductor also charges the capacitor and supplies current to the load. While the switch is closed, the voltage across the inductor has the plus-minus polarity shown in Fig. 39-17b. As the current through the inductor increases, more energy is stored in the magnetic field.

When the pulse goes low, the switch opens. At this instant, the magnetic field around the inductor starts collapsing and induces a reverse voltage across the inductor, as shown in Fig. 39-17c. This reverse voltage is called the *inductive kick*. Because of the inductive kick, the diode is forward-biased, and the current through the inductor continues to flow in the same direction. At this time, the inductor is returning its stored energy to the circuit. In other words, the inductor acts like a source and continues supplying current for the load.

Current flows through the inductor until the inductor returns all its energy to the circuit or until the switch closes again, whichever comes first. In either case, the capacitor will also supply load current during part of the time that the switch is open. This way, the ripple across the load is minimized.

The switch is being continuously closed and opened. The frequency of this switching can be from 100 kHz to 3 MHz. The current through the inductor is always in the same direction, passing through either the switch or the diode at different times in the cycle.

The diode is usually a Schottky type so that it can switch fast enough. The average output value is related to the duty cycle and is given by

$$V_{out} = DV_{in} \qquad \textbf{(39-12)}$$

The larger the duty cycle, the larger the dc output voltage.

When the power is first turned on, there is no output voltage and no feedback voltage from the R_1-R_2 voltage divider. Therefore, the comparator output is very large and the duty cycle approaches 100%. As the output voltage builds up, however, the feedback voltage V_{FB} reduces the comparator output, which reduces the duty cycle. At some point, the output voltage reaches an equilibrium value at which the feedback voltage produces a duty cycle that gives the same output voltage.

Because of the high gain of the comparator, the virtual short between the input terminals of the comparator means that

$$V_{FB} \cong V_{REF}$$

From this, we can derive this expression for the output voltage:

$$V_{out} = \frac{R_1 + R_2}{R_1} V_{REF} \qquad \textbf{(39-13)}$$

After equilibrium sets in, any attempted change in the output voltage, whether caused by line or load changes, will be almost entirely offset by the negative feedback. For instance, if the output voltage tries to increase, the feedback voltage reduces the comparator output. This reduces the duty cycle and the output voltage. The net effect is only a slight increase in output voltage, much less than without the negative feedback.

Similarly, if the output voltage tries to decrease because of a line or load change, the feedback voltage is smaller and the comparator output is larger. This increases the duty cycle and produces a larger output voltage that offsets almost all the attempted decrease in output voltage.

Boost Regulator

Figure 39-18a shows a **boost regulator,** another basic topology for switching regulators. The switch is usually a MOSFET but could be a BJT. A boost regulator always steps the voltage up. The theory of operation is similar to that for a buck regulator in some ways but very different in others. For instance, when the pulse is high, the switch is closed

and energy is stored in the magnetic field of L, as previously described.

When the pulse goes low, the switch opens. Again, the magnetic field around the inductor collapses and induces a reverse voltage across the inductor, as shown in Fig. 39-18b. Notice that the input voltage now adds to the inductive kick. This means that the peak voltage on the right end of the inductor is

$$V_p = V_{in} + V_{kick} \qquad \textbf{(39-14)}$$

The inductive kick depends on how much energy is stored in the magnetic field. Stated another way, V_{kick} is proportional to the duty cycle.

With a stiff input voltage, a rectangular voltage waveform appears at the input to the *capacitor-input filter* of Fig. 39-18c. Therefore, the regulated output voltage approximately equals the peak voltage given by Formula (39-14). Because V_{kick} is always greater than zero, V_p is always greater than V_{in}. This is why a boost regulator always steps the voltage up.

Aside from using a capacitor-input filter rather than a choke-input filter, the regulation with boost topology is similar to that with buck topology. Because of the high gain of the comparator, the feedback almost equals the reference voltage. Therefore, the regulated output voltage is still given by Formula (39-13). If the output voltage tries to increase, there is less feedback voltage, less comparator output, a

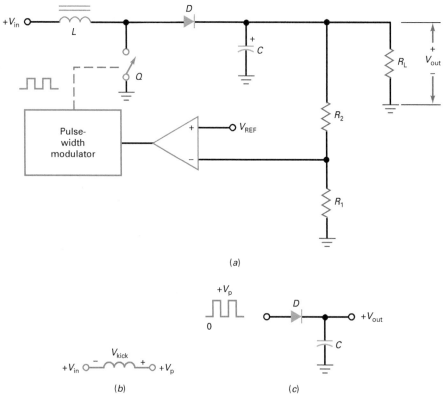

(a)

(b)

(c)

Figure 39-18 (a) Boost regulator. (b) Kick voltage adds to input when switch is open. (c) Capacitor-input filter produces output voltage equal to peak input.

smaller duty cycle, and less inductive kick. This reduces the peak voltage, which offsets the attempted increase in output voltage. If the output voltage tries to decrease, the smaller feedback voltage results in a larger peak voltage, which offsets the attempted decrease in output voltage.

Buck-Boost Regulator

Figure 39-19a shows a **buck-boost regulator,** the third most basic topology for switching regulators. A buck-boost regulator always produces a negative output voltage when driven by a positive input voltage. When the PWM output is high, the switch is closed and energy is stored in the magnetic field. At this time, the voltage across the inductor equals V_{in}, with the polarity shown in Fig. 39-19b.

When the pulse goes low, the switch opens. Again, the magnetic field around the inductor collapses and induces a kick voltage across the inductor, as shown in Fig. 39-19c. The kick voltage is proportional to the energy stored in the magnetic field, which is controlled by the duty cycle. If the duty cycle is low, the kick voltage approaches zero. If the duty cycle is high, the kick voltage can be greater than V_{in}, depending on how much energy is stored in the magnetic field.

In Fig. 39-19d, the magnitude of the peak voltage may be less than or greater than the input voltage. The diode and the capacitor-input filter then produce an output voltage equal to $-V_p$. Since the magnitude of this output voltage can be less than or greater than the input voltage, the topology is called *buck-boost.*

An inverting amplifier is used in Fig. 39-19a to invert the feedback voltage before it reaches the inverting input of the comparator. The voltage regulation then works as previously described. Attempted increases in output voltage reduce the duty cycle, which reduces the peak voltage. Attempted decreases in output voltage increase the duty cycle. Either way, the negative feedback holds the output voltage almost constant.

IC Switching Regulators

There are hundreds of different IC regulators made by companies like Analog Devices, Linear Technology, Maxim Integrated Products, Texas Instruments, and others. Most of the previously mentioned topologies are available in IC form. Each regulator IC contains the PWM circuits, error amplifier, and related components. All these regulators are supported by external discrete components that cannot be easily integrated. They include the inductor, storage/filter capacitors, and the Schottky diode. In lower-voltage, lower-current (less than several amps) devices, the series MOSFET

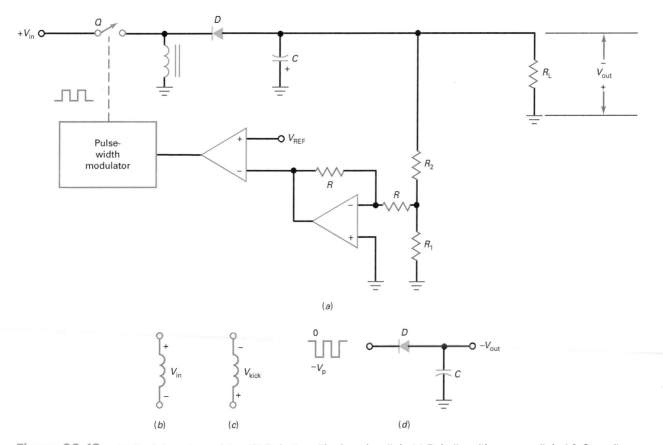

Figure 39-19 (a) Buck-boost regulator. (b) Polarity with closed switch. (c) Polarity with open switch. (d) Capacitor-input filter produces output equal to negative peak.

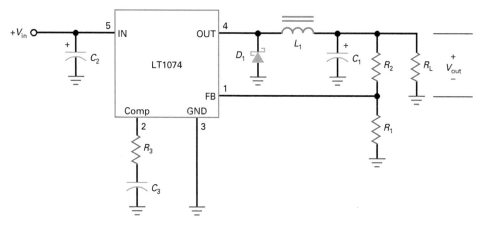

Figure 39-20 Buck regulator IC.

or bipolar switch is inside the IC. For higher voltages and currents, an external MOSFET or BJT is used.

As an example, a generic buck regulator IC circuit is shown in Fig. 39-20. It contains most of the components discussed earlier, such as a reference voltage (2.21 V), a MOSFET or BJT switch, an internal oscillator, a pulse-width modulator, and a comparator. It runs at a switching frequency of 100 kHz, can handle input voltages from +8 to +40 Vdc, and has efficiency of 75 to 90% for load currents from 1 to 5 A.

Figure 39-20 shows the IC connected as a buck regulator. Pin 1 (FB) is for the feedback voltage. Pin 2 (comp) is for frequency compensation to prevent oscillations at higher frequencies. Pin 3 (GND) is ground. Pin 4 (out) is the switched output of the internal switching device. Pin 5 (in) is for the unregulated dc input voltage.

In Fig. 39-20, D_1, L_1, C_1, R_1, and R_2 serve the same functions as described in the earlier discussion of a buck regulator. But notice the use of a Schottky diode to improve the efficiency of the regulator. Because the Schottky diode has a lower knee voltage and faster switching speed, it wastes less power. The data sheet of the IC recommends adding a capacitor C_2 from 200 to 470 μF across the input for line filtering. Also recommended are a resistor R_3 of 2.7 kΩ and a capacitor C_3 of 0.01 μF to stabilize the feedback loop (prevent oscillations) in the internal error amp.

The circuit is incredibly simple. The IC includes everything except the components that cannot be integrated (choke and filter capacitors) and those left for the user to select (R_1 and R_2). By selecting values for R_1 and R_2, you can get regulated output voltage from about 2.5 to 38 V. Since the reference voltage is 2.21 V, the output voltage is given by

$$V_{out} = \frac{R_1 + R_2}{R_1}(2.21 \text{ V}) \qquad \textbf{(39-15)}$$

Most of the newer IC switching regulators operate at higher frequencies, and 1, 2, and 3 MHz are popular because the inductor size can be much smaller as can the capacitor values. Frequencies as high as 6 MHz are used. Another variation is the integrating of two, three, or even more switching regulators in the same package. These ICs are designed for battery and portable products with multiple power supplies that must be smaller and more compact.

39.8 DC-DC Converters

A *dc-dc converter* is a circuit that changes one dc value to another. Many modern electronic products require multiple dc supply voltages to operate the different ICs. Laptop computers and cell phones are excellent examples. Rather than use one completely independent power supply for each voltage, some electronic equipment today employs a single dc supply and then uses either regulators or dc-dc converters to produce the other desired voltages.

Figure 39-21 shows a common power supply architecture found in modern equipment. A single ac-to-dc power supply (or a battery) supplies a common power bus of unregulated dc voltage. This bus is distributed to the different circuits, modules, or subsystems over a printed-circuit board copper pattern or via connectors. Then regulators or dc-dc converters convert the bus voltage to the desired end voltages.

Another example is a complex IC whose internal circuits need multiple voltages. Rather than require the designer to use multiple external supplies, the IC designer builds in the needed dc-dc converters on chip to produce the extra voltages so that the IC can operate from a single supply.

In reality, the switching regulators discussed in the previous section are one common form of dc-dc converter, as they do definitely transform one dc level into another but introduce regulation as well. There are also several other forms of dc-dc converters you may encounter in your work, including the charge pump, forward converter, and flyback converter. Following is an introduction to these basic types.

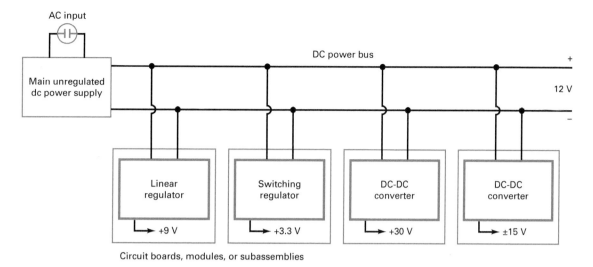

AC input

DC power bus

Main unregulated
dc power supply

12 V

+

−

| Linear regulator | Switching regulator | DC-DC converter | DC-DC converter |

+9 V   +3.3 V   +30 V   ±15 V

Circuit boards, modules, or subassemblies

Figure 39-21 Bus architecture power supply.

Charge Pump

A charge pump is a circuit that uses switches and capacitors to generate a higher dc voltage than the input or a voltage of a different polarity. Capacitors store charges that are transferred to other capacitors through the switches. The techniques used are similar to the operation of voltage multiplier circuits discussed back in Chap. 30. The basic charge pump circuit is illustrated in Fig. 39-22. Capacitors are used to store voltages and Schottky diodes are used for the switches.

The dc input voltage to be converted V_{dc} is applied to the input along with rectangular clock pulses that switch between V_{dc} and ground. A CMOS inverter can to be used to supply such voltages as shown. When the clock input is zero, C_1 is grounded while V_{dc} is applied to C_2, and C_1 then charges to V_{dc} through D_1 which is forward-biased.

When the clock switches, the lower end of C_1 is connected to V_{dc} from the clock* signal. The voltage across C_1 is then in series with V_{dc} from the clock, making a voltage of $2V_{dc}$. The lower end of C_2 is grounded via the inverted clock* input so C_2 charges to $2V_{dc}$. On the next clock reversal, the voltage on C_2 is transferred in series with V_{dc} from

the clock, so C_3 charges to $3V_{dc}$. Clock speeds vary from 10 kHz to over 100 kHz.

For most applications the output needs to be only two or three times the input dc voltage. If only a two times multiplication is needed, D_3 and C_3 can be eliminated. If more than $3V_{dc}$ is needed, extra diode and capacitor sections can be added. Note that the output voltage is actually less than $3V_{dc}$ by an amount equal to the sum of the diode voltage drops. Each diode drop is in the 0.2- to 0.4-V range of the typical Schottky diode.

In modern designs, the diodes are replaced by MOSFETs, connected as diodes or as conventional switches. The output is unregulated and may have to be followed by a regulator. Capacitor sizes are usually large, meaning that in most designs the capacitors are too large to integrate, so external capacitors are needed. Electrolytic capacitors are common. Most charge pump converters are used with light (low-current) loads. For high-current needs other forms of dc-dc converters should be used.

Forward Converter

A typical forward converter is shown in Fig. 39-23. A MOSFET, M_1, "chops" the input dc voltage into a high-frequency square wave that is applied to the primary of the transformer. Switching frequencies range from a few hundred kilohertz to over 1 MHz. The transformer can then step up or step down the voltage to the desired level. A standard full-wave rectifier with D_1 and D_2 converts the output to dc pulses that are filtered into a constant dc by L_1 and C_1. DC feedback from the output via R_1 and R_2 operates the PWM circuit that varies the duty cycle of the switching wave to the MOSFET. The dc output is regulated.

A wide range of variations of the forward converter are used. All feature the external transformer, Schottky diodes, inductor, and filter capacitor. For higher-current applications a push-pull version of this circuit with two MOSFETs

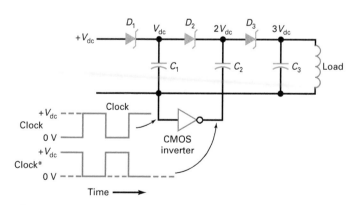

Figure 39-22 A charge pump dc-dc converter.

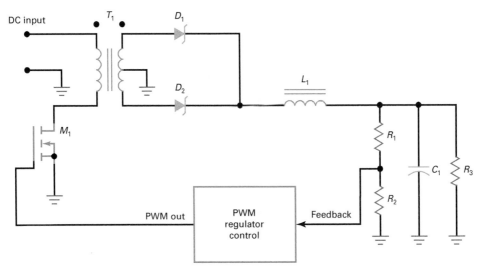

Figure 39-23 A dc-dc forward converter.

switching the input dc between the two sides of a center-tapped primary winding is used.

Flyback Converter

A flyback converter is illustrated in Fig. 39-24. Its design is similar to a forward converter; however, a special transformer stores energy in its magnetic field to induce higher pulses of voltage into the secondary windings. In this design, three secondary windings provide three different output voltages. Note the upper voltage, which is 35 kV. Flyback power supplies have been used for years in the older TV sets and video monitors that use cathode ray tubes (CRTs) that require a high-operating voltage. Only the lower-output voltage is regulated by the PWM circuit.

While a dc-dc converter is commonly made with a PWM IC and external MOSFETs, inductors, capacitors, and diodes, they are also available as a complete component. These converters are available in a wide range of voltages and power capabilities. Just apply input voltage and the desired output voltage is available. These packaged dc-dc converters are standardized and are referred to as "bricks." See Fig. 39-25 which shows the basic brick that can handle a power of up to 500 W. Smaller bricks of ½, ¼, ⅛ and ¹⁄₁₆ size and power ratings are also available. These dc-dc converters are fully isolated, meaning they use an internal transformer to separate and isolate the input and outputs so that separate grounds may be used.

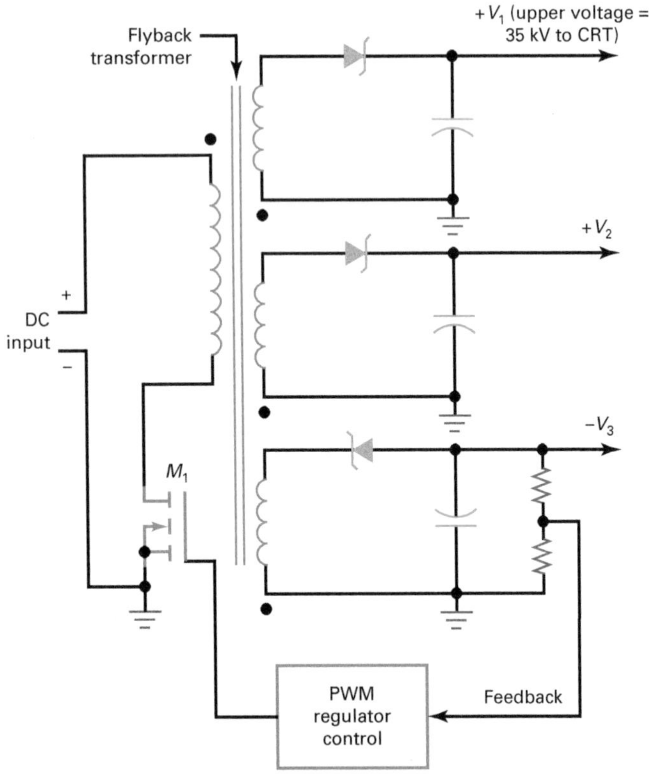

Figure 39-24 A flyback dc-dc converter.

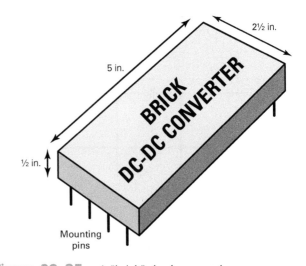

Figure 39-25 A "brick" dc-dc converter.

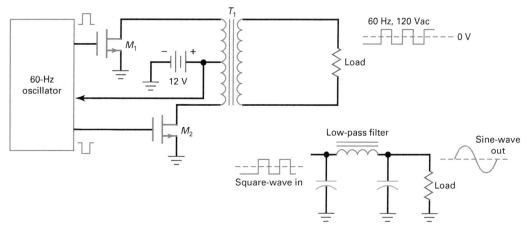

Figure 39-26   A basic inverter circuit.

39.9 Inverters

An *inverter* is an electronic circuit that converts dc to ac. It converts dc power into ac power. It is a special case of a switch-mode power supply. Power MOSFETs, or special semiconductors called *thyristors,* are used to perform the switching of high voltages and currents. The most common example of an inverter is one that converts 12-V dc battery power to standard 120-V, 60-Hz ac. Inverters are widely used, and their use is increasing with the green movement and the many alternative energy sources.

Inverter Concepts

Figure 39-26 shows the basic idea of an inverter. A battery powers an oscillator and driving circuit that operates two power MOSFETs. The oscillator generates a 60-Hz square wave, shapes it, and changes its level so that it can drive the gates of the MOSFETs. The MOSFETs are switches that alternately connect one side of the primary winding of a transformer to the 12-V battery. When M_1 conducts, M_2 is off and current flows in the upper part of the primary winding. The resulting square wave of voltage is stepped up by the transformer and appears as a larger square wave across the secondary winding.

When Q_2 conducts, Q_1 cuts off. Current flows in the lower part of the transformer. The voltage is stepped up by the transformer to produce the higher output voltage. In this case, because of the reversal of the current flow in the primary, the secondary voltage has the opposite polarity. The result is a 60-Hz ac square wave of the desired voltage level.

The 60-Hz, square-wave output can be used as is for some applications, but others require a waveform that is a pure sine wave or at least closer to one. Figure 39-27 shows a popular alternative pulse waveform that can substitute for a sine wave. The 60-Hz oscillator generates the two pulse trains to drive the MOSFET gates. In either case, the pulse waveforms can be filtered with a low-pass filter, as shown in Fig. 39-26 to produce a near perfect 60-Hz sine wave.

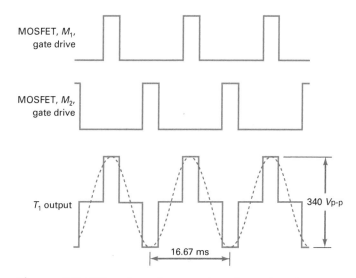

Figure 39-27   Alternative waveform for an inverter.

Some Alternative Inverter Circuits

Early inverters used bipolar transistor in the circuit of Fig. 39-28. A special transformer provided the voltage step up, as well as positive feedback voltage windings (L_2 and L_3) for the transistors. Note that L_1 and L_4 are the primary and secondary windings, respectively, making the inverter self-oscillating. Both Q_1 and Q_2 alternately conduct, connecting the battery to the transformer windings. Voltages induced in the feedback windings turn the transistors off and on as the circuit oscillates.

The frequency of oscillation is a function of the transformer characteristics, including core size and type and the saturation flux density of the core. The core is designed to be easily saturated and, as a result, has a large distinctive hysteresis curve. The transformer switches from a saturated state in one magnetic direction to saturation in the opposite magnetic direction. This produces a square-wave output at the design frequency, usually 60 Hz.

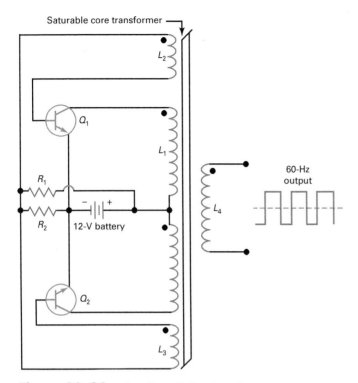

Figure 39-28 A self-oscillating inverter.

Another approach to inverter implementation is to use PWM. A pulse-width modulator can replace the oscillator driving circuit of the inverter in Fig. 39-26 to produce PWM in the secondary. The switching oscillator will switch at several kilohertz or higher. The output would look something like that in Fig. 39-29. When the output pulses are averaged in the load, the result is a near sine-wave response.

Inverter Applications

Inverters are more common than you may think. Here are a few of the more widely used examples.

Uninterruptible Power Supply An uninterruptible power supply (UPS) is a common product that is widely used to keep computers and other critical equipment working during a power failure. A computer system can lose critical data during an ac power outage. Network servers are particularly vulnerable to a power failure and connections are lost. It can take hours or days to return a network to fully operational

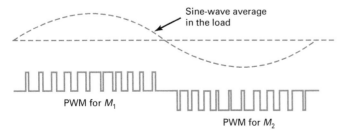

Figure 39-29 A PWM inverter.

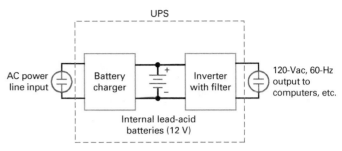

Figure 39-30 An uninterruptible power supply.

condition. Therefore the goal is to keep it from losing power by plugging the computers into a UPS.

Figure 39-30 shows the block diagram of the typical UPS. The UPS has a built-in 12-V battery with its own internal charger that keeps the battery fully charged from the ac line. The battery, in turn, operates an inverter that generates standard ac line voltage from which the computers and other equipment operate. If the ac power should fail, the battery keeps on supplying the inverters and power continues. When the power returns, the battery is recharged. The size and capacity of the battery determine how long the output can continue, which is typically several hours. UPSs with a wide range of capacity and duration options are available.

Generic Inverter You can also buy an all-purpose inverter as a stand-alone unit. Typical models come with a plug that is designed to plug into the 12-V outlet in most cars. The output is two or more standard three-prong ac outlets, just like those conventional wall outputs in any home. Units are available with power output ranges from about 150 W to several kilowatts. They can easily power electric lights, radios, TV sets, audio equipment, or almost anything else that operates from the 120-V line. Larger kilowatt inverters are also available.

Solar Inverter An inverter is the key component in solar power systems. It converts the dc voltage from solar panels to the standard ac power line. Home solar systems use multiple inverters, which are available in several sizes to fit the solar panel capacity and the power load of the house. Special inverters are also used to connect unused solar power back to the grid.

Motor Drives AC motors turn at a speed determined by the frequency of the ac voltage that powers them. To vary the speed, you need an ac variable-frequency drive. This is a special circuit that takes standard 60-Hz input and rectifies it into dc that is then used to operate a variable-frequency inverter to power the motor, so its speed can be varied. AC drives are a common piece of equipment in industrial equipment using motors. Most electric cars use ac motors for power. The battery packs in these cars power special inverters or ac drives to control the car's speed.

1. Voltage regulators normally use
   a. negative feedback.
   b. positive feedback.
   c. no feedback.
   d. phase limiting.

2. During regulation, the power dissipation of the pass transistor equals the collector-emitter voltage times the
   a. base current.
   b. load current.
   c. zener current.
   d. foldback current.

3. Without current limiting, a shorted load will probably
   a. produce zero load current.
   b. destroy diodes and transistors.
   c. have a load voltage equal to the zener voltage.
   d. have too little load current.

4. A current-sensing resistor is usually
   a. zero.
   b. small.
   c. large.
   d. open.

5. A capacitor may be needed in a discrete voltage regulator to prevent
   a. negative feedback.
   b. excessive load current.
   c. oscillations.
   d. current sensing.

6. If the output of a voltage regulator varies from 15 to 14.7 V between the minimum and maximum load current, the load regulation is
   a. 0%.
   b. 1%.
   c. 2%.
   d. 5%.

7. If the output of a voltage regulator varies from 20 to 19.8 V when the line voltage varies over its specified range, the source regulation is
   a. 0%.
   b. 1%.
   c. 2%.
   d. 5%.

8. The output impedance of a voltage regulator is
   a. very small.
   b. very large.
   c. equal to the load voltage divided by the load current.
   d. equal to the input voltage divided by the output current.

9. Compared to the ripple into a voltage regulator, the ripple out of a voltage regulator is
   a. equal in value.
   b. much larger.
   c. much smaller.
   d. impossible to determine.

10. A voltage regulator has a ripple rejection of -60 dB. If the input ripple is 1 V, the output ripple is
    a. -60 mV.
    b. 1 mV.
    c. 10 mV.
    d. 1000 V.

11. Thermal shutdown occurs in an IC regulator if
    a. power dissipation is too low.
    b. internal temperature is too high.
    c. current through the device is too low.
    d. any of the above occur.

12. If a linear three-terminal IC regulator is more than a few inches from the filter capacitor, you may get oscillations inside the IC unless you use
    a. current limiting.
    b. a bypass capacitor on the input pin.
    c. a coupling capacitor on the output pin.
    d. a regulated input voltage.

13. The 78XX series of voltage regulators produces an output voltage that is
    a. positive.
    b. negative.
    c. either positive or negative.
    d. unregulated.

14. The LM7812 produces a regulated output voltage of
    a. 3 V.
    b. 4 V.
    c. 12 V.
    d. 78 V.

15. A series regulator is an example of a
    a. linear regulator.
    b. switching regulator.
    c. shunt regulator.
    d. dc-dc converter.

16. To get more output voltage from a buck switching regulator, you have to
    a. decrease the duty cycle.
    b. decrease the input voltage.
    c. increase the duty cycle.
    d. increase the switching frequency.

17. An increase of line voltage into a power supply usually produces
    a. a decrease in load resistance.
    b. an increase in load voltage.
    c. a decrease in efficiency.
    d. less power dissipation in the rectifier diodes.

18. A power supply with low output impedance has low
    a. load regulation.
    b. current limiting.
    c. line regulation.
    d. efficiency.

19. A zener-diode regulator is a
    a. shunt regulator.
    b. series regulator.
    c. switching regulator.
    d. zener follower.

20. The efficiency of a voltage regulator is high when
    a. input power is low.
    b. output power is high.
    c. little power is wasted.
    d. input power is high.

21. A switching regulator is considered
    a. quiet.
    b. noisy.
    c. inefficient.
    d. linear.

22. The zener follower is an example of a
    a. boost regulator.
    b. shunt regulator.
    c. buck regulator.
    d. series regulator.

23. The efficiency of a linear regulator is high when the
    a. headroom voltage is low.
    b. pass transistor has a high power dissipation.
    c. zener voltage is low.
    d. output voltage is low.

24. The dropout voltage of standard monolithic linear regulators is closest to
    a. 0.3 V.
    b. 0.7 V.
    c. 2 V.
    d. 3.1 V.

25. In a buck regulator, the output voltage is filtered with a
    a. choke-input filter.
    b. capacitor-input filter.
    c. diode.
    d. voltage divider.

26. The regulator with the highest efficiency is the
    a. shunt regulator.
    b. series regulator.

27. In a boost regulator, the output voltage is filtered with a
    a. choke-input filter.
    b. capacitor-input filter.
    c. diode.
    d. voltage divider.

28. The buck-boost regulator is also
    a. a step-down regulator.
    b. a step-up regulator.
    c. an inverting regulator.
    d. all of the above.

29. A key disadvantage of switch-mode power supplies is its
    a. lower efficiency.
    b. higher efficiency.
    c. noise generation.
    d. complexity.

30. A low-dropout regulator
    a. is a switching regulator.
    b. has very low overhead.
    c. consumes more power than a standard linear regulator.
    d. is not available In IC form.

31. In pulse-width modulation, the following is correct.
    a. The pulse duty cycle is varied.
    b. The pulse frequency is varied.
    c. The pulse amplitude is varied.
    d. Both *b* and *c*.

32. Switching power supplies operated up to which pulse frequencies?
    a. Less than 50 kHz.
    b. 100 kHz to 300 kHz.
    c. 500 kHz to 1 MHz.
    d. Over 3 MHz.

33. A PWM signal has a pulse amplitude of 5 V and a duty cycle of 35%. What is the average output voltage?
    a. 1.75 V.
    b. 2.5 V.
    c. 3.5 V.
    d. 4.25 V.

34. A charge pump is a type of
    a. regulator.
    b. dc-dc converter.
    c. linear regulator.
    d. inverter.

35. The main purpose of the switching transistor in a SMPS is to
    a. vary the series load current.
    b. convert ac into dc.

c. average the load current.

d. switch dc into a square wave.

36. A "brick" is a type of

a. regulator.

b. dc-dc converter.

c. linear regulator.

d. inverter.

37. An inverter converts

a. dc to ac.

b. ac to dc.

c. dc to dc.

d. ac to ac.

38. The output of an inverter may be a

a. square wave.

b. pulse wave.

c. sine wave.

d. any of the above.

39. In a self-oscillating inverter, the frequency of operation is set by the

a. transistors.

b. *RC* network.

c. transformer.

d. an external circuit.

40. PWM can be used to create a sine-wave output.

a. True.

b. False.

CHAPTER 39 PROBLEMS

SECTION 39.2 Supply Characteristics

39.1 A power supply has $V_{NL} = 15$ V and $V_{FL} = 14.5$ V. What is the load regulation?

39.2 A power supply has $V_{HL} = 20$ V and $V_{LL} = 19$ V. What is the line regulation?

39.3 If line voltage changes from 108 to 135 V and load voltage changes from 12 to 12.3 V, what is the line regulation?

39.4 A power supply has an output resistance of 2 Ω. If the minimum load resistance is 50 Ω, what is the load regulation?

SECTION 39.4 Series Regulators

39.5 In Fig. 39-5, $V_{in} = 20$ V, $V_Z = 4.7$ V, $R_1 = 2.2$ kΩ, $R_2 = 4.7$ kΩ, $R_3 = 1.5$ kΩ, $R_4 = 2.7$ kΩ, and $R_L = 50$ Ω. What is the output voltage? What is the power dissipation in the pass transistor?

39.6 What is the approximate efficiency in Prob. 24-8?

39.7 In Fig. 39-7, the zener voltage is changed to 6.2 V. What is the approximate output voltage?

SECTION 39.5 Linear IC Regulators

39.8 What is the load current in Fig. 39-31? The headroom voltage? The power dissipation of the LM7815?

39.9 What is the output ripple in Fig. 39-31?

39.10 If $R_1 = 2.7$ kΩ and $R_2 = 20$ kΩ in Fig. 39-11, what is the output voltage?

39.11 The LM7815 is used with an input voltage that can vary from 18 to 25 V. What is the maximum efficiency? The minimum efficiency?

SECTION 39.7 Switching Regulators

39.12 A buck regulator has $V_{REF} = 2.5$ V, $R_1 = 1.5$ kΩ, and $R_2 = 10$ kΩ. What is the output voltage?

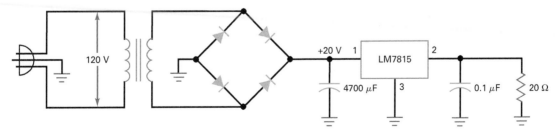

Figure 39-31 Example.

39.13 If the duty cycle is 30% and the peak value of the pulses to the choke-input filter is 20 V, what is the regulated output voltage?

39.14 A boost regulator has $V_{REF} = 1.25$ V, $R_1 = 1.2$ kΩ, and $R_2 = 15$ kΩ. What is the output voltage?

39.15 A buck-boost regulator has $V_{REF} = 2.1$ V, $R_1 = 2.1$ kΩ, and $R_2 = 12$ kΩ. What is the output voltage?

SECTION 39.8 DC-DC Converters

39.16 A dc-dc converter has an input voltage of 5 V and an output voltage of 12 V. If the input current is 1 A and the output current is 0.25 A, what is the efficiency of the dc-dc converter?

39.17 A dc-dc converter has an input voltage of 12 V and an output voltage of 5 V. If the input current is 2 A and the efficiency is 80%, what is the output current?

39.18 If the load regulation is 5% and the no-load voltage is 12.5 V, what is the full-load voltage?

39.19 If the line regulation is 3% and the low-line voltage is 16 V, what is the high-line voltage?

39.20 A power supply has a load regulation of 1% and a minimum load resistance of 10 Ω. What is the output resistance of the power supply?

Chapter 40

Thyristors

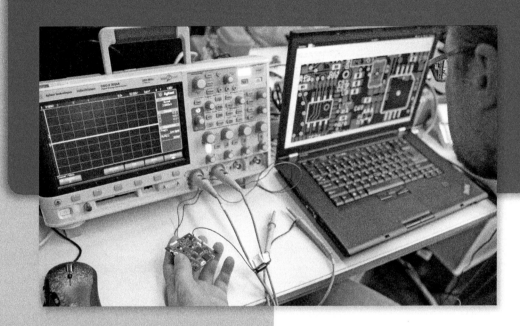

Learning Outcomes

After studying this chapter, you should be able to:

> *Describe* the four-layer diode, how it is turned on, and how it is turned off.

> *Explain* the characteristics of SCRs.

> *Demonstrate* how to test SCRs.

> *Calculate* the firing and conduction angles of *RC* phase control circuits.

> *Explain* the characteristics of triacs and diacs.

> *Compare* the switching control of IGBTs to power MOSFETs.

> *Describe* the major characteristics of the photo-SCR and silicon-controlled switch.

The word thyristor comes from the Greek and means "door," as in opening a door and letting something pass through it. A thyristor is a semiconductor device that uses internal feedback to produce switching action. The most important thyristors are the silicon-controlled rectifier (SCR) and the triac. Like power FETs, the SCR and the triac can switch large currents on and off. Because of this, they can be used for overvoltage protection, motor controls, heaters, lighting systems, and other heavy-current loads. Insulated-gate bipolar transistors (IGBTs) are not included in the thyristor family, but are covered in this chapter as an important power-switching device.

40.1 The Four-Layer Diode

Thyristor operation can be explained in terms of the equivalent circuit shown in Fig. 40-1a. The upper transistor Q_1 is a *pnp* device, and the lower transistor Q_2 is an *npn* device. The collector of Q_1 drives the base of Q_2. Similarly, the collector of Q_2 drives the base of Q_1.

Positive Feedback

The unusual connection of Fig. 40-1a uses *positive feedback*. Any change in the base current of Q_2 is amplified and fed back through Q_1 to magnify the original change. This positive feedback continues changing the base current of Q_2 until both transistors go into either saturation or cutoff.

For instance, if the base current of Q_2 increases, the collector current of Q_2 increases. This increases the base current of Q_1 and the collector current of Q_1. More collector current in Q_1 will further increase the base current of Q_2. This amplify-and-feedback action continues until both transistors are driven into saturation. In this case, the overall circuit acts like a closed switch (Fig. 40-1b).

On the other hand, if something causes the base current of Q_2 to decrease, the collector current of Q_2 decreases, the base current of Q_1 decreases, the collector current of Q_1 decreases, and the base current of Q_2 decreases further. This action continues until both transistors are driven into cutoff. Then, the circuit acts like an open switch (Fig. 40-1c).

The circuit of Fig. 40-1a is stable in either of two states: *open* or *closed*. It will remain in either state indefinitely until acted on by an outside force. If the circuit is open, it stays open until something increases the base current of Q_2. If the circuit is closed, it stays closed until something decreases the base current of Q_2. Because the circuit can remain in either state indefinitely, it is called a *latch*.

Closing a Latch

Figure 40-2a shows a latch connected to a load resistor with a supply voltage of V_{CC}. Assume that the latch is open, as shown in Fig. 40-2b. Because there is no current through the load resistor, the voltage across the latch equals the supply

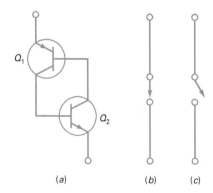

Figure 40-1 Transistor latch.

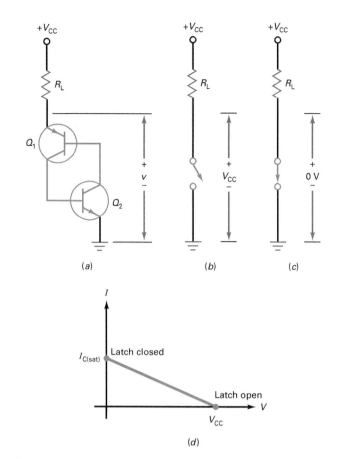

Figure 40-2 Latching circuit.

voltage. So the operating point is at the lower end of the dc load line (Fig. 40-2d).

The only way to close the latch of Fig. 40-2b is by **breakover**. This means using a large enough supply voltage V_{CC} to break down the Q_1 collector diode. Since the collector current of Q_1 increases the base current of Q_2, the positive feedback will start. This drives both transistors into saturation, as previously described. When saturated, both transistors ideally look like short circuits, and the latch is closed, (Fig. 40-2c). Ideally, the latch has zero voltage across it when it is closed, and the operating point is at the upper end of the load line (Fig. 40-2d).

In Fig. 40-2a, breakover can also occur if Q_2 breaks down first. Although breakover starts with the breakdown of either collector diode, it ends with both transistors in the saturated state. This is why the term *breakover* is used instead of *breakdown* to describe this kind of latch closing.

Opening a Latch

How do we open the latch of Fig. 40-2a? By reducing the V_{CC} supply to zero. This forces the transistors to switch from saturation to cutoff. We call this type of opening **low-current dropout** because it depends on reducing the latch current to a value low enough to bring the transistors out of saturation.

The Schockley Diode

Figure 40-3*a* was originally called a **Schockley diode** after the inventor. Several other names are also used for this device: **four-layer diode,** *pnpn diode,* and **silicon unilateral switch (SUS).** The device lets current flow in only one direction.

The easiest way to understand how it works is to visualize it separated into two halves, as shown in Fig. 40-3*b*. The left half is a *pnp transistor,* and the right half is an *npn transistor.* Therefore, the four-layer diode is equivalent to the latch of Fig. 40-3*c*.

Figure 40-3*d* shows the schematic symbol of a four-layer diode. The only way to close a four-layer diode is by breakover. The only way to open it is by low-current dropout, which means reducing the current to less than the **holding current** (given on data sheets). The holding current is the low value of current where the transistors switch from saturation to cutoff.

After a four-layer diode breaks over, the voltage across it ideally drops to zero. In reality, there is some voltage across the latched diode. Figure 40-3*e* shows current versus voltage for a 1N5158 that is latched on. As you can see, the voltage across the device increases when the current increases: 1 V at 0.2 A, 1.5 V at 0.95 A, 2 V at 1.8 A, and so on.

Breakover Characteristic

Figure 40-4 shows the graph of current versus voltage of a four-layer diode. The device has two operating regions: cutoff and saturation. The dashed line is the transition

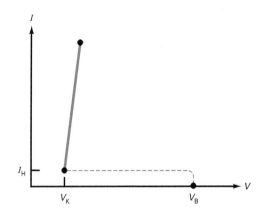

Figure 40-4 Breakover characteristic.

path between cutoff and saturation. It is dashed to indicate that the device switches rapidly between the OFF and ON states.

When the device is at cutoff, it has zero current. If the voltage across diode tries to exceed V_B, the device breaks over and moves rapidly along the dashed line to the saturation region. When the diode is in saturation, it is operating on the upper line. As long as the current through it is greater than the holding current I_H, the diode remains latched in the ON state. If the current becomes less than I_H, the device switches into cutoff.

The ideal approximation of a four-layer diode is an open switch when cut off and a closed switch when saturated.

40.2 The Silicon-Controlled Rectifier

The **SCR** is the most widely used thyristor. It can switch very large currents on and off. Because of this, it is used to control motors, ovens, air conditioners, and induction heaters.

Triggering the Latch

By adding an input terminal to the base of Q_2, as shown in Fig. 40-5*a*, we can create a second way to close the latch. Here is the theory of operation: When the latch is open, as shown in Fig. 40-5*b*, the operating point is at the lower end of the dc load line (Fig. 40-5*d*). To close the latch, we can couple a *trigger* (sharp pulse) into the base of Q_2, as shown in Fig. 40-5*a*. The trigger momentarily increases the base current of Q_2. This starts the positive feedback, which drives both transistors into saturation.

When saturated, both transistors ideally look like short circuits, and the latch is closed (Fig. 40-5*c*). Ideally, the latch has zero voltage across it when it is closed, and the operating point is at the upper end of the load line (Fig. 40-5*d*).

Gate Triggering

Figure 40-6*a* shows the structure of the SCR. The input is called the *gate*, the top is the *anode*, and the bottom is the *cathode*. The SCR is far more useful than a four-layer

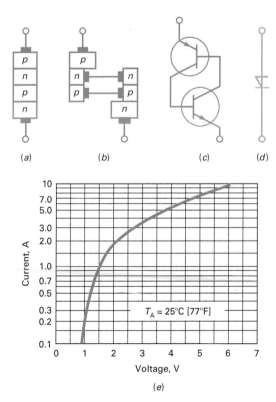

Figure 40-3 Four-layer diode.

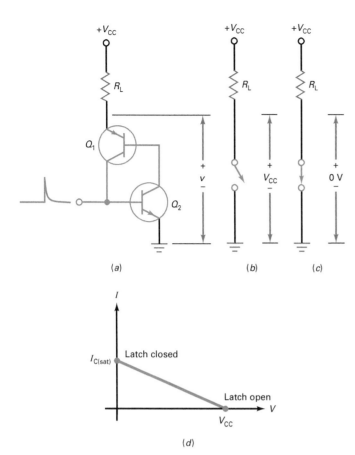

(a)   *(b)*   *(c)*

(d)

Figure 40-5   Transistor latch with trigger input.

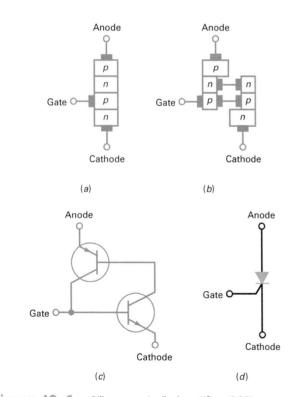

(a)   *(b)*

(c)   *(d)*

Figure 40-6   Silicon-controlled rectifier (SCR).

Figure 40-7   Typical SCRs.

diode because the gate triggering is easier than breakover triggering.

Again, we can visualize the four doped regions separated into two transistors, as shown in Fig. 40-6b. Therefore, the SCR is equivalent to a latch with a trigger input (Fig. 40-6c). Schematic diagrams use the symbol of Fig. 40-6d. Whenever you see this symbol, remember that it is equivalent to a latch with a trigger input. Typical SCRs are shown in Fig. 40-7.

Since the gate of an SCR is connected to the base of an internal transistor, it takes at least 0.7 V to trigger an SCR. Data sheets list this voltage as the **gate trigger voltage, V_{GT}.** Rather than specify the input resistance of the gate, a manufacturer gives the minimum input current needed to turn on the SCR. Data sheets list this current as the **gate trigger current, I_{GT}.**

Some typical trigger voltage and current values are:

$$V_{GT} = 1.0 \text{ V}$$
$$I_{GT} = 9.0 \text{ mA}$$

This means that the source driving the gate of a typical 2N6504 series SCR has to supply 9.0 mA at 1.0 V to latch the SCR.

Also, the breakover voltage or blocking voltage is specified as its peak repetitive OFF state forward voltage, V_{DRM}, and its peak repetitive OFF state reverse voltage, V_{RRM}. Depending on which SCR of the series is used, the breakover voltage ranges from 50 V to 800 V.

Required Input Voltage

An SCR like the one shown in Fig. 40-8 has a gate voltage V_G. When this voltage is more than V_{GT}, the SCR will turn on and the output voltage will drop from $+V_{CC}$ to a low value. Sometimes, a gate resistor is used as shown here. This resistor limits the gate current to a safe value. The input voltage needed to trigger an SCR has to be more than:

$$V_{in} = V_{GT} + I_{GT}R_G \qquad \textbf{(40-1)}$$

In this equation, V_{GT} and I_{GT} are the gate trigger voltage and current for the device. For instance, the data sheet of a 2N4441 gives $V_{GT} = 0.75$ V and $I_{GT} = 10$ mA. When you

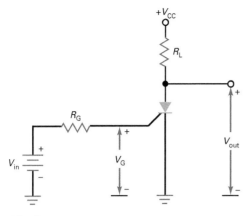

Figure 40-8   Basic SCR circuit.

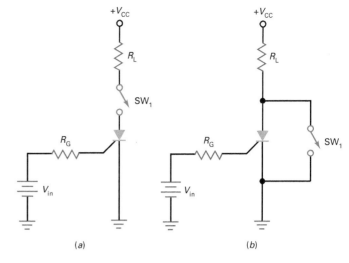

(a)

(b)

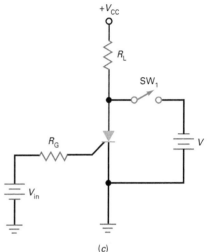

(c)

Figure 40-9   Resetting the SCR.

have the value of R_G, the calculation of V_{in} is straightforward. If a gate resistor is not used, R_G is the Thevenin resistance of the circuit driving the gate. Unless Formula (40-1) is satisfied, the SCR cannot turn on.

Resetting the SCR

After the SCR has turned on, it stays on even though you reduce the gate supply, V_{in}, to zero. In this case, the output remains low indefinitely. To reset the SCR, you must reduce the anode to cathode current to a value less than its holding current, I_H. This can be done by reducing V_{CC} to a low value. The data sheet for the 2N6504 lists a typical holding current value of 18 mA. SCRs with lower and higher power ratings generally have lower and higher respective holding current values. Since the holding current flows through the load resistor in Fig. 40-8, the supply voltage for turnoff has to be less than

$$V_{CC} = 0.7 \text{ V} + I_H R_L \qquad \textbf{(40-2)}$$

Besides reducing V_{CC}, other methods can be used to reset the SCR. Two common methods are current interruption and forced commutation. By either opening the series switch, as shown in Fig. 40-9a, or closing the parallel switch in Fig. 40-9b, the anode-to-cathode current will drop down below its holding current value and the SCR will switch to its OFF state.

Another method used to reset the SCR is forced commutation, as shown in Fig. 40-9c. When the switch is depressed, a negative V_{AK} voltage is momentarily applied. This reduces the forward anode-to-cathode current below I_H and turns off the SCR. In actual circuits, the switch can be replaced with a BJT or FET device.

Power FET versus SCR

Although both the power FET and the SCR can switch large currents on and off, the two devices are fundamentally different. The key difference is the way they turn off. The gate voltage of a power FET can turn the device on and off. This

is not the case with an SCR. The gate voltage can only turn it on.

Figure 40-10 illustrates the difference. In Fig. 40-10a, when the input voltage to the power FET goes high, the output voltage goes low. When the input voltage goes low, the output voltage goes high. In other words, a rectangular input pulse produces an inverted rectangular output pulse.

In Fig. 40-10b, when the input voltage to the SCR goes high, the output voltage goes low. But when the input voltage goes low, the output voltage stays low. With an SCR, a rectangular input pulse produces a negative-going output step. The SCR does not reset.

Because the two devices have to be reset in different ways, their applications tend to be different. Power FETs respond like push-button switches, whereas SCRs respond like single-pole, single-throw switches. Since it is easier to control the power FET, you will see it used more often as an interface between digital ICs and heavy loads. In applications in which latching is important, you will see the SCR used.

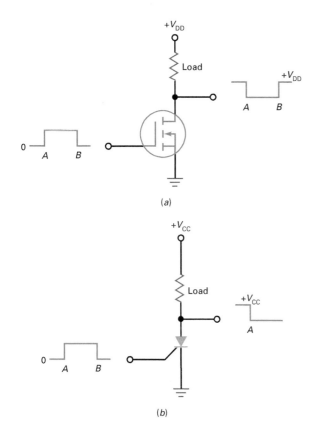

Figure 40-10   Power FET versus SCR.

EXAMPLE 40-1

In Fig. 40-11, the SCR has a trigger voltage of 0.75 V and a trigger current of 7 mA. What is the input voltage that turns the SCR on? If the holding current is 6 mA, what is the supply voltage that turns it off?

Answer:
With Formula (40-1), the minimum input voltage needed to trigger is

$$V_{in} = 0.75 \text{ V} + (7 \text{ mA})(1 \text{ k}\Omega) = 7.75 \text{ V}$$

With Formula (40-2), the supply voltage that turns off the SCR is

$$V_{CC} = 0.7 \text{ V} + (6 \text{ mA})(100 \text{ }\Omega) = 1.3 \text{ V}$$

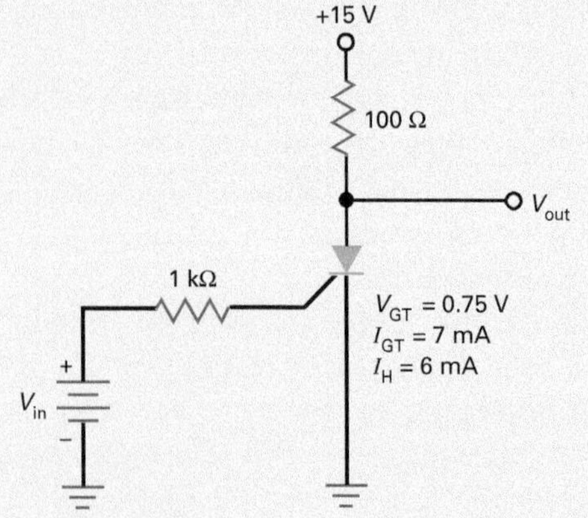

Figure 40-11   Example.

40.3 The SCR Crowbar

If anything happens inside a power supply to cause its output voltage to go excessively high, the results can be devastating. Why? Because some loads such as expensive digital ICs cannot withstand too much supply voltage without being destroyed. One of the most important applications of the SCR is to protect delicate and expensive loads against overvoltages from a power supply.

Prototype

Figure 40-12 shows a power supply of V_{CC} applied to a protected load. Under normal conditions, V_{CC} is less than the breakdown voltage of the zener diode. In this case, there is no voltage across R, and the SCR remains open. The load receives a voltage of V_{CC}, and all is well.

Now, assume that the supply voltage increases for any reason whatsoever. When V_{CC} is too large, the zener diode breaks down and a voltage appears across R. If this voltage is greater than the gate trigger voltage of the SCR, the SCR fires and becomes a closed latch. The action is similar to throwing a *crowbar* across the load terminals. Because the SCR turn-on is very fast (1 μs for a 2N4441), the load is quickly protected against the damaging effects of a large overvoltage. The overvoltage that fires the SCR is

$$V_{CC} = V_Z + V_{GT} \qquad \textbf{(40-3)}$$

Crowbarring, though a drastic form of protection, is necessary with many digital ICs because they can't take much overvoltage. Rather than destroy expensive ICs, therefore, we can use an SCR crowbar to short the load terminals at the first sign of overvoltage. With an SCR crowbar, a fuse or *current limiting* (discussed later) is needed to prevent damage to the power supply.

Integrated Crowbar

The simplest solution is to use an IC crowbar, as shown in Fig. 40-13. This is an integrated circuit with a built-in zener diode, transistors, and an SCR. Different models are available to protect different voltage supply levels. In some cases the crowbar circuit is built into other power supply ICs.

The prototype crowbar is all right if the application is not too critical about the exact supply voltage at which the SCR

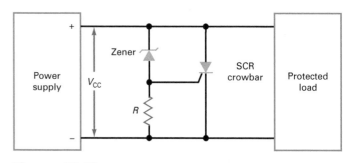

Figure 40-12   SCR crowbar.

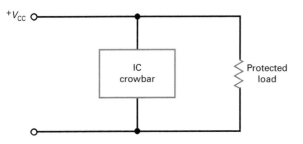

Figure 40-13 IC crowbar.

turns on. For instance, the 1N752 has a tolerance of ±10%, which means that the breakdown voltage may vary from 5.04 to 6.16 V. Furthermore, the trigger voltage of a 2N4441 has a worst-case maximum of 1.5 V. So, the overvoltage can be as high as

$$V_{CC} = 6.16 \text{ V} + 1.5 \text{ V} = 7.66 \text{ V}$$

Since many digital ICs have a maximum rating of 7 V, a simple crowbar cannot be used to protect them.

40.4 SCR Phase Control

SCRs are available in a wide range of current and voltage ratings. Typical current ranges are from 1.5 A to over 1.5 kA with voltage ratings exceeding 2 kV. Most of these devices are used to control heavy industrial loads using phase control.

RC Circuit Controls Phase Angle

Figure 40-14*a* shows ac line voltage being applied to an SCR circuit that controls the current through a heavy load. In this circuit, variable resistor R_1 and capacitor C shift the phase angle of the gate signal. When R_1 is zero, the gate voltage is in phase with the line voltage, and the SCR acts like a half-wave rectifier. Resistor R_2 limits the gate current to a safe level.

When R_1 increases, however, the ac gate voltage lags the line voltage by an angle between 0 and 90°, as shown in Fig. 40-14*b* and *c*. Before the trigger point shown in Fig. 40-14*c*, the SCR is off and the load current is zero.

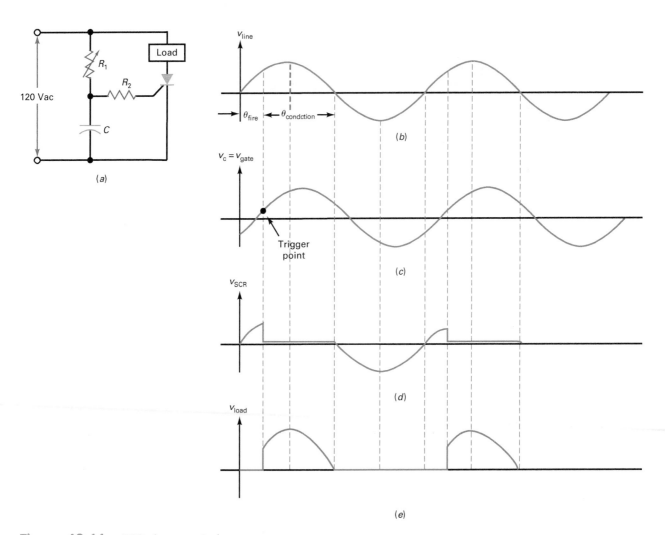

Figure 40-14 SCR phase control.

At the trigger point, the capacitor voltage is large enough to trigger the SCR. When this happens, almost all the line voltage appears across the load and the load current becomes high. Ideally, the SCR remains latched until the line voltage reverses polarity. This is shown in Fig. 40-14c and d.

The angle at which the SCR fires is called the **firing angle,** shown as θ_{fire} in Fig. 40-14a. The angle between the start and end of conduction is called the **conduction angle,** shown as $\theta_{conduction}$. The RC phase controller of Fig. 40-14a can change the firing angle between 0 and 90°, which means that the conduction angle changes from 180 to 90°.

The shaded portions of Fig. 40-14b show when the SCR is conducting. Because R_1 is variable, the phase angle of the gate voltage can be changed. This allows us to control the shaded portions of the line voltage. Stated another way: We can control the average current through the load. This is useful for changing the speed of a motor, the brightness of a lamp, or the temperature of an induction furnace.

By using circuit analysis techniques studied in basic electricity courses, we can determine the approximate phase-shifted voltage across the capacitor. This gives us the approximate firing angle and conduction angle of the circuit. To determine the voltage across the capacitor, use the following steps:

First, find the capacitive reactance of C by

$$X_C = \frac{1}{2\pi fc}$$

The impedance and phase angle of the RC phase-shift circuit are

$$Z_T = \sqrt{R^2 + X_C^2} \qquad \textbf{(40-4)}$$

$$\theta_Z = \angle -\arctan\frac{X_C}{R} \qquad \textbf{(40-5)}$$

Using the input voltage as our reference point, the current through C is

$$I_C \angle \theta = \frac{V_{in}\angle 0°}{Z_T\angle - \arctan\dfrac{X_C}{R}}$$

Now, the voltage value and phase of the capacitor can be found by

$$V_C = (I_C \angle \theta)(X_C\angle -90°)$$

The amount of delayed phase shift will be the approximate firing angle of the circuit. The conduction angle is found by subtracting the firing angle from 180°.

EXAMPLE 40-2

Using Fig. 40-14a, find the approximate firing angle and conduction angle when $R = 26$ kΩ.

Answer:
The approximate firing angle can be found by solving for the voltage value and its phase shift across the capacitor. This is found by

$$X_C = \frac{1}{2\pi fc} = \frac{1}{(2\pi)(60\ \text{Hz})(0.1\ \mu\text{F})} = 26.5\ \text{k}\Omega$$

Because capacitive reactance is at an angle of −90°, $X_C = 26.5$ kΩ∠−90°.

Next, find the total RC impedance Z_T and its angle by

$$Z_T = \sqrt{R^2 + X_C^2} = \sqrt{(26\ \text{k}\Omega)^2 + (26.5\ \text{k}\Omega)^2} = 37.1\ \text{k}\Omega$$

$$\theta_Z = \angle -\arctan\frac{X_C}{R} = \angle -\arctan\frac{26.5\ \text{k}\Omega}{26\ \text{k}\Omega} = -45.5°$$

Therefore, $Z_T = 37.1$ kΩ∠ −45.5°.

Using the ac input as our reference, the current through C is

$$I_C = \frac{V_{in}\angle 0°}{Z_T\angle\theta} = \frac{120\ \text{Vac}\angle 0°}{37.1\ \text{k}\Omega\angle -45.5°} = 3.23\ \text{mA}\angle 45.5°$$

Now, the voltage across C can be found by

$$V_C = (I_C\angle\theta)(X_C\angle -90°)$$
$$= (3.23\ \text{mA}\angle 45.5°)(26.5\ \text{k}\Omega\angle -90°)$$
$$V_C = 85.7\ \text{Vac}\angle -44.5°$$

With the voltage phase shift across the capacitor of −44.5°, the firing angle of the circuit is approximately −45.5°. After the SCR fires, it will remain on until its current drops below I_H. This will occur approximately when the ac input is zero volts.

Therefore, the conduction angle is

$$\text{Conduction } \theta = 180° - 44.5° = 135.5°$$

The RC phase controller of Fig. 40-14a is a basic way of controlling the average current through the load. The controllable range of current is limited because the phase angle can change from only 0 to 90°. With op amps and more sophisticated RC circuits, we can change the phase angle from 0 to 180°. This allows us to vary the average current all the way from zero to maximum.

Critical Rate of Rise

When ac voltage is used to supply the anode of an SCR, it is possible to get false triggering. Because of capacitances inside an SCR, rapidly changing supply voltages may trigger the SCR. To avoid false triggering of an SCR, the rate of voltage change must not exceed the *critical rate of voltage rise* specified on the data sheet. For instance,

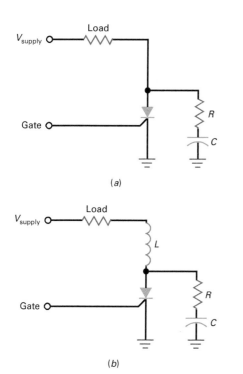

(a)

(b)

Figure 40-15 *(a)* RC snubber protects SCR against rapid voltage rise. *(b)* Inductor protects SCR against rapid current rise.

the 2N6504 has a critical rate of voltage rise of 50 V/μs. To avoid false triggering, the anode voltage must not rise faster than 50 V/μs.

Switching transients are the main cause of exceeding the critical rate of voltage rise. One way to reduce the effects of switching transients is with an *RC snubber,* shown in Fig. 40-15*a*. If a high-speed switching transient does appear on the supply voltage, its rate of rise is reduced at the anode because of the *RC* time constant.

Large SCRs also have a *critical rate of current rise.* For instance, the C701 has a critical rate of current rise of 150 A/μs. If the anode current tries to rise faster than this, the SCR will be destroyed. Including an inductor in series with the load (Fig. 40-15*b*) reduces the rate of current rise to a safe level.

40.5 Bidirectional Thyristors

The two devices discussed so far, the four-layer diode and the SCR, are unidirectional because current can flow in only one direction. The **diac** and **triac** are *bidirectional thyristors.* These devices can conduct in either direction. The diac is sometimes called a *silicon bidirectional switch (SBS).*

Diac

The diac can latch current in either direction. The equivalent circuit of a diac is 2 four-layer diodes in parallel, as shown in Fig. 40-16*a*, ideally the same as the latches in Fig. 40-16*b*.

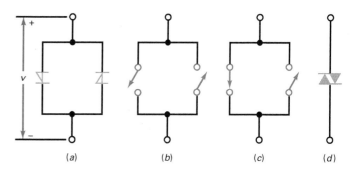

Figure 40-16 Diac.

The diac is nonconducting until the voltage across it exceeds the breakover voltage in either direction.

For instance, if *v* has the polarity indicated in Fig. 40-16*a*, the left diode conducts when *v* exceeds the breakover voltage. In this case, the left latch closes, as shown in Fig. 40-16*c*. When *v* has the opposite polarity, the right latch closes. Figure 40-16*d* shows the schematic symbol for a diac.

Triac

The triac acts like two SCRs in reverse parallel (Fig. 40-17*a*), equivalent to the two latches of Fig. 40-17*b*. Because of this, the triac can control current in both directions. If *v* has the polarity shown in Fig. 40-17*a*, a positive trigger will close the left latch. When *v* has opposite polarity, a negative trigger will close the right latch. Figure 40-17*c* is the schematic symbol for a triac.

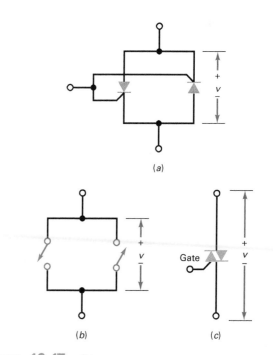

(a)

(b)   *(c)*

Figure 40-17 Triac.

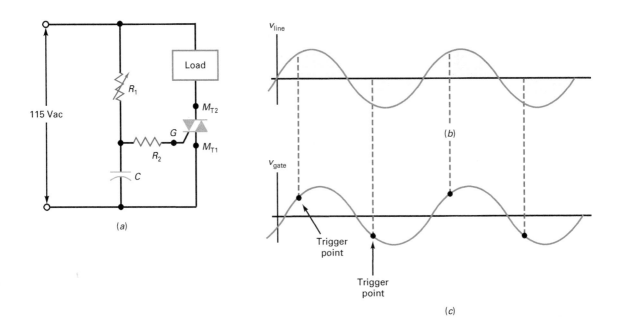

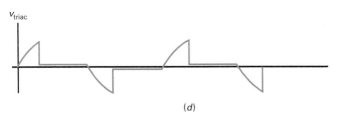

Figure 40-18 Triac phase control.

There are a wide range of commercially available triacs. Because of their internal structure, triacs have higher gate trigger voltages and currents than comparable SCRs. Gate trigger voltages range from 2 to 2.5 V, and the gate trigger currents are from 10 to 50 mA. The maximum anode currents are commonly from 1 to 15 A.

Phase Control

Figure 40-18a shows an *RC* circuit that varies the phase angle of the gate voltage to a triac. The circuit can control the current through a heavy load. Figure 40-18b and c shows the line voltage and lagging gate voltage. When the capacitor voltage is large enough to supply the trigger current, the triac conducts. Once on, the triac continues to conduct until the line voltage returns to zero. Figure 40-18d and e show the respective voltages across the triac and the load.

Although triacs can handle large currents, they are not in the same class as SCRs, which have much higher current ratings. Nevertheless, when conduction on both half-cycles is important, triacs are useful devices especially in industrial applications. They are also the main component in common commercial light dimmers.

Triac Crowbar

Figure 40-19 shows a triac crowbar that can be used to protect equipment against excessive line voltage. If the line voltage becomes too high, the diac breaks over and triggers the triac. When the triac fires, it blows the fuse. A potentiometer R_2 allows us to set the trigger point.

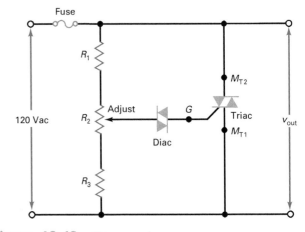

Figure 40-19 Triac crowbar.

EXAMPLE 40-3

In Fig. 40-20, the switch is closed. If the triac has fired, what is the approximate current through the 22-Ω resistor?

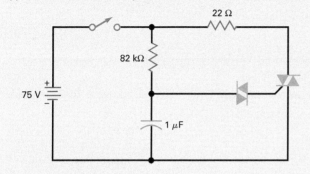

Figure 40-20   Example.

Answer:
Ideally, the triac has zero volts across it when conducting. Therefore, the current through the 22-Ω resistor is

$$I = \frac{75\ V}{22\ \Omega} = 3.41\ A$$

If the triac has 1 or 2 V across it, the current is still close to 3.41 A because the large supply voltage swamps the effect of the triac on voltage.

40.6 IBGTs

Basic Construction

Power MOSFETs and BJTs can both be used in high-power switching applications. The MOSFET has the advantage of greater switching speed, and the BJT has lower conduction losses. By combining the low conduction loss of a BJT with the switching speed of a power MOSFET, we can begin to approach an ideal switch.

This hybrid device exists and is called an **insulated-gate bipolar transistor (IGBT).** The IGBT has essentially evolved from power MOSFET technology. Its structure and operation closely resemble a power MOSFET. Figure 40-21 shows the basic structure of an *n*-channel IGBT. Its structure resembles an *n*-channel power MOSFET constructed on a *p*-type substrate. As shown, it has gate, emitter, and collector leads.

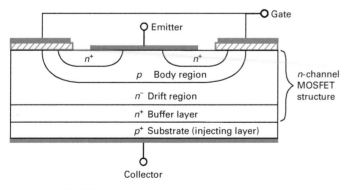

Figure 40-21   Basic IGBT structure.

IGBT Control

Figure 40-22*a* and *b* show two common schematic symbols for an *n*-channel IGBT. Also, Fig. 40-22*c* shows a simplified equivalent circuit for this device. As you can see, the IGBT is essentially a power MOSFET on the input side and a BJT on the output side. The input control is a voltage between the gate and emitter leads. The output is a current between the collector and emitter leads.

The IGBT is a normally off high-input impedance device. When the input voltage, V_{GE}, is large enough, collector current will begin to flow. This minimum voltage value is the gate threshold voltage, $V_{GE(th)}$. A typical $V_{GE(th)}$ for one type of IGBT is 5.0 V when $I_C = 60$ mA. The maximum continuous collector current is shown to be 60 A. Another important on characteristic is its collector-to-emitter saturation voltage, $V_{CE(sat)}$. The typical $V_{CE(sat)}$ value, shown on the data sheet, is 1.5 V at a collector current of 10 A and 2.5 V at a collector current of 60 A.

IGBT Advantages

Conduction losses of IGBTs are related to the forward voltage drop of the device, and the MOSFET conduction loss is based on its $R_{DS(on)}$ values. For low-voltage applications, power MOSFETs can have extremely low $R_{D(on)}$ resistances. In high-voltage applications, however, MOSFETs have increased $R_{DS(on)}$ values resulting in increased conduction losses. The IGBT does not have this characteristic. IGBTs also have a much higher collector-emitter breakdown voltage as compared to the V_{DSS} maximum value of MOSFETs. This is important in applications using higher-voltage inductive loads. As compared to

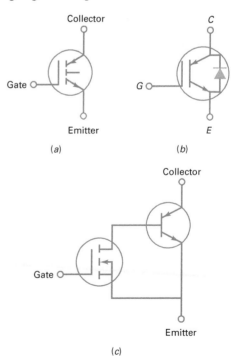

Figure 40-22   IGBTs. (*a*) and (*b*) Schematic symbols. (*c*) Simplified equivalent circuit.

BJTs, IGBTs have a much higher input impedance and have much simpler gate drive requirements. Although the IGBT cannot match the switching speed of the MOSFET, new IGBT families are being developed for high-frequency applications.

IGBTs are, therefore, effective solutions for high-voltage and current applications at moderate frequencies. Some hybrid and electric vehicles use IGBTs in the inverters that convert the battery dc into ac to operate the synchronous motors.

CHAPTER 40 SYSTEM SIDEBAR

IGBT Switching Power Converters

You have seen how switching transistors can be used to form regulators, dc-dc converters, and inverters. Insulated-gate bipolar transistors (IGBTs) are also used for this purpose, especially when very high power is required. A good example is a dc-to-ac power inverter to drive an ac motor. It is often desirable to operate a three-phase induction motor from dc.

The speed of an ac motor is determined by the applied ac frequency. To vary the speed of an ac motor, you need to vary the frequency of the voltage applied to it. One such device is an ac drive that develops variable three-phase ac voltages to operate the motor. Another case is the use of a three-phase induction motor in an electric vehicle. Varying the speed of the motor requires a source of variable-frequency ac.

Figure S40-1 shows the basic block diagram of ac power drive. If the input is dc from a battery, as it would be in an electric vehicle, the dc power bus operates a group of IGBTs that

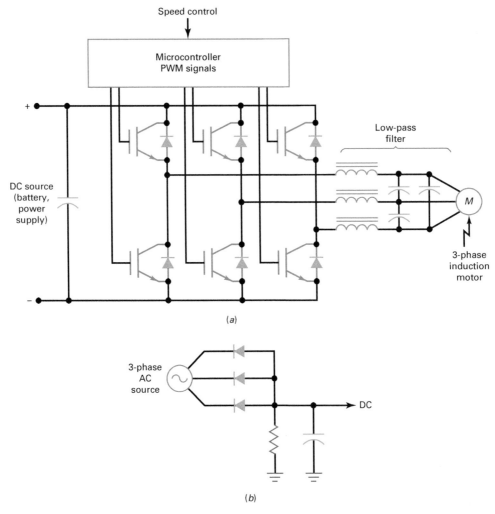

Figure S40-1   (a) I-GBT motor driver. (b) Optional ac source.

are switched off and on to form pulses. The pulses are varied in width or time duration, and these are filtered in low-pass filters to smooth them into a signal closely resembling a sine wave. The digital controller is usually an embedded microcontroller programmed to produce pulse-width modulation (PWM). The controller also knows which IGBTs to turn off or on at the correct time to provide not only the correct pulse width but also whether the pulse is positive or negative.

If the original power source is also ac, for example, from the 60-Hz power line, a three-phase rectifier can be used to convert the ac into dc that is then used as discussed above.

Figure S40-2 shows the pulses developed by the IGBTs. The pulses are fixed in amplitude. They are narrow at the lower-amplitude points on the sine wave and maximum duration for the highest-amplitude points of the sine wave. The pulse frequency is constant, but the time duration varies. Pulse frequency is usually several kilohertz. Note in Fig. S40-1 that low-pass filters made up of inductors and capacitors average the pulses into a continuous sine wave.

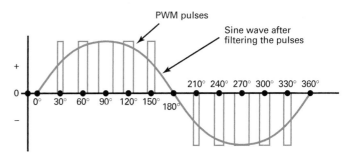

Horizontal axis is time but labeled in degrees.

Figure S40-2   Variable-width pulses, when filtered, form a sine wave.

This method can also use power MOSFETs or even SCRs. The concept is the same. However, these devices are mostly used at lower power levels. The main advantage of IGBTs is that higher voltages and greater power levels can be used. IGBTs are used in systems delivering hundreds or even thousands of watts of power to an ac motor.

❓ CHAPTER 40 REVIEW QUESTIONS

1. A thyristor can be used as
   a. a resistor.
   b. an amplifier.
   c. a switch.
   d. a power source.

2. Positive feedback means that the returning signal
   a. opposes the original change.
   b. aids the original change.
   c. is equivalent to negative feedback.
   d. is amplified.

3. A latch always uses
   a. transistors.
   b. negative feedback.
   c. current.
   d. positive feedback.

4. To turn on a four-layer diode, you need
   a. a positive trigger.
   b. low-current dropout.
   c. breakover.
   d. reverse-bias triggering.

5. The minimum input current that can turn on a thyristor is called the
   a. holding current.
   b. trigger current.

   c. breakover current.
   d. low-current dropout.

6. The only way to stop a four-layer diode that is conducting is by
   a. a positive trigger.
   b. low-current dropout.
   c. breakover.
   d. reverse-bias triggering.

7. The minimum anode current that keeps a thyristor turned on is called the
   a. holding current.
   b. trigger current.
   c. breakover current.
   d. low-current dropout.

8. A silicon controlled rectifier has
   a. two external leads.
   b. three external leads.
   c. four external leads.
   d. three doped regions.

9. An SCR is usually turned on by
   a. breakover.
   b. a gate trigger.
   c. breakdown.
   d. holding current.

10. SCRs are
    a. low-power devices.
    b. four-layer diodes.
    c. high-current devices.
    d. bidirectional.

11. The usual way to protect a load from excessive supply voltage is with a
    a. crowbar.
    b. zener diode.
    c. four-layer diode.
    d. thyristor.

12. An *RC* snubber protects an SCR against
    a. supply overvoltages.
    b. false triggering.
    c. breakover.
    d. crowbarring.

13. When a crowbar is used with a power supply, the supply needs to have a fuse or
    a. adequate trigger current.
    b. holding current.
    c. filtering.
    d. current limiting.

14. The diac is a
    a. transistor.
    b. unidirectional device.
    c. three-layer device.
    d. bidirectional device.

15. The triac is equivalent to
    a. a four-layer diode.
    b. two diacs in parallel.
    c. a thyristor with a gate lead.
    d. two SCRs in parallel.

16. Any thyristor can be turned on with
    a. breakover.
    b. forward-bias triggering.
    c. low-current dropout.
    d. reverse-bias triggering.

17. The trigger voltage of an SCR is closest to
    a. 0 V.
    b. 0.7 V.
    c. 4 V.
    d. Breakover voltage.

18. Any thyristor can be turned off with
    a. breakover.
    b. forward-bias triggering.
    c. low-current dropout.
    d. reverse-bias triggering.

19. Exceeding the critical rate of rise produces
    a. excessive power dissipation.
    b. false triggering.
    c. low-current dropout.
    d. reverse-bias triggering.

20. A four-layer diode is sometimes called a
    a. unijunction transistor.
    b. diac.
    c. *pnpn* diode.
    d. switch.

21. An SCR can switch to the ON state if
    a. its forward breakover voltage is exceeded.
    b. I_{GT} is applied.
    c. the critical rate of voltage rise is exceeded.
    d. all of the above.

22. The maximum firing angle with a single *RC* phase control circuit is
    a. 45°.
    b. 90°.
    c. 180°.
    d. 360°.

23. A triac is generally considered most sensitive in
    a. quadrant I.
    b. quadrant II.
    c. quadrant III.
    d. quadrant IV.

24. An IGBT is essentially a
    a. BJT on the input and MOSFET on the output.
    b. MOSFET on the input and MOSFET on the output.
    c. MOSFET on the input and BJT on the output.
    d. BJT on the input and BJT on the output.

25. The maximum ON-state output voltage of an IGBT is
    a. $V_{GS(on)}$.
    b. $V_{CE(Sat)}$.
    c. $R_{DS(on)}$.
    d. V_{CES}.

CHAPTER 40 PROBLEMS

SECTION 40.2 The Silicon-Controlled Rectifier

40.1 The SCR of Fig. 40-23 has $V_{GT} = 1.0$ V, $I_{GT} = 2$ mA, and $I_H = 12$ mA. What is the output voltage when the SCR is off? What is the input voltage that triggers the SCR? If V_{CC} is decreased until the SCR opens, what is the value of V_{CC}?

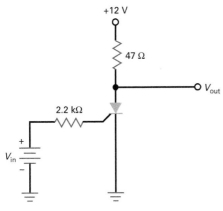

Figure 40-23

40.2 What is the peak output voltage in Fig. 40-24 if R is adjusted to 500 Ω?

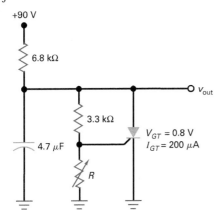

Figure 40-24

40.3 If the SCR of Fig. 40-23 has a gate trigger voltage of 1.5 V, a gate trigger current of 15 mA, and a holding current of 10 mA, what is the input voltage that triggers the SCR? The supply voltage that resets the SCR?

40.4 In Fig. 40-24, R is adjusted to 750 Ω. What is the charging time constant for the capacitor? What is the Thevenin resistance facing the gate?

40.5 The resistor R_2 in Fig. 40-25 is set to 4.6 kΩ. What are the approximate firing and conduction angles for this circuit? How much ac voltage is across C?

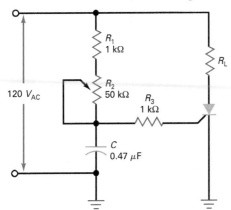

Figure 40-25

40.6 Using Fig. 40-25, when adjusting R_2, what are the minimum and maximum firing angle values?

40.7 What are the minimum and maximum conduction angles of the SCR in Fig. 40-25?

SECTION 40.3 The SCR Crowbar

40.8 Calculate the supply voltage that triggers the crowbar of Fig. 40-26.

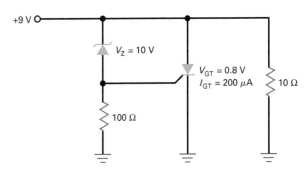

Figure 40-26

40.9 If the zener voltage in Fig. 40-26 is changed from 10 to 12 V, what is the voltage that triggers the SCR?

SECTION 40.5 Bidirectional Thyristors

40.10 The diac of Fig. 40-27 has a breakover voltage of 20 V, and the triac has a V_{GT} of 2.5 V. What is the capacitor voltage that turns on the triac?

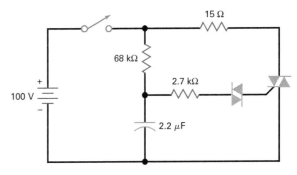

Figure 40-27

40.11 What is the load current in Fig. 40-27 when the triac is conducting?

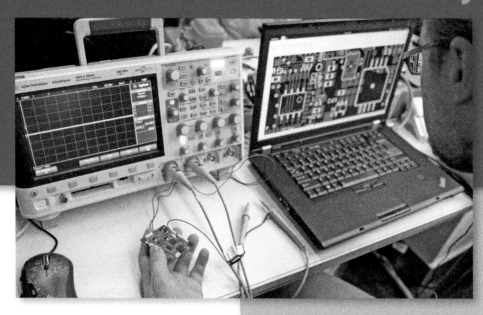

Chapter 41

Electronic System Troubleshooting

Looking at what most technicians (and engineers) do as part of their jobs, one thing stands out: troubleshooting. That, of course, is the art and science of finding and correcting problems in electronic systems, equipment, and circuits. There is some scientific basis for troubleshooting as logical reasoning, and essential principles can be applied to trace and identify troubles. But the process is also an art since it relies to a great extent on personal knowledge, knowledge of the specific system or equipment, and real experience. It is something that can be taught up to a point, but it is also a skill that grows with experience and practice. This chapter examines the troubleshooting process and some of the equipment used to facilitate and expedite the process.

Learning Outcomes

After you complete this chapter, you should be able to:

> *Define* virtual instrumentation.

> *List* at least five key electronic instruments, other than the multimeter and the oscilloscope, that are critical to troubleshooting.

> *Describe* a generic procedure that will aid in almost any electronic troubleshooting.

> *Troubleshoot* a common electronic power supply.

> *Troubleshoot* a typical linear circuit.

> *Troubleshoot* a common digital circuit.

> *Describe* an approach to finding the problem in a complex electronic system.

41.1 Electronic Instrumentation

Electronic instrumentation refers to the test equipment used by engineers and technicians to test and measure electronic devices and systems. This equipment is used not only to troubleshoot electronic products but to design them as well. We focus on troubleshooting in this chapter, but the principles are the same for design and development.

The most widely used electronic test instruments are the digital multimeter and the oscilloscope. Measuring voltage, current, and resistance is basic to any troubleshooting endeavor. The multimeter is a low-cost instrument that gives excellent accuracy and precision. The oscilloscope is also a critical instrument that is essential to troubleshooting. It is the eyes of the technician in looking at signals to see if they are correct. It also allows viewing things such as noise and interference that would otherwise be difficult to detect. The multimeter and the oscilloscope were covered in previous chapters but will be referred to here as appropriate.

Other instruments can ease and speed up troubleshooting. As electronic equipment has become more complex due to the increased integration of circuits and systems, it becomes more difficult to ferret out the problems. Furthermore, the use of higher frequencies for wireless communications and the higher data rates of digital systems has made other instruments more appropriate and necessary. The most important of these we cover here are the logic analyzer, the spectrum analyzer, and signal generators.

This chapter also introduces the concept of virtual instrumentation. *Virtual instruments (VIs)* are more software than hardware and have as their base a personal computer. Almost all basic instrument functions can be implemented in virtual instrument form. An introduction to the software called LabVIEW shows that special virtual instruments can also be developed for dedicated or unique applications.

The Digital Approach to Instrumentation

Most modern test instruments come in three basic forms, stand-alone bench instruments, rack-mount instruments, and virtual instruments. The bench instruments are as described, a box that occupies space on the bench of a design engineer or a service technician. The rack-mount instruments are usually the bench instruments that have been repackaged for mounting in cabinets or open racks, common to test facilities used in manufacturing and final product test. Virtual instruments use a laptop or desktop PC as the core, along with a box or rack of modules that perform tests for specific applications. You will encounter all three in your work.

Another form of test system is the automated test system, a collection of different test instruments packaged together along with a central control computer that is programmed to perform automatic test procedures for a specific product or system. Automatic test equipment (ATE) systems are widely used in manufacturing to perform fast automatic tests on specific products. Testing millions of cell phones is an example. Other test systems are used to troubleshoot and validate complex systems such as the electronics on a fighter aircraft.

The important thing to realize is that most modern instruments are implemented like digital oscilloscopes. They have the form shown in Fig. 41-1. Input signals to be

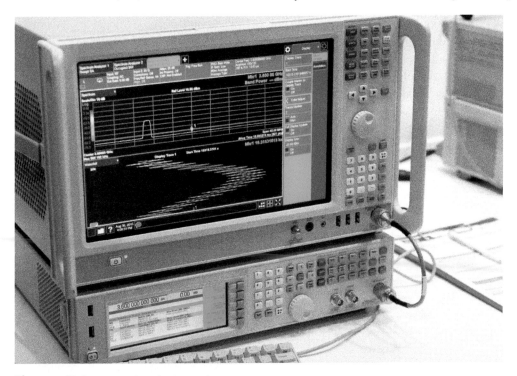

Figure 41-1 A modern logic analyzer.

measured or tested are first applied to attenuators and or amplifiers to scale the level of the input signal to the needs of the scope circuitry. The appropriately sized signal is then applied to a high-speed analog-to-digital converter (ADC) and digitized into a stream of binary numbers that are then stored in a memory. This digital representation of the signal is then analyzed by software on an internal computer, usually a special dedicated, high-speed processor. The software defines the measurements and implements a visual display, usually on a color LCD. Controls are provided to adapt to different signal levels, speeds, frequencies, and other conditions. Essentially, all modern test instruments have this basic form.

Specialized Electronic Instruments

Most electronic tests and measurements are still conducted with multimeters and oscilloscopes. But depending on the type of equipment you are working on, other special instruments will actually be more useful. Two of the most important are the logic analyzer and the spectrum analyzer. Signal generators are also key to testing and measuring. These are described here along with a short list of other useful special instruments.

The Logic Analyzer   A logic analyzer is similar to an oscilloscope in that it generally displays digital logic signals on an LCD screen with respect to time. While an oscilloscope permits display and measurement of amplitude, shape, frequency, time, and other characteristics of both analog and digital signals, a logic analyzer is designed only for digital signals. A typical unit is shown in Fig. 41-2. The logic analyzer samples the signal but only looks at the threshold voltage between a binary 0 and binary 1 logic level.

Instead of the usual dual- or four-trace display of most oscilloscopes, a logic analyzer allows you to look at 16, 32, or up to about 136 different signals simultaneously. In complex digital systems, it is the relationship between signals that is important. A logic analyzer lets you check the signals for proper synchronization and timing. It allows you to detect race conditions by which one circuit may be too fast or too slow to produce the desired effect.

The logic analyzer is particularly valuable in testing complex circuits implemented with field-programmable gate array (FPGA) integrated circuits. These ICs have hundreds or thousands of digital circuits on them, and most of their interconnections are programmable. Their operations are extensive with many simultaneously occurring signals. A logic analyzer is the ideal instrument to detect and isolate problems. Logic analyzers can troubleshoot any digital circuit, including those with an embedded microcontroller at its heart.

The digital signals are connected to the logic analyzer by way of special probes or connectors that are designed to minimize the capacitance, inductance, and resistance of the connection. This permits signals of many megabits or gigabits per second to be analyzed. The input signals are then sampled and digitized as if they were in an oscilloscope, and the resulting binary data are stored in a high-speed memory. The data in the memory can then be analyzed and displayed.

One common display is multiple, simultaneous time domain binary signals in multiple traces on the LCD screen. Alternatively, the binary data representing the signals can be displayed as either binary or hexadecimal data. In some logic analyzers, the data can be analyzed to determine its binary meaning and displayed as actual communication protocol words and frames, making them easier to analyze.

Spectrum Analyzer   A spectrum analyzer is an oscilloscope-like instrument that displays signals in the frequency domain rather than the time domain. One typical unit is shown in Fig. 41-3. Input signals are analyzed and displayed as the individual frequency components that make up a complex signal. The display shows frequency on the horizontal axis and power on the vertical axis.

The spectrum analyzer shows the Fourier theorem harmonics of a signal and any sidebands generated by modulation. Figure 41-4 gives several examples of how different signals are presented. In Fig. 41-4a a sine wave is shown as just a single vertical line at its frequency with the length of the line representing the amount of signal power. The square wave in Fig. 41-4b is presented as a fundamental frequency sine wave in addition to its many odd harmonic sine waves. An amplitude-modulated signal is presented as the carrier sine wave along with its upper- and lower-sideband components. Note the random variation at the bottom of the display. This is the noise signal power in the system.

A spectrum analyzer is more widely used than an oscilloscope, especially in testing radio-frequency (RF) circuits and

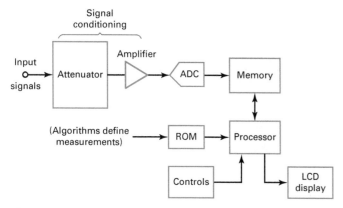

Figure 41-2   General block diagram for all modern test instruments.

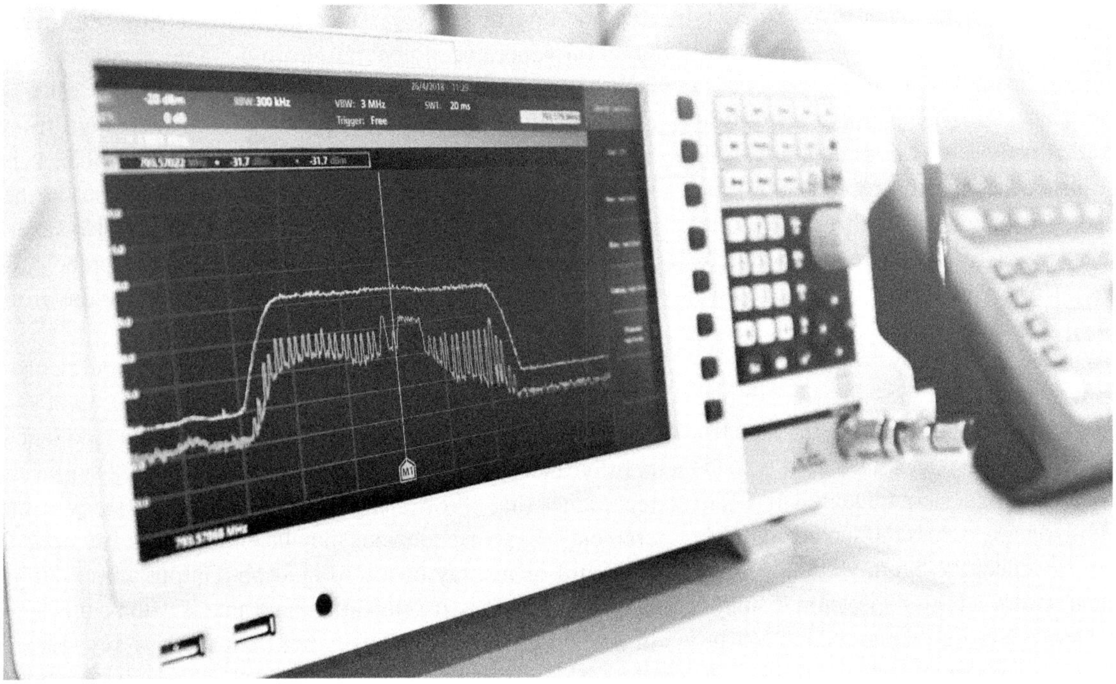

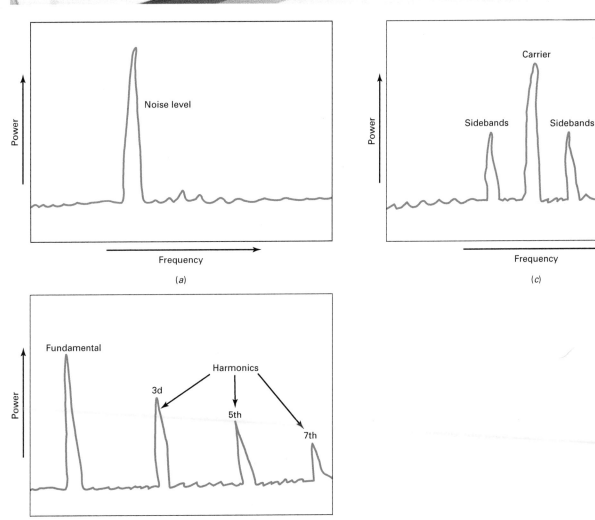

Figure 41-3
A spectrum analyzer shows the power level of signals in the frequency domain.

Figure 41-4 Common spectrum analyzer displays.
(*a*) Sine wave. (*b*) Square wave. (*c*) Amplitude modulation.

communications equipment. This instrument gives a better view of what is going on with a signal in addition to showing noise and interfering signals on adjacent frequencies.

There are three basic ways of implementing a spectrum analyzer. The first and older method is analog in nature. The spectrum analyzer is structured as a radio receiver with its input signal being applied to a mixer along with a local oscillator signal. The mixer produces a difference frequency called in *intermediate frequency (IF)*. At the IF, a bandpass filter defines a channel in the spectrum to be displayed. The local oscillator frequency is then varied linearly from a low to a high frequency, so that input signals over the bandwidth of the channel are presented to a detector that recovers the signals and determines their power level. The detector output is then presented to a display, a CRT in older equipment. The CRT electron beam is swept horizontally across the screen, and the detected signals are used to deflect the beam vertically. The result is the classic frequency versus power spectrum display.

The other two types of spectrum analyzer are digital instruments. The input signal is amplified and then digitized in an analog-to-digital converter. The input signal is sampled at fast rate, resulting in proportional binary numbers for each sample. The binary numbers are stored in a memory. The binary data are then accessed by a special processor that subjects the binary signal data to a special algorithm called the *discrete Fourier transform (DFT)* or the *fast Fourier transform (FFT)*. This mathematical algorithm is a digital signal processing (DSP) technique for extracting frequency and amplitude data. It is a Fourier analysis of the binary data. The processed data are sent to a digital-to-analog converter (DAC) where an analog signal representing the various signal powers over a frequency range is created. This signal is then sent to an LCD, where it is plotted vertically as the display is swept horizontally.

This type of spectrum analyzer is generally known as a *vector signal analyzer (VSA)*. It provides a detailed and accurate frequency spectrum output. Its main limitation is that the display is not in real time. That is, the signal is not displayed at the time the signal is being input. It takes time to capture, digitize, store, and analyze the signal before display. This delay may or may not be a disadvantage as determined by the measurement application.

To display the signal in real time requires a more advanced spectrum analyzer called a *real-time signal analyzer (RSA)*. It digitizes the signal as a VSA but introduces the DSP signal analysis before the data are stored. Thanks to very fast sampling ADCs and high-speed digital processors, the Fourier analysis takes place as the signal is occurring and is digitized. The processed data are stored in a memory for further analysis and display. The RSA not only displays the input in real time but also is fast enough to capture short-duration pulses, transients, and random signals or nonperiodic events that elude the VSA.

Signal Generators You have already been exposed to the function generator, a general-purpose signal generator that generates sine, square, and triangular waves of different frequencies. Typical units produce signals in the range from less than 1 Hz to 20 MHz. The amplitude is variable. It is a good instrument for testing circuits and equipment in the audio-frequency range and low radio frequencies. However, there are many other special generators for different purposes. Two of these are the RF signal generator and the arbitrary waveform generator.

An RF generator produces sine waves at radio and microwave frequencies. The range is from about 30 MHz up to and over 20 GHz. The amplitude is variable, and the frequency is fully selectable. These generators use frequency synthesis techniques that allow changing the frequency in steps or increments. Exact frequencies may be entered on a front panel keyboard or by way of a tuning knob. Output amplitude is varied with a resistive step attenuator that is often calibrated in decibal-milliwatts (dBm) rather than volts. A key feature of an RF generator is that the sine-wave output may be modulated to simulate wireless signals. Modulation can be AM, FM, PM, or some other popular modulation.

An *arbitrary waveform generator (AWG)* is a generator that you can program to produce any shape wave at any frequency. The desired waveform is a digital representation in binary form stored in a memory. See Fig. 41-5. The AWG does not digitize these signals. This is done by synthesizing them in a computer and then loading them into memory. The binary words representing the waveform are read out and applied to a digital-to-analog converter where they are transformed into an analog signal unique to the equipment being tested. A clock and frequency synthesizer determine the rate at which the data are read out and that determines the frequency of the output signals. Most AWGs have prestored waveforms like sine, square, triangular, sawtooth, and other common shapes that can be called up for output. The user can use one of several types of software to generate the binary bit patterns representing special waveforms, including those of special modulated signals. Like standard signal generators, the AWG uses frequency synthesis to set the output frequency. An amplifier or an attenuator conditions the DAC output to the desired level at the output.

Virtual Instrumentation

The test instruments discussed so far are general-purpose instruments designed to fit a wide range of measurement needs. Such instruments incorporate many functions and features so that a broad range of measurement applications may be addressed in either design and development or in service and troubleshooting. This versatility makes standard bench instruments overly expensive for simpler, more specialized tests. Often there are needs in manufacturing or in

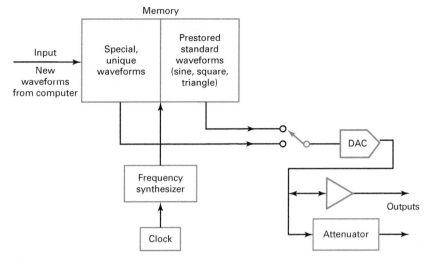

Figure 41-5 An arbitrary waveform generator (AWG).

process control for chemicals where only a few dedicated and special measurements are needed. Standard instruments are overkill. As a result specialized instruments were developed just for those specific cases. This then lead to the creation of virtual instruments, instruments that are a combination of hardware and software and that solve a very specific measurement problem.

Virtual instrumentation is user-defined measurement solutions based on customizable software and modular hardware combined with a standard desktop or laptop computer. Such virtual instruments perform three basic functions: Capture input signals, process or analyze them with software, and either display and/or store the result for human consumption and analysis. Modern virtual instrumentation systems are usually made up of hardware such as signal-conditioning circuits, like amplifiers and filters, and high-speed analog-to-digital converters that digitize the signals to be measured. The captured signal data in digital form are stored on a computer, and software processes the data to make measurements on such signal characteristics as amplitude, frequency, timing, or shape. Computer software algorithms are applied to analyze the data. That may include a Fourier analysis using the FFT, described earlier, to determine frequency content. Statistical analysis is another example.

Once the data are analyzed, they are further processed for display. Software again plays a role in providing a way to present the data output on the computer LCD. It may be a simulated oscilloscope display, spectrum analyzer display, or just digital numerical outputs presented as they might appear on a standard digital multimeter. Even analog dial meters can be simulated. The data can be stored on a hard disk drive, flash memory, or other storage device.

Figure 41-6 shows a common VI package. It consists of a box or chassis containing a uniquely packaged personal

Figure 41-6 The PXI-5154 digitizer, a virtual instrument. The chassis houses a special personal computer in the leftmost module along with standard PXI modules used for signal conditioning and digitizing.

Source: Courtesy National Instruments.

computer on the left and multiple plug-in modules for signal conditioning (amplification, filtering, etc.) and analog-to-digital conversion. The LCD display shows the program used for processing and the output results in the lower left.

The most difficult part of virtual instrumentation is creating the software. Early VI systems used custom programs created in BASIC, C, Pascal, or assembly language. Only experienced programmers were capable of creating the

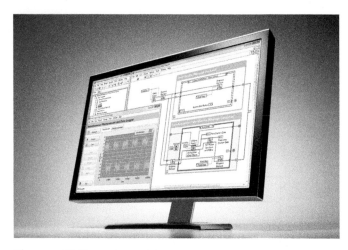

Figure 41-7 LabVIEW has a graphical user interface of blocks to program the desired instrument functions.
Source: Courtesy National Instruments.

desired test and measurement functions. Today, general-purpose software has been developed to make it easy for any engineer or technician to create custom measurement programs.

The most famous and widely used VI software is called LabVIEW, a complete development environment developed by National Instruments. The LabVIEW development platform is based on a graphical user interface that a programmer deploys to produce a virtual instrument. Icons representing measurement and processing functions are "wired" together on the screen to produce the desired VI, including the display format. Figure 41-7 shows one example. The individual blocks may be hardware in the chassis or segments of software code created by LabVIEW. The software then organizes the simulation programs evoked into a complete instrument customized to the application. LabVIEW is easy to learn and fast to use by not only nonprogrammers but also engineers who understand the instrumentation needed for their application.

Virtual instrumentation is widely used in all sectors of industry for manufacturing, development testing, and generic measurements. VI leverages the fast powerful personal computers and high-speed data converters and software simulation techniques to quickly develop customized instrumentation at a reasonable price.

41.2 An Introduction to Troubleshooting

Troubleshooting is the general term applied to locating defects in electronic equipment. As indicated earlier, a great deal of the work performed by electronic technicians is the repair of electronic equipment. For example, an instrument may come off the production line in a manufacturing plant and, when tested, found to be defective. This instrument has

to be repaired before it can be shipped to the customer. Troubleshooting finds the defect.

Another example is a piece of equipment that fails after a considerable period of normal use. While electronic equipment is generally reliable, most equipment fails sooner or later. When this occurs, an electronics technician must track down the problem and replace the bad components or make other repairs as needed.

Another example is the troubleshooting of problems in a newly designed circuit. A design engineer, in creating a new circuit, frequently makes a mistake in the design process, or the prototype circuit has been wired incorrectly. This calls for troubleshooting to determine why the circuit is not operating as designed. All these example situations make use of various common troubleshooting procedures that will help you track down and fix almost any fault.

There are eight general steps to follow in electronic troubleshooting:

1. Verify the reported problem.
2. Note specific symptoms.
3. Verify that power is applied.
4. Do a visual inspection.
5. Isolate the problem to a specific stage, module, IC, or subassembly through signal tracing or signal injection.
6. Isolate the problem to a specific component or module.
7. Repair the problem.
8. Test the unit.

Assume the user of the equipment notes a malfunction and reports it. Your first job is to be sure that the equipment is indeed defective by verifying the complaint. Often when the equipment is operated incorrectly or used other than intended, the impression may appear to be a hardware defect when in reality the equipment is operating perfectly. These operator problems are typically called "cockpit troubles" and do not really involve an equipment failure. Reeducation or training of the operator is the solution.

Another common factor today is that the problem is software and not the hardware. Since everything in electronics is controlled by a computer or embedded controller, software is involved. In products with embedded controllers, the software is in a ROM, so is called firmware. It rarely goes bad, but in other products and systems involving an attached computer, software must be considered in troubleshooting. The problem may actually be the user's inability to use the computer or software correctly.

Once you have determined that the problem does exist, you should look for specific symptoms. Just what is the equipment doing or not doing? In some cases the equipment may be totally dead. In other cases it may be operating but not correctly. It may not be meeting stated specifications or performing some of its designated functions. List these symptoms, as they will give you clues as to where to look for the problem.

The third step in the troubleshooting procedure is to verify that the unit is receiving power. Obviously during the first two steps if the unit is working, then it must be receiving power. However, some of the symptoms may reveal that the unit may not be receiving one or more of its power supply voltages. Remember that most electronic equipment receives its main power from the ac power lines. A power supply in the unit converts this into one or more dc voltages to operate the circuits. In portable or mobile equipment, the dc power is from a battery. They battery may be bad and need replacing, or it just may not have been recharged. Your job is to measure all the dc voltages to be sure that they are all present and of the correct values.

If the power supply voltages are not present, you can assume that the problem lies somewhere in the power supply. At this point, you can begin more detailed troubleshooting procedures on the power supply. If all the power supply voltages are good, then the problem lies elsewhere and you can go on with the next step.

A visual inspection is the fourth troubleshooting step. Look at all the front panel controls and displays. Verify that all are either set correctly or reading out the proper values. Check all input and output cables and connectors. Wires frequently break, and connectors can fail to make a good connection. Even the ac cord could be unplugged. And if a computer and software are involved, be sure all the software conditions have been fulfilled to make the hardware work. Be sure the software settings are correct.

Next look inside. Check to see that printed-circuit boards are properly seated in their connectors. Look for burned components such as resistors. Search for broken wires or other signs of trouble. A burned smell is a clue to a defective part. Touching the various components might reveal an unusually hot IC or transistor that might signal trouble. Use all your senses during troubleshooting.

Signal Tracing and Injection

The next step is to use the common troubleshooting techniques of signal tracing and signal injection. Most electronic equipment can be viewed as shown in Fig. 41-8. It accepts inputs which are electronic signals of some sort. The equipment processes these signals in multiple stages and provides some output. Each stage is some kind of electronic circuit, such as an amplifier, filter, mixer, oscillator, modulator, processor, memory, or whatever. It could be a single IC or multiple discrete components or a complete module or printed-circuit board. The first thing you do is check to

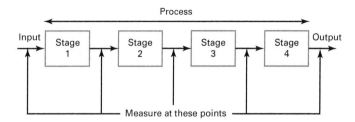

Figure 41-8 Multiple stages of circuits or modules process one or more inputs into outputs.

see that the equipment is getting the proper inputs. To determine if the equipment is operating correctly, naturally you will measure the outputs of each stage to see if they are as they are supposed to be. Again, this is done by measurement with a meter, scope, or spectrum analyzer. If the inputs are correct but the output is incorrect, then obviously there is something wrong in between. The idea is to trace the input signals through the various stages of processing in an attempt to isolate the problem to one of them.

Signal tracing is a technique of using a multimeter, an oscilloscope, or possibly some other instrument to make measurements inside the equipment as the signal is processed. Figure 41-8 shows a hypothetical piece of equipment with four stages. Assume that the input voltage is measured and found to be correct. The output voltage does not exist. The multimeter or oscilloscope is then connected to the output of each of the various stages in the equipment. As soon as you have found the point where the signal no longer exists, you can assume that the stage prior is defective. For example, assume that you measure the correct output voltages at stages 1 and 2. The output of stage 3 is zero. Obviously, stage 3 is the defective stage. At this point you can begin more detailed troubleshooting procedures on stage 3. The big issue with using this approach is to know just what level or type of signal is expected at each input and output. Only knowledge of the equipment through documentation or training can provide this.

Assume that the output of stage 1 in Fig. 41-8 is measured and found to be zero volts. Which of the following is the most likely cause of the trouble?

In this example, stage 1 could be defective if no output signal is measured. This assumes that the correct input signal is present. However, stage 1 may be perfectly good and give no output if the input signal is zero. Or the output may be zero if the input to stage 2 is shorted to ground.

In some cases signal injection can be used to determine the source of trouble. For example, a multimeter or oscilloscope can be connected to the output of stage 4 in Fig. 41-8. Then connect an electronic signal similar to that typically processed by the equipment to the input of stage 4. A signal or function generator is a good source for such a signal. If the output is present, then stage 4 is good. Next, the signal is injected at stage 3. Again, if the correct output is measured, stage 3 is also good. The signal is injected to the input of stage 2. Assume at this point that no output signal occurs. Obviously, since stages 3 and 4 are good, stage 2 must be defective. Detailed troubleshooting procedures can then be initiated on stage 2.

The question you might have regarding signal injection and signal tracing is how do you know what the various stages of the equipment are and what their inputs and outputs should be? The answer is that this information is usually found in the equipment documentation. Documentation consists of operating manuals and service manuals, which usually contain block diagrams, schematics, specifications, parts lists, and other detailed information on the equipment. Today, documentation for complex equipment and systems may be so extensive it is available only on a CD or online at a website. Most electronic equipment is extremely complex and simply cannot be serviced without this documentation. It is foolish to even attempt to service equipment without the manufacturer's service manuals. Only then will you know at what points in the circuit to check and what the inputs and outputs should be.

After the problem has been isolated to a particular stage, detailed measurements are made on that stage to determine the exact problem.

41.3 Circuit Troubleshooting

An example of detailed circuit troubleshooting is illustrated in Fig. 41-9, which shows a simple class A amplifier. If this is the stage that has been identified as being defective, then you will need to use a multimeter and/or oscilloscope to make further tests. A quick initial test to check the circuit is to measure the dc voltage between collector and ground. Most class A amplifiers are designed so that this voltage is approximately one-half the dc supply voltage ($V_{CC}/2$). If the collector voltage is in the correct range, the transistor stage is probably operating correctly. If no signal is getting through, you should look for broken connections or components in the signal path that may be defective. For example, capacitor C_1 or C_2 may be open.

As an example, assume the collector voltage of a class A stage is found to be 6.7 V. The supply voltage is +12 V. This stage is probably biased correctly.

If you should measure the collector voltage in Fig. 41-9 and find that it is not equal to approximately one-half the supply voltage, then you may get a clue to the problem. For example, if the collector voltage is equal to the supply voltage, V_{CC}, then no current is flowing through the transistor. All you are seeing is V_{CC} through R_1. This usually means that the transistor is defective (that is, open). Alternatively, it could mean that the bias network made up of R_1, R_2, and R_4 is defective. For example, R_1 may be open, preventing base current from turning Q_1 on. Or R_4 may be open. If the collector voltage is very low or near zero volts, the transistor may be shorted.

This same troubleshooting technique can also be used on oscillators. Most oscillators are just class A amplifiers with feedback. The same test should reveal the problem. Regardless of the type of stage, you should measure the dc voltages on any transistors and compare them to the voltages specified in the service manual. If a transistor is conducting, its emitter-base voltage drop will be about 0.7 V. Any differences between measured and specified values will give you a clue as to what the defective component might be. If the dc voltages are correct, then look for signal components such as defective capacitors or transformers that could be causing the problem.

The basic technique for troubleshooting a single stage is to measure the dc voltages and compare them to those in the service literature. Then use signal tracing or injection to isolate the problem. If the circuit is an integrated circuit, all you can do is measure to be sure it has the correct power supply voltages applied. Then you can check to see that the input and output signal voltages are as specified. If not, your only choice is to replace the IC.

A common MOSFET circuit is shown in Fig. 41-10. This transistor is used to turn a solenoid or relay coil off or on. If the input is +12 V, the MOSFET conducts and connects the coil to +12 V, and the solenoid or relay is energized. If the input is zero, the MOSFET will be off, and the solenoid or relay will not be turned on. A good first step is to check to see that +12 V is being supplied to the circuit. Then provide an input of either +12 V or zero to see if the solenoid or relay is working. If not, the most likely causes are a bad (open) transistor or an open coil. If the latter is suspected, disconnect the supply voltage and one side of the solenoid coil, and measure its resistance. A good coil will have a low resistance. An open coil will have an infinite resistance.

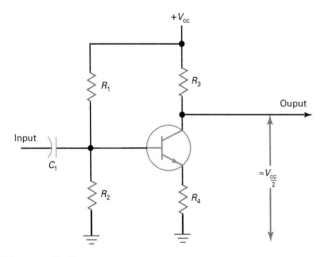

Figure 41-9   A class A amplifier.

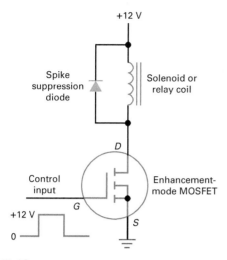

Figure 41-10   A MOSFET switch-driver circuit.

Another problem may be the spike suppression diode across the coil. This diode is sometimes contained within the transistor itself connected between the source and drain. If this diode is shorted there, the supply voltage will be connected directly to the transistor, damaging it. If the diode is open, it will not suppress the large inductive kick voltage when the coil is deenergized. The spike would then damage the transistor.

Another circuit is shown in Fig. 41-11. This op-amp circuit is typical of many in all types of equipment. Start the troubleshooting by measuring the power supply voltages, in this case both the positive and negative 5 V supplied to all three op amps. You may even want to use an oscilloscope to verify that the ripple on the power supplies is low and barely visible.

Next, be sure you have the correct inputs, and measure the outputs of IC_1 and IC_2 with a multimeter or oscilloscope. Next, IC_1 has a gain of R_2/R_1 and phase inversion, so verify that. Since IC_2 is a follower, its output should be the same as its input. No output could mean an open R_1 on IC_1 or a bad

IC. A saturated output or positive or negative voltage near the supply voltage could mean an open R_2 on IC_1 or a bad IC. Keep in mind that both op amps could be inside the same IC package, so you will need to know the IC pin numbers to find the inputs and outputs.

Next, measure the output of IC_3. This summer adds the two signals from IC_1 and IC_2. A saturated output could mean R_5 is open. An incorrect output could mean an open R_3 or R_4.

Digital circuits are a bit different to troubleshoot. The discrete logic circuits of the past are rarely found today. Most digital circuits are either embedded microcontrollers or field-programmable gate arrays (FPGAs). Both are single-chip ICs containing many circuits. You do have access to the inputs, outputs, and power pins, so that is what you use. Refer to Fig. 41-12.

As usual, the first step is to be sure you have the correct dc power, usually 3.3 or 5 V. If so, check for the presence of a clock signal that drives all the other circuits. Find the clock oscillator, and verify its output on an oscilloscope for correct logic levels and frequency. The clock circuit may be internal to the processor or FPGA IC.

Next, verify the inputs and outputs according to documentation or another source. Look at these on an oscilloscope to be sure the logic levels are correct and that the rise and fall times of the pulses are correct or within range. You can also use a logic analyzer here to view multiple signals at once to see that the sequence of pulses is correct.

Input/output circuits called interfaces are a common source of problems as they connect to external cables or equipment. Use an oscilloscope to verify signal integrity. Some scopes have signal capture capability for specific interface standards like USB, RS-232, I²C, or SPI. These make troubleshooting faster and simpler if available.

Finally, in a digital circuit with an embedded controller, remember that the problem could be in the firmware. Check any external memory ICs for proper operation.

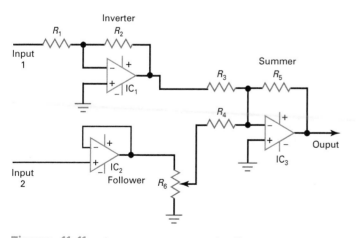

Figure 41-11   A common op-amp circuit.

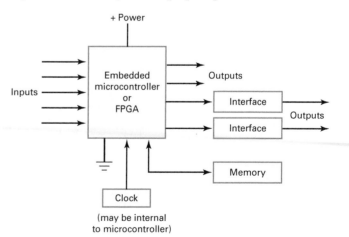

Figure 41-12   A common digital circuit configuration.

41.4 Power Supply Troubleshooting

One of the most common problems in electronic equipment is a power supply failure. The power supply is a key element of any piece of electronic equipment, as the unit will not work without its correct dc voltages. Since the entire power for the equipment comes from the power supply, naturally it is under more stress than probably any other circuit in the equipment. For that reason, power supplies tend to fail more than almost any other circuit.

Figure 41-13 shows a schematic diagram of a typical power supply. It uses an input ac filter, a power transformer, a full-wave bridge rectifier, a capacitor filter, and several IC voltage regulators and dc-dc converters. Assume that the initial testing reveals that the power supply seems to be defective, as some initial measurements indicate that one or more of the output voltages are zero. Here's how you would go about troubleshooting a power supply.

Start by using a multimeter to measure the dc output voltage at each regulator and dc-dc converter. The technique you will use is signal tracing. In this case the signal is the dc output voltage. Assume that you measure the output voltage of the linear regulator and find it to be zero. Next, you should check the input to the regulator circuit at point A. Assume that you measure a voltage at the input to the regulator that is slightly above the desired output voltage. With this measurement, you can conclude that the regulator IC is defective and can be replaced.

Check each of the other outputs the same way. One common problem is excessive ripple voltage on the dc output. This ripple can get into the circuits being powered by the power supply and cause all sorts of unexpected problems. This is especially true in switching power supplies with ripple voltages of 100 kHz to 6 MHz or so. An oscilloscope is the best instrument to use here. A common problem is a defective filter capacitor near the regulator or dc-dc converter IC.

Let's assume that you did not measure the correct input voltage to the linear regulator. You can conclude then that the regulator is probably okay. The fact that you do not measure a voltage at its input on the dc power bus means that some earlier portion of the power supply is defective. The input to the regulator, of course, is the power bus or the output of the rectifier and the filter. For example, you may measure a lower than normal dc voltage. In this case, the filter capacitor may be open, causing the average dc voltage to be low. An open filter capacitor means that there is pulsating dc at point A. You could check this with an oscilloscope to see the ripple.

If the input to the regulator (point A) is zero, it could mean that the filter capacitor is shorted. It could also mean that the rectifier(s), power transformer, or other components are defective. A bad rectifier is a very common problem. But there could be other troubles too.

The next measurement you should make is to check for the presence of ac voltage across the secondary of the power transformer. If the correct voltage is present, then you can assume that the rectifier is bad. The rectifier unit may be a packaged bridge unit where individual diodes are not accessible. In this case, you replace the whole bridge rectifier device. If individual diodes are used, you can test each diode to determine if it is shorted or open. Then replace the defective diodes. You can use your multimeter as described earlier to do this.

If you find that the ac voltage is not present across the transformer secondary, the problem may be one of several. The transformer itself may be defective, such as an open secondary or open primary winding. The way to

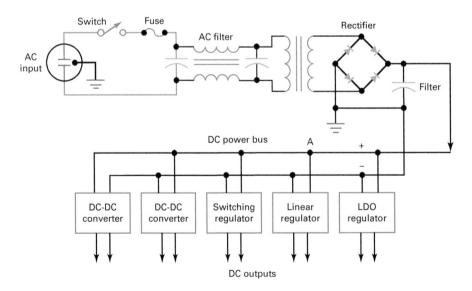

Figure 41-13   One possible power supply configuration.

isolate this problem is to check for the presence of voltage across the primary winding. If voltage is present across the primary but not across the secondary, then the transformer is most likely bad. Use your ohmmeter to check the continuity of both primary and secondary windings to see if one is open. Be sure that the power is off for this measurement.

If no ac voltage appears at the primary winding, again there may be one of several possible defects. These are a defective ON-OFF switch, a blown fuse, an open in the filter coils, or the lack of ac power. The lack of ac power could simply mean that the circuit breaker in the ac power service box is blown and no ac power is at the outlet into which the equipment is plugged. It could also mean that the ac plug or line cord is defective. The most likely cause is a blown fuse or defective switch. In this power supply, the filter unit may also be bad, either an open inductor or shorted capacitor. Most filters are a packaged unit that must be replaced as a whole.

Once your signal tracing has determined that a component may be defective, there are two steps you can use to determine whether the component is good or bad. The first approach is to test the component in some way. This is usually done by disconnecting it from the circuit and checking it with an ohmmeter or a special tester. The continuity tests on the transformer, fuse, and switch are examples. This approach is not viable for integrated circuits.

An alternate approach is to substitute a good component for the one suspected of being defective. In many applications, replacement parts are commonly available and can be substituted to verify whether or not a component is bad. If the replacement component does not solve the problem then, of course, troubleshooting should continue until the defective unit is found.

Component Failure

Components may fail in a variety of ways. Table 41-1 lists the most common electronic components in the order they

Table 41-1 Common Component Failures in Order of the Most Likely to Fail

1. Fuses, indicator lights
2. Switches and relays
3. Cables (wires) and connectors
4. Diodes and transistors
5. Integrated circuits
6. Capacitors
7. Resistors
8. Inductors and transformers
9. Printed-circuit boards

are most likely to fail. Take a moment to look over this information.

In testing circuits, verify the operation of the more likely components first as it will usually identify the problem with less testing.

The repair on a piece of equipment is concluded when the defective component or other problem is identified and repaired. Then the correct operation of the equipment should be determined. A thorough test should be made of all equipment operation and functions prior to returning the equipment to regular operation.

41.5 Alternative Troubleshooting Techniques

Repair vs. Replace

While it is generally possible to troubleshoot and repair almost any piece of electronic equipment, we have reached a stage of technology and economy that in some cases repair is impractical or too expensive. There are three reasons for this:

1. Electronic equipment and components are very inexpensive, particularly when they are made and purchased in large quantities. Foreign-manufactured equipment is often so cheap that it is less expensive just to buy a new unit rather than repair the defective unit.
2. The time it takes to troubleshoot and repair a piece of equipment and the cost of that time is very high. Technician repair rates may be $50 to $200 per hour.
3. Often very expensive test instruments and special repair equipment are required to fix a unit. They may not be available or affordable.

Because of these reasons, it is often cheaper and faster to buy a new unit than repair a defective unit. Inexpensive consumer electronic devices are a good example. If a clock radio, portable MP3 player, or cordless telephone goes bad, you can usually buy a new one for less than the price of a repair, if indeed you can actually find someone to repair the unit. It is still possible to repair more expensive items like TV sets and personal computers and the like at a reasonable price.

Even so, the way the repair is made is not to troubleshoot the problem down to the component level. Instead, the troubleshooting is done by substituting larger subassemblies rather than individual components. Consider a personal computer. Instead of fixing a bad hard disk or DVD drive, it is simply replaced with a new one because the cost of a new one is less than repairing the old one. In a modular TV set the repair is made by replacing major subassemblies or components. This is often as easy as plugging in a replacement PC board for a defective one. Not only does this approach lower repair costs, but also it is considerably faster. In our fast-paced society, speed is often more important than cost. Downtime is unacceptable. Replacing a defective unit is faster and cheaper. In a business or industrial setting, equipment downtime costs the enterprise money for being out of service. Think of the lost income on a $3 million dollar MRI machine in a hospital not to mention the delay in patient care. In manufacturing, a broken piece of machinery causes production line backups and lost manufacturing income. Repair speed is critical. This factor must be seriously considered in how to make a repair.

Troubleshooting and repair are still possible with some equipment, but you should always try to determine the fastest and least expensive solution to a defective unit.

Think in Block Diagrams

Get used to thinking at a higher level. Instead of going to schematics first, go to a block diagram. Use the block diagrams in the documentation to help you zero in on a problem. Once you isolate the problem to a major block, you can then dig deeper with the schematics if necessary. Chances are your block diagram approach will lead you to an IC, module, PC board or other replaceable subassembly that you can remove and replace to fix the problem.

Divide and Conquer

Divide and conquer is a generic technique that is handy especially if you are using signal tracing. The idea is to divide the problem circuit or device into two parts and then test one part to determine if it is okay. If it is, check the other part. It may be bad. If one part is defective, divide it into two parts and repeat the process. By continuing to repeat this "binary" process, you can eventually pinpoint the problem.

Substitution

A widely used troubleshooting technique is replacing a suspected defective part with a known good one. This technique is one of the fastest methods of putting a piece of equipment or system back in operation. The substitution approach works at the component level, module level, and equipment level.

A circuit example is when you suspect from your troubleshooting efforts that a transistor is bad. If you have an exact replacement, do the substitution. It will usually involve unsoldering the suspected device then reinstalling the new device.

The virtue of this approach is that it is fast and low in cost. Most new transistors are very inexpensive, and they may be readily available from local stock or by mail order. Exact replacement devices are essential. Don't use substitutes or so-called equivalents.

One downside to this approach is that you may not have identified what caused the transistor to fail in the first place. Replacing the defective part could lead to immediate failure of the new part if the original cause has not been detected. The real problem may be a shorted capacitor in a nearby circuit, overvoltage from a power supply problem, or a transient occurring randomly. If the replaced part becomes defective again, then you must begin the real troubleshooting again.

In larger systems, the equipment is often made up of plug-in modules or printed-circuit boards. In your troubleshooting you may have traced the signal to a specific module that seems to be defective. If a replacement module is available as it often is in big critical systems, a substitution may be the fastest way to get the system online again.

One big issue here is cost. If the module is inexpensive, substitution becomes the fastest, less expensive way to make the repair. An example of this is a defective power supply in a personal computer. You may have determined the absence of one or more dc supply voltages. Repairing a complex switch-mode power supply is difficult and challenging. It is rarely done. The reason for this is the very low cost of a replacement supply. This is a commodity product that costs less than $50 in most cases. Direct replacement is the best solution.

If the module or subsystem is very expensive (many hundreds or thousands of dollars), it is not likely that a direct replacement is available. You may be able to acquire one, but you must make sure of your diagnosis. If that module is not the problem, you have wasted a huge amount of money and time. And a replacement could be subject to whatever caused the original fault anyway.

Is It Hardware or Software?

In systems controlled by a computer or equipment operated by an internal embedded microcontroller, software becomes one aspect to consider in troubleshooting. Did the software fail, or was it the hardware? Sometimes the exact cause is difficult to determine because the two are so entwined.

One clue is that if the system was previously operating normally, ask what changed between the time of normal operation and the failure. If there was no change in the software, as would be the case of an internal embedded controller, then it is probably not a software problem. Most embedded controller software is stored in ROM or a flash memory chip. Although more commonly called firmware, it does not change. One possibility is the failure of the program storage memory. This may be an internal memory on the microcontroller chip or a defective external memory IC.

Another common source of problems in equipment with embedded microcontrollers is the interface. The interface is that circuit or circuits that connect the equipment to external devices like sensors or other pieces of equipment. Interfaces often encounter excessive voltages or currents. Interfaces also usually involve connectors and cables, a major source of problems.

If the failure occurs in a system running on a personal computer or other computer that uses external software sources that may be loaded from a CD-ROM, a flash drive, online source, or a hard disk, then any newly installed software is probably the problem. Sometimes, software upgrades, patches, or enhancements could be the problem. Start with the assumption that there was a change in software, and work from there.

Don't Forget the Feedback

Many if not most equipment or systems have one or more feedback paths. The feedback monitors system or product outputs with sensors or other circuits and sends a signal back to the input or controller to modify the processing in some way. Feedback is useful as it is stabilizing and permits circuits and equipment to be self-adjusting and well mannered. Without that feedback all sorts of weird things can happen.

First, be sure the feedback is present. Go to the block diagram or other documentation, determine what it is and where it is, and then make measurements or visual inspections to verify its operation.

Second, if you are having troubleshooting difficulty, the feedback may actually be obfuscating the results of your tests. One way to get a better check on signal flow or other operations is to disable the feedback, and then make you tests. Once you prove that the signal chain is okay, you can reconnect the feedback. Keep in mind that there may be more than one feedback path.

Diagnostics

If the equipment or system is computer-based, in some cases, diagnostic software has been developed to help locate problems. Ask or otherwise determine if such troubleshooting software is available. If so, use it. It is a speedy way to a solution.

External Help

There may be sources of help from outside sources. Some companies may have a help desk or phone line or web-based help resources. Go to the website of the equipment manufacturer, and search for such resources. Then use them.

Sometimes a fee is involved, but it may be justified if time is of the essence. Don't overlook internal help, such as a friend, colleague, supervisor, or some other person who may know the equipment or have troubleshooting experience. When time and money are on the line, don't be embarrassed to ask for help.

Another point is the availability of field service techs or engineers who work for the company that makes the equipment. These individuals are trained and experienced in the equipment and can facilitate a repair in record time. Normally they are only available on the larger more complex systems (MRI machines or other medical gear, semiconductor manufacturing equipment, etc.). And they are expensive. Yet the expense may be worth it if failure of the equipment is causing larger losses.

Documentation

Documentation was mentioned earlier, but it bears repeating. If good documentation in the form of manuals or computer-based references (CDs, online, etc.) is available, be sure to use them. Most equipment documentation includes troubleshooting information.

Training and Experience

Troubleshooting is a learned skill. You learn by actually doing. The more experience you can get, the better you will be at ferreting out trouble and fixing it fast. Such skill is always acknowledged in a work environment.

Another way to get better at troubleshooting quickly is to take any training courses related to the equipment you will be working with. Many companies will automatically train you on the equipment. In other cases, you may have to find and ask for such training from the equipment manufacturer. Most training is classroom or lab-based, but some online self-training may also be available. Since such training is usually expensive, you will need to justify the cost and your time. Generally, you may be able to justify the training by analyzing how much money and time will be saved by faster troubleshooting and repair.

A Final Word

The troubleshooting advice in this chapter is by necessity broad and general. The reason for that is there are just too many different types of electronic equipment and not all techniques work well on all equipment. You literally have to learn to repair a specific piece of equipment from someone else, from training, or from the documentation with your own experimentation. Troubleshooting is specific to the circuit or equipment in question. You will learn troubleshooting mainly by the experience of using one or more of the techniques outlined here.

41.6 Systems Troubleshooting

Today, most electronics is in the form of systems. A system is just a collection of components, circuits, modules, equipment, or whatever to make a complete solution to a need or problem. Some examples are a personal computer, a home entertainment system, or a large complex piece of equipment like an MRI machine. These are very complex systems and require a different kind of thinking to troubleshoot. The first rule is that you probably will never troubleshoot to the component level in the real world. It is too time-consuming and expensive to look for individual component failures in any electronic product. There may be exceptions, of course, but in most cases you want to fix the system quickly at lowest cost. Therefore, forget about circuits, and start thinking at a higher level. Think in terms of the major components of the system you are working with. The best way to do this is to think in terms of block diagrams of major components.

Next, think of tracing signals through the system from major block to block. Using the processes described earlier for troubleshooting, try to determine where the signal is lost and where the trouble may lie. Keep in mind that most systems have feedback signals from outputs to inputs that affect system operation. Such feedback sometimes thwarts the normal process of logical signal tracing. Disconnecting the feedback often helps the troubleshooting process.

Following are several system examples that will illustrate processes you can use.

Garage Door Opener

A garage door opener is a small simple system but a good illustration of how to troubleshoot in a systems environment. Refer to Fig. 41-14 for a block diagram of a common garage door opener. The major components are the mechanical subsystem that actually moves the door up and down, the motor and its driver circuits that operate the mechanical subsystem,

the control subsystem that includes push buttons to initiate an operation, and the sensors that determine if obstacles are present. The control system may be simple logic circuits or an embedded controller. Finally, there is the wireless subsystem that uses a UHF radio to remotely operate the system.

The door can be opened or closed by the push button or the wireless remote. The infrared (IR) sensor detects the light from an IR source like an LED. The IR sensor detects if there is an object under the door and prevents the door from closing as a safety measure. If the IR light is blocked, the door will not close.

The motor driver turns the motor off and on and reverses it direction of rotation for the opening and closing operations. The driver gets its inputs from the control circuits. A power supply operates all the circuits. A light usually operated from the ac line turns on when the door is opened. A timer in the control circuitry turns the light off after a specific period of time after the door closes.

When a failure occurs, you use the same process described earlier.

1. Verify the reported problem.
   Door won't open or close. Light won't turn on or off.

2. Note specific symptoms.
   Is the issue with existing wired push buttons or wireless or both?

3. Verify that power is applied.
   Is ac power on or is the battery in the wireless unit dead?

4. Do a visual inspection.
   What has changed since the last operation? Is anything visually missing, wrong, or broken? Are the mechanical components working as they should? Can you hear the motor? Are objects blocking the IR sensor?

5. Isolate the problem to a specific stage, module, IC, or subassembly through signal tracing or signal injection.
   If wireless is not working, try the wired push buttons. Is there any blockage between the IR source and sensor?

6. Isolate the problem to a specific component or module.
   If power is on and the motor is not working, it may be a bad motor or motor drive circuit. Was any wiring changed or broken? Are the infrared sensors blocked in some way?

7. Repair the problem.
   Replacing the whole unit is always an option here as it may be the fastest and cheapest solution.

8. Test the unit.

Garage door openers on the whole are very reliable. The most common problems are things like dead batteries in the wireless remote, burned out lightbulbs and blocked sensors from boxes, bikes, or toys in the garage.

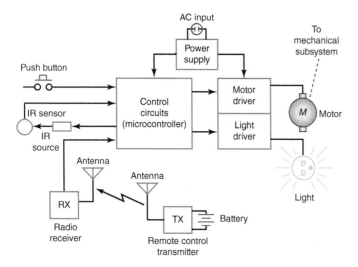

Figure 41-14 A garage door opener.

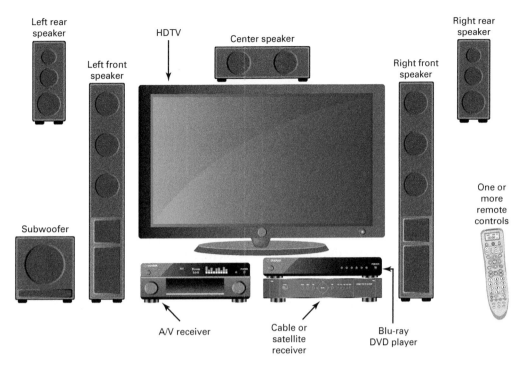

Figure 41-15 Common electronic home entertainment system with HDTV and surround sound.

Home Entertainment System

A home entertainment center typically consists of a large-screen HDTV and a variety of ancillary devices. The HDTV with its audio output is the main component, and the other equipment, such as a higher-quality receiver and audio subsystem, serves as input devices to enhance the overall user experience.

Figure 41-15 shows one common arrangement we can use as an example. The main component is a large-screen HDTV. It may have an LCD, plasma, LED, or projection screen, but it does have stereo output. The audio is limited by the very small built-in speakers (not shown), so often an auxiliary audio subsystem is added to give improved-quality audio.

The other components consist of a cable TV top box or receiver having multiple channels of programming that typically serves as the main content input to the TV. Alternatively, the main input may come from a satellite TV receiver (DirectTV or DishNetwork), or it may come from an Internet provider (AT&T's U-verse video/audio programming or Netflix Roku).

Other auxiliaries include a DVD/Blu-ray disc player and a digital video recorder (DVR) that records movies or other content to a hard disk drive, which can be accessed later at a more convenient time. The DVR is often internal to the cable receiver or satellite receiver. Older systems may include a VCR with its VHS cassette tape. Combined VCR and DVD players are also common.

Another input source is the A/V receiver. This box is essentially a set of high-fidelity audio amplifiers that implement a surround system with multiple speakers. For example, a 5.1 surround system has five major speakers and a subwoofer for bass. Figure 41-16 shows the most common arrangement. There are three front-facing speakers: center (C), left front (LF), and right front (RF). Then behind the listener are two surround speakers on the left rear (LR) and right rear (RR). The ".1" reference in the 5.1 designation means that an additional woofer handles the

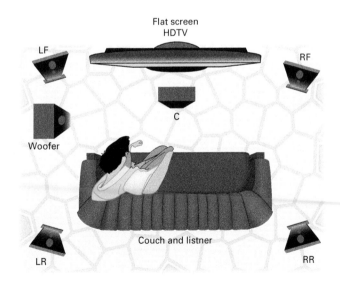

Figure 41-16 A typical 5.1 surround sound system showing speaker locations.

bass signals. The speakers require wiring back to the A/V receiver which contains all the power amplifiers that drive the speakers.

The A/V receiver also includes AM and FM radios and, in some cases, digital HD radio or satellite radio (Sirius or XM). The A/V receiver also has extensive front panel controls, as well as a remote control unit, allowing many choices for inputs and outputs, so a variety of configurations are selectable. With such an arrangement, the user has a variety of video and audio entertainment options.

Wiring A key part of such a system is the wiring, which is how all the components are interconnected. The block diagram in Fig. 41-17 shows one possible arrangement. Most connections are with high-definition multimedia interface (HDMI) cables or with cables having RCA-type "phono" video connectors. HDMI is a special cable that transmits digitized video and audio from one piece of equipment to another. The RCA connectors were originally developed to carry audio signals from a phonograph turntable to an amplifier. They are still widely used to carry audio as well as video signals.

Speaker wiring is by zip cord or twisted-pair cable. Some systems use a special digital audio fiber-optical

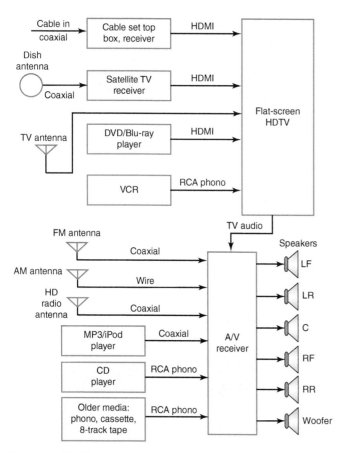

Figure 41-17 Wiring in a typical home entertainment center.

cable. Keep in mind that a variety of electric interfaces and cables are used. The secret to understanding such a system is to keep in mind the various inputs and outputs on each piece of equipment. Each device has its own inputs and outputs, and those are, in turn, routed to other equipment for display or output.

Remote Controls Once all the equipment is connected, the desired mode must be selected. Do you want to watch TV or video or just listen to music? What is the source of the programming? What is the destination of the output? Are you watching cable or satellite TV, or do you want to view a DVD? Do you want standard TV audio, or would you like the audio to be routed to the A/V receiver for higher-quality sound?

The setup and operation of the system is generally through the remote control units of each device. Usually each piece of equipment has its own remote control. One for the TV, the cable or satellite box, A/V receiver, DVD player, or other device. What you must learn is how to use the remotes to set the desired inputs and outputs to each device. This whole procedure is complex and the cause of many problems in home entertainment systems. In some cases, a single master remote may be an option.

Troubleshooting The process of successful troubleshooting in a home entertainment system comes in two forms: initial installation and problems with a working system. Here are suggestions to deal with both situations.

For installation, it is essential to have all the equipment manuals, which have plenty of details about how to wire various configurations. Details on operations with the remote control are also given. Read the manuals, wire the system, and then test it. Testing involves experimenting with different input sources and watching for the desired outputs. Keep in mind that you often need to set the inputs and outputs on two or more pieces of equipment to affect the desired end result. Each piece of equipment should be tested. The most common problems are wiring errors in equipment interconnection and incorrect remote control selections. Cables may be plugged into the wrong connectors, and remote controls may have dead batteries or be set incorrectly. Through a process of "playing around" with the system and meticulous attention to the connection and operational details in the manuals, most problems can be resolved.

Troubleshooting a previously operating system requires a different approach. Use the procedure outlined earlier, by which you first verify the complaint or problem and then go on to experiment with the remote controls, trying various combinations of inputs and outputs. Be sure to ask the question, What is different? Or what changed? If the system was working, when did it fail, and what happened between the time it was working and the time that it failed? A common problem is that a new user came along and tried to use the system, thereby changing input/output settings.

Next, think up individual tests for each box in the system. If there is no picture on the TV screen, is it completely dark or is there "snow" or just a blank screen? If the screen is completely dark, it could mean TV set failure. But if the screen shows light, chances are the TV inputs have been changed or lost. Try different input settings to see if you can get a picture via the cable or satellite receiver. If not, try a DVD input. Change input settings with the remote as necessary.

If sound is the problem, try to play a normal CD. Or try an AM or FM radio station. If the TV won't play through the sound system, check input/output settings.

Troubleshooting is often just a trial-and-error process. Just be sure to thoroughly verify the connections and settings before considering that a box is defective. More commonly it is not. Most problems seem to be initial interconnections problems or incorrect settings via the remote controls.

? CHAPTER 41 REVIEW QUESTIONS

1. The most common job of an electronics technician is
   a. design.
   b. circuit analysis.
   c. testing.
   d. troubleshooting.

2. Which test instrument presents its output in the frequency domain?
   a. Spectrum analyzer.
   b. Oscilloscope.
   c. Multimeter.
   d. Logic analyzer.

3. Which test instrument presents its output in the time domain?
   a. Spectrum analyzer.
   b. Oscilloscope.
   c. Multimeter.
   d. Signal generator.

4. The spectrum analyzer shows what characteristic on the vertical scale of the output?
   a. Voltage.
   b. Current.
   c. Power.
   d. Impedance.

5. A signal generator that produces sine, square, and triangular waves from about 1 Hz to 6 MHz is called a(n)
   a. RF signal generator.
   b. AWG.
   c. function generator.
   d. frequency synthesizer.

6. Most modern test instruments are
   a. analog.
   b. digital.

7. What is the first process that an input signal encounters in a modern test instrument?
   a. Filtering.
   b. Frequency conversion.
   c. Digital-to-analog conversion.
   d. Analog-to-digital conversion.

8. In a modern test instrument, how is the measurement actually made?
   a. An embedded microcontroller processes the digital input.
   b. Analog circuits interpret the signal and display it.
   c. The signal is directly displayed for the user to measure.
   d. Level comparators decide the signal amplitude.

9. A virtual instrument generally uses what to process the input signal?
   a. Embedded controller.
   b. Personal or laptop computer.
   c. An FPGA or DSP chip.
   d. No processing is done.

10. What makes the measurement in a virtual instrument?
    a. Hardware.
    b. Software.

11. What is the name of the software that is used to create virtual instruments?
    a. The C language.
    b. Java.
    c. LabVIEW.
    d. MathCAD.

12. Which of the following is *not* part of an AWG?
    a. DAC.
    b. Frequency synthesizer .
    c. Memory.
    d. ADC.

13. It is often more economical and faster to replace a product or subassembly than to repair it.
    a. True.
    b. False.

14. The first step in most troubleshooting processes is to
    a. isolate the problem.
    b. do a visual inspection.
    c. verify that the problem exists.
    d. check for power.

15. You should not attempt to troubleshoot a piece of equipment without
    a. the documentation.
    b. an oscilloscope.
    c. multimeter.
    d. training.

16. Electronic components are generally more likely to fail than mechanical components.
    a. True.
    b. False.

17. A good first step in troubleshooting a circuit is to
    a. test all transistors.
    b. check for overheated ICs.
    c. verify that the correct dc voltages are present.
    d. begin signal tracing.

18. In Fig. 41-8, the input to stage 1 is good and the output from stage 4 is not correct. The problem lies in
    a. stage 1.
    b. stage 2.
    c. stage 4.
    d. any of the stages.

19. In Fig. 41-8, there is no output from stage 1. The problem is
    a. a defective stage 1.
    b. no input signal.

    c. either *a* or *b*.
    d. not having enough information to determine.

20. Most detailed troubleshooting should not begin unless
    a. you have the documentation.
    b. you are familiar with the equipment.
    c. you have been trained on the equipment.
    d. all of the above.

21. The most likely component to fail in a system is a(n)
    a. diode.
    b. integrated circuit.
    c. capacitor.
    d. resistor.

22. Which is often the primary goal of troubleshooting a piece of equipment?
    a. Low cost.
    b. Minimum downtime.
    c. Continued long product life.
    d. No need for experienced repairers.

23. For complex digital circuit troubleshooting, the best instrument is probably the
    a. oscilloscope.
    b. spectrum analyzer.
    c. logic analyzer.
    d. signal generator.

24. In systems troubleshooting, it is best to approach the problem with
    a. block diagram analysis.
    b. schematic diagram analysis.
    c. testing all individual components.
    d. replacement of complete subsystems.

25. A quick but expensive and often effective troubleshooting approach is
    a. complete unit replacement.
    b. component or module substitution.
    c. signal tracing to the component level.
    d. dc voltage measurement.

 CHAPTER 41 PROBLEMS

SECTION 41.3 Circuit Troubleshooting

41.1 In Fig. 41-18, a +5-V signal is applied to the input of the transistor. The LED does not light.

A measurement at the collector of the transistor reveals a voltage of +5 V. What are the most likely problems?

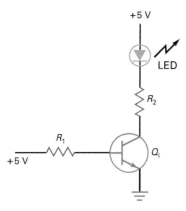

Figure 41-18 Circuit for prob. 41-1.

41.2 In Fig. 41-19, the circuit is not performing as it should. This schematic is all you have. Determine what the output is supposed to be. Measuring the output, you get a dc voltage value of –12 V. What is the problem?

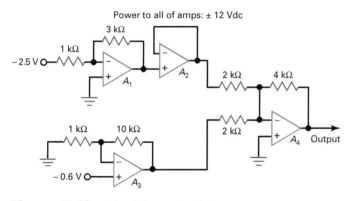

Figure 41-19 Circuit for prob. 41-2.

SECTION 41.4 Power Supply Troubleshooting

41.3 Consider the circuit of Fig. 41-20. What is the desired output? If the output is a series of 60-Hz pulses across D_2, what is the most likely problem?

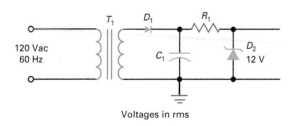

Figure 41-20 Circuit for prob. 41-3.

41.4 In Fig. 41-21, the circuits are operating properly except for a 500-kHz signal riding on the 8-V dc output. What may be the problem?

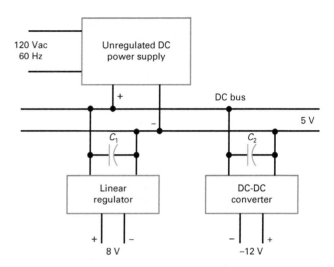

Figure 41-21 Circuit for prob. 41-4.

SECTION 41.5 Alternative Troubleshooting Techniques

41.5 One of the PC board modules in a piece of equipment has failed. A replacement board costs $85. The repair can be made by just plugging in a new board. On the other hand, a technician can repair the board in an estimated 2 hours if the $6 defective IC is available. The technician cost is $45 per hour. What is the recommended solution?

41.6 If documentation is not available for the device or system you are troubleshooting, what may be a way to get that?

SECTION 41.6 System Troubleshooting

41.7 In the garage door example given in Fig. 41-14, the door will open but not close. What may be the problem?

41.8 In the home entertainment center of Fig. 41-17, the TV set is working but you cannot play a movie DVD. Give several reasons why you cannot do this.

41.9 The automatic lawn sprinkler system, shown in Fig. 41-22, has four zones of operation. Water sprinkler heads in each zone turn on in sequence for a specific amount of time.

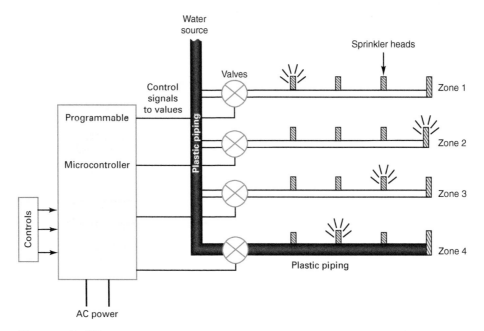

Figure 41-22 An automatic programmable lawn sprinkler system.

A microcomputer-based controller is used to program the system with the zones, days, times, and durations of the watering. When the system is operating, the microcontroller provides the signals to turn on the water valves in each section. Zone 3 is not working. Give several possible causes of the problem.

Photo Credits

Chapter 1
Part Opener: Pedro Castellano/Getty Images RF.

Chapter 2
Figures 2-1, 2-10a: © Mark Steinmetz; **2-16a–b, 2-17a–c:** Courtesy of Mitchel E. Schultz.

Chapter 3
Figure 3-1b: © Rueangsin Phuthawil/123RF; **3-2, 3-6, 3-16, 3-19:** © Mark Steinmetz; **3-3a–b, 3-7b:** The McGraw-Hill Companies, Inc./Cindy Schroeder, photographer; **3-8b:** Martyn F. Chillmaid/ Photo Researchers, Inc.; **S3-2:** Courtesy of R. Stanley Williams © 2012 Hewlett-Packard Development Company, L.P. Reproduced with permission.

Chapter 5
Figure S5-1: Steven Puetzer/Photodisc/Getty Images RF.

Chapter 9
Figures 9-3a–d, 9-5a–g, 9-5i, 9-15a–c: The McGraw-Hill Companies, Inc./Cindy Schroeder, photographer; **9-4, 9-6a–b, 9-10, 9-12, 9-13, 9-14:** © Mark Steinmetz; **9-5h:** © George Doyle & Ciaran Griffin/Stockbyte/Getty Images; **S9-1:** © Jon Boyes/ Photographer's Choice/Getty Images.

Chapter 10
Figures 10-1, 10-2, 10-6, 10-7, 10-12: © Mark Steinmetz; **10-8:** The McGraw-Hill Companies, Inc./Cindy Schroeder, photographer; **10-10:** The McGraw-Hill Companies, Inc./Doug Martin, photographer.

Chapter 11
Figures 11-4a–b: © Richard Megna—Fundamental Photographs, NYC; **11-8, 11-31:** © Mark Steinmetz.

Chapter 12
Figure 12-1a: Courtesy of MCM Electronics; **12-1b:** © Volodymyr Krasyuk/Shutterstock; **12-12:** Courtesy of Simpson Electric Company/www.simpsonelectric.com; **12-13:** © Siarhei Dzmitryienka/Shutterstock; **12.14:** ©Industrykb/Alamy Stock Photo; **S12-2:** Courtesy of National Instruments.

Chapter 14
Part Opener: © Phil Boorman/The Image Bank/Getty Images.

Chapter 15
Figure 15-1b: The McGraw-Hill Companies, Inc./Cindy Schroeder, photographer; **15-4b, 15-5b, 15-6, 15-7, 15-8, 15-9c, 15-10:** © Mark Steinmetz; **S15-3:** Nick Rowe/Getty Images RF.

Chapter 18
Figures 18-4a, 18-11, 18-14a: © Mark Steinmetz; **18-4b:** The McGraw-Hill Companies, Inc./Cindy Schroeder, photographer; **18-4c, 18-23:** Courtesy of Design Criteria, Inc.

Chapter 24
Figure 24-30: © Mark Steinmetz; **24-34:** Stockbyte/Getty Images RF; **24-36:** © Dreamframer/Shutterstock; **S24-1:** © belchonock/123RF.

Chapter 26
Figures 26-1a: © Andrew Lambert Photography/Science Source; **26-8:** © Dmitry Strizhakov/Shutterstock; **26-1b, 26-9, 26-21:** © Agilent Technologies, Inc. 2012. Reproduced with Permission, Courtesy of Agilent Technologies, Inc.; **26-22:** Courtesy of National Instruments; **26-23:** Courtesy of Digilent Inc.

Chapter 27
Part Opener: © Adam Berry/Getty Images.

Chapter 29
Figure 29-19c: Steven Puetzer/Photodisc/Getty Images RF.

Chapter 30
Figure 30-8e: © Brian Moeskau/Moeskau Photography.

Chapter 33
Figures 33-47c, 33-48b: © Brian Moeskau/Moeskau Photography.

Chapter 38
Figure 38-14: Courtesy Fox Electronics, an Integrated Device Technology Company.

Chapter 41
Figures 41-1: © Aleksey Dmetsov/Alamy Stock Photo; **41-3:** © Sergey Ryzhov/123RF; **41-6, 41-7:** Courtesy of National Instruments.

Index

Wire(s)
  ac power, 414–415
  ground, 416–418
  resistance of, 135
  standard gage sizes, 128–129
  troubleshooting, 223
  types of, 130
Wire cable, 130–131
Wire conductors, 130–131
Wire size (twisted pair cables), 440
Wire-wound resistors, 34–35
  color coding and marking of, 38
  power ratings for, 42
  variable, 40
Wireless
  in homes, businesses, and industry, 414
  jobs in, 10

Wiring
  in boat electrical systems, 226
  defective, 416
  in electronic systems, 657–660
  in homes, 409–413
Work
  power vs., 51
  practical units of, 52
Wound-field dc motors, 184–185

X

X axis, 50
XOR gates, 476, 477

Y

Y axis, 50
Yahoo, 8

Z

Zener diodes, 507–511
  loaded zener regulators, 509–511
  specifications, 511
Zener followers, 747–748
Zener regulators, 508–511, 746–748
Zener resistance, 511
Zero α, 136
Zero-crossing detectors, 666–668
Zero-ohm resistors, 39
Zero-ohms adjustment, 200
Zero-power resistance, 39
Zip cord cable, 414